Encyclopedia of

Library and Information Sciences, Fourth Edition

Volume 1

Encyclopedias from the Taylor & Francis Group

Print	Online

Agriculture

Encyclopedia of Agricultural, Food, and Biological Engineering, 2nd Ed., 2 Vols.　Pub'd. 10/21/10
K10554 (978-1-4398-1111-5)　　K11382 (978-1-4398-2806-9)

Encyclopedia of Animal Science, 2nd Ed., 2 Vols.　Pub'd. 2/1/11
K10463 (978-1-4398-0932-7)　　K10528 (978-0-415-80286-4)

Encyclopedia of Biotechnology in Agriculture and Food　Pub'd. 7/16/10
DK271X (978-0-8493-5027-6)　　DKE5044 (978-0-8493-5044-3)

Business and Computer Science

Encyclopedia of Computer Science & Technology, 2nd Ed., 2 Vols.　Pub'd 12/21/2016
K21573 (978-1-4822-0819-1)　　K21578 (978-1-4822-0822-1)

Encyclopedia of Information Assurance, 4 Vols.　Pub'd. 12/21/10
AU6620 (978-1-4200-6620-3)　　AUE6620 (978-1-4200-6622-7)

Encyclopedia of Information Systems and Technology, 2 Vols.　Pub'd. 12/29/15
K15911 (978-1-4665-6077-2)　　K21745 (978-1-4822-1432-1)

Encyclopedia of Library and Information Sciences, 4th Ed.　Publishing 2017
K15223 (978-1-4665-5259-3)　　K15224 (978-1-4665-5260-9)

Encyclopedia of Software Engineering, 2 Vols.　Pub'd. 11/24/10
AU5977 (978-1-4200-5977-9)　　AUE5977 (978-1-4200-5978-6)

Encyclopedia of Supply Chain Management, 2 Vols.　Pub'd. 12/21/11
K12842 (978-1-4398-6148-6)　　K12843 (978-1-4398-6152-3)

Encyclopedia of U.S. Intelligence, 2 Vols.　Pub'd. 12/19/14
AU8957 (978-1-4200-8957-8)　　AUE8957 (978-1-4200-8958-5)

Encyclopedia of Wireless and Mobile Communications, 2nd Ed., 3 Vols.　Pub'd. 12/18/12
K14731 (978-1-4665-0956-6)　　KE16352 (978-1-4665-0969-6)

Chemistry, Materials and Chemical Engineering

Encyclopedia of Chemical Processing, 5 Vols.　Pub'd. 11/1/05
DK2243 (978-0-8247-5563-8)　　DKE499X (978-0-8247-5499-0)

Encyclopedia of Chromatography, 3rd Ed.　Pub'd. 10/12/09
84593 (978-1-4200-8459-7)　　84836 (978-1-4200-8483-2)

Encyclopedia of Iron, Steel, and Their Alloys, 5 Vols.　Pub'd. 1/6/16
K14814 (978-1-4665-1104-0)　　K14815 (978-1-4665-1105-7)

Encyclopedia of Plasma Technology, 2 Vols.　Pub'd 12/12/2016
K14378 (978-1-4665-0059-4)　　K21744 (978-1-4822-1431-4)

Encyclopedia of Supramolecular Chemistry, 2 Vols.　Pub'd. 5/5/04
DK056X (978-0-8247-5056-5)　　DKE7259 (978-0-8247-4725-1)

Encyclopedia of Surface & Colloid Science, 3rd Ed., 10 Vols.　Pub'd. 8/27/15
K20465 (978-1-4665-9045-8)　　K20478 (978-1-4665-9061-8)

Engineering

Dekker Encyclopedia of Nanoscience and Nanotechnology, 3rd Ed., 7 Vols.　Pub'd. 3/20/14
K14119 (978-1-4398-9134-6)　　K14120 (978-1-4398-9135-3)

Encyclopedia of Energy Engineering and Technology, 2nd Ed., 4 Vols.　Pub'd. 12/1/14
K14633 (978-1-4665-0673-2)　　KE16142 (978-1-4665-0674-9)

Encyclopedia of Optical and Photonic Engineering, 2nd Ed., 5 Vols.　Pub'd. 9/22/15
K12323 (978-1-4398-5097-8)　　K12325 (978-1-4398-5099-2)

Environment

Encyclopedia of Environmental Management, 4 Vols.　Pub'd. 12/13/12
K11434 (978-1-4398-2927-1)　　K11440 (978-1-4398-2933-2)

Encyclopedia of Environmental Science and Engineering, 6th Ed., 2 Vols.　Pub'd. 6/25/12
K10243 (978-1-4398-0442-1)　　KE0278 (978-1-4398-0517-6)

Encyclopedia of Natural Resources, 2 Vols.　Pub'd. 7/23/14
K12418 (978-1-4398-5258-3)　　K12420 (978-1-4398-5260-6)

Medicine

Encyclopedia of Biomaterials and Biomedical Engineering, 2nd Ed.　Pub'd. 5/28/08
H7802 (978-1-4200-7802-2)　　HE7803 (978-1-4200-7803-9)

Encyclopedia of Biomedical Polymers and Polymeric Biomaterials, 11 Vols.　Pub'd. 4/2/15
K14324 (978-1-4398-9879-6)　　K14404 (978-1-4665-0179-9)

Concise Encyclopedia of Biomedical Polymers and Polymeric Biomaterials, 2 Vols.　Pub'd. 8/14/17
K14313 (978-1-4398-9855-0)　　KE42253 (978-1-315-11644-0)

Encyclopedia of Biopharmaceutical Statistics, 3rd Ed.　Pub'd. 5/20/10
H100102 (978-1-4398-2245-6)　　HE10326 (978-1-4398-2246-3)

Encyclopedia of Clinical Pharmacy　Pub'd. 11/14/02
DK7524 (978-0-8247-0752-1)　　DKE6080 (978-0-8247-0608-1)

Encyclopedia of Dietary Supplements, 2nd Ed.　Pub'd. 6/25/10
H100094 (978-1-4398-1928-9)　　HE10315 (978-1-4398-1929-6)

Encyclopedia of Medical Genomics and Proteomics, 2 Vols.　Pub'd. 12/29/04
DK2208 (978-0-8247-5564-5)　　DK501X (978-0-8247-5501-0)

Encyclopedia of Pharmaceutical Science and Technology, 4th Ed., 6 Vols.　Pub'd. 7/1/13
H100233 (978-1-84184-819-8)　　HE10420 (978-1-84184-820-4)

Routledge Encyclopedias

Encyclopedia of Public Administration and Public Policy, 3rd Ed., 5 Vols.　Pub'd. 11/6/15
K16418 (978-1-4665-6909-6)　　K16434 (978-1-4665-6936-2)

Routledge Encyclopedia of Modernism　Pub'd 5/11/16
　　Y137844 (978-1-135-00035-6)

Routledge Encyclopedia of Philosophy Online　Pub'd. 11/1/00
　　RU22334 (978-0-415-24909-6)

Routledge Performance Archive　Pub'd. 11/12/12
　　Y148405 (978-0-203-77466-3)

Encyclopedia of

Library and Information Sciences, Fourth Edition

Volume 1

From: *Academic Libraries* To: *Careers and Education in Records and Information Management*

Encyclopedia Edited By

John D. McDonald

and

Michael Levine-Clark

CRC Press
Taylor & Francis Group
Boca Raton London New York

CRC Press is an imprint of the
Taylor & Francis Group, an **informa** business

First published 2018 by CRC Press

Published 2019 by CRC Press
Taylor & Francis Group
6000 Broken Sound Parkway NW, Suite 300
Boca Raton, FL 33487-2742

First issued in paperback 2020

ISBN 13: 978-1-4665-5259-3 (HB Set)
ISBN 13: 978-0-8153-8623-0 (Vol. 1) (hbk)

ISBN 13: 978-0-3675-7010-1 (PB Set)
ISBN 13: 978-0-3675-7016-3 (Vol. 1) (pbk)

**Visit the Taylor & Francis Web site at
http://www.taylorandfrancis.com**

**and the CRC Press Web site at
http://www.crcpress.com**

Encyclopedia of Library and Information Sciences, Fourth Edition

Brief Contents

Volume VI (cont'd.)

Volume VII

Volume VII (cont'd.)

Volume VII (cont'd.)

Encyclopedia of Library and Information Sciences, Fourth Edition

Editors-in-Chief

John D. McDonald
Analytics and Assessment, EBSCO Information Services

Michael Levine-Clark
University of Denver Libraries, Denver, Colorado

Editorial Advisory Board

Contributors

June Abbas / *School of Library and Information Studies, University of Oklahoma, Norman, Oklahoma, U.S.A.*

Richard Abel / *Portland, Oregon, U.S.A.*

Eileen G. Abels / *College of Information Science and Technology, Drexel University, Philadelphia, Pennsylvania, U.S.A.*

Tia Abner / *American Medical Informatics Association (AMIA), Bethesda, Maryland, U.S.A.*

Donald C. Adcock / *Dominican University, River Forest, Illinois, U.S.A.*

Kendra S. Albright / *School of Library and Information Science, University of South Carolina, Columbia, South Carolina, U.S.A.*

Mikael Alexandersson / *University of Gothenburg, Gothenburg, Sweden*

Joan M. Aliprand / *Cupertino, California, U.S.A.*

Jacqueline Allen / *Dallas Museum of Art, Dallas, Texas, U.S.A.*

Romano Stephen Almagno / *International College of St. Bonaventure, Rome, Italy*

Connie J. Anderson-Cahoon / *Southern Oregon University Library, Ashland, Oregon, U.S.A.*

Karen Anderson / *Archives and Information Science, Mid Sweden University, ITM, Härnösand, Sweden*

Rick Anderson / *University of Utah, Salt Lake City, Utah, U.S.A.*

Silviu Andrieş-Tabac / *Institute of Cultural Heritage, Moldova Academy of Sciences, Chişinău, Republic of Moldova*

Peng Hwa Ang / *Wee Kim Wee School of Communication and Information, Nanyang Technological University, Singapore*

Hermina G.B. Anghelescu / *School of Library and Information Science, Wayne State University, Detroit, Michigan, U.S.A.*

Leah Arroyo / *American Association of Museums, Washington, District of Columbia, U.S.A.*

Terry Asla / *Senior Lifestyles Researcher, Seattle, U.S.A.*

Shiferaw Assefa / *University of Kansas, Lawrence, Kansas, U.S.A.*

Ilse Assmann / *Radio Broadcast Facilities, SABC, Johannesburg, South Africa*

Maija-Leena Aulikki Huotari / *University of Oulu, Oulu, Finland*

Henriette D. Avram / *Library of Congress, Washington, District of Columbia, U.S.A.*

Sven Axsäter / *Department of Industrial Management and Logistics, Lund University, Lund, Sweden*

Murtha Baca / *Getty Research Institute, Los Angeles, California, U.S.A.*

Roger S. Bagnall / *Institute for the Study of the Ancient World, New York University, New York, New York, U.S.A.*

Nestor Bamidis / *GSA-Archives of Macedonia, Thessaloniki, Greece*

Franz Barachini / *Business Innovation Consulting—Austria, Langenzersdorf, Austria*

Rebecca O. Barclay / *Rensselaer Polytechnic Institute, Troy, New York, U.S.A.*

Judit Bar-Ilan / *Department of Information Science, Bar-Ilan University, Ramat Gan, Israel*

Alex W. Barker / *Museum of Art and Archaeology, University of Missouri, Columbia, Missouri, U.S.A.*

John A. Bateman / *University of Bremen, Bremen, Germany*

Marcia J. Bates / *Department of Information Studies, Graduate School of Education and Information Studies, University of California, Los Angeles (UCLA), Los Angeles, California, U.S.A.*

Philippe Baumard / *School of Engineering, Stanford University, Stanford, California, U.S.A., and University Paul Cézanne, Aix-en-Provence, France*

David Bawden / *City, University of London, London, U.K.*

Jennifer Bawden / *Museum Studies Program, Faculty of Information Studies, University of Toronto, Toronto, Ontario, Canada*

David Bearman / *Archives & Museum Informatics, Toronto, Ontario, Canada*

William K. Beatty / *Northwestern University Medical School, Chicago, Illinois, U.S.A.*

A.R. Bednarek / *University of Florida, Gainesville, Florida, U.S.A.*

Clare Beghtol / *Faculty of Information Studies, University of Toronto, Toronto, Ontario, Canada*

Lori Bell / *Alliance Library System, East Peoria, Illinois, U.S.A.*

Danna Bell-Russel / *Library of Congress, Washington, District of Columbia, U.S.A.*

William Benedon / *Benedon & Associates, Encino, California, U.S.A.*

Anna Bergaliyeva / *Kazakhstan Institute of Management, Economics and Strategic Research (KIMEP), Almaty, Kazakhstan*

Sidney E. Berger / *Phillips Library, Peabody Essex Museum, Salem, Massachusetts, U.S.A.*

Andrew J. Berner / *University Club of New York, New York, New York, U.S.A.*

Sean F. Berrigan / *Policy, Library and Archives Canada, Ottawa, Ontario, Canada*

John W. Berry / *NILRC: Network of Illinois Learning Resources in Community Colleges, Dominican University, River Forest, Illinois, U.S.A.*

Michael W. Berry / *Department of Electrical Engineering and Computer Science, University of Tennessee, Knoxville, Tennessee, U.S.A.*

Suresh K. Bhavnani / *Center for Computational Medicine and Bioinformatics, University of Michigan, Ann Arbor, Michigan, U.S.A.*

Tamara Biggs / *Chicago History Museum, Chicago, Illinois, U.S.A.*

Frank Birkebæk / *Roskilde Museum, Roskilde, Denmark*

Ann P. Bishop / *Graduate School of Library and Information Science, University of Illinois at Urbana-Champaign, Urbana, Illinois, U.S.A.*

Julia Blixrud / *Association of Research Libraries, Washington, District of Columbia, U.S.A.*

Gloria Bordogna / *Italian National Research Council, Institute for the Dynamics of Environmental Processes, Dalmine, Italy*

Steve Bosch / *Administration Department, University of Arizona, Tucson, Arizona, U.S.A.*

Kimberly S. Bostwick / *Ecology and Evolutionary Biology, Cornell University Museum of Vertebrates, Ithaca, New York, U.S.A.*

Natalia T. Bowdoin / *University of South Carolina Aiken, Aiken, South Carolina, U.S.A.*

Patrick J. Boylan / *Department of Cultural Policy and Management, City University, London, U.K.*

Amy E. Brand / *CrossRef, Lynnfield, Massachusetts, U.S.A.*

Judy Brooker / *Australian Library and Information Association, Deakin, Australian Capital Territory, Australia*

Terrence Brooks / *iSchool, University of Washington, Seattle, Washington, U.S.A.*

Vanda Broughton / *School of Library, Archive and Information Studies, University College London, London, U.K.*

Cecelia Brown / *School of Library and Information Studies, University of Oklahoma, Norman, Oklahoma, U.S.A.*

Jos de Bruijn / *Digital Enterprise Research Institute, University of Innsbruck, Innsbruck, Austria*

Steve Bryant / *BFI National Archive, Herts, U.K.*

Alan Bryden / *International Organization for Standardization, Geneva, Switzerland*

Jeff E. Bullard / *Free Library of Philadelphia, Philadelphia, Pennsylvania, U.S.A.*

Kathleen Burns / *Beinecke Rare Book and Manuscript Library, Yale University, New Haven, Connecticut, U.S.A.*

Brenda A. Burton / *Library, Kirkland & Ellis LLP, Chicago, IL, U.S.A.*

E. Burton Swanson / *Anderson School of Management, University of California, Los Angeles, Los Angeles, California, U.S.A.*

Donald I. Butcher / *Canadian Library Association, Ottawa, Ontario, Canada*

Kevin Butterfield / *Wolf Law Library, College of William and Mary, Williamsburg, Virginia, U.S.A.*

Alex Byrne / *University of Technology, Sydney—Sydney, New South Wales, Australia*

Brian Byrne / *Discipline of Psychology, School of Behavioural, Cognitive and Social Sciences, University of New England, Armidale, New South Wales, Australia, Australian Research Council Centre of Excellence in Cognition and its Disorder, Australia, and National Health and Medical Research Council Centre of Research Excellence in Twin Research, Australia*

Bernadette G. Callery / *School of Information Sciences, University of Pittsburgh, Pittsburgh, Pennsylvania, U.S.A.*

Paul D. Callister / *Leon E. Bloch Law Library, University of Missouri-Kansas City School of Law, Kansas City, Missouri, U.S.A.*

Perrine Canavaggio / *International Council on Archives, Paris, France*

Sarah R. Canino / *Dickinson Music Library, Vassar College, Poughkeepsie, New York, U.S.A.*

Robert Capra / *School of Information and Library Science, University of North Carolina, Chapel Hill, North Carolina, U.S.A.*

Nicholas Carroll / *Hastings Research, Inc., Las Vegas, Nevada, U.S.A.*

Ben Carterette / *Department of Computer and Information Sciences, University of Delaware, Newark, Delaware, U.S.A.*

Vittorio Castelli / *T.J. Watson Research Center, IBM, Yorktown Heights, New York, U.S.A.*

Jane Rosetta Virginia Caulton / *Library of Congress, Washington, District of Columbia, U.S.A.*

Richard Cave / *Formerly at the Public Library of Science, San Francisco, California, U.S.A.*

Roderick Cave / *Loughborough University, Loughborough, U.K.*

Marcel Caya / *Department of History, University of Quebec at Montreal (UQAM), Montreal, Quebec, Canada*

Frank Cervone / *Purdue University Calumet, Hammond, Indiana, U.S.A.*

Leslie Champeny / *Alaska Resources Library and Information Services (ARLIS), Anchorage, Alaska, U.S.A.*

Lois Mai Chan / *School of Library and Information Science, University of Kentucky, Lexington, Kentucky, U.S.A.*

Sergio Chaparro-Univazo / *Graduate School of Library and Information Science, Simmons College, Boston, Massachusetts, U.S.A.*

Mary K. Chelton / *Graduate School of Library and Information Studies, Queens College Flushing, New York, U.S.A.*

Hsinchun Chen / *Department of Management Information Systems, University of Arizona, Tucson, Arizona, U.S.A.*

Jianhua Chen / *Computer Science Department, Louisiana State University, Baton Rouge, Louisiana, U.S.A.*

Eric R. Childress / *OCLC, Dublin, Ohio, U.S.A.*

Michael A. Chilton / *Department of Management, Kansas State University, Manhattan, Kansas, U.S.A.*

TzeHuey Chiou-Peng / *Spurlock Museum, University of Illinois at Urbana-Champaign, Urbana, Illinois, U.S.A.*

Hyun-Yang Cho / *Department of Library and Information Science, Kyonggi University, Suwon, South Korea*

Jae-Hwang Choi / *Department of Library and Information Science, Kyungpook National University, Daegu, South Korea*

Carol E.B. Choksy / *School of Library and Information Science, Indiana University, Bloomington, Indiana, U.S.A.*

Su Kim Chung / *University Libraries, University of Nevada–Las Vegas, Las Vegas, Nevada, U.S.A.*

James Church / *University Libraries, University of California, Berkeley, Berkeley, California, U.S.A.*

Barbara H. Clubb / *Ottawa Public Library, Ottawa, Ontario, Canada*

Arlene Cohen / *Pacific Islands Library Consultant, Seattle, Washington, U.S.A.*

Barbara Cohen-Stratyner / *New York Public Library for the Performing Arts, New York, U.S.A.*

Edward T. Cokely / *Center for Adaptive Behavior and Cognition, Max Planck Institute for Human Development, Berlin, Germany*

Arthur H. Cole / *Harvard University, Cambridge, Massachusetts, U.S.A.*

John Y. Cole / *Center for the Book, Library of Congress, Washington, District of Columbia, U.S.A.*

Patrick Tod Colegrove / *DeLaMare Science & Engineering Library, University Libraries, University of Nevada, Reno, Reno, Nevada, U.S.A.*

Edwin T. Coman, Jr. / *University of California, Riverside, California, U.S.A.*

Nora T. Corley / *Arctic Institute of North America, Montreal, Quebec, Canada*

Sheila Corrall / *Department of Information Studies, University of Sheffield, Sheffield, U.K.*

Erica Cosijn / *Department of Information Science, University of Pretoria, Pretoria, South Africa*

Richard J. Cox / *School of Computing and Information, University of Pittsburgh, Pittsburgh, Pennsylvania, U.S.A.*

Barbara M. Cross / *Records and Information Management, Sony Pictures Entertainment, Culver City, California, U.S.A.*

Kevin Crowston / *School of Information Studies, Syracuse University, Syracuse, New York, U.S.A.*

Adrian Cunningham / *National Archives of Australia (NAA), Canberra, Australian Capital Territory, Australia*

Judith N. Currano / *University of Pennsylvania, Philadelphia, Pennsylvania, U.S.A.*

Susan Curzon / *University Library, California State University–Northridge, Northridge, California, U.S.A.*

Ingetraut Dahlberg / *Bad Koenig, Germany*

Nan Christian Ploug Dahlkild / *Royal School of Library and Information Science, Copenhagen, Denmark*

Jay E. Daily / *University of Pittsburgh, Pittsburgh, Pennsylvania, U.S.A.*

Kimiz Dalkir / *Graduate School of Library and Information Studies, McGill University, Montreal, Quebec, Canada*

Prudence W. Dalrymple / *Drexel University College of Computing & Informatics, Philadelphia, Pennsylvania, U.S.A.*

Marcel Danesi / *Department of Anthropology, University of Toronto, Toronto, Ontario, Canada*

Xuan Hong Dang / *Computer Vision and Image Understanding, Institute for Infocomm, A* STAR, Singapore*

Yan Dang / *Department of Management Information Systems, University of Arizona, Tucson, Arizona, U.S.A.*

Evelyn Daniel / *School of Information and Library Science, University of North Carolina at Chapel Hill, Chapel Hill, North Carolina, U.S.A.*

Richard A. Danner / *School of Law, Duke University, Durham, North Carolina, U.S.A.*

Regina Dantas / *Museu Nacional, HCTE, Universidade Federal do Rio de Janeiro, Rio de Janeiro, Brazil*

Daniel C. Danzig / *Consultant, Pasadena, California, U.S.A.*

Robert Allen Daugherty / *University Library, University of Illinois at Chicago, Chicago, Illinois, U.S.A.*

Charles H. Davis / *Indiana University, Bloomington, IN, U.S.A., and School of Library and Information Science, Indiana University, Bloomington, Indiana, U.S.A.*

Gordon B. Davis / *Carlson School of Management, University of Minnesota, Minneapolis, Minnesota, U.S.A.*

Mary Ellen Davis / *American Library Association, Chicago, Illinois, U.S.A.*

Peter Davis / *International Centre for Cultural and Heritage Studies, Newcastle University, Newcastle upon Tyne, U.K.*

Sheryl Davis / *University Library, University of California, Riverside, Riverside, California, U.S.A.*

Ronald E. Day / *School of Library and Information Science, Indiana University, Bloomington, Indiana, U.S.A.*

Cheryl Dee / *School of Library and Information Science, University of South Florida, Tampa, Florida, U.S.A.*

Robert DeHart / *Department of History, Middle Tennessee State University, Murfreesboro, Tennessee, U.S.A.*

Brenda Dervin / *School of Communication, Ohio State University, Columbus, Ohio, U.S.A.*

Brian Detlor / *Information Systems, McMaster University, Hamilton, Ontario, Canada*

Don E. Detmer / *American Medical Informatics Association (AMIA), Bethesda, Maryland, U.S.A.*

Stella G. Dextre Clarke / *Information Consultant, Oxfordshire, U.K.*

Catherine Dhérent / *National Library of France, Paris, France*

Anne R. Diekema / *Gerald R. Sherratt Library, Southern Utah University, Cedar City, Utah, U.S.A.*

Susan S. DiMattia / *DiMattia Associates, Stamford, Connecticut, U.S.A.*

Gloria Dinerman / *The Library Co-Op, Inc., Edison, New Jersey, U.S.A.*

Jesse David Dinneen / *School of Information Studies, McGill University, Montreal, Quebec, Canada*

Bernard Dione / *School of Librarianship, Archivists Information Science (EBAD), Cheikh Anta Diop University, Dakar, Senegal*

Dieyi Diouf / *Central Library, Cheikh Anta Diop University of Dakar, Dakar, Senegal*

Keith Donohue / *National Historical Publications and Records Commission, Washington, District of Columbia, U.S.A.*

Ann Doyle / *X̱wi7xwa Library, First Nations House of Learning, University of British Columbia, Vancouver, British Columbia, Canada*

Carol D. Doyle / *Government Documents Department and Map Library, California State University, Fresno, California, U.S.A.*

Marek J. Druzdzel / *School of Information Sciences and Intelligent Systems Program, University of Pittsburgh, Pittsburgh, Pennsylvania, U.S.A., and Faculty of Computer Science, Bialystok Technical University, Bialystok, Poland*

Kathel Dunn / *National Library of Medicine, Bethesda, Maryland, U.S.A.*

Luciana Duranti / *School of Library, Archival and Information Studies, University of British Columbia, Vancouver, British Columbia, Canada*

Joan C. Durrance / *School of Information, University of Michigan, Ann Arbor, Michigan, U.S.A.*

Maria Economou / *Department of Communication and Cultural Technology, University of the Aegean, Mytilini, Greece*

Gary Edson / *Center for Advanced Study in Museum Science and Heritage Management, Museum of Texas Tech University, Lubbock, Texas, U.S.A.*

Mary B. Eggert / *Library, Kirkland & Ellis LLP, Chicago, IL, U.S.A.*

Daniel Eisenberg / *Florida State University, Tallahassee, Florida, U.S.A.*

Innocent I. Ekoja / *University Library, University of Abuja, Abuja, Nigeria*

Sarah Elliott / *International Centre for Cultural and Heritage Studies, Newcastle University, Newcastle upon Tyne, U.K.*

David Ellis / *Department of Information Studies, Aberystwyth University, Wales, U.K.*

Jill Emery / *Portland State University Library, Portland, Oregon, U.S.A.*

Zorana Ercegovac / *InfoEN Associates, Los Angeles, California, U.S.A.*

Timothy L. Ericson / *School of Information Science, University of Wisconsin-Milwaukee, Milwaukee, Wisconsin, U.S.A.*

Elena Escolano Rodríguez / *National Library of Spain, Madrid, Spain*

Leigh S. Estabrook / *Graduate School of Library and Information Science, University of Illinois at Urbana- / Champaign, Champaign, Illinois, U.S.A.*

Mark E. Estes / *Alameda County Law Library, Oakland, California, U.S.A.*

Beth Evans / *Library, Brooklyn College, City University of New York, Brooklyn, New York, U.S.A.*

Joanne Evans / *Centre for Organisational and Social Informatics, Monash University, Melbourne, Victoria, Australia*

Dominic J. Farace / *Grey Literature Network Service, TextRelease/GreyNet, Amsterdam, The Netherlands*

David Farneth / *Special Collections and Institutional Records, Getty Research Institute, Los Angeles, California, U.S.A.*

Sharon Fawcett / *Office of Presidential Libraries, National Archives and Records Administration, College Park, Maryland, U.S.A.*

Dieter Fensel / *Institute of Computer Science, University of Innsbruck, Innsbruck, Austria, and National University of Ireland, Galway, Galway, Ireland*

Thomas L. Findley / *Leo A. Daly/Architects & Engineers, Omaha, Nebraska, U.S.A.*

Karen E. Fisher / *Information School, University of Washington, Seattle, Washington, U.S.A.*

Nancy Fjällbrant / *Chalmers University of Technology Library, International Association of Technological University Libraries, Gothenburg, Sweden*

Julia Flanders / *Brown University, Providence, Rhode Island, U.S.A.*

Nancy Flury Carlson / *Westinghouse Electric Corporation, Pittsburgh, Pennsylvania, U.S.A.*

Roger R. Flynn / *School of Information Sciences and Intelligent Systems Program, University of Pittsburgh, Pittsburgh, Pennsylvania, U.S.A.*

Helen Forde / *Department of Information Studies, University College London, London, U.K.*

Douglas J. Foskett / *University of London, London, U.K.*

Susan Foutz / *Institute for Learning Innovation, Edgewater, Maryland, U.S.A.*

Christopher Fox / *Department of Computer Science, James Madison University, Harrisonburg, Virginia, U.S.A.*

Carl Franklin / *Consultant, Columbus, Ohio, U.S.A.*

Jonathan A. Franklin / *Gallagher Law Library, University of Washington, Seattle, Washington, U.S.A.*

Thomas J. Froehlich / *School of Library and Information Science, Kent State University, Kent, Ohio, U.S.A.*

Steve Fuller / *Department of Sociology, University of Warwick, Coventry, U.K.*

Crystal Fulton / *School of Information and Communication Studies, University College Dublin, Dublin, Ireland*

Carla J. Funk / *Medical Library Association, Chicago, Illinois, U.S.A.*

Jonathan Furner / *Department of Information Studies University of California, Los Angeles, Los Angeles, California, U.S.A.*

Dennis Galletta / *Katz Graduate School of Business, University of Pittsburgh, Pittsburgh, Pennsylvania, U.S.A.*

D. Linda Garcia / *Communication Culture and Technology, Georgetown University, Washington, District of Columbia, U.S.A.*

Holly Gardinier / *Honnold/Mudd Library, Libraries of The Claremont Colleges, Claremont, California, U.S.A.*

Sally Gardner Reed / *Association of Library Trustees, Advocates, Friends and Foundations (ALTAFF), Philadelphia, Pennsylvania, U.S.A.*

Janifer Gatenby / *Online Computer Library Center (OCLC), Leiden, The Netherlands*

Ramesh C. Gaur / *Kalanidhi Division, Indira Gandhi National Centre for the Arts (IGNCA), New Delhi, India*

Lee Anne George / *Association of Research Libraries, Washington, District of Columbia, U.S.A.*

David E. Gerard / *College of Librarianship Wales, Cardiganshire, Wales, U.K.*

Malcolm Getz / *Department of Economics, Vanderbilt University, Nashville, Tennessee, U.S.A.*

Mary W. Ghikas / *American Library Association, Chicago, Illinois, U.S.A.*

Nicholas Gibbins / *School of Electronics and Computer Science, University of Southampton, Southampton, U.K.*

Gerd Gigerenzer / *Center for Adaptive Behavior and Cognition, Max Planck Institute for Human Development, Berlin, Germany*

Tommaso Giordano / *Library, European University Institute, Florence, Italy*

Lilian Gisesa / *Kenya National Archives, Nairobi, Kenya*

Edward A. Goedeken / *Iowa State University, Ames, Iowa, U.S.A.*

Warren R. Goldmann / *National Technical Institute for the Deaf, Rochester Institute of Technology, Rochester, New York, U.S.A.*

David Gordon / *Milwaukee Art Museum, Milwaukee, Wisconsin, U.S.A.*
David B. Gracy II / *School of Information, University of Texas at Austin, Austin, Texas, U.S.A.*
Karen F. Gracy / *School of Library and Information Science, Kent State University, Kent, Ohio, U.S.A.*
Renny Granda / *Universidad Central de Venezuela, Caracas, Venezuela*
Paul Gray / *School of Information Systems and Technology, Claremont Graduate University, Claremont, California, U.S.A.*
Jane Greenberg / *Metadata Research Center, School of Information and Library Science, University of North Carolina at Chapel Hill, Chapel Hill, North Carolina, U.S.A.*
Karen Greenwood / *American Medical Informatics Association (AMIA), Bethesda, Maryland, U.S.A.*
Jill E. Grogg / *Libraries, University of Alabama, Tuscaloosa, Alabama, U.S.A.*
Melissa Gross / *School of Information, Florida State University, Tallahassee, Florida, U.S.A.*
Andrew Grove / *Guest Faculty, Information School, University of Washington, Seattle, Washington, U.S.A.*
Dinesh K. Gupta / *Department of Library and Information Science, Vardhaman Mahaveer Open University, 3 Kota, India*
Laurel L. Haak / *Open Researcher and Contributor ID, Inc. (ORCID), U.S.A.*
Kate Hagan / *American Association of Law Libraries, Chicago, Illinois, U.S.A.*
Kathleen Hall / *Leon E. Bloch Law Library, University of Missouri-Kansas City School of Law, Kansas City, Missouri, U.S.A.*
Virginia M.G. Hall / *Center for Educational Resources, The Sheridan Libraries, Johns Hopkins University, Baltimore, Maryland, U.S.A.*
Wendy Hall / *Intelligence, Agents, Multimedia Group, University of Southampton, Southampton, U.K.*
Stuart Hamilton / *International Federation of Library Associations and Institutions, The Hague, The Netherlands*
Maureen L. Hammer / *Knowledge Management, Batelle Memorial Institute, Charlottesville, Virginia, U.S.A.*
Jong-Yup Han / *Research Information Team, KORDI, Seoul, South Korea*
Debra Gold Hansen / *School of Library and Information Science, San Jose State University, Yorba Linda, California, U.S.A.*
Derek L. Hansen / *University of Maryland, College Park, Maryland, U.S.A.*
Eugene R. Hanson / *Shippensburg State College, Shippensburg, Pennsylvania, U.S.A.*
Jane Hardy / *Australian Library and Information Association, Deakin, Australian Capital Territory, Australia*
Julie Hart / *American Association of Museums, Washington, District of Columbia, U.S.A.*
Hiroyuki Hatano / *Surugadai University, Saitama, Japan*
Robert M. Hayes / *Department of Information Studies, University of California, Los Angeles, Los Angeles, California, U.S.A.*
Caroline Haythornthwaite / *Graduate School of Library and Information Science, University of Illinois at Urbana- / Champaign, Champaign, Illinois, U.S.A.*
Penny Hazelton / *Gallagher Law Library, University of Washington, Seattle, Washington, U.S.A.*
P. Bryan Heidorn / *Graduate School of Library and Information Science, University of Illinois at Urbana-Champaign, Champaign, Illinois, U.S.A.*
Helen Heinrich / *Collection Access and Management Services, California State University–Northridge, Northridge, California, U.S.A.*

Doris S. Helfer / *Collection Access and Management Services, California State University–Northridge, Northridge, California, U.S.A.*

Markus Helfert / *School of Computing, Dublin City University, Dublin, Ireland*

Jean Henefer / *School of Information and Communication Studies, University College Dublin, Dublin, Ireland*

Steven L. Hensen / *Rare Book, Manuscript and Special Collections Library, Duke University, Durham, North Carolina, U.S.A.*

Pamela M. Henson / *Archives, Smithsonian Institution, Washington, District of Columbia, U.S.A.*

Peter Hernon / *Graduate School of Library and Information Science, Simmons College, Boston, Massachusetts, U.S.A.*

Dorothy H. Hertzel / *Case Western Reserve University, Cleveland, Ohio, U.S.A.*

Francis Heylighen / *Free University of Brussels, Brussels, Belgium*

Randolph Hock / *Online Strategies, Annapolis, Maryland, U.S.A.*

Theodora L. Hodges / *Berkeley, California, U.S.A.*

Sara S. Hodson / *Huntington Library, San Marino, California, U.S.A.*

Judy C. Holoviak / *American Geophysical Union, Washington, District of Columbia, U.S.A.*

Aleksandra Horvat / *Faculty of Philosophy, University of Zagreb, Zagreb, Croatia*

Ali Houissa / *Olin Library, Cornell University, Ithaca, New York, U.S.A.*

Pamela Howard-Reguindin / *Library of Congress Office, Nairobi, Kenya*

Han-Yin Huang / *International Centre for Cultural and Heritage Studies, Newcastle University, Newcastle upon Tyne, U.K.*

Kathleen Hughes / *American Library Association, Chicago, Illinois, U.S.A.*

Betsy L. Humphreys / *National Library of Medicine, Bethesda, Maryland, U.S.A.*

Charlene S. Hurt / *University Library, Georgia State University, Atlanta, Georgia, U.S.A.*

Sue Hutley / *Australian Library and Information Association, Deakin, Australian Capital Territory, Australia*

John P. Immroth / *University of Pittsburgh, Pittsburgh, Pennsylvania, U.S.A.*

Peter Ingwersen / *Royal School of Library and Information Science, University of Copenhagen, Copenhagen, Denmark*

Vanessa Irvin / *Library and Information Science Program, Information and Computer Sciences Department, University of Hawaii at Mānoa, Honolulu, Hawaii, U.S.A.*

Karla Irwin / *University Libraries, University of Nevada–Las Vegas, Las Vegas, Nevada, U.S.A.*

October R. Ivins / *Ivins eContent Solutions, Sharon, Massachusetts, U.S.A.*

Kalervo Järvelin / *School of Information Science, University of Tampere, Tampere, Finland*

Jean Frédéric Jauslin / *Federal Department of Home Affairs (FDHA), Swiss Federal Office of Culture, Bern, Switzerland*

V. Jeyaraj / *Hepzibah Institute of Conversion, Chennai, India*

Scott Johnston / *McPherson Library, University of Victoria, Victoria, British Columbia, Canada*

Trevor Jones / *Mountain Heritage Center, Western Carolina University, Cullowhee, North Carolina, U.S.A.*

William Jones / *Information School, University of Washington, Seattle, Washington, U.S.A.*

Jay Jordan / *OCLC Online Computer Library Center, Inc., Dublin, Ohio, U.S.A.*

Corinne Jörgensen / *School of Information Studies, Florida State University, Tallahassee, Florida, U.S.A.*

Gene Joseph / *Aboriginal Library Consultant, Langley, British Columbia, Canada*

Daniel N. Joudrey / *School of Library and Information Science, Simmons College, Boston, Massachusetts, U.S.A.*

Heidi Julien / *Library and Information Studies, State University of New York–Buffalo, Buffalo, New York, U.S.A.*

Janet Kaaya / *Department of Information Studies, University of California, Los Angeles, California, U.S.A.*

Philomena Kagwiria Mwirigi / *Kenya National Library Service (KNLS), Nairobi, Kenya*

Athanase B. Kanamugire / *Library Consultant, Dhahran, Saudi Arabia*

Paul B. Kantor / *School of Communication and Information, Rutgers University, New Brunswick, New Jersey, U.S.A.*

Sofia Kapnisi / *International Federation of Library Associations and Institutions, The Hague, the Netherlands*

Nelson Otieno Karilus / *Kenya National Library Service (KNLS), Nairobi, Kenya*

Amy M. Kautzman / *University of California, Berkeley, Berkeley, California, U.S.A.*

Karalyn Kavanaugh / *Account Services Manager, EBSCO Information Services, Birmingham, Alabama, U.S.A.*

Caroline Kayoro / *Kenya National Library Service (KNLS), Nairobi, Kenya*

Andreas Kellerhals / *Federal Department of Home Affairs (FDHA), Swiss Federal Archives, Bern, Switzerland*

John M. Kennedy / *Indiana University, Bloomington, Indiana, U.S.A.*

Kristen Kern / *Portland State University, Portland, Oregon, U.S.A.*

Christopher S.G. Khoo / *School of Communication and Information, Nanyang Technological University, Singapore*

Tapan Khopkar / *University of Michigan, Ann Arbor, Michigan, U.S.A.*

Irene Muthoni Kibandi / *Kenya National Library Service (KNLS), Nairobi, Kenya*

Ruth E. Kifer / *Dr. Martin Luther King, Jr. Library, San Jose State University, San Jose, California, U.S.A.*

Seong Hee Kim / *Department of Library and Information Science, Chung-Ang University, Seoul, South Korea*

Pancras Kimaru / *Kenya National Library Service (KNLS), Nairobi, Kenya*

Karen E. King / *Washington, District of Columbia, U.S.A.*

William R. King / *University of Pittsburgh, Pittsburgh, Pennsylvania, U.S.A.*

Susan K. Kinnell / *Consultant, Santa Barbara, California, U.S.A.*

Laurence J. Kipp / *Harvard University, Cambridge, Massachusetts, U.S.A.*

Thomas G. Kirk, Jr. / *Earlham College Libraries, Earlham College, Richmond, Indiana, U.S.A.*

Breanne A. Kirsch / *Library, Emerging Technologies, University of South Carolina Upstate, Spartanburg, South Carolina, U.S.A.*

Vernon N. Kisling, Jr. / *Marston Science Library, University of Florida, Gainesville, Florida, U.S.A.*

Adam D. Knowles / *San Diego, California, U.S.A.*

Rebecca Knuth / *Library and Information Science Program, University of Hawaii, Honolulu, Hawaii, U.S.A.*

Michael Koenig / *College of Information and Computer Science, Long Island University, Brookville, New York, U.S.A.*

Jesse Koennecke / *Cornell University Library, Cornell University College of Arts and Sciences, Ithaca, New York, U.S.A.*

Jes Koepfler / *Museum Studies Program, Faculty of Information Studies, University of Toronto, Toronto, Ontario, Canada*

Amelia Koford / *Blumberg Memorial Library, Texas Lutheran University, Seguin, Texas, U.S.A.*

Toru Koizumi / *Library, Rikkyo University, Tokyo, Japan*

Josip Kolanović / *Croatian State Archives, Zagreb, Croatia*

Sjoerd Koopman / *International Federation of Library Associations and Institutions, The Hague, the Netherlands*

Donald Kraft / *Department of Computer Science, U.S. Air Force Academy, Colorado Springs, Colorado, U.S.A.*

Allison Krebs / *University of Arizona, Tucson, Arizona, U.S.A.*

Judith F. Krug / *Office for Intellectual Freedom, American Library Association, Chicago, Illinois, U.S.A.*

D.W. Krummel / *Emeritus, Graduate School of Library and Information Science, University of Illinois at Urbana-Champaign, Champaign, Illinois, U.S.A.*

Carol Collier Kuhlthau / *Department of Library and Information Science, Rutgers University, New Brunswick, New Jersey, U.S.A.*

Krishan Kumar / *Former Head, Department of Library and Information Science, University of Delhi, New Delhi, India*

Sanna Kumpulainen / *Library, Tampere University of Technology, Tampere, Finland*

Michael J. Kurtz / *National Archives at College Park, U.S. National Archives and Records Administration, College Park, Maryland, U.S.A.*

Zhenhua Lai / *Department of Management Information Systems, University of Arizona, Tucson, Arizona, U.S.A.*

Mounia Lalmas / *Department of Computing Science, University of Glasgow, Glasgow, U.K.*

Heather M. Lamond / *Massey University Library, Palmerston North, New Zealand*

F.W. Lancaster / *Graduate School of Library and Information Science, University of Illinois at Urbana-Champaign, Urbana, Illinois, U.S.A.*

Ronald L. Larsen / *School of Information Sciences, University of Pittsburgh, Pittsburgh, Pennsylvania, U.S.A.*

Ray R. Larson / *School of Information, University of California—Berkeley, Berkeley, California, U.S.A.*

Jesús Lau / *Library Services Unit USBI Veracruz (USBI VER), University of Veracruz, Veracruz, Mexico*

Judith V. Lechner / *Department of Educational Foundations, Leadership, and Technology, Auburn University, Auburn, Alabama, U.S.A.*

Christopher A. Lee / *School of Information and Library Science, University of North Carolina at Chapel Hill, Chapel Hill, North Carolina, U.S.A.*

Janet Lee / *University of Denver, Denver, Colorado, U.S.A, and Regis University, Denver, Colorado, U.S.A.*

Catherine Leekam / *Museum Studies Program, Faculty of Information Studies, University of Toronto, Toronto, Ontario, Canada*

Kjell Lemström / *Department of Computer Science, University of Helsinki, Helsinki, Finland*

Timothy F. Leslie / *Department of Geography and Geoinformation Science, George Mason University, Fairfax, Virginia, U.S.A.*

Noémie Lesquins / *Scientific Mission (DSR), National Library of France, Paris, France*

Rosalind K. Lett / *Information-2-Knowledge, Atlanta, Georgia, U.S.A.*

Allison V. Level / *Colorado State University, Fort Collins, Colorado, U.S.A.*

Michael Levine-Clark / *Penrose Library, University of Denver, Denver, Colorado, U.S.A.*

Anany Levitin / *Department of Computing Sciences, Villanova University, Villanova, Pennsylvania, U.S.A.*

Marjorie Lewis / *Canaan, New York, U.S.A.*

Elizabeth D. Liddy / *School of Information Studies, Syracuse University, Syracuse, New York, U.S.A.*

Silje C. Lier / *Software & Information Industry Association, Washington, District of Columbia, U.S.A.*

Jane E. Light / *Dr. Martin Luther King, Jr. Library, San Jose Public Library, San Jose, California, U.S.A.*

Paul M. Lima / *Canadian Heritage Information Network (CHIN), Gatineau, Quebec, Canada*

Louise Limberg / *Swedish School of Library and Information Science, University of Borås and University of Gothenburg, Borås, Sweden*

Shin-jeng Lin / *Department of Business Administration, Le Moyne College, Syracuse, New York, U.S.A.*

Sarah Lippincott / *Educopia Institute, Atlanta, Georgia, U.S.A.*

Peter Johan Lor / *School of Information Studies, University of Wisconsin-Milwaukee, Milwaukee, Wisconsin, U.S.A., and Department of Information Science, University of Pretoria, Pretoria, South Africa*

Beth Luey / *Fairhaven, Massachusetts, U.S.A.*

Joseph Luke / *Kazakhstan Institute of Management, Economics and Strategic Research (KIMEP), Almaty, Kazakhstan*

Claudia Lux / *Central and Regional Library of Berlin (ZLB), Berlin, Germany*

Marianne Lykke / *Information Interaction and Architecture, Royal School of Library and Information Science, Aalborg, Denmark*

Elena Macevičiūtė / *Faculty of Communication, Vilnius University, Vilnius, Lithuania, and Swedish School of Library and Information Science, University of Borås, Borås, Sweden*

Juan D. Machin-Mastromatteo / *Universidad Central de Venezuela, Caracas, Venezuela*

Barbara A. Macikas / *American Library Association, Chicago, Illinois, U.S.A.*

Leslie Madsen-Brooks / *Boise State University, Boise, Idaho, U.S.A.*

William J. Maher / *Archives, University of Illinois at Urbana-Champaign, Urbana, Illinois, U.S.A.*

Thomas Mann / *Library of Congress, Washington, District of Columbia, U.S.A.*

Sylva Natalie Manoogian / *Department of Information Studies, University of California, Los Angeles, Los Angeles, California, U.S.A.*

Daniel Marcu / *Information Sciences Institute, University of Southern California, Marina del Rey, California, U.S.A.*

James W. Marcum / *Fairleigh Dickinson University, Madison, New Jersey, U.S.A.*

Francesca Marini / *School of Library, Archival and Information Studies, University of British Columbia, Vancouver, British Columbia, Canada*

Johan Marklund / *Department of Industrial Management and Logistics, Lund University, Lund, Sweden*

Dian I. Martin / *Small Bear Technical Consulting, LLC, Thorn Hill, Tennessee, U.S.A.*

Susan K. Martin / *Lauinger Library, Georgetown University, Washington, District of Columbia, U.S.A.*

Paul F. Marty / *College of Communication and Information, Florida State University, Tallahassee, Florida, U.S.A.*

Dan Marwit / *Lee H. Skolnick Architecture + Design Partnership, New York, New York, U.S.A.*

Laura Matzer / *Arizona Museum for Youth, Mesa, Arizona, U.S.A.*

Robert L. Maxwell / *Special Collections and Metadata Catalog Department, Brigham Young University, Provo, Utah, U.S.A.*

Hope Mayo / *Houghton Library, Harvard University, Cambridge, Massachusetts, U.S.A.*

Sally H. McCallum / *Network Development and MARC Standards Office, Library of Congress, Washington, District of Columbia, U.S.A.*

Gavan McCarthy / *eScholarship Research Centre, University of Melbourne, Melbourne, Victoria, Australia*

Ian McGowan / *Former Librarian, National Library of Scotland, Edinburgh, U.K.*

Roger McHaney / *Department of Management, Kansas State University, Manhattan, Kansas, U.S.A.*

I.C. McIlwaine / *University College London, School of Library, Archive and Information Studies, London, U.K.*

Sue McKemmish / *Centre for Organisational and Social Informatics, Monash University, Melbourne, Victoria, Australia*

Marie E. McVeigh / *JCR and Bibliographic Policy, Thomson Reuters - Scientific, Philadelphia, Pennsylvania, U.S.A.*

Linda Mboya / *National Museums of Kenya, Nairobi, Kenya*

Judith Adams Meadows / *State Law Library of Montana, Helena, Montana, U.S.A.*

K. van der Meer / *Faculty of Electrical Engineering, Mathematics and Computer Science, Delft University, the Netherlands; Information and Library Science, IOIW, Antwerp University, Belgium; and D-CIS, Delft, The Netherlands*

Bharat Mehra / *School of Information Sciences, University of Tennessee, Knoxville, Tennessee, U.S.A.*

Margaret Ann Mellinger / *OSU Libraries & Press, Oregon State University, Corvallis, Oregon, U.S.A.*

Elizabeth E. Merritt / *American Association of Museums, Washington, District of Columbia, U.S.A.*

David Millman / *Academic Information Systems, Columbia University, New York, U.S.A.*

Jack Mills / *North-Western Polytechnic, London, U.K.*

Kevin L. Mills / *National Institute of Standards and Technology, Gaithersburg, Maryland, U.S.A.*

Staša Milojević / *Department of Information Studies, University of California, Los Angeles, Los Angeles, California, U.S.A.*

Marla Misunas / *Collections Information and Access, San Francisco Museum of Modern Art, San Francisco, California, U.S.A.*

Joan S. Mitchell / *OCLC Online Computer Library Center, Inc., Dublin, Ohio, U.S.A.*

Yoriko Miyabe / *Rikkyo University, Tokyo, Japan*

Diane Mizrachi / *University Libraries, University of California–Los Angeles, Los Angeles, California, U.S.A.*

William Moen / *Texas Center for Digital Knowledge, University of North Texas, Denton, Texas, U.S.A.*

Abdul Moid / *University of Karachi, Karachi, Pakistan*

Hermann Moisl / *Center for Research in Linguistics, University of Newcastle upon Tyne, Newcastle upon Tyne, U.K.*

Ole Magnus Mølbak Andersen / *Danish State Archives, Copenhagen, Denmark*

Mavis B. Molto / *Utah State University, Logan, Utah, U.S.A.*

Philip Mooney / *Heritage Communications, Coca-Cola Company, Atlanta, Georgia, U.S.A.*

Reagan W. Moore / *San Diego Supercomputer Center, University of North Carolina at Chapel Hill, Chapel Hill, North Carolina, U.S.A.*

Mersini Moreleli-Cacouris / *Department of Library Science and Information Systems, Technological Educational Institute (TEI) of Thessaloniki, Sindos, Greece*

Paul K. Moser / *Department of Philosophy, Loyola University Chicago, Chicago, Illinois, U.S.A.*

Clara C. Mosquera / *Library, Kirkland & Ellis LLP, Chicago, IL, U.S.A.*

David J. Muddiman / *Leeds Metropolitan University, Leeds, U.K.*

Nancy C. Mulvany / *Bayside Indexing Service, Fort Collins, Colorado, U.S.A.*

Sue Myburgh / *School of Communication, University of South Australia, Adelaide, South Australia, Australia*

Elli Mylonas / *Brown University, Providence, Rhode Island, U.S.A.*

Jeremy Myntti / *J. Willard Marriott Library, Salt Lake City, Utah, U.S.A.*

Jacob Nadal / *ReCAP: The Research Collections and Preservation Consortium, Princeton, New Jersey, U.S.A.*

Diane Nahl / *Information and Computer Sciences Department, University of Hawaii, Honolulu, Hawaii, U.S.A.*

Robert Nardini / *Vice President, Library Services, ProQuest Books, La Vergne, Tennessee, U.S.A.*

Arnold vander Nat / *Department of Philosophy, Loyola University Chicago, Chicago, Illinois, U.S.A.*

Charles M. Naumer / *Information School, University of Washington, Seattle, Washington, U.S.A.*

Sophie Ndegwa / *Kenya National Library Service (KNLS), Nairobi, Kenya*

Dixie Neilson / *University of Florida, Gainesville, Florida, U.S.A.*

Sarah Beth Nelson / *School of Information and Library Sciences, University of North Carolina at Chapel Hill, Chapel Hill, North Carolina, U.S.A.*

Stuart J. Nelson / *National Library of Medicine, Bethesda, Maryland, U.S.A.*

Stephanie Nemcsok / *Museum Studies Program, Faculty of Information Studies, University of Toronto, Toronto, Ontario, Canada*

Ken Neveroski / *College of Information and Computer Science, Long Island University, Brookville, New York, U.S.A.*

Jennifer Ng / *Museum Studies Program, Faculty of Information Studies, University of Toronto, Toronto, Ontario, Canada*

Melissa Niiya / *Portland Public Schools, Portland, Oregon, U.S.A.*

Angela Noseworthy / *Museum Studies Program, Faculty of Information Studies, University of Toronto, Toronto, Ontario, Canada*

Barbara E. Nye / *Ictus Consulting, LLC, Pasadena, California, U.S.A.*

Charles Nzivo / *Kenya National Library Service (KNLS), Nairobi, Kenya*

Dennis O'Brien / *Maps and Wayfinding, LLC, Mystic, Connecticut, U.S.A.*

Karen Lynn O'Brien / *American Library Association, Chicago, Illinois, U.S.A.*

Kieron O'Hara / *Intelligence, Agents, Multimedia Group, University of Southampton, Southampton, U.K.*

Elizabeth O'Keefe / *Morgan Library and Museum, New York, U.S.A.*

Denise I. O'Shea / *Fairleigh Dickinson University, Teaneck, New Jersey, U.S.A.*

Douglas W. Oard / *College of Information Studies, University of Maryland, College Park, Maryland, U.S.A.*

Maria Oldal / *Morgan Library and Museum, New York, U.S.A.*

Lorne Olfman / *School of Information Systems and Technology, Claremont Graduate University, Claremont, California, U.S.A.*

Bette W. Oliver / *Austin, Texas, U.S.A.*

Annette Olson / *Biological Resources Division, U.S. Geological Survey, Reston, Virginia, U.S.A.*

Hope A. Olson / *School of Information Studies, University of Wisconsin-Milwaukee, Milwaukee, Wisconsin, U.S.A.*

Lawrence J. Olszewski / *OCLC Library, Dublin, Ohio, U.S.A.*

Kok-Leong Ong / *School of Information Technology, Deakin University, Burwood, Victoria, Australia*

Tim Owen / *Chartered Institute of Library and Information Professionals (CILIP), London, U.K.*

John C. Paolillo / *School of Informatics and School of Library and Information Science, Indiana University, Bloomington, Indiana, U.S.A.*

Eun Bong Park / *Library Service Department, National Library of Korea, Seoul, South Korea*

Soyeon Park / *Department of Library and Information Science, Duksung Womens University, Seoul, South Korea*

Gabriella Pasi / *Department of Informatics, Systems and Communication, University of Studies of Milano Bicocca, Milan, Italy*

Norman Paskin / *Tertius Ltd., Oxford, U.K.*

Christiane Paul / *Whitney Museum of American Art, New York, U.S.A.*

Ellen Pearlstein / *Information Studies and UCLA / Getty Program in the Conservation of Ethnographic and Archaeological Materials, University of California, Los Angeles, Los Angeles, California, U.S.A.*

Kathleen de la Peña McCook / *School of Library and Information Science, University of South Florida, Tampa, Florida, U.S.A.*

Steve Pepper / *Department of Linguistics, University of Oslo, Oslo, Norway*

Manuel A. Pérez-Quiñones / *Department of Software and Information Systems, University of North Carolina, Charlotte, North Carolina, U.S.A.*

Paul Evan Peters / *University of Pittsburgh, Pittsburgh, Pennsylvania, U.S.A.*

Jakob Heide Petersen / *Danish Agency for Libraries and Media, Copenhagen, Denmark*

Mary Jane Petrowski / *American Library Association, Chicago, Illinois, U.S.A.*

Katharine J. Phenix / *Northglenn Branch, Rangeview Library District, Northglenn, Colorado, U.S.A.*

Robert B. Pickering / *Gilcrease Museum, and Museum Science and Management Program, University of Tulsa, Tulsa, Oklahoma, U.S.A.*

Janice T. Pilch / *Rutgers University Libraries, Rutgers University, New Brunswick, New Jersey, U.S.A.*

Thomas E. Pinelli / *Langley Research Center, National Aeronautics and Space Administration (NASA) Hampton, Virginia, U.S.A.*

Daniel Pitti / *Alderman Library, Institute for Advanced Technology in the Humanities, University of Virginia, Charlottesville, Virginia, U.S.A.*

Elena Ploşniţă / *Science Department, National Museum of Archaeology and History of Moldova, Chisinau, Republic of Moldova*

Gabriela Podušelová / *Slovak National Museum, Bratislava, Slovak Republic*

Danny C.C. Poo / *School of Computing, Department of Information Systems, National University of Singapore, Singapore*

Martine Poulain / *Department of Libraries and Documentation, National Institute for the History of Art (INHA), Paris, France*

Tammy Powell / *National Library of Medicine, Bethesda, Maryland, U.S.A.*

Stephen Prine / *Library of Congress, Washington, District of Columbia, U.S.A.*

Mary Jo Pugh / *Editor, American Archivist, Walnut Creek, California, U.S.A.*

Ajit K. Pyati / *University of Western Ontario, London, Ontario, Canada*

Aimée C. Quinn / *Government Publications Services, Brooks Library, Central Washington University, Ellensburg, Washington, U.S.A.*

Jennie Quiñónez-Skinner / *University Library, California State University–Northridge, Northridge, California, U.S.A.*

Debbie Rabina / *School of Library and Information Science, Pratt Institute, New York, New York, U.S.A.*

Katalin Radics / *Research Library, University of California—Los Angeles, Los Angeles, California, U.S.A.*

Carl Rahkonen / *Harold S. Orendorff Music Library, Indiana University of Pennsylvania, Indiana, Pennsylvania, U.S.A.*

Jocelyn Rankin / *Centers for Disease Control and Prevention Library, Atlanta, Georgia, U.S.A.*

Samuel J. Redman / *Department of History, University of California, Berkeley, Berkeley, California, U.S.A.*

Thomas C. Redman / *Navesink Consulting Group, Little Silver, New Jersey, U.S.A.*

Barbara Reed / *Recordkeeping Innovation, Sydney, New South Wales, Australia*

Marcia Reed / *Getty Research Institute, Los Angeles, CA, U.S.A.*

CarrieLynn D. Reinhard / *Department of Communication, Business, and Information Technologies, Roskilde University, Roskilde, Denmark*

Harold C. Relyea / *Congressional Research Service, Library of Congress, Washington, District of Columbia, U.S.A.*

Steve Ricci / *Department of Information Studies/Film and Television, University of California–Los Angeles, Los Angeles, California, U.S.A.*

Ronald E. Rice / *Department of Communication, University of California–Santa Barbara, Santa Barbara, California, U.S.A.*

John V. Richardson, Jr. / *Department of Information Studies, University of California, Los Angeles, Los Angeles, California, U.S.A.*

Soo Young Rieh / *School of Information, University of Michigan, Ann Arbor, Michigan, U.S.A.*

Kevin S. Rioux / *Division of Library and Information Science, St. John's University, Queens, New York, U.S.A.*

Julian Roberts / *Wolfson College, University of Oxford, Oxford, U.K.*

Lyn Robinson / *City, University of London, London, U.K.*

Diane Robson / *University Libraries, Media Library, University of North Texas, Denton, Texas, U.S.A.*

Michael Rodriguez / *Michigan State University Libraries, East Lansin, Michigan, U.S.A.*

Juraj Roháč / *Department of Archival Science and Auxiliary Historical Sciences, Comenius University in, Bratislava, Slovak Republic*

Mark Roosa / *Pepperdine University, Malibu, California, U.S.A.*

Jonathan Rose / *Department of History, Drew University, Madison, New Jersey, U.S.A.*

Howard Rosenbaum / *School of Library and Information Science, Indiana University, Bloomington, Indiana, U.S.A.*

Catherine Sheldrick Ross / *Faculty of Information and Media Studies, University of Western Ontario, London, Ontario, Canada*

Shannon Ross / *Canadian Heritage Information Network (CHIN), Gatineau, Quebec, Canada*

Richard Rubin / *School of Library and Information Science, Kent State University, Kent, Ohio, U.S.A.*

Lynne M. Rudasill / *University of Illinois at Urbana-Champaign, Champaign, Illinois, U.S.A.*

Michael Rush / *Beinecke Rare Book and Manuscript Library, Yale University, New Haven, Connecticut, U.S.A.*

Mariza Russo / *Faculty of Administration and Accounting Sciences (FACC), Federal University of Rio de Janeiro, Rio de Janeiro, Brazil*

Athena Salaba / *Kent State University, Kent, Ohio, U.S.A.*

Romelia Salinas / *California State University, Los Angeles, Los Angeles, California, U.S.A.*

Airi Salminen / *Department of Computer Science and Information Systems, University of Jyväskylä, Jyväskylä, Finland*

Michael J. Salvo / *Department of English, Purdue University, West Lafayette, Indiana, U.S.A.*

Robert J. Sandusky / *University Library, University of Illinois at Chicago, Chicago, Illinois, U.S.A.*

Tefko Saracevic / *School of Communication and Information, Rutgers University, New Brunswick, New Jersey, U.S.A.*

Chris Sauer / *Said Business School, University of Oxford, Oxford, U.K.*

Rejéan Savard / *School of Library and Information Science, University of Montreal, Montreal, Quebec, Canada*

Reijo Savolainen / *School of Information Sciences, University of Tampere, Tampere, Finland*

Barbara Schaefer / *Geneseo, New York, U.S.A.*

Silvia Schenkolewski-Kroll / *Department of Information Science, Bar-Ilan University, Ramat Gan, Israel*

Lael J. Schooler / *Center for Adaptive Behavior and Cognition, Max Planck Institute for Human Development, Berlin, Germany*

Joachim Schöpfel / *Department of Library and Information Sciences (IDIST), GERiico Laboratory Charles de Gaulle University Lille 3, Villeneuve d'Ascq, France*

Catherine F. Schryer / *Department of English Language and Literature, University of Waterloo, Waterloo, Ontario, Canada*

Marjorie Schwarzer / *Museum Studies Department, John F. Kennedy University, Berkeley, California, U.S.A.*

Jo Ann Secor / *Lee H. Skolnick Architecture + Design Partnership, New York, New York, U.S.A.*

Sara Selwood / *Department of Cultural Policy and Management, City University, London, U.K.*

Frank B. Sessa / *University of Pittsburgh, Pittsburgh, Pennsylvania, U.S.A.*

Mark Sgambettera / *Bronx County Historical Society, Bronx, New York, U.S.A.*

Ayman Shabana / International Institute, University of California, Los Angeles, Los Angeles, California, U.S.A.

Nigel Shadbolt / *School of Electronics and Computer Science, University of Southampton, Southampton, U.K.*

Kalpana Shankar / *School of Informatics, Indiana University, Bloomington, Indiana, U.S.A.*

Debora Shaw / *School of Library and Information Science, Indiana University, Bloomington, Indiana, U.S.A.*

Conrad Shayo / *Department of Information and Decision Sciences, California State University—San Bernardino, San Bernardino, California, U.S.A.*

Elizabeth Shepherd / *Department of Information Studies, University College London, London, U.K.*

Beverly K. Sheppard / *Institute for Learning Innovation, Edgewater, Maryland, U.S.A.*

Ross Shimmon / *Faversham, U.K.*

Snunith Shoham / *Department of Information Science, Bar-Ilan University, Ramat Gan, Israel*

Lyudmila Shpilevaya / *New York Public Library, New York, New York, U.S.A.*

David Shumaker / *School of Library and Information Science, Catholic University of America, Washington, District of Columbia, U.S.A.*

Judith A. Siess / *Information Bridges International, Inc., Champaign, Illinois, U.S.A.*

John Edward Simmons / *Museologica, Bellefonte, Pennsylvania, U.S.A.*

Anestis Sitas / *Aristotle University of Thessaloniki, Thessaloniki, Greece*

Roswitha Skare / *Institute of Culture and Literature, UiT The Arctic University of Norway, Tromsø, Norway*

Katherine Skinner / *Educopia Institute, Atlanta, Georgia, U.S.A.*

Lee H. Skolnick / *Lee H. Skolnick Architecture + Design Partnership, New York, New York, U.S.A.*

Mette Skov / *Department of Communication and Psychology, Aalborg University, Aalborg, Denmark*

Bobby Smiley / *Vanderbilt University, Heard Libraries, Nashville, Tennessee, U.S.A.*

Linda C. Smith / *School of Information Sciences, University of Illinois at Urbana-Champaign, Champaign, Illinois, U.S.A.*

Lois Smith / *Human Factors and Ergonomics Society, Santa Monica, California, U.S.A.*

Lori Smith / *Linus A. Sims Memorial Library, Southeastern Louisiana University, Hammond, Louisiana, U.S.A.*

Patricia A. Smith / *Colorado State University, Fort Collins, Colorado, U.S.A.*

Scott A. Smith / *Langlois Public Library, Langlois, Oregon, U.S.A.*

A. Patricia Smith-Hunt / *Science Library, Preservation Services, University of California, Riverside, Riverside, California, U.S.A.*

Karen Smith-Yoshimura / *Online Computer Library Center (OCLC), San Mateo, California, U.S.A.*

Diane H. Sonnenwald / *University College Dublin, Dublin, Ireland*

Nour Soufi / *Library Cataloging and Metadata Center, University of California, Los Angeles, Los Angeles, California, U.S.A.*

Barbara M. Spiegelman / *Churchill Associates, Pittsburgh, Pennsylvania, U.S.A.*

Robert P. Spindler / *Department of Archives and Manuscripts, Arizona State University, Tempe, Arizona, U.S.A.*

Joie Springer / *Information Society Division, UNESCO, Paris, France*

Suresh Srinivasan / *National Library of Medicine, Bethesda, Maryland, U.S.A.*

Guy St. Clair / *Knowledge Management and Learning, SMR International, New York, New York, U.S.A.*

Cheryl L. Stadel-Bevans / *National Archives and Records Administration, College Park, Maryland, U.S.A.*

Jill Stein / *Institute for Learning Innovation, Edgewater, Maryland, U.S.A.*

Marcia K. Stein / *Museum of Fine Arts, Houston, Houston, Texas, U.S.A.*

Jela Steinerová / *Department of Library and Information Science, Comenius University in, Bratislava, Slovak Republic*

Dick Stenmark / *Department of Applied IT, IT University of Gothenburg, Gothenburg, Sweden*

Andy Stephens / *OBE, Board Secretary, Head of International Engagement, The British Library, London, U.K.*

Margaret Stieg Dalton / *School of Library and Information Studies, University of Alabama, Tuscaloosa, Alabama, U.S.A.*

Katina Strauch / *Addlestone Library, College of Charleston, Charleston, South Carolina, U.S.A.*

Robert D. Stueart / *Graduate School of Library and Information Science, Simmons College, Boston, Massachusetts, U.S.A.*

Paul F. Stuehrenberg / *Yale Divinity Library, New Haven, Connecticut, U.S.A.*

Brian William Sturm / *School of Information and Library Sciences, University of North Carolina at Chapel Hill, Chapel Hill, North Carolina, U.S.A.*

Anna Suorsa / *University of Oulu, Oulu, Finland*

Brett Sutton / *Aurora University, Aurora, Illinois, U.S.A.*

Sarah Sutton / *Mary and Jeff Bell Library, Texas A&M University-Corpus Christi, Corpus Christi, Texas, U.S.A.*

Destinee Kae Swanson / *Adams Museum & House, Inc., Deadwood, South Dakota, U.S.A.*

H.L. Swanson / *GSOE, University of California, Riverside, California, U.S.A.*

Miriam E. Sweeney / *School of Library and Information Studies, University of Alabama, Tuscaloosa, Alabama, U.S.A.*

Shelley Sweeney / *University of Manitoba, Winnipeg, Manitoba, Canada*

Jean Tague-Sutcliffe / *Graduate School of Library and Information Science, University of Western Ontario, London, Ontario, Canada*

Masaya Takayama / *National Archives of Japan, Tokyo, Japan*

Sanna Talja / *Department of Information Studies and Interactive Media, University of Tampere, Tampere, Finland*

G. Thomas Tanselle / *Vice President, John Simon Guggenheim Memorial Foundation, New York, New York, U.S.A.*

Ivan Tanzer / *Museum Studies Program, Faculty of Information Studies, University of Toronto, Toronto, Ontario, Canada*

Melissa Terras / *UCL Department of Information Studies, UCL Centre for Digital Humanities, University College London, London, U.K.*

Mike Thelwall / *School of Computing and Information Technology, University of Wolverhampton, Wolverhampton, U.K.*

Lynne M. Thomas / *Rare Books and Special Collections, Northern Illinois University, DeKalb, Illinois, U.S.A.*

Lawrence S. Thompson / *University of Kentucky, Lexington, Kentucky, U.S.A.*

Jens Thorhauge / *Danish Agency for Libraries and Media, Copenhagen, Denmark*

Anne Thurston / *International Records Management Trust, London, U.K.*

Michael Tiemann / *Open Source Initiative, Chapel Hill, North Carolina, U.S.A.*

Christinger Tomer / *School of Information Sciences, University of Pittsburgh, Pittsburgh, Pennsylvania, U.S.A.*

Elaine G. Toms / *Faculty of Management, Dalhousie University, Halifax, Nova Scotia, Canada*

Jack Toolin / *Whitney Museum of American Art, New York, U.S.A.*

Jennifer Trant / *Archives & Museum Informatics, Toronto, Ontario, Canada*

Barry Trott / *Williamsburg Regional Library, Williamsburg, Virginia, U.S.A.*

Alice Trussell / *Hale Library, Kansas State University, Manhattan, Kansas, U.S.A.*

John Mark Tucker / *Abilene Christian University, Abilene, Texas, U.S.A.*

James M. Turner / *School of Library and Information Sciences, University of Montreal, Montreal, Quebec, Canada*

Louise Tythacott / *Centre for Museology, University of Manchester, Manchester, U.K.*

George Tzanetakis / *Department of Computer Science, University of Victoria, Victoria, British Columbia, Canada*

Franklyn Herbert Upward / *Centre for Organisational and Social Informatics, Monash University, Melbourne, Victoria, Australia*

Richard Urban / *Graduate School of Library and Information Science, University of Illinois, Champaign, Illinois, U.S.A.*

Rachel E. Vacek / *University of Michigan, Ann Arbor, Michigan, U.S.A.*

Ron Van den Branden / *Centre for Scholarly Editing and Document Studies, Royal Academy of Dutch Language and Literature, Gent, Belgium*

Sydney C. Van Nort / *The City College of New York, The City University of New York, New York, U.S.A.*

Edward Vanhoutte / *Centre for Scholarly Editing and Document Studies, Royal Academy of Dutch Language and Literature, Gent, Belgium*

Rebecca Vargha / *Information and Library Science Library, University of North Carolina at Chapel Hill, Chapel Hill, North Carolina, U.S.A.*

Jana Varlejs / *School of Communication, Information and Library Studies, Rutgers University, New Brunswick, New Jersey, U.S.A.*

Jason Vaughan / *Library Technologies, University of Nevada, Las Vegas University Libraries, Las Vegas, Nevada, U.S.A.*

Dale J. Vidmar / *Southern Oregon University Library, Ashland, Oregon, U.S.A.*

Diane Vizine-Goetz / *OCLC Online Computer Library Center, Inc., Dublin, Ohio, U.S.A.*

Ellen M. Voorhees / *Information Technology Laboratory, National Institute of Standards and Technology, Gaithersburg, Maryland, U.S.A.*

Sharon L. Walbridge / *Libraries Washington State University, Pullman, Washington, U.S.A.*

Stephanie Walker / *Brooklyn College, City University of New York, Brooklyn, New York, U.S.A.*

Virginia A. Walter / *Department of Information Studies, University of California, Los Angeles, Los Angeles, California, U.S.A.*

Mark Warschauer / *School of Education, University of California, Irvine, CA, U.S.A.*

Nigel M. Waters / *Department of Geography and Geoinformation Science, George Mason University, Fairfax, Virginia, U.S.A.*

Kathryn M. Wayne / *Art History/Classics Library, University of California, Berkeley, California, U.S.A.*

Frank Webster / *City University, London, U.K.*

Jeff Weddle / *School of Library and Information Studies, University of Alabama, Tuscaloosa, Alabama, U.S.A.*

Judith Weedman / *School of Library and Information Science, San Jose State University, Fullerton, California, U.S.A.*

Stuart L. Weibel / *Office of Research and Special Projects, OCLC Research, Dublin, Ohio, U.S.A.*

Jennifer Weil Arns / *School of Library and Information Science, University of South Carolina, Columbia, South Carolina, U.S.A.*

Bella Hass Weinberg / *Division of Library and Information Science, St. John's University, Queens, New York, New York, U.S.A.*

Volker M. Welter / *Department of the History of Art and Architecture, University of California, Santa Barbara, Santa Barbara, California, U.S.A.*

Caryn Wesner-Early / *ASRC Aerospace & Defense, US Patent & Trademark Office, Alexandria, Virginia, U.S.A.*

Lynn Westbrook / *School of Information, University of Texas at Austin, Austin, Texas, U.S.A.*

Howard D. White / *College of Computing and Informatics, Drexel University, Philadelphia, PA, U.S.A., and College of Information Science and Technology, Drexel University, Philadelphia, Pennsylvania, U.S.A.*

Layna White / *San Francisco Museum of Modern Art, San Francisco, California, U.S.A.*

Michael J. White / *Engineering and Science Library, Queen's University, Kingston, Ontario, Canada*

Sarah K. Wiant / *School of Law, Washington and Lee University, Lexington, Virginia, U.S.A.*

Stephen E. Wiberley, Jr. / *University of Illinois at Chicago, Chicago, Illinois, U.S.A.*

Gunilla Widén-Wulff / *Information Studies, Åbo Akademi University, Åbo, Finland*

Bradley J. Wiles / *Hill Memorial Library, Louisiana State University, Baton Rouge, Louisiana, U.S.A.*

Mary I. Wilke / *Center for Research Libraries, Chicago, Illinois, U.S.A.*

Barratt Wilkins / *Retired State Librarian of Florida, Tallahassee, Florida, U.S.A.*

Peter Willett / *Department of Information Studies, University of Sheffield, Sheffield, U.K.*

Kate Williams / *University of Illinois at Urbana-Champaign, Champaign, Illinois, U.S.A.*

Kirsty Williamson / *Caulfield School of IT, Monash University, Caulfield, Victoria, Australia and School of Information Studies, Charles Sturt University, Wagga Wagga, New South Wales, Australia*

Concepción S. Wilson / *School of Information Systems, Technology and Management, University of New South Wales, Sydney, New South Wales, Australia*

Ian E. Wilson / *Librarian and Archivist of Canada 2004–2009, Ottawa, Ontario, Canada*

Kristen Wilson / *North Carolina State University Libraries, Raleigh, North Carolina, U.S.A.*

Thomas D. Wilson / *Publisher/Editor in Chief, Information Research, U.K.*

Catherine C. Wilt / *PALINET, Philadelphia, Pennsylvania, U.S.A.*

Charles Wilt / *Association for Library Collections and Technical Services (ALCTS), Chicago, Illinois, U.S.A.*

Niels Windfeld Lund / *Institute of Culture and Literature, UiT The Arctic University of Norway, Troms , Norway*

Michael F. Winter / *Shields Library, University of California, Davis, California, U.S.A.*

Erica Wiseman / *Graduate School of Library and Information Studies, McGill University, Montreal, Quebec, Canada*

Steve W. Witt / *University of Illinois at Urbana-Champaign, Champaign, Illinois, U.S.A.*

Blanche Woolls / *iSchool, San Jose State University, San Jose, California, U.S.A.*

Louisa Worthington / *Public Library Association, Chicago, Illinois, U.S.A.*

Jadwiga Woźniak-Kasperek / *Institute of Information and Book Studies, University of Warsaw, Warsaw, Poland*

Judith Wusteman / *School of Information and Communication Studies, University College Dublin, Dublin, Ireland*

Iris Xie / *School of Information Studies, University of Wisconsin–Milwaukee, Milwaukee, Wisconsin, U.S.A.*

Yiyu Yao / *Department of Computer Science, University of Regina, Regina, Saskatchewan, Canada, and International WIC Institute, Beijing University of Technology, Beijing, China*

Janis L. Young / *Library of Congress, Washington, District of Columbia, U.S.A.*

Priscilla C. Yu / *University Library, University of Illinois at Urbana-Champaign, Urbana, Illinois, U.S.A.*

Jana Zabinski / *American National Standards Institute, New York, New York, U.S.A.*

Lisl Zach / *iSchool, Drexel University, Philadelphia, Pennsylvania, U.S.A.*

Olga Zaitseva / *Kazakhstan Institute of Management, Economics and Strategic Research (KIMEP), Almaty, Kazakhstan*

Marcia Lei Zeng / *School of Library and Information Science, Kent State University, Kent, Ohio, U.S.A.*

Yi Zeng / *International WIC Institute, Beijing University of Technology, Beijing, China*

Višnja Zgaga / *Museum Documentation Center, Zagreb, Croatia*

Jun Zhang / *Pitney Bowes, Shelton, Connecticut, U.S.A.*

Yulei Zhang / *Department of Management Information Systems, University of Arizona, Tucson, Arizona, U.S.A.*

Kai Zheng / *Department of Health Management and Policy, University of Michigan, Ann Arbor, Michigan, U.S.A.*

Ning Zhong / *Department of Life Science and Informatics, Maebashi Institute of Technology, Maebashi-City, Japan, and International WIC Institute, Beijing University of Technology, Beijing, China*

Maja Žumer / *University of Ljubljana, Slovenia*

Vladimir Zwass / *Computer Science and Management Information Systems, Fairleigh Dickinson University, Teaneck, New Jersey, U.S.A.*

Encyclopedia of Library and Information Sciences, Fourth Edition

Contents

Volume I

Volume I (*cont'd.*)

Volume I (*cont'd.*)

Volume II

Volume II (*cont'd.*)

Volume III

Volume III (*cont'd.*)

Volume III (*cont'd.*)

Volume IV

Volume IV (*cont'd.*)

Volume V

Volume V (*cont'd.*)

Volume VI

Volume VI (*cont'd.*)

Volume VI (*cont'd.*)

Volume VII

Volume VII (*cont'd.*)

Introduction to the Encyclopedia of Library and Information Sciences, Fourth Edition

How to Use This Encyclopedia

Entries are arranged alphabetically in this encyclopedia (see end papers for alphabetical list). The editors of this edition (ELIS-4) have decided to forego the Topical Table of Contents that was provided in ELIS-3 by editors Marcia Bates and Mary Niles Maack. At the time of publication of ELIS-3, the Topical TOC was crucial for readers to get a sense of how subjects were grouped and an understanding of the field or subfield through the clustering of categorical entries in the print edition. ELIS-4 is envisioned as a primarily online reference work where a Topical TOC does not serve the same purpose. The print edition is served well by the main TOC as well as the detailed index, while entries in the online version are easily discoverable through title, author, keyword, and full text searches.

In sum, relevant entries can be found by

1. Entry title (alphabetical arrangement of entries in the encyclopedia or listing in the end papers)
2. Specific name or keyword, including the index at the end of each volume

If the first name or keyword searched is not found, try several more variations—either different words or a different order of words. Most topics are described in several ways in the literature of a discipline, and the first term or phrase that comes to mind may not be the one used here.

Scope of the Encyclopedia

The title of the third edition, *Encyclopedia of Library and Information Sciences*, ended with the letter "s" because the encyclopedia was broadened to cover a spectrum of related and newly emerging information disciplines, including archival science, document theory, informatics, and records management, among others. The fourth edition continues this trend but with an extensive focus on the aspects of library and information sciences that have been heavily impacted by the adoption and reliance on online information distribution. This focus is reflected in the inclusion of numerous new entries such as digital preservation, altmetrics, web-scale discovery services, demand-driven acquisitions, and global open knowledgebases. Alongside these entries based on entirely new topics, the expanded use of the Internet for information has led to new treatment of traditional LIS topics such as resource description and access (RDA) that reflects the adoption of new standards for cataloging.

ELIS-4 also seeks to build upon the description of professional practice to round out the theoretical perspective that previous editions covered very well. Both current editors are academic research librarians and thus, focused heavily on addressing gaps in the encyclopedia related to academic research information while still relying heavily on the structure established by editors of ELIS-3. For example, ELIS-3 introduced country profiles and ELIS-4 builds upon that with new entries for New Zealand and a third on Brazil, in addition to revisions for Slovakia, Netherlands, Canada, Belarus, Kazakhstan, and Brazil among others. This edition also expands the number of entries for named cultural and information entities that did not appear in previous editions, such as the National Library of Medicine, North American Serials Interest Group (NASIG), the International Association of Scientific, Technical and Medical Publishers (STM), and ASLIB, as well as entities like the HathiTrust that have been established since the last edition was published. A number of new entries describing important information conferences such as the Acquisitions Institute at Timberline, the Charleston Conference, and Electronic Resources in Libraries (ER&L) also help round out the encyclopedia and further the description of the current state of academic research librarianship.

ELIS-4 also continues the tradition of designating important entries of historical or theoretical importance as "ELIS Classics." These are entries by major figures in the library and information sciences or those that describe core concepts in LIS theory, practice, or education that appeared in earlier editions of the encyclopedia. The current editors preserved the approximately 40 previous "ELIS Classics" and designated 13 previous entries as new "ELIS Classics."

There are more than 550 entries, of which more than 20 are new, another 93 are revisions to prior entries that have been brought up to date by their authors or by new authors, about 30 are ELIS Classics, and about 400 are reprinted from an earlier edition since they have remained relevant to the present. It is important to note that the editors also had to make some choices related to retiring entries that were no longer relevant—due to the passage of time and the development of the field, the technologies and theories described in those entries were deemed to be out of scope for the new edition and thus not revised or reprinted.

Encyclopedia Authors

As in past editions, the authors writing for the encyclopedia are major researchers, librarians and practitioners, and leaders in the fields and subfields in the disciplines in which they are writing. Noted scholars are well represented, and a number of authors are former leaders in LIS associations, including the American Library Association (ALA), the Association for College and Research Libraries (ACRL), the International Federation of Library Associations and Institutions (IFLA), the American Society for Information Science and Technology (ASIS&T), and the American Association of Library and Information Science Education (ALISE). In addition, there are many contributors who are current or former directors of major institutions. As in past editions, the editors are very proud of the range and diversity of authors who have written these entries for the encyclopedia and we thank them for sharing their expertise with the current and future readers and researchers in the field.

Finally, the editors for ELIS-4 have grappled with the challenges of entry generation that was noted by previous editors in nearly every edition: that not all ideas, topics, and potential entries were able to be completed for publication in this edition. While we made a valiant attempt to include entries identified by ELIS-3 editors but not secured for publication in that edition, we sometimes could not find authors willing to take those topics on. Similarly, we were sometimes unable to secure revisions to entries from new authors when previous authors were unable to perform that task. To the greatest extent possible, we endeavored to replace authors when entries were deemed important enough to appear in ELIS-4 but initial or previous authors had to decline or defaulted. No doubt, the editors of ELIS-5 will also pick up the mantle and attempt to round out the encyclopedia with entries for anything that ELIS-4 missed. As noted by editors Bates and Niles Maack in ELIS-3, this problem of missing topics was also acknowledged by Allen Kent, editor of the first edition of ELIS. Kent stated in 1973, "I have prepared this presentation to make sure the lessons of Diderot-d'Alembert are recalled in terms of encyclopedia-making as an exercise in the art of the possible."

Background and Development of the Encyclopedia

The first edition of ELIS, under the editorship principally of Allen Kent and Harold Lancour, was published between 1968 and 1982. The 33 volumes of the first edition were published in alphabetical sequence during those years. After the "Z" volume appeared in 1982, a number of supplements were published at roughly the rate of two per year, up to and including volume 73, which appeared in 2003. Miriam Drake was appointed editor for the second edition, which appeared in 2003, both online and in paper. The second edition came out at one time in four large-format volumes, with a supplement in 2005 [3]. Kent and Lancour covered a wide range of librarianship, information science, and some computer science topics. Drake, an academic library director, emphasized academic libraries, and the ELIS-2 volumes contained many profiles of major academic libraries and professional library associations.

The third edition, under the editorship of Marcia Bates and Mary Niles Maack, reflected a growing convergence among the several disciplines that concern themselves with information and the cultural record. As information science educators and noted researchers in the field, their focus was on growing the encyclopedia in the theoretical fields of information sciences as well as drawing together the associated information and cultural disciplines such as archival sciences and museum studies within the overall field of LIS.

For this edition, we have focused on developing the encyclopedia to reflect the changing nature of information production and consumption through online and digital forms. We have also endeavored to fill in gaps in the description of important people, places, and theories in the information sciences, and further enhanced the description of important concepts related to the provision of research information and the field's major institutions.

We continue to see the audience for the encyclopedia just as previous editors have: as principally consisting of 1) the educated lay person interested in one or more of its topics, 2) students learning about a topic, and 3) professionals and researchers in the several fields who want to learn about something new, or to be refreshed on a familiar topic.

We honored the previous editors by reengaging their superb Editorial Advisory Board with significant new additions of experts known to the current editors. (See listing in the front matter.) These leaders and experts from as many disciplines as are in the encyclopedia provided excellent guidance and feedback for the editors as they began the process of new topic generation, evaluation of previous entries, and offering to author or review numerous entries throughout the process of publication.

All new and revised entries were reviewed by one or more outside expert reviewer as well as one or more of the editors. Referees provided invaluable feedback to authors, including noting errors or omissions as well as making suggestions on additional aspects of the topic to cover. While we made every reasonable attempt through this process to check the accuracy of every entry and every fact, undoubtedly readers will find some topics explained more thoroughly or accurately than others. Indeed, due to the time frame from the beginning of the generation of the fourth edition and the time of publication, readers will reasonably note that some topics have been quickly superseded due to this passage of time, so the date of acceptance of the entry will be noted on each entry since several years may have passed since the writing of the entry and the publication of this edition.

Acknowledgments

This edition of the encyclopedia was possible only through the countless hours that the editors, John McDonald and Michael Levine-Clark, spent reviewing the previous encyclopedia entries, outlining the topics that were missing or that were newly emerging in the field, and identifying appropriate expert authors to write those new entries. In addition, the editors devoted extensive time to corresponding with previous authors encouraging them to revise their entries, and finding replacement authors for important entries that needed revisions but whose original authors were unavailable.

Both editors wish to acknowledge the expertise of each other and their knowledge of our field, their extensive network of contacts, and their ability to work closely together to ensure the success of this encyclopedia. Neither of them could have completed this project alone.

They acknowledge and thank the Taylor & Francis Group editors, Claire Miller and Rich O'Hanley, as well as Susan Lee, who passed away at the early stages of the preparation of this edition, and more recently, Alexandra Torres, who supported and kept the editors and authors on track over the course of the years of work on this edition of the encyclopedia.

The editors thank the authors who wrote and revised entries, and the huge number of reviewers who refereed the entries. Without their dedication, expertise, and willingness to share their knowledge with others, there would be no encyclopedia. They also wish to thank the Editorial Advisory Board for their advice, suggestions of topics and authors, their hours spent writing or reviewing for the final edition. They also wish to thank the previous editors, Marcia Bates and Mary Niles Maack, whose organization and structure for ELIS-3 provided an excellent blueprint for ELIS-4.

Encyclopedia of Library and Information Sciences, Fourth Edition

Volume 1

Pages 1–722

Academic–American

Approval–Archival

Archives–Art

Artificial–Association

Artificial–Back

Bibliographic–Bibliothèque

Binding–British

British–Careers

Academic Libraries

Susan Curzon
Jennie Quiñónez-Skinner
University Library, California State University–Northridge, Northridge, California, U.S.A.

Abstract
This entry provides an introduction to academic libraries throughout the world including an overview of the mission, history, governance, external influences, collections, services, organizational structure, personnel, administration, and facilities of academic libraries. The issues and future of academic libraries are also discussed.

INTRODUCTION

Academic libraries are libraries that belong to institutions of higher education, including publicly funded, federal, state, provincial, or national universities or colleges, privately funded universities or colleges, 2-year community or junior colleges that can be publicly or privately funded, tribal colleges, professional schools, and special focus institutions that offer a single or small set of programs. These institutions grant a range of degrees from the associate of arts to the baccalaureate and from the master's degree to the doctorate. This entry will provide an overview of academic libraries, including such topics as their mission, history, governance, external influences, collections, services, organizational structure, personnel, administration, and facilities.

The mission of academic libraries is to support the educational and research activities of their parent institution through the provision of collections, services, and user education. Providing support for teaching, learning, and research is the focus of academic libraries. Academic libraries are vital to the success of the academic enterprise of their institutions whose research and teaching agenda are greatly enriched by libraries whose collections are broad, rich, and diverse.

Academic libraries are found in almost all post-secondary educational environments. While there is no absolute count of the number of higher education institutions, there are at least 18,000 institutions of higher education throughout the world today as indexed by the *International Handbook of Universities*.[1] There is no complete and final count of the number of academic libraries internationally. However, OCLC has a helpful spreadsheet revealing many academic libraries country by country.[2] In the United States, there are 3700 academic libraries today according to the National Center for Education Statistics.[3]

While the core mission of academic libraries remains consistent throughout institutions of higher education, academic libraries vary considerably from each other in size, in resources, in collections, in services, in complexity, and in numbers of library facilities serving the institution. The variation is largely a product of the mission, the history, the size, and the funding of its parent institution. Looking at some examples of variations around the globe, there is the Harvard University Library founded in 1638[4] to serve this institution's extensive research agenda and internationally recognized programs. It is the largest academic library in the United States by volumes held[5] and has over 70 separate library units.[6] Another example of variation is the University of Botswana formally established in 1982.[7] Its library has a main and three branch libraries—two of which are located in other cities.[8] Yet another example of variation is the University of the South Pacific, which has libraries located throughout its 12 member countries (Cook Islands, Fiji, Kiribati, Marshall Islands, Nauru, Niue, Samoa, Solomon Islands, Tokelau, Tonga, Tuvalu, and Vanuatu).[9] This is in accordance with the University's vision of being "an exemplar of tertiary education in the Pacific region..."[10]

HISTORICAL CONTEXT

From the early card catalogs to Online Public Access Catalogs (OPACs), from print indexes to periodicals via electronic article databases hosted online, the organization and access of academic library materials has changed dramatically over the past century. For hundreds of years, the predecessors of today's academic libraries relied on inventory lists, similar to contemporary archival finding aides, to keep track of books, scrolls, and tablets.[11] During the nineteenth century, the rotary printing press, public funding for colleges and universities, and the emergence of the profession of librarianship laid the groundwork for the meeting of technology and user-centered services, which changed libraries in the twentieth century from a book focus to a user focus. As two academic librarians described in a popular online blog, the role of academic

Encyclopedia of Library and Information Sciences, Fourth Edition DOI: 10.1081/E-ELIS4-120053333

libraries will be to meet the "Three Cs," which are creation, collaboration, and contemplation."[12] New collaborative study spaces, creative media studios, or the silent mediation created in a reading room are continuing the rich history of academic library service to patrons as their needs change over throughout the years.

During World War II and throughout the Cold War, government funding for natural and social science research expanded to include area studies, such as Latin American studies. These new disciplines were integral in the development of new collections and the creation of new resources.[13] The children of the World War II generation, the baby boomers, entered college during a time of great social and political change. Just as the student population increased, so did the collections in academic libraries, which now grew to 100,702,000 volumes.[14] The Civil and Women's Rights movements diversified the campus population and challenged the traditional research cannons. For libraries, as the population changed, so did library collections change with new subject areas, and so did new issues emerge in user services and education. With a large number of first-generation college students, the role of the academic library was not just to preserve a collection, but to continue a larger discussion within the profession, lasting into the twenty-first century about how to teach users to search independently and more effectively with a wealth of information now available in the library and online. For academic libraries, two great shifts in technology occurred during the twentieth century. The first shift occurred with the development of Machine-Readable Catalog (MARC) standards. In conjunction with MARC, the Anglo-American Cataloguing Rules provided the format for descriptive cataloging while the Library of Congress Subject Headings were used to describe the subject matter of a record.[15] Automation of library records allowed computers to read information about a book using data elements, such as code. The first OPAC was launched in 1970 at Ohio State University and facilitated the sharing of library information between institutions, and in 1981 the Ohio State project would be known as the Online Computer Library Center (OCLC).[16]

The second great shift, the 1990s, was the age of the Internet. For centuries, libraries were the primary source for information and materials, but the Internet broadened the meaning of library science and the potential for sharing information. Now, academic libraries were competing with information sources not just created by scholars, but sources created by everyday people who were challenging the library and faculty as the gatekeepers of knowledge. Libraries have responded by opening their collections to the public and increasing the sharing of resources to other libraries. Many archival collections have a portion of their materials, including photographs and finding aids, available online. Similarly, the OPACs are accessible through library homepages giving users throughout the world the ability to search the holdings at any particular institution,

at any time. More than a quarter century after OCLC was formed, WorldCat launched a public tool for searching over 2 billion items in over 10,000 libraries worldwide.[17] In the years ahead, libraries in higher education face many hurdles: funding cutbacks, increased costs for electronic serials and books, and a changing student population with user-education needs. The demand for 24/7 access, cloud storage, and changes in publishing for books and distribution for films has impacted both the institution and library. The field itself is in a state of flux and self-discovery as academic libraries rethink their place in the larger world of information. Much of the strength of library resources lies in their print collection and knowledge of staff. Finding ways to package and promote services to a new generation of students and faculty will require opening the field to opportunities for increasing the sharing of resources, for cooperatively creating online tutorials and instructional components for the growing online classes. Our success will depend on the ability of the librarians and administrators to adapt to social, economic, and political changes that have impacted academic libraries throughout the twentieth century.

EXTERNAL INFLUENCES

Academic libraries, as a key component of the academic enterprise, are influenced by the same forces influencing education today. While these forces vary from country by country, such important issues of student enrollment and retention, student preparation for college, time to graduation, competitive salaries for campus personnel, trends in academic disciplines, distance instruction, changes in educational legislation, assessment, the rising cost of tuition, technology-based instruction, and adequate funding are universal and impact the library. The library must assist with solving the institution's challenges, exist within the confines of its budget, and react to its changing needs. Other external influences related to accreditation, legislation, intellectual property, and the actions of organizations and associations can also impact the library very directly.

Accreditation is an important external influence in some countries. Within the United States, for example, there are regional commissions that accredit the universities because "in a quality assurance environment, accreditation is the most visible 'report card' regarding the quality of institutional effectiveness and performance."[18] Academic libraries are often evaluated as part of the accreditation process on the range of services, the depth of collections in a variety of subject fields, the outreach to the academic departments, the integration of the library into the academic enterprise, and the information literacy focus. The core question for accrediting agencies is, can this library support the stated educational and research mission of the institution?

To assist with this process, the Association of College and Research Libraries (ACRL) in the United States "promulgates standards and guidelines to help libraries, academic institutions, and accrediting agencies understand the components of an excellent library."[19] An extensive list of standards, guidelines, and policies is provided,[20] including the *Standards for Libraries in Higher Education,* which are "designed to guide academic libraries in advancing and sustaining their role as partners in educating students, achieving their institutions' missions, and positioning libraries as leaders in assessment and continuous improvement on their campuses. Libraries must demonstrate their value and document their contributions to overall institutional effectiveness and be prepared to address changes in higher education. These *Standards* were developed through study and consideration of new and emerging issues and trends in libraries, higher education, and accrediting practices."[21]

Academic libraries can also be impacted and controlled by legislation or governmental regulations. For example, the American Library Association has raised concerns about provisions in the USA PATRIOT ACT, which may "endanger constitutional rights and privacy rights of library users."[22] Any legislation that impacts the user's borrowing and searching history is a sea change as user privacy has been protected historically in most academic libraries.

Another example of an external influence is intellectual property, which has grown ever more complex with emerging technologies. As the American Library Association states, "Copyright issues are among the most hotly contested issues in the legal and legislative world; billions of dollars are at stake. Legal principles and technological capabilities are constantly challenging each other and every outcome can directly affect the future of libraries."[23]

Of course, in the global economy, intellectual property issues are not simply an issue for one country alone. For example, countries that are members of the World Trade Organization know of the "Agreement on Trade Related Aspects of Intellectual Property Rights, Including Trade in Counterfeit Goods," which "sets out the obligations of member governments to provide procedures and remedies under their domestic law to ensure that intellectual property rights can be effectively enforced, by foreign right holders as well as by their own nationals."[24]

Another example of a global influence is the agreement between the European Union and the U.S. Department of Education with the European Union-United States Atlantis Program. Among other activities, "the Atlantis Program will fund collaborative efforts to develop programs of study leading to joint or dual undergraduate or graduate degrees."[25] Such cooperation is very desirable globally but may call upon academic libraries for increasing or new services.

PROFESSIONAL ASSOCIATIONS

Academic libraries are also participants in and influenced by professional associations. Such associations further the success of academic libraries by providing a forum for the exchange of ideas, by influencing public policy relating to libraries and information and higher education, by providing opportunities to network with colleagues, by creating opportunities for cooperation and dialogue, and by providing leadership on emerging issues.

Library associations for academic libraries are all across the world. For example, there is the Association of Caribbean University and Research Libraries located in Puerto Rico; the International Association of Technological University Libraries located in Ireland; the Canadian Association of Research Libraries/Association des bibliothèques de recherche du Canada located in Ottawa or the Association of Libraries of Czech Universities located in Prague.[26]

Academic libraries also participate in the International Federation of Library Associations and Institutions (IFLA). IFLA has an Academic and Research Libraries Section, which "provides members with the opportunity to exchange experiences and ideas with colleagues worldwide and is a vehicle for collaborative action."[27]

In the United States, the American Library Association has a division called the ACRL. ACRL is "dedicated to enhancing the ability of academic library and information professionals to serve the information needs of the higher education community and to improve learning, teaching, and research."[28]

Another example is the Association of Research Libraries (ARL) whose 125 member libraries are research-extensive universities in the United States and Canada. ARL is focused on "providing leadership in public and information policy to the scholarly and higher education communities, fostering the exchange of ideas and expertise, facilitating the emergence of new roles for research libraries, and shaping a future environment that leverages its interests with those of allied organizations."[29]

There are other associations and organizations that set standards, which have an impact on libraries. For example, the National Information Standards Organization (NISO) "is a nonprofit association accredited as a standards developer by the American National Standards Institute, the national clearinghouse for voluntary standards development in the United States."[30] NISO is "where content publishers, libraries, and software developers turn for information industry standards that allow them to work together. Through NISO, all of these communities are able to collaborate on mutually accepted standards—solutions that enhance their operations today and form a foundation for the future."[31]

GOVERNANCE AND HIERARCHY

The governance of academic libraries is subject to the governance structure of the colleges or universities of which they are a part. This governing structure establishes the budget as well as the policies and the regulations with which the library must also comply.

The governance of institutions does vary but usually the institution reports to a lay governing board, often called a Board of Trustees, whose members are either elected or appointed from such groups as alumni, faculty, community members, major donors, or political or religious affiliates. Occasionally, there is a nonvoting student representative. A lay governing board governs within the charter and by-laws of the institution that lays out their rights and responsibilities and that is legally binding. One example of a lay governing board is from the University of Queensland in Australia, which is "governed by a 22-member senate representing University and community interests."[32] Another example comes from the University of British Columbia in Canada, which has a 21-person board of governors including 11 persons appointed by the lieutenant-governor of the province.[33]

The head of the institution, frequently called a president or chancellor, reports directly to the lay governing board. In turn, the chief academic officer often called a provost, a vice chancellor, or vice president, reports to the president or chancellor.

The head of the library, who often uses such titles as university librarian (UL), dean, director, and occasionally vice-president or vice-provost, often reports to the chief academic officer because libraries are part of the academic enterprise or the activities of teaching, learning, and research, all of which are the domain of the chief academic officer.

Additionally, the library might also have faculty advisors appointed or elected by the academic senate of the institution. The academic senate, which not all institutions of higher education have, comprises faculty members elected by their peers to participate in the shared governance of the institution in matters pertaining to teaching, curriculum, research, and faculty personnel matters. These faculty senators work directly with the UL advising on such matters as policy, services, and budget as well as providing the UL with input on the changing needs of the faculty and students. These faculty advisors might also act as advocates for the library particularly in supporting the library's financial needs.

ORGANIZATIONAL STRUCTURE

The organizational structure, that is, the way that the service units within the library are organized hierarchically, varies substantially in academic libraries depending upon size, history, and need. The following is a general description of a common organizational structure with discussion on some variations. Generally speaking, many academic libraries are divided into Public Services and Technical Services, although there are many variations on the names. All services or units report into library administration.

Library administration usually comprises the Office of the University Librarian as well as management and staff personnel associated with planning, policymaking, budget, personnel, facilities, security, and fundraising. Often, the UL has one or more assistant or associate university librarians (AULs) responsible for various service units such as Public Services. Administration collectively is responsible for leadership and all its many dimensions, for the provision of resources and for decision-making. Administration also has a regulatory role to make sure that the library is in compliance with policies, regulations, and budgetary matters.

Public Services, which usually reports to an AUL or equivalent, is responsible for connecting the user to resources through the provision of direct services that make the library understandable, accessible, and easy to use. Public Services comprise service units such as reference or research services, user education such as information literacy or media literacy, circulation, interlibrary loan, media, government documents, maps, and other specialized services. Sometimes, outreach services to students is located here and sometimes in Administration. The reference services unit and the user education unit can often be separate services from Public Services.

Technical Services, which also reports to an AUL or equivalent, comprises the acquisition, the cataloging, and the processing of resources. It is the job of Technical Services to order, receive, and process resources in an expeditious fashion so that the resources are available to the user as soon as possible. Bindery and repair of resources are usually located in Technical Services. Many Technical Services units today have contract personnel because some of the traditional responsibilities of Technical Services, personnel are being outsourced such as the copy cataloging, bar coding, and labeling of books.

Collection development/bibliography, or the selection of resources, can be located in Technical Services or in Public Services, but it could also be a separate unit if the library is large enough. Personnel in this unit, usually librarians, are responsible for selecting all information resources, including print, electronic, and media. Librarians with collection development responsibility work closely with faculty in the academic departments so that the collection continually meets the changing curriculum and current research needs of the institution.

Library Systems, or the unit that is responsible for library technology, can exist as one unit managing the integrated library system, the web development, other online systems, and the computers and computer labs. Management personnel, such as a systems administrator or director of technology, usually are responsible for Library Systems. However, these duties can also be split into different areas. For example, the personnel who support the integrated library system might report to Technical Services. The web developer might report to Administration. Personnel who manage the computers and computer labs might report to Administration or to Public Services.

Special collections of rare materials, archives, maps, oral histories, and other resources are usually a separate unit particularly in libraries of considerable size. Special collections are usually headed up by a curator or highly specialized librarian versed in both the subject matter and in the preservation and conservation of the materials. The curator may report to the AUL of Public Services or to the UL. Often, exhibits come under this unit as exhibits are an important way of making the academic community and beyond aware of these unique resources. Increasingly, digital archives or institutional repositories are embedded within special collections. However, these emerging fields are not yet stable in their operations, are usually interdepartmental by nature, and so could report anywhere in the library.

Large institutions of higher education usually have more than one library on the campus. In fact, large research universities can have dozens of libraries on campus. For example, Oxford University in England has over a 100 separate libraries serving 38 colleges and other units.[34]

Sometimes, libraries in these large institutions are united under one UL, but often not. For example, the heads of medical or law libraries often report directly to the dean of the medical or law school rather than the UL. The head of an academic departmental library, such as a business library, could report to the dean of the college of business rather than the UL.

Sometimes, academic departments launch their own small library without regard for the professional management of libraries, for an ongoing budget, for staffing, for access, or for continuing acquisitions. Many of these little libraries end up shuttered after interest wanes, and materials go missing. Occasionally, an academic department eventually will ask the academic library for help, which presents an opportunity to rescue resources, some of which may be important donations, and to strengthen a partnership with this academic department.

Institutions of higher education that have developed satellite campuses to reach students in remote geographic areas usually have satellite libraries on those campuses. The heads of the satellite libraries might also report to the UL.

Regardless of the reporting line, it is increasingly important for the libraries in one institution to cooperate through a single technical integrated library system, by access to shared resources, by a common portal, common policies and procedures, and by shared Technical Services support. This cooperative and integrated approach benefits greatly the students and faculty.

The important issue is not how the library is organized but how efficient and effective its structure is and how well its systems and processes are integrated for the benefit of the user.

USER PRIVILEGES

The users of academic libraries are primarily students and faculty of the parent institution. Academic libraries also extend library privileges to visiting scholars and other researchers. Publicly funded colleges and universities may provide some limited library privileges to members of the greater community. Members of the Friends of the Library, a fundraising and support group, alumni, other significant friends, and donors also usually receive some limited library privileges. Sometimes, there is reciprocity between nearby academic libraries to extend library privileges to each other's faculty and occasionally students.

Library privileges in academic libraries are not consistent across groups of users. Along a continuum of increasing privileges, faculty members usually receive the most privileges as befitting their pivotal role in a teaching and research enterprise. After faculty come graduate students in terms of range of privileges and then undergraduate students. University staffs also receive privileges, which can vary widely but is often similar to the privileges of undergraduate students. Employees of the library may sometimes have special personal use privileges.

The range of library privileges include the ability to checkout resources including media; to use interlibrary loan; to request document delivery; to access online resources; to access course reserves; to use the computers; to browse closed stack areas; to use private study carrels and group study rooms; and to access rare and valuable materials.

RESOURCES AND SERVICES

Each generation of library users has had different expectations for service and access to their library's collection, resources, and services. In response, academic libraries are continually adapting to their user's needs and lifestyles.

The purpose of library services is to connect the user to resources and support research, instruction, and lifelong learning. However, the development of services throughout the years is based on the idea that the library should be user-friendly, straightforward, and rapid in response. Enhanced services and expert staff continue to add value to the library. The academic library is also a "community center" with experts to assist students, faculty, and surrounding neighborhoods.[35]

Resources and Expertise

As information is more accessible online, experts in copyright, digital content management, and digital preservation are necessary for campuses with robust research and creative projects. Scholarly and instructional content created by faculty and staff must be organized and preserved for the success of scholarly communication. More libraries are restructuring based on areas of expertise: instructional design, digital preservation, web services, and copyright. Resources include the instruction, research consultations, digital preservation and archiving, web services and information technology, rare books and special collections, and finally the ongoing role of librarians in copyright and accessibility issues. Many libraries inform patrons about services and resources via strategic outreach and marketing plans that often include social media.

Many academic libraries have a designated outreach librarian who informs the campus about new services and resources. Outreach is an education service provided by the library to raise awareness, educate patrons by bringing library services or information about services directly to them. Typically, the outreach librarian is tasked with the important work of making the services and resources of the library visible to new students and faculty. New faculty orientation and freshman welcome week are two opportunities for publicizing new services and collections to the campus community. Some outreach librarians also work directly with primary and secondary education in their service areas. A local teacher might want to bring his/her class to campus to visit an exhibit in special collections, tour the facilities, or request an information literacy session for a class. With school libraries underfunded, the college or university library often partners to ensure students have access to materials and prepared to enter a higher education institution.[36] Most librarians have an outreach responsibility to their job description as a department or college liaison that is critical to disseminating information about services and resources to the campus.

Digital collections of a library are often materials from special collections and archives. It is important to note that digital collections are not entirely special collections and a great deal of work is happening to preserve the intellectual works of faculty. Staff with expertise in digital libraries, repositories, and preservation work directly with faculty on campus to make sure that copies of their scholarly publications, digital learning objects created for courses, or even creative projects are available online for future generations. This is a relatively new service provided by libraries and requires staff with technical expertise and an outreach plan to educate faculty about the benefits of contributing their works to a repository on their campus.

Copyright advisement is necessary on campuses with research or instruction foci. Staff at the library guide and advise faculty (and administrators) regarding fair use and equitable access of materials for instruction or research. Faculty seek advice when signing a contract for publication with a particular publisher and have questions about their contracts and what they are able to do with their work. Professors teaching an online class often need assistance locating materials in the collection that can be shared via a learning management system (LMS). For works in archives, copyright becomes increasingly important as more pressure to digitize collections results in difficult questions about who owns the rights to unique and rare materials donated to the library. As more types of information, media for example, are easily shared and streamed, this continues to be an important issue for libraries and campus administration.

An area of continued growth is instruction, instructional design, and information literacy services provided by librarians for undergraduate and graduate classes. Librarians at a college or university teach users how to find, access, evaluate, and understand what information sources are needed (e.g., articles, books, media, and websites). Information literacy programs at academic libraries consist of a combination of in-person instruction and online learning objects to enhance student learning. The development and mainstreaming of instruction programs share a common purpose—provide students with the training to help them complete their assignments and develop critical thinking skills for their life during and post college. The popularity of these programs increased with major information literacy initiatives started, for example, by the California State University librarians who brought a program of information literacy to 23 campuses.[37] In today's academic libraries, bibliographic/library instruction often takes place once a semester in a computer lab or online via a LMS, both types of instruction require collaboration with instruction faculty on campus and librarians to create meaningful information literacy learning outcomes for a class. Throughout the year, librarians teach a variety of sessions that cover a wide range of subject from general information literacy tasks to more specialized research skills.

Digital services and online instruction design positions at libraries provide faculty and students with resources such as customized tutorials, library guides for course projects, quizzes and assessment of learning outcomes, and new ways of accessing primary sources. Library staff and faculty collaborate to scaffold information literacy standards throughout the undergraduate curriculum. Often, undergraduate first-year experience classes and introductory writing classes have some type of information literacy requirement. The expertise of the staff and their ability to connect faculty and students to resources is a service to the campus, and while not contained by a traditional desk, their outreach to users in the digital space, LMS, is critical to the mission of the library.

Reference

The reference desk provides immediate one-on-one research assistance usually from a librarian. A reference interaction can be as simple as pointing out the circulation desk to a lost library visitor or can involve more complicated research questions. Most reference questions require extensive knowledge of the library's collection and a solid subject specialization in a particular discipline. The role of the reference desk as a service point is a topic of continued debate in library science and among administrators. Statistics from ARL show a decrease in reference transactions between the years 1991 and 2004, by 24%.[38] The statistics from ARL were further explored in a 2012, which indicated that more spending on electronic resources actually increased reference transactions.[39] More libraries are moving to data collection systems, such as LibAnswers, to capture the different types of questions, duration of time spent helping the patron, and additional rich data that describes the interaction between library (or staff) and patron (undergraduate, graduate, faculty) that could be helpful in evaluating and improving library services in the future. Reference services are also handled through chat, e-mail, and texting services. Patrons can also request one-on-one research consultations with a librarian for more time-extensive questions. While some libraries have moved away from the reference desk, most offer a hybrid of services and see the desk at a critical site for outreach to patrons. Many visitors at the reference desk are first time library users and might not have found an online reference or office appointments.

Public Service

The entire library can best be understood as a network of services and resources that work together through careful coordination of library staff (and faculty) to provide seamless access, organization, and context to the information used by faculty and students. The traditional public service sites include the circulation desk, interlibrary loan, course reserves, and specialized departments, such as music and media collections.

　Circulation: The circulation unit in the library is a key access service point. As library leadership looks to brand service points, changing the name of the circulation desk is happening as renovations and new outreach plans are created at academic libraries. At the circulation desk, users can checkout library materials, renew books, return materials, and pay fines; these functions are some of the more visible and traditional tasks associated with the circulation desk. Circulation personnel usually staff the library for all hours that the library is open. In addition to overseeing student employees, they shelve books, supervise group and individual study rooms, collect books from campus book drops, manage holds and fines, tag damaged items for repair, assist students with disabilities, troubleshoot printing and copying, oversee weeding/shaping of the print collections, and collect statistics when necessary. Staff in circulation serves as backup to other service desks, often managing problems or troublesome users and

responsible for contacting campus police or other emergency services as needed.

　Interlibrary loan: Interlibrary loan broadens the user's access to a collection, beyond the holdings of their own library. There are three items that are typically requested via interlibrary loan (depending on institution): books, periodical articles, and dissertations or theses. Many libraries will also lend media and microfiche. Articles are usually delivered to users as a PDF unless there are strict copyright rules or the publication is not available online. As more libraries purchase e-book subscriptions, the loan of books is changing. Most ILL requests for books and dissertations are delivered physically through the mail or intersystem delivery services (i.e., University of California).[40]

　Course reserves: As the cost of textbooks increases, students and faculty continue to look to the library for affordable options to support instruction on campus. Academic libraries manage course reserves where faculty can place books, media, and articles on a limited check out for their students. Some course reserve programs coordinate with instructors and order a copy of a textbook for a library. Others are working with faculty to create a collection of readings (journal articles and book chapters) that are available through electronic course reserves. Electronic reserves includes selection of readings that are scanned and only students associated with the course are able to access with their login and password. Staff that work in course reserves are familiar with copyright and fair use because so much of their day-to-day work involves balancing fair use with the rights of users to have access to a particular subject. One challenge for academic libraries are the growing number of online courses and streaming rights for film that differ in cost greatly from what an individual can purchase and the pricing for an institution.

　Systems and technology services: Most academic libraries provide computers for student use including laptop or tablet checkouts. Computer use is a common service associated with library systems, but their services extend beyond the hardware. Partnerships with campus information technology provide better quality services to users. One example is the integration of more cloud-based software by IT departments that are used by library patrons for research and projects. Systems and information technology departments within the academic library are also developing new services and technologies through a variety of initiatives to develop mobile sites, dynamic library websites, and Content Management Systems (LibGuides), blogs, wikis, indexed-based discovery systems, institutional repositories streaming servers, digital archives, and any other software running off the servers that enhances the experience of library users. With hundreds of databases and thousands of e-books, the staff with technology expertise are necessary in making sure the patron's experience searching for sources and using resources at the seamless. Patrons continue to expect the same type of online experience at the library as they do with commercial services, the library systems/technology staff work to leverage resources (hardware/software) to meet these expectations within reason.

Imaging services: More campuses are reducing the amount of paper and printing generated on their campus; as they look for more alternatives to print, there are some instances when students need to print a paper, unofficial transcripts, or documents for presentation. These photocopy, printing, and even faxing needs still exist. Most libraries are unable to subsidize image services and opt to outsource to a vendor rather than provide them directly. A partnership is formed and both the library and vendor share in the profits (if any) from printing.

TECHNICAL SERVICES

Technical Services is responsible for acquiring materials for the collection in support of the mission of the research and educational priorities. Traditional functions included in Technical Services operations are acquisitions, cataloging, preservation, and conservation. After the bibliographer selects materials for the collection, an acquisitions unit, within Technical Services, is in charge of orders, payments, tracking budgets, and processing received materials. Academic libraries purchase books and other media that are printed by many different publishers, but dealing with so many publishers individually would be a logistical headache for the library's staff. To simplify purchases, there are vendors, sometimes known as jobbers, who act as the intermediary between the academic library and publisher. Many libraries work with several different vendors as opposed to thousands of different publishers. Vendors are not the only source of materials, but they help consolidate the purchasing process for a large number of items within the collection. Once materials are received (physically or digitally), the item moves to cataloging. Cataloging is the process by which entries are made into the library's catalog of holdings. These entries include descriptive information about the book, subject headings, and call numbers. An academic library in the United States generally uses the Library of Congress Subject Headings to describe content of a book, periodical genre, media, and, sometimes, special collections; one item can have several different subject headings. A call number is generated from the 21 main classes found in the Library of Congress.

Other classification schemes include the Dewey Decimal System, not often used in academic libraries, and the U.S. Superintendent of Documents classification system for federal documents. Cataloging at an academic library can take two different formats: copy cataloging and original cataloging. While copy cataloging, or the use of existing cataloging from a bibliographic utility, has helped institutions standardize their metadata, the more specialized and rare materials require original cataloging by a subject specialist. Next, the library's staff prepares each item for the stacks by placing checkout slips, ownership stamps, classification labels, security strips, and protective covers as needed on the item. With more vendors offering shelf ready materials, many libraries opt to pay upfront for these activities. Most libraries have a split with some items coming shelf ready and some being processed in-house. Some items will take a different path through Technical Services. For example, print journal subscriptions will be checked in with close attention being paid to verifying that the journal is on schedule or initiating the claiming process to the publisher if not. Electronic resources are very complex and require staff that specializes in their management. Activities include establishing and checking links to databases or e-books, ensuring that materials are still accessible, troubleshooting, and the monitoring of license agreements. Conservation and preservation are two additional functions of Technical Services. For print collections, the definition of conservation is the "noninvasive physical or chemical methods employed to ensure the survival of manuscripts, books, and other documents" and preservation can be a "broad range of activities intended to prevent, retard, or stop deterioration of materials or to retain the intellectual content of materials no longer physically intact."[41] Both definitions describe conservation and preservation as it relates to print resource, but libraries are working together now to address some of the unique preservation issues involved in digital collections.[42] An academic library's efforts in caring for its resources can vary widely from a small in-house repair effort to a substantial unit with specialized staff handling valuable and rare materials. As more collections are available online, and this is especially true for serials, the function and role of Technical Services will continue to develop. Two issues of interest for the future are outsourcing and the development of digital collections. First, outsourcing for cataloging and processing is an increasing option for academic libraries. Reaction to outsourcing is mixed. On one side, sending materials offsite decreases some of the staffing and work at the local level reducing costs. On the other side, outsourcing may adversely affect the quality of local cataloging. The second issue for academic libraries and Technical Services is the continuing development in electronic resources and digital content. This introduces concerns such as the role of licensing agreements, assessing copyright for online journals and article databases, and the tracking and cataloging of these resources. One significant issue is that the idea of ownership is blurred with electronic resources. Once the subscription stops, the library has lost its investment unless back files have also been purchased. Librarians are now working for guarantees that the materials available today will be available and accessible for patrons in the years to come. This is referred to as perpetual access. This is problematic when there is no guarantee of the lifespan of a vendor or publishing company, and therefore, there are no guarantees that access will be the same. The role of Technical Services is the management of these resources and ensuring that both print and electronic resources are available to users.

ADMINISTRATION

Most academic libraries are multimillion dollar operations and as such require the same sound management principles that any business requires to succeed. Planning, leadership, policymaking, customer service, personnel management, assessment, marketing and outreach, communication, and budgeting are all vital components and must be managed with the same skill as in any other business.

Planning is vital for success in order to keep academic libraries viable with their academic community particularly by keeping up with the dynamic changes within the information and technological environment. Most academic libraries have a strategic plan to point the way forward. These plans are usually renewed every 5 years and are linked to the university's broader strategic plan. In one example, the University of Tasmania clearly states that "the UTAS strategic plan, Open to Talent, frames the context for the Library's plan."[43]

Customer service is a vital component of academic libraries. A focus on the needs of the students and faculty is essential as the library is an important player in the teaching, learning, and research process. Students and faculty rely upon the library to provide them with essential knowledge resources. Some libraries, in order to understand user needs, have participated in LibQual, which is offered by the ARL and which is used to "solicit, track, understand, and act upon users' opinions of service quality."[44] These and other assessment methods have helped libraries to continually understand their users and to make needed changes in resources and services as user needs alter.

Another example is marketing and outreach. In this increasingly complex electronic environment, libraries have many competitors for the users searching for information in spite of the depth of scholarly resources that academic libraries provide. In order to encourage usage and maintain visibility and viability, marketing and outreach are vital. For example, most academic libraries issue information about new acquisitions or new services through social media, on the library's web pages or through informative, glossy newsletters that are often electronic now.

Libraries also offer programs to attract the academic and surrounding community such as speaking engagements with well-known authors or scholars and exhibit openings. These events frequently have a fundraising purpose also.

Outreach can take many other forms. Most libraries participate in new student orientations and new faculty orientations and offer workshops on library research. Some libraries participate in outreach efforts before the students enroll in college by offering advanced placement privileges for high school students.

Importantly, it is vital that the library, through its UL and its librarians, have close relationships with the colleges and the academic departments. While the UL works closely with college deans, librarians are often assigned academic departments and work closely with department chairs to make sure that the library is fulfilling the needs of that academic discipline.

Increasingly academic libraries are required to participate in fundraising to supplement the budget. Fundraising can take many forms including obtaining rare collections, endowments, support for library services, collections and programs, funding for buildings, equipment and furniture, and so on. In order to have successful fundraising, the library needs to have a director of development and other fundraising staff in order to work closely with the donors. It is a significant effort, which can have significant results but which requires a long-term effort.

In order to achieve economies of scale and expand their fiscal reach, some academic libraries are engaging in cooperative buying. A good example is the California State University. In their *Libraries of the Future* report, "The CSU has managed the cooperation of all 23 campuses to negotiate significant discounts in the cost of academic content, software licenses, and vendor training and support services."[45]

Academic libraries that are in a consortium can also take advantage of membership in the International Coalition of Library Consortia (ICOLC), which comprises of 200 library consortia around the world. The ICOLC is "dedicated to keeping participating consortia informed about new electronic information resources, pricing practices of electronic information providers and vendors, and other issues of importance...."[46]

Vigilant, skilled, and supportive library management, dedicated supervisors and staff, close attention to the users, comfort in the face of continual change, and a deep commitment to quality service are some of the most important components in the successful administration of an academic library.

ACADEMIC LIBRARY PERSONNEL

Academic libraries are staffed by a mix of personnel. The proportion of librarians in libraries is relatively small to the total number of personnel. Most categories of staffing fall into librarians (i.e., persons with a degree in library or information science), library assistants (i.e., specialists in the work of libraries), clerical, student employees, and administrators. There are also other specialists in the fields of accounting, personnel, building management, and, of course, technology (See *Academic Librarians*).

LIBRARY BUILDINGS

Academic library buildings range greatly in size depending upon their purpose, history, size of collection, size of student population, and available funding. Regardless of size, many similar issues confront academic libraries today in terms of their buildings. The most challenging of these issues is the impact of digital information and therefore a common belief that library buildings are either not needed or can be greatly reduced when the physical collection is not growing at the rate it once did. A more sophisticated view is that the role of the library is expanding and changing and therefore demanding new uses of its facilities in addition to its traditional uses.

As Leighton and Weber point out, "the library is recognized more than ever as a locus for educational effort in a social setting. The building must meet the need for the 'library as place' within academe...."[47] The authors go on to say that "...this intangible characteristic is often a major force in shaping the academic objective of the library building to meet the institutional goals for the years ahead."[47]

One of the most compelling trends over the last two decades is the creation of an information commons that largely refers to a place where students have access to both information and technology resources. This concept emerged because students needed a place where they could use the latest technology, research and write their paper online, and/or create a multimedia presentation. Importantly, an information commons provides the students with a sense of place and shared community. The layout of the room often encourages collaboration and group work.

For a good example, see the University of Southern California (USC) in Los Angeles, which was one of the early pioneers in creating an information commons. The USC provides a full range of services, including computers, research and writing support and consultation, technical support, adaptive technologies, printing, and collaborative workrooms.[48]

Study facilities have also changed significantly in recent years. There is a push for more student-friendly study areas largely out of concern that students will no longer come to the library because of the remote access provided to digital resources. In order to encourage studying and learning in the library, more attention in the library today is directed at learning spaces that are more comfortable with better seating and wireless access. Additionally, students now need more collaborative and group study space in order to work on group projects assigned by their professors or simply to study together. These rooms or areas designated for collaborative study often have flexible seating, with technology supportive of group work.

In keeping with the need to attract students, libraries are also relaxing the rules about drinks and sometimes food in the library. Academic libraries are also adding coffee shops to lure students in and also to benefit from the proceeds. For example, the University of Waterloo's Dana Porter Library in Canada has "Browsers," which is a coffee shop located on the main floor of the library and which provides students with an area of relaxation for studying, being with friends and enjoying a cup of coffee.[49]

What is still unknown is whether or not more comfortable spaces, increasing group study rooms, and providing cafes have actually increased students' study time or improved their study habits.

Libraries in the United States today also need to be very conscientious to come into compliance with the Americans with Disabilities Act (ADA) of 1990, which gives civil rights protections for persons with disabilities. The ADA requires that buildings be both usable and accessible to a person with a disability including both customers and employees. A library today needs to make sure that it is removing barriers to access, designing for accessibility, or providing alternative access.[50] Many libraries also go a step further by providing students with disabilities with special rooms and/or special assistive technology equipment such as adjustable furniture, Kurzweil reading machines, speech recognition software, magnifiers, and screen readers.[51]

As print collections grow beyond the capacity of the library to retain, some libraries are weeding but more and more are developing alternative storage places. The world's first automated storage and retrieval system was developed at California State University Northridge's Oviatt Library. This system contains over 13,000 bins occupying an 8000 ft^2 facility that is 40 ft high and allows for over one million items to be stored. This library uses the storage facility for lesser used materials and older periodicals. The system interfaces with the library's catalog and the user commands the system to retrieve the item needed.[52] Approximately 20 libraries in the United States have this system.

Some academic libraries use offsite storage. For example, the University of California maintains the Southern Regional Library Facility, which is offsite storage for lesser used but still needed collections. Storage is not its only responsibility. This service sends books for interlibrary loan, guaranteeing 24 hour service for its member UC libraries, and also is engaged in preservation and digitization.[53]

Another example is the Harvard University Library's the Harvard Depository, which was designed to be modular so that more storage units can be added as resources increase. This depository is offsite, secure, and accessible as well as climate controlled. It is a cost-effective method to store resources in an environment that preserves resources for the long term.[54]

Libraries now and in the future must be able to be flexible in their spaces, provide robust technology, and create collaborative study spaces and a more appealing environment. This is all coupled with the need to still provide for the care and housing of the collection for the wide range of services and of course a good environment for the users and the staff. This is particularly challenging in older buildings that may not be scheduled for renovation or rebuilding. The American Library Association has a useful web page for any academic library that is building a new building or renovating an older one. This page contains a wide range of resources.[55]

CONCLUSION

The continually changing world of digital information continues to have a dramatic impact on academic libraries in terms of services, resources, facilities, and the skill set needed for personnel. As the premiere support service to teaching, learning, and research, academic libraries will continue to thrive and play an important role but their future will require vigilance. Academic libraries need to become easier to use in order to compete effectively against popular search engines and other online services and companies that are aggressively adding information sources including scholarly ones. The online presence of the library, including its services and its information resources, will need unflagging attention so that students and faculty have the 24 hour easy access that they have come to expect in a digital world. The rare collections of academic libraries must be digitized continually to provide access to unique resources. The environment will always need to be scanned for significant changes in technology and user needs so that the library's strategies can shift in keeping with the times. Marketing will play an important role in order for the voice of libraries to be heard above the crowded world of information. An enhanced focus on national and local policies affecting libraries including intellectual property is vital.

Given their strong sense of purpose, the richness of their resources, and the dedication of their staff, academic libraries will maintain their all-important mission of supporting teaching, learning, and research.

REFERENCES

1. *International Handbook of Universities*, Palgrave MacMillan: London, U.K., 2013; 5532 pp.
2. OCLC. Global Library Statistics-Country Data, 2016; https://www.oclc.org/global-library-statistics.en.html (accessed June 19, 2016).
3. National Center on Education Statistics. (N.D.) Library Statistics Program. http://nces.ed.gov/surveys/libraries/academic.asp (accessed June 19, 2016).
4. Harvard University. History, 2016; http://www.harvard.edu/history (accessed June 17, 2016).
5. American Library Association. The Nation's Largest Libraries: A Listing by Volumes Held, October, 2012; http://www.ala.org/tools/libfactsheets/alalibraryfactsheet22 (accessed June 17, 2016).
6. Harvard University. Harvard at a Glance, June 17, 2016; http://www.harvard.edu/harvard-glance (accessed June 17, 2016).
7. University of Botswana. (N.D.) History. http://www.ub.bw/content/id/1366/History/ (accessed June 17, 2016).
8. University of Botswana. (N.D.) Main Library about us. http://www.ub.bw/content/id/1687/pid/1447/sd/1/sec/6/dep/91/About-Us/ (accessed June 17, 2016).
9. University of the South Pacific. About the University. USP: An Introduction, February 12, 2013; http://www.usp.ac.fj/index.php?id=usp_introduction (accessed June 19, 2016).
10. University of the South Pacific. USP's Vision, Mission and Values, February 14, 2013; http://www.usp.ac.fj/index.php?id=usp_mission (accessed June 19, 2016).
11. Boden, Dana W.R. A history of the utilization of technology in academic libraries. (1993): *ERIC*. ED 373806. Web. 20 June 2016.
12. Carlin, J. and Barb M. Books? Or No Books? Envisioning the Academic Library of the Future." The Huffington Post, 2, June 2014. Web. 20 June 2016. http://www.huffingtonpost.com/jane-carlin-and-barb-macke/academic-library-of-the-future_b_5078456.html.
13. Rafael, V.L. The cultures of area studies in the United States. Soc. Text **1994**, (41), 91–111.
14. Carpenter, K.E. Libraries. In *Encyclopedia of American Cultural and Intellectual History*; Cayton, M.K., Williams, P.W., Eds.; Charles Scribner's Sons: New York, 2001; Vol. 3, 372–373. .
15. Rau, E.P. Managing the machine in the stacks: Operations research, bibliographic control and library computerization, 1950-2000. Lib. Hist. **2007**, *23* (2), 151–168. Academic Search Elite, EBSCOhost (accessed September 10, 2014).
16. Things we used in libraries and when they were invented. American Libraries 2006, *37* (1), 46. Academic Search Premier (accessed September 10, 2014).
17. History of the OCLC Research Library Partnership. *OCLC*. Online Computer Library Center (OCLC) Research, n.d. Web. 20. http://www.oclc.org/research/partnership/history.html (June 2016).
18. Bangert, S. Understanding accreditation. In *Proven Strategies for Building an Information Literacy Program*, Curzon, S.C., Lampert, L.D., Eds.; Neal Schuman Publishers: New York, 2007; 335. 216 pp.

19. Association of College and Research Libraries. (N.D.) Guidelines, Standards and Frameworks. http://www.ala.org/acrl/standards (Accessed June 19, 2016).

20. Association of College and Research Libraries. (N.D.) Guidelines, Standards and Frameworks, http://www.ala.org/acrl/standards/standardsguidelinestopic (accessed June 19, 2016).

21. Association of College and Research Libraries. (N.D.) Standards for Libraries in Higher Education, http://www.ala.org/acrl/standards/standardslibraries (accessed June 19, 2016).

22. American Library Association. Government Relations. (N.D). USA Patriot Act. http://www.ala.org/advocacy/advleg/federallegislation/theusapatriotact (accessed June 19, 2016).

23. American Library Association. (N.D.) Copyright. http://www.ala.org/advocacy/copyright/ (accessed June 19, 2016).

24. World Trade Organization. Legal Texts: The WTO Agreements, 2016; http://www.wto.org/english/docs_e/legal_e/ursum_e.htm#nAgreement (accessed June 19, 2016.).

25. U.S. Department of Education. (N.D) Programs, http://www2.ed.gov/programs/fipseec/index.html (accessed June 19, 2016).

26. American Library Association. (N.D). Library Associations around the World. http://www.ala.org/offices/iro/intlassocorgconf/libraryassociations (accessed June 19, 2016).

27. International Federation of Library Associations and Institutions. About the Academic and Research Libraries Section, October 9, 2015; http://www.ifla.org/about-the-academic-and-research-libraries-section (accessed June 19, 2016).

28. American Library Association. (N.D.) Association of Research Libraries Fact Sheet-History. http://www.ala.org/news/mediapresscenter/presscenter/onlinemessagebook/acrlfactsheet (accessed June 19, 2016).

29. Association of Research Libraries. (N.D.) Mission Statement. http://www.arl.org/about#mission (accessed June 19, 2016).

30. American Library Association. Association for Library Collections and Technical Services. NISO Standards, January 25, 2001; http://www.ala.org/alcts/resources/guides/serstdsbib/niso (accessed June 19, 2016).

31. National Information Standards Organization. NISO. Welcome to NISO, 2016; http://www.niso.org/home/ (accessed June 19, 2016).

32. The University of Queensland. About UQ. Governance, 2016; http://www.uq.edu.au/about/governance (accessed June 19, 2016) .

33. University of British Columbia. (N.D.) University Act. http://www.bclaws.ca/EPLibraries/bclaws_new/document/ID/freeside/00_96468_01#part6 (accessed June 19, 2016).

34. University of Oxford. (N.D). Libraries. http://www.ox.ac.uk/research/libraries/index.html (accessed June 19, 2016).

35. Fennell, J. All in together: Building community through outreach services at Frazer Library. New Members Round Table (NMRT): News. American Library Association., Nov. 2008. Web. 20. June 2016. http://www.ala.org/nmrt/news/footnotes/november2008/outreach_services_finnell 38(2).

36. Walters, W. E-books in academic libraries: Challenges for sharing and use. J. Lib. Infor. Sci. **2014**; *46* (2), 85–95.

37. Rockman, I.F. Integrating information literacy into the learning outcomes of academic disciplines: A critical 21st century issue. Coll. Res. Libr. News. 2003, *64* (9), 612–615. OmniFile, WilsonWeb (accessed September 10, 2014).

38. Zabel, D. Trends in reference and public services librarianship and the role of RUSA: part one. Ref. User Ser. Quarter. *45* (1), 7–10. Academic Search Premier EBSCO (accessed September 10, 2014).

39. Dubnjankovic, A. Electronic resource expenditure and the decline in reference transaction statistics in academic libraries. J. Acad. Lib. **2012**; *38* (2), 94–100. Science Direct (accessed September 10, 2014).

40. Essinger, C., Irene K. Outreach: What works? Collaborative Librarianship **2013**; *5*(1), 52–58. General OneFile (accessed August 30, 2014).

41. Johnson, P. *Fundamentals of Collection Development and Management*; American Library Association: Chicago, IL, 2004, 312 pp.

42. Landgraf, G. Task force to address Digital Preservation. Am. Lib. **2007**; *38* (10), 29. *Expanded Academic ASAP*. Web. 20 June 2014.

43. University of Tasmania. (N.D.) UTAS Library Strategic Plan. *2013-2015*. http://www.utas.edu.au/__data/assets/pdf_file/0020/332552/Library-Strat-Plan-Poster1.pdf (accessed June 19, 2016).

44. Association of Research Libraries. What is LibQual+R, *2016*; http://www.libqual.org/ (accessed June 19, 2016).

45. California State University. Libraries of the Future, February 1, 2013; p. 4; http://www.calstate.edu/library/documents/LOFT_Recommendations_Report_v1_0.pdf. (Accessed June 19, 2016) from.

46. International Coalition of Library Consortia. (N.D.) About ICOLC. http://icolc.net/about-icolc (accessed June 19, 2016).

47. Leighton, P.D. David C.W. Planning Academic and Research Library Buildings, 3rd Ed.; American Library Association: Chicago, IL, 2000; p. 25.

48. University of Southern California. Leavey Library. About this Library, 2016; http://www.usc.edu/libraries/locations/leavey/ic (accessed June 19, 2016).

49. University of Waterloo. Library. (N.D.) Dana Porter Library. http://www.lib.uwaterloo.ca/locations/browsers.html (accessed June 19, 2016).

50. U.S. Department of Justice. 2010 ADA Standards for Accessible Design. https://www.ada.gov/regs2010/2010ADAStandards/2010ADAstandards.htm (accessed June 20, 2016).

51. California State University, Northridge. Oviatt Library. Services for Users with Disabilities, January 8, 2014; http://library.csun.edu/Services/UsersWithDisabilities (Accessed June 20, 2016).

52. California State University, Northridge. Oviatt Library. Automated Storage and Retrieval System, October 28, 2015; http://library.csun.edu/About/ASRS (accessed June 20, 2016).

53. University of California. (N.D.) Southern Regional Library Facility. http://www.srlf.ucla.edu/ (accessed June 20, 2016).

54. Harvard University Library. (N.D.) The Harvard Depository. http://hul.harvard.edu/hd/pages/facility.html (accessed June 20, 2016).

55. American Library Association. (N.D.) Building Library and Library Additions: A Selected Annotated Bibliography. http://www.ala.org/tools/libfactsheets/alalibrary-factsheet11 (accessed June 20, 2016.

Accessibility

Lori Bell
Alliance Library System, East Peoria, Illinois, U.S.A.

Abstract

Although libraries have come a long way in making their spaces, places, and Web sites in compliance with the Americans with Disabilities Act (ADA), there are still issues and challenges in providing services and materials to individuals who have a physical challenge or disability. This entry looks at traditional services and materials libraries have offered and also new technologies libraries should consider for serving this population. Because the ADA was passed almost 20 years ago, it did not really address the Internet and the World Wide Web. The ADA is a civil rights law to prevent discrimination because of disability. Disability is defined as "a physical or mental impairment that substantially limits a major life activity." The ADA requires places and services for the public to be accessible, but there is no specific reference to the Internet as a public place of accommodation for people with disabilities. This is where Section 508 of the U.S. Code comes in. This entry will discuss specific ways in which libraries are complying with the ADA and Section 508 and where progress needs to be made.

SERVICES WHICH HELP WITH ACCESSIBILITY

For libraries that are Carnegie-type buildings or that cannot afford the changes necessary to their physical libraries to make them accessible, there are services they can offer that help make the library accessible and show the library as a concerned agency that wishes to serve everyone.

Making a Building Accessible

There are numerous ways the bricks and mortar library can make itself accessible. Entire books have been written on this topic, but basic building accommodations can include an elevator, an automatic door, a large restroom which can accommodate a wheelchair, a curb cut for smooth travel by a wheelchair, Braille signs around the building, and much more. In addition to building accommodations, below are services libraries can offer to improve accessibility.

Homebound Book Delivery

Home book delivery is a time-intensive but effective service for people who cannot make it to the library. Librarians or volunteers choose books of interest to the homebound on a biweekly or monthly basis and then physically deliver the books to the reader's home. They pick up the books the reader is finished with and return them to the library. Sometimes libraries require a note signed by a doctor or medical worker to verify they need the service. Since this is a time-intensive service, many libraries rely on volunteers to deliver the books. The volunteer picks the selected books up at the library and delivers them to the home of the person needing the service.

Books by Mail Service

A few libraries offer books by mail services. Some offer all the books available in the library and others have a special collection, for instance, of paperbacks which are less expensive to mail. The books are delivered to the reader via U.S. mail and returned in the same way. Many books by mail services are for people who live in rural areas and do not have access to a library. Some libraries offer books by mail to the homebound instead of relying on volunteers. The librarian selects the books for the reader and mails them with a postage paid return envelope the reader can use to return the books. Many libraries offering books by mail to the homebound require some verification from a medical doctor or agency that the individual is in need of the service.

Deposit Collections

Many libraries offer deposit collections to agencies like nursing homes, senior centers, and even day care centers. This way, residents have access to a large number of books for a certain period of time without having to visit the library. For many seniors in nursing homes and adaptive living centers, this brings the library to them.

Large Print Books

Many libraries have large print books with size 14 or 16 point or higher fonts for people with vision impairments. Large print publishing really only began as recently as 1964 in England. Although many people are eligible for the talking book program if they have problems reading regular print, many are not ready to give up reading print.

Encyclopedia of Library and Information Sciences, Fourth Edition DOI: 10.1081/E-ELIS4-120045388

Large print books have increased in popularity and now most are published at the same time as the traditional print book. Most public libraries, even small ones, now carry large print books.

Braille Books

Some libraries have collections of Braille books for the blind. Braille is a system widely used for blind people to read and write and was invented by Louis Braille in 1821. Because of the space required for Braille, these books are very thick and one book can consist of a number of volumes. Most libraries refer the blind to the talking book program which carries a large Braille collection and is sent to readers through the mail.

Audio/Recorded Books

Many libraries offer collections of audio/recorded books in different formats for readers. These are enjoyed by everyone, not just people with disabilities. Audiobooks became popular on cassette in the 1980s and that is when public libraries started offering them. The talking book program started offering books on cassette in the 1970s to qualified readers. Audiobooks then became available on CD, and most recently as downloadables from the Internet. Readers qualified for the talking book program can supplement their book selections at the public library with different formats of audio/recorded books.

ADAPTIVE HARDWARE AND SOFTWARE FOR ADULTS

With shrinking budgets and the challenge of providing more materials in more formats, it is difficult for libraries to select which accessibility options to offer that will provide the greatest increase in accessibility to the greatest number of physically challenged library users. The majority of public libraries serve populations of under 25,000, which means many do not have professional librarians or decent budgets. Although making many services accessible can be simple and common sense, other accommodations can be expensive and cost prohibitive even for libraries who want to provide accessible materials and services. Even if grants are available, there is the problem that people with mobility and vision problems do not come to the library because they think the library does not have enough for them that is accessible. Because this population does not come to the library, the library staff thinks there is no need for the accommodations.

Another challenge for library staff is finding the latest information on adaptive software and hardware. Many states like Illinois have agencies which provide information about and access to adaptive equipment. The Illinois Assistive Technology Project has a toll-free number for agencies and individuals with questions. The National Assistive Technology Project Partnership maintains a contact list of state offices funded by the Assistive Technology Act of 1998.

Closed-Circuit Television

Some libraries offer closed-circuit television designed for people with low vision. CCTVs can enlarge text up to 50 times and can change the color of background and print. CCTVs may be useful for the elderly or persons of any age with vision problems who come in to balance their checkbook, read their bills, or to read a newspaper on their own.

Typing and Voice Recognition Software

For many with mobility impairments or arthritis, writing and typing may be impossible. There are resources that can help including check-writing guides, pencil grips, slip-on typing aids, and touch and type sticks. Voice recognition software may help those with mobility and vision problems. There have been major recent improvements in this technology. Voice recognition software used to require individual users to take a good amount of time to train the software to understand his or her own voice. Programs such as Dragon Naturally Speaking allow a user to dictate commands to the computer in order to open programs, create reports, e-mail and documents, and so forth.

Screen Reading Software

Screen reading software is used by the visually impaired to vocalize the written text on a computer screen, including the "alt" text embedded in Web pages that describes visual images and icons. It is available in a variety of synthetic voices and will read what is on a computer screen to a visually impaired user. The software will read whatever is on the screen, from windows, to word processing to Web sites.

Mouse Challenges

People with visual impairments cannot use a mouse because they cannot see the cursor. Those with mobility impairments may not be able to move the mouse effectively. Software with keystroke commands and which will work with function keys help with these challenges. There are also large mice, large track balls, switches, a head mouse, joybox, and other options.

Signing for Public Meetings

Some states require that libraries offer signing for hearing impaired patrons during public meetings. When these meetings are advertised publicly, these services must be offered. Those with hearing impairments planning to attend the meeting notify the library a certain amount of time in advance so that the library can hire an interpreter.

TDDS and TTYS

Before the digital age, some libraries had TDDS/TTYS—telecommunications devices for the deaf. A hearing impaired person would phone the library and use a typing device to transmit information. When the library received the call from this person, they would view the message from the patron as it printed out on paper and staff would communicate with the hearing impaired user by typing back. For libraries without TDDS/TTYs, most states have a relay service which can transmit information either way for those with hearing impairments trying to communicate with the hearing population. Now, with the Internet, both populations can use communication software to talk with one another.

Talking Books

The National Library Service (NLS) for the blind and physically handicapped is a national program through the Library of Congress that offers books in Braille and audio recording for individuals who cannot comfortably read regular print because of a vision, physical, or certain learning disabilities. Most libraries offer large print and audiobooks in a variety of formats such as CD, downloadable, and Playaway. The NLS has more than 60,000 titles and provides a free cassette machine for eligible readers. More than 750,000 readers in the United States enjoy the talking book program. The NLS is moving to digital flash cartridges and downloadable digital audiobooks and has quit producing cassette machines, but it will be several years before the program is totally transitioned.

Other Digital Book Formats

The Recording for the Blind and Dyslexic continues to offer recorded textbooks for the visually impaired. Benetech developed Bookshare.org which provides print-disabled people in the United States with over 40,600 books and 150 periodicals which are converted to Braille, large print, or digital formats for text to speech audio. Originally, members had to pay a modest startup and annual registration fee for access to this rapidly growing library. Thanks to special funding from the Office of Special Education programs, memberships for schools and qualifying students is free.

Audiobooks and Playaways

Public libraries have had audiobooks on cassette since the 1980s. In the twenty-first century, the popularity of audiobooks continues to rise while the sale of print remains flat. In addition to audiobooks on cassette, libraries make them available on CD. Digital audiobooks are now available via

download through vendors like netLibrary and OverDrive. A new company called Findaway developed the Playaway in 2005 which allowed libraries to buy a self-contained audiobook. Patrons who do not want to download and then upload on to an MP3 player or mobile device can borrow a self-contained unit with digital quality and the portability of a handheld player. Again, with shrinking budgets, it is a challenge for libraries to purchase audiobooks in many formats.

Audio Description

Audio description of plays and movies for the visually impaired provides narration during places in the performance that have no dialog or sound effects. Many libraries and talking book centers offer audio described videos which can be enjoyed by the sighted and sight impaired alike. The narrative is professionally written and recorded and is not intrusive. The Illinois State Library has funded an effort by the Alliance Library System to provide audio description for historical digital images on Web sites. Without audio description, digital images are almost totally inaccessible except for metadata tags which do not fully describe an image. This effort is very much in its infancy and there are few agencies providing audio description for even a few of their images.

Outreach Efforts

Users and nonusers of your core service population who would benefit tremendously from accessibility initiatives often are difficult to identify and reach out to. Often they may be trying to use the online systems, collections, and services offered by your library, because their personal return on investment in their search for information may be better online than in person. When designing your outreach efforts to this population, use this fact to your advantage. Because many print-impaired users are extraordinarily adept at using computers and networks, use the power of the Web and the Internet to reach out to this underserved population.

Web Sites

Section 508 is a law requiring federal agencies in the United States to make electronic and information technology accessible to people with disabilities and requires organizations to comply with the standard for content accessibility outlined in the statute if they wish to do business with the federal government. There are books and training programs which should help Web designers to learn how to create accessible Web sites. There are programs like BOBBY which take a look at the coding on Web sites to make sure it is accessible and provide information on what about the site is not accessible. Yet, as Web sites contain more and more media and high-end graphics, more Web sites are not accessible to the visually impaired. Accessibility means more and easier access for everyone, yet most sites are not accessible or easy to use.

Virtual Reference

Twenty-first century libraries are offering more and more services on the Web. Virtual Reference is no exception. In 2008, OCLC Question Point and Tutor.com are the primary services libraries use to offer Web-based chat reference services. At first, these programs required separate downloads by librarian and client and had high overhead and hardware requirements. The programs are simplified, but that does not mean they are now accessible. Several talking book libraries worked with OCLC to create an accessible VR program called InfoEyes for use by the visually impaired. Visually impaired users were able to use Question Point for e-mail questions and a Web conferencing program by Talking Communities, a company led by a visually impaired gentleman, for live chat sessions.

Web Conferencing

Again, as more and more business is conducted on the Web, the visually impaired and others with disabilities comprise a large number. Talking Communities has created an accessible Web conferencing platform which is used by talking book libraries to provide live, Web-based reference services. Online Programming for All Libraries (OPAL) uses the accessible conferencing software to provide accessible library programs of all kinds. Talking book libraries, academic and public libraries use this easy-to-use conferencing software to offer book discussions, guest lectures, training programs, and all kinds of

library programs to their patrons on the Web. Not all Web conferencing programs are accessible. Elluminate is one of the few other programs that is accessible.

CONCLUSION

Although technology improves opportunities for accessibility, many technologies create a more complex environment for a person with a disability. Libraries need to keep up as much as possible with technology or who to ask to continue to improve library services for people with disabilities.

BIBLIOGRAPHY

1. Bell, L.; Peters, T. Digital library services for all. Am. Libr. September **2005**, *36*(8), 46–49.
2. Blansett, J. Digital discrimination. Libr. J. August 15, **2008**, *133*(13), 26–29.
3. Hutchinson, N.G. Beyond ADA compliance: Redefining accessibility. Am. Libr. **2001**, *32*(6), 76.
4. Mates, B. Wakefield, D. Dixon, J. *Adaptive Technology on the Internet: Making Electronic Resources Accessible to All*, American Library Association: Chicago, IL, 2000.
5. Peters, T.; Bell, L. Hello IM goodbye TTY. Comput. Libr. **2006**, *26*(55), 18–21.
6. Peters, T.; Bell, L. Is web conferencing software ready for the big time?. Comput. Libr. February **2006**, *26*(2), 32–35.
7. Peters, T.; Bell, L. Choosing and using text-to-speech software. Comput. Libr. February **2007**, *27*(2), 26–29.
8. Schmetzke, A. Web accessibility at university libraries and library schools. Libr. Hi Tech **2001**, *19*(1), 35–49.

Accreditation of Library and Information Studies Programs in the United States and Canada

Karen Lynn O'Brien
American Library Association, Chicago, Illinois, U.S.A.

Abstract
The entry details the purposes, definition, historical and operational contexts, and future prospects for accreditation of library and information studies programs by the American Library Association.

INTRODUCTION

Accreditation is a quality assurance process guided by standards, policies, and procedures. The process entails the assessment of educational quality and the continued enhancement of operations through the development and validation of standards. Accreditation is also a condition wherein an accredited status signifies that the education provided is of quality.[1]

PURPOSE OF ACCREDITATION

Accreditation provides for quality assessment and enhancement, with quality defined as the effective utilization of resources to achieve appropriate educational objectives. The purpose of accreditation is to assure the public, members of a profession, employers, students, and the educational community that an institution or program 1) has clearly defined educationally appropriate objectives, 2) maintains conditions under which their achievement can reasonably be expected, 3) is in fact accomplishing them substantially, and 4) can be expected to continue to do so.[2] Accreditation does not result in the ranking of programs or institutions, but rather respects the uniqueness of each, while ensuring that all meet appropriate standards.

The fundamental assumptions of accreditation are that 1) self-regulation in education is preferable and in the long run is more effective than governmental regulation (in accordance with the 10th Amendment), 2) a system of quality assurance can be effective only to the extent that it recognizes and builds upon constituent willingness to engage in the process, 3) self-regulation is possible if expectations are clear, 4) most institutions will self-regulate if it is believed that they might otherwise be identified by their peers as doing something wrong, and 5) only a few institutions deliberately engage in behavior that they know is not in the public interest.[3]

The American Library Association (ALA) *Standards for Accreditation of Master's Programs in Library and Information Studies* express the purpose of accreditation as protecting "the public interest" and providing guidance "for educators." The ALA "Committee [on Accreditation] offers a means of quality control in the professional staffing" for "library and information services."[4]

CONTEXT

Accreditation is a voluntary, nongovernmental, collegial process of self-study and external review. It involves a framework of international, national, regional, vocational, specialized, and professional accrediting organizations. In the United States, there are 67 accreditors recognized as reputable by the U.S. Department of Education (USDE) or the Council for Higher Education Accreditation (CHEA) and some accreditors by both.[5]

"Regional" and "national" organizations accredit whole institutions, while specialized and professional agencies accredit programs and free-standing institutions. These organizations, while based in the United States, may accredit institutions and programs of study outside the United States. The ALA, for instance, also accredits programs at the master's degree level in library and information studies (LIS).

ALA is recognized by the CHEA, whose requirements state that accreditors "[must maintain] appropriate and fair policies and procedures that include effective checks and balances."[6] ALA *Accreditation Process, Policies, and Procedures*, developed by the ALA Office for Accreditation (OA) and approved by the ALA Committee on Accreditation (COA), guide the process. The OA conforms to good accreditation practice as set forth by the CHEA and the Association of Specialized and Professional Accreditors (ASPA).

As a member of the ASPA, the ALA follows the *ASPA Code of Good Practice*, which states that a member must exhibit "a system of checks and balances in its standards development and accreditation procedures. [A member must avoid] relationships and practices that would provoke

Encyclopedia of Library and Information Sciences, Fourth Edition DOI: 10.1081/E-EISA-120053574

questions about its overall objectivity and integrity."[7] The ALA process includes appeal policies and procedures that provide a system of checks and balances to protect the integrity and fairness of the process for the institution in which programs reside.[1]

The ALA accredits the professional degree in librarianship, established by the ALA governing council as the master's degree. The ALA is recognized by the CHEA as the accreditor for LIS programs offered under the degree-granting authority of regionally accredited institutions located in the United States, Puerto Rico, and by agreement with the Canadian Library Association, in Canada. The ALA is one of the 20 ASPA member agencies that accredit educational entities in both in the United States and Canada.[8] CHEA recognition requires that an accreditor undergo review every 5 years to confirm eligibility and demonstrate compliance with its standards.

The ALA falls outside the scope of the USDE recognition, which by statute does not concern itself with accreditors that do not perform a Title IV "gate-keeping role" for federal loan and grant funds. That gate-keeping role falls to the institutions in which LIS programs reside, which are regionally accredited. The ALA standards require a program in the United States to reside within a regionally accredited institution.[4]

Canadian institutions of higher education, however, operate under provincial legislation, with older institutions likely to have an act specific to themselves. As the number of postsecondary institutions seeking degree-granting powers has increased through the 1990s to the present, provinces have devised their own mechanisms for considering these requests; for example, the Province of Alberta operates the Campus Alberta Quality Council, and Ontario has the Higher Education Quality Council of Ontario. However, while the processes for obtaining provincial approval to offer degree programs may have similarities to accreditation processes (incorporating self-studies and external reviews), the language in Canada is not that of "accreditation" of each institution.

Individual academic programs in Canada are subject to accreditation processes, as is the case with library and information science education. In recognition of cross-border mobility between the two nations, Canadian accreditation processes often recognize the American program in some way. The Canadian Medical Association accredits graduate medical education, and its accreditation committee includes two observers from the Accreditation Council for Graduate Medical Education, from the United States. The Canadian Architectural Certification Board comprises only Canadians, but acknowledges a close working relationship with NAAB (National Architectural Accrediting Board).

The COA has 12 members, stipulated by the ALA Committee on Organization as five practitioners and five academics who are ALA members, and two public-at-large members, all appointed by ALA presidents-elect. The ALA members serve a 4-year nonrenewable term,

while the public-at-large members are appointed to serve a 2-year term, renewable once.

HISTORY

The origins of accreditation are traced to the progressivism movement of the mid-nineteenth through mid-twentieth centuries, which saw the rise of higher education in the United States beginning in 1862 with the passage of the Morrill Act, which came to be known as the Land Grant College Act.[9]

Accreditation of postsecondary education developed in parallel with library studies education. Through efforts led by Melvil Dewey, the first library school was founded in 1883 as the School of Library Economy at Columbia College (now Columbia University). Similar schools elsewhere followed, and in 1900, accreditation of library studies was born when the ALA established the Committee on Library Training that formulated the first standards of quality for library education.[10]

The first comprehensive study of library education was undertaken in 1921 with Carnegie Corporation funding by librarian and economist Dr. Charles C. Williamson. His report *Training for Library Service* stressed the need to draw a strong distinction between professional and clerical work in libraries, recommending that library education schools require applicants to hold an undergraduate degree.[11] It was not until 1951, however, that the master's degree was established by ALA policy 54.2 as the professional degree.

The ALA Board of Education for Librarianship (BEL) was established in 1924 to develop "minimum standards for library programs." The Carnegie Corporation used these quantitative training measures as a basis for endowing existing and newly formed schools, including the Graduate Library School at the University of Chicago, the first library school to offer a graduate educational program in librarianship.[12]

As library education moved into more traditional academic contexts within university settings, the next revision of the standards in 1933 was more descriptive of the principles and practices of the profession and less prescriptive and training oriented. This revision was a cooperative effort between the BEL and the Association of American Library Schools [AALS; now the Association for Library and Information Science Education (ALISE)]. At that time, there were 25 LIS schools in the United States and Canada.

The next revision was undertaken by the BEL, the AALS, and the ALA Library Education Division and approved by the ALA Council on July 15, 1951. These standards incorporated the premise that the professional degree be a minimum of 5 years of study beyond secondary school and should lead to a master's degree.

In 1955, the COA was established as the decision-making body for accrediting professional degree programs in LIS. The work of the COA is administered by the OA, formerly the Office for Education. The OA also coordinates reviews of school library media teacher education programs.

The COA established a standards revision subcommittee in 1970 that produced the 1972 *Standards for Accreditation*, which notes that its requirements and recommendations emphasize "qualitative rather than quantitative considerations… and that, therefore the standards lend themselves to some variation in interpretation… The standards are indicative but not prescriptive." The 1972 standards guided programs until the ALA Council adopted the *Standards for Accreditation of Master's Programs in Library and Information Studies 1992*, which became effective January 1, 1993.

Between the 1972 and 1992 standards, a grant from H.W. Wilson funded the ALISE to convene a conference of 17 organizations including ALA to discuss whether the accreditation process could be broadened to provide for governance and operation through a collaborative, mutually supported mechanism.[12] That same year, the U.S. Department of Education funded the COA to explore implementation of a collaborative accreditation structure. As a result of that exploration, the COA prepared a report *Accreditation: A Way Ahead*, released April 1986.

In 1999, members of the ALA, the ALISE, the American Association of Law Libraries, the Medical Library Association, and the Special Libraries Association convened as a steering committee for the first Congress on Professional Education (COPE I) to focus on "initial preparation for librarianship." Their report included 30 recommendations.[13] Following those recommendations, COA revised its process policies and procedures, including a streamlined appeals process and a policy requiring that a program be placed on conditional accreditation status before accreditation could be withdrawn. Those revisions are published as *ALA Accreditation Process Policies and Procedures*, effective December 15, 2006. New editions followed in 2012 and 2015.

In 2002, a COA subcommittee reviewed the standards, surveyed constituents, and provided a report with recommendations. That report led to a draft of proposed changes that were released for comment on November 17, 2006. A revised version of the *Standards* was approved for adoption by the ALA Council on January 15, 2008. A 5-year COA standards review (2009–2014) resulted in a revised edition approved for adoption by the ALA Council on February 2, 2015.

STANDARDS AND PROCESS

Revisions of the ALA standards over the years reflect the evolution of accreditation in general from quantitative to qualitative. The ALA standards emphasize improvement by stating that the intent of the standards is "to foster excellence through a program's development of criteria for evaluating effectiveness, developing and applying qualitative and quantitative measures of these criteria, analyzing data from measurements, and applying analysis to program improvement. The Standards stress innovation, and encourage programs to take an active role in and concern for future developments and growth in the field." In each of the standards, emphasis is on a program demonstrating "how the results of evaluation are systematically used" for improvement and "plan for the future." Programs must demonstrate use of "the results of their evaluations for broad-based, continuous program planning, development, and improvement."[4]

The standards define the field as "concerned with recordable information and knowledge, and the services and technologies to facilitate their management and use. Library and information studies encompasses information and knowledge creation, communication, identification, selection, acquisition, organization and description, storage and retrieval, preservation, analysis, interpretation, evaluation, synthesis, dissemination, and management."[4]

A 5-year review of the 2008 *Standards* conducted by the ALA COA in public discussion with stakeholders, virtually through webinar and website comment collection, and in face-to-face meetings, resulted in a revision adopted by the ALA Council on February 2, 2015, at the ALA Midwinter Meeting. The 2015 standards specify five areas for determining program quality: Standard I. Systematic Planning; Standard II. Curriculum; Standard III. Faculty; Standard IV. Students; and Standards V. Administration, Finances, and Resources.

Programs seeking initial or continued ALA accreditation status address how they are meeting the *Standards* by following the established *Accreditation Process Policies and Procedures*. A primary function of the OA is to administrate the review process. The process entails preparation of a self-study that is verified by an external review panel at least every 7 years as part of the comprehensive review process, an annual statistical report, and a biennial narrative report addressing changes since the last comprehensive review.

FUTURE PROSPECTS

Interest in accreditation as a quality assurance mechanism remains intense, especially with the ongoing development of online program delivery. Currently, 29 of the 59 institutions offering an ALA-accredited program are providing a 100% online program. Regional accreditors depend on specialized accreditors and vice versa. ALA accreditation requires programs to reside within regionally accredited institutions.

Most developed countries outside the United States rely on governmental entities such as ministries of education to assure educational quality. Visitors to the ALA and attendees at accreditation conferences from outside the United States express interest in the collegial peer-review nongovernmental nature of accreditation as practiced in the United States. Visitors to the OA have expressed admiration of the clarity and brevity of the ALA *Standards* and the well-developed review process.

It is the self-regulatory nature of accreditation as practiced in the United States that appears to promise a greater potential for the kind of agility sought to meet demand for education on a global scale. Based on the U.S. Census Bureau Population Division International Database, the Center for Higher Education projects that global demand for higher education will continue to grow exponentially from 48 million learners in 1997 to 159 million by 2025. Programs are likely to proliferate in an effort to meet this demand, and with that, quality assurance will become all the more important.[14]

Master of library and information studies programs in the United States and Canada continue to come forward seeking ALA accreditation. In January 2014, an eighth Canadian program was granted initial accreditation. Yet another program domestically was initially accredited in 2015 to bring the total number to 64. Two programs in the United States are currently in Candidacy status under review for initial ALA accreditation.[15]

REFERENCES

1. Committee on Accreditation. *Accreditation Process Policies and Procedures, 4th Ed.; American Library Association: Chicago, IL, 2015.*

2. Chernay, G. *Accreditation and the Role of the Council on Postsecondary Accreditation*; COPA: Washington, DC, 1990.

3. Young, K.F. New pressures on accreditation. J. High. Educ. **1979**, *50*, 132–144.

4. Committee on Accreditation. *Standards for Accreditation of Master's Programs in Library and Information Studies.* American Library Association: Chicago, IL, 2015.

5. Council for Higher Education Accreditation. *Database of Institutions and Programs Accredited by Recognized U.S. Accrediting Organizations.* Council for Higher Education Accreditation: Washington, DC. http://www.chea.org/search/default.asp (accessed February 29, 2016).

6. Council for Higher Education Accreditation. CHEA Policy and Procedures. *Recognition Standard 12D*; Council for Higher Education Accreditation: Washington, DC, June 28, 2010.

7. Association of Specialized and Professional Accreditors. *ASPA Code of Good Practice.* Association of Specialized and Professional Accreditors: Chicago, IL, April 2013.

8. Association of Specialized and Professional Accreditors. *ASPA Accreditation Profile Database.* Association of Specialized and Professional Accreditors: Chicago, IL, 2015.

9. Young, Chambers, Kells and Associates. *Understanding Accreditation: Contemporary Perspectives on Issues and Practices in Evaluating Educational Quality*; Jossey-Bass: San Francisco, CA, 1983.

10. Hayes, R.M. Accreditation. Libr. Trends. **Spring 1986.**

11. Vann, S.K. *The Williamson reports of 1921 and 1923*; The Scarecrow Press, Inc.: Metuchen, NJ, 1971.

12. Stieg, M.F. *Change and Challenge in Library and Information Science Education*; American Library Association: Chicago, IL/London, U.K., 1992.

13. Ghikas, M. *Education and Accreditation*; American Library Association: Chicago, IL, July 2003.

14. Peace Lenn, M. *The World is Flat: Globalization of Professional and Specialized Accreditation: A Workshop*; Center for Quality Assurance in International Education, National Center for Higher Education: Washington, DC, September 10, 2005.

15. Outlook. *Prism*; American Library Association: Chicago, IL, Spring 2014.

Acquisitions Institute at Timberline Lodge

Scott A. Smith
Langlois Public Library, Langlois, Oregon, U.S.A

Abstract

This entry describes the history and program content of the Acquisitions Institute at Timberline Lodge. The institute is an annual international library conference, held in mid-May in Oregon, focusing on issues pertaining to acquisitions, collection development, and technical services, primarily as relates to academic libraries.

DESCRIPTION OF THE ACQUISITIONS INSTITUTE AT TIMBERLINE LODGE

The Acquisitions Institute at Timberline Lodge is an annual international library conference held in mid-May at Timberline Lodge, Oregon. The conference focuses on topics relating to library acquisitions, collection development, and technical services. The scope of the institute is not, in theory, confined to any particular type of library, but in practice nearly all presenters and delegates are from academic libraries or commercial vendors who serve the academic market. Most attendees are from institutions in North America, but librarians from Africa, Australia, the Middle East, New Zealand, and South America have also attended. Although sometimes referred to as "the Timberline Institute" or "the Timberline conference," the formal name of the meeting is "The Acquisitions Institute at Timberline Lodge."

HISTORY AND EVOLUTION OF THE INSTITUTE

The institute has been held under two different names and in two different locations. The origins of the conference can be traced to a conversation between two librarians and a bookseller at the annual American Library Association meeting held in San Francisco in June, 1987. By then the Charleston Conference had become well established as a smaller, more focused meeting relating to acquisitions. Richard Brumley, then Head of Acquisitions at the California Polytechnic State University, San Luis Obispo; Tom Leonhardt, then Dean of Libraries at the University of the Pacific in Stockton; and Scott Alan Smith, then with Blackwell's, pondered the question as to whether acquisitions policies and procedures in the West differed from acquisitions practice in the East. The three agreed the topic was worthy of exploration and over the course of the next few years began working on what was envisioned, initially, as a one-time event.

As plans for the meeting coalesced, Leonhardt noted that the University of the Pacific had an off-site conference facility located in the Sierras known as the Feather River Inn. (The Inn was built in 1914 and acquired by the university in 1977. It was sold to a private equity company in 2005.) After initial discussions with the Inn in 1989, plans were made to hold a meeting the following May. The first Feather River Institute was held on May 18–20, 1990. The keynote speaker was Melvin Voight, University Librarian Emeritus at the University of California, San Diego. His keynote address was titled "Historical perspective on acquisitions/collection development." Other presentation titles that first year included "Lonesome dove: doing business in the West"; "Boom or bust"; "How the West was won"; "Gunfight at the OK Corral: Holding your own with collection development"; and "Young guns: library education today" (a complete list of keynoters, presenters, and institutional affiliations is included as an appendix).

During the wrap-up session, the group agreed that in fact there is nothing unique about acquisitions in the West. The meeting was deemed productive and successful, however, and the delegates felt the idea of making this an annual event had merit. Eight more Feather River Institutes were held through 1998 (there was no conference in 1993).

In part due to the rustic and informal nature of the Inn, the Institute had a decidedly relaxed and laid-back style. Far from being unprofessional, however, this somewhat unconventional atmosphere fostered a candid and fertile environment for presentations, discussions, and debates among and between librarians, publishers, book vendors, serials agents, and others associated with the library community.

One of the unusual characteristics of the conference was the fact that, from the beginning, there were no exhibits. Nearly a third of the attendees were vendors, but they came as colleagues, not as company representatives motivated to sell product. A strict policy of "noncommercial behavior" was established early on and has been observed with but a few transgressions. Most vendors expressed genuine gratitude for the opportunity to engage

Encyclopedia of Library and Information Sciences, Fourth Edition DOI: 10.1081/E-ELIS4-120049542

in substantive conversations on topics germane to the library world without the burden of normal conference exhibit demands.

Moreover, to the consternation of some attendees, name badges do not reflect institutional affiliation. This is deliberate; the organizers feel this helps stimulate conversations and initially break the ice.

Some of the vendors and publishers who have attended the institute over the years include Alibris, Blackwell's, Brill Academic, University of California Press, Cambridge University Press, Casalini, University of Chicago Press, Credo Reference, EBSCO, Elsevier, Gale/Cengage Learning, Harrassowitz, Innovative Interfaces, Kluwer Academic Publishers, Martinus Nijhoff, University of Minnesota Press, now publishers, Oxford University Press, Puvil, SAGE, Springer, Stanford University Press, Swets, Syndetic Solutions, Taylor & Francis, University of Washington Press, John Wiley & Sons, and YBP Library Services.

During the Feather River years, outside sponsorship for the institute was not necessary. With the move to Timberline, it was deemed appropriate to develop an ongoing network of corporate supporters; many of the aforementioned vendors have been steadfast contributors to the institute.

By 1999, the planning committee consisted of Scott Alan Smith (now library director at the Langlois Public Library, Oregon), Richard Brumley (by then Head of Acquisitions and Collection Development at Oregon State University; now retired), and Nancy Slight-Gibney (then Head of Acquisitions; now Director, Library Resource Management and Assessment, University of Oregon). For a variety of reasons, the committee decided to investigate a different venue for the conference. After considering a number of possible sites, Smith selected Timberline Lodge. The conference was renamed The Acquisitions Institute at Timberline Lodge, and the first institute at Timberline was held on May 20–23, 2000. With the exception of 2006, when no conference was held, the institute has met annually.

Prior to the move to Timberline, the conference organization had been very informal. By the time the decision to relocate had been made, it was apparent the group had outgrown its initial structure, and more traditional organizational models were considered. Today, the institute is a 501 (c) 3 nonprofit corporation, EIN 71-1001116; Oregon nonprofit corporation 361549-95.

TIMBERLINE LODGE—THE FACILITY

Timberline was built in 1936–1937 as a project of the Works Progress Administration. The lodge is sited at 6000 ft on the south slope of Mt. Hood, approximately an hour east of Portland. With the exception of six massive timber columns that came from what is now the Gifford Pinchot National Forest on the Washington side of the Columbia Gorge, most of the building materials were gathered within a mile of the lodge.

During the depths of the Great Depression, the WPA and the Civilian Conservation Corps sought to give unemployed, desperate people a living wage and, just as importantly, a sense of self-worth and dignity. The results at Timberline are magnificent and enduring.

Gilbert Stanley Underwood, who had designed lodges at Bryce Canyon and Yosemite, was retained as consulting architect. Marjorie Hoffman Smith served as the interior designer, and Ray Neufer supervised the woodworking shop. Under Smith's direction, workshops were established in Portland to weave cloth for the lodge; Neufer oversaw workers who evolved into craftsmen, building unique furniture that is still in use today.

On September 28, 1937, Franklin D. and Eleanor Roosevelt visited Oregon and dedicated the Bonneville Dam and Timberline Lodge. In 1975, a conference wing was added—named the C. S. Price Wing, after one of Oregon's foremost artists and whose work is featured in the lodge. In 1977, the lodge was declared a national landmark, and in 1981 the Wy'East Day Lodge was opened to help accommodate the heavy ski, snowboard, and snowshoe traffic the facility serves year round. Today more than two million people visit the lodge every year.

THE ACQUISITIONS INSTITUTE TODAY

Smith selected Timberline as the new home for the conference because it is uniquely Oregonian, and its size and configuration readily meet the objectives of the conference. The conference planners intend to keep the meeting small. The immediacy and collegiality of the conference are values that would likely be sacrificed were the institute to grow too large.

The schedule deliberately allows time outside of the formal program to allow delegates to network. There are no breakout sessions; everyone attends all presentations together. The goal of the planning committee is to give all attendees the best conference experience possible.

Program content has evolved over the years, reflecting changes in libraries and publishing. Today the focus of many presentations is access of content rather than the acquisition and management of print collections.

THE PLANNING COMMITTEE

As of February, 2014, the Planning Committee consists of

Kristina DeShazo, Acquisitions and E-Resources Librarian, Oregon Health and Science University

Stacey Devine, Assistant Head, Acquisitions and Rapid Cataloging, Northwestern University

Nancy Slight-Gibney, Associate Professor, Director, Library Resource Management, University of Oregon

Scott Alan Smith, Library Director, Langlois Public Library

APPENDIX—PREVIOUS CONFERENCE PROGRAMS

Feather River Institute, May 18–20, 1990
Conference title: Acquisitions in the West.
Keynote address: Melvin Voight, University Librarian
Emeritus, University of California, San Diego
Historical perspectives on acquisitions/collection development
Lonesome Dove: Doing business in the West.
Adrian Alexander, The Faxon Company
Marcia Anderson, Arizona State University
Dana D'Andraia, University of California, Irvine
Steve Pugh, Yankee Book Peddler
How the West was won.
Marion Reid, California State University, San Marcos
Marilyn Myers, Arizona State University West Campus
Melvin Voight, University Librarian Emeritus, University of California, San Diego
Gunfight at the OK Corral: Holding your own with collection development.
Henry Yaple, Whitman College
Young guns: library education today.
Bill Fisher, San Jose State University

Feather River Institute, May 29–June 1, 1991
Conference title: Acquisitions conference.
Whether it is better to be loved or feared: Librarianship as Machiavelli might have described it.
Jan Maxwell, University of Oregon
Personnel management: a vendor's perspective.
Bob Mastejulia, Baker & Taylor
Acquisitions personnel management—a librarian's perspective.
Richard Brumley, Cal Poly, San Luis Obispo
Toward greater objectivity: formal production standards for processing units in libraries.
Andrew Shroyer, University of California, Santa Barbara
Selection and development of library professionals as sales representatives.
Adrian Alexander, The Faxon Company
The Changing role of the library supplier.
Don Satisky, Blackwell North America

Collection development and automation: threat or promise?
Margo Sasse and Patricia Smith, Colorado State University
A knowledge-based expert system application in library acquisitions: monographs.
Pam Zager, Iowa State University
Acquisitions or access: impact and implications of electronic publishing.
Bill Fisher, San Jose State University

Feather River Institute, May 28–31, 1992
(No unique conference title in 1992.)
Haven't I seen you somewhere before?: a history of acquisitions.
Karen Schmidt, University of Illinois, Urbana-Champaign
Acquisitions principles and the future of acquisitions.
Joe Barker, University of California, Berkeley
The Tenets of acquisitions?: library acquisitions: practice and theory.
Joyce Ogburn, Yale University
How the Richard Abel Company changed the way we work.
Ann O'Neill, University of North Carolina, Chapel Hill
A Brief history of the Richard Abel Company.
Dora Biblarz, Arizona State University
Ethics: vendor perspectives.
Mary Devlin, The Faxon Company
Acquisitions ethics: the evolution of models for hard times.
Mary Bushing, Montana State University
Firing an old friend, painful decisions: the ethics between librarians and vendors.
Roger Presley, Georgia State University
Coordinating collection development in a hostile environment: the impact of reduced funding on organization for collection development.
Wanda Dole, SUNY Stony Brook
History of library-vendor relations since 1945.
Bill Fisher, San Jose State University

There was no conference in 1993.

Feather River Institute, May 19–22, 1994
Conference title: Dealing with change, or changing the deal.
Developing a vision for acquisitions.
Karen Schmidt, University of Illinois
The Upside of downsizing.
Karen Cargille, University of California, San Diego
Peter Stevens, University of Washington
Library organizational culture in strategic planning.
Joe Barker, University of California, Berkeley

Postmodern acquisitions: organizing to change.
    Sharon Propas, Stanford University
Darwinism in technical services: natural selection in
    an evolving information delivery environment.
    Christian Boissonnas, Cornell University
Does TQM really help anyone?
    Bill Fisher, San Jose State University
Awards from the front lines: participative recogni-
    tion of peer performance.
    Steve Marquardt, University of Wisconsin, Eau
    Claire
Academic librarians: acculturation and efficacy.
    Mary Bushing, Montana State University
Case studies of ethics and business practices.
    Mary Devlin, The Faxon Company
    Meta Nissley, California State University, Chico
Profiling the internet: can vendors survive in
    cyberspace?
    Stephen Pugh, Yankee Book Peddler
    Rick Lugg, Yankee Book Peddler

Feather River Institute, May 18–21, 1995
    Conference title: Controversies and debates in
        acquisitions.
    The Economics of monographs acquisitions: results
        of a time/cost study.
        Pam Rebarcak, Iowa State University
    Building bridges between acquisitions and collec-
        tion development: communication models for
        the electronic environment.
        Margaret Axtmann and Barbara Stelmasik,
        University of Minnesota
    Prioritizing firm order costs and vendor services.
        Scott Alan Smith, Blackwell's
        Richard Brumley, Oregon State University
    Debate: Any technical services operation can sustain a
    25% budget reduction without significant erosion of
        service.
    Reorganization revisited: acquisitions as an endan-
        gered species.
        Eleanor Cook, Appalachian State University
    Sleeping with the enemy: the love/hate relationship
        between acquisitions and collection development.
        Karen Cargille, University of California, San
        Diego
        Douglas Cargille, California State University,
        San Diego
    I Love me, I love not: schizophrenic behavior among
        acquisitions/collection development librarians.
        Terry Allison, California State University, San
        Marcos
    What Does acquiring require, or, Is there a librarian
        in the acquisitions department?
        Kirk Russell, Brigham Young University
    Borderline ethics and business practices: case studies.
        Mary Devlin, Blackwell's
        Meta Nissley, California State University, Chico

Library management: the latest fad, a dismal sci-
    ence, or just plain work?
    Bill Fisher, San Jose State University
Debate: Consolidation of purchases with a single
    vendor makes good business sense - yes or no.

Feather River Institute, May 16–19, 1996
Conference title: Death and rebirth in library acquisitions.
History of forgotten library technologies.
    Karen Schmidt, University of Illinois, Urbana-
    Champaign
When bad things happen to good vendors: lessons for
    librarians.
    Adrian Alexander, Swets
    Mike Markwith, Swets
Rearranging the universe: reengineering, reinventing,
    and recycling.
    Sharon Propas, Stanford University
The Fugitive: gifts, exchanges, and booksellers to
    acquire the elusive monograph.
    Tom Leonhardt, University of Oklahoma
Panel debate: Libraries are businesses.
T2: Theory in acquisitions revisited.
    Joyce Ogburn, Old Dominion University
Applying serials and acquisitions skills on the "other
    side".
    Nancy Chaffin, Arizona State University West
Realities of the team model in library technical services.
    Rita Echt, Michigan State University
Evolution of acquisitions in the electronic information
    realm.
    Chris Hector, University of Waikato
Analysis of acquisitions and collection development
    web pages.
    Steve Johnson, Clemson University
Acqnet & Acqweb: past, present, and future.
    Eleanor Cook, Appalachian State University

Feather River Institute, May 15–18, 1997
    Conference title: Oh, the Change is going to do me
        good...
    Library schools and library education: change
        agents for the profession.
        Bill Fisher, San Jose State University
    What's for dinner? Continuing education after the
        MLS.
        Ann O'Neill, University of South Carolina
    Unhurried change.
        Larry Ostler and Terry Dahlin, Brigham Young
        University
    Acquisitions packages and licensing agreements:
        changing the way we do business.
        Scott Wicks, Cornell University
    Licensing information: where can we go from here?
        Steve Bosch, University of Arizona
        Kittie Henderson, EBSCO
        Brian Schottlaender, UCLA

Acquisitions and collection development in an era of full-text and package deals.
  Nancy Persons, Sonoma State University
  Scott Wicks, Cornell University
Changing roles, changing services: partnerships in the new information age.
  Mike Markwith, Swets
Developing a relationship with your new subscription agent/understanding your new client: 1001 Thai burritos on the road to great two-way communication.
  Sandy Barstow, University of Wyoming
  Barbara Woodford, EBSCO
Dynamic partnerships in the library marketplace.
  Kit Kennedy, Blackwell's Periodicals
  Anne McKee, Blackwell's Periodicals
Got any spare change?
  Ron Ray, University of the Pacific
The Books are shelf ready, are you?
  Rick Lugg, Yankee Book Peddler
  Albert Joy, University of Vermont
The Undead in library technical services: activities and attitudes that have exhausted their life, but which refuse to die.
  Ron Ray, University of the Pacific.
Integrating solutions: examining the collection management process using OCLC's Electronic Collections Online as a model.
  Elizabeth Cooley, OCLC
  Chip Nilges, OCLC

Feather River Institute, May 14–17, 1998
  Conference title: "As I walked through the wilderness of this world" - Acquisitions: the Pilgrim's progress.
  The Five blind men and the elephant: differing perceptions of the introduction of an approval plan.
    Martin Cohen, McGill University
    James Galbraith, Blackwell's
  Slicing the pie: implementing and living with a journal allocation formula.
    Robert Soregnfrei, Colorado School of Mines
  How far have we come: benchmarking time and costs for monograph purchasing.
    Nancy Slight-Gibney, University of Oregon
    When write is wrong: a look at the professional literature in library and information science.
    Bill Fisher, San Jose State University
  Publishers on the web: from Addison to Ziff.
    Virginia Scheschy, University of Nevada, Reno
  Who's number one: evaluating acquisitions departments.
    Peter Stevens, University of Washington
  Libraries and vendors working together for a change.

Keith Schmiedl, Coutts
Libraries "in Dutch": their response to the Elsevier-Kluwer merger.
  Carol MacAdam, Swets

There was no conference in 1999.

The Acquisitions Institute at Timberline Lodge, May 20–23, 2000
  Conference title: Acquisitions and Collection Development in the twenty-first century.
  Core competencies for the acquisitions librarian.
    Bill Fisher, San Jose State University
  Aiming for continuous improvement: performance measurement in a re-engineered technical services.
    Glenda Smith, Griffith University
  Physical setting and organizational success.
    Mary Ellen Kenreich, Portland State University
  Consortial purchases: are they really worth it?
    Deborah Carver, University of Oregon
  Be careful what you wish for; or, all for nothing—nothing for all.
    Julia Gammon, University of Akron
    Rick Lugg, R2 Consulting
  Making one size fit all: a legislative mandate for Minnesota State Colleges and Universities.
    Diane Richards, Minnesota State University, Mankato
  E-Journals: The OhioLINK experience.
    Carol Pitts Dietrich, Ohio State University
  California State University journal access core collection.
    Marc Langston, California State University, Chico
  There is more to OP ordering than searching the internet.
    Sharon Propas, Stanford University
    James Bryant, Backer-Bryant Academic Library Services
  A New social contract for acquisitions and collection development.
    Martin Cohen, McGill University
  Book selection responsibilities for the reference librarian.
    Robert Sorgenfrei, Colorado School of Mines
    Christopher Hooper-Lane, Colorado School of Mines
  Collection development for the digital age.
    Sandra Kerbel, University of Virginia
  E-Book panel discussion.
    Bonnie Allen, Oregon State University
    Carol Pitts Dietrich, Ohio State University
    Scott Alan Smith, Blackwell's

The Acquisitions Institute at Timberline Lodge, May 19–22, 2001
  Conference title: Acquisitions and collection development: The Beat goes on—cool jazz or hard rock?

Keynote address: Deregulation and the Academy: Opportunities and strategies under the new rules for resource gathering.
Frank D'Andraia, University of Montana

Collections and systems: a new organizational paradigm for collection development.
John Webb, Washington State University

E-Journal databases—a long term solution?
Nathalie Schultz, Griffith University

Evolution of the supply chain in library bookselling.
Jackie Coats, University of Washington
Matt Nauman, Blackwell's Book Services
Brian Elliott, Alibris
Jeff Dixon, Amazon.com
Robert Rooney, Taylor & Francis

E-Books, the latest word.
Bonnie Allen, Oregon State University
Len Liptak, ebrary
Katharine Phenix, netLibrary
Robert Rooney, Taylor & Francis

Accounting for access (costs and benefits).
Karen Schmidt, University of Illinois
Nancy Slight-Gibney, University of Oregon

Putting theory into practice: needs analysis.
Karen Rupp-Serrano, University of Oklahoma

Finding the right balance: campus involvement in the collections allocation process.
Lisa German, University of Illinois
Karen Schmidt, University of Illinois

Impact of organizational structure on acquisitions and collection development.
Bill Fisher, San Jose State University

On the nature of our business: A forum.
Richard Brumley, Oregon State University

The Acquisitions Institute at Timberline Lodge, May 18–21, 2002

Conference title: Acquisitions and collection development on the mountain.

Keynote address: Vessels and voyagers: some thoughts on reading and writing, books and libraries.
Dr. Richard Hume Werking, U.S. Naval Academy

Print to electronic journal conversion: criteria for maintaining duplicate print subscriptions.
Sandy Campbell, University of Alberta

Joint remote storage projects.
Carol MacAdam, JSTOR

The California State University E-Book pilot project.
Marc Langston, California State University, Chico
Nancy Noda, San Francisco State University

Publishing and library bookselling: trends in publishing and information distribution.
Adam Chesler, Kluwer Academic Publishers
Matt Nauman, Blackwell's Book Services

Robert Rooney, Taylor & Francis
Jannette Schuele, YBP Library Services
Jonathan Weiss, Oxford University Press

Collection assessment: what do you do with the data?
Peggy Cooper, Boise State University
Marita Kunkel, Oregon Institute of Technology
Faye Chadwell, University of Oregon

Electronic resources librarians.
Bill Fisher, San Jose State University

Going back to the future: Deconstructing a subject based acquisitions structure.
Karl Debus-Lopez, University of Wisconsin

Automating selection through cataloging: workflow redesign.
Jackie Coats, University of Washington
Joe Kiegel, University of Washington

Acquisitions and collection development in the age of content.
Christa Easton, Stanford

Digitization projects in acquisitions.
Jackie Coats, University of Washington
Lisa Spagnolo, University of Washington

The Acquisitions Institute at Timberline Lodge, May 17–20, 2003

Conference title: Library collections and information access: Great notion or cuckoo's nest?

Keynote address: Collection development in interesting times.
Michael Gorman, California State University, Fresno

Publishers and publishing: panel discussion.
Bonnie Allen, Moderator, Oregon State University
Pat Soden, University of Washington Press
Laura Driussi, University of California Press
Sloane Lederer, Cambridge University Press
Amy Yodanis, Blackwell Publishing

Now you see it, now you don't.
Bill Fisher, San Jose State University

Copyright law for librarians.
Robert Ogden, Gursky & Ederer, LLP

Gearing up for the next insanely great thing: training for new skill sets and new mindsets.
Marita Kunkel, Oregon Institute of Technology
Kit Kennedy, Swets Blackwell

External forces impacting twenty-first century library collection development—the transformation of decision-making power.
Carolyn Henebry, University of Texas at Dallas
Ellen Safley, University of Texas at Dallas

Managing electronic resources in a time of shrinking budgets.
Emily Miller-Francisco, Southern Oregon University

Collection assessment: panel discussion.

Nancy Slight-Gibney, Moderator, University of Oregon

Laurel Kristick, Oregon State University

Cyril Oberlander, Portland State University

Diane Richards, Minnesota State University, Mankato

Mark Stengel, California State University, San Marcos

Dan Streeter, Portland State University

Collection management as risk management.

Mary Casserly, University at Albany, State University of New York.

The Journal stop: a complete serials information system.

Peggy Cooper, Boise State University

Dan Lester, Boise State University

The Case for acquisitions standards in integrated library systems.

Katherine Farrell, Princeton University

Marc Truitt, University of Notre Dame

Bringing library content to the palm of users' hands.

Denise Koufogiannakis, University of Alberta

Pam Ryan, University of Alberta

Outsourcing your overflow (let the vendor do the work for you).

Jeff Earnest, National University, San Diego

Bud Sonka, National University, San Diego

The Acquisitions Institute at Timberline Lodge, May 15–18, 2004

Conference title: New discoveries on Lost Lake.

Keynote address: Golden rods, tub files, encumbrances, and how to put the cat among the pigeons—or, what I learned in acquisitions, 1973–1987.

Henry Yaple, Whitman College

Defining functional requirements for acquisitions records: vendor metadata.

Katherine Farrell, Princeton University

Marc Truitt, University of Houston

Panel discussion: Acquisitions standards.

Martha Gettys, VTLS

Katherine Farrell, Princeton University

Dan Miller, Blackwell's Book Services

Jannette Schuele, YBP Library Services

Marc Truitt, University of Houston

The Impact of technology on reading behavior.

Bill Fisher, San Jose State University

A Leap in the dark: a pilot project for an e-only engineering collection.

Laurel Kristick, Oregon State University

Margaret Mellinger, Oregon State University

Migration to electronic journals: one library's journey.

Gwen Bird, Simon Fraser University

The Scholarly journal: an editor's perspective.

Pat Wheeler, Oregon State University

Panel discussion: publishers and publishing.

Bonnie Allen, Moderator, Oregon State University

Daviess Menefee, Elsevier

Peter Milroy, University of British Columbia Press

Niko Pfund, Oxford University Press

Changing horses midstream: restructuring collection development.

Stefanie Wittenbach, University of California, Riverside

Cooperative collection development in an interdisciplinary subject.

Sarah George, Illinois Wesleyan University

Serials decision database: selection, management and evaluation of print and online journals.

Diane Carroll, Oregon Health & Science University

From data to decisions: using surveys and statistics to make collection management decisions.

Julie Blake, St. Cloud State University

Susan Schleper, St. Cloud State University

The Ethics of academic collection development in a politically contentious era.

Wendy Highby, University of Northern Colorado

The Acquisitions Institute at Timberline Lodge, May 14–17, 2005

Conference title: The View from the mountain.

Keynote address: Wines of fear, wines of conviction.

Lewis Miller, Butler University

Retail bookselling and libraries.

Phil Wikelund, Great Northwest Bookstore

Publishers and publishing: panel discussion.

Bonnie Allen, Oregon State University

Ellen Endres, Brill Academic

Greg Giblin, Wiley

Living with e-books.

Carolyn Henebry, University of Texas at Dallas

Ellen Safley, University of Texas at Dallas

New opportunities for technical services: reconsidering Priorities, reshaping processes.

Kathy Carter, University of Alberta

Marilyn McClary, University of Alberta

Kit Wilson, University of Alberta

Problem analysis and appreciative inquiry for acquisitions management.

Cynthia Coulter, University of Northern Iowa

The Impact of technology on learning.

Bill Fisher, San Jose State University

Making serials purchasing decisions using pay per view usage statistics.

Beth Bernhardt, University of North Carolina, Greensboro

The Contours of collection development in a consortial setting.

Faye Chadwell, University of Oregon
Cara List, University of Oregon

There was no conference in 2006.

The Acquisitions Institute at Timberline Lodge, May 19–22, 2007
Publishers and publishing: panel discussion.
David Jackson, Stanford University Press
Medhi Khosrow-Pour, IGI Global
Pascal Schwarzer, Springer Verlag
Mitigating loss of content in times of shrinking collections budgets.
Bonnie Allen, University of Montana
Library support for open access publishing.
Faye Chadwell, University of Oregon
Developing a print repository in a consortial environment.
Linda DiBiase, University of Washington
Susan Hinken, University of Portland
Mark Watson, University of Oregon
Thinking critically about critical thinking: the impact of information literacy on teaching and learning.
Bill Fisher, San Jose State University
Black, white, or grey: organizing acquisitions and collection development.
Paula Popma, California State University, Fresno
Kimberly Smith, California State University, Fresno
It's no longer the tip of the iceberg: navigating change in technical services.
Bill Kara, Cornell University
Anna Korhonen, Cornell University
Post-modern acquisitions revisited.
Sharon Propas, Stanford University
Building a high quality flexible acquisitions workflow in a Web 2.0 world.
Adam Wathen, Kansas State University
New liaisons, new ideas, new energy.
Peggy Cooper, Boise State University
Memo Cordova, Boise State University
Melissa Kozel, Boise State University
Richard Stoddart, Boise State University
Metadata planning: keys to communication.
Sheila Bair, Western Michigan University

The Acquisitions Institute at Timberline Lodge, May 17–20, 2008
Keynote address: Achieving the impossible? Reflections on transformational change in the system of scholarly communication.
Ray English, Oberlin College
Strategically building collections: attempts to coordinate budgeting, tracking, data gathering, and policy making.
Adam Wathen, Kansas State University
The Lois Hole Campus Alberta Digital Archive: building collections for a diverse community

Kit Wilson, University of Alberta
"Core" and "long tail" materials in collection development consortia.
Katalin Radics, UCLA Collections 2.0
Margaret Mellinger, Oregon State University
New liaison, new ideas, new energy: one year later.
Peggy Cooper, Boise State University
Melissa Kozel, Boise State University
Richard Stoddart, Boise State University
From fraught to unfettered: developing tools to help librarians improve e-book selection.
Sarah Polkinghorne, University of Alberta
Denise Koufogiannakis, University of Alberta
Collection development in an open access world.
Karen Estlund, University of Oregon
John Russell, University of Oregon
Making sure we have well qualified staff to do the work of acquisitions.
Leo Agnew, University of Missouri, Columbia
Mike Arnold, Blackwell's Book Services
Karen Daring, University of Missouri, Columbia
Recruiting librarian to careers in technical services.
Deborah Thomas, University of Tennessee
Bill Fisher, San Jose State University
Collections connection: streamlining the interface between collection development and acquisitions.
Katherine Treptow Farrell, Princeton University
Adventures in wonderland: subject librarians and assessment in collection development.
Allison Level, Colorado State University
Michelle Wilde, Colorado State University

The Acquisitions Institute at Timberline Lodge, May 16–19, 2009
Keynote address: Open data: encouraging open source and integration between vendors and the library developer community.
Terry Reese, Oregon State University
Going next gen: OCLC's WorldCat Local and Acquisitions.
Keith Powell, University of California, Irvine
Lisa Spagnolo, University of California, Irvine
But how do we pay for it? Realigning the budget to support new access models.
Elizabeth Mengel, Johns Hopkins University
R. Cecilia Knight, Grinnell College
Patricia Tully, Wesleyan University
Streaming video: issues, challenges, and new horizons.
Mike Hill, Brigham Young University
Rebecca Schroeder, Brigham Young University
Julie Williamsen, Brigham Young University
Managing personnel resources in tough times: things we've learned to do (and to avoid next time).
Karen Darling, University of Missouri, Columbia
Priorities and policies for gift operations: best practices for handling gift materials.

Michael A. Arthur, University of Central Florida
Kelli Getz, University of Houston
Squeezed between the e's: electronic resources, print monographs, economic woes.
Lisa Barricella, East Carolina University
Joseph Thomas, East Carolina University
Herding cats: providing oversight, support, and coordination to library liaisons at a small academic library.
Karen Jensen, University of Alaska, Fairbanks
Collecting together: developing the Crater Lake Digital Research Collection.
Anne Hiller Clark, Oregon Institute of Technology
Lia Vella, Oregon Institute of Technology
All the water in the ocean—all the books on the sea: collaborative collection development in Oregon's Marine Laboratory Branch Libraries and beyond.
Barb Butler, University of Oregon
Janet Webster, Oregon State University

The Acquisitions Institute at Timberline Lodge, May 15–18, 2010
Welcoming remarks: Camila Alire, President, American Library Association
Keynote address: Katina Strauch, College of Charleston
E-books: taking the next step.
Diane Carroll, Washington State University
Evidence based decision making: data mining for serials cancellations.
Dean Walton, University of Oregon
Developing a new allocation model for the library budget.
Sandra Shropshire, Idaho State University
Assessing the Evergreen Acquisitions Module from an academic library perspective.
Karen Clay, Eastern Oregon University
Whither journals? or wither journals?
Faye Chadwell, Oregon State University
Kittie Henderson, EBSCO
Daniel Morgan, Elsevier
Mary Nugent, Taylor & Francis
Twittering for books: patrons using social media for library collection requests.
Joan Petit, Portland State University
Buy request: just in time versus just in case as OSU Libraries.
Uta Hussong-Christian, Oregon State University
Kerri Georgen-Doll, Oregon State University
OARS: toward automating the ongoing subscription review.
Jonathan H. Harwell, Georgia State University
The Serials decision database as a research tool.
Diane Carroll, Washington State University
Joel Cummings, Washington State University
What to do with reference collections?

Paul Frantz, University of Oregon
Cheryl Middleton, Oregon State University
Overcoming the shrinking budgets blues—creative and collaborative acquisitions strategies at Cornell
University Library in a time of financial constraints.
William J. Kara, Cornell University
Boaz Nadav-Manes, Cornell University
The Consortial balancing act: how your consortium impacts collection development.
Alice Crosetto, University of Toledo
Round book in a square hole: providing access to artists' books.
Cara List, University of Oregon

The Acquisitions Institute at Timberline Lodge, May 14–17, 2011
Welcoming remarks: Molly Raphael, President-Elect, American Library Association
Keynote address: The Opening hand.
Joseph Janes, University of Washington
This is not a test…. Patron-driven acquisitions.
Marcia Anderson, Arizona State University
Ginny Sylvester, Arizona State University
The e-book ecosystem: what are the challenges?
Panel discussion.
Douglas Armato, University of Minnesota Press
Emily McElroy, Oregon Health and Science University
Michael Zeoli, YBP Library Services
Using academic course data to assess library collection strength.
April Henson, Naropa University
Mark Kille, Naropa University
Project GIST: The Getting it System Toolkit; tools to enhance and streamline workflows.
Cyril Oberlander, SUNY College at Geneseo
GIST for you!
Kristina DeShazo, Oregon Health and Science University
Rethinking the role of collection development librarians.
Kerry Scott, University of California, Santa Cruz
Demand-driven acquisitions: patrons riding shotgun.
Linda DiBiase, University of Washington
Alison Bobal, Oregon State University
Collection use from the other side: assessing print collection usage by humanities faculty.
Charlene Kellsey, University of Colorado, Boulder
Jennifer Knievel, University of Colorado, Boulder
Blended libraries: becoming one family.
Cheryl Adams, Utah State University
Lori Brassaw, Utah State University—College of Eastern Utah

Angela Dresselhaus, Utah State University

Betty Rozum, Utah State University

Moving beyond 10Ks: collecting and providing access in a business library.

Nathan Rupp, University of Michigan

Maintaining an LGBT studies collection in an academic library.

Rachel Wexelbaum, St. Cloud State University

Science video journals in academic research and education.

Moshe Pritsker, JoVE

A Patron-driven acquisition project in a consortial environment.

Catherine Davidson, York University

Tony Horava, University of Ottawa

Jumping without a parachute: transitioning to on-demand purchasing for journal articles.

Camila Gabaldon, Western Oregon University

Implementing an e-preferred approval plan: assessing duplication.

Carmelita Pickett, Texas A&M University

Demonstrating the value of scholarly collections through ROI and other methods.

Kira Cooper, Elsevier

Regina Mays, University of Tennessee

Carol Tenopir, University of Tennessee

The Future of collection development and acquisitions.

Panel discussion.

Faye Chadwell, Oregon State University

Peggy Cooper, Boise State University

Mel DeSart, University of Washington

Stephen Pugh, Oranjarra Partners

Scott Alan Smith, Kent State University

The Acquisitions Institute at Timberline Lodge, May 19–22, 2012

Welcome message: Molly Raphael, American Library Association.

Keynote address: The Changing landscape of academic libraries.

Susan Gibbons, Yale University

Using service design thinking to reinvent acquisitions.

Steven Sowell, Oregon State University

All textbooks in the library! The Portland Community College experience.

Tony Greiner, Portland Community College

Scholarly publishers, research data and implications for academic library collections.

Jeremy Kenyon, University of Idaho

Nancy Sprague, University of Idaho

Patrons: the new subject selectors?

Kerri Georgen-Doll, Oregon State University

E-books: they're a heartache, nothing but a heartache.

Stacey Marien, American University

Kari Schmidt, American University

E-books at the tipping point? How faculty view e-books for teaching and research.

Laurel Kristick, Oregon State University

Margaret Mellinger, Oregon State University

The Value of purchasing e-book collections from a large publisher: a usage-based analysis of Oxford University Press e-books.

Jennifer W. Bazeley, Miami University, Ohio

Aaron K. Shrimplin, Miami University, Ohio

Beyond ROI: transitioning assessment into action.

Denise Pan, University of Colorado, Denver

Leslie Williams, University of Colorado, Health Sciences

Additional authors not attending:

Yem Fong, University of Colorado, Boulder

Gabriella Wiersma, University of Colorado, Boulder

Discussion among all participants about the impact of Web-scale discovery service.

Alice Eng, University of North Florida

Stephanie Weiss, University of North Florida

Managing electronic resources with open source software: perspectives from a small and medium sized academic library.

Roen Janyk, Okanagan College

Sandra Wong, Simon Fraser University

Stanford University Libraries and everyday electronic materials.

Sharon Propas, Stanford University

Electronic sandbox: cross-organizational electronic resource management.

Jesse Koennecke, Cornell University

Colleen Major, Columbia University

Boaz Nadav-Manes, Cornell University

Greening your acquisitions from A to Z.

Marie Bloechle, University of North Texas

How suite it is: rehabbing acquisitions at UNC.

Caroline Norton, University of Northern Colorado

Rick Kerns, Creighton University

Managing and preserving unique library collections.

Scott Devine, Northwestern University

Show and tell: new artists' books at the University of Oregon's Architecture and Allied Arts Library.

Cara List, University of Oregon

The Acquisitions Institute at Timberline Lodge, May 18–21, 2013

Keynote address: Beyond measure: valuing libraries.

Chris Bourg, Stanford University

Should we buy Angry Birds? And other questions for collection development and acquisitions librarians in a world of iPads and Apps.

Juleah Swanson, Ohio State University

All that counts: evaluation of an aggregator package from three libraries.

James Bunnelle, Lewis & Clark College

Jill Emery, Portland State University

Emily McElroy, Oregon Health and Science University

New models for publishing, packaging, and purchasing: implications for libraries and suppliers. Panel discussion.
John Dove, Credo Reference
Kittie Henderson, EBSCO
Zac Rolnik, now publishers
Michael Zeoli, YBP Library Services

Maine shared collections strategy.
Sara Amato, University of Maine

Two versions on MI-SPI: a collaborative approach to deselection and collection management.
Randy Dykhuis, Midwest Collaborative for Library Services
Pamela Grudzien, Central Michigan University

Monographs: titles cited in faculty-authored works.
Amy Lana, University of Missouri, Columbia

Citation analysis as a tool for collection development and instruction.
Karen Kohn, Arcadia University

Americans with Disabilities Act compliance and library acquisitions.

Angela Dresselhaus, University of Montana

Library transitions: collections and access before, during and after a library renovation.
Eric Hanson, University of Portland
Susan Hinken, University of Portland

Collaborative endeavor between TRLN, OUP, and YBP.
Lenny Allen, Oxford University Press
Ann-Marie Breaux, YBP Library Services
Nancy Gibbs, Duke University

Consortial patron driven acquisitions.
Marc Langston, California State University, Chico
Jodi Shepherd, California State University, Chico

Adding a use factor measure to a materials allocation plan for books in an academic library.
Lisa Barricella, East Carolina University

What is unique about unique? An Assessment to uncover and value distinct collections within the main library collections.
Laurel Kristick, Oregon State University
Rick Stoddart, Oregon State University
Andrea Wirth, Oregon State University

African Librarianship

Natalia T. Bowdoin
University of South Carolina Aiken, Aiken, South Carolina, U.S.A

Janet Lee
University of Denver, Denver, Colorado, U.S.A, and
Regis University, Denver, Colorado, U.S.A

Abstract

Africa is a continent of contrasts, comprised of 54 countries that vary in size from the expansive Sudan to small island nations. Diversity of languages and cultural traditions abound both among the nations within this second largest continent and within each nation. Each nation has diverse human populations, mineral resources, biological diversity, and political and social histories. Several have written records and libraries that date back centuries such as the Library of Alexandria in Egypt, and the Ge'ez scripts housed in the monasteries of Ethiopia and Eritrea. One possible unifying factor among the African nations, with the exception of Ethiopia and Liberia, is the legacy of colonialism, which greatly permeates life today despite independence in the 1960s. African libraries were modeled after European libraries, frequently created to serve the colonials themselves or to exert intellectual or religious control over indigenous populations. New approaches and views of African librarianship began to be articulated following the end of the colonial era, calling for a revisioning of African librarianship reflecting on the social and cultural values of Africans.

INTRODUCTION

Africa is the second largest continent, measuring approximately 5000 miles (8000 km) from north to south and approximately 4600 miles (7400 km) from east to west.[1] Excluding the associated dependent island nations and contested states such as Somaliland and Western Sahara, Africa comprises 54 countries, each recognized by the United Nations.[2,3] Although the second largest continent, Africa constitutes 13.2% of the world population in the year 2000.[4] Despite a small population density, it is an extremely diverse continent with regard to human populations, mineral resources, biological diversity, and political and social histories. Its people use approximately 2000 indigenous African languages that represent a third of the world's linguistic diversity,[5] as well as five international or colonial languages (English, French, Portuguese, Spanish, and Arabic).

A comprehensive definition and discussion of African librarianship is therefore extremely difficult, if not impossible. For purposes here, however, the main historical events and evolving trends, perceptions, and practices that have contributed to African librarianship over time will be highlighted. The developments in the past decade that indicate new directions and paths for African librarianship will also be discussed.

AFRICAN LIBRARIES: THE HISTORICAL CONTEXT

Precolonization

The continent of Africa was home to some of the oldest libraries and collections of documents in history, The Library of Alexandria in Egypt, being the most famous example from classical antiquity. The Alexandrian Library was founded by the Ptolemies in the third century BC. While physically located on the African continent, the library was principally composed of Greek literature and was modeled on the organization of libraries in Athens.[6] It has been argued that for the Greeks of the Ptolemaic empire, their literature represented the culture that legitimized their rule over indigenous populations of the region. Similarly, their literature united them against the Egyptians as an intellectual community.[7] By these accounts, then, imperialism and cultural hegemony played an early role in library development on the African continent.

Early indigenous examples of African libraries include the monasteries of Ethiopia and Eritrea where Ge'ez script texts were housed and preserved for centuries. Similarly, the Mali and Songhai empires during the thirteenth to sixteenth centuries in West Africa contained resource centers with many scholarly texts. Sankoré University in

Encyclopedia of Library and Information Sciences, Fourth Edition DOI: 10.1081/E-ELIS4-120049506

Timbuktu (in present day Mali) remained a vitally important center of higher learning during the middle ages.[8] Another example of early African print culture is the East African history, *Kitab al-Sulwa* (known in English as *The Kilwa Chronicle*) written in Arabic by an unknown author in the early to mid-sixteenth century.[9] While few libraries that housed print resources existed on the African continent before European penetration and colonization, the practice of oral tradition played an extremely strong and similar role to traditional print libraries across Africa for centuries. Anaba A. Alemna points out that, "There is ample evidence to indicate that oral librarianship is appropriate within the African context, given that African traditional societies did have information specialists whose official and social functions resided in the narration of local history in public places."[10] While libraries and graphic records were by no means strangers to the African soil before colonialism, for much of the continent, the greatest venues for the preservation and transmission of knowledge and cultural values were distinctly in the realm of oral tradition and the arts.[11]

European Colonization and Librarianship

Despite the early examples of African libraries and librarianship mentioned above, the story of African librarianship has been heavily influenced by, and in many respects entwined with, the story of European colonization and imperialism, most particularly in sub-Saharan Africa.[12] Apart from Ethiopia, which had only a brief occupation by Italian forces for five years from 1936 to 1941, and Liberia, founded in 1822 by the American Colonization Society (ACS) as a colony for American free blacks who elected to immigrate to Africa until it gained its independence in 1847,[13] European colonization had a significant and enduring influence on all African nations that carried over into library development and librarianship. African colonization was carried out by Belgium, Great Britain, France, Germany, Italy, Portugal, and Spain. The apex of European colonization of Africa is represented by the Berlin Conference of 1884–1885 that effectively eliminated African self-governance and autonomy. Most African countries emerged from colonial rule in the early 1960s.

Colonial administrators and missionaries brought their European-modeled libraries with them during the period of African colonization. In some cases, these libraries were created exclusively to serve the colonials themselves, but in other cases they were used to exert a type of intellectual or religious control over the indigenous African populations. The Western-modeled libraries were stocked with materials that the colonial powers believed to be important for local populations. Indigenous knowledge and local systems of information sharing were disparaged.[14] For many local African populations, the colonial libraries were often viewed as propaganda tools for colonial administrators.[15] Collections in these libraries often glorified the Western powers or denigrated the primitiveness of indigenous populations. The African libraries that sprang up during the colonial period reflected Western ideals of librarianship in multifaceted ways—in their physical facilities, their materials, and in their ideology. The colonial powers and religious missionaries believed strongly that they could import the philosophy of librarianship that had developed in Europe and the United States to the continent without modification or respect for existing models of knowledge transmission and preservation. It was essentially expected that African populations would adapt to this model of librarianship rather than libraries adapting to the needs and desires of the African populations.[16] Western colonialists viewed the written word and the book as the pinnacle of knowledge and imposed this view on others. Colonial languages were favored and indigenous ones were dismissed. There was also no recognition that information needs might vary between communities, or between local populations and European populations. The imposition of such foreign structures resulted, for the most part, in the alienation of African communities from library services. While impressive buildings were sometimes erected to house these libraries, over time the collections and services tended to be seriously neglected.[17] This was particularly pronounced in public and school libraries, though academic libraries were in no way immune to this situation. As the economies and stability of many African countries declined after the optimism of the 1960s and 1970s, all types of libraries suffered setbacks. Books were often in short supply as a result of the national economic situations, lack of infrastructure in the book trade, and restrictions on imports.[18] Afeworki Paulos writes "But in general, the level of knowledge production of Africa's cultural output remained very low in post-colonial Africa. . .Further, as early as 1980s, university libraries in Africa were in serious financial difficulty. . .In the 1980s and the 1990s, library resources in most African universities no longer met the basic needs of users. The library was marginalized in the life of the university and it led to the use of 'alternative methods of accessing and acquiring information'."[14] Public libraries often housed collections that had little to no relevance to local populations while academic libraries in universities often contained outdated materials with little research value.[16]

Independence and Postcolonization Period—African Librarianship Responds

Despite gaining independence, the majority of African countries in the early 1960s maintained dependency relationships with their former colonial powers and such influence persisted for institutions, including libraries, across the continent. African libraries continued their colonial heritage in ways that were often stifling. Even after independence, many of the pioneer librarians were expatriates from Europe.[16] By the 1960s, there were three dominant European languages on the continent: English, French, and Portuguese. The role and importance of libraries in those countries and cultures associated with these languages differed greatly depending on whether the African nation was colonized by Britain, France, or Portugal. Portugal, for example, had a weak library tradition itself, and this was reflected in the lack of strong library institutions in the Portuguese colonies Angola, Cape Verde, Guinea-Bissau, and Mozambique. Similarly, the French library tradition was primarily intended for the scholar and bibliophile rather than the ordinary citizen. French libraries symbolized the heritage of a civilization rather than serving as an instrument for further progress. In the former French colonies, its few libraries had hardly any qualified staff and extremely limited public access to collections of published information. Sturges and Neill have suggested however that a positive aspect of this may be that the former Francophone countries had more freedom within which to invent new responses to information needs and provisions than the former British colonies had at independence.[19]

In contrast to France and Portugal, Britain passed on a legacy of library development that saw the library, particularly the public library, as a key component of a modern and complete nation state. Although this model of the library in the African colonial countries originally was almost exclusively racially restrictive, by the mid-twentieth century, just before independence, the library was being promoted as a means to expand literacy. The 1953 UNESCO Seminar at Ibadan, Nigeria, was an important turning point in library development serving as the affirmation of the belief that the expansion of public libraries in Africa would help increase literacy. Of the 29 participants, however, only eight were Africans. Also significant, several of the participants were not librarians but had other interests with regard to literacy and illiteracy. One tangible result of this seminar was the establishment of the West African Library Association. Another recommendation was the adoption of the UNESCO Public Library Manifesto that called for the public library to be a democratic institution, open for free and equal use by all members of the community. This would become the basis on which national public library service should be built in Africa. At the time of the seminar and adoption of the manifesto, public library service in Africa differed widely in achieving its ideals.[20]

The Anglo-American library tradition that was embodied in the UNESCO Manifesto had an effect on libraries throughout Africa after the UNESCO Seminar, particularly in the Anglophone countries. This tradition had firmly placed national library services as the foundation for all library services. However, there was limited success in providing what many would expect of a national library in terms of a national legal deposit collection, national bibliography, interlibrary loan services, and other cooperative programs. Instead, national libraries have

generally functioned as a centralized national public library service coordinating regional branches and postal loans or mobile libraries in some locations, and to mixed degrees of success. In contrast to these public libraries, the special library sector in Africa has been stronger and longer established. These include geological, agricultural, and other special libraries tied to government ministries, NGOs, and parastatals. Similar to public libraries, academic libraries were in a relatively weak state at the time of independence. The majority of academic libraries were established after the Second World War during a period of significant expansion in African educational efforts. Despite this expansion, the library was often placed in a central location, both physically and in the rhetoric espoused by these new African universities, the reality was that a great majority of these institutions could not afford to keep up a high-quality library collection to support teaching and research.[21] This had, and continues to have, a serious impact, on scholarship and teaching in most African countries.

In spite of the imposition and primarily negative legacy left by the colonial powers on library development in Africa, new approaches and views of African librarianship began to be proposed and articulated following the end of the colonial era in the 1960s. These views occasionally came from Europeans themselves though the majority of the calls for change came from within the continent. For example, one early European promoter of a revisioning of African librarianship came from Ronald Benge who fleshed out his ideas in his book *Cultural Crisis and Libraries in the Third World* after recognizing that Africa had its own social and cultural values and that librarianship in Africa should reflect and build on these values rather than simply following European models of the profession.[22]

Many African librarians also took up this cause. One of the strongest examples of this anticolonial call for a nonhegemonic library tradition can be found in Adolphe Amadi's book *African Libraries: Western Tradition and Colonial Brainwashing* published in 1981. Amadi detailed the negative effects and history of the colonial influence on African libraries and called for a radically different approach to information provision.[23] Another key contributor in this area was Kingo J. Mchombu whose article "On the Librarianship of Poverty" published in 1982, made the case that the conditions of poverty must be addressed in any discussion of information work in underdeveloped countries. In this way, Mchombu directed concerns in African librarianship to neglected populations living in the rural areas. He elaborated four principles that he saw as being necessary to make information work socially relevant in African countries:

1. That the chief factor determining information work in developing countries should be poverty rather than affluence.

2. That information work in developing countries differs markedly from information work in developed countries.
3. That it is possible to gather a body of knowledge on how best to meet this challenge.
4. That information workers must play an active role in the process of socioeconomic development.[24]

Mchombu's piece became a rallying cry for a whole generation of African librarians (and some of their European allies) who accepted his challenge to produce a new library and information model for Africa, particularly among rural communities. Mchombu argued that information services in developing countries should not blindly follow those of developed countries and that information workers need to develop an aggressive attitude and participate fully in the social struggle for national development. Mchombu concluded, unequivocally, "…Information workers must look for solutions to their problems within their own societies rather than depending on foreign aid."[25] During the second half of the twentieth century, many African librarians and information scholars carried forth Mchombu's challenge in a variety of ways. Some focused their efforts on the need for greater research on community needs and existing patterns of information transfer. Others examined ways that information services could repackage information into forms that were more culturally and socially acceptable by local populations. Others emphasized the importance of utilizing the oral tradition itself as an information resource within African libraries and the issues of availability of resources in local African languages. Related to this, many stressed the importance of recognizing the information needs and abilities of nonliterates.[26] By the closing of the twentieth century, there were many voices from within Africa as well as some from outside that had articulated, researched, and written about a substantial reordering of priorities for African librarianship than those presented in the model that had been imposed on the continent by the colonial powers.

Liberalization of Higher Education in Africa and Its Impact on Libraries

Liberalization of higher education, beginning in the 1990s, had a significant impact on higher education and on libraries within those institutions of higher education. Damtew Teferra noted, "The number of institutions in sub-Saharan Africa has increased from half a dozen in the 1960s—when most of the nations in the sub-region declared independence—to over 300 in 2003."[27] The total of institutions of higher education that grant four-year undergraduate and/or postgraduate degrees and that are officially recognized, licensed, or accredited by national or regional bodies in 2013 number over 600.[28] This rapid growth placed a heavy burden on the nations involved since these states provided upward of 90% of support for

the education of qualified students.[29] Paul Tiyambe Zeleza and Adebayo Olukoshi note that the "liberalization of African Universities has increasingly manifested itself in the growth of private universities and the privatization of programmes and funding sources."[30] Mahlubi Mabizela further reflects on the effects of global pressures, "The contemporary or new generation private HE is largely a direct consequence of the hegemonic neo-capitalist and neo-liberal post cold war social context."[31]

Public or private, this growth in the number of institutions of higher education has placed both added challenges and opportunities for the libraries serving those institutions. Budget cuts curtailed the acquisition of journals, core textbooks, and encyclopedias in a number of libraries, acknowledged the Association of African Universities.[32] Furthermore, Muyoyeta Simui states, "Inadequate funding for the purchase of new books and subscriptions to journals and other information resources rank amongst the most severe obstacles faced by African university libraries."[33] According to Simui, factors affecting funding range from "growing populations exerting new and expanded demands on government, debt burden, structural adjustment programmes whose conditions are unfavourable to social services like education to changes in government policy."[34] ICT (Information and Communication Technologies) has the potential to be the great equalizer as the continent adopts these technologies. Teferra cites L.A. Levey in stating, "despite numerous hurdles, the major factor that catalyzed the development of ICT in the continent (especially e-mail) was the great thirst and enormous need for academic and scholarly communication in the land of forbidding cost and extremely poor communication services."[35] Libraries are responding to these challenges through consortia, use of open source products, and the promotion, use, and creation of open access (OA) journals and directories, all of which are discussed below. One library that exemplifies the modern African university library and the possibilities for these challenges to be overcome is Kenyatta University in Nairobi.[36] Services provided by the library include a robust online catalogue (KOHA, Open Source Integrated Library Management software), institutional repository, e-books, electronic journals, chat reference, a full complement of professional staffing, adaptive technology, 370,000 volumes of books and bound periodicals, current journals, audio visuals magazines and dailies, and 74 computers for student use. The library is a member of the following organizations: Kenya Library and Information Services Consortium (KLISC) and the Kenya Library Association (KLA).[37]

NEW DEVELOPMENTS IN AFRICAN LIBRARIANSHIP

Several new developments since the end of the twentieth century and the beginning of the twenty-first century have influenced visions for a new way forward for African librarianship. These new visions revolve around the themes of unity, pan-African responses, and a revisioning and redefining of information ethics and Library and Information Science (LIS) education for the African context.

Pan-African Responses: African Library Summits and Emergence of New Organizations

Beginning in the late twentieth century, a greater effort was being made to have a pan-African response to library and information challenges in Africa. One such example of this is the biennial conference of the Standing Conference of Eastern, Central and Southern African Library and Information Associations (SCECSAL). SCECSAL derived its origins from the East African Library Association (EALA), which was founded in 1957. When the EALA dissolved in 1970, members agreed to continue to hold a biennial conference and invited other countries in the subregion. SCECSAL is now one of the largest and fastest growing library and information professional associations in Africa. Membership to SCECSAL is open to national library and information associations from the following countries: Angola, Botswana, Burundi, Congo Republic, Democratic Republic of Congo, Djibouti, Eritrea, Ethiopia, Kenya, Lesotho, Malawi, Madagascar, Mauritius, Mozambique, Namibia, Rwanda, Seychelles, Somalia, South Africa, South Sudan, Sudan, Swaziland, Tanzania, Uganda, Zambia, and Zimbabwe. The first SCECSAL conference was held in Dar es Salaam, Tanzania, in 1974. There have since been a total of 20 SCECSAL conferences to date, which have rotated within the countries represented in the organization. Themes of SCECSAL conferences have included a wide range of topics from "Enhancing democracy and good governance through effective information and knowledge services" to "From Africa to the World—the globalization of indigenous knowledge systems."[38]

Given the nature of shared commonalities, histories, and challenges in African librarianship, attempts at continent-wide solutions and attempts at unified pan-African LIS have been growing. A major development in this area has been three all-African LIS summits. The first was the African Library Summit held in Muldersdrift, Gauteng, South Africa, from May 11 to 13, 2011. The theme of the conference was "The Future of African Librarianship." The summit was hosted by the University of South Africa (UNISA) Library, the IFLA(International Federation of Library Associations and Institutions) Regional Office for Africa, and the IFLA Africa Section.[39] The summit brought together 158 LIS policy makers, senior managers, educators, and researchers from 24 African countries and four countries outside Africa. During the deliberations, participants compiled thematic frameworks for the future of African library development. The summit delegates

recommended that a task team investigate forming a continental African Library Association or a Federation of African Library Associations, and that national library associations establish chartering bodies to regulate and accredit the LIS profession and education programs. Issues of indigenization versus internationalization of LIS education and curricula were also debated and discussed.[40]

Following the first African Library Summit, the African Public Libraries Summit was held in Johannesburg, South Africa, September 19–21, 2012. This summit, intended solely for the public library sector, was organized around the theme "Informing Africa, Developing Africa" and was funded by the Bill & Melinda Gates Foundation and IFLA. This summit was a direct result of the first African Library Summit where a framework for public libraries had been discussed and developed. Its framework encouraged public libraries to "...examine the broadest context and develop strategies to reach the underserved and disadvantaged. They should innovate and provide a wide range of services. Public libraries should be open, flexible, multipurpose community spaces. They should add value, tap into opportunities through one voice in Africa and aggressively align and advocate within Government strategies to participate within the national development process."[41] In addition to public library directors and decision makers from 44 African countries and 8 countries outside the continent, the summit was also attended by government ministers, permanent secretaries, director generals, members of the African Union Commission, United Nations Economic Commission for Africa, and a variety of other commission members and development practitioners and researchers.[42] The summit was much more of a "think tank" environment than a traditional academic conference.

Following the African Public Libraries Summit, a second African Library Summit was held in Pretoria, South Africa, July 2–5, 2013. This time the theme was "The Horizon and Beyond." This second all-continent, all-sectors, summit was hosted by the UNISA Library, the IFLA Regional Office for Africa, and the IFLA Africa Section.[43] This second summit attracted 280 participants from 35 countries representing greater participation from across Africa than that of the first summit. The three days of the summit focused on the themes of leadership, innovation, and cooperation. In addition to common sessions, breakout sessions took place in which more focused conversations and planning occurred by sector and topic including African library associations; training of LIS practitioners; public and community information practitioners; school libraries; academic, research, and special libraries; and LIS in Francophone Africa and Lusophone Africa. The conclusion of the summit saw the launch of three important new initiatives: the African Library and Information Association and Institutions (AfLIA), the Public Library Network, and the Conference of African National Libraries (CANL).

AfLIA's vision is to be the association of choice for strategic growth of the library and information profession in Africa, while its mission is to be the trusted African voice for the library and information community and to drive equitable access in information and knowledge for all. AfLIA's membership currently does not include an option for individuals but rather it is an organization that represents other associations or institutions from across the continent.[44] Similarly, the Public Library Network is an effort to assist the public library community in Africa by expanding on the networking and knowledge sharing that occurred at the Public Library Summit. It is also intended to enable African public librarians to push for the placement of public libraries firmly on the development agenda with a unified voice.[45] And finally, the CANL emerged as an independent organization of chief executives of national libraries. Its goals are to facilitate discussion and promote understanding and cooperation on matters of common interest across the continent; facilitate the exchange of information, training opportunities, sharing of resources, experiences, and expertise and promoting research; improve regional cooperation and visibility; set standards against which performance can be measured; be a representative body, recognized at national, regional, continental, and international levels; provide guidelines whenever needed; enable and promote communication; and reach out to countries without national libraries.[46]

The three all-Africa summits have been an important step in moving African librarianship forward into the twenty-first century with a pan-African strategy for improving information provision and adapting libraries to the specific needs of African communities and contexts. What remains to be seen is if these efforts can truly be pan-African in scope. While there was representation from Francophone and Lusophone countries at the three summits, the vast majority of participants were from the Anglophone regions of Africa. Unless this changes, the countries that were former colonies of France and Portugal may continue to lag behind the rest of Africa in terms of strengthened libraries, library associations, and professional development opportunities in the LIS field.

Information Communication Technology

Improved access to ICTs, including mobile-cellular phones, has made a substantial impact on information access and provision worldwide, and Africa is no exception. Despite this, Africa lags behind the rest of the world in a number of areas. Estimated mobile-cellular penetration in Africa in 2013 was 63% compared to developing countries overall (89%) and developed countries (128%). Similarly, in 2013, Africa had the lowest Internet penetration rate at 16%, only half the penetration rate of Asia and the Pacific and far below the Americas and Europe. An estimated 7% of households in Africa had Internet access

in 2013. In Africa, as in other developing areas, fixed-broadband services continue to be extremely expensive, accounting for 30.1% of average monthly incomes. There was a substantial gap in fixed-broadband penetration rates with less than 1% in Sub-Saharan Africa compared to 27.2% in developed countries. Although mobile broadband penetration was also lower in Africa in 2013 at 11% than in other parts of the world, Africa is the region with the highest growth rates over the past three years in this arena. While mobile broadband is considerably cheaper than fixed-broadband services, mobile broadband services remain prohibitively expensive in Africa where the price of a computer-based plan with 1 GB of data volume is on average 50% of gross national income per capita.[47]

Scholarly Publishing

Due to these ICT issues and the traditionally high cost of access to peer-reviewed scholarship in journals and databases, researchers in Africa and other developing countries have been least able to pay to access vital scientific information. Additionally, research output from Africa has had a very difficult time reaching an international audience. Both of these results have had a significant and damaging impact on the progression of scientific and scholarly progress on the continent.[48] Starting in the 1990s, several initiatives began that were designed to increase the availability of scholarly information in Africa and thus address the deteriorating situation in which most African institutions and researchers were finding themselves. These projects were primarily driven and funded by donors and nongovernmental or nonprofit institutions. Some examples of these programs are the Programme for the Enhancement of Research Information (PERI) from the International Network for the Availability of Scientific Publications (INASP), Electronic Information for Libraries (EIFL), JSTOR's Africa Access Initiative, The Essential Electronic Agricultural Library (TEEAL) from Cornell University, and projects related to the United Nations such as the Health InterNetwork Access to Research Initiative (HINARI), Access to Global Online Research in Agriculture (AGORA), and Online Access to Research in the Environment (OARE).[49] While most of these efforts have primarily been focused on increasing the flow of scholarly and scientific information from outside Africa to the continent, there have also been efforts to reverse this information flow and promote the publications and research output of African researchers and institutions to the rest of the world. Examples of this include NISC in South Africa, a bibliographic database publishing company that compiles licenses and aggregates bibliographic databases with a mission to promote the publications and research output of Africa. Another example is African Journals Online (AJOL), a nonprofit organization whose mandate is to promote access to African research output and support African scholarly publishing.[50]

Open Access Initiatives: Open Access Journals and Institutional Repositories

OA initiatives have also been seen as a potential leveling mechanism through which African scholars and institutions may gain much needed access to scholarly publications and may broaden their reach with their own research outputs. OA literature has been defined as "digital, online, free of charge, and free of most copyright and licensing restrictions."[51] OA literature is also subject to the same rigorous peer-review process as traditionally published scholarly literature. Proponents of OA recognize the importance of the flow of scientific literature for human development. While several different economic models have developed for covering the costs of OA literature production, it is relatively common for OA publishers to charge article processing charges (APCs) for the articles that are selected after the peer-review process. This has the potential to continue the imbalances between developed and developing countries in the process of scholarly dissemination. For that reason, many OA publishers, such as BioMed Central, support research output from developing countries by providing automatic waivers for APCs for all research published from low-income or lower-middle-income countries.[48] Not only are African scholars publishing in OA journals published in other parts of the world, they are also creating their own OA publications. These are being created and published on a number of different platforms. Some are created using Open Journal Systems (OJS), an open source software for the management of peer-reviewed academic journals created by the Public Knowledge Project.[52] Additionally, many OA journals have been created using the AJOL system which provides free hosting for over 400 peer-reviewed journals from 30 African countries.[53] Other platforms include AOSIS Open Journals (formerly known as OpenJournals Publishing), which utilizes the OJS system, and South African Bibliographic and Information Network (Sabinet) Open Access Collection.[54,55]

African higher education institutions are also getting increasingly involved in the creation of OA institutional repositories, which are a significant and important development as an alternative OA route (the so-called green road of OA) designed to help disseminate scholarship from the continent to the larger scientific community. Similarly, digitization projects have gained momentum across the continent, which are helping to preserve rare and unique documents and collections. The software most commonly used in African countries for both these initiatives, institutional repositories and digital preservation projects, is the DSpace software, an open source product.[56]

Open Source Initiatives

Another outcome of improved access to ICTs is the increase in library automation, oftentimes replacing card catalogues with open source products. Open source software is software that can be freely used, changed, and shared. Many people can contribute to the development of the software, and it is distributed under licenses that comply with the Open Source Definition.[57] Open source software provides an inexpensive alternative to commercial solutions. Source code is available to libraries and can be modified or developed to the library's needs. Implementation costs are also more affordable than commercial software products. Open source products do require technical expertise to operate and maintain requiring skilled professionals in house.[58] Koha has been adopted by a number of libraries including Rwanda National Public Library, Strathmore University Library (Kenya), Addis Ababa University Libraries (Ethiopia), and Bibliothèque de l'Université de Kinshasa Bibliothèque Centrale (Congo).[59]

Consortiums and Networks

As African libraries become more automated, there are increased opportunities for resource sharing through the development of consortiums and consortial agreements and through networks. Typically, consortia are formed of similar types of organizations to meet a common end, be it resource sharing, a common project such as a unified catalogue, collaborative purchasing, or the sharing of expertise. Ngozi Blessing Ossai, Delta State University in Nigeria, states "Globally, the development of new technologies with a concomitant exponential increase in the amount of information available has made the building and expansion of library consortia an imperative."[60] Sabinet, a private company since 1996,[61] was established as a nonprofit network in 1983. It is an aggregator of southern African publications as well as providing access to South African Union. It provides access to African Electronic Journals (SA ePublications), South African Union Catalogue (SACat), South African National Bibliography (SANB), Index to South African Periodicals (ISAP), the Union Catalogue for Theses and Dissertations (UCTD), and African Digital Repository.[62] Library consortiums served by Sabinet include SANLiC—South African National Library and Information Consortium (SANLiC) NPC, CALICO—Cape Library Consortium, SEALS—South East Academic Library System, and ESAL—Eastern Seaboard Alliance of Libraries.[63]

Of the 200 participating consortia in ICOLC (International Coalition of Library Consortia), two are from Africa: Consortium of Ethiopian Academic and Research Libraries and the SANLiC.[64] The EIFL, founded in 1999, partners with libraries and library consortia in more than 60 developing and transitioning countries in Africa, Asia, Europe, and Latin America.[65] Of the 23 African nations listed, 17 have consortia including Botswana, Cameroon,

Egypt, Ethiopia, Ghana, Kenya, Lesotho, Malawi, Mali, Nigeria, Senegal, Sudan, Swaziland, Tanzania, Uganda, Zambia, and Zimbabwe.[66]

Privately Funded Public Libraries

The scarcity of publicly funded libraries has given rise to social entrepreneurs who give up both time and capital to build libraries and develop programs. Two such endeavors include the Busia Community Library and its partner organization, Maria's Libraries (Kenya), and the Axumite Heritage Library in historic Axum, Ethiopia. The Busia Community Library is the 1988 vision of community leader Maria Wafula and serves a population of 500,000 along the Kenya-Uganda border.[67] In 2008, it formally affiliated with the new umbrella library network, Maria's Libraries.

The Axumite Heritage Library is a project of the Axumite Heritage Foundation, a subsidiary of the Ethiopian Community Development Council (ECDC). The library is housed in the former Governor's palace (the 'Inda Nebri'id) in Axum, Ethiopia, and opened its doors as a temporary facility in 2002. The Axumite Heritage Foundation was established to oversee the creation of the new Axumite Heritage Library, the museum, and eventually the Institute of Axumite Studies. The construction of a permanent facility is nearby on the ground of the former Governor's palace.[68]

Innovative Outreach to Rural Communities

Scarcity of library services in rural areas has resulted in innovative outreach to members of the community. These include the use of camels for transportation of materials in Kenya,[69] cloth pocket libraries in Ethiopia,[70] bicycles in Kenya,[71] and Kindles in Ghana, Kenya, Malawi, Rwanda, Tanzania, and Uganda.[72]

Growth and Changes in Library and Information Science Education

The original training of professional librarians in Africa took place primarily overseas in institutions located in the former colonial powers or through correspondence courses from those same countries and was often criticized for being too remote from the reality of the African geographic, cultural, social, economic, and political environment. African countries began to develop library education programs of their own to meet their own needs for professional training and to counter some of these criticisms.[73] With the exception of South Africa, which began LIS education as early as 1938, the majority of LIS programs only began to develop in the 1960s.[74] The first was the Institute of Librarianship at Ibadan University followed by the Library School in Accra. Shortly afterward, regional library schools were founded such as those in Dakar, Senegal, for Francophone Africa and the East African School of Librarianship at Makerere University in Uganda, which was designed to cater to Eastern African countries.[73] Today, the majority of African countries have at least one library school or training program, and several countries (such as Kenya, Nigeria, and South Africa) have multiple schools for LIS education. While there has been great expansion in access to LIS education since independence from the colonial powers, a fair amount of recent reflection on the aims, quality, and relevance of LIS programs in Africa has been occurring, particularly since the turn of the twenty-first century. While a plethora of research publications and sessions at the pan-African conferences mentioned above give testimony to the fact that collaboration and a unified vision and standards for LIS education in Africa are strongly desired, substantial disparities between programs and countries exist. For example, although students and faculty across the continent have a strong desire to learn and use ICTs in their LIS educational programs, LIS institutions in South Africa have a much greater capacity and are more effective at exploiting ICTs in their programs than other countries in sub-Saharan Africa.[75]

CONCLUSION

African librarianship has evolved significantly since the postcolonial era beginning in the mid-twentieth century when it was conceived as a distinctly different type of librarianship with unique perspectives, challenges, and needs. As Kingo Mchombu articulated toward the end of the second African Library Summit in 2013, African librarianship is not a static phenomenon, rather, it is one that has been changing over many years. While the idea of the "barefoot librarian" who worked in African villages, received an honorarium for homegrown library and information work, and participated in literacy work (among other services) was put forward by Mchombu and other African librarians in the 1970s and 1980s, he noted that that concept was a particular response to the information and political environment of the time. Although these environments and landscapes have now changed, the role of the library as a tool for human development and elevation is just as important in the African context as it was in the immediate postcolonial period. However, Mchombu has pointed out that the relationship between African librarianship and emerging technologies is now of great interest, importance, and challenges. African libraries and librarians cannot afford to ignore the new challenges brought about by increasing ICTs. The primary challenge is how to make the technology work for African communities and libraries rather than allowing those same communities to be overpowered and devoured by such technologies. The ongoing question to be negotiated in African librarianship is how to use ICT as enabling tools without these same technologies becoming "the master" that disregards the unique needs and environment in which the majority of African populations live. As Mchombu stated, "African society is changing rapidly as well, which we must take full advantage of. The stakes are much higher now. Failure is not an option but we must realize we are riding the tiger—we cannot afford to get down and there's no way back."[76] The future and precise meaning of African librarianship is being written and revised, daily, across the continent.

REFERENCES

1. Encyclopædia Britannica Online Academic Edition, s. v. "Africa," http://www.britannica.com/EBchecked/topic/7924/Africa (accessed August 26, 2013).
2. Composition of macro geographical (continental) regions, geographical sub-regions, and selected economic and other groupings, revised February 11, 2013, http://unstats.un.org/unsd/methods/m49/m49regin.htm#africa.
3. World Atlas: Countries listed by continent, http://www.worldatlas.com/cntycont.htm.
4. U.S. Census Bureau, International Data Base, http://www.census.gov/ipc/www/idb/ (accessed June 2010).
5. BBC, Are indigenous languages dead? January 6, 2006, http://news.bbc.co.uk/2/hi/africa/4536450.stm (accessed August 15, 2013).
6. Paulos, A. Library resources, knowledge production, and Africa in the 21st century. Int. Inform. Libr. Rev. **2008**, *40* (4), 251–256.
7. Blum, R. In *Kallimachos: The Alexandrian Library and the Origins of Bibliography*; Hans H. Wellisch: Madison, WI, 1991; 97.
8. Amadi, A.O. The emergence of a library tradition in pre- and post-colonial Africa. Int. Libr. Rev. **1981**, *13*, 65–72.
9. Saad, E. Kilwa dynastic historiography: A critical study. History Africa. **1979**, *6*, 177–207.
10. Alemna, A.A. *Issues in African Librarianship*; Type Co. Ltd.: Accra, Ghana, 1996; 3.
11. Kaungamno, E.E.; Ilomo, C.S. *Books Build Nations. Library Services in West and East Africa*. Transafrica Book Distributors: London, U.K., 1979, *1*.
12. Sturges, P.; Neill, R. In *The Quiet Struggle: Information and Libraries for the People of Africa*, 2nd Ed.; Mansell: London, U.K., 1998.
13. Oyebade, Adebay, "Liberia" Oxford Bibliographies, http://www.oxfordbibliographies.com/view/document/obo-9780199846733/obo-9780199846733-0112.xml (accessed December 9, 2013).
14. Afeworki, P. Library resources, knowledge production, and Africa in the 21st century. Int. Informat. Libr. Rev. **2008**, *40*, 251–256.
15. Odi, A. The colonial origins of library development in Africa: Some reflections on their significance. Libr. Culture **1991**, *26* (4), 598.
16. Mchombu, K.J. Which way African librarianship? IFLA J. **1991**, *17* (1), 26–38.
17. Sturges, P.; Neill, R. *The Quiet Struggle: Information and Libraries for the People of Africa*, 2nd Ed; Mansell: London, U.K., 1998; 93–95.
18. Libraries., Middleton, J.; Miller, J.C. In *New Encyclopedia of Africa*, 2nd Ed.; Charles Scribner's Sons: Detroit, MI, 2008; Vol. 3, 302–304. Retrieved from http://go.galegroup.com/ps/i.do?id=GALE%7CCX3049000375&v=2.1&u=regis_main&it=r&p=GVRL&sw=w&asid=a8cbd9a1e2e6a4a1403b3e629a239045.
19. Sturges, P.; Neill, R. *The Quiet Struggle: Information and Libraries for the People of Africa*, 2nd Ed.; Mansell: London, U.K., 1998; 81–84.
20. Olden, A. *Libraries in Africa: Pioneers, Policies, Problems*; Scarecrow Press: Lanham, MD, 1995; 11–14.
21. Sturges, P.; Neill, R. *The Quiet Struggle: Information and Libraries for the People of Africa*, 2nd Ed.; Mansell: London, U.K., 1998; 82–83.
22. Sturges, P.; Neill, R. *The Quiet Struggle: Information and Libraries for the People of Africa*, 2nd Ed.; Mansell: London, U.K., 1998; 128–129.
23. Amadi, A.O. In *African Libraries: Western Tradition and Colonial Brainwashing*; Scarecrow Press: Metuchen, NJ, 1981.
24. Mchombu, K.J On the librarianship of poverty. Libri **1982**, *32* (3), 241.
25. Mchombu, K.J. On the librarianship of poverty. Libri **1982**, *32* (3), 250.
26. Poppeliers, N.T. Cultural rights and library development and discourse in sub-saharan Africa: Is the colonial legacy still alive. In *Beyond Article 19: Libraries and Economic, Social, and Cultural Rights*; Julie, B.E.; Stephan, P.E., Eds.; Library Juice Press: Duluth, MN, 2010.
27. Teferra, D. Higher education in sub-Saharan Africa. *International Handbook of Higher Education*. Springer International Handbooks of Education, **2006**, *18*, 557.
28. International universities and colleges http://www.4icu.org/Africa (accessed December 15, 2013).
29. Teferra, D. Higher education in sub-Saharan Africa. *International Handbook of Higher Education*. Springer International Handbooks of Education, **2006**, *18*, 559.
30. Paul, T.Z.; Adebayo, O. *African Universities in the Twenty-First Century*; Council for the Development of Social Science Research in Africa: Dakar, Senegal, 2004; 7.
31. Mabizela, M. Private surge amid the dominance in higher education: the African perspective. JHEA/REA **2007**, *5* (2&3), 22.
32. Teboho, M. Policy responses to global transformation by African higher education systems. In *African Universities in the Twenty-First Century*; Paul, T.Z.; Adebayo, O., Eds.; Council for the Development of Social Science Research in Africa: Dakar, Senegal, 2004; 37.
33. Muyoyeta, S. The provision of scholarly information in higher education in Zambia. In *African Universities in the Twenty-First Century*; Paul, T.Z., Adebayo, O., Eds.; Council for the Development of Social Science Research in Africa: Dakar, Senegal, 2004; 400.
34. Muyoyeta, S. The provision of scholarly information in higher education in Zambia. In *African Universities in the Twenty-First Century*; Paul, T.Z., Adebayo, O., Eds.;

Council for the Development of Social Science Research in Africa: Dakar, Senegal, 2004; 401.

35. Teferra, D. Higher Education in Sub-Saharan Africa. *International Handbook of Higher Education*. Springer International Handbooks of Education, **2006**, *18*, 566.

36. Kenyatta University Library, http://library.ku.ac.ke/.

37. Historical Development of Kenyatta University Library, http://library.ku.ac.ke/wp-content/uploads/2012/09/HISTORICAL-DEVELOPMENT-OF-KENYATTA-UNIVERSITY-LIBRARY.pdf.

38. SCECSAL. SCECSAL: Standing Conference of Eastern, Central and Southern Africa Library and Information Associations, http://www.scecsal.org/documents/history.html.

39. Library and Information Association of South Africa (LIASA), African Library Summit 2011, http://www.liasa.org.za/node/458 (accessed August 15, 2013).

40. Dick, A.L. Library associations: A leadership role? Inform. Develop. **2012**, *28* (1), 11–12.

41. Makhanya, M. The African Public Library Summit, September 19–21, 2012: Welcome Address by Professor Mandla Makhanya, Principal and Vice Chancellor, UNISA.

42. Mulindwa, G.K. Launch of Public Libraries Network. Presentation at African Library Summit 2013 : The Horizon and Beyond, University of South Africa (UNISA) Muckleneuk Campus, Pretoria, South Africa, July 2–5, 2013.

43. African Library Summit 2013, http://www.unisa.ac.za/default.asp?Cmd=ViewContent&ContentID=28174/ (accessed August 15, 2013).

44. Helena, A.H. Launch: African Information Association. Presentation at African Library Summit 2013, Pretoria, South Africa, July 2–5, 2013. The Horizon and Beyond. University of South Africa (UNISA) Muckleneuk Campus.

45. Mulindwa, G.K. Launch of Public Libraries Network. Presentation at African Library Summit 2013, Pretoria, South Africa, July 2–5, 2013. University of South Africa (UNISA) Muckleneuk Campus.

46. Tsebe, J. Launch: Conference of African National Libraries (CANL). Presentation at African Library Summit 2013: The Horizon and Beyond, Pretoria, South Africa, July 2–5, 2013. University of South Africa (UNISA) Muckleneuk Campus.

47. International Telecommunication Union. The World in 2013: ICT Facts and Figures, http://www.itu.int/en/ITU-D/Statistics/Documents/facts/ICTFactsFigures2013.pdf (accessed October 10, 2013).

48. McKay, M. Improving access to scholarly research in Africa: Open access initiatives. Serials **2011**, *24* (3), 251–254.

49. Masinde, S.; Okoh, T. Challenges of broadening access to scholarly e-resources in Africa : The JSTOR example. Afr. Res. Document. **2011**, *117*, 49–57.

50. Murray, S. Access to African published research: the complementary approaches of NISC SA and African Journals OnLine. IFLA Conf. Proc. 2007, 1–2, 15.37.

51. Suber, P. *Open Access*; MIT Press: Cambridge, MA, 2012; 4.

52. PKP: Public Knowledge Project, http://pkp.sfu.ca (accessed December 15, 2013).

53. African Journals Online (AJOL). About AJOL, http://www.ajol.info/index.php/ajol/pages/view/AboutAJOL (accessed December 15, 2013).

54. AOSIS Open Journals. History, http://www.openjournals.net/index.php/about-us/history (accessed December 15, 2013).

55. Sabinet. African Electronic Journals, http://www.sabinet.co.za/journals (accessed December 15, 2013).

56. Bowdoin, N.T. Cultures of access: Differences in rhetoric around open access repositories in Africa, Europe, and the United States and their implications for the open access movement. In *The Global Librarian*; Caroline, F.; Jason, K., Eds.; Metropolitan New York Library Council and the Greater New York Metropolitan Area Chapter of the Association of College and Research Libraries: New York, 2013.

57. Open Source Initiative, http://opensource.org/ (accessed December 10, 2013).

58. Kamble, V.T. Open source library management and digital library software. J. Libr. Inform. Technol. **2012**, *32* (5), 388.

59. KohaUsers/Africa, http://wiki.koha-community.org/wiki/KohaUsers/Africa (accessed December 10, 2013).

60. Ossai, N. Consortia Building among Libraries in Africa, and the Nigerian Experience. Collabora. Libr. **2010**, *2*(1), collaborativelibrarianship.org/index.php/jocl/article/viewFile/46/50.

61. South African Companies and Intellectual Property Commission, http://www.cipro.co.za/ccc/EntDet.asp?T1=%DD%47%2C%8C%56%C0%1D%B8%1E%BA%3A%88%BA%AD%25&T2=SABINET%20ONLINE (accessed December 10, 2013).

62. Olivier, E.; Fourie, I. African Electronic Journals (SA ePublications). Charleston Advisor **2013**, *14* (4), 33–38. doi:10.5260/chara.14.4.33, http://charleston.publisher.ingentaconnect.com/content/charleston/chadv/2013/00000014/00000004/art00008;jsessionid=4j2rv4pvfae96.victoria (accessed December 10, 2013).

63. Sabinet Library consortium, http://www.sabinet.co.za/library_consortia (accessed December 10, 2013).

64. ICOLC (International Coalition of Library Consortia) Participating Consortia (accessed December 10, 2013).

65. EIFL. Who we are, http://www.eifl.net/who-we-are (accessed December 10, 2013).

66. EIFL Where we work. http://www.eifl.net/where-we-work (accessed December 10, 2013).

67. Turner, J. Beyond access member profile: Busia community library, Maria's Libraries, http://beyondaccess.net/2012/06/22/beyond-access-member-profile-busia-community-library-marias-libraries/ (accessed December 10, 2013).

68. Axumite Heritage Foundation, http://axumiteheritage-foundation.org/about-us.asp (accessed December 10, 2013).

69. Mobile libraries as effective solutions to reading access and reading promotion in remote communities in Marlene Asselin and Ray Doiron's Linking Literacy and Libraries in Global Communities Ashgate Publishing Limited: Surrey, BC, 2013; 108.

70. Segenat Foundation: Cloth pocket libraries have arrived, http://segenatfoundation.org/2012/05/10/cloth-pocket-libraries-have-arrived/ (accessed December 10, 2013).

71. Kaplan, E. Library services in Kenya: A process of reinvention, International Leads v. 27 #. March 1, 2013; 3. http://www.ala.org/irrt/sites/ala.org.irrt/files/content/intlleads/leadsarchive/201303.pdf.

72. Worldreader, http://www.worldreader.org/about-us/faq/ (accessed December 10, 2013).

73. Sturges, P.; Richard, N. *The Quiet Struggle: Information and Libraries for the People of Africa*, 2; Mansell: London, U.K., 1998; 108–109.

74. Ocholla, D.; Bothma, T. Trends, challenges and opportunities of LIS education and training in Eastern and Southern Africa. New Libr. World **2007**, *108* (1/2), 58–78.

75. Ocholla, D.N. An overview of information and communication technologies (ICT) in the LIS schools of Eastern and Southern Africa. Edu. Inform. **2003**, *21*, 181–194.

76. Mchombu, K.J. African librarianship. Presentation at African Library Summit 2013: The Horizon and Beyond, Pretoria, South Africa, July 2–5, 2013. University of South Africa (UNISA) Muckleneuk Campus.

BIBLIOGRAPHY

1. Aina, L.O. Towards an ideal library and information studies (LIS) curriculum for Africa: Some preliminary thoughts. Edu. Inform. **2005**, *23*, 165–185.

2. Benge, R.C. *Cultural Crisis and Libraries in the Third World*; Clive Bingley: London, U.K., 1979.

3. Dick, A.L. *African library summits 1 and 2: An overview. Presentation at African Library Summit 2013: The Horizon and Beyond,* Pretoria, South Africa, July 3–5, 2013. University of South Africa, Pretoria, Republic of South Africa.

4. Issak, A. *Public Libraries in Africa: A Report and Annotated Bibliography*; INASP: Oxford, U.K., 2000.

5. Iwuji, H.O.M. Librarianship and oral tradition in Africa. Int. Libr. Rev. **1990**, *22* (1), 53–59.

6. Lor, P.J.; Britz, J. Knowledge production from an African perspective: International information flows and intellectual property. Int. Inform. Libr. Rev. **2005**, *37*, 61–76.

7. Maack, M.N. Books and libraries as instruments of cultural diplomacy in Francophone Africa during the Cold War. Libr. Culture. **2001**, *36* (1), 58–86.

8. Ocholla, D.; Bothma, T. Trends, challenges and opportunities for LIS education and training in Eastern and Southern Africa. New Libr. World **2007**, *108* (1/2), 55–78.

9. Sitzman, G.L. *African Libraries*; Scarecrow Press: Metuchen, NJ, 1988.

10. Sturges, P. The poverty of librarianship: An historical critique of public librarianship in Anglophone Africa. Libri **2001**, *51*, 38–48.

11. Wise, M. *Aspects of African Librarianship: A Collection of Writings*; London, U.K.: Mansell, 1985.

Altmetrics

Richard Cave
Formerly at the Public Library of Science, San Francisco, California, U.S.A.

Abstract

Altmetrics provide a new way to measure the impact of research. New channels of evaluation for scholarly work have become available by tracking its activity through usage, captures, mentions, social media, and citations. This allows researchers to gain a faster, broader picture of research as it is viewed, shared, and critiqued.

INTRODUCTION

Traditionally, citations in scholarly literature were used as a proxy to establish the quality of scholarly output with higher citation counts generally interpreted to represent higher scientific impact of the scholarly output. Used in this way, citations have limitations in that they are a substitution for quality and that they can be influenced by superficial factors. The Journal Impact Factor (IF) is another way frequently used to judge the impact of scholarly output, as it counts the average number of citations to recent articles published in a journal. But the Journal Impact Factor is highly controversial as it has a number of limitations such as skewed distributions within journals and manipulation by a journal's editorial policy.[1] There is a need to assess research on its merits rather than on the basis of the journal in which it is published.

Altmetrics was a response to the limitations of these traditional metrics. As the volume of academic literature has exploded in the past decade, citation counts are not sufficient for measuring the impact of research. Given the large amount of online sharing of research articles and the discussions on the web about research, there are new ways to filter science for importance. Altmetrics provide a way to measure scientific impact through the conversations, bookmarks, and shares happening daily on websites around the world.

According to the Altmetrics Manifesto, "altmetrics is the creation and study of new metrics based on the Social Web for analyzing and, information scholarship."[2] The term "altmetrics" was coined by Jason Priem, a graduate student at the School of Information and Library Science at University of North Carolina-Chapel Hill. Altmetrics is also a general-purpose term for "alternative metrics" (Fig. 1).

THE USE OF ALTMETRICS

The use of altmetrics aims to capture a broader view of scholarly research by tracking the activity of this research as it is viewed, shared, critiqued, cited, and built upon. Some metrics are primarily used to track scholarly activity, while other metrics are used to show the public interest in scholarly research. Altmetrics measures not just articles but datasets, code, blogs, comments, and "nanopublications."

Altmetrics are gathered from a number of sources through public APIs. The data that is gathered is generated more rapidly than traditional metrics such as citations, so researchers do not have to wait years to show their worth. By providing this information, altmetrics can be used to show the impact of scientific research not just by researchers through citations but also the impact of the research to the public through the Social Web.

As outlined in "Overview of the Altmetrics Landscape,"[3] altmetrics are generally categorized in the following ways:

- *Usage*: The measurement of the use of a research object. This includes the number of HTML views of an article on a publisher's website, the number of times that the article PDF and XML have been downloaded, and how many times its supplemental data has been accessed.
- *Captures*: The number of times an article has been stored as a bookmark or in a reference library. This includes how often an article has been bookmarked on CiteULike and how many times an article has been shared on Mendeley readers/groups.
- *Mentions*: The discussions about a research article. This includes blog posts, news stories, Wikipedia articles, comments, and reviews about a research article.
- *Social Media*: The use of an article in the Social Web. This includes tweets about an article, the number of Facebook likes, and how many times an article has been shared on social networks such as LinkedIn.
- *Citations*: The number of citations to an article that shows the long-term contribution of an article to scholarly literatures. This includes article citations from Scopus, CrossRef, and Pubmed Central (Fig. 2).

Encyclopedia of Library and Information Sciences, Fourth Edition DOI: 10.1081/E-ELIS4-120049483

rfishbase: exploring, manipulating and visualizing FishBase data from R.

(2012) Boettiger, Lang, Wainwright. *Wiley Blackwell (Blackwell Publishing).* Journal of fish biology

highly saved by scholars	**94 - 99 percentile** ⓘ of articles published in 2012	**56** Mendeley readers
highly discussed by public	**94 - 99 percentile** ⓘ of articles published in 2012	**1** Topsy influential tweet
highly discussed by public	**97 - 100 percentile** ⓘ of articles published in 2012	**17** Topsy tweets
saved by scholars	**86 - 98 percentile** ⓘ of articles published in 2012	**1** CiteULike bookmark
cited by scholars	**79 - 95 percentile** ⓘ of articles published in 2012	**1** PubMed Central pmc citation
cited by scholars	**67 - 87 percentile** ⓘ of articles published in 2012	**3** Scopus citations
saved by public	**96 - 99 percentile** ⓘ of articles published in 2012	**1** Delicious bookmark

Fig. 1 ImpactStory percentiles showing that an article was highly saved by scholars in Mendeley and highly discussed by the public through Tweets.
Source: CC-By license 2.0 ImpactStory.com 2013. As shown on article "rfishbase: exploring, manipulating and visualizing FishBase data from R." Reprinted with permission.

633

Score in context

Puts article in the top 5% of all articles ranked by attention

show more...

Mentioned by

- 72 tweeters
- 18 Facebook users
- 61 news outlets
- 1 video uploaders
- 9 science blogs
- 25 Google+ users
- 1 Redditors

Readers on

- 28 Mendeley
- 0 CiteULike

Fig. 2 Altmetric score with the number of mentions across different social media outlets.
Source: Altmetric.com 2013. Reprinted with permission.

Altmetrics can be customized to address the needs of researchers, publishers, institutional decision-makers, and funders. Researchers can track the impact of their research, filter research to their individual needs, and share the research with others. They can use altmetrics to show the impact of their scholarly output beyond a journal's impact factor. Research institutions can access up-to-date metrics on the research published by their faculty members and use this information for tenure and promotion decisions. Publishers can gain an understanding on how their publications are accessed and distributed. Funders can effectively track the impact of their grant awardees and measure the wider engagement of the research that they funded.

The altmetrics movement builds on previous work by other researchers. Johan Bollen has studied usage on articles for over a decade. He started MESUR (http://mesur.informatics.indiana.edu/) in 2006 as a large-scale usage data collection for scientific research. The MESUR database currently contains over one billion article-level usage events obtained from publishers, institutions, and aggregators.[4] Eigenfactor is a rating of the total importance of a scientific journal. It was created in 2007 by Jevin West and Carl Bergstrom of the University of Washington as an academic research project to rank journals based on a vast network of citations.[5]

There are a number of organizations already collecting altmetrics and aggregating the data including ImpactStory (http://impactstory.org/), Mendeley (http://www.mendeley.com/), Altmetric (http://www.altmetric.com/), Plum Analytics (http://www.plumanalytics.com/), and Public Library of Science (PLOS) (http://article-level-metrics.plos.org).

plos.org create account sign in

PLOS | MEDICINE Browse For Authors About Us Search 🔍

advanced search

🔓 OPEN ACCESS

ESSAY

| 919,988 | 1,146 | 4,143 | 7,035 |
| VIEWS | CITATIONS | SAVES | SHARES |

Why Most Published Research Findings Are False

John P. A. Ioannidis

Published: August 30, 2005 • DOI: 10.1371/journal.pmed.0020124

Article | About the Authors | Metrics | Comments | Related Content

Download PDF ▼

Print Share

Viewed ❓

Total Article Views		HTML Page Views	PDF Downloads	XML Downloads	Totals
919,988	PLOS	656,843	118,252	2,293	777,388
Aug 30, 2005 (publication date) through Dec 16, 2013ᵃ	PMC	121,726	20,874	n.a.	142,600
	Totals	778,569	139,126	2,293	919,988

17.87% of article views led to PDF downloads

Cumulative Views: 1,000k, 500k, 0k; Months 1 8 15 22 29 36 43 50 57 64 71 78 85 92 99

☐ Compare average usage for articles published in 2005 in the subject area: ❓

Computational biology ▾ | Show reference set

ᵃAlthough we update our data on a daily basis, there may be a 48-hour delay before the most recent numbers are available. PMC data is posted on a monthly basis and will be made available once received.

Cited ❓

SCOPUS	crossref	PMC	ISI Web of SCIENCE	Europe PubMed Central
1146	628	248	1028	457

Google Search

Saved ❓

citeulike	Connotea	MENDELEY
372	18	3771

Discussed ❓

Research Blogging	ScienceSeeker	WordPress.com	Wikipedia	twitter
10	1	20	9	953

facebook	reddit	Comments	Trackbacks	Google Search
6082	41	31	3	

Recommended ❓

F1000 Prime
12

Information on PLOS Article-Level Metrics
Questions or concerns about usage data? Please let us know

Related PLOS Articles

When Should Potentially False Research Findings Be Considered Acceptable?

Most Published Research Findings Are False—But a Little Replication Goes a Long Way

Minimizing Mistakes and Embracing Uncertainty

Subject Areas ❓

Clinical research de...
Gene prediction
Genetic epidemiology
Genetics of disease
Randomized controll...
Research design
Research laboratories
Schizophrenia

ADVERTISEMENT

🐦 Archived Tweets

12 Dec
Toronto Stare™
@TorontoStare
RT @AlexFGoldberg: "Why Most Published Research Findings Are False" A potentially paradoxical study? http://t.co/ILUApgWLEf

11 Dec
Recommended Tweets™
@RecommTweet
RT @AlexFGoldberg: "Why Most Published Research Findings Are False" A potentially paradoxical study? http://t.co/ILUApgWLEf

Load More ▾

Comments

A small group research
Posted by samgul

Surely the answer is Bayes theorem?
Posted by mickofemsworth

A Critique of Ioannidis JPA (2005) Why Most Published Research Findings Are False. PLoS Med 2(8) II: proposals
Posted by vetter

Fig. 3 PLOS Article-Level Metrics displaying the usage statistics, the number of citations, the number of bookmarks, the discussions, and the recommendations for an article.
Source: Creative Commons Attribution License 3.0 PLOS 2013. As shown on the article "Why Most Published Research Findings Are False." Reprinted with permission.

Publishers of scientific research have enabled altmetrics on their articles, open source applications are available for platforms to display altmetrics on scientific research, and subscription models have been created that provide altmetrics. There are also a number of applications using altmetrics, including ReaderMeter (http://readermeter.org/), PeerEvaluation (http://www.peerevaluation.org/), Research Scorecard (http://researchscorecard.com/), and PLOS Article-Level Metric (ALM) Reports (http://almreports. plos.org/).

Altmetrics are sometimes conflated with ALMs. ALMs are an attempt to measure impact at the article level. ALMs include many of the same data points as altmetrics but also include usage statistics and traditional metrics such as the number of citations to an article. ALMs are typically associated with the publisher PLOS, who introduced ALMs on all PLOS articles in 2009[6] (Fig. 3).

CONCLUSION

Altmetrics provide a way for researchers to measure the impact of research by tracking its activity through usage, captures, mentions, social media, and citations. Altmetrics can be customized for use by researchers, publishers, institutional decision-makers, and funders. Altmetrics can also be used to understand a researcher's influence beyond a journal's impact factor. A handful of companies have started to collect and aggregate altmetrics and a number of publishers have enabled altmetrics on their articles. The idea of altmetrics has recently gained a lot of traction, but there is still work to be done to promote the use of altmetrics to measure scholarly impact. Altmetrics will continue to evolve along with the ways that the scholarly community views, shares, and critiques scholarly activity.

REFERENCES

1. San Francisco Declaration on Research Assessment (DORA), December 2012. http://www.ascb.org/dora/.
2. Priem, J.; Taraborelli, D.; Groth, P.; Neylon, C., 2010, Altmetrics: A manifesto, (v.1.0), October 26, 2010. http://altmetrics.org/manifesto.
3. Cave, R. Overview of the almetrics landscape. Proceedings of the Charleston Library Conference, Charleston, SC, 2012. DOI: 10.5703/1288284315124.
4. Bollen, J.; Van de Sompel, H.; Rodriguez, M.A. Towards usage-based impact metrics: first results from the MESUR project, JCDL 2008, Pittsburgh, PA, June 2008 (arXiv:0804.3791v1).
5. Bergstrom, C.T.; West, J. D.; Wiseman, M. A., The eigenfactor metrics, J. Neurosci. **2008**, *28* (45): 11433–11434.
6. Article-level metrics. Retrieved from https://www.plos.org/article-level-metrics (accessed 10/30/2013).

American Association of Law Libraries (AALL)

Kate Hagan
American Association of Law Libraries, Chicago, Illinois, U.S.A.

Abstract

The American Association of Law Libraries (AALL) was founded in 1906 by a group of 24 law librarians who heeded the call of A.J. Small (Curator, Law Department of the Iowa State Law Library) to form a national organization to create a professional identity for those who dealt with the burgeoning body of legal publications.

That first meeting brought together participants from state libraries, bar associations, and law school libraries, among others. Diversity among venues for the profession continues to infuse and strengthen the association, which had grown to 5041 members by September 30, 2007. This entry focuses on the recent history and the state of the organization. For a detailed review of AALL's founding and history, see earlier editions of the *Encyclopedia of Library and Information Science*, which cover the period from AALL's founding through the late 1990s, or the chronology in the *AALL Directory and Handbook*, which is updated yearly.

INTRODUCTION

The American Association of Law Libraries (AALL) was founded in 1906 by a group of 24 law librarians who heeded the call of A.J. Small (Curator, Law Department of the Iowa State Law Library) to form a national organization to create a professional identity for those who dealt with the burgeoning body of legal publications.

That first meeting brought together participants from state libraries, bar associations, and law school libraries, among others. Diversity among venues for the profession continues to infuse and strengthen the association, which had grown to number 5041 members by the year 2007.

This entry focuses on the recent history and the state of the organization and the profession. For a detailed review of AALL's founding and history, see the first edition of the *Encyclopedia of Library and Information Science*, which covers the period from AALL's founding to the late 1960s. For a detailed review of AALL at the turn of the new century, see the second edition of the *Encyclopedia of Library and Information Science*, or the chronology in the *AALL Directory and Handbook*, which is updated yearly.

AALL MISSION AND VISION

Since its founding in 1906 and later incorporation in 1935, AALL has been a leader in promoting the value of law libraries, fostering the profession of law librarianship, and providing leadership in the field of legal information and policy. Since 1990, AALL and law librarians have become even more vocal, more visible, and more effective in representing their needs and interests beyond the profession itself.

Members are AALL's core and set its priorities. The current 15-member staff handles administration and carries out the directives of the membership. William H. Jepson was appointed the first executive director in 1981, followed by Judith Genesen in 1989, Roger H. Parent in 1993, Susan E. Fox in 2003, and Kate Hagan in 2007.

AALL 2005–2010 Strategic Directions

In July 2005, the AALL Executive Board adopted a new strategic plan (called "Strategic Directions") for the years 2005–2010 at the Annual Meeting in San Antonio. These strategies for the future will strengthen AALL's core purpose and values and will give AALL the direction and flexibility needed to grow as an organization, while placing primary focus on members.

Core purpose of AALL

The American Association of Law Libraries strengthens the profession of law librarianship and supports the individual efforts of our members.

Core values of AALL

AALL values:

- Lifelong learning and intellectual growth.
- The role of the law librarian in a democratic society.
- Equitable and permanent access to legal information.
- Continuous improvement in the quality of justice.
- Community.

Encyclopedia of Library and Information Sciences, Fourth Edition DOI: 10.1081/E-ELIS4-120044743

GOALS AND OBJECTIVES

Goal I: Leadership

Law librarians will be recognized and valued as the foremost leaders and experts in legal information, research, and technology.

Objectives:

* Provide tools to increase members' abilities to position themselves as essential to the mission of their organizations.
* Provide leadership training opportunities.
* Expand mentoring programs and opportunities.
* Increase AALL participation in organizations within the legal and library communities.

Goal II: Education

Law librarians will have the education and training they need to meet and leverage the challenges of the changing information environment.

Objectives:

* Expand the scope of educational offerings to meet the ever-changing needs of members.
* Develop partnerships to increase the range of educational offerings.
* Use a wide range of delivery means and opportunities to provide education beyond the Annual Meeting.
* Increase the number of library school programs for law librarianship and increase awareness of law librarianship as a profession.
* Increase the number and amount of grants and scholarships.

Goal III: Advocacy

Law librarians will influence the outcome of legal information, technology policy, and librarianship issues of concern to AALL members.

Objectives:

* Increase resources available for advocacy efforts.
* Continue to expand international role.
* Improve grassroots participation in advocacy efforts.

MEMBERSHIP AND STRUCTURE

A September 2007, tally of AALL's 5041 members showed a division into the following types of library categories: 36% (a total of 1826 members) worked in academic law libraries; 11% (578 members) in state, court, and county law libraries; 37% (1865) in private law firm and corporate law libraries; and 15% (772) in other organizations.

The American Association of Law Libraries members worked for more than 2000 different institutions, including 1073 law firms, 233 law schools, 179 corporations, 200 state or municipal government agencies, 150 state or municipal courts, and 52 federal government agencies. Other employers included vendors, independent self-employed librarians, other public or academic libraries, and associations.

In 2007, AALL members' libraries had a total combined information budget —hard copy and electronic—of more than $1.3 billion; the average information budget per library was $1.1 million. Libraries were spending growing percentages of their budgets on electronic resources, varying considerably by library type: academic libraries, 18%; government, 20%; and private law firms, 62%.

SPECIAL INTEREST SECTIONS

The American Association of Law Libraries' special interest sections (SIS) help members focus on issues of personal interest or concern to their daily work. At a forum held in 2000, the leaders of the SISs defined themselves as "a self-selecting group of members with a common interest, which serves as a forum, contributes educational value, serves as a resource for expertise, advocates, and provides leadership group opportunities."

These SISs hold meetings, produce publications, and provide a vehicle for networking and the exchange of information. Individual memberships in SISs remained high through the years. For example, in December 2007, AALL's members held 7172 memberships in its SISs:

Academic Law Libraries: 1220
Computing Services: 501
Legal History and Rare Books: 243
Legal Information Services to the Public: 313
Foreign, Comparative, and International Law: 423
Government Documents: 323
Micrographics and Audiovisual: 82
Online Bibliographic Services: 321
Private Law Libraries: 1512
Research Instruction and Patron Services: 800
Social Responsibilities: 205
State, Court, and County Law Libraries: 541
Technical Services: 688

AALL Entries

In addition to the SISs, members early on recognized the value in allying along geographic lines. Some 31 entries now provide members a forum for their AALL and region-specific interests.

Atlanta Law Libraries Association
Association of Law Libraries of Upstate New York
Arizona Association of Law Libraries
Chicago Association of Law Libraries
Colorado Association of Law Libraries
Dallas Association of Law Librarians
Greater Philadelphia Law Library Association
Houston Area Law Librarians
Law Libraries Association of Alabama
Law Library Association of Greater New York
Law Library Association of Maryland
Law Librarians Association of Wisconsin, Inc.
Law Librarians of New England
Law Librarians of Puget Sound
Law Librarians Society of Washington, DC, Inc.
Mid-America Association of Law Libraries
Minnesota Association of Law Libraries
Michigan Association of Law Libraries
New Jersey Law Librarians Association
New Orleans Association of Law Librarians
Northern California Association of Law Libraries
Ohio Regional Association of Law Libraries
San Diego Area Law Libraries
Southern California Association of Law Libraries
Southeastern Entry of the American Association of Law
    Libraries
South Florida Association of Law Libraries
Southern New England Law Librarians Association
Southwestern Association of Law Libraries
Virginia Association of Law Libraries
Western Pacific Entry of the American Association of
    Law Libraries
Western Pennsylvania Law Library Association

AALLNET

Since the late 1990s, AALL's Web site, AALLNET, has
served to foster the mission and Strategic Directions of
the Association. The American Association of Law
Libraries NET promotes law librarians, law librarianship,
and the Association itself to as wide an audience as
possible.

It is a valuable tool, providing members with archived
information about the Association's past, up-to-date information about AALL and law librarianship, and information to help law librarians who wish to continue their
professional education.

Through the years, AALLNET has continually evolved
in order to provide its members with valuable tools,
including:

- In 2004, the AALL general election was held electronically for the first time. This gave members the ability
  to electronically submit a ballot anytime of the day at
  anytime of the week during the election.

- In 2005, the same online election system was modified
  so that AALL entries and SIS may conduct their elections electronically.
- In 2006, AALLNET launched a calendar of events.
  Sustained by the contributions of AALL members,
  information about events pertaining to law librarianship can be found online.
- In 2007, audio and video recordings of professional
  development programs, which were the result of continuing professional education (CPE) grants, were
  made available in the Members Only Section.

FOSTERING MEMBER GROWTH

The American Association of Law Libraries is committed
to building the profession of law librarianship in numerous
ways, including providing financial aid for library school
and law school education, and helping new professionals
become successful through student memberships and
mentoring programs.

CPE

The American Association of Law Libraries provides
increasing support to the professional development of its
members. In the fall of 2005, the Association held an Education Summit to connect AALL members who expected
their Association to provide more opportunities for continuing education between AALL's Annual Meetings. This
need expressed by members aligns with AALL's three
Strategic Directions—leadership, education, and advocacy—the core service goals of AALL from 2005 to 2010.
A number of initiatives grew out of the Education Summit.

In January 2006, AALLNET launched a calendar of
events, which includes listings of professional education
events throughout the country and abroad that are of interest
to law librarians. Built largely by contributions of AALL
members, the calendar lists upcoming and archived events
that are sponsored by a wide variety of entities.

By May 2006, AALLNET included a Speakers Directory, which provides names of speakers suggested by
AALL members. Searchable by either name or area of
speaker expertise, this database showcases the expertise
of AALL members and makes known the names of
speakers who have presented to law librarian groups.

Eager to participate in providing continuing education
tailored to meet the needs of law librarians, Michael
Chiorazzi, director and professor of law at the University
of Arizona College of Law Library, and Mark Estes, director of library services at Holme Roberts Owen LLP,
quickly developed programs for AALL members.
Chiorazzi and Estes received approval and funding for
their proposals from the AALL Executive Board at its
Spring 2006 meeting.

Longtime supporter of AALL professional education activities, BNA Inc., which had generously sponsored the Education Summit, agreed to largely fund Estes' program on law firm library management.

Then Vice President Sally Holterhoff formed the AALL CPE Committee. The stage was set to revitalize AALL's role in providing more high-quality professional education opportunities for law librarians.

The new approach to determining content for continuing education programs is to be member driven, and the CPE Committee's role is to work with staff to make these events happen as quickly and seamlessly as possible for everyone.

In 2007, the Association added new content and resources to AALLNET, the Association's Web site, including streaming media. Now posted as free, members-only educational resources are more than 40 video and audio programs, including *How to Use Competitive Intelligence to Win Clients*, *The Creative Conflict Resolution Toolbox*, and *The Future of Cataloging*.

Bringing select Annual Meeting content to members was also part of new effort to repackage and extend learning opportunities to members who could not attend the Annual Meeting.

These resources allow geographically-dispersed members to have access to online content. These resources also allow members to see and/or hear specific speakers and use them for a future program. The content serves as a springboard for education sessions. One example includes members who utilized one of the videos for staff training and discussion.

The American Association of Law Libraries continues to provide quality educational programs at the AALL Annual Meeting and Conference. Nearly 1700 registrants attended the 100th AALL Annual Meeting and Conference in New Orleans, July 14–17, 2007. The theme, "Rise to the Challenge," reflected both the challenges facing the host city in the aftermath of Hurricane Katrina and those currently faced by law librarians. Thanks to the hard work and flexibility of AALL staff and members, the Annual Meeting schedule provided a balance of educational programming, membership entity meetings, and networking time within a four-day time frame (one day shorter than previous Annual Meetings), beginning with the Opening Event on Saturday evening and concluding with the Closing Banquet on Tuesday night.

Presented in New Orleans were 66 conference programs and four workshops, which had been selected by the 2007 Annual Meeting Program Committee. These, plus 23 special interest section-sponsored programs, covered a wide range of topics.

Scholarships and Grants

Each year, AALL offers a number of different scholarships and grants to both students and practicing law librarians to either complete their education or further continue it.

The American Association of Law Libraries has four different scholarship funds from which approximately 20 scholarships are given each year. These scholarships are funded by a combination of an AALL endowment and contributions from both individual members and legal publishers. These scholarships are given to students that have chosen law librarianship as their career to help with their education expenses.

The American Association of Law Libraries also awards more than 20 grants each year to attend that Annual Meeting and/or educational workshops. These grants are supported by both individual member contributions and donations from legal publishers. The AALL grant program was started in 1952. Since the inception, more than 1000 law librarians have been given funding assistance to attend the Annual Meeting and educational programs.

A CPE Grant Program has also been established by AALL. The Strategic Direction of education is to endure that the law librarians receive the education and training needed to meet and leverage the challenges of the changing information environment. The purpose of the grant program is to encourage program development and promote sharing among AALL entities. The grants program provides funding to AALL Headquarters, entries, SIS, member institutions, caucuses, and individual AALL members to assist in providing ongoing quality continuing education programming outside of the AALL Annual Meeting, which can be distributed to a wider audience.

New Members: Conference of Newer Librarians

Mentoring new members has been an Association priority. A program developed by the Mentoring Committee helps newer members—or those needing career guidance—establish personal contact with experienced law librarians, who act as resources and advisers on the profession and the Association. The Conference of Newer Librarians is now a standing committee and will conduct an annual program to take place at each AALL Annual Meeting.

Ethics and Diversity

The American Association of Law Libraries members recognize their professional roles in the broader legal and social environments.

In 1999, AALL's members endorsed updated "Ethical Principles" produced by the AALL Special Committee on Ethics, to replace the Association's prior "Code of Ethics." The principles offer standards for law librarians to follow when serving diverse clientele, building business relationships, and fulfilling their professional responsibilities.

A symposium on diversity is held annually, supporting AALL's commitment toward diversity in the profession. The sessions at each Annual Meeting examine law librarian's roles in providing legal information to all citizens, including cultural minority groups. The symposia help participants appreciate how ethnic, cultural, and lifestyle differences affect law librarianship, and provide guidance on how to address these issues in daily work.

AALL LEADERSHIP IN SHAPING THE LEGAL INFORMATION ENVIRONMENT

The American Association of Law Libraries continues to advance its unique role as a twenty-first century leader in the legal information arena. It provides the enhanced credibility that a national presence on many fronts can bring to both the Association and to its membership. Due to the impact of the rapid pace of technological change on libraries, as well as the opportunities and challenges of the digital age, AALL's policy work is among the top priorities of the Association.

AALL's Washington Affairs Office

The American Association of Law Libraries' Washington Affairs Office has become a driving force in policy debates affecting members and is a leader on information policy issues. Its staff adheres to the AALL Government Relations Policy approved in March 2001. Washington Affairs Representative Robert Oakley (director of the law library and professor of law, Georgetown University Law Center) was a national and international expert on copyright law until his untimely death in November 2007. The American Association of Law Libraries' Washington Affairs Office staff led the Association's efforts to ensure that law librarians' views are voiced and considered in the formation of state, national, and international policy, focusing on copyright and fair use, legal publishing, the digital information environment, and public access to government information.

The Washington Office actively represents member interests to executive, judicial, and legislative policy makers through formal comments and testimony. The staff also participates in important coalitions of information professionals, including the Library Copyright Alliance (the library community's united voice on copyright issues such as fair use and preservation) and OpenTheGovernment.org (a broad-based coalition that seeks to advance the public's right to know and to reduce secrecy in government). In addition, the Washington affairs representative serves as AALL's liaison to a number of government agencies and organizations.

The American Association of Law Libraries' Washington presence means that the Association can act quickly to advocate member interests as issues emerge, whether they relate to legal publishing, legislative, or policy concerns. The office also alerts members to important federal and state issues, and encourages grassroots advocacy campaigns that involve members of AALL and its entries in promoting the Association's policies and the profession of law librarianship.

As a national leader in information policy, in 2003 AALL published the groundbreaking *State-by-State Report on Permanent Public Access to Electronic Government Information* that researched and reported what, if anything, state governments were doing to meet the enormous challenges of ensuring permanency and public accessibility of government information on the Web. This report raised national awareness and encouraged states to take steps to ensure permanent public access to electronic state government information. As a result, to date nine states have enacted legislation requiring permanent public access to their electronic government information.

In order to focus on an equally important requirement of the life cycle of electronic government information, in 2007 AALL published a second groundbreaking report, the *State-by-State Report on Authentication of Online Legal Resources*. It examined the results of a state survey that investigated whether government-hosted legal resources on the Web are *official* and capable of being considered *authentic*. In researching the question of how trustworthy state-level primary legal resources on the Web are, the report concluded that a significant number of state online resources are *official* but none are authenticated or afford ready authentication by standard methods. State online primary legal resource are, therefore, not sufficiently trustworthy.

The report also found that many states are discontinuing the print versions of legal sources, especially administrative materials, and substituting online versions that are neither official nor authenticated. The American Association of Law Libraries' *Authentication Report* raised concerns that must be addressed by all levels of government, both as high-level policy decisions and as technical issues.

The American Association of Law Libraries convened a National Summit on Authentication of Digital Legal Information in Chicago in April 2007. Approximately 50 delegates from the judiciary, legal community, state governments, and interested organizations, all of whom share AALL's concern about ensuring the authenticity of digital legal information, participated in discussions about the *Authentication Report* findings. The American Association of Law Libraries continues to work with these new allies to explore legal and technological solutions to ensure that state online legal resources are authenticated and trustworthy.

The following illustrate the broad range of policy issues addressed by AALL. The Association:

Advocated for adequate funding for the U.S. Government Printing Office (GPO) and the National

Archives and Records Administration to enable these agencies to meet the challenges of the digital age in providing access to and preserving electronic federal government information;

- Promoted public access to government information through the Federal Depository Library Program and GPOs Federal Digital System that will bring the program into the twenty-first century, particularly in regards to digital authentication;
- Collaborated with other library associations to advocate for balance in federal copyright law as a member of the Library Copyright Alliance;
- Participated in filing several amicus briefs in *Greenberg v. National Geographic Society*, a case that alleged copyright infringement by National Geographic Society due to the publication of 100 years of past issues of its magazine on CD; AALL believed that digitizing past issues of the magazine is a permitted revision of the collective works;
- Participated in a coalition to promote openness in government and fight secrecy, including through legislation such as the OPEN Government Act of 2007 (P.L. 110–175);
- Worked with our entries to engage them in our advocacy work and built a successful member grassroots advocacy program that is critical to our legislative and policy successes.

UNIVERSAL CITATION GUIDE

In 1995, the Task Force on Citation Formats issued its report addressing the controversial issue of vendor- and medium-neutral citations to legal authorities. At a special meeting in July, the executive board adopted the report's recommendations regarding public domain case citation form.

The American Association of Law Libraries became the first national organization to recommend a new standard for citing primary law, case, constitution, statutory, and administrative law in response to member expectations for vendor-neutral citations and the increasing use of electronic information sources. The new format—using standardized abbreviations and other detailed specifications—was published in AALL's *Universal Citation Guide* in 1999. This publication, culminating three years' work by AALL's Committee on Citation Formats, was published by the State Bar of Wisconsin. The work allows users to cite court opinions as issued, regardless of the format in which they appear. The American Bar Association Special Committee on Citation Issues endorsed AALL's citation format and recommended that it become a national standard. In the late 1990s, several states adopted public domain citation for their bodies of case law and recognized AALL's leadership. As of January 2008, 15 states had adopted a medium and vendor-neutral citation system.

A second edition of the guide, published and distributed by William S. Hein & Co., Inc., was released in 2004. The new edition updates the material presented in the original guide.

PUBLISHER RELATIONS

The American Association of Law Libraries' advocacy work with legal publishers is headed up by the Committee on Relations with Information Vendors (CRIV). This committee facilitates communication among information vendors, legal publishers, and Association members. It monitors legal publishing developments, presents the annual new product award, and investigates complaints from members.

The CRIV Sheet, the committee's newsletter that comes out three times per year, tracks industry developments and presents substantive articles about complaint resolutions and legal publishing issues. On its Web site, the CRIV lists CRIV Tools, such as checklists, sample forms, or sample letters, to aid acquisitions and serials librarians in the performance of routine tasks or to assist them in solving commonly occurring problems.

The committee also conducts periodic site visits to legal publishers to learn more about their plans and operations and to represent the interests and concerns of members. In recent years, the CRIV has visited Commerce Clearing House, Inc. (CCH), Lexis Publishing, William S. Hein and Company, Aspen Publishing, West Group, and the Bureau of National Affairs, Inc. Each year at the Association's Annual Meeting, the CRIV sponsors a forum for legal publishing leaders and members.

The CRIV gained considerable importance in the Association during the 1990s, as information vendors increasingly merged to form megapublishers. In 1999–2000, AALL created a Special Committee on Fair Business Practices to prepare guidelines to replace the recently expired Federal Trade Commission's *Guides for the Law Book Industry*. In 2002, the AALL Executive Board approved the *AALL Guide to Fair Business Practices for Legal Publishers* and created a Fair Business Practices Implementation Task Force. The task force steered the distribution and promotion of the new guide as the accepted standard in the industry and monitored its ongoing interpretation, revision, and evaluation. A second edition of the guide was published and distributed to AALL members in 2007.

The *AALL Guide to Fair Business Practices for Legal Publishers* describes standards for the business practices of publishers that most directly affect law librarians and covers the full range of their interactions, from advertising and solicitation to purchases and customer support. It consists of five general principles, each of which is accompanied by subprinciples and examples. It does not explicitly require any methods of operation because it is not intended to interfere with particular business models. As indicated in the introduction, the guide "is designed to allow legal publishers to take advantage of evolving technology and to foster innovation while adhering to principled business practices that will ensure fair and appropriate treatment for customers."

In 2002, recognizing the growing role of license agreements in law library acquisitions, AALL created a Special Committee on Licensing Principles for Electronic Resources. The committee was charged to review the Principles for Licensing Electronic Resources endorsed by the AALL Executive Board in July 1997 and revise and update them as needed.

In 2004, the AALL Executive Board approved the new Principles for Licensing Electronic Resources. The revised principles provide expanded guidance in the areas of access, interlibrary loan, archiving, usage statistics, and dispute resolution. The organization of the document facilitates its use as a checklist for reviewing license agreements, so it is a valuable tool for librarians and vendors alike. The principles are not intended to dictate specific licensing terms, but rather are intended to serve as guidelines and best practices for parties involved in the licensing process.

Association for Library and Information Science Education
Association of American Law Schools
Association of Legal Administrators
British and Irish Association of Law Librarians
Canadian Association of Law Libraries
Center for Computer-Assisted Legal Instruction
Federal Depository Library Council
Friends of the Law Library of Congress
Institute of Museum and Library Services
International Association of Law Libraries
International Federation of Library Associations
Legal Information Preservation Alliance
Legal Marketing Association
Mayflower II Conference
National Center for State Courts
National Commission on Libraries and Information Science
National Equal Justice Library
National Information Standards Organization
Self Represented Litigation Network
Special Libraries Association
Special Libraries Association Legal Division

ENCOURAGING VALUED PARTNERSHIPS

The American Association of Law Libraries' many accomplishments are the result of collaborative efforts with other sister associations and of activities generously supported by legal publishers. These partners work with AALL to advance the interest of law libraries in many important areas.

In July 2008, AALL will again cosponsor a Joint Study Institute, "Harmonization and Confrontation: Integrating Foreign and International Law into the American Legal System" to be held at Georgetown University Law School. The American Association of Law Libraries joins four other law library associations, the Australian Law Librarians' Association, British and Irish Association of Law librarians, Canadian Association of Law Libraries, and the New Zealand Law Librarians' Association, in sponsoring the event.

The American Association of Law Libraries also provides funds for designated leaders or the executive director to represent the Association at other national organizations that share common objectives. This presence leads to greater law librarian involvement in the library and legal communities, such as consulting on library standards and sponsoring joint educational programs. The AALL supports involvement with such organizations as

American Bar Association
American Library Association

PUBLICATIONS

AALL Spectrum

In September 1996, the AALL launched its new magazine, *AALL Spectrum*, which is now issued nine times each year, monthly except for January and August with a combined September/October. The full-color magazine replaced the former newsletter and positioned the Association toward new editorial directions. It focuses on broad issues of interest to members and others in the larger legal and library communities. It also covers Association news and legislative reports from the Washington Affairs Office.

AALL Spectrum includes three regular inserts. *The CRIV Sheet* is written and coordinated by the AALL CRIV three times per year to provide general information on vendor/library relations and to foster a constructive dialogue between the legal publishing/legal information industry and the law library community. *Members' Briefing* also appears three times per year and is written by member subject experts who summarize important issues and present practical ideas for handling them in the workplace. Lastly, each June issue includes a city survival guide with important travel information for AALL members attending the Annual Meeting. Other regular features of *AALL Spectrum* include practice-oriented columns on "Practicing Law Librarianship," "Public Relations," and "Teaching Legal Research," as well as popular interactive departments like "Member to Member" and "Views from You."

Law Library Journal

Law Library Journal (*LLJ*) has been the official publication of the association since it was first published in 1908 in a combined format with the *Index to Legal Periodicals*. Although the *Index to Legal Periodicals* has long since become a separate publication, the *LLJ* has been continuously published on a quarterly basis ever since. As the premier publication in the law library profession, it features scholarly articles on law, legal materials, and librarianship, as well as practice-oriented articles, book reviews, AALL Executive Board election information, and proceedings of the business sessions of the AALL Annual Meeting. In addition to its traditional print format, *LLJ* is currently available in a variety of electronic sources, including four online databases. In recent years, the *LLJ* has had three editors: Richard A. Danner, volumes 77–86, 1984–1994; Frank G. Houdek, volumes 87–99, 1995–2007; and Janet Sinder, volumes 100-present, 2008-present. The editor is assisted by the *Law Library Journal/ AALL Spectrum* Editorial Board and Advisory Committee.

AALL Directory and Handbook

Published annually, the *AALL Directory and Handbook* is distributed to all dues-paying members of the Association. It provides an alphabetical listing of all members, as well as listings of all libraries where members are employed. The handbook section contains Association policies and guidelines, committee charges and rosters, and other pertinent Association information. The directory was first published as *List of Law Libraries in the United States and Canada* in 1940 for the Association through the courtesy of CCH. It has been continuously published by CCH, now Wolters Kluwer Law & Business, ever since.

AALL Biennial Salary Survey

The American Association of Law Libraries is a definitive source for information and statistics about the salaries paid to law librarians and about trends in law library expenditures and services. Every 2 years, since 1995, AALL has gathered meaningful information about members' libraries through the *AALL Biennial Salary Survey and Organizational Characteristics*, which provides such statistics as ratios of attorneys to librarians and expenditures for hard copy versus electronic information. The survey's results help librarians manage their own libraries and leverages AALL's influence with publishers and opinion leaders in the legal community.

Index to Foreign Legal Periodicals

The *Index to Foreign Legal Periodicals* is a multilingual index to articles and book reviews appearing in approximately 540 journals and annuals published worldwide. It provides a subject, author, and geographic approach to comparative and foreign law and the law of all jurisdictions throughout the world, except the United States, United Kingdom, Canada, and Australia. It also provides in-depth coverage of public and private international law.

In addition to periodical publications, the *Index to Foreign Legal Periodicals* analyses about 80 individually published collections of legal essays, Festschriften, Melanges, and congress reports each year. Beginning in 1960, the *Index to Foreign Legal Periodicals* was first published by the Institute of Advanced Legal Studies of the University of London. In 1984, the *Index to Foreign Legal Periodicals* moved to its present location in the Garret W. McEnerney Law Library at Boalt Hall School of Law at the University of California, Berkeley, where it utilizes the library's international and foreign law collection, the largest and most comprehensive in the western United States.

The *Index to Foreign Legal Periodicals* is published quarterly with an annual bound cumulation by the University of California Press. Each single issue consists of approximately 2600 articles and 300 book reviews. In addition to the hard bound volume, the index is also available electronically via Ovid Technologies. Ovid offers complete online coverage of the index from 1985 to the present.

Subscription information and a complete list of all serials currently indexed, as well as links to the index's electronic publishing partners, are available at the Index Web site (http://www.law.berkeley.edu/library/iflp).

AALL Price Index for Legal Publications

The *AALL Price Index for Legal Publications* is an annual table-based index that documents the mean cost of titles and percentage increases over previous years for monographs, serial publications, legal periodicals, looseleaf services, commercially-published court reporters, and legal continuations. The *Price Index* concludes with an appendix of all products and legal vendors that were surveyed. Pricing information is compiled each year by the AALL Price Index for Legal Publications Committee, which is comprised of seven AALL members.

AALL Publications Series

This series of monographs, sponsored by the Association, began in 1960 with the assistance of the Fred B. Rothman Company as publisher. It continues today in partnership with W.S. Hein and Company Inc. as publisher. It now includes more than 60 individual titles and contains a wealth of information for law librarians, including Practicing Reference: Thoughts for Librarians and Legal Researchers, United States Tribal Courts Directory, and Werner's Manual for Prison Law Libraries. Potential manuscripts for the series are reviewed and evaluated by the AALL Publications Committee, which manages the series.

American Association of Museums (AAM)

Elizabeth E. Merritt
American Association of Museums, Washington, District of Columbia, U.S.A.

Abstract
Founded in 1906, American Association of Museums (AAM) is an organization whose mission is "to enhance the value of museums to their communities through leadership, advocacy and service." It is the only organization representing the entire scope of museums and the professionals and nonpaid staff who work for and with museums. As of 2008, it represents more than 15,000 individual museum professionals and volunteers, 3000 institutions, and 300 corporate members.

HISTORY

The American Association of Museums (AAM) was organized in 1906 to develop standards and best practices for museums, gather and share knowledge, and provide advocacy on issues of concern to the entire museum community. AAM is the only organization representing the entire scope of museums and the professionals and nonpaid staff who work for and with museums. As of 2008, it represents more than 15,000 individual museum professionals and volunteers, 3000 institutions and 300 corporate members. Individual members span the range of museum occupations, including directors, curators, registrars, educators, exhibit designers, public relations officers, development officers, security managers, trustees, and volunteers. Every type of museum is represented, including art, history, science, military, maritime, and youth museums, as well as aquariums, zoos, botanical gardens, arboretums, historic sites, and science and technology centers.

The association was established at a meeting of representatives of 21 U.S. museums held at the American Museum of Natural History on May 15, 1906. It filed for incorporation in 1920, and in 1923 established a headquarters in Washington, D.C., at the Smithsonian Castle. Notable accomplishments include

- 1925: Publication of *Code of Ethics for Museum Workers*.
- 1930: Publication of the comprehensive *Museums in America*.
- 1968: AAM accreditation committee chartered.
- 1969: *America's Museums: The Belmont Report* assessed the future and the needs of American museums, developing a solid case for federal support.
- 1970: AAM designates the first accredited museums.
- 1973: *Museum Studies: A Curriculum Guide for Universities and Museums* offers guidelines and minimum standards for museum studies programs.

- 1980: Museum Assessment Program (MAP) begins.
- 1994: *Museum Counts*, the first comprehensive survey of U.S. museum finances and practice.
- 1992: *Excellence and Equity*, a landmark report outlining museums' educational and public service roles.
- 1995: *Museums in the Life of a City: Strategies for Community Partnerships*, which laid the groundwork for a more expansive role for museums in community service.
- 2002: *Mastering Civic Engagement: A Challenge to Museums*, a call to action to museums to realize fully their potential as active, visible players in the lives of their communities.
- 2006: To mark AAM's centennial, the publication of *Riches, Rival and Radicals: 100 Years of Museums in America*, a history of the evolution of museums in the United States.
- 2008: *National Standards and Best Practices for U.S. Museums*, the current, field-wide operating standards in all areas of museum operations, published for the first time.
- 2009: The founding of AAM's Center for the Future of Museums, the field's think tank and forum for discussions on how best to prepare for and shape the future of the nation's museums.

STRUCTURE AND GOVERNANCE

AAM is a 501(c) 3 tax-exempt organization whose mission is "to enhance the value of museums to their communities through leadership, advocacy and service." It is governed by a board of directors consisting of 3 officers of the association (chair, vice-chair, and immediate past chair) and 18 board members-at-large elected by the membership. Its bylaws establish various committees, notably the International Council of Museums-United

Encyclopedia of Library and Information Sciences, Fourth Edition DOI: 10.1081/E-ELIS4-120044036

States (ICOM-US), which helps the association represent museums within the United States to ICOM as well as ICOM's interests within the United States. The International Council of Museums (ICOM) is dedicated to improving and advancing the world's museums and the museum profession as well as preserving cultural heritage while respecting the culture from which it comes. To join ICOM, museum professionals within the United States join their national committee, ICOM-US.

AAM actively seeks the opinions and counsel of its members and is advised by a variety of committees and associations throughout the museum field. Per the association's bylaws, these advisers include three groups. The Standing Professional Committees, open to AAM members and representing specific disciplines such as curators, development, membership, education, media and technology, PR, and marketing and small museum administrators. The Council of Regional Associations informs AAM of policy deliberations that may have an impact in the museum field; it comprises the New England Museum Association, Mid-Atlantic Association of Museums, Southeastern Museums Conference, Association of Midwest Museums, Mountain-Plains Museums Association, and Western Museums Association. The Council of Affiliates, made up of associations of professional categories such as state and local history, conservation, museum stores, volunteers, computers, and trustees, in addition to discipline-specific museum associations representing, for example, zoo, aquariums, colleges, children, science, and Jewish and African American history and culture.

In addition, AAM has created Professional Interest Committees, which enable people and institutions with common interests or needs to develop professional associations. They are open to AAM members. Committee interests range from those of Asian Pacific Americans, Latinos, Native Americans, and gays and lesbians to professionals in the areas of historic house museums, traveling exhibitions, and visitor services.

ACTIVITIES AND PRIORITIES

The association also presents forums for museum professionals to share information and experience. Principal among these is the annual meeting and MuseumExpo™, open to participants from around the world. It is the world's largest conference of museum professionals, offering educational opportunities through its sessions and events at local museums as well as job-hunting forums and the MuseumExpo's extensive showcase of museum products and services.

A key AAM priority is advocating for the value of museums, especially before Congress and within the federal government. The association addresses legislative and regulatory issues that affect museums, particularly increased

funding, and strives to keep advocates informed about issues in Washington through regular communications to AAM's Museum Advocacy Team network and its member newsletter, *Aviso*.

AAM's projects and initiatives serve as a venue and catalyst for new ideas to address issues facing the field, resulting in a variety of products and services. Examples include International Museum Day, Museums and Diversity, the Nazi-Era Provenance Internet Portal, and Small Museums initiatives, and collaborations such as the Non-Profit Listening Post Project; Museums, Archives and Libraries and Partners in Tourism. AAM also seeks to serve new audiences by identifying groups such as Emerging Museum Professionals—those in the profession for less than 10 years. The association offers "EMPs" person-to-person and online networking and mentoring, fellowships, blogs, events, and other resources to guide their career path.

AAM provides a number of services to the field at large. Many are available to museums that are not members. These include the following:

The Accreditation Program serves as the field's primary vehicle for quality assurance and self-regulation, formally recognizing museums' commitment to excellence, accountability, high professional standards, and continued institutional improvement. It accomplishes this by employing a standardized process of self-study and peer review to assess how well each museum achieves its stated mission and goals and meets the standards and best practices generally accepted in the museum field. Any museum can apply for accreditation. Reaccreditation is required after 10 years.

The Museum Assessment Program (MAP), established and administered in collaboration with the Institute of Museum and Library Services, helps maintain and improve operations through a confidential, consultative process, providing guidance in meeting priorities and goals and understanding how a museum compares to standards and best practices. The four categories of assessment are institutional, collections management, public dimension, and governance.

AAM's online bookstore provides publications on a full range of topics relevant to the field. A notable recent example is *AAM Standards and Best Practices for U.S. Museums*, which spells out voluntary national standards and best practices—benchmarks against which museums measure their own performance. A key goal is to help policy makers, media, philanthropic organizations, donors, and members of the public to assess museums' achievements.

The Professional Education Program offers tools for finding solutions to some of the most critical issues in the field. Seminars and workshops are held throughout the year across the United States, in addition to webinars and Web conference programs, addressing the needs of museum professionals at all levels of their careers—emerging, mid-level, executive, and beyond. Topics address exhibitions, and legal issues, visitor experience, executive leadership, technology, interpretation, and historical administration.

SERVICES TO MEMBERS

Beyond its service to the entire field, AAM offers its members the following benefits: It helps them save money, become more influential in their communities and profession, and better serve their public. Its resources include free subscriptions to publications; discounts on books and annual meeting registration, access to peer groups, full access to jobHQ and the Information Center, professional education seminars, books, free and discounted admission to participating member museums and affinity partners; and more. Advantages include the following.

JobHQ allows members to tailor their job searches, receive e-mail alerts when new jobs are posted and post resumes. Employers can list jobs 24 hr a day, track and measure online recruitment activity, and search a database of highly qualified candidates.

AAM's publications department produces members-only benefits such as *Museum*, an award-winning, bimonthly magazine; *Aviso*, a monthly e-newsletter that reports on museums in the news, federal legislation affecting museums, upcoming seminars and workshops, federal grant deadlines and AAM activities and services; bookstore discounts; and Web Exclusives: special online articles and extras, including Practically Speaking (hands-on information); "Day in the Life," first-hand accounts on how active professionals do their jobs; and audio clips of interviews and excerpts.

The Information Center offers fast, easy access to extensive online resources to AAM members and staff of AAM member museums: fact sheets, glossaries, and hundreds of links that have been individually reviewed and described for their usefulness to museums. Staff of AAM member museums also receive confidential, customized reference services within two business days, as well as access to more than 900 sample documents in more than 50 categories, from a variety of types and sizes of museums.

The association's awards and competitions celebrate excellence in the museum field. Two popular competitions open to museum staff around the world are the Publications Competition (Pub Comp), honoring 16 categories from books, catalogs and posters to marketing materials and annual reports, and the Brooking Paper on Creativity in Museums. Other awards include the Nancy Hanks Memorial Award for Professional Excellence, the Award for Distinguished Service to Museums, and more. The Museums and Community Collaborations Abroad (MCCA) is a new grant program designed to strengthen international connections through innovative, museum-based exchanges; funding is provided through a partnership with the U.S. Department of State Bureau of Educational and Cultural Affairs. AAM also provides fellowships for attendance at its annual meeting and professional education seminars and webinars.

CONCLUSION

As AAM embarks on its second century, it is guided by new leadership, structure, and vision—for itself and the field at large. Its new Strategic Framework contains ideas and goals that will drive the association's behavior and activities into the future and guide decisions. The upcoming Center for the Future of Museums will bring in thought leaders from a variety of professions to identify and shape new directions. AAM welcomes input from all who care about museums and their critical role in American society. For additional information or to contact the association, please visit the AAM Web site at http://www.aam-us.org.

American Association of School Librarians (AASL)

Donald C. Adcock
Dominican University, River Forest, Illinois, U.S.A.

Abstract

The American Association of School Librarians (AASL), a type-of-library division of the American Library Association (ALA), serves the needs of over 10,000 school library media specialists. Its mission is to advocate excellence, facilitate change, and develop leaders in the school library media field. This entry discusses the history of AASL, describes its organization and governance, and highlights its major accomplishment.

INTRODUCTION

The American Association of School Librarians (AASL), a type-of-library division of the American Library Association (ALA), is interested in the general improvement and extension of school library services for children and young people.

The American Association of School Librarians has specific responsibility for: planning of programs of study and services for the improvement and extension of library services in elementary and secondary schools as a means of strengthening the educational program; evaluation, selection, interpretation and utilization of media as it is used in the context of the school library program; stimulation of continuous study and research in the school library media field and to establish criteria of evaluation; synthesis of the activities of all units of the American Library Association in the areas of mutual concern; representation and interpretation of the need for the function of school libraries to other educational and lay groups; stimulation of professional growth, improvement of the status of school librarians, and encouragement of participation by members in appropriate type-of-activity divisions; conduct activities and projects for improvement and extension of service in the school library media center when such projects are beyond the scope of type-of-activity divisions, after specific approval by the ALA Council.[1]

DESCRIPTION

Vision

The American Association of School Librarians is

- A proactive organization that addresses issues, anticipates trends, and sets the agenda for the profession.
- An advocate for the indispensable role of school library media programs with school library media specialists, for best practices in school librarianship, and for the core values and ethics of the library profession.
- An open, friendly, welcoming organization that embraces cultural and ethnic diversity.
- An inclusive professional home for all school library media specialists and a partner in mutual interests with educators, technologists, researchers, vendors, and other librarians.
- An essential resource for school library media specialists seeking professional development, leadership opportunities, communication with peers, and the most current information, research, and theory in the field.
- A flexible, responsive organization that models effective management practices.[2]

Mission

The mission of the American Association of School Librarians is to advocate excellence, facilitate change, and develop leaders in the school library media field.

Goals

AASL works to ensure that all members of the school library media field collaborate to

- Provide leadership in the total education program.
- Participate as active partners in the teaching/learning process.
- Connect learners with ideas and information.
- Prepare students for lifelong learning, informed decision-making, a love of reading, and the use of information technologies.[3]

Encyclopedia of Library and Information Sciences, Fourth Edition DOI: 10.1081/E-ELIS4-120043844

Values

AASL will

- Make a difference for members of the field/profession by addressing the important issues.
- Have a national presence and a national identity.
- Encourage diversity in its membership.
- Be organized to clearly address the identified essential functions.
- Have a structure to allow members to be represented in the decision-making process.
- Proved involvement opportunities that will be varied and geographically dispersed.
- Be fast, focused, friendly, flexible, and fun.[4]

EARLY BEGINNINGS

Although the formation of the AASL can be traced back to the ALA Midwinter Meeting in 1914, it did not achieve division status until 1951. In 1914, the Normal and High School Librarians Roundtable petitioned the ALA Council to form a school libraries section. The petition was granted and the section held its first meeting in June of 1915 and elected Mary E. Hall as its president. Two other groups outside the ALA were also interested in promoting school libraries. The National Council of Teachers of English had a Library Section from 1913 until 1919. In July 1896, the National Education Association (NEA) established a Library Department, and, in September of that same year, the ALA Executive Board appointed the Committee on Cooperation with the NEA to foster cooperative efforts with the NEA. However, the school library group affiliated with the ALA became the sole national professional association for school librarians when the Committee on Cooperation with NEA evolved into the School Libraries Committee, which existed from 1931 until 1935. The School Libraries Committee merged with the School Librarians Section in December 1935. The School Libraries Section became one of two sections within the Division of Libraries for Children and Young People in 1941 and changed its name to the American Association of School Librarians in 1944. In 1947, the number of sections within the Division of Libraries for Children and Young People was increased to three and now consisted of the Children's Library Association, Young People's Reading Round Table, and the AASL. In 1951, the AASL became a separate division of ALA.[5] Margaret K. Walraven was elected the first president of the AASL as a division of the American Library Association and Mildred Batchelder became the first executive secretary of the association.[6]

ORGANIZATION AND GOVERNANCE

Board of Directors

The policy-making body of the AASL is its board of directors, which meets at the ALA Midwinter Meeting and Annual Conference. The officers of the association are president, vice-president/president elect, treasurer, and immediate past president. Members of the board are the division officers, a representative of each section, and the AASL representative to the ALA Council, regional directors chosen from nine geographical regions, two members at large, and the executive director. Board members elected from the geographical regions are elected for a 4-year term. They serve 2 years as members of the Executive Committee of the Affiliate Assembly and are designated observers of the AASL Board of Directors. During the second 2 years of their term, they are voting members of the AASL board. The association's executive committee consists of the officers, the division representative to the ALA Council, one member elected by the board of directors, and the executive director.

Executive Committee

The executive committee of the AASL consists of the president, president-elect, and treasurer, immediate past president, the AASL division councilor and one member elected on an annual basis by the board of directors. The executive director serves as an *ex-officio* nonvoting member. The executive committee acts as the budget committee of the association and is restricted to acting on policy and fiscal issues requiring action between meetings of the board.[7]

Affiliate Assembly

The Affiliate Assembly is a more formalized version of the old State Assembly, which had its origins in the AASL Council. The council originated with the practice Ruth Ersted began during her presidency in 1947 and was continued by Frances Henne in 1948–1949. They had invited state leaders to attend AASL Executive Board meetings to establish a forum to communicate with the state leaders and get feedback from them.[6] The assembly is composed of two representatives from state or regional school library media associations affiliated with the AASL. It was established to provide a mechanism for the affiliates to communicate the concerns of their members to the AASL Board of Directors and to report the actions of AASL to their members. To be eligible for affiliation, the president of an organization must be a member of AASL and 25 members or 10% of the association members, whichever is smaller, must also be members of AASL. The assembly meets once a year at the ALA Annual Conference to conduct business and holds a caucus at the ALA Midwinter Meeting for the leaders of the affiliates to discuss mutual problems or attend training sessions to improve the leadership skills of the affiliate representatives. The assembly has an executive committee that meets at the ALA Midwinter Meeting and Annual Meeting and consists of its officers and the regional directors-elect.[8]

Sections

The AASL Bylaws provide for the formation of sections, which consist of at least 25 members who represent a distinct special field of activity within the profession. Each section has a chair, a chair-elect, a secretary, and a representative to the AASL Board of Directors. Currently, there are three sections.

The Independent Schools Section provides a means for the discussion of and action on the problems relating to all phases of nonpublic school librarianship.

The Supervisors Section provides a means for discussion of and action on the problems relating to all phases of school library supervision.

The Educator of School Library Media Specialists, whose purpose is to exchange ideas, review and study curricula, and develop research activities for educators in colleges and universities whose programs focus on school library media education and training.[8]

Committees

The programs of the association are conducted by its standing committees, special committees, and task forces. Standing committees are established to carry on the continuing work of the association. The AASL bylaws provide for three standing committees: budget, nominations, and bylaws. The president may establish additional standing or special committees or task forces at any time with the approval of the board of directors. Special committees are established for a 2-year term to deal with a specific issue. A task force is established with a specific charge to study a specific issue and recommend solutions to the board of directors. A task force exists until it has completed its charge or is dismissed by the board of directors. Currently, there are 38 standing and special committees and three task forces.[7]

PUBLICATIONS

AASL publishes one print journal, *Knowledge Quest* (KQ), one electronic journal *School Library Media Research*, and an electronic newsletter, *AASL Hotlinks*. Each journal has an editor appointed by the president with the consent of the board of directors and an advisory board consisting of members of the association.

Knowledge Quest, published bimonthly September through June, is devoted to offering substantive information to assist building-level library media specialists, supervisors, library educators, and other decision makers concerned with the development of school library media programs and services. Articles address the integration of theory and practice in school librarianship and new developments in education, learning theory, and relevant disciplines. *Knowledge Quest* is mailed to all members and is available to nonmembers by subscription.

Provided at no cost, *KQ on the Web* (http://www.ala.org/aasl/kqweb/), expands *Knowledge Quest* via the Internet. It contains "KQ Extra" items supporting one feature article and columns published in *Knowledge Quest*. These Web-only items are referenced within the relevant print article or column and direct the reader to the online content. Abstracts of feature articles and information on becoming a member of AASL or subscribing to *Knowledge Quest* are also provided.[9]

School Library Media Research (http://www.ala.org/aasl/SLMR/), a refereed research journal published electronically, is the successor to *School Library Media Quarterly Online*. The purpose of *School Library Media Research* is to promote and publish high-quality original research concerning the management, implementation, and evaluation of school library media programs. The journal also emphasizes research on instructional theory, teaching methods, and critical issues relevant to school library media. *School Library Media Research* is currently available at no charge.[10]

AASL Hotlinks is the text-only monthly e-mail newsletter of the AASL. *AASL Hotlinks* is primarily composed of brief summaries with links to more in-depth content, and includes previews of upcoming association activities and continuing education programs, news from AASL and all of ALA, valuable Web resources, highlights of new products and services, summaries of new articles from AASL's print and online journals. This text-based newsletter is sent to AASL members with valid e-mail addresses on file in the ALA member database. Additional "pass along" distribution occurs through AASL members, who are encouraged to share each issue with colleagues and administrators in their school and district, and AASL's state and regional affiliate organizations, which can redistribute the newsletter to their own members.[11]

In addition to these two journals, AASL publishes pamphlets, brochures, small monographs, and resource guides designed to assist school library media specialists in establishing and operating school library media programs. There is a nominal charge for some of these publications, but others are available at no charge on the AASL home page (http://www.ala.org/aasl).

AWARDS, GRANTS, AND SCHOLARSHIPS

Generous contributions made to AASL by a number of sponsors enable AASL to fund the following awards, grants, and scholarships.

AASL Collaborative School Library Media Award

Established in 2000, the $2500 award, sponsored by Highsmith, Inc. recognizes and hopes to encourage collaboration and partnerships between school library media specialists and teachers in meeting educational goals outlined in *Information Power: Building Partnerships for Learning*[12] through joint planning of a program, unit, or event in support of the curriculum and using media center resources. Applicants must be AASL personal members.

ABC-CLIO Leadership Grant

Established in 1986, the grant, up to $1750 donated by ABC-CLIO, is given to school library media associations that are AASL affiliates for planning and implementing leadership programs at the state, regional, or local levels.

Beyond Words: Dollar General School Library Relief Fund

Dollar General, in collaboration with the American Library Association (ALA), the American Association of School Librarians (AASL) and the National Education Association (NEA), sponsors the school library disaster relief fund for public school libraries in the states served by Dollar General. The fund will provide grants of $5,00–$15,000 to public schools whose school library program has been affected by a natural disaster (tornado, earthquake, hurricane, flood, etc.). Grants are to replace or supplement books, media, and/or library equipment in the school library setting.

Distinguished School Administrators Award

Established in 1985, the $2000 award sponsored by ProQuest, honors a school administrator who has made worthy contributions to the operations of an exemplary school library media center and to advancing the role the school library media center in the educational program. Nominations must be made by AASL personal members.

Distinguished Service Award

Established in 1978, the $3000 award donated by Baker and Taylor recognizes an individual member of the library profession who has, over a significant period of time, made an outstanding national contribution to school librarianship and school library development. Nominations must be made by AASL personal members.

Frances Henne Award

Established in 1986, the $1250 award sponsored by the Greenwood Publishing Group recognizes a school library media specialist with 5 years or less experience in the profession who demonstrates leadership qualities with students, teachers, and administrators to attend an AASL conference or ALA Annual Conference for the first time. Applicants must be AASL personal members.

Information Technology Pathfinder Award

Established in 1985 as the Microcomputer in the Media Center Award, this award of $1000 to the school library media specialist and $500 to the library is sponsored by the Follett Software Company. The award recognizes a school library media specialist who demonstrates vision and leadership through the use of technology. The award is given in two categories—elementary (K–6) and secondary.[7–12] Applicants must be AASL personal members.

Innovative Reading Grant

Established in 2006, the $2500 grant sponsored by Capstone Publishers supports the planning and implementation of a unique and innovative program for children, which motivates and encourages reading, especially struggling readers.

Intellectual Freedom Award

Established in 1982, the award, $2000 to the recipient and $1000 to the media center of the recipient's choice, is sponsored by ProQuest and recognizes a personal member of AASL who has upheld the principles of intellectual freedom. Applicants must be AASL personal members.

National School Library Media Program of the Year Award

Established in 1963, the award, $10,000 ($30,000 total) in three categories, sponsored by Follett Library Resources, recognizes school districts and single schools for exemplary school library media programs that are fully integrated into the school's curriculum. Winners receive a crystal obelisk and $10,000 each, one district award and two single schools.

President's Crystal Apple Award

Established in 1992, the crystal apple award is given at the discretion of the AASL president to an individual or group who has had significant impact on school libraries and students.

School Librarian's Workshop Scholarship

The scholarship, $3000 sponsored by Jay W. Toor, president, Library Learning Resources, recognizes a full-time student preparing to become a school library media specialist at the preschool, elementary, or secondary level. The recipient must pursue graduate level education in an ALA-accredited library school program or in a school library media program that meets the ALA curriculum guidelines for the National Council for Accreditation of Teacher Education.[13]

CONFERENCES AND EVENTS

National Conference and Exhibition

The first AASL national conference was held at the Commonwealth Convention Center in Louisville, Kentucky, in 1980. The daunting task of planning the first conference was undertaken by Rebecca Bingham, a former president of AASL and director of Library Media Services for Jefferson County (Kentucky) Public Schools, as chairperson of the conference planning committee. To bring programs and commercial exhibits closer to its members, AASL has held a national conference in a different geographical region every 2 years since that first conference. The number of attendees, the scope of the programs, and the number of exhibits have increased with each conference.

Regional Institutes

AASL licenses regional institutes to organizations who pay AASL a fee for a one-time presentation. AASL provides the organization with promotional information that it can use to develop publicity about the institute, the content for the institute and the honorarium, and all travel expenses of the presenter it provides. The organization sets its own registration fee and is responsible for all local planning, logistics, and publicity.

Fall Forum

In the fall of the years when AASL does not hold its national conference, the association provides a multiday national institute. The forum provides several speakers making presentations that deal with different aspects of a central topic. Time is provided for forum participants to interact with the presenters and each other in small discussion groups. Topics in the past have included "Assessing Student Learning in the Library Media Center" and "Collaboration and Reading to Learn @ Your Library."[14]

GUIDELINES AND STANDARDS

School Library Programs

A preliminary report of the Commission on Library Organization and Equipment of the National Education Association and the North Central Association of Colleges and Secondary Schools chaired by C.C. Certain, principal of Cass Technical High School in Detroit, Michigan, was published in 1917.[15] The final report, *Standard Library Organization and Equipment for Secondary Schools of Different Sizes*,[16] became the first school library standards and was published by the ALA in 1920. Mary E. Hall, librarian at Girl's High School in Brooklyn, New York, and Hannah Logasa, librarian at the University of Chicago School of Education, both early leaders in the development of the concept of school libraries, were members of the commission. In 1925 a joint committee of the National Education Association and the American Library Association, chaired by C.C. Certain, coauthored *Elementary School Library Standards* that established standards for elementary school libraries.[17] The next set of standards for school libraries was published in 1945 by the ALA; *School Libraries for Today and Tomorrow*[18] was prepared by the Committees on Post-War Planning of the ALA, Division of Libraries for Children and Young People and its section, the AASL under the chairmanship of Mary Peacock Douglas. In 1960, the ALA published the third set of school library standards, *Standards for School Library Programs*.[19] The AASL, by now a division of the ALA, prepared these standards in cooperation with the Association for Colleges for Teacher Education. *Standards for School Media Programs*,[20] published in 1969, were the first standards jointly developed by AASL and the Department of Audiovisual Instruction [now Association for Educational Communications and Technology (AECT)]. The standards, *Media Programs: District and School*,[21] published in 1979, were jointly developed by the AASL and the AECT and published by the ALA.

In 1988, AASL and AECT jointly prepared *Information Power: Guidelines for School Library Media Programs*. Leaders of the two associations realized that the two associations lacked authority to establish national standards but could establish national guidelines. Therefore, for the first time, the national document focusing on the development of school library programs was called guidelines rather than standards. In 1998, AASL and AECT prepared a new edition of the guidelines entitled *Information Power: Building Partnerships for Learning*,[12] which was published by the ALA. In 2009, the American Association of School Librarians authored a new set of guidelines, *Empowering Learners: Guidelines for School Library Media Programs*, published by the ALA.

As minimal as the "Certain standards" may seem today, they were the first published national standards that provided some guidance to school administrators attempting to organize a school library. The 1945 standards established the model that still impacts school library media programs today and established AASL as the authority in the field. The 1960 standards recognized the impact of the school library media program on the educational goals of the school and the teaching role of the school library media specialist. They were also the first to discuss the importance of the collaboration between classroom teachers and the school library media specialist. The 1969 standards attempted to emphasize the inclusion of resources other than print items in school library collections and the perceived new role of the school librarian by eliminating the word "library" and using the terms media, media specialist, and media center. The 1975 standards placed a heavy emphasis on quantitative standards but also provided a set of guiding principles for each area of the standards. It also, as the title implies, addressed the role of the district, as well as the local school, in the development of school media programs. The emphasis in the 1988 standards was the role of the school library media specialist and library media program in the achievement of the school's curricular goals. The term "library" was reinstated and used along with "media" (e.g., library media specialist and library media program). There was also a strong emphasis on collaboration with the various members of the school community. The major addition to the new edition of the 1988 standards, published in 1998, is the *Information Literacy Standards for Student Learning*.[21] In 2009 the Association published a new set of guidelines, *Empowering Learners: Guidelines for School Library Media Programs*. These new guidelines reflect the heightened influence of technology and evidenced-based learning on school library media programs and the necessity of producing successful learners skilled in multiple literacies.[22]

Other National Guidelines and Standards

In 1998, AASL collaborated with the Alliance for Curriculum Reform in the development of the National Study of School Evaluation's (NSSE) standards-based guide for program evaluation of library media services. This publication, *Program Evaluation: Library Media Services*,[23] is one volume of the study's Indicators of Quality Program Evaluation Series. The NSSE serves as the research and development branch of the regional accrediting commissions, which accredit the educational programs of elementary, middle, and secondary schools and colleges and universities. Its board of directors is composed of representatives from each of the six regional accrediting associations in this country. The director of the project to develop the guide extended a special thanks to the members of the AASL staff, Julie Walker and Barbara Herrin, and members, Betty Marcoux, Carol Newman, and Barbara Stripling, for their extensive contribution to the development of the guide.[24]

The AASL, with 11 other national associations, was a partner of the International Society for Technology in Education (ISTE) in their development of the *National Educational Technology Standards for Teachers* and *National Educational Technology Standards for Students*. The standards were developed as a part of the ISTE initiative, National Educational Technology Standards, funded by the National Aeronautics and Space Administration in consultation with the U.S. Department of Education, the Milken Exchange on Education Technology, and Apple Computer. The student standards were designed to provide teachers, technology planners, and other educational decision makers a framework to establish learning environments supported by technology. The framework incorporates educational technology skills into relevant curricular areas.[25] The intent of the teacher standards is to establish a set of performance-based standards for the institutions preparing teachers in order for teachers to be able to provide a technology-supported learning environment for students.[25]

MAJOR GRANT PROJECTS

Knapp Project

In 1962, the Knapp Foundation provided a grant in excess of $1 million to AASL to develop model elementary and secondary school library programs through improving library resources and personnel. Peggy Sullivan from Montgomery County, Maryland, was appointed project director. The Knapp School Libraries Project was to be developed in three phases with the funding spread over a 5-year period. The purpose of the project was to establish model programs throughout the country that could be visited by librarians, educators, and community members in order that they might replicate these model school libraries in their communities.[6]

In 1968, the Knapp Foundation made a second grant of over $1 million to AASL for the Knapp School Library Manpower Project. The grant was made in response to the identified need for qualified school librarians in the field. Robert N. Case was appointed the project director and Anna Mary Lowrey was appointed the associate director. The purpose of the program was threefold: 1) perform a task and job analysis of professional and nonprofessional school library staff members; 2) identify a manpower pool for recruitment to the profession; and 3) identify experimental and demonstration educational programs for the development of school librarians.[6]

DeWitt Wallace *Reader's Digest* Library Power Project

From 1988 through 1998, the DeWitt Wallace *Reader's Digest* Fund provided major funding, through a grant to the ALA, for the AASL to coordinate the National Library Power Program. Ann Carlson Weeks served as the national coordinator of the project from its beginning until she left AASL; Donald Adcock was named to fill that position until the conclusion of the project. This $41 million initiative provided 19 communities with 3-year grants to improve school library media programs. Funding to the local communities was made to local education funds rather than directly to the schools. The AASL coordinated the national program, provided administrative and technical assistance to the 19 Library Power sites, and collaborated with the Public Education Network, who provided technical assistance to the local education funds. The initiative was based on *Information Power: Guidelines for School Library Media Programs*. Its purpose was to "create a national vision and new expectations for public elementary and middle school library programs and to encourage new and innovative uses of the library's physical and human resources." Library Power provided funds for professional development for teachers, administrators, and school library media specialists; to renovate school library media centers; to match local funds for library books and other library resources; and to hire project staff. To participate in the program, the local school agreed to provide a full-time, certified school library media specialist, keep the library open throughout the school day with schedules that provided for open access to the library, support release time for staff to attend Library Power professional development activities, and cover labor costs for renovation and remodeling of school library media facilities. Library Power projects were located in the following cities: Atlanta, Georgia; Baton Rouge, Louisiana; Berea, Kentucky; Cambridge, Massachusetts; Chattanooga, Tennessee; Cleveland, Ohio; Dade County (Miami), Florida; Denver, Colorado; Lincoln, Nebraska; Lynn, Massachusetts; Mon Valley (McKeesport), Pennsylvania; Nashville, Tennessee; New Haven, Connecticut; New York, New York; Paterson, New Jersey; Philadelphia, Pennsylvania; Providence, Rhode Island; Tucson, Arizona; and Wake County (Raleigh), North Carolina.[26] In 2001, The AASL and the Public Education Fund collaborated with the ALA to publish *The Information-Powered School*, edited by Sandra Hughes-Hassell and Anne Wheelcock. Each chapter of the publication deals with a specific aspect of Library Power and was written by a member of one of the 19 sites. The intent of the publication is to show how other schools can use the lessons learned by the Library Power participants to implement any part, or all, of Library Power.

CONCLUSION

As it has throughout its history, the AASL will continue being the professional school library organization that provides a national voice that advocates providing each student in elementary and secondary schools access to a school library media program with access to the inviting environment of a school library media center and the human and intellectual resources that allow them to become effective users of information. It will continue to be the voice that calls attention to the importance of providing each school with a qualified school librarian to work with other members of the school community to provide students with the skills necessary to locate, select, evaluate, and use information effectively. The AASL will continue to collaborate with other national educational organizations to advocate for the improvement of the quality of elementary and secondary education and the preparation of those who work with children and youth in our schools. Additional Information about the American Association of School Librarians and its programs, activities, publications, and organization can be found at http://www.ala.org/aasl.

REFERENCES

1. *ALA Handbook of Organization 2006–2007*. American Library Association: Chicago, IL, 2006; 62.
2. AASL Vision Statement. Available at http://www.ala.org/aasl/aboutaasl/aaslvision/aaslvisionstatement.cfm.
3. AASL Mission and Goals. Available at http://www.ala.org/aasl/aboutaasl/mission and goals/Aaslmissiongoals.cfm.
4. AASL Values. Available at http://www.ala.org/aasl/aboutaasl/aaslvalues/aaslvalues.cfm.
5. Pond, P. Development of a professional school library association: American Association of School Librarians. School Media Q. **1976**, *5* (1), 12–14 Fall.
6. Koch, C.W. *A history of the American Association of School Librarians, 1950–1971*, Southern Illinois University: Carbondale, IL, 1976; 150. Ph D dissertation.
7. American Association of School Librarians. Bylaws. Available at http://www.ala.org/aasl/aboutaasl/aaslgovernance/aasldocuments/aaslbylaws.cfm American Association of School Librarians. Bylaws.
8. *ALA Handbook of Organization 2006–2007*. American Library Association: Chicago, IL, 2006; 66.
9. American Association of School Librarians. *About Knowledge Quest*. Available at http://www.ala.org/ala/aasl/aaslpubsandjournals/kqweb/aboutkq/aboutkq.cfm (accessed May 2007).
10. American Association of School Librarians. *School Library Media Research*. Available at http://www.ala.org/ala/aasl/aaslpubsandjournals/slmrb/schoollibrary.cfm (accessed May 2007).
11. American Association of School Librarians. *AASL Hotlinks: The Official Monthly E-mail Newsletter of AASL*. Available at http://www/ala.org/aaslTemplate.cfm?section=aaslhotlinks (accessed May 2007).
12. American Association of School Librarians and Educational Communications and Technology. *Information Power: Building Partnerships for Learning*, American Library Association: Chicago, IL, 1998.

13. American Association of School Librarians. *AASL Awards, Grants and Scholarships*, http://www.ala.org/aasltemplate. cfm?Section=aaslawards (accessed May 2007).

14. American Association of School Librarians, *AASL Conferences and Events*. Available at http://www.ala.org/ aasltemplate.cfm?Section=conferencesandevents (accessed May 2007).

15. Certain, C.C. *Standard Library Organization for Accredited High Schools of Different Sizes: First Preliminary Report*, National Education Association: Washington, DC, 1917.

16. Certain, C.C. *Standard Library Organization for Accredited High Schools of Different Sizes*, American Library Association: Chicago, IL, 1920.

17. Certain, C.C. *Elementary School Library Standards*, American Library Association: Chicago, IL, 1925.

18. Committee on Post-war Planning for the American Library. *Association School Libraries for Today and Tomorrow*, American Library Association: Chicago, IL, 1945.

19. American Association of School Librarians. *Standards for School Library Programs*, American Library Association: Chicago, IL, 1960.

20. American Association of School Librarians and Department of Audiovisual Instruction. *Standards for School Media Programs*, American Library Association: Chicago, IL, 1969; National Education Association: Washington, DC, 1969.

21. American Association of School Librarians and Association for Educational Communications and Technology. *Media Programs: District and School*, American Library Association: Chicago, IL, 1975; Association for Educational Communications and Technology: Washington, DC, 1975.

22. American Association of School Librarians. *Empowering Learners: Guidelines for School Library Media Programs*. American Library Association: Chicago, IL, 2009.

23. Fitzpatrick, K.A. Edwards, B. *Program Evaluation: Library Media Services*, National Study of School Evaluation: Schaumburg, IL, 1998; viii.

24. Fitzpatrick, K.A. Edwards, B. *Program Evaluation: Library Media Services*, National Study of School Evaluation: Schaumburg, IL, 1998; xiii.

25. International Society for Technology in Education. *National Educational Technology Standards for Teachers*, The Society: Eugene, OR, 2000; 29.

26. National Library, Power Program. *Transforming Teaching and Learning for Children*; National Library Power Program: Chicago, IL, 1994.

American Library Association (ALA)

Mary W. Ghikas
American Library Association, Chicago, Illinois, U.S.A.

Abstract
The American Library Association (ALA) was founded in 1876 in Philadelphia, at a conference held at the Pennsylvania Historical Society. The defined purposes of the ALA are fundamentally public service purposes–primarily public and educational, rather than professional. ALA is, therefore, characterized as a nonprofit corporation under Section 501(c)(3) of the U.S. Internal Revenue Service code and has, from its formation, been open to "any person, library, or other organization interested in library service and librarianship. . .upon payment of the dues provided for in the Bylaws."

INTRODUCTION

The American Library Association (ALA) was founded in 1876 in Philadelphia, at a conference held at the Pennsylvania Historical Society. The purposes described in the enabling resolution would later be expressed succinctly in Article II of the ALA Constitution: "The object of the ALA shall be to promote library service and librarianship."[1] This founding conference grew from an 1853 meeting in New York, initiated by Charles B. Norton, publisher; Hastings Grant, librarian of the New York Mercantile Library; Charles Jewett, librarian of the Smithsonian Institution; and others. A committee was appointed in 1853 and charged to draft a constitution and bylaws—and to call a second meeting.[2]

Civil war and its aftermath delayed that second conference. Then, with the United States moving toward its centennial, various library leaders suggested a congress of librarians. On April 22, 1876, a letter from James Yates, a public librarian from England, was printed in Frederick Leypoldt's "Library and Bibliographical Notes" column in R.R. Bowker's *Publishers Weekly*, commenting on the fact that no congress of librarians had been called, though other scientific, educational, and professional groups were convening. The call was picked up and pushed forward by Melvil Dewey, then librarian of Amherst College. Publishers Weekly advertised both the impending conference and the birth of the *American Library Journal* in September 1876.[3]

The conference arrangements committee for the Philadelphia conference included: Justin Winsor, Boston Public Library; Lloyd P. Smith, Philadelphia Library Company; and William F. Poole, Chicago Public Library. One hundred and three persons attended the conference, including 13 women. Justin Winsor was elected president (1876–1885) and Melvil Dewey was elected secretary.[2]

In 1879, the ALA was officially incorporated under the laws of the commonwealth of Massachusetts, by Justin Winsor, C.A. Cutter, Samuel S. Green, James L. Whitney, Melvil Dewey, Fred B. Perkins and Thomas W. Bicknell. The language of the charter echoed the founding resolution:

> for the purpose of promoting the library interests of the country by exchanging views, reaching conclusions, and inducing cooperation in all departments of bibliothecal science and economy; by disposing the public mind to the founding and improving of libraries; and by cultivating good will among its members. . . .

In 1942, the Charter was revised, restating the purposes of the Association to include "promoting library interests throughout the world," and adding that the Association might fulfill its mission "by such other means as may be authorized from time to time by the Executive Board or Council of the ALA."

ALA VALUES, PRIORITIES, AND MISSIONS

In 1986, the ALA Council approved this mission statement:

> The mission of the American Library Association is to provide leadership for the development, promotion, and improvement of library and information services and the profession of librarianship in order to enhance learning and ensure access to information for all.[4]

The 1995 report of the ALA Organizational Self-Study Committee (OSSC), comparing this mission to the 1879 charter, as amended in 1942, noted: "The fact that they are similar is one of the enduring strengths of ALA."[5]

Over the years, ALA has defined its directions and priorities within the broad contexts of social, economic, and political change. A 1986 statement of organizational and operational priorities notes: "Any organization as

Encyclopedia of Library and Information Sciences, Fourth Edition DOI: 10.1081/E-ELIS4-120044705

large, diverse, and dynamic as ALA must periodically reassess priorities...."[6]

In 1959, the ALA Council adopted the first *ALA Goals for Action*; a revision was approved in 1967. The 1984 *ALA Annual Report* noted formation of a Process Planning Group to guide ALA's "first comprehensive long-range planning process"—Strategic Long-Range Planning (SLRP). The following year, the ALA Office for Research and Statistics surveyed over 450 members of the leadership and general membership. On January 9, 1985, ALA Council approved the process outline. A planning session at the 1985 ALA Annual Conference was attended by more than 400 individuals.

From this process, six "priority areas" were defined in 1986:[7]

- *Access to information*: ALA will promote efforts to ensure that every individual has access to needed information at the time needed and in a format the individual can utilize, through provision of library and information services.
- *Legislation/funding*: ALA will promote legislation at all levels that will strengthen library and information services. Means will be developed for facilitating the effective competition of libraries for public funds as well as for funds from the private sector.
- *Intellectual freedom*: ALA will promote the protection of library materials, personnel, and trustees from censorship; the defense of library personnel and trustees in support of intellectual freedom and the Library Bill of Rights; and the education of library personnel, trustees, and the general public to the importance of intellectual freedom.
- *Public awareness*: ALA will promote the role of librarians and the use of libraries and their resources and services as well as the awareness of their importance to all segments of society.
- *Personnel resources*: ALA will promote the recruitment, education, professional development, rights, interests, and obligations of library personnel and trustees.
- *Library services, development, and technology*: ALA will promote the availability of information tools and technologies which assist librarians in providing services responsive to the changing needs of society.

These six priority areas were supported by three "organizational support goals," which defined organizational philosophies regarding roles and relationships, finances, and human resources.

These priorities remained in the ALA Policy Manual until 2006.[8] Over the next two decades following 1986, ALA approved additional statements of goals and programmatic priorities. There is striking continuity in the successive goals and statements, across several generations of leadership.

On February 6, 1995, the Association approved ALA Goal 2000, which called for the ALA to be "as closely associated with the idea of the public's right to a free and open intellectual society—intellectual participation—as it is with the idea of intellectual freedom." Originally presented to the ALA Executive Board in October 1994, by then executive director Elizabeth Martinez, the plan's tactical recommendations included expansion of the ALA Washington Office, establishment of an Office of Information Technology Policy and establishment of an independent charitable foundation called the Fund for America's Libraries.[9]

Arising from analysis of Goal 2000, the ALA Executive Board defined five "key action areas"—arenas in which the Association must be active in order to fulfill its mission and ALA Goal 2000. Beginning January 13, 1998, at the Midwinter Meeting in New Orleans, the ALA Council approved these "key action areas" annually, through 2005, as the Association's current "programmatic priorities": diversity, intellectual freedom, equity of access, twenty-first century literacy, and education and continuous learning.

As 2000 approached, ALA leaders began development of a new statement of vision and goals. In July 2000, the ALA Council approved ALAction 2005, including the following goals:

- By 2005, ALA will have increased support for libraries and librarians by communicating clearly and strongly why libraries and librarians are both unique and valuable.
- By 2005, ALA will be recognized as the leading voice for equitable access to information resources in all formats for all people.
- By 2005, ALA will be a leader in the use of technology for communication with, democratic participation by, and for shared learning among its members.
- By 2005, ALA will be a leader in continuing education for librarians and library personnel.

With the approach of 2005, the Association again looked at its strategic directions. This time, the ALA Executive Board and ALA Management undertook the most ambitious effort, since the 1980s, to secure member input: focus groups at the American Association of School Librarians (AASL) (Fall 2003) and Public Library Association (Spring 2004) National Conferences and at the 2004 ALA Midwinter Meeting and Annual Conference, including one focus group specifically related to academic libraries; open ALA Forums at chapter conferences, including student chapters; an email questionnaire completed by 17,000 members; analysis of planning documents from ALA divisions and other groups; telephone interviews with nonmembers; collection of a broad array of external environmental scan information; and, widespread dialogue with leadership.

In Fall 2004, the ALA Executive Board, ALA Division presidents, ALA Round Table representatives, and senior ALA and Division staff met to develop the first stage of a 2010 plan. The process continued at the 2004 joint meeting of the ALA Executive Board and executive committees of the 11 divisions. Following review of a draft plan by ALA divisions, round tables, committees at others at the 2005 ALA Midwinter Meeting, a proposed plan— ALA Ahead to 2010—was distributed for review and approved by the ALA Council at the 2005 ALA Annual Conference.[10]

The approved plan includes six major goal areas, each with several specific objectives:

Goal Area I: *Advocacy/Value of the Profession*: ALA and its members are the leading advocates for libraries and the library profession.

Goal Area II: *Education*: Through its leadership, ALA ensures the highest quality graduate and continuing education opportunities for librarians and library staff.

Goal Area III: *Public Policy and Standards*: ALA plays a key role in the formulation of national and international policies and standards that affect library and information services.

Goal Area IV: *Building the Profession*: ALA is a leader in recruiting and developing a highly qualified and diverse library work force.

Goal Area V: *Membership*: Members receive outstanding value for their ALA membership.

Goal Area VI: *Organizational Excellence*: ALA is an inclusive, effectively governed, well-managed, and financially strong organization.[11]

In Fall 2005, the ALA Executive Board recommended revised programmatic priorities—or key action areas—to increase alignment with the new plan. The revised programmatic priorities are diversity, equitable access to information and library services, education and lifelong learning, intellectual freedom, advocacy for libraries and the profession, literacy and organizational excellence. The new priorities have been presented to and approved by the ALA Council annually since January 2006.[12]

MEMBERSHIP AND ORGANIZATIONAL CHANGE

The defined purposes of the ALA are fundamentally public service purposes—primarily public and educational, rather than professional. ALA is, therefore, characterized as a nonprofit corporation under Section 501 (c) (3) of the U.S. Internal Revenue Service code and has, from its formation, been open to "any person, library, or other organization interested in library service and librarianship...upon payment of the dues provided for in the Bylaws."[13]

As of August 31, 2007, ALA membership included the following personal membership categories: Regular, Student, Library Support Staff, Trustees and Friends, Associate Members (including those not employed in library and information services or related activities), Retired, Non-Salaried, and International, as well as Life Members and Continuing Members. Within each category, dues are flat rate—not on a salary-based sliding scale, with the exception that a new (first-time) member pays a reduced rate during the first and second years of membership. In 1986, the membership year changed from the calendar year to the "anniversary" year, marked from the month in which the member initially joined the Association. Organizational membership (based on budget size) and corporate membership (patron or contributor) is also available.

Basic ALA membership does not include membership in any ALA division or round table, which is separately selected with (as of August 31, 2007) approximately 67% of ALA membership belonging to one or more division or round table. In aggregate, almost as many division and round table membership are purchased (approximately 61,000) as ALA personal memberships (approximately 62,000). All division and round table members must first be members of the ALA.

As early as 1918, when ALA membership was 3380 (smaller than many current ALA divisions), the question "is ALA too big?" was raised.[2] In 1989, a major milestone was reached when Wendy Sinnott became the 50,000th member of ALA. Ten years later, 1998–1999, ALA experienced its largest 1-year growth—adding 3459 members in 1 year, for a total of 58,777.

On August 31, 2007, the ALA had 3480 organizational members, 266 corporate members, and 60,983 personal members—a total of 64,729. ALA's membership retention rate has been notable, ranging from 85% to 87% over a 10-year period.

ORGANIZATIONAL DEVELOPMENT: GROWTH AND DEMOCRATIZATION

ALA has sought, from its earliest history, to maintain an organizational structure characterized by democratic processes, individual member participation, and effective operations—though the understanding of these has varied significantly.

The original ALA constitution called for members to elect an executive board of five members, which in turn had the power to select additional members and to elect officers from the executive board so constituted. Nominating and election practices resulted in a largely self-perpetuating and homogenous governance body.[3] In 1892, the constitution was revised to add the ALA Council. Recommendations in regard to library policy and creation of new sections within ALA both required Council action. This first council consisted of 10 councilors, elected by the membership. At the same time, the constitution established direct election of officers.[14] The 1899 constitution did not clearly delineate the division of power between the council and the executive board. With the 1909 constitutional revision, the council grew to include the executive board (*ex officio*), all ex-presidents of the Association, and 50 additional members, one-half chosen by Council and one-half chosen by the membership. No questions of policy could be voted by the membership without prior consideration by Council. The "business affairs" of the Association were entrusted to the executive board.[14] In 1919, further revisions made the Executive Board the governing body of the Association—a move viewed with some suspicion by many members.[14] Further constitutional revisions would continue to alter the composition of both the Council and Executive Board and to redefine the distribution of powers.

PERIODIC SCRUTINY

In 1928, at the West Baden, Indiana, conference, the ALA Council approved a special committee report which recommended:

> A periodic scrutiny of Association activities within 3 years and not less frequently than every third year thereafter, by a committee appointed by the president, such scrutiny to include as complete consideration of the effectiveness and results of the various activities as is warranted and practicable, with a view of suggesting to Council possible changes of policy.[15]

Over the next two decades, four such committees were appointed.

The First Activities Committee report (December 1930) covered library extension, adult education, education for librarianship, foundation projects, financing, publishing, membership, personnel practices, and ALA headquarters. The Second Activities Committee report (December 1934) noted the ALA Executive Board's increasing tendency to systematically review the Association's activities and recommended that committees be appointed to survey the Association's activities only every 6 years, unless particular circumstances warranted more frequent survey.

Five years later, in 1939, the Third Activities Committee produced a report calling for a more democratic organization, particularly for a revised Council that would be completely elective, based on specific interests and on geography,[14] as well as establishment of a sliding scale of membership dues, based on individual salary or (for institutional members) library budget. Also included in the report were recommendations to give more authority to sections within the Association and call them divisions,

and to give those divisions an allotment of 20% of the dues paid by their members to ALA.

In 1949, against a background of fiscal concerns, the report of the Fourth Activities Committee was presented to the ALA Council in two parts. Its recommendations would lead, by 1951, to an increase in dues—referred to a mail vote of the membership and approved.[16]

TAKING THE MODERN SHAPE: CRESAP, MCCORMICK, AND PAGET

For ALA's next "periodic scrutiny," authorized by the ALA Executive Board in July 1954, the management consulting firm of Cresap, McCormick, and Paget was employed.

The final report, based on intensive study of the ALA membership organization, headquarters organization, and fiscal policies and practices, was discussed by Membership and unanimously approved by the ALA Council in July 1955, in Philadelphia. Among the major recommendation of this report were the following:

- Establishment of divisions by type of library and type of activity.
- Assignment of greater responsibility to Council.
- Closer relationship between the Council and the Executive Board.
- Appointment of divisional executive secretaries by the ALA executive secretary, with the approval of the division concerned.
- Simplification of the dues structure.

The special committee appointed to guide implementation of the management report adopted several guiding principles: (1) a closer relationship between the organization and its 20,000 members, accompanied by greater membership participation; (2) decentralization of responsibility and authority; (3) strengthening of the Council as the governing body of the organization; and (4) allowance for more planning, control, and flexibility in managing the financial resources of the organization.[14]

By the late 1950s, the present shape of the ALA became visible. The divisional framework was restructured and type-of-library and type-of-activity divisions were designated, with some decentralization of responsibility. The ALA Council was clearly designated as the policy-making body of the Association. The Executive Board was elected by Council from among its members, to act for Council. Council representation became more democratic—with some councilors representing geographic areas, some divisions, some elected at-large. Member oversight of the budget was assigned to the Program Evaluation and Budget Committee, its membership then made up of past officers of the 12 divisions, plus the ALA treasurer, immediate past president and president-elect.[14]

REVOLUTION AND REFORM: ACONDA/ANACONDA

Arising from the social turmoil of the 1960s, forces for continuing change within ALA reached the stage of "revolution" at the 1968 annual conference in Kansas City and the 1969 conference in Atlantic City. From Kansas City, the momentum was maintained by the "Congress for Change" in Washington, DC, as well as New York organization of "Librarians for 321.8" and "Libraries to the People"—culminating at the Atlantic City conference. Requested changes ranged from asking Council candidates for a statement of views, to taking activist roles in intellectual freedom, library education, and recruitment for the profession.

Incoming president William Dix suggested appointment of a committee to examine the objectives and program of ALA. The Activities Committee on New Directions for ALA (ACONDA) was appointed and charged "to recognize the changes in the interests of ALA members," to provide "leadership and activities relevant to those interests," to reinterpret and restate "the philosophy of ALA to provide a meaningful foundation which is capable of supporting a structure and program which reflects the beliefs and priorities of the profession." It was also to reexamine priorities and organizational structure.[17]

The committee's final report, presented at the 1970 conference in Detroit, suggested a broad array of changes, including changes to the composition of Council, the nominating process, Association communications, the number and types of divisions, and ALA organizational structure. The resulting furor resulted in Membership Meetings which lasted 15 hr—and did not cover all the recommendations.[18]

Council approved the first three ACONDA recommendations in July 1970:

1. ALA's "overarching objective" would remain improvement of library service and librarianship. It would continue to be an organization of libraries and librarians.
2. An Office for Social Responsibility—changed by Council to Office for Library Service to the Disadvantaged—would be established.
3. The Intellectual Freedom Office would be expanded.

To aid in completely articulating the remaining recommendations, Council charged an Ad Hoc Council Committee (later dubbed ANACONDA) to work in conjunction with ACONDA. Their work was considered by Membership and Council through 1971. At the Midwinter Meeting in 1971, ACONDA/ANACONDA reiterated ACONDA's strong recommendation that ALA needed restructuring, with ACONDA pointing out issues of overlapping interests and duplication of effort.

ACONDA criticized the makeup of Council, recommending a smaller council elected on a district basis. ANACONDA recommendations also attempted to deal with problems of democratization and reorganization.[14] At the 1971 Annual Conference, Council passed a new policy statement implementing bylaws provisions for Council membership. All Council terms would end at the close of the 1972 Chicago conference and a new Council would be seated for the first time at the 1973 Midwinter Meeting. With this change, ALA past presidents were no longer automatically members of Council.[19]

ACONDA/ANACONDA would have an impact on subsequent strategic planning within the Association, leading to establishment of a Planning Committee "to provide the thorough planning necessary to achieve the long-range goals of the association."[20] The strategic planning role was later moved to the ALA Executive Board following the work of the Official Self-study Committee. ACONDA/ANACONDA also resulted in adoption of statements of goals and objectives, though there was a lack of clarity regarding priority.[21]

In 1977, the Future Structures Committee would note "there is no established means by which the individual member's priorities can be tapped to provide a firm basis on which to assign the association's resources," and recommended that a member survey be undertaken. That survey was conducted in 1979, with results released in 1980. The survey results, along with ACONDA/ANACONDA reports, recommendations from the 1st White House Conference on Library and Information Services and other reports, formed the basis for recommended priorities, considered by Council in 1981.[20]

CONTINUED CHANGE: THE HOLLEY COMMITTEE TO OSSC

In 1981, the Special Committee to Review Program Assessment Processes and Procedures (SCRPAPP), chaired by Edward G. Holley, identified Association needs in the area of planning and budgeting: for clearly understood common purposes; a clear set of goals, objectives, and priorities; balance between overall association interests and unit (i.e., division or round table) interests; more effective use of resources—including staff and member time, as well as dollars; resource allocation based on priorities; and a process for ongoing planning and evaluation.[22]

The Executive Board's management role was one focus of SCRPAPP (commonly known as "the Holley Committee").[23] The report noted: "There is no formal mechanism within ALA which brings together within a total management system the authority and the responsibility for planning, resource allocation, and program assessment." The special committee recommended that the ALA Executive

Board assume responsibility for overall financial planning and control. That recommendation would be echoed in the mid-1990s in Part I of the report of the OSSC.[24]

In January 1985, the ALA Council approved an outline for SLRP. The Council-approved document pointed to ALA's "tradition of flexible patterns of participation by individuals and groups who address association-wide and unit goals and an enormous range of issues facing libraries and librarians."[25] It articulated roles and responsibilities: for Council and its sub-entities—the Committee on Program Evaluation and Support (COPES) and the Planning and Budget Assembly (PBA), for the Executive Board and its sub-entities—the Directions and Program Review Committee and Process Planning Committee, for consultants, and for staff. The SLRP was to be characterized by broad communication and participation, and opportunities to address "critical issues," would result in clear goals and objectives, and would be evaluated continuously. In 1986, ALA adopted the resulting "priority areas" which remained in ALA's policy manual until 2006.[8] Indeed, the continuity between these priorities and subsequently adopted programmatic priorities is striking.

ORGANIZATIONAL SELF-STUDY: 1992–1995

In June, 1992, then president-elect Marilyn Miller, with Council approval, appointed an ALA Self-study Committee, chaired by past-president William Summers. The proposal to the Executive Board and Council noted:

> There has been neither an internal nor an external study of ALA since the 1955 ALA Management Survey conducted by Cresap, McCormick, and Paget. Since the reorganization prompted by the Cresap study, the ALA structure has remained virtually unchanged except for those modifications we have made to respond to democratization and to accommodate new units.

The proposal further noted that the "present pattern of division fiscal responsibility is a given," that "management by coalition based on coalescing principles will continue to be the predominant management force...," and that the then-present (1986) ALA goals and objectives should stand.[26]

Phase I of the Self-study was a management study, conducted by Consensus Management Group and focused primarily on the roles and responsibilities of the ALA Council, ALA Executive Board, and ALA Executive Director. Phase II of the OSSC Report, in January 1995, contained recommendations relating to management and governance of the association. Among the recommendations implemented were provision for membership ballot initiatives; an increase (to 1% of membership) in the number of members required to petition for a mail vote to overturn an action of Council; elimination of the ALA

Planning Committee, with responsibility for strategic planning clearly designated as an Executive Board role; a restructuring of fiscal oversight processes, with the COPES becoming the Budget Analysis and Review Committee and including significant representation from the Executive Board; a decrease in the term of members of the Council and Executive Board from four to three (excluding the ALA Treasurer), whose term remained at four until subsequently reduced to 3 years in 2003, when ALA members ratified the action taken by Council in June 2002.

Among the OSSC management and governance recommendations either not considered or rejected were term limits for members of ALA Council and elimination of the Membership Meeting in favor of a Membership Forum. In initially proposing the latter, the committee noted:

> The Membership Meetings at Annual Conference have been a cause of concern for a number of years. Attendance at the meetings is very low, representing a small proportion of the members attending the conference and a very low proportion of the total membership. Yet at these meetings, questions are posed and actions recommended, which have far reaching consequences for the Association.[27]

In subsequently deciding not to seek approval of that recommendation, the committee noted that membership had recently (1994) approved an increase in quorum size for a Membership Meeting. The issue remained a significant one within the Association. At the 2001 Annual Conference, Council requested establishment of a Special Presidential Task Force on Membership Meeting Quorum. Recommendations of that task force led to reduction of the quorum to one-half of 1% and removed the authority of the Membership Meeting to set aside actions of Council. In 2003, a Membership Meeting Committee (standing) was established to plan the agenda for Membership Meetings, identify topics of potential interest for discussion, and encourage participation. In 2005, the quorum for the Membership Meeting was reduced to 75 personal members.[28]

FOLLOWING CONTROVERSY

The OSSC also made extremely controversial structural recommendations. As a result of the heated discussions within both the ALA Membership Meeting and ALA Council, in 1995 the ALA Council established an Ad Hoc Task Force on Structure Revision (SRTF), chaired by Sarah M. Pritchard.

Its report to Council, February 1997, noted:

> The SRTF is not recommending major structural revision of the divisions and other membership groups since we

believe that, to be effective, such change must come from the direct collaboration and initiative of those groups rather than as a top-down design.... Nor do we believe the general governing concept of a delegate council and an executive board needs to be changed. These structures have a long tradition in the ALA and extensive member support; the effort to change them might well distract us from our primary professional goals for years on end with no guarantee that a different structure would be any more effective. We propose to concentrate instead on improving, enhancing and modernizing the operations we now have.

Acknowledging other ongoing work within the Association (including ALA Goal 2000 and establishment of new committees such as the ALA Conference Committee), the SRTF determined to focus its efforts on the structure of Council.[29] The SRTF presented its final report in June 1997. Among the resulting changes was the addition of six round tables representatives to the ALA Council.

ORGANIZATIONAL DEVELOPMENT: ALA DIVISIONS AND THE ALLIED PROFESSIONAL ASSOCIATION

It was clear early in ALA's history that provision would have to be made for specialization, to accommodate various institutional perspectives and the growing diversity of librarianship itself. In 1889, the college and reference librarians formed the first special section and held their first separate meeting apart from the general body of the Association. In 1890, the trustees formed a special section, followed by the catalogers in 1900.

With acceptance of the recommendation of the Third Activities Committee (December 1939), ALA's growing sections gained increased authority within the Association and became divisions. In 1952, grants were made to divisions from ALA endowment capital to enable them to have executive secretaries, based on the ALA Executive Board's conviction that staff would enable divisions to develop and maintain more effective programs for their members and the entire Association. The resulting division membership gains were to fund staff following expiration of the grants.

DIVISIONS AFTER 1955

The 1955 Cresap, McCormick, and Paget report defined type of library and type of activity divisions, headed by executive secretaries appointed by the ALA executive secretary. Among the specific provisions were that each division encompasses a field of activity clearly distinct from that of other divisions, that basic ALA dues include one type-of-activity and one type-of-library division, and that

the "60–40" allocation of dues be replaced by an allotment proportional to number of members, with additional funding based on proposals to the Association's budget committee. This report shaped the basic divisional structure in operation today. The final reorganization eventually resulted in the formation of 12 divisions and 6 round tables.

By 1975, with ALA membership at 33,491, 73% of ALA members selected at least one division. By 1976, with membership at 34,491, it was 64%. In 1982, with total ALA membership at 38,330, 65% of ALA members selected at least one division. That percentage proved to be reasonably stable. In 1984, with 39,000 ALA members, 66% selected at least one division. By January 2007, 67% selected at least one division or one round table, with 7 of 10 being Division memberships and 3 of 10 round table memberships.

THE DUES SCHEDULE TRANSITION DOCUMENT

Based on the recommendations of a succession of reviews, by the 1970s ALA had moved to the principle of funding divisions from income derived from division dues and other division activities. This required a basis on which to separate those areas of division financial responsibility from those of the ALA "General Fund." That basis was provided by the "Dues Schedule Transition Document," adopted by the ALA Council in January 1976.

The "Dues Schedule Transition Document" ended inclusion of division membership within the ALA basic dues; members now paid additional dues to each division joined. Divisions became largely self-supporting at this point, assuming financial responsibility for cost of staff, publications, and programs. The ALA General Fund assumed responsibility for a defined set of "indirect cost" items, including office space, administrative services (including both fiscal and human resources services) and other similar kinds of expenses. In 1975, ALA divisions spent just under $600,000 for programming. By 1980, division expenditure on programming reached $1.3 million and by 1983, a short 3 years later, exceeded $2 million. With the passage of time, the several revisions of the ALA dues structure, and the move into the new headquarters building, by April 1982 it was time for a new statement.[30]

THE OPERATING AGREEMENTS

Responsibility for preparation of a new statement outlining the fiscal relationship between ALA and its divisions was given to the COPES. Initial discussions between COPES and divisions began in 1981 and resulted in a draft operating agreement, which was presented to ALA Council in July 1982, and approved as an "Operating

Agreement between ALA and its Membership Divisions." This operating agreement articulated its primary purpose as

> to define those services which divisions receive from ALA at no cost and those for which charges are made to divisions. In addition, the document seeks to establish a cooperative framework in which the inevitable questions of organizational relationship can be addressed and resolved.[31]

The document approved called for revision in not more than 5 years.

In 1984, executive directors of ALA divisions met with COPES to discuss the relationship between the divisions and ALA. In December 1986, a revised draft was presented to COPES. Members of COPES, division representatives, ALA officers and staff met in Lisle, IL, November 11–13, 1988, to discuss revision of the 1982 Operating Agreement. The current "Policies of the American Library Association in Relation to Its Membership Divisions" was approved by the ALA Council on June 28, 1989.[32] It was implemented through a series of "management practices," providing the flexibility to accommodate changes in technology and other conditions.

The 1989 Operating Agreement was based on four "fundamentals:"

1. ALA and all its units abide by the ALA Constitution, Bylaws, and policies adopted by the ALA Council.
2. ALA is one legal, financial, and organizational entity, with indivisible assets.
3. All staff are employees of the ALA, and subject to ALA personnel policies.
4. All ALA units use services provided by ALA (including such services as personnel, membership services, telephone, insurance, purchasing, fiscal services, legal services, archives and general administrative services) and are housed in property owned or leased by ALA.[31]

Just prior to the 1988 Lisle meeting, addressing a concurrent meeting of the division executive committees, then ALA executive director Thomas Galvin said:

> In the words of the ALA Business Plan, "ALA as a whole and its individual units exist in an environment of mutual programmatic and financial interdependence." ALA is only as programmatically strong as its most fragile unit, only as financially robust as its most impoverished entity....[33]

DIVISIONS AND THE SELF-STUDY

In their 1995 Phase II Report, the OSSC defined "guiding principles" as a context for organizational change and also noted current issues or concerns:[5]

1. "ALA provides holistic participation in the Association." OSSC noted overlap among divisions (and other units) as they responded to emerging issues and sought to address the various concerns of their members.
2. "ALA's governance structure fosters quick, concerted, and coordinated national action." Here the committee noted that members "experience frustration" when an opportunity to act on an issue quickly is lost because of the need to consult with many different units.
3. "ALA speaks externally with one voice." OSSC noted the need to provide opportunities for internal disagreement, examination of different perspectives, and development of consensus, in order to "ensure that ALA speaks with one voice in its external communications."
4. "Governance structures facilitate decision-making and consume a minimal amount of the Association's resources." The committee recommended a "flatter" and "more streamlined" governance structure.
5. Membership in ALA is a satisfying, positive professional experience.

Believing that responses to a 1994 "concept paper" distributed by the OSSC indicated support for these guiding principles, the OSSC proposed a broad range of organizational changes, including expanding networks with state and regional groups, consolidating organizational units, regrouping or clustering existing units, replacing some units with different ones, implementing an active "sunrise" philosophy, adding new offices and creating new groupings such as "policy councils." These recommendations proved controversial—and were not adopted. The Membership Meeting at which the proposal was discussed (Chicago, 1995) was one of the few during the late nineties to achieve a quorum.

FORMATION OF THE ALA-ALLIED PROFESSIONAL ASSOCIATION

One of the recurring overtones in ALA's long history is the pull—for resources and attention—between a focus on libraries and a focus on the individual library practitioner. This normal condition—stemming from ALA's broad mission and scope—was exacerbated by the increasing body of law and regulation governing tax-exempt organizations.

In July 1996, boards of two ALA divisions—the Public Library Association (PLA) and Library Administration and Management Association (LAMA)—voted to support a proposal to establish a "certified public library administrator program." Based on interest expressed by other divisions, PLA and LAMA, joined by the Association of Specialized and Cooperative Library Agencies (ASCLA), recommended that ALA create a special committee to develop a "framework" certification proposal for ALA Executive Board approval. Until 2001, activity proceeded along several tracks: (1) the definition of such a "framework" by the ALA Committee on Education;[34] (2) the continued development of a "standard for professional practice" by PLA and LAMA, now joined by the ASCLA; and (3) the definition of a legally permissible strategy to enable certification. In June 2001, the ALA Council approved the following statement:

> Toward fulfillment of its declared mission to provide leadership for the development, promotion, and improvement of the profession of librarianship, the American Library Association establishes an allied professional association to certify individuals in areas of specialization beyond the ALA-recognized masters degree.[35]

Overlapping this strand of ALA development were the issues of the salaries and status of librarians. In Spring 1999, the report of the 1st Congress on Professional Education recommended that ALA "address the need to improve salaries." In Spring 2001, then-president Nancy Kranich, in response to issues raised by members, appointed a Task Force on the Status of Librarians, chaired by Tom Wilding, charged to articulate status issues, recommend strategies to address them and outline the issues that must be resolved in order to follow those strategies. In Fall 2001, ALA president-elect Maurice J. Freedman appointed a presidential initiative committee, chaired by Patricia Glass Schuman—the Better Salaries Task Force. In December 2001, the task force outlined Options for Action, recommending that the scope of the allied professional organization be broadened to enable it to undertake a variety of activities related to status and salaries.

In Fall 2001, the ALA Executive Board appointed a working group, including the chairs of both the Status Task Force and Better Salaries Task Force, to coordinate development of a proposal to broaden the scope of the Council-authorized allied professional organization.

At the 2002 Midwinter Meeting, the ALA Council approved preliminary Bylaws for an ALA Allied Professional Association, broadening its scope "to promote the mutual professional interests of librarians and other library workers." An ALA–APA Transition Team, chaired by Nancy Kranich, was appointed in March 2002, reporting to Council in May 2002. The ALA–APA Board of Directors met for the first time at the 2002 Annual Conference. In July 2002, the ALA Allied Professional Association was incorporated in the State of Illinois. It subsequently received tax-exempt status under Section 501 (c) (6) of the U.S. Internal Revenue Code. In November 2002, the ALA–APA Board of Directors accepted a business plan for the ALA–APA and, based on that plan, requested a $250,000 establishment loan from the ALA.[36]

ALA STRUCTURE TODAY

ALA Council

The governing body of the ALA is the ALA Council. Council is empowered by the ALA Constitution to "determine all policies of the Association." Its decisions are binding upon the Association, except when "set aside by a majority vote by mail in which one-fourth of the members of the Association have voted."[37] A majority vote of the Council may also refer a question of policy to the Association for vote at either a Membership Meeting or by mail. Only personal members of ALA may serve on the Council.

As of August 31, 2007, the ALA Council includes 186 members. There are 100 councilors at large, elected by ALA membership. Members of each of ALA's 11 divisions elect one councilor. Members of the each round table whose membership is equal to or greater than 1% of the total ALA personal membership, as of August 31 of the year preceding the election, elect one councilor; members of the remaining round tables jointly elect one councilor. Finally, there are 53 chapter councilors, representing the 50 states, the District of Columbia, Guam and the Virgin Islands. Members of the ALA Executive Board, including the treasurer and immediate past president, are members of the ALA Council. Officers of the ALA Council are the ALA president and president-elect. The ALA executive director is the secretary of Council. Councilors serve a 3-year term. ALA Council must meet at least twice a year, once at the annual conference and once not less than 3 months prior to the annual conference, by current practice at the ALA Midwinter Meeting.

ALA Executive Board

The ALA Executive Board includes the president, president-elect, immediate past president, treasurer—all elected by the ALA membership—and eight members elected by the ALA Council from among its members. The ALA executive director is a non voting member of the Executive Board. Members of the Executive Board serve a 3-year term.

The ALA Executive Board "acts for the Council in the administration of established policies and programs." The Executive Board "manages within this context the affairs of the Association," delegating management of day-to-day operations to the ALA Executive Director. The Executive Board also makes recommendations to Council regarding policy.[38]

ALA Divisions

ALA has 11 membership divisions, each focusing on a type of library or type of service. Divisions offer programs, publications, institutes, and workshops; and provide other services to their members. Three divisions—the American Association of School Librarians, the Association of College and Research Libraries, and the Public Library Association—currently hold biennial national conferences. Division members must be members of the ALA. Division staff members are ALA staff members.

Each division has, by policy:

- A statement of responsibility developed by its members and approved by ALA Council.
- A set of goals and objectives established by its members, which drive its activities.
- An Executive Director and other personnel as necessary to carry out its programs.
- Responsibility for generating revenue to support staff and carry out its programs.
- A separate Board of Directors, elected by its members, and responsible to ALA Council.

ALA Bylaws [Article VI, Section 2 (b)] give divisions authority "to act for the ALA as a whole" within their Council-designated areas of responsibility. An operating agreement—ALA Policy 6.4.1 "Policies of the American Library Association in Relation to its Membership Divisions"—outlines a collaborative and intertwined relationship between ALA and its divisions. The policy notes that ALA:

> ...is unique among American associations in the manner in which it is structured. It is one association, with indivisible assets and a single set of uniform administrative, financial, and personnel policies and procedures. It is governed by one Council, from which its Executive Board is elected, and is managed by an Executive Director who serves at the pleasure of that Board. It is also the home for eleven Divisions. ...[32]

Each division is represented on the ALA Council, in the PBA and on other association-wide assemblies.

ALA Round Tables

In accordance with ALA Bylaws, Council may authorize organization as an ALA Round Table by any group of not less than 100 members of the Association "who are interested in the same field of librarianship not within the scope of any division. ..."[39] Interested members must petition for the formation of a round table and that petition must include a statement of the proposed purpose. As of August 31, 2007, ALA had 17 round tables.

Round tables may charge dues, issue publications (with the approval of the ALA Publications Committee) and affiliate with regional, state or local groups with the same interests. Round tables do not have their own staff. ALA

staff are assigned "liaison" responsibility for round tables. Round tables may not "commit the Association by any declaration of policy."[40] Round tables have been represented on the ALA Council since 1999–2000. In Spring 2006, ALA Membership ratified a change to Article VI of the Bylaws, to provide that any round table with personal membership equal to or greater than 1% of ALA's total personal membership may elect one councilor. The remaining round tables, those with less than 1% of ALA's total personal membership, jointly elect one councilor.[41] As of June 2008, nine round tables were individually represented in Council, with a 10th, elected in the Spring 2008 election, to be individually represented beginning in 2008–2009.

Standing Committees

ALA Bylaws provide for standing, special, interdivisional and joint committees. Standing committees may be committees of the Association or committees of Council. By Council action (July 1969), the Executive Board and Council may also create ad hoc committees, to pursue a designated project for a specific period and not continuing beyond the life of that project. ALA currently has approximately 35 standing committees, including both Association and Council committees, as well as five joint committees—committees jointly created by the ALA and one or more external organization. Many standing committees have associated subcommittees, assemblies and other subordinate bodies. Approximately, 350 ALA members serve on these standing committees.

Member Initiative Groups

Membership Initiative Groups (MIGs) were initially authorized in 1979 at the Annual Conference in Dallas, with implementation guidelines approved in 1980. In 1985, ALA Council approved a recommendation from the ALA Committee on Organization (COO) that MIGs be continued and given organizational status in the Bylaws. Council and members subsequently approved a Bylaws amendment [Article VIII, Section 2 (b) (ii)] enabling COO to authorize MIGs. MIGs may not assume any of the delegated responsibilities of other units and may not speak for the Association. Originally conceived as a short-term organizational vehicle to provide for prompt, organized membership activity on topics of mutual interest, MIGs were limited to a 3-year life, after which they must either cease to exist or "apply for a place within the ALA structure." In 2008, recognizing shifts in the way members sought to organize themselves for work, the COO authorized Member Initiative Groups to repetition for successive 3-year terms of existence. Since 2000, there have been four Member Initiative Groups formed: Libraries Fostering Civic Engagement, Information

Commons, Virtual Communities and Libraries, Games and Gaming.[42]

Chapters

Under Article V of ALA's Bylaws, Council

> may establish a chapter...in any state, province, territory, or region in which a majority of ALA members residing within the area involved and voting on the issue favors such action; provided, however, that the total number of persons voting on the issue shall not be less than ten percent of the total number of ALA members residing within the area.[43]

Council representation is through the state chapter, unless chapters in a region choose to take representation through the regional chapter, in which case the regional chapter would elect one representative from each state or provincial chapter included in the region. As of August 31, 2000, all state chapter representation was through state chapters, not regional chapters. Chapters may admit members who are *not* members of the ALA.

It is important to note that many states have more than one library association. Particularly, school library media specialists frequently have separate associations. These separate school library media associations are generally affiliated with the AASL, a division of the ALA, but are not represented on the ALA Council. AASL has a separate Affiliate Assembly. Other ALA divisions also have state or regional affiliates.

Affiliated Organizations

The ALA Council may, by vote, affiliate with the Association or any of its subdivisions "any national or international organization having purposes similar to those of the Association or its subdivision."[44] Likewise, it may affiliate the Association with such an organization. No subdivision of the Association may separately affiliate with an organization to which the Association as a whole is affiliated. Affiliated organizations are not represented in the ALA governance structure. Representatives of affiliated organizations meet informally at ALA conferences.

BEYOND STRUCTURE: INTERNATIONAL CONNECTIONS AND WEB-BASED NETWORKING

International Connections

From the earliest period of its history, ALA has seen its role and the interests of its members in an international context. James Yates, librarian of the Free Library of Leeds (England) was an invited delegate at the 1876 founding conference. A year later, following the 1877 conference in New York City, 12 American librarians went to England, where they were present at the organization of the Library Association of the United Kingdom. International contact continued to expand, with many international librarians attending the 1904 conference in St. Louis.[2]

In 1917, the ALA Executive Board appointed the Committee on Mobilization and War Service Plans (later the War Service Committee). Herbert Putnam served as general director of the program. ALA undertook to supply books and periodicals to military personnel at home and overseas. The initial campaign raised $1 million for camp libraries, as well as including a book drive.[2] In August 1918, ALA opened a library for U.S. military personnel in Paris; others followed. Following the Armistice, there was significant pressure from Americans in France to keep these facilities open. ALA promised to leave the books and equipment and to provide a $25,000 "endowment." On May 20, 1920, the American Library in Paris was incorporated as a private, nonprofit organization under the laws of the state of Delaware.[45]

This spirit of missionary librarianship was further evidenced in the establishment of the Biblioteca Benjamin Franklin in 1942,[2] as well as later projects such as the Keio Library School in Tokyo and the Institute of Librarianship in Ankara, both in the 1970s. In 1988, a delegation of Soviet librarians attended the ALA Annual Conference for the first time. In the United States, they signed a protocol opening the way to exchange of current national bibliographic data. ALA's international participation would accelerate by the beginning of the twenty-first century, through visits, workshops, and bi-national conferences around the globe.

In 1942, the ALA Council approved establishment of a "board" on International Relations. In October 1942, a grant from the Rockefeller Foundation made possible the establishment of the ALA International Relations Office in Washington, DC, in space donated by the Library of Congress, to "lend effective aid to libraries and advance the interests of librarianship."[2] The office operated for the next 5 years using further grants, serving as secretariat to the International Relations Board. The scope of operations was broad, with funding from both the Rockefeller Foundation and the U.S. State Department to supply library science literature to libraries overseas, to distribute bibliographies, to assist in administration of library schools in Latin American countries, to administer the program for bringing foreign librarians to the United States to study and observe American libraries, to purchase American journals for distribution to newly forming libraries in the post-World War II period, and other projects.

The International Relations Board became ALA's International Relations Committee in 1956. That same year, another grant from the Rockefeller Foundation and a supplemental grant from the Council on Library Resources made possible the establishment of the

International Relations Office, operating both at the Chicago headquarters and in Washington, DC. Its original purpose was to offer assistance, direct or indirect, to the underdeveloped countries in Asia, Africa, Latin America, and the Near East, particularly promoting education for librarianship in those countries. The scope of the Office was later broadened. An international relations office existed 1943–1949 and again 1956–1972, when it temporarily fell victim to ALA financial stresses in the 1970s.[2]

In 1986, ALA received a grant from USIA for the Library/Book Fellows Program, which was to continue until 1998—sending over 100 U.S. librarians to over 70 countries, and bringing 40 librarians from 35 countries to the United States. In 1989, the chair of the ALA International Relations Committee, E.J. Josey, reported to the ALA Executive Board on the outcomes of a November 1988 planning meeting convened to reassess ALA's international role, develop a strategy for identifying that role within ALA's organizational priorities, and achieve funding to support that role. Establishment of an International Relations Office within ALA was proposed—and in the late 1990s the ALA International Relations Office was established, using ALA operating funds rather than grant funds. It continues today.[46]

The first steps toward an international library organization had been taken at ALA's 50th anniversary conference in 1926. The International Library Committee met in Edinburgh in 1927, the 50th anniversary of the Library Association, with ALA as one of the founding members. In 1952, the Committee became the International Federation of Library Associations.[2]

Six U.S. based associations within IFLA—the American Association of Law Libraries, the ALA, the Art Libraries Society/North America, the Association of Research Libraries, the Medical Library Association, and the Special Libraries Association—together with U.S. IFLA members and the National Commission on Libraries and Information Science, joined together to sponsor the 1985 IFLA Conference in Chicago, drawing delegates from more than 86 countries. In 2001, IFLA again met in the United States, setting a new attendance record with approximately 5500 attendees in Boston.

ALA in the Networked World

By 2007, the impact of rapid expansion in use of social networking tools was visible within the Association. In Spring 2008, an informal census of "web 2.0" deployment revealed 70 blogs (doubled since Spring 2007), 125 wikis (quadrupled since Spring 2007)—both supplementing, and in some cases supplanting, the more than 948 discussion lists in use by ALA groups. Some 15 individuals a day were joining the ALA group in Facebook—which totaled 3800 by September 2008—and other ALA-related groups had formed on other social networking sites.[47] ALA staff and members had held events—celebrating traditions such

as Banned Books Week and National Library Week—in Second Life and ALA leaders were meeting and speaking in Second Life. A series of videos created by American Libraries for National Library Week (2008) received over 78,000 views, with comments appearing on YouTube, Blip, AL Focus and other sites. In 2007, the ALA Council authorized formation of a special ad hoc Task Force on Electronic Member Participation, chaired by Janet Swan Hill, to explore ways of integrating "virtual" participation into the organizational fabric of ALA. The special committee is due to report in January, 2009.

ALA CONFERENCES, SERVICES, AND OPERATIONS

To support its mission and its members, the ALA provides core services and products generally common to all associations, including publishing, conferences, standards and guidelines, research and statistics, and awards and scholarships. These and other functions are managed through a headquarters operation.

ALA Publishing

Professional literature on librarianship was minimal prior to the mid-nineteenth century. In 1886, the ALA formed the publishing section for publication of catalogs, indexes, and other bibliographical tools not easily produced by individual libraries.[2] Early ALA publications were printed and distributed by other organizations, including Melvil Dewey's own Library Bureau, Houghton Mifflin and the U.S. Government Printing Office.[48] In 1900, a Publishing Board was created, replaced by an Editorial Committee in 1921, and reestablished in 1966, retaining the Editorial Committee in an advisory capacity. As the Association grew, both the editorial and technical work of publishing became increasingly the work of staff. The Association's publication program is now guided by the ALA Publishing Committee, a standing committee of Council.

ALA's present publishing business began in 1886 with starting capital of $486. In 1902, Andrew Carnegie provided increased impetus with the gift of a $100,000 endowment for "the preparation and publication of reading lists, indexes, and bibliographical aids."[2] The program made available to libraries small publications that might not have been economically feasible without such support. That endowment, now the Carnegie–Whitney endowment, still supports such lists. By 1922, ALA Publishing sales had reached almost $24,000. By 1966, ALA's publishing department had sales in excess of $990,000. By 2000, ALA Publishing had published over 1520 titles—not including titles published through individual divisions of the Association.

In 1904, at the conference in Atlantic City, discussion began which would lead to the establishment of *ALA Bulletin* in 1907, competing with both *Library Journal* and *Public Libraries*, begun by Dewey in 1896. While the new *ALA Bulletin* was given the "Proceedings" of the Association to publish, it was not designated ALA's "official organ." That designation was retained by *Library Journal* until 1908, when the *ALA Bulletin* was so-designated.[14] The *ALA Bulletin* was published through December 1969—a total of 63 volumes. In January 1970 volume one of a new journal—*American Libraries*—appeared and continues today as ALA's official organ, with a standing advisory committee. *American Libraries* committed to four-color coverage in January 1980 for its report on the White House Conference.[49]*American Libraries* has continued to change with the Association, expanding to include *AL Direct*, a weekly online newsletter, in 2006 and opening *AL Focus*, a video and discussion site, in 2007.

In 1905, the Carnegie endowment made possible the establishment of *The Booklist*—a book review journal. The number of subscriptions rapidly grew to 3000, with state library commissions purchasing multiple copies.[14] Originally published eight times a year, *Booklist* is now published 22 times a year and has a print subscription of over 20,000. *Booklist* reviews are licensed by numerous electronic publishers, in both CD and online formats. In 1994, *Booklist* spawned a companion journal, *Book Links*, which has over 18,000 subscribers, as of August 2001. *Book Links* is focused on connecting books, libraries, and classrooms. In 2006, 100 years after its initial establishment, *Booklist* opened a new era with the release of *Booklist Online*.

As of August 2008, ALA Publishing's journals include *American Libraries*, *Booklist*, *Booklist Online*, *Book Links*, and *ALA TechSource*. ALA Editions publishes approximately 30 new or revised titles a year, with a backlist of over 300 titles. The *Guide to Reference*, long a staple—under a variety of titles—of the ALA Editions catalog, *released* as an online title in 2008 (Appendix).

ALA Graphics provides posters and other promotional items supporting libraries, literacy, and reading. ALA Graphics has also played a key role in ALA's advocacy for libraries. In 1921, ALA commissioned Chicago Tribune cartoonist John T. McCutcheon (the 1932 Pulitzer Prize winner) to draw a poster advocating use of libraries. Copies of the poster were made available to libraries—at $5.00/100 posters, $3.50/1000 bookmarks.[50]

In 1981, ALA inaugurated its celebrity "Read" posters—now a staple offering in the ALA Graphic catalog. The first "star" to be featured was not a person but a cartoon—Mickey Mouse. ALA produced its first set of celebrity "Read" posters in 1984–1985. Posters featured such stars as Bette Midler, Paul Newman, Mikhail Baryshnikov, and Michael J. Fox. Of that first set, the most popular would be a poster of "Bill Cosby and friends"—with over 26,000 copies sold by the end of the decade.[51] Current posters feature Tony Hawk, Rachel Ray, Corbini Bleu and Common, among others. In addition to the celebrity "Read" posters, ALA Graphics also markets a "Read" CD, enabling libraries across the country to produce their own posters—featuring staff and local celebrities.

ALA Conferences

Annual conference: Growth and development

The first ALA conference was held in 1876. With the exception of 1878, 1880, and 1884, the Association then held annual conferences until the war years of 1943–1945. Six of these meetings were international conferences, held in European cities in conjunction with European associations. Four conferences were held in Canada—but were not considered "international."[2] The fifth Canadian conference, in Montreal in 1960, was a joint ALA—Canadian Library Association conference. In June 2003, the American and Canadian Library Associations once again held a joint conference in Toronto. With registration underway—and on track for a record attendance—an outbreak of SARS (Severe Acute Respiratory Syndrome) was reported in Toronto. Both ALA and CLA staff and leaders would spend the countless hours over the next months in consultation with public health agencies, medical practitioners, and others—and, in the end, the ALA, along with CLA, would be the first major convention and exhibit to keep its commitment to meet in Toronto following the SARS outbreak.

In 1903 (Boston), conference attendance surpassed 500 for the first time—setting a new record at 1018 at a time when total ALA membership was 1152. By 1938 (New York), attendance had reached the 5000 mark. Over 12,000 gathered to celebrate ALA's Centennial in 1976. In 1990, *American Libraries* reported: "Back in January, portents of a possibly impossibly large Annual Conference were evident."[52] With ALA membership at the 50,000 mark, famous speakers scheduled (including Tom Wolfe, Ray Bradbury and Senators Paul Simon and Patrick Leahy), ALTA celebrating its 100th anniversary, attendance totaled 19,982—surpassing the previous year's attendance in San Francisco by 2776. Total attendance 10 years later, at the 2000 Annual Conference in Chicago, was 24,913. In June 2007, ALA set a new Annual Conference participation record, in Washington, DC, at 28,499, including registrants, exhibitors, staff, press, and guests.

Throughout ALA's history, conferences have served the purposes of governance, education, business, and entertainment. In 1891, following ALA's first conference in San Francisco, ALA president Samuel S. Green, elected to replace an ailing Dewey, suggested that future conferences be held in locations with fewer non-conference enticements.[14] That meeting was attended primarily by westerners, with only 40 ALA members making the then-lengthy trek from the east coast.

Member complaints about meeting conflicts at conferences are also almost as old as the Association. In 1889, the newly formed College and Reference Section held its first meeting apart from the general body of conference attendees. Within a few years, members were complaining about missing meetings because of conflicts. Growth in membership and growth in demand for establishment of separate units within the Association to address special needs led to a continuing escalation in the number of meetings during the annual conference—as well as to increasing numbers of pre- and post-conference meetings. By 1967 (San Francisco), there were 987 meetings at the annual conference. By 2000 (Chicago), there were approximately 2500 meetings, programs, and events. That number has remained fairly stable.

Exhibits also began early. The Chicago conference of 1893, in conjunction with the World's Fair, included a model library, architectural exhibits, the Leyden Books, and the Rudolph continuous indexer.[14] By the 1967 San Francisco conference, there were 375 commercial exhibits. By the 2001 San Francisco conference there were 1600 booths.

The Midwinter Meeting

In 1908, ALA began holding an annual business meeting—the Midwinter Meeting—in Chicago. Over the next four decades, that meeting expanded steadily in number of registrants and types of meetings. Beginning in 1952, the Midwinter Meeting was limited to the meetings of the ALA Council, the Executive Board and the business meetings of all ALA units, thus beginning a continuing cycle of expansion and restriction. In 1958, Council acted to further restrict the Midwinter Meeting, eliminating "programs, general business, or Membership Meetings of the divisions, sections, or round tables." Following that restriction, the Midwinter Meeting again began to broaden.

In response to member pressures, in the Spring 1965, a committee of the Executive Board and Council, chaired by Katherine Laich, was appointed to study and make recommendations regarding the Midwinter Meeting. The committee's recommendations, at the 1966 Midwinter Meeting, were adopted without discussion. While the Midwinter Meeting would continue to be primarily "for carrying out the business of the Association through meetings of the Council, boards, and committees," and while there would continue be "be no programs, general business, or Membership Meetings of the divisions, sections, or round tables," the Executive Board now had the power to authorize "a limited number of program meetings, institutes, conferences, or workshops. . . ."[53]

The report also allowed the ALA Executive Board to "authorize the occasional movement of the Midwinter Meeting to an appropriate center other than Chicago. . ." and authorized ALA staff "to experiment with a limited number of commercial exhibits at the Midwinter Meeting,

whether in Chicago or elsewhere." By March 1987, a headline in American Libraries declared: "A burgeoning ALA gobbles up 3800 hr of meetings in six days as it programs for information access and a library-card campaign."[54] That Midwinter Meeting in Chicago drew 6337 registrants (a 12% increase over the previous year) and a "house full of exhibitors." In 2001, 13,989 gathered for the 2001 Midwinter Meeting in Washington, DC, including 10,249 registrants and exhibits visitors, as well as 3819 exhibitors, plus staff, press, and guests.

The educational function of the Midwinter Meeting has continued to grow, both based on and encouraging strong regional attendance. Early in the twenty-first century, ALA Divisions and other groups began offering "Pre-Midwinter Institutes" in conjunction with the ALA Midwinter Meeting. The number of such institutes grew over the next several years. The Midwinter Meeting has been particularly characterized by a rich, and growing, array of discussion groups, interest groups, and other "communities of interest." Approximately, 200 such groups currently convene. In 2007 (Seattle), the ALA Executive Board authorized a speaker series—currently called the "Sunrise Speaker Series"—designed to give first-time and regional attendees a "taste" of the Annual Conference.

Conferences: Change and controversy

ALA's strong support for key values has, at times, affected conference site selection.

The ALA Annual Conference was held in Richmond, Virginia, in 1936. African-American members of the Association were faced with segregated lodging and eating facilities. Faced with protests from members, the ALA Council, on a recommendation from the new Committee on Racial Discrimination in 1937, adopted a policy that ALA would not meet in cities where access to conference hotel and meeting rooms was not available on equal terms. It would be another 20 years before the Association would meet in the South again. In 1977, the ALA, having gone on record in support of the Equal Rights Amendment in 1974, joined a "conference boycott" of states that had not approved the ERA. It would be 7 years before an ALA Annual Conference or Midwinter Meeting returned to Chicago when the 104th Annual Conference would set a new record with 10,152 paid attendees, breaking the 1980 record set in New York (9479). Issues related to gay rights would later result in the shifting a Midwinter site from Cincinnati to Philadelphia, as well as a Council resolution calling on the Association not to complete conference site contracts with jurisdictions that discriminate by law, with the ALA Executive Board to determine the appropriate action in each case, based on information to be provided by ALA Conference Services.[55]

If ALA's strong values have sometimes led the Association to boycott sites, values have also been a key factor in staying with sites under difficult circumstances. Late in the 2005 hurricane season, Katrina devastated New Orleans, scheduled site of the June 2006 ALA Annual Conference. That fall, following a brief visit to the city by key ALA staff, and extensive consultation, the ALA Executive Board agreed with a management recommendation that the 2006 ALA Annual Conference would proceed in New Orleans. Again, as in Toronto, ALA would be the first major conference and exhibition to make that commitment. In June, in addition to gathering for a conference, almost 1000 ALA members fanned out to volunteer sites across the city to assist in recovery efforts.

ALA's demographic profile and commitment to participation was reflected in a 1985 recommendation by the Committee on the Status of Women in Librarianship, endorsed by the ALA Executive Board, to ask Annual Conference Local Arrangements to provide a list of recommended child care services in host cities. Today, "Camp ALA" is a regular part of each ALA Annual Conference.[56]

Reflecting rising national awareness of physical fitness, in 1984, at the Annual Conference in Dallas, ALA held its first Annual Conference Fun Run. 1st ALA Fun Run. Runners completed a 3 mi course in approximately 16.35.[49] The ALA Fun Run ended following the 2004 Annual Conference, falling victim to rising costs and competition for scarce resources. Health concerns made a return to the ALA Annual Conference, however, in 2008, with ALA President Loriene Roy's Wellness Fair.

Awards and Scholarships

ALA grants and awards are presented for distinguished service to the profession, for publication and for research. Grants and awards, including some of the best known awards (such as the Caldecott and Newbery Medals), are given by ALA and by its Divisions and Round Tables. The ALA Awards Committee coordinates the work of the ALA awards juries and provides a broad policy and procedural framework within which the awards committees of various ALA units function. "Awards" initially encompassed all types of awards, including scholarships. In 2005, with the growth in the number of awards and scholarships, the ALA Scholarships and Study Grants committee was established, dividing the responsibility with the ALA Awards Committee.

"Check off" contributions to ALA scholarship endowments, through the annual ALA membership renewal process, began in 1976, with contributions going either to the David H. Clift scholarship or the Minority Scholarship. In 1977, the latter was renamed the Louise Giles scholarship in honor of a past president of Association of College and Research Libraries (ACRL). Over the next 3 decades, scholarship programs would increasingly—though not exclusively—focus on increasing diversity. The first Louise Giles Minority Scholarship was awarded by ALA in 1977. By 1991, the Library and Information Technology Association, a division of ALA, began offering a minority scholarship (LITA/OCLC, LITA/LSSI). In 2000, the ASCLA Century Scholarship was established, to recruit people with disabilities to the profession.

In 1997, the ALA Executive Board approved the Spectrum Initiative, a 3-year, $1.35 million, recruitment and scholarship initiative providing 50 $5,000 scholarships/year to students from four traditionally under-represented racial/ethnic groups, as well as an ongoing Spectrum Institute (valued at $1500/scholar), designed to build connections and community, and to facilitate the transition to professional leadership.[57] The Spectrum Initiative subsequently developed into the "Spectrum family" of scholarships, including, as of 2008 the Giles, Albert, Teeple, Gordon, and Turock Scholarships, as well as the Spectrum scholarships, [Scholarship] Bash Scholarships, and a scholarship from the Medical Library Association. By 2008, the total number of Spectrum Scholars, past and current, had reached 560. Growth in the Spectrum program was significantly aided by the receipt of two awards (2004, 2007) from the federal Institute of Museum and Library Services (IMLS), under the Laura Bush 21st Century Librarian program. In 2006, IMLS awarded a grant to the University of Pittsburgh, working with ALA, for a Spectrum Doctoral Fellowship. The first Spectrum Doctoral Fellowships were awarded through ALA in 2006.

Standards and Guidelines

Over the course of its history, ALA has promulgated standards and guidelines for college, school, state, public, hospital, and institutional libraries. ALA is not a regulatory body and these standards have been promulgated for use by autonomous institutions. Standards have, nevertheless, been seen as valuable in providing a guide for self-measurement. Libraries have used ALA standards in seeking increased support, improved legislation. Standards and guidelines are reviewed by ALA's Standards Committee for consistency with ALA policy. A current list of standards and guidelines is available on the ALA Web site.

ALA Offices

ALA's programmatically focused Offices largely mirror the member-determined priority areas of the Association. As of August 31, 2008, ALA Offices included the following:

Chapter Relations Office, Development Office, Diversity Office, International Relations Office, Office for Accreditation, Office for Government Relations (in Washington, DC), Office for Human Resource Development and Recruitment, Office for Information Technology Policy (in Washington, DC), Office for Intellectual Freedom, Office for Library Advocacy, Office for Literacy and Outreach Services, Office for Research and Statistics, Office for ALA Governance, Public Information Office, Public Programs Office, Washington Office (in Washington,

DC). ALA maintains a headquarters Library and Knowledge Management Center—serving members and staff, as well as the public. Most offices work with standing advisory committees of members.

Headquarters offices

For the first 30 years of its life, ALA conducted all its activities, except publishing, without a permanent headquarters office and, generally, without paid staff. In 1906, ALA leased office space on Newbury Street in Boston—a lease that would be terminated by the end of 1907. By 1909, following a protracted debate over location, ALA took a small step toward establishment of a permanent headquarters, using space provided by The Chicago Public Library in its Central Library.[2]

By 1924, the Association was outgrowing its quarters at the Chicago Public Library and part of the headquarters staff moved across the street to quarters provided by the John Crerar Library.[2] Then, in May 1929, the entire ALA staff, 50 people, moved into rented quarters at 520 North Michigan Avenue, Chicago. In July 1946, the Association moved into the former Cyrus H. McCormick mansion at 50 East Huron, which the Association had purchased and remodeled. Despite debate—but with addition—ALA continues at that site.

With headquarters established in Chicago, in 1944 ALA Council approved a Library Development Fund drive, to raise money from ALA membership for the temporary support of an ALA Washington Office.[58] In 1946, ALA opened its federal office in Washington, DC; the first director was Paul Howard. The fund would continue to support the Washington Office until 1949, when closure was threatened due to the Fund depletion. An alternative funding proposal—including both ALA operating and endowment funds, as well as contributions from state library associations and individuals—was proposed by Ralph Lindquist, then chair of ALA's Federal Relations Committee, and ALA's Washington Office continued.

Soon after the Chicago headquarters moved into the McCormick mansion, in 1949, the Fourth Activities Committee recommended that all ALA headquarters activity be in Washington, DC. (ALA already housed government-relations staff there.) While this recommendation was rejected at the time, the issue of headquarters location remained a significant one to many association leaders. In 1957, the Executive Board Subcommittee on Headquarters Location recommended the sale of the Chicago property and relocation of its headquarters to Washington, DC, by the beginning of 1959, citing greater convenience to the majority of members and officers, proximity to other national associations with which ALA maintained working relationships, the economy and administrative efficiency of combining the Chicago and Washington operations, and legal matters. While the ALA Council approved the recommendation (82–34), the ALA

membership exercised its right to set aside Council action, petitioned for a mail vote by the membership, and reversed the Council decision (5739–2199).

In 1963, during the Annual Conference in Chicago, ALA dedicated its own building, at 50 E. Huron Street in Chicago, erected on the site of the old McCormick mansion. The staff then numbered 164. ALA had, however, entered a period of rapid programmatic growth. A short 2 years later, headquarters space was again a problem, the publishing department was moved to rented space, and the new headquarters building was remodeled (1966). It was obvious that additional space would be required, which again brought up the question of moving to Washington, DC. Again, in 1967, a subcommittee of the Executive Board recommended that the ALA headquarters be moved to Washington, DC. Again, Council supported the Board's recommendation and, again, membership petitioned for a mail vote and overturned the Council action.

Following lengthy debate and some lean years, in 1974 ALA President Jean Lowrie signed the agreement which would lead to ALA's new headquarters building, adjoining the 50 E. Huron facility. In November 1977, ALA entered into a complex development agreement that would exchange approximately 26,000 ft^2 of undeveloped property owned by ALA—and used by ALA staff for parking—for six floors of office space, lobby, and related facilities.[59]

In addition to the Chicago headquarters, ALA also owns and maintains offices in Washington, DC for staff involved in government relations, as well as the Office for Information Technology Policy. The staff of *CHOICE* magazine, a publication of the ACRL, a division of ALA, is located in a Middletown, CT office. In 2008, ALA began the process of purchasing office space for *CHOICE*.

CONCLUSION

Reporting at the end of FY2007, ALA Treasurer Rod Hersberger presented a complex, but healthy, organization, with total FY2007 revenues over $50 million and a consolidated balance sheet showing net assets in excess of $36 million. As of the end of fiscal year 2008, the total staff of the ALA had grown to about 280.[60] Various reports on the state of the association would present a strong organization facing significant challenges—from rapidly changing technology to a faltering economy, from challenges to privacy and intellectual freedom to the complexities of copyright in a networked global society.

Looking back over ALA's history, however, provides grounds for optimism. From the late nineteenth century to the early twenty-first century, you see an organization that has held steadfastly to its core mission "to promote library service and librarianship."[1] To enable that essential continuity, it has been willing to periodically review and change its structure, develop and modify policies to address an increasingly diverse and globalized society and take advantage of a succession of new technologies. Current programmatic priorities—advocacy, diversity, education and lifelong learning, equitable access to information, intellectual freedom, literacy—run like bright threads through the long and complex history, knitting together the organization, policy, and public stance. That persistence of mission and message is essential and it provides a foundation for organizational change.

APPENDIX

American Library Association: Membership Statistics, August 2008[a]

American Library Association	66,624
ALA Divisions	
American Association of School Librarians (AASL)	9,013
Association for Library Collections & Technical Services (ALCTS)	4,749
Association for Library Service to Children (ALSC)	4,222
Association for Library Trustees and Advocates (ALTA)	1,173
Association of College and Research Libraries (ACRL)	12,782
Association of Specialized & Cooperative Library Agencies (ASCLA)	896
Library Leadership, Administration and Management Association (LLAMA)[b]	5,104
Library and Information Technology Association (LITA)	4,018
Public Library Association (PLA)	11,934
Reference and User Services Association (RUSA)	4,960
Young Adult Library Services Association (YALSA)	5,702
ALA Round Tables	
Continuing Library Education Network and Exchange Round Table (CLENERT)	432
Ethnic and Multicultural Information Exchange Round Table (EMIERT)	647
Exhibits Round Table (ERT)	450
Federal and Armed Forces Libraries Round Table (FAFLRT)	387
Gay, Lesbian, Bisexual, Transgendered Round Table (GLBTRT)	954
Government Documents Round Table (GODORT)	1,039
Intellectual Freedom Round Table (IFRT)	1,709
International Relations Round Table (IRRT)	1,797
Library History Round Table (LHRT)	571
Library Instruction Round Table (LIRT)	1,643
Library Research Round Table (LRRT)	1,486
Library Support Staff Interest Round Table (LSSIRT)	622
Map and Geography Round Table (MAGERT)	374
New Members Round Table (NMRT)	1,906
Social Responsibilities Round Table (SRRT)	2,213
Staff Organizations Round Table (SORT)	195
Video Round Table (VRT)	284

[a]Total membership, including Personal, Organizational, Corporate.
[b]2007–2008 name change—previously Library Administration and Management Association.

REFERENCES

1. ALA Constitution, Article II: Object.
2. American Library Association, *Encyclopedia of Library and Information Science*, Marcel Dekker, Inc.: New York, 1968; Vol. 1, 267–269, 273, 279, 280, 288 [informaworld].
3. Wiegand, W.A. *The Politics of an Emerging Profession: The American Library Association, 1876–1917*, Greenwood Press: New York, 1986; 3–7, 17.
4. *ALA Handbook of Organization: Policy Manual 1.2*; Mission, American Library Association: Chicago, IL, 2007–2008.
5. American Library Association, Internal document, 1994–1995 CD2.2, ALA Organizational Self-Study Committee, Phase II Report, 11, 12, 36.
6. *ALA Handbook of Organization: Policy Manual, Section 1: Organization and Operational Policies*; American Library Association: Chicago, IL, 2000.
7. *ALA Handbook of Organization: Policy Manual, Policy 1.3: Priority Areas and Goals*; American Library Association: Chicago, IL, 2000.
8. American Library Association, Internal document, 2005–2006 CD#17.
9. *ALA Sets "Goal 2000"*; American Library Association: Chicago, IL, October 26, 1994. NEWS, Public Information Office.
10. American Library Association, Internal document, 2004–2005 CD31.1 and 2004–2005 EBD12.50.
11. American Library Association, Internal document, ALAhead to 2010 (brochure).
12. American Library Association, Internal document, 2007–2008 EBD12.5.
13. ALA Constitution, Article III.
14. Thomison, D. *A History of the American Library Association. 1876–1972*, American Library Association: Chicago, IL, 1978; 36–39 54, 57, 59, 74, 138, 200, 201, 229, 230.
15. Compton, C.H.; Countryman, G.A.; Meyer, H.A.B. ALA activities committee report. ALA Bull. **1930**, *24*(12), 607–680.
16. Shaw, R.R.; Coney, D.; Ersted, R.; Logsdon, R.H.; Rutzer, R. Final report of the fourth activities committee: memo to members. ALA Bull. **1949**, 17–43, 43.
17. Farley, J.; Breivik, P.S.; Rudd, A.S.; Garen, R.J.; Hensley, C.; Mutschler, H.; Wrigut, R. Internal document. Reorganizations, restructurings, and self-studies of the American Library Association (From 1965–present). 1995–1996 CD#40 (Midwinter).
18. American Library Association, Internal document, 1995–1996 CD#40.
19. Up the down concourse. Am. Libr. **1971**, 813–814.
20. Recommended ALA priorities for the 1980s. Am. Libr. **1981**, *12*(4), 190.
21. American Library Association, Internal document, 1975 Reference File.
22. ALA strategic long-range planning: an outline. Am. Libr. **1985**, *16*(2), 125.
23. ALA report. Am. Libr. **1981**, *12*(3), 126.
24. Executive board: seeking "total management". Am. Libr. July–August, **1981**, 406.
25. American Library Association, ALA strategic long-range planning (SLRP), 1985 CD#11d, with ALA Council Minutes, January 7–9, 1985, exhibit 15.

26. American Library Association, Internal document, 1991–1992 CD#44 (Rev.), Proposed ALA Self-Study.

27. American Library Association, Internal document, 1993–1994 CD#50 (Rev.), ALA Organizational Self-Study, Phase I Report, 1.

28. American Library Association, Internal document, 2000–2001 CD#41, 2001–2002 CD#44, 2001–2002 CD#44.1.

29. American Library Association, Internal document, 1996–1997 CD#4. Report of the Ad Hoc Task Force on Structure Revision to the ALA Council at the Midwinter Meeting, February 1997, 1–3.

30. ALA report. Am. Libr. **1982**, *13*(4), 257.

31. American Library Association, Internal file, Operating Agreement.

32. *ALA Handbook of Organization: Policy Manual, Policy 6.4.1: Policies of the American Library Association in Relation to its Membership Divisions*; American Library Association: Chicago, IL, 2000.

33. American Library Association, Internal document. Remarks by ALA Executive Director to Concurrent Meetings of ALA Division Executive Committees, Lisle, IL, October 21, 1988.

34. American Library Association, Internal document, 2000–2001 CD50.1.

35. American Library Association, Internal document, 2002–2003 CD2.2.

36. American Library Association, Internal document, 2002–2003 ALA-APACD2.2.

37. ALA Constitution Article VI.

38. ALA Constitution Article VII.

39. ALA Bylaws Article VII.

40. ALA Bylaws, Article VII, Section 2.

41. ALA Bylaws, Article IV.

42. American Library Association, Internal document. COO Action/Information Reporting Form, Membership Initiative Groups (Ghikas).

43. ALA Bylaws Article V Editors: This is listed as Chapter V in the prev Ency version.

44. *ALA Handbook of Organization: Policy Manual, Policy 9.2*; American Library Association: Chicago, IL, 2000; ALA Constitution: Chicago, IL, Article X.

45. Bone, L.E., The American Library in Paris: Fifty Years of Service. Am. Libr. **1970**, *1*(3), 279.

46. American Library Association, Internal document. White Paper: ALA International Relations Office.

47. American Library Association, Internal document. Presentation on ALA 2.0: New Technologies in an Association Context, Ghikas, M.W., for the Association Forum of Chicagoland, Spring 2008.

48. *World Encyclopedia*; American Library Association: Chicago, IL, 2000; 50.

49. The 80s. Am. Libr. **1989**, *20*(11), 1049–1058.

50. AL/ 92 112, Karen Schmidt, UIUC ALA Archives.

51. AL Wrapup 1989 1058.

52. Just another record-breaker. Am. Libr. **1990**, 638–643.

53. Highlights of the midwinter meeting. ALA Bull. **1966**, *60* (3), 239–246.

54. An information-access omnivore. Am. Libr. **1987**, *18*(3), 216.

55. Kniffel, L. Am. Lib. *25*(4), 367.

56. ALA acts: Membership, council and executive board summaries. Am. Libr. **1985**, *16*(7), 510–511.

57. American Library Association, Internal document. Celebrating 10 Years of the Spectrum Scholarship Program (brochure).

58. Holley, E.G. Schremser, R.F. *The Library Services and Construction Act: An Historical Overview from The Viewpoint of Major Participants*, JAI Press: Greenwich, CT, 1983; Vol. 18, 5–6, Foundations in Library and Information Science.

59. American Library Association, Internal memo, Ernest Martin to William Gordon, November 19, 1999.

60. American Library Association, Internal document. Treasurer Report, FY2007.

American Medical Informatics Association (AMIA)

Don E. Detmer
Tia Abner
Karen Greenwood
American Medical Informatics Association (AMIA), Bethesda, Maryland, U.S.A.

Abstract

The American Medical Informatics Association (AMIA) incorporated 20 years ago through the merger of three existing informatics groups: the Symposium on Computer Applications in Medical Care (SCAMC), the American Association for Medical Systems and Informatics (AAMSI), and the American College of Medical Informatics (ACMI). AMIA has nearly 4000 members from over 55 countries worldwide, including informatics professionals, students, institutions, and corporations. Together they represent all basic, applied, and clinical interests in informatics. AMIA is the trusted source of knowledge about the effective use of informatics to transform health care within three domains: clinical informatics (including health care, research, and personal health management), public health and population informatics, and translational bioinformatics. AMIA holds several meetings each year—the Summit on Translational Bioinformatics in March, the Spring Congress in May, an Academic Forum meeting in the summer, and two fall meetings, the Health Policy Meeting in September and its signature event, the Annual Symposium held in November.

AMIA—THE PROFESSIONAL HOME FOR BIOMEDICAL AND HEALTH INFORMATICS

American Medical Informatics Association (AMIA) is the premier professional organization in the United States dedicated to the development and application of informatics in the support of health and patient care, public health, teaching, research, administration, and related policy. AMIA incorporated 20 years ago as the American Medical Informatics Association through the merger of three existing informatics groups, the Symposium on Computer Applications in Medical Care (SCAMC), the American Association for Medical Systems and Informatics (AAMSI), and the American College of Medical Informatics (ACMI). AMIA has nearly 4000 members from over 55 countries worldwide. Members include informatics professionals, students, institutions, and corporations and together, they represent all basic, applied, and clinical interests in informatics. AMIA is governed by a Chairman and Board of Directors and calls upon its members to serve in volunteer roles on task forces, committees, and working groups. AMIA office is led by the President and CEO who works with a professional staff to manage AMIA's programs and services.

AMIA is the trusted source of knowledge about the effective use of informatics to transform health care within three domains: clinical informatics (including health care, research, and personal health management), public health and population informatics, and translational bioinformatics. AMIA is frequently called upon as a source of informed and scholarly opinion on policy issues relating to the national health information infrastructure, uses and protection of personal health information, and public health considerations and international dimensions of informatics education and research. AMIA publishes a leading industry cited scholarly journal, *the Journal of the American Medical Informatics Association (JAMIA)*, proceedings of its meetings, the e-News Weekly, an electronic newsletter, and digest of syndicated news as well as AMIA's Standards Standard bulletin, a periodic review of international health-related standards.

In 2005, the Board of Directors, members and AMIA staff undertook a strategic planning exercise to review and where needed redefine the vision, mission, and goals of the association. Now, 4 years later, AMIA has cemented its standing in the health care community as the professional home for biomedical and health informatics. AMIA continues to work on new projects to advance health and health care with the development of educational offerings and programs, public policy and advocacy initiatives, scientific meetings and symposia, white papers and reports, and ongoing membership activities.

AMIA holds several meetings per year—the Summit on Translational Bioinformatics in March, the Spring Congress in May, an Academic Forum meeting in the summer, and two fall meetings, the Health Policy meeting and the Annual Symposium. Summit programming includes informatics research emerging from the analysis of molecular and clinical data and measurement; relating and representing phenotypes and disease; delineating disease through the study of organisms, evolution, and taxonomy; and computational approaches to finding molecular

Encyclopedia of Library and Information Sciences, Fourth Edition DOI: 10.1081/E-ELIS4-120044717

mechanisms and therapies for disease. The congress program engages attendees to develop a common view of the opportunities and challenges facing informatics. Congress topics include the development of effective electronic health records and the evaluation of their impact on care delivery; the incremental definition of personal health records and the new challenges associated with emerging approaches to Personal Health Records (PHR) development and use; the technologies, policies, research, and social structures required to create a stronger public health informatics infrastructure; and the opportunities, challenges, and progress associated with accelerating the field of clinical research informatics.

AMIA's Annual Symposium is the premier scientific meeting for new research and development in biomedical and health informatics. It is the place to hear about leading edge scientific work, to learn about evolving standards and policies for the management of biomedical information, and to understand how cutting-edge information and communication technologies can best be developed and deployed. The symposium provides a wide range of formats for education and discussion and includes half-day tutorials, expert panels, papers, posters, and invited talks. Over 2000 biomedical and health informatics professionals attend this signature event each year. AMIA's education, training, and academic-related opportunities are rigorous and independently evaluated, and we continue to develop new programs and services. Our 10×10 program to train 10,000 health care professionals in applied health and medical informatics by the year 2010 continues to grow. In 2007, AMIA launched a 5-year cooperative agreement with the Centers for Disease Control and Prevention. The purpose of the program is to strengthen the breadth and depth of the public health workforce by providing a training in public health informatics and encourage new explorations of innovations at the intersection of public health and health informatics. AMIA also continues to support nursing informatics through the Alliance for

Nursing Informatics, the TIGER initiative, and our own educational endeavors. AMIA received the exciting news in 2007 that it was selected to host NI2012, the 11th triennial Congress on Nursing Informatics. This is the premier meeting for nurses and informatics experts representing global communities and working with informatics applications in nursing care, administration, research and education, and will bring together thought leaders of the nursing informatics community.

In 2008, AMIA launched the Academic Forum and it now has 45 members. The Forum provides a unique platform for surveying and analyzing activities in academic units dedicated to biomedical and health informatics, and for recommending best practices related to education, scholarship, faculty development, salary structures and retention, and other selected issues relating to the academic enterprise.

One of AMIA's newest agenda items is to develop through its membership in and collaboration with the International Medical Informatics Association (IMIA), an international distance learning continuing education program in informatics for low-resource environments. Working under a global umbrella program known as 20/20 e-Health Capacity Building, AMIA is creating *Health Informatics Building Blocks* to improve basic clinical informatics skills.

Also relating to the association's e-Health Capacity strategy, AMIA received a $1.2 million grant from the Bill & Melinda Gates Patnership to support the development of a blueprint to create a robust global health informatics education and research infrastructure in low-resource settings. Through this Global Partnership Program (GPP), AMIA will lead a team of experts to develop scalable approaches to e-health education, and help address the need for a global a informatics workforce and scholarly network.

More information about AMIA may be found on its Web site: http://www.amia.org.

American National Standards Institute (ANSI)

Jana Zabinski
American National Standards Institute, New York, New York, U.S.A.

Abstract
The American National Standards Institute (ANSI) is a private, nonprofit organization that oversees the standards and conformity assessment system in the United States.

INTRODUCTION

The American National Standards Institute (ANSI) is a private, nonprofit organization that oversees the standards and conformity assessment system in the United States, and represents the needs and views of U.S. stakeholders in standardization forums around the globe. ANSI's membership comprises government agencies, organizations, corporations, academic and international bodies, and individuals. The Institute represents the interests of more than 125,000 companies and 3.5 million professionals.

The American National Standards Institute approves standards that are developed by accredited organizations comprised of representatives of standards developing organizations (SDOs), government agencies, consumer groups, companies, and others. The Institute also accredits organizations that carry out product or personnel certification in accordance with requirements defined in international standards. (See Fig. 1.)

Standards ensure that the characteristics and performance of products are consistent, that the same definitions and terms are used, and that products are tested the same way. For example, standards ensure that libraries can share electronic card catalog information, and patrons can search those catalogs from connected computers with Internet access.

HISTORY

By the twentieth century, the need for coordination among U.S. standards-setting groups became evident. In October 1918, three government agencies and five private sector organizations joined together to form a coordination body known as the American Engineering Standards Committee, the predecessor of what is now known as ANSI.

Today, the U.S. standardization community is comprised largely of nongovernmental SDOs and consortia; these groups are primarily supported by industry participation.

OVERVIEW OF THE U.S. STANDARDIZATION SYSTEM

Standardization encompasses a broad range of considerations—from the actual development of a standard to its promulgation, acceptance, implementation, and demonstration of compliance. A primary facilitator of commerce, standardization has become the basis of a sound national economy and the key to global market access.

Voluntary consensus standards serve as the cornerstone of the distinctive U.S. standardization system. These documents arise from an open process that depends upon data gathering, a vigorous discussion of all viewpoints, and agreement among a diverse range of stakeholders. Thousands of individual experts representing the viewpoints of consumers, companies, industry and labor organizations, and government agencies at the federal, state, and local level voluntarily contribute their knowledge, talents, and efforts to standardization activities.

"Voluntary" refers only to the manner in which the standard was developed; it does not necessarily refer to whether compliance to a consensus standard is optional or whether a government entity or market sector has endorsed the document for mandatory use.

AMERICAN NATIONAL STANDARDS

The Institute oversees the creation, promulgation and use of thousands of norms and guidelines that directly impact businesses in nearly every sector: from manufacturing and construction to agriculture, food service, software engineering, energy distribution, and more. Likewise, ANSI-accredited standards developers span the full gamut of industry sectors and services.

Though ANSI itself does not develop standards, the Institute facilitates the development of American National Standards, also known as ANS, by accrediting the procedures of standards developing organizations. ANSI accreditation signifies that the procedures used by these bodies meet the Institute's essential requirements for openness,

Encyclopedia of Library and Information Sciences, Fourth Edition DOI: 10.1081/E-ELIS4-120044428

Fig. 1 ANSI Logo.

balance, consensus, and due process. Currently, more than 200 active SDOs are accredited under the *ANSI Essential Requirements: Due process requirements for American National Standards*. Approximately 10,500 American National Standards carry the ANSI designation.

Hallmarks of the American National Standards process involve

- Consensus by a group that is open to representatives from all interested parties.
- Broad-based public review and comment on draft standards.
- Consideration of and response to comments.
- Incorporation of submitted changes that meet the same consensus requirements into a draft standard.
- Availability of an appeal by any participant alleging that these principles were not respected during the standards-development process.

For example, the National Information Standards Organization (NISO) develops and maintains standards for information systems, products, and services relating to bibliographic and library applications. NISO's standards span both traditional and new technologies, and address the full range of information-related needs, including retrieval, storage, and preservation.

Another ANSI-accredited standards developer, the Data Interchange Standards Association (DISA) writes standards that support e-commerce and business-to-business data exchange, from order processing to electronic payment. DISA also serves as the secretariat for the ANSI-Accredited Standards Committee (ASC) X12, which develops e-business exchange standards in XML and X12 EDI formats.

ANSI INVOLVEMENT IN INTERNATIONAL STANDARDS ACTIVITIES

In the international arena, ANSI promotes the use of U.S. standards abroad, advocates U.S. policy and technical positions in international and regional standards and conformity assessment organizations, and encourages the adoption of international standards as national standards where appropriate. The Institute is the official U.S. representative to the two major international standards organizations,

the International Organization for Standardization (ISO) and, via the U.S. National Committee (USNC), the International Electrotechnical Commission (IEC). ANSI is also a member of the International Accreditation Forum (IAF).

Through ANSI, the U.S. has immediate access to the ISO and IEC standards development processes. ANSI and the USNC frequently carry U.S. standards forward to ISO and IEC where they are adopted in whole or in part as international standards. U.S. positions are developed by U.S. Technical Advisory Groups (TAGs) that have been accredited by ANSI or approved by the USNC. Participation in a U.S. TAG is open to all affected stakeholders.

On behalf of ANSI, NISO administers the U.S. TAG to ISO Technical Committee 46 (TC 46), *Information and Documentation*. TC 46 develops international standards relating to records management and museum documentation, as well as publishing, archiving, and indexing.

ASC X12 serves as the U.S. TAG administrator to ISO TC 154, *Processes, Data Elements and Documents in Commerce, Industry and Administration*, which supports international standardization activities in the field of industrial data. ISO TC 154 standards address business administration processes and information interchange between individual organizations.

CONFORMITY ASSESSMENT

On the other side of the standardization coin is conformity assessment, a term used to describe the evaluation of products, processes, systems, services, or personnel to confirm adherence to the requirements identified in a specified standard. In general, conformity assessment includes sampling and testing, inspection, supplier's declaration of conformity, certification, and management system assessment and registration. It can also include accreditation of the competence of those activities by a third party and recognition (usually by a government agency) of an accreditation program's capability.

Conformity assessment forms a vital link between standards that define product characteristics or requirements and the products themselves. It can verify that a particular product meets a given level of quality or safety, and it can provide explicit or implicit information about the product's characteristics, the consistency of those characteristics, and/or the performance of the product.

The American National Standards Institute's role in the conformity assessment arena includes accreditation of organizations that certify that products and personnel meet recognized standards. The ANSI-American Society for Quality National Accreditation Board (ANAB) serves as the U.S. accreditation body for management systems certification, primarily in areas such as quality (ISO 9000 family of standards) and/or the environment (ISO 14000 family of standards). ANSI also is involved in several international and regional organizations to promote multilateral recognition of conformity assessments across borders to preclude redundant and costly barriers to trade.

The American National Standards Institute's accreditation programs themselves are created in accordance with international guidelines as verified by government and peer review assessments.

STANDARDS PANELS

Through its standards panel program, ANSI provides standards-based solutions to national and international priorities. Each of the Institute's panels engages a broad range of stakeholders in the coordination and harmonization of standards and conformity assessment activities relevant to the panel's area of focus.

In 2004, the ANSI Homeland Security Standards Panel (ANSI-HSSP) supported a special project on private-sector emergency preparedness that had been requested by the 9/11 Commission. The panel continues to provide ongoing support for the Department of Homeland Security and other agencies.

At the request of the Office of Science and Technology Policy in the Executive Office of the President, ANSI launched the Nanotechnology Standards Panel (ANSI-NSP) to facilitate the development of standards for nanotechnology nomenclature and terminology; materials properties; and testing, measurement, and characterization procedures.

The Healthcare Information Technology Standards Panel (HITSP) is under contract with the Department of Health and Human Services to assist in establishing a national health IT network for the United States.

The Identity Theft Prevention and Identity Management Standards Panel (IDSP) is supporting all citizens in its efforts to facilitate the identification and development of standards to secure and protect personal information.

Launched in May 2007, the ANSI Biofuels Standards Panel (ANSI-BSP) is a cross-sector coordinating body established to promote the development and compatibility of standards and conformity assessment programs to support the large-scale commoditization of biofuels.

CONCLUSION

The American National Standards Institute provides the forum through which all affected stakeholders may cooperate in establishing, improving, and recognizing consensus-based standards and certification programs that are dynamically responsive to national needs. ANSI continues to be fully involved in its support of the goals of U.S. and global standardization and remains committed to enhancing of the quality of life for all global citizens.

BIBLIOGRAPHY

1. American National Standards Institute official Web site. Available at http://www.ansi.org.

American Society for Information Science and Technology (ASIST)

Charles H. Davis
Debora Shaw
School of Library and Information Science, Indiana University, Bloomington, Indiana, U.S.A.

Abstract

The American Society for Information Science and Technology (ASIST) was founded as the American Documentation Institute in 1937. In response to the evolution of the field, the society has grown and changed its structure and focus. This entry describes the Society's history, purpose, structure (regional chapters plus topical special interest groups), governance, meetings, and publications. This entry is based on the History of ASIST section of the Society's Web site, http://www.asis.org.

INTRODUCTION

The American Society for Information Science and Technology, ASIST, is an association of researchers and information professionals leading the search for new and better theories, techniques, and technologies to improve access to information. Among its membership are some 4000 specialists from such fields as computer science, linguistics, management, psychology, librarianship, engineering, law, medicine, chemistry, and education; individuals who share a common interest in improving the ways society stores, retrieves, analyzes, manages, archives, and disseminates information, coming together for mutual benefit.

American Society for Information Science and Technology provides educational and conference programs, highly regarded journals, and other publications, as well as professional services for information systems developers, online professionals, information resource managers, librarians, record managers, and others who bridge the gaps between research and application, and between developer and user. The Society's flagship publications are the highly regarded *Journal of the American Society for Information Science and Technology* and *Annual Review of Information Science and Technology*. The American Society for Information Science and Technology is a nonprofit 501(c)3 professional association organized for scientific, literary, and educational purposes.

HISTORY

1937—Beginnings in Documentation

The American Society for Information Science and Technology was founded on March 13, 1937, as the American Documentation Institute (ADI), a service organization made up of individuals nominated by and representing affiliated scientific and professional societies, foundations, and government agencies. Its initial interest was in the development of microfilm as an aid to learning. Watson Davis, the Institute's first president (1946–1943), was the editor of *Science News Letter* and director of Science Service, a nonprofit organization founded to promote public appreciation of science. The American Documentation Institute compiled an impressive record of achievement in its early years: development of microfilm readers, cameras, and services; fostering negotiations and research that resulted in the so-called "gentleman's agreement" covering the photo-duplication of copyrighted materials; establishment of programs for the storage and reproduction of auxiliary publications in support of journal editors; operation of an Asian scientific literature service during World War II; support of Interlingua, an early rival of Esperanto, to foster international science communications; and cosponsorship of the 1958 International Conference on Scientific Information.[1–4]

1950s—Transition to Modern Information Science

As the number of people engaged in developing new principles and techniques in the many areas of documentation and information services increased, the Bylaws were amended in 1952 to admit individual as well as institutional members. Thus, ADI became the national professional society for those concerned with all elements and problems of information science. With the 1950s came increasing awareness of the potential of automatic devices for literature searching and information storage and retrieval. As these concepts grew in magnitude and potential, so did the variety of professional interests.[1,4,5]

Encyclopedia of Library and Information Sciences, Fourth Edition DOI: 10.1081/E-ELIS4-120044410

1960s—The Information Explosion

During the 1960s, membership increased sevenfold as the problems created by the "information explosion" became of national concern. Reflecting this change in its total range of activities, as well as the emergence of information science as an identifiable configuration of disciplines, the membership voted to change the name of the ADI to the American Society for Information Science (ASIS). The name change took effect January 1, 1968, and emphasized the fact that the membership of ASIS was uniquely concerned with all aspects of the information transfer process, and that the Society was a national professional organization with regional chapters for those concerned with designing, managing, and using information systems and technology. Harold Borko's frequently cited article, "What Is Information Science?," helped to define the field:

> Information science is that discipline that investigates the properties and behavior of information, the forces governing the flow of information, and the means of processing information for optimum accessibility and usability. It is concerned with the body of knowledge relating to the origination, collection, organization, storage, retrieval, interpretation, transmission, and utilization of information.[6]

One of the Society's flagship publications, the *Annual Review of Information Science and Technology (ARIST)*, was launched in 1966, under the editorship of Carlos A. Cuadra and with support from the National Science Foundation.

1970s—The Move to Online Information

The transitions from batch processing to interactive interaction and from mainframe to mini- and micro-computers accelerated in the 1970s. Traditional boundaries among disciplines began to fade; many library schools and some departments of computer science added "information" to their titles. In 1974, the Society's *Newsletter* became the *Bulletin of the American Society for Information Science*, a mainstay membership publication of the Society. During this decade, ASIS made the midyear meeting an annual event focusing on a single topic of current interest and sponsored a bicentennial conference (1976) on the role of information in the country's development. From 1970 to 1974, ASIS administered the Educational Resources Information Center (ERIC) Clearinghouse on Library and Information Sciences; the Society was also an active participant in the planning and implementation of the 1979 White House Conference on Library and Information Services.

1980s—Personal Computers Change the Market

By the 1980s, large databases such as Grateful Med at the National Library of Medicine and user-oriented services such as Dialog and CompuServe, were for the first time accessible by individuals from their personal computers. The American Society for Information Science added to and revised its special interest groups (SIGs) to respond to the changes establishing groups on office information systems and personal computers as well as international information issues and rural information services. By the end of the decade, SIGs had been formed and reformed to represent various interests, including nonprint media, social sciences, energy and the environment, and community information systems, and ASIS had its first non-North American chapters.

1990s and 2000s—The Human/Social Perspective

The social implications of information and communication technologies became increasingly apparent in the 1990s and the Society responded. Ann Prentice's "environmental scan" identified key topics that were expected to reshape work and the workforce, including public access to and ownership of information, internationalization, and increasing expectations of information technology. Themes that appeared regularly in the Society's *Bulletin* and at national meetings included information management, globalization, intellectual property, computer networks (National Research and Education Network, the Internet), and digitization of text and visual information. Interest in representation of information found new applications in, for example, markup languages, taxonomy, metadata, and information architecture. Understanding of information retrieval was broadened and deepened by such developments as hypertext, the World Wide Web, search engines and browsers, and the addition of new domains to the Text Retrieval Conferences.

The Society experienced changes in its own use of information technology. Merri Beth Lavagninio introduced the ASIS-L listserv in 1992. The Society created its first Web site in 1995;[7] it was chosen as a Yahoo! "Pick of the Day" in 1996. After considerable deliberation about making the Society's publications available electronically, a digital library was unveiled in 2006. In 2007, the *Bulletin of the American Society for Information Science and Technology* became an electronic-only publication.

The Society's work reflected the increasing interest in social and historical aspects of information science in other ways as well. In 1991, it renamed and reoriented one of the SIGs to focus on the History and Foundations of Information Science, and in 1992, ASIS adopted its first professional guidelines. Conferences cosponsored with the Chemical Heritage Foundation on the History and Heritage of Scientific and Technical Information Systems were held in 1998 and 2002.

Reflecting both internal and external influences, the Society again changed its name in 2000, becoming the American Society for Information Science and Technology. President Eugene Garfield championed the change, calling attention to the importance of practical, information-technology focused programs. The Society continued its annual meetings, but discontinued the midyear meeting in favor of "summits" featuring topics such as information architecture, knowledge management, and digital archives for engineering and scientific research.

PURPOSE

ASIST's mission is to advance the information sciences and related applications of information technology by providing focus, opportunity, and support to information professionals and organizations.

The Society seeks to stimulate participation and interaction among its members by affording them an environment for substantive professional exchange. It encourages and supports personal and professional growth through opportunities for members to extend their knowledge and skills, develop and use professional networks, pursue career development goals, and assume leadership roles in the Society and in the broader information community. The American Society for Information Science and Technology increases the influence of information professionals among decision makers by focusing attention on the importance of information as a vital resource in a high technology age; in this way the organization seeks to promote informed policy on national and international information issues. It supports the advancement of the state-of-the-art and practice by taking a leadership position in the advocacy of research and development in basic and applied information science.

To accomplish these goals, ASIST edits, publishes, and disseminates publications concerning research and development; convenes annual meetings providing a forum for papers, discussions, and major policy statements; holds smaller local and special interest meetings, as well as special symposia; and acts as a sounding board for the promotion of research and development and for the education of information professionals.

CHAPTERS AND SIGS

To give scope and focus to the diverse interests of its members, special interest groups (SIGs) were established in 1966. These groups are, in effect, small professional organizations within ASIST that encourage the discussion and development of both theoretical and practical issues. The introduction and widespread adoption of the World Wide Web encouraged and supported five virtual SIGs: Bioinformatics; Blogs, Wikis, and Podcasts; Critical Issues; Information Architecture; and Metrics. As of 2009, the other 17 SIGs are Arts and Humanities; Classification Research; Digital Libraries; Education for Information Science; History and Foundations of Information Science; Human–Computer Interaction; Information Architecture; Information Needs, Seeking and Use; Information Policy; International Information Issues; Knowledge Management; Library Technologies; Management; Medical Informatics; Scientific and Technical Information Science; Social Informatics; and Visualization, Images, and Sound.

All ASIST members are automatically enrolled as members of SIG/CON, which seeks to "explore the fundamental notions of information science, and expose them for what they are."[8] The official SIG/CON Web site describes the initial meeting: "Those who stumbled into the first SIG/CON session found it to be one of the most informative and intellectually challenging sessions of the conference, a tradition which continues to this day."[8] Dr. Llewellyn C. Puppybreath III, founder and perpetual chair of the group, has yet to appear at any of its meetings. The breadth of topics covered is evident in this list of titles from presentations: "Baloonean Logic" by Ev Brenner, "Conical Classifications" by C. David Batty, "Titular Colonicity" by Candy Schwartz, and "Early Binary Separation: Implications for Information Retrieval (relevance judgments by identical twins separated at birth)" by Joseph Janes.

At a local and regional level, ASIST has, in 2009, 18 chapters in major cities and regions in the United States and around the world: Arizona, Carolinas, Central Ohio, Europe, Florida, Indiana, Los Angeles, Metropolitan New York, Michigan, Minnesota, New England, New Jersey, Northern Ohio, Pacific Northwest, Potomac Valley (Washington, D.C. area), Southern Ohio, Taipei, and Wisconsin. These chapters provide local forums for discussion and information exchange for members in a given locality.

In addition, 39 student chapters encourage the discussion and are in development of research and education, and are a medium for channeling student interest toward issues of concern to the profession: Catholic University of America, Drexel University, European Student Chapter, Florida State University, Indiana University, Long Island University, Louisiana State University, North Carolina Central University, Ohio Virtual, Pratt Institute, Rutgers University, San Francisco Bay Area, Simmons College, St. John's University, State University of New York-Albany, State University of New York-Buffalo, State University of New York-Oswego, Taipei University, University of British Columbia, University of California at Los Angeles, University of Denver, University of Hawaii, University of Illinois at Urbana-Champaign, University of Iowa, University of Kentucky, University of Maryland, University of Michigan, University of North Carolina Chapel Hill, University of North Texas, University of Pittsburgh, University of South Carolina, University of South Florida, University of Tennessee, University of Texas-Austin, University of Toronto, University of Washington, University of Wisconsin-Madison, University of Wisconsin-Milwaukee, and Wayne State University.

GOVERNANCE

The ASIST Board of Directors consists of the president, president-elect, immediate past president, treasurer, a parliamentarian, directors chosen by the SIGs and chapters, and six directors-at-large. The American Society for Information Science and Technology maintains headquarters under an executive director at 1320 Fenwick Lane, Suite 510, Silver Spring, Maryland 20910. The Society's Web address is http://www.asis.org.

The Executive and Budget and Finance committees are standing committees of the Board of Directors; standing committees of the Society are Awards and Honors, Constitution and Bylaws, Information Science Education, International Relations, Leadership, Membership, Nominations, Publications, and Standards. An annual business meeting of the membership is held during the ASIST annual meeting.

The presidents of the Society have been:

ADI
Watson Davis 1937–1943
Keyes D. Metcalf 1944
Waldo G. Leland 1945
Watson Davis 1946
Waldo G. Leland 1947
Vernon D. Tate 1948–1949
Luther H. Evans 1950–1952
E. Eugene Miller 1953
Milton O. Lee 1954
Scott Adams 1955
Joseph Hilsenrath 1956
James W. Perry 1957
Herman H. Henkle 1958
Karl F. Heumann 1959
Cloyd Dake Gull 1960
Gerald J. Sophar 1961
Claire K. Schultz 1962
Robert M. Hayes 1963
Laurence B. Heilprin 1964–1965
Harold Borko 1966
Bernard M. Fry 1967
ASIS
Robert S. Taylor 1968
Joseph Becker 1969
Charles P. Bourne 1970
Pauline A. Atherton 1971
Robert J. Kyle 1972
John Sherrod 1973
Herbert S. White 1974
Dale B. Baker 1975
Melvin S. Day 1976
Margaret T. Fischer 1977
Audrey N. Grosch 1978
James M. Cretsos 1979
Herbert B. Landau 1980

Mary C. Berger 1981
Ruth L. Tighe 1982
Charles H. Davis 1982–1983
Donald W. King 1984
Bonnie C. Carroll 1985
Julie A. C. Virgo 1986
Thomas H. Hogan 1987
Martha E. Williams 1988
W. David Penniman 1989
Toni Carbo 1990
Tefko Saracevic 1991
Ann E. Prentice 1992
José-Marie Griffiths 1993
Marjorie M.K. Hlava 1994
James E. Rush 1995
Clifford Lynch 1996
Debora Shaw 1997
Michael Buckland 1998
Candy Schwartz 1999
Eugene Garfield 2000
ASIST
Joseph A. Busch 2001
Donald Kraft 2002
Trudi Bellardo Hahn 2003
Samantha Hastings 2004
Nicholas Belkin 2005
Michael Leach 2006
Edie Rasmussen 2007
Nancy Roderer 2008
Donald Case 2009

PUBLICATIONS

Publication affords a major channel of communicating current significant research reports, development work, and professional activities, both to the membership and to the interested public. It also serves as an historical record of achievement. In addition to a substantial monograph series, ASIST produces four continuing publications.

The *Journal of the American Society for Information Science and Technology* is a highly ranked and frequently cited source for reports of research and development in information science. The *Journal of Documentary Reproduction* (edited by Vernon D. Tate and published by the American Library Association, 1938–1942) is considered the predecessor of the Society's journal. In 1950, it was reborn as *American Documentation*. With the Society's changes of name it became the *Journal of the American Society for Information Science* (in 1970) and subsequently the *Journal of the American Society for Information Science and Technology* (in 2001). The *Journal* serves as a forum for new research in information transfer and communication processes in general, and in the context of recorded knowledge in particular. Concerns include the generation, recording, distribution, storage, representation, retrieval, and dissemination of information, as well as its social impact and management of information agencies. The *Journal of the American Society for Information Science and Technology* consistently ranks at or near the top in assessments of influence in the field. It has gradually increased from four issues per year (1950–1969) to 14 issues per year (1998 to date). The *Journal* has been edited by Vernon D. Tate (1950–1951), Mortimer Taube (1952), Jesse H. Shera (1953–1960), Luther H. Evans (1961), James D. Mack (1962–1963), Arthur W. Elias (1964–1976), Charles T. Meadow (1977–1984), Donald H. Kraft (1985–2008), and Blaise Cronin (2009 to date).

The Annual Review of Information Science and Technology was first published in 1966. It surveys the landscape of information science and technology, providing the reader with an analytical, authoritative, and accessible overview of recent trends and significant developments. The initial volumes of this series were supported by a grant from the National Science Foundation. *ARIST* has been edited by Carols A. Cuadra (1966–1975), Martha E. Williams (1976–2001), and Blaise Cronin (2002–to date).

The proceedings of the annual meetings collect significant contributed papers. These proceedings are available on disc to meeting registrants and electronically through the ASIST digital library (which also includes the *Journal*, *Annual Review*, and *Bulletin*).

The *Bulletin of the American Society for Information Science* (formerly the Society's *Newsletter*) began publication in 1974. It is a bimonthly news magazine focused on developments and issues affecting the field, pragmatic management reports, opinion, and news of people and events in the information science community. Irene Travis was named editor in 1997; in 2007 the *Bulletin* became an electronic-only publication.

MEETINGS

Both the annual meeting, held in the fall of the year, and the special symposia provide important forums for communication among people of diverse interests. The annual meetings bring together researchers and professionals from virtually every field: attorneys, bankers, electronic and print publishers, government officials, businesspersons, librarians, engineers, social scientists, information managers, computer scientists, and many others. The topical summits feature presentations from research and industry leaders, opportunities for intense personal interaction, and awards recognizing excellence in all aspects of the field.

Special interest groups also organize and present specialized meetings. The SIG on Classification Research has conducted the highly regarded Workshop of Classification Research each year since 1990. The SIG on Information Needs, Seeking, and Use holds a research symposium in conjunction with ASIST annual meetings.

Table 1 Recipients of the Award of Merit and Watson Davis Award.

Year	Award of Merit recipients	Watson Davis Award recipients
1964	Hans Peter Luhn	
1965	Charles P. Bourne	
1966	Mortimer Taube	
1967	Robert Fairthorne	
1968	Carlos A. Cuadra	
1969	[not awarded]	
1970	Cyril W. Cleverdon	
1971	Jerold Orne	
1972	Phyllis Richmond	
1973	Jesse H. Shera	
1974	Manfred Kochen	
1975	Eugene Garfield	
1976	Laurence Heilprin	James M. Cretsos
		Laurence B. Heilprin
		Lois F. Lunin
		Gerald J. Sophar
		Herbert S. White
1977	Allen Kent	Arthur W. Elias
		Irene Farkas-Conn
		Simon M. Newman
		Roy D. Tally
1978	Calvin N. Mooers	Mary C. Berger
		Joe Ann Clifton
		Charles H. Davis
1979	Frederick Kilgour	Frank Slater
1980	Claire K. Schultz	Gerard O. Platau
1981	Herbert S. White	Jan Krcmar
1982	Andrew A. Aines	Edmond Sawyer
1983	Dale B. Baker	Toni Carbo
		Margaret T. Fischer
1984	Joseph Becker	Robert Tannehill, Jr.
1985	Robert L. Chartrand	Lawrence Woods
		Michel Menou
1986	Bernard M. Fry	N. Bernard Basch
1987	Donald W. King	George Abbott
		Barbara Flood
1988	F. Wilfrid Lancaster	Bonnie C. Carroll
1989	Gerard Salton	Madeline Henderson
		G. Daniel Robbins
1990	Pauline Atherton Cochrane	Julie Virgo
1991	Roger K. Summit	Mickie Voges-Piatt
1992	Robert S. Taylor	Marianne Cooper
1993	Robert M. Hayes	Debora Shaw
1994	Harold Borko	Audrey Grosch
1995	Tefko Saracevic	Martha Williams
		Donald Kraft
1996	Jean Tague-Sutcliffe	Marjorie M.K. Hlava
1997	Dagobert Soergel	Karla Petersen
1998	Henry Small	Judy Watson
1999	José Marie Griffiths	Jessica L. Milstead
2000	Donald R. Swanson	Candy Schwartz
2001	Patrick G. Wilson	Julie Hurd
2002	Karen Spärck Jones	Thomas Hogan
2003	Nicholas J. Belkin	Nancy Roderer
2004	Howard D. White	Joseph Busch
2005	Marcia Bates	Michael Buckland
2006	Blaise Cronin	Trudi Bellardo Hahn
		Steve Hardin
2007	Donald H. Kraft	Paula Galbraith

AWARDS

The American Society for Information Science and Technology's Award of Merit is the Society's highest honor. It is bestowed on individuals who have made noteworthy contributions to the field of information science, including the expression of new ideas, the creation of new devices, the development of better techniques, and outstanding service to the profession of information science. The Watson Davis Award commemorates the Society's founder. It is given to members who have shown outstanding continuous contributions and dedicated service to ASIST. The recipients of these awards are listed in Table 1.

CONCLUSION

The American Society for Information Science and Technology has responded to more than seven decades of change in our understanding and use of information technologies. It has evolved from an organization of institutions to an individual membership society. ASIST's meetings reach academic, industry, and professional audiences and the society's long-standing publications increase their already high impact as they move to electronic formats.

REFERENCES

1. Farkas-Conn, I.S. *From Documentation to Information Science: The Beginnings and Early Development of the American Documentation Institute-American Society for Information Science*, Greenwood Press: New York, 1990.
2. Richards, P.S. *Scientific Information in Wartime: The Allied-German Rivalry, 1939–1945*, Greenwood Press: Westport, CT, 1994.
3. Schultz, C.K. ASIS: Notes on its founding and development. B. Am. Soc. Inform. Sci. **1976**, *2*(8), 49–51.
4. Schultz, C.K.; Garwig, P.L. History of the American Documentation Institute: A sketch. Am. Doc. **1969**, *20*(2), 152–160.
5. Hahn, T.B.; Buckland, M.K. *Historical Studies in Information Science*, Information Today: Medford, NJ, 1998.
6. Borko, H. What is information science? Am. Doc. **1968**, *19*(1), 3.
7. American Society for Information Science and Technology [Homepage of the American Society for Information Science and Technology]. Available at http://www.asis.org/.
8. SIG/CON [what? how? why?]. Available at http://web.simmons.edu/~schwartz/con/sig-con.html.

Approval Plans

Robert Nardini
Vice President, Library Services, ProQuest Books, La Vergne, Tennessee, U.S.A.

Abstract

An approval plan is an acquisitions method under which a library receives regular shipments of new print books or activation of new ebooks selected by a vendor. The selections are based on a formalized profile of local collection interests, and a library has the right to reject what it decides not to buy. The dealer also provides electronic new title announcements which the library may use to place orders for titles fitting the profile less perfectly. Approval plans were first implemented in the 1960s and following a period of controversy, by the 1990s had become the predominant way larger academic libraries in the United States and Canada acquired new books. By 2010, the rise of ebooks, new acquisitions models, and new financial pressures had caused many libraries to stop or to reduce the size of their approval plans. The underlying profile mechanisms first developed to support approval plans, however, are still widely used for new title announcements and to support "Patron-Driven" or "Demand-Driven" acquisitions programs, as well as to generate new title selection records for use by librarians.

INTRODUCTION

An approval plan is an acquisitions method under which a library receives regular shipments of new print books or activation of new ebooks selected by a vendor. The selections are based on a formalized profile of local collection interests, and a library has the right to reject what it decides not to buy. Approval plans were first implemented in the 1960s, and following a period of controversy, by the 1990s had become the predominant way larger academic libraries in the United States and Canada acquired new books. By 2010, the rise of ebooks, new acquisitions models, and new financial pressures had caused many libraries to stop or to reduce the size of their approval plans. The underlying profile mechanisms first developed to support approval plans, however, are still widely used to support "patron-driven" acquisition (PDA) or "demand-driven" acquisition (DDA) programs, as well as to generate new title selection records for use by librarians.

HISTORY OF APPROVAL PLANS

The forerunners of the modern approval plan were begun after World War II, when mass-buying programs such as the Farmington Plan were organized to enable North American academic libraries to acquire books from areas of the world where war had disrupted the book trade or where buying books was otherwise difficult. Various types of domestic gathering plans or blanket order plans, as they were known, also appeared in the postwar years, when support from the federal government increased book budgets to the point where library staffs strained to spend their money through title-by-title firm orders.

In the early 1960s, Richard Abel, the manager of Portland, Oregon's Reed College Book Store, which had grown an extensive business with academic libraries, began a company dedicated to this market. His insight that vendor advance buying in anticipation of library orders could preempt the orders themselves, if books were shipped to libraries automatically, was the core of the approval plan concept. Within a few years, approval plans were widespread among North American academic libraries, growth enabled by automation of many of the company's processes.

The scale of Abel's operation outpaced his finances, however, and in 1974 the firm failed. To alarmed librarians and publishers, this put the future of the approval plan in doubt. But the venerable U.K. firm Blackwell's company bought the remains of the Richard Abel Company; by then, many libraries had come to depend on approval plans and several competitors offered their own programs. The idea not only survived the Abel demise but approval plans continued to spread.[1–3]

Acceptance did not occur without dissent, however, because many librarians vigorously opposed approval plans. The most important forum for debate was a series of four conferences held from 1968 to 1979, organized by Western Michigan University's Peter Spyers-Duran.[4–7] Some librarians argued against handing over to commercial interests the professional activity of book selection, fearing that lapses of undependable or unstable vendors and omissions by vendors focused on moneymaking would leave damaging gaps in library collections. Vendor concentration on mainstream, profitable books would produce library collections that were too much alike, without the collective richness resulting from local selection in support of local needs.

Encyclopedia of Library and Information Sciences, Fourth Edition DOI: 10.1081/E-EISA-120053689

Vendors were aware, as these critics pointed out, that far more often than not libraries would keep a book shipped on approval, thus putting many marginal titles into collections. At the same time, it was difficult to predict whether a needed title would be shipped at all. A significant monitoring effort would be necessary to prevent gaps, to minimize unwanted titles, to eliminate duplication, and to stay within budget. The expense of this work, and the costly handling of returns, would erase any savings a library might gain with an approval plan.

None of these objections ever disappeared, but arguments in favor of approval plans proved stronger during this era. Efficiencies won from approval plans allowed libraries to reallocate staff to other duties, even to operate with less staff. Discounts, passed on from vendors able to buy from publishers in volume, helped to stretch budgets. The ability to acquire new books soon after publication, while sought by patrons and still safely in print, and to make selection decisions based on the book itself, rather than on reviews or publisher advertisements, were other reasons to establish an approval plan. Librarians administering budgets valued evenness in spending across the year, and librarians overseeing collections liked the subject balance approval plans ensured, even without strong local subject expertise.

Although Spyers-Duran and others debated critics of approval plans, the argument already had been won in the venue that truly mattered, library operations themselves. A survey published in 1977, only 3 years after the Abel Company's failure, found that 79% of respondents had approval plans.[8] In 1979, Oryx Press saw enough demand to publish a handbook, *Practical Approval Plan Management*.[9] A survey of Association of Research Library (ARL) members published in 1982 documented that 85% of these large libraries maintained approval plans, about the same percentage recorded by an earlier survey of ARL libraries in 1969.[10] ARL surveys published in 1988 and 1997 remarked on the stability of the concept through good and bad financial times and found that more than 90% of ARL libraries used approval plans.[11,12]

In 1999, a survey of nearly 300 libraries found high satisfaction with approval plans.[13] Because many of these libraries reported annual approval plan expenditures under $100,000, this survey documented what had long been evident in the literature, that use of approval plans had spread beyond research libraries. In fact, approval plans became nearly as common among medium-sized as among large academic libraries in North America, as well as among law, health sciences, art, and other specialized academic libraries. A study published in 2008 concluded that use of approval plans by college libraries had not declined.[14]

transform approval plans. This was the launch of netLibrary, an ebook platform heavily marketed to academic libraries. Other platforms offered by other companies followed, and the idea that ebooks might be included in approval plans was a natural result. Then, in 2005, Google announced a mass digitization project in cooperation with five research libraries, raising questions about the value of local collections. By then, the out-of-print book market had largely moved to the Web and it was no longer difficult to obtain most print books when and if needed, even long after publication, undermining what had always been an approval plan advantage. Finally, the financial crisis of 2008 overturned assumptions about the stability of library funding for collections.

Academic libraries responded to these changes by experimenting with new arrangements. In 2007, the University of Vermont's library stopped approval shipments for three large publishers. Instead, patrons were invited to request orders for these print books by using an online form resulting from an OPAC search, the library promising 3-day delivery if desired.[15] Nova Southeastern University, in Fort Lauderdale, Florida, reported an ebook approval plan begun in 2009, and in 2010 Texas A&M University reported implementation of an "e-preferred" approval plan.[16,17]

Academic librarians had only sporadically expressed interest about usage of their print collections. But the rise of ebooks and other electronic resources, where usage measures were common, combined with financial pressures after 2008, and with concerns over building space claimed by growing print collections, led many librarians to look critically at approval plans. A study completed in 2009 by librarians at Penn State University and the University of Illinois at Urbana-Champaign raised as a problem the high percentage of unused books acquired on their approval plans.[18] The findings of a Cornell University Library task force published in 2010 did not directly question approval plans, but raised the same concern over books that did not circulate.[19]

These concerns were widely shared, leading to a surge in the use of PDA or DDA programs.[20–23] Often, approval plan profiles were modified to produce MARC records for use as patron selection vehicles, instead of or in addition to the traditional outputs of automatic book shipments and selection slips intended for librarians' use. The practice became widespread quickly enough for the chair of the Cornell 2010 task force to remark in 2011 that a summary was difficult to achieve, and to ask if U.S. academic libraries had not reached a "tipping point" toward the new method of acquiring books.[24]

TRANSFORMATION OF APPROVAL PLANS

In that same year, however, 1999, the first of four major events took place that directly or indirectly would

PROFILES

While many libraries have reduced or eliminated their approval plans, one enduring component of approval plans has been the "profile" recording local collection interests and guiding the vendor in identification of pertinent titles. Vendors have their own systems of documentation, but all work with customers to create and then revise over time a profile that will be at the heart of an approval plan, or a profile devoted entirely or in large part to selection records for librarians, or today, a PDA/DDA program. Often, libraries establish multiple profiles with a vendor, by subject, publisher, library location, print/ebook format, or some combination of these. Libraries may also establish their own combinations of approval plans, librarian selection plans, and DDA programs, all working in parallel.

Profiles have several basic components that can be configured to produce the optimal combination of automatic purchases, library or patron selection records, and exclusions. Subject parameters are recorded by using standard library classification systems supplemented by a vendor's own controlled vocabulary of subject terms. Vendors also designate categories of nonsubject descriptors for dozens of bibliographic categories such as textbook, reprint, country of origin, language, binding, and price. Publisher parameters are established as well. A list of authors important to a library might be included in an approval plan profile, and some vendors make it possible for libraries to incorporate book awards or significant review notices into a profile. Libraries may also embed their own descriptors for budget/fund and location in a profile, so that these local terms for fiscal and selection responsibility are assigned to each title generated by the profile, enabling ongoing analysis and accounting by both the vendor and library.

The advent of ebooks introduced a new set of components into profiles. A preference for ebook or print can be supported, as can an agreed upon number of days a library is willing to wait for their preferred format when it is not the first available. Libraries can also incorporate into their profiles preferences for the ebook platforms offered by different ebook aggregators and publishers, and for the various usage models defined by the license types those companies offer.

VENDORS

Vendors provide an array of services closely associated with their approval plans. Library approval plans often incorporate some level of technical services, where the vendor might provide cataloging records and, for print books, physical processing, which can include the application of spine labels, ownership stamps, security devices, and barcodes, as well as binding services. When physical processing is provided, libraries waive their right of return and no longer truly have an "approval" plan. These libraries, after having achieved a very low rate of return—perhaps 2% or 3% or even lower—calculate that the cost of returns outweighs any savings. Librarians may,

however, already have "approved" shipment of either print or ebooks using their vendor's online interface. These online services are integral to approval plans, providing access to profiles, review of shipments, ordering and selection support, and management reporting. Vendors can also provide title metadata and library holdings information to OCLC's WorldCat (Online Computer Library Center) and to discovery service providers.

Many vendors proved unable to provide the level of support approval plans require, one factor leading to consolidation in the industry. The 1999 acquisition of Academic Book Center by Blackwell's, the 2006 acquisition of Coutts Information Services by Ingram, the 2009 acquisition of Blackwell North America by Baker & Taylor, the 2012 merger of the French vendors Aux Amateurs de Livres and Librairie Internationale Touzot, the 2013 acquisition of the United Kingdom's Blackwell Library Services by Dawson Books, and the respective 2015 acquisitions first of YBP Library Services by EBSCO, and then of Coutts by ProQuest, were the most significant of these changes in the vendor landscape.

While many smaller vendors were also acquired or went out of business, other vendors continue to offer specialized approval plans. Among those providing European books are Casalini Libri, East View Information Services, Harrassowitz, and Puvill Libros.[25] Other vendors include Rittenhouse Book Distributors and Matthews Book Company, for medical books; Theodore Front Musical Literature, for music scores; Worldwide Books, for art publications; and Midwest Library Service, who markets to smaller academic libraries.

CONCLUSION

Approval plans, which began as mechanisms for the mass acquisition of print books, have evolved into an interwoven set of systems for academic libraries and their patrons to discover titles in both print and electronic format, and for libraries to acquire the most locally pertinent in an efficient manner. While questions have been raised about the value of the collections academic libraries built through approval plans, due to the efficiencies they offer approval plans continue in use by many large academic libraries in the United States and Canada. There is no question, finally, that approval plans and the profiles now supporting PDA and DDA programs succeeded in what has always been one central goal, that is, to enable librarians to focus their work on activities other than book selection.

REFERENCES

1. *Papa Abel Remembers: The Tale of a Band of Booksellers*; Smith, S.A., Strauch, K., Eds.; Against the Grain Press: Charleston, SC, 2013.

2. Newlin, L. The rise and fall of Richard Abel and Co., Inc. Sch. Publish. **1975**, *7*(1), 55–61.

3. O'Neill, A. How the Richard Abel Co., Inc., changed the way we work. Libr. Acquis. Pract. Theory **1993**, *17*(1), 41–46.

4. *Approval and Gathering Plans in Academic Libraries*; Spyers-Duran, P., Ed.; Libraries Unlimited: Littleton, CO, 1969.

5. *Advances in Understanding Approval and Gathering Plans in Academic Libraries*; Spyers-Duran, P., Gore, D., Eds.; Western Michigan University: Kalamazoo, MI, 1970.

6. *Economics of Approval Plans*; Spyers-Duran, P., Gore, D., Eds.; Greenwood Press: Westport, CT, 1972.

7. *Shaping Library Collections for the 1980s*; Spyers-Duran, P., Mann, T. Jr., Eds.; Oryx Press: Phoenix, AZ, 1980.

8. McCullough, K.; Posey, E.D.; Pickett, D.C. *Approval Plans and Academic Libraries: An Interpretive Survey; A Neal-Schuman Professional Book*; Oryx Press: Phoenix, AZ, 1977.

9. Cargill, J.S.; Alley, B. *Practical Approval Plan Management*; Oryx Press: Phoenix, AZ, 1979.

10. *Approval Plans in ARL Libraries*; SPEC Kit 83. Association of Research Libraries: Washington, DC, 1982.

11. *Approval Plans*; SPEC Kit 141. Association of Research Libraries: Washington, DC, 1988.

12. *Evolution and Status of Approval Plans*; SPEC Kit 221. Flood, S., Ed.; Association of Research Libraries: Washington, DC, 1997.

13. Brown, L.A.; Forsyth, J.H. The evolving approval plan: how academic libraries evaluate services for vendor selection and performance. Libr. Collect. Acquis. Tech. Serv. **1999**, *23*(3), 231–277.

14. Jacoby, B. Status of approval plans in college libraries. Coll. Res. Libr. **2008**, *69*(3), 227–241.

15. Spitzform, P. Patron-driven acquisition: collecting as if money and space mean something. Against Grain **2014**, *23*(3), 20–24.

16. Buckley, M.; Tritt, D. Ebook approval plans: integration to meet user needs. Comput. Libr. **2011**, *31*(3), 15–18.

17. Pickett, C.; Tabacaru, S.; Harrell, J. E-approval plans in research libraries. Coll. Res. Libr. **2014**, *75*(2), 218–231.

18. Alan, R.; Chrzastowski, T.E.; German, L.; Wiley, L. Approval plan profile assessment in two large ARL libraries: University of Illinois at Urbana-Champaign and Pennsylvania State University. Libr. Resour. Tech. Serv. **2010**, *54*(2), 64–76.

19. Walker, K.; Entlich, R.; Green, G.; et al. Report of the Collection Development Executive Committee Task Force on Print Collection Usage; Cornell University Library: Ithaca, NY, November 22, 2010. Available online atstaffweb.library.cornell.edu/system/files/CollectionUsage TF_ReportFinal11-22-10.pdf (accessed April 2015).

20. *Patron-Driven Acquisitions: history and Best Practices*; Swords, D.A., Ed.; De Gruyter Saur: Berlin, Germany/Boston, MA, 2011.

21. *Patron-Driven Acquisitions: Current Successes and Future Directions*; Nixon, J.M., Freeman, R.S., Ward, S.M., Eds.; Routledge: New York, 2011.

22. Tyler, D.C.; Falci, C.; Melvin, J.C.; Epp, M.; Kreps, A.M. Patron-driven acquisition and circulation at an academic library: interaction effects and circulation performance of print books acquired via librarians' orders, approval plans, and patrons' interlibrary loan requests. Collect. Manage. **2012**, *38*(1), 3–32.

23. Goedeken, E.A.; Lawson, K. The past, present, and future of demand-driven acquisitions in academic libraries. Coll. Res. Libr. **2015**, *76*(2), 205–221.

24. Walker, K. Patron-driven acquisition in U.S. academic research libraries: at the tipping point in 2011? Bibliothek Forschung Praxis **2012**, *36*(1), 126–130.

25. Iorio, M.C. The Casalini Libri approval plan: origins, contexts and future prospects. Ital. J. Libr. Arch. Inf. Sci. **2015**, *6*(1), 121–146.

Approval–Archival

Arab Federation for Libraries and Information (AFLI)

Nour Soufi
Library Cataloging and Metadata Center, University of California, Los Angeles, Los Angeles, California, U.S.A.

Abstract

This entry provides an overview of the founding and the history of the Arab Federation for Libraries and Information (AFLI). The federation's objectives, membership, and organizational structures are discussed as well as its financing and conferences. The Arab Federation for Libraries and Information holds training and workshops in different Arab countries, publishes a newsletter "Sada al Ittihad" and publishes the proceedings of the yearly conferences in cooperation with the Association or the institution in the country where it is held. It also publishes a journal named IALAM twice a year. The Arab Federation for Libraries and Information also encourages and gives awards to the outstanding professionals in the field of library and information science.

INTRODUCTION

In the last 30 years of the twentieth century, many library schools were established at various Arabic universities. As a result, large numbers of students graduated in the specialty and needed jobs at different levels in libraries and in documentation and information centers. This growth in the number of trained professionals lead to the establishment of national level library associations in Arab countries. The oldest of these are the Egyptian Library Association, the Jordanian, and Iraq Library Associations.

From the point of view of library professionals these associations were not adequate to meet a deep-felt need for cooperation among Arab librarians. A broader organization was needed in order for this new generation of professionals to formulate mutual goals and work together for their accomplishment. Several Arab libraries recommended establishing an Arab federation for libraries and for information professionals. For example, the Round Table of the Regional Studies to Improve Libraries in the Arab Countries, which was held in Beirut in 1959, recommended the establishment of national associations as well as a regional association.[1] The Seminar on Library and Information Science, which was held in Baghdad in 1977, recommended that the Iraqi Libraries Association to establish an Arab Libraries Association; then it recommended the same project at the seminar in Tunisia in 1979, but no progress was made in implementing the project at that time. The Jordanian Library Association requested the same project in 1984. The Jordanian Library Association then sent the proposal to the Arab Associations and Institutions involved in libraries and library sciences, and requested that they study it, and give their opinions. This project was revisited as well during the Seminar of Cataloging the Arab Books, which was held under the sponsorship of the Higher Institute of Documentation in Tunisia in 1984. Despite all of these efforts, there were no positive outcomes.[2] The second Arab Seminar held by the Higher Institute of Documentation (in Tunisia from April 5 to 7, 1985), focused on the theme "the benefits drawn from the services of the libraries and the Arabic documentation centers." Participants in the second seminar also stressed the importance of creating of an Association for Arab librarians, and they identified a need to create a directory of professionals and libraries in the Arab world.[3]

FOUNDING

The project remained a dream for librarians, professionals, and information specialists in Arab libraries until January 19, 1986, when the 3rd Arab seminar was held in Qayrawan, Tunisia from January 16 to 20, 1986, where there was a round table to study this important project. Professor Dr. Abdul Jalil al Tamimi, the former director of the Higher Institute of Documentation in Tunisia, and at that time the owner of the Tamimi Institution of the Scientific Research, played a key leadership role during this seminar. In attendance were professionals in library science from Tunisia, Algeria, Saudi Arabia, Sudan, Syria, Iraq, Kuwait, Libya, Egypt, and Yemen. These professionals drafted some general policies during a series of four meetings and officially founded what they called "the Arab Federation of the Arab Librarians and the Specialists in Information." This name was later changed to "the Arab Federation for Libraries and Information (AFLI)." The first elected Executive Board included seven members; the chairperson was Dr. Wahid Qdoura, Professor at the Higher Institute of Documentation in Tunisia. However, the key leader who struggled to create the Arab

Encyclopedia of Library and Information Sciences, Fourth Edition DOI: 10.1081/E-ELIS4-120043622

Federation was Dr. Abdul Jalil Al Tamimi formerly director of the same institute in Tunisia.

Beginning in 1986, the AFLI began to be active under the leadership of this Institute, which fully supported it. After a few years, the Tamimi Institution of the Scientific Research took over the responsibility of leading AFLI. Step by step, AFLI developed successfully due to the perseverance of information professionals and librarians. Its crucial role is to advocate for the development of libraries and to increase public awareness of the impact of libraries on cultural development in general.[1]

OBJECTIVES

Like every federation, association, and institution, AFLI has its own objectives.

These objectives are to:

1. Strengthen cooperation among the societies and institutions in the Arab world.
2. Introduce and preserve the audiovisual and written Arab heritage everywhere.
3. Support all efforts to promote the profession and librarianship.
4. Conduct and encourage scientific research and studies in the field of libraries and information science, as well as hold specialized seminars, symposiums, and conferences.
5. Endeavor to improve the standard of preservice and in-service training institutes for librarians and specialists in information.
6. Work on the unification of information and library terminology.
7. Attempt to issue statutes and regulations concerning libraries and information institutions.
8. Participate in the publication of specialized guides and in the development of the profession.
9. Encourage the establishment of national societies of libraries and information specialists in those Arab countries where such societies have not yet been established.
10. Issue a specialized professional periodical as the organ of the federation.
11. Cooperate with those international and Arab organizations concerned with the same objectives of the federation.[4]

MEMBERSHIP

There are four categories of membership in AFLI:

1. Category A includes library and information associations in the Arab countries.

2. Category B includes libraries, departments, and institutes of library and information science.
3. Category C includes qualified librarians or those who possess experience in the field of libraries and information for no fewer than 5 years.
4. Category D includes the honorary members who have made outstanding contributions to the profession and to AFLI.[5]

ORGANIZATIONAL STRUCTURES

The Arab Federation for Libraries and Information organization consists of the following:

1. General Assembly
   The General Assembly is the highest committee in the federation. It includes all the members of AFLI. It has the authority to assign general policy, and to define the program of its activities by considering suggestions which have been made by the executive board and members. It also decides the federation budget, and it examines and makes appropriate decisions on the following matters:
   - "The reports and recommendations submitted by the committees on the duties assigned to them by the executive board.
   - "The reports of the secretary general and the treasurer on the federation activities during the general assembly's terms.
   - "Matters related to the field of librarianship and all items in the agenda.
   - "Suggestions on amendments to the statute and executive acts.
   - "Elects the federation chairman and members of the executive board.
   - "Approves the federation budget, and defines the annual subscription fees."[5]
2. The Executive Board
   The general assembly elects an executive board every 3 years; it consists of seven members, divided into five groups as follows: Chairman, Vice Chairman, Secretary General, Treasurer, and three additional members. The executive Board has seven essential functions, which are:
   - Prepare the agenda of the general assembly, as well as the federation budget and the program of its activities.
   - Follow up on implementation of the program approved by the general assembly.
   - Transfer some of its powers as appropriate, to the chairman of any of its members.
   - Form the committees that it deems necessary to be responsible for the implementation of its functions.

- Invite individuals or representatives of institutions to attend its meetings as consultants, if it considers this necessary for the undertaking of its functions.
- Consider applications for membership of the federation.
- The chairman, or any member of the execution board assigned by the chairman, is authorized to sign contracts or bonds for selling or buying properties, or for paying rent.[5]

Through 2007 there have been eight executive boards, chaired by:

- Temporary executive board: Chairman: Prof. Dr. Shaaban Khalifah (Egypt).
- 1st executive board: Chairman: Prof. Dr. Wahid Qdoura (Tunisia).
- 2nd executive board: Chairman: Prof. Dr. Wahid Qdoura (Tunisia).
- 3rd executive board: Chairman: Prof. Dr. Abdul Jalil al Tamimi (Tunisia).
- 4th executive board: Chairman: Prof. Dr. Abdul Jalil al Tamimi (Tunisia).
- 5th executive board: Chairman: Prof. Dr. Wahid Qdoura (Tunisia).[1]
- 6th executive board: Chairman: Prof. Dr. Mabruka Mhiriq (Libya).
- 7th executive board: Chairman: Dr. Saad al Zahri (Saudi Arabia).[6]
- 8th executive board: Chairman: Prof. Dr. Hassan A. Alsereihy (Saudi Arabia).[6]

3. The General Secretariat

The General Secretariat works on editorial assignments, and implements the General Assembly and Executive Board's decisions.

4. The Working Committees

Working committees are formed as needed. For example, standing working committees have been formed for bibliography, budget, knowledge production, training, terminologies standards, and international relations.

FINANCE

The Arab Federation for Libraries and Information obtains its financial resources and support from the subscriptions of members; from any profits of its conferences, services, and activities; and from the donations, grants, and other income. Financial decisions must be approved by the Executive Board.[5]

LOCATION

The permanent location of AFLI is in Tunisia at the Tamimi Institution of the Scientific Research. This foundation was established in 1989, by the greatly appreciated efforts of Professor Dr. Abdul Jalil al Tamimi, in Zaghwan, a city in Tunisia about 55 km from the capital of Tunis. The Tamimi foundation building consists of a library that contains more than 18,000 titles. The Foundation also produces three journals, the *Magazine of Maghrebi History*, the *Arab Magazine of Archives, Documentation and Information* and the *Arab Historical Review for Ottoman Studies* (AHROS).[7]

CONFERENCES

The Arab Federation for Libraries and Information organizes yearly scientific seminars and conferences. The first six seminars were held in Tunisia. Dr. Abdullatif Soufi (former professor in Library Science at the University of Constantine in Algeria) then suggested that AFLI start holding its conferences in different Arab countries, and the organization began doing so starting with its seventh seminar. This was an important challenge and step that led to AFLI being known to a wider range of librarians and professionals in all Arab countries. From that point onward, membership tripled, until it grew to about 450 attendees per conference. These seminars and conferences were held as follows:[1]

- 1st Arab Seminar: held in Hamamat (Tunisia) from February 23 to 26, 1988. Titled "Indexing and classification in the Arab information centers."
- 2nd Arab Seminar: held in the capital of Tunisia from January 18 to 21, 1989. Titled "Information and telecommunications technology in the Arab world: the challenges of the future."
- 3rd Arab Seminar: held in Zaghwan (Tunisia) from October 20 to 23, 1991. Titled "Information in the development service in the Arab world."
- 4th Arab Seminar: held in Tunisia from December 4 to 6, 1993. Titled "University libraries under the service of scientific research and education in the Arab world."
- 5th Arab Seminar: held in Zaghwan (Tunisia) from October 21 to 23, 1994. Titled "the position of studies in libraries and information science in our Arab world: perspectives of the future."
- 6th Arab Seminar: held in Tunisia from October 24 to 26, 1995. Titled "National and public libraries, their roles in the elaboration of national systems of information."
- 7th Arab Conference: held in Amman (Jordan) from November 2 to 6, 1996. Titled "Edition and bibliographic control in Arab scholarly publishing."
- 8th Arab Conference: held in Cairo (Egypt) from November 1 to 4, 1997. Titled "Information technologies in the libraries and information centers in Arab countries: current and future trends."

- 9th Arab Conference: held in Damascus (Syria) from October 21 to 26, 1998. Titled "The unified Arab strategies of information in the era of telecommunications and other studies."
- 10th Arab Conference: held in Nabeul (Tunisia) from October 8 to 12, 1999. Titled "Electronic publishing and information services in Arab world: perspectives on future trends."
- 11th Arab conference: held in Cairo (Egypt) from August 12 to 16, 2000. Titled "Towards building a strategy to enter Arabic printed literature into cyber space."
- 12th Arab conference: held in Sharjah (United Arab Emirates) from November 5 to 8, 2001. Titled "Arab libraries in the third millennium: infrastructure, technology, and manpower capabilities."
- 13th Arab Conference: held in Beirut (Lebanon) from October 29 to November 1, 2002. Titled "Electronic information directory: knowledge and efficiency."
- 14th Arab Conference: held in Tripoli (Libya) from December 14 to 18, 2003. Titled "Knowledge engineering in Arab world."[8] During 2004 to 2005, AFLI had some problems concerning changing the location of AFLI from Tunisia, where AFLI was born, to Cairo. Because of these problems, AFLI held two conferences: the first in 2004 in Egypt and the second in Tunisia in early 2005. After this, the conflicts were resolved.
- 15th Arab Conference: held in Alexandria (Egypt) from December 27 to 30, 2004. Titled "Arab libraries and the educational development in a challenging world."[9]
- 16th Arab Conference: held in Hammamat (Tunisia) from March 2 to 5, 2005. Titled "Libraries and documentary information services: contribution to the emergence of the knowledge society."[10]
- 17th Arab Conference: held in Algiers (Algeria) from March 19 to 22, 2006. Titled "Access to information for all: partnership between librarian and archivists."[11]
- 18th Arab Conference: held in Jeddah (Saudi Arabia) from November 17 to 20, 2007. Titled "Librarianship: current challenges and the future and its role in the free access to scientific information."
- 19th Arab Conference: held in Cairo (Egypt) from November 24 to 26, 2008. Titled "Cooperation among Arabic Information Institutions in the digital era."
- 20th Arab Conference: will be held in Casablanca (Morocco) from December 9 to 11, 2009. Titled "Towards a new generation of Information systems and Information specialists: a future look."

TRAINING AND WORKSHOPS

From time to time, AFLI organizes workshops and continuing education for librarians, and also sponsors workshops in cooperation with the national library associations and information centers (like the Arabian Advanced Systems), in order to develop librarians' knowledge and capabilities in the use of information technologies, and the use of new systems. Any member from any Arab country who wishes to learn more about these technologies may attend. The Arab Federation for Libraries and Information also cooperates with the International Federation for Libraries and Information.[12]

AFLI PUBLICATIONS

The Arab Federation for Libraries and Information cooperates with other Arab institutions, like libraries, universities, and other federations, in regards to publishing the proceedings of the yearly conferences. These publications include abstracts and papers which were presented at the conferences. The proceedings are distributed free of charge to members who present papers at the conferences. Others may purchase a copy. The Arab Federation for Libraries and Information also publishes a periodical, "Sada Al-Ittihad" that reports on AFLI's activities during the current year, provides information on its members accomplishments, and lists current research in the field, as well as published books, conferences, and seminars. The Arab Federation for Libraries and Information also provides additional activities at every conference, such as book exhibitions, new technology exhibits, etc.[1]

RELATIONSHIPS AND COOPERATION

The Arab Federation for Libraries and Information maintains relationships with other library associations both in Arab countries and internationally. It organizes its conferences in cooperation with national library associations (depending on the country where the conference is being held), such as the Egyptian Library Association, Jordanian Library Association, and with Arab universities and with other organizations such as The League of Arab States, The Arab Club of Information in Damascus, etc. The Arab Federation for Libraries and Information tries to support these relationships by setting the yearly conferences in different Arab countries, in order to allow the largest number of librarians in each country to contribute to and attend its conferences.[2]

HONORS AND AWARDS

Beginning in 2000, AFLI started to give awards to its outstanding members, who had contributed with distinction to excellence in the field of Library and Information Science, either through scholarly contributions or by working on committees. Until now, AFLI has presented four outstanding member awards, as follows:

1. Prof. Dr. Shaaban Abdelaziz Khalifah (Egypt), Professor in the Egyptian universities, received the award at the tenth AFLI conference in 1999 in Tunisia.[1]
2. Prof. Dr. Abdullatif Soufi (Algeria), Professor in Library Science in the University of Mentouri Constantine, Algeria, received the award at the 13th AFLI conference in 2002 in Beirut.
3. Prof. Dr. Abduljalil Al-tamimi (Tunisia), Owner of the Tamimi Institution of scientific research, and former president of AFLI, received the award at the 14th AFLI conference in 2005 in Libya.
4. Prof. Dr. Wahid Qdoura (Tunisia), Professor at the Higher Institute of Documentation (Tunisia), received the award at the 16th AFLI conference in 2006 in Algeria.[6]
5. Prof. Dr. Mohamed Fathi Abdulhadi (Egypt), Professor at the Institute of Library and Information Sciences (Cairo University), received the award at the 18th AFLI conference in 2007 in Jeddah (Saudi Arabia).
6. Prof. Dr. Abubakr El Houshe and Prof. Dr. Mabruka Mhiriq (Libya), Professors at the Institute of Library Science (Libya), received the award at the 19th conference in 2008 in Egypt.

CONCLUSION

The Arab Federation for Libraries and Information, is the only such federation in the Arab countries. It brings together professionals and specialists in the Arab libraries and information institutions for purposes of communication and provides a forum for discussing the current situation and the future trends of their profession. It supports cooperation and encourages research. The Arab Federation for Libraries and Information supports professional development and assists in the training of librarians in the Arab universities and institutions. It also works on codifying terminologies and encourages formation of national library associations. In addition, AFLI strengthens the relationships between these national library associations and international associations in the profession. Finally, AFLI continues to hold its yearly conferences in different Arab countries, and grant honors and awards to outstanding professionals.

REFERENCES

1. Soufi, N. *Arab Federation for Libraries and Information (AFLI): Current Aspects and Future Trends*, Constantine, Algeria, 2002. B.A. Thesis presented to the Library and Information Science Department.
2. *Arabic Encyclopedia in Library and Information Science*, 2nd Ed.; Shaaban Abd al-Aziz, K., Ed.; Dar al Masriyah al-Lubnaniyah: Lebanon, 1999.
3. Soufi, A. *Glimpses in the History of Books and Libraries*, Tlas Publisher: Damascus, Syria, 1987.
4. AFLI, Edition and bibliographic control to the Arab intellectual production Proceedings of the Seventh Arabic Conference Jordan November, 2–6, 1996 AFLI: Jordan, 1997.
5. http://213.150.161.217/website_afli/index.htm.
6. http://www.afli.cybrarians.info/state.htm.
7. Tamimi, A.J. *Tamimi Institution of the Scientific Research*, Papyrus publication: Tunisia, 1997; 2.
8. http://213.150.161.217/website_afli/previous_conference_14.htm.
9. http://www.librariannet.com/mo2tamrat/27_12_2004.asp.
10. http://213.150.161.217/website_afli/previous_conference.htm.
11. http://213.150.161.217/website_afli/main.htm.
12. AFLI, *Sada al Ittihad*, AFLI Newsl, December 2005.

Archival Appraisal and Acquisition

Barbara Reed
Recordkeeping Innovation, Sydney, New South Wales, Australia

Abstract

Archival appraisal is the analytic process involved in determining how long records are required, thus driving acquisition strategies, and determining the archival record left for the future. Developing as a major archival function during the twentieth century, appraisal has been subject to intensive professional discussion as archivists search for theoretically sound approaches and techniques. The advent of the digital environment has created great challenges for appraisal theory and practice. This entry outlines the purpose and principles underlying appraisal, discusses the development of traditional appraisal practices, identifies professional concerns with traditional appraisal practices, and discusses methodologies and approaches proposed as alternatives to traditional appraisal methods to address digital recordkeeping challenges.

INTRODUCTION

Appraisal is the set of recordkeeping processes involved with determining how long to keep records, and what ultimately will become the archival representation for the future. Appraisal is possibly the most contested of archival functions and the most discussed in the professional literature. Definitions of what appraisal encompasses and methods of achieving reliable and replicable appraisal outcomes are in discussion.

Perhaps most strongly associated with the public (or government) record tradition, appraisal also has a manifestation in acquisition strategies employed by collecting archives, that is, those archives that target specific records for acquisition into focused collections held in a dedicated repository maintained independently by a third party as distinct from the record-creating body/person. These traditions are not the same, one focusing on a selection process, a winnowing of the extant records into a kernel for continuing preservation, and the other focusing on identifying records which are of sufficient significance to justify expenditure of a third party entity collecting and therefore implicitly undertaking the expense and responsibility for continuing preservation of the records. While there is a relationship, one is a selection process and one is a collection process. These processes therefore have different techniques associated with appraisal.

In recent years, the advent of digital recordkeeping has challenged recordkeeping processes in general, and appraisal techniques developed to suit the paper-based world have been subject to extensive discussion. The paper-based, reactive, and physical examination-dependent techniques of the past are widely acknowledged to be unscalable to meet the challenge presented by the digital environment. The struggle to rearticulate appraisal techniques and develop an appraisal theory that is coherent, timely, and effective in the digital environment, is an ongoing process.

This entry

- Outlines the purpose and principles underlying appraisal.
- Discusses the development of traditional appraisal practices.
- Identifies professional concerns with traditional appraisal practices.
- Discusses methodologies and approaches proposed as alternatives to traditional appraisal methods to address digital recordkeeping challenges.

WHAT IS APPRAISAL?

Traditionally, archivists define appraisal as the intellectual process of determining the value of records. By determining the value of records, it is then possible to arrange for their orderly disposal/disposition by organizing their transfer to a secondary storage facility or to an archival institution for destruction or retention as appropriate. Usually formal documentation is devised, following the evaluative process, which lays out the classes of records requiring similar disposal action. It is, in simple terms, the process which determines which records should be kept and which should be destroyed. It is therefore the most important archival activity around which all other tasks revolve as it determines what is acquired by an archival institution.

Appraisal evaluates records in their recordkeeping, organizational, and wider social context; in terms of their origin, their functional, evidential and research

Encyclopedia of Library and Information Sciences, Fourth Edition DOI: 10.1081/E-ELIS4-120044327

characteristics; their relationship with other records; their age, volume, and physical format and the long-term cost of preserving them in appropriate archival storage or their destruction once they have fulfilled their administrative and legal requirements. Traditionally, appraisal occurred when records were no longer of current administrative use.

APPRAISAL TRADITIONS

Prior to the mid-twentieth century, appraisal, or the determination of which records should be retained as archives, largely evolved in practice. There was little articulated theory of appraisal, but rather codification of specific practices. The practices evolved dependent on the nature of the archival system in place.

European Traditions

In European archival theory, appraisal was initially the responsibility of the record-creating body. The identification of which documents were superseded by others or which were destroyed was embedded in procedures undertaken as routine by record-creating bodies. In more modern times categories of records able to be destroyed were formalized by regulations devised by Committees established for this purpose. Within this tradition, some forms were seen as requiring more secure protection and were removed for safe keeping outside the records creating body. From the 1920s European archivists gradually established the principle that they should be consulted before any records were destroyed.[1] Archivists were involved in determining what should be retained once the material was no longer needed for administrative purposes.

The European recordkeeping tradition argues that the presence of records in the archives repositories creates a web of relationships between the records which in itself constitutes meaning. The archivists' role was not to select, but to protect these records and their context. Archivists actively involved in selecting within these designated records risk destroying the archival bond and alienating the characteristics of authenticity and reliability.[2] This was incorporated into English-speaking traditions through the work of Sir Hilary Jenkinson in his *A Manual of Archival Administration*, first published in 1922. Jenkinson stressed the impartiality of the archivist and physical and moral duties to the record. He warns against active involvement of archivists in destruction which can equate to the archivist and/or historian having a role in records creation and at risk of imposing a personal judgment of value. Such actions negate the principle of impartiality of the archives and involves the archivist in the contested territory of determining the historical record. Jenkinson's view places the appraisal processes including disposal decisions with the individual agency/office of origin.[3]

Guidance on appraisal criteria was established as early as 1901 within the German tradition, encompassing principles such as respect for age, that records created for temporary purposes generally should be able to be destroyed, that the nature of the administration (both function and structure) should determine the relative importance of records for continuing retention.[4]

Further into the twentieth century, archivists began to be more actively involved in setting appraisal standards. The efforts of Prussian archivists to reject the laissez-faire *fingerspitzengefuhl* (finger in the wind) methods of determining appraisal were reported in 1937,[5] resulting in more defined appraisal criteria based on an analysis of the relative functions and positioning of the creating body.

A more function/structure approach was developed to combat subjectivity in appraisal decision making. The Prussian archivists were concerned primarily with the "quality" of the record for historical research. In 1926, Karl Otto Muller proposed a division of records according to administrative level—central, intermediate, and local authorities—with the proposition that central agencies of government would create more valuable records than local ones. This subsequently developed into the 1950s proposition that rather than looking at the record level, the agency as a whole could be defined as being "more or less worthy of permanent preservation" and that agencies could be ranked according to their importance. The administrative level of the agency and the degree of independence in decision making were factors in the ranking. Georg Wilhelm Sante:

> Archivists must make their selection with a view to the function of the departments and the significance attached to that function. They must begin by analyzing the functions of the individual agencies, and only thereafter can the records produced by these agencies be appraised.
> —(1956–1957). Quoted in Booms.[6]

By 1957 in Germany there was a call to "cease the practice of simply disposing of valueless records and to adopt instead the principle of selecting valuable records." This interesting concept reverberates through a number of traditions of appraisal. Ole Kolsrud in discussing the 1981 Wilson Report reviewing the Public Record Office U.K.'s appraisal system writes "it suggested that the time had come to concentrate more on preservation than destruction" and similarly in Australia, "The Australian Archives records disposal policy in 1979 shows a shift in emphasis from the need to justify the destruction of records to a requirement involving the critical appraisal of material to justify permanent retention" in Dianne Easter "Records Disposal in the Australian Archives" *Archives Conference 1979. Papers Presented to the Second Biennial Conference of the Australian Society of Archivists, Sydney May 18–20, 1979*, Supplementary Volume, p. 37. The German practice also strongly argued that records could not be appraised in isolation but must be placed in their administrative context.[7] These ideas were not particularly well translated into practice at the time. However these ideas clearly influenced Schellenberg, through the translation of the lectures and papers of Adolf Brenneke, cited in the Introduction to *Modern Archives*. These ideas similarly had an influence on the work of Hans Booms in the 1970s and indirectly through Booms on Terry Cook in the 1990s.

In Britain, the demands for paper salvage efforts associated with World War II led to the first articulation of appraisal principles in relation to government records, stressing again that those responsible for the conduct of the business should decide which records should be preserved for business purposes, and that an additional set of criteria would be added to that to serve the requirements of historical or general uses. The historical or general uses were to show history of the organization concerned, to answer technical questions regarding its operations and to meet possible scholarly needs for the information that is incidentally or accidentally contained in the records.[7]

Post World War II, a huge proliferation of records occurred as a result of the rapid expansion of government functions, the diverse scope of their activities, and the ability of machines to duplicate records. Large-scale destruction of records became inevitable. The costs of preserving more than a small fraction of the records produced was prohibitive.[8]

American Traditions

T.R. Schellenberg codified the American practice in 1956 in his highly influential book *Modern Archives*. This work built on practices already in place within the (then) National Archives and Records Service, articulating criteria for appraisal work based on the distinction between "primary" (or values to the originating agency) and "secondary" (values for other agencies and private users). Records having secondary values were broadly categorized as having "evidential" and "informational" value. Schellenberg defines these categories quite broadly. Evidential value does not reference the Jenkinsonian notion of the sanctity of evidence in archives that is derived from unbroken custody, nor of the value that inheres in public records because of the merits of the evidence they contain. Rather, evidential value is the value that depends on the importance of the matter evidenced, that is the organization and functioning of the agency that produced the records. Informational value is "the information they [records] contain on persons, corporate bodies, problems, conditions, and the like, with which the government body dealt." Schellenberg's text went on to consolidate what constituted examples of both of these value criteria.

Following Schellenberg, much of the writing on appraisal during the last half of the twentieth century focused on developing criteria through which to assess value. Taxonomies of values were developed: evidential, financial, administrative, research values are still the most common means of evaluating records in practice. Appraisal criteria based on these taxonomies of value became the norm, and for many archivists today, continue to be the basis for appraisal decision making.

APPRAISAL TOOLS

Collecting or Acquisition Policies

During the 1980s, particularly in America, collecting archives borrowed from the library tradition of collection development.[9] A response to limited and contracting resources and to an expanding universe of responsibility saw collecting archives attempt to rationalize and define boundaries for their collecting activity.

The collecting policy, or acquisition policy, establishes the scope of the collecting remit—defining what will be accepted and under what conditions. They enable decision making against an established framework to determine whether specific material identified or offered to a collecting archives will be accepted for continuing preservation as archives.

Collecting policies typically cover

- Information about the legal status of the repository and its authority to collect.
- Information about scope or limitations of the policy (e.g., geographical area, subject name, chronological period or genre/media).
- Information about the process of collection, such as methods of acquisition—donations, loans, and conditions associated with ownership.
- Information about the rights to dispose or deaccession material.
- Information about access to the material.[10]

Initially a tool used by collecting archives, by the late 1990s many government archives had adopted this tool to broadly outline the rationale for decision making on what records would be accepted into archival custody.

Disposal/Retention Schedules

Early twentieth century appraisal practice in European countries favored the development of lists of records proposed for destruction. These lists were examined by archivists and authorized prior to destruction of records. As archival practices became more codified, archival institutions were increasingly granted authority over the appraisal process or the business of determining which records to keep through archival legislation which generally provided the mandate for their responsibility to determine which records would be selected as archives.

As a development from the simple listing and authorizing of specific records for destruction, formalized disposal/retention schedules were developed. These tools were issued by an archival authority to enable the ongoing authorization of destruction or transfer to archival storage of records nominated within the schedule. Particularly associated with the work of the U.S. National Archives and Records Administration (and its predecessors), the technique of issuing disposal authorities is still the most common form of records disposal/disposition authorization.

Disposal/retention schedules list categories of records which are assigned a retention span based on a time period. Categories in such schedules which share a retention period are known as disposal classes. Disposal authorities can be specific to one agency, or part of an agency, or may cover records that are common/general to many agencies. They provide, for the period of their currency, a continuing authority for agencies to destroy records which fall within the disposal classes, without any further authorization from the authorizing body.

Responsibility for preparing disposal/retention schedules was variously placed with archival authorities or with records staff in agencies. Compiling disposal schedules embodies appraisal decision making. The process of developing a disposal schedule typically involves identification of the records to be covered, an analysis of the related function/process/transaction, an analysis of recordkeeping systems and information flow, compilation of information about the records being targeted (including volume, system of arrangement, physical condition, frequency of use etc.), development of broad classes of records sharing a retention period, consideration of appraisal criteria and, more recently, justification of the retention period proposed against mandates such as operating legislation and internal and external stakeholder requirements.

Generally agencies are responsible for the determination of the primary value (i.e., the values for the originating agency itself) and archivists in an archival institution are responsible for the determination of secondary values (i.e., values for other agencies and private users).

Systemic Reviews

Practice in the U.K. post-Jenkinson era modified his position on appraisal as reflected in the practices of the Public Record Office (PRO). Following the recommendations of The Grigg Report of 1954,[11] the PRO implemented a system of regulated life cycle reviews which comprised the basis of their practices from the 1950s to the 1990s. The practice was based on a network of departmental record officers responsible for working sequentially with records creators and Public Record Office Inspecting Officers to undertake a series of reviews at different points of time each with a predetermined point of view.

At the point when records were no longer required for current business, they were moved to remote storage. Five to seven years after their movement to remote storage they were reviewed by administrators 'to determine whether the organization needs the records for its own administrative purposes ... interpreted broadly to encompass issues such as continued accountability.' While originally there was an assumption that records would have no research value, modifications to the methodology over time introduced a consideration of research or informational value. Once the first review was completed, the remaining records continued to be stored in remote storage until 25 years after the creation of the record. A second review by Departmental Officers accompanied by a Public Record Office Inspecting Officer was undertaken at this time on a file by file basis with the primary consideration the research value of the records. The records determined to possess this value were transferred to the Public Record Office.[12]

Sampling

Case-based records, or particular instance papers as they were known in the British tradition, caused problems for archivists with their voluminous nature and their basically repetitive structure—for example, taxation files, personnel files, or individuals' medical files, in which each document records the same basic process conducted for many individual instances with differing circumstances reflecting individual experiences. The technique of sampling such series became the accepted practice. Sampling could be either random (using genuine statistical methods), illustrative (selecting cases that illustrated the business process), chronological (selecting the complete records of a given period of time), or sequential (selecting every 10th case).[13]

Implementing appraisal decisions, whether embodied in a collecting policy, in disposal/retention schedules or in sampling plans, involved a physical process of examining records and determining whether or not they and their history of compilation were such that they fitted into specific categories. Depending on which categories, the records would be designated either archives and destined for retention in archives or allocated a retention period after which they could be (with appropriate documentation) destroyed.

UNEASE WITH APPRAISAL PRACTICE

Appraisal has always been a contested process. Described as evaluation, it involves application of individual judgment, which, even when bounded by rules or criteria, will always be subject to differing opinions. Using the same rules and processes, will two appraisers always reach the same conclusions on value?

Lack of documentation, or inadequate documentation, of appraisal decision making in the past, and records taken into archival custody in last minute emergency rescues have all raised the controversial specter of reappraisal.[14] When, in what circumstances and how should records determined to be archives be deaccessioned?

The techniques of implementing appraisal as articulated in these methodologies was physically focused— records were examined and evaluated against the appraisal criteria, collection policy, or retention/disposal schedule. The determination of value came after the record was finished—at the end of its administrative usefulness. By the mid-1980s the flood of material requiring coordinated appraisal swamped the capacity of any archival organization or collecting body to deal with the physical universe within its jurisdiction or collecting domain. As David Bearman put it, archivists were failing in their mission using these methods by orders of magnitude,[15] and this assessment was made before the tsunami of digital records, which are unable to be physically inspected, had much impact on mainstream professional practice.

The results of archival appraisal were implicitly recognized to be the creation of a "representative" reflection of our society. But representative of what? Whose view was to be represented? Was it to be representative of all recorded memory, of activities or members of society, or of those aspects determined at a particular time to be important to social activity?

At much the same time, the growth of social history was focusing much historical scholarly activity on the experience of individuals, reflected in the archival records amid the case files or particular instance papers. Traditional archival appraisal practices using samples to deal with this type of record were being questioned.

From the archival institutions themselves came concerns with the foundation of appraisal on taxonomies of values, approaching appraisal from the bottom up and being overwhelmed by the vastness of the records to appraise and being unable to formulate methods which will work with the digital environment of the modern office.

Perhaps most of all, these unsettling questions that arose during the last decades of the twentieth century pointed to the lack of capacity to evaluate the success or otherwise of the appraisal decision-making processes. To what extent were archivists accountable for their appraisal decision making and how did they report on and justify their appraisal practices?

These questions arose from professional practice at the same time as the postmodernism philosophy was extending into humanities disciplines provoking a reconsideration of the reading of texts. The postmodernist discourse also raised a greater awareness of the real and symbolic exercise of social power in decision making about the representation of the past.

ALTERNATIVE APPROACHES TO APPRAISAL

Documenting Society

Inherent in the notion of evaluation of records is the notion of framing or shaping the form that the archival record will take for future generations. The evaluation appraisal criteria introduced by Schellenberg contrasts with the more passive notions of accumulation or elimination of the valueless which is the characteristic advocated by the European and Jenkinsonian approach. But the subjectivity that this approach introduces requires models for active documenting and these models are generally lacking. There is however, a history of attempts to address this issue, and the influence of the collecting traditions become particularly important in this regard.

Hans Booms, writing in German in 1971 (translated into the English professional literature in 1987), rejects the notion that administrative structure and positioning is the appropriate basis for appraisal. He argues for a much more socially situated approach to "the formation of a documentary heritage" focusing on the content of the records to reflect social processes. This requires a comprehensive view of the total societal development process and an interpretation of the way in which society has actually developed and forms a documentation model from which the archivist forms the documentary heritage.

East German archivists were actively experimenting with appraisal during the 1970s and beyond. Their practice was based on the proposition that "the value of archives is determined by the social importance of the events, activities and subjects it refers to"[16] and that

> documents got their value through their importance for fulfilling of the manifold tasks a socialist society sets itself for carrying through the historic mission of the working class. Furthermore it was declared that the function and the place of an administrative body defines essentially the information potential and relevance of its documents ... and thereby their value.

East German attempts to implement Booms' ideas during the 1980s resulted in the development of a list of some 500 events which ought to be documented. Angelika Menne Haritz warns of the problem in this approach which "fits the records into a politically desirable image of history,"[17] references great difficulty in implementation and warns

> As important decisions are delegated to authorities outside the profession, archivists are reduced merely to executing guidelines that we cannot investigate, even if they cause us to act as instruments for political purposes we would not support as individuals.[17]

Documentation Strategies

Within the collecting archives domain, the concept of documentation strategies was articulated initially by Helen Samuels in the mid-1980s. This appraisal strategy articulates active and targeted documenting according to a collaborative cross institutional strategy. In essence a documentation strategy

> is a plan formulated to assure the documentation of an ongoing issue, activity, or geographic area. The strategy is ordinarily designed, promoted, and in part implemented by an ongoing mechanism involving records creators, administrators (including archivists), and users. The documentation strategy is carried out through the mutual efforts of many institutions and individuals influencing both the creation of the records and the archival retention of a portion of them. The strategy is refined in response to changing conditions and viewpoints.[18]

Reports on one of the projects which attempted to faithfully follow the theoretical proposal concluded that the methodology is "to supplement rather than replace the traditional methods of archival appraisal."[19] The idea of the documentation strategy is in the tradition of the documentation plan articulated by Hans Booms and also has echoes in the collaborative collecting of library institutions. Now seen as something of an unachievable aim, the concepts have had some influence, most notably in focusing attention on under represented areas of collection and cross institutional cooperation.

Macro Appraisal

Influenced by Hans Booms, and the writings of sociologists such as Anthony Giddens, Terry Cook has articulated the concept of macro appraisal, the first coherent appraisal theory. Macro appraisal attempts to capture the significant activities of government and its interaction with the society in which it exists. That is, it attempts a more holistic and strategic approach to appraisal of government records. Macro appraisal adds a layer of appraisal methodology that sits above the appraisal of individual agencies.[20] Archivists

> seek to understand why records were created rather than what they contain, how they were created and used by their original users rather than how they might be used in the future, and what formal functions and mandates of the creator they supported rather than what internal structure or physical characteristics they may or may not have.[21]

The theoretical focus of macro appraisal is the

> "image" of society—that is to say, not on the objective reality of society per se, which can never be known absolutely, but rather on the mechanisms or loci in society

where the citizen interacts with the state to produce the sharpest and clearest insights into societal dynamics and issues .... It is at these points of sharpest interaction of the structure, function, and client that the best documentary evidence will be found.[21]

Essentially the focus is on "documenting the process of governance rather than of governments governing and corporations operating."

The critical feature of macro appraisal is the need for detailed analytic work in advance of consideration of the records of an agency. It is a "top-down" approach that provides a solid and justifiable framework for decision making about records of individual functions. It enables prioritization of scarce resources to focus on those functions of government identified which require the most intensive documentation into the future as part of the archives. It is a media-neutral strategy, equally applicable to records in all formats, although implementation strategies for various media may need different approaches. It removes the reliance on the physical examination, moving appraisal to an intellectual process made within a stated framework.

Macro appraisal theory has influenced practice internationally,[22–24] and can be seen as a part of a broader professional refocusing on the notions of provenance as a core archival tenet.

Minnesota Method—Functional Analysis for the Collecting Domain

The notion of functional analysis moved further into discussions of acquisition strategies for collecting archives with the development of the "Minnesota method," a combination of collection analysis, documentation strategy, appraisal, and functional analysis.[25] Originally articulated in the late 1990s,[26] this approach developed to address the reality that collection or acquisition policies for most collecting archives were unachievably large. This approach involved defining and analyzing the particular area in the collecting policy, determining the documentary universe for that area, defining criteria for prioritizing and applying the criteria, defining the functions most appropriate to the collecting area and the level of documentation to support those functions.

Functions-Based Appraisal

If macro appraisal with its inherent requirement for a formalized and well-defined analytic framework established in advance of any appraisal decision making was too grand an undertaking for many archival institutions, the revitalized emphasis on functions as the basis of appraisal practice was not.

The integration of functions as a core recordkeeping tenet was reinforced in the International Standard on

Records Management, published in 2001. Function and activity is defined as the basis for both records classification in current records systems and as the basis for decision making on retention or destruction periods. The explicit linking of the two recordkeeping processes enabled the implementation of appraisal decisions at the time of records creation, thus providing both an intellectual coherence to recordkeeping processes which could be established in advance of records creation, and proactively addressing the process of decision making on length of retention.

REVIEW OF TRADITIONAL APPRAISAL METHODS

By the late 1990s and early years of the twenty-first century all archival institutions were acknowledging that the traditional methods of appraisal and implementing appraisal decisions were not adequately coping with the volume of records to be appraised, the resources available, or the nature of the increasingly digital record in the workplace. Various strategies have been adopted by individual archives.

The most developed implementation of macro appraisal is at the (now) Library and Archives Canada, where framework documents for the approach include an Acquisition Policy, Appraisal Strategy, Government Wide plan, and institutional profiles. The outcomes of the approach are Multi Year Disposition Plans "which formally set down records disposition project targets over a period of time according to criteria and priorities mutually agreed between the National Archives and the client at a senior management level, and Implementation Timetables. Revisited and modified in 2003–2004, macroappraisal continues to underlie the practices of Library and Archives Canada and has proven successful in enabling a coherent approach to the growing volume of government records and encompassing records in all media.[20]

At the same time that the Canadians were implementing macroappraisal strategies, the Dutch government were also undertaking an appraisal process focused on analyzing the functions of government in order to intellectually identify those functional areas of government producing the records required as archives.[27]

The National Archives (TNA) United Kingdom revised its appraisal framework in the late 1990s, continuing to use the notion of graduated time-based reviews established by the Grigg Report framework of 1954 for paper-based records, but varying the time intervals. New tools, Operational Selection Policies, apply the selection criteria established in the Acquisition Policy to records of individual government agencies or cross institutional themes. These are issued either as ongoing authorities or for records covering a specified time period. Combined with a focused acquisition strategy, TNA has a more proactive strategy for digital records and data sets.[28]

The U.S. National Archives and Records Administration (NARA) is similarly experimenting with approaches to appraisal. In 2004 its strategy for flexible scheduling was introduced to overcome the reality that NARA's previous methodology for appraisal was too resource-intensive for agencies to implement. The revisions to the process sought to transfer much greater responsibility to agencies to determine both the levels and the groups of records which would be treated as disposal classes. It encouraged agencies to use flexible scheduling, big buckets, and records retention bands.

> Flexible scheduling provides for concrete disposition instructions that may be applied to groupings of information and/or categories of records. Flexibility is in defining the record groupings, which can contain multiple records series and electronic systems.

> A "big bucket" or large aggregation schedule is a type of flexible schedule. Flexible scheduling using 'big buckets' or large aggregations is an application of disposition instructions against a body of records grouped at a level of aggregation greater than the traditional file series or electronic system that can be along a specific program area, functional line, or business process.

Record retention bands "Specify a time band that provides both minimum and maximum retention periods for records. Example: Destroy when records are no less than 3 years old but no more than 6 years old."[29] The new policy direction was accompanied by four strategies aimed at coordinating records scheduling to business areas and lines of business defined in the government's information architecture framework.[30]

In Australia, the National Archives of Australia established strictly regulated processes based on the "Developing and Implementing a Recordkeeping System" (DIRKS) methodology, a version of which had been incorporated into the International Standard on Records Management (ISO 15489). This approach adopted the framework of records continuum thinking and sought to establish partnerships between agencies and archives in establishing consistent recordkeeping frameworks for the whole of government's records, for however long they are required. This approach explicitly linked the creation of business classification schemes to appraisal decision making. Proposals on the appropriate retention period for records were made by agencies to the Archives, based on extensive documentation and risk analysis. The Archives then reviewed individual agency proposals for consistency with their published appraisal directions. Revised substantially in 2007 the degree of agency justification and documentation for records recommended for short-term retention has been minimized to focus on a more balanced risk-based approach, focusing on those records determined to have continuing value.

These approaches, while all responding to the individual needs of the jurisdiction, have features in common. They all stress a more coherent function basis for appraisal decision making. There is less emphasis on identifying and documenting individual records series, and greater emphasis on identification and documentation of functions that create records. The resulting disposal/disposition instruments are media-neutral enabling agencies to vary the format and media of records relating to specific functions rather than having authorizations tied to specific records, series, or formats. The principles of risk management have been acknowledged in these approaches, with greater latitude for agency decision making in the realm of appraisal, particularly for those records deemed not to have continuing value.

In contrast to some appraisal decision making of the past which was generally recognized to be inadequately documented, the practices of the major archival institutions have redressed this by specifying requirements for detailed appraisal justification documentation. In doing so, requirements may have swung too far and the documentation burden on those submitting appraisal recommendations have been recognized to be unduly onerous, resulting in revisions to adopt more risk-tolerant strategies and documentation. At the same time, archival institutions have themselves become more open and transparent about their appraisal decision making, through published appraisal or acquisition policies, publishing of disposal/retention schedules on Web pages and calls for public comments on disposal/retention schedules prior to their authorization.

DIGITAL RECORDKEEPING

Appraisal theory and practice has embraced requirements for digital recordkeeping by moving away from targeting specific records towards appraising functions and consequently the records (in any form) that they create. Embodying appraisal decision making into tools which formally authorize destruction will continue to be required and disposal/disposition tools are being subtly transformed to encompass changes to suit the digital environment. However, there are continuing debates about whether these are the correct tools to manage appraisal decision making into the future, with concerns about how implementable our traditional levels of appraisal decision making are in constantly changing workplaces.

The digital environment forces archival practices to change. Digital records are fragile and difficult to sustain, involving layers of technology dependence each requiring management to prevent records from being unusable and uninterpretable through technology change. Most pressing in the digital world is the requirement to ensure that robust records are made in the first place.

The Australian records continuum approach to recordkeeping has redefined appraisal to encompass the decision-making processes associated with determining what records need to be made and captured to document a business activity. "Appraisal is an analytic process which determines: (1) which records should be captured into recordkeeping systems and (2) how long such records should be maintained."[31] This redefinition was included into the Australian Standard on Records Management, AS 4390, published in 1996, which formed the basis of the International Standard on Records Management. While the idea of proactively determining what records to capture was transferred into the International Standard, the broadening of the concept of appraisal to encompass this was not adopted.

The digital environment requires proactive involvement of archivists (recordkeepers) in appraisal decision making. The older model of leaving appraisal decision making until the record is not required for continuing business use, is unviable in the digital world where records will not survive technology change and obsolescence without active intervention. Rather, digital records require continuing intensive management in systems necessary to ensure that they will survive to serve personal, corporate, or societal memory in the form of archives.

Records continuum approaches have redefined appraisal as a continuous process, occurring at multiple sites of intervention. In this reconceptualisation, we appraise

- Transactions and the actors initiating them for whether documents should be created.
- Activities and those involved in them for how evidential the process of records capture needs to be.
- Functions in an organizational context for the place of records in the memory of the organization and its archive.
- In a broader institutional or societal context for their purpose and their place in collective memory.

Appraisal becomes an ongoing process with multiple points of specific intervention—not just about selecting "archival records," but intimately concerned with managing robust records from the time of their creation/capture within environments that will sustain them for as long as they are required. Appraisal in this view is not only the domain of the archivist operating externally to agency recordkeeping frameworks, but rather, one of a team of recordkeeping professionals operating at various points of active intervention in the existence of records, affected by social, legal, and sometimes political factors.

This view is acknowledged within Australian recordkeeping practice and supported by coherent recordkeeping guidance issued to agencies. The archival tools and techniques have yet to transform significantly to suit the digital world.

For many archives, engagement with the digital world has been slow. Dealing with a set of paper-based practices that are essentially reactive has enabled many archivists to avoid engagement with digital realities. The necessary retooling of professional practice has yet to be incorporated into many work practices. Digital repositories and the capacity to take custody of digital archives have lagged the digital library community's uptake of the challenge. Appraisal decision making for the digital world must be accompanied by robust accompanying tools such as agreed metadata standards to enable records to move between system boundaries on a continuing basis, while retaining all contextual data essential to maintaining records, rather than digital objects.

Other information professionals are engaging with discussions on appraisal. The sophistication of the engagement ranges from provocative responses of the information technology community that information storage is cheap, we should just keep everything, to sophisticated interdisciplinary approaches to address particular needs, such as research data in the higher education sphere or scientific data. For example, see "An Australian e-Research Strategy and Implementation Framework."[32] For collaborative work in scientific data, see "Appraisal of Scientific Data."[33] Much of the discussion is arising and being conducted in the digital preservation discipline rather than in the archival discipline. The potential uses and capacity to manipulate born digital records in new and continually evolving ways substantially challenges archival practices such as sampling of case-based material. Archivists are increasingly collaborating with other information colleagues in addressing appraisal of complex digital records and data sets. See for example, http://www.interpares.org;[34] The work of the Digital Curation Centre at Glasgow University, particularly Harvey;[35] and The collaboration between the Public Record Office and the UK Data Archive for managing UK government data sets.

The digital recordkeeping environment has also raised profound challenges for collecting archives, responsible traditionally for taking custody of material created and managed by others after their administrative or business uses have ended. The challenge of digital collecting has been taken on by some major institutions supported by digital library repositories. Recent collaborative projects, such as the "Paradigm" project, have resulted in appraisal models for digital personal papers,[36] and the British Library "Digital Lives" project is investigating use of computers to store personal memory, aiming to devise strategies for collecting such material into the future.[37]

CONCLUSION

Appraisal is the professional recordkeeping function of applying a conscious process to determine what "records to keep." It developed as a key archival function in the twentieth century as a result of the overwhelming bulk of modern documentation and initially took the form of an evaluative process based on physically examining records and evaluating them against a set of criteria.

By the late twentieth century, the emergence of digital records, fluid organizational structures, and the concern that anything can be of value to someone, it became clear that existing appraisal thinking and processes were not able to be scaled up to meet demands. New thinking and experimentation in appraisal theory and methodologies emerged. Largely based upon the archival notion of provenance, these newer approaches stress the intersection of function and structure, advocate a greater acceptance of risk management and involve greater accountability in appraisal decision making. Many of these new methods have been incorporated into proactive recordkeeping standards to address recordkeeping from the time of creation for as long as the records are required.

While a ferment of appraisal thinking and experimentation has been underway, the practice of archivists is often still lagging, often remaining in the older paper-based criteria model. Appraisal has been the subject of constant professional discussion, as archivists strive towards methodologies that meet objectives, are coherent, timely and cost effective.

REFERENCES

1. Iacovino, L. Appraisal of public archives. In *Encyclopedia of Library and Information Science*; Kent, A., Hall, C.M., Eds.; Marcel Dekker, Inc: New York, 1993; Vol. 52 (Supplement 15), 1–23.
2. Duranti, L. The concept of appraisal and archival theory. Am. Arch. **1994**, *57*(2), 332.
3. Jenkinson, Sir H. *A Manual of Archive Administration*; revised 2nd Ed. Percy Lund, Humphries & Co Ltd: London, U.K., 1966, Part III Modern archives 148.
4. Schellenberg, T.R. *Modern Archives Principles and Techniques*; The University of Chicago Press: Chicago, IL, 1956; 133–160 Part III Archival management, Chapter 12, Appraisal standards.
5. Kolsrud, O. The evolution of basic appraisal principles—some comparative observations. Am. Arch. **1992**, *55*(1), 30–33.
6. Booms, H. Society and the formation of a documentary heritage: issues in the appraisal of archival sources (originally published in German in 1972). Archivaria **1987**, *24*, 95–97.
7. Schellenberg, T.R. *Modern Archives Principles and Techniques*; The University of Chicago Press: Chicago, IL, 1956; 137–160 Part III Archival management, Chapter 12, Appraisal standards.
8. Hurley, C. Appraisal of records created by the new technologies, in the national and international environment Proceedings of the 6th Biennial Conference of the Australian Society of Archivists Inc Australian Society of Archivists: Perth, Western Australia, Australia, 1987; 93–94.

9. Abraham, T. Collection policy or documentation strategy: Theory and practice. Am Arch. **1991**, *54*(1), 44–53.

10. Kitching, C.; Hart, I. Collection policy statements. J. Soc. Arch. **1995**, *16*(1), Appendix.

11. Gowing, M. British modern public records: A vital raw material. Arch. Manusc. **1981**, *9*(2), 15–27.

12. Honer, E.; Graham, S. Should users have a role in determining the future archive? The approach adopted by the Public Record Office, the U.K. National Archive, to the selection of records for permanent preservation. Liber Quart. **2001**, *11*, 382–399.

13. Hull, F. *The use of Sampling Techniques in the Retention of Records, a RAMP Study with Guidelines*; UNESCO: Paris, France,1981.

14. Rapport, L. No grandfather clause: Reappraising accessioned records. Am. Arch. **1981**, *44*, 143–150.

15. Bearman, D. Archival methods. Archiv. Mus. Inform. Tech. Rep. **1989**, *3*(1).

16. Menne-Haritz, A. Appraisal or documentation: Can we appraise archives by selecting content. Am. Arch. **1994**, *57*(3), 528–542.

17. Menne-Haritz, A. Appraisal or documentation: Can we appraise archives by selecting content. Am. Arch. **1994**, *57*(3), 536.

18. Samuels, H.W. Who controls the past. Am. Arch. **1986**, *49*(2), 115.

19. Cox, R.J. A documentation strategy case study: Western New York. Am. Arch. **1989**, *52*(2), 193.

20. Loewen, C. 'Cart or horse? macro appraisal and the contemporary drivers of archival appraisal.' Association of Canadian Archivists 2006 Conference Proceedings Association of Canadian Archivists Session 22, https://www.members-archivists.ca/conference_proceedings.asp (accesed November 2008).

21. Cook, T. Mind over matter: Towards a new theory of archival appraisal. In *The Archival Imagination. Essays in Honour of Hugh A Taylor*; Craig, B., Ed.; Association of Canadian Archivists: Ottawa, Ontario, Canada, 1992; 38–70.

22. Roberts, J. One size fits all? The portability of macroappraisal by a comparative analysis of Canada, South Africa, and New Zealand. Archivaria **2001**, *52*, 47–68.

23. Cunningham, A. *Some functions are more equal than others: The National Archives' macro-appraisal project July 2005*, Available at http://www.arkivrad.no/foredrag/2007/cunninghamjul05.pdf (accessed November 2008).

24. National Archives of Scotland, Records selection policy. http://www.nas.gov.uk/recordKeeping/governmentRecords SelectionPolicy.asp.

25. Society of American Archivists Glossary 'Minnesota Method,' Available at http://www.archivists.org/glossary/term_details.asp?DefinitionKey=2703.

26. Greene, M.A. Daniels-Howell, T.J. Documentation with 'an attitude': A pragmatist's guide to the selection and acquisition of modern business records. In *The Records of American Business*; O'Toole, J., Ed.; Society of American Archivists: Chicago, IL, 1997.

27. Hol, R.C.; de Vries, A.G. Pivot down under. Archiv. Manuscripts **1998**, *26*(1), 78–101.

28. The National Archives, Appraisal Policy. Background paper—'The Grigg system' and beyond. 2004. Available at http://www.nationalarchives.gov.uk/recordsmanagement/selection/appraisal.htm.

29. National Archives and Records Administration NARA Bulletin 2008-04, April 30, 2008. Available at http://www.archives.gov/records-mgmt/bulletins/2008/2008-04.html.

30. National Archives and Records Administration. Strategic directions: Flexible scheduling, January 2004. Available at http://www.archives.gov/records-mgmt/initiatives/flexible-scheduling.html.

31. AS 4390.5-1996, Australian Standard on Records Management, 1996, Part 5, Appraisal and Disposal, 3.

32. *An Australian e-Research Strategy and Implementation Framework*; Australian Government, 2006, Final Report of the e-Research Coordinating Committee. Chair: Dr Mike Sargent, http://www.dest.gov.au/sectors/research_sector/policies_issues_reviews/key_issues/e_research_consult/.../eResearch_Final_Report_pdf.htm (accessed November 2008).

33. Appraisal of Scientific Data. Papers of the ERPANET Workshop Lisboa, December 2003. Available at http://www.erpanet.org/events/2003/lisbon/index.php.

34. The interdisciplinary work of the InterPARES project. Available at http://www.interpares.org.

35. Harvey, R. Appraisal and selection. *Digital Curation Manual*; 2006; http://www.dcc.ac.uk/resource/curation-manual/chapters/appraisal-and-selection/ (accessed November 2008).

36. Personal Archives Accessible in Digital Media (paradigm project) 2005–2007. *Workbook on Digital Private Papers*; 2007; htp://www.paradigm.ac.uk/workbook/index.html (accessed November 2008).

37. British Library 'Digital Lives' project. Available at http://www.bl.uk/digital-lives/about.html.

BIBLIOGRAPHY

1. Booms, H. Society and the formation of a documentary heritage: Issues in the appraisal of archival sources. Archivaria **1987**, *24*, 69–107 (originally published in German in 1972).

2. Cook, T. Mind over matter: Towards a new theory of archival appraisal. In *The Archival Imagination. Essays in Honour of Hugh A Taylor*; Craig, B., Ed.; Association of Canadian Archivists: Ottawa, Ontario, Canada 1992; 38–70.

3. Duranti, L. The concept of appraisal and archival theory. Am. Arch. **1994**, *57*(2), 328–344.

4. Samuels, H.W. Who controls the past. Am. Arch. **1986**, *49*(2), 109–124.

5. Schellenberg, T.R. *Modern Archives Principles and Techniques*; The University of Chicago Press: Chicago, IL, 1956; 133–160 Part III Archival management, Chapter 12, Appraisal standards.

Archival Arrangement and Description

Joanne Evans
Sue McKemmish
Centre for Organisational and Social Informatics, Monash University, Melbourne, Victoria,
Australia

Barbara Reed
Recordkeeping Innovation Pty Ltd, Sydney, New South Wales, Australia

Abstract

Arrangement and description are processes traditionally associated with the transformation of an individual or corporate archive into a broader archival framework in order for it to function, alongside other such archives, as accessible collective memory. They are integral to maintaining the authenticity, integrity, and reliability of records in an archival environment so that they may continue to function as evidence of the activities and contexts in which they were created and used. This entry outlines the principles underlying archival arrangement and description, discusses the construction of arrangement and description traditions, and explores their transformation in response to the recordkeeping challenges of the twenty-first century.

INTRODUCTION

Arrangement and description are processes traditionally associated with the documentation and processing of records into archival systems at, or near, the time of their transfer to archival institutions. Archival systems document, manage, and enable the ongoing accessibility of records of continuing value under the control of an archival institution. They have been shaped by a number of social, political, cultural, economic, and technological factors in different countries and at different times. This has given rise to different traditions for arrangement and description that reflect the nature of records, the contexts in which they are created and maintained, and the archiving contexts into which they are transferred. However, despite their differences, all archival arrangement and description practices are fundamentally a practical application of underlying archival principles of *respect des fonds*, provenance, and original order.

In the late twentieth century, the uptake of digital information and communications technologies in society in general, and in archival institutions in particular, has had a significant impact on arrangement and description practices. On the one hand, descriptive processes have been codified into various national (a selection of guides and manuals for arrangement and description produced or sponsored by national professional associations for archivists is provided at the end of the entry) and international standards (see entries on "International Standards for Archival Description," p. 2950, and "Encoded Archival Description (EAD)," p. 1699) to enable cross-institutional sharing of archival documentation and automated processing and to assist in retrieval of archival data. On the other hand, the fundamental change in the nature and structure of records and on business and recordkeeping processes brought about by computing technologies has exposed the practices of arrangement and description as reflective of a physical focus based strongly in the prevailing paper-based paradigm of their origin.

This entry

- Outlines the purpose and principles underlying archival arrangement and description.
- Discusses the construction of arrangement and description traditions.
- Explores how these traditions are changing to address the recordkeeping challenges of the twentieth and twenty-first centuries.

It is limited to canvassing archival practices in English-speaking countries as this reflects the experience of the authors.

ARRANGEMENT AND DESCRIPTION PRINCIPLES

The articulation of archival principles fundamental to arrangement and description arose during the nineteenth century. Their development represented a reaction to the practices employed during the previous period of consolidation of archival records in specifically designated repositories geared to serve the purposes of historical research. Those practices were intent on guiding the researcher to the most "interesting" documents typically identified by subject. The physical arrangement of the material, then, was ordered to reflect that of the guide.[1] In reaction to this practice, which put a subjective, imposed order on records, the principle of respect des fonds was articulated. Subsequently translated into the English principles of provenance and original order, these have formed the basis of all ensuing archival arrangement and descriptive practices.

Respect des fonds, Provenance, and Original Order

Respect des fonds is the nineteenth-century French dictum that the authenticity and integrity of records depends, at least in part, on tracing their origin and history. To preserve records as evidence, the functional structural context of their creation must be explicitly revealed. *Respect des fonds* aims to maintain the integrity of records in relation to their documentary, provenancial, functional, and jurisdictional contexts. The *fonds* or archive is defined as

> ... the whole of the documents of any nature that every administrative body, every physical or corporate entity, automatically and organically accumulated by reason of its function or of its activity. (p. 40)[2]

It has both exterior dimensions, relating to the external jurisdictional and functional structures in which records are created or accumulated, and interior dimensions, relating to its internal recordkeeping structures. The principle of *respect des fonds* is about preserving the interior and exterior dimensions of a corporate or individual archive within an archival environment so that its authenticity and integrity is maintained, allowing the records to function as part of collective memory.[2]

In English-speaking communities, *respect des fonds* has been translated into the principles of provenance and original order. The principle of provenance is about preserving "the relationship between records and the organizations or individuals that created, accumulated, and/or maintained and used them in the conduct of personal or corporate activity" (p. 11).[3] The principle of original order is about maintaining the ordering and grouping of records used by a creating organization or individual so that they reliably reflect the creator's business and recordkeeping activities. Keeping records in an archival environment according to these principles maintains their authenticity, integrity, and reliability (pp. 15–16).[4]

Table 1 gives definitions of *respect des fonds*, provenance, and original order from key publications associated with arrangement and description traditions in North America, the United Kingdom, and Australia. It illustrates that the definitions of these principles tend to be expressed in terms of their application in practice, with that practice, by and large, reflecting the physicality of paper recordkeeping environments.

Arrangement and description practices have been constructed around applying the principles in paper

Table 1   Definitions of arrangement and description principles.

Respect des fonds	One axiom holds that records created or accumulated by one records creator must be kept together and not intermixed with the records of other creators, often referred to as *respect des fonds*. (*Rules for Archival Description*, p. xv)
	Respect des fonds is the principle that the records created, accumulated, and/or maintained and used by an organization or individual must be kept together in their original order, if it exists or has been maintained and not be mixed or combined with the records of another individual or corporate body. (*Arranging and Describing Archives and Manuscripts*, p. 33)
	Respect for the principle of provenance that the archives of an agency or person are not mixed or combined with those of other agencies or people. (*Keeping Archives*, p. 479)
Principle of provenance	*Provenance* is defined as "the relationship between records and the organizations or individuals that created, accumulated, and/or maintained and used them in the conduct of personal or corporate activity." The principle of provenance states that records should be maintained according to their origin and not "intermingled" with those of another provenance that is those created by another person or agency. They should be kept together on the basis of the organization that created, used, or accumulated the group of records and in the groupings of records the person or organization created. (*Arranging and Describing Archives and Manuscripts*, p. 15)
	In archival theory, the principle of provenance requires that the archives of an agency or person not be mixed or combined with the archives of another, that is, the archives are retained and documented in their functional and/or organizational context. (*Keeping Archives*, p. 476)
Principle of original order	A second axiom follows from the observance of *respect des fonds*: the way archives are described depends on their arrangement. Implicit in the archivist's observance of *respect des fonds* is the assumption that the way a creator "automatically and organically created and/or accumulates records" will affect the way archivists arrange a fonds. (*Rules for Archival Description*, p. xv)
	The principle of original order involves keeping records in the order by which they were kept by the person or organization that created, accumulated, assembled, or maintained them. The groupings, file system, subdivisions, and other physical structuring provided by the records creator should be

(Continued)

Table 1 Definitions of arrangement and description principles. *(Continued)*

maintained. Keeping records in the order the creator kept them provides information about the context and use of those records. (*Arranging and Describing Archives and Manuscripts*, pp. 15–16)

The order in which records and archives were kept in active use, that is, the order of accumulation as they were created, maintained, and used. The principle of original order requires that the original order be preserved or reconstructed unless, after detailed examination, the original order is identified as a totally haphazard accumulation making the records irretrievable (but not an odd, unorderly, or difficult arrangement). (*Keeping Archives*, p. 475)

recordkeeping environments and also tend to be defined in terms of their application in particular traditions, as illustrated in Table 2.

ARRANGEMENT AND DESCRIPTION PRACTICES

Arrangement and description are processes traditionally associated with the transformation of an individual or corporate archive into a broader archival framework in order for it to function, alongside other such archives, as accessible collective memory. They are integral to maintaining the authenticity, integrity, and reliability of records in an archival environment so that they may continue to function as evidence of the activities and contexts in which they were created and used. As archival functions, their principal aim is to establish, capture, and sustain the physical and intellectual controls that enable an archive to be comprehensible and accessible for future use beyond the environment in which it originated.

Archival arrangement and description processes differ from other cataloging, classification, and organizing practices in that they are not seeking to impose order and structure but to reveal the inherent relationships between the records and their contexts of creation and use. The focus of arrangement and description is on explicating the relationships of records to the business and recordkeeping processes of their creators, not on classifying their subject matter or format. While cross-referencing into classification frameworks utilized in other information management traditions may be useful for accessibility, and physical arrangement based on format may be more efficient for repository management, neither one should dictate the way in which records in archival institutions are principally arranged and described. One of the potential advantages of digital and networked technologies is that

Table 2 Definitions of arrangement and description.

Arrangement	Arrangement involves analyzing the records to see who or what created them, how and why they were created, what functions and activities they document, when they were created, and what their physical nature is. (Basic RAD, http://scaa.sk.ca/rad/section1.htm)
	Arrangement is the process of organizing materials with respect to their provenance and original order, to protect their context, and to achieve physical and intellectual control over the materials. (*Arranging and Describing Archives and Manuscripts*, p. 11)
	The intellectual and physical process of putting archives and records into order in accordance with accepted archival principles, particularly those of provenance and original order. If, after detailed examination, the original order is identified as a totally haphazard accumulation making the records irretrievable (but not an odd, unorderly, or difficult arrangement), the archivist may (after documenting this original order) impose an arrangement that presents the records objectively and facilitates their use. (*Keeping Archives*, p. 464)
	The intellectual and physical processes and results of analyzing and organizing documents in accordance with archival principles. (*ISAD(G)* 2nd edition, p. 10)
Description	Description is the process of explaining that arrangement so that people—researchers, administrators, whomever—who want to use the records know where to look to find the answers to their questions. Arrangement, therefore, is the process of studying the records to discover how they relate to the entities that created them. Description is the way of sharing that knowledge with everyone else. (Basic RAD, http://scaa.sk.ca/rad/section1.htm)
	Description is the creation of an accurate representation of a unit of archival material by the process of capturing, collating, analyzing, and organizing information that serves to identify archival material and explain the context and record system(s) that produced it. (*Arranging and Describing Archives and Manuscripts*, p. 13) The process of recording information about the nature and content of the records in archival custody. The description identifies such features as provenance, arrangement, format, and contents and presents them in a standardized form. (*Keeping Archives*, p. 467)
	The creation of an accurate representation of a unit of description and its component parts, if any, by capturing, analyzing, organizing, and recording information that serves to identify, manage, locate, and explain archival materials and the context and records systems which produced it. This term also describes the products of the process. (*ISAD(G)* 2nd edition, p. 10)

they may more efficiently enable such cross-referencing and inventory management without compromising archival principles.

Record Group Traditions

In the late nineteenth/early twentieth century, when archivists were mainly dealing with closed, complete, paper, records series from the records systems of relatively stable and long-lived administrations, reconstructing the physical arrangement of records in the archival repository "inherently reflect[ed] the functions, programmes, and activities of the person or institution that created them" (p. 35).[2]

The principle of *respect des fonds* was implemented through the segregation within archival repositories of records from different creators, with the physical arrangement of these groups representing "the external structures in which each particular archive had been formed, thus also reflecting the functions and activities of the records creator" (p. 166).[5] Internal filing structures among record series within these groupings were then also physically recreated, maintaining original order and thus also preserving the interior dimensions of the *fonds*. Once arranged, the resultant record groups were then described in archival finding aids to produce a surrogate of the "physical manifestation of the archive or *fonds* on the repository shelf."

These arrangement and description practices reflect the recordkeeping and archiving contexts of their time, characterized by relatively stable and long-lived administrative structures, simpler and hierarchical organizational structures, and archival institutions mainly handling records of past administrations. In these circumstances, taking records into custody once they were closed, identifying the boundaries of a *fonds*, nominating a single records creator, and capturing the documentary, provenancial, functional, and jurisdictional relationships of the *fonds* through physical arrangement was relatively straightforward. This gave rise to the record group approach to description, involving the successive depiction of whole-part relationships between levels of physical arrangement.

In the record group approach, a representation of the *fonds* is captured by successive enumeration of its hierarchy from the top down, that is, the overall *fonds* is described and contextualized, followed by each of the series making up the *fonds*, and within each series the files making up the series, and (resources permitting) within each file the items making up the file. Implicit in this is the view that there are only one-to-one relationships between the levels of arrangement and their provenancial and functional contexts, and between the levels of arrangement themselves. The arrangement also preserves the filing order of the records systems, thereby representing the activities of the records creator. With the

record group approach, description takes place after arrangement, proceeds from the general to the specific, and emphasizes the importance of multilevel rules to avoid redundancy and repetition in the production of paper-based finding aids. The arrangement hierarchy implicitly captures relationships of the records to the recordkeeping contexts of the creator of the *fonds*, with physical arrangement reconstructing these relationships within the archival institution.

With the increasing complexity of recordkeeping processes and contexts throughout the twentieth century, these arrangement and description practices have been put under stress.

> As the dynamics of social institutions changed, administrative structures become more volatile, organizations became larger, more complex and more variably structured, archival institutions came to be concerned with modern records (*fonds* in the process of becoming, rather than already formed in any sense), more than one records creator came to be associated with a single *fonds*, and they and any number of successor organizations might be responsible for transferring to archival custody subsets of records belonging to the same *fonds* at different points in time, archival systems based on the old rules of arrangement and description became increasingly dysfunctional as archivists struggled to solve insurmountable problems associated with scalability, defining the boundaries of the *fonds*, and creating a one-to-one relationship between a *fonds* and a records creator (p. 167).[5]

This has given rise to alternative approaches to arrangement and description based on decoupling physical and intellectual control and viewing the *fonds* as a logical rather than a physical construct. In this approach, arrangement and descriptive processes focus on capturing multiple, virtual relationships between records, and their contexts of creation and use. Relationships in the records are virtually reconstructable rather than physically reconstructed by arrangement. Finding aids that reconstruct the *fonds* are an output from the system rather than being the unit of description. Description documents contexts rather than cataloging an arrangement. According to this approach, an archival system is in essence a register of recordkeeping systems, as well as itself being a form of recordkeeping system.

Context Control Systems

Registry systems

The recordkeeping tradition known as registration systems, or registries, was widespread throughout Northern Europe (particularly in Prussia) and the United Kingdom. It formed an important part of the control tools associated with the administration of empire, consequently spreading in the English-speaking world to Australia, Canada, and other colonies. Registry systems, as their title suggests, center on the concept of registration and are based on the notion of controlling documentation from the time it is received by or authored within an organization. In the German recordkeeping tradition, the Registratur is "the office that controls the flow of incoming and outgoing correspondence and preserves, arranges, and services those parts of it that remain with the agency until they can be transferred to archival custody."[6] Documents are registered (controlled) into the system as soon as they enter the organization, or as they are dispatched, being always managed conceptually as records. A number of indexing schemes complement the registration process—typically an alphabetic name index and a topic or subject index. In the German tradition, "since the eighteenth century, there has existed a common agreement that the registry ... must arrange its holdings in accordance with the main functions of the administrative unit it serves" (p. 91).[6] Registry systems thereby apply archival functions to records that are still current.

Registry systems associate incoming documents with any previous transactions of relevance, in many cases physically gathering together the associated documents. The system then directs the records around the organization to those that need to take relevant action, tracking the work that is done, or needs to be done, in proto-workflow systems. The records travel through the organization gathering the documentation associated with the work as it is carried out, finally ending up once action is completed in the registry. Here, all correspondence being dispatched outside the organization is also registered in separate sequences and linked to the records related to it. The registry then stores the records of completed business in various sequences, commonly either the original sequential registration number or by any later registration number given to a subsequent document arising from action. If the latter option is adopted, sequences of records related to a business action are gathered together. These are the antecedents of the "file" that grew out of the inability to cope with the demands of registering every document in the paper explosion of the mid-twentieth century.

Registry systems separate out the physical from the intellectual, relying on multiple indexes constructed around the records to provide entry points for retrieval to serve multiple purposes. They focus on controlling documents from the moment they are received or created, with records acting as enablers and drivers of business action, sequencing activity. In paper registry practices, indelible bonds between documents and actions are created through physical markings on records, physical arrangement, and in registry control systems, for example, card indexes that also serve as retrieval tools. These practices embody the principle of original order. The physical manifestation of the record sequences can, and do, change depending on the documents brought together for action. Registry systems allow for the actual storage arrangements of the documents to be quite separate from the constructs brought together for the actioning of work.

This recordkeeping tradition is not universal and was not the norm in twentieth-century American administration, the source of much of the first wave of office automation software. The disappearance of registries and records functions at the time of initial automation of work processes was based on the presumption that these paper-based behemoths epitomized outdated practices. Undoubtedly, understandings of the importance of records and recordkeeping were swamped by the routine undertaking of day-to-day activity in automated systems. Out-of-control, chaotic recordkeeping resulted as the old registry systems broke down and oftentimes were not replaced with any commensurate organizational systems.

With the increased application of computing technologies to the processes of document and records formation and structuring, the concepts behind many of the practices embedded in registry systems are finding new life in electronic recordkeeping best practice. Indeed, all IT systems work on registration of some kind even if it is not explicit. Simple sequential numbers as locators, rather than the bearers of embedded meaning; the development of classification systems for current records based on the work that is done in organizations, as manifested in activities and functions; the provision of multiple indexing points as aids to searching and retrieval; and workflow grouping of sequences of actions reflected in records are some examples. These characteristics are implicit or explicit in the registry systems of the nineteenth and twentieth centuries. Reworking of these concepts can be found in best practice statements for recordkeeping in the electronic world, where they can be delivered in ways attuned to automation and minimal manual intervention (see entry on "International Organization for Standardization (ISO)," p. 2917).

Series systems

In Australia, the circumstances leading to the establishment of the Commonwealth Archives Office in 1961 were quite different from those that formed the context for the establishment of national archives in Europe and the United States. In the Commonwealth of Australia (formed at federation in 1900), recordkeeping problems were essentially those presented by twentieth-century records, with all their complexity and volume. Australian recordkeeping was plagued with difficulties arising from the penchant for administrative change. Agencies were created and abolished with quite amazing regularity,[7] and this trend increased dramatically in the second half of the twentieth century. This resulted in records series being split, merged, and resplit as functions moved through multiple agencies. The notion of documenting provenance in terms of the last creating agency was not sustainable as it could create a substantially distorted view of the provenance of the records.

The so-called Series System was developed to address these and other problems.[8] Its designers were very conscious of requirements to maintain conceptual compliance with the core archival principles of *respect des fonds*, provenance, and original order. It also drew heavily from the recordkeeping systems in place to manage current records. To these, it added a further layer of control— a registration system for recordkeeping systems imposed by an external (archival) authority. In this system, record series are allocated a simple sequential number—a registration number that bears no inherent meaning. A separate registration process documents the records-creating agencies, with a different simple sequential number. Thus, two types of entities are independently registered and documented. They are then linked in as many ways as are relevant to each other. Using the notion of previous and subsequent relationships inherited from the earlier registration systems, each record series is linked to any and all relevant agencies. These relationships become the pivotal mechanism in allowing complex depiction of provenance in a continually cumulating archival documentation system. This approach enables representation of multiple roles in relation to records, thus documenting these roles over time and as they change. Thus, the system maintains all the traditional requirements for *respect des fonds* through enabling a very detailed representation of provenance, both at a specific point in time and over time.

The "Series" System addressed a number of problematic areas in Australian attempts to apply traditional archival description. The nexus between the physical arrangement of records in a repository and their intellectual arrangement was broken. As in a registry system managed through multiple registers and indexes, it was no longer necessary to locate the records derived from a single agency together in one place, or even all parts of a record system together. The arrangement was an intellectual construct. While agencies and records series were the two entities and aggregation layers defined by Scott and his colleagues for the purposes of archival description at the Commonwealth Archives Office, other entities and layers of aggregation can, and have been, introduced and managed in different implementations using exactly the same concepts. Multiple, layered, and complex relationships between agencies (and other entities, such as function) and records can be constructed and expressed to reflect the volatility of the administrative structures of the government. Roles and rights associated with these relationships can also be clearly delineated—for agencies, these were initially defined as "creating," "controlling," and "transferring" when linked to records series.

Similarly, this system is not dependent on records being in custody. The circumstances of recordkeeping in volatile administrative contexts mean that the earlier the documentation of agencies and recordkeeping systems are documented, the easier it is to get an accurate depiction.

This is ideally undertaken at the time of change of government through the management mechanisms of the machinery of government announcements (e.g., administrative arrangement orders). Documentation of recordkeeping systems can, in theory, be carried out by anyone involved in recordkeeping regardless of their designation as an archivist or records manager. Description occurs before any arrangement and can encompass all of the records— both documents and files, those extant and those that have been destroyed at various times. The system is not restricted to depicting the records held in archival custody as if they are the whole of the records. The "archief" of Muller, Feith, and Fruin can therefore be much more faithfully represented in this system of archival documentation than in the traditional representation of a *fonds* as a catalog of surviving records.

The "Series" System implemented in the Commonwealth Archives Office was also based on pragmatism. In being pragmatic, it achieved a high adherence to the principle of original order, using the contemporary documentation of records' creators rather than creating independent lists of records once received into archival custody. The fledgling archives could not hope to process voluminous twentieth-century records to the level of detail of calendars and many other traditional archival finding aids recommended by European antecedents. Instead, the system was based on the controls that were in place at the time the records were created. If the agency relied on the registers and indexes that they created to manage the records, so too would the archives and the archives users. Additional layers of description could be added to provide additional interpretative material where necessary, but whole-scale reprocessing need not be undertaken. With hindsight, these wholly pragmatic decisions are now a fundamental component of ensuring the authenticity of records.

All these characteristics of the "Series" System— connection to the whole of the records; the treatment of agencies and records as independent entities; documenting multiple relationships over time to maintain links between agencies and records as they change; multiple points of registration; inheritance of the records controls created to manage records in a different context—can be seen reverberating through the rich conceptual thinking embedded in the Records Continuum Model (see entry on "Records Continuum Model," p. 4447).

DIGITAL AND NETWORKED ENVIRONMENTS

As already noted, the world of digital recordkeeping and networked environments has brought substantial challenges to traditional arrangement and descriptive practices. The complexity of modern recordkeeping, the heterogeneity of modern records, the exigencies of repository management, and the advent of computing technologies mean that increasingly the connection between records in an individual or corporate archive and their originating context is constructed as an intellectual rather than a physical exercise. A major challenge for archivists in the face of increasing complexity and virtualization of recordkeeping with the introduction of copying, digital, and networking technologies has been to come to a greater conceptual understanding of the principles underlying arrangement and description, so that practices can be constructed to meet past, present, and future recordkeeping realities as required. In Canada, Cook and other Canadian archivists have written about the need to view the *fonds* as an intellectual rather than physical construct and of the need for arrangement and description practices to enable representation of "a dynamic interconnection between the abstract description of the records creator(s) and of the concrete description of the actual records (series, files, items)" (p. 74).[2]

Recent definitions of arrangement emphasize that it is an analytical process, in which the records of an individual or corporate entity and their contexts are examined in order to establish by whom, how, and why they were created and used.[9] Arrangement and description are also increasingly perceived as intertwined and interdependent functions rather than sequential activities, with interconnection to other functions such as appraisal, preservation, and access.[4] There has been recognition of the need to capture appropriate records of arrangement and description activities, along with the notion that these processes may occur throughout the records existence rather than just at the point of archival transfer. This challenges the idea of description as merely a cataloging activity that explicates archival arrangement.

The past few decades have also seen an increasing interest in standardization of archival arrangement and description practices in order to capitalize on the capabilities of computing technologies for archival processing, access, and retrieval. This has resulted in the codification of arrangement and description practices (to varying degrees) in a number of manuals, guides, and other kinds of publications that have then been shared within communities of practice. These publications provide a set of rules for implementing arrangement and description processes that reflect the type of jurisdictional and functional arrangements, organizational structures, and recordkeeping processes of their places and times. They are also shaped and limited by available recordkeeping and archiving technologies. The International Council of Archives (ICA) has supported the development of ISAD(G), ISAAR (CPF), ISDIAH, and ISF as a suite of international archival descriptive standards, aimed at developing better understanding of arrangement and description principles in order to construct more unified practices. The ICA's latest project, the Expert Group on Archival Description (EGAD), is seeking to develop an overarching conceptual model of archival description to facilitate the standardization of descriptive systems and allow for the combination of descriptive information from different sources to foster collaboration among memory institutions. Such a higher-level abstract model is necessary for semantic interoperability across descriptive schemas not just within the archival community, but also with those used in library, museum, and other information systems. Conceptual descriptive models allow for common and distinctive areas to be identified, enabling the integration of description from different perspectives and sources in ways that preserve and respect meanings. However, while these kinds of standardization activities may enable greater sharing of information about archival holdings across institutions, the practices either implied or codified in the standards are mainly those relating to existing cumulated holdings of archives and reflect essentially paper-based practices. They do not address the challenges of digital and networked records and recordkeeping.

Standards for Arrangement and Description in Digital Environments

In a digital world, records do not exist as fixed artifacts. The relationships between records, once preserved by their physical ordering, can no longer be represented by physical arrangement. This provides a substantial challenge to the notion of original order as traditionally articulated and applied. All records are accessed and viewed through the mediation of a machine. What order the machine stores them in is unknown. The best that can be asserted is that the digital components of a specific record are managed within a known application. Protocols of presentation embedded in that application structure the ways the records are presented to a viewer. Records are virtual constructs—what we see on a machine in a form and format that looks familiar to a paper-based mind is only a representation of the record, rendered to appear in that given form and format. Records are brought together on the fly. They can be composed on multiple fragments and drawn from many different underlying parts of a system. Breaking traditional thinking from a physical to a virtual mindset is essential if we are to maintain the ability to present records in context to users of the future.

In a virtual world, location is irrelevant. An electronic record can live anywhere from the time it is initially created—it can live in an organization; it can be shared by multiple organizations; it can be stored by a third party; it can be lodged with an archival institution from the time it is locked into a defined format. Any of these physical locations are possible and may coexist. Thus, the archival endeavor of description is to ensure that when represented to a user (any user, at any time) the record is meaningful, able to be rendered on the technology of choice, and possesses the characteristics that make it reliable, authentic, and trustworthy.

Approval–Archival

The need for digital preservation processes to focus on capabilities to represent records is a key finding from the International Research on Permanent Authentic Records in Electronic Systems (InterPARES) Project. The project reached the conclusion that preservation of records in digital environments is about preserving the ability to reproduce records, rather than the electronic records per se.[10] Reproducing records involves assembling preserved digital components into the proper arrangement to represent their intrinsic documentary form and extrinsic archival bond, essentially to maintain and render relationships to transactional contexts in order to enable the transmission of authentic, reliable records through time and space. Preservation and description functions are thus inextricably linked.

Along with the undoubted challenges of ensuring that digital records survive in readable format long enough to reach the portals of an archival institution comes awareness that reconstructing provenance and other contextual linkages for such records is unachievable when applied retrospectively. This combined set of challenges has pushed the locus of descriptive activity into recordkeeping environments more contemporary with the creation of the record. To achieve this, understandings about what descriptive processes for recordkeeping entails must be broadened out and embedded into all recordkeeping environments, commencing from the initial point of a record's capture. A corollary is the requirement to standardize these understandings in the form of professionally accepted practice (see entries on "International Standards for Archival Description," p. 2950, and "International Organization for Standardization (ISO)," p. 2917). Using such means, the locus of the application of the descriptive standards can be changed and the record can be created with many of the components required to ensure the survival of the record, connected to the business activities it documents, through time, and across multiple locations.

Moving away from description as a professional practice associated only with the work of archivists within an archival institution has required a thorough review of the purpose and nature of the work of arrangement and description. An Australian archivist, Hurley,[11] argues that the traditional role of all archival finding aids is to maintain the linkages of records, now in archival custody, to the circumstances of their creation and management. This posits a very critical role for finding aids as a mechanism for maintaining the authenticity of records through exposition of the processes of formation of records, rather than just a presentation of explanatory material for users to facilitate access.

Reconceptualizing description as an archival function that ensures authenticity of records over time provides an immediate link back to the creating environment as there too the authentic nature of records is a key concern. The recordkeeping endeavor then becomes a shared one across multiple sites of implementation. The requirement to

define the characteristics and components that will ensure that records survive over time is an inherent part of the records-creating process. Enhancing the context of creation with other layers of descriptive information as the record moves through multiple points of creation and management, and weaving different relationships as it travels through multiple uses, reuses, and sites of interaction become components in a cumulative process of records formation.

In digital environments of increasing complexity, grounding recordkeeping theory in a strong understanding of the rationale of records becomes vital. Concepts such as cascading regions of recordkeeping actions that reflect different points in registration processes, and the primacy of relationships and notions of aggregations become essential to effective intervention in the design and development of systems to create and maintain records over time. The ability to depict complex and dynamic relationships becomes the key to maintaining records authenticity.

Recordkeeping Metadata Models

Recent international recordkeeping metadata initiatives (see entry on "International Organization for Standardization (ISO)," p. 2917) are a direct result of conceptualizing the world of recordkeeping in this way. ISO 23081 defines recordkeeping metadata as

> data describing the context, content, and structure of records and their management through time. As such metadata is structured or semi-structured information that enables the creation, registration, classification, access, preservation, and disposition of records through time and within and across domains ... metadata can be used to identify, authenticate, and contextualize records and the people, processes, and systems that create, manage, maintain, and use them and the policies that govern them.[12]

It specifies different layers of aggregation and different sites of implementation of recordkeeping metadata in order to enable the continual construction and weaving of relationships as the record appears in multiple sequences of action. This approach is predicated on the proposition that records need to be actively defined, and that at the initial point of capture of a record object, a set of metadata about its content, structure, and immediate context of action also needs to be captured and persistently linked to it. Moreover, record objects can only exist through time if they are persistently linked to an accumulating set of metadata, providing broader layers of context, and recording the business, recordkeeping, and archiving actions that manage them. This recordkeeping metadata is itself a record and so must incorporate metadata about its "recordness," that is, who created it, when, etc.

With this view archival description is another layer of recordkeeping metadata that maintains evidential contextualization when records are represented outside of their originating environments in transfer to an archival institution. Recordkeeping metadata about the recordkeeping metadata enables intellectual separation of that which is created by the originating and archival contexts.

Emerging metadata standards and models, including the ISO, take as their frame of reference a conceptual model of recordkeeping metadata in business and socio-legal contexts developed by the Australian SPIRT Recordkeeping Metadata Project (Fig. 1). This model extends the Australian Series System concepts of context, drawing on records continuum thinking relating to a record's complex and dynamic social, functional, organizational, procedural, and documentary contexts of creation, management, and use in and through spacetime. It is also informed by the insights of Hurley and Cook. The model defines layers of descriptive metadata about the following:

- Records at any layer of aggregation
- The people or agents involved in creating, managing, and using records
- The organizational and social structures in which they interact
- The activities and transactions in which the people or agents are engaged, and related social and business purposes and functions
- The recordkeeping functions, activities, and transactions that capture, manage, and make accessible the records of these activities and transactions
- The complex, multiple, and dynamic relationships between all of these entities
- The warrants or mandates that govern all of this social and organizational activity, including metadata about social mores, laws, business policies, and rules

Other research initiatives are also investigating requirements for recordkeeping metadata and frameworks for its

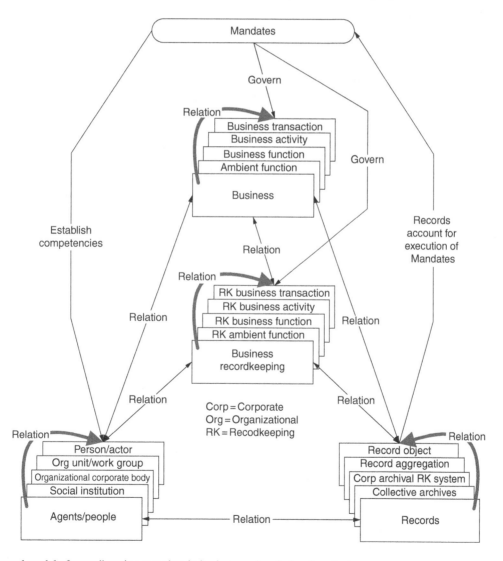

Fig. 1 Conceptual model of recordkeeping metadata in business and socio-legal contexts.
Source: From McKemmish et al.[13]

management in electronic environments. The Clever Recordkeeping Metadata Project, involving researchers at Monash University in partnership with the National Archives of Australia, State Records Authority of NSW, and the Descriptive Standards Committee of the Australian Society of Archives, explored the automated capture and re-use of recordkeeping metadata,[14] while the InterPARES 2 Description Research Team prototyped a Metadata Schema Registry for describing and analyzing extant metadata schemas and archival descriptive standards against emerging recordkeeping warrants.[15]

ARCHIVAL DESCRIPTIVE SYSTEMS OF THE FUTURE

Recordkeeping metadata initiatives implemented in digital networked environments have the potent to enable simultaneous multiple views of recordkeeping realities. While most implementations of archival description document, and therefore privilege, the role of a singular records creator, representing their context rather than that of other parties to the transactions that the records evidence, and draw on functional classification schemes and thesauri developed at the level of the corporate or individual archive, the frameworks could enable alternative readings of our society, to ensure that the voices of other parties to transactions in which records are created, captured, and used are heard.

Hurley's writings have extended the conceptual basis and principles of the "Series" System to encompass the challenges of describing context and records entities and their complex, multidimensional relationships in the virtual world of the beginning of the twenty-first century. His further development of the conceptual framework for the Series System includes two layers of contextual entities, higher-level ambient entities representing organizational structures, jurisdiction-based or functional groupings, or functions themselves, corresponding to the French view of the external dimension of the *fonds*, and lower-level German-like provenance entities representing records-creating units and activities.

This extension of the Series System and emerging recordkeeping metadata models enable archivists to go beyond Peter Scott's original vision of sequential multiple provenance to address what Hurley has named parallel and multiple simultaneous provenance.[16]

Recognising that the documentation created within the New Zealand national archives system largely reflects the cultural views of the Pakeha majority, but living in a society in which bi-culturalism is more than mere rhetoric, Hurley began to question how the views of the Maori could be accommodated in systems defined by Pakeha standards, and to seek a set of alternative, equally valid ways of viewing and documenting the records. (p. 192)[5]

Postmodern ideas are informing the critique and refiguring of archival processes and systems. Ketelaar has also posited a different approach to archival descriptive practice with reference to the concept of "communities of records" as developed by Bastian, and associated ideas about shared ownership and joint heritage. Bastian defines a community of records as "the aggregate of records in all forms generated by multiple layers of actions and interactions between and among the people and institutions within a community."[17] According to this view, "the records of a community become the products of a multi-tiered process of creation that begins with the individual creator but can be fully realized only within the expanse of an entire community of records"; thus, "all layers of society are participants in the making of records, and the entire community becomes the larger provenance of the records." Drawing out the implications of these conceptual approaches, Ketelaar[18] points to the matrix of mutual rights and obligations of all the parties involved and how they would extend to all aspects of recordkeeping and archiving, including description. Archival descriptive systems of the future need the capabilities to negotiate such matrices of mutual rights and obligations and manage integrated archiving and recordkeeping processes to fulfill these multiple purposes.

Frameworks for capturing layers of rich contextual metadata, along with multiple contexts of creation, management, and use are emerging. The advent of Web 2.0 technologies has seen archival institutions experiment with adding user annotation, tagging, and transcription capabilities to their online discovery systems. While labeled as participatory archives, many would argue that more radical transformation is needed to enact more dynamic and democratic models of archival description. Huvila has identified three characteristics of a participatory archives approach, namely "decentralized curation," "radical user orientation," and "contextualization of both records and the entire archival process" and describes it as "not a complementary layer, but a primary knowledge repository about records and their contexts."[19] In a participatory paradigm, the role of an archivist shifts from crafting descriptions to providing frameworks for creating, capturing, and connecting recordkeeping metadata, across, not just within, archival repositories. Such frameworks should also have the potential to make archival description of value and use in other information management contexts, for example, digital humanities, big data management, information governance, etc., where accountability, traceability, and authority of information, reuse, sharing, retention, and disposal are required.[20]

Participatory archival approaches reposition what have traditionally been viewed as the subjects of records and recordkeeping processes as participatory agents with legal and moral recordkeeping rights and responsibilities.

By expanding the definition of record creators to include everyone who has contributed to a record's creative process or been directly affected by its action, notions of co-creation and parallel or simultaneous multiple provenance reposition 'records subjects' as 'records agents'. They support a broader spectrum of rights, responsibilities, and obligations relating to the ownership, management, accessibility, and privacy of records in and through time. (p. 132)[21]

Documenting and servicing those rights requires abandoning the concept of singular records creation, ownership, and custody that dominates extant archival description practices and standards and embracing models that enable representation of a plurality of perspectives. Furthermore, in confronting the power of records and archival frameworks to empower, disempower, harm, and protect, Gilliland has also identified an ethical platform for participatory description, based on five facets—acknowledging, respecting, enfranchising, liberating, and protecting—and with individual and community rights to name, self-identify, respond, disclose, take down, and be secure.[22]

CONCLUSION

Arrangement and description processes, as well as the understanding of these practices, are undergoing transformation as archivists attempt to deal with records created and kept in newer information technologies, business systems, and social environments. This entry has explored the principles underlying arrangement and description processes, along with how these processes have been shaped by the recordkeeping and archiving contexts in which they are constructed. It has also outlined emerging metadata management frameworks to inform the development of the recordkeeping and archiving systems of the future.

REFERENCES

1. Duranti, L. Origin and development of the concept of archival description. Archivaria **1993**, *35* (Spring), 47–54.
2. Cook, T. The concept of the archival fonds: theory, description and provenance in the post-custodial era. In *The Archival Fonds: From Theory to Practice*; Eastwood, T., Ed.; Bureau of Canadian Archivists: Ottawa, Ontario, Canada, 1992; 31–86.
3. International Council of Archives. *ISAD(G): General International Standard Archival Description*, 2nd Ed.; International Council of Archives: Ottawa, Ontario, Canada, 2000.
4. Roe, K.D. *Arranging and Describing Archives and Manuscripts*; Society of American Archivists: Chicago, IL, 2005.
5. McKemmish, S.; Reed, B.; Piggott, M. The archives. In *Archives: Recordkeeping in Society*; McKemmish, S., Reed, B., Piggott, M., Upward, F., Eds.; Centre for Information Studies, Charles Sturt University: Wagga Wagga, New South Wales, 2005; 159–195.
6. Posner, E. The role of records in German administration. In *Archives and the Public Interest. Selected Essays by Ernst Posner*; Munder, K., Ed.; The Society of American Archivists: Chicago, IL, 1967, 2006; 87–97. First published in 1941.
7. Scott, P.; Finlay, G. Archives and administrative change some methods and approaches (part 1). Arch. Manuscr. **1978**, *7* (3), 115–127.
8. Scott, P. The record group concept: a case for abandonment. Am. Arch. **October 1966**, *29*, 493–504.
9. O'Brien, J. Basic RAD: an introduction to the preparation of fonds- and series-level descriptions using the Rules for Archival Description; October 1997. http://scaa.sk.ca/basic_rad.html (accessed August 2007).
10. Duranti, L. *The Long-term Preservation of Authentic Electronic Records: Findings of the InterPARES Project*; Archilab: San Miniato, Italy, 2005. http://www.interpares.org/book/index.htm (accessed August 2007).
11. Hurley, C. The making and keeping of records: (1) what are finding aids for? Arch. Manuscr. **1998**, *26* (1), 58–77.
12. *ISO 23081: Information and Documentation-Records Management Processes-Metadata for Records, Part 1: Principles*; International Standards Organisation: Geneva, Switzerland, 2004.
13. McKemmish, S.; Acland, G.; Ward, N.; Reed, B. Describing records in context in the continuum: the Australian Recordkeeping Metadata Schema. Archivaria **1999**, *48* (Fall), 3–43.
14. Evans, J.; McKemmish, S.; Bhoday, K. Create once, use many times: the clever use of recordkeeping metadata for multiple archival purposes. Arch. Sci. **2005**, *5* (1), 17–42.
15. Gilliland, A.; Rouche, N.; Lindbergh, L.; Evans, J. Towards a 21st century metadata infrastructure supporting the creation, preservation and use of trustworthy records: developing the InterPARES 2 metadata schema registry. Arch. Sci. **2005**, *5* (1), 43–78.
16. Hurley, C. Parallel provenance: (1) what, if anything, is archival description? Arch. Manuscr. **2005**, *33* (1), 146–175.
17. Bastian, J. *Owning Memory: How a Caribbean Community Lost Its Archives and Found Its History*; Libraries Unlimited: Westport, CT/London, U.K., 2003.
18. Ketelaar, E. Sharing: collected memories in communities of records. Arch. Manuscr. **2005**, *33* (1), 44–61.
19. Huvila, I. Participatory archive: towards decentralised curation, radical user orientation, and broader contextualisation of records management. Arch. Sci. **2007**, *8* (1), 15–36.
20. Reed, B. Metadata? A contestable concept. *Presentation by Barbara Reed for 'Paradigm Shift', a seminar in honour of Hans Hofman*; National Archives of the Netherlands: The Hague, the Netherlands, January 27, 2014. http://rkroundtable.org/2014/02/05/metadata-a-contestable-concept/.
21. McKemmish, S.; Piggott, M. Toward the archival multiverse: challenging the binary opposition of the personal and corporate archive in modern archival theory and practice. Archivaria **2013**, *76*, 111–144.
22. Gilliland, G. Acknowledging, respecting, enfranchising, liberating and protecting: a platform for radical archival description, paper. In *Radical Archives Conference*, New York, April 2014.

BIBLIOGRAPHY

A selection of manuals and guides for arrangement and description.

Historical

1. Hensen, S.L. *Archives, Personal Papers, and Manuscripts: A Cataloging Manual for Archival Repositories, Historical Societies, and Manuscript Libraries*, 2nd Ed.; Society of American Archivists: Chicago, IL, 1989.
2. Jenkinson, H. *A Manual of Archive Administration*, 2nd Ed.; Percy Lund, Humphries & Co. Ltd: London, U.K., 1966.
3. Miller, F.M. *Arranging and Describing Archives and Manuscripts*; Society of American Archivists: Chicago, IL, 1990.
4. Muller, S.; Feith, J.A.; Fruin, R. *Manual for Arrangement and Description of Archives*; Originally published in Dutch in 1898, 1940 translation of second edition by Leavitt, A.H. SAA Archival Classics Series. Society of American Archivists: Chicago, IL, 2003.
5. Schellenberg, T.R. *Modern Archives: Principles and Techniques*; University of Chicago Press: Chicago, IL, 1956. SAA Archival Classics Series. Society of American Archivists: Chicago, IL, 2003.

Australia

1. Australian Society of Archivists Committee on Descriptive Standards. *Describing Archives in Context: A Guide to Australasian Practice*; Australian Society of Archivists: Canberra, Australian Capital Territory, Australia, 2007.

Canada

1. Canadian Committee on Archival Description. *Rules for Archival Description*. Revised Version November 2007; Canadian Council on Archives, http://www.cdncouncilarc hives.ca/archdesrules.html (accessed December 2007).

United Kingdom

1. Procter, M.; Cook, M. *Manual of Archival Description*, 3rd Ed.; Gower Publishing Ltd: Hampshire, U.K., 2000.

United States of America

1. Roe, K.D. *Arranging and Describing Archives and Manuscripts*; Society of American Archivists: Chicago, IL, 2005.
2. Society of American Archivists. *Describing Archives: A Content Standard*; Society of American Archivists: Chicago, IL, 2007.

Archival Documentation

Gavan McCarthy
eScholarship Research Centre, University of Melbourne, Melbourne, Victoria, Australia

Joanne Evans
Caulfield School of Information Technology, Monash University, Melbourne, Victoria, Australia

Abstract

Archival documentation is the term used to refer to those documents that record the activities of an archival program. As the term archives has come to mean different things in different contexts, the term archival program has come into common use to represent that wide range of tasks and roles undertaken by people when endeavoring to ensure that information is preserved as evidence. This entry does not attempt to prescribe the specific forms of records that archival documentation should create, as circumstances will vary widely from jurisdiction to jurisdiction and from country to country. Rather, the entry outlines the fundamental issues and purpose of archival documentation in order to help the reader judge the adequacy of existing systems and plan for improved forms of documentation. The guiding principle of archival documentation is that it helps tell the story of the materials managed by an archival program.

INTRODUCTION

Archival documentation is the term used to refer to those documents that record the activities of an archival program. As the term archives has come to mean different things in different contexts, the term archival program has come into common use to represent that wide range of tasks and roles undertaken by people when endeavoring to ensure that information is preserved as evidence. The description and management of materials in an archive is described elsewhere (see entry on "Archival Arrangement and Description," p. 130) and is not covered in detail here. There is an underlying assumption in this entry that the materials being managed by an archival program are created elsewhere. This could be in another place within the same organization or, as with a collecting archive, from people or organizations from the outside world. In all cases the archival program needs to create adequate records that document its activities, as these provide the essential operational framework in which the archival materials are managed.

This entry does not attempt to prescribe the specific forms of records that archival documentation should create, as circumstances will vary widely from jurisdiction to jurisdiction and from country to country. Rather, the entry outlines the fundamental issues and purpose of archival documentation in order to help the reader judge the adequacy of existing systems and plan for improved forms of documentation.

The guiding principle of archival documentation is that it helps tell the story of the materials managed by an archival program. These materials, once managed by an archival program, are generally referred to as archives and in this sense the term refers to a body of records generated by an entity, i.e., a person, an organizational unit or some other agent, in the course of their activities. The aim of the archival program is to ensure that the materials for which it is responsible can be meaningfully interpreted by users at any time in the life of the records, from the moment of receipt into the archival program, through the period of custody and curation, and, if it eventuates, the removal of the materials from the archival program. In order to fulfill this aim archival documentation must be created that records the actions of the archival program with respect to all the materials under its control.

Archival documentation, a critical component in the capture of contextual information surrounding archival materials, is created, kept, and made available:

1. To support the effective and efficient management of an archival program.
2. To maintain the authenticity, integrity, and reliability of records under the control of the archival program.
3. To make the records under the control of an archival program accessible and meaningful to its users.

It encompasses the documentation of records within an archival program, as well as documentation of the archival program itself. Its various forms reflect the complex and heterogeneous nature of archival material, along with the physical, virtual, multi-entity, and multilayered dimensions of records, including the recordkeeping systems in which they are created, managed, and used. All archival activities—initial survey and appraisal; acquisition,

Encyclopedia of Library and Information Sciences, Fourth Edition DOI: 10.1081/E-ELIS4-120044326

transfer of custody and ownership; preservation and conservation; arrangement and description; location management; reference, access conditions, and copyright; and public programming and outreach—are sources of archival documentation and indeed each of these functions relies on information from the others to work effectively.

Most extant archival programs are based around the handling of paper records, generally generated from paper recordkeeping systems. Archivists are facing a challenging time in determining the documentation requirements for records created and managed in digital and networked environments. Consequently, archival programs need to manage the paper/digital hybrids of the last few decades as well as new forms of "born digital" records. Practices and processes need to not only deal with the technicalities of digital records, but also the changing nature of digital and networked recordkeeping systems, and in particular the new narratives or stories that the digital technologies make possible for materials managed by archival programs.

This entry will

- Establish principles governing archival documentation.
- Detail the archival documentation associated with the different phases of a traditional archival program.
- Discuss the impact of digital and networked technologies on documentation requirements and practices.

ARCHIVAL DOCUMENTATION PRINCIPLES

To tell the story of archival materials throughout their life in an archival program, the program itself will need to establish systematic, thorough, and integrated record creation and keeping systems. The emergence, in the latter decades of the twentieth century, of digital database technologies to assist in this process has significantly altered what archival programs can achieve in this area. However, it is not effectively possible to build a computerized database system that can deal with all the variables that may need to be documented in the life of archival materials. Indeed, this is why more generalized recordkeeping systems appeared and still remain an essential part of an archival documentation system. So, it is more than likely that comprehensive archival documentation will be found in a range of systems within an archival program. It is important that these systems are interrelated and cross-reference each other in clearly articulated ways so that future archivists and users can trace and track specific information about the materials as required. The careful use of unique identifiers within archival program documentation systems will aid in the unambiguous interconnection of related information.

Archival materials are assembled and managed on the understanding that they will ultimately be used without recourse to the originating entity or indeed the people responsible for their transfer into an archival program. It is important that specific information held only in the minds of those people is captured and systematically documented so that both the archivists of the future and those wishing to use the materials understand why the materials are where they are and why they are in the form or structure in which they have been preserved in the archive.

Although it is generally understood by archivists, it is not more commonly understood by others outside the profession that, unlike libraries (as a rule), the logical and actual arrangement of materials in an archival program is not determined by a preexisting classification system. Indeed, the description and arrangement of archival materials is predominantly determined by the actions of the entities responsible for the creation and initial purpose of the materials and the recordkeeping systems they created to manage them while they were in the first phase of their life. In accepting materials into an archival program, archivists are sensitive to the provenance of the materials, and to the relationships between materials determined by that shared provenance. Archival documentation must ensure that these relationships are preserved. Very closely aligned with that guiding principle is a respect for the actual juxtaposition of materials as determined by the filing structures, systems, and arrangement of the material at the time they are transferred to the archival program. What should emerge from an engagement with archival materials is a sense of the structure and function of the materials in that first, operational phase of their life. What should also be clear is what has happened to those materials since they entered the archival program. The hand of the archivist should not be invisible but clear and evident to all users of the materials.

The challenges facing those responsible for archival documentation are significant as each case, each new set of materials, each relationship with a transferor and indeed each user will be highly individualized and arise from its own context. Therefore, what remains fundamental to successful archival documentation is the application of consistent and well-recorded processes for handling the materials at all stages in their life.

THE PHASES OF AN ARCHIVAL PROGRAM

The main phases of an archival program involve

- Establishment of the archival program (policy, mandate, scope, governance, rights management, workplace and storage buildings, location control, digital repository systems, and documentation systems and processes).
- Entry or acceptance of materials into the program (survey, accession, mapping and registration, transfer of custody, transfer of ownership, registration of rights, and appraisal).

- Archival processing incorporating
  - Provenance identification, registration, and description.
  - Inventory identification, registration, and description.
  - Collection/series identification, registration, and description.
  - Preservation actions (type and nature of containers for physical objects, type and nature of the digital repository(ies) for digital objects).
- Location, conservation, and access management (storage locations, movements between locations, conservation actions, registration of users, and documentation of use).
- Review of holdings (reappraisal, deaccessioning).

The passage of materials through an archival program is often a discontinuous process. All phases of the program must be meticulously and carefully managed and documented so that all those involved in the various phases of the program have ready access to explicit information that documents where the materials are in the program. However, before materials can commence their journey the archival program itself has to exist.

Establishment

Setting up an archival program is a serious business and with it comes obligations to individuals and organizations that may be moral, ethical, and legal. Consequently, it is essential that there are documents that record the establishment of the program and include the following: the purpose, mandate and authority; reporting requirements; the governance framework; the collecting policy; the access policy; a statement articulating the appraisal process; the roles of the people in the program; privacy policies; copyright management processes; and where required systems that deal with legally enforced freedom of information rights.

Associated with these fundamental administrative records are the operational administrative records that deal with finance, human resources, procurement, and day-to-day management. Although they form an important element of the records of an archival program they have more in common with general administrative practice rather than archival documentation so will not be dealt with here.

However, the records that tell the story of why the archival program was established and who is responsible for its ongoing life are important for users of the records. The reasons why some records may have been kept and others not, and why records may have been documented in a particular way, may be found in the policies of the program or indeed in the policies of the mandating organization. In the contemporary era it is not uncommon for changes in operational and organizational frameworks to occur in frequent although ad hoc phases. It is important that changes in policy and operational frameworks are clearly documented as they occur. In dealing with this sort of change over longer periods of time, it is worthwhile producing comprehensive annual reports that restate clearly the operational and policy framework under which the archival program is operating, making special note of when there are changes, what they are, and the implications for the program.

An archival program requires places for people to work and systems for the storage and preservation of the materials. The extent of the provision of these practical aspects of the program will be determined by the budget for the program and the technologies available at the time. Therefore records documenting the characteristics, design, commissioning, and upgrading of buildings including repository storage systems (physical and digital) are essential.

Allied with these records are the records dealing with the design, establishment, and commissioning of the systems to manage archival documentation. It is likely that these will evolve over time as technologies change and more productive tools become available, allowing better integration and interconnection. In an ideal world, archival documentation systems would be fully self-explaining, but the world is far from ideal so external systems documentation and explanation is necessary. It will be critical for most of these records to be kept for at least as long as the materials in the archival program exist. In other words, the plan should be to keep them indefinitely.

Entry into the Archival Program—The Survey

With the exception of the records mentioned above, the materials managed by an archival program come from outside the program. In order for the materials to become part of an archival program they must be identified and registered. The crossing of the boundary into the archival program is probably the most significant moment in the life of a record. If it survives the day-to-day rough and tumble of life in the operational zone and is kept for whatever reason, it may become a candidate for an archival program. It is not uncommon for life in the operational zone to be quite dynamic and recordkeeping systems to be highly localized and idiosyncratic. The archival program rarely has any influence over what happens in this zone but there are exceptions—it is possible that archival program requirements can influence the identification, management, and appraisal of key records, for example, records of governance, minutes of boards, and similar materials. However, once the archival program boundary has been crossed things need to change dramatically for the record.

Although it is accepted that the ability to interpret archival materials requires knowledge of the context of their life in the operational zone, the best that the archival program can usually achieve is a snapshot of how they were positioned at the time they crossed the archival program boundary. To prepare for this transition an initial survey of the materials in situ needs to be undertaken. New digital documentary tools such as digital cameras, sound recorders, and video recorders can bring significant productivity gains to this process. Archival program management systems should allow for the systematic integration of these items into the documentation process. This will assist in enabling the traceability of the items from this context-as-found through to their preservation location within the archive. Images of materials on shelves, in offices, with the people who use them, or even in garages after years of neglect help capture their story in ways not possible when only using words. Interviews with staff, family, and others associated with the materials, especially stories around the people and circumstances of the projects and activities associated with the materials, can be invaluable.

Transfer and Accessioning

The accession, or the formal registration of the materials as found, is the traditional entry point for records into an archival program. As a rule, the formal custody arrangements of the materials will change during this aspect of the archival program. Therefore, appropriate instruments documenting, authorizing these changes, and articulating the terms of transfer should be utilized. In some cases an exchange of letters may be sufficient, whereas in other cases, forms and other legal documents may be appropriate. For many materials the accession registration may be the first concrete anchor point where their story is documented. The systematic collection of data about the materials at this point is critical. Meticulous attention to detail in process and description is an imperative. For example, data that documents terms and conditions of ownership and custody, date of registration, location of registration, detailed information about the transferee, storage unit labeling, description of contents, notes about the condition of the materials and thoughts about future processing issues, and the provenance of the materials are critical elements of the archival documentation of this process.

The ability to produce a range of accession reports that reproduce the information gathered during the accession process can be most useful. These should be made available to all authorized stakeholders as evidence of the work undertaken and the condition or state of the records on entry into the archival program. All stakeholders should have the opportunity to contribute knowledge to the archival program. This point in the process provides an important opportunity to assess the descriptions of the materials, identify records where there may be ambiguities of interpretation, and check the allocation of key terms and other forms of controlled classification. Each accession or transfer should be allocated a unique identifier within the registration system.

During the accession process information can be gathered about how the incoming material relates to material already managed by the archival program. Is the new material an addition to an existing series or collection or is it something completely new? Documentation that captures this information and analyses it in a systematic way will help with the eventual description, management, and use of the materials. The funding of archival programs is known to be problematic. As a result there can be significant gaps in time between the accession registration and further archival processing or documentation. These gaps are sometimes measured in decades. Indeed, there are archival programs that undertake no further archival processing and use their accession registers as their primary form of control and access.

An accession once registered, consistently documented, and housed in uniquely identified containers, can be responsibly relocated into appropriate storage without comprising the integrity of the records and minimizing the risk of materials going missing. The systematic documentation of the sequence of any new locations is therefore a new requirement of the archival program. The completion of this phase is only the beginning of the archival program but if this phase is undertaken in a painstaking and careful manner, the following phases, rather than being problematic, can be responsibly and efficiently managed.

Once in the Archival Program—Description, Housing, and Management

Once formally established within an archival program subsequent archival processes which require documentation include more detailed identification of materials, the registration and description of entities associated with provenance, the identification and registration of collections, series, record groups, box lists, and inventory items. Again, the allocation of unique identifiers to each of these entities will enable the systematic interlinking of elements of the documentation systems.

The selection of the levels and elements at which an archival program chooses to arrange and describe the materials varies from archival program to archival program. The issues surrounding arrangement and description are covered elsewhere and are therefore not covered here. Most importantly, what is required is clear documentation of the processes, practices, and protocols used by the archival program. There is no right answer or indeed correct arrangement of archival materials but what must be evident to the user are the actions taken by archives staff in preparing the materials for use.

The International Council on Archives has prepared standards that describe the elements of description that should be gathered as part of archival documentation at this phase of an archival program: ISAD(G)—the International Standard for Archival Description (General) and ISAAR(CPF)—the International Standard for Archival Authority Records (Corporations, Persons and Families).

Some countries have developed more detailed local rules for archival documentation during this phase of the archival program and these should be consulted as necessary.

Appraisal—It Happens All the Time

Of all the traditional archival functions, appraisal and the selection of materials for long-term retention is probably one of the most difficult to systematize and therefore document. Appraisal and selection is a process that can and is often undertaken in all phases in the life of materials and is an essential function of the archival program. From the initial survey to the review of materials based on use and changing values, appraisal and selection can be invoked. This may lead to either the noninclusion of materials or the removal of materials from an archival program. The appraisal of materials can also include the identification of any sensitivity associated with the materials due to personal, cultural, or legal circumstances. Archival materials are rarely the province of just one stakeholder, so it is necessary that these issues are identified and clearly documented, even if these sensitivities and resultant restrictions on use remain uncertain. The clear and unequivocal documentation of these actions and processes wherever undertaken is critical.

Uses of Archival Materials—The Role of Feedback

Archival programs may produce a range of guides, finding aids, online databases, and other documentary instruments that enable users to locate, request, and utilize the materials managed by the archival program. These are all important elements of the framework of archival documentation. However, these public instruments generally do not reveal all the information pertinent to the story of the materials, some of which may not be suitable, for privacy or other reasons of sensitivity, for the general public. As a rule access to archival materials, even those that do not have access restrictions, requires users to register with the archival program and agree to abide by a set of local rules. These rules may cover such things as copyright and moral rights obligations. In an ideal setting all the information resulting from this user documentation and any other feedback or additional information the users might provide about the materials they are viewing should be systematically incorporated into the descriptive documentation of the materials. If well done this could be used to enrich guides and finding aids, identify communities of users and aid in appraisal reviews.

IMPACTS OF DIGITAL AND NETWORKED TECHNOLOGIES

From the use of digital capture documentation tools such as cameras, and video and sound recorders in the archival survey and accession process, to the use of imaging and digitization technologies of the materials themselves for enhanced access, security, and preservation, these technologies are not only transforming archival practice but also reconfiguring archival documentation.

The impact of digital and networked technologies on archival programs has been profound and seems likely to continue to drive significant transformation. Not only are most materials destined for archival programs now produced in the first instance in digital forms but there is increasing pressure to radically increase the rate of digitization of analogue materials. However, this is just the beginning of the data revolution facing archives programs. Further manifestations of digitized archival materials where content is extracted and marked up, for example in XML schemas, are being created and these new records must also be documented and integrated into the archival program.

The advent of sophisticated digital repository systems makes possible the effective archival preservation of digital materials. These systems rely extensively on the use of unique identifiers to manage digital objects as well as the related metadata elements required for management and preservation. Digital preservation is a vast topic that is covered elsewhere but there a few points worth reiterating, especially the use of nonproprietary formats, the ability to store the objects on a range of media, the automated checking of integrity and the management of file type migration to enable use through time and changing technologies.

Therefore, it is critical that system description and documentation is well maintained and that data disaster recovery processes are regularly tested. Like the storage and location management systems for paper materials, these systems form only part of a full archival program. The challenge for the archivist is to ensure that all components remain integrated.

Digital systems are becoming increasingly interconnected via the World Wide Web and its underlying services. It is an area that is still to be fully explored by archivists but users are driving the need for better interconnection and cross-referencing of the materials managed by different archival programs. This will add yet another dimension to the archival documentation matrix. The quality and the information architectures utilized by archival programs in their documentation systems will determine how quickly and efficiently these frameworks of connectivity may emerge.

However, despite the challenges posed by rapid technological change, archival programs are shaped by the needs of the human beings and their relationships with archival materials—as the creators of records, the subjects of records, and as the users of records. It is this ongoing narrative, this interplay between people and the evidence of their world, and the stories that make their lives meaningful, that lay the foundations for archival documentation. Archival documentation is driven by epistemological, ethical, and practical imperatives that can all be traced back to the telling of the story of records. Without the stories of the records the stories in the records become increasing difficult for future users to decode.

SUMMARY

Archival documentation involves

1. Records of establishment, including governance, policies, reporting, and administration with a focus on capturing change in these parameters over time.
2. Records of archival systems, including the design, building, and commissioning of archival record-keeping systems (including both database and other systems) and the way they are interconnected and integrated, e.g., contact files/database, collection files/database, accession registers, consignment forms, and disposal records.
3. Records of survey and transfer, including making contact, surveying, appraising, consigning, accessing, physical and custodial transfer, in other words formal registration in the archival program.
4. Records of rights, sensitivities, and access conditions.
5. Records of arrangement and description including further appraisal and assessment, structural analysis, identification of related materials, and processing tasks.
6. Records of location, preservation, and conservation.
7. Records of access and use, including the guides, finding aids and user interfaces to archival documentation; the registration of users and their usage (including publications); annotation, qualification, addition of new information; and management access for restricted or sensitive material.
8. Records of the interconnections and cross-references with other archival programs.

BIBLIOGRAPHY

The following references represent a small and highly selective subset of the literature that either talks directly or indirectly about archival documentation. This encyclopedic entry has not attempted to be scholarly in the citation of sources as an element of presentation and style but the authors acknowledge the broad range of experience and literature they have been able to draw on in writing this entry.

1. Australian Society of Archivists Committee on Descriptive Standards, *Describing Archives in Context: A Guide to Australasian Practice*, Australian Society of Archivists Inc.: Canberra, Australian Capital Territory, Australia, 2007.
2. *Keeping Archives*, 3rd Ed.; Bettington, J., Eberhard, K., Loo, R., Smith, C., Eds.; Australian Society of Archivists Inc.: Canberra, Australian Capital Territory, Australia, 2008.
3. Cook, T. Electronic records, paper minds: The revolution in information management and archives in the post-custodian and post-modernist era. Arch. Manus. **1994**, *22* (2), 300–328.
4. Cox, R. *Managing Institutional Archives: Foundational Principles and Practices*, Greenwood Press: Westport, CT, 1992.
5. Dryden, J. *Respect for Authority: Authority Control, Context Control and Archival Description*, The Haworth Information Press: Binghamton, NY, 2007.
6. Jenkinson, H. *A Manual of Archive Administration*, 2nd Ed. P. Lund, Humphries: London, U.K., 1965. reissue.
7. McKemmish, S. Introducing archives and archival programs. *Keeping Archives*, 2nd Ed. D.W. Thorpe and the Australian Society of Archivists Inc.: Port Melbourne, Victoria, Australia, 1993.
8. In *The Records Continuum: Ian McLean and Australian Archives: First Fifty Years*; McKemmish, S., Piggott, M., Eds.; Ancora Press and Australian Archives: Canberra, Australian Capital Terrritory, Australia, 1994.
9. Nesmith, T. Reopening Archives: Bring new contextualities into archival theory and practice. Archivaria **2005**, *60*, 259–274.
10. Schellenberg, T.R. *Modern Archives: Principles and techniques*, F W Cheshire: Melbourne, Victoria, Australia, 1956.
11. Thomas, S. Gittens, R. Martin, J. Baker, F. *Paradigm: Workbook on Personal Digital Archives*, Bodleian Library, University of Oxford: Oxford, U.K., 2007. Online at: http://www.paradigm.ac.uk/workbook/index.html (accessed January 2009).
12. Yakel, E. *Starting an Archives*, Society of American Archivists and The Scarecrow Press: Chicago, IL, 1994, and Metuchen, NJ.

Archival Finding Aids

Su Kim Chung
Karla Irwin
University Libraries, University of Nevada–Las Vegas, Las Vegas, Nevada, U.S.A.

Abstract

Archival finding aids serve as the primary means by which users can obtain intellectual access to collections of primary source materials be they archives or manuscripts. Although the term may reference the general guide to a repository's collections as a whole, it is more commonly thought of as the guide to an individual collection within a repository. In addition to describing the physical contents of an archival collection, the finding aid comprises other identifying, contextual, and administrative information about the materials. The online environment has led to significant changes in the archival finding aid in the past decade with the development of the encoded archival description, *Describing Archives: A Content Standard*, archival collection management systems, links to digital content, and the addition of Web 2.0 features that seek to add interactive qualities to online finding aids. The future is certain to see continued development and refinement of the finding aid as more materials are born digital or are digitized and ways are developed to describe them in the context of a user-friendly finding aid.

INTRODUCTION

In its broadest conception, the archival finding aid can be considered as any type of descriptive tool that provides intellectual access to the materials held in an archival or manuscript repository on a collective or individual level. Thus, this definition holds that a summary guide to the various record groups and/or collections held in a repository or a document providing box, folder, or item-level description to an individual collection in that repository are both considered finding aids. Whether "published or unpublished, manual or electronic," its purpose is "to establish physical and/or intellectual control over records and/or archival materials"[1] and serve as the means by which users "gain access to and understand the materials."[2]

Although the finding aid can take a variety of formats and is referred to by various names, including inventories, calendars, registers, guides, and container lists, it is generally thought of by most archival practitioners as the primary descriptive tool by which they and users can access materials in an individual archival or manuscript collection. Thus, this entry will focus on the finding aid primarily in its function as an access tool to an individual collection. It will provide a brief introduction to the archival finding aid by considering its function, structure, and content, and efforts at standardization that have evolved with the presentation of finding aids in the online environment. The entry will also discuss recent developments that may affect the creation and presentation of finding aids in the future. Suggestions for further resources in which to find more detailed information and examples of finding aids can be found at the end of the entry.

DEFINING THE ARCHIVAL FINDING AID

Whether a guide that focuses broadly on the totality of records or manuscripts in a repository, or a document that describes an individual collection within that repository, the finding aid is the central means of access and description for archival or manuscript materials. In this manner, it serves much the same function as an index or table of contents does in a book. Thus, it reflects the intellectual and physical organization of a collection and provides descriptive information on these materials at a collection, series, folder, or item level. The level of description and organization reflected in a finding aid can vary and depends on the collection, the processor, and the policies of an institution.

Although they serve much the same purpose, archival finding aids have in the past been referred to by different names depending on whether they were found in an archives or a manuscript repository. In the former, they were traditionally called inventories, while in the latter, they were known as registers. Calendars, a name typically reserved for the most lengthy, detailed finding aids created to list documents at an item level, are more likely to be found in a manuscripts repository where such detail may be helpful for particular formats of personal papers such as correspondence. This difference became less of a concern in recent years, as noted by O'Toole, with the move toward standardization of archival description in the late 1970s. According to O'Toole, "archivists acknowledged this to be a distinction without a difference, since all repositories wanted to know and record the same things about their collections in the effort to understand them and make them usable."[3] Specifically, he credits the

Encyclopedia of Library and Information Sciences, Fourth Edition DOI: 10.1081/E-EISA-120053420

Approval–Archival

publication *Inventories and Registers: A Handbook of Techniques and Examples* as the work that demonstrated "graphically how much alike the supposedly different approaches were."[4]

ARRANGEMENT AND DESCRIPTION IN THE FINDING AID

Whatever its nomenclature, the finding aid can only be created after the essential business of arrangement and description of materials, be they archives or manuscripts, has taken place. Thus, most guides to arranging and describing archives and manuscripts[5–7] consider the finding aid embodies the concept of archival description. During the physical processing of a collection, for instance, as the archivist rehouses and arranges the archival material at hand, he or she takes notes and focuses on what information will eventually need to be placed in the finding aid in order to make both the content and context of the collection clear to users.

The archivist's decisions about how the materials in a collection will be intellectually and physically arranged are also reflected in the finding aid. In fact, before any work is done on the organization of individual items or folders, the archivist must determine the appropriate level of arrangement for the collection as a whole. This arrangement may be at a record group, series, or subseries level depending on the type of materials being processed. Although rare, the correspondence or diaries of a notable person may be described at an item level while it is more likely that the records of an institution or organization may only be described at the series or box level. Following the concept of original order (respect des fonds), some materials, such as the archives of an organization or business, should retain something of the natural record group or series arrangement they had as functioning records in a working business or organizational environment; they can then be easily arranged in series and subseries due to the fact that their previous functions are easy to discern.

Other materials such as personal papers and manuscripts, however, may lack any type of obvious arrangement. As a result, the archivist will generally create the arrangement for personal papers during the course of processing by grouping them into series and sometimes subseries based on elements such as formats, time periods or the roles, occupations or interests of the person whose papers are being processed. The series and subseries arrangement provides the necessary hierarchical structure for the physical and intellectual organization of the materials that will then be conveyed in the description presented in the finding aid.

Beyond the larger concepts of arrangement, finding aids, as guides to individual collections of records or papers, typically consist of a number of both administrative and descriptive elements designed to provide

intellectual and physical control over the materials being described. Additional descriptive elements typically included in an archival finding aid serve as a means to provide contextual information such as size, date span, access restrictions, and organizational or individual history. Such information forms part of what Miller[5] described in 1990 as an "archival description system" that should be included in the creation of all finding aids. This system consists of various elements: information *about* the records and their creators, their intellectual content, use of the records including access and relationships to other records, physical characteristics, and gaining access to the records.

Some 15 years later, Roe's[7] version of these necessary elements is somewhat similar but with differences that reflect the development of the encoded archival description (EAD) standard for the structuring of finding aid content. She describes the elements needed for archival description within the finding aid as comprised of the following: core identifying information, contextual information, physical characteristics, administrative information needed to support use, informational content, and access/index points. Whatever these subtle differences, however, the elements described by Miller and Roe both comprise the descriptive information that a finding aid *should* ideally contain even if the structure and presentation of these elements may differ in terms of the detail and scope decided on by the archivist (processor) or the policies dictated by the archival repository.

Within these general categories of descriptive information are the detailed elements that will form part of the finding aid. For instance, the core identifying information may contain details such as the title, dates, and quantity. Contextual information consists of the biographical sketch of an individual or the administrative history of an organization or business. The section on physical characteristics may contain details on the quantity/extent of records and the actual organization and arrangement of the collection. Administrative information regarding use may describe any restrictions or copyright issues that apply to the records. Finally, there is the heart of the collection—the series, folder, or even item-level listing that describes the content of the collection at the highest level of detail.

CREATING THE FINDING AID

Once the archivist has compiled this information, he/she transforms it into the finding aid. Structurally, a finding aid is divided into two parts—the front matter and the container list. In combination, the two parts illustrate the arrangement and description of the collection of records, archives, or manuscript material as described earlier. The front matter contains the identifying, contextual, administrative, and physical information about the collection, while the container list is comprised of the content—the listing of all box and/or folder (sometimes item) titles in the collection. See Fig. 1 for a sample finding aid that follows this format.

For finding aids that describe materials that are particularly rich in content, such as literary manuscripts, diaries, correspondence, photographs, or maps, some archivists may choose to include an item-level inventory in the container listing. Typically, because of the extra effort and resources needed to create finding aids at this level of detail, archivists only produce item-level description when it is anticipated that a collection will be used heavily.

Until the advent of word processing software, finding aids were usually created on typewriters and the resulting print copies were only available within, or by user request

Collection Name:	Desert Breeze Hotel Casino Collection
Collection Number:	MS 105
Creator:	Desert Breeze Hotel
Extent:	5 linear feet, (5 boxes)
Dates:	1948-1979 (bulk 1955-1965)
Prepared by:	Cecilia Sinatra, 1990

Organizational History

The Desert Breeze Hotel Casino opened in 1949 on the Las Vegas Strip. For thirty years it was located on Las Vegas Boulevard just north of the legendary Flamingo Hotel Casino resort. The hotel was built in what was a cutting-edge design at the time and is now referred to as mid-century modern design. It was known for its lush landscaping and the many ponds and waterfalls that dotted the property. The Desert Breeze was also renowned for the top caliber of entertainment it featured in its showrooms and for its famous line of chorus girls – the Breezettes. Prominent singers from the opera and jazz worlds as well as Hollywood singing stars graced its stages frequently as did vaudeville performers and comedians.

A scandal befell the hotel in 1965 when it was found that individuals connected with organized crime in New York City had been skimming the casino profits for millions of dollars over a period of five years.
Despite the bad press, the Desert Breeze bounced back and was still regarded as one of the top hotels in Vegas in the early 1970s. After many successful years, however, the Desert Breeze Hotel began to fall into disrepair in the late 1970s. In 1979, the Desert Breeze Hotel Casino was sold by its owners and imploded to make way for the Tropical Sands Hotel Casino Resort.

Scope and Content

The Desert Breeze Hotel Casino Collection is comprised solely of materials from the public relations department of the hotel. The materials document the hotel's efforts at promoting itself within the Las Vegas community and nationwide. The public relations department was charged with initiating publicity stunts in order to promote the hotel and with advertising the casino in a variety of media including print, radio, and television. The collection includes correspondence from the head of the public relations department to a variety of newspapers and public relations figures across the country as well as press releases and press clippings on the hotel. Scrapbooks created by the public relations are also included in the collection.

Arrangement

The collection is arranged into two series:
- I.   Correspondence, 1948-1978
- II.   Publicity, 1948-1979.

Series I: Correspondence
Extent: 2 linear feet, (2 boxes)
Date: 1948-1978
Scope and Content
This series is comprised of correspondence from the public relations department of the Desert Breeze Casino. The majority of correspondence is from the head of the Desert Breeze Hotel public relations department, Sid Smith, who worked in this capacity from 1948-1972. The remaining correspondence is from various individuals who worked in the Public Relations department after Smith's departure. It also contains Christmas cards from various celebrities and entertainers who either stayed at the hotel or performed in the showroom.
Arrangement
Material is arranged chronologically.

Box	Folder	Contents
1	1	Incoming Correspondence, 1948-1960
	2	Outgoing Correspondence, 1948-1960
	3	Incoming Correspondence, 1960-1975
	4	Outgoing Correspondence, 1960-1970
	5	Incoming Correspondence, 1975-1978
2	1	Christmas Cards, 1948-1955
	2	Christmas Cards, 1956-1960
	3	Christmas Cards, 1961-1965
	4	Christmas Cards, 1966-1975

Fig. 1   Sample finding aid (fictional collection).

Series II: Publicity
Extent: 3 linear feet, (3 boxes)
Date: 1948-1979
Scope and Content
This series is comprised of materials generated by the public relations department of the Desert Breeze Hotel Casino and articles clipped from local and national newspapers.
Arrangement
The series is arranged according to format: press clippings, press releases and scrapbooks.

Box	Folder	Contents
3	1	Press clippings: local, 1948-1960
	2	Press clippings: local, 1965-1975
	3	Press clippings: national, 1948-1960
	4	Press clippings: national, 1961-1975
4	1	Press releases, 1948-1960
	2	Press releases, 1966-1970
	3	Press releases, 1972-1975
	4	Press releases, 1976-1979
5	1	Scrapbook, 1950-1955
	2	Scrapbook, 1956-1960
	3	Scrapbook, 1961-1965
	4	Scrapbook, 1966-1972

Fig. 1   (*Continued*).

from, the repository in which they were created. Initially, only those institutions with the resources and staff to produce published versions of their finding aids were able to see them in wider circulation as these versions could be purchased by other individuals or libraries. In 1983, the development of the *National Inventory of Documentary Sources in the United States* (*NIDS-US*), a compilation of finding aids on microfiche (and later CD-ROM) for collections located at the Library of Congress, federal and state repositories, and select historical societies and academic libraries, did provide a wider degree of access to some finding aids. Even then, however, this access was limited to those who had access to a library that held *NIDS-US*.[8] Fortunately, this situation has changed dramatically as a result of the Internet and the development of the EAD standard for the encoding of finding aids. Archival and manuscript repositories throughout the United States and the world now regularly place their finding aids online, providing a greater degree of access for a much wider range of users than was previously possible.

Developing Archival Standards for Finding Aids

Prior to EAD standardization, efforts in the archives community were centered around efforts to make the *summary descriptions* of collections available through the national bibliographic utilities (at the time the Online Computer Computer Library Center, Inc. and the Research Libraries Group) that had enabled libraries to make the existence and location of published works known—first in print catalogs and then over networked computing systems. In adapting the MARC standard for archival materials (through the work of the National Information Systems Taskforce of the Society of American Archivists), the USMARC Archival and Manuscripts Control (MARC AMC) format was created as a means to make it possible for archives and manuscripts to share summary descriptions of individual collections within their repositories in bibliographic catalogs. As Pitti[9] has noted, however, MARC AMC was a content encoding standard only and not a standard for the content of the records themselves. Only with the publication of the *Archives, Personal Papers, and Manuscripts* (APPM) cataloging rules by Steven L. Hensen in 1983 (revised in 1989) were archive and manuscript repositories able to fully utilize the MARC AMC format and create catalog records of their individual collections to place into the national bibliographic utilities.

Although this was a giant step in the direction of providing greater access to archival materials at the collection level, the development of EAD brought this access to the next level by providing the means to place complete finding aids online. Although early computer applications such as SPINDEX I, II, and III (1960s–1970s) had provided greater access to collections by enabling the automation of finding aids, they did not have the delivery mechanism of the online environment.[2] The EAD standard addressed the structure of a finding aid and enabled a visual representation in the online environment that would reflect the hierarchical organization and structure of the finding aid. Developed by Daniel Pitti and a committee of other archival description experts in 1995–1996, EAD is a data content standard that enables users to markup (encode) the contents of a finding aid using the Extensible Markup Language (XML) standard, thus providing a structure for describing an entire collection and its various components. Tags surrounding different elements (such as the scope and content note, biography, or organizational history or related materials) in a finding aid encoded using EAD enable the contents of a finding aid to be searched at various hierarchical levels of description from the broadest collection level to the more granular series, subseries, or folder level. Beyond the individual finding aid, however, EAD tags also enable searching on a larger scale within union catalogs of finding aids or in a potential federated search of multiple finding aids systems.

Although finding aids have traditionally varied in style across repositories, the structural requirements of EAD have encouraged the existence of more standardized finding aid content and presentation in many repositories. EAD has been widely adopted in the archival community and has become a true international standard.[10] However, adoption is still unattainable for some and despite the increased access afforded by EAD barriers to implementation still remain. These range from a lack of staff, resources, and technology expertise to a concern with revising existing finding aids to fit EAD structures to questioning its propriety as a standard or how it might fit in within a repository (see for example: Tatem,[11] Meissner,[12] Roth,[13] Marshall,[14] Yakel and Kim,[15] and Yaco.[16])

EAD/XML editing software such as Archivists' Toolkit, Archon, and, most recently, ArchivesSpace have removed some barriers by minimizing the time and expertise needed to convert older finding aids and place EAD encoded finding aids online. Archon and ArchivesSpace provide all-in-one software that can be utilized to encode finding aids and output them in several formats such as HTML, XML, or PDF.[17]

ArchivesSpace is also being designed to support another recent XML encoded standard: Encoded Archival Context for Corporate Bodies, Persons, and Families (EAC-CPF). Adopted by the Society of American Archivists in 2001 and maintained with the Berlin State Library, the EAC-CPF standard is a means to improve creator description. EAC-CPF supports linkages between creators (agents) to show the relationship among record-creating bodies with additional links to record description and other related entities. According to Weimer, EAC-CPF will "eliminate the need for redundant administrative/biographical information in various finding aids; enable the linking of access by creator to records within a single repository, or across multiple repositories; and ease the maintenance of variant and related names."[18] Users gain a better understanding of records through information about the context of how agents created and used the material.[19] The number of EAC-CPF implementers is set to grow in the coming years with increased education and training on the subject, but use thus far is modest. Only 5% of respondents in a recent survey claim use of the standard at their repository.[20]

Another standard that had a significant effect on finding aids occurred with the release of *Describing Archives: A Content Standard* (DACS). Whereas EAD was created to reflect and standardize the hierarchical *structure* of a finding aid, DACS serves as a standard for describing both archival catalog records and finding aid content. Following both internal SAA committee reviews and input from the archival community, it was officially approved by the Society of American Archivists as an SAA standard in 2004.

DACS superseded the APPM standard, while also supplementing and extending the skeletal rules of description found in Chapter 4 of the *Anglo-American Cataloging Rules* (*AACR2*). Its rules for describing archival content are compatible with both the EAD and MARC formats. It also serves as the US implementation of international descriptive standards (primarily ISAD(G) and ISAAR (CPF)). What is significant about DACS with regard to finding aids is that it allows for the description of archival and manuscript material at any level of description from the collection to the item level, leaving behind the bibliographic approach that was inherent in AACR2 and followed to some degree in APPM. In 2013, SAA released the second edition of DACS with the intent of harmonizing its rules to the standards of ICA, ISAD, and Resource, Description, and Access (RDA) framework.[10] RDA was published in 2012 as the primary standard for bibliographic description in libraries, replacing *AACR2*.[21]

Current and Future Finding Aid Developments

With more and more repositories using EAD as the means to present their finding aids online, there has been a corresponding increase in the concern over the usability of these archival finding aids in the online environment. In addition, with online finding aids reaching a more varied audience unfamiliar with the function and structure of these guides, archivists have realized the need to develop a greater awareness of these elements as they design finding aids for the online environment. Thus, both user studies and usability testing of online finding aids have increased. Archival journals such as the *American Archivist, Archival Issues,* and the *Journal of Archival Organization* have all featured case studies on the usability tests carried out by various repositories in different stages of implementing EAD (see for example: Roth,[13] Prom,[22] Yakel,[23] Coats,[24] and Daines and Nimer[25]). Conclusions are varied, but all suggest the need to pay closer attention to issues of terminology, navigation, display, and structure. In particular, there has been criticism about the simple transfer of the print finding aid into the web environment with little consideration of how this structure should be adjusted, either cosmetically or structurally, to take advantage of the features offered by the online environment. User expectation of finding aids and ease of navigation is also a concern.

The continued development of online finding aids is also making possible other innovations such as the linking of digital objects representing selected collection materials to their description in the finding aid. Light and Hyry emphasize the potential of enriching online finding aids with web-based annotations, suggesting that this enhanced description could promote discovery by augmenting existing access or by offering alternative descriptive language for researchers. In this enriched form, they argue, the finding aid could truly become "the center for the accumulation of knowledge about a collection, instead of residing in the experience and knowledge of seasoned reference archivists."[26] Speaking from a postmodern perspective, Light and Hyry (2002) also advocate for placing colophons in finding aids as a means to inform researchers about the (processing) archivist's role in representing and interpreting a collection—a means to acknowledge his/her editorial contribution and "acknowledge our mediating role in shaping the historical record."[27] It appears adoption of Light and Hyry's suggestion has been minimal thus far. In a 2014 survey, a low percentage of respondents answered "yes" to whether or not they included commentary and annotations to their institution's finding aids, expressing concern with time required to address feedback and changes, as well the quality of user feedback.[28,29]

Recently, as user-generated content on the web (blogs, RSS feeds, user feedback, podcasts, and wikis) has become more pervasive, some archives have begun to look at these technologies as a means to transform their online finding aids and provide them with these more interactive qualities. The "Next Generation Finding Aids" project at the University of Michigan's School of Information that began in 2005 is one prominent example; in seeking to expand the capability of EAD, its goals have been to use these Web 2.0 technologies (collaborative filtering and social navigation mechanisms in this case) to "challenge the traditional finding aid structure" by making the archival and research experience "collaborative and participatory."[30,31] Calisphere by the California Digital Library is an accessible interface for K-12 audiences and an example of the relationship between online finding aids and images.[32,33] With archival repositories continually looking for ways to make their websites and finding aids responsive to user needs, research continues into the means by which Web 2.0 can enrich and transform the capabilities of online finding aids.

Another development in the archival world that has affected the creation of finding aids is the 2005 Greene–Meisner study, which addresses a problem common to many archival repositories: the existence of large processing backlogs. The study suggests a number of physical and intellectual methods for the minimal processing and description of archival collections that can reduce processing times and the subsequent creation of finding aids. While acknowledging the critical importance of archival description, Greene and Meisner's rubric, succinctly expressed as "More Product, Less Process," acknowledges the critical importance of archival

description while arguing that "it needn't be long-winded, laborious or minutely detailed to be effective."[34] To this end, they suggest that not all collections require the same level of description, and that even within collections the same level of description may not be needed for every series. Archival repositories are implementing some or all of these guidelines in processing their collections (see for example: McCrea,[35] Weideman,[36] and Meissner and Greene[37]). Even DACS offers guidelines on the description of minimally described collections in the form of "minimum" requirements as opposed to "optimum."[38]

CONCLUSION

The archival finding aid, as the primary means by which users obtain intellectual access to collections of primary source materials, plays an essential role in a researcher's information-seeking behavior within an archives or manuscripts repository. Although the term may reference the general guide to a repository's collections as a whole, it is more commonly thought of as the guide to an individual collection within a repository. In addition to describing the physical contents of an archival collection, the finding aid is comprised of other identifying, contextual, and administrative information about the materials. The online environment has led to significant changes in the archival finding aid with the development of the EAD and DACS descriptive standards, links to digital content, archival collection management systems, and the potential addition of Web 2.0 features that seek to add interactive qualities to online finding aids. The future is certain to see continued development and refinement of the finding aid as more materials are born digital or are digitized and ways are developed to describe them in the context of a user-friendly finding aid. Ultimately, however, the task of the finding aid, whether in the print or online environment, is to provide adequate access and information on archival collections, and it is important not to lose sight of that fact in the face of whatever advantages technology may provide in the future.

REFERENCES

1. Bellardo, L.J.; Bellardo, L.L.; Society of American Archivists. *A Glossary for Archivists, Manuscript Curators, and Records Managers*; Society of American Archivists: Chicago, IL, 1992.

2. Pearce-Moses, R. *A Glossary of Archival and Records Terminology*; Society of American Archivists: Chicago, IL, 2005.

3. O'Toole, J.M. *Understanding Archives and Manuscripts*; Society of American Archivists: Chicago, IL, 1990; 43.

4. O'Toole, J.M. *Understanding Archives and Manuscripts*; Society of American Archivists: Chicago, IL, 1990; 46.

5. Miller, F.M. *Arranging and Describing Archives and Manuscripts*; Society of American Archivists: Chicago, IL, 1990.

6. Fox, M.J.; Wilkerson, P.L. *Introduction to Archival Organization and Description: Access to Cultural Heritage*; Getty Information Institute: Los Angeles, CA, 1999.

7. Roe, K. *Arranging & Describing Archives & Manuscripts*; Society of American Archivists: Chicago, IL, 2005.

8. Burke, F. The national inventory of documentary sources in the United States (NIDS-US). Ref. Libr. **1997**, *56*, 67–81.

9. Pitti, D.V. Encoded archival description: the development of an encoding standard for archival finding aids. Am. Arch. **1997**, *60* (3), 268–283.

10. Gracy, K.F.; Lambert, F. Who's ready to surf the next wave? A study of perceived challenges to implementing new and revised standards for archival description. Am. Arch. **2014**, *77* (1), 98.

11. Tatem, J.M. EAD: obstacles to implementation, opportunities for understanding. Arch. Issues **1998**, *23* (2), 155–169.

12. Meissner, D.E. First things first: reengineering finding aids for implementation of EAD. Am. Arch. **1997**, *60* (4), 372–387.

13. Roth, J.M. Serving up EAD: an exploratory study on the deployment and utilization of encoded archival description finding aids. Am. Arch. **2001**, *64* (2), 214–237.

14. Marshall, J.A. The impact of EAD adoption on archival programs: a pilot survey of early implementers. J. Arch. Organ. **2002**, *1* (1), 35–55.

15. Yakel, E.; Kim, J. Adoption and diffusion of encoded archival description (EAD). J. Am. Soc. Inf. Sci. Technol. **2005**, *56* (13), 1427–1437.

16. Yaco, S. It's complicated: barriers to EAD implementation. Am. Arch. **2008**, *71* (2), 456–475.

17. Yaco, S. It's complicated: barriers to EAD Implementation. Am. Arch. **2008**, *71* (2), 471–472.

18. Weimer, L. Pathways to provenance: *DACS* and creator descriptions. J. Arch. Organ. **2008**, *5* (1–2), 37.

19. Gracy, K.F.; Lambert, F. Who's ready to surf the next wave? A study of perceived challenges to implementing new and revised standards for archival description. Am. Arch. **2014**, *77* (1), 98.

20. Gracy, K.F.; Lambert, F. Who's ready to surf the next wave? A study of perceived challenges to implementing new and revised standards for archival description. Am. Arch. **2014**, *77* (1), 111.

21. Gracy, K.F.; Lambert, F. Who's ready to surf the next wave? A study of perceived challenges to implementing new and revised standards for archival description. Am. Arch. **2014**, *77* (1), 113.

22. Prom, C.J. User interactions with electronic finding aids in a controlled setting. Am. Arch. **2004**, *67* (2), 234–268.

23. Yakel, E. Encoded archival description: are finding aids boundary spanners or barriers for users? J. Arch. Organ. **2004**, *2* (1/2), 63–78.

24. Coats, L.R. Users of EAD finding aids: who are they and are they satisfied? J. Arch. Organ. **2004**, *2* (3), 25–39.

25. Daines, J.G.; Nimer, C.L. Re-imagining archival display: creating user-friendly finding aids. J. Arch. Organ. **2011**, *9* (1), 4–31.

26. Light, M.; Hyry, T. Colophons and annotations: new directions for the finding aid. Am. Arch. **2002**, *65* (2), 216–230. p. 228.

27. Light, M.; Hyry, T. Colophons and annotations: new directions for the finding aid. Am. Arch. **2002**, *65* (2), 216–230. p. 217.

28. Gorzalski, M. Examining user-created description in the archival profession. J. Arch. Organ. **2014**, *11* (1–2), 9.

29. Gorzalski,, M. Examining user-created description in the archival profession. J. Arch. Organ. **2014**, *11* (1–2), 19–20.

30. http://quod.lib.umich.edu/p/polaread/ (accessed November 7, 2014).

31. Krause, M.G.; Yakel, E. Interaction in virtual archives: the polar bear expedition digital collections next generation finding aid. Am. Arch. **2007**, *70* (2), 282–314.

32. http://www.calisphere.universityofcalifornia.edu/ (accessed November 7, 2014).

33. Daniels, M.G.; Yakel, E. Seek and you may find: successful search in online finding aid systems. Am. Arch. **2010**, *73* (2), 562.

34. Greene, M.A.; Meissner, D. More product, less process: revamping traditional archival processing. Am. Arch. **2005**, *68* (2), 208–263. p. 246.

35. McCrea, D.E. Getting more for less: testing a new processing model at the University of Montana. Am. Arch. **2006**, *69* (2), 284–290.

36. Weideman, C. Accessioning as processing. Am. Arch. **2006**, *69* (2), 274–283.

37. Meissner, D.; Greene, M.A. More application while less appreciation: the adopters and antagonists of MP LP. J. Arch. Organ. **2010**, *8* (3–4), 174–226.

38. Rush, M.; Holdzkom, L.; Backman, P.; Santamaria, A.; Leigh, A. Applying DACS to finding aids: case studies from three diverse repositories. Am. Arch. **2008**, *71* (1), 215.

BIBLIOGRAPHY

1. Bellardo, L.J.; Bellardo, L.L. *A Glossary for Archivists, Manuscript Curators, and Records Managers*; Society of American Archivists: Chicago, IL, 1992.

2. Burke, F. The national inventory of documentary sources in the United States (NIDS-US). Ref. Libr. **1997**, *56*, 67–81.

3. Coats, L.R. Users of EAD finding aids: who are they and are they satisfied? J. Arch. Organ. **2004**, *2* (3), 25–39.

4. Daniels, M.G.; Yakel, E. Seek and you may find: successful search in online finding aid systems. Am. Arch. **2010**, *73* (2), 535–568.

5. Fox, M.J.; Wilkerson, P.L. *Introduction to Archival Organization and Description Access to Cultural Heritage*; Getty Information Institute: Los Angeles, CA, 1999.

6. Gorzalski, M. Examining user-created description in the archival profession. J. Arch. Organ. **2013**, *11* (1–2), 1–22.

7. Gracy, K.F.; Lambert, F. Who's ready to surf the next wave? A study of perceived challenges to implementing new and revised standards for archival description. Am. Arch. **2014**, *77* (1), 96–132.

8. Greene, M.A.; Meisner, D. More product, less process: revamping traditional archival processing. Am. Arch. **2005**, *68* (2), 208–263.

9. Hensen, S.L. *Archives, Personal Papers, and Manuscripts: A Cataloging Manual for Archival Repositories, Historical Societies, and Manuscript Libraries*, 2nd Ed.; Society of American Archivists: Chicago, IL, 1989.

Approval–Archival

10. Krause, M.G.; Yakel, E. Interaction in virtual archives: the polar bear expedition digital collections next generation finding aid. Am. Arch. **2007**, *70* (2), 282–314.

11. Light, M.; Hyry, T. Colophons and annotations: new directions for the finding aid. Am. Arch. **2002**, *65* (2), 216–230.

12. Meissner, D.E. First things first: reengineering finding aids for implementation of EAD. Am. Arch. **1997**, *60* (4), 372–387.

13. Miller, F.M. *Arranging and Describing Archives and Manuscripts*; Society of American Archivists: Chicago, IL, 1990.

14. O'Toole, J.M. *Understanding Archives and Manuscripts*; Society of American Archivists: Chicago, IL, 1990.

15. Pearce-Moses, R. *A Glossary of Archival and Records Terminology*; Society of American Archivists: Chicago, IL, 2005.

16. Pitti, D.V. Encoded archival description: the development of an encoding standard for archival finding aids. Am. Arch. **1997**, *60* (3), 268–283.

17. Prom, C.J. User interactions with electronic finding aids in a controlled setting. Am. Arch. **2004**, *67* (2), 234–268.

18. Roe, K. *Arranging & Describing Archives & Manuscripts*; Society of American Archivists: Chicago, IL, 2005.

19. Roth, J.M. Serving up EAD: an exploratory study on the deployment and utilization of encoded archival description finding aids. Am. Arch. **2001**, *64* (2), 214–237.

20. Rush, M.; Holdzkom, L.; Backman, P.; Santamaria, A.; Leigh, A. Applying DACS to finding aids: case studies from three diverse repositories. Am. Arch. **2008**, *71* (1), 210–227.

21. Society of American Archivists. Committee on Finding Aids. In *Inventories and Registers: A Handbook of Techniques and Examples: A Report*, Society of American Archivists: Chicago, IL, 1976.

22. Society of American Archivists. *Describing Archives: A Content Standard*; Society of American Archivists: Chicago, IL, 2013.

23. Tatem, J.M. EAD: obstacles to implementation, opportunities for understanding. Arch. Issues **1998**, *23* (2), 155–169.

24. Weimer, L. Pathways to provenance: *DACS* and creator descriptions. J. Arch. Organ. **2008**, *5* (1–2), 33–48.

25. Yaco, S. It's complicated: barriers to EAD implementation. Am. Arch. **2008**, *71* (2), 456–475.

26. Yakel, E. Encoded archival description: are finding aids boundary spanners or barriers for users? J. Arch. Organ. **2004**, *2* (1/2), 63–78.

Archival Management and Administration

Michael J. Kurtz
National Archives at College Park, U.S. National Archives and Records Administration, College Park, Maryland, U.S.A.

Abstract

This entry provides a detailed overview of the key elements involved in managing archival institutions in the twenty-first century. Focus is on effective leadership exercised in the paradigm of organizational complexity and relationships. Core management functions of organizing work; planning; and managing financial and human resources are described in the context of dramatic societal change and the unique role which archives fulfill in society. The transformative role of information technology in revolutionizing archival work processes and providing access to potentially unlimited numbers of users forms a key underpinning of contemporary archival management.

INTRODUCTION

Archival institutions in the first years of the twenty-first century are in the midst of a profound transformation. The revolution in information technology and the advent of the digital era have challenged traditional concepts of how records are created and used; how archives manage and preserve their collections; and, most importantly, provide access. Access to records and archives is a hallmark of a democratic society. Thus, whether in the public or private sectors the demand for transparency and accountability has a dramatic impact on the management of archival institutions.

Archival institutions in most instances face stiff competition in obtaining needed recognition and resources. In the era of Web and online access the continued relevance of amassing collections and managing them in the traditional manner is under continual challenge. Archival managers are faced with a workplace in transition. New technologies, new work processes, and a diverse and demanding workforce pose particular challenges for archival managers.

Though there are professional and management issues specifically relevant to the archival setting, at its heart, management as Mary Parker Follett, an early twentieth century management theorist, put it, "is the art of getting things done through other people."[1] This succinctly captures the essence of management work. Managers must lead in organizing work and aiding workers in being productive. From the perspective of management guru Peter Drucker, management is the key organ of any institution, responsible for the performance and survival of the organization.[2]

Management is basically about people, what they do, and the organizations in which they work. From this perspective, archives are like any other organization. The basic management tasks of judiciously applying available resources (i.e., money, people, space, technology, and equipment) to mission-related work are carried out in a variety of organizational frameworks and settings.

This entry presents a management paradigm, organizational complexity, as a particularly useful method to understand the web of relationships—personal, professional, and organizational—around which the work of complex modern organizations revolves.[3] Though many archival organizations may be relatively small in size, almost all are located in larger, hierarchical organizations such as libraries, museums, historical societies, and at all levels of government. In addition to purely internal relationships all archival entities are constantly involved in a web of complex external relationships including resource allocators, donors, stakeholders, customers, the media, and oversight bodies.

Administrators who seek to succeed in leading archival institutions must be able to understand the complex systems and subsystems that undergird all organizations. Mastery of organizational complexity enables the administrator to lead, organize work, manage resources, and communicate in a variety of forums essential to achieving the mission of the archives. Archival management, as is management in any other setting, is complicated and often confusing without a clear foundation in theory as well as practice.

MANAGEMENT THEORY AND PRACTICE

Modern management theory is principally derived from the world of business and industry. Beginning with the Industrial Revolution in the mid-nineteenth century, theorists such as Belgian industrialist Henri Fayol and American engineer Frederick W. Taylor identified the basic management functions which formed the basis for organizing work in the emerging factory system. These

Encyclopedia of Library and Information Sciences, Fourth Edition DOI: 10.1081/E-ELIS4-120044286

functions—planning, organizing, budgeting, directing, and controlling—also formed the basis for managing the increasingly large bureaucratic organizations that began dominating modern society's landscape.

The development of the human and social sciences has profoundly affected management theory and practice moving far beyond Taylor and Fayol's "scientific" approach which viewed work and the organization from a "machine" perspective emphasizing the division of work into specialized tasks or functions with a command and control management structure. Concepts from psychology and sociology, in particular, such as group behavior, group dynamics, and collaborative work processes have found their way into contemporary management and organizational theory.

Theorists and management practitioners draw from a variety of concepts and theories in order to better understand and lead organizations situated in a rapidly evolving knowledge-based economy. Contingency theory, resource dependence theory, institutional theory, and organizational learning, among others, provide different insights and approaches into understanding the nature of work and organizations, and the critical impact of societal change on organization culture.

An ongoing workplace revolution is underway which is transforming society, organizations, and the role of managers. This definitely includes archives and archival managers at all levels of the organization. The management functions identified by Fayol and Taylor and modified by their successors remain required in the workplace. Leadership in management continues as an essential function required for organizational success. Yet, how these functions are carried out is undergoing great change. Information technology has revolutionized communications within organizations and between organizations and the external environment. The modern workforce is educated, diverse, and demands autonomy in the workplace. Managers are no longer commanding or directing staff and passing decisions down and information up the hierarchical ladder. Rather, in a world increasingly based on the effective acquisition and utilization of knowledge swiftly absorbed and communicated, the new paradigm for managers is one of complex organizational relationships. Picture for a moment a web of relationships connecting the managers with staff, other parts of the parent organization (the usual situation for archives), and numerous groups external to the organization. Successful management is based on understanding the complicated and changeable nature of organizational relationships and developing relationships to further the archival mission.

Particularly in any organization which is totally knowledge-based such as archives, managers cannot operate self-contained units. The varying ways in which managers relate to staff, to other parts of the institution, and to the complex world of customers (i.e., patrons, researchers, constituent groups, and the general public) demands a new set of skills. The ability to manage diverse relationships is the ultimate measure of a manager's success.

Archival organizations operate in a variety of settings including the public sector, not-for-profit organizations, religious denominations, and corporations, among others. All have archival entities embedded in them. But archives in all of these settings increasingly operate within a reengineered and automated workplace. Managers can no longer just assign and monitor work, report to their supervisors, and operate in traditional ways. The flood of information coming in via the Internet, the Web, and the intranet is transforming the work environment. Staff members are able to communicate and perform their jobs in this automated milieu across the organization, the city, the state, the nation, and the world. Staff members are not really constrained to operate and communicate within traditional hierarchical channels. Management theories and techniques such as total quality management, business process reengineering, or knowledge management emphasize the critical importance of staff empowerment and investment in the institution's mission if long-term success is to be achieved. Corporate restructuring and tight budget times for all levels of government place archival organizations and staff members in potential peril.

Role of the Manager

The manager's principal contribution to achieving success for the archives is in managing multiple relationships. The archival and program staff must be "managed" in new ways. The manager more and more assumes the role of coach and mentor. The educated, Internet-savvy, and mobile workforce of today requires leaders who assist staff in developing individually and as a group. Teamwork, whether formal, ad hoc or a mix, dominates the workplace. The manager as coach must aid the staff in gaining the skills and the autonomy needed to make decisions about the work at hand.

The information-age manager must aid staff in learning decision-making and consensus-building skills so they can make as many work-process-related decisions as possible. This empowerment of staff, whether it takes place on the factory floor or the archives processing room, reflects widespread technological and social changes that have dominated the past 50 years. This new relationship is one where tasks previously considered as solely the prerogative of management are shared. The manager in the role of leader and coach shares responsibility and authority with those directly involved in a particular work process or project. This shared approach can take place in a variety of guises grouped under the term teamwork. Generally teams range from those tasked with ad hoc special projects to those working on recurring processes such as accessioning, archival processing, or reference service. Teams, to a greater or lesser extent, participate in such management responsibilities as planning, assigning work, and performance evaluation. The relationship of shared responsibility and commitment with the staff is vital for the archival unit to perform at full effectiveness.

The successful manager must relate with the staff as more than the coach. Organizations can seem fragmented to staff as they and their functions undergo reengineering, restructuring, and outsourcing. The manager must become a builder aiding the staff to communicate with each other and with those outside the unit who share similar interests and needs. The manager must be flexible, open-minded, and knowledgeable of techniques such as communities of practice and hot groups. All this fosters creativity and the sharing of expertise and breaks out of the confines of rigid hierarchical systems built around stove-piped work processes and bureaucratic structures.

The manager must also build and sustain relationships beyond the archival unit. This is a critical aspect of the manager's full-time job in networking and relationship building. All archival managers operate in a competitive, cost-conscious, and technologically driven organizational climate. In a number of settings, such as universities, corporations, and the public sector, there are other parts of an organization that can compete with the archives as the provider of information and even records. Building relationships with other parts of the institution and with constituent groups outside the institution is extremely important. The manager must market the archives for its value, service, and contributions to the bottom line.

Building Partnerships

The demand for quality work and competitive efficiency is relentless. Regardless of the sector of the economy, archival programs seen as ineffective, too expensive, or irrelevant will disappear, most likely through elimination or consolidation with another unit. The basic mindset that the archival manager must have for external relationship building is the concept of partnership. Forming partnerships with corporations, relevant constituent groups, or other organizations with a compatible mission, services, or holdings is essential in meeting the archival program's objectives. Partnership aids in exploiting opportunities and coping with threats. For example, a partnership with a nonprofit organization or a corporation could be essential in digitizing and making available a particularly important collection. Getting the collection digitized and online could raise the archives' profile and earn additional support. To achieve these results, the archival manager must master the communication tools of negotiation and persuasion in reaching out to the external environment.

All this means that the manager must constantly demonstrate that the archives are so valuable, so essential for organizational success, that the archives not only survive but also grow. Organizations (like individuals) either move forward or decline. The status quo is never truly maintained. The manager's major responsibility is to ensure that the archives move forward. Management is not an additional duty or a step up the organizational ladder. It is a calling with results for success and penalties for failure. The roles of the manager include that of mentor, coach, negotiator, and partner. But first and foremost the manager is a leader.

LEADING PEOPLE

Our understanding of the nature of leadership has evolved tremendously from the days of the nineteenth century British historian, Thomas Carlyle, who propounded the "Great Man" theory of history that claimed that leadership is reserved to those with certain personality or character traits. In other words, leadership is innate and always male. Contemporary theorists using a variety of different models offer what is called the "Networked Talent Model" for understanding the nature and role of leadership in modern, knowledge-based organizations.[4]

From this perspective all members of the organization have leadership responsibility. Leadership roles will vary according to the individual's duties in the organization. Executives and managers must articulate a leadership philosophy and philosophy to guide the organization. But all staff members must have the opportunity (and responsibility) for input into the leadership philosophy and organizational vision. In a team-based environment, increasingly the norm in knowledge-based organizations, management and staff must work together to develop the most effective and efficient work processes to meet customer needs. This collaboration must result in products and services that are cost-effective and at an appropriate level of quality.

Everyone must exhibit leadership through commitment to the mission, the vision, values, products and services of the archives. Staff members must understand their place in achieving the mission and their responsibility for their own performance and that of the unit. A team-based work environment with an at least relatively flat organizational structure is most conducive to the "Networked Talent Model." With archives facing turbulent internal and external environments a collaborative management and work process is the only practical way to proceed.

While leadership is at all levels and throughout the organizations, managers have particular leadership tasks. Management as a team must begin the process leading to the articulation of a leadership philosophy for the unit. There is a wide range of possibilities ranging from the direst autocracy to a participatory style that puts a premium on widespread input, collaboration, and consensus. With the workforce in general (and archives in particular) ever more diverse, educated, and independent the participatory leadership style is the most appropriate for long-term organizational success.

In leading people the manager has an obligation to serve as a mentor for others in the organization. The archival manager is responsible not only for his or her own self-knowledge and development, but also for the development of leadership and management skills by all others on the staff. Most importantly, this can be done by example. An organized, competent, and effective manager who focuses on the mission of the organization and of the staff becomes a model of professional leadership. Attention to big-picture issues as well as to routine management chores such as timely and accurate personnel performance appraisals demonstrates how a concerned manager operates.

Mentoring includes assisting employees in developing individual development plans for professional growth and advancement. Where appropriate and in keeping with the perspective of the "Networked Talent Model" these plans should have leadership and management components. The mentoring manager should identify training courses or educational programs that will aid employees in developing leadership and management skills. Training assignments in other parts of the organization, e.g., may be part of a mentoring effort, as well as allowing staff members to "shadow" managers to observe firsthand the duties and responsibilities of management.

The archival manager should work with the institution's Human Resources Department to utilize existing mentoring programs or to develop one if needed. Leadership and management competency is not a gift. It is a set of skills developed over time through education, training, and experience. Part of any good manager's legacy is leaving behind competent leadership to assume the responsibilities of management.

Communication

There is one critical ingredient necessary for the manager to succeed as leader, coach, mentor, team-builder, negotiator, and partner. This is the ability to effectively communicate. It is almost a cliché to blame everything on "poor communication." Yet in many situations it is a fundamental problem. What most managers do not understand is that though we communicate everyday and the process seems normal, it is in fact a complex phenomenon involving the senses, emotions, and logic. Linguists have long-studied the two-way process called communication. Whenever communication occurs, either between individuals or in a group setting, there is similar abundance of sensory stimuli as well as emotional factors, sometimes labeled biases or blinders, so that what is said or heard is only a highly condensed or abstract version. Lack of communication is a fertile area for workplace difficulties.

There are several elements that comprise communication: verbal, nonverbal, and listening. Words can be spoken or written and the communication expressed through words is more than a linear progressive unfolding of logic. Words reflect cultural backgrounds, have different meanings in different contexts, and can be difficult to interpret if the words or thoughts are too abstract. Nonverbal messages (i.e., tone of voice, body language, and facial expressions), in fact, are estimated to account for more than 90% of communication between people, while verbal communication is estimated at 10%. Listening, a skill that can be supported and strengthened through training, is another critical element of communication. Effective listening is acutely impacted by sensory overload, feelings and emotions which, in turn, affect what is heard or understood.

Clearly, the manager alert to the manifold issues affecting communication should expect more often than not to be misunderstood. When it is all said and done, the basic task the manager must achieve is creating and sustaining a communications network. The manager's task is to appropriately create, use, and disseminate information. From the perspective of the organizational complexity and relationship paradigm the manager and staff must identify the basic needs of the archival unit as well as the parent institution; identify the specific information needs of staff and public; and create a communications process that enables effective decision-making and policy dissemination.

There are many tools and strategies that can be used to build an effective communications network. Well-structured meetings and presentations with an agenda, "rules of the road," and definite start and end points are essential to every organization. Judiciously using tools such as e-mail, the Web, the Internet, and intranet can keep management and staff involved with and informed about policies, procedures, and administrative information. This can contribute to organizational knowledge, assist in democratizing the workplace, and enhance productivity. Never underestimate the power of informal hallway conversations to further the communication goals of the organization. But above all communication is about trust. If the manager is perceived as fair, approachable, and competent by the unit, the manager will survive the times of stress and turmoil arising from inevitable communication problems.

Leadership and management are clearly complex tasks. Among the many relationships the manager must build and sustain is the one with the Human Resources Department. A key element of the administrative and communications process is to have fair, transparent, and clear personnel rules and processes. The manager must work closely with the Human Resources unit in analyzing work processes and staffing needs; and in structuring positions, writing position descriptions, and recruiting staff appropriate to the needs of the organization. All these steps involve administrative, legal, and regulatory requirements that must be met. A close working relationship with Human Resources specialists is critical to managing a fair personnel system.

In today's society the manager must be alert to the reality and demands of a diverse workforce. This directly impacts on the need for a personnel system perceived as fair, and an effective communications process.

How policies, e.g., are crafted and written can be problematic. The needs of a workforce transformed by profound social and cultural change must be harmonized with the requirements of the archives' mission and services to its customers. All of this is best undertaken with a leadership philosophy that is participatory and inclusive, and with the sure understanding that everything communicated will be filtered through sensory stimuli and emotions, as well as logic.

MANAGING INFORMATION

Management theory and practice, leadership, and communications are the foundation for the other tasks of archival management. Foremost among these is managing the key asset of information. Within this broad rubric is strategic and operational planning, information technology, and the specific skills involved with project management. Every archival organization, regardless of its institutional setting, is engaged in a fierce competition for resources and relevancy. Careful planning supports an effective strategy-making process which enables management to meet the requirements of the archives' mission and to obtain the financial, personnel, and other resources needed for success. Careful strategic thinking, planning, and decision-making are critical for survival, much less success. Done effectively, planning is at the heart of mission achievement.

Planning

Planning is how archives attempt to create its future situation by setting goals and strategies to reach those goals. This involves looking ahead with the information available to estimate what is likely to occur using a variety of tools and techniques. The effectiveness of archival planning depends on the quality of the planning process itself as well as the quality of the strategic thinking and decision-making which planning supports. Archives must develop plans and strategies to meet their mission, goals, and objectives and relate them to the direction and purpose of the parent institution. Planning serves as the focal point that integrates all the organization's activities needed to meet the mission and goals of the archives. In broad terms, planning is both strategic and operational.

Strategic planning is a concept and practice deeply embedded in corporate, governmental, and academic institutions. There are three ways to create strategy: learning, vision, and planning. All three must be utilized and integrated in any process used in developing a strategic plan. The actual nature of the strategic planning process will be determined by the type of organization (i.e., hierarchical and centralized; hierarchical and decentralized; flat, less hierarchical), and by the management's leadership philosophy. For example, organizations which put a premium on a participatory management philosophy will use a team approach

to strategic planning. Staff members and management from all levels work as a team to develop strategies and plans, with all employees provided a chance for input.

Whatever model or approach is adapted, the first step is analyzing the archives' strengths, weaknesses, opportunities and threats (SWOT analysis). This analysis looks at external factors such as funding, technology, and demographic changes, as well as internal factors such as organizational culture, quality of staff, and technical and facility infrastructure. It is also critically important to have a trained, outside facilitator to assist in guiding the strategic planning process. In addition, outside experts and constituent and user groups must provide input so that the environmental SWOT provides as accurate as possible perception of the environment.

The SWOT environment is part of the learning methodology in creating strategy. Mission and vision are also needed to define effective strategies. This is defining the organization's reason for existence at the highest level. In other words, what makes the archives unique, those whom the organization serves, and the holdings which the archives acquire, preserve, and provide access. The vision puts the mission in broader societal terms and purposes. Also, a statement of institutional values needs to be developed as part of the process. Values are those that the archival organization seeks to inculcate in its employees and in the organization's culture.

After the SWOT analysis is completed and the mission—vision—values are crafted, strategic goals are identified. These should be overarching, responsive to the articulated mission and vision, reflective of the opportunities and challenges facing the archives, and what the organization must achieve to be successful. Strategies must be determined which lay out the various avenues that should be pursued in order to fulfill the strategic goals. Strategies may include forging new institutional partnerships, exploiting technology in a different manner, or creating additional sources of funding. Because strategic plans cover anywhere from a 5 to 10-year time frame, it is necessary to identify objectives or tasks on a yearly basis as measurable milestones toward achieving the strategic goals. Objectives are assigned to units in the archives, with performance plans for managers tied, at least in part, to achieving the objectives.

In addition to the strategic plans, the archival manager must also ensure that a yearly operational plan is developed which covers all program activities and functions of the archives. The operating plan accounts for the hours to be expended for the year ahead by all categories of employees (including volunteers). The plan should also account, by function, for the units of work to be accomplished by the staff. Each program activity should be broken into major subcategories that constitute the tasks to be carried out. The program functions, activities, and subcategories that constitute the tasks will vary from archives to archives but must cover all the records of life-cycle activities carried out by the archives.

Policies and Procedures

A key activity related to planning is the development of policies and procedures that provide the structure and content needed for successful operations. A policy is a governing principle that provides overall guidance to the way in which archive conduct their business. Policy guidance, which is general in nature, is refined and standardized through the use of specific archival procedures. An access policy, e.g., outlines who may use archival collections, when they are open for research, and how researchers can gain access to closed collections. Procedures outline how staff members are to deal with researchers when they visit or communicate with the archives. For example, access procedures would cover such issues as how written requests for information (letter, fax, e-mail) should be handled; how the research room will be managed, addressing issues such as where personal belongings of researchers are to be placed; and what specific steps should be taken when researchers request access to restricted collections. To implement all this, manuals should be prepared that codify the policies and procedures. This is critical both for staff training purposes and for reference when questions arise from staff members or researchers.

Any planning effort is only as good as the results. It is vitally important to have a management performance measurement system in place to monitor organizational performance. Managers and staff together are responsible for setting measurable objectives and setting up systems to measure the quantity and quality of work necessary to achieve program goals and objectives. Quality is a particularly important issue. With limited resources all archives must meet customer expectations, process collections, and preserve holdings. Work must be done well the first time. Quality work is the only answer. Quality standards should be incorporated into the archives' procedures. Further, quality of work must be included in the performance measurement system. Quality can be measured. For example, standards can be developed that addresses the extent of research to be undertaken into a reference inquiry or the elements that must be included in a quality response. Managers and teams can use auditing or sampling techniques to review reference responses for adherence to quality standards. Quality standards should be incorporated into standardized procedures, and then into individual and team performance standards.

Project Management

An important component in the effective use of organizational information and knowledge is project management. The Project Management Institute defines a project as "a temporary endeavor undertaken to create a unique product or service."[5] Successful project management requires effective teamwork, good planning, and the creative use of available information. Projects are really building blocks in the design and execution of an organization's strategic goals and objectives. Projects often involve a matrix management where horizontal teams pulled together from across organizational boundaries are grafted on, for a time, to the vertical hierarchical structure. Projects by definition are temporary, finite, and goal-oriented.

Project management involves the overall context in which the project takes place (i.e., organizational environment, role of stakeholders), scope management and planning, human resources management, and integrating the project into the products, services, and work processes of the organization. Because projects and project teams are temporary, cross-functional, and operate under tight time and logistical constraints, a clear understanding of roles and responsibilities within the organization is critical for success. Project managers and team members must be carefully selected based upon their skills and expertise and the requirements of the project. The project manager must be able to negotiate and obtain the resources and support to complete the project. Senior management is responsible for ensuring that the needed support is provided by other parts of the archives.

Careful planning and execution of all aspects of the project life-cycle forms the basis for project accomplishment. Every project goes through four stages: conception, definition, acquisition, and operation. Although projects impact on existing systems of operation and organization already in place, the project itself as it goes through the four phases of the life-cycle is also a system of people and resources integrated, organized, and managed to achieve the goals of the project. The project manager and the team must understand the project as a system and manage all the parts and relationships.

As an example of how a project can impact an archival system, consider how most archives permit copies of records to be made for research purposes. To carry this out, most institutions will have some sort of order fulfillment system. The parts of such a system include archivists, conservators (for fragile records), accountants (financial controls) and, in many cases, information technology staff if the system is automated. Researchers are also part of the system. A project to devise a new order fulfillment system or even to change one component must involve looking at the big picture of how the whole system operates and the interrelationship of the parts.

There are also a number of methodologies available for systems or project management. These include systems analysis, systems engineering, and systems management. The basic principles in any methodology involve looking at an entire work process and how systems relate to one another. When analyzing a system or process, the project manager and the team must understand how a change in one part of the system affects other parts (systems thinking). For example, a change in the pricing structure may impact the number of copies made and this will have an impact on other parts of the system. Another principle utilizes problem solving (systems analysis) as the basic project tool or methodology. Problem solving includes clarity about requirements and project objectives, setting criteria for making decisions, and examining alternative strategies to achieve the desired goals of the project. Thus, if the project manager and the team do not take a holistic approach to the project, the chances of achieving the desired result are markedly diminished.

In every contemporary organization the effective acquisition, use, and distribution of information is profoundly affected by information technology. Information technology is basically a tool; like other tools, designed to assist in planning, decision-making, program support, and customer service. All management tools or instruments must be designed and implemented to carry out the archives' mission (effectiveness) in the most business-like manner (efficiency). All tools and instruments must be evaluated on the basis of these two fundamental criteria. This is particularly critical given the cost and complexity involved with information technology.

Archival managers cannot expect to become experts in the information technology arena. But the archival manager does need to bring archival skills and knowledge about records and records systems together with information technology. For example, the ways in which records are appraised and scheduled, described, and made available are increasingly affected by information technology through use of the Web and distinctive markup language (hyper text markup language, HTML).

To manage in this complex environment the archival manager must engage in systematic information technology analysis, planning and decision-making. The use of information technology must flow from the archives' mission, goals, objectives, and business needs. The core of effective information technology planning (as with all other planning) focuses on customers, products, and related work processes. The International Council on Archives have developed a planning and decision-making process for the use of information technology in archives.[6] In addition to identifying the customers in the archives' strategic plan, other steps include identifying products and sources required by the customers, and business processes designed to produce the products and services. Archivists are expected to understand these business requirements, and to work with information technology experts in determining appropriate information technology architecture; hardware, software, and communication tools; and the application, acquisition, and development of the information technology support system.

MANAGING MONEY, FACILITIES, AND RELATIONSHIPS

As if mastery of management theory and practice, leading people, organizing work, and managing information were not enough, the archival manager must also master the management of financial resources, the operations of archival facilities, and building effective public relations. As in every other aspect of management, finance requires a careful review and understanding of an organization's operating environment. The manager must understand how the archives receive its funds, the sources of the funding, and the procedures involved in obtaining the funding. On the other side of the financial ledger are expenses. How do the archives expend the resources allocated to it? What authorities must grant permission before the archives commit resources to personnel, facilities, equipment, or other costs? The answers to these questions represent the financial control function of management.

Managers must fully integrate financial planning with plans for the acquisition and use of personnel, facilities, and other resources. The plan—and its attendant costs—must be matched against the revenues likely to be available to the organization. All revenue projections should be realistic in that they reflect possibilities that might actually occur. For example, an organization applying for grant money to accomplish a project is realistic in calculating the availability of the requested funds, especially if it has successfully competed for grants in the past. As part of the financial planning all fixed resources must be prioritized. The archives must know which operations have the first claim in money coming in, which comes next, and so on. In most instances, current operations—being able to sustain ongoing efforts—come first. Expansion and new projects must compete for whatever is left. This emphasis, though it may seem to discourage innovation and experimentation, is entirely appropriate for archives, which are essentially conserving institutions.

Budgets

Careful budgeting is another key element of financial management. Budgets are necessary for planning and controlling funds, as products of the planning process, and as tools by which managers control the expenditure of resources. Effective budgeting is part of a multiyear (three to five) process in advance of planned activities. There are various types of budgets such as line-item budgets and program budgeting that are appropriate for both small archival operations as well as more complex organizations. Whatever type of budgeting is used, archival managers must bear responsibility for cost controls. Records of expenditures must be kept in whatever format the institution requires. Expenditures need to be tracked on at least a quarterly basis to ensure that the basic tenets of the financial plan are being met.

Resource Development and Outreach

The archival manager's financial management responsibilities may include participating in fund raising and

Archival Management and Administration

development for the archives. This involves planning and identifying program needs beyond what can be met through the organization's regular funding mechanism. This involves seeking external funding and working with the development office of the archives or the parent organization to seek grants or initiate capital or endowment campaigns. Federal, local, and state governments provide a variety of grants through a number of grant-giving agencies. In addition, there is extensive philanthropic giving by foundations.

Involvement with managing archival facilities seems the most mundane and tedious of tasks. Yet, it is one of the most critical management tasks. Archivists must be able to work with facility managers, conservators, architects, engineers, and contractors to ensure that the archival holdings are kept in appropriate storage conditions; that supplies and equipment obtained are archivally sound from a preservation perspective; and that adequate space is designed and obtained for the archives. All archivists are faced with planning for the best use of their facilities as well as for the purchase of supplies and equipment. Such decisions have a significant impact on the total program and should always be carefully considered. Through applied common sense backed by thorough research and information gleaned from conservation experts, the archivist should be able to make decisions which lead to positive results and an improved archival program.

All that has gone before in describing the elements of archival management rests upon the need for an effective public relations program. Public relations are the communication or dialog which archives have with individuals or groups inside and outside their institutional setting intended to convey information about services and goals. This can be done in a variety of forms: up to date and informative Web sites, press releases, communications between staff and researchers, or annual reports prepared for the parent organization.

Public relations also include special events designed to invite specific groups to ceremonial openings or other events. Communication must be a part of a well-planned program designed to explain the archives' mission and involve people in its programs. Public relations is an integral part of archival management. It must be a deliberate effort and should be included in the strategic planning process. Plans for public relations should evolve from the archives' mission statement that defines not only program goals but also the constituencies to which the archives are responsible. A variety of communication tools should be used to approach these groups—open houses, lecture and film series, exhibits, press releases, newsletters, listserves, e-mail lists, and fax lists. With limited resources the public relations effort must target the audiences it wishes to reach and the methods that will most effectively communicate with particular groups.

Public relations are critical in developing and maintaining support for the archival program. With the information overload felt by individuals in modern society, people will not go out of their way to learn about new programs or new ideas. Information about archives must be brought to people in familiar and understandable ways. Failure to build support for the archives through a public relations program may lead not only to a lack of funding but also to the demise of the program itself. Successful archival managers must find ways for their programs not only to survive, but also to prosper, and this can only be done through a planned and thorough public relations program.

CONCLUSION

Contemporary archival management is complex. The pressure for high performance grows steadily more intense, with sustained, effective leadership as perhaps the single most important ingredient for organizational success. Leadership at all levels of the archival enterprise must build an organizational commitment to the mission, vision, and values of the archives, with a pledge to meet customer needs by providing quality products and services in a timely and cost-effective manner. Archives are cornerstones of memory, community, and the historical record. Those engaged in the work of archival management play a key role in maintaining a democratic society rooted in the right of access to records and information.

REFERENCES

1. Evans, G.E.; Ward, P.L.; Rugaas, B. *Management Basics for Information Professionals*; Neal-Schuman Publishers Inc.: New York, 2000; 5.
2. Kurtz, M.J. *Managing Archival and Manuscript Repositories*; Society of American Archivists: Chicago, IL, 2004; 1.
3. Kurtz, M.J. *Managing Archival and Manuscript Repositories*; Society of American Archivists: Chicago, IL, 2004; 35–45. [The core concepts for this essay are contained this referenced monograph and in the Evans, Ward, and Rugaas monograph.].
4. Commonwealth Center for High Performance Organizations Inc. *Building High Performance Organizations in the Twenty-First Century*; Commonwealth Center for High Performance Organizations Inc.: Charlottesville, VA, 1988–2006; 1–40.
5. Kurtz, M.J. *Managing Archival and Manuscript Repositories*; Society of American Archivists: Chicago, IL, 2004; 89–100. [The Project Management Institute provides literature and training to support effective project management. For information see Pinto M.J., Ed.; *Project Management Handbook*; Jossey-Bass, San Francisco, CA, 1998.].
6. Kurtz, M.J. *Managing Archival and Manuscript Repositories*; Society of American Archivists: Chicago, IL, 2004; 104–111.

Archival Reference and Access

Mary Jo Pugh
Editor, American Archivist, Walnut Creek, California, U.S.A.

Abstract

This entry introduces reference services in archival and manuscript repositories and concepts of access to archival holdings. Archival holdings and reference services differ significantly from those typically provided in library settings. This entry is distilled from my book, *Providing Reference Services for Archives and Manuscripts*. Archives are tools; like all tools, they are kept to be used. Archives allow people to communicate information and evidence through time. Reference services in archival and manuscript repositories assist users, and potential users, in using archival holdings and locating information they need. This entry discusses the nature of archival holdings, institutional settings, and the range of uses and users of archival records. It explores information seeking and its relationship to archival reference services. The essay discusses the intellectual, interpersonal, and administrative elements of providing reference services in archives and manuscript repositories. It concludes with a brief consideration of outreach, management, and evaluation.

INTRODUCTION

Reference services" in archives provide "intellectual," "legal," and "physical" access. "Intellectual access" is provided through arrangement, description, and reference services. Traditionally, reference services have been mediated. Increasingly, however, they are provided through the repository website or online finding aids and are unmediated or mediated though e-mail, tutorials, or Web 2.0 tools. In a mediated setting, interpersonal dynamics between reference archivists and users are critical to successful transactions.

Second, the word "access" can also be used in a narrower, "legal," sense, meaning permission or authority to use archival records. Third, "physical access" means providing opportunities to examine records, traditionally in person in a repository, but also through copies, digital surrogates and loans.[1]

ARCHIVAL REFERENCE

People create documents to communicate across space and through time. Records provide "evidence" of the actions that created them, long after the actions are completed. Records also provide "information" about the people, events, and objects that were the subject of these actions.[2–4]

The mission of the archival profession is to identify records that have continuing usefulness, preserve them, and make them accessible through time. The use of archives depends, in part, on the mission of the parent organization. Three common rationales for maintaining archives are:

- Continuing administrative use by the creating institution.
- Public use of government records in a democracy to provide oversight of government actions, ensure accountability, and preserve the rights of its citizens.
- Research by users from a wide variety of disciplines.

Reference services were slower to develop an identity than other functions of archival administration, perhaps because early archivists assumed that reference services fall at the end of a continuum that begins with the creation of the records in the originating office (shaped by records management), followed by appraisal, accessioning, arrangement and description, and concluding with reference services and outreach activities. Driven by the need to manage and protect large quantities of records, archivists initially thought of reference services as a series of administrative decisions made after records had been appraised, accessioned, arranged, and described. Since the 1970s, however, archives have become user-centered and recognize the importance of reference services. This interest has led to new research into the users of archives, which has increased with electronic access to archives. The archival profession moved from a custodial role, in which the archivist's primary duty was to protect repository collections by limiting use, to an activist role in which the archivist's primary duty is to promote the wider use of archives.

Uses and Users of Archives

Many people use archives, both directly and indirectly, to find information or evidence for a purpose outside the repository. Practical needs bring most people to archives;

Encyclopedia of Library and Information Sciences, Fourth Edition DOI: 10.1081/E-ELIS4-120043645

most are less likely to seek entertainment or recreation than they might in many libraries (Fig. 1).

THE USES OF ARCHIVES

Primary Uses of Records

Organizations and individuals create records initially to carry out administrative, fiscal, economic, social, legal, or other activities.[5] These uses may continue as long as the organization continues. Architectural records are created to build structures and they are needed for as long as the structure is maintained.

Secondary Uses of Records

Some records, however, can be used by a wide variety of users beyond the creator. These secondary uses may be of three types. "Informational" use describes the continuing usefulness of the information in records about people, events, objects, or places. "Evidential" use is the continuing usefulness of the evidence (or information) that records provide about their creators, their activities, or the

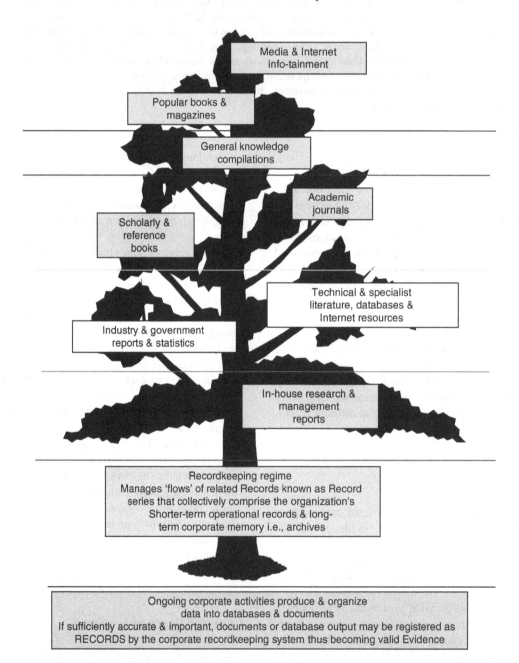

Fig. 1 The information family tree.
Source: From Understanding society through its records, by Pederson, A., http://john.curtin.edu.au/society/evidence/tree.html.

actions that generated the records. File contents and filing structures can be as important for documenting the flow of information as for retrieving it. "Intrinsic" use describes records as artifacts—as symbols, or tangible links to the past. Although uses are practical, users often find the age, authenticity, or beauty of original documents as important as the information or evidence they contain. Authentic documents often move and even awe people.

Direct Use of Archives

Direct use occurs when someone obtains information from a record or uses a record as evidence of the activity that it documents. Direct use can be measured when someone reads a document in the repository, obtains a copy by mail, or receives information by telephone, e-mail, or letter. Loaning documents is another type of direct use. Placing records on the Internet means that many remote users can consult them directly, but this direct use is harder to measure.

Indirect Use of Archives

Identifying and measuring indirect use is more difficult still. Indirect users may never enter an archives or archival Web site, but benefit nevertheless from archival information by using the many and varied products arising from direct use. Though a relatively small number of researchers actually uses archives, their work has a "multiplier effect," transmitting information that affects how others think about themselves and their past.[6] Books, newspaper and periodical articles, dissertations and theses, genealogies, speeches, term papers, films, television documentaries, slide shows, exhibits, Web sites, legal briefs, environmental impact statements, and other policy documents are some of the products that convey information and evidence derived from archival holdings.

UNDERSTANDING INDIVIDUAL NEEDS

People come to archival repositories because they need to solve problems that require historical information or evidence of past activities. Direct users have a variety of needs depending on several variables.

Research Purpose

Research purpose is in part determined by whether the user's information problem is motivated by work requirements or by personal interest. Most people come to archives because their work requires it, whether on behalf of a business, government, association, university, class assignment, or client. They are motivated by external rewards. Other researchers come to archives because

they seek information for enjoyment, recreation, or personal growth.

Intended Use

The intended use—report, film, exhibit, family history, book, article, legal brief, advertisement, or environmental impact statement—affects the types of information sought and the questions asked. It is helpful to distinguish between the form or genre of information and the media on which it is stored. Reference archivists need to understand the form of evidence that researchers require for their intended use. For example, minutes of meetings may be found as paper summaries, paper transcripts, audio-recordings, video recordings, or electronic documents.

Type of Question

Factual researchers typically ask closed-ended questions, seeking a particular document or specific information about a particular person, place, object, or event. Their questions are precise and focused, asking, "Who, What, Where, When, How many, or How much?" Although precise, such questions may require considerable time and ingenuity to answer.[7,8] Interpretive researchers ask more abstract questions, intended to increase general understanding or competence. Such users read comprehensively through bodies of records to tell a story, develop a narrative, or test a hypothesis. Posing open-ended questions, they seek to answer broad questions of motivation, causality, and change. They ask the question, "Why?" William L. Joyce formulates a similar distinction.[9]

Joyce.[9] Experience and Preparation

Many users, regardless of education, occupation, or nature of inquiry, are not well versed in archival research. They may need help in conceptualizing the research process, analyzing the process of finding relevant documentation, and devising search strategies. Once launched on research, they may need assistance in understanding the context of the records, or practical help deciphering handwriting, understanding dating conventions and abbreviations, or interpreting slang and references to contemporary event.

IDENTIFYING USER GROUPS

Identifying Vocational User Groups

Staff of the parent organization are an important user population. Archives, the corporate memory of the institution, are preserved so that the parent institution can understand its history and the sources of current policy and can maximize its return on its investment in information resources. Archives are used to ensure continuity, build on experience, and identify solutions for current problems. Since administrators of the parent institution are simultaneously creators and users of archives, access and reference policies may give priority to their requests. Archivists are the employees most likely to know an institution's history and organizational changes and can identify precedents, policies, and documents relevant to current issues.

Staff from other organizations also use archival resources in their work. These users represent many professions, such as lawyers, legislators, engineers, landscape architects, preservationists, urban planners, architects, film and television producers, picture researchers, journalists, and publishers. These professionals are direct users linking archival sources to many indirect users. They use archival information on behalf of groups, clients, firms, governments, or professional associations.

Many assume that *scholars* are the primary constituency for archives, but in most repositories their numbers are significantly smaller than other user groups. As direct users of archives, however, scholars transfer information from archival sources to indirect users of archives. Many disciplines use archival holdings —history, geography, political science, demography, sociology, literature, medicine, epidemiology, among other. Whatever their field, scholars are more likely to be interpretive researchers.

Students are a major constituency of many repositories, particularly university archives and manuscript libraries on or near college campuses. Secondary school classes have become significant users of archival material as some state curriculum standards require this component. In addition, the National History Day competition brings requests from students as young as middle school, following instructions to use primary sources in their research. Using archives can make history come alive for students, whether young or old. Providing access to archives empowers students to learn and discover on their own. Most students, like the general public, respond enthusiastically to the authenticity of original source materials.

For *teachers* archives are laboratories for research in the humanities and social sciences and archivists and teachers work together to instruct students in research. Archivists have new opportunities to collaborate with K–12 teachers in local schools and with universities who teach potential teachers to improve teachers' knowledge, understanding, and appreciation of American history.[10–17]

Identifying Avocational User Groups

Avocational users are "recreational" or "leisure" users. Archives play a significant role in educating the general public, providing opportunities for lifelong learning for a public that is better educated and living longer. Many people who found classroom history boring or even alienating search eagerly for a personal, usable past.

Genealogists: Many Americans seek meaningful connections with the past through family history. Genealogy, one of the most popular hobbies in America, appeals to all economic, racial, and ethnic groups. Genealogists have been called "historians of their own family,"[18,19] and they often share the results of their research with other family members. Because the direct impact of genealogy beyond individuals is not easily measured and because of its past association with elitist societies, its value has been discounted or dismissed as idle curiosity. Many genealogists are expert researchers who have completed many projects using primary sources and have wide-ranging and sophisticated interests and research skills.

Local historians are examples of avocational historians. Like genealogists, their interests reflect a modern society's need for a usable past. Community celebrations and anniversaries often stimulate interest in local history. These may focus on the community as a whole or on its many parts: churches, civic groups, businesses, schools, labor unions, ethnic groups, hospitals, or other institutions. Homeowners or business owners seek information about the history of their house or company.

Hobbyists find meaning in the past by focusing on selected historical objects and events. For almost any common object or event, a group is devoted to its preservation and history: trains, ships, lighthouses, bottles, circuses, automobiles, carousels, prospecting, and military engagements.

INFORMATION SEEKING AND ITS RELATIONSHIP TO ARCHIVES

Information retrieval is the identification, preservation, and description of information, anticipating user queries. In contrast, information seeking focuses on the needs, characteristics, and actions of the information seeker. Information surrounds us all (Fig. 2). People use their mental models for events, experiences, and domains of knowledge to run scenarios for contemplated actions. Information seekers in organizations tend to turn first to people, then to readily available personal collections and electronic resources to answer information needs. Libraries are seldom the first choice, archives almost never are.

Seeking Information from People

Face-to-face oral communication remains a preferred means of communication, because much human communication remains nonverbal and nonverbal elements carry meaning. Personal interaction also provides the opportunity to clarify meaning. Seeking information from others is so ingrained in our daily behavior that we take it for granted. People seek information from other people because such information is filtered and, more importantly,

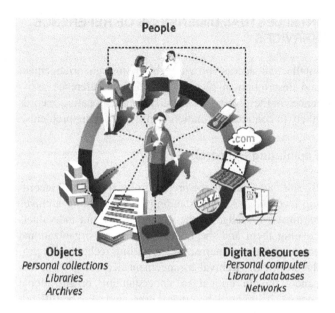

People

Objects
Personal collections
Libraries
Archives

Digital Resources
Personal computer
Library data bases
Networks

Fig. 2 Information seekers are surrounded by information.
Source: From Pugh.[1]

authenticated by experience. Information givers responsible for their actions are given credence.

Seeking Information from Documents

Common information-bearing objects are found in personal collections, records, electronic networks, and libraries. Research suggests that people will use accessible information, regardless of whether it is the best. The "principle of least effort" is well documented.[20]

Personal collections

People tend to turn first to personal collections. They compile some tools personally, such as calendars and lists, reach for familiar books close at hand, such as procedures manuals and reference books. Most of these information sources are organized and retrieved by physical location and by form, such as size, shape, and color. People select sources close at hand, not because they are lazy, but because these sources are authenticated by personal experience, annotated to reflect recent changes, and accommodate time constraints in the press of daily business.

Records

People also turn to their working files, such as reports, research notes, and correspondence. This observation also helps archivists understand why staff members are often reluctant to send their records to the archives. Organizations, whether corporate, governmental, non-profit, religious, or community, embed evidence of activities and

corporate knowledge in documents and capture them in records systems.

Electronic resources

Personal reference files are now created and retained mostly in personal computers. As computers are linked in local area networks within organizations or through the Internet to other organizations, users now rely on online tools and information resources. These convenient tools also add powerful modes of retrieving and repurposing information.

Libraries

To turn to libraries, people must first conceive that a book or article might be published on the subject of interest, that is, the user must have a mental image of finding the information in a publication. Library users have effective search techniques beyond the catalog, and studying actual information-seeking behaviors offers insight into the assumptions and expectations of information seekers as they approach archives.

Many users prefer browsing, which provides full-text retrieval. By scanning the table of contents and bibliography, users can quickly determine the relevance of a publication. Library classification systems collocate material by subject, and propinquity reinforces the serendipity of discovery. Browsing emphasizes to the physical nature of information seeking. Users accustomed to this powerful technique may be frustrated by closed stacks and gray boxes arranged by provenance in archives.

Chasing footnotes is another common behavior. A footnote is information in context. The nuances of a subject reference can be very precise when a footnote barks at the heels of a sentence, in a way that subject headings in a catalog cannot. Further, the footnote also evaluates a source. Footnotes are a means of authenticating an argument, but they are also a means for people to communicate about sources of information.[21] Although we think that users are looking for all bits of information, these search techniques indicate that they are often looking for authenticated, evaluated information in context, the kind of evidence found in archives.

Information Seeking in Organizations

When seeking information for their daily work, staff members typically rely first on their own memory or on readily accessible records that document their knowledge and actions, whether analog or digital.[22–25] For information beyond their memory, files, and scope of activity, staff consult other people. People trust information selected and authenticated by the person responsible for the action. An employee summons knowledge of the organization and its functions to infer what person or department is responsible for activity and its associated information. Employees (and their clients or customers outside the organization) find information by asking, "Who is responsible? Who keeps that information? Who would know or need to know about this problem?"

Often this information-seeking behavior is so ingrained that people do not think about these processes, and the search for information is so obvious that the process is transparent. Information seeking is grounded in a world of personal networks and increasingly in networked electronic environments.

Information Seeking in Archives

Information seeking in institutional archives builds on these patterns in the parent institution. Although people leave, the organization continues. Records documenting significant actions with continuing consequences are transferred to organizational archives so that later incumbents or others seeking evidence of past actions, can find them. Some staff members know that records of their predecessors are in the archives. Other staff ask, "who would have created this information in the organization in the past and where are the records now?" As time passes, however, the reference archivist is needed to assist staff members in asking these questions and locating information.

An important question is, "How do people find archives?" Researchers still find their way to archival repositories through informal personal networks, that is, by word of mouth. Electronic networks are, however, becoming increasingly important.[26,27] Archivists must be part of the personal and electronic networks of their organizations or communities.

Reference Services in Archives

Reference services are the activities by which archivists bring users and records together to meet user needs. Reference services assist users in person in the research room or remotely by telephone, mail, fax, or e-mail systems. Public programs also provide reference services. On the Internet, reference services assist users who discover them there (Fig. 3).

- Intellectual elements
  - Arrangement
  - Description
  - Reference services
- Human dimension: Interpersonal elements
- Administrative elements

Fig. 3 Fundamental elements of reference services in archives. **Source:** From Pugh.[1]

INTELLECTUAL DIMENSIONS OF REFERENCE SERVICES

Intellectual access, provided both through arrangement and description of records and through reference assistance, is the process of identifying and locating records likely to contain information useful for solving problems.

Facilitating Research

Because archives are the products of activities and recording technologies, understanding and evaluating archives requires knowledge of the organizations and individuals creating them and the historical context, organizational communication patterns, and recording technologies producing them. Archival arrangement according to provenance, though critical to understanding organizational context, is unfamiliar to many users and may be difficult for users outside the organization or even for later users from the organization.

Facilitating research requires continuing interaction between archivist and user throughout the research project and may not be completed until the last question regarding copyright or citation is answered. In addition to using records and descriptive tools within the repository, reference archivists often facilitate research by referring researchers to other sources beyond the repository.

Undertaking Research

Undertaking research is a second important intellectual role. Reference archivists undertake research to learn about the parent organization and records creators, to understand the functions and forms of records, to place the finding aids and records in context, to locate information in the records for others who cannot do so, and to evaluate records or information. They also undertake systematic analysis of the use of their holdings to improve access and services.

Educating Users

Educating users is a third essential intellectual function. Instructing users in the research room is an important part of daily work. Few users have experience with primary sources, and most are unprepared for the complexity of archival sources, finding aids, and archival practice. Most have no experience integrating and understanding the undigested mass of information so often found in primary sources. Archivists impart an understanding of archival theory and practice, to provide "archival intelligence" which is an "intellectual framework for understanding both the organizational principles behind archival records as well as the multiplicity, interconnectedness, and appropriate selection of one type of access tool over another."[28]

Reference archivists also participate in more formal educational programs. Orientation sessions, class presentations, workshops, publications, tutorials, and online presentations extend the ability of the reference archivist to help users to be more efficient in exploiting archives. Increasingly, these educational programs are presented through a repository's Web site.

INTERACTING WITH USERS TO PROVIDE INTELLECTUAL ACCESS

The extent of reference interaction varies from simple to complex, depending on the nature of the repository, the kinds of finding aids available, the research problem, and users' research skills. Not all users require, or even desire, all elements of the reference process, nor do all repositories have sufficient staff to carry them out. The reference interaction has both intellectual and administrative elements and is powerfully affected by interpersonal dynamics between user and archivist (Fig. 4). The quality of reference service depends on acknowledging and successfully resolving the complicated dynamics of reference transactions. An effective reference interaction enables a user to exploit archival resources fully, while meeting the administrative requirements of the repository.

Question Negotiation

As an intellectual exchange, the reference interaction consists of three activities: "query abstraction," "resolution," and "refinement," through three phases—initial interview, continuing assistance, and follow-up activities.[29] Query abstraction and query resolution usually take place during the initial interview. Since query refinement takes place during research, interaction between user and archivist may be needed throughout the research project. Interaction may be extended by phone or e-mail follow-up activity after the researcher leaves (Fig. 5).

Initial interview
    Query abstraction
    Query resolution
    Search strategy
  Continuing interaction
    Query refinement
Exit interview

Fig. 4 Intellectual dimensions of reference services in archives.
Source: From Pugh.[1]

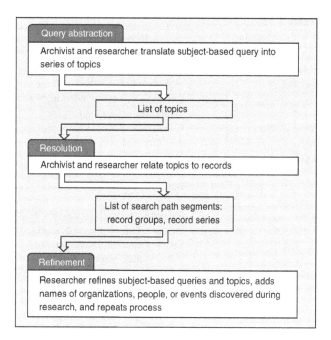

Fig. 5 Question negotiation.
Source: From Pugh.[1]

Initial Interview

The initial interview, an intellectual interchange between archivist and user, is the archivist's opportunity elicit what information the user needs for the inquiry, and the amount, variety, level, and complexity of the source materials needed to resolve it.

Query abstraction

The initial interview begins with question abstraction, in which the seeker and the archivist identify the topic, delimited by time, place, and the researcher's intended use. The archivist tries to glean enough information to translate the user's natural language into the retrieval language of the finding aid system. Reference staff and researchers ask questions to determine which archival and manuscript collections will be most useful. Asking such questions helps to educate researchers about the process of finding information in archives (Table 1).

Query resolution

In the second stage, the archivist and the user analyze the problem to form a search strategy. The archivist assesses available sources, identifies records, and suggests an order in which the researcher might use them. Based in part on an assessment of the probability of finding useful information, the archivist helps the user identify some sources as highly relevant, others as possibly useful, and others as of marginal interest. The arrangement of a series may dictate the sequence for the research strategy. For example, to use

Table 1 Question negotiation.

Question	Information sought
Who would have needed this information or evidence?	Names of persons, organizations, places, and events. Research subjects must be linked with the activities of specific individuals and organizations and their records.
Why would someone have needed this information or evidence?	Functions that would have required this information or evidence suggest organizational entities or individuals whose records might be useful.
What time period is of interest?	Date range
How might this information been recorded?	Recording technologies available at the time. Forms of records likely to have relevant information: textual, audio, visual, graphic, digital.
Where would those records be now?	Archival access tools to locate likely sources.
What product will result from the research project?	Assists in identifying the most useful types of sources and alerts the archivist to potential legal ramifications, such as copyright.
How much time is available for this research?	Research for a speech to be given tomorrow differs that for a dissertation or book.
Is the user is asking on behalf of someone else?	A question is altered as it passes though an administrative assistant or research assistant. An intermediary may not know why information is needed or the full ramifications of the project. Archivists can give much better service talking directly to the person with the question. In organizational archives, speaking to the actual user, often higher in the organizational chart, is an important means for advocating for the archives.

a series arranged by name, the researcher must have the names of individuals.

The archivist's response depends in large measure on the level of description and the types of finding aids available. Sometimes the archivist identifies terms for searching catalogs and databases. For other queries the archivist identifies likely record groups and series. In other cases, the archivist locates and explains the use of finding aids.

The resolution stage of question negotiation is particularly important in institutional archives. The archivist translates a question about a subject into a question about the related organizational functions. Knowledge of institutional history is necessary to ascertain which agencies were responsible for those functions in the period under investigation and likely to have recorded the needed information. The archivist also uses knowledge of the forms in which information was recorded to infer which series might bear on the subject, to suggest, for example, that

annual reports will be more fruitful than minutes. To refer users to relevant information sources outside the repository, archivists may also use national reference sources and their knowledge of holdings of other repositories or of records retained by creating departments within the parent organization.

Locating information in archives is often inferential, based on what is known about the records, their creators, and the circumstances of their creation. Reference archivists link subject queries to the records and the functions that created them, extending the user's information seeking in the creating organization into the past. Archivists play a vital role in this process because of their understanding of the universe of documentation and how a user's questions fit within that universe.

By explaining their reasoning to users, archivists help researchers build their research skills, to help them understand record creation, finding aids, and the process leading to a particular search strategy. Fig. 6 suggests the complicated relationships among archivists, finding aids, records, and users in resolving questions. Archivists strive to make users as independent as possible by helping them to think archivally—that is, functionally and hierarchically. As teachers, archivists help users to think, "Who would have been likely to record the information I am seeking, how would it have been recorded and filed, and where are the records now?"

Continuing Interaction

In the third, or query refinement stage, interaction between archivist and user continues to refine the problem and search strategy. Research in archives is iterative. As users work through archival materials, they discover new aspects of their topic, including the names of other organizations and individuals whose activities bear on the subject of research.

Researchers also discover questions about records as they work through them. Provenance-related questions, or external evidence, about the source, creation, or custody of records help users judge the authenticity of documents, understand bias or interpretation, or explain gaps in the records. Users also may need technical assistance—deciphering handwriting, identifying archaic words or references, resolving problems with dating, or using difficult file structures. Archivists also may help users understand how best to use formats new to them, such as architectural drawings, photographs, maps, or electronic records.

Exit Interview

Ideally, the reference interaction is closed by an exit interview as long and thorough as the initial interview, but this is seldom realized. Because users do not always announce their departure, it is wise to request and, if possible,

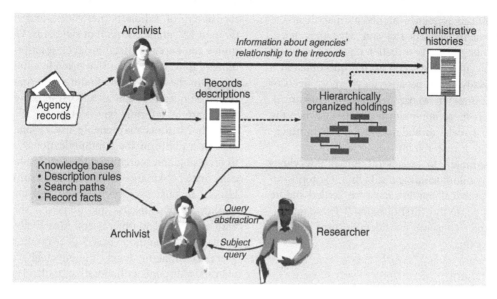

Fig. 6   Inside the black box.
Source: From Pugh.[1]

schedule an exit interview during the initial interview. Although continuing interaction during research helps to ensure that researchers have seen all pertinent materials, an exit interview provides an opportunity to review the sources used and to discover if additional materials warrant another visit.

The exit interview is an opportunity to capture the knowledge and expertise of researchers and use it to enhance collection descriptions. Researchers often bring, or discover, additional information about collections or creators. An exit interview also is an opportunity to evaluate reference services, assess finding aids, report on arrangement and preservation, or suggest leads for acquisition. Now Web 2.0 interactive finding aids allow researchers to add information to finding aids, and surveys and other tools allow them to evaluate services.

HUMAN DIMENSIONS OF REFERENCE SERVICES

Reference Interactions in Person

All phases of reference interaction—initial interview, continuing assistance, and exit interview—are affected by interpersonal dynamics between user and archivist. Reference encounters in libraries are usually short and voluntary, each devoted to a single question. In contrast, reference transactions in archives are more likely to be substantive, obligatory, and continuing.[30,31] Because reference archivists frequently mediate among users, finding aids, and records, understanding the human dimension is critical to providing intellectual access to archives. Beyond assisting individual users, reference

archivists also provide services to groups of users and potential users through public programs.

Good communication between users and archivists is critical. Archivists and users may have very different expectations about the reference interaction. Discrepancies in expectations may cause confusion, disappointment, or failure to use archival holdings effectively. Users and archivists may differ in their expectations of the conditions under which archival materials are to be used. Expecting archives to be like libraries, users may feel rebuffed if they cannot be accommodated without an appointment and disappointed that documents do not circulate. Archives must ensure the integrity of unique materials, users may find registration and security procedures intrusive.

Users and archivists may also have different expectations about the appropriate role of the archivist and the amount of time required to locate needed information.[32,33] Most archivists expect to provide instruction in using finding aids and guidance to records, but users may expect archivists to furnish information directly. Research in archives requires an effective partnership between archivists and users. Archivists strive to make every user feel welcome and to treat all users fairly. Sensitivity, clarity, and a genuine spirit of public service are needed to ensure successful interpersonal relationships in archival reference services. Nonverbal signs and symbols are significant components of interpersonal communication. Positive body language makes users feel more comfortable approaching a staff member with a question.

The first question, the initial inquiry

Reference archivists quickly learn that the first question is usually not the real question, or at least not a full statement of need. Experienced reference archivists marvel at the gap between what researchers say and what they need. Understanding this gap leads to better reference service. People often feel vulnerable when asking questions, at a psychological disadvantage because they expose need or ignorance to a stranger whose response is unknown. Active listening is more important than talking when the reference archivist seeks to understand the full ramifications of a question or research project.

Another impediment to successful initial interviews may be repository administrative needs. In most repositories, numerous practical matters must be settled before researchers even get to the "first question." These administrative details can consume considerable time and are likely to be distracting.

Cues for researchers

Researchers need clear cues to identify the reference archivist. Layout and signage can facilitate the interpersonal exchange by making it clear that the reference archivist is to be approached and marking clear pathways to the reference workstation.

Interpersonal dynamics of continuing interaction

If a good relationship is forged during the initial interview, interpersonal communications are less likely to be a problem in continuing interaction and the exit interview. Staff members cannot control the behavior of users, but they do have control over their responses to users. They strive to be pleasant and courteous to all users, regardless of their demeanor. If staff members cannot provide requested information or services, they explain such limitations fully and politely.

Reference Interaction with Remote Users

Off-site users direct inquiries to the archives through letters, telephone calls, electronic mail, and fax.[34–36] For some users, such inquiries initiate or continue research conducted in person, but for many users they are the sole means of interacting with the repository. Remote inquiries often request archivists to do research for them. Usually archivists have neither the time nor the resources to do so. Policies and priorities must be clearly and convincingly communicated; they must also be reviewed regularly to recognize changing publics, user needs, and technologies.

Reference Services on the Web

Many elements of reference services are moving to the Internet and the World Wide Web, which expand both archival services and researcher expectations. For collecting repositories, information about researcher needs may require assessment of researcher patterns in the repository, of scholarly networks, and of kindergarten through 12th grade curriculum standards. Collecting repositories can export collection-level descriptions to national bibliographical networks that provide a link in the MARC record to the Web-based finding aid. Analysis of e-mail reference questions can also assist in designing interfaces for electronic information systems.

Web 2.0 tools may enable use of interactive finding aids as a platform for communications.[37–41] In addition to browsing and searching one or more finding aids, users can create bookmarks and follow link paths of other users. Users can post user profiles and add comments to the finding aid and discuss questions and share information with archivists and other users. Max Evans introduces the concept of "commons-based peer-production" in which users determine the level of intellectual access to archival materials through collaborative tools. Users might add value to the finding aid by identifying, indexing, and digitizing and posting documents, using such tools as Flickr.

Archivists in institutional archives find opportunities through the organizational network electronic information systems to integrate archival information resources into the information ecosystem of the organization. An institutional archives can use the intranet to become a clearinghouse for all organizational information, beyond information about the records it holds. The archives can integrate information about organizational functions, activities, and forms of records, whether paper, visual, audio, or digital, and regardless of whether records are in offices, records center, or archives. The organization chart could be used as an information interface. Better yet, using the business processes or functions of the organization provides a more useful map of the organization. The provenance method of information retrieval can be made explicit and can empower the information seeker.

LEGAL ACCESS

In a narrower sense, access is "the permission to locate and retrieve information for use (consultation or reference) within legally established restrictions of privacy, confidentiality, and security clearance."[42] Access in this sense is "Who gets to see what and when." The repository must face these issues before users request to use records. Access policies are determined in consultation with all interested parties, applying relevant laws, in the context of the mission and resources of the repository and are a foundational document for the repository.

Providing reference services and determining access to records are separate functions, yet they are linked in access policies, first because reference assistance is critical to obtaining access in the broader intellectual sense, and second because reference archivists usually administer physical access and supervise researchers using documents. Reference archivists are responsible for administering and explaining to users any restrictions on access to records. Access policies are administered, but not determined, at the reference desk.

Records created for personal or internal use may contain private or confidential information that cannot be disseminated immediately. Legal and ethical issues may condition the use of archives: most importantly, privacy, confidentiality, and freedom of information. Archivists strive to provide fair, equitable access to records in their care but must also protect information affecting creators or third parties until it is no longer sensitive. Further, the copyright law, which allows for access to documents, may govern the secondary uses that can be made of them, such as publication.

Archivists have a dual responsibility to the creators and subjects of the records on one hand and to scholarship and the public good on the other. Archivists seek to make as much information as possible available to users as soon as possible, but recognize that some information must be withheld to protect legitimate interests of privacy and confidentiality. Archivists maintain a delicate balance between encouraging use and protecting the rights of creators and third parties. Without some restrictions on access to recent and sensitive materials, individuals may be harmed. Furthermore, without such protection, creators may destroy records rather than transfer them to archives, or they may not record certain kinds of information at all.

Policies for access in its larger intellectual sense and its narrower legal sense must be in place before users actually examine records and obtain information from them. Access policies protect records from harm and some information from premature disclosure, while making as much information available to researchers as possible. An access policy mediates among the competing demands of privacy, confidentiality, public right to know, and equality of access. Some of these concepts are embodied in law, others in deeds of gift, and still others in ethical norms. Access policies also allocate repository resources for reference services as equitably possible. Policies and procedures help the archivist to balance user needs against the staff resources available to meet them while also protecting archival holdings from harm.

Understanding the concepts underlying the laws, regulations, and policies governing access to archival materials is important when designing repository access policies. Four concepts are particularly relevant: privacy, confidentiality, right to know, and equality of access. These guiding concepts are expressed in various forms. The first three are expressed in privacy and freedom of information laws affecting records generated or accumulated by institutions such as federal, state, and local governments, colleges and universities, businesses, and other corporate bodies. Agreements regarding privacy and confidentiality are embodied in contracts in deeds of gift or

transmittal documents. Ethical norms also govern access policies.

Information to be protected is identified and segregated during acquisition and processing, so that the reference archivist does not have to judge the motives or wisdom of individual researchers at the reference desk.

Archivists strive to provide equality of access to repository holdings, although they find they often must educate both records creators and records users about appropriate access to sensitive information. In general, repositories grant access to all users to all materials, with the exception of those covered by law or other restrictions. Insofar as possible, records open to one user are open to all users. No exclusive use of materials is allowed. When interested users approach the repository, archivists strive to treat them equally. Equality of access is now the governing principle for use of records in most repositories.

In addition to advocating equal access for all users, archivists strive to give equal service to all users. Archivists in both archives and manuscript repositories are unlikely, however, to have sufficient resources to be able to serve all users equally and must therefore devise policies to administer reference services fairly. In some settings limitations on services must be acknowledged (Fig. 7).

PHYSICAL ACCESS

Access also means "physical" access, the opportunity to examine documents. To find information in records, someone must consult documents, whether in analog or digital form. Providing physical access means that repositories maintain regular and sufficient hours of operation and provide space to study records.

Most information in archives remains in tangible forms in unique aggregations, although information is increasingly "born digital," that is, created and maintained in electronic form. Records must be protected from tampering and from accidental or careless damage. The evidential value of archives depends on preserving their integrity, proven through an unbroken line of responsible custody. Archival documents are often unique, and the aggregations of documents in one record group are unique. If the document or the order of documents is lost, it cannot be recovered. Records need physical protection from all types of hazards. Repositories protect records from environmental threats such as fires, floods, and other natural catastrophes, and extend the life of records by creating a stable environment. Archivists protect records while making them available to users.

Planning for security and preservation must be integrated with planning the services to meet users' needs. Providing physical access primarily means setting administrative policies, but these policies have important consequences for the ability of users to perform their

1. User communities. Identify the communities of users to be served by the repository.
2. Resources and restrictions. State generally the types of records held by the repository. State the types of information that may need to be restricted. Identify applicable laws and institiutional information policies that apply to information in the repository, and append them to the policy. Indicate how restrictions will be applied.
3. Intellectual access and reference services. Describe the finding aids, levels of reference services, and the relationship between the two. If necessary, specify distinctions in service levels. Describe searching services, copying services, and services for remote users, whether by phone, email, fax or mail.
4. Fees. Indicate fees for services.
5. Physical access and conditions of use. Describe how records will be made available for researach. Include rules for using materials and policy statements for researchers.
6. Use of information. Establish policies to respond to requests for permission to publish from holdings. Indicate forms for citations. Determine terms for staff user of holdings.
7. Loan of materials. Specify conditions under which matierials will be loaned.

Fig. 7 Elements of an access policy.
Source: From Pugh.[1]

work efficiently. The access policy should specify the physical conditions under which research materials may be used.

Further, potential users find it difficult to come to a repository to carry on their research. Archivists must respond to requests from individuals unable to visit personally for information from their holdings and for research services. Remote users request archivists to locate documents, search them, and provide information from them. Policies must address ways of providing information and documents for remote users. It may be necessary to spell out in some detail how much research reference archivists can do for remote users. It is permissible to provide more research time for the parent institution than for other users.

Access for people with disabilities is a matter of public law.

Security

Security is a basic function for all staff members, not just those in reference. Three components interact to provide security for documents: secure storage, description of holdings, and protection of materials while in use. Records are most at risk when they are in active use, and reference archivists or their assistants are generally responsible for ensuring that they are protected during use.[43]

Preservation

Records also need protection to minimize the wear and tear inherent in handling, copying, loaning, and exhibiting them. Providing physical access to archives and manuscripts means negotiating a balance among the needs of users, records, and staff. While meeting these needs of current researchers, archivists must also consider those of future users by protecting archives from theft or abuse and from wear and tear.

PROVIDING INFORMATION FROM ARCHIVES: COPIES, DIGITAL SURROGATES, OR LOANS

Copies

Physical access may also be provided through copies and loans. When users begin to use materials, they often request copies of them. Responding to requests for copies is a significant administrative component of reference service. Copying raises questions of technology, preservation, and copyright. To use information found in archives, users note it, copy it, or borrow the documents in which it is found. They request copies for research use, legal use, publication, and exhibition. Individuals also often request information from holdings rather than about holdings. Some requests for such information can be provided by the reference archivist, but most such requests are answered by providing copies.

Digital Surrogates or Electronic Records

Today, copying often means creating digital surrogates disseminated electronically through the Internet or other digital means. Repositories are placing archival documents online, first in exhibition-like presentations, but more recently, through mass digitization of collections or series of records, mediated through interactive finding aids. "Born digital" documents are also provided electronically.

Loans

Although most archival materials do not circulate, repositories do loan records. The three most common requests for loans are administrative use, research use, and exhibition. Ensuring preservation and security of records are critical to making loans, which are recorded in a loan agreement.[44–46]

OUTREACH

Reference archivists are integral to outreach programs in their repository. The daily activities of providing reference services profoundly affect public relations. For better or worse, the reference archivist contributes daily to the public perception of the repository and the profession.

Developing Personal Networks in the Parent Organization

For institutional archivists, developing networks of people and constituencies is as important as developing electronic networks. Archivists recognize and build on the information-seeking patterns that staff in organizations use to find information through other people. Developing networks within the parent institution has two important benefits. First, staff members may add the archives to their mental map of information resources and think to call when they have information needs. Second, building such networks reinforces the usefulness of archives; staff will understand why the archival program is important to the organization.

Developing Networks Outside the Parent Institution

Building outside networks is another important aspect of outreach. Archivists should participate in conferences of scholarly associations in relevant subject areas; visit with records creators, donors, and collectors; develop relationships with teachers and faculty; and prepare public programs describing the repository's records and how they could assist current work.

PUBLIC PROGRAMS

Some users' needs can be met more effectively through public programs for groups of users than through personal reference interaction.

Managing Reference Services and Evaluating the Use of Archives

To meet users' needs, protect records, and use staff effectively, repositories organize, administer, and evaluate reference services. Staff qualified to provide reference services must be recruited, trained, supervised, and organized in reporting relationships conducive to meeting user needs. Managing reference services requires planning policies and procedures. Effective management of reference services also depends on measuring and evaluating the use of archives through routine data collection and user studies.

ADMINISTRATIVE COMPONENTS OF REFERENCE SERVICES

Administering the daily tasks of reference services, or managing the staff providing them, may be the most time-consuming aspect of archival reference services. These tasks include receiving, identifying, registering, and orienting users; locating, retrieving, and reshelving materials; and supervising research use, copying, and loans.

ORGANIZING REFERENCE SERVICES

Repositories typically organize reference staff in one of three models: (1) a curatorial organization, in which reference services are integrated with arrangement and description, often in a subject or format area; (2) a rotating organization, in which all staff members take turns providing reference services in rotation; and (3) a functional organization, in which reference services are organized as a separate department. The effectiveness of each of these arrangements depends on the staff, the size and complexity of the holdings, and the nature of the finding aids. In most organizations, the functional model works best, so that reference archivists have a distinct identity, accountable for ensuring service for users and advocating for their needs.

STAFF QUALIFICATIONS

No matter what pattern or combination of patterns is employed, service to users must be the foundation of all archival programs. Reference staff must be capable of meeting both the intellectual and personal needs of users. Archivists have a genuine spirit of public service, but all reference staff must be dedicated to public service, whether professional, paraprofessional, or clerical. A repository's image is largely shaped by the services provided to users.

Reference staff members must also be knowledgeable about repository finding aids and holdings, understand the organizational history, functions, and record forms of the records creators, and possess or develop subject knowledge of the activities documented by repository holdings. Since appropriate referrals to other information sources are an important element of good service, knowledge of the information universe is also necessary, as is familiarity with general reference tools and online sources. Reference archivists must be familiar with laws and ethics relating to access and copyright.

MANAGING REFERENCE SERVICES

Managing reference services requires planning, establishing reference and access policies, implementing those policies in procedures, and administering these policies and procedures. It also requires advocacy and communication. Planning for reference services should be integrated into repository planning, both for strategic planning and annual planning. Establishing policies for reference services rests on repository policies embodied in the access policy. A procedures manual that outlines the details of reference procedures saves time and helps ensure consistent reference practices.

Time management is difficult, but because of that very difficulty, it is vital. In the research room the time constraints of staff and users intersect and sometimes conflict. No repository can be all things to all users or potential users. Each repository acknowledges its limitations of staff and budget and establishes policies that serve the greatest number in terms of its overall mission. At the same time, the repository can work to identify unmet needs and seek resources to meet them. Priorities must be set as rationally as possible; it is unfair both to users and to reference archivists to make such decisions on an ad hoc basis.

Reference archivists are advocates for the needs of users in repository planning, reporting regularly to repository management on use and user needs. Providing meaningful information to management is necessary to obtain adequate space and equipment, a sufficient number of well-trained staff members, usable finding aids, appropriate public programs, and useful publications. Reference archivists also communicate user needs to other staff members, especially those in technical services and acquisitions.

MEASURING AND EVALUATING THE USE OF ARCHIVES

Archivists measure and evaluate the use of archives at three levels: at the reference function or department, the repository, and the profession.[47] Evaluating the effectiveness of the reference function means measuring the uses and users of records. Qualitative assessments of service outputs evaluate such attributes as promptness, thoroughness, courtesy, care, and adequacy of response. Evaluating performance of individual reference staff members requires a set of objectives for reference performance linked to behaviors that can be observed. Meaningful evaluation of reference services compares performance with some standard—either repository objectives or professional standards.

At the repository level, understanding the use of the holdings is necessary to plan descriptive programs, set processing priorities, develop acquisition strategies, and plan public programs. Quantitative and qualitative measures justify the value of archives to resource allocators. Evaluating repository performance requires assessment of such qualitative factors as accessibility, quality of finding aids, comfort, quality of holdings, and accuracy of information provided. Quantitative measures can evaluate timeliness, costs, and other aspects of user satisfaction.[48]

At the broadest level, understanding and communicating the use of archives contributes to greater public understanding of the value of preserving them and the need to support archival and manuscript repositories.[49]

CONCLUSION

The archival system is now less predicated on interaction between the user and the archivist. Researchers can directly access finding aids and increasingly locate documents online. Remote inquiries come through e-mail, often from "accidental users," and are increasing exponentially. Web 2.0 technologies allow researchers to annotate, update, and index archival finding aids, and the next generation finding aid may serve as the basis for a community of learners, including archivists.

Information-seeking behaviors are changing rapidly as information is increasingly "born digital," recorded in electronic forms, especially in networked environments. A search engine on the World Wide Web may yield a list of archival repositories or archival holdings, available to the public in ways undreamed of a decade ago, but it will list them among hundreds, or thousands, or tens of thousands of sites. Understanding the information needs and research methods of significant users of archives is even more important now as archivists compete for the attention of users.

Perhaps a greater concern is the new popular illusion that all information is available online, especially when this illusion is coupled with the normal consequence of the "Principle of Least Effort." This combination militates against searching for information that is not online, even though most archival records will remain in analog forms for the foreseeable future. Changing researcher expectations mean that archivists have to develop innovative ways to connect with them. Although archivists are increasingly linked to users via telecommunication networks, personal networks remain the most important means by which users find archival repositories. Archivists connect with those personal networks to ensure that archives will be on the mental maps of users. Users must think of archives as an information resource.

A reference archivist is a guide not only to finding aids and records, but to the structures and forms of the information landscape of the repository and beyond. The reference archivist is not a barrier, nor a gatekeeper, but rather a partner, a facilitator, and a guide in the joint quest to find meaning in the past to understand the present and plan for the future.

ACKNOWLEDGMENTS

I am grateful to the Society of American Archivists, and Teresa Brinati, Director of Publishing, for allowing me to repurpose text from my book. Elizabeth Yakel, Paul Conway, Kathy Marquis, and Nancy Bartlett provide rigorous, yet always constructive, criticism and stimulate my thinking about reference services. The Bentley Historical Library at the University of Michigan provides an intellectual home.

REFERENCES

1. Pugh, M.J. *Providing Reference Services for Archives and Manuscripts*; Society of American Archivists: Chicago, IL, 2005.
2. Levy, D.M. *Scrolling Forward: Making Sense of Documents in the Digital Age*; Arcade Publishing: New York, 2001.
3. Yates, J. *Control through Communication: The Rise of System in American Management*; Johns Hopkins University Press: Baltimore, MD, 1989.
4. Wosh, P.J. Going postal. Am. Archivist **1998**, Spring *61*, 220–239.
5. Schellenberg, T. The appraisal of modern public records. In *Modern Archives Reader: Basic Readings on Archival Theory and Practice*; Daniels, M.F., Walch, T., Eds.; National Archives and Records Service: Washington, DC, 1984; 57–70.
6. Miller, P.P. *Developing a Premier National Institution: A Report from the User Community to the National Archives*; National Coordinating Committee for the Promotion of History: Washington, DC, 1989; 9.
7. Mick, C. Human factors in information work. ASIS **1980**, *17*, 21–23.
8. Bearman, D. User presentation language in archives. Arch. Mus. Inform. **1989–1990**, Winter *3*, 3–7.
9. Joyce, W.L. Archivists and research use. Am. Arch. **1984**, Spring *47*, 124–133.
10. Kobrin, D. *Beyond the Textbook: Teaching History Using Documents and Primary Sources*; N.H. Heinemann: Portsmouth, U.K., 1997.
11. Hendry, J. Primary sources in K-12 education: opportunities for archives. Am. Arch. Spring/Summer **2007**, *70*, 114–129.
12. National Archives and Records Administration. *Teaching with Documents: Using Primary Sources from the National Archives*, National Archives: Washington, DC, 1989.
13. Greene, M.A. Using college and university archives as instructional materials. Midwest. Arch. **1989**, *14*, 31–38.
14. Robyns, M.C. The archivist as educator: integrating critical thinking skills into historical research methods instruction. Am. Arch. **2001**, Fall/Winter *64*, 363–384.
15. Gilliland-Swetland, A.J. An exploration of K–12 user needs for digital primary source materials. Am. Arch. **1998**, *61*, 136–157.
16. Gilliland-Swetland, A.J.; Kafai, Y.; Landis, W. Integrating primary sources into the elementary school classroom: A case study of teachers' perspectives. Archivaria **1999**, *48*, 89–116.
17. Osborne, K. Archives in the classroom. Archivaria **1986–1987**, *23*, 16–40.
18. Boyns, R. Archivists and family historians: Local authority record repositories and the family history user group. J. Soc. Arch. **1999**, *20*, 61.
19. Yakel, E.; Torres, D.A. Genealogists as a "community of records.". Am. Arch. Spring-Summer **2007**, *70*, 93–113.
20. Mann, T. *Library Research Models*; Oxford University Press: New York, 1993; 91.
21. Gordon, A.D. A portrait of research in legal history. *Public Services Issues with Rare and Archival Law Materials*, Haworth Press: New York, 2001; also published as Legal Reference Services Quarterly, 2001; *20*, 12.
22. Yakel, E. Thinking inside and outside the boxes: archival reference services at the turn of the century. Archivaria **2000**, Spring *49*, 140–160.
23. Sprague, M.W. Information-seeking patterns of university administrators and nonfaculty professional staff members. J. Acad. Libr. **1994**, *19*, 378.
24. Brown, W.; Yakel, E. Redefining the role of college and university archives. Am. Arch. **1996**, Summer 282.
25. Cross, R.L. *A relational view of information seeking*; Boston University: Boston, MA, 2001; Dissertation.
26. Conway, P. *Partners in Research: Improving Access to the Nation's Archive*; Archives and Museum Informatics: Pittsburgh, PA, 1994.
27. Southwell, K.L. How researchers learn of manuscript resources at the western history collections. Arch. Issues **2002**, *26*, 91–109.
28. Yakel, E.; Torres, D.A. AI: Archival intelligence and user expertise. Am. Arch. Spring-Summer **2003**, *66*, 51–78.
29. American Management Systems, *Methodology for Developing an Expert System for Information Retrieval at the National Archives and Records Administration*, National Archives and Records Administration: Washington, DC, 1986.
30. Long, L.J. Question negotiation in the archival setting: the use of interpersonal communication techniques in the reference interview. Am. Arch. Winter **1989**, *52*, 40–51.

31. Malbin, S.L. The reference interview in archival literature. Coll. Res. Libr. January **1997**, *58*, 69–80.

32. Craig, B. Old myths in new clothes: expectations of archival users. Archivaria Spring **1998**, *45*, 122.

33. Cook, T. Viewing the world upside down: reflections on the theoretical underpinnings of archives public services. Archivaria Winter **1990–1991**, *31*, 123–134.

34. Tibbo, H.R. Interviewing techniques for remote reference: electronic versus traditional environments. Am. Arch. Summer **1995**, *58*, 294–310.

35. Bell, M.M. Managing reference e-mail in an archival setting: tools for the increasing number of reference queries. Coll. Res. Libr. News February **2002**, *63*.

36. Martin, K.E. Analysis of remote reference correspondence at a large academic manuscripts collection. Am. Arch. Spring/Summer **2001**, *64*, 17–42.

37. Krause, M.G.; Yakel, E. Interaction in virtual archives: the polar bear expedition digital collections next generation finding aid. Am. Arch. Fall/Winter **2007**, *70*, 282–314.

38. Evans, M.J. Archives of the people, by the people, for the Peopl. Am. Arch. Fall/Winter **2007**, *70*, 387–400.

39. Yakel, E.; Kim, J. Midwest state archives on the web: a content and impact analysis. Arch. Issues **2003–2004**, *28*(1), 47–62.

40. Yakel, E. Inviting the user into the virtual archives, OCLC systems & services. Int. Digit. Libr. Perspect. **2007**, *22*(3), 159–163.

41. Samouelian, M. Embracing Web 2.0: Archives and the newest generation of web applications. Am. Arch. **2009**, Spring/Summer *72*, 42–71.

42. Pearce-Moses, R. *A Glossary of Archival and Records Terminology*; Society of American Archivists: Chicago, IL, 2005.

43. Trinkhaus-Randall, G. *Protecting Your Collections: A Manual of Archival Security*; Society of American Archivists: Chicago, IL, 1995.

44. American Association of Museums, Registrars Committee, *Standard Facility Report*, 2nd Ed., 1998; Washington DC AAM 2008.

45. Rare Books and Manuscript Section, Guidelines for the loan of rare and unique materials. CRL News **1993**, May *54*, 267–269.

46. Rare Books and Manuscript Section, Guidelines for borrowing and lending special collections materials for exhibition. CRL News **1990**, May *51*, 430–434.

47. Dearstyne, B. What is the use of archives? A challenge for the profession. Am. Arch. **1987**, Winter *50*, 76–87.

48. Dearstyne, B. What is the use of archives? A challenge for the profession. Am. Arch. **1987**, Winter *50*, 80.

49. Cox, R.J. Researching archival reference as an information function: observations on needs and opportunities. RQ **1992**, Spring 387–397.

BIBLIOGRAPHIC ESSAY

These citations offer sources not cited in the notes. Few monographs are devoted to reference services in archives. The first American manual was Sue Holbert, *Archives & Manuscripts: Reference and Access* (Chicago, IL: Society of American Archivists, 1977) followed by the first edition of Mary Jo Pugh, *Providing Reference Services for Archives and Manuscripts* (Chicago, IL: Society of American Archivists, 1992). The second edition in 2005 has a lengthy bibliographic essay. Hugh Taylor offered a broad overview in *Archival Services and the Concept of the User: A Ramp Study* (Paris: UNESCO, 1984). Haworth Press published two useful collections of essays: Lucille Whalen, Ed. *Reference Services in Archives* (New York: Haworth Press, 1986) and Laura B. Cohen Ed. *Reference Services for Archives and Manuscripts* (New York: Haworth Press, 1997). An excellent summary of references services in academic repositories that offers models for all settings is Elizabeth Yakel, "Managing Expectations, Expertise, and Effort While Extending Services to Researchers in Academic Archives," in *College and University Archives*, Christopher J. Prom and Ellen D. Swain, Eds. (Chicago, IL: SAA, 2008, 261–286).

The *American Archivist* published a special section on users and archival research in Volume 66 (Spring/Summer 2003): Helen R. Tibbo, "Primarily history in America: How U.S. historians search for primary materials at the dawn of the digital age," 9–50; Wendy M. Duff and Catherine A. Johnson, "Where is the list with all the names? Information-seeking behavior of genealogists," 79–95; and Barbara L. Craig, "Perimeters with fences? Or thresholds with doors? Two views of a border," 96–101.

Richard Lytle identified and analyzed the two methods of access in "Intellectual access to archives: Provenance and content indexing methods of subject retrieval," *American Archivist* 43 (Winter 1980): 64–75. Pugh elaborated these insights in "The illusion of omniscience: Subject access and the reference archivist," *American Archivist* 45 (Winter 1982): 33–44. Elizabeth Yakel argues for a new vision for reference services in the new millennium, particularly stressing the multiple roles organizational archivists might play as knowledge brokers, "Knowledge management: The archivist's and records manager's perspective," *Information Management Journal* 34 (July 2000): 24–30.

Two essays in *Encoded Archival Description on the Internet* (Binghamton, New York: Haworth Press, 2001) assess the impact of EAD on reference services: Richard V. Szary, "Encoded finding aids as a transforming technology in archival reference service," and Anne J. Gilliland-Swetland, "Popularizing the finding aid: Exporting EAD to enhance online discovery and retrieval in archival information systems by diverse user groups." Michelle Light and Tom Hyry, "Colophons and annotations: New directions for the finding aid," *American Archivist* 65 (Fall/Winter 2002): 216–230 suggest that annotations be used to enhance description by reference archivists and users. For further exploration of the potential directions of archival reference services on the Web, see Christopher Prom, "User interactions with electronic finding aids in a controlled setting," *American Archivist* 67 (Fall/Winter 2004: 234–268, and Richard V. Szary,

"Encoded finding aids as a transforming technology in archival reference service," in *College and University Archives*, Christopher J. Prom and Ellen D. Swain, Eds. (Chicago, IL: SAA, 2008, 245–260). Elsie Freeman Finch, Ed., *Advocating Archives: An Introduction to Public Relations for Archivists* (Lanham, MD: Scarecrow Press, 1994) provides overviews of outreach in archives.

Navigating Legal Issues in Archives by Menzi Behrnd-Klodt (Chicago, IL: Society of American Archivists, 2008) introduces legal issues. Michel Duchein, *Obstacles to the Access, Use and Transfer of Information from Archives: Ramp Study* (Paris: UNESCO 1983); and Gabrielle Blais, *Access to Archival Records: A Review of Current Issues: A Ramp Study* (Paris: UNESCO 1995) provide international perspectives. Excellent collections of essays are found in *Libraries, Museums, and Archives: Legal Issues and Ethical Challenges in the New Information Era*, Tomas A. Lipinski, Ed. (Lanham, MD: Scarecrow Press, 2002) and *Public Services Issues with Rare and Archival Law Materials*, Michael Widener, Ed. (New York: Haworth Press, 2001). Also useful are *Archives and the Public Good: Accountability and Records in Modern Society* (Westport, CT: Quorum Books, 2002) and Heather MacNeill, *Without Consent: The Ethics of Disclosing Personal Information in Public Archives* (Lanham, MD: Society of American Archivists and Scarecrow Press, 1992). Peter B. Hirtle, "When works pass into the public domain in the United States: Copyright term for archivists," posted on-line by Cornell University at http://www. copyright.cornell.edu/Training/Hirtle_Public_Domain.htm, is a useful summary in table form. Also important is his "Archives or assets?" *American Archivist* 66 (Fall/Winter 2003): 235–248, in which he challenges archivists to consider the financial barriers they place in the use of archival resources.

The Society of American Archivists publishes the *American Archivist* (1938–), and the Association of Canadian Archivists publishes *Archivaria* (1975–). The Midwest Archival Conference publishes *Archival Issues*. *Archives and Manuscripts is* published in Australia. Two recent journals are published by commercial presses, *Archival Science* (2001–) and the *Journal of Archival Organization* (2002–). Blogs offer discussion of new trends. See for example, http://www.archivesnext.com/, accessed December 10, 2008.

Approval–Archival

Archival Science

Elizabeth Shepherd
Department of Information Studies, University College London, London, U.K.

Abstract

Archival science is the academic and professional discipline concerned with the theory, methodology, and practice of the creation, preservation, and use of records and archives. It encompasses the creation, preservation, and use of records in their functional context, whether organizational or personal, and the wider social, legal, and cultural environment within which records are created and used. In the United Kingdom the discipline is known as archives and records management, although in other cultural contexts, archives management and records management are sometimes separated, and sometimes bound more closely together, for example, by proponents of the records continuum. Organizations and individuals create records in the conduct of their current business, to support administration, to ensure accountability, and for cultural purposes, to meet the needs of society for collective memory and the preservation of individual and community identity and history. This entry serves as an introduction to the field of archival science. The first section gives a broad overview of the nature of archives and records. The main section provides some definitions and an introduction to the core concepts and principles which are central to the discipline, including records, archives, and activities; attributes of records; the records lifecycle; the records continuum; provenance and original order; postmodernist ideas around archive fever. The entry concludes with a discussion of the professional discipline of archives and records management, an overview of the key functions of archives and records services, and a brief history of archival science in the Western world.

INTRODUCTION

Archival science is the academic and professional discipline concerned with the theory, methodology, and practice of the creation, preservation, and use of records and archives. It encompasses the creation, preservation, and use of records in their functional context, whether organizational or personal, and the wider social, legal, and cultural environment within which records are created and used. In the United Kingdom the discipline is known as archives and records management, although in other cultural contexts, archives management and records management are sometimes separated, for example in Germany and in North America, where separate education programs and professional bodies have developed for each subfield. Sometimes archives and records management are bound more closely together than in the United Kingdom, for example, by proponents of the records continuum in Australia. Organizations and individuals create records in the conduct of their current business, to enable actions to be taken and decisions to be made. Records used for business purposes support administration, regulation, public or professional services, economic activity, and dealings between individuals and organizations. Records can be used to ensure accountability, to make people and businesses account for their actions and obligations and when there is a need to prove that organizations have complied with legal or regulatory requirements or recognized best practice. Records enable organizations to meet legal, regulatory, and financial requirements, and to protect their assets and rights. They help to support the expectations of a democratic society for transparency and they protect citizen's rights. They enable governments to deliver electronically enabled services to citizens (e-government) and facilitate citizen participation through the provision of information and digital interaction. Records support lifelong learning, education, and training for people of all ages, from school children to senior citizens. Some records are preserved for posterity as archives, ensuring that they are available for historical, demographic, sociological, medical, scientific, genealogical, or biographical research, for cultural purposes, and to meet the needs of society for collective memory and the preservation of individual and community identities and histories. Increasingly, writers including Andrew Flinn and Sue McKemmish are interested in the role of archives outside traditional organizations and the ways in which archives provide evidence of individuals and communities.

This entry serves as an introduction to the field of archival science and many of the topics which are covered briefly here can be found discussed in more detail in related entries. The first section gives a broad overview of the nature of archives and records. The main section provides some definitions and an introduction to the core concepts, theories, and principles which are central to the discipline, including records, archives, and activities;

Encyclopedia of Library and Information Sciences, Fourth Edition DOI: 10.1081/E-ELIS4-120043644

attributes of records; the records lifecycle; the records continuum; provenance and original order; postmodernist ideas around archive fever. The entry concludes with an overview of the professional discipline of archives and records management, a discussion of the key functions of archives and records services, and a brief history of archival science.

OVERVIEW OF ARCHIVAL SCIENCE

The Nature of Archives and Records

Archives and records can be created or maintained in any media and format, including paper, parchment, and digital, maps, plans, and technical drawings, volumes and files, photographs and prints, moving images, films, audio and videos, voicemail, other materials such as clinical samples and specimens, Web sites and computer-generated records of all kinds. Records are traditionally text-based but often also contain nontextual material such as sound, images, or three-dimensional objects. Recent literature has challenged some of the more traditional ideas about the nature of the record and its attributes, to encompass ideas of orality, community archives and fluidity. Some writers, such as Terry Cook and Ciaran Trace, emphasize the social context of records creation and suggest that records cannot be neutral: this is supported by the work by Verne Harris in a South African context.

In the traditional view, a record is not defined by its physical format, but rather by its purpose and use. There is significant discussion in the literature about the nature of a record or archive, but a simple definition is that a record is "recorded evidence of an activity" and that archives are those records "recognized as having long-term value." Extracts from Shepherd and Yeo[1] used throughout this entry are reproduced with permission of the publisher. Records management is the "field of management responsible for the efficient and systematic control of the creation, receipt, maintenance, use and disposition of records" according to the International Standard.[2] Archives management is "the selection, preservation, maintenance, arrangement, description and access to materials of continuing value for the long term."[3]

CORE CONCEPTS AND THEORIES

Records, Archives, and Activities

If we accept the definition of a record as "recorded evidence of an activity," we need to understand what an activity is. An activity is "an action or set of actions undertaken by an individual, a group of individuals or a corporate body, or by employees or agents acting on its behalf, and resulting in a definable outcome."[1] An activity has an identifiable starting point in time and also an end point. Some activities, such as drafting and note taking, only involve one party (or actor). In an organizational context most significant activities involve two or more actors: some form of transaction or communication takes place between them. One or more of the actors may be external to the organization (such as a customer or supplier). Other transactions may be purely internal (for example, between a manager and a staff member, or between one department and another). Where more than one party is involved in the activity each party may create its own record; alternatively the second party may receive and retain the record transmitted to it by the creator, while the creator retains a copy of it. Records may be created either in the course of the activity, or afterwards in a conscious act of record keeping. A transaction may be effected in whole or part by the creation or transmission of records such as letters, invoices, or purchase orders. On the other hand, minutes of meetings, staff training records, equipment maintenance records, personal diaries, and the registration of births, marriages, and deaths, are created in order to provide evidence of activities which are already complete.

Records maintained digitally are known as digital records (or electronic records). These terms refer to records created or received digitally which are then maintained in digital form, as opposed to records which are created using word-processing or other software and are then printed out on paper. Such records are sometimes called "born digital," to distinguish them from paper records which are digitized, i.e., turned into digital images of the original. In most organizations, records are a mixture of formats and media, and include both paper-based and digital records. Most organizations operate hybrid records systems which accommodate records and archives in any medium.

More recently the concept of the record has been extended to include records from different cultures and traditions, including stories, oral records of all kinds, and other cultural objects including dance, song, and painting, which provide evidence of community histories and identities.

Attributes of Records

If a record is to function effectively as "evidence of an activity," it must be compliant with any external requirements in the environment where the organization operates. Requirements for records may derive from legislation, regulation, mandatory standards, codes of best practice and ethics, or community expectations. Many traditional aspects of fixity in records have now come into question, but a summary of the established view is given here.

Records must possess content, context, and structure. The content of a record must reflect the facts about the activity. For a reliable record these should be accurate (the facts should be correct) and complete (everything of significance should be recorded). The context of a record must be supported by information about the circumstances in which it was created and used. Records cannot be fully understood without adequate knowledge of the activity which gave rise to them, the wider function of which that activity forms part, and the administrative context, including the identities and roles of the various participants in the activity. Contextual information must be captured in the records themselves or in the systems that are used to maintain them. The structure of records and records systems must reflect the relationships between their constituent parts. In a business letter, for example, there is a formal structural relationship between the details of the addressee, the date, the body of the text divided into paragraphs, and the signature at the end. There are also structural relationships between the individual letters in a file or folder, and between records in a series. The structure of a record forms a link between content and context. Structure organizes the content in such a way as to denote context, and thus contributes to a user's understanding of the record.

Records should also have the qualities of authenticity, integrity, usability, and reliability. The authenticity and integrity of records need to be guaranteed over time, so that users can be confident that records are genuine and trustworthy and that no illicit alterations have been made to them. Records need to be usable: they must be accessible to authorized users and provide sufficient evidence of the context of their creation to support a user's understanding of their significance. Records created within an organization should also be reliable and accurate in their content.

For this to be achieved, records must be created and maintained systematically. Sufficient measures are needed to ensure that records are what they purport to be. If this is not clear the weight of a record as evidence of an activity will be diminished. Records may be destroyed when no longer required, but for as long as they are kept they should remain unaltered. Moreover their usability must be preserved so that they can be retrieved, consulted, and interpreted when they are needed.

Records can be used as sources of information as well as evidence. The information in a textual record may be in the form of raw data (name, address, etc.) or may be derived from narrative. Records typically contain information relating to the parties involved in an activity and to the contents or subject matter of the activity itself, but may also contain information relating to other matters such as the political, organizational, or social environment within which the activity occurred. Users may refer to a record to obtain information of this kind, quite independently of any need to use the record as evidence. Users are also likely to employ other methods of obtaining information, and may not consciously distinguish between records and other information sources.

Archives and records management systems have, therefore, to meet both informational and evidential needs.

Records Lifecycle

The records lifecycle is a concept which is in common use in archival science. It seeks to show that records are not static, but have a life similar to that of biological organisms: they are born, live through youth and old age, and then die. The idea was developed in the 1950s in the United States by T.R. Schellenberg, who wrote about "the life span" of records which included their current use and final destiny.[4] Since the 1950s many variants on the records lifecycle concept have been developed. Most models aim to show a progression of actions at different times in the life of a record: typically, its creation, capture, storage, use, and disposal. Some writers show this as a linear progression, while others describe a loop or a circle, see Fig. 1.

A different model of the concept suggests that records pass through three ages (or stages), as shown in Fig. 2 a current or active stage, when they are used for business; a semi-current or semi-active stage, when their business value is reduced; and a noncurrent or inactive stage, when they have little or no business value but may be used for secondary, research purposes. Writers who use this version generally also emphasize the physical movement of paper records to alternative storage at each phase of their life, together with the destruction of unwanted records at each stage. There are varying views on where archives fit into this picture, for instance, are they all encompassed within the noncurrent records stage or can archives be identified at other stages in the lifecycle? Records that have continuing significance to the organization, for example, as evidence of its ongoing rights and obligations, can be identified as archives while they are still current.

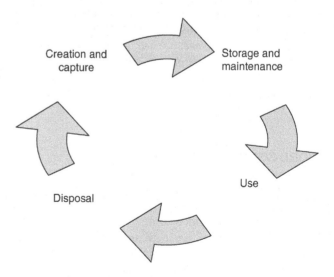

Fig. 1 "Progression of actions" lifecycle model.
Source: E. Shepherd; G. Yeo.[1]

Current records are records regularly used for current business of an organisation and which continue to be maintained in their place of origin.

↓

Semicurrent records are records required infrequently for current business which should be transferred from offices to a records centre awaiting their ultimate disposal.

↓

Noncurrent records are records no longer needed for current business.

Fig. 2 "Three ages" lifecycle model.
Source: Adapted from International Council on Archives, *Dictionary of Archival Terminology*, Walne, P., Ed., 1988.

Cultural values can also be identified at an early stage, as well as later.

In the 1980s and 1990s, the lifecycle concept was subject to much adverse criticism. Critics noted that some records do not "die," but are retained indefinitely because of their continuing value. The division between stages of the lifecycle in the "three ages" model is seen as artificial: for example, records which have been thought to be noncurrent may have a renewed period of currency if the activity which gave birth to them is revived. Lifecycle models do not appear to allow for iteration or repetition of stages or actions, or for stages to be omitted, though in practice this frequently happens. It has also been argued that the lifecycle concept perpetuates an artificial

distinction between records kept for business purposes and records kept for historical or cultural reasons and between the two professions of archivist and records manager. Critics of the lifecycle models also suggest they are too focused on records as physical entities and on operational tasks, especially those associated with the custody of paper records. Digital records rely on logical rather than physical structure, and the tasks associated with the physical storage of paper are largely irrelevant to their management.

Records Continuum

The records continuum concept was developed in the 1980s and 1990s, particularly by academics in Australia. As Fig. 3 shows, managing records is seen as a continuous process. In the continuum there are no separate parts or stages, but one element of the continuum passes seamlessly into another. Dimensions in the continuum are said not to be time-based, but rather represent different perspectives on the management of records. The circles move out from the creation of records of business activities, to ensuring that records are captured as evidence, and to their inclusion in formal systems for records management within the organization; while the fourth dimension looks out toward the needs of society for collective memory. In contrast to the older view that records are kept for organizational purposes during the early stages of their lives, and only later come to meet the needs of a wider society as

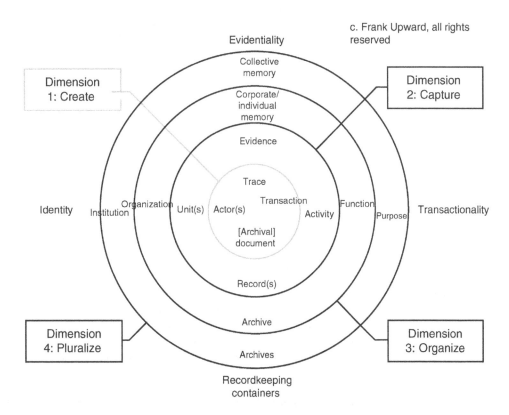

Fig. 3 Records continuum model. Copyright F. Upward.

archives, the continuum takes the view that records can function simultaneously as organizational and collective memory from the time of their creation.[5]

Many archivists and records managers have rejected the lifecycle concept altogether and promote the continuum concept as offering a truer insight. In fact the two concepts may not be incompatible. The continuum is flexible and able to reflect a range of issues surrounding the role of records in contemporary organizations and society. However when planning and executing a program of archives and records management in an organization, formulations based on the lifecycle concept can provide a useful practical framework and can help to identify stages and actions which have to be taken.

Provenance and Original Order

The establishment of public national archives following the French Revolution led to a separation between record keepers in government administration and archivists in national archives. Initially archivists were content to receive records selected by the administration, seeing their role, in a purely custodial light, as the preservation, arrangement, and description of records, and the provision of access to them. For these purposes, archivists developed a coherent body of theory based on the innate characteristics of records, summed up in the principles of provenance and original order. These are called in French, *le respect des fonds*, and in German, *Provenienzprinzip*.

An early coherent explanation of these European principles was given in the Dutch *Manual for the arrangement and description of archives* in 1898 by Samuel Muller, Johan Feith, and Robert Fruin.[6] They defined archives as "the whole of the written documents, drawings and printed matter, officially received or produced by an administrative body of one of its officials. . ." and said that archives "must be kept carefully separate" and not mixed up with the archives of other creators, thus preserving the principle of provenance. In addition, they said that the arrangement of such archives "must be based on the original organization of the archival collection, which in the main corresponds to the organization of the administrative body that produced it" and not placed into artificial arrangements based on chronology, geography, or subject, which is a statement of the principle of original order.

It is worth noting that at this time the U.K. Public Record Office finding aids were based on subject arrangements rather than on strict archival principles of provenance. For example, Scargill-Bird's *Guide* appeared in 1891: the 3rd edition in 1908 changed from an alphabetical subject arrangement to one respecting groups and classes. Giuseppi's *Guide* (begun in 1914 and published in 1923–1924) was more fully provenance-based.

Hilary Jenkinson of the Public Record Office was the author of the first significant U.K. text on archives, *A Manual of Archive Administration* published in 1922.[7]

Jenkinson stated that the Primary duty of the archivist was to safeguard the archives in his custody and his Secondary duty was to provide for the needs of historians and researchers. Jenkinson subdivided Primary duty into, first, the physical defense (preservation and conservation) and, secondly, the moral defense (unbroken chain of custody, preservation of original order), and stated that "The good Archivist is perhaps the most selfless devotee of Truth the modern world produces." Jenkinson's Moral defense embraced his belief that records accumulate naturally as a result of administration and are untainted evidence of acts. They require an unbroken chain of custody to maintain their authenticity. As a result, the archivist should not interfere with the order in which documents were received or with original bindings or filing or undertake appraisal, since the archivist's role was to keep and not to select archives. Only the administrator could select archives for preservation.

Jenkinson's understanding of the principles of arrangement were that "the only correct basis of arrangement is exposition of the administrative objects which the archives originally served" and that the archivist should "establish or reestablish the original arrangement." He noted that the immediate source of acquisition of the archive should not necessarily be regarded as its true provenance and "whatever else we do we must not break up the Archive Group." In this he adopts the French *respect des fonds*.

These early theories up to 1930s were developed by archivists who worked in national archives mainly with medieval and early modern records from stable administrations. They described archives in order to make them accessible through calendars and lists.

The principles rely on an idea of the fonds, which is the whole of the records, regardless of form or medium, organically created and/or accumulated and used by a particular person, family, or corporate body in the course of the creator's activities. The fonds, or in English the archive group, is an organic whole, which can be subdivided. One main level of subdivision is the series, that is archives arranged in accordance with a filing system or maintained as a unit because they result from the same accumulation or filing process, or the same activity; have a particular form; or because of some other relationship arising out of their creation, receipt, or use. These levels of description are made manifest in late twentieth century standards for archival description, in particular the International descriptive standard, ISAD(G), which gives general guidance for the preparation of archival descriptions, which themselves identify and explain the context and content of archives.[8]

Traditionally understood as referring to the administrative origin of records, the term provenance is now often reinterpreted to include an understanding of the functions and activities which underlie records creation and maintenance.

The principle of retaining the original order relates to records in the paper world: it preserves context by protecting the physical structure of archives, for example, the physical juxtaposition of a record to other records, typically in a file or cabinet. To extend this concept into the digital world, where storage is random and the physical juxtaposition of records has no significance, it is necessary to reinterpret it in terms of the intellectual relationships between one record and another. These long-standing principles are still valid, but postmodernist reinterpretations of provenance include understanding the role of the archivist as a cocreator and wider ideas such as societal provenance, developed in particular by Canadian writers.[9]

Archive Fever

Philosophers and cultural theorists, such as Derrida, Foucault, structuralist Levi-Strauss, and Ricoeur, in recent decades experienced uncertainty about positivist approaches which fix identities and meaning, and they questioned whether concepts and phenomena have distinctive and unchanging identities and whether it is possible to accept scientifically verifiable facts or truths. Since 1970s various forms of constructivist and relativist thinking, loosely labeled "postmodernism," have become prevalent. Postmodernist writers commonly argue that there are no meanings independent of human experience, only interpretations which arise within particular social or cultural contexts. They propose legitimate parallel readings of the same concepts, concepts which are not fixed and can vary over time, cultures, and languages. Archivists began to pay attention to the trend in the 1990s, especially after the publication in English of Derrida's *Archive Fever* in 1996.[10] Relativism affects how we think of and conceptualize archives and records and the role of archivists. Derrida suggested that the archivist's role is to do with context and creation of meaning, setting boundaries around context (provenance), and giving title and order through hierarchization (original order). Archival writers, including Terry Cook and Tom Nesmith, suggest that archivists, who create, transmit, keep, arrange, destroy, and use records, are cocreators of the record, and that archival activities are a part of the provenance and context of the record, which is an evolving and not fixed concept.[11] Archivists may have custody of the record for centuries, far longer than the original creator, and thus shape the record and its provenance more than any other actor. These ideas also raise the issue of the representativeness or diversity of archives.

Key Functions of Archives and Records Services

Most organizations have a need for an archives and records management service to ensure that records and archives are created, preserved, and made accessible to the right people in a timely fashion for agreed purposes. Many larger organizations have a formal archives and records service which employs professional and support staff to ensure that a quality service is provided. Although there has been recent debate about the nature of an archives and records service in a digital environment, a number of key functions can be identified, which may typically be available.[1,3]

Organizational analysis and records creation

Records managers and archivists need first to understand the records which are produced, the organizational activities which generate records, and the systems used to control them. In addition, they should have a thorough understanding of the organization itself. To gain this understanding, they must analyze the role and responsibilities of the organization, study its structures and working methods, and discover how these have changed during its life. They also need to identify the broader issues which influence the way the organization operates, including its corporate culture and the interests and expectations of stakeholders both within the organization and externally. The next task is to acquire a deeper understanding of its functions and the activities which are performed to support them, using techniques such as functional and systems analysis. The knowledge gained from these investigations can then be used to assess how each of these factors affects the organization's needs for evidence and information, and the degree to which its needs are met by the existing records and records systems. The archivist and records manager can then design and implement a program which fits the organization's requirements.

Much organizational activity leads naturally to the creation of records, but few organizations seek to record everything that they do. Written communications generate records, but most spoken communications leave no record unless a written note is made, and manual and physical tasks can also pass unrecorded. However, new ways of working and new technologies often make it simpler to create records where none were created in the past. With the growth of e-mail, many messages that might once have been verbal are now written; voicemail has made it easier to capture evidence of telephone communications when records are required. Archivists and records managers advise an organization on its records creation strategy, which helps to ensure that records meet the appropriate standards of authenticity, integrity, usability, and reliability, and that they are captured into a secure and effective records system so that these qualities will remain intact over time. Archivists may be responsible for the procedures for identifying records that need to be captured and for managing the systematic capture of records in both paper and digital form.

Appraisal

Appraisal is the process by which an organization identifies its requirements for maintaining records. Records managers have developed appraisal techniques primarily to support decisions about retention: which records can be destroyed at an early stage, and which merit longer-term or indefinite retention? According to the Australian records management standard, appraisal seeks "to determine which records need to be captured" into a records management system as well as "how long the records need to be kept" (AS 4390.1-1996, clause 8.1).[12] Appraisal can also be used to support other decisions, such as which records require special measures for protection or security. In an archival context, appraisal helps to determine which records have long-term value and are worthy of permanent preservation as archives. Appraisal decisions must take account of the organization's requirements for records for business use and accountability. Decisions about retention also acknowledge cultural interests, and the interests of external users, to ensure the preservation of corporate or societal memory. In practice, these wider interests will carry greater weight in some organizations than others. In the private sector, an organization may choose to keep only those records needed for its own purposes. In a democracy, public sector bodies and some private ones can be expected to take a broader view. Appraisal systems need three components: methodologies for making appraisal decisions, documentation of decisions, and operational measures for their implementation. Much theoretical and methodological work has been done on appraisal, from Schellenberg's taxonomy of values (primary and secondary, evidential and informational), through to macro and functional appraisal approaches developed by Terry Cook and others, which are discussed in the related entry.[13]

Acquisition

One of the outcomes of appraisal is archival acquisition. Acquisition is the process by which an archives service adds to its holdings by accepting archives as a transfer, donation, loan, or purchase. There are two simple models of archival acquisition. First, by managing the archives of the employing organization (either seamlessly with the records management service or separately from it) in order to document the organization which created them: these are institutional archives. Second, by acquiring archives from outside the employing organization, in order to preserve records about some activity external to the organization that maintains the archives: these are collecting repositories. Collecting repositories generally have an acquisition mandate which is defined by one or more themes, such as a particular geographical area (a region or state), a subject (such as literature or women's studies), a specific individual, group, event, or era (such as a prime minister or president, trades unions, or a particular war or conflict), the media of the records (such as film archives), or a time period. Many archives services in fact adopt both of these two models alongside each other, combining an

institutional mission with a collecting responsibility. The archivist should publish a clearly defined statement of acquisition policy which indicates the mandate, such as the subject areas within which records are sought and acquired and the media for which storage and access facilities are available. As well as being a reference tool for staff and potential depositors, the acquisition policy makes acceptance or rejection decisions less subjective, it provides a basis for cooperation with other archives services and for future service development. A well-crafted acquisition policy, backed up by an active acquisition strategy, helps to ensure that the archive is diverse and representative of the community it seeks to document.

Preservation

Preservation is a key activity, whether the intention is to maintain records for a short time for business purposes, or permanently as archives. With paper records, the main emphasis is on storing the physical media and protecting them from loss or damage. Digital records require a different approach: the physical carriers are likely to be short-lived, but the records must be maintained over time and probably across several generations of storage media. While preservation of paper records is costly in terms of space occupancy, electronic storage media are compact and relatively inexpensive, but costs are incurred in ensuring that records remain accessible. Older paper and parchment archives may require active intervention from a specialist conservator to ensure their physical preservation. There is a whole separate discipline of preservation management, archival conservation, and materials science which concerns itself with the preservation of archives, and archivists must work closely with their preservation colleagues to plan and implement a preservation strategy and program.

Arrangement and description

Archivists and records managers need to classify and describe records and document the context of their creation, so as to exercise intellectual control over records and facilitate their management and use over time. For records managers, a key element in this is the classification scheme, which provides links (both intellectual and physical) between records which originate from the same activity or from related activities, helps to determine where a record should be placed in a larger aggregation of records, and assists users in retrieving and in interpreting records. In a paper-based system, the classification scheme determines the identity of the file in which each item is housed and the place of each file within the system as a whole. Classification schemes play an equally important role in digital records systems, where paper files do not exist, and in hybrid systems, where paper and digital records exist side by side. Classification schemes are based on an analysis of functions, processes, and activities and are based on the business activities which generate records. They document the structure of a records system and the relationships between records and the activities which generate them. The archival activity of arrangement and description builds on the descriptive systems devised to manage records while they were in current use. Arrangement is rooted in the archival principles of provenance and original order, referred to earlier in the entry. Archivists usually create a range of finding aids, which might include descriptive catalogs, inventories or lists of each fonds, repository guides which list all the different types of records held by a single archive, subject guides which bring together archives relating to a specific research topic (such as family history or gardening), together with access and index points (which enable a user to ask about a specific person or topic) and authority records which provide contextual or biographical information. Increasingly finding aids are made available online in shared systems and may be cross searched by users. Catalogs or inventories are usually the main hierarchical representation of an archival fonds and are constructed following one or more of the available archival description standards. The key international description standard, ISAD(G) and its supplements, issued by the International Council on Archives, sits alongside national guidelines and rules. These include the Canadian *Rules for Archival Description*, in the United States *Describing Archives: a content standard*, and in the U.K. *Manual of Archival Description*.[14–16]

Access and use

In order to design effective description and retrieval systems, archivists need an understanding of user requirements and search behaviors, since the purpose of description is to enable users to access archives and records. At the time of their creation, records are normally accessible only to the creator workgroup. When they are captured into a records management system, they normally become more widely accessible. Once they have been designated as having value as archives, and especially when they have been transferred to archival custody, wider access to them needs to be provided, including to users from outside the organization. In countries where citizens have rights of access under freedom of information or other legislation, access requests must be handled by someone with appropriate knowledge of the legal obligations. Increasingly archivists are concerned to provide equality of access for all communities, across different legislative, constitutional and cultural frameworks, seeing access to public archives as a right in a democratic society. Most countries have developed standards for access to archives: for example in the United Kingdom, the Public Services Quality Group published a *Standard for Access to Archives* which helps archivists to think about equity, openness, responsiveness and

effectiveness in their access services and to capture these in a written access policy.[17] Archives provide reference services, which typically include onsite access to archives and to finding aids and other explanatory and supplementary materials, and online access to digital copies of archives, searchable finding aids of various kinds and other services. Archives services increasingly engage in other advocacy and outreach activities. Archives which are visually attractive are a prime resource for marketing and public relations, for instance in advertising or promotional materials. Exhibitions and publications are popular and most archives now have educational programs and online learning resources.

This is necessarily a rather brief overview of the key activities offered by archives and records management services: other entries deal with many of these in much greater detail.

The Professional Discipline

Archival science is a professional discipline and those who practice the discipline to provide archives and records management services are called archivists and records managers. Other job titles include manuscript curator, special collections archivist, access to information manager and data specialist. Archivists and records managers can work in any organizational context. Government agencies and authorities at every level (international, national, federal, county, local, city), universities and schools, hospitals, museums, businesses of every kind including multinational corporations, charities and not-for-profits, professional organizations, families and individuals all create and use records and archives and, potentially, offer work to archivists and records managers. Archives and records management services may form part of library and information services, cultural and leisure services, history and local studies, risk and compliance, access and data privacy, legal and chief executive's department, corporate and facilities services, information and communication technology departments or public relations. Sometimes archives services are administered separately from records management, although within a single organization; some organizations only offer one or the other; and sometimes they are a single seamless service.

Entry to professional work is generally through specialist higher level university education, in many countries at Masters or doctoral level, followed by a period of professional development in post, guided by the requirements of national professional bodies. For instance in the United Kingdom, the normal requirement is for a Masters level qualification (for example, a 1-year M.A. in archives and records management) from a university program recognized by the professional body, the U.K. Society of Archivists, preceded by a period of work experience in an archives or records management service, and followed by a portfolio of achievement for continuing professional development. In the United States, entry-level positions generally require an undergraduate and a graduate degree. Although archivists take a variety of undergraduate majors, most receive graduate degrees in history or library science, which may include archival courses and a practical work placement. The Society of American Archivists (SAA) is North America's largest national archival professional association. It serves the educational and informational needs of its members and provides leadership to ensure the identification, preservation, and use of records of historical value. The largely separate profession of records management in the United States is supported by ARMA International, a not-for-profit professional association for those interested in managing records and information. ARMA supports the certification process for experienced professionals in records and information management which leads to the Certified Records Manager (CRM). In Canada, the Association of Canadian Archivists (ACA) provides leadership to the archival profession and a number of universities offer Masters level qualifications in archival science. In Australia, the profession is served by both the Records Management Association of Australasia (RMAA) and by the Australian Society of Archivists (ASA).

BRIEF HISTORY OF ARCHIVAL SCIENCE IN THE WESTERN WORLD

This section examines, briefly, some of the key events and issues arising in archives and records development in England, continental Europe, the United States of America, Canada, and Australia. Archives have been created and preserved by literate societies for several millennia, and some writers trace the history of archives back to ancient Babylonian clay tablets and Egyptian papyruses.[18] However, the modern history of archival science is generally agreed to have emerged after the French revolution of 1789 and the Napoleonic period, which resulted in the upheaval of law, administration and government across Europe. After 1794 French citizens were given the right to have access to public archives, which had previously been closed: as Duchein said, "the notion that research in archives was a civic right was increasingly recognized" in Europe.[19] A shifting of purpose can be seen in many countries, including the Netherlands after 1795, when the "administrative-legal" function was lost and a new "historical-antiquarian" interest emerged. In the nineteenth century, new national archives were established which, Duchein suggests, maintained an historical stance at least initially. These included as the Archivo Historico Nacional in Spain in 1866, the appointment of the first national archivist in the Netherlands in 1802, and the PRO in England in 1838. Duchein identifies two distinct traditions in Europe from this period. The first was in those countries, including Germany and much of

central Europe, which adopted a "registratur" system "by which each administrative document is registered with a registry number corresponding to a methodical schedule known as the Aktenplan." The second tradition emerged in other countries, including France, Belgium, the Netherlands and England, where archivists had to arrange and classify records after their transfer to the archives. This need to classify led to the development of the theoretical debate over the arrangement of archives, initially by subject-based classification schedules which were adopted in France, Prussia, Austria, and England, and then the emergence of the provenance-based approach, especially notable in the Netherlands, which respected the fonds or creating body. The principle of provenance, defined in 1841 by Natalis de Wailly, and respect for original order first articulated in Prussia in 1880, provided the distinctive European approach to archives from the late nineteenth century, and were expressed in the influential texts by the Dutchmen, Muller, Feith, and Fruin in 1898 and by Hilary Jenkinson in 1922.

Although no comprehensive study of the modern development of the archival profession across Europe has been published, there is an emerging body of research into the history of archives and archivists across cultures and in specific countries, mostly unpublished so far, but including studies of the profession in France, England, and in the Netherlands.[20] In late eighteenth century England, for example, public records were scattered between sixty buildings in London and Westminster. The Record Commission, which was active between 1800 and 1837, reported that "the first and most obvious defect in the present system is that records are deposited in different and widely scattered buildings." It recommended a single central repository for public records and laid the foundations for the Public Record Office Act 1838. In 1851 a new building was begun in Chancery Lane, London to provide a home for the public records, initially records of the courts, but from 1852 including central government department administrative records. A new Public Records Act 1958 made significant changes, including the introduction of the Grigg two-stage appraisal process. The history of the Public Record Office up to 1969 has been comprehensively discussed by John Cantwell, a former Assistant Keeper.[21,22] In 1869, a Royal Commission for Historical Manuscripts was appointed, initially only for 5 years, to address questions of private records. Its purpose was to identify and describe the records of private individuals and families held in the great country houses. The two bodies, the Public Record Office and the Historical Manuscripts Commission, were independent of each other, although they shared premises and staff for many decades, but they were brought together in an administrative arrangement in 2003 to form The National Archives and are now both housed on a single site in Kew, Surrey. Outside the Public Record Office, local, university, business and specialist archives developed in England during the twentieth century. The 1880s saw burgeoning interest in local record publications, the foundation of local antiquarian and record societies and a growth in genealogy. The study of local history developed. After the establishment of county council authorities in 1889, a number of different models of local archives provision developed: some justices and clerks of the peace preserved the records of quarter sessions, city and borough authorities maintained their records and public libraries acquired manuscripts alongside printed materials. In a few places, privately run antiquarian and archaeological societies, trusts and museums collected archives in the absence of, or sometimes in conflict with, official bodies. Bedfordshire can claim the earliest established county record office, appointing its Records Committee in 1898 and establishing an archive in 1913, led by G H Fowler. Gradually through the twentieth century, each county area established an archives service which usually preserved the official records of the local government and acquired the records of local churches, families, societies and businesses. Independent university, business and specialist archives also developed in the later twentieth century. In the United Kingdom, there is no single piece of national archives and records legislation. Instead separate legislation regulates the records of various types of institutions (such as central or local government, the established church, manors), and for different jurisdictions within the United Kingdom (Scotland, Northern Ireland, Wales) and responsibility is divided among government departments.

The history of archival science in the United States has been recounted by, among others James O'Toole and Richard Cox.[23] Their account of the history of archives and the archival profession in the United States referred back to the European (Old World) antecedents and explored the American traditions. They propounded the view, explored by a number of writers, that the American tradition was two-fold: the public archives tradition alongside a historical manuscripts tradition. The historical manuscripts tradition emerged from state historical societies, beginning with Massachusetts in 1791, which collected mainly private papers useful for the study of the past, alongside museum artifacts and other objects. Notable names in this movement included Jared Sparks, the American historian and educator, who edited the letters of George Washington and of Benjamin Franklin. O'Toole and Cox suggest that in the nineteenth century much greater scholarship was devoted to historical manuscripts, while the public records were relatively neglected, kept simply as an adjunct to the current business of government.

The establishment of the Public Archives Commission in 1899 was "a catalyst in the formation of state archives." Gradually, state archives departments were established, starting in 1901 with Alabama: the state archives network was eventually completed in 1978 with New York. The 1930s were the decade of real progress, according to

O'Toole and Cox. In 1933 construction finally began on a building to house federal records and the following year legislation was passed which established the National Archives as an independent federal agency and the first Archivist of the United States, Robert Connor, was appointed. Adopting ideas of provenance from the European writers, American archivists, notably Margaret Cross Norton, Illinois State archivist, and T.R. Schellenberg of the National Archives, developed ideas about the links between record creators and record groups, the administrative value of records and their place in the public archives. These ideas became very influential, not only in the United States.

In the 1950s and 1960s, records management in the United States developed as a distinct activity, driven by the huge increase of records in wartime. The distinctiveness of the new activity was perceived by some to constitute a separate profession. This perception was marked by the separation of the committee concerned with selection and appraisal from the Society of American Archivists to form a new organization, the American Records Management Association (ARMA) in 1956. Standardization of archival description emerged earlier in the United States than in England, perhaps partly as a result of the dominance of manuscript archivists, often working in a library context, in America. American archivists led the way with the publication of U.S. MARC-AMC (MAchine Readable Cataloguing—Archival and Manuscripts Control) in the mid-1980s and a decade later, Encoded Archival Description (EAD), based on an international standard markup language (SGML), which allowed archivists to construct searchable finding aids in an online environment. According to O'Toole and Cox, "for all practical purposes, this movement in the direction of standardization resulted in a merger of the distinct public archives and historical manuscripts traditions." It led to the emergence of more unified professional practice, especially in areas such as archival description.

The Canadian archival tradition evolved differently from those in Europe and America. In Canada, government archives preserved not only the official records of the state, but also private and other records which related the history of the area, in a model not dissimilar to English local record offices, except that in Canada, "total archives" were embraced within the official mandate of many publicly funded cultural agencies. Millar suggested that Canadian archives developed in three phases.[24] In the first phase, the collection and copying of records relating to Canada from many sources was funded by government. The Literary and Historical Society of Quebec, established in 1824, pioneered the work, followed by Nova Scotia in the 1850s. In 1872 Douglas Brymner was appointed to preserve the archives of the Dominion of Canada. The Public Archives Act 1912 enshrined the collecting approach and allowed the Public Archives to acquire "public records, documents and other historical material of every kind." During the second phase, 1900 to the 1970s, the growing sense of national identity and the need for the public sector to manage its institutional records meant that many Canadian archivists pursued the preservation of both institutional records and those from private sources: what became know as the "total archives" approach. The appointment of W. Kaye Lamb as Dominion Archivist in 1948 led to the transformation of the archival profession in Canada. In 1967, the Public Archives joined the National Library in a new building, where acquisition programs expanded to include the preservation of film and television archives and machine readable archives. Provincial archives gradually emerged across the country, with institutions in place in all provinces by 1968, although the services offered and the funding to support activities varied widely. Some, but not all, provincial archives had a role in the records management of their province.

The third phase of development, characterized by Millar as running from the late 1970s, saw significant changes. The Canadian Council of Archives was established in 1985 to support the development of a coordinated archival system, including identifying national priorities, facilitating improved communications across regions, and advising on mechanisms for consistent and sustainable funding and support. In 1975, a new Association of Canadian Archivists was established, launching a new scholarly and professional journal, *Archivaria*, and developing plans for archival education in cooperation with a university.

In Australia, soon after federation in 1901, there were proposals to establish a Commonwealth Archives. The Commonwealth Parliament's Library took an interest and in the 1920s the Commonwealth National Library brought Australian history collections together. In 1917 the Australian War Records Section was established, which developed in 1925 into the Australian War Memorial, with both museum and archival functions. More extensive national archival developments began in the 1940s, with the establishment of the War Archives Committee in 1942 and the appointment of government archivists in half of the Australian states. In 1944, Ian Maclean was appointed as Archives Officer in the Commonwealth National Library, the precursor to the Commonwealth Archives Office, founded in 1961, which finally separated the archives from the national library.

During the 1950s and 1960s, many archivists in Australia were strongly influenced by practices from the United States, partly because of the famous visit by T.R. Schellenberg in 1954, to give the series of lectures which were the basis of his book published in 1956 as *Modern Archives*; another influence was the visit by Ian Maclean to the United States in 1958 to study American practices. The Commonwealth Archives Office became Australian Archives in 1974 and the Commonwealth Archives Act was passed in 1983.

In 1975 the Library Association recommended winding up its Archives Section and setting up an independent Australian Society of Archivists, which took over the publication of the journal, *Archives and Manuscripts*. This marked the Australian archivists' emancipation from the library profession. It also marked the beginning of the intellectual contributions made by Australian archivists, confirming the importance of the work by Peter Scott on the record series and encouraging research into the changing role of the archivist, the influence of postmodernism on the profession and the development of the records continuum.

Archival Education and Research

In nineteenth century European countries, the potential loss of archival skills of palaeography and diplomatics led to the establishment of specialist independent schools to teach historical sciences: the Scuola del Grande Archivio in Naples in 1811, the Archivalische Unterrichtsinstitut in Munich in 1821, and the Ecole des Chartes in Paris, also in 1821. These schools taught archivists as a specialist professional group, generally separately from librarians. In some countries including France from 1850, only graduates from these schools could be employed in the national archives.

In other countries, the history of archival science as an independent academic discipline has a more recent genesis. In the United Kingdom, university teaching of palaeography, librarianship, local history, and diplomatics developed in the period between the two World Wars. These eventually coalesced into archival education in the immediate postwar period when training schools for archivists were established at the Universities of Liverpool and London in 1947, followed by two courses in the University of Wales (Bangor 1954 and Aberystwyth 1955), together with a practical scheme at the Bodleian Library, Oxford, which ran from 1947 to 1980. Most of these universities offered a 1-year graduate program leading to a Diploma in archive studies. The syllabuses typically comprised subjects including palaeography, archaic languages, diplomatics, English constitutional and administrative history, the study of archival materials, and practical work in a repository. For the next 25 years or so a steady stream of classically trained archivists emerged from the remarkably uniform university schools. Later, new programs emerged, such as archives diploma at University College Dublin from 1973 and the Society of Archivists correspondence course (1980–2000). In the 1990s new subjects, such as records management (at the University of Northumbria) and digital preservation (at the University of Glasgow) were offered. The academics teaching these programs were also active in research, although in the early years, research interests were in diplomatics, palaeography, and historical sciences, mainly undertaken by lone scholars. In the 1990s the archives and records academics began to move from a scholarly, historical approach to research toward a social science model, which enabled research projects and collaborative and cross disciplinary research to develop. In 2008, there were several active research groups in the United Kingdom, with interests as diverse as community archives and identities, digital preservation and humanities computing, the history of archives and records, and records compliance and management.

American archival education developed rather late by European standards, although there had been calls for such education since the early twentieth century. In the 1930s, the Society of American Archivists established a Committee on Training of Archivists to discuss archival education in the wake of the establishment of the National Archives, which created a need for qualified staff. This committee recommended that archivists should be educated in history or political science, together with a practical training in an archive. The National Archives evolved its own training for staff. Most archivists came into the profession through historical studies. In the 1970s some education programs began in American universities, either in history departments where it often became associated with public history programs, or in library schools. Generally, archives studies formed a small part of the curriculum, which was heavily weighted toward the predominant interests of the parent department. The programs usually included an internship or practical experience project and sometimes required a thesis. During the 1970s and 1980s, there were intense debates between library and history programs about the proper place of archival education. In 1977 the Society of American Archivists published graduate education guidelines in an attempt to produce some uniformity in the curriculum for the profession, although these recommendations endorsed the approach of small numbers of credits for archival studies within wider programs of library science or history. By the late 1990s most programs were in information science schools. A growing number of archival educators were appointed in American universities, estimated by Cox to stand at about 30 faculty by the end of the twentieth century. This group of archival academics undertook teaching and research and enabled an academic discipline to develop, through dissemination, publication, and doctoral research, which began to establish a sound knowledge foundation for the archival profession.

Canadian archival education took a different route from its neighbor, the United States. A range of graduate programs focusing on archival studies were established, first at the University of British Columbia (UBC) in 1981 and then at the University of Manitoba in 1991. Other programs included an archival studies specialization in the Library School at the University of Toronto, which expanded into a broader information studies degree, and courses in the oldest library school in Canada, at McGill University in Montreal, which established archival studies as a distinct part of its information studies curriculum in 2002. Many of the Canadian schools now have strong research programs, including the INTERPARES projects at UBC.

In Australia, the 1970s saw developments in archival education. Since 1963, the University of New South Wales had offered archives options in its Diploma of Librarianship, but no separate archives qualification existed. In 1973, a new Diploma in Archives Administration began. In the 1990s, archival education in Australia developed significantly, with the Records Continuum Research Group at Monash University, in Victoria and the distance learning program offered at Edith Cowan University in Perth. Each offered something quite distinctive to the Australian, and international, educational scene. Extensive research began into records issues, especially by the Monash University research group, which brought leading thinkers together and generated significant research projects, such as the recordkeeping metadata projects. The publications of these researchers marked a distinctive Australian contribution to the profession.

CONCLUSION

This entry serves as an introduction to the field of archival science and many of the topics which are covered briefly here can be found discussed in more detail in related entries. The first section gave a broad overview of the nature of archives and records. The main section provided some definitions and an introduction to the core concepts and principles which are central to the discipline, including records, archives, and activities; attributes of records; the records lifecycle; the records continuum; provenance and original order; postmodernist ideas around archive fever. The entry concluded with on overview of the professional discipline of archives and records management, a discussion of the key functions of archives and records management services, and a brief history of archival science.

REFERENCES

1. Shepherd, E.J.; Yeo, G. *Managing Records: A Handbook of Principles and Practice*; Facet Publishing: London, U.K., 2003.
2. ISO 15489-1:2001. *Information and Documentation—Records Management—Part 1: General*; International Standards Organization.
3. Williams, C. *Managing Archives: Foundations, Principles and Practice*; Chandos: Oxford, U.K., 2006.
4. Schellenberg, T.R. *Modern Archives: Principles and Techniques*; FW Cheshire: Melbourne, Victoria, Australia, 1956.
5. Upward, F. In search of the continuum: Ian Maclean's Australian experience. In *The Records Continuum: Ian*

Maclean and Australian Archives First Fifty Years; McKemmish, S., Piggott, M., Eds.; Ancora Press in association with Australian Archives: Melbourne, Victoria, Australia, 1994.

6. Muller, S.; Feith, J.; Fruin, R. *Manual for the Arrangement and Description of Archives*; Netherlands Association of Archivists, 1898. English translation published Wilson: New York, 1940.

7. Jenkinson, H. *A Manual of Archive Administration*; Clarendon Press: Oxford, U.K., 1922.

8. International Council on Archives *General International Standard Archival Description ISAD(G)*, 2nd Ed.; International Council on Archives: Seville, Spain, 2000.

9. Cook, T. What is past is prologue: A history of archival ideas since 1898 and the future paradigm shift. Archivaria **1997**, *43*, 1763.

10. Derrida, J. *Archive Fever: A Freudian Impression* (trans. E. Prenowitz); University of Chicago Press: Chicago, IL, 1996.

11. Nesmith, T. Seeing archives: Postmodernism and the changing intellectual place of archives. Am. Arch. **2002**, *65*(1), 2441.

12. AS 4390-1996. *Records Management*; Standards Australia: Sydney, New South Wales Australia, 1996.

13. Cook, T. Mind over matter: Towards a new theory of archival appraisal. In *The Archival Imagination: Essays in Honour of Hugh A Taylor*; Craig, B., Ed.; Association of Canadian Archivists: Ottawa, Ontario, Canada, 1992.

14. Canadian Council of Archives. *Rules for Archival Description*; CCA: Ottawa, Ontario, Canada, 1990–2007.

15. Society of American Archivists. *Describing Archives: A Content Standard (DACS)*; SAA: Chicago, IL, 2004. revised 2007.

16. Procter, M.; Cook, M. *Manual of Archival Description*, 3rd Ed.; Gower: Aldershot, U.K., 2000.

17. National Council on Archives, Public Services Quality Group. *Standard for Access to Archives*; National Council on Archives: London, U.K., 2003. Available at http://www.ncaonline.org.uk/materials/access_standard.pdf (accessed May 1, 2008).

18. Posner, E. *Archives in the Ancient World*; Harvard University Press: Cambridge, MA, 1972. Reprinted by Society of American Archivists, 2003.

19. Duchein, M. The history of European archives and the development of the archival profession in Europe. Am. Arch. **1992**, *55*(1), 14–25.

20. Shepherd, E. *Archives and Archivists in 20th Century England*; Ashgate: Aldershot, U.K. forthcoming.

21. Cantwell, J. *The Public Record Office 1838–1958*; HMSO: London, U.K., 1991.

22. Cantwell, J. *The Public Record Office 1959–1969*; Public Record Office: Surrey, U.K., 2000.

23. O'Toole, J.M.; Cox, R. *Understanding Archives and Manuscripts*; Society of American Archivists: Chicago, IL, 2006.

24. Millar, L. Discharging our debt: The evolution of the total archives concept in English Canada. Archivaria **1998**, *46*, 103–146.

BIBLIOGRAPHY

1. Bettington, J., Australian Society of Archivists, Ed. *Keeping Archives, 3rd Edn.*; Australian Society of Archivists: Virginia, Queensland, Australia, 2008.

2. Cox, R.J. *Archives and Archivists in the Information Age*; Neal-Schuman Publishers: New York/London, U.K., 2005.

3. Craig, B.L. Ed. *The Archival Imagination: Essays in Honour of Hugh A Taylor*; Association of Canadian Archivists: Ottawa, Ontario, Canada, 1992.

4. Jimerson, R.C. Ed. *American Archival Studies: Readings in Theory and Practice*; Society of American Archivists: Chicago, IL, 2000.

5. Kennedy, J.; Schauder, C. *Records Management: A Guide to Corporate Record Keeping*, 2nd Ed.; Addison Wesley Longman: Melbourne, Victoria, Australia, 1998.

6. McKemmish, S.; Piggott, M. Eds. *The Records Continuum: Ian Maclean and Australian Archives First Fifty Years*; Ancora Press: Melbourne, Victoria, Australia, 1994.

7. McKemmish, S.; Piggott, M.; Reed, B.; Upward, F., Eds. *Archives: Recordkeeping in Society*; Charles Sturt University Centre for Information Studies: Wagga Wagga, New South Wales, Australia, 2005.

8. Nesmith, T., Ed. *Canadian Archival Studies and the Rediscovery of Provenance*; Society of American Archivists and Association of Canadian Archivists. Scarecrow Press: Metuchen, NJ/London, U.K., 1993.

9. O'Toole, J.M.; Cox, R. *Understanding Archives and Manuscripts*; Society of American Archivists: Chicago, IL, 2006.

10. Shepherd, E.; Yeo, G. *Managing Records: A Handbook of Principles and Practice*; Facet: London, U.K., 2003.

11. Williams, C. *Managing Archives: Foundations, Principles and Practice*; Chandos: Oxford, U.K., 2006.

Archives

Adrian Cunningham
National Archives of Australia (NAA), Canberra, Australian Capital Territory, Australia

Abstract

This entry provides an overview and comparative analysis of the varied manifestations and roles of archival institutions across the world. The entry argues, and illustrates by example (many of which are from Australia), that archival institutions are contingent and mutable creations with ever-shifting, always-contested forms and missions that reflect the dynamic nature of human experience, aspiration, and activity. A secondary argument of the entry is that archival institutions are simultaneously reflectors of and active shapers of their time and place—institutions whose form and function change as the dynamics of societal power relations evolve and transform around them.

INTRODUCTION

Common dictionary definitions of "archives" state that they are either places where historical records are kept or the organizations responsible for collecting and storing such documents. Beneath such seemingly simple, straightforward, and innocuous definitions lies a fascinating and far more interesting terrain of complex, contestable and dynamic views of archives and their role in society. It is the aim of this entry to explore this interesting terrain—to dig beneath the deceptively unproblematic conventional static view of archives as being merely "dusty old stuff." In short, there is much more at stake in archives than might first meet the eye.

One of the features that has characterized all human societies since time immemorial has been an instinct for collective cultural self-preservation. While culture is contestable and ever-evolving, human beings nevertheless like their cultures and cultural achievements and experiences to endure across generations. This cultural persistence is made possible through the preservation of stories, both orally and in writing and through dance, rituals, art, music and performance. The keeping of many of these valuable cultural "records" is fostered and institutionalized in an "archive(s)." The forms, functions, and mandates of archival programs and institutions have varied and continue to vary enormously depending on the nature of the society in which they exist and the objectives of those who own or have control of the archives.

This entry provides an overview and comparative analysis of the varied manifestations and roles of archival institutions across the world. One of the aims of the entry is to illustrate by example, some of which come from international archival history while others come from the author's Australian context, just what mutable creations archival institutions really are. Recognition of this seemingly obvious fact is argued as a counterpoint to any tendency that other authors may have to argue in favor of universal laws and immutable truths about the nature of the archival institution. While common themes, objectives, and issues can be identified through such a comparative analysis, the main argument of this entry is that there is no universal law governing the form and mission of archival institutions. All archival institutions fulfill their mission by, as a minimum, controlling and preserving the records that constitute the archive, but the nature of the mission served can and does vary from case to case. The ever-shifting, always-contested form and mission of the archive reflects the dynamic nature of human experience, aspiration, and activity in all its infinitely rich variety.

The secondary aim of this entry is to illustrate not only that all archival programs and institutions are the contingent products of their time and place, but also that they are active shapers of their time and place. In the words of Verne Harris, archives "at once express and are instruments of prevailing relations of power."[1] Indeed, as we shall see, it is the nature of the prevailing power relations and the particular roles archives play as contested sites of power struggle that determine the forms and functions of archival programs—forms and functions that can and do change as the dynamics of societal power relations evolve and/or transform around them.

ARCHIVES AND HUMAN IMPULSES: THE INSTITUTIONALIZATION AND PLURALIZATION OF THE RECORD

Records are made as a means of conducting and/or remembering activities. They are created for pragmatic or symbolic purposes—as enablers and evidence of experience and activity, as aids to memory and/or as artifacts. Some of these records are consciously retained for future reference as archives in order to transmit the activity and

Encyclopedia of Library and Information Sciences, Fourth Edition DOI: 10.1081/E-ELIS4-120044329

experience through time. As authors such as James O'Toole and Sue McKemmish have argued, human beings throughout the ages have demonstrated impulses to save and to bear witness.[2,3] Human beings are the sum of their memories. The nature of their interaction with other humans, indeed their very identity, is determined by their memories. While all memory is cognitive, literate individuals learn to rely at least to some extent on the written word to document, express, and supplement cognitive processes. In turn, these cognitive processes give meaning to the archives for, as Jacques Derrida says, the archive does not speak for itself—users inscribe their own interpretations into it.[4]

When these impulses move beyond the purely personal and take on a broader collective or societal purpose the archives so retained take on a more formal character. One manifestation of this phenomenon is that the records can become part of an archival program or institution. This institutionalization of the record, which Derrida calls *domiciliation*, marks the passage of information from the private to a collective domain (p. 2).[4] For instance, when a novelist decides to offer for sale her personal papers to a publicly funded manuscript collecting program and when that program assesses the papers and agrees that they would constitute a worthwhile addition to its collection, the records are simultaneously institutionalized and made public property. The novelist is acting on an individual impulse to pluralize her private records for the benefit of current and future generations, while the broader community is acting (through its publicly funded manuscript collecting program) on an impulse to preserve and provide public access to the evidence of a novelist's creative processes and private life.

There are a wide variety of reasons why records may be institutionalized in this way:

- Organizations need to retain their archives in order to meet their legal obligations, to protect and advance their rights and entitlements, and to retain corporate memory of the decisions and activities of the collective over time to support future decision making and organizational continuity.
- Communities, including entire nations, retain archives as a means of remembering and connecting with their pasts, their origins. There are many complex and subtle variations driving this kind of institutionalization of memory. Eric Ketelaar describes archives in this sense as "time machines"—"a bridge to yesteryear."[5] Others describe the need to capture and retain ancestral voices or to listen to the whispers of the souls of long ago.[6,7] In serving this role archival institutions have much in common with other cultural and memory institutions such as museums.
- Similarly, communities and nations often establish archives to inform, enlighten, educate, and sometimes to entertain. Related to this is the collective need to

support and control storytelling about the pasts and origins of the community. Often archives are retained as a means of expressing, asserting, and preserving a unifying group consensus on the nature of its identity, as forged through a shared history—or alternatively to support competing articulations of group identity and plurality.

- Organizations and communities retain archives for their symbolic significance. Objects stored in the archives can themselves be invested with and convey enormous symbolic significance.[8,9] The creation of a national archives can be symbolically significant as a form of solidification and memorialization in the context of nation building.[10] The heavy symbolism of the archives and its contents can in turn cause the archives to be a site of mythmaking and myth perpetuation.[11] Powerful rulers or administrators often establish archives as symbolic monuments to their own power and as a means of controlling and directing mythmaking activities concerning their achievements. See, for instance, Verne Harris on the role of the South African Archives Service as "an important vehicle for Afrikaaner nationalist historiography, with the legitimation of white rule and the exclusion of oppositional voices being key objectives in the selection policy."[12]
- Powerful rulers create archives not only as symbolic monuments to their greatness, but also to legitimize, reinforce, and perpetuate their power. The deeds, treaties, and founding documents in such an archive can legitimize power in a legalistic and evidential sense, while the information on individual subjects and their relationships and activities in such an archive can provide the information such rulers need to control their dominions and perpetuate their power. Moreover, because archives exercise control over selective memory, they are a source of power that is of enormous utility to autocratic rulers. When endeavoring to control the past, deciding what should be forgotten is just as important as deciding what should be remembered. As Antoinette Burton says, "the history of the archive is a history of loss."[13]
- Conversely, in democratic societies archives are meant to enable democratic accountability by providing access to the evidence of governance that can empower citizens against potential maladministration, corruption, and autocracy. In addition to, or perhaps instead of, protecting the rights and entitlements of rulers and governments, such archives are meant to protect the rights and entitlements of the governed. In the words of John Fleckner, such archives are bastions of a just society where "individual rights are not time bound and past injustices are reversible," where "the archival record serves all citizens as a check against a tyrannical government."[14]

As it will be seen, these reasons for the existence of archival institutions are not mutually exclusive. Most such institutions exist for a combination of these reasons. Indeed, many archival institutions struggle either consciously or subconsciously with the ambiguities, complementarities, and contradictions associated with serving these multiple purposes, whether the purposes are served explicitly or implicitly. The ongoing crisis of identity of government archives in democratic countries is a major theme of this entry. Are archives a part of government or a check on it? Do government archives exist to serve the legal and administrative needs of government and/or the people, or do they exist primarily as cultural and memory institutions? How do archival institutions balance the often-competing demands of public and private interests and the differing imperatives of public and private records and their uses? What is the interplay of symbolic roles with these other functions and mandates? Most importantly, what factors influence responses to these dilemmas in practice and what are the consequences of the different responses?

Of course, being brought under the control of an archival institution is not the only form of institutionalization that can be experienced by records. Registrar style arrangements in public administrations, identification by auditors, records commissions and/or documentation programs are all examples of alternative forms of institutionalization, some of which will be explored later in this entry.

INSTITUTIONAL FORM AND FUNCTION SINCE THE DAWN OF TIME

Ernst Posner has argued that the first archives were created by the Sumerians in the middle of the fourth millennium B.C. These records took the form of clay tablets with cuneiform characters. The archives were used to support commercial activity and property ownership. Later ancient societies such as the Hittites, Assyrians, and Mesopotamians all kept archives, although one can only speculate today on just how institutionalized these archives were and what form, if any, such institutions took? In at least some of these societies, archives were kept in temples and courts for religious, legal, administrative, commercial, and genealogical purposes.[15]

During the second and third millennium B.C. the Egyptians developed an extensive system of archives to support their empire, as did the later Persian Empire. These archives existed primarily to serve the legal, administrative, and military purposes of the rulers. An early indication of the perceived role of archives as tools of political oppression occurred in Egypt around 2200 B.C. when, during a revolt, an angry mob destroyed a records office "as the custodians of hated property rights."[15] Persian archives often incorporated the captured archives of defeated governments to help establish control over the newly occupied territories.

Archives in China can be traced back almost as far as the Sumerians. These records were inscribed on bones and tortoise shells for religious, administrative, and symbolic purposes. By 700 B.C. bamboo, silk, and stone tablets were in use, with records of military value being stored in secure buildings. While the Egyptians used papyrus, the Chinese began using plant fiber paper after 200 B.C. The Chinese also demonstrated an early interest in the use of archives to control the writing of history. In the first century A.D. the Han Dynasty established a Bureau of Historiography.

The Greek city-state of Athens began housing its archives in the Metroon, the temple of the mother of the gods next to the courthouse, by around 400 B.C. This archive contained laws, decrees, minutes, financial and diplomatic records, contracts, records of court proceedings, and manuscripts of plays by Sophocles, Euripides, and others. In what was perhaps the first example of an archival institution fulfilling the function of public access to records and consistent with the democratic principles of Athenian government, private citizens could obtain copies of the records in the archives.

The power that resides in the archives is illustrated in the etymology of the word archives, which can be traced to this time. The Greek *archeion* referred to the office of the magistrate or *archon* and the records kept by that office. *Archontes* wielded executive power, which in large part was legitimized by the legal documents in the *archeion*. Similarly, the Greek *arkho* meant to command or govern. The Latin *archivum* was likewise the residence of the magistrate and the place where records of official legal and administrative significance were kept.

Rome's first public archives was founded about 509 B.C. in the Aerarium, or treasury, of the temple of Saturn and housed laws, decrees, reports, and financial records. Like the Metroon, the laws housed in this archives could be consulted by all citizens. When the Aerarium was destroyed by fire in 83 B.C. it was replaced by the Tabularium, a large stone building. In later imperial Rome the Tabularium adopted a narrower mission as the archives of the Senate. It was supplemented by imperial archives and a network of provincial, municipal, military, and religious archives. Various emperors, most notably Justinian I, were keen advocates of archives. The Justinian Code of 529 A.D. was not only written with the assistance of archives, it also included a section on the role of archives and archivists. This code emphasized the importance of archives as a public place of deposit and as guarantors of the integrity and authenticity of the records housed therein.

Most medieval European archives were maintained in ecclesiastical settings, often in "muniment rooms." By the middle of the sixth century a papal archives had been established. In the eighth century, the Venerable Bede was able to make use of archives to write his landmark history of the church in England. Following the collapse of the Roman Empire a number of municipal archives persisted in Italy and France until the ninth and tenth centuries. Venice and Florence established archives during the eleventh and thirteenth centuries respectively. It was common practice for royal archives in Europe to have no fixed location, but instead to travel with the King's household. Toward the end of the twelfth century, however, there were some moves toward the establishment of a central government archives in England. A century later Exchequer rolls began to be housed in the Tower of London.[16] In time, this archives was expanded to include all of Britain's Chancery records. In 1323, the first inventory of English archives was completed and served as a model for similar initiatives elsewhere in western Europe. In 1346, the archives of the kingdom of Aragon were created.[17]

In 1524, the archives of the crown of Castille was established by Charles V at Simancas near Valladolid. The archive was greatly expanded by Philip II, who regarded archives as vital for controlling, administering, and legitimizing an empire and who also viewed archives as symbols of power and prestige. The Simancas archive is now regarded as the classic prototype of a centralized "national archives." Two hundred years later the Archives of the Indies was established in Seville for the same reasons. When Cortes conquered the Americas, it was considered essential to not only burn the archives of the conquered Incas and Aztecs,[18] but also ensure legitimate documentation of the occupation by a legally appointed Notary, whose records were eventually deposited in archives back in Spain.[9] Between the sixteenth and early eighteenth centuries royal archives repositories were established in France, Sweden, Denmark, and China. The combined effect of the advent of the printing press and the emergence of the modern administrative state generated a significant growth in records creation and, as a consequence, archives holdings.

The creation, control, and use of archives became increasingly important in the context of religious, legal, and political power struggles such as the Reformation and parliamentary reform movements, when opposing factions used records to support their arguments. The Renaissance had created demand for access to information for the purpose of supporting scholarly enquiry as opposed to the more common political, financial, legal, administrative, and symbolic purposes. Nevertheless, access to archives was strictly controlled by their owners, usually monarchs or churches, who very often kept them inaccessible to all except themselves and their functionaries.

THE FRENCH REVOLUTION AND THE NINETEENTH CENTURY

The French Revolution provides perhaps the clearest example of the mutable nature and purpose of archives

and their tendency to inspire extremes in human emotion. Between 1789 and 1793, much of the archives of the *Ancien Régime* were attacked and destroyed by mobs or in state-sponsored bonfires and paper recycling campaigns, with the aim of obliterating what the revolutionaries regarded as symbols of their erstwhile oppression. While such actions might sometimes have had the practical benefit of destroying the evidence of feudal debts and obligations, by and large they were cathartic acts of retribution and ritual cleansing of the body politic.

In the midst of this destruction of old archives there coexisted a desire to create new archives, out of which emerged a new archival system for the new society. A legislative repository was provided for by the new Assembly just two weeks after the fall of the Bastille. In September 1790 a law was passed establishing a new National Archives that was to be open to the public and which was to report to the Assembly. By 1794 the desire to destroy the documentary evidence of the *Ancien Régime* had been replaced by a desire to preserve and manage those records as nationalized public property, reinvented for the purpose of symbolically highlighting the glory of the new Republic in contrast to the sinful decadence and oppression of the old regime. A decree issued in June 1794 granted the National Archives jurisdiction over the records of government agencies, provinces, communes, churches, universities, and noble families, thus creating the world's first centrally controlled national archival system. The same decree also proclaimed the right of public access to these records, thus establishing the first modern instance of archives fulfilling a legal role as protectors of the rights and entitlements of the people and as instruments of accountability and transparency in government. The creation of national archives as both symbols of nation building in the midst of turbulent change and ideological—indeed almost mythological—assertions of legitimacy by new orders is a pattern that has been repeated often since. The fate of the archives of the *Ancien Régime* testify to the fact that no archives can assume an eternal mandate—in the words of Judith Panitch, they are forever "subject to the judgment of the society in which they exist."[19,20]

Another aspect of the impact of the French Revolution on archives is worth exploring at this point. Luciana Duranti has argued that the 1794 decree created for the first time a dichotomy between administrative and historical archives—the distinction between the archives of the Republic and the archives of the *Ancien Régime*. Duranti considers this an unfortunate development in that it represents a usurpation of the administrative and legal functions of archives by social and cultural functions—a usurpation that has echoes in various places and times since the Revolution.[21] Other commentators, however, beg to differ. Judith Panitch, for instance, argues that in the 1790s the notion of French archives as sites of "historical or cultural scholarship had yet to take hold." While they had acquired the new function of public access for the new purpose of accountability, their essential role as legal, administrative, and symbolic institutions remained unaltered.[22]

Nevertheless, Duranti is correct in highlighting the distinction between the administrative/legal and cultural/historical roles of archives—a source of contestation that shall be explored in more detail later—even if the cultural role of French archives did not become apparent until some decades after the Revolution. Duranti's portrayal of one role as being innately superior to another is, however, a position that is far more difficult to sustain, as we shall see. Nor, as we have already seen, is it true that the world had to wait until the late eighteenth century to witness an example of an archives that was established for cultural and historical purposes. While such phenomena were indeed unusual, they were not unprecedented—see for example the case of the Han Dynasty Bureau of Historiography referred to above.

The creation of a centralized national archives in France provided a model for archival development in a number of other countries such as Finland, Norway, the Netherlands, and Belgium during the nineteenth century. Similarly, in Sweden, Denmark, and Prussia central archives evolved out of pre-existing royal or administrative repositories. Forty eight years after the creation of the French national archives, the English followed suit, but for very different reasons and in much less dramatic circumstances. Between 1800 and 1837 a variety of committees and commissions of inquiry had highlighted the scattered and poorly controlled and preserved state of public records in that country. These efforts culminated in the passage of the Public Records Act in 1838 and the eventual establishment of the Public Record Office during the 1850s by a government that was concerned to ensure the proper care and preservation of records that guaranteed the legal rights and entitlements of English people. Lawmakers in Westminster were no doubt aware of the fact that their counterparts in Scotland had beaten not only themselves but also the French in establishing a national archives, when the principal collection of Scottish public records had been assembled in Edinburgh's General Register House as early as 1784.

By the middle of the nineteenth century the growth in historical scholarship based on the use of written sources was becoming an important factor in the evolution of European archival institutions. Selected series of historical documents were published, such as the "Roll Series" and the "Calendars of State Papers" in England. In 1869, the Historical Manuscripts Commission was established in the United Kingdom to identify, describe, and promote the preservation and use of significant historical records that were not otherwise catered for under the Public Records Act. The Commission, which existed until April 2003 when it was amalgamated with the Public Record Office to form a re-branded National Archives, is probably the best example of a state-sponsored documentation program for the nationally distributed holdings of historically significant private records.

ARCHIVAL INSTITUTIONS IN THE TWENTIETH CENTURY: THE POSTCOLONIAL ERA

Globalization, the spread of modern bureaucracies and the worldwide interest in history and cultural/national identity together provided the impetus for the emergence of archival systems around the world during the twentieth century. Soon after the Bolshevik revolution the Soviet Union established a highly centralized archival system as both a reflection and enabler of centralized state power. In contrast to democratic states, access to archives in totalitarian states was not a guaranteed right of the citizen.

In Asia, Latin America, Africa, and the South Pacific, European colonial powers were responsible for the creation of administrative archives that in turn formed the basis for national archives once independence was achieved.[23,24] For instance, the National Archives of Malaysia was established in 1957 and was based on the model of the Public Record Office in London.[25] In Vietnam the French colonial administrators established an Archives in 1917. This was eventually superseded by the State Archives Department in 1962, which in turn was consolidated by the 1982 Decree on the Protection of National Archives Documents. The scope of this decree is, however, limited to government records.[26] In many such territories archival development has also benefited from a strong precolonial archival tradition. Thailand, which was never colonized by a European power, inherited an impressive system of royal legal/administrative and cultural archives of palm leaf manuscripts and bark paper stretching back many hundreds of years. This system was overlaid with a more western approach, including the adoption of a registry system, during the late nineteenth century. In the early twentieth century records retention schedules were introduced and a National Archives was established in 1952 with responsibility for preserving the historical records of government administration.[27]

A common feature of archival institutions in the postcolonial developing world is that institutions established with the best of intentions on a European model have often struggled to fulfill expectations in the harsh economic and political reality of independent governance. Just as these emerging nations have struggled to consolidate inherited democratic institutions, so too have inherited archival institutions often failed to establish themselves as robust organic components of the culture and governance of postcolonial societies. (See, for instance, the summary of current issues and developments from the Proceedings of the 9th Conference of the Pacific Regional Branch of the International Council on Archives.)[28,29] Not only have administrators often been inclined to view archives as at best luxuries and at worst irrelevant western white elephants, but also citizens living in predominantly oral cultures have often been slow to develop attachments with institutions primarily associated with preservation of the written word for use by western academics.[30] Indeed, an interesting variant on the traditional archival institutional model in non-Western territories has been the emergence of alternative forms of memory institutionalization such as so-called "keeping places"[31] and memory institutions that deal primarily with orality in preference to written records.

Archival institutions in these territories are having to develop flexible new conceptions of indigenous knowledge ownership, control, and access in response to a rejection of the inappropriate aspects of Eurocentric archival theory, which support the systematic marginalization and dispossession of the indigenous by dominant global discourses. The more successful of these have been able to demonstrate the potential of archives to support the rediscovery of suppressed cultural identities and the redressing of past injustices.[32] Indeed, the often-tenuous place of archival institutions in oral societies tells us much about the mutable and contingent nature of such institutions. While many have endeavored to increase their relevance by instituting oral history programs, others argue that such activities fail to comprehend the difficulties involved in converting fluid orality into fixed material custody without destroying the very thing that the archives is trying to capture. In the words of Verne Harris there exists "A reluctance to engage indigenous conceptualizations of orality not as memory waiting to be archived, but as archive already."[33,34]

ARCHIVAL INSTITUTIONS IN NORTH AMERICA

Contrary to the more usual pattern of legal/administrative archives gradually acquiring a cultural/historical role (or alternatively being supplemented by the establishment of separate cultural/historical records programs), in North America the primary impetus for the creation of archival institutions were cultural/historical imperatives. The Public Archives of Canada was established in 1872, only 5 years after confederation, following a petition to government by the Quebec Literary and Historical Society. Of particular concern to Canadian historians was the desire to have access to records of Canadian historical interest held in Britain and France. Driving the cultural/historical interest was a perceived need to build national unity and identity through the study of the origins of the Canadian people. Although the Public Archives of Canada lacked both proper facilities and a legislative mandate during its early decades, these shortcomings were rectified in 1906 with the construction of an archives building and in 1912 with the passage of archival legislation. This legislation was informed exclusively by the need to preserve records for historical rather than for legal/administrative purposes.

From its outset the Canadian archival endeavor encompassed both public and private records, a concept later articulated as "total archives."[35,36] The total archives

concept reflects a long-standing social consensus that public funds should be used to preserve a wide range of Canadian documentary heritage, regardless of its origins and format, and that this preservation effort should be pursued via a planned national system. As a result Canada has not experienced the emergence of separate (and sometimes warring) archival tribes or traditions for public records and historical manuscripts, as has been the case in the United States and Australia. With Canadian national identity constantly at risk of being swamped by the more dominant identity of its southern neighbor, recognition of the need to take coordinated action to preserve something distinctly Canadian has compelled generations of Canadians to take a holistic approach to the preservation and management of their archival heritage. The National Archives of Canada (since 2004, Library and Archives Canada) has, at least in theory, always given equal priority to the preservation of records originating in the private sector and to the preservation of records originating in the public sector. Nevertheless, it was not until the 1950s that the then Public Archives of Canada began to exert authority over public records and perform the legal/administrative role that provided the original basis for its counterpart institutions in Europe.[37,38]

Like Canada, it was cultural/historical concerns that led to the creation of a national archives in the United States. Unlike Canada, which wasted little time in establishing a central archival program, the United States had to wait until 1934—over 150 years after the Declaration of Independence—before its national archives was established. This represented the culmination of many decades of agitation by historians, most notably the American Historical Association. By that time the desire to rescue, preserve, and provide access to historical records had manifested itself in the emergence of the so-called "historical manuscripts tradition." This tradition, which dated back to the earliest years of the nation, had been shaped by an antiquarian collecting instinct and had become institutionalized in organizations such as state historical societies and the Library of Congress, where the main focus was the collecting, researching, and publishing of the private papers of prominent individuals.

Why did the United States take so long to establish a national archives? In 1939, Ernst Posner argued that a major contributing factor was American ambivalence, if not hostility, toward state bureaucratic power. So, while manuscript collecting endeavors pursued for scholarly purposes were considered laudable, proposals to create an archival institution as an integral part of the state bureaucracy was something that was regarded, at least subconsciously, with suspicion.[39] Americans had to wait until the latter part of the twentieth century before there was a clear articulation of the role of archives as guarantors of the democratic rights of citizens and as means of holding public officials to account—a role that, while recognized as an ideal, is yet to be fully realized both in practice and in public perception.

Just as the Public Archives of Canada was established for historical purposes and had to wait until the 1950s to acquire an administrative/legal role, so too the U.S. National Archives had to wait until 1950, with the passage of the Federal Records Act, before it acquired a role as a supporter and enabler of public administration. Public records archivists such as Margaret Cross Norton from Illinois pursued a campaign to articulate and assert an administrative and legal accountability role for archives in the face of the primacy of the historical/cultural role.[40] While Norton had very good reasons for pursuing these efforts, they had the unfortunate effect of creating a polarization of the American archival community—a polarization that persists to this day in a profession which seems unable to attain a comfortable and balanced view of the dual role of archival institutions.[41,42]

The American experience highlights in sharp relief the tensions and contradictions that have emerged in the roles of archives worldwide since the nineteenth century and which represent contested ground everywhere. Arguably, the polarization is more pronounced in the United States because there is more at stake. Archives in Europe were initially established for legal and administrative purposes, thus conferring on them a valuable legitimacy in the eyes of government that has enabled them to acquire a cultural/historical role from a position of strength. In contrast, at the time of their establishment American archives had no such legal/administrative legitimacy and have had to struggle ever since to attain such a role and the government support and funding that it could attract. The addition of democratic accountability to the legal/administrative role by proponents such as Margaret Cross Norton merely helped the struggle to work against itself. So, while the general public may be suspicious of the legal/administrative role of archives and supportive of the accountability role, the reverse is very often the case from the perspective of those that control the corridors of power in government. Once again we see the mutable nature of archives and the fact that they are forever subject to the judgment of the societies in which they exist.

PUBLIC RECORDS INSTITUTIONS IN AUSTRALIA

The emergence of archival institutions in Australia mirrors in many respects the experience of the United States. As in the United States, the establishment of government archival authorities was the result of advocacy from historians, with a unique Australian contribution coming from leading librarians. In most jurisdictions the government archives programs started their lives as units within the government research libraries, which themselves inherited control of colonial-era records from heterogeneous administrative locations.[43,44]

The decision to establish a Commonwealth Government archives was made during World War II, 40 years after Federation, and was informed primarily by the desire to "ensure the availability of material for the preparation of a history of the war."[45,46] The Archives Division was initially located administratively within the Commonwealth Parliamentary Library, later the National Library. Despite these cultural/historical origins, early Commonwealth archivists were heavily influenced by the legal/administrative tradition as embodied in the writing of English archivist Sir Hilary Jenkinson. While this was partly a reflection of the British origins of the Australian bureaucracy, it also reflects the simple fact that Jenkinson provided the only archival handbook in the English language to which neophyte archivists could turn for guidance.[47,48]

Two North American visitors to Australia exerted a significant subsequent influence on archival development in Australia. T.R. Schellenberg from the U.S. National Archives and W. Kaye Lamb from Canada were both firmly of the historical/cultural tradition. Schellenberg toured Australia in 1954 on a Fulbright Fellowship at the invitation of National Librarian Harold White.[49] Perhaps Schellenberg's most enduring legacy in the Australian profession was his philosophy regarding the appraisal of voluminous modern public records. While Schellenberg's cultural/historical message may have registered well with Harold White, the government archivists themselves were determined to stick to the Jenkinsonian path—a determination that manifested itself in a desire to break free from their cultural/historical roots and what they saw as the ill-informed control of librarians. Doubtless also by this time, the North American debates surrounding the campaign of Margaret Cross Norton were resonating in Australia. Despite their philosophical differences on the role of archival institutions, the local Jenkinsonians nevertheless also found in Schellenberg an ally in their arguments in favor of separation from the Library. A Committee of Inquiry recommended separation of the Archives Division from the Library, influenced partly by the differences between the two professional disciplines and partly by the view that the role of the Archives was to manage government records for the benefit of the government and its departments—a very different role than that asserted by Harold White of building a systematic record of national life and development.[50]

In 1961, the Commonwealth Archives Office separated from the National Library. The newly independent Office then proceeded to bury itself deep into the Federal bureaucracy and largely turn away from any cultural/historical role—indeed from the rest of the profession in Australia, a situation that persisted into the 1990s. The most positive aspect of this bureaucratic focus was, through the efforts of Ian Maclean and Peter Scott, the development of an innovative and enduring Australian school of thought on the management and intellectual control of current records.[51–53] Nevertheless, throughout the 1960s and 1970s the Australian Archives, as it became known in

1974, proved to be ambivalent about pursuing an active role in support of democratic accountability in the manner that opinion leaders such as Sir Paul Hasluck and Margaret Cross Norton had advocated. It appeared to define its role solely in terms of supporting the administrative and legal requirements of the Commonwealth Government.[54,55]

At this time the Government was contemplating the need for legislation to govern the work of the national archives. Canadian Dominion Archivist Kaye Lamb was invited to investigate and recommend a way forward. Lamb's 1973 report was critical of the lack of support and assistance provided to researchers by the Archives and recommended legislation that gave the organization a broad cultural and administrative mandate and a leadership role at the centre of a national archival system.[56] Lamb's recommendations were eventually enacted in legislation with the passage of the Archives Act in 1983—itself part of a suite of administrative law reform bills including a Freedom of Information Act. This law created the Archives as a quasi-independent entity with an appointed Advisory Council.

While the advent of legislation gave the Archives a mandate to pursue a range of activities, during the 1980s only those activities relating to a narrowly defined administrative/legal role were pursued with any vigor. Although lack of resources and institutional inertia help explain this situation, changes initiated by George Nichols as Director-General during the 1990s resulted in the Archives pursuing a vibrant and proactive role as both a cultural institution[58] and an agent of democratic accountability.

In State Government jurisdictions archival institutions have suffered from a variety of malaises including underfunding, lack of public visibility, and control by the sometimes stifling hand of librarians. More recently, however, new public records legislation in a number of jurisdictions has given State and Territory archives greatly enhanced powers as cultural institutions and as semi-independent agents of democratic accountability. In many cases these positive developments have been assisted by spectacular examples of failed public administration, assisted at least in part by a lack of regulation regarding public recordkeeping.[58–60] Time will tell if the State and Territory Archives are given the resources and the true independence both to reinvent themselves organizationally and pursue their new mandates with the vigor that they deserve.

THE COLLECTING TRADITION IN AUSTRALIA

An Australian equivalent of the Historical Manuscripts Tradition, in the form of a network of State and Commonwealth Government research libraries, was the first to apply serious endeavor to the business of identifying, preserving, and making available valuable archival materials.

These efforts grew out of the antiquarian work of private collectors and historical societies and, until the 1940s were characterized by a desire to preserve documents relating to the origins of European settlement in Australia. The Mitchell Library, opened in Sydney in 1910, was founded on the bequest of the prodigious collector David Scott Mitchell. Similar, though less extensive, collections of personal papers preserved along side other categories of historical source material were subsequently established in State Libraries in each of the other jurisdictions.[61]

During the 1950s and 1960s there was an almost exponential expansion in the institutional collecting of private archives in Australia. At the forefront of this expansion was the National Library in Canberra,[62,63] but also significant was the emergence of new collecting programs such as the University of Melbourne Archives[64] and the Australian National University's Archives of Business and Labour, which have specialized in collecting the records of Australia's leading businesses and trade union organizations.[65] In large part this expansion reflected a similar expansion in the study and teaching of Australian history in Australia's rapidly expanding university system. According to Stuart Macintyre, the number of full-time professors and lecturers in history in Australian tertiary institutions expanded from fewer than 20 in 1939 to more than 700 in 1973.[66]

The emergence of new collecting programs and the simultaneous expansion and shift in collecting emphasis by the existing programs stemmed from a desire to serve the needs of this expanding researcher clientele. Both phenomena reflected a more self-confident, nationalistic, and prosperous Australia, a nation that was keen to apply the methods of scientific history to the task of understanding and articulating a national identity. There was also continuity between the old amateur collecting paradigm and the new professional, institution-building paradigm. Both types of collecting constituted, in the words of James Clifford, "a form of western subjectivity" and a "crucial process of Western identity formation."[67] Human beings are storytelling creatures. Every society develops mechanisms for the formation and persistence of collective memory for storytelling purposes. Archives, when they pursue a cultural/historical role, provide one such mechanism that will very often secure and retain the support of the society in which they exist.

Just as the growth of social history, with its commitment to uncovering the lived experience of ordinary people, provided a major boost to the collection and preservation of textual archives, so too did it foster the growth of associated documentation and preservation programs such as oral history, film and sound archives and, more recently, computer data archives. The National Library commenced gathering and preserving sound recordings and transcripts of oral history interviews with prominent Australians during the 1950s as an adjunct to its manuscript collecting program. Initially based on the

Columbia University model of an oral history program, the Library's efforts have since expanded to encompass thematic social history projects such as the history of the timber industry and documentation of the HIV/AIDS epidemic. The Library's oral history holdings now include some 30,000 hr of original recordings. The program also incorporates a component of field recordings of folkloric tales and music, similar to that pioneered by the U.S. Library of Congress.[68] Over 400 similar, though much smaller, oral history collections have been established in libraries and archives such as the Northern Territory Archives Service.[69,70]

The National Library also pioneered the collecting of another category of Australia's documentary heritage in the form of films and sound recordings. As early as 1935 the Library established (as a result of a Cabinet decision) a "National Historical Film and Speaking Record Library." After some years of lobbying by the film and sound industry and others, the National Film Archive and Sound Recording Section were eventually separated from the Library in 1984 as a new institution called the National Film and Sound Archive (NFSA), with its headquarters in Canberra. Although it currently lacks a strong legislative mandate and the kind of bureaucratic independence that is enjoyed by both the National Library and the National Archives, the NFSA enjoys a solid international reputation for professional excellence in the field of film and sound preservation and is an active participant in international professional forums such as the International Association of Sound Archives (IASA) and the International Federation of Film Archives (FIAF). For the most part the NFSA restricts its collecting activity to the output of the Australian private sector media industries, with the various public records archives in each jurisdiction being responsible for the control and preservation of archival value audio visual records produced in the public sector.[71–74]

Another strand in the collecting archives scene can be found in the area of data archives. The leading archive of this type in Australia, the Social Science Data Archives, was established at the Australian National University in 1981. Its brief is to collect, preserve, and make available for use computer readable data and statistical sets emanating from social, political, and economic research projects in the disciplines of the social sciences. The creators and depositors of these data sets include academics, government and private organizations, and individuals. Such initiatives are consistent with an international tradition of data archiving in the sciences and social sciences, as represented by the activities of the International Federation of Data Organisations and institutions such as the U.K. Data Archive, based at the University of Essex. In recent years data archives have made great use of the World Wide Web to help identify and disseminate qualitative and quantitative statistical data sets for secondary use research and learning.

Unlike their public records counterparts, collecting archives programs have been little troubled by existential dilemmas caused by the need to either combine or choose between the often conflicting roles of supporting a bureaucracy, enabling democratic accountability and supporting cultural/historical endeavor. The collecting archives' sole *raison d'être* is a cultural/historical one. In this, they have much in common with museums, which are themselves significant players in the collecting and preserving of archival materials in Australia.[75] For instance, one of the country's best known museums, the Australian War Memorial in Canberra, also acts as a collecting archive and as a repository of Commonwealth records relating to Australia's various military endeavors.[76]

The expansion in funding for archival collecting activities during the 1950s and 1960s provides ample evidence of the value Australian society placed on the pursuit of their role of documenting society by means of collecting.[77] Today the funding and relative fortunes of the collecting programs may not appear as impressive as they were during this earlier period of rapid growth. This is in part a reflection of declining support for academic history coupled with changes in historiographical philosophy, which nowadays places less emphasis on finding "truth" in source documents. Certainly the quantity of holdings, in terms of items and shelf meters, in Australia's various collecting archives today is dwarfed by that of the public records institutions. Nevertheless, the collecting archives enterprise in Australia today remains impressively robust and continues to enjoy good public support and a reasonable, though sometimes uncertain, level of funding, including the provision of taxation incentives for donors in the form of the Commonwealth Government's Cultural Gifts Program.

In the field of science and technology the work of collecting archives is also complemented by the longstanding contributions of the University of Melbourne's e-Scholarship Research Centre, formerly the Australian Science and Technology Heritage Centre (first established as the Australian Science Archives Project in 1985). This noncollecting archival documentation program locates, identifies, describes, and publicizes significant archival material relating to scientific endeavor in Australia and, when necessary, arranges for the preservation of such records by a suitable collecting organization (see the Center's Website at http://www.esrc.unimelb.edu.au/).

The distributed national collection of private records in Australia is, however, not without problems. Graeme Powell has highlighted the lopsided nature of these holdings—the fact that certain areas of activity such as politics and literature are very well represented, while other significant areas of human endeavor in Australia are grossly underrepresented.[78,79] Doubtless a contributing factor to this is the ad hoc acquisitions policies of the various collecting programs, a problem accentuated by a lack of a centrally coordinated national archival system,[80] such as exists in many other countries.

BUSINESS ARCHIVES IN AUSTRALIA

At the beginning of this entry it was argued that many archives combine multiple roles. Good examples of this can be found in Australian business and university archives. Mention has already been made in our examination of collecting archives of the University of Melbourne Archives and the Australian National University's Archives of Business and Labour (now called the Noel Butlin Archives Centre). Both archives were established with the main aim of collecting the records of Australian businesses, primarily to serve the needs of economic historians. In the case of the Australian National University this occurred at the instigation of historian Noel Butlin, while in Melbourne the initiative came from one of the most interesting and significant groups ever to exert an influence on recordkeeping in Australia, the Business Archives Council of Australia (BACA).

BACA was established at the University of Sydney in 1954 and was modeled on the British Records Association. It brought together historians, archivists, librarians, and businessmen with the aim of promoting awareness among the business community of the importance of and methods for preserving valuable business records in order to support the pursuit of business history. Initially the aim of BACA was to encourage good recordkeeping within companies, including the establishment of in-house archives. The collecting of business records by other organizations was very much an afterthought. A Victorian Branch of BACA was established in 1957 at the suggestion of the ubiquitous and indefatigable Harold White. White used his contacts with senior Melbourne librarians, history professors, and captains of industry to assemble a formidable alliance of business archives advocates. Inevitably, White was attracted to this endeavor by the prospect of eventually enriching the National Library's collection with the records of some of Australia's leading businesses. This collecting aim was, however, a longer term objective. To achieve that objective White recognized that the collecting archives had to both establish a positive working relationship with the potential donors and also provide immediate assistance to businesses to ensure that the most valuable records were properly identified and managed by their creators in the short term. By emphasizing the business benefits of the latter to his corporate audience White hoped to also achieve the former—a positive working relationship that could lead to later donations of records from the corporate sector.[81]

The Victorian Branch of BACA was based at the University of Melbourne, where leading historians like John La Nauze and Geoffrey Serle decided that Melbourne should emulate the collecting efforts that Noel Butlin was energetically pursuing in Canberra. Before long, Harold White found himself in collecting competition with not one, but two university archives. Eventually White agreed that the National Library would leave the collecting of business archives to the universities, but the concession was not granted without a bitter struggle. In the meantime White, Frank Strahan (inaugural University of Melbourne Archivist), and others had pursued a campaign under the auspices of BACA to educate businessmen in the fundamentals of professional records management, using the argument that good recordkeeping is good for business. This was a rare and relatively successful example of archivists with overriding cultural/historical objectives adopting and pursuing objectives associated with supporting the business needs of organizations (pp. 77–85).[81]

The innovative adoption of a dual historical/administrative role by archivists operating in a private sector records environment proved to be quite successful, especially considering the relatively low level of total resources that constituted the combination of BACA and the two university archives. A survey of the state of Australian business archives conducted by Simon Ville and Grant Fleming in the late 1990s found a remarkably high retention rate of archives of Australia's top companies for the first 60–70 years of the twentieth century—preserved either by in-house archives such as the Archives of Westpac Bank or BHP Billiton or by collecting archives.[82,83] It is not unreasonable to attribute this success, at least in part, to the work of BACA and the major collecting archives. Sadly, they and others have noted an alarming decrease in the preservation of significant Australian business records dating from the early 1970s after BACA had effectively ceased to operate, a situation that may also reflect a decline in interest in economic history in the tertiary education sector.[84,85]

EDUCATIONAL AND RELIGIOUS ARCHIVES IN AUSTRALIA

The significance of the role of two university-based collecting archives, the Noel Butlin Archives Centre and the University of Melbourne Archives, has already been discussed in some detail. Similar collecting programs have also been established at university archives in Wollongong and Newcastle and at the University of New England in Armidale and Charles Sturt University in Wagga Wagga, each of which adopt a regional emphasis in their collecting programs.[86,87] The University of Queensland's Fryer Library has for many years collected records associated with literary endeavor in Australia as, more recently, has the Australian Defence Force Academy's Library in Canberra. These university-based collecting archives are generally the product of the energies of enthusiastic and committed individuals responsible for the creation and continued existence of collections that relate to their research interests. In theory they exist primarily to support the research and teaching activities of the university concerned, although in practice their continued existence often relies on patronage and is, as a consequence, tenuous.

Equally precarious is the existence of in-house university archives responsible for the universities' own records. While most of Australia's 38 publicly funded universities have records officers and/or records managers, they do not always have an in-house archives. Those in-house archives that do exist find themselves in a variety of structural reporting arrangements, with some reporting to the university librarian and others associated with the administrative/chancellery arm of the university. In many cases the in-house archives that do exist are severely under-resourced. As a result of recent legislative changes in some State jurisdictions, university recordkeeping requirements are now more formally mandated, with the relevant State Archives exercising overall responsibility for setting recordkeeping standards and authorizing the destruction of university records.

Some university archives combine a collecting and an in-house function. Don Boadle has argued that this marriage is not always a successful one, with university administrators and in some cases the archivists themselves being unable to articulate a clear and compelling mission for a unified archival program. Indeed, in some cases it could be argued that the combined function has resulted in an identity crisis that suggests the worst of both worlds rather than the best of both worlds.[88–90] As with many of their State and Commonwealth Archives colleagues, these archivists have struggled to achieve consensus on the value of simultaneously pursuing legal, administrative, cultural, and accountability roles, much less consensus on how such an integrated vision might actually be pursued and achieved in an environment where support from senior university administrators is often ambivalent at best. Moves toward a national or even State-based strategy in this area will, if past experience is any guide, struggle in the face of the tendency by universities to guard their independence and autonomy with jealousy and vigor.

Finally, there exist a very large number of in-house archival programs in Australia's various independent (nongovernment) schools and religious organizations. These programs are often staffed by part-time and/or volunteer staff and struggle to attain resourcing commensurate with the scope of their operations.[91] Church archives house records of dioceses, parishes, and/or religious orders. In recent years their work has been thrust into the spotlight because of the role churches played in the now controversial and emotive issues of child migration and the Aboriginal "stolen generations." The access policies of these archives have been greatly tested in the context of the churches coming to terms with their role in these unfortunate episodes of Australian history and their relationships with aggrieved individuals who were separated from their families and often badly mistreated in the hands of church officials.[92,93]

ARCHIVES AS A PLACE AND VIRTUAL ARCHIVES

Throughout the ages one of the regularly recurring functions of archival institutions is to provide a secure place for the safekeeping of valuable records to guarantee the ongoing legal authenticity of those records. This is especially common for archives that serve solely or primarily a legal/administrative role, where control and possession of the records is recognized as a source of power. Luciana Duranti has highlighted the importance of this function in archives stretching back to the days of the Justinian Code and the Tabularium in Ancient Rome, while Michel Duchein has identified the same issue as being important to archives in Flanders and Hungary. One of Sir Hilary Jenkinson's more influential contributions to the archival discourse is the related notion of the need to guarantee an uninterrupted transmission of custody from records creator to archival institution—the physical and moral defense of the record. Duranti has argued that when records "cross the archival threshold" they are attested to be authentic and henceforth guaranteed to be preserved as such by an archives that is independent from the records creating office and for which the preservation of the authenticity of its holdings is its *raison d'être*.[17,94,95]

While this is a common theme in the history of archival institutions, it is not a universal one. Duchein has argued that there are many countries in which the notion has never existed, including France "where the fact of its being preserved in a public archival repository does not give a document any guarantee of authenticity."[17] Similarly, while the preservation of authenticity is undoubtedly an objective of most collecting/historical archives programs, it cannot be said to be their *raison d'être*. More recently, archivists who agree with Duranti and Jenkinson about the absolute importance of guaranteeing the authenticity of records have disagreed with Duranti's argument that this can only be achieved by means of archival institutions taking physical custody of the records. To these critics adequate control of records to guarantee authenticity in the digital age can be achieved without the need for archives to provide a physical place of safekeeping. In the digital age the very physicality of records is superseded by a virtual concept or "performance" where the idea of a record having a set physical location becomes meaningless. These critics also object to the notion of records crossing an "archival threshold" at some point in time after their creation. They argue that the "archival bond" and subsequent guarantees of authenticity should commence at the point of records creation which, by definition cannot be physically in the archives (See the entry on

"Diplomatics," p. 1593 for an explanation of the concept of the "archival bond."). If the archival bond is achieved and guaranteed at the point of records creation the decision when or whether to perform a physical act of custodial transfer to an archives becomes a minor administrative consideration, not a matter of central significance.[96–99]

Another strand to this topic is the architectural use of archival buildings to make symbolic statements about the role and significance of archives in society. Many archival buildings throughout the ages have architectural features suggestive of solidity, impenetrability, durability, and authority. Indeed, such featurism is so common as to be almost a cliché—something which itself speaks volumes about perceptions of archival institutions. Recent, more imaginative architectural representations of the form and function of archives, such as the Gatineau Preservation Centre in Canada, have attempted to convey an image of archives as "the epitome of liberal-humanist and objective-scientific activity," but perhaps unwittingly reflect instead the ultimately indeterminate and mutable nature of the archival pursuit.[100]

One feature of the "archives as a place" debate has been the perhaps naïve assertion by the post-custodialists[101] that technological change has made it possible, indeed essential, for digital records to be archivally captured, described and controlled in such a way as to guarantee the authenticity and integrity of the records from the instant of creation onwards. A good explanation of post-custodial archival thinking was provided by Terry Cook in 1994 when he said ". . . our traditional focus on caring for the physical things under our institutional custody will be replaced or (at the very least) enhanced by a focus on the context, purpose, intent, interrelationships, functionality, and accountability of the record and especially its creator and its creation processes. All this goes well beyond simple custody, and thus has usefully been termed post-custodial." Perhaps the closest archivists have yet come to achieving this vision is with the "VERS encapsulated objects" of the Victorian Electronic Records Strategy (VERS).[102] The fact remains, however, that the assertion remains an unproved—though appealing—hypothesis.

Ultimately, different archives will make their own choices as to how important guarantees of authenticity are and, if they are considered vital, which strategies they feel will give them the best chance of achieving that objective. Certainly, the post-custodialists argue for a more proactive and virtual "archives without walls" as an antidote to the traditional passive custodial view, although there is no reason why a custodial approach could not also be combined with a more proactive role. Jeannette Bastian has recently argued that (distributed) custody and authenticity should not be ends in themselves, but rather means to a more important end—that of facilitating use of the archives by those who stand to benefit from such activity.[103] As Frank Upward has argued, "the externalities of place are becoming less significant day-by-day . . . the

location of the resources and services will be of no concern to those using them. . . ."[104]

In the online world the development of virtual archives is not only desirable, but also essential for continued relevance and survival. Users will wish to be assured of authenticity, but will not care less about the existence of or necessity for places of custody. It will be interesting to watch as archival institutions respond to new user expectations of 24-hr online virtual access and the opportunities offered by digitization programs for transforming and democratizing archival access programs.[105]

CONCLUSION

We have seen how archives in different times and in different places take different forms, pursue different strategies and different combinations of objectives. We have seen that these differences can be explained with reference to the political and cultural environment in which archives exist and the objectives of those who own, control, or are responsible for the existence of the archives. We have seen that archives are not passive, objective, and "neutral repositories of facts," but rather active and subjective participants in and shapers of political and cultural power relations.[106] The power to decide which records constitute the archives and, conversely, which records do not, coupled with the power to determine who can have access to the archives are powers that can be used for good or for bad. Usually, they are used by those who control the archives to support and bolster their own political and economic position, views, and mythologies. In democracies, however, there is at least the potential for archives to act as enablers of citizen empowerment and democratic accountability and transparency—a potential that unfortunately remains more latent than actual in most cases. While public records archives in democratic nations should be substantially independent from the executive government of the day, so that they can play a proactive role in ensuring the accountability of that government, most such archives are in fact a part of executive government and enjoy only limited independence. This suggests that the main role of these archives is to serve the legal, administrative, and culture constructing objectives of the government, rather than any truly democratic purpose. Even in democracies, governments do not surrender lightly the power of archival consignation.

We have also seen that the political and social purposes of archives are never eternal. These purposes are always being contested, reconsidered, reinvented, and transformed, reflecting the judgments of and changing power relations in society. This was seen most dramatically as a result of the French Revolution, but there are other notable examples, such as the transformation of archives in post-Apartheid South Africa, in post-Communist Eastern Europe and in post-Colonial Asia and the Pacific. We have also seen that archives that were originally established to serve the narrow legal, economic, and administrative purposes of a ruling elite can also be transformed and reinvented to serve broader cultural objectives of society—objectives that are themselves eternally subject to contestation and reinvention.[24,107–110]

The consequences of these lessons depend on your values and perspectives. From this author's perspective the best archives are those that serve broad social, cultural, and democratic accountability purposes. In reality, we have seen that such archives often struggle to emerge from or succeed within a governance environment that does not place high value on such purposes. In more tightly controlled and less democratic environments archives that serve the narrow political, legal, economic, and symbolic objectives of the ruling elite will generally enjoy greater funding, support, and patronage. As we have seen, however, circumstances can change. Any archives are better than no archives and those which today serve a narrow set of power interests may tomorrow be reinvented to serve broader social and democratic interests. The key is for archivists to understand the roles that they play and to remain ever alert and sensitive to the political and social dynamics in which their archives operates. Archivists should always be ready to take advantage of changing circumstances that may permit their archives to serve more pluralistic, socially inclusive, and democratically empowering roles.

Any discussion of the role of archives as enablers of wider democratic transparency and accountability cannot ignore the issue of archival institutions themselves being accountable and transparent. If the archives should help to hold a government accountable, to whom is the archives accountable? The answer should be the wider community. Unfortunately, however, most archives have a long distance still to travel before they can claim to practice what they preach with regard to transparency and accountability for their own politically charged decisions and activities. Too often archives and archivists are guilty of making their decisions behind closed doors and justifying their actions with spurious claims of sacrosanct professionalism and scientific objectivity. The first steps for archives in a democracy to become truly effective enablers of transparency and accountability are for this role to be effectively communicated to and understood by the wider community and for the archives itself to become fully transparent and accountable for its own operations and decisions.

Depending on their social and political circumstances and on their own choices and actions, archives can pursue missions that can either hinder or assist society in being civil, pluralistic, open, just, and democratic.

REFERENCES

1. Harris, V. The archival sliver: Power, memory, and archives in South Africa. Arch. Sci. **2002**, *2*, 63.
2. O'Toole, J.M. *Understanding Archives and Manuscripts*; Society of American Archivists: Chicago, IL, 1990; 13–15.
3. McKemmish, S. Evidence of me.... Arch. Manus. **1996**, *24*(1), 28–45.
4. Derrida, *J. Archive Fever;* University of Chicago Press: Chicago, IL, 1996; 68.
5. Ketelaar, E. Is everything archive. In Seminar Presentation. Monash University: Melbourne, Australia, August 5, 2002.
6. Wehner, M.; Maidment, E. Ancestral voices: Aspects of archival administration in Oceania. Arch. Manus. **1999**, *27*(1), 22–31.
7. Steedman, C. *Dust*; Manchester University Press: Manchester, U.K., 2001; 70.
8. O'Toole, J. The symbolic significance of archives. Am. Arch. **1993**, *56*(Spring), 234–255.
9. O'Toole, J. Cortes's notary: The symbolic power of records. Arch. Sci. **2002**, *2*, 45–61.
10. Le Goff, J. *History and Memory*; Columbia University Press: New York, 1992; 87–88.
11. Echevarria, R. *Myth and Archive: A Theory of Latin American Narrative*; Cambridge University Press: Cambridge, U.K., 1990.
12. Harris, V. The archival sliver: Power, memory, and archives in South Africa. Arch. Sci. **2002**, *2*, 74.
13. Burton, A. Thinking beyond the boundaries: Empire, feminism and the domains of history. Soc. Hist. **2001**, *26*(1), 66.
14. Fleckner, J. Dear Mary Jane: Some reflections on being an archivist. Am. Arch. **1991**, *54*(Winter), 13.
15. Posner, E. *Archives in the Ancient World*; Harvard University Press: Cambridge, MA, 1972; 71–85.
16. Clanchy, M.T. *From Memory to Written Record, England 1066–1307*, 2nd Ed. Basil Blackwell: Oxford, U.K., 1993; 72.
17. Duchein, M. The history of European archives and the development of the archival profession in Europe. Am. Arch. **1992**, *55*(Winter), 15.
18. Martin, H.-J. *The History and Power of Writing*; University of Chicago Press: Chicago, IL, 1988; 26.
19. Panitch, J.M. Liberty, equality, posterity?: Some archival lessons from the case of the French Revolution. Am. Arch. **1996**, *59*(Winter), 101–122.
20. Posner, E. Some aspects of archival development since the French revolution. Am. Arch. **1940**, *3*(July), 161–162.
21. Duranti, L. Archives as a place. Arch. Manus. **1996**, *24*(2), 248–249.
22. Panitch, J.M. Liberty, equality, posterity?: Some archival lessons from the case of the French Revolution. Am. Arch. **1996**, *59*(Winter), 118.
23. Alexander, P.; Pessek, E. Archives in emerging nations: The anglophone experience. Am. Arch. **1988**, *51*(Winter and Spring), 121–129.
24. Stoler, A.L. Colonial archives and the arts of governance. Arch. Sci. **2002**, *2*, 87–109.
25. Nor, Z.H. The national archives of Malaysia—Its growth and development Archives in the Tropics: Proceedings of the Australian Society of Archivists Conference, Townsville, May 9–11, 1994 Australian Society of Archivists: Canberra, Australian Capital Territory, Australia, 1994; 94–98.

26. Professional identity of the archivist in Vietnam Archivists—The Image and Future of the Profession: 1995 Conference Proceedings; Piggott, M., McEwen, C., Eds.; Australian Society of Archivists: Canberra, Australian Capital Territory, Australia, 1996; 142–148 Pham thi Bich Hai.

27. Prudtikul, S. Records and archives management in Thailand Paper Presented at the Annual Congress of the International Federation of Library Associations Bangkok, Thailand, 1999. Available at http://www.tiac.or.th/thailib/ifla/ifla99_20.htm.

28. Proceedings of the 9th Conference of the Pacific Regional Branch of the International Council on Archives, Palau, 2001; 12.

29. Papadopoulis, S. The image and identity of Africa's Archivists Archivists—The Image and Future of the Profession: 1995 Conference Proceedings; Piggott, M., McEwen, C., Eds.; Australian Society of Archivists: Canberra, Australian Capital Territory, Australia, 1996; 149–156.

30. Wehner, M.; Maidment, E. Ancestral voices: Aspects of archives administration in Oceania. Arch. Manus. **1999**, 27(1), 22–41.

31. McIver, G.; Best, Y.; Hutchinson, F. Friends or enemies? Collecting archives and the management of archival materials relating to Aboriginal Australians Archives in the Tropics, Proceedings of the Australian Society of Archivists Conference, Townsville, Australia, May 9–11, 1994 Australian Society of Archivists: Canberra, Australian Capital Territory, Australia, 1994; 135–140.

32. 'Archives and Indigenous Peoples' theme issue of *Comma, Int. J. Arch.,* **2003**.

33. Harris, V. *Exploring Archives: An Introduction to Archival Ideas and Practice in South Africa*; 2nd Ed. National Archives of South Africa: Pretoria, South Africa, 2000; 92–93.

34. Hatang, S. Converting orality to material custody: Is it a noble act of liberation or is it and act of incarceration?. ESARBICA J. **2000**, 19.

35. Millar, L. The spirit of total archives: Seeking a sustainable archival system. Archivaria **1999**, 47(Spring), 46–65.

36. Millar, L. Discharging our debt: The evolution of the total archive concept in English Canada. Archivaria **1998**, 46 (Fall), 103–146.

37. Wilson, I. A noble dream: The origins of the public archives of Canada. Archivaria **1982/83**, 15(Winter), 16–35.

38. Atherton, J. The origins of the Public Archives Records Centre, 1897–1956. Archivaria **1979**, 8(Summer), 42.

39. Posner, E. Archival administration in the United States. *Archives & The Public Interest: Selected Essays by Ernst Posner*; Mundsen, K. Ed.; Public Affairs Press: Washington, DC, 1967; 114–130.

40. Norton, M.C. *Norton on Archives*; Society of American Archivists: Chicago, IL, 1975.

41. Berner, R.C. *Archival Theory and Practice in the United States: A Historical Analysis*; University of Washington Press: Seattle, WA, 1983.

42. Gilliland-Swetland, L.J. The provenance of a profession: The permanence of the public archives and historical manuscripts traditions in American archival history. Am. Arch. **1991**, 54(Spring), 160–175.

43. Doust, R. The administration of official archives in New South Wales 1870–1960. Master of Librarianship Thesis, University of New South Wales: Sydney, New South Wales, Australia, 1969.

44. Snow, R.H. The developments that led to the establishment of the Public Record Office of Victoria Archives and Reform—Preparing for Tomorrow: Proceedings of the Australian Society of Archivists Conference, Adelaide, July 25–26, 1997; Crush, P., Ed.; Australian Society of Archivists: Canberra, Australian Capital Territory, Australia, 1998; 256–262.

45. Piggott, M. Beginnings. In *The Records Continuum: Ian Maclean and Australian Archives First Fifty Years*; McKemmish, S., Piggott, M., Eds.; Ancora Press in Association with Australian Archives: Melbourne, Victoria, Australia, 1994; 8.

46. The commonwealth's first archives bill 1927. Arch. Manus. **2001**, 29(1), 98–109 For a description of an earlier failed attempt to establish a Commonwealth Government archives see Ted Ling.

47. Maclean, I. An analysis of Jenkinson's 'manual of archive administration' in the light of Australian experience. In *Essays in Memory of Sir Hilary Jenkinson*; Hollaender, A.E.J., Ed.; Society of Archivists: London, U.K., 1962; 1281–1352.

48. Hasluck, S.P. A narrow and rigid view of archives. Arch. Manus. **1981**, 9(2), 3–10.

49. Piggott, M. The visit of Dr. T.R. Schellenberg to Australia 1954: A study of its origins and some repercussions on archival development in Australia. Master of Archives Administration thesis, University of New South Wales, Sydney, New South Wales, Australia, 1989.

50. Piggott, M. An important and delicate assignment: The paton inquiry, 1956–57. Aust. Acad. Res. Libr. **1990**, 21(4), 213–223.

51. Upward, F. In search of the continuum: Ian Maclean's "Australian Experience" essays on recordkeeping. In *The Records Continuum: Ian Maclean and Australian Archives First Fifty Years*; McKemmish, S., Piggott, M., Eds.; Ancora Press in Association with Australian Archives: Melbourne, Victoria, Australia, 1994; 110–130.

52. Scott, P.J. The record group concept: A case for abandonment. Am. Arch. **1966**, 29(4).

53. Wagland, M.; Kelly, R. The series system—A revolution in archival control. In *The Records Continuum: Ian Maclean and Australian Archives First Fifty Years*; McKemmish, S., Piggott, M., Eds.; Ancora Press in Association with Australian Archives: Melbourne, Victoria, Australia, 1994; 131–149.

54. Golder, H. *Documenting a Nation*; Australian Archives: Canberra, Australian Capital Territory, Australia, 1994; 37.

55. Smith, C. A hitchhikers guide to Australian archival history. In *Archival Documents: Providing Accountability Through Recordkeeping*; McKemmish, S., Upward, F., Eds.; Ancora Press: Melbourne, Victoria, Australia, 1993; 197–210.

56. *Development of the National Archives: Report by* Dr W Kaye Lamb September 1973, AGPS: Canberra, Australia, 1975; Australia. Parliament.

57. Hyslop, G. For many audiences: Developing public programs at the national archives of Australia. Arch. Manus. **2002**, 30(1), 48–59.

58. Schwirtlich, A.-M. Overview of archival legislation in Australia Archives at Risk: Proceedings of the Australian

Society of Archivists Annual Conference, Brisbane, 1999 Australian Society of Archivists: Canberra, Australian Capital Territory, Australia, 2001; 95–99.

59. Ling, T. Acts of atonement: Recent developments in Australian archival legislation. J. Soc. Arch. **2002**, *23*(2), 209–221.

60. Hurley, C. From dust bins to disk-drives and now to dispersal: The State Records Act 1998 (New South Wales). Arch. Manus. **1998**, *26*(2), 390–409.

61. Powell, G. The collecting of personal and private papers in Australia. Arch. Manus. **1996**, *24*(1), 62–64.

62. Thompson, J. Let time and chance decide: Deliberation and fate in the collecting of personal papers. In *Remarkable Occurrences: The National Library of Australia's First 100 Years 1901–2001*; Cochrane, P., Ed.; National Library of Australia: Canberra, Australian Capital Territory, Australia, Australia, 2001; 105–122.

63. Powell, G. Modes of acquisition: The growth of the manuscript collection of the National Library of Australia. In *Library for the Nation*; Biskup, P., Henty, M., Eds.; Australian Academic and Research Libraries (special issue): Canberra, Australian Capital Territory, Australia, Australia, 1991; 74–80.

64. Swancott, L. Origins and development of the University of Melbourne archives. Arch. Manus. **1999**, *27*(2), 40–47.

65. Moore, P.; Maidment, E. The archives of business and labour, 1954–1982. Labour Hist. **1983**, *44*, 107–112 See also a very interesting discussion of the tensions inherent in the Australian collecting archives tradition by Michael Piggott in his address "Alchemist Magpies: Collecting Archivists and their Critics", Public lecture at the Australian National University, Canberra, Australian Capital Territory, Australia, September 16, 2008. Available at http://www.anu.edu.au/discoveranu/content/podcasts/alchemist_magpies_collecting_archivists_and_their_critics/.

66. Macintrye, S. The writing of Australian history. In *Australians: A Guide to Sources*; Borchardt, D.H., Crittenden, V., Eds.; Syme & Weldon Associates: Sydney: Fairfax, VA, 1987; 22.

67. Clifford, J. *The Predicament of Culture: Twentieth-Century Ethnography, Literature, and Art*, Harvard University Press: Cambridge, MA, 1988; quoted in Griffith, T. *Hunters and Collectors: The Antiquarian Imagination in Australia*; Cambridge University Press: Cambridge, U.K., 1996; 25.

68. York, B. Impossible on less terms: The oral history collection. In *Remarkable Occurrences: The National Library of Australia's First 100 Years 1901–2001*; Cochrane, P., Ed.; National Library of Australia: Canberra, Australian Capital Territory, Australia, Australia, 2001; 183–197.

69. *Australia's Oral History Collections: a National Directory*. National Library of Australia, Available at http://www.nla.gov.au/ohdir/about.html.

70. Good, F. Technology and oral history at the Northern Territory Archives Service Archives in the Tropics: Proceedings of the Australian Society of Archivists Annual Conference, Townsville, May 1994; Edwards, J., Ed.; Australian Society of Archivists: Canberra, Australian Capital Territory, Australia, Australia, 1994; 60–66.

71. Cochrane, P. Rescuing "The Sentimental Bloke.". In *Remarkable Occurrences: The National Library of Australia's First*

100 Years 1901–2001; Cochrane, P., Ed.; National Library of Australia: Canberra, Australian Capital Territory, Australia, Australia, 2001; 78–80.

72. Shirley, G. Activism towards a national film archive. Cinema Papers July.1984..

73. Edmondson, R. Sacrilege or synthesis? An exploration of the philosophy of audiovisual archiving. Arch. Manus. **1995**, *23*(1), 18–29.

74. Edmondson, R. A case of mistaken identity: Governance, guardianship and the ScreenSound saga. Arch. Manus. **2002**, *30*(1), 30–46.

75. Smith, B. Archives in museums. Arch. Manus. **1995**, *23*(1), 38–47.

76. Schwirtlich, A.-M. The Australian war memorial and commonwealth records, 1942–1952. In *The Records Continuum: Ian Maclean and Australian Archives First Fifty Years*; McKemmish, S., Piggott, M., Eds.; Ancora Press in Association with Australian Archives: Melbourne, Victoria, Australia, 1994; 18–34.

77. Fairbanks, S. Social warrants for collective memory: Case studies of Australian collecting archives. Master of Arts (archives and records) thesis, Monash University, Melboune, Victoria, Australia, 1999.

78. Powell, G. The collecting of personal and private papers in Australia. Arch. Manus. **1996**, *24*(1), 62–77.

79. Ayres, M.-L. Evaluating the archives: 20th century Australian literature. Arch. Manus. **2001**, *29*(2), 32–47.

80. Cunningham, A. From here to eternity: Collecting archives and the need for a national documentation strategy. Lasie **1998**, *29*, 32–45.

81. Fairbanks, S. Social warrants for collective memory: Case studies of Australian collecting archives. Master of Arts (archives and records) thesis, Monash University, Melbourne, Victoria, Australia, 1999; 61–67.

82. Ville, S. Fleming, G. Locating Australian corporate memory. Bus. Hist. Rev. **1999**, *73*, 256–264.

83. Terwiel, D.; Ville, S.P.; Fleming, G.A. *Australian Business Records: An Archival Guide*; Department of Economic History, Australian National University: Canberra, Australian Capital Territory, Australia, Australia, 1998.

84. Dan, K.; Smith, B. Where have all the [business] archives gone? Archives at Risk: Proceedings of the Australian Society of Archivists Annual Conference, Hicks, S., Ed ; Brisbane, Queensland, Australia, July1999.

85. Pritchard, C. Survey of business records. Arch. Manus. **1987**, *15*(2), 139–148.

86. Boadle, D. Origins and development of the New South Wales regional repositories system. Arch. Manus. **1995**, *23*(2), 274–288.

87. Boadle, D. Documenting 20th century rural and regional Australia: Archival acquisition and collection development in regional university archives and special collections. Arch. Manus. **2001**, *29*(2), 64–81.

88. Boadle, D. Australian university archives and their prospects. Aust. Acad. Res. Libr. **1999**, *30*(3), 153–170.

89. Boadle, D. Australian university archives and their management of the records continuum 1953–1997 Archives and Reform: Proceedings of the Australian Society of Archivists Annual Conference, Adelaide, 1997; Crush, P., Ed.; Australian Society of Archivists: Canberra, Australian Capital Territory, Australia, Australia, 1998; 247–255.

90. Boadle, D. Archives at the edge? Australian university archives and the challenge of the new information age Place, Interface and Cyberspace: Archives at the Edge—Proceedings of the Australian Society of Archivists Conference, Fremantle, August 6–8, 1998 Australian Society of Archivists: Canberra, Australian Capital Territory, Australia, Australia, 1999; 73–82.

91. Riley, J. Integrating archival programs into the core business of the independent school. Arch. Manus. 1997, 25(1), 50–61.

92. Australia. Parliament. Senate legal and constitutional committee. Healing: A Legacy of Generations, The Report of the Inquiry into the Federal Government's Implementation of the Recommendations Made by the Human Rights and Equal Opportunity Commission in Bringing Them Home, Canberra, Australian Capital Territory, Australia, Australia, 2000.

93. Thorpe, K. Indigenous records: How far have we come in bringing the history back home?. Arch. Manus. 2001, 29(2), 28.

94. Duranti, L. Archives as a place. Arch. Manus. 1996, 24(2), 242–255.

95. Jenkinson, H. A Manual for Archive Administration; The Clarendon Press: London, U.K., 1922.

96. Cook, T. Electronic records, paper minds: The revolution in information management and archives in the post-custodial and post-modernist era. Arch. Manus. 1994, 22(2), 300–328.

97. Upward, F.; McKemmish, S. Somewhere beyond custody. Arch. Manus. 1994, 22(1), 136–149.

98. Ham, F.G. Archival strategies for the post-custodial era. Am. Arch. 1981, 44(Summer), 207–216.

99. Hedstrom, M. Archives as repositories: A commentary. Archival Management of Electronic Records, Archives and Museum Informatics Technical Report No. 13, Pittsburgh, PA, 1992.

100. Koltun, L. The architecture of archives: Whose form, what functions?. Arch. Sci. 2002, 2(3–4), 239–261.

101. Cook, T. Electronic records, paper minds: The revolution in information management and archives in the post-custodial and post-modernist era. Arch. Manus. 1994, 22(2), 308.

102. Victorian Electronic Records Strategy; Available at http://www.prov.vic.gov.au/vers/vers/default.htm.

103. Bastian, J.A. Taking custody, giving access: A postcustodial role for a new century. Archivaria 2002, 53, 76–93.

104. Upward, F. Structuring the records continuum part one: Post-custodial principles and properties. Arch. Manus. 1996, 24(2), 282.

105. 'Online archives unlock the past', The Australian, November 6, 2007. http://www.australianit.news.com.au/story/0,24897,22707582–5013037,00.html.

106. Schwartz, J.M.; Cook, T. Archives, records, and power: The making of modern memory. Arch. Sci. 2002, 2, 1–19.

107. Harris, V. The archival sliver: Power, memory, and archives in South Africa. Arch. Sci. 2002, 2, 63–86.

108. Kahlenberg, F.P. Democracy and federalism: Changes in the national archival system in a united Germany. Am. Arch. 1992, 55(Winter), 72–85.

109. Ress, I. The effects of democratization on archival administration and use in Eastern Middle Europe. Am. Arch. 1992, 55(Winter), 86–93.

110. Grimstead, P.K. Beyond Perestroika: Soviet-area archives after the August Coup. Am. Arch. 1992, 55(Winter), 94–124.

BIBLIOGRAPHY

1. Fairbanks, S. Social warrants for collective memory: Case studies of Australian collecting archives. Master of Arts (archives and records) thesis, Monash University, Melboune, Victoria Australia, 1999. Examines the activities that informed the development of key collecting archives institutions in Australia during the 1950s and 1960s, exploring the warrants and mandates that gave impetus to these archival programs.

2. Gilliland-Swetland, L.J. The provenance of a profession: The permanence of the public archives and historical manuscripts traditions in American archival history. Am. Arch. 1991, 54(Spring), 160–175 Traces the origins and history of the split between public records institutions and those working in the so-called "Historical Manuscripts Tradition" in the United States and considers the implications of the existence of these very different views of the role and functions of archival institutions.

3. Panitch, J.M. Liberty, equality, posterity? Some archival lessons from the case of the French revolution. Am. Arch. 1996, 59(Winter), 30–47 Using the pivotal example of the French Revolution, this article illustrates how archival institutions are the mutable and contestable products of their time.

4. Piggott, M. Beginnings. In The Records Continuum: Ian Maclean and Australian Archives First Fifty Years; McKemmish, S., Piggott, M., Eds.; Ancora Press in Association with Australian Archives: Melbourne, Victoria, Australia, 1994; 1–17 Surveys the activities that culminated in the establishment in the 1940s of the organization that is today called the National Archives of Australia.

5. Posner, E. Archives in the Ancient World; Harvard University Press: Cambridge, MA, 1972. The definitive text on the development of archival institutions in the ancient world.

Archivists and Collecting

Richard J. Cox
School of Computing and Information, University of Pittsburgh, Pittsburgh, Pennsylvania, U.S.A.

Abstract
Archival collecting is seen by archivists in some countries principally within the context of well-established archival appraisal practices and principles. The author argues for a specifically North American approach to collecting, where the notion of collecting has been the strongest, most well-developed, and most debated. However, many traditional aspects of collecting have been and are being challenged.

INTRODUCTION

Archival collecting—the acquisition of organizational records and personal papers for housing, management, and use in archival and historical records repositories—has a long and tangled history. In many cases, archivists equate appraisal, the identification of records with continuing or enduring value, with collecting (and vice versa). Some archivists believe that records do not become archives without the blessing of archivists through an appraisal process or, at the least, the physical housing (by whatever means) within archives and historical manuscripts repositories. Others place records on a continuum, and they equate a record as archival based on a continuing value because of the importance of the warrant or function generating the record. In the midst of this comes the much older idea of collecting, bringing records into centralized repositories for convenience of research use.

These are complicated matters, cutting into cultural, political, professional, and personal concerns. There are differences with archival collecting within American, European, and other nations. Europeans, operating within a more centralized and controlled regime, tend not to think about collecting in the way North Americans approach this matter. In the younger nation of Australia, the collection of personal and private papers has not been a primary emphasis, at least when compared to the focus on institutional and governmental records.[1] Many archivists on both continents adhere to a view of appraisal advocated more than half a century ago by Hilary Jenkinson, who believed that the creators of records made selection decisions and that this ensured an objectivity that is impossible to achieve by any other means.[2] That American archivists have stressed more pragmatic ideas about collecting can be seen in the response to the Duranti essay (where she argues for a more theoretical basis) by Boles and Greene.[3] American archivists, however, have focused on a serendipitous form of collection, as well as more systematic appraisal approaches, reflecting their less regimented national archival system.[4] Some would argue that the greatest emphasis, both practical and theoretical, on *collecting* has come from the North Americans, and as a result, this article concentrates on the topic of the archivist and collecting from a distinctly North American perspective.

BASIC DEFINITIONS

Some observers have connected the ability to document the past with the strength of the archival and historical manuscripts repositories that have been formed. In an assessment of how well America was caring for its historical records, Gerald George (then director of the National Historical Publications and Records Commission, the national funding arm of the U.S. National Archives) wrote: "In the ongoing struggle with capricious fortune, the nation's collective success in documenting its history is directly connected to the strength of archival and historical institutions, the security of their facilities, the scope of their holdings, their ability to provide access, and the aggressiveness of their collecting—or their ability to persuade others to preserve materials accessible to historians."[5] The issue of collecting is what this article is about, but the degree of aggression or activity George emphasizes seems to me to be far from the main issue to be worried about. Collecting has played an important role in forming what archivists do, but it has also wrecked havoc with systematic or standardized approaches to the management of archival records.

For the past century, the nature of U.S. archives and historical manuscripts programs that might collect records has remained fairly constant. Fifty years ago, one archivist provided a typology of collecting programs in this fashion: "Basically there are four types of collecting agencies: 1) the private collector; 2) historical agencies devoted to the preservation of research materials—including manuscripts—primarily for cultural or historical purposes; 3) archival agencies serving their own governments, businesses, or organizations; and 4) agencies that combine

Encyclopedia of Library and Information Sciences, Fourth Edition DOI: 10.1081/E-ELIS4-120053554

both manuscript-collecting and archival functions."[6] This aptly describes the nature of archival programs until recently. New types of community and indigenous archives have gained prominence, part of a remarkable international movement.[7] While such programs are more diverse and greater in number today, they still fall largely within these categories, and just as they did a century ago, there are tensions caused by the existence of these various collecting groups. Some of these tensions are reflected within the newer community and indigenous programs.

In the United States, the development of archives and historical records programs has been connected to two distinct traditions, which have been built around the manner in which they acquire records. The historical manuscripts tradition relates to collecting repositories such as historical societies and colleges and universities. The public archives tradition is connected to the local, state, and federal archives in which records are acquired primarily through regular transfers, usually as part of a records management program.[8] Even public archives have "collected," generally in their formative years as they discover older records, deal with the recovery of older records found by agencies, and evaluate records delivered to them unannounced and without any relationship to formal records retention and disposition schedules.[9] The movement for the U.S. National Archives was largely focused on gathering and centralizing the archives of the nation.[10–15] Many state governments worked in the early nineteenth century to copy records related to their colonial past in the archives of Europe, constituting a primitive kind of archival collecting.[16] In addition, most states labored to gather, copy, and preserve their oldest records.[17] For a long time, one could not find a formal definition of archival "collecting" within the main professional glossary, but it is still the case that the physical acquisition of records is what most archivists have their sights set on. There are definitions for "acquisition" and "acquisition policy," but there is no formal definition for the idea of collecting.[18] The expanded Society of American Archivists (SAA) glossary includes a brief entry for "collecting archives," a slight improvement.[19] These definitions are remarkably lifeless when considering the dynamic processes of acquiring records, including the personal, psychological, cultural, economic, and other facets making up such processes.

PIONEER COLLECTORS AND COLLECTING

The collecting emphasis in the American archival profession stems substantially from the role of nineteenth-century historical societies in forming the foundation of the modern archives profession. These historical societies were part of a major shifting by Americans to develop a usable past for their young nation, developing in tandem with historical novels, museums, autograph collecting, and other such

activities and trends. One historian even characterized this period as one beset by "documania."[20–23] The historical societies were also modeled after English and European organizations that were established in the Renaissance, generally with similar antiquarian and nationalistic motives to their American descendants.[24] The equating of archives with their collecting has long been associated with nationalism, as in some nations' war memorials, including not just the erection of monuments but the collecting of records related to the conflict.[25] It is also the case that the U.S. National Archives appealed to veterans' groups looking for a means to commemorate their service, especially as recounted in Victor Gondos' book on Jameson and the National Archives cited earlier.[10]

It is not accidental, therefore, that the earliest historical societies were established in the first years of the American Republic. The Massachusetts Historical Society was the first historical society, formed in 1791, and it was followed by a few others in the subsequent decades. Between 1820 and 1860, historical societies were established in nearly every hamlet, major urban center, and certainly every state. By the end of the nineteenth century, hundreds of these repositories existed, each generally subscribing to a "collect-everything" mission. As many have pointed out, these were the main protectors of America's historical records before the advent of government, institutional, and university and college archives. Historical societies often primarily collected to reflect the interests of their board members, and they often acquired items in a highly competitive way. Admittedly these institutions operated long before there was an archival profession, professional standards, and distinctions between the business of libraries and archives, but their highly charged desire to collect everything and anything set a tone that still permeates large sections of the modern archival profession. Despite the fact that we can discern some narrowing in collecting foci by these institutions as they ran out of space, faced increasing competition from other repositories, identified particular constituencies, and hired more professionally trained staff, there is still no question that by any standard their acquisition desires remained broad and sometimes quite eclectic.[26,27] In some ways, the general American's intense interest in the past, including collecting, continues today, and it serves as a reminder that archivists, despite training and vocational choice, are part of the general populace as well.[28]

One of the devices used by these historical societies was the issuance of wanted lists, circulated to individuals and communities in order to advise them of important and valuable archival records and historical manuscripts. The practice continued well into the twentieth century, when organizers of historical pageants often put out calls for historical documents and artifacts.[29] Beyond such pageants, modern historical societies, such as the Minnesota Historical Society's issuing of a brief pamphlet in 1951, effectively did the same. This pamphlet carried an age-old sentiment: "Between the records and those who would destroy stands the collector. He knows that it is impractical or impossible to save everything, for only a fraction of all the records that have been created have enough historical value to warrant the investment of time and money necessary to preserve them. But, out of the mass, the collector must select the essentials, the manuscripts that will insure remembrance of our past."[30] In this statement, the collector was the state or local repository; a century ago it would have been *both* institution and individual.

The historical societies served many roles, but it was primarily as a collecting agency—even if somewhat unfocused—that these organizations made their most important contributions to the development of a documentary heritage. For example, long before government archives were established in the early twentieth century, historical societies often acquired public records that might have otherwise been lost due to neglect or willful destruction.[30] Still, it is also obvious that these organizations at times inflicted damage on public records by removing them from their records context, treating them as isolated personal papers rather than organizational records, and removing the responsibility from governments to care for their archives. In this way, they reflected the activities of earlier English and European collectors who gathered up stray public records, personal papers, artifacts, and other materials in the pursuit of documenting the past for present use while arguing for their more systematic care.[31]

One of the roles of these historical societies was to acquire printed ephemera and nontraditional record sources. Over a century ago, it was likely that ephemeral material would be swept up just as eagerly as other more substantial records. Many of the most valuable ephemera collections reside in these older institutions. Now archivists are developing methodologies to help them identify criteria for dealing with ephemera and other nontraditional records.[32] Sometimes what seem to be fairly routine business records take on more useful importance to repositories and society where their research value is recognized beyond their business utility. The obsolete files of publishers and literary agents have become sought after, for example, because of what they reveal about particular writers as well as the publishing industry.[33] Archival collecting can become messy and undisciplined when missions are set so broadly, even when individual researchers attest to the potential and actual value of such collections. Without some criteria or parameters, the older antiquarian tradition of collection for collecting's sake becomes the guiding principle. The question remains what has become the primary objective in archival work; it is easy to state an archival mission or mandate but harder to explain it in plain, understandable language. The digital era has opened up new challenges and opportunities for documenting aspects of society and its organizations seemingly unobtainable just a decade ago.[34] Will collecting continue?

COLLECTING AND THE FORMATION OF THE EARLY MODERN PROFESSION

Acquisition, collecting, appraisal, and selection have long been thought to be critical functions of the modern archival profession. The profession is usually considered to have started in the late nineteenth century in Europe and in the early twentieth century in North America.[35] German archivist Karl Otto Muller, working as the American archival profession was forming, is considered by some to be the first to state that selection was the primary matter for archivists to resolve, and he argued that "What is not worth preserving, should never be allowed into an archive." In the years since then, archivists worldwide have developed a variety of models, methods, and mandates for guiding archival appraisal and acquisition, ranging from allowing the records creator to make the decision to elaborate cooperative planning approaches.[36] Always working against the systematic approach to appraisal and selection have been the more romantic stories of the great collectors, the pursuit of important collections, and the defying of odds in saving significant records and sources. Such stories are often a main attraction for individuals to become professional archivists and manuscripts curators.[37–39]

The romance of collecting fits well with what occurred during the Second World War. The obsession that is collecting can be seen most dramatically in tragic events in which one people's misfortunes become an opportunity for another's collecting.[40] For centuries, conquering nations have carried off the archives of other nations (when they haven't destroyed them), although the reasons are often complicated; motivations for destroying a people's identity are often stronger than those for making a profit or acquiring an object.[41] The Second World War was particularly troublesome in this regard, as the Nazis carted off art and historical treasures, including many historical manuscripts. These documents and objects came onto the market, and many wound up in the hands of both private collectors and public repositories. In the 1980s and 1990s, as records of the warring nations were declassified and as the generation of Holocaust survivors began to die off, the looting of Europe became more public and controversial with shifting attitudes about events such as the Holocaust.[42–45] The question as to why it took 50 year for some of these issues to become of sufficient public and media scrutiny is also complicated and has much to do with broader political, nationalistic, ethnic, and cultural events.[46] The end of the Cold War also has brought new attention to matters such as these, especially with calls for the return of displaced archival treasures.[47] The increased public scrutiny on the events of the Second World War also brought more intense light on the troublesome aspects of the international trade in antiquities, art, and archives.[48,49]

SHIFTING NOTIONS OF ARCHIVAL COLLECTING

There have been noteworthy changes in archival collecting in the past half-century in the United States and around the world, some caused by societal developments and others generating from within the profession itself. Philip Mason, a leading labor archivist and archival educator, sees a substantial change after 1950, when "emphasis on elite groups continued to dominate the collecting priorities of established archival institutions but increasingly attention was devoted to new areas." As Mason sees it,

> the major change in the focus of archives came about through the efforts of the newly established archival programs, especially those affiliated with colleges and universities. Unfettered by traditional collecting practices and the free from conservative governing boards, the new archives branched out into new subject fields. Some chose a geographical area or region as their focal point of collecting, whereas others selected a subject theme around which to develop an archives.[50]

In fact, the origins of many nationally and internationally known repositories developed because of perceived gaps in collecting or due to the need to locate archives and records closer to the researchers.[51–53] This is a very different emphasis from what propelled such pioneers as Jeremy Belknap (founder of the Massachusetts Historical Society) in the early years of the historical societies; when there were few repositories, researchers resorted to building their own collections, and critical records were viewed as "monuments" in claiming a past for the young nation.[27]

After a century of the wide-scale founding of state and local historical societies, the twentieth century became a period of more focused collecting by repositories, either to document specific groups or to support the research of scholars.[54,55] Those programs that did not refocus their collecting efforts often discovered that the financial burden of collecting had become substantial, requiring them to either rethink their mission or to face escalating costs that threatened to put them out of business.[56] Nonetheless, archivists persistently have collected virtually every type of record representing every discipline, topic, and trend, such as science,[57] publishing,[58] law,[59] ethnic groups,[60,61] and labor.[50,62] Most of these collecting areas developed because historians and other scholars identified gaps in or needs for the collecting of sources missing from repositories for their research. While these efforts have often had noteworthy results, they have also contributed to the general randomness of collecting and often minimized more rigorous standards for appraisal in favor of "we better acquire these records just in case they are valuable" or because "they might be of use to some researcher, at some time." Such views are particularly vulnerable as the future brings immense changes in records technology, the use of records, and how society understands and approaches such fundamental matters as evidence and information. Within the archival profession, this can be seen in the massive recent writing on the nature of records and record keeping. For citations to the extensive literature on this topic, refer to.[63–66]

THE FUTURE OF COLLECTING

If anyone wonders where archival collecting might lead, a reading of Thomas Mallon's fictional article about the National Archives in the year 2099 will inspire thoughts. Archives II, the newer building of the National Archives in College Park, Maryland, has been turned into condos since all the records have been digitized and the space is no longer needed. The National Archives collects artifacts, or as one of the characters in the short story mutters, "junk is the Archives' business."[67] This is a major shift from what has actually occurred in recent years, when there has been evidence of a shift to criteria used by public archivists to the practices of collecting institutions. In general, collecting was driven by the desire to acquire almost anything or to serve the needs of researchers. Meanwhile, public records staff have begun to shift to more critical matters, such as accountability. Some have argued that such matters are also relevant for collecting repositories, such as in environmental issues and political freedom.[68] Despite two centuries of collecting and continued reflection on what it means to acquire records for archives, archivists and their predecessors have spent little time in gathering data about why people donate records to repositories. This is a fruitful area for research,[69] as is almost any aspect of archival appraisal and collecting.[70,71]

While the emphasis of collecting in the nineteenth century was on the papers of individuals and families rather than institutions (and when institutional records that were acquired were treated as personal papers), many believe today that the emphasis has shifted too far to stress government records, especially as they have had to deal with electronic records. Some have even argued that the archivists who collect personal papers have been victims of a "marginalization."[1,72,73] Some of these commentators have stressed that the changing record-keeping technologies necessitate the need for both collecting and organizational repositories to cooperate as well as to develop similar approaches to appraising records.[74] Other archivists have gone even farther with this concern, worrying that the shifts to contending with information (the emphasis of electronic records management) will have archivists forget about the artifacts of record keeping. One archivist contends that archives do not store information but they hold "artifacts in which information inheres." Archives are, according to such a perspective, like a "documents museum."[75] There has also been increasing tension between archivists, who seem to emphasize responsibility for organizational records, and researchers (especially historians), who worry about where privately created records are being acquired.[76–79] Such debates indicate that archivists are both worried about what they will be able to collect and whether or not they will be relegated to lesser roles as places to store old and obsolete documents.

COLLECTING INSTITUTIONS AND INDIVIDUAL COLLECTORS

The idea of professional collecting does butt up against the avocation of collecting by private individuals. Despite the connection of collecting to various stages in the history of the archival profession and the formation of a genuine documentary heritage, in the past few decades, there has been increasing tension between private collectors (individuals, not institutions) and the role of repositories from government archives to collecting programs in universities and historical societies. One case involved the state government of North Carolina seeking legal action against an autograph dealer who offered in his catalog records that had been in the state government's hands a long time before.[80] A famous incident arose in the 1950s regarding a question of ownership (were the manuscripts the property of the federal government, which commissioned the expedition, or of descendants of the explorers?) of the records of the Lewis and Clark expedition to the Pacific Ocean in the early nineteenth century.[81] Another replevin case in the early 1980s occurred when one repository, the Western Reserve Historical Society, deaccessioned a group of surveyors' drawings and notes related to Louisiana and adjacent states and put the records up for auction. Pursued by the Historic New Orleans Collection, a private repository, and the Louisiana State Museum, a public agency, the state ultimately attempted to seize a portion of the records as having belonged to the government. While the contention that these were public records appeared a bit specious, the various parties managed to reach a compromise whereby the entire collection was acquired by the Historic New Orleans Collection.[82]

From the archivist's viewpoint, the concept of collecting and ownership is generally very clear: "The archives and institutional collections that are now established have taken the place of the private collector as the major agent for the preservation of our documentary heritage," or "Replevin actions for public archival estrays sustain a historical view and belief that official records belong to the people as represented by their governments."[83] Private individual collectors have, of course, a very different take on the matter. While private collectors had played a critical role in the nineteenth century, they are now seen with the multitude of established programs as threats to security, access, and other matters.[84] More recently, archivists have begun to do research revealing

that replevin is a far more complex issue, with many questions and aspects still needing to be resolved.[85,86]

The tension between private collectors and archival repositories should not be surprising. In many cases, individuals and institutions compete for the same items. Sometimes, collectors use institutions such as libraries as ways of building their own private holdings (as libraries unsuspectingly weed out rare and valuable materials).[87] Just as often, archival programs acquire the collections of private collectors. This has been more documented in terms of literary collections, such as Watson.[88] In rare but important cases, private collectors establish important repositories for their collections, blurring the distinction between personal and institutional collecting as professional staffs become involved in the new institutions.[89] At the very least, many archivists, manuscripts curators, and special collections librarians relished collecting with the same passion as private collectors. University librarian Lawrence Thompson noted that "Strict objectivity, total indifference to partisan issues on the site of collecting, and an unmitigated passion for identifying the truth are the hallmarks of the collector's trade."[90] Collectors were identified as being both individuals and institutions.

The problems represented in this aspect of collecting are due to how popular the autograph trade has become in an investment hedge—important enough to be featured in articles such as those appearing in Forbes.[91,92] The continuing interest in personal autograph collecting can be seen in the ongoing proliferation of how-to books on the subject.[93–96] Despite this fascination with autographs, we do not have a good social history of such collecting, but rather we possess isolated studies on autograph albums as genres[97,98] and the autograph business.[99] There are no particularly scholarly studies of individual autograph collectors. The journal *Manuscripts*, the publication of the Manuscript Society, an organization dominated by individual collectors and dealers, contains many articles on individual autograph collectors. The rising monetary value of manuscripts also contributes to other problem activities, such as forgery, although the history of forgery is also connected to other developments in record keeping (such as the transition to written records as the primary form of evidence and changing trends in scholarship).[100,101] Some forgery cases, such as the Hitler diaries debacle or the celebrated case concerning the forgeries of Mormon documents in the 1980s, have become international events and leave no doubt that the primary motive was to make money.[102–104]

Even today, conflict between the appraisal of government records and the collecting of records exists, evident in litigation concerning the U.S. National Archives' appraisal of FBI records and the maintenance of electronic mail concerning the Iran–Contra affair.[105,106] Such cases are less problematic than the ethical aspects of collecting. Older versions of the Society of American Archivists' code of ethics suggests that any archives or historical manuscripts repository ought to have a collecting policy and adequate resources for maintaining the records acquired, work against excessive donor restrictions, use legal documents in making acquisitions, and seek acquisitions in a fair manner.[107] Most repositories do have these elements in collecting or acquisition policies, but the specter of theft and other problems haunts these repositories and the profession. Again, the more human aspects of collecting often overwhelm the mundane but necessary elements of maintaining archives. Thoughts of publicity or of beating another repository in acquiring a particular collection can lessen the influence of even ethical issues.

Despite the intense discussion and development of new appraisal methods and planning over the past several decades, the fact remains that many collecting programs still acquire the collections of private collectors. Many of the core holdings of historical manuscripts programs are actually private collections later donated to them. The stress between private and institutional collectors is somewhat modified by this relationship. Private collectors often create trends in collecting or have the financial resources to build collections; it is often up to institutional repositories to house them and make them available for researchers.[108] It is precisely at this intersection between private collector and institutional repository that the most dramatic aspects of the psychological nature of collecting are seen.

THE PSYCHOLOGY OF COLLECTING AND THE COLLECTING IMPULSE

In some cases, archivists have exhibited some of the most blatant desires for collecting as a human and competitive instinct. Howard Gotlieb, a long-time archivist at Boston University and the chaser of records of celebrities and other noteworthies, was described as collecting to preserve a moment in time; thus, he pleads that each member of his ever-growing "family" save every scrap of paper upon which a little or just one single word has been written or typed. He claims that each manuscript or type-script piece of paper is like an irreplaceable painting, as it is the only one extant. Gotlieb is so persnickety about this matter that he, on occasion, sends empty boxes to collectees to use as receptacles for any paper discards that they may have. "We want everything," explains Gotlieb.[109]

Such sentiments are supported by the many volumes of memoirs of collectors and dealers who describe the thrill of the chase, ecstasy in finding treasures in dung heaps, and the pleasures associated with outdoing individual and institutional competitors.[110,111] This is a far cry from those archivists who struggle to understand whether a systematic analysis of organizational records can lead to the type of documentation needed for purposes of organizational memory and future research use. For example, Canadian archivists B. Corbett and E. Frost[112] examine whether or not records retention/disposition schedules lead to a fuller archival documentation of government agencies.

Collecting has been connected to all sorts of human instincts and interests from the personal[113] to societal consumerism.[114] A significant challenge to the archivist and collecting is the archivist's impulse to collect. Most archivists remain fearful of the prospect of losing potentially valuable records if they do not step in to save them.[115] Writing about the more than 1000 human rights organizations in North America, Bruce Montgomery stated that the lack of research about human rights "stems largely from the paucity of archival evidence in research universities and other educational institutions, and illustrates the danger to the historical record when significant materials on leading worldwide social and political movements remain uncollected."[116] While some of this stems from personal and psychological impulses to collect, some also extends from the importance of assembling archives in new areas. As an archivist of dance writes,

> The emergence of new archives can only offer greater nourishment and benefit to any tradition . . . It is also a fact that the growth of an archive has a direct impact on a community and a country's sense of self or their personal self-esteem. It is a force which guards its heritage, while at the same time providing educational materials on traditions of the past and customs of the present for its members.[117]

A part of coming to terms with the psychological dimensions of collecting has been a substantial rethinking of archival appraisal and collecting in the past two decades.

RETHINKING ARCHIVAL APPRAISAL AND COLLECTING

In the 1970s and 1980s, concerns rose on how effective collecting was led by efforts to systematize the development of archival collecting policies. Here is where archival appraisal and collecting approaches crossed over, both hindered by a sense that the records (and other materials) that archivists and manuscripts curators were gathering were inadequate for a full or balanced documentation of society.[118] There was some borrowing from librarians' experience with the development of collecting policies, stressing planning, selecting, evaluation, and cooperative activities.[119] The fullest articulation of what goes into a collecting policy for manuscripts was also developed in this time period, emphasizing a statement of purpose, types of programs supported by the collection, clientele served by the collection, priorities and limitations of the collection, cooperative agreements affecting the policy, resource sharing, deaccessioning, procedures for supporting the collecting policy, and procedures for monitoring the development of the collection.[120] By the early 1990s, archivists such as Barbara Craig were stating this about appraisal (and by the implication, acquisition, and collecting): "We do not create the past like an artist creates a work of art; rather we aim to control the past, or more accurately, to control the documentation of the past, like a systems methodizer, balancing aims, objectives, resources and demand."[121]

In some ways, the opening salvo in this reconsideration of collecting by archivists was fired by F. Gerald Ham in his 1974 presidential address to the Society of American Archivists. Ham, then state archivist for Wisconsin, lamented that archivists and manuscripts collectors seemed to be prey to dealing with the constantly changing foibles of research trends and practices. Ham argued that archivists needed to take the lead by acquiring records that would both support current research and open up new avenues for historical research.[122] Ham's eloquent message brought both short-term and long-term responses. Typical of the short-term response was that of Lester Cappon, drawing on 50 years' experience, who argued that

> On the archival edge the archivist as collector is confronted with certain dilemmas, not inherently new in the twentieth century, that are insoluble, to some degree, but open to accommodation. On the one hand, the quantity of certain records demands drastic measures wisely to save and to destroy; on the other hand, the paucity or lack of certain records attributable to telephone communication, tempts him to fill the void by creating records for the service of scholarship.[123]

The challenges facing archivists in collecting and appraising are, of course, difficult and complicated.

Because of dissatisfaction with what has long seemed to be random or skewed collecting, the decades of the 1970s and 1980s were a time in which archivists devoted considerable attention to developing a system for acquisition. Archivists began to question the usual approaches, by which they sought to appraise records, and this ultimately affected the standard means by which archivists collected. There were some efforts to develop a more objective appraisal process, although the development of models and the use of models differed wildly.[124] Archival documentation strategies emerged as a model of such an approach, promising to bring together archivists, records creators, and the users of archives into a forum that would support clearer planning and literal collecting.[125,126] Archival macroappraisal approaches, stressing functions of records creation, also emerged in this period, although they have been more directed to organizational realms such as government and corporations.[127–129] Since these professional discussions, there has been some contentious debate about the foundation of archival appraisal and acquisition, generally revolving around notions of archival

science versus historical approaches to appraising and collecting.[130,131] The documentation strategy model or concept (it has been termed various things) was attractive enough that it was discussed as a foundation for documenting certain aspects of society, such as adult education,[132] higher education,[133] medicine and health care,[134] and geographic regions.[135] While the core ideas of documentation strategy have continued to generate debate,[136,137] these ideas have deeply influenced the archivist's means for thinking about their appraisal and collecting approaches.[138,139]

This transformation of thinking about appraisal and collecting led to a consideration of more standardized approaches and a reevaluation of cherished ideas and values. For a long time, the assumption and practice of collecting records has been to focus on their informational value (i.e., their value beyond the immediate uses to the creators of the records—the evidential value). Some case studies suggest that such distinctions are quite artificial.[140] On the other hand, some archivists have come to see the degree of use as the only means by which appraisal or acquisition can be effectively evaluated.[141] At the least, there is some room for connecting archival appraisal, records management scheduling, and the more systematic collecting approaches.[142]

The most recent move to reconsider archival appraisal generally stemmed from the realization that records were voluminous and increasing in volume, leading to more calls for discipline, restraint, and intelligence in collecting.[143] As some might argue, as the nature and breadth of records systems increased, appraisal (and collecting) has simply become more difficult.[144,145] The challenges of volume have been commented on by every archival thinker or theoretician for the past century, however. There must be other motives for such reevaluation. While government records have become subject to reappraisal[146,147] (the systematic reevaluation of records acquired by archival programs), for example, the records and manuscripts acquired by collecting repositories have not been the topic of such reexamination. Since the notion of reappraisal or deaccessioning surfaced 30 year ago, there still seems to have been little effort to incorporate this approach into repositories. Little research has been done, but what has been done suggests reappraisal and deaccessioning are not being used. An example of such research is Hempe.[148] Collection analysis (closely related to reappraisal), the systematic reevaluation of holdings against perceived or assumed collecting objectives, has had a mildly better history, with some intense case studies and follow-up examinations.[149,150] The lack of attention to such analysis suggests the importance of developing careful collecting criteria in the first place and that systematic approaches still have not reinvented the appraisal or collecting practice. It may be other factors that transform the basic archival attitudes toward collecting.

Archives-Art

BARRIERS TO COLLECTING

Many problems persist in the modern collecting of personal papers, such as privacy regarding those still living.[151–153] In fact, personal privacy has become so significant an issue that popular novels have focused on questions about the role of archivists and manuscript collectors in preserving manuscripts that reveal the private lives of public figures.[154] Life imitates art, as the daughter of writer Bernard Malamud laments anyone having access to her late father's papers, contemplating why such personal papers might be better off destroyed.[155] Some of these problems have persisted for a long time, such as the recognition of the complicated and laborious relationship between repositories and the donors of archives and manuscripts.[156] The most famous (or infamous) case was that of access to the Sigmund Freud papers at the Library of Congress, placed on deposit and controlled by Freud disciples.[157] Such challenges extend beyond the nature of the materials, requiring that hard questions be asked. It is no wonder that many archives and historical manuscripts programs have shied away from confronting (and collecting) controversial topics.[158]

Those programs that have tried to deal with controversial topics by acquiring the records of such organizations as the Ku Klux Klan have found it difficult to contend with public reaction and media attention.[159,160] Indeed, we could ask today whether it is possible to collect at all and avoid controversy. Nearly every aspect of history, historical interpretation, and historical sources have been caught up in controversies, ranging from national history standards in education to exhibitions in major museums.[161–163] The American archival profession seems timid in dealing with such issues, suggesting that acquiring records posing the slightest problems may be overlooked in favor of collecting less contested and often older materials. While there are continuing challenges to access to government records, it is difficult to find evidence of a dynamic archival profession arguing for a more liberal interpretation of classification regulations. The best recent volume on the importance of records and the challenges of access to government records is Theoharis[164] and the best recent example is the work of the National Security Archive,[165] showing a weakened archival profession not willing to try to enforce existing records legislation or to work for strengthened legislation.

The records of American businesses have presented a particularly difficult aspect of collecting, as archivists realized that the businesses themselves were rarely establishing institutional archives and that the records were voluminous in comparison to what other types of records represented.[152] Archival programs with a national scope in documenting businesses often had to reevaluate just how they were conducting their mission. Harvard University's Baker Library shifted from a national scope to one focused on the New England region, and even then it had to develop more aggressive criteria in reducing the size of its holdings—at a time long before the archival profession began to develop theoretical foundations for such work.[166–168] In the mid-1990s, a national conference on the appraisal of American business records tried to deal with the myriad challenges posed by corporate records, but whether or not it really helped get any closer to developing workable solutions is unclear. The American archival profession still sees itself as collectors of such records, while the quantity and complexity (especially the uses of electronic information technology) of these records defy the possibility of acquisition.[169] Calls for the profession to work on establishing institutional or corporate archives have gone largely unheeded. The specter of an archival fascination with acquisition seems to suggest the problems with archivists who view themselves as collectors.[170,171]

Cooperation has also been recognized as a critical but elusive element for collecting and appraising. Government archivists pushed for cooperation, especially in sharing scheduling information, so as to bring control to the housing of voluminous qualities of records. In 1988, the Intergovernmental Records Project, uniting in 1990 with Research Libraries Group's Government Records Project to share descriptive information, started as a pioneering effort at some sort of joint appraisal effort.[172] Cooperation has not been a general success, however, even though it is obviously needed in the archival community. When surveys accumulate data about collecting practice, interest in and activity supporting cooperative approaches generally come up as the weakest link.[173] Some would contest this conclusion. In the early 1980s, Philip Mason had this description about labor archives: "The issue of competition between labor archives is no longer the problem that it was 20 or 30 years ago when the "scarcity" theory of collecting dominated the field. Archivists now recognize that not only are there enough important and valuable collections for all archives to share, but indeed, the existing programs cannot begin to collect and preserve the available union sources. Cooperation has replaced competition among labor archivists."[50] If this is the case, such cooperation is a rarity.

There have been other documented instances in which two or more repositories have worked together to deal with voluminous or complex records.[174] These have been either rare occurrences or rarely documented, however. The establishment of statewide regional networks in the late 1960s seems to be the apogee of such archival cooperation, but they have been well documented as being more effective in nurturing use than in coordinating collecting and acquisitions. See Cameron, Ericson, and Kenney[175] for the most balanced assessment of these networks. Some have argued that the networks helped manuscript collecting, such as Fogerty.[176] Forty years ago, archivists realized that they needed to be organized in some manner that would enable more focused collecting and that would reduce competition. In the United States, a number of states set up regional systems, usually under the direction of the state archives.[177] There is another tension here, however. For generations, archival repositories seemed to thrill in the competitive hunt for personal papers, family manuscripts, and similar materials. Increasingly there has been recognition that there must be strong and able cooperation.[178]

There have been other persistent barriers to collecting. Prior to such ideas of systematic analysis and collecting, archivists tended to focus on approaches promising flexibility and experimentation from a conviction of the inability to assess the overall documentation universe.[179] In the not-too-distant past, many archivists stressed more good public relations and cooperative efforts rather than a high degree of analysis that could be called appraisal.[180] The amazing growth in thematically oriented collecting programs also created new programs for such institutions as colleges and universities. The fact that a single faculty member with a diverse career and research interests might be the target of two or more such programs generated the need for more cooperation between repositories.[181] This is a different problem from trying to develop such criteria as tenure status, reputation, institutional service, and community role as the basis for collecting decisions.[182]

Some aspects of appraisal and collecting require multi-institutional partnerships, such as moving image records, that somehow move the process far beyond just a collecting function, although thousands of repositories worldwide probably acquire motion pictures when they are offered.[183] Still images—photographs—have been even more problematic. They are generated in the millions by government, corporations, artists and commercial photographers, and private citizens, yet they are eagerly collected, often violating provenance, without much regard to rigorous standards (although such standards have been around for quite a while).[184] Some record materials, such as sound recordings, require so many accommodations for technical issues of maintenance and use that their collecting is dictated by many different factors than the normal or traditional paper-based records systems.[185]

It has long been recognized that one of the greatest challenges is the lack of a national system in appraising and acquiring records, leading to some early calls for better communication about appraisal decisions.[186] Many other countries have recognized problems with the lack of national policies that provide some guidance for coordinated work such as appraisal and collecting.[187,188] There have been some considerations of the legal implications of such problems, with the usual laments about weak national, regional, and local legislation supporting the acquisition and maintenance of records.[189]

As a result of such problems as these barriers, some archivists have called for sensitivity in appraising large accumulations of government records such as case files for underdocumented elements of society.[190] This concern brings up the question of what the collecting mission of any archives should be. Examining the collecting of Native American visual records, Canadian archivist Jim Burant laments that too many archivists are "driven" to meet researcher's needs "rather than attempting to take a more all-embracing approach to documenting and reflecting society." Burant continues that "as an archivist, as a native Algonquin, and as a human being, I feel that public institutional archives must reflect as broadly as possible the nature, fabric, and conflicts of the society from which they spring."[191] Others have made similar pleas for a social sensitivity in appraising and collecting,[190,192] but the uninhibited human nature of collecting works against any kind of uniform approach for dealing with such matters.

COMPLICATED TIMES

There is little question that how archivists perceive their holdings, their collections, and, consequently, their mission has become more complicated. Archivists are now far more sensitive to issues like privacy, accountability, public memory, truth, politics, identity, and the ethical implications of their actions. Some of these matters affect deeply how archivists think of collecting, a function they once took for granted.[193] The mission is changing. Especially important to the transformation of the archival mission has been the increased role of indigenous and community groups in taking responsibility for their own records. This works against established, traditional archives sweeping up records as they have for decades, while increasing the possibility of preserving and protecting a greater and more diverse portion of our documentary heritage.[194,195] Along with this transformation has come some particularly difficult stress points, such as those witnessed with the debate about the proposed Native American Archival Materials and, more recently, the claims about social justice being a purpose for archives.[196–198] The notion of collecting is being challenged, or, at least, the assumptions about collecting are being questioned.

CONCLUSION

Archival collecting is not just a mindless exercise in sweeping up old records or sitting back and waiting for the important records to appear for maintenance by an archives; rather, collecting is a process enmeshed with political, theoretical, psychological, and historical elements. More research is needed about the nature of archival collecting. More understanding by society about how archives and historical manuscripts repositories are formed is needed as well. The image of an archivist as an Indiana Jones–type character hunting out the treasures of the past in exciting pursuits is romantic but inaccurate; rather, archivists are flawed humans trying to develop clear and reliable methods for identifying records that should be acquired by archives. Much of merit has been accomplished by archivists and manuscripts curators gathering records, but more reflection and experimentation needs to be done on this topic. It seems that the new archival hunters and gatherers will be using very different techniques to sleuth about in the sophisticated record-keeping technologies of the twenty-first century.

REFERENCES

1. Powell, G. Collecting of personal and private papers in Australia. Arch. Manus. May **1996**, *24*, 62–77.
2. Duranti, L. The concept of appraisal and archival theory. Am. Arch. Spring **1944**, *57*, 328–344.
3. Boles, F.; Greene, M.A. Et Tu Schellenberg? Thoughts on the dagger of American appraisal theory. Am. Arch. Summer **1996**, *59*, 298–310.
4. Cox, R.J. The Federal Government's interest in archives of the United States. In *Archival Science on the Threshold of the Year 2000*; Bucci, O., Ed.; University of Macerata: Macerata, Italy, 1993; 207–241.
5. George, G. The State of the American Record: A Report on How Well Americans Are Documenting Their History. In *National Historical Publications and Records Commission*, Washington, DC, 1994; 8.
6. Duniway, D.C. Conflicts in collecting. Am. Arch. January **1961**, *24*, 55–63.
7. Bastian, J.A.; Alexander, B. *Community Archives: The Shaping of Memory*; Facet: London, U.K., 2009.
8. Berner, R.C. *Archival Theory and Practice in the United States: A Historical Analysis*; University of Washington Press: Seattle, WA, 1983.
9. Duniway, D. How does one collect archives? The Oregon experience. Ind. Arch. **1967–1968**, *17*, 50–57.
10. Gondos, V. Jr. J. *Franklin Jameson and the Birth of the National Archives 1906–1926*; University of Pennsylvania Press: Philadelphia, PA, 1981.
11. Jones, H.G. *The Records of a Nation: Their Management Preservation, and Use*; Atheneum: New York, 1969.
12. McCoy, D.R. *The National Archives: America's Ministry of Documents 1934–1968*; University of North Carolina Press: Chapel Hill, NC, 1978.
13. Walch, T. *Guardian of Heritage: Essays on the History of the National Archives*; National Archives and Records Administration: Washington, DC, 1985.
14. Dozois, P. Beyond Ottawa's reach: The federal acquisition of regional government records. Archivaria Winter **1991–1992**, *33*, 57–65.
15. MacLeod, D. Quaint specimens of the early days: Priorities in Collecting the Ontario Archival Record, 1872–1935. Archivaria Summer **1986**, *22*, 12–39.
16. Hawes, L.M.; Britt, A.S., Jr. *The Search for Georgia's Colonial Records*; Georgia Historical Society: Savannah, GA, 1976; 18.
17. Jones, H.G. *For History's Sake: The Preservation and Publication of North Carolina History 1663–1903*; University of North Carolina Press: Chapel Hill, NC, 1996.
18. Bellardo, L.J.; Bellardo, L. *Lady A Glossary for Archivists, Manuscript Curators, and Records Managers*; Society of American Archivists: Chicago, IL, 1992; 1.
19. Pearce-Moses, R. *A Glossary of Archival and Records Terminology*; Society of American Archivists: Chicago, IL, 2005.
20. Callcott, G.H. *History in the United States 1800–1860: Its Practice and Purpose*; Johns Hopkins University Press: Baltimore, MD, 1970.
21. Jones, H.G. Historical Consciousness in the Early Republic: The Origins of State Historical Societies, Museums, and Collections, 1791–1861 North Caroliniana Society, Inc. and North Carolina Collection: Chapel Hill, IL, 1995.
22. Van Tassel, D.D. *Recording America's Past: An Interpretation of the Development of Historical Societies in America 1607–1884*; University of Chicago Press: Chicago, IL, 1960.
23. Whitehill, W.M. *Independent Historical Societies: An Enquiry into Their Research and Publication Functions and Their Financial Future*; Boston Athenaeum: Boston, MA, 1962.
24. Evans, J. *A History of the Society of Antiquaries*; University Press for the Society of Antiquaries: Oxford, U.K., 1956.
25. Thomson, A. *Anzac Memories: Living with the Legend*; Oxford University Press: New York, 1994.
26. Tucker, L.L. From Belknap to Riley: Building the Collection of the Massachusetts Historical Societ. In *Witness to America's Past: Two Centuries of Collecting by the Massachusetts Historical Society*; Massachusetts Historical Society: Boston, MA, 1991; 15–23.
27. Tucker, L.L. *Clio's Consort: Jeremy Belknap and the Founding of the Massachusetts Historical Society*; Massachusetts Historical Society: Boston, MA, 1990.
28. Rosenzweig, R.; Thelen, D. *The Presence of the Past: Popular Uses of History in American Life*; Columbia University Press: New York, 1998.
29. Bodnar, J. *Remaking America: Public Memory, Commemoration, and Patriotism in the Twentieth Century*; Princeton University Press: Princeton, NJ, 1992.
30. Kane, L.M. *A Guide for Collectors of Manuscripts*; Minnesota Historical Society: St. Paul, MN, 1951. Service bulletin no. 1, 1.
31. Bickford, C.P. Public records and the private historical society: A Connecticut example. Gov. Publ. Rev. **1981**, *8A*, 311–320.
32. Wright, C.J. Ed. *Sir Robert Cotton as Collector: Essays on an Early Stuart Courtier and His Legacy*; British Library: London, U.K., 1997.
33. Burant, J. Ephemera, archives, and another view of history. Archivaria Fall **1995**, *40*, 189–198.
34. Lee, C.A.I. *Digital: Personal Collections in the Digital Era*; Society of American Archivists: Chicago, IL, 2011.
35. Lohf, K.A. Treasures for Alma Mater: How Columbia University acquired the papers of major New York Publishers and literary agents. Manuscripts Spring **1977**, *29*, 103–109.
36. Cook, T. What is past is prologue: A history of archival ideas since 1898, and the future paradigm shift. Archivaria Spring **1997**, *43*, 17–63.

37. Kolsrud, O. Developments in archival theory. In *Encyclopedia of Library and Information Science*; Marcel Dekker, Inc: New York, 1978; Vol. 61, supplement 24.

38. Buchanan, D. *The Treasure of Auchinleck: The Story of the Boswell Papers*; McGraw-Hill: New York, 1974.

39. Cahill, T. *How the Irish Saved Civilization: The Untold Story of Ireland's Heroic Role from the Fall of Rome to the Rise of Medieval Europe*; Anchor Books, Doubleday: New York, 1995.

40. Cox, R.J. Archives, war, and memory: Building a framework. Libr. Arch. Secur. **2012**, *25*, 21–57.

41. O'Toole, J.M. The symbolic significance of archives. Am. Arch. Spring **1993**, *56*, 234–255.

42. Feliciano, H. *The Lost Museum: The Nazi Conspiracy to Steal the World's Greatest Works of Art*; Harper Books: New York, 1997.

43. Honan, W.H. *Treasure Hunt: A New York Times Reporter Tracks the Quedlinburg Hoard*; Fromm International: New York, 1997.

44. Nicholas, L.H. *The Rape of Europa: The Fate of Europe's Treasures in the Third Reich and the Second World War*; Vintage Books: New York, 1994.

45. Simpson, E. *The Spoils of War: World War II and Its Aftermath: The Loss, Reappearance, and Recovery of Cultural Property*; Harry, N., Ed.; Association with the Bard Graduate Center for Studies in the Decorative Arts: New York, 1997.

46. Novick, P. *The Holocaust in American Life*; Houghton Mifflin: Boston, MA, 1999.

47. Kecskemeti, C. Displaced European archives: Is it time for a postwar settlement? Am. Arch. Winter, *55*, 132–138.

48. Cuno, J. *Who Owns Antiquity? Museums and the Battle Over Our Ancient Heritage*; Princeton University Press: Princeton, NJ, 2008.

49. Rothfield, L. *The Rape of Mesopotamia: Behind the Looting of the Iraq Museum*; University of Chicago: Chicago, IL, 2009.

50. Mason, P.P. Labor archives in the United States: Achievements and prospects. Labor Hist. Fall **1982**, *23*, 487–497.

51. Atherton, L.E. Western historical manuscripts collection: A case study of a collecting program. Am. Arch. January **1963**, *26*, 41–49.

52. Kane, L. Collecting policies of the Minnesota Historical Society, 1849–1952. Am. Arch. April **1953**, *16*, 127–136.

53. Fox, E.M. The genesis of Cornell University's collection of regional history. Am. Arch. April **1951**, *14*, 105–116.

54. Goggin, J.; Carter, G. Woodson and the collection of source materials for Afro-American history. Am. Arch. Summer **1985**, *48*, 261–271.

55. Grigg, S. A world of repositories, a world of records; redefining the scope of a national subject collection. Am. Arch. Summer **1985**, *48*, 286–295.

56. Guthrie, K.M. *The New-York Historical Society: Lessons from One Nonprofit's Long Struggle for Survival*; Jossey-Bass: San Francisco, CA, 1996.

57. Anderson, P.G. Appraisal of the papers of biomedical scientists and physicians for medical archives. Bull. Med. Libr. Assoc. October **1985**, *73*, 338–344.

58. Pratt, D.; Kenneth, A. Lohf: Collecting for Columbia. Columbia Libr. Col. November **1992**, *42*, 2–12.

59. Trimble, M. Archives and manuscripts: New collecting areas for law libraries. Law Libr. J. Summer **1991**, *83*, 429–450.

60. Lee, C.C. Collecting organizing and using Chinese-American resources: An archival approach. JLIS October **1990**, *16*, 24–42.

61. Anderson, R.J. Building a multi-ethnic collection: The research library of the Balch Institute for Ethnic Studies. Ethnic Forum **1985**, *5*, 7–19.

62. Mason, P.P. The archives of labor and urban affairs, Walter P. Reuther Library, Wayne State University. Labor Hist. Fall **1982**, *23*, 534–545.

63. Cox, R.J. The record: Is it evolving? Records Retriev. Rep. March **1994**, *10*, 1–16.

64. Cox, R.J. Archives as a multi-faceted term in the information professions. Records Retriev. Rep. March **1995**, *11*, 1–15.

65. Cox, R.J. The record in the manuscript collection. Arch. Manus. May **1996**, *24*, 46–61.

66. Cox, R.J. The importance of records in the information age. Records Mgmt. Q. January **1998**; *32*, 36–46, 48–49, 52.

67. Mallon, T. Love among the records. GQ **November 1999**, *69*, 336, 346.

68. Cook, M. Appraisal and access: We should expect changes driven by the media and by public awareness. Records Mgmt. J. April **1998**, *8*, 3–9.

69. Church, L.T. *What Motivates African-Americans to Donate Personal Papers to Libraries and How Their Giving Decisions Affect the Quantity and Quality of Collections Procured for Archives*; Master of Science in Library Science, University of North Carolina at Chapel Hill: Chapel Hill, NC, 1998.

70. Marshall, J.A. Documentation strategies in the twenty-first century? Rethinking institutional priorities and professional limitations. Arch. Iss. **1998**, *23* (1), 59–74.

71. Cox, R.J. *No Innocent Deposits: Forming Archives by Rethinking Appraisal*; Scarecrow: Lanham, MD, 2004.

72. Cunningham, A. Beyond the pale? The 'flinty' relationship between archivists who collect the private records of individuals and the rest of the archival profession. Arch. Manus. May **1996**, *24*, 20–26.

73. Hurley, C. Beating the French. Arch. Manus May **1996**, *24*, 12–18.

74. Cunningham, A. From here to eternity: Collecting archives and the need for a national documentation strategy. LASIE March **1998**, *29*, 32–45.

75. Heald, C. Are we collecting the 'right stuff?'. Archivaria Fall **1995**, *40*, 182–188.

76. McDonald, R.A.J. Acquiring and preserving private records—A debate. Who is preserving private records? Archivaria Fall **1994**, *38*, 155–157.

77. Fall Hives, C. Thinking globally, acting locally. Archivaria Fall **1994**, *38*, 157–161.

78. McDonald, R.A.J. Acquiring and preserving private records: Cultural versus administrative perspectives. Archivaria. **Fall 1994**, *38*, 162–163.

79. Blouin, F.X., Jr.; Rosenberg W.G. *Processing the Past: Contesting Authority in History and the Archives*; Oxford University Press: New York, 2011.

80. Price, W.S., Jr.; N.C. v. B.C. West, JR. Am. Arch. Winter **1978**, *41*, 21–24.

81. Cutright, P.R. *A History of the Lewis and Clark Journals*; University of Oklahoma Press: Norman, OK, 1976.

82. Brady Schmidt, P. Compromise resolves fate of documents: Replevin avoided. Manuscripts Fall **1985**, *37*, 275–282.

83. O'Neil, J.E. Replevin: A public archivist's perspective. College Res. Libr. January **1979**, *40*, 26–30.

84. Cox, R.J. Collectors and archival, manuscript, and rare book security. Focus Sec. Mag. Libr. Arch. Mus. Sec. April **1995**, *2*, 19–27.

85. Dow, E.H. *Archivists, Collectors, Dealers, and Replevin: Case Studies in Private Ownership of Public Documents* Scarecrow: Lanham, MD, 2012.

86. Mattern, E. The Replevin process in Government Archives: Recovery and the Contentious Question of Ownership. PhD dissertation, University of Pittsburgh: Pittsburgh, PA, 2014.

87. Swan, J. Sound archives: The role of the collector and the library. Wilson Libr. Bull. February **1980**, *54*, 370–376.

88. Watson, J.G. Carvel Collins's Faulkner: A newly opened archives. Libr. Chron. UT Austin **1990**, *20* (4), 89–97.

89. Dickinson, D.C.; Harry, E. *Huntington's Library of Libraries*; Huntington Library: San Marino, CA, 1995.

90. Thompson, L.S. Incurable mania. Manuscripts Summer **1966**, *18*, 18–25.

91. Brown, C. Buyer beware, seller too. Forbes March 1, **1993**, *151*, 124.

92. Queenan, J.; Kripalani, M. The handwriting's off the wall. Forbes October 2, **1989**, *144*, 186.

93. Berkeley, E., Jr. Ed. Autographs and Manuscripts: A Collector's Manual; Klingelhofer H.E., Rendell, K.W., Eds.; New York, 1978. sponsored by the Manuscript Society.

94. Benjamin, M.A. *Autographs: A Key to Collecting*; Walter R. Benjamin Autographs: New York, 1966.

95. Hamilton, C. *Collecting Autographs and Manuscripts*, rev. ed.; Modoc: Santa Monica, CA, 1993.

96. Rendell, K.W. *History Comes to Life: Collecting Historical Letters and Documents*; University of Oklahoma Press: Norman, OK, 1995.

97. Weston, P. Vincent Novello's autograph album: Inventory and commentary. Music Lett. August **1994**, *75*, 365–380.

98. Green, T.A.; Devaney, L. Linguistic play in autograph book inscriptions. West. Folklore January **1989**, *48*, 51–58.

99. Cappon, L.J.; Walter, R. Benjamin and the autograph trade at the turn of the century. Proc. Mass. Hist. Soc. January–December **1966**, *78*, 20–37.

100. Bozeman, P. Forged documents: Proceedings of the 1989 Houston Conference, Organized by the University of Houston Libraries; Oak Knoll Books, New Castle, DE, 1990.

101. Grafton, A. *Forgers and Critics: Creativity and Duplicity in Western Scholarship*; Princeton University Press: Princeton, NJ, 1990. Hamilton, C. *Great Forgers and Famous Fakes: The Manuscript Forgers of America and How They Duped the Experts*, Crown, New York, 1980.

102. Hamilton, C. *The Hitler Diaries: Fakes That Fooled the World*; University Press of Kentucky: Lexington, KY, 1991.

103. Harris, R. *Selling Hitler*; Penguin: New York, 1986.

104. Naifeh, S.; Smith, G.W. *The Mormon Murders: A True Story of Greed, Forgery, Deceit, and Death*; New American Library: New York, 1988.

105. Steinwall, S.D. Appraisal of the FBI files case: For whom do archivists retain records? Am. Arch. Winter **1986**, *49*, 52–63.

106. Wallace, D. *The Public's Use of Federal Record Keeping Statutes to Shape Federal Information Policy: A Study of the PROFS case*. PhD dissertation; University of Pittsburgh: Pittsburgh, PA, 1977.

107. Wilsted, T. Observations on the ethics of collecting archives and manuscripts. Provenance **1993**, *11*, 25–38.

108. Basbanes, N.A. *A Gentle Madness: Bibliophiles, Bibliomanes, and the Eternal Passion for Books*; Henry Holt: New York, 1995.

109. Yoken, M. Collecting the twentieth century: Curator Howard Gotlieb. Wilson Libr. Bull. April **1986**, *60*, 25–29.

110. Everittt, C.P. *The Adventures of a Treasure Hunter: A Rare Bookman in Search of American History*; Little, Brown: Boston, MA, 1952.

111. Williams, R. *Adventures of an Autograph Collector*; Exposition Press: New York, 1952.

112. Corbett, B.; Frost, E. The acquisition of federal government records: A report on records management and archival practice. Archivaria Winter **1983–1984**, *17*, 201–232.

113. Muensterberger, W. *Collecting: An Unruly Passion: Psychological Perspectives*; Princeton University Press: Princeton, NJ, 1994.

114. Belk, R.W. *Collecting in a Consumer Society*; Routledge: New York, 1995.

115. Ericson, T.L. At the 'rim of creative dissatisfaction'; Archivists and acquisition development. Archivaria Winter **1991–1992**, *33*, 66–77.

116. Montgomery, B.P. Collecting human rights evidence: A model for archival collection development. In *Encyclopedia of Library and Information Science*; Marcel Dekker, Inc.: New York, **1999**, *64*, 51–61.

117. Oswald, G. One approach to the development of a dance archive: The dance collection in the library and museum of the performing arts (The New York Public Library at Lincoln Center). In *Libraries, History, Diplomacy, and the Performing Arts: Essays in Honor of Carleton Sprague Smith*; Katz, I.J., Ed.; Pendragon Press, in Cooperation with the New York Public Library: Stuyvesant, NY, 1999; 77–84. Festschrift Series no. 9.

118. Henry, L.J. Collecting policies of special-subject repositories. Am. Arch. Winter **1980**, *43*, 57–63.

119. Reed-Scott, J. Collection management strategies for archivists. Am. Arch. Winter **1984**, *47*, 23–29.

120. Phillips, F. Developing collecting policies for manuscript collections. Am. Arch. Winter **1984**, *47*, 30–42.

121. Craig, B.L. The acts of the appraisers: The content, the plan, and the record. Archivaria Summer **1992**, *34*, 175–180.

122. Ham, F.G. The archival edge. Am. Arch. January **1975**, *38*, 5–13.

123. Cappon, L.J. The archivist as collector. Am. Arch. October **1976**, *39*, 429–435.

124. Boles, F.; Marks Young, J. *Archival Appraisal*; Neal-Schuman: New York, 1991.

125. Alexander, P.N.; Samuels, H.W. The roots of 128: A hypothetical documentation strategy. Am. Arch. Fall **1987**, *50*, 518–531.

126. Samuels, H. Improving our disposition: Documentation strategy. Archivaria Winter **1991–1992**, *33*, 125–140.

127. Bailey, C. From the top down: The practice of macro-appraisal. Archivaria Spring **1997**, *43*, 89–128.

128. Brown, R. Macro-appraisal theory and the context of the public records creator. Archivaria Fall **1995**, *40*, 121–172.

129. Brown, R. Records acquisition strategy and its theoretical foundation: The case for a concept of archival hermeneutics. Archivaria Winter **1991–1992**, *33*, 34–56.

130. Cook, T. 'Another brick in the wall': Terry Eastwood's masonry and archival walls, history and archival appraisal. Archivaria Spring **1994**, *37*, 96–103.

131. Eastwood, T. Nailing a little jelly to the wall of archival studies. Archivaria Spring **1993**, *35*, 232–252.

132. Keenan, T. Documenting adult education: Toward a cooperative strategy; Syracuse University Kellogg Project Technical Report Series Report no. 2, October 1989.

133. Samuels, H.W. *Varsity Letters: Documenting Modern Colleges and Universities*; Scarecrow Press: Metuchen, NJ, 1992.

134. Krizack, J.D. Ed. *Documentation Planning for the U.S. Health Care System*; Johns Hopkins University Press: Baltimore, MD, 1994.

135. Cox, R.J. *Documenting Localities: A Practical Model for American Archivists and Manuscripts Curators*; Scarecrow Press: Metuchen, NJ, 1996.

136. Johnson, E. Our archives, our selves: Documentation strategy and the re-appraisal of professional identity. Am. Arch. April **2008**, *71*, 190–202.

137. Malkmus, D. Documentation strategy: Mastodon or Retro-success. Am. Arch. No.2 **2008**, *71*, 384–409.

138. Cook, T., ed. *Controlling the Past: Documenting Society and Institutions*; Essays in Honor of Helen Willa Samuels Society of American Archivists: Chicago, IL, 2011.

139. Robyns, M.C. *Using Functional Analysis in Archival Appraisal: A Practical and Effective Alternative to Traditional Appraisal Methodologies*; Rowman and Littlefield: Lanham, MD, 2014.

140. Davis, R.C. Getting the lead out: The appraisal of silver-lead mining records at the University of Idaho. Am. Arch. Summer **1992**, *55*, 454–463.

141. Greene, M. 'The surest proof': A utilitarian approach to appraisal. Archivaria Spring **1998**, *45*, 127–169.

142. Cox, R.J. Records management scheduling and archival appraisal. Records Info. Mgmt. Rep. April **1998**, *14*, 1–16.

143. Ham, F.G. Archival choices: Managing the historical record in an age of abundance. Am. Arch. Winter **1984**, *47*, 11–22.

144. Eastwood, T. How goes it with appraisal? Archivaria Autumn **1993**, *36*, 111–121.

145. Bearman, D. *Archival Methods*; Archives and Museum Informatics: Pittsburgh, PA, 1989.

146. Rapport, L. No grandfather clause: reappraising accessioned records. Am. Arch. Spring **1981**, *44*, 143–150.

147. Powell, S. Archival reappraisal: the immigration casefiles. Archivaria Winter **1991–1992**, *33*, 104–116.

148. Hempe, A. *Deaccessioning Practices in Selected North Carolina Archives*; University of North Carolina at Chapel Hill: Chapel Hill, NC, 1996.

149. Endelman, J.E. Looking backward to plan for the future: collection analysis for manuscript repositories. Am. Arch. Summer **1987**, *50*, 340–355.

150. Weideman, C. A new map for field work: Impact of collections analysis on the Bentley Historical Library. Am. Arch. Winter **1991**, *54*, 54–60.

151. Hodson, S.S. Private lives: Confidentiality in manuscripts collections. Rare Books Manus. Libr.**1991**, *6* (2), 108–118.

152. MacNeil, H. *Without Consent: The Ethics of Disclosing Personal Information in Public Archives*; Society of American Archivists and the Scarecrow Press: Metuchen, NJ, 1992.

153. Whyte, D. The acquisition of lawyers' private papers. Archivaria Summer **1984**, *18*, 142–153.

154. Cooley, M. *The Archivist: A Novel*; Little, Brown: Boston, MA, 1998.

155. Smith, J.M. *Private Matters: In Defense of the Personal Life*; Addison-Wesley: Reading, MA, 1997.

156. Kaiser, B.J. Problems with donors of contemporary collections. Am. Arch. April **1969**, *32*, 103–107.

157. Malcolm, J. *In the Freud Archives*; Knopf: New York, 1984.

158. Lamoree, K.M. Documenting the difficult or collecting the controversial. Arch. Iss. **1995**, *20* (2), 149–154.

159. Boles, F. 'Just a bunch of bigots': A case study in the acquisition of controversial material. Arch. Iss. **1994**, *19* (1), 53–65.

160. Hackbart-Dean, P. A Hint of scandal: Problems in acquiring the papers of Senator Herman E. Talmadge—A case study. Provenance **1995**, *13*, 65–80.

161. Nash, G.B.; Crabtree, C.; Dunn, R.E. *History on Trial: Culture Wars and the Teaching of the Past; A*; Knopf: New York, 1997.

162. Stearns, P.N. *Meaning Over Memory: Recasting the Teaching of Culture and History*; University of North Carolina Press: Chapel Hill, NC, 1993.

163. Cox, R.J. Archival anchorites: Building public memory in the era of the culture wars. Mutlicult. Rev. June **1998**, *7*, 52–60.

164. Theoharis, A.G., Ed. *A Culture of Secrecy: The Government Versus the People's Right to Know*; University Press of Kansas: Lawrence, KS, 1998.

165. Blanton, T., Ed. *White House E-Mail: The Top Secret Computer Messages the Reagan/Bush White House Tried to Destroy*; New Press: New York, 1995.

166. Burckel, N.C. Business archives in a university setting: Status and prospect. College Res. Libr. May **1980**, *41*, 227–233.

167. Altman, E.C. A history of Baker Library at the Harvard University Graduate School of Business Administration. Harvard Libr. Bull. April **1981**, *39*, 169–196.

168. Bartoshesky, F. Business records at the Harvard Business School. Bus. Hist. Rev. Autumn **1985**, *59*, 475–483.

169. Lathrop, F. Toward a national collecting policy for business history: The view from Baker Library. Bus. Hist. Rev. Spring **1988**, *62*, 134–143.

170. O'Toole, J.M., Ed. *The Records of American Business*; Society of American Archivists: Chicago, IL, 1997.

171. Cox, R.J. *Managing Institutional Archives: Foundational Principles and Practices*; Greenwood Press: New York, 1992.

172. Allen, M.B. Intergovernmental records in the United States: Experiments in description and appraisal. Info. Dev. April **1992**, *8*, 99–103.

173. Harris, R. *Bridges over Troubling Waters: Collection Development Patterns in Archival Holdings*; University of North Carolina: Chapel Hill, NC, July 1994.

174. Couch, N. Collection division as an acquisition method: A case study. Acquist. Libr. **1992**, *8*, 23–31.

175. Cameron, R.A.; Ericson, T.; Kenney, A.R. Archival cooperation: A critical look at statewide archival networks. Am. Arch. Fall **1983**, *46*, 414–432.

176. Fogerty, J.E. Manuscript collecting in archival networks. Midwest Arch. **1982**, *6* (2), 130–141.

177. Kyvig, D.E. Documenting urban society: A regional approach. Drexel Libr. Q. October **1977**, *13*, 76–91.

178. Cumming, J. Beyond Intrinsic value towards the development of acquisition strategies in the private sector: The experience of the Manuscript Division, National Archives of Canada. Archivaria Fall **1994**, *38*, 232–239.

179. Anderson, R.J. Managing change and chance: Collecting policies in social history archives. Am. Arch. Summer **1985**, *48*, 296–303.

180. Grabowski, J.J. Fragments or components: Theme collections in a local Setting. Am. Arch. Summer **1985**, *48*, 304–314.

181. Wolff, J. Faculty papers and special-subject repositories. Am. Arch. Fall **1981**, *44*, 346–351.

182. Honhart, F.L. The solicitation, appraisal, and acquisition of faculty papers. College Res. Libr. May **1983**, *44*, 236–241.

183. Kula, S. *The Archival Appraisal of Moving Images: A RAMP Study with Guidelines*; UNESCO: Paris, France, 1983; PGI-83/WS/18.

184. Leary, W.H. *The Archival Appraisal of Photographs: A RAMP Study with Guidelines*; UNESCO: Paris, France, 1985; PGI-85/WS/10.

185. Harrison, H. *The Archival Appraisal of Sound Recordings and Related Materials: A RAMP Study with Guidelines*; UNESCO: Paris, France, February 1987; PGI-87/WS/1.

186. Evans, M.J. The visible hand: Creating a practical mechanism for cooperative appraisal. Midwest. Arch. **1986**, *11* (1), 7–13.

187. Kitching, C.; Hart, I. Collection policy statements. J. Soc. Arch. Spring **1995**, *16*, 7–14.

188. Harvey, P.D.A. Archives in Britain: Anarchy or policy? Am. Arch. Winter **1983**, *46*, 22–30.

189. Hall, K. Archival acquisitions: Legal mandates and methods. Archivaria Summer **1984**, *18*, 58–69.

190. Cook, T. 'Many are called, but few are chosen': Appraisal guidelines for sampling and selecting case files. Archivaria Summer **1991**, *32*, 25–50.

191. Burant, J. The acquisition of visual records relating to native life in North America. Provenance **1992**, *10*, 1–26.

192. Laberge, D. Information knowledge, and rights: The preservation of archives as a political and social issue. Archivaria Winter **1987–1988**, *25*, 44–49.

193. Avery, C.; Holmlund, M., eds. *Better Off Forgetting? Essays on Archives, Public Policy, and Collective Memory*; University of Toronto Press: Toronto, Ontario, Canada, 2010.

194. Daniel, D.; Levi, A.S., eds. *Identity Palimpsests: Archiving Ethnicity in the U.S. and Canada*; Litwin Books: Sacramento, CA, 2013.

195. Caldera, M.A.; Neal, K.M., eds. *[Through the] Archival Looking Glass: A Reader on Diversity and Inclusion*; Society of American Archivists: Chicago, IL, 2014.

196. Fox, L.; Bhasin, A.; Arriaga, S.K., eds. *Tribal Libraries, Archives, and Museums: Preserving Our Language, Memory, and Lifeways*; Scarecrow Press: Lanham, MD, 2011.

197. Greene, M.A. A critique of social justice as an archival imperative: What is it we're doing that's all that important? Am. Arch. Fall-Winter **2013**, *76*, 302–334.

198. Jimerson, R.C. Archivists and social responsibility: A response to Mark Greene. Am. Arch. Fall-Winter **2013**, *76*, 335–345.

Archives–Art

Area and Interdisciplinary Studies Literatures and Their Users

Lynn Westbrook
School of Information, University of Texas at Austin, Austin, Texas, U.S.A.

Abstract

Interdisciplinary scholars and their literature pervade modern academia, reaching into the humanities, social sciences, and natural sciences. Their publication formats include standard monographs and journals, as well as a rich array of gray literature. Online searching remains, however, problematic with library services and collections struggling to meet their information needs. Using both formal and informal information networks, interdisciplinary scholars construct and maintain their own knowledge domains. Variations in the information experiences and literatures of interdisciplinary work in the humanities, social sciences, and natural sciences continue to require focused support mechanisms.

INTRODUCTION

The stock phrase "increasingly interdisciplinary" describes so much of modern academia that many fields within the humanities, the natural sciences, and the social sciences now expect disciplinary boundary crossing as part of their intellectual landscape.[1,2] Interdisciplinary studies are characterized by their deep integration of the theories, methods, perspectives, and basic concepts of two or more disciplines with a resultant body of scholarship that moves beyond the boundaries of any single discipline. Area studies, a subset of interdisciplinary studies, concentrate holistically on the culture, nature, concerns, and problems of a particular geopolitical or cultural region, thereby generating that essential integration of disciplines. The resultant information landscape of any interdisciplinary and area studies (IAS) field is inherently rugged and unstable. The goal of this entry is to synthesize current understanding regarding IAS information structures and the related experiences of scholars who use and contribute to those structures. Included in this synthesis are materials on IAS scholars' experiences in using available publication formats, finding primary sources and gray literature, online searching, using library services and collections, using informal information systems, and building and maintaining their knowledge domain. A brief research agenda and bibliography provide support for additional study.

UNDERSTANDING IAS

Definitions

At its simplest, interdisciplinary work may be defined as the deep integration of the theories, methods, perspectives, and basic concepts of two or more disciplines with a resultant body of scholarship that moves well beyond the boundaries of any single discipline.[3,4] However, academic disciplines, like ecosystems, are in almost constant flux and the tidy categorization of a field as "interdisciplinary" may ignore roots which still impact the direction of its growth. (For an in-depth explication of these complex issues and forces, see Palmer).[5] Therefore, understanding the context, measures, artifacts, and drivers of interdisciplinarity supports any effort to strengthen relevant information infrastructures.

Disciplinarity is not, of course, a binary proposition; single-discipline and IAS fields are not the sole forms. Other boundary crossing[6–8] approaches include multidisciplinarity, transdisciplinarity, and information probing. Multidisciplinary research entails the use of two or more disciplines in examination of a research question without any integration of approach, method, theory, or tools; parties bring their own expertise to the table, and the resulting research is quilted together by connecting disparate segments rather than integration. Disciplines are respected, boundaries maintained, and applications focus on the research problem. Transdisciplinary research transcends disciplinary boundaries entirely, building on an overarching perspective to synthesize research. While interdisciplinary fields are indeed academic disciplines with all the infrastructure and information ecologies appertaining thereto, transdisciplinary research consciously eschews academic disciplinary boundaries. It does not integrate or absorb them but, rather, utilizes whatever is needed at any given moment with no intent to form a new discipline.[1,9] "Information probing"[10] is an individual's effort to test the waters of another field by examining information resources, data, theories, and/or principles in light of a particular information need. Such probing may lead to active participation in transdisciplinary/multidisciplinary/interdisciplinary work but it may also be a onetime experience.

Encyclopedia of Library and Information Sciences, Fourth Edition DOI: 10.1081/E-ELIS4-120043493

Measures and Artifacts of Interdisciplinarity

Measures and artifacts of interdisciplinarity have been studied for almost 50 years in an effort to understand the boundaries of intellectual domains and the consequences of changes in those boundaries.[11–14] The bibliometric concept of "scatter" underpins much of this work by measuring scholars' output (for example, journals and monographs) as well as the meta-structures of the field (for example, periodical indexes and digital libraries). In "high scatter" fields, scholars publish in a wide range of journals and monographic series so that numerous periodical indexes and digital libraries are required in any comprehensive search of the literature. In "low scatter" fields, scholars publish primarily in a handful of forums which are readily indexed in a few key tools. IAS fields are, inherently, "high scatter" in that they publish in, are cited by, and choose to cite in a wide range of forums as they pull from various disciplines. The interdisciplinarity of particular journals can actually be measured through social network analysis using "betweenness centrality" as an indicator to denote where a journal lies in relationship to the disciplinary location of other journals.[15,16]

Apart from bibliometric analyses, the academic infrastructure has a powerful influence on hiring faculty, providing the intellectual freedom of tenure, and building research support. In any field, including those of IAS, that infrastructure develops in relationship to life cycles in research specializations which develop and redefine their intellectual boundaries. The typical pattern is an initial period of slow growth which moves into an explosion of research that gradually tapers off to a steady state.[17] In much of academia, that steady state has been reached when faculty work within a funded infrastructure that actively supports the tenure of effective scholars. Since IAS fields, however, are formed outside that academic norm, they must choose to either construct their own infrastructure or deliberately live outside of it. On any given campus, an IAS faculty may work to gain program, research center, or school status with line-item appointments for permanent faculty, an appropriate tenure system, and research support. Many will start with faculty who hold formal line appointments in multiple departments, run projects based in multi-jurisdictional research centers, hold appointments in departments other than those in which their doctoral work was completed, regularly teach and/or hold research concentrations in multiple departments, and build collaborative research connections across academic units.[18]

Drivers of Interdisciplinarity

The drivers of IAS impact users' experiences and the development of the literature. The histories of modern IAS fields are far too varied to permit a complete list of their origin patterns. For example, those of the natural sciences alone include subsets of, variations on, and additions to the patterns explicated below.[19] Nevertheless, these three broad patterns directly influence the information ecologies within which IAS scholars must function[12] and, therefore, merit delineation.

In many IAS fields, isolates built and continue to develop their own academic infrastructure around their own discipline because their original "home-discipline" colleagues deemed their work too activist, unimportant, or offensive for support. Those who hold political and economic power could, and do, simply deny tenure, refuse research support, and reject curricular change. The growth of women's studies exemplifies the isolates' drive for disciplinary autonomy. The work of Florence Howe and Catharine Stimpson, among many others, eventually resulted in the development of women's studies courses, degrees, and graduate programs as well as such infrastructure elements as the Feminist Press and journals such as *Signs*, *Feminist Studies*, and *Women's Studies*.[20] These scholars construct their own publications streams, academic organizations, and networks within, across, and beyond campuses.

Isolate driven fields often live with a certain degree of tension between scholarship and activism that adds a particular layer of complexity to their entire information stream. As new materials are created, published, and indexed both within and outside of academia on a wide range of issues, such fields must "rescue" works which have been ignored as unimportant, archive the nonacademic results of social change, and otherwise integrate social outcomes with scholarly endeavors. In Black Studies, for example, the intellectual output of both scholarship and activism is used by individual adherents of both approaches.[21]

In contrast with the isolates approach, most area studies are built on a broad but robustly integrated, sociopolitical approach that is rooted in the nexus between culture and history. Although politically unpopular at times and/or driven by governmental needs relating to foreign policy (for example, Middle Eastern Studies), area studies focus on the cultural region as a whole. For example, American Studies grew from the pre-World War II development of a widely held conceptualization of U.S. values, traditions, and ideals. Nourished by such books as Ruth Benedict's *Patterns of Culture* (1934) and championed by such intellectually pluralistic individuals as President Roosevelt's vibrant Librarian of Congress, Archibald MacLeish, the American Studies movement, from its inception, genuinely involved art, history, literature, journalism, sociology, and political science.[22] These scholars redeploy, reconfigure, and redevelop existing infrastructure to meet new demands; they work from an intellectual space which is more within than apart from the academic norm.

More recently, deliberately developed interdisciplinary research groups and institutes have formed around broad research areas (for example, biomedicine), social/physical problems (for example, global warming), or technology (for example, the Internet or the Advanced Photon Source at the Argonne National Laboratory) with the intention of weaving together an effectively chosen array of disciplines. For example, the Rockefeller Institute/University, with its 23 Nobel laureates, 5 McArthur fellows, and 12 National Medal of Science winners, produces stellar biomedical research without utilizing a traditional academic disciplinary organizational structure.[23] Scholars in this type of intentionally constructed, interdisciplinary organization utilize existing publication and indexing structures but augment them through the development of their own networks for both formal and informal information dissemination.

All of these approaches share a critical, common element, i.e., a focus on the research questions. As Karl Popper noted 45 years ago, "We are not students of some subject matter, but students of problems. And problems may cut right across the borders of any subject matter or discipline."[24] Scholars, with few exceptions, focus on their work rather than the information structures supporting that work. To a certain extent, IAS scholars have had to build their own information dissemination mechanisms (for example, new journals and monograph series in women's studies), construct their own information networks (for example, conferences and associations in American Studies), and design their own information access systems (for example, data archives and clearinghouses in biomedicine). These actions provide a degree of self-definition and demarcate arenas of intellectual responsibility. Library and information studies (LIS) scholars and practitioners must take those factors into account when building support structures to help IAS scholars find the information they need.

ISSUES IN THE STRUCTURE AND FINDING OF INFORMATION

Six issues repeat in the numerous analyses of IAS information structures, activities, and concerns. They are publication formats, primary sources and gray literature, online searching, library services and collections, informal information systems, and knowledge-domain maintenance. Obviously they overlap and intersect to some extent but each issue provides a valuable perspective on IAS work.

Publication Formats

As IAS fields coalesce into established programs with all the support structures of academia,[25] their professional associations and university presses begin to produce journals, monographs, conference proceedings, and book series in support of their scholarship. Journals may be focused on the field directly (for example, *Internet Mathematics* and *Neuropsychopharmacology*) while others will deliberately include connections to the latest IAS research

within the publications of sister disciplines (for example, *Cell*'s regular section highlighting materials recently published in interdisciplinary journals such as *Neuron* and *Immunity*).[26]

Gray literature, discussed more fully below, is defined as research and professional materials published outside of mainstream formats and generally overlooked by indexing tools. It is often integral to the development of interdisciplinary fields, particularly in the early stages of their development. In some underfunded fields, gray literature remains crucial and actively calls on IAS scholars' time. Archaeology, for example, produces reams of primary material that is intended to support additional studies as well as reports that are meant to, in part, inform the public. Much of it, however, is in the form of gray literature found through Web-accessed databases that are voluntarily populated by scholars.[27]

All publication formats are embroiled in the broader concerns of information-technology development. For those nascent and less well funded IAS fields, the lure of information technology is complex. In archaeology, for example, moving decades of paper records to a digital format is alluring, but migration is extremely expensive for the original effort, much less any subsequently required migration.[29] Digital libraries (for example, Archnet) and databases (for example, the National Archaeological Database) develop from professional organizations and subgroups taking on the responsibility of utilizing information technology, but the autonomy of scholars combines with their focus on producing, rather than disseminating, research to prevent comprehensive digitization programs. The Internet's ability to permit scholarly communities to define their own boundaries has long been recognized as a significant support for IAS communities[28] but significant work in building on that support still requires substantial organization, connectivity, and resources.

Primary Sources and Gray Literature

In addition to the general publication format issues already highlighted, IAS scholars face particularly complex problems in their access to primary sources and gray literature. The barriers to creation, publication, indexing, and collecting are profound. The solutions, often strengthened by user-controlled information technology, are neither comprehensive nor standardized.

Certain materials are simply not kept because they are not recognized as worth keeping by those with the power to do so.[29] Original writing, art, music, archives, and instructional materials are, therefore, lost and not always susceptible to rescue. Again, women's studies provide concrete examples of voices that are not privileged, such as Florence Howe's 1971 effort to teach a course on "Women and Identity" at Goucher, which was rejected as "offensive."[30]

Many important collections have been developed locally, particularly in women's studies, but they often lack substantive off-site access.[31] Creating digital libraries of these primary source materials is far too costly and labor-intensive for most IAS fields to support on the basis of internal infrastructure and funding. Even the simple effort to create online finding aids for primary source material and gray literature, including archives, is beyond the scope of possibility. Historians, as well as others, still want the type of item-level record that is not really feasible in most archives.[32] In fields where much of the gray literature has moved to the Internet, such as social policy research, the access mechanisms have not kept pace, which has led to the paradoxical outcome that users have difficulty in culling out the unwanted material.[33]

Nevertheless, strategies and techniques to maximize access continue to develop. Organizations, libraries, and even individuals have long been gathering oral histories, letters, and other unconventional documents as a way to provide voice to the unheard, dismissed, and marginalized of society and academia.[32] Designing electronic portals to institutional, shared, and library-managed repositories of otherwise unpublished material on a global scale requires substantial will, cooperation, and effort but has been explored in various areas including the problem-based arena of geology's work on karst formations[34] and the global environmental change community.[35] Constructing the controlled vocabulary that best facilitates access to these collections adds a layer of complexity but can also involve scholars in reflective monitoring of their field's conceptual developments.[36] Electronic finding aids available on the Web and constructed by librarians who know both their collections and their users are increasingly used by historians.[35]

Online Searching

The three interconnected problems in online searching for IAS information are vocabulary failures, inadequate indexing, and multi-database use. Consistently applied and universally understood terminology tends to be more sought than available. Terms do not map across disciplines[37] or even across the databases within a single discipline.[38] For example, the "elusive" nature of terminology in anthropology combines with the wide range of its subject matter to make information seeking particularly difficult. Post-coordinate indexing, like keyword searching, gives the user control over conceptualizing connections among ideas and approaches in an IAS field[39] but it also requires more thought, self-awareness, and skill than would an effectively pre-coordinated controlled vocabulary. Even apart from disciplinary jargon, words such as "culture" carry very different meanings in various disciplines so that a kind of intellectual "translation" is needed to actually mine the databases of disparate fields. IAS scholars often find the adaptation to multiple intellectual vocabularies quite problematic.[40] The advent of keyword searching on the natural language used by authors in the titles, abstracts, and even full text of their works opens the floodgates of information access but provides none of the mapping or translation required to hone a search. Reliance on keyword alone overlooks the difficulties inherent in navigating multiple perspectives using unstable language. IAS scholars must, therefore, develop deep online search skills that enable them to use both controlled vocabulary and natural language in databases and search engines.

Even well-developed skills are useless, however, if they cannot be applied. Search skills can only be applied when indexes are available and when those indexes cover the needed material. The cost of indexing journals, conference proceedings, collected works, and monographic series is substantive, and commercial companies follow the market, investing in that effort only after the market demand makes their effort financially viable. Particularly in the newer IAS fields, problems with indexing are common so that finding the best research in specific journals is not always easy. In women's studies, for example, the majority of the journals were inadequately indexed over a decade ago,[41] a situation that is neither uncommon nor readily resolved although a great number of new resources are available including some which are free.[42] In both economics and psychology coverage of the many subdisciplines within these interdisciplinary fields varies widely.[43] Of 11 indexes covering quaternary studies, relevancy rankings on broad searches ranged from 49% to 99%.[44]

Between language problems and indexing inadequacy, most IAS scholars need to search several databases efficiently.[45] Using a single database is almost always insufficient for a comprehensive IAS search[46] but using multiple databases requires mastering so many search skills and techniques that this, too, poses significant problems. In social work, for example, both *Social Work Abstracts* and *Social Service Abstracts* are required for a substantive search since these two indexes complement each other, rather than overlap.[47] In later-life migration studies there is so little overlap among the 12 most relevant databases, both disciplinary and multidisciplinary, that scholars need to search all of them when recall is paramount.[48]

Fortunately, both database availability and search options are improving for IAS scholars. New databases are developing to address specific fields and to deliberately include an array of fields,[50] although their use is not uniformly productive for all users.[49] Citation searching, a favorite information seeking technique for people in many social science and humanities-rooted IAS fields, is increasingly available via ISI Web of Knowledge[SM].[32] Meta-database searching[32,50] is available via vendors' standardized interfaces that allow end users to search several databases simultaneously. Although still developing,

Google Scholar already covers 27% more core articles in one IAS field than the next most nearly complete database, Social Sciences Citation Index.[51]

More specialized refinements are still limited but their potential is encouraging. Visual search interfaces, although still rudimentary, show promise of particular interest to IAS scholars who must search multidisciplinary databases. The ability to both identify and refine subtopics may be enhanced by the visual search mode.[52] An ontological approach to mapping knowledge domains can support scholars' exploration of concepts, theories, research questions, and relationships with, perhaps, greater effectiveness than a subject approach to controlled vocabulary.[34] Similarly, the social network approach of tagging provides a user-driven avenue for identifying worthwhile information; such control might be particularly useful in developing and/or rapidly changing IAS fields.

Using Library Services and Collections

As a natural by-product of their effort to locate, integrate, and manage information from multiple disciplinary perspectives, IAS scholars have more reference questions than do single-discipline scholars, as has long been understood.[13] In addition to more numerous questions, IAS scholars need holistic, continuous reference service that is integrated into their research streams. Librarians should actively partner with these researchers, helping them identify connections, language, viable resources, authoritative sources, and accessible experts with whom to network.[53] Developing the detailed understanding of IAS scholars' work that makes such in-depth service effective can be developed by bibliometrically mapping the intellectual environment in which the scholars work.[54]

In addition to vigorous reference services, libraries provide rich collections of valued information in multiple formats. Overly narrow libraries undercut IAS connections to sister disciplines[55,56] but even the most open collection management policies are stymied by practicalities of the IAS information stream. Frequently monographic print-runs are small so that scholars, as well as librarians, note that materials are out of print quickly.[57] Some IAS clusters are so small and informal that their needs can be unmet by efforts to meet the needs of the majority.[58] Establishing purchasing structures which deliberately account for the needs of IAS research through development of funding lines and procedures is essential. For example, more than half of the monographs reviewed in major history journals are actually classed outside of "history" in the LC system.[59]

The serials crisis has long made adding new titles, without cutting old ones, almost impossible, but IAS fields commonly generate and require new journals in which to publish the research that is not acknowledged or valued in scholars' "home" disciplines. Some libraries lack budget lines, procedures, approval plans, and staff who are properly trained for building print and digital collections to support IAS fields[60] although systems for properly identifying core IAS journals offer some support.[61,62] In racial and ethnic studies libraries frequently lack the depth of journal collections that most fully support a developing program; for example, one study indicates that less than a third of Association of Research Libraries collections include half of the Latino research titles.[63]

In addition to standard collection management and reference service concerns, the large number of IAS fields that use government documents requires government document librarians to review their priorities and patterns. Integrating government documents collections into the main collections exemplifies the extreme of collection management and service concerns. Librarians who merge collections, properly bolstered by reference and information literacy services, must balance the potential for greater access via browsing and searching[64] with the potential loss of access due to variable formats and heavy reliance on the most complex of corporate author systems. As government publications continue to migrate to electronic format, accessibility may increase use in some IAS fields.[65]

As already noted, classification and subject-heading choices privilege certain perspectives in library catalogs. Perhaps the most obvious connection between the librarian's codification and the discipline's self-definition lies in Melville Dewey's insistence that the nascent field of "Home Economics" be classed as a "Useful Art" (600s) rather than a form of Sociology (300s).[66] Efforts to correct, update, expand, and refine Dewey, LC, and other organizational systems to enhance access across the range of IAS fields are long-standing. Examples of such efforts include the following: *A Women's Thesaurus*,[67] *Women in LC's Terms*,[68] an empirical analysis of subject headings on women's studies with amelioration recommendations,[69] an entire IAS adaptation process for classifications,[70] and NASA's Global Change Mastery Directory.[71]

Informal Information Systems

In all of academia, the informal information systems embedded in social networks, conference structures, and collegial connections are often more a part of the scholar's daily information life than is the library. In IAS fields, however, the disciplinary influences on language mentioned earlier play a role in the development and efficacy of these informal information systems. Interpersonal communications, regardless of channel or context, require some means of reaching a shared understanding of terminology. "Fundamental terminology should be established early and reviewed regularly. This may mean that the minority chooses to speak the majority tongue."[72] Time to learn how to communicate across disciplines, attitudes of willingness to engage in developing a shared understanding, and the real and/or virtual proximity needed to make regular contact convenient are essential for deep collaborations of a genuinely interdisciplinary nature.[73]

Evidence of these patterns is found in the norms of interdisciplinary research centers, which have been shown to foster interpersonal connections across disciplines more than within disciplines. Indeed, one study determined that an average of 84% of these connections develop after individuals join the center. Although center directors and select senior faculty act as nodes for cross-disciplinary communication, the greatest range of connections to different disciplines is found among graduate students and postdoctorates who actually build the links among faculty.[74]

In both social science and natural science interdisciplinary, research collaboratories, the informal information networks among the scholars yield more than simple factual information. Exchanges are productive in terms of learning new processes, research methods, and technology skills, as well as in terms of generating new ideas.[75,76]

The proliferation of discussion lists supports the informal seeking and exchange of information as well as the general discussion of ideas; small groups of people, however, tend to be most active in these endeavors.[77] Some disciplines, for example, music, utilize unstructured, interpersonal connections but prefer in-person contact to electronic channels.[78] The development of Web-based communities (WBCs) supports both IAS communication and development of customized resources. Although the lack of research on WBC generates more questions than answers, the potential for active, global information support of an informal nature is well worth further examination.[79]

Knowledge Domain Maintenance

As might be expected within these parameters, many IAS scholars have long had to work extremely hard to both construct and maintain their domain knowledge. The difficulties of keeping up with the growth in their fields are part of their academic life.[80] Engineers, for example report that the fragmentation of their literature and the sheer volume of publications combine to create a sense of information overload, generating the conviction that they are not finding all they need for their work.[81] Obviously various current-awareness tools and alerting services, such as *Current Contents*, can be helpful; one study noted that about two-thirds of faculty in natural resources made use of them.[82] In women's studies, one exemplary complaint of scholars is that keeping up "in the way that discipline-specific folks do is impossible" and those who do gather material effectively may well echo their colleague's statement that "I don't have a sense how to organize the materials...."[67]

One aspect of this continuous effort to gather useful information is the increasingly common problem of discerning quality. Surviving the academic promotion and tenure process is vital for most humanities and social science scholars who want to continue their work, as well as those natural science scholars who want to work within academic rather than governmental or corporate environments. As IAS fields mature, recognized and generally acknowledged standards of publication quality develop, which support consistent application of promotion and tenure requirements.[83] In nascent or rapidly changing fields, however, those widely accepted benchmarks are lacking, and this adds another layer of uncertainty to the information seeking. The ability to rapidly recognize indicators of quality scholarship is a hallmark of the expert but even expertise in a new IAS field may be insufficient to support efficient analysis across the multiple channels and formats of information available.

DISCIPLINARY OVERVIEW

Within the context of all six of the information concerns delineated above, IAS scholars share, to some degree, a primary information-seeking trait, i.e., an open-minded approach to work in other disciplines. This willingness to consider other academic perspectives, theories, language, assumptions, research, and applications colors their information experiences. In the IAS fields of the humanities, social sciences, and natural sciences, that open-minded approach leads to serendipitous information encounters,[84] the use of multiple disciplinary vocabularies, a drive to maintain currency in multiple areas, and a strong need for effective information management skills.[85] Within that overarching context, an examination of information experiences more common to each broad arena is presented below.

Humanities

In the humanities, scholars still prefer "footnote chasing, colleague networks and their own files over online systems" with the lack of retrospective access still a significant problem. Working independently, these researchers build their own personal collections through browsing and expect to travel in order to access distant collections. Making use of personal bibliographic control software and image-capturing tools, some scholars build their own databases and collections which reflect their own analytic perspective, for example, chronological or thematic.[86]

Analysis of several IAS fields within the humanities reveals more focused concerns. Medieval studies, for example, includes women's studies, art history, archaeology, and sociology, as well as a growing focus on history. Nevertheless, the International Medieval Congress papers remain conceptually housed in literature[87] as do monographic publications in cultural studies.[88] Databases of visual collections remain clumsy and counterintuitive for art historians so that the vast majority of their work is still done in print formats.[89,90]

Unfortunately, scholars with roots in the humanities generally lag behind those of the physical and social sciences in development of Internet and database search skills. Those who most need effective databases are the ones least likely to have them. Almost 86% of music scholars make use of materials in "criticism, literature, art history, philosophy, anthropology, or psychology" with informal information seeking among colleagues a quite common part of the research process.[91]

Social Sciences

A reliance on journals, informal networking with colleagues, and citation-tracing is still the hallmark of the social sciences[92] although an increasing number of substantive databases support keyword searching across disciplines. The widely accepted information-seeking model of social scientists applies to many of the IAS scholars with its emphasis on six basic patterns (starting, chaining, browsing, differentiating, monitoring, and extracting)[93] but the four additional patterns employed by scholars studying stateless nations (accessing, networking, verifying, and information managing)[94] are of particular interest.

In the social sciences, foci often integrate in clusters of sister disciplines. Demography, for example, clusters more closely around the triad of sociology, family studies, and economics than it does around its next most common cluster of public health and medicine.[95] Interdisciplinarity is more of a continuum than a pervasive characteristic in some social science arenas; for example, those historians who take a problem-centered approach utilize works from sister disciplines far more frequently than do more traditional historians.[96]

Access to these clusters of information is found in various journals covered by both discipline-specific indexes (for example, for history, America, History, and Life) and discipline-spanning indexes (for example, for public administration, Expanded Academic Index).[97] Full-text journals are in demand but not always readily available (for example, European archaeologists have a few online publications but none provide substantial coverage of work on American sites).[98] Personal libraries remain the primary information source for faculty in social work/family studies, who read their few online publications on more of a monthly basis. Younger faculty, who are, presumably, still building their personal libraries, make more extensive use of interlibrary loan than do senior faculty.[99]

A few areas have been studied in sufficient depth to permit more focused patterns to emerge. The advent and development of electronic resources have moved scholars in history to a heavier use of indexes but have not fundamentally altered their basic practices of using print materials, following citations closely, browsing, and using book reviews.[100] Their reliance on archival material requires them to orient their information-seeking to the structure of each separate archive, including the appropriate finding aids and collections, before building the contextual knowledge that allows them to identify relevant material.[101] Anthropologists rely on journals, visual materials (for example, maps), their own personal data and libraries, and interlibrary loan.[102] In women's studies, faculty value librarian support in the early stages of their literature review process, need detailed access to primary sources, find indexes and monograph collections inadequate, struggle to keep current on their research issues from multiple fields, and want to work with well-educated, enthusiastic librarians who evaluate Internet resources rigorously.[63,103]

In its relationship to other social science IAS fields, LIS is uniquely situated in that it is itself both interdisciplinary and information-seeking. LIS produces resources that are used by other fields seeking information, and it actively connects to other interdisciplinary, information-seeking fields.[104] LIS both studies and exists within an environment of interdisciplinarity. Widely integrated, LIS has been established as an interdisciplinary field for the past 40 years with its boundaries continuing to broaden.[105] One study found that almost half of its cites came from outside LIS while only 18% of computer science cites did so.[106]

Natural Sciences

In the natural sciences, interdisciplinarity is more embedded and extensive than might, at first, appear. Although cross-disciplinary citations are decreasing in the last 20 years for the biological sciences, they do continue there and are actively increasing in the areas of agriculture[107] and chemistry.[108] Chemistry cites non-chemistry journals almost half the time.[109] Bradford's law of scatter describes the dispersion of forestry journal citations in which 50% of the citations come from 35 journals and the other 50% from 1234 journals.[110]

In one sense, IAS scholars working in the natural sciences are prone to even greater vocabulary problems than their counterparts in the humanities and social sciences. Terminology, while often precisely defined, still varies substantially in disciplinary contexts; for example, "gene" had five different meanings in one conversation.[111] While most researchers expect to come to a shared understanding of broad terms such as "culture," fewer would expect that "gene" requires similar attention.

Journal literature is far more commonly used than any other format but not simply because it contains more recent materials than do monographs. In neuroscience, for example, over 80% of citations are to journals but 30% of those journals are 15 years old or older.[112] The fundamentals of the field are established enough to make older material of use. Multidisciplinary science journals are heavily cited in support of new work.[113]

In geography, the focus on more scientific aspects of the field involves a greater use of journals while books are used in those areas of geography that focus more on the social aspects of the field.[114] Initial research indicates that almost half of geography publications entail at least two different disciplines.[115] Using approval plans to effectively collect appropriate works in geography requires substantial analysis but can be worthwhile.[116]

In the less generously funded fields, such as archaeology, comprehensive databases remain elusive[29] although efforts to consolidate local systems continue. For example, the ARGOS multiunion catalog of 14 libraries in 10 languages[117] was not fully successful but later developed into Ambrosia, a more functional catalog of two libraries in two languages.[118]

The inherent tension between problem-centered researchers and structure-centered administrators poses information access problems for scientists who construct, or try to construct, their working environments on the basis of integration rather than administration. Many must work in teams, making their reliance on these administrative structures more critical than that of the more independent scholars in the social sciences and humanities. Administrative support for interdisciplinary research involves administrators in data archiving, organization, mining, and maintenance.

Although long-recognized as early adapters to electronic information resources, scientists in interdisciplinary fields can be somewhat unskilled and even naïve in using Google, databases, and the Internet in general. Not only do they have little sense of the differences (for example, comprehensiveness and search options) between Google and databases, but they are not always well informed about the authority of the resulting documents.[119]

CONCLUSION

Research on interdisciplinary studies integrates the theories, methods, perspectives, and basic concepts of two or more disciplines to create a body of scholarship that moves beyond the boundaries of any single discipline. The resultant information ecology requires scholars and librarians to maximize access to journals, gray literature, and primary sources. Problems with variable terminology, inadequate indexing, online searching, thin research collections, and subject access support structures are part of the landscape. Information technology provides both the risk of costly retrospective conversions and the promise of enhancing informal information networks. Digital libraries, Google scholar, multidisciplinary databases, and digital archives are developing throughout IAS fields although

the pace of the growth is directly connected to the funding available in various fields or to the possibility of fiscal feasibility.

In the coming generation of IAS scholarship it should be possible to address several research questions in some depth. The varying nature of information support structures in IAS fields involves inherent tensions in the more common adaptations to information technology. Fields in which individuals produce heavy loads of original data (for example, archaeology) and fields which are still struggling to find baseline funding in academia (for example, women's studies) benefit from the grassroots autonomy provided by the Internet, but they have little experience in the information architecture and usability work needed to maximize that freedom. Research is still needed on the best practices, adaptation techniques, informatics, and usability criteria of IAS fields.[32] Digital library and archive development, as well as nurtured growth of WBC, requires carefully planned design[120] to be followed by evaluation.

From the perspective of information retrieval, an LIS-centric analysis of IAS fields provides raw data that can be used to better construct databases, communication infrastructure, gray literature repositories, and focused digital libraries. The evolving nature of IAS, influenced by academic power structures and external funding source priorities,[12] provides a moving target for such analysis at the microlevel. The analytic lens might be directed toward homogeneities of approach to knowledge generation.[121] For example, while characterizing the information experiences and preferences of Hispanic Studies scholars provides certain insights, a more productive path might be derived from analysis of ethnic/racial studies communities and biomedical studies communities. What do those research approaches and environments have in common and what information architecture designs best meet their needs?

Concurrently, however, additional research is needed on the most effective means of supporting IAS scholarship in terms of directly addressing the disciplinary issues and opportunities available to these scholars who deliberately embrace research in multiple fields. The IAS scholars' mental models of academic information-seeking certainly differ from those scholars whose work carefully stays within limited disciplinary boundaries. How do controlled vocabularies and indexing systems best support the intellectual exploration of multiple perspectives? Parsing large concepts more finely, for example, may lead to more effective searching in gender studies.[122] Using Bradford's work on scatter, libraries might maximize serials access to support needs across disciplines by identifying indexes which serve as nexus points among disciplinary views.[123] Digital libraries have the potential to serve as creative, information-rich, highly connective, and dynamic, scholarly spaces in which the complex needs of IAS scholars can be actively anticipated.[124]

The growing meta-awareness of IAS approaches continues to generate deliberate efforts to integrate widely different disciplinary methods, theories, and principles. The relationship, for example, between art and nanotechnology has been proposed as a new approach to consciousness.[125] In terms of knowledge discovery and theory generation, what information infrastructures are most supportive? As new IAS fields are created, how can LIS best anticipate information needs and become integrated into the information resource development processes?

IAS scholarship addresses the living framework of intellectual life in a global, complex, and problematic world. Bringing to bear the most appropriate tools, these fields address social, technological, scientific, and human concerns at the point of need. LIS is uniquely positioned both within and in support of IAS. The opportunity is also a responsibility.

REFERENCES

1. Klein, J. *Humanities, Culture, and Interdisciplinarity*; State University of New York Press: Albany, NY, 2005.
2. Committee on Facilitating Interdisciplinary Research, Committee on Science, Engineering, and Public Policy, National Academy of Sciences, National Academy of Engineering, and Institute of Medicine; *Facilitating Interdisciplinary Research*; The National Academies Press: Washington, DC, 2005.
3. Klein, J. *Interdisciplinarity: History, Theory, and Practice*; Wayne State University Press: Detroit, MI, 1990.
4. Klein, J. Interdisciplinary needs: The current context. Libr. Trends **1996**, *45* (2), 134–154.
5. Palmer, C. *Work at the Boundaries of Science: Information and the Interdisciplinary Research Process*; Kluwer Academic Publishers: Boston, MA, 2001.
6. Gieryn, T. Boundary-work and the demarcation of science from non-science. Am. Sociol. Rev. **1983**, *48*, 781–795.
7. Fisher, D. Boundary work and science. In *Theories of Science in Society*; Cozzens, S., Gieryn, T., Eds.; Indiana University Press: Bloomington, IN, 1990; 98–119.
8. Pierce, S. Boundary crossing in research literatures as a means of interdisciplinary information transfer. J. Am. Soc. Inform. Sci. Technol. **1999**, *50* (3), 271–279.
9. Moran, J. *Interdisciplinarity*; Routledge: London, 2002.
10. Palmer, C. Structures and strategies of interdisciplinary science. J. Am. Soc. Inform. Sci. **1999a**, *50* (3), 242–253.
11. Klein, J. *Crossing Boundaries Knowledge, Disciplinarities, and Interdisciplinarities*; University Press of Virginia: Charlottesville, VA, 1996a.
12. Mote, L. Reasons for the variations in the information needs of scientists. J. Doc. **1962**, *18* (4), 160–175.
13. Hurd, J. Interdisciplinary research in the sciences: implications for library organization. Coll. Res. Libr. **1992**, *53* (4), 283–297.
14. Pikoff, H. Improving access to new interdisciplinary materials. Libr. Resour. Tech. Ser. **1991**, *35* (2), 141–147.
15. Leydesdorff, L. Betweenness centrality as an indicator of the interdisciplinarity of scientific journals. J. Am. Soc. Inform. Sci. Technol. **2007**, *58* (9), 1303–1319.
16. Leydesdorff, L.; Schank, T. Dynamic animations of journal maps: Indicators of structural changes and interdisciplinary developments. J. Am. Soc. Inform. Sci. Technol. **2008**, *59* (11), 1810–1818.
17. Crane, D. *Invisible Colleges: Diffusion of Knowledge in Scientific Communities*; University of Chicago Press: Chicago, IL, 1972; 172.
18. Palmer, C. Structures and strategies of interdisciplinary science. J. Am. Soc. Inform. Sci. **1999a**, *50* (3), 243.
19. Palmer, C. *Work at the Boundaries of Science: Information and the Interdisciplinary Research Process*; Kluwer Academic Publishers: Boston, MA, 2001; Chap. 3.
20. Rosenberg, R. Women in the humanities: Taking their place. In *The Humanities and the Dynamics of Inclusion since World War II*; Hollinger, D., Ed.; Johns Hopkins University Press: Baltimore, MD, 2006; 256–258.
21. Weissinger, T. Black studies scholarly communication: A citation analysis of periodical literature. Collect. Manag. **2002**, *27* (3/4), 45–56.
22. Zenderland, L. Constructing American studies: Culture, identity, and expansion of the humanities. In *The Humanities and the Dynamics of Inclusion since World War II*; Hollinger, D., Ed.; Johns Hopkins University Press: Baltimore, MD, 2006; 276–278.
23. Committee on Facilitating Interdisciplinary Research, Committee on Science, Engineering, and Public Policy, National Academy of Sciences, National Academy of Engineering, and Institute of Medicine; *Facilitating Interdisciplinary Research*; The National Academies Press: Washington, DC, 2005; 35–38; 176.
24. Popper, K. *Conjectures and Refutations: The Growth of Scientific Knowledge*; Routledge and Kegan Paul: New York, 1963; 88.
25. Abbott, A. *System of Professions*; University of Chicago: Chicago, IL, 1988.
26. Committee on Facilitating Interdisciplinary Research, Committee on Science, Engineering, and Public Policy, National Academy of Sciences, National Academy of Engineering, and Institute of Medicine; *Facilitating Interdisciplinary Research*; The National Academies Press: Washington, DC, 2005; 140.
27. Seely, A. Digging up archeological information. Behav. Soc. Sci. Libr. **2005**, *24* (1), 1–20.
28. Clark, P. Disciplinary structures on the Internet. Libr. Trends **1996**, *45* (2), 226–238.
29. Bates, M. Learning about the information seeking of interdisciplinary scholars and students. Libr. Trends **1996**, *45* (2), 155–164.
30. Rosenberg, R. Women in the humanities: Taking their place. In *The Humanities and the Dynamics of Inclusion since World War II*; Hollinger, D., Ed.; Johns Hopkins University Press: Baltimore, MD, 2006; 257.
31. Denda, K. Fugitive literature in the cross hairs: An examination of bibliographic control and access. Collect. Manag. **2002**, *27* (2), 75–86.
32. Anderson, I. Are you being served? Historians and the search for primary sources. Archivaria **2004**, *58* (Fall), 81–116.

33. Hartman, K. Social policy resources for social work: Grey literature and the Internet. Behav. Soc. Sci. Libr. **2006**, *25* (1), 9.

34. Chavez, T.; Perrault, A.; Reehling, P.; Crummett, C. The impact of grey literature in advancing global karst research: An information needs assessment for a globally distributed interdisciplinary community. Grey J. **2007**, *3* (3), 126–137.

35. Downs, R.; Chen, R. Cooperative design, development, and management of interdisciplinary data to support the global environmental change research community. Sci. Technol. Libr. **2003**, *23* (4), 5–19.

36. Chavez, T.; Perrault, A.; Reehling, P.; Crummett, C. The impact of grey literature in advancing global karst research: An information needs assessment for a globally distributed interdisciplinary community. Grey J. **2007**, *3* (3), 131–132.

37. Weisgerber, D. Interdisciplinary searching: Problems and suggested remedies. J. Doc. **1993**, *49* (3), 241–244.

38. Smith, L. Systematic searching of abstracts and indexes in interdisciplinary areas. J. Am. Soc. Inform. Sci. **1974**, *25* (5), 343–353.

39. Kotter, W. Improving subject access in anthropology. Behav. Soc. Sci. Libr. **2002**, *20* (2), 1–4.

40. Spanner, D. Border crossings: Understanding the cultural and informational dilemmas of interdisciplinary scholars. J. Acad. Libr. **2001**, *27* (5), 354, 355–356.

41. Gerhard, K.; Jacobson, T.; Williamson, S. Indexing adequacy and interdisciplinary journals: The case of Women's Studies. Coll. Res. Libr. **1993**, *54* (2), 125.

42. Dickstein, R.; Hovendick, K. Women's studies databases. In *Encyclopedia of Library and Information Science*; Marcel Dekker: New York, 2004; 409–418.

43. Frandsen, T.; Nicolaisen, J. Interdisciplinary differences in database coverage and the consequences for bibliometric research. J. Am. Soc. Inform. Sci. Technol. **2008**, *59* (10), 1577.

44. Joseph, L. Comparison of retrieval performance of eleven online indexes containing information related to quaternary research, an interdisciplinary science. Ref. User Serv. Q. **2007**, *47* (1), 60.

45. Weisgerber, D. Interdisciplinary searching: Problems and suggested remedies. J. Doc. **1993**, *49* (3), 231–254.

46. Joseph, L. Comparison of retrieval performance of eleven online indexes containing information related to quaternary research, an interdisciplinary science. Ref. User Serv. Q. **2007**, *47* (1), 58.

47. Flatley, R.; Lilla, R.; Widner, J. Choosing a database for Social Work: A comparison of Social Work Abstracts and Social Service Abstracts. J. Acad. Libr. **2007**, *33* (1), 55.

48. Walters, W.; Wilder, E. Bibliographic index coverage of a multidisciplinary field. J. Am. Soc. Inform. Sci. Technol. **2003**, *54* (14), 1305–1312.

49. Fister, B.; Gilbert, J.; Fry, A. Aggregated interdisciplinary databases and the needs of undergraduate researchers. Portal **2008**, *8* (3), 273–292.

50. Bates, M.; Wilde, D.; Siegfried, S. Research practices of humanities scholars in an online environment: The Getty Online Search Project report no. 3. Libr. Inform. Sci. Res. **1995**, *17* (Winter), 5–40.

51. Walters, W. Google Scholar coverage of a multidisciplinary field. Inform. Process. Manag. **2007**, *43* (4), 1121.

52. Fagan, J. Usability testing of a large, multidisciplinary library database. Inform. Technol. Libr. **2006**, *25* (3), 148.

53. Westbrook, L. Information needs and experiences of scholars in women's studies: Problems and solutions. Coll. Res. Libr. **2003**, *64* (3), 192–209.

54. Dilevko, J.; Dali, K. Improving collection development and reference services for interdisciplinary fields through analysis of citation patterns: An example using tourism studies. Coll. Res. Libr. **2004**, *65* (3), 234.

55. Hurd, J. Interdisciplinary research in the sciences: Implications for library organization. Coll. Res. Libr. **1992a**, *53* (4), 295.

56. Hurd, J. The future of university science and technology libraries: Implications of increasing interdisciplinarity. Sci. Technol. Libr. **1992b**, *13* (Fall), 29.

57. Westbrook, L. Information needs and experiences of scholars in women's studies: Problems and solutions. Coll. Res. Libr. **2003**, *64* (3), 198.

58. Allen, B.; Sutton, B. Exploring the intellectual organization of an interdisciplinary research institute. Coll. Res. Libr. **1993**, *54* (6), 513.

59. Hickey, D.; Arlen, S. Falling through the cracks: Just how much "history" is history? Libr. Collect. Acquis. Tech. Serv. **2002**, *26*, 97.

60. Gerhard, K. Challenges in electronic collection building in interdisciplinary studies. Collect. Manag. **2000**, *25* (1/2), 51–65.

61. Kushkowski, J.; Gerhard, K.; Dobson, C. A method for building core journal lists in interdisciplinary subject areas. J. Doc. **1998**, *54* (4), 477–488.

62. Kelsey, P.; Diamond, T. Establishing a core list of journals for forestry. Coll. Res. Libr. **2003**, *64* (September), 357–377.

63. García, S.V. Racial and ethnic diversity in academic library collections: Ownership an access of African American and U.S. Latino periodical literature. J. Acad. Libr. **2000**, *26* (September), 314.

64. Cheney, D. Government information collections and services in the social sciences: The subject specialist integration model. J. Acad. Libr. **2006**, *32* (3), 303–312.

65. Hogenboom, K. Has government information on the Internet affected citation patterns? A case study of population studies journals. J. Gov. Inform. **2003**, *29* (6), 392–401.

66. Fields, A.; Connell, T. Classification and the definition of a discipline: The Dewey Decimal Classification and Home Economics. Libr. Culture **2004**, *39* (3), 245.

67. Capek, M., Ed. *A Women's Thesaurus: An Index of Language Used to Describe and Locate Information By and About Women*; Harper & Row: New York, 1987.

68. Dickstein, R.; Mills, V.; Waite, E., Eds. *Women in LC's Terms: A Thesaurus of Library of Congress Subject Headings Relating to Women*; Oryx Press: Phoenix, AZ, 1988.

69. Gerhard, K.; Su, M.; Rubens, C. An empirical examination of subject headings for women's studies core materials. Coll. Res. Libr. **1998**, *59* (2), 129–137.

70. Kublik, A.; Clevette, V.; Ward, D.; Olson, H. Adapting dominant classifications to particular contexts. Cataloging Classif. Q. **2003**, *37* (1/2), 13–31.

71. Major, G. Beyond bibliography: A dynamic approach to the cataloging of multidisciplinary environmental data for global change research. Sci. Technol. Libr. **2003**, *23* (4), 21–36.

72. Epstein, S. Making interdisciplinary collaboration work. In *Interdisciplinary Collaboration: An Emerging Cognitive Science*; Derry, S., Schunn, C., Gernsbacher, M., Eds.; Lawrence Erlbaum Associates, Publishers: Mahwah, NJ, 2005; 249.

73. Epstein, S. Making interdisciplinary collaboration work. In *Interdisciplinary Collaboration: An Emerging Cognitive Science*; Derry, S., Schunn, C., Gernsbacher, M., Eds.; Lawrence Erlbaum Associates, Publishers: Mahwah, NJ, 2005; 249–251.

74. Committee on Facilitating Interdisciplinary Research, Committee on Science, Engineering, and Public Policy, National Academy of Sciences, National Academy of Engineering, and Institute of Medicine; *Facilitating Interdisciplinary Research*; The National Academies Press: Washington, DC, 2005; 157–162.

75. Haythornthwaite, C. Learning and knowledge networks in interdisciplinary collaborations. J. Am. Soc. Inform. Sci. Technol. **2006**, *57* (8), 1079.

76. Spanner, D. Border crossings: Understanding the cultural and informational dilemmas of interdisciplinary scholars. J. Acad. Libr. **2001**, *27* (5), 354.

77. Berman, Y. Discussion groups on the Internet as sources of information: The case of social work. Aslib Proc. **1996**, *48* (February), 31.

78. Brown, C. Straddling the humanities and social sciences: The research process of music scholars. Libr. Inform. Sci. Res. **2002**, *24* (1), 83.

79. Neelameghan, A. Patterns of interdisciplinary interactions and formations in web-communities. Inform. Stud. **2006**, *12* (1), 61–68.

80. Packer, K.; Soergel, D. The importance of SDI for current awareness in fields with severe scatter of information. J. Am. Soc. Inform. Sci. **1979**, *30* (3), 125–135.

81. Ackerson, L. Challenges for engineering libraries. Sci. Technol. Libr. **2001**, *21* (1/2), 43–52.

82. Quigley, J.; Peck, D.; Rutter, S.; Williams, E. Making choices: Factors in the selection of information resources among science faculty at the University of Michigan. Issues Sci. Technol. Libr. **2002**, *34* (Spring), http://www.istl.org/02-spring/refereed.html (accessed December 2007).

83. Klein, J. Afterword: The emergent literature on interdisciplinary and transdisciplinary research evaluation. Res. Eval. **2006**, *15* (1), 79.

84. Foster, A.; Ford, N. Serendipity and information seeking: An empirical study. J. Doc. **2003**, *59* (3), 321–340.

85. Spanner, D. Border crossings: Understanding the cultural and informational dilemmas of interdisciplinary scholars. J. Acad. Libr. **2001**, *27* (5), 355–358.

86. Palmer, C.; Neumann, L. The information work of interdisciplinary humanities scholars: Exploration and translation. Libr. Q. **2002**, *72* (1), 89–100.

87. Herubel, J. Disciplinary affiliations and subject dispersion in Medieval studies: A bibliometric exploration. Behav. Soc. Sci. Libr. **2005**, *23* (2), 67–83.

88. Michalski, D.; Taub, A. Measuring interdisciplinarity: A three tiered analysis of cultural studies. Behav. Soc. Sci. Libr. **2001**, *20* (1), 90.

89. Elam, B. Readiness or avoidance: e-Resources and the art historian. Collection Building **2007**, *26* (1), 4–6. http:// www.emeraldinsight.com.ezproxy.lib.utexas.edu/ Insight/viewPDF.jsp?Filename=html/Output/Published/ EmeraldFull TextArticle/Pdf/1710260101.pdf (accessed December 2007).

90. Rose, T. Technology's impact on the information-seeking behavior of art historians. Art Doc. **2002**, *21* (2), 39–40.

91. Brown, C. Straddling the humanities and social sciences: The research process of music scholars. Libr. Inform. Sci. Res. **2002**, *24* (1), 82–83.

92. Folster, M. Information seeking patterns: Social sciences. Ref. Libr. **1995**, *49/50*, 83–93.

93. Ellis, D. A behavioral approach to information retrieval system design. J. Doc. **1989**, *45* (September), 171–212.

94. Meho, L.; Tibbo, H. Modeling the information-seeking behavior of social scientists: Ellis's study revisited. J. Am. Soc. Inform. Sci. Technol. **2003**, *54* (6), 570–587.

95. Liu, Z.; Wang, C. Mapping interdisciplinarity in demography: A journal network analysis. J. Inform. Sci. **2005**, *31*, 314.

96. Buchanan, A.; Herubel, J. Interdisciplinarity in historical studies. Libres **1994**, *4* (3), 1–13. http://dhsws1.humanities.curtin.edu.au/libres/LIBRES4N2/BUCHANAN.txt (accessed December 2007).

97. Tucker, J. Database support for research in public administration. Behav. Soc. Sci. Libr. **2005**, *24* (1), 47–60.

98. Seely, A. Digging up archeological information. Behav. Soc. Sci. Libr. **2005**, *24* (1), 9.

99. Mayfield, T.; Thomas, J. A tale of two departments: A comparison of faculty information-seeking practices. Behav. Soc. Sci. Libr. **2005**, *23* (2), 55–56, 59.

100. Dalton, M.; Charnigo, L. Historians and their information sources. Coll. Res. Libr. **2004**, *65* (5), 400–425.

101. Duff, W.; Johnson, C. Accidentally found on purpose: Information-seeking behavior of historians in archives. Libr. Q. **2002**, *72* (4), 472–496.

102. Hartmann, K. Social policy resources for social work: Grey literature and the Internet. Behav. Soc. Sci. Libr. **2006**, *25* (1), 1–11.

103. Westbrook, L. *Interdisciplinary Information Seeking in Women's Studies*; McFarland: Jefferson, NC, 1999.

104. Beghtol, C. Within, among, between: Three faces of interdisciplinarity. Can. J. Inform. Libr. Sci. **1995**, *20* (2), 30–41.

105. McNicol, S. LIS: The interdisciplinary research landscape. J. Libr. Inform. Sci. **2003**, *35* (1), 23–30.

106. Herring, S.D. The value of interdisciplinarity: A study based on the design of Internet search engines. J. Am. Soc. Inform. Sci. **1999**, *50* (4), 362.

107. Zhang, L. Discovering information use in agricultural economics: A citation study. J. Acad. Libr. **2007**, *33* (3), 403–413.

108. Ortega, L.; Antell, K. Tracking cross-disciplinary information use by author affiliation. Coll. Res. Libr. **2006**, *67* (5), 458.

109. Hurd, J. Interdisciplinary research in the sciences: Implications for library organization. Coll. Res. Libr. **1992a**, *53* (4), 283.

110. Kelsey, P.; Diamond, T. Establishing a core list of journals for forestry. Coll. Res. Libr. **2003**, *64* (September), 366.

111. Palmer, C. *Work at the Boundaries of Science: Information and the Interdisciplinary Research Process*; Kluwer Academic Publishers: Boston, MA, 2001; 73.

112. Burright, M.; Hahn, T.; Antonisse, M. Understanding information use in a multidisciplinary field: A local citation analysis of neuroscience research. Coll. Res. Libr. **2005**, *66* (3), 206.

113. Ackerson, L.; Chapman, K. Identifying the role of multidisciplinary journals in scientific research. Coll. Res. Libr. **2003**, *64* (6), 478.

114. Robinson, W.; Poston, P. Literature use by geography scholars. Behav. Soc. Sci. Libr. **2006**, *25* (1), 27.

115. Allen, R. Interdisciplinary research: A literature-based examination of disciplinary intersections using a common tool, Geographic Information System (GIS). Sci. Technol. Libr. **2001**, *21* (3/4), 196.

116. Bartolo, L.; Wicks, D.; Ott, V. Border crossing in a research university: An exploratory analysis of a library approval plan profile of geography. Collect. Manag. **2002**, *27* (3/4), 29–44.

117. Roccos, L. Archaeological research online—finally!. Comput. Libr. **2000**, *20* (10), 37.

118. Ambrosia. http://193.92.187.46:8990/F (accessed December 2007).

119. Kuruppu, P.; Gruber, A. Understanding the information needs of academic scholars in agricultural and biological sciences. J. Acad. Libr. **2006**, *32* (6), 620.

120. Green, A.; Gutman, M. Building partnerships among social science researchers, institution-based repositories and domain specific data archives. OCLC Syst. Serv. **2007**, *23* (1), 35–53.

121. Palmer, C. Aligning studies of information seeking and use with domain analysis. J. Am. Soc. Inform. Sci. Technol. **1999b**, *50* (12), 1140.

122. López-Huertas, M.; Ramírez, I. Gender terminology and indexing systems: The case of woman's body, image and visualization. Libri **2007**, *57* (1), 39–40.

123. von Ungern-Sternberg, S. Bradford's law in the context of information provision. Scientometrics **2000**, *49* (1), 161–186.

124. Palmer, C.; Neumann, L. The information work of interdisciplinary humanities scholars: Exploration and translation. Libr. Q. **2002**, *72* (1), 85–117.

125. Ascott, R. Technoetic pathways toward the spiritual in art: A transdisciplinary perspective on connectedness, coherence and consciousness. Leonardo **2006**, *39* (1), 65–69.

BIBLIOGRAPHY

1. Denda, K. Beyond subject headings: a structured information retrieval tool for interdisciplinary fields. Libr. Resour. Tech. Serv. **2005**, *49* (4), 266–275.

2. Epstein, S. Making interdisciplinary collaboration work. In *Interdisciplinary Collaboration: An Emerging Cognitive Science*; Derry, S., Schunn, C., Gernsbacher, M., Eds.; Lawrence Erlbaum Associates, Publishers: Mahwah, NJ, 2005; 245–263.

3. Green, A.; Gutman, M. Building partnerships among social science researchers, institution-based repositories and domain specific data archives. OCLC Syst. Serv. **2007**, *23* (1), 35–53.

4. Klein, J. *Crossing Boundaries Knowledge, Disciplinarities, and Interdisciplinarities*; University Press of Virginia: Charlottesville, VA, 1996a.

5. Palmer, C. *Work at the Boundaries of Science: Information and the Interdisciplinary Research Process*; Kluwer Academic Publishers: Boston, MA, 2001.

6. Walters, W. Google Scholar coverage of a multidisciplinary field. Inform. Process. Manag. **2007**, *43* (4), 1121–1132.

ARMA International, Inc.

Carol E. B. Choksy
School of Library and Information Science, Indiana University, Bloomington, Indiana, U.S.A.

Abstract

ARMA International is a not-for-profit professional association dedicated to the management of records and information. Its primary focus is education and advocacy for its membership. As a professional association, it is the primary resource for persons needing education or networking in records management.

BACKGROUND

ARMA International (http://www.arma.org) is a not-for-profit professional association dedicated to the management of records and information. It provides educational resources such as seminars, conferences, and publications to records and information management (RIM) professionals and others responsible for managing organizations' information assets, creates standards and guidelines related to records management, and advocates for records management with the governments of the United States, Canada, and the European Union. The mission of the association

> is to provide education, research, and networking opportunities to information professionals to enable them to use their skills and experience to leverage the value of records, information, and knowledge as corporate assets and as contributors to organizational success.

Headquartered in the Kansas City metro area, ARMA International's 11,000-plus members include records managers, attorneys, information technology managers, archivists, consultants, and others involved in various aspects of managing records and information. The association also supports approximately 125 chapters, which provide additional programs at the local and regional levels. Most of the chapters are in the United States and Canada with chapters in the Caribbean, Europe, and Japan and a growing membership outside of North America.

For many years, ARMA International has been known worldwide for its active participation in setting standards and best-practice guidelines in the field. It is the only association in the United States that develops RIM-specific standards approved by The American National Standards Institute (ANSI), which coordinates the U.S. voluntary standardization system. ARMA International is also actively involved in the International Organization of Standards (ISO), the world's largest developer and publisher of international standards. ARMA played a key role in developing and launching the only international records management standard, ISO 15489. The association also works closely with other organizations, including The Sedona Conference, on groundbreaking guidelines.

ARMA is also well-known for advocacy in the United States, Canada, and more recently, the EU. ARMA testified in the U.S. Congress and organized a letter-writing campaign to ensure the Paperwork Reduction Act of 1980 was passed. In 2004, ARMA's association president testified in Washington, DC before a federal panel of judges on recommended changes to the Federal Rules of Civil Procedure that went into effect in 2006.

HISTORY

Business Records Management

Although the practice of records management has been traced back to the Sumerian civilization in 5000 B.C.,[1] the profession is comparatively a much newer concept. In the late nineteenth century, many companies in the Unites States discovered the burgeoning number of typed and gel-pressed documents and records was making effective and efficient work nearly impossible. Several different social, political, and technological trends in the nineteenth century brought about the rise in records management independent of archives. The rise of the notion that society could be engineered and that government should intervene on behalf of citizens led to a greater need to document activities.

> The narrow scholarly aspect of many of these traditional archives was to make it difficult for them to adjust to the information revolution unleashed by modern administrative and managerial practices. Across Europe and North America in the late nineteenth century, spurred by increasing state intervention in society, central administrations were growing in relative terms, and creating increasing masses of paper. In the private sector similar developments were under way which led, especially in the United States, to the emergence of new forms of records management.[2]

Encyclopedia of Library and Information Sciences, Fourth Edition DOI: 10.1081/E-ELIS4-120044712

Archives–Art

Experts in Scientific Management applied their techniques to the office environment improving what we now call the "active phase" of the life cycle. Referred to as "Filing and Transferring" nearly everything we understand in the practice of records management was present by 1920.

In the United States, by World War I, the Warren Filing Association was founded in Chicago.

In 1914 Irene Warren, librarian at the University of Chicago, brought together a group of people interested in records-handling and began the Warren Filing Association. This was associated with the Warren School of Filing, which she had also founded for the training of filing clerks and supervisors.[2]

In 1920, the Filing Association of New York was founded in New York City. The Chicago Filing Association was established in 1932 primarily for file personnel.

Government Records Management

After the Civil War, the U.S. federal government recognized that agencies exercised little, or no management of records through several congressional studies like the Hays Commission study of 1877 recommending fireproof buildings for storage of veteran's records. An 1888, Senate study of recordkeeping processes produced many descriptions of problems of records having been kept long past any useful life and recommended the destruction of such records. In 1889, the General Records Disposal Act was passed but made very little difference to the burgeoning number of documents. In 1913, the Taft Report recommended using some of the methods adopted by business for handling, managing, and destroying information. By World War II, the number of U.S. federal records had grown past a size such that the National Archives, established in 1934, could possibly manage. A number of archivists and records officers within the federal government established several concepts such as the life cycle and retention schedules that became the foundations of modern U.S. records management.[3]

CREATION OF ARMA

The Chicago Filing Association changed its name to the Records Management Association of Chicago and was incorporated in 1952, reflecting the contribution made by the U.S. federal government to the field. In the early 1950s, the New York association was renamed the Records Management Association of New York and was incorporated in 1955. In that same year, 12 records managers in New York formed the Association of Records Executives and Administrators (AREA).

At that time, some believed that the other records-related organizations were centered too much on filing and

retrievals and not enough on the management aspects of records. The Association of Records Executives and Administrators membership increased steadily, and other chapters were chartered in the mid-1960s.

ARMA became international in the 1960s when Canadian chapters joined including Montreal in 1968, Toronto in 1969, Vancouver in 1970, Ottawa in 1971, and Edmonton in 1975.[2]

Shortly after its inception, ARMA began developing resources for records management professionals. In 1956, it launched a member newsletter and held its first annual conference. Four years later, the association debuted its first "official" publication, *Records Review*, and published the technical resource *Rules for Alphabetical Filing*, which eventually became an ANSI standard. In 1958, ARMA became part of the International Records Management Federation, a "steering committee" between ARMA, the Records Management Association of Australia, and the Records Management Association of South Africa. The International Records Management Federation's goal was to help develop records associations in countries in need of such organizations.

ARMA continued to extend its reach and its activities in the 1970s. Then in 1975, it merged with AREA and became ARMA, the Association of Records Managers and Administrators. As the association's influence both in North America and internationally grew, the association added "International" to its logo and became known as ARMA International in 1987. The board of directors decided in 1995 to officially do business as ARMA International, dropping the "Association of Records Managers and Administrators" in general use; this remains the official business name of the association today.

In 1997, ARMA signed an agreement with the International Records Management Trust and the International Council on Archives (ICA) to foster the development of records management standards and education globally. The agreement ended in 1999, but the relationship between ARMA and the ICA has continued, particularly in the area of standards development, such as competencies.

In 2005, ARMA extended its administrative reach to include Europe and European issues such as Data Security. This initiative was so popular among European members that they requested, and the Board of Directors granted, the creation of a member unit in Europe where cross-jurisdictional issues could be discussed.

DEFINING MOMENTS FOR THE ASSOCIATION AND THE PROFESSION

During the last 50-plus years, there have been a variety of noteworthy occurrences helping to fuel the growth and recognition of RIM in the business community. A major boost came from passage of the Paperwork Reduction Act in 1980. This legislation addressed the need to manage information as a resource that is planned, managed, and included in the budget.

When several high-level e-mails became public in the Microsoft antitrust case (*United States v. Microsoft*) in the late 1990s, many executives expressed surprise that their e-mails could become part of litigation and be made public.

The effects of ISO 15489, the standard on records management unveiled at the 2001 ARMA conference in Montreal, Canada, made records management a global practice. Until the release of the standard, records management was primarily an Anglophone activity, the concept of "records" being an English-only word.[4] Non-U.K. European companies struggled to conform to this new standard.

Nothing could equal the effect of the infamous Enron Corporation scandal. The handling of information played a critical supporting role in the drama that unfolded in 2001. It was played out in *U.S. v. Arthur Andersen, LLP*, which is generally regarded as the most widely publicized criminal case involving the destruction of business records. Andersen, one of the nation's largest public accounting firms, was accused of destroying accounting records for audits it had performed for Enron after the U.S. Securities and Exchange Commission had begun its investigation into Enron. Andersen was eventually convicted of obstruction of justice, but the damage was done before the verdict was even announced. Customers fled. The Enron case and Andersen's alleged role in it ultimately led to the rapid and bloody demise of the powerhouse accounting firm. Andersen was later acquitted, but too late for the thousands of employees and partners who had worked so hard and had lost, in some cases, a lifetime of work.

Later that year, Congress reacted to the Enron scandal and other widely publicized reports of corporate accounting irregularities by passing the far-reaching Sarbanes-Oxley Act of 2002. The act made regulatory compliance a higher priority than ever before for publicly traded companies, regardless of whether they were U.S.-based or simply traded on the U.S. stock exchange. For the first time, chief executive officers began to feel the very real threat—including a jail sentence and hefty financial penalties—that could result from mismanaging their corporate information assets. More and more, executives began turning to their legal, information technology, and RIM professionals to ensure their companies did not go the way of Andersen and others in the headlines. An important part of Arthur Andersen's sudden decline occurred because employees had been told to, and did, destroy e-mails relating to their engagement with Enron. The employees as well as the persons instructing them did not know that the e-mails could easily be restored from back-up tapes. Many in information technology were surprised to discover that e-mails could be relevant to a case of regulatory compliance and that back-ups could be viewed as business records and be subpoenaed for litigation.

That same surprise was compounded when a judgment against Morgan Stanley (*Coleman v. Morgan Stanley*) was issued for $1.4 billion in part because the brokerage firm had not been able to deliver all the appropriate e-mails in the discovery phase of the trial. The judgment was reversed on appeal, but it sent shockwaves through the information technology community as well as the legal community. Another high-profile case that was not reversed, *UBS Warburg v. Zubulake*, cost the global banking giant, UBS Warburg $38 million. Producing a few relevant e-mails each of the seven times the judge in the case ordered the company to produce what Ms. Zubulake knew and could prove they had, UBS Warburg became a victim of its own poor record-keeping. The e-mails involved in this case, however, were not "records" in the sense that governments define them, as having to do with the business of the organization. The e-mails the judge was compelling UBS Warburg to produce involved the authors' beliefs in Ms. Zubulake's "fitness" to be a broker.

While the judgment of $38 million was far smaller than in *Coleman v. Morgan Stanley*, the *UBS Warburg v. Zubulake* case came only a few months before the 2006 revisions to the Federal Rules of Civil Procedure. One of the effects of those revisions was to clarify what could be produced for litigation from specific format references like, "documents" and "databases," to all "electronically stored information." Many companies as well as government entities struggled, and are still struggling, to ensure the organization can produce all information, whether electronically stored or not, relevant to a discovery subpoena. As case after case of lost or destroyed electronically stored information comes to light, all organizations continue to struggle to implement life-cycle management from creation to final disposition.

ASSOCIATION STRUCTURE

ARMA's decision-making body is the Board of Directors. Lead by the President, the Board makes strategic and governance decisions. Membership is primarily geographical, with most members living near a Chapter belonging to that Chapter. Each Chapter is incorporated and has a board of directors as well as bylaws. The role of Chapters is primarily to provide education, and leadership and networking opportunities. Chapters are geographically organized into Regions. Regions are lead by a Region Manager, and several Coordinators whose job it is to advise the Chapters and give help where necessary. ARMA also has Committees that perform various duties including reaching out to membership, education, awards, public policy advice, and governance. ARMA creates task forces when a specific, limited action is required. The

Board makes decisions by first gathering information from its members' wants, needs, and preferences. The association also has an annual strategic planning meeting in which information gathered from members as well as reports from analysts and ideas from attendees are sorted through to determine what ARMA should be doing over the next few years. ARMA's records management competencies, published in October 2007, were the result of the first strategic planning meeting and are now the basis for all education in ARMA. Attendees to strategic planning includes all members of the Board, a member of ARMA's Company of Fellows, a member of the ARMA International Education Foundation, a regent from the Institute of Certified Records Managers (ICRM), and a Region Manager.

The Company of Fellows is made up of individuals selected by the Awards Committee. On an annual basis, the Awards Committee reviews nominations for Fellow of ARMA International (FAI), scores the nominations and determines whether any rise to the high standard set for this recognition to the furtherance of the profession. Individuals selected for this honor are permitted have the designation FAI after their name.

The ARMA International Educational Foundation was created as an independent foundation with its own Board of Trustees in 1996 for the purposes of funding research and education in the area of records management.

MEMBERSHIP

ARMA has more than 11,000 members globally. The backgrounds of members is extremely diverse as most people who become records managers are simply handed the job. At the last survey, more than two-thirds of members had some college or a college degree. A decade before, only half had some college. The reason for the change is the aging and retirement of a demographic group that could obtain gainful employment in business administration with only a high school degree. As this population has aged and retired, job descriptions for records managers have changed and now require a college degree. Most members do not have the title "records manager" as titles vary from level to level, organization to organization, and industry to industry. In addition, records managers have never reported to a single area or level of responsibility. There is some evidence that records managers are reporting to an executive instead of a manager and are reporting to an enterprise-wide function like compliance or legal rather than to an administrative function like facilities over the past decade. This lack of consistency makes identifying the person in charge of an organization's records very difficult and makes recruiting new members very difficult. Before the profile of records management and ARMA rose in the late 1990s, the only way to recruit new members was by word of mouth.

Anyone may join ARMA, including members and employees of associations considered competitors. Many members of ARMA joke that there are actually two qualifications for joining ARMA, one is a love of jigsaw puzzles, crossword puzzles, or Sudoku, and the other is to have a controlling, bordering on obsessive-compulsive, personality. Those two characteristics: a predilection for pattern recognition over logical thought, and a desire to put the pattern in order are very common personality characteristics for records managers. This does not mean that straight-line logical thinking is uncommon among ARMA members, but pattern-recognition is a commonly preferred mode of thought.

PUBLICATIONS

ARMA publishes the *Information Management Journal*, the world's only publication devoted solely to records management and the issues that impact records management professionals, quarterly and *InfoPro Online*, focused on association news, monthly. ARMA publishes three monthly newsletters that impact the practice and profession, the *Washington Policy Brief*, devoted to news from the U.S. federal government, the *European Policy Brief*, devoted to news from the European Union government, the *Canadian Policy Brief*, devoted to news from the Canadian federal government, and the *Information Management NewsWire*, devoted to issues outside the government. ARMA also publishes standards, textbooks, and practitioners monographs on issues including, technology, risk management, compliance, and industry issues.

Most Chapters have a newsletter or a Web site where articles and news of local interest is also published.

STANDARDS

Most notable of ARMA's standards is the Alphabetic standard mentioned above, which is a guide to creating file systems for active records. Other ARMA International Standards and Guidelines include, Glossary of RIM Terms, Guideline for Evaluating Offsite Records Storage Facilities, ISO 15489 Information Documentation—Records Management—Part 1: General, ISO 15489 Information Documentation—Records Management—Part 2, ISO/S 23081-1—Information and Documentation—Records Management Processes, Procedures, and Issues for Managing Electronic Messages as Records, Records Center Operations, Records Management Responsibility in Litigation Support, Retention Management for Records and Information, Revised Framework for Integration of electronic document management (EDMS) & electronic records management (ERMS), Requirements for Managing Electronic Messages as Records, The Digital Records Conversion Process: Program Planning, Requirements,

Procedures, Vital Records Programs: Identifying, Managing, and Recovering Business Critical Records, Working Collaboratively in an Electronic World.

ARMA creates and partners with other organizations and associations to create standards that impact records management professionals. Frequent partners include the Association for Imaging and Information Management (AIIM) and Society of American Archivists (SAA). ARMA is now partnering with the Object Management Group to create a technology framework for government records management that will retire DOD 5015.2.

COMPETENCIES AND CERTIFICATION

ARMA's independent certifying body is the ICRM. The ICRM creates, revises, and administers a six-part test leading to the designation of CRM, Certified Records Manager. Recognized globally, the CRM is currently held by more than 800 records managers. To deal with the ever-burgeoning duties and importance of records managers, ARMA created a set of competencies for the records management profession. Validated in the United States and Canada, the competencies, created by ARMA were created independently of the CRM. A recent joint task force mapped the CRM body of knowledge to the ARMA competencies and found nearly 100% agreement between the competencies and the ICRM test outline, further validation of both. Both the ICRM test outline and the ARMA competencies also reflect the broadly-based need to work with electronic records and all other forms of electronically stored information including e-mail and instant messaging.

EDUCATION

Nearly all of ARMA's educational offerings are now mapped to the competencies. Education is delivered to ARMA's members primarily through Chapter meetings. Approximately one-sixth of the membership attends the annual conference, the purpose of which is primarily education, but also networking. Conference attendees can choose from nearly a hundred sessions offered by practitioners and consultants in the meeting rooms and from nearly the same number of technology advisory sessions provided by vendors on the exposition floor. The accompanying exposition has technology vendors, shelving and archiving vendors, storage vendors, as well as consultants. Monthly webinars permit the reach of education to extend to anyone with an Internet connections. Additional conferences, such as the "E-Discovery and Beyond" conference have arisen as the need for more specialized education arises.

ARMA'S ROLE IN INFORMATION AND RECORDS MANAGEMENT

ARMA is a major hub in a network of networks. It functions as a professional network, an educational network, a sales network, and a social network. Its most important role is as a professional network. At various types of meetings: geographically or regionally focused, industry focused, nationally and internationally focused, members meet to listen to presentations by practitioners, consultants, educators, and vendors on issues relevant to their work lives. Because the speakers are vetted by records management professionals, or by staff well-educated in records management issues, ARMA meetings of all types are the only place where one can go and be assured of hearing a relevant, high-quality presentation. Conversations break out among participants with similar problems as a result of the presentations. Problems may be industry-related, geographical, life cycle, media, migration or conversion, vendor, business, resource, political, etc. Members share their experiences, both good and bad, making ARMA meetings the best place to learn the ins and outs of being a records manager. In this way, ARMA is truly a Community of Practice in Wenger's sense of the term.

ARMA is a vendor network. Vendors can become members of ARMA as membership is not limited to practitioners. However, only individuals can become members. Companies can join ARMA by paying for membership for its employees. Vendors have all the same privileges as members, but are asked not to use ARMA meetings to make sales pitches. ARMA chapters often request a vendor to speak at a meeting about their product or subject. For example, shredding companies are asked to speak about data security and privacy. ARMA meetings and vendor expositions at ARMA conferences, seminars, and chapter meetings are good venues for vendors to get their messages across to members in a low-key environment. ARMA practitioners and consultants are usually not decision-makers for vendor products, but they are recommenders. For all types of soft and hard products, ARMA members are key links in the sales of records management products and services. Because ARMA addresses the risk side of document life-cycle management, the rising profile of records management in litigation made ARMA a natural place for attorneys interested in electronic discovery to receive and provide education.

As the profile of records management has risen, more organizations ask information management vendors to address the risk side of information. Vendors of electronic content management (ECM), EDM, and imaging were targets of this request. To remain relevant these vendors, including IBM, FileNet, OpenText, Interwoven, and Stellent purchased or developed ERM modules. The trend in all industries toward consolidation of companies also lead these companies to buy each other: IBM bought FileNet, Oracle bought Stellent, OpenText bought Hummingbird, etc.

Archives–Art

Consolidation within this multibillion dollar a year industry created reduced income for the trade association that represents most of those companies, AIIM, by reducing the number of large company members that pay much more than other members. Before these vendors realized their products were deficient in life-cycle management and were actually creating more expensive problems for organizations than they were solving (discovery being a seven-figure problem where productivity increases were only six-figure problems), they all showed at the AIIM show, which was attended primarily by information technology professionals responsible for managing documents rather than data. At about the same time, consolidation was rapidly increasing and records management was becoming a higher profile problem, AIIM sold its trade show to Questex (an organization that services associations primarily with publications and conferences), which combined the AIIM show with the On Demand show (output management), creating an annual shopping mall for information technology vendors managing documents. The Association for Imaging and Information Management has also ventured into records management education, but geared primarily toward the information technology professionals who were previously the primary decision-makers for ECM, EDM, and imaging technology. The education provided is at a very elementary level and is primarily focused on what a consumer would need to know in order to purchase ECM, EDM, and ERM applications.

The change in focus of purchasers from creation and distribution of information to full life-cycle management of information meant that vendor members showing at both the AIIM and the ARMA shows were spending twice as much money as they would if all potential purchasers were at one show. Another trend is the shift in decision-making regarding information management technology from the information technology department to the legal department. Because attorneys were more inclined to attend the ARMA show than the AIIM show, vendors found themselves forced to purchase space at both vendor shows in order to reach the information technology professionals, records management professionals, and legal professionals who would together make the decision to purchase document-related information technology. The expense and employee time required to have a booth at both shows, has lead vendors to request AIIM and ARMA to merge. While this suggestion has been made twice by AIIM and once by vendors to ARMA, there has been no movement as no model for merging a trade association where dollars are votes with a professional associations where individual members are votes has been suggested. The Association for Imaging and Information Management continues to be a valued partner of ARMA, providing education about document-related information technology products, and working with ARMA on document-related information technology standards.

For the same reasons and causes, ARMA is an education network. Until the past decade, ARMA was one of the few places one could go to get education in records management. Only recently have institutions of higher education offered courses in records management. Until that time records management was taught primarily at the community college level by practitioners and consultants. Two universities had faculty who taught records management, one in a department of business and one in library science, but only one of those had a program in records management. Neither program had a degree in records management. Most of these programs, both at the community college and university level ended when the faculty member or practitioner moved on or retired. Consequently, the only place a student of records management could be assured of a steady supply of education was through ARMA meetings, seminars, and conferences. Practitioners and consultants have been and still are the main source of education in records management. Product vendors also provide education through ARMA either in a regular program meeting or seminars advertised through ARMA publications and newsletters. Such sessions are focused primarily on content their products address, though some address broader industry issues.

Another problem with education in records management, as noted above, is that records management could as easily be associated with business and with library and information science programs. There is also a very strong legal component. This lack of fit into a normal university curriculum makes records management a very poor fit in any particular school, much less a single department. With the widespread use of computers another component has arisen that is normally associated only with Ph.D. studies in information technology, enterprise information architecture (EIA). EIA has three components: technology, information, and people, but departments of information technology usually only deal with one and one-half of those: technology and databases—documents are not addressed in the standard IT curriculum. Records management deals with how people use information, regardless of whether it is hard copy or electronic, i.e., regardless of whether humans are interacting with computers or paper; how people use documents—the best-known form of records; and with how governance impacts how people use and abuse information. The lack of understanding within most of the information technology community about EIA means that most software applications that handle information are created to function within a small department and to assist in the creation and distribution of information, not in the actually life-cycle management of information. Microsoft products are an excellent example of this, as are all ECM, EDM, ERM, and imaging applications. Schools of library and information science have begun to address these issues over the past decade, but still wrestle with the problem that records management addresses business objects whereas library science, archival science, and information science address cultural objects. Despite all the changes in records management education over the past decade, perhaps the strongest network for getting education in records management is still ARMA meetings and seminars, as well as seminars advertised by education providers and vendors through ARMA publications.

As noted above in the publications section, ARMA International is the only place where one can find education in and discussions of cross-jurisdictional issues. Canadians and U.S. residents take this for granted and both share information and knowledge about political, economic, social, and technology issues without thinking. Other Anglophone countries have records management societies, but those associations are focused on a single country. With 50 states and as many different laws and regulations, it is the norm for U.S. members and Canadian members working for companies doing business in the U.S. to need cross-jurisdictional discussions. Canadian ARMA members have worked for many years to try to educate U.S. members on the need to pay attention to European Union issues. This finally sunk in over the past 5 years when European members told ARMA representatives that ARMA meetings were the only place where they could discuss cross-jurisdictional issues. Happy with their countries' records management associations, they needed a place to discuss EU and pan-European issues such as data security.

Conversations that begin at ARMA meetings continue off-line often to become warm friendships. A friend who can commiserate with the challenges and frequent lack of respect records managers have experienced is a special friend. For example, a records management Listserv completely independent of ARMA sprang up in the mid-1990s that discusses subjects related to records management. Rambunctious at times, the Listserv provides a way to remain connected while at work. Sometimes the discussion gets off-topic, and many posts are made in jest. The conversation includes frequent contributions from members in Australia as well as in Canada, and the U.S. Most members of the Listserv are also ARMA members, and there are frequent discussions of the association. Many Listserv members meet at a party arranged at the annual conference. People place names and postings to faces and grow closer. While social networking is not the reason one becomes a member of ARMA, it is one of the reasons members remain members.

Because ARMA is the only place one can go to get a thorough understanding of records management, both as a profession and for complete, formal education in records management, the association has credibility and magnitude that is unparalleled worldwide. As the past decade has seen a rise in visibility for records management, membership has become even more diverse, including lawyers, accountants, information technology professionals, and compliance professionals. Instead of weakening the various types of networks, this has strengthened the various networks, both expanding and strengthening them.

REFERENCES

1. Penn, I.A.; Pennix, G.B.; Coulson, J. *Records Management Handbook*, 2nd Ed.; Gower: Aldershot, U.K., 1994.
2. Higgs, E. The role of tomorrow's electronic archives. In *History and Electronic Artefacts*; Higgs, E., Ed.; Clarendon Press: Oxford, U.K., 1998; 184–194.
3. Choksy, C. *Domesticating Information: Managing Documents Inside the Organization*; Scarecrow Press: Lanham, MD, 2006.
4. Walne, P. *Dictionary of Archival Terminology*; K. G. Saur: Munich, Germany, 1984.

BIBLIOGRAPHY

1. Gill, S. *File Management and Information Retrieval Systems: A Manual for Managers and Technicians*; 2nd Ed. Libraries Unlimited: Englewood, CO, 1988.

Armenia: Libraries, Archives, and Museums

Sylva Natalie Manoogian
Department of Information Studies, University of California, Los Angeles, Los Angeles, California, U.S.A.

Abstract

Throughout the millennial history and worldwide dispersion of the Armenian nation, libraries have played and continue to play a critical role. They have been instrumental in collecting and preserving the written works which embody the creativity of the Armenian mind, and making widely accessible recorded knowledge about the history, language, and literature of one of the oldest peoples on earth. This entry presents an overview of the establishment and development of libraries, museums, and archives, with special emphasis on those institutions functioning today in the Republic of Armenia.

INTRODUCTION

Throughout the millennial history and worldwide dispersion of the Armenian nation, libraries have played and continue to play a critical role. They have been instrumental in collecting and preserving the written works that embody the creativity of the Armenian mind; and making widely accessible recorded knowledge about the history, language, and literature of one of the oldest peoples on earth.

This entry presents the establishment and development of libraries, museums, and archives, with special emphasis on those institutions functioning today in the Republic of Armenia (Fig. 1). Following an historical overview and a brief discussion of Armenian libraries in the Diaspora, the entry introduces the academic, public, children's, school, and special libraries of Armenia within the regional context of the South Caucasus during the Soviet era and following the declaration of Armenia as an independent republic in 1991. Profiles of Armenia's major libraries are followed by a discussion of the Armenian Library Association, its activities, its role in continuing education, and its involvement in international and regional cooperation. A brief overview of Armenia's national archives and museums concludes the entry. Appendix 1 provides facts about Armenia, and Appendix 2 contains a chronology of Armenian libraries, archives, and museums.

History

The earliest information about Armenian libraries comes to us from the fifth-century historian, Ghazar Parpet'si, who tells us about the establishment at the Catholicosate of Ejmiatsin of a scriptorium ("gratun") which contained Armenian and Greek manuscripts. Surviving time, the elements, and hostile armies, copies of these and other texts can be viewed today in the "Matenadaran," the manuscript library in Yerevan, Armenia, named after the visionary monk Mesrop Mashtots, who created the Armenian alphabet ca. 405–406 A.D.

Throughout the centuries, Armenian libraries continued to develop and flourish in centers of education, almost always as part of Armenian monasteries, and sometimes as separate facilities. Manuscripts were also stored in nearby caverns and in wall recesses. Entire libraries started coming into being. It is said that in 1170 the Seljuks destroyed a library in Siwnik consisting of 10,000 Armenian manuscripts. These encompassed religious subjects as well as philosophy, music, and calligraphy. The first Armenian academic libraries were established during the eighteenth and nineteenth centuries, away from the Armenian homeland, in Venice, Vienna, Constantinople, Tiflis (Tbilisi), and Moscow.

LIBRARIES OF EURASIA AND CENTRAL ASIA DURING THE SOVIET ERA

Prior to the 1991 break-up of the Soviet Union, its vast territory was administratively structured into 15 union republics (SSRs), which in turn contained 20 autonomous republics (ASSRs), 10 autonomous areas (or National Okrugs), 6 territories (Krais), and 130 regions (Oblasts), eight of which carried the qualifier "autonomous." The union republics and the various autonomous divisions were created to foster, or at least to contain, national or ethnic groups with distinct cultures, languages, or dialects. One may wonder about the origin of all these peoples, with their more than one hundred languages and diverse ethnic backgrounds. In nearly all cases, their established cultures predate both the Soviet and earlier tsarist expansions, which gradually drew them into a sprawling, contiguous Imperium.

Encyclopedia of Library and Information Sciences, Fourth Edition DOI: 10.1081/E-ELIS4-120043538

Fig. 1 Map of Republic of Armenia, 1991.
Source: From CIA World Factbook 2008. https://www.cia.gov/library/publications/the-world-factbook/geos/countrytemplate_AM.html (accessed July 17, 2009).

During the seven decades of the Soviet regime, the closed societal nature of the USSR promoted an essentially Russian culture, with Russian as the language of administration, law, and military affairs, as it had been during the Tsarist Empire. Government policy toward the minorities, decreed by Lenin and carried on by Stalin and his successors, nominally allowed for the equality of the national languages with Russian, the right to self-determination and development, and the minorities guaranteed, and strongly encouraged, access to socialist federation within the union. There had been numerous long-term and short-term adjustments to the nationalities policy throughout the years, but difficulty always characterized the Soviet Union's attempts to ease or promote the merging of its people.

Libraries of the Soviet Union, as reflections of the cultural and societal institutions of the various countries and their peoples, were centralized and carefully controlled by the state, with the USSR Council of Libraries in Moscow as the ultimate authority on their governance, their resources, and services. Hierarchies of libraries were firmly established. Library directors were spared most of the decision-making. Norms for library work were prescribed with the same kind of central planning that decided in Moscow how many hectares would be planted in Armenia.

Librarians were told how many books of each kind must circulate in their library and filled out their reports accordingly. The usefulness of Soviet libraries was never

contested, nor were libraries especially threatened by closing. Librarians enjoyed a certain stature as "cultural workers." Their libraries, especially the large, important ones, were highly esteemed. Soviet librarianship was proud of huge collections, of the universal literacy in their state, and of excellent bibliographic control. The unmasking of Soviet libraries began during the last years of "glastnost," when librarians began to complain in their professional journals about the formalism of work, the necessity to inflate statistics, the collections full of redundant material that no one wanted to read. Some even began to question the premise that the librarian's most important job was to be a propagandist for the Communist party.

This was the general climate when the Soviet Union disintegrated in August 1991, with participants in the International Federation of Library Associations and Institutions (IFLA) Congress in Moscow as eye-witnesses. Instead of one rigid and homogeneous system, there suddenly were 15. In each case, librarianship had to reform and carry on, to choose what should be kept of the old ways, and to discard what should be thrown out with old party slogans.

The 1988 Earthquake

On December 7, 1988, a killer earthquake devastated the northern region of Armenia. The tremor affected territory of almost 30,000 sq. km. In a few short seconds, it took tens of thousands of lives, crushed by falling debris, and left half a million people injured, maimed, and homeless. It toppled the buildings of many social and cultural institutions, including libraries. Soviet President Mikhail Gorbachev cut short his visit to the United States to return to the earthquake site. Armenian communities world-wide sprang to their feet and worked feverishly to send humanitarian aid to their brothers and sisters in the homeland.

Three weeks later, at the American Library Association Midwinter Meeting in Washington, D.C., a resolution by its International Relations Committee was presented to the decision-making ALA Council, under the presidency of William F. Summers. On January 10, 1989, the Council responded swiftly and unanimously by establishing a $5000 emergency fund and establishing the Armenian Earthquake Disaster Committee to help restore libraries and library services in the Armenian earthquake belt.

October 7, 1994, marked the inauguration of the Library Association of Armenia. Ten years later, the decennial conference of ArLA took place in Yerevan, with the participation of 119 delegates and 64 guests.

LIBRARIES IN THE REPUBLIC OF ARMENIA

The book has been considered a national treasure since the creation of the Armenian alphabet at the beginning of the fifth century of our era. Armenia's libraries have long been the mainstay of its rich culture. They are considered as centers of education, surviving time, the elements, and hostile armies. Most of the contemporary libraries of Armenia were established during Soviet rule, from 1921 to 1991. Based on the Soviet formula of one library per 1000 persons, in 1975, there were approximately 3300 libraries: one-third of them public; one-third, school; and the rest, technical, trade union and others. A reorganization of the libraries took place between 1975 and 1980, as a result of which independent city, regional, and rural libraries were unified into 42 centralized library systems, serving three and a half million people. In 1989, the total number of libraries numbered 2883.

The Armenian Book Chamber was established on December 27, 1922. This institution registers and records all published works in the Republic, including books and brochures, periodical publications, posters, postcards, music scores, and maps. The Chamber has been publishing the current national bibliography, *Tpagrut'ian taregirk* = The Yearbook of Printing, since 1925.

With the collapse of the Soviet Union, the challenges of dramatic changes from relative plenty to economic instability and political volatility hit hard. Abridgement of services and reorganization were disturbing. While librarians thought and spoke in terms of their team efforts, unlike a team, they were discouraged from individual initiative and certainly from collaborative networking. The concept of librarians as controllers of information, rather than as facilitators of its flow, would be slow to change.

The work of libraries since the collapse of the Soviet Union has been difficult. Insufficient finances, the disintegration of the book acquisition system, deregulation of the price of books, social hardships, and the energy crisis have disrupted and brought to a halt library functions and services. Some libraries were closed, and the number of users decreased. Presently, efforts are under way to overcome the challenges, and ground is being prepared to move library life into the new technological age.

Academic Libraries

Academic libraries are divided into two groups. The first group includes the Armenian Academy of Sciences library network, consisting of a main library and 31 specialized libraries in its various institutes. The second group of academic libraries includes the libraries of the Republic's 12 state higher educational institutions. The largest of these is the Yerevan State University Library, which functions as the methodological and coordinating center for about 60 libraries affiliated with institutions of higher education in Armenia.

Since the Armenian Republic's declaration of independence in 1991, a few private and non-governmental universities were established and formed their own libraries. Notable among these is the Papazian Library of the American University of Armenia, a WASC-accredited affiliate of the University of California system.

Public Libraries

A well-organized network of public libraries is spread throughout Armenia. These libraries function under the auspices of the Ministry of Culture, Sports, and Youth, and are also accountable to their local governmental bodies. Their services are free of charge to the public. However, in the last decade, public libraries have adopted fee-based services, such as photocopying, computer training, and bibliographic searching.

Notable in the most recent developments have been increased efforts in grant-writing toward the establishment special library areas focusing on English language materials to raise awareness about the United States. These "American Corners" serve as regional resource centers for information and programs highlighting American culture, history, current events, and government.

Children's and School Libraries

Information services to young readers are provided in three types of libraries: separate children's libraries; Centralized Library System children's sections; and school libraries. The Yerevan Centralized Children's Library System, formed in 1979, consists of a Central Library and 14 branch libraries with a total of 350,000 volumes in 1994. This library system caters to 40,000 children annually. Every school has its own library. In the Republic there are 1349 school libraries, with over 18.4 million items, of which 10 million are textbooks. The methodological center of the children's and school libraries is Khnko-Aper Apor National Children's Library, established in 1932.

Special Libraries

In this category are scientific-research, medical institute, museum, and military libraries. Notable among these is the Republic's Scientific-Technical Library, under the Health Ministry. Established in 1946, this library has begun the task of creating and computerizing a network of 19 medical libraries, which primarily serve physicians, surgeons, and medical researchers. The Library of the Eghishe Charents State Museum of Literature and Art, established in 1954, contains books, ps, manuscripts, photographs, audiotapes, musical instruments, works of art and other realia, connected to the renowned literary giant.

PROFILES OF ARMENIA'S MAJOR LIBRARIES

National Library of Armenia

The National Library of Armenia was established in 1919, and subsequently named the Alexander Miasnikian State Library. The 1920s were a crucial period for the formation and growth of the Library's collections, with the acquisition of items provided by the Lazarian Institute in Moscow, the Nersisian School of Tbilisi, the Armenian Benevolent and Armenian Publishers Societies of the Caucasus, the Yerevan Men's Gymnasia (established in 1831), and private contributions made by prominent scholars and intellectuals. In July 1990, it was renamed the National Library of Armenia (NLA) (Fig. 2). The largest library of Armenian books in the world, as of 2000, the collections numbered some 6.2 million items in Armenian, Russian and other world languages, available for on site use and loan to some 40,000 registered readers.

The first legislation regarding the national Deposit of Publications was adopted in February of 1924. Since 1925, the NLA has gained depository rights to a copy of every item published in Armenia. Presently, an amended and improved version of this law, reviewed and ratified in 1959, is in force. In addition to this function, the NLA also serves as the methodological center for all the public libraries in the Republic.

The Library's Old, Rare and Archival Literature Collection contains single copies of every Armenian publication. Its oldest book is the first printed Armenian book *Urbatagirk* (Friday Prayer Book, or, *A Book on Remedies against Witchcraft*), published in Venice in 1512 by Hakob Meghapart, the first Armenian printer. The Library also has a separate, specialized department for performing arts, cartography, music, and literature in world languages.

Beginning in March 2004, the NLA has begun issuing a monthly newspaper, *Hogevor Hayrenik (Spiritual Homeland)*.

American University of Armenia Papazian Library

Soon after Armenia's declaration of independence in September 1991, the American University of Armenia, an affiliate of the University of California, opened its doors in the capital city of Yerevan. Offering a graduate program in engineering and business, the mission of the university was to prepare a new generation of leaders of the new democracy.

Established with a $1 million endowment by Swiss-Armenian benefactors Pinna and Simon Arman Papazian, the University's library has evolved at a rapid pace, implementing contemporary Western-style methods of management and technology.

Since its humble beginnings in 1991, the Library has grown in capacity and readership, currently housing some 28,000 volumes in open stacks and serving over 9000 members—students, faculty, and staff of the University, as well as students, scholars from other institutions, and the community at large. With the American Embassy's USIS library collocated on the same site, a state-of-the-art library would be created to serve as a hands-on, Western-style prototype of what might be possible throughout the country and its existing more than 3000 multitype libraries. At first, the library would occupy a large space on the first floor of the University's building, formerly the site of Communist education headquarters.

What began as a single unit has now evolved into a small information resource network, together with the University's Legal Research Center and the Krikor Soghikian Public Health Reference Room.

Khnko-Aper National Children's Library

Named after a noted author and teller of children's stories, the Khnko-Aper Library was established in 1932 and celebrated its 70th anniversary in October 2003 (Fig. 3). Its mandate is to collect and develop Armenian children's literature. At the entrance to the library building is a sign, which urges children, "Every day, read one page from a new book." Open

Fig. 2 National Library of Armenia, main entrance, Yerevan (Note: Sylva Natalie Manoogian was contributor and translator into English of Armenian text).
Source: From SNM photos, permission granted 2009.

Fig. 3 Khnko-Aper National Children's Library, Yerevan.
Source: From SNM photos, permission granted 2009.

to children from the age of 4 up until the 9th form, the
Library's collections number nearly one half million items.
Adults are allowed only if accompanied by a child.

National Academy of Sciences of Armenia Library (NAS RA)

The National Academy of Sciences of Armenia Library
(NAS RA), which was established in November 1943 as
an affiliate of the USSR Academy of Sciences, unifies
scientific and research institutes, subsidiary services and
the governing body of this state scientific organization.
The Presidium of the Academy includes more than 50
scientific institutions and other organizations in five scientific divisions: Mathematical and Technical Sciences;
Physics and Astrophysics; Natural Sciences; Armenology
and Social Sciences; Chemistry and Earth Sciences.

The NAS RA Web site currently lists 71 full members,
54 corresponding members, 4 honorary members, 117 foreign members, 59 honorary doctors, and 180 deceased
members.

The total number of the workers in the system of the
Academy is 4395.

More than four million items can be found in the
Academy's libraries, with more than half in the main
library. Of these approximately one fourth are in languages other than Armenian and Russian. Annually, this
information resource center serves some 20,000 users. The
Library is organized to serve the needs of its members and
is also part of the Consortium of Libraries engaged in
automation.

Republican Scientific-Medical Library (RSML)

In 1990, the RSML of the Ministry of Health of Armenia,
in collaboration with the Fund for Armenian Relief, created a vision of a national library network supported by
information technology. This vision incorporated four
goals: (1) to develop a national resource collection of
biomedical literature accessible to all health professionals;
(2) to develop a national network for access to bibliographic information; (3) to develop a systematic mechanism for sharing resources; and (4) to develop a national
network of health sciences libraries. The RSML has
achieved significant progress toward all four goals and
has realized its vision of becoming a fully functional
national library, now providing rapid access to the biomedical literature and training health professionals and
health sciences librarians of Armenia in information system use.

Yerevan State University Library

The University's library, with its 10 divisions, is one
of the largest libraries in Armenia, with 12 specialized reading rooms in a number of faculties, housing
collections constantly enhanced by the further addition of academic and scholarly literature, and a division of antiquarian and rare books dating from the
beginning of the seventeenth century. Over 14,000
readers, students, lecturers, postgraduates, researchers,
and other affiliates of the University have access to
the University collection of approximately two million
books.

Among the University's 100 partners in the international exchange of books are the United States Library of Congress, the National Library of the Russian Federation, and university and academic libraries in the United States, France, Germany, Italy, Argentina, and elsewhere. In 1989, the Library received an award as a first-rate scholarly library.

LIBRARY EDUCATION

Degree programs in library science were established in Yerevan State University in the 1950s. Since 1969, librarians are being trained at the Yerevan Armenian Pedagogical Institute, in a 4-year program leading to a diploma in library science, with 25 graduates each year. The curriculum includes both general subjects and technical courses in library and information science. The Holy See of Ejmiatsin Library College has been offering secondary library training to its seminarians since 1931.

The Armenian Library Association (ArLA) has also taken responsibility for continuing professional education for the Republic's librarians and other library personnel, with funding from various sources, such as IREX, OSI, and the Carnegie Corporation, among others. In addition to continuing education courses, consideration is being given to the establishment of a graduate program of information studies offering master's and doctoral degrees at the American University of Armenia.

ARMENIAN LIBRARY ASSOCIATION, 1994–2008

The first steps toward creating a professional library association in Armenia were taken toward the end of the 1980s. During the same time period, the first library associations were also being established in Moscow and St. Petersburg. In the Baltic States librarians re-established their library associations, which had been forbidden during the Soviet years.

Based on a strong, millennial tradition of the book as a national icon, the libraries of Armenia had continued to thrive during the 70-year Soviet era, with a total of 2883 multitype (public, academic, higher education, technical, health, school, and children's) libraries in 1989.

The tiny land-locked republic of the Caucasus was still reeling from the aftershocks of the earthquake and the effects of hastened democratization, exacerbated by the chokehold of an economic blockade from neighboring Azerbaijan to the east and Turkey to the west.

Libraries had stayed open during the winter, even when schools and businesses were closed. Of Armenia's 7000 library workers, 53 librarians, representing 10 of Yerevan's leading multitype libraries, participated in a short course, entitled "Strategic Planning for Armenia's Libraries," offered through the Extension Department of the American University of Armenia, from May 23 to June 9, 1993.

The librarians participating in the course were organized into a planning group, representing the major libraries of Armenia, took active part in the following goals:

- To develop the constitutional structure of the Armenian Library Association
- To convene the inaugural conference of the Association in 1994
- To develop strategic plans for the improvement and automation of its 3000 multitype libraries.

On October 7, 1994, the Armenian Library Association conference was convened in the National Library of Armenia. The Inaugural Conference participants were 119 delegates from all parts of the Republic and 64 guests, including four representatives of the American Library Association. Representing all major research, academic technical, special, public, and children's libraries of the Republic, and with delegates from nearly all of the nation's regional libraries, the participants were 100 women and 19 men.

The initiators of Armenia's new Library Association were the National Library of Armenia, the Academy of Sciences Library, the Republican Scientific-Medical Library, the Republican Scientific-Technical Library, the Khnko-Aper Republican Children's Library, the Avetik Isahakyan Yerevan City Central Library, and the Republic of Armenia Ministry of Culture.

The conference approved the Association's constitution and elected a 21-member administration. Nerses Hayrapetyan, currently Director of the U.S. Embassy's Information Resource Center, was elected the first president of the Library Association. Upon examining the status of library work in the Republic, the inaugural meeting approved the mission statement submitted to the library workers, the intellectuals, and the public, which states:

> We, the participants in this inaugural meeting, are applying to our co-workers and to all those supporters of library work, asking them for their enthusiastic support of the goals of the Association. The free development of our national culture and knowledge depends on the expression of our individual intellectual strength. Libraries, as centers for the preservation of national spiritual values, moral upbringing, cultural and historical traditions, must assume the place of utmost importance in the sphere of illumination and education.

See the Web site of the Armenian Library Association (ArLA) for organizational constitution and by-laws.[3]

The Association was officially registered with the Republic of Armenia Ministry of Justice in March 1995 under the name "Library Association of Armenia." On March 6, 1998, by decision No. 9/1–3 of the Ministry of

Justice, the organization was reregistered under a new name, "Armenian Library Association (ArLA)."

During the 10th anniversary of ArLA, October 7 was designated as National Library Day, with the motto: "New Century; New Libraries."

The ArLA and the National Library of Armenia play key roles as regulative, normative, and cultural-cognitive institutions. They are the leaders in the Consortium of Armenian Libraries, currently involved in automation of the nation's information institutions, and also as collaborators with the libraries of the neighboring South Caucasus republics of Azerbaijan and Georgia.

Association Activities

Since its establishment, the Association has had the purpose of utilizing the specialized abilities of its members for the development of library work in Armenia, to improve access to services and resources by means of modern methods and computer technology.

Toward this end, Association members have enthusiastically participated in approving government documents concerning library work, developing library legislation. The Association has organized and implemented numerous and various types of library programs (Table 1), as follow:

- The Association also implements various types of large and small programs. Among these should be mentioned the subscription for Armenia's libraries, through the support of the South East Florida Library Information Network (SEFLIN), to the EBSCO publications company's EIFLDirect program
- Taking into account the low probability of Internet access in Armenia, approval was given to secure CD and DVD subscriptions for libraries. From January to June 2000, through a "Technical means for participants in the EIFLDirect program" grant, nine regional libraries acquired DVD players.
- In cooperation with the International Research and Exchanges Board (IREX), the Armenian board of the Internet Access and Training Program is currently preparing to mount individual pages on the World Wide

Web for the Armenian Library Association and regional public libraries.

The Association and Continuing Education

One of the main priorities for the Association is the organization of continuing education courses for Armenian librarians (Table 2).

The Association cooperates with Armenia's only institution of higher education for the preparation of library specialists, the faculty of library science and bibliography of the Khachatur Abovian Armenian Pedagogical University. The Association has organized lectures by American library specialists for the student body.

Many of the Association's members participate in international scholarly conferences, seminars, and retrainings. Numerous representatives of the Association are participating in the largest conferences convened throughout the Commonwealth of Independent States, the annual conferences in Crimea. In December 1998, in cooperation with the Russian Library Association, St. Petersburg Library Association member, Karine Arabyan participated in the planning of the seminar, "The role of special organizations in the work of shaping the library into a modern facility." Several members of the Association have also participated in the Open Society Institute Central European University, Bucharest, summer school program.

International and Regional Cooperation

One of the goals of the Association is to participate in international library cooperation (Table 3). Toward that end, in 1995, the Association applied to the governing bodies of IFLA (International Federation of Library Associations and Institutions), indicating its interest in membership in that most important international library organization. In August 1996, during the 62nd IFLA congress in Beijing, the Armenian Library Association was elected as a national association member.

Table 1 Armenian Library Association library programs.

Date	Participants	Program
May 10, 1996	ArLA, National Library of Armenia, National Book Palace	Round table: "The current status of the receipt of obligatory copies of documents of the Armenian Republic and visions of cooperation within the Commonwealth of Independent States."
July 10, 1996	ArLA, National Library of Armenia, Yerevan State University Scientific Library, Khnko-Aper Children's Library, Republican Scientific-Medical Library, and others	Round table: "Status of Armenia's libraries"
October 1994–present	ArLA	Annual conference

Table 2 The Armenian Library Association and continuing education.

Date	Participants	Program
March to June 1998	78 library workers from nine libraries in Yerevan	4-session course: "Education of Library Workers"
October 21–24, 1998	18 representatives from nine regional libraries of the Republic	3-day seminar
March to April 1999	60 library workers	3-week training course, in collaboration with Ararat Republican Regional Library

The Association is registered in four of IFLA's working group sections: national libraries, scientific-technical libraries, national libraries, and information technologies.

Regional relations with Georgian and Azeri library workers began in 1996–1997. During those years, after the break-up of the former Soviet Union, the first immediate linkages among the three nations' library workers took place.

As a result of such collaborative efforts, several librarians from Armenia have received fellowships to study library and information science in the United States so that they can return to the homeland and train their colleagues in new ideas and techniques for improved library service. They have taken courses at the University of Illinois-Urbana/Champaign, Mortenson Center; University of Pittsburgh; and Louisiana State University, among others. They have also served internships in several U.S. libraries, including the Library of Congress, Los Angeles Public Library, and the Bakersfield Public Library, in California.

ARMENIAN LIBRARIES OF THE DIASPORA

With the loss of statehood and the tragedy of the first genocide of the twentieth century, Armenians were compelled to leave their homeland on the Armenian plateau, east of the Euphrates River. Because of their affinity for global exploration and adaptability to new environments, they settled in all corners of the world, carrying with them and re-establishing their cultural institutions in émigré colonies. They built churches, founded schools and libraries, and established printing presses, thus bridging the past and present of their unique heritage.

In the United States, as early as 1875, the quest to preserve the language and instill the history of Armenia led various groups to organize reading rooms and lycea, usually in a church, library, or office building. There, Armenian and English-language periodicals and books were made available. College students and visiting intellectuals spoke to reading-room subscribers in Armenian on matters of current political and literary interest. In 1890, before the city of Fresno, California, had a public library, the Fresno Armenian Library Union was organized, an example emulated by Armenians in the California cities of Fowler and Los Angeles. The desire for a public library was so ardent that even the tiny Armenian community in Kenosha, Wisconsin, raised funds at an Armenian wedding in 1905 for an Armenian reading room.

The number of multitype and specialized Armenian libraries in the Diaspora today is significant. A listing of no fewer than 302, surveyed for their holdings of books printed between 1512 and 1800, can be found in a compilation prepared by three bibliographic scholars and published under the title *The Armenian Book in 1512–1800* (Yerevan: 1988). Notable among them are the libraries of the Mekhitarist Order in Vienna, Austria, and San Lazzaro in Venice Italy; the Monastery Library of Isfahan, Iran; and the Calouste Gulbenkian Library of the Armenian Patriarchate of Jerusalem.

In its latest edition, the *Armenian American Almanac* (Los Angeles: 1995) devotes an entire section to libraries in the United States and Canada which house Armenian collections, categorizing them as organizational and research, public, school, college and university, church and seminary, special and private collections.

NATIONAL ARCHIVES OF ARMENIA

The Web site of the National Archives of Armenia (http://www.armarchives.am) provides a detailed chronology of significant events from the establishment of the Archives

Table 3 ArLA and regional and international cooperation.

Date	Participants	Program
May 5–6, 1997	ArLA with 15 Georgian librarians and two from the United States	Regional conference:The Library as an information center: A new model" took place in Yerevan
May 5–8, 2001	Representatives of ArLA, Library Associations of Azerbaijan, Georgia, and American Library Association	Regional conference, Tiblisi, Georgia: "Strengthening Library Associations in the South Caucacus"
September 27–30, 2002	Representatives of ArLA, Library Associations of Azerbaijan, Georgia, and American Library Association	Regional conference, Yerevan, Armenia: "The Role of Academic Libraries in Fostering Civil Society"

Archives-Art

in 1921–2006. Also included are summaries of organizational structure, archives legalization, description of holdings, and information about archives of state bodies and organizations in the Republic of Armenia and the Autonomous Republic of Nagorno-Karabakh. The *Law of the Republic of Armenia Concerning Archive Keeping* was accepted on June 8, 2004 and its 10-chapter content can be accessed online.

MUSEUMS OF ARMENIA

A visitor to Armenia will be amazed by the beauty of its natural environment, often described as "a museum of antiquities," with its mountains, caves, rivers and lakes, and ancient monasteries and fortresses. "Armeniapedia. org" offers brief descriptions of some 24 museums in the capital city of Yerevan and environs. "TACentral.com" provides a more extensive list of museums in Yerevan and 52 museums throughout the Republic.

The Armenian State Teachers University named for Khachatur Abovian offers Master's Courses in Museology and Protection of Monuments.

The following table lists the Yerevan museums by name and summarizes the regional museums by quantity (Table 4).

Mesrop Mashtots Institute ("Matenadaran") of Ancient Manuscripts

After its confiscation by the Soviets, the scriptorium of the Holy See of Ejmiatsin, which housed an extensive collection of Armenian and Greek manuscripts, became the basis for the "Matenadaran," the manuscript library in Yerevan, Armenia, named after the visionary monk Mesrop Mashtots, who created the Armenian alphabet ca. 405–406 A.D.

The "Matenadaran" is one of the oldest and richest book depositories in the world and, as such, it is listed on the UNESCO Memory of the World Register Nominated Documentary Heritage (Fig. 4). Its collection of some 17,000 manuscripts includes almost all the areas of ancient and medieval Armenian culture and sciences—history, geography, grammar, philosophy, law, medicine, mathematics-cosmography, theory of calendar, alchemy-chemistry, translations, literature, chronology, art history, miniature, music and theatre, as well as manuscripts in Arabic, Persian, Greek, Syrian, Latin, Ethiopian, Indian, Japanese and others. In this cultural heritage repository, many originals, lost in their original languages and known only because of their translations into, have been saved from disappearance.

Genocide Museum and Institute

Principal activities of the Armenian Genocide Institute and Museum include: Scholarly exhibitions of historical-

Table 4 Museums of Armenia.

Yerevan	Tags
Armenian Craft Museum	Arts and crafts
Minas Avetisyan Museum	Armenian painter and stage designer
Yeghishe Charents Museum	Armenian poet
Children's Art Gallery	Center for aesthetic education
Contemporary Art Museum	Contemporary painting, drawing, sculpture
Erebuni Fortress Museum	Archaeology
Genocide Memorial and Museum	Armenian genocide, 1915–1923
Geological Museum	Metals and minerals; fauna and flora
Avetik Isahakyan House-Museum	Armenian poet
Aram Khachaturian Museum	Armenian composer
Matenadaran Manuscript Museum	17,000 manuscripts in Armenian and other languages
Middle East Museum	Arts of the Mediterranean
Museum of Modern Art	Contemporary arts
Sergei Parajanov House-Museum	Armenian cinematographer
Martiros Sarian House-Museum	Armenian landscape painter
Alexander Spendiarian House-Museum	Armenian composer and composer
State History Museum	
State Picture Gallery	
Hovhannes Tumanian Museum	National poet and folklorist
Wood Carving Museum	Folk art
Aragatsotn region (Ashtarak)	Four House-Museums
Ararat region	One House-Museum and one Art Gallery
Armavir region (Ejmiatsin)	11 Museums and Art Galleries
Gegharkunik region (Sevan)	Four Museums and Art Galleries
Lori region (Vanadzor)	Eight Museums and Art Galleries
Kotaik region (Tsaghkadzor)	Five Museums
Shirak region (Giumri)	Five Museums
Siunik region (Sisian-Goris-Meghri)	Nine Museums
Vayots Dzor region (Yeghegnadzor)	Five Museums

documentary materials, archival documents, photos on the Armenian Genocide in 1915–1923; guided tours of the Museum in English, French, German, Russian and Armenian; continuing research. The Genocide Museum and Institute cooperates with various scholarly institutions in Germany, France, United States, Austria, Italy, Russia, and other countries. In 1996 the Museum organized an international conference in collaboration with Italian and

Fig. 4 Mesrop Mashtots Institute of Ancient Manuscripts, Yerevan.
Source: From SNM photos, permission granted 2009.

German scholars devoted to the memory of Armin Wegner, the noted German writer and journalist, a witness of the Genocide. In 1998, the Museum sponsored an international conference dedicated to the 140th anniversary of the birth of Johannes Lepsius, a great friend of the Armenian people. The Museum also conducted international conferences in 1999 devoted respectively to Henry Morgenthau, United States Ambassador to Turkey during the Genocide, and Anatole France, famous French author and supporter of the Armenian cause. On April 23, 2001 it convened an international conference on the occasion of the 86th anniversary of the Armenian Genocide, attended by scholars from all over the world.

Children's Art Gallery and Aesthetic Center[5]

"The World through the Eyes of Children" is the name of an exhibit that opened in the Artists' Union of Armenia in 1968 on the initiative of art expert Henrik Igitian. Soon after, the Children's Art Gallery was founded in 1970 and was later reorganized as the Museum of Children's Art. The Gallery has a permanent exhibit and a showroom where individual and thematic exhibitions are organized. It accepts children's paintings from all over the world and exchanges some of its own with institutions abroad. In 1975 the British Prime Minister Harold Wilson donated 100 paintings by British children

to the Gallery. The Gallery has about 200,000 exhibits (paintings, sculptures, graphic works, ceramics, appliqué, carpets, rugs, and embroidery) from about 120 countries of the world. Artwork from the Museum's stock has been sent to exhibitions staged in the United States, France, Italy, Czech Republic, Slovakia, Japan, and elsewhere, and shown at the Children's Paintings Exhibition "From Fuji to Ararat" displayed in 1997 in Yerevan at the gallery of the Armenian Painters Union, where 400 paintings from 160 countries were displayed. With the founding of the National Aesthetic Education Center in 1978, the Gallery became the Armenian Museum of Children's Artwork (with branches in Gyumri, Vanadzor, Goris, Kapan, Sevan, Spitak, and Meghri). The Center also has painting and decorative applied arts studios, a symphony orchestra, theaters, and a film studio. The Children's Art Gallery was the first museum of children's artwork in the world.

CONCLUSION

This overview of Armenia's libraries, archives, and museums aims to introduce these information institutions against a millennial historical backdrop, within a sociocultural and organizational context, in order to demonstrate their importance in preserving, promoting, and perpetuating Armenian patrimony, cultural heritage, and ethnic identity.

Armenia was a full participant in the World Summit on the Information Society in Geneva (2003), with President Robert Kocharian and his staff representing the Republic. All of the documents from the Summit regarding Armenia are available on-line in English and Armenian. The Republic was also represented at the Tunis World Summit in 2005.

The Armenian Library Association, with its National Library Day motto: "New Century; New Libraries," promises to sustain itself and grow as an advocate for the Armenian Republic's libraries and librarians. Through its participation in the international arena, it has become an active participant in the globalization of its libraries. Future plans for library, archival, and museum specializations include research on educational reform in the South Caucasus, with an aim to establish an integrated graduate information studies program and training center for the region.

APPENDIX 1

 Republic of Armenia = Hayastani Hanrapetut'yun

Government

National: Republic with a unicameral National Assembly (Parliament—131 seats); 5-year term for President; 4-year term for Parliament; universal suffrage; 18 and older (CIA World Fact Book, 2008)

Official Languages: Armenian is the primary official language (97.7%); Yezidi, 1%; Russian, 0.9%; other; 0.4% [2001 census) (CIA World Fact Book, 2008)

Capital City and Administrative Divisions: Yerevan and 11 *marz*-es (provinces)—Aragatsotn, Ararat, Armavir, Geghark'unik', Kotayk', Lorri, Shirak, Syunik', Tavush, Vayots' Dzor (CIA World Fact Book, 2008).

Legal System: Based on civil law system. Has not accepted compulsory ICJ jurisdiction (CIA World Fact Book, 2008).

Key Dates: September 21, 1991 (Independence from Soviet Union) (CIA World Fact Book, 2008)

People and Economy

Total Population: 2,968,586 (July 2008 est.); Rank (World Bank): 131 (2007 est.) (CIA World Fact Book, 2008)

Major Ethnic Groups (%): Armenian (97.9%); Yezidi (Kurd) (1.3%); Russian (0.5%); other (0.3%) (2001 est.) (CIA World Fact Book, 2008)

Religious Affiliation (%): Armenian Apostolic (94.7%); other Christian (4%); Yezidi (monotheist with

elements of nature worship) (1.3%) (CIA World Fact Book, 2008)

Gross domestic product (GDP in 000s US $ and rank): $17.15 billion and 111

(2007 est.) (CIA World Fact Book, 2008)

GDP Per Capita (US $ and rank): $4,900 and 111 (2007 est.) (CIA World Fact Book, 2008)

Education

Adult Literacy Rate: 99.6% M; 99.0% F (2005 regional avg.) (CIA World Fact Book, 2008)

Preprimary School Enrollment: NA/29% M; NA/27% F (2005 regional avg.) (CIA World Fact Book, 2008)

Primary School Enrollment: 101% M; 100% F (2005 regional avg.) (CIA World Fact Book, 2008)

Secondary School Enrollment: 91% M; 87% F (2005 regional avg.) (CIA World Fact Book, 2008)

Tertiary School Enrollment: 25% M; 27% F (2005 regional avg.) (CIA World Fact Book, 2008)

Higher Education Enrollment: (number of institutions—16 State, 72 non-State Universities; 42,505 and 19,755 students, respectively)

Information Infrastructure and Institutions

Number of Books Published: 516 (1999) (UNESCO – UIS, 2008)

Circulation of Daily Newspapers (per 1000 inhabitants): 7.60 (2004) (UNESCO - UIS, 2008)

Radios (per 1,000 inhabitants): 850,000 (1997) (CIA World Fact Book, 2008)

TV Sets (per 1,000 inhabitants): 825,000 (1997) (CIA World Fact Book, 2008)

Films (number produced and seats per 1,000 inhabitants): 38.70 (1995) (UNESCO - UIS, 2008)

Internet Hosts (total): 8,270 (2007) (CIA World Fact Book, 2008)

Internet Users (total): 172,800 (2006) (CIA World Fact Book, 2008)

APPENDIX 2

Chronology of Armenian libraries, archives, and museums

Dates	Event
301 A.D.	Armenia accepts Christianity as state religion
405–406 A.D.	Mesrop Mashtots creates the Armenian alphabet, launching the Golden Age of Armenian literature (5th century), and establishment of monastic libraries and centers of education

(Continued)

Chronology of Armenian libraries, archives, and museums
(Continued)

Dates	Event
1512	Publication of first Armenian printed book by Hakob Meghapart, San Lazzaro, Venice, Italy
18th–19th centuries	Establishment of academic libraries in Diaspora
1875	Inception of reading rooms and lycea in United States Armenian communities
• 1890	Establishment of Fresno Armenian Library Union
• 1905	Armenian reading room established in Kenosha, Wisconsin
1895	Hamidian massacres (Ottoman Empire)
1909	Massacres at Adana, Cilician Armenia
1915–1923	Armenian genocide, Western Armenia
1917	Eastern Armenia becomes part of the USSR
1919	Establishment of National Library of Armenia
1921	Establishment of National Archives of Armenia
December 27, 1922	Establishment of Armenian Book Chamber
1932 November	Establishment of Khnko-Aper National Children's Library
1943 November	Establishment of National Academy of Sciences of Armenia
1950s	Establishment of degree programs in library science, Yerevan State University
December 7, 1988	Earthquake in northern region of Armenia
August 1991	Collapse of Soviet Union
September 21, 1991	Armenia declares independence; American University of Armenia established
October 7, 1994	Inaugural meeting of Armenian Library Association
May 5–8, 2001	Regional conference: "Strengthening Library Associations in the South Caucasus"
September 27–30, 2002	Regional conference: "The Role of Academic Libraries in Fostering Civil Society"
October 7, 2004	Decennial of Armenian Library Association and declaration of National Library Day

BIBLIOGRAPHY

Printed Sources

1. *ALA World Encyclopedia of Library and Information Services*, 2nd Ed American Library Association: Chicago, IL, 1986. (Useful for its overview article on national libraries, although Armenia was not included in the list.).

2. *Armenia at the Crossroads: Democracy and Nationhood in the Post-Soviet Era*; Blue Crane Books: Watertown, MA, 1991. (Edited by Gerard J. Libaridian, these are essays, interviews, and speeches by the leaders of the national democratic movement in Armenia.).

3. Armenian Library Association. *Haykakan Gradaran: Taregirk = Armenian Library: Annual*, ALA: Yerevan, Armenia 2002. (The official journal of the Armenian Library Association, this is the sixth volume of the serial, containing an overview of the progress in librarianship since the establishment of the Association in 1994.).

4. Armenian Tourism Association, *Sustainable Travel and Tourism in Armenia*, ATA: Yerevan, Armenia, 2000. (Funded by the Eurasia Foundation, this full-color brochure includes photographs, maps, and descriptions of sites worthy of visiting, in order to promote the tourism industry in Armenia. Further information is available at http://www.sustravel.am.).

5. *Bibliography of New Armenian Periodical Press, 1987–1996*; Girk: Yerevan, Armenia, 1999. (A chronological survey of new periodicals from the Republic of Armenia, Nagorno Karabakh, the Commonwealth of Independent States, and the Baltic states. The introductory essay is in Armenian, followed by an English version. Citations are in the vernacular of their publications.).

6. Hayrapetyan, N. Libraries of Armenia. *ALA World Encyclopedia of Library and Information Services*; 1995. prepared for (At the time of its preparation, in 1993, the author of this article was in residence at the Mortenson Center, University of Illinois, Urbana Champaign. It was during this period that the guidelines and bylaws for the Armenian Library Association were developed. International Federation of Library Associations and Institutions. IFLA/FAIFE World Report: Libraries and Intellectual freedom—Armenia. Available from http://www.ifla.org.sg/faife/report/Armenia.htm).

7. Hayrapetyan, N. *Amerikyan gradaranagitutyun ev teghekatuner = American Library Studies and Information*; Yerevan, Armenia, 2001. (Based on the author's research during his tenure as a Mortenson Scholar, the work presents resources and services offered in academic and public libraries in the United States.).

8. Libaridian, G.A. *Modern Armenia: People, Nation, State*, Transaction Publishers: New Brunswick, NJ, 2004. (Compiled by a key governmental advisor in the early years of independence, this brand-new publication reviews Armenian politics and political thinking from the mid-19th century to the present.).

9. National Library of Armenia. *Bilingual Guide*, NLA: Yerevan, Armenia, 2000. (An IREX-funded project, of which I was the administrator, this full-color, illustrated overview of the 85-year-old principal state library for the Republic of Armenia made its international debut at the IFLA conference in Jerusalem.).

10. Sochocky, C. Undoing the legacy of the Soviet era. Am. Libr. **1994**, July/August *25*(7), 684–686 (This article highlights the challenges for librarians caused by the end of Soviet totalitarian control.).

11. Zargaryan, T. *T'vayin gradaranner = Towards an Information Society: Digital Libraries*, Author: Yerevan, 2002. (In this survey of librarianship from its historical beginnings, the author, director of information technology at Yerevan State University, explores the importance of automation to libraries and their new role in an information society.).

Online Sources

1. American Corners Website (http://www.americancorners.am/en/).

2. Armenian Library Association Website (http://www.ala.
   nla.am). The site includes English and Armenian versions
   of the association's background, structure, activities, cooperation, publications, bylaws, news, archives, photos, information about National Library Week, links to other library
   sites, and contact information.

3. Armenian Library websites: National Library of Armenia
   (http://www.nla.am); American University of Armenia
   Papazian Library (http://www.aua.am/aua/library); Khnko-
   Aper National Children's Library (http://Khnko-Aper.iatp.
   irex.am/0e.htm); Mashtots Matenadaran Ancient Manu-
   scripts Collection (http://www.matenadaran.am); National
   Academy of Sciences of Armenia Library (RA NAS)
   (http://www.sci.am); Republican Scientific-Medical
   Library (RSML) (http://www.medlib.am/); Yerevan State
   University Library (http://www.ysu.am).

4. Kazaryan, R. Continuing professional education activities
   of Armenian Library Association. (http://www.ais.org.ge/
   conf99/proceedings_eng/kazaryan.htm). Remarks by one
   of the Deputy Directors of the National Library of Armenia

and one of the founding members of the Armenian Library
Association at a regional conference funded by the Open
Society Institute in 1999.

5. National Archives of Armenia Website (http://www.
   armarchives.am/en). Includes structure, legislation, chronology, collections, bulletin, conferences and exhibitions,
   archives of state bodies and organizations, and news.

6. Republic of Armenia. *Archival Affairs Administration
   Website* (http://www.arminco.com/hayknet/archiv-e.
   htm). Describes the structure and system of the Archival
   Affairs administration and the regional archives of the
   Republic.

7. Tour Armenia Website (http://www.tacentral.com/culture).
   Lists Museums of Armenia by location and provides phone
   numbers.

8. World Summit on the Information Society and Armenia:
   Geneva 2003–Tunis 2005. (http://www.undpi.am/wsis).

9. WSIS documents in Armenian and English versions,
   including: Declaration of principles, plan of action, and
   frequently asked questions.

Art Galleries

Daniel C. Danzig
Consultant, Pasadena, California, U.S.A.

Abstract

The transition of the support structure for the arts from one of patronage to a primarily market-based environment provided an opportunity for art connoisseurs to establish private commercial art galleries. With the growth of the art market, the gallery system has developed into the primary enterprise for distributing art, as well as the dominant venue for introducing and presenting new art. Art galleries reflect the diversity in the art world and are similarly differentiated in terms of high and low, popular and elitist, avant-garde, and commercial. This entry defines the ideal-typical art gallery and describes its operations and role in the distribution of art. The qualities that distinguish galleries are situated within the cultural hierarchies of the art world as a way to better understand the various roles of art galleries within an integrated, but stratified system for the production and consumption of art.

INTRODUCTION

In the broadest sense, an art gallery is any architectural space utilized for the display of art or artifacts. The Western concept of the gallery originated during the Renaissance when noble residences featured colonnades, balconies, and eventually entire rooms for displaying art, antiquities, and cabinets of curiosities. These galleries were primarily private affairs, functioning as symbols of wealth, status, education, and taste for the benefit of the aristocracy and cultured elite. Only in the eighteenth century did public spaces begin to be designated as exhibition galleries for art, most often within institutions such as academies, universities, museums, libraries, and other civic organizations. (For a history of the establishment of museums see Bazin[1] and Impey and MacGregor.[2]) Thus, a large museum is generally a collection of galleries, with each exhibition space often dedicated to a specific period, discipline, genre, artist, or theme.

But art galleries are more than exhibition spaces: they are also organizations in and of themselves, and when we think of art galleries today, particularly in the United States, we generally refer to privately owned, commercial businesses operating as retail enterprises that buy and sell works of art.

As with any creative endeavor, the arts are measured in degrees of competency, excellence, and significance, and art galleries reflect these distinctions. Galleries can be located on a continuum between art created for the sake of aesthetics and expression at one end, and art produced for its marketability on the other. The paradox of the contemporary art world lies somewhere along that stretch of multiformity, a balance, (some would say compromise), between artistry and success.[3] That is, the art world is a tiered system that distinguishes galleries by those that handle fine art and high culture from those that promote commercial art and popular culture. Fine arts galleries are closely allied with museums, curators, auction houses, and others in the international art market, while commercial galleries lean toward interior design and decoration. The majority of galleries are a composite, yet the ideal-typical gallery conforms largely to the top-tier model: a large, minimalist white room, single works of art, and dramatic lighting characterize the common notion of an art gallery. (For an interpretation of the aesthetics of display in art galleries and exhibit design in museums see O'Doherty[15] and Staniszewski.[4])

This entry begins with a definition of the quintessential contemporary art gallery and its role in the presentation and distribution of contemporary art. Although the model gallery is not unusual, a variety of specializations exist among them, while others appeal to quite different markets altogether and operate with dissimilar business models and organizational structures. Differentiation among galleries, particularly for those involved in the international contemporary art market, is situated within the cultural hierarchies of the art world as a way to better understand the various roles of art galleries within an integrated, but stratified system for the production and consumption of art. A brief history of the development of the modern art market places the establishment of art galleries in context. The diversity within the field is described to provide an introduction to the range of variance.

DEFINITION

Art galleries are, basically, retail establishments exhibiting and selling original works of art. More importantly, however, art galleries facilitate the distribution of art through a cooperative enterprise, providing art

Encyclopedia of Library and Information Sciences, Fourth Edition DOI: 10.1081/E-ELIS4-120043708

Archives–Art

collectors a mediated and authoritative source for acquiring works of art, while at the same time integrating artists into society by helping them transform their vision, talent, and investment in materials into financial support for their creations, by converting aesthetic value into economic worth.[5] The following description more specifically represents what is, perhaps, the archetypal art gallery within this system, albeit a top-tier one selling art by living artists in the international contemporary art market.

Operations

An art gallery is a small business, generally a sole owner/director supported by a few staff.[6] The dealer or gallerist, as the owner is known, represents a select group, or stable of artists, usually around 20, and exhibits their work publicly. Galleries present an annual schedule of six to ten changing exhibitions, including solo shows that feature the work of an individual artist, or group exhibitions of multiple artists organized by artistic theme or affiliation.

Exhibitions are an opportunity to not only present each artist's new work, but also, more importantly, to promote the artists to art critics and museum curators. Dealers then use their confirmation of artistic merit to cajole collectors to purchase artwork. Ideally then, growing collector interest, particularly among museums, bolsters the artist's reputation, which in turn leads to appreciating value for their work, and ultimately results in a sustained artistic career. Financially, the gallery generally assumes all the expenses associated with presenting and publicizing the art and artist, and in return receives a percentage of the sale price. The standard commission is 50%, although individual business arrangements vary. Dealers may also purchase their artist's work outright—artist stipends are less common today—and then resell it, making an additional profit if the value increases over time.

As promoters, dealers must be able to convince others that their artists have the potential to become important in the art historical context, and consequently collectible. As part of the promotional process for building an artist's reputation, dealers advertise in the major art magazines, e.g., *Art in America*, *Artforum*, *ArtNews*, (these influential trade journals are an important resource for current events and a leading indicator of trends in the international contemporary art market) and mail announcements of exhibit openings to a large, but targeted audience. Dealers then begin to establish a demand in the market, or a following for their artists by encouraging critics—who hopefully will write positive reviews and topical articles—and curators—who may include their artists in museum exhibitions and/or collections—to also recognize and acclaim an artist's importance.

A collector's reputation can also affect the reputation of the artists they collect. Dealers are able to increase the value, both economic and aesthetic, of an artist's work by placing it in a high profile, well-respected collection, a

museum for example. Dealers actively cultivate their collectors, be they individuals, corporations, or museums by networking and providing value-added services such as art education, investment counseling, appraisals, and financing in order to maintain this relationship. Unlike a typical commercial business where all buyers are welcome, generally on a first-come-first-served basis, art dealers may even restrict sales, if demand allows, of important artwork to their best clients in an effort to maximize the benefits. Establishing prices for art works and increasing their value over time is a complicated formula involving more than just basic supply and demand, but also, as with other luxury goods, more intangible factors as status, reputation, provenance, and fashion. (For more on the economics of art see Robertson.[26])

Role

As the primary distributor for art, galleries function in a somewhat similar capacity as both publishers and booksellers do for literature. However, the role of the art dealer is often not limited to publicity and sales, but may also include discovering and premiering new artists, and directly brokering the relationship between artist and collector. In this capacity, an art dealer is perhaps more like a literary agent. Dealers constantly wade through a sea of new artists, often received as unsolicited submissions to the gallery. The dealer evaluates an artist's work, rejects those they regard as unworthy of exhibition and unlikely to advance, and accepts for a trial run those whom they feel have potential. The artists who fail to garner critical notice, or lucrative sales, are released, while those that succeed are retained and promoted. Each dealer has individual artistic preferences, or aesthetic points of view, and chooses artists to represent accordingly.

Museums rarely exhibit work by young, untested artists, so galleries function as the proving ground for emerging artists—although rarely lauded for providing a public service, art galleries significantly enhance cultural life by presenting free public exhibitions of art, much of it for the first time. Since galleries are the dominant venue for introducing and presenting new art, dealers also assume the role of gatekeeper,[6] first, by initially deciding which works will be exhibited and which will not, and second, by being as selective as possible about the type of collectors to whom they are willing to sell, particularly when it comes to their best artists. One way to evaluate an art dealer, as art critic Lawrence Alloway[7] sees it, is to "…measure dealers…by their power to sell art, infiltrate museums, and generate art criticism."

Art dealers who follow this model for seeking out the rising stars of the art world are more interested in choosing unrecognized artists they feel are breaking new artistic ground, ones that are truly avant-garde, than in artists who may be more of sure thing financially. While some dealers prefer to represent these emerging artists, others favor artists who are more established and already recognized, and consequently their work is more likely to sell. Still other galleries do not even represent living artists, but only historical artwork. These differences in focus or specialization among galleries are only some of the ways in which they vary. Regardless of which type of gallery a dealer develops, the style is often a reflection of the owner's individual tastes, goals, and values, as is the case with most small independent businesses. (For more on the idiosyncrasies of individual dealers see Klein.[8]) So while the description above represents the common idea of an art gallery, it does not sufficiently describe all art galleries.

DISTINCTION

Perhaps the most common way to distinguish one type of gallery from another is by the quality of the artwork exhibited. This method is, however, highly subjective and invariably leads to applying something like a high- vs. low-culture framework. (For an investigation of high vs. low culture see Gans[9] and Bourdieu.[10]) The top tier of galleries adopting this criterion present only museum-quality art, i.e., art that meets rigorous standards regarding the artist's education, aesthetics, art historical significance, etc., and which could conceivably be exhibited in an art museum.[3] A second criterion for the quality of art and the galleries that handle it would be investment potential. These two benchmarks distinguish elite galleries from more commercially-oriented galleries, frame shops with art displays, and similar retail stores offering more popular, or less aesthetically challenging decorative art and graphics, mass-produced posters, functional crafts, and tourist mementos.

The 2002 US Economic Census[11] reported more than 6000 galleries in the United States; however, the number handling museum-quality art is much smaller. The national art magazine *Art in America* publishes an annual guide to artists, arts organizations, and art services. In 2007 it provided free listings for just over 4000 commercial galleries.[12] The classified exhibition directory, *Gallery Guide*, included just over 3000 paid gallery listings the same year.[13] The number of galleries in invitational organizations is much smaller: Art Dealers Association of America lists only 170 members, all of which are top ranked and most of which are located in New York. The four major national associations of art and antique galleries in the United States (see Resources, below), represent approximately 400 members, or less than 10% of the total number of establishments. These gallery groups operate in all three economic markets, and handle the great majority of the most sought-after museum, and investment-quality art represented in major museum exhibitions and collections.

Among these 400 or so elite galleries, distinctions become more complex and involve the dealers' access to,

and levels of, participation in social networks, and their degree of professional reputation. Artist Martha Rosler[14] recounts a conversation with a dealer who pointed to his coffee table and said, "See this table? That's the art world. And you're either on it or you're not." Only a small group of galleries meet the top criteria, and have achieved influence and power within the small network of players in the fine arts.

THE ART WORLD

Since the middle of the twentieth century, art galleries have operated as part of what is broadly referred to as the art world. Art critic Lawrence Alloway[7] describes the art world as a system: an integrated network of artists, dealers, critics, museum curators, historians, collectors, and staff affiliated with for-profit and nonprofit galleries, museums, academies, publications, auction houses, government cultural agencies, etc. Within this system, art moves through various spheres or stages as its audience changes. Artworks are first created in the artist's studio, where the initial audience includes the artist's friends. The audience expands to a specialized, interested public when the work is presented at a gallery exhibit. Works that are sold then reach the collector's circle. Museum exhibits, and publication in catalogs, magazines, and newspapers, expand the audience to a broader public.

Sociologist Howard Becker[5] describes the relations between artists and audiences in terms of "patterns of cooperation." He too sees the art world as an interdependent system or network of social relationships that define the production and consumption of art. Rather than seeing artists in the cliché of isolated, romantic mythically gifted individual creators toiling alone in their studios, they are part of integrated spheres of colleagues, styles, movements, regions, etc., their efforts being shaped by interactions among various participants. For Becker, art worlds are composed of all those whose cooperation and coordinated collective activities, utilizing a set of shared conventions, are required to produce, distribute, and consume art, rather than any essential quality or inherent characteristics of the art work itself. Membership in the art world is, at least in the commercial sector, self-determined, since there are no regulations that control entry into the field, and no required training, certification, or licensing for art dealers.

Therefore, there is no singular art gallery, beyond the basics of an architectural space for exhibiting art and a business engaged in selling it. While general occupations and affiliations may be easily attributed to the art world and membership within those categories seem clearly determined, for example, any business calling itself an art gallery is an art gallery, the sheer number of galleries and artists encourages not only specialization and market segmentation, but hierarchies as well that help distinguish the relative position of participants and organizations.

For many, the extent to which art galleries engage in elitism is problematic. Artist and art critic Brian O'Doherty's[15] important series of critical essays on contemporary galleries as exhibition space captures what has become for many, particularly artists, a persistent problem with galleries, chiefly that the "gallery space still gives off negative vibrations." Three perceptions contribute to the bad connotations, according to O'Doherty, first "esthetics are turned into a kind of social elitism—the gallery space is exclusive…[second] esthetics are turned into commerce—the gallery space is expensive…[and third] without initiation…art is difficult."

For better or worse, fine art continues to function as, or at the very least is still perceived symbolically to be a luxury commodity that conveys status. As a result, it is often classified in a stratified pyramid of cultural production and consumption that defines works of art in terms of high and low, popular and elitist, avant-garde and commercial.[3] Determining the value of that status is often quite subjective and it is generally established by complicated interactions among various, but unequal art world constituents.[10]

FROM PATRONAGE TO MARKET

Although there has been an arts trade for as long as there have been artists and patrons, the practice of art as commerce is a product of the modern era.[16] Patronage had been the dominant system for supporting the arts until the seventeenth century. An art patron, that is, the church, court, or aristocrat, would commission works directly from artisans, who usually worked as part of a guild. The patron often dictated the subject, if not the style and method, and effectively controlled the market, such as there was one. The Medici family of Florence, for example, was a significant art patron during the Renaissance. Again, these artworks were symbols of the patron's prestige and their sponsorships required a certain amount of knowledge about the subject, usually history or religion, as well as about aesthetics and craftsmanship. Eventually, however, structural changes that fostered the development of bourgeois society made possible shifts in the production and consumption of art from the aristocracy to a developing upper-middle class, from commissioned patronage by the elite to public participation in a commercial market.[17]

For these new admirers, art was mostly acquired through inheritance, as gifts or travel mementos, or at estate auctions, rather than directly through commissions. Art entered the marketplace as early as the seventeenth century, with the Dutch art trade being one of the earliest examples.[18] Shops and merchants of various types began to sell paintings and prints, but the art was more often intended as interior design pieces along with furniture, decor, and curios, recognized more for their craftsmanship and not yet generally regarded as "high" art. Similarly, as the demand for art grew, especially among an aspiring new class of merchants, tastes changed as well, with a growing preference for landscapes, portraiture, and domestic scenes, as opposed to the grand historical narratives and religious allegories commissioned by the state or church. To meet this demand, the number of artists increased dramatically, and so did various academic and artist associations, which began organizing expansive public art exhibitions. The most renowned during the seventeenth and eighteenth centuries was the Salon, the official art exhibition of the Académie des Beaux-Arts in Paris, France, with equally important exhibitions organized by the Royal Academy of Arts in London, which provided an opportunity for both burgeoning groups to connect. As art schools and academies were established and artists trained and professionalized, art movements with specified philosophies, styles, and manifestos developed, and consequently, interpretation, criticism, and the historicization of art. Soon, what was and was not art became formalized, specialized, and debatable. For those who were unfamiliar with art history and these new artistic trends—the cultural avant-garde—the need, or at least the opportunity for an arbiter between artists and this new audience developed. And it was not long before connoisseurs entered the world of art in order to help negotiate an exchange and make a profit for themselves along the way.[19–21]

The developing moneyed classes of the nineteenth century, in the United States in particular, wanted to establish their social status by building their cultural capital in addition to their increasing economic fortunes. However, most had not inherited a history of arts patronage, and therefore had little experience acquiring an art collection. This became an opportunity for entrepreneurs to insert themselves as intermediaries and aspirant brokers in art between now-independent artists and the nouveaux patrons-cum-collectors. With the novice collectors' aspirations in mind, the galleries tended, early on, to mostly feature paintings by well-established, prominent European masters, rather than unknown provincial American artists. The dealers recognized that consecrated works would more readily convey the sophistication of their collectors. The Duveen Brothers was a prominent gallery at this time. But as the interest in collecting art grew and art galleries flourished, Old Masters and other prominent art became harder to acquire relative to the growing demand, so dealers increasingly began to feature art work by living artists in addition to the classics. At the same time, artists were asserting their creative freedom to produce innovative work and were gradually being recognized as legitimate. By the late nineteenth century, art galleries exhibiting and offering original works of art exclusively as fine art, rather than merely decorative items, were becoming well-established. Two important galleries that were established at this time, and continue to operate today, are Knoedler and Co. and Wildenstein and Co.[22]

The rise of art dealers and their establishment of galleries marked a significant change in the development of a new art market as they assumed the role of middleman. But art dealers gradually undertook a much more significant role than merely facilitating transactions. In this new capacity, gallery owners expanded their function in the art trade, becoming impresarios of the fine arts who represented artists, presented exhibitions, and built audiences. Alfred Stieglitz was one of the earliest art dealers of this type, opening his New York gallery in 1905. Other important galleries were started by Betty Parsons and Sidney Janis, both in the late 1940s, and Leo Castelli in 1957. (Today's prominent dealers include Arnold Glimcher and his PaceWildenstein gallery, Mary Boone, and Larry Gagosian.)[22]

Much as theater producers sought new plays, art dealers sought to discover and premiere new artists, only selling art instead of tickets. Ultimately, they assumed the role of gatekeepers, deciding which art works to display, and which artists to promote, and became tastemakers, courting critics and favorable reviews, and cultivating collectors by advising them on what to buy. The gallery eventually developed as the primary outlet for an artist's work. While artists have periodically initiated alternatives to the dealer–artist relationship, e.g., nonprofit cooperative galleries run by artists, for the most part, artists need to be represented by a gallery, similar to actors having an agent, in order to regularly exhibit, promote, and sell their art. This relationship became firmly established with the fall of the Parisian art market after World War II and development of the contemporary avant-garde art movement in New York and its assumption as the art world capital.[6]

Diversity in the Market

The transition of the support structure for the arts from one of patronage to a primarily market-based environment provided an opportunity for art dealers to facilitate the distribution of art. With this system established, the overall expansion of the United States in the second half of the twentieth century was set to easily accommodate a significant boom in the arts. As a growing population of professionally educated artists began producing more and more art, a commensurate number of entrepreneurial connoisseurs joined the gallery system as dealers to negotiate its distribution. And as the market expanded and segmented, with artists creating work in endless styles and media, art dealers differentiated their galleries from one another to meet equally diverse tastes.

According to the 1970 census, the number of individuals classified as professional artists ("Professional artists" refers to the census occupational classification of painters, sculptors, craft-artists, and artist printmakers. This is a distinct category from performing artists, authors, etc.) was 87,000.[3] By 2000, this figure had increased to 231,000.[23] The number of art galleries grew at an even faster rate. The 1982 US Economic Census reported 1563 establishments; just 10 years later 4543 galleries were

counted, and the 2002 report totaled 6328 galleries.[11,24] These totals represent significant growth in the sector. In New York City alone, for example, there were approximately 150 galleries in 1946; by 1975 the total had increased to 761.[3] By 2007, New York had more than 1000 commercial galleries.[12]

Just as the entertainment business is often referred to as an industry, e.g., the film industry, the buying and selling of works of arts is a well-established commercial enterprise, with art dealers and their galleries playing a central role. This industry is part of the U.S. retail trade sector, is classified by the North American Industry Classification System as *453920-Art Dealers*, and defined as "establishments primarily engaged in retailing original and limited edition art works. Included in this industry are establishments primarily engaged in displaying works of art for retail sale in art galleries."[11] The category includes auction houses selling art, in addition to art dealers and art galleries selling art to retail customers. These are distinct from retail establishments selling reproduction art or art supplies, art galleries displaying art not for retail sale (e.g., museums and other similar arts organizations), or those selling art without a physical location (e.g., electronic home shopping or direct sales). The sector is responsible for generating annual sales in the United States of more than $4 billion in 2002,[11] up from $3 billion in 1997, and $2 billion in 1992,[24] art galleries account for approximately 70% of this market,[25] half of which is generated by galleries in New York, California, and Florida.[24] Auction houses are responsible for the remaining 30% of sales in the sector.[25] There are about 100 art auction houses in the United States,[11,13] but the market is dominated by two international companies, United States-based Sotheby's and United Kingdom-based Christie's, who together account for approximately 80% of auction sales.[25]

For economists, and others concerned with art as a commodity, the art market is broadly divided into the primary, secondary, and tertiary markets. The primary, or source, market refers to contemporary art works being sold for the first time, and this generally means the work of young, emerging artists, although unknown artists from previous generations whose work has never been sold are occasionally discovered. The secondary market deals in the resale of artworks, either the work of recognized artists who have built a following among collectors, or other established historical pieces. The sale of fine art by auction houses is classified as the tertiary market. Art dealers participate in this market when they buy and sell artwork at auction.[26]

Similarly, art galleries can be segmented by the same market categories based upon the type of artwork for sale, i.e., emerging or established artists, or auctionable art. Given the quantity of art produced, the majority is only ever sold in the primary market, if it is sold at all. There is very little demand for most art beyond its initial purchase, since very few artists gain enough of a reputation to build a secondary market. Auction houses, particularly the large international ones, are becoming increasingly selective about the art they accept on consignment for sale; therefore an artist's reputation and an artwork's provenance are increasingly important.

As with individual businesses in any commercial sector, galleries specialize in a number of different ways beyond these broad market categories, according to the types of art they handle and other factors. The art market and galleries can also be divided by artistic discipline, in art historical terms, or by category,[26] in economic terms (e.g., paintings, prints, or photographs); by period and region, or sector, respectively (e.g., Italian Renaissance, Twentieth Century American, or Chinese Contemporary); by style and movement, or type (e.g., Impressionism, Modernism, or Pop Art); by genre (e.g., portraits, landscapes, or sporting); or by media (e.g., ceramics, film and video, or textiles). For example, one gallery might concentrate on anything from the more general category of contemporary art by emerging artists, another in world folk art, while others might focus narrowly on American nineteenth century marine paintings, or black-and-white photography. The various concentrations are nearly endless. (See Table 1 for a listing of categories and sectors in the art and antiques market used by the two major international auction houses, Sotheby's and Christie's. These are closely aligned with gallery specializations.)

Galleries may also focus on a particular type of customer. The connoisseur and institutional collector was the clientele of choice for the galleries described earlier. However, galleries cater to many other types of art buyers as well. Some galleries may appeal to tourists by featuring themed art on local attractions, history, or culture, or by promoting only local artists (e.g., American Indian ceramics in Santa Fe, New Mexico, or plein-air paintings

Table 1 Structure of international art market by category and sector.

Discipline or category	Period and region or sector
Ancient and Ethnographic Arts	Aboriginal Art; African and Oceanic Art; American Indian Art; Antiquities; Islamic Art; Pre-Columbian Art
Asian Art	Chinese Ceramics and Works of Art; Chinese Classical and Modern Paintings; Chinese Contemporary Art; Indian and Southeast Asian Art; Japanese Art; Korean Art; Southeast Asian Paintings
Books and Manuscripts	Atlases, Maps, Natural History, and Travel; Continental Books and Manuscripts including Science; English Literature, History, Private Press, and Children's Books; Printed and Manuscript Music; Printed and Manuscripts Americana; Western Manuscripts; Medieval Illuminated Manuscripts
Ceramics and Glass:	English and Continental porcelain and pottery from the early eighteenth to nineteenth century; glass and paperweights from the fifteenth to the nineteenth centuries; and, Chinese export porcelain.
Collectibles and Memorabilia	Arms, Armor, and Militaria; Automobiles; Collectibles and Memorabilia; Sporting Guns
Fashion	Lace, costume, textiles, fans, needlework, quilts, and samplers dating from the sixteenth to the twentieth centuries, as well as haute couture from the 1920s onward.
Furniture and Decorative Arts:	Nineteenth Century Furniture, Sculpture, and Decorative Works of Art; Twentieth Century Decorative Arts and Design; American Furniture, Decorative Works of Art and Folk Art; Arcade: Furniture, Decorative Works of Art, and Carpets; Decorative Arts and Jewelry Department; English Furniture and Decorations; European Furniture; European Sculpture and Works of Art; French and Continental Furniture, Decorations and Tapestries; Garden Statuary and Architectural Items; General Furniture and Decorations; House Sales; Judaica; Rugs and Carpets
Jewelry	Modern and period jewels, signed and unsigned, precious stones, as well as jewelry from specific periods, styles, or jewelers.
Musical Instruments	Stringed, fretted, and wind, as well as early keyboard instruments
Paintings, Drawings, and Sculpture	Nineteenth Century European and British Paintings; Twentieth Century British Art; American Paintings, Drawings, and Sculpture; Australian Art; British Drawings, Watercolors, and Portrait Miniatures; British Paintings 1500–1850; Canadian Art; Contemporary Art; Fine Arts (the following collecting areas: Old Masters and Nineteenth Century European Art, Impressionist and Modern Art, American Paintings, Drawings, and Sculpture, and Contemporary Art); German and Austrian Art; Greek Paintings and Sculpture; Impressionist and Modern Art; Irish Art; Israeli and International Art; Latin American Art; Marine Paintings and Nautical Works of Art; Old Master Drawings; Old Master Paintings; Orientalist Paintings; Scandinavian Paintings; Scottish Art; Spanish Paintings; Sporting Art; Swiss Art; Victorian and Edwardian Art
Photographs	Photographs; Picture Library
Prints	Old Master, Modern, and Contemporary Prints
Silver, Russian, and Vertu	American Silver; English and European Silver and Vertu; Russian Paintings, Works of Art, Faberge, and Icons
Stamps, Coins, and Medals	Coins; Postage Stamps
Watches and Clocks	Clocks, Watches, Barometers, and Mechanical Music
Wine	Fine and Rare Wine

Source: Compiled from the Specialist and Sales Departments of the auction houses Sotheby's and Christie's.

in Laguna Beach, California). Other galleries may cater to interior designers or corporate art curators, for example, by providing reasonably priced prints that might integrate easily in a home or office setting, or by offering professional discounts.

Art galleries often locate near one another in order to take advantage of a shared audience. Clusters of galleries may exist in specific neighborhoods that are designated as art districts, especially in large urban centers. New York has the highest number of galleries in the United States and several recognized art districts, such as SoHo, Chelsea, and 57th Street, and its galleries' influence continue to dominate the international art market. However, arts communities in other locations (e.g., Los Angeles, Chicago, San Francisco, Washington, D.C., and the metropolitan areas of Florida and Texas) have proliferated and grown in influence recently. Successful galleries also operate in smaller cities that have large populations of artists, or the so-called art colonies, which in turn attract art collectors (e.g., Santa Fe and Taos, New Mexico). Many small local markets throughout the country also support artists and galleries, albeit in proportionally fewer numbers, and with lesser influence in the wider national and international markets.[3]

CONCLUSION

Private commercial art galleries have developed as the primary enterprise for distributing art to collectors and appreciators of art. The art market in general, and the market for museum-quality art in particular, has expanded quite measurably to become a not-insignificant part of the retail sector. The diversity and quantity of art being produced combined with increasing public participation in the arts has fostered an equal expansion of galleries to meet the supply and demand. Consequently, art dealers have worked to distinguish themselves in a number of ways, through specialization, market segmentation, and status in order to succeed.

The independence of art galleries as entrepreneurial businesses operating in a relatively unregulated marketplace, combined with no real alternative system for introducing and presenting new artists, enables art dealers to exercise significant control over what art is, and is not exhibited. This gatekeeping role is further stratified by a closed network that distinguishes and privileges elite galleries within the art world. This closed network, comprised of a fairly small group of dealers, collectors, curators, and auctioneers, enables a recurring number of conflicts of interest and ethical dilemmas, particularly among interactions with the public nonprofit side of the art world. Although these galleries represent a small percentage of the total number of establishments they are most often presented as the model art gallery and the focus of the majority of press and research on the international art market.

Most artists have maintained a love/hate relationship with art dealers and their galleries, and periodically organize for themselves alternative venues and systems for exhibiting and selling their art with limited success. The relationship is somewhat similar to that between musicians and record companies. The recent trend in that medium to self-produce and distribute may be an indication of new alternative approaches for artists to the art gallery.

RESOURCES

Professional Associations

The Art Dealers Association of America, http://www.artdealers.org

International Fine Print Dealers Association, http://www.ifpda.org

The Art and Antique Dealers League of America, http://www.artantiquedealersleague.com

The National Antique & Art Dealers Association of America, http://www.naadaa.org

The Private Art Dealers Association, http://www.pada.net

CINOA: Confédération Internationale des Négociants en Oeuvres d'Art, or International Confederation of Traders in Works of Art (an international organization of art dealer's associations), http://www.cinoa.org

The Fine Art Dealers Association, http://www.fada.com

Association of International Photography Art Dealers, http://www.aipad.com

New Art Dealers Alliance NADA, http://www.newartdealers.org

Magazines and Journals

Art in America, http://www.artinamericamagazine.com
Artforum, http://artforum.com
ARTnews, http://artnewsonline.com
FlashArt, http://www.flashartonline.com
The Art Newspaper, http://www.theartnewspaper.com
Art + Auction, http://www.artinfo.com/artandauction
American Art Collector, http://www.americanartcollector.com
The Burlington Magazine, http://www.burlington.org.uk
Art History Journal of the Association of Art Historians, http://www.blackwellpublishing.com
Art Bulletin, http://www.collegeart.org/artbulletin

Directories and Archives

Gallery Guide monthly exhibition listings, http://www.artinfo.com/galleryguide

artguide, *Artforum* magazine's monthly exhibition listings, http://www.artforum.com/guide

e-flux, an electronic listing of exhibitions, http://www.e-flux.com

Art-Collecting.com, Online Art Gallery Guides and listings of fine art galleries located in the United States

Smithsonian Institution's Archives of American Art: A Guide to Commercial Art Gallery Records in the Archives of American Art, http://www.aaa.si.edu/guides/pastguides/artgall/artgalac.htm

Art in Context is a Web-based reference library providing public access to information concerning artists around the world. The site maintains an extensive international gallery directory, http://www.artincontext.org/gallery

The Art Reference Resource Center a Web site maintained by the Albright-Knox Art Gallery Library that includes a listing of galleries throughout the world. http://www.albrightknox.org/resourcecenter

Fairs, Festivals, and Biennials

Art Basel, http://www.artbasel.com

Art Basel Miami Beach, http://www.artbasel-miamibeach.com

The Armory Show, The International Fair of New Art, http://www.thearmoryshow.com

Art Chicago, http://www.artchicago.com

Art Forum Berlin, http://www.art-forum-berlin.com

Art Los Angeles, http://www.artfairsinc.com

Venice Biennale, http://www.labiennale.org

Whitney Biennial, http://www.whitney.org/biennial

São Paulo Art Biennial, http://bienalsaopaulo.globo.com

Documenta, Kassel, http://www.documenta.de

ARCO Madrid, http://www.arco.ifema.es

Blogs, Galleries, and Auctions Online

ArtsJournal, http://www.artsjournal.com

Modern Art Notes, http://www.artsjournal.com/man/

CultureGrrl, http://www.artsjournal.com/culturegrrl

art:21, http://blog.art21.org/

Rhizome, http://www.rhizome.org

Saatchi Gallery Online, http://www.saatchi-gallery.co.uk

Artnet, http://www.artnet.com

REFERENCES

1. Bazin, G. *The Museum Age*, Universe Books: New York, 1967, Translated by Jane van Nuis Cahill.
2. In *The Origins of Museums*; Impey, O., MacGregor, A., Eds.; Clarendon Press: Oxford, U.K., 1985.
3. Plattner, S. *High Art Down Home: An Economic Ethnography of a Local Art Market*, The University of Chicago Press: Chicago, IL and London, U.K., 1996.
4. Staniszewski, M.A. *The Power of Display: A History of Exhibition Installations at the Museum of Modern Art*, The MIT Press: Cambridge, MA, 1998.
5. Becker, H.S. *Art Worlds*, University of California Press: Berkeley, Los Angeles, CA and London, U.K., 1984.
6. Crane, D. *The Transformation of the Avant-Garde: The New York Art World, 1940–1985*, The University of Chicago Press: Chicago, IL and London, U.K., 1987.
7. Alloway, L. *Network: Art and the Complex Present*, UMI Research Press: Ann Arbor, MI, 1984; 193.
8. Klein, U. *The Business of Art Unveiled: New York Art Dealers Speak Up*, Peter Lang: Frankfurt, Germany, 1994.
9. Gans, H. *Popular Culture and High Culture: An Analysis and Evaluation of Taste*, Basic Books: New York, 1971.
10. Bourdieu, P. *Distinction: A Social Critique of the Judgment of Taste*, Harvard University Press: Cambridge, MA, 1984. Translated by Richard Nice.
11. United States Bureau of the Census, *2002 Economic Census*, Department of Commerce, Bureau of the Census: Washington, DC, 2004.
12. *Art in America: Annual Guide to Galleries, Museums, Artists*, Art in America, Inc.: New York, 2007; 95, 75–278 (7).
13. http://www.artinfo.com/artmarket/ Gallery Guide.
14. Rosler, M. Money, power, contemporary art. Art Bull. **1997**, *79*(1), 20.
15. O'Doherty, B. *Inside the White Cube: The Ideology of the Gallery Space*, The Lapis Press: Santa Monica, CA, 1986; 76.
16. *Art Markets in Europe, 1400–1800*; North, M., Ormrod, D., Eds.; Ashgate: Aldershot, U.K., 1998.
17. Bürger, P. *Theory of the Avant-Garde*, University of Minnesota Press: Minneapolis, MN, 1984. Translated by Michael Shaw.
18. North, M. *Art and Commerce in the Dutch Golden Age*, Yale University Press: New Haven, CT and London, U.K., 1997.
19. White, H.C. White, C.A. *Canvases and Careers: Institutional Change in the French Painting World*, John Wiley & Sons: New York, 1965.
20. Crow, T.E. *Painters and Public Life in Eighteenth-Century Paris*, Yale University Press: New Haven, CT, 1985.
21. Moulin, R. *The French Art Market: A Sociological View*, Rutgers University Press: New Brunswick, NJ, 1987. Translated by Arthur Goldhammer.
22. Goldstein, M. *Landscape with Figures: A History of Art Dealing in the United States*, Oxford University Press: Oxford, U.K. and New York, 2000.
23. National Endowment for the Arts. *Research Division Report #48: Artists in the Workforce: 1990–2005*, NEA: Washington, DC, 2008. (http://www.nea.gov/research/ResearchReports_chrono.html).
24. National Endowment for the Arts. *Research Division Note # 65: Retail Art Dealers Continue Strong Growth in the Economic Census of 1992*, NEA: Washington, DC, 1998. http://www.nea.gov/research/ResearchNotes_chrono.html.
25. *IBISWorld Industry Report: Art Dealers in the US: 45392*, IBISWorld United States: New York, 2008.
26. Robertson, I. The international art market. In *Understanding International Art Markets and Management*; Robertson, I., Ed.; Routledge: London, U.K. and New York, 2005; 13–36.

Art Librarianship

Kathryn M. Wayne
Art History/Classics Library, University of California, Berkeley, California, U.S.A.

Abstract

Art librarianship is a profession that unites academic backgrounds in the visual arts, art history, architecture, and allied fields together with an advanced degree in library science and information management. The collections art librarians manage continue to be mostly comprised of print resources although technological advances, such as digitization of images and the creation of online databases and catalogs, have significantly impacted art research methodology. Art librarians work in several library settings such as academic, museum, public, and visual resources. Depending on the size and scope of the collection, responsibilities can include management, collection development, reference, bibliographic instruction, research, and cataloging. Art librarians work closely together through worldwide professional organizations, contribute to the literature through articles and book reviews, and write essential reference works and standards for the field.

HISTORICAL INTRODUCTION

Since the early twentieth century, art librarians have been building comprehensive collections in order to meet the needs of art historians, scholars, and curators in public, academic, and museum institutions, both large and small. Surprisingly, no one monograph exists that provides comprehensive descriptions of prominent libraries such as the National Art Library at the Victoria and Albert Museum in London, the Metropolitan Museum of Art's Libraries, Harvard's Fine Arts Library, the New York Public Library, the Canadian Centre for Architecture's Library in Montreal, to name a few. Sarah Scott Gibson traces the development of art librarianship and origins of art libraries in her article "*The Past as Prologue*: *The Evolution of Art Librarianship*."[1] The article is part of an issue with the theme "The State of Art Librarianship." A seminal work on art librarianship that continues to be relevant is *The Art Library Manual*: *A Guide to Resources and Practice*.[2] Edited by Philip Pacey, it covers all types of art libraries and resources housed, for example artists' books, photographs, exhibition catalogs, books, and periodicals. Pacey describes his other title, *A Reader in Art Librarianship*, as being

> about how and why art librarianship has developed into a distinct branch of librarianship; about what it takes to be an effective art librarian; about how art librarians differ from other librarians; and about how and why art librarians have developed their own associations.[3]

Another early work appeared in an issue of *Library Trends*,[4] edited by Guy Marco and Wolfgang Freitag and entitled "Music and Fine Arts in the General Library." The contributors represent a "*Who's Who*" in art librarianship in the mid-1970s. Two publications include *A History of Art Libraries in Canada*[5] covering a chronological history of Canadian art libraries from 1632 to present and biographies of Canadian art librarians no longer living. *Art Museum Libraries and Librarianship*,[6] edited by Joan Benedetti, provides a series of essays contributed by art librarians currently working in museums worldwide, with a bibliography covering additional sources on museum librarianship. Ann Abid's introduction provides a brief history of the major museum libraries in the United States, from the late 1800s to present.

Another way to identify art collections and their librarians, is through directories, both historical and current. Beginning in 1923, the *American Library Directory*[7] was one of the first directories to record the existence of art libraries, included under the heading of "Special Libraries." Within this listing are some of the early American museum libraries such as the Cleveland Museum of Art Library, the Ryerson Library at the Art Institute of Chicago, and the Cincinnati Art Museum Library. It was not until the late 1960s and early 1970s that the first professional art librarian organizations were formed in Europe and North America: the Art Libraries Society (now ARLIS/UK and Ireland), the Art Libraries Society of North America (ARLIS/NA), and the International Federation of Library Association's (IFLA) Art Libraries Section. Each had the goal of creating specialized art directories that would provide researchers with a reference that described art library collections. After publishing three preliminary lists, the IFLA Section of Art Libraries published its *Directory of Art Libraries*[8] under the direction of Jacqueline Viaux, the first chairwoman of the IFLA group which in 1977 was called the Round Table of Art Librarians. Its North American counterpart, *The Directory of Art Libraries and Visual Resource Collections in North*

Encyclopedia of Library and Information Sciences, Fourth Edition DOI: 10.1081/E-ELIS4-120043242

Archives–Art

America[9] was compiled by Judith Hoffberg, the founder of ARLIS/NA. Today, the *International Directory of Art Libraries*[10] provides almost 3000 listings to art, architecture, and archeology libraries and over 4000 art librarians worldwide. The print edition is complemented by an online directory of the same title.[11] Histories of individual art library collections can also be found in the periodical literature, for example Susan Allen's "*Toward an International Art Library: The Growth of the Research Library at the Getty Research Institute 1979–2002.*"[12]

The role of the art librarian in the first half of the twentieth century was to serve the information needs of faculty, students, curators, collectors, and art professionals, and to focus on collecting, organizing, cataloging, maintaining, and preserving the largest collections with the greatest historical significance. At the same time, the modern art library began to proliferate in national libraries, museums, and academic institutions throughout Europe and the United States. Access to these library collections was provided onsite to users via a card catalog. Many of the prominent art libraries published catalogs of their holdings in printed format, such as Harvard University's Fine Arts Library[13] and the Victoria and Albert Museum.[14] Art librarians purchased these catalogs and, in turn, were able to provide their patrons with bibliographic access to some of the largest art library collections.

In the 1980s, art librarians focused on converting card catalog records into machine readable format and the first computerized local catalogs became available. The Internet now makes it possible to search holdings of almost all major art libraries worldwide, either by looking at an individual online public access catalog (OPAC), such as the *Getty Research Institute Research Library Catalog*[15] or *Watsonline*, the catalog of the libraries of the Metropolitan Museum of Art.[16] Over 9000 library catalogs worldwide can be searched via the Online Computer Library Center's *Worldcat*.[17] The database *CAIRNS* (Cooperative Information Retrieval Network for Scotland)[18] includes 52 library catalogs. *The European Library*[19] online catalog provides bibliographic access to the holdings of 47 national libraries across Europe, including access to online books and images. There are still many art libraries in the United States and Europe that have not completely converted all of their card catalog records into their OPACs and others have created databases that are not accessible via the Internet.

TWENTY-FIRST CENTURY TRENDS

The new millennium has brought changes to art librarianship through continuing technological advancements. One of the most significant trends is the digitization of analog image and slide collections. Paula Hardin provides a model for implementing this new technology in her chapter *Integrating the Digitization of Visual Resources into Library Operations*.[20] Another area that has been impacted by digitization is special collections material. Various rare collections from art libraries all over the world are being scanned and text and images will become available for those resources through finding aids. Historical maps, architectural drawings, manuscripts, and rare books are examples of formats that lend themselves well to the two-dimensional scanning process. The Institute of Museum and Library Services' *IMLS Digital Collections Registry* is currently funding the digitization of archives in several museums and libraries.[21] The third area that has received much attention is the mass book digitization projects supported by Google, Microsoft, and Yahoo begun in 2006 through partnerships with over 27 libraries. Books in all disciplines published prior to 1923 and considered to be in the public domain, are being scanned and their full text, along with illustrations, is becoming accessible to the public.

One of the major differences between arts and humanities users versus those in social sciences and science disciplines is their preference for high-quality color reproductions in beautifully published books, often oversized. Today's art historians and curators still deem the printed book the most desirable format. Even when a special collection is available digitally, many scholars will travel to the library to examine the original object, the paper, the ink, and the binding; a similar process to that of viewing an original work of art. One study, *Art History and its Publications in the Electronic Age*,[22] discusses the state of scholarly publishing for the discipline of art history.

Despite the overall increase in digitized resources, today's art librarians are still collecting primarily print-based collections. Only a small percentage of art journals are being published electronically in full-text format. *JSTOR: the Scholarly Journal Archive*[23] is a database for archiving academic journals. It includes full text for approximately 64 art and art history-related journals, excluding the last 5 years. Major print-based art periodical indexes such as the *Bibliography of the History of Art*,[24] *Art Full Text and Art Index Retrospective*,[25] *ARTbibliographies Modern*,[26] and the *Avery Index to Architectural Periodicals*[27] are now available as Web-based searchable databases. *Oxford Art Online*[28] includes *Grove Art Online*, the only scholarly art encyclopedia available online with full text and images. After many years of development, the Mellon Foundation has published *ARTstor*,[29] the largest digital image database available complementing academic art and architecture teaching and scholarship. The *Imageline* project is a proposal "to assess the feasibility of establishing an international web gateway providing access to sources of images."[30]

EDUCATIONAL REQUIREMENTS

Professional art librarian positions require an accredited master's degree in library science (M.L.S.) or library and information science (M.L.I.S.). In the United States, there are numerous American Library Association-accredited library degree programs but currently only three that grant a dual master's in library science and art history: Pratt School of Information and Library Science in New York City; the School of Library and Information Science, Indiana University; and the University of North Carolina Chapel Hill School of Information and Library Science. In the United Kingdom, the Chartered Institute of Library and Information Professionals[31] accredits undergraduate and postgraduate library and information courses in England, Wales, and Scotland. Many librarians obtain a master's degree in library science after completing either an undergraduate or graduate degree in studio art, art history, architecture, or allied discipline.

The publication *Core Competencies & Core Curricula for the Art Library and Visual Resources Professions* "identifies the fundamental knowledge, behaviors, and skills currently essential to most professional positions within the art information field."[32] The document *Criteria for the Hiring and Retention of Visual Resources Professionals* provides standards for visual resources positions.[33] Position listings for art librarians can be found through professional organizations' listservs as well as periodicals such as *Library Journal* and the *Chronicle of Higher Education*.

ART LIBRARY SETTINGS

With an accredited library degree, a professional librarian interested in pursuing a career in an art library has a variety of settings from which to choose: academic (art, architecture, art, and design school), museum, public, and visual resources. The size and staffing of libraries in all of these settings can vary greatly, from a small one-person library to a large internationally prominent collection with numerous staff.

Academic

Academic art libraries can cover one or more of the fine arts disciplines. Art history, architecture, landscape architecture, archaeology, urban design and planning, and studio art curriculums often serve undergraduates and graduate students, both masters and Ph.D.s. The focus of an academic art and architecture collection is generally on history and theory while studio art and design collections may be broader in scope and include practical manuals on drawing, painting, etc. Fine arts libraries are typically housed in a separate facility and are commonly referred to as branch or subject specialty libraries. At some universities, these collections may be integrated into a main library complex. Although collections can vary from large (150,000 + volumes) to very small (10,000 + volumes), the types of resources most often collected include print

and online reference sources, databases, periodical indexes, books, periodicals, exhibition catalogs, catalogue raisonnés, visual resources such as print reproductions, slides and digital image databases, and media (CDs, DVDs, video). Collections of local theses and dissertations may also be housed. Special collections in the art library can include unique archival and primary source material such as artists' books, ephemera, and artists' papers. Microform sets such as the Courtauld Institute of Art's *Conway Library* or the *Museum of Modern Art Artists Files* provide access to unique archival material. Comprehensive academic art collections house resources published in all languages with an emphasis on Western European languages.

No published ranking of academic art libraries exists; however, the Association of Research Libraries provides rankings and statistics for the top academic libraries in the United States.[34] The accreditation bodies and requirements for the accrediting process related to academic art libraries are described in detail in *The Library and the Accreditation Process in Design Disciplines: Best Practices*.[35]

Museum

In the museum library setting, art librarians primarily serve the needs of museum curators and the institution's exhibition and educational outreach programs. Museum library collections usually include books, periodicals, exhibition catalogs, catalogue raisonnés, reference sources, periodical indexes, and sales catalogs from auction houses such as Christie's and Sotheby's. Subscriptions to online auction catalog databases are also integral to the museum library. Special collections often include primary sources such as artists' books, artists' files, artists' archives, and oral histories, along with the museum's historical correspondence and administrative archives. These collections are focused on the interests of the museum's art collections by period, e.g., Museum of Modern Art and/or cultural emphasis, e.g., Asian Art Museum of San Francisco. The publication *Art Museum Libraries and Librarianship*[6] provides a current overview of this field. (See also the entry Art Museum Librarians in the second edition of the *Encyclopedia of Library and Information Science*.) In the United States, museums go through an accreditation process sponsored by the American Association of Museums and the museum library participates in this review.

Public

Many public libraries support art collections that might complement a strong local community interest in the arts. These collections tend to include popular art books and magazines, how-to guides for artists, along with access to one or more of the major online art periodical indexes. Public art librarians serve patrons with little or no initiation to the field, or with a general interest in art, along with researchers and professionals, such as architects who may not have a college or university library available in the area. Although most public art librarians do not have the funding to build extensive collections, some of the public libraries established in the late nineteenth or early twentieth century, such as the New York Public Library and the British Library, house comprehensive research-level art collections similar o those found in academic libraries. Heather Rowland's chapter in the book *Information Sources in Art, Art History and Design*,[36] provides a history of art resources in public libraries in the United Kingdom along with current trends in collection management and reference sources.

Visual Resources

Visual resource curators manage pictorial material such as slides, digital images, and prints (reproductions and photographs). These collections are often found in the art library settings described above, and in university or art school settings can be large, stand-alone operations that report directly to an academic department rather than the institution's library. Collections are often comprised of images photographed or scanned in-house from books and journals, purchased image collections, and gifts. Curators are responsible for cataloging these collections with many professionals choosing to follow national standards by using the *VRA Core*[37] and the Getty Research Institute's Getty Vocabulary Program databases[38] such as the *Art and Architecture Thesaurus*, *Union List of Artist Names* and the *Getty Thesaurus of Geographic Names*. For additional information see the entries "Visual Resources Association (VRA)," p. 5572, and "Visual Resources Management, in Cultural Institutions," p. 5578, in this edition of the *Encyclopedia of Library and Information Sciences*.

PROFESSIONAL RESPONSIBILITIES

Depending upon the scope of the position, size of library and setting, art librarians can be responsible for one or all of the following areas: collection development, reference, bibliographic instruction and research, cataloging, and management of staff, budget, and library facilities.

Collection Development

One of the most important roles the art librarian fulfills is determining which resources should be added or deleted from the collection. A collection development policy can provide an overview of the scope of collections such as areas, period and languages covered, collecting level, and

types of materials collected. The publication *Collection Development Policies: for Libraries & Visual Collections in the Arts*[39] provides art librarians with examples of collection development policies along with guidelines for writing a policy.

Both historical and recently published art bibliographies are helpful tools for building core reference and monograph collections. One title, *Guide to the Literature of Art History 2*[40] complements an earlier edition, *Guide to the Literature of Art History*,[41] providing the most comprehensive art bibliography published. The titles *Fine Arts: A Bibliographic Guide to Basic Reference Works, Histories, and Handbooks*[42] and *Art Books: A Basic Bibliography of Monographs on Artists*[43] provide a core list of important art books and catalogue raisonnès for well-known artists.

Book review publications that focus on art resources include *The Art Book: Issues, News and Reviews*,[44] *Bookforum: The Book Review for Art and Culture*,[45] the College Art Association's *CAA Reviews*[46] and *ARLIS/NA Reviews*.[47] The *Art Libraries Journal (ALJ)*[48] provides reviews of books about, or relevant to, art librarianship, such as major art-related bibliographies and reference works. Other library journals such as the Association of College and Research Library's *Choice Magazine*[49] provide reviews for all disciplines including the fine arts.

Art librarians may opt to use art-based approval plans to obtain recently published museum and gallery exhibition catalogs along with trade and university press titles. Books are shipped automatically or selected title-by-title following pre-determined parameters that are based on the library's collection development policy. Art approval plans are available from several vendors located in the United States and Europe. Some of the most widely known vendors that supply European imprints include Erasmus Boekhandel,[50] Casalini Libri,[51] Touzot,[52] and Harrassowitz.[53] Worldwide Books[54] provides international coverage of exhibition catalogs as well as current trade publications. Two approval vendors that cover titles published in the United States and United Kingdom include Yankee Book Peddler[55] and Blackwell.[56] Art librarians with limited budgets might prefer to place orders directly to publishers of interest. Titles can be ordered through *Amazon.com* along with European counterparts such as *Amazon.co.uk* (United Kingdom), *Amazon.fr* (France), *Amazon.de* (Germany), *Bol.it* (Italy), and *FNAC.com* (Spain).

Art books, especially exhibition catalogs, are generally published in small runs, and can quickly go out-of-print. A list of art-related antiquarian and out-of-print art book dealers worldwide can be found by searching the International League of Antiquarian Booksellers' Web page.[57] Online ecommerce search engines such as *Bookfinder*, *Abebooks*, and *AddAll* provide extensive lists of new, used, rare, and out-of-print art books linking to individual booksellers worldwide.

Reference, Bibliographic Instruction, and Research

Art librarians serve primary clientele, local art professionals, and the general public with questions related to myriad art topics. The increase of online art resources combined with the continued publication of printed resources, has made the research process more complex for patrons to locate, navigate, and evaluate relevant publications which in turn has expanded the art librarians' role in both reference service and bibliographic instruction. Reference "points" for most art librarians can include a reference desk with scheduled hours, online chat, email, telephone, or by appointment.

Bibliographic instruction can include tours, general library orientations, in-class research seminars on a particular subject, or locally published research guides that might describe one source comprehensively or provide annotated bibliographies for an art-related topic. The guide, *Library Instruction for Students in Design Disciplines*,[58] provides descriptions of a variety of art-related courses taught by art librarians in academic institutions throughout the United States.

Art librarians also create library Web sites providing links to important art resources available, both licensed and free. Blogs, or Web logs of text that might describe recent acquisitions or new trends, often are included as well. In addition, art librarians are beginning to utilize Google's *You Tube* as another means to promote research to their users—for example the tutorials created by librarians and faculty at Otis College of Art and Design.[59] Librarians are also utilizing social networking destinations such as *MySpace*, *Facebook* and *Twitter* to reach their audiences.

Art librarians are responsible for original research and publication and continue to be the major contributors to the art library journal literature, primarily *Art Documentation* and the *Art Libraries Journal*. Article content can be searched using the periodical index *Library Literature*. Combined, these two journals, along with newsletters published by the Art Libraries Society and other art library organizations, provide a comprehensive history of the field of art librarianship.

Cataloging

The cataloging of all types of art materials requires subject background and coursework based on the *Anglo-American Cataloguing Rules* (AACR2)[60] and *MARC21 Concise Format for Bibliographic Data*.[61] One-person art libraries, museum libraries, and visual resources collections usually catalog their own collections onsite, where academic institutions generally catalog books in a central cataloging unit housed within the technical services department. In the academic setting, art catalogers might cover the fine arts as well as additional humanities disciplines, such as theater, dance, and music. Art catalogers communicate across institutions and through professional organizations such as the Visual Resources Association (VRA) and the ARLIS/NA Cataloging Section[62] about current cataloging trends and art-related issues. The *Cataloguer's Toolbox*[63] created by Memorial University, Newfoundland, provides a variety of international cataloging resources.

Management

The head of the art library is responsible for numerous administrative activities such as hiring, training, supervising and evaluating staff (professional, paraprofessional, students, interns, and volunteers), long-range planning, annual reports, statistics, and budgets.

The management of facilities is an area requiring increased attention, as print collections often outgrow available stack space. Art librarians must carefully determine whether to withdraw or store materials off-site if that option is available. Security is another critical issue as a result of the increased value of rare art books and special collections.

Six art library facilities appear in a special issue of *Art Libraries Journal*.[64] Two examples of art libraries designed for the twenty-first century are the Fleet Library at the Rhode Island School of Design (Fig. 1) and the

Fig. 1 Interior, Fleet Library, Rhode Island School of Design, Office dA.
Source: Courtesy of John Horner, 2006.

Fig. 2   Exterior, Docklands Campus Library and Business School, University of East London.
Source: Courtesy of David Barbour/Building Design Partnership.

Docklands Campus Library and Business School at the University of East London (Fig. 2). The key to a successful art library project is the articulate communication of needs from the art librarian to the architect during the programming phase. The handbook *Planning Academic and Research Library Buildings*[65] is a key reference along with *Facilities Standards for Art Libraries and Visual Resources*[66] (currently under revision) and IFLA's *Libraries as Places: Buildings for the Twenty-first Century*.[67] Macken's article, *The Art Library as Place*, examines "the impact of changing technology on the physical space of the academic art library and the relevance of current trends in space planning to the art library of the future."[68]

PROFESSIONAL ORGANIZATIONS

There are several organizations that support the field of art librarianship. Art librarians often participate in a variety of professional organizations and affinity groups depending upon their specialty.

ARLIS/UK and Ireland

The Art Libraries Society United Kingdom and Ireland (ARLIS/UK and Ireland)[69] was founded in 1969 to promote all aspects of the librarianship of the visual arts, including architecture and design. Approximately 350 members work in academic settings, museum libraries, and public libraries, and include booksellers, publishers,

and professional practice libraries. Publications include a scholarly quarterly journal *Art Libraries Journal*, and a bimonthly newsletter *ARLIS News-Sheet* providing news, bibliographic information, and scholarship for art librarians internationally. An annual conference is held at a university location in the United Kingdom or Ireland. Members can communicate via *ARLIS-LINK*,[70] its online discussion listserv.

ARLIS/NA

The Art Libraries Society of North America (ARLIS/NA)[71] is the primary professional association for art librarians in the United States, Canada, and Mexico. Inspired by ARLIS/UK and Ireland, it was founded by Judith Hoffberg and nine other charter members in 1972.[72] The stated mission in the ARLIS/NA Strategic Plan 2006–2009 is "to foster excellence in art and design librarianship and image management."[73] Its 1000 members include art and architecture librarians working in art libraries and visual resources collections as well as students, book publishers, dealers, and gallery owners. Members can participate on a local level by joining one of the Society's chapters. An annual conference includes workshops, tours, and juried sessions that respond to a programmatic theme. Publications include a journal, *Art Documentation*, an *Occasional Papers* series and *Bibliographic Notes*, an online column that includes recent articles and books covering topics devoted to art librarians and art librarianship. ARLIS/NA hosts an open, moderated listserv, *ARLIS-L*.[74]

ART LIBRARIES SOCIETIES WORLDWIDE

ARLIS/ANZ (Australia and New Zealand): http://arlisanz.anu.edu.au/.

ARLIS/NL (Netherlands) – Overleg Kunsthistorische Bibliotheken Nederland: http://www.okbn.nl/.

ARLIS Norden (Scandinavia): http://www.arlis.org.uk/link/link.html.

ARLIS Flanders/Belgium: http://www.vvbad.be/okbv/engels.

VRA

The Visual Resources Association: The International Association of Image Media Professionals (VRA)[75] grew out of a long history of involvement with other art-related organizations, the College Art Association (CAA), the Art Libraries Society of North America (ARLIS/NA), and the Mid-America College Art Association (MACAA). The VRA was formally established in 1982 and has over 800 members. The purpose of the VRA is to support and educate image professionals on the creation, description, and distribution of both analog and digital images, primarily for art- and architecture-related collections and to develop standards for the profession. The association also has regional chapters. The *VRA Bulletin* is the association's quarterly journal featuring conference reports and scholarly articles. *Images Newsletter* is published bi-monthly as an electronic journal and contains member and organization-related news. The *VRA Special Bulletins* are occasional publications covering current trends and topics of interest to the profession. The *VRA-L* online discussion list is open to members only.

IFLA

Although the International Federation of Library Associations and Institutions (IFLA) was formed in 1927, its Art Libraries Section evolved in 1982 through the efforts of French art librarian Jacqueline Viaux.[76] The Section "endeavors to represent libraries and organizations concerned with all formats of textual and visual documentation for the visual arts, including fine arts, applied arts, design, and architecture."[77] J. Margaret Shaw's article[78] provides an account of the Art Libraries Section's 25-year history, including its officers, and a list of its publications and projects. A strategic plan[79] outlines its mission and goals. The Art Libraries Section meets annually at the IFLA World Library and Information Congress. Serial publications include the *International Directory of Art Libraries* and the electronic journal, *IFLA Art Libraries Section Newsletter*. The IFLA Publication series covers numerous topics on librarianship, some focused on art librarianship, e.g., *Multilingual Glossary for Art Librarians*.[80] An online discussion list, *IFLAART*, is a closed, moderated list, with approximately 250 subscribers.

ADDITIONAL ART-RELATED LIBRARY AND AFFINITY ORGANIZATIONS

American Association of Museums (AAM): http://www.aam-us.org/.

Established in 1906, AAM has developed standards and best practices, promoted gathering and sharing of knowledge, and provided advocacy on issues of concern to the entire museum community. Every type of museum is represented, including art.

American Library Association (ALA), Association of College and Research Libraries (ACRL), Arts Section: http://www.ala.org/ala/acrl/aboutacrl/acrlsections/arts/arts.cfm.

A section of ACRL, a division of ALA, which provides an umbrella organization for librarians and specialists working in or interested in the fields of visual and performing arts.

Arbeitsgemeinschaft der Kunst- und Museumsbibliotheken (AKMB) (Germany): http://www.akmb.de/.

The Association of Art and Museum Libraries supports education, training, and cooperation for art librarians working in Germany. Affiliated with ARLIS/NA.

Association of Architecture School Librarians (AASL): http://www.architecturelibrarians.org/.

Established in 1979 at the annual meeting of the Association of Collegiate Schools of Architecture, AASL's goal is to enhance various aspects of academic architectural librarianship. Affiliated with ARLIS/NA.

Association des Bibliothécaires de France (ABF): http://www.abf.asso.fr/.

Founded in 1906 and recognized as a public utility in 1969, ABF is the oldest library association in France. The Groupe des Bibliothèques d'art is one of its several working groups. Affiliated with ARLIS/NA.

College Art Association (CAA): http://www.collegeart.org/.

Founded in 1911, CAA's members are committed to the practice of art, teaching, and research. Affiliated with ARLIS/NA.

Japan Art Documentation Society (JADS): http://www.soc.nii.ac.jp/jads/eng/index.html.

Founded in 1989, with the object of promoting the development of art documentation, members include librarians, curators, art history students, media representatives, and computer specialists. Affiliated with ARLIS/NA.

Kunstfaggruppen (Denmark): http://grupper.bf.dk/kunst/.

Organization dedicated to all aspects of art librarianship, including education and training.

Museum Computer Network (MCN): http://www.mcn.edu/.

Affiliated with ARLIS/NA. See the entry "Museum Computer Network (MCN)," p. 3711, in this edition of the *Encyclopedia of Library and Information Sciences*.

Museum Documentation Association (MDA) (U.K.): http://www.mda.org.uk/.

Archives-Art

Organization devoted to working with professionals in museums, libraries, and archives to promote best practices, accreditation guidelines, standards, and training.

Redarte-SP (Sao Paulo, Brazil): http://www.goethe.de/ins/br/sap/prj/red/ptindex.htm.

Network of information services related to art that brings together professionals who work in libraries, museums, and archives. The group develops projects that promote the broad dissemination of art information within the city of Sao Paulo.

Society of American Archivists (SAA): http://www.archivists.org/.

Founded in 1936, SAA is North America's oldest and largest national archival professional association which provides leadership to ensure the identification, preservation, and use of records of historical value. Affiliated with ARLIS/NA.

Society of Architectural Historians (SAH): http://www.sah.org/.

International organization that promotes the study and preservation of the built environment worldwide. Affiliated with ARLIS/NA.

Special Libraries Association (SLA), Museums, Arts, and Humanities Division: http://units.sla.org/division/dmah/.

Vitruvio: http://www.vitruvio.cpaupage.com/.

Organization for architecture, art, design, and urbanism librarians working in academic settings in Argentina and Latin America.

Established in 1929, the Division includes librarians and information specialists from all types of museums from historical societies, institutions, and other organizations having special departments or special collections devoted to the arts, decorative and performing arts, architecture, and humanities.

CONCLUSION

Over the past two centuries, art librarians have played a major role in building art libraries that support the world's finest academic and public institutions. The role of the art librarian will continue to expand as technological advances improve and affect the way resources are published, delivered, and received in the arts and humanities. The art librarian of the future will need to respond to the technological needs of future generations and at the same time give consideration to the preservation of rare, out-of-print materials. Some efforts, such as the Institut National d'Historie de l'art in Paris,[81] that combine several important art libraries into one facility, may be a trend for future collaborations worldwide. Perhaps the words Pacey wrote in 1985 still hold true,

> Now as never before the collective power of art librarians must be exercised, on the one hand to safeguard what has been achieved hitherto, on the other to seek and to share

new solutions to old problems, and at all times to exercise art library power for the benefit of civilization.[82]

ACKNOWLEDGMENTS

The author offers special thanks to colleagues Sheila Klos, Dumbarton Oaks Research Library; Barbara Rominski, San Francisco Museum of Modern Art Research Library; Lynn Cunningham, Visual Resource Collection, University of California Berkeley; Christine Sundt, University of Oregon (retired); Judith Anne Preece, University of East London, Docklands campus; and Carol Terry, Rhode Island School of Design.

REFERENCES

1. Gibson, S.S. The past as prologue: The evolution of art librarianship. Drexel Libr. Quart. **1983**, *19*(3), 3–17.
2. In *Art Library Manual: A Guide to Resources and Practice*; Pacey, P., Ed.; Bowker: London,U.K., 1977.
3. In *A Reader in Art Librarianship*; Pacey, P., Ed.; IFLA Publications 34; K. G. Saur: New York 1985; ix.
4. Music and fine arts in the general library. Libr. Trends, **1975**, *23*(3), Marco, G.A.; Freitag, W.M., Eds.
5. *A history of Art libraries in Canada: Essays in the history of art librarianship in Canada*. 2006, ARLIS/Canada. Available at http://www.arliscanada.ca/.
6. In *Art Museum Libraries and Librarianship*; Benedetti, J., Ed.; Scarecrow Press [Ottawa]: Art Libraries Society of North America: Lanham, MD, 2007, Occasional Paper 16.
7. *American Library Directory: A Classified List of Libraries in the United States and Canada, With Personnel and Statistical Data*. R.R. Bowker: New York, 1923.
8. Viaux-Locquin, J. *IFLA Directory of Art Libraries, Repertoire de Bibliotheques d'art de l'IFLA, Addressbuch der Kunstbibliotheken von IFLA, Directorio de Bibliotecas de Arte de la IFLA*, Garland Reference Library of the Humanities; Garland Publication: New York, 1985; Vol. 510.
9. Hoffberg, J.A.; Hess, S.W. *Directory of Art Libraries and Visual Resource Collections in North America*, Neal-Schuman Publishers: Santa Barbara, CA, 1978. Distributed by ABC-Clio: New York.
10. In *International Directory of Art Libraries*; Hill, T.E., Ed.; IFLA Publications 82; K.G. Saur: Munchen, Germany, 1997.
11. International Directory of Art Libraries, Available at http://artlibrary.vassar.edu/ifla-idal/ (accessed July 2009).
12. Allen, S.M. Toward an international art library: The growth of the research library at the Getty Research Institute, 1979–2002. Art Libr. J. **1988**, *27*(4), 25–31.
13. Harvard University, *Catalogue of the Harvard University Fine Arts Library, the Fogg Art Museum*, G.K. Hall: Boston, MA, 1971.
14. Catalogue of Exhibition Catalogues, G.K. Hall: Boston, MA, 1972. National Art Library Catalogue, Victoria and Albert Museum, London, England.

15. Getty Museum, Research Library Catalog. Available at http://library.getty.edu/cgi-bin/Pwebrecon.cgi?DB=local&PAGE=First (accessed July 2009).

16. Watsonline, The Catalog of the Metropolitan Museum of Art Libraries. Available at http://library.metmuseum.org/screens/opacmenu.html *Watsonline*, The Catalog of the Metropolitan Museum of Art Libraries.

17. Worldcat, Available at http:/worldcat.org (accessed July 2009).

18. Cooperative Information Retrieval Network for Scotland, *CAIRNS*. Available at http://cairns.lib.strath.ac.uk/CAIRNSService/ZCatSrch.cfm?uMiniID=11 Cooperative Information Retrieval Network for Scotland, *CAIRNS*.

19. The European Library. Available at http://www.theeuropeanlibrary.org/portal/index.html (accessed July 2009).

20. Hardin, P. Integrating the digitization of visual resources into library operations. In *The Twenty-First Century Art Librarian*; Wilson, T.L., Ed.; Haworth Information Press: Binghamton, NY, 2003; 45–55.

21. IMLS Digital Collections and Content. Available at http://imlsdcc.grainger.uiuc.edu/ (accessed July 2009).

22. Ballon, H.; Westermann, M. Available at http://cnx.org/content/col10376/1.1/ (accessed July 2009).

23. JSTOR: The Scholarly Journal Archive. Available at http://www.jstor.org/ (accessed July 2009).

24. Bibliography of the History of Art. Available at http://www.getty.edu/research/conducting_research/bha/ (accessed July 2009).

25. Art Full Text and Art Index Retrospective. Available at http://www.hwwilson.com/Databases/artindex.htm (accessed July 2009).

26. ARTbibliographies Modern.Available at http://www.csa.com/factsheets/artbm-set-c.php (accessed July 2009).

27. Avery Index to Architectural Periodicals. Available at http://www.getty.edu/research/conducting_research/avery_index/ (accessed July 2009).

28. Oxford Art Online. Available at http://www.oxfordartonline.com/ (accessed July 2009).

29. ARTstor. Available at http://www.artstor.org (accessed July 2009).

30. Kosovac, B. *IFLA* Imageline Scope and Feasibility Report, 2002. Introduction, 1.1 (Background). Available at http://www.ifla.org/VII/s30/pub/imageline_report.pdf Vancouver, BC (accessed July 2009).

31. Chartered Institute of Library and Information Professionals. Available at http://www.cilip.org.uk/qualificationschartership/Whereto study.

32. In *Core Competencies & Core Curricula for the Art Library and Visual Resources Professions*; Ball, H., Ed.; Art Libraries Society of North America: Kanata, ON, Canada, 2006; 7.

33. ARLIS/NA and the VRA Boards of Directors, Criteria for the hiring and retention of visual resources professionals, 1995; August and June revised October 2002. Available at http://www.collegeart.org/guidelines/resources.html (accessed July 2009).

34. Association of Research Libraries, ARL Statistics. Available at http://fisher.lib.virginia.edu/cgi-local/arlbin/arl.cgi?task= setuprank.

35. Brown, J.; Glassman, P.; Henri, J.J. *The Library and the Accreditation Process in Design Disciplines: Best Practices*, Art Libraries Society of North America: Kanata, ON, Canada, 2003; Occasional Paper 14.

36. Rowland, H. Public libraries. In *Information Sources in Art, Art History and Design*; Ford, S., Ed.; K.G. Saur: Munchen, Germany, 2001; 22–33.

37. VRA Core. Available at http://www.vraweb.org/projects/vracore4/index.html.

38. Getty Research Institute, Getty Vocabulary Program. Available at http://www.getty.edu/research/conducting_research/vocabularies/.

39. Whiteside, A.B. *Collection Development Policies for Libraries and Visual Collections in the Arts*, Art Libraries Society of North America: Laguna Beach, CA, 2000; Occasional Paper 12.

40. Marmor, M.; Ross, A. *Guide to the Literature of Art History 2*, American Library Association: Chicago, IL, 2005.

41. Arntzen, E.; Rainwater, R. *Guide to the Literature of Art History*, American Library Association: Chicago, IL, 1980.

42. Ehresmann, D.L. *Fine Arts: A Bibliographic Guide to Basic Reference Works, Histories, and Handbooks*, 3rd Ed. Libraries Unlimited: Littleton, CO, 1990.

43. *Art Books: A Basic Bibliography of Monographs on Artists*, 2nd Ed.; Freitag, W.M., Ed.; Garland Publication: New York, 1997; Vol. 1264. Garland Reference Library of the Humanities.

44. *The Art Book: Issues, News and Reviews*; Blackwell Publishing and the Association of Art Historians: Oxford, U.K.

45. *Bookforum: The Book Review for Art, Fiction and Culture*; Artforum International: New York.

46. *CAA Reviews, College Art Association: New York.* Available at http://www.caareviews.org/ (accessed July 2009).

47. ARLIS/NA Reviews. Art Libraries Society of North America: Kanata, ON, Canada. Available at http://www.arlisna.org/resources/reviews/index.html (accessed July 2009).

48. *Art Libraries Journal*, Suffolk, U.K ARLIS/UK & Ireland.

49. Choice Reviews Online. Association of College and Research Libraries: Chicago, IL. Available at http://www.ala.org/ala/mgrps/divs/acrl/publications/choice/choicereviewsonline/cro.cfm (accessed July 2009).

50. Erasmus Boekhandel B.V., Amsterdam, the Netherlands. Available at http://www.erasmusbooks.nl/.

51. Casalini Libri, Fiesole, Italy. Available at http://www.casalini.it/.

52. Jean Touzot Librairie Internationale, Paris, France. Available at http://search.touzot.fr/.

53. Harrassowitz, Weisbaden, Germany. Available at http://www.harrassowitz.de/.

54. Worldwide Books, Ithaca, New York. Available at http://www.worldwide-artbooks.com/.

55. Yankee Book Peddler, Contoocook, New Hampshire. Available at http://www.ybp.com/.

56. Blackwell, Oxford, United Kingdom; Lake Oswego, Oregon. Available at http://www.blackwell.com/.

57. International League of Antiquarian Booksellers. Available at http://www.ilab.org/.

58. In *Library Instruction for Students in Design Disciplines: Scenarios, Exercises, and Techniques*; Brown, J.M., Ed.; Art Libraries Society of North America: Kanata, ON, Canada, 2002; Occasional Papers 13.

59. Otis College of Art and Design Library Tutorials. Available at http://youtube.com/otiscollege.

60. Anglo-American Cataloguing Rules (AACR2). Available at http://www.aacr2.org/ (accessed July 2009).

61. MARC21 Concise Format for Bibliographic Data. Available at http://www.loc.gov/marc/bibliographic/ (accessed July 2009).

62. ARLIS/NA Cataloging Section. Available at http://www.arlisna.org/organization/sec/cataloging/index.html.

63. Cataloguer's Toolbox. Available at http://staff.library.mun.ca/staff/toolbox/ (accessed July 2009).

64. Special issue: Art libraries in new and converted buildings. Art Libr. J. **2007**, *32*(4), 3–36.

65. Leighton, P.D.; Weber, D.C. *Planning Academic and Research Library Buildings*, 3rd Ed. American Library Association: Chicago, IL, 2000.

66. In *Facilities Standards for Art Libraries and Visual Resources Collections*; Irvine, B.J., Ed.; Libraries Unlimited, Art Libraries Society of North America: Englewood, CO, 1991.

67. Bisbrouck, M.F. Libraries as places: Buildings for the 21st century Proceedings of the Thirteenth Seminar of IFLA's Library Buildings and Equipment Section Together with IFLA's Public Libraries Section Paris, France, 28–August 1,2003; IFLA Publications 109; K. G. Saur: Munchen, Germany, 2004.

68. Macken, M.E. The art library as place: The role of current space planning paradigms within the academic art and architecture library. Art Doc. **2006**, *25*(2), 18.

69. ARLIS/UK and Ireland Home Page. Available at http://www.arlis.org.uk/.

70. ARLISL/UK and Ireland Listserv, ARLIS-LINK. Available at http://www.jiscmail.ac.uk/lists/arlis-link.html.

71. ARLIS/NA Home Page. Available at http://www.arlisna.org.

72. Judith, A. Hoffberg: The early years of ARLIS/NA. Art Doc. **2008**, *27*(1), 41–51.

73. ARLIS/NA Strategic Plan 2006–2009. Available at http://www.arlisna.org/organization/admindocs/planning/stratplan06–09.pdf.

74. ARLIS/NA Listserv, ARLIS-L. Available at http://www.arlisna.org/about/arlisl.html.

75. VRA Home Page. Available at http://www.vraweb.org/.

76. Markson, E. A tribute to Jacqueline Viaux (1913–1998). Art Doc. **1999**, *18*(1), 41–44.

77. IFLA Art Libraries Section, Home Page. Available at http://www.ifla.org/en/art-libraries.

78. Shaw, J.M. Twenty-five years of international art library co-operation: The IFLA art libraries section. Art Libr. J. **2007**, *32*(3), 4–10.

79. IFLA Art Libraries Section, Strategic Plan, 2006–2007. Available at http://www.ifla.org/en/publications/strategic-plan-18.

80. IFLA Section of Art Libraries, *Multilingual Glossary for Art Librarians: English with Indexes in Dutch, French, German, Italian, Spanish and Swedish*, 2nd rev. and enl. Ed. K.G. Saur: Munchen, Germany, 1996. IFLA Publication No. 75. New Providence, NJ.

81. Institut National d'Historie de l'Art. Available at http://www.inha.fr/.

82. In *A Reader in Art Librarianship*; Pacey, P., Ed.; IFLA Publications 34; Saur: Munchen, New York, 1985; 17.

Art Museums

David Gordon
Milwaukee Art Museum, Milwaukee, Wisconsin, U.S.A.

Abstract

The urge to collect things of beauty and significance goes deep into history. Art museums safeguard art for future generations. Works of art have power, and that power has been coopted throughout history by those who wish to assert authority, position, and wealth. The evolution of the display of art for the public reflects social, economic, and political developments, and can best be understood historically. This entry traces that history from ancient times on to the Renaissance, the Enlightenment, and the nineteenth century and on to modern times. Art museums provide an opportunity for civic pride, and for an architectural statement. The past three decades have seen a boom in new museum building. They are popular gathering places, and in catering (literally) for their growing audiences with cafés, special events, functions, and stores, they have to keep a proper balance between commercial activities and their core mission. Good governance is essential. Success should be measured in ways other than just by number of visitors. Temporary exhibitions should be worthwhile and not just crowd-pleasers. An issue of concern in collection management is ensuring that works of art have not been looted or expropriated. The popularity of art museums in a digital age is likely to continue as screen captives escape to look at real objects.

INTRODUCTION

An art museum is an institution that collects, preserves, and presents art for the public run by a professional staff driven by a mission to encourage a love and appreciation of art and working according to considered standards and procedures. While other museums show objects to explain, art museums show objects to inspire, nourish, and transport: explanation is an aid rather than the point. The spiritual aspect of art museums, and their often imposing buildings, gives them a kind of standing as secular cathedrals. As such they define the values, aspirations, and civilization of their cities, communities, and countries.

Not all art museums have "museum" in their title; sometimes they are called "gallery" or "institute" or "collection." Art museums come in all sorts of sizes and can be categorized in different ways. In the United States there are over 4000 of them, mostly small, with around 200 of the leading ones with budgets of over $2 million qualifying their directors for membership in the Association of Art Museum Directors (AAMD). A few art museums are encyclopedic, covering wide swathes of visual culture in depth. Some are general, with several substantial collections. Many specialize in, for example, sculpture, portraits, photography, textiles, decorative arts, Asian art, arts and crafts, and regional art; or in one artist. There are museums that are completely or predominantly the collection of one collector. There are museums that are parts of universities or other institutions. Most museums collect, but some simply mount exhibitions: the German word *Kunsthalle* (literally art hall) is apt but does not have an

English equivalent. One thinks of a museum as a building, but there are also open-air sculpture parks. Museums that are not art museums but historical or ethnographic or house museums have art works and objects in them, but in order to tell a story or provide context rather than to focus on aesthetics. Museums can also be classified by ownership: government, municipality, trustee, private individual.

Art museums with their cafés and stores are visitor attractions in the exploding area of cultural tourism. Some have become icons of economic regeneration and urban renaissance. Architecturally ambitious new museums and new additions have been springing up all over the world. As they have become larger, more complicated and more expensive to run, museums have confronted issues surrounding their collecting policies, focus, governance and management, and the potentially competing claims of increasing admissions and deepening scholarship. As not-for-profit organizations, relying increasingly on donations, they need to retain the high measure of public trust that has been reposed in them.

Museums can best be understood in a historical context. The word stems from the Greek word *mouseion*, meaning "temple of the Muses." The range of what is considered collectible art has widened over time, usually following controversy about what is in the canon. The history of the art museum and its antecedents mirrors the history of culture. It illustrates the urge to collect things of beauty, significance, and interest made by imaginative and skillful humans; to use objects to seek affinity with other beings, human or supernatural; to share them with other

Encyclopedia of Library and Information Sciences, Fourth Edition DOI: 10.1081/E-ELIS4-120044672

connoisseurs and people of taste; to demonstrate the "superiority" of western culture against the mere artifacts of native peoples—and latterly to celebrate those artifacts as works of art in their own right and manifesting a culture equal to that of the West; to show off works of art as evidence of wealth; and to make them available to the general public as part of a mission to improve, set standards and educate that public.

HISTORY

There has been considerable debate about whether the history of museums is linear and evolutionary or discontinuous. In the preface to the book they edited, Impey and Macgregor[1] state:

> With due allowance for the passage of years, no difficulty will be found in recognizing that, in terms of function, *little has changed*; along with libraries, botanical and zoological gardens, and research laboratories, museums are still in the business of "keeping and sorting" the products of Man and Nature and in promoting understanding of their significance.

The emphasis was added by Berelowitz,[2] who takes a contrary view, influenced by Michel Foucault, that each period should be considered within its world view; there are huge disjunctions between them; and historians should avoid the temptation of presenting a story of continuous betterment from primitive to civilized. The brief account of the history of museums that follows takes a middle view: some things have remained constant, like keeping and sorting, but the meaning and context in which this has taken place has varied enormously.

Museums in Antiquity

While there is archaeological evidence of ancient peoples collecting things without obvious practical use, it was in the Hellenistic period that a taste for art and for the collecting of art developed. Greek temples, or *mouseions*, were dedicated to the various Muses of the arts and sciences. Votive offerings, often war booty, were stored in *thesauroi*, or treasuries, and were a visible display of power and influence. The temples became places of learning as well as offering. Whereas Plato's lyceum taught through argument and induction, Aristotle's drew deductions from an observation of things that were collected and categorized and stored. The lyceum needed a *mouseion*. Ptolomy I Soter (ca. 367–283 B.C.E.) incorporated objects as well as texts in his great library of Alexandria.

In 212 B.C.E., General Marcus Claudius Marcellus brought back treasures from Syracuse and the Romans discovered the wonders of Greek art and became avid collectors. Generals put their booty on display to the public to demonstrate their successes and as a show of power. The collecting habit spread to the rich. Roman artists copied Greek originals and adopted their style, increasing supply. The statesman and General Marcus Agrippa (63 BC–12 BC) urged that statues and pictures should be viewable by the public instead of being stashed away in villas. "This was the first explicit declaration of the value of an art collection as a cultural heritage and of the right of the public to share in its enjoyment."[3]

Christianity understood the iconic power of art and attacked pagan art. Many ancient collections were destroyed, although in the Eastern Empire Constantine brought antique statues to his new capital.

Art Museums in the Middle Ages

In the Middle Ages churches were also museums—the only places where the public could see art. Art was no longer votive, or to celebrate victory in battle, or to flaunt individual wealth, but primarily devotional. Cathedral workshops and monasteries became centers of artistic production and recipients of gifts. Soaring Gothic arches and the glitter of gold objects emphasized to worshippers where authority resided. As society developed under the influence of trade and prosperity, courts and the bourgeoisie started collecting as well as the church, and works of art again became appreciated for themselves rather than for their symbolic value; artists gradually began to be valued as creators and not just artisans.

Art Museums in the Renaissance

Fittingly it was in Florence, heart of the Renaissance, the explosion of creativity in literature, sculpture, painting and architecture, and during its "High" phase in the sixteenth century, that the first art museum was constructed. Known as the *Uffizi* because state offices were on the ground floor, the palace was designed in 1560 by Giorgio Vasari (1511–1574) with the second floor *galleria* (hence probably "gallery" as a place for a collection of paintings) purpose-built for the display of art. The Medici installed their art collection and it was opened to the public (at first by appointment only) in 1591. Vasari, the first art historian and author of the *Lives of the Artists*, stressed the importance of the antique as an inspiration to modern artists and gave credit to the idea of genius and to historical development. He was a founder of the first official academy of art, the Accademia del Disegno, incorporated in 1563. While in the modern age art is associated with museums, art of a level rarely exceeded since had already been on public display in the form of Ghiberti's bronze doors in the Florence baptistery (completed in 1452), Brunelleschi's dome for the cathedral (1436), and Michelangelo's *David*, put on display to widespread awe in 1504. Renaissance man witnessed the "rebirth" in the streets.

Art Museums in the Enlightenment

How did the idea of a place of contemplation and study open to the public develop? From several roots.

Art museums developed originally from what we would now classify more as science museums. Private collections of objects from the natural world, science collections, antiquities, manuscripts, and books put together in a spirit of inquisitiveness, go back to the Middle Ages, but spread from the Renaissance onward. Variously known as *Wunderkammer*, or as "cabinets of curiosity," they were open not to the general public but to like-minded seekers of a universal pattern in what to today's eyes would seem a jumble. In 1683 Elias Ashmole presented his collection to the University of Oxford and the Ashmolean Museum became the first private collection to enter the public domain. The "use of the term "Museum" was a novelty in English: a few years later the *New World of Words*" (1706) defined it as a Study, or Library; also a College, or Publick Place for the Resort of Learned Men."[4] The physician, naturalist, and collector Sir Hans Sloane (1660–1753) bequeathed his large collection to the nation and the British Museum was created by Act of Parliament in 1753. It was a new type of institution in that it was governed by a body of trustees responsible to Parliament; its collections belonged to the nation, with free admission for all. Entry was given to "all studious and curious Persons," linking public enjoyment with education.[5]

Art museums also developed from the spread of art collecting around Europe in the seventeenth and eighteenth centuries. Art dealers and auction houses sprang up. Rome was the artistic capital of Europe until the Popes decided that art was too much of a distraction, and collections moved elsewhere. Rubens advised Duke Vicenzo I Gonzaga of Mantua about collecting. After the Duke's death a large part of his collection was sold in 1627 to Charles I of England, an avid collector. The Spanish collected the more Catholic art of Flanders, and Philip IV sent Velázquez to Italy in 1649 to buy Italian art. The Dutch collected from around Europe as well as their home-grown artists enjoying a golden age. The French saw collecting as an expression of royal authority. The British nobility returned from their Grand Tours laden with art and artifacts to fill their country houses. Lord Burlington brought back the drawings of Palladio and changed the style of architecture of those houses. Art emerged from the cabinet into the long gallery, was arranged, and labeled.

Third, with the Enlightenment came the establishment of art academies (after Florence's, came France in 1648, Venice in 1750, and London in 1768), with their collections to inspire and teach students, and their exhibitions. The annual shows of works by Academicians at the Royal Academy of Arts were selling exhibitions open to the public for a fee and were hugely popular. Appreciation of art became more widespread. Collectors who did not want their collections broken up gave to the growing number of art and archaeology academies.

A fourth factor in the development of art museums was egalitarianism. While several art collections were opened to the public voluntarily by rulers (the Gemäldegalerie in Kassel by William VIII of Hesse in 1760, the Schloss Belvedere gallery in Vienna by Joseph II ca. 1781) it was the French Revolution that decisively put art into the public domain. With the fall of Louis XVI on August 10, 1792, the royal collection was declared the property of the nation, and the National Assembly moved fast to assemble choice works from the royal and church collections together in the Louvre, where Louis had been constructing a Grand Gallery until the Revolution broke out in 1789. The Louvre was also home to the Academy. A year later, to celebrate the anniversary, the Louvre was opened as part of a Festival of National Unity, and those who visited the Museum "would have come away with a ... sense of Revolutionary triumph over despotism."[6] The Museum was for education; it was also for propaganda. The two were linked. The Academy's master-pupil relationship was deemed elitist and it was crushed. The eclectic display of 1793 was deemed to hark back to the *Ancien Régime* and was replaced by 1801 by the more rational, enlightened arrangement by school and period. What mattered was that the "Louvre contained the greatest collection of Western art ever assembled under one roof, and nothing was to prevent that fact from being self-evident to the beholder."[6] In the same year the government decreed the establishment of 15 other museums around France.

Art Museums in the Nineteenth Century

The nineteenth century saw a boom in museum building in Europe and later in the United States. Monarchs wanted to appear more democratic by opening up their collections to the public. Nation-states and cities within them emphasized their power and prestige by establishing important collections in impressive buildings. The rising middle classes wanted to enjoy the polite arts as well as the useful arts of manufacturing and transportation. Educational reformers wanted to improve the lot of the working classes and saw art as a civilizing influence.

At the beginning of the century Napoleon brought vast quantities of art back from his conquests for the Louvre (renamed Musée Napoléon) and the provincial museums, but he also established museums in the lands that he conquered, for example, the Galleria dell'Accademia in 1807 in Venice as part of the art school established 17 years earlier, and the Museo del Prado in 1809. After his defeat in 1815, the art was seized back. In Berlin the returned art was put on display by Frederick William III, King of Prussia, who called for a national museum, the Altes Museum, possibly inspired by visits to the Louvre during the peace negotiations. Karl Friedrich Schinkel, professor of architecture and Berlin's city planner, designed the purpose-built museum and worked with an art historian, Gustav Friedrich Waagen, on the display of only original works of art and only of masterpieces. They argued that the state's provision of aesthetic reverence would promote social unity and defuse the fervor for dissent. The Altes Museum opened in 1830.

Royal collections were in one way or another opened up in most of Europe. Not so in Britain. For one thing, the magnificent collection of Charles I had been auctioned off by Oliver Cromwell in 1649 to no popular protest other than about the poor prices realized. Subsequent monarchs, particularly George III, rebuilt it to some extent but had no intention of letting the public have access. In 1777 John Wilkes, a member of Parliament, argued for the purchase by the nation of Robert Walpole's superb Houghton Collection, but to no avail. When Catherine the Great moved it to Russia, a commentator on this occasion protested: "The riches of a nation have generally been estimated according to as it abounds in works of art. . . ."[7] In 1811 the Dulwich Picture Gallery opened as the first public art gallery in Britain. The collection was given to Dulwich College in the absence of a national gallery. Arguments for a national collection grew, although the Royal Academy of Arts wanted national to be defined as British art only. In 1824 the connoisseur George Beaumont donated his collection to the nation, Parliament voted funds and the National Gallery was born. In competition with schools and sewers, it remained underfunded in spite of the arguments of Radical thinkers that access to works by Old Masters would improve the level of taste and thus make British textiles more competitive with the high-end products of France and Germany and that free entry would wean the working class off drink.[7] In the second half of the century, municipal museums spread around the country. In response to the demands to show British art, and as a condition of Sir Henry Tate's gift of his collection, the Tate Gallery was spun out of the National Gallery in 1894 (but had its remit expanded to international modern art in 1917).

Art Museums in the United States

While some museums in the United States were established in the first half of the century (Wadsworth Atheneum, Hartford, CT, 1842), most were formed in the economic boom that followed the Civil War, in the great growing cities of the North and in many of the smaller cities as well. The Metropolitan Museum of Art in New York was founded in 1870, the Museum of Fine Art in Boston in the same year, the Philadelphia Museum of Art in 1876, the Art Institute of Chicago in 1879, the Los Angeles County Museum of Art in 1910, and the Cleveland Museum of Art in 1916. These museums were established by wealthy donors who wanted prestige for

their cities and for themselves. Whereas in Europe access to museums was symptomatic of the bourgeois struggles to end aristocratic privilege, in the United States museums were overt displays of privilege.[8] The titans of industry and commerce collected on a grand scale, often aided by the great dealer Sir Joseph Duveen: Altman, Frick, Johnson, Kress, Morgan, Mellon, Walters, and Widener gave their works away to existing museums or to new ones that preserved their heritage and gave prominence to their names. When the Metropolitan Museum expanded in 1925, the *New York Times* commented that the Met was "not so much an institution for the instruction and pleasure of the people as a sort of joint mausoleum to enshrine the fame of American collectors."[8] In a salute to them, the Met organized in 2007 an exhibition of all its seventeenth-century Dutch art, *Rembrandt and his Time*. The art was presented neither chronologically nor by artist but by the date when the works were acquired. Andrew Mellon, more self-effacing than most, donated his collection to create the National, rather than Mellon, Gallery of Art in 1937.

The art museum was to serve many purposes, besides the obvious ones of providing enlightenment, education, and pleasure: first, to prove that the dynamic American economy knew the value of higher things. At a dedication in Chicago in 1913, the sculptor Loredo Taft said "art and culture had arrived to crown commercial life as was crowned commercial life of Athens and Florence and Venice."[8] Since Europe was the model, it was predominantly European art that was collected in preference to American art which was often considered inferior. "During the late nineteenth and early twentieth century, history's largest transfer of cultural wealth from one hemisphere to another took place."[9] Second, since the church, separated from the state, could not provide homogeneity of values for the disparate immigrant hordes, impressive museums were to be organs of acculturation, providing a set of core values. Third, galleries with displays of decorative arts modeled on those in the influential Victoria and Albert Museum in London (founded in 1857 as the South Kensington Museum and renamed in 1899) would provide the lower middle classes with aspirational models for their households and inspire manufacturers to produce goods of superior design.

Art Museums in the Modern Period

Two world wars and dire economic conditions put a damper on museum building until the 1950s. But from then on the momentum of building new museums and museum extensions gathered pace. Collections grow and need to be shown to the public rather than being kept in storage. Funding shifted from almost complete reliance on a few wealthy patrons, who, while still very much in evidence in the United States, were joined by corporations, foundations, and museum members, with agendas of broadening audiences and enhancing access.

Art museums have become less stuffy and are gathering places for the community, with gourmet restaurants, high-end stores, and sought-after venues for weddings and receptions. They have become more popular—too popular for some who fear the spread of commercialism.

Notable new museums or extensions of the second half of the twentieth century that were architecturally adventurous included the extension to the Kröller-Müller Museum in Otterloo, Holland, designed by Henry van de Velde (1953); the Milwaukee County War Memorial, Wisconsin, United States, by Eero Saarinen (1957) containing the Milwaukee Art Center; the Louisiana Museum of Art, Humlebaek, near Copenhagen, Denmark, by Jorgen Bo and Vilhelm Wohlert (1958); the Whitney Museum of American Art, New York, by Marcel Breuer (1966); the Hayward Gallery, London, by London County Council architects, part of a cultural center on London's South Bank (1968); the Neue Nationalgalerie, Berlin, West Germany, by Ludwig Mies van der Rohe (1968); the Kimbell Art Museum, Fort Worth, TX, by Louis Kahn (1972); the Sainsbury Centre for the Visual Arts, Norwich, England, by Norman Foster (1977); the East Wing of the National Gallery of Art, Washington, D.C., by I.M. Pei (1978), and Pei's pyramid for the Louvre (1989); Neue Staatsgalerie, Stuttgart, West Germany, by James Stirling and Michael Wilford (1984) and their Clore Gallery extension to London's Tate Gallery (1986); the Menil Collection, Houston, TX, by Renzo Piano (1987); the Museum of Contemporary Art, Barcelona, by Richard Meier (1995) and his Getty Center in Los Angeles (1997); the Quadracci Pavilion extension to the Milwaukee Art Museum by Santiago Calatrava (2001) (Fig. 1), the de Young Museum, San Francisco, by Herzog and de Meuron (2005); the Bloch Pavilion at the Nelson-Atkins Museum of Art, Kansas City, MO, by Stephen Holl (2007); the Hamilton Building at the Denver Art Museum, Colorado, by Daniel Libeskind (2007); the Broad Contemporary Art Museum at the Los Angeles County Museum of Art, by Renzo Piano (2008), and his extension for the Art Institute of Chicago (2009). Museums provide rare opportunities for architects to show talents not sought in commercial real estate development.

The explosion of modern art from the early part of the twentieth century onward presented art museums with a dilemma: to collect only art that had stood the test of time or to collect challenging contemporary art. The solution most commonly adopted was to lag well behind the times, thus creating a gap for new museums for new art in new buildings.

The unrivalled pioneer and a model for others, the Museum of Modern Art (MoMA) in New York was set up under its formidable director Alfred H. Barr in 1929 and moved into its first purpose-built structure on West 53rd Street, designed by Philip Goodwin and Edward Stone, in 1939. The Museum embraced all aspects of visual culture with departments of architecture and design, film (and later video and digital), and photography. The latest and largest of several expansions, by Yoshio Taniguchi, opened in 2004, and has made MoMA a must-visit attraction.

Fig. 1 Milwaukee's Santiago Calatrava – designed Quadracci Pavillion extension: The museum as art object.
Source: Photograph by Fritz Jusak, courtesy of the Milwaukee Art Museum.

The industrialist Solomon R. Guggenheim began to show his extensive collection of "nonobjective art" from the 1930s and, advised by Hilla Rebay, he continued to collect until his death in 1949. Ten years later (and six months after the death of the architect, Frank Lloyd Wright), the museum bearing his name opened on Fifth Avenue. The startling round shape and spiral gallery of the Guggenheim signaled the new adventurousness in architecture, and set off a continuing debate on whether dramatic new museum buildings outshine the art inside them.

In France, discussions were held about creating a modern art museum in Paris in the 1930s, but the war intervened. Making up for lost time, the Pompidou Center in the Palais Beaubourg was opened in 1977, housing a library and contemporary music center as well as a huge collection of modern art. The young architects who won an international competition, Richard Rogers (English) and Renzo Piano (Italian), introduced new concepts: large spaces unimpeded by columns, services on colorful display instead of being hidden, a feeling of openness and transparency. It was and has remained a huge popular success.

In 1997 the Guggenheim Bilbao, an asymmetrical sculpture clad in titanium designed by American architect Frank Gehry, opened to international notice and visits. The museum was paid for by the Basque government as part of a comprehensive economic development plan to reinvigorate a failing industrial area. Since the opening of Guggenheim Bilbao, other cities have been opening new museums and museum additions hoping for a "Bilbao effect," but without the accompanying investment in subways, airports, and infrastructure. The art came from the extensive collection of the Guggenheim Foundation, and other Guggenheim branches were opened in Las Vegas (with Russia's Hermitage in 2001; it closed in 2008), Berlin (with Deutsche Bank, 1997) and another vast one also designed by Frank Gehry is planned for Abu Dhabi. The Peggy Guggenheim Collection opened in Venice in 1951 and is part of the international family.

In 2000 the international modern and contemporary art collections of Tate (in rebranding having lost "the" and "Gallery") were split off into Tate Modern and installed in a disused power station remodeled by the Swiss architects Herzog and de Meuron who had won an international competition. The cavernous Turbine Hall is transformed each year with an installation by a contemporary artist. The combination of the drama of the building, an enticing art program, the site on a revivified stretch of river, and, importantly, free general admission (with payment for special exhibitions) has made Tate Modern one of the most popular art museums in the world, with an annual attendance of some 5 million.

Museums focusing variously on modern or contemporary art sprung up in San Francisco (1935), Chicago (1967), and Los Angeles (1979). The Boston Institute of Contemporary Art opened a bold building by Diller and Scofidio in 2006, and the New Museum of Contemporary Art opened a subtle and complex building by the Japanese practice SANAA on New York's Bowery in 2007. Some institutions such as the Institute of Contemporary Art in London (1947) and the Walker Art Center in Minneapolis (originally 1927; focus on contemporary in the 1940s) have spaces for dance, theater, and film.

Many traditional museums like the Metropolitan Museum have long collected modern art as the passage of time venerates it into art history, but in recent years, seeing the popularity of contemporary art, they have been less and less willing to cede a monopoly of the new to specialist museums. The large and elegant extension to the Art Institute of Chicago is called the Modern Wing and contains modern art, contemporary art, photography, design and architecture, and sets a challenge to Chicago's forbidding Museum of Contemporary Art, designed by Josef Paul Kleihues (1996).

Some of the underlying reasons for the building boom mirror those of the nineteenth century: pride in city, economic prosperity, a cause and consequence of higher education. A new driving force is the wish to stimulate the creative economy as the manufacturing economy moves to low-wage countries and attract those industries whose workers are looking for quality of life and not just quantity of salary.

The spate of museum building has fostered criticism that the star architect or "starchitect" usually commissioned often create grandiose structures that deflect attention from the art to their own creations. Victoria Newhouse, in *Towards a New Museum* is the most influential of such critics.[10]

ETHICS AND GOVERNANCE

Museums are repositories of values as well as of objects. Those who control the museum should regard themselves as stewards and trustees, not owners: they have to act unselfishly for the long-term good of the institution. The museum should safeguard and protect the art it collects for future generations, and that includes cataloging and documentation. The museum should be run by the museum's director and staff with integrity and in accordance with the highest ethical principles and with professional standards. Written policies should cover key areas such as acquisition, deaccessioning, and loans. The museum should have a mission statement and a long-range strategy or plan that ensures long-term financial stability. While overseeing management, trustees should not become involved in it. Trustees should avoid conflicts of interest.

Museums should be run in a businesslike way, but they are not businesses. They often do things that puzzle the business mind, like putting on an expensive exhibition that is artistically worthwhile even though it will not draw in a large audience. Of course such a course of action is only sensible provided the resources are available to pursue it, but the very notion can raise the business eyebrow.

In the United States, the usual pattern is to have a large board to encompass wealthy individuals, collectors, captains of local industry and commerce, civic leaders, ethnic minorities, and local politicians. Much work is typically done in committees. Museums can readily tap local expertise to create strong committees of finance, audit, investment, remuneration and human resources, and fund-raising. Committees for acquisition, exhibitions, and education have mixed motives: oversight and advice, but also to engage members so that they are more likely to be generous to the museum.

Museums are porous. They are influenced by trustees, by members, by support and special interest groups, by donors, by local politicians, by community groups, and of course by the public, who have access to each other and to the director and to the staff. It is up to the director to ensure that the museum is led rather than being led. It is up to the chairman of the board of trustees to ensure that the director is supported in leading. It is up to the trustees to ask the tough questions to ensure that the direction in which the director is leading is sensible and fiscally responsible.

In Britain, boards are smaller, and at Tate and the National Portrait Gallery their members include artists. In the rest of Europe, museums are typically under the control of the state or of a local municipality with perhaps an advisory board of lay people.

MONEY AND OTHER MEASUREMENTS

Broadly, art museums in the United States are financed by the private sector, in Europe by the state, and in England by both. It must be remembered that in the United States the federal government is a "silent partner." As part of a long-standing policy of encouraging the not-for-profit sector, charitable donations of money and of art can be deducted by individuals from their taxable incomes, and so their giving is subsidized. Britain under a Labour government has moved towards the American practice of making it attractive to give money to museums and other charities.

This is an age of measurement and evaluation. What are the "outputs" and "deliverables" by which a board can judge its management and a community its museum?

The annual operating statement of a museum should demonstrate that revenues and expenses are in balance. There are three categories of operating revenue: contributed (donations, grants from foundations, sponsorship, government support); earned (admission fees, typically 5–10% of revenues), membership (sometimes categorized as contributed), store, café, rental, exhibition fees); and transfer from endowment funds (usually set at between 4% and 6% of the value of endowment assets). The higher the endowment in relation to the other sources of revenue, the greater the likelihood of fiscal stability; the smaller it is, the more the museum has to depend on courting donors or earning more. Expenses are usually categorized by function—curatorial, education, development, marketing, administration—and by type—of which salaries and benefits is the largest, typically about half of operating costs. A persistent operating deficit indicates a mismatch between ambition and means and should send out a warning to the board to take action.

The number of people who come to visit a museum is clearly an important indicator of how it is connecting to the community. If a museum is heavily dependent on admission revenue and store and café revenue that goes up and down with foot traffic, then the number of tickets sold is obviously of great concern. However, if a museum is tempted to put on exhibitions intended purely to draw a crowd, then it is substituting for the high-minded mission of bringing art to the people the less noble one of bringing people to the museum any way it can. There is a possible cost in credibility and integrity. The Solomon R. Guggenheim Museum put on an exhibition on the *Art of the Motorcycle* in 1998, sponsored by the car and motorcycle manufacturer BMW, that was criticized on just those grounds.

A better way of making museums more accessible is to abolish admission fees. During the Thatcher government of the 1980s, the major British museums were set free of control by the Treasury (ministry of finance) and encouraged to raise more money from the private sector and charge for entry if they wished. Many did. In 2001 the Labour government abolished admission fees: the result has been a dramatic increase in visitorship. The profile of visitors did not change, but since there were many more of them, the number of less-affluent visitors did go up in absolute terms. The Walters Museum of Art in Baltimore and the Baltimore Museum of Art were both given extra money by the city to allow free admission in 2006, and the Indianapolis Museum of Art, which has a large endowment, went back to free admission in 2007.

In an influential paper entitled "Metrics of Success for Museums,"[11] Maxwell Anderson, now director at the Indianapolis Museum of Art, argues for looking at measures of quality such as the number of scholarly publications, the size of the library, and loans made to other museums in an attempt to free museums from chasing numbers of bodies. It remains true that the most important function of a museum—the experience of being moved by or challenged by or uplifted by a work of art—is problematic to measure.

EXHIBITIONS

Most museums put on exhibitions of works of art brought especially together from one or many lenders. While not a new phenomenon, the size and frequency of exhibitions grew from the 1970s and museums carved out space devoted specifically to them. Some of these exhibitions are labeled "blockbusters," a term of praise in the cinema but of mild abuse in museums. Exhibitions such as *Monet in the Twentieth Century*, mounted by the Royal Academy of Arts in London and the Museum of Fine Arts in Boston in 1998, and *Picasso/Matisse*, mounted by Tate and MoMA in 2000, brought huge crowds. The Royal Academy set a trend for such exhibitions by being open all night on the last Saturday of the run. The Monet exhibition drew attention to the twentieth century art of a great artist considered as a nineteenth century figure and the Picasso and Matisse exhibition explored the creative rivalry between two great twentieth century artists. They were great visual experiences but the exhibitions added to scholarship and understanding. Popularity was the by-product. However, given the appeal of Impressionist and Postimpressionist artists, for example, there have been exhibitions of them where popularity was the goal, and this has generated debate about the role of all temporary exhibitions, blockbuster or otherwise.

The case against blockbuster shows begins with the argument that exhibitions detract attention and curatorial and other resources away from the museum's own collection. The public is led to expect a linear progression of art and argument and then finds the collection dull by comparison. Fragile works of art traveling to the several venues of a tour have to endure the hazards of transportation. Marketing departments are let loose in ways that undermine the mission of the museum. Museums follow the false gods of money and attendances. On a proper allocation of expenses, including overhead, they do not really make a profit. The attendant exhibition store and special merchandise bring commercialism into the museum.

The case for blockbusters is that such exhibitions bring a focus on an artist or theme that would be impossible in any other way. They enable scholarship that would otherwise not get done. They do bring a sense of event and excitement to the museum that does, indeed, usually bring in a higher audience—an audience for whose time and attention museums are in competition with a plethora of other activities. The appropriate exhibition can bring in minorities that do not usually come to museums: *The Quilts of Gee's Bend*, an exhibition of quilts by African-American women, attracted the local black population wherever it was shown. Exhibitions are exciting for the public and for curators. They foster international collaborations.

Museums have reacted to the criticism by making more of an exhibition of the "permanent" collection. Tate Britain rotates the display of its collection annually. The Metropolitan Museum's exhibition of Dutch art referred to above came entirely from the Met's own collection. However, it remains a true criticism of many museums

that scholarship and publication of the collection have taken second place to exhibition-making.

GLOBALIZATION

The Guggenheim Foundation, rich in collection and poor in endowment, has a strategy of international expansion on the Guggenheim Bilbao model: to lend its curatorial and architectural expertise, traveling exhibitions, and collection in return for a fee. Lest this be seen as the exclusive preserve of American institutions following the profit-maximizing strategies of global corporations, the government-financed Louvre has agreed to do exactly the same thing with a rich Gulf oil state. In March 2007, France and Abu Dhabi agreed to the Abu Dhabi Louvre, with the Louvre benefitting to the tune of approximately $1 billion. The architect is Jean Nouvel, thus providing France an opportunity for one of its star architects. Protesters argued that the Louvre was selling its soul, exporting culture for cash and contributing to the Disney-fication of artistic experience. The full meaning of these international exploits has yet to sink in. Since museums are busy digitizing images of their works of art to make them internationally available, why not follow up by making some of the real works themselves viewable in conditions set by museum professionals? Is it not better to have the works on display than in storage? But if the works are from storage, is the museum sending its best works out? The director of the Louvre, Henri Loyrette, is keen on enhancing the international brand of the Louvre, and has sent part of the collection on long-term loan for a fee to the High Museum in Atlanta. Is it too commercial to use the consumer-goods concept of branding for a museum—or is he simply being realistic about the way the world now works? The Louvre is a national museum, financed by the state, and should one argue if the state wishes to add a cultural dimension to its diplomacy?

OWNERSHIP OF ART

In late 2007 the Russian authorities threatened to cancel a blockbuster exhibition of art from four Russian collections at the Royal Academy of Art in London. The exhibition was already installed in Dusseldorf, Germany. Germany has a law that prevents the filing of legal claims for the restitution of art in loan exhibitions; Britain had delayed the implementation of such a law. The Russians, fearing a suit from an heir of the Shchukin family, whose art has been confiscated after the Russian Revolution, said they were too worried to lend (the deterioration in relations between Britain and Russia also played a role: art is a weapon of diplomacy). The law was brought forward in Britain, and the crisis defused. The United States has a similar law in place, though with loopholes. While sensible to protect cross-country traveling exhibitions in this way, the incident did highlight one of the major issues facing museums in recent years: giving back art to its rightful owners. This issue has four main strands.

The first is confiscation under totalitarian regimes in the twentieth century. Russia's current government does not accept that confiscation during the revolution is an act for which restitution needs to be made. During the Nazi era, art owned by Jews was either confiscated, or fleeing Jews were forced to sell for low prices. After several decades of prevarication, and under pressure from the American Department of State, the German museums and their municipal owners have been cooperating in giving back to heirs art where there is good evidence of ownership. The Austrians have been less helpful. Museums in Europe and the United States are under pressure to research the provenance of Nazi-era acquisitions. In 1998 the AAMD issued guidelines encouraging American museums to research their collections and if they suspect they have such works to try and trace the heirs, and if they cannot find them, to publish any details known about the work.

The second strand is antiquities that might have been looted. The Italian and Greek authorities have since the early part of the twenty-first century been active in pursuit and in 2007 the Getty Museum and the Met were obliged to give back important objects once they were confronted with good evidence that these objects had been looted. Art museums are under pressure not to acquire objects whose provenance back to 1970—the date of a UNESCO resolution—is not clear. Some feel that the imposition of this arbitrary date will mean the disappearance of such objects on the market into the hands of private collectors and that it would be better to acquire them for the public realms and display them and publish what is known on the Web.

The third strand concerns objects collected by museums during their ethnographic phases that turn out to be sacred to the peoples that once owned them. The American government passed the Native American Graves Protection and Repatriation Act (NAGPRA) in 1990 to provide a process to return certain Native American cultural items to living descendants and affiliated Indian tribes. Australia and New Zealand have similar policies.

The fourth is what might turn out not to be the special case of the Elgin Marbles, bought for the British Museum by Lord Elgin from the Ottoman Empire, the then occupying power of Greece. The Elgin Marbles (also known as the Parthenon Marbles) are ever more actively being sought by the now sovereign Greek government that in 2007 opened a museum to house them on their return. The British Museum, the Metropolitan Museum, and the Louvre, the three great encyclopedic museums of the world, argue that it is in the interests of everyone for antiquities marking the birth of Western civilization to be dispersed rather than given back to governments who happen now to govern the territories from which they once derived.

CONCLUSION

This entry has concentrated on art museums as institutions rather than on the art within their walls.

We are in an age that values imagination and creativity. We are also in the digital age. We are in an age of a profusion of visual images. Art sparks the imagination. In a screen-based society, people increasingly appreciate the single, authentic object. Contemplation of art in museums provides a welcome respite from the rush of imagery. The boom in art museum expansion is likely to continue after the 2009 recession is over, provided that museums remain true to themselves.

In an influential book of essays by leading museum directors entitled *Whose Muse*,[12] James Wood, then director of the Art Institute of Chicago and subsequently president of the J. Paul Getty Trust eloquently sums up the implicit contract of museums with the public:

> Ultimately the American art museum's authority must flow from its ability to be one of the most tangible and accessible forums for the experience of excellence, the affirmation of tolerance, the appreciation of personal expression, and the pursuit of the individual happiness embodied in our Constitution. Guaranteeing the integrity of this forum has never been more important, and to the degree that we succeed, our museums will indeed be worthy of the public trust.

REFERENCES

1. Impey, O.; MacGregor, A. Introduction. *The Origins of Museums: The Cabinet of Curiosity in Sixteenth- and Seventeenth-Century Europe*; Clarendon Press: Oxford, U.K., 1985; 1.
2. Berelowitz, J. From the body of the prince to Mickey Mouse. Oxford Art J **1990**, *13*(2), 70–84.
3. *Encyclopedia of World Art*; McGraw Hill: New York, 1965; 379–380.
4. Ashmolean website. Available at http://www.ashmolean.org.
5. British Museum website. Available at http://www.britishmuseum.org.
6. McLellan, A. The Museé du Louvre as revolutionary metaphor during the terror. Art Bull. June **1988**, *70*(2), 305. College Art Association.
7. Conlin, J. *The Nation's Mantelpiece*; Pallas Athene: London, 2006; 23–66.
8. Duncan, C. Art museums and galleries. In *The Oxford History of Western Art*; Kemp, M., Ed.; Oxford University Press: Oxford, 2000; 405.
9. Abt, J. Museums. In *Grove Dictionary of Art*; Turner, J., Ed.; Oxford University Press: Oxford, 1996; 359.
10. Newhouse, V. *Towards the New Museum*; 2nd Ed. Monacelli Press: New York, 2006.
11. Anderson, M.L. *Metrics of Success in Art Museums*; Getty Leadership Institute: Los Angeles, CA, 2004.
12. Cuno, J., Ed. *Whose Muse*; Princeton University Press: Princeton, NJ, 2004; 127 and Harvard University Press: Cambridge, MA.

BIBLIOGRAPHY

1. Carrier, D. *Museum Skepticism: A History of the Display of Art in Public Galleries*; Duke University Press: Durham, 2006 and London.
2. Cuno, J., Ed. *Whose Muse;* Princeton University Press: Princeton, NJ, 2004; 127 and Harvard University Press: Cambridge, MA.
3. Duncan, C. *Civilizing Rituals Inside Public Art Museums*; Routledge: Oxford, 1995.
4. Mauries, P. *Cabinets of Curiosities*; Thames and Hudson: London, 2001.
5. McLellan, A. *The Art Museum from Boullée to Bilbao*; University of California Press: Los Angeles, CA, 2008.
6. Meyer, K. *The Art Museum: Power, Money, Ethics*; William Morrow: New York, 1979.
7. Schubert, K. *The Curator's Egg: The Evolution of the Museum Concept from the French Revolution to the Present Day*; Christie's Books: London, 2000.
8. Spalding, J. *The Poetic Museum: Reviving Historic Collections*; Prestel: London, 2002.
9. Taylor, F.H. *The Taste of Angels: A History of Art Collecting from Rameses to Napoleon*; Little, Brown: Boston, MA, 1948.
10. Thompson, K. *Treasures on Earth: Museums, Collections and Paradoxes*; Faber and Faber: London, 2002.

Artificial Intelligence

Jianhua Chen
Computer Science Department, Louisiana State University, Baton Rouge, Louisiana, U.S.A.

Abstract

Artificial intelligence (AI) is a multidisciplinary subject, typically studied as a research area within Computer Science. AI study aims at achieving a good understanding of the nature of intelligence and building intelligent agents which are computational systems demonstrating intelligent behavior. AI has been developed over more than 50 years. The topics studied in AI are quite broad, ranging from knowledge representation and reasoning, knowledge-based systems, machine learning and data mining, natural language processing, to search, image processing, robotics, and intelligent information systems. Numerous successful AI systems have been deployed in real-life applications in engineering, finance, science, health care, education, and service sectors. AI research has also significantly impacted the subject area of Library and Information Science (LIS), helping to develop smart Web search engines, personalized news filters, and knowledge-sharing and indexing systems. This entry briefly outlines the main topics studied in AI, samples some typical successful AI applications, and discusses the cross-fertilization between AI and LIS.

INTRODUCTION

This entry is about artificial intelligence (AI),[1–4] a multidisciplinary subject, typically studied within Computer Science. Ever since the dawn of civilization, humans have constantly asked questions regarding mechanisms of human intelligence. Human's abilities to think, reason, learn, act to achieve goals, adapt to changing environment, etc., which are central to intelligence, fascinated philosophers, scientists for centuries. There is a long history of human endeavor in unveiling the mystery of human intelligence and building artificial systems capable of doing smart things like humans do. The early works in understanding human intelligence focused on studying how humans "know" the world around them and how the human thinking and reasoning are performed. As early as 2300 years ago, Aristotle, a great Greek philosopher, studied the laws of thought and proper ways of reasoning. In his work "Prior Analytics,"[5] Aristotle defines syllogism, a kind of logical argument, which allows deduction of a valid conclusion from two given premises. For example, from the premises that "All men are mortal" (major premise) and "Socrates is a man" (minor premise), one can infer by syllogism that "Socrates is mortal." Over the long time after Aristotle, logicians such as Freg, Russell, Leibniz, Godel, Tarski, and others, have fully developed formal logic systems such as propositional logic and predicate logic, which formalize the thinking and reasoning process of humans. Moreover such formal logic systems open up the possibility of being implemented on computational systems.

Endeavors of constructing mechanical/electronic artifacts to do calculation, concept manipulation, reasoning, and game playing can be found in many eras of human history. Such efforts contribute significantly to the foundations of AI. See Entry 1.1[1,2] for more discussions on the foundations of AI. Around twenty-sixth century B.C., the Chinese invented the abacus, the first mechanical tool in human history for performing arithmetic calculations (Entry 1.1.1).[2] Similar calculating equipments were also discovered in Roman relics, in India, and Egypt from ancient times. In 1623, Wilhelm Schickard, a German mathematician, created a calculating clock for addition and subtraction. Soon after in 1642, the famous calculating machine Pascaline was created by Blaise Pascal, a great French philosopher and mathematician. Pascaline is capable of addition and subtraction with carries and borrows. Pascal noted[6] "The arithmetical machine produces effects which approach nearer to thought than all the actions of animals." Gottfried Wilhelm Leibniz, a great German philosopher and mathematician, believed that human reasoning could be reduced to mechanical calculations of some kind, and thus one could use the calculation results to find out who is right and who is wrong in cases of conflicting opinions. He wrote[7]

> The only way to rectify our reasonings is to make them as tangible as those of the Mathematicians, so that we can find our error at a glance, and when there are disputes among persons, we can simply say: Let us calculate [calculemus], without further ado, to see who is right.

He envisioned that a machine could be devised for automatic derivation of scientific knowledge by deductive inference. In the late 1950s and early 1960s, amid the initial enthusiastic development of AI, Arthur Samuel developed[8] a computer program that learns to play the game of Checkers, which could learn to improve its

Encyclopedia of Library and Information Sciences, Fourth Edition DOI: 10.1081/E-ELIS4-120043680

game-playing skills by playing against a copy of itself, playing with human players, and storing good moves from Master game books. In 1997, IBM's Deep Blue Chess program,[9] with a combination of parallel processing, fast search, and AI ideas, scored a historical win against world Chess champion Kasparov.

As can be seen from the brief descriptions above, the philosophical roots of AI can be traced back to over 2300 years ago. The recent past 200–300 years have witnessed a rapid development in mathematics and science. The formalization of mathematics and science has laid the intellectual foundations of AI. AI as a multidisciplinary area draws on the development in diverse disciplines in addition to philosophy and mathematics, including economics, psychology, linguistics, control theory and cybernetics, and neurosciences. In particular, the birth of the electronic computer in the 1940s was instrumental and crucial to making AI a viable distinctive scientific discipline within Computer Science. The availability of digital computers in late 1940s made it possible for researchers at that time to write computer programs for playing games, performing logical reasoning, and problem-solving. Researchers could then empirically study the computer's performance and analyze whether the computer demonstrated some kind of intelligence. In 1956, at Dartmouth College in Massachusetts, a two-month summer workshop was held[1] and attended by 10 prominent researchers of AI, including John McCarthy, Marvin Minsky, Claude Shannon, Arthur Samuel, Allen Newell, and Herbert Simon. The workshop was a milestone that signified the birth of AI—a name suggested by McCarthy and agreed by all the attendees of the workshop.

In the early 1950s, researchers and the general public were all fascinated by the possibilities made prominent by the advent of the electronic computer era. People asked numerous questions about whether computers could be intelligent, e.g., do things that used to require human intelligence, what is intelligence, what would it take for us to consider a computer to be intelligent. Objecting views were raised by many to the idea that indeed a computer could be intelligent given sufficient storage memory and processing power. Alan Turing, a great British mathematician and considered by many as the founding father of Computer Science/AI, proposed in 1950[10] the famous Turing test. Turing proposed to replace the question "Can machines think" by the question of whether a digital computer can pass the Turing test. In the Turing test, a human interrogator converses in natural language with a computer and a human participant, that are located in rooms separated from the interrogator. The questions from the interrogator and the answers from the computer/human participant are transmitted via online typed messages (similar to today's computer-relayed talk or instant messaging). After conversations for 5 min, the human interrogator needs to identify which one is the computer/human participant. According to Turing, a computer should be considered as "intelligent" if it passes the Turing test, i.e., if it fools the human interrogator over 30% of the time in many repeated trials. The central idea behind the Turing test is that a system is deemed intelligent if it can behave like humans. This conceptualization of intelligence (behaving like humans) makes it easier to discern intelligence because one does NOT need to know the inner workings of a system to judge whether the system is intelligent or not—it is sufficient to just look at the system's behavior. Turing predicted that by the year 2000, man could program computers with large storage capacities (109 units) so well that the computers would easily pass the test with an average human interrogator. Although his prediction was not realized, the discipline of AI certainly has achieved great advancement over the 58 years from the proposal of the Turing test.

After over 50 years of development, AI has become an industry and a gradually maturing subject. Theories of AI—computational theories of intelligence, have advanced significantly, with many flourishing research topic areas developed and numerous successful AI systems deployed in real-world applications. Today, we enjoy the great benefits of modern computer and information technology in our daily lives, many with important AI components. We have smart online shopping tools that can recommend suitable products catering to the specific preferences of customers; personalized message filtering tools that help to sort out spam e-mails; robots that perform (or assist doctors to perform) medical procedures with great precision; intelligent online information tools that allow us to know what is going on in the world with the click of a mouse, and to obtain, create, and share knowledge efficiently. We have seen the great miracle of electronic computers and AI that our great intellectual pioneers have envisioned, and much much more. Today, AI is more exciting than ever as a research area, playing an increasingly important role in the age of information technology. See Buchanan[11] for a brief account of AI history.

The rest of the entry is organized as follows. In the next entry, the major topics of AI will be briefly described. Entry on "Artificial Intelligence and Application" presents some sample applications of AI, along with a brief discussion of the impact of AI on Library and Information Science (LIS), and is followed by the conclusion. Readers interested to know more about the AI subject can consult leading text books[1–3] and other online sources.[4]

TOPICS OF STUDIES IN AI

In this entry we survey some representative topics of AI study. Since AI is a very big and broad field, it is impossible to make a complete coverage of all topics of AI within the limited space of one entry. The omission of a topic in this entry is by no means an indication that the topic is not important.

Heuristic Search and Problem Solving

Heuristic search is a topic studied since the early days of AI. Researchers realized long ago that many AI problems could be viewed as a search problem. The concept of problem-solving as state space search was introduced in the 1950s.[12] In problem-solving, each possible scenario related to the task at hand is formulated as a state, and the entire collection of all possible states is called the state space, which contains the initial state of the problem, and the desired solution state (often called goal state). A state S has a number of neighboring states $N(S)$ which could be reached from state S in one step (by applying some state-transition operators). A state space could then be modeled as a (possibly weighted) graph with nodes representing states and edges connecting neighboring states. Given this view, solving a problem amounts to searching the state space (graph) to find a sequence of states that leads from the initial problem state to the goal state. For example, in computer checkers game playing, each feasible board configuration constitutes a state in the search space. The initial state corresponds to the initial game board configuration and a board configuration in which the player (the computer) has won the game is one of the many goal states. A move of a piece in board configuration S transits it into a board configuration which is a member of $N(S)$. The problem of playing the checkers game successfully against an expert human player is reduced to finding a sequence of moves (states) in response to the human player's moves such that the final state is a winning state for the computer.

For most search problems with real-world application, the search space is huge (or even infinite). Any blind exhaustive search method (such as breadth-first search, depth-first search) would suffer from combinatorial explosion that renders such search impractical. How to efficiently search the state space becomes a critical issue if we want to build useful AI applications. Heuristic search methods have been developed by AI researchers[13] to efficiently search the state space and overcome combinatorial explosion. Typical heuristic search methods are based on using a heuristic evaluation function h to guide the search process. The best-known heuristic search algorithm is perhaps the $A*$ algorithm[14] for searching a weighted graph for shortest path from the starting node ns to the destination node nd. The algorithm maintains a list of open nodes and a list of closed nodes. Each open node n corresponds to a current partial path ns, ..., n which could be extended into a path from ns to nd via n. For each open node, a measure $f(n) = g(n) + h(n)$ is used to estimate the length of the shortest path from ns to nd via n, where $g(n)$ is the actual length of the path ns, ..., n and $h(n)$ is the heuristic estimate for the length of the shortest

path n, ..., nd. The $A*$ algorithm always selects from the open list the node n with the lowest $f(n)$ value, considers the neighbors of node n for expanding the current partial path in search for the shortest path from ns to nd. It has been shown[14,15] that if the heuristic function h is admissible, namely $h(n)$ is always an underestimate of the actual length of the shortest path from node n to nd, then the $A*$ algorithm is optimal in that it is guaranteed to find the shortest path from ns to nd. Various improvements of the $A*$ algorithm have been proposed in the literature, including memory bounded $A*$ (MA) and simple memory bounded $A*$ (SMA), etc.

In using heuristic search for problem solving for specific applications, the design of the heuristic function h is a nontrivial task. One has to carefully analyze the specific problem at hand, formulate the search problem, and choose the h function by considering characteristics of the problem to be solved.

Knowledge Representation and Automated Reasoning

An intelligent agent must "know" the world around it, have knowledge about how to achieve goals (namely, what actions are needed to bring about a desired outcome), and can infer useful information from what it already knows in making intelligent decisions. Therefore, an intelligent agent should be a knowledge-based agent, with the ability to represent knowledge and perform automated reasoning from its knowledge.

Knowledge representation research (KR)[16] studies and develops formal systems for representing knowledge in a knowledge base, whereas automated reasoning research[17] focuses on finding efficient algorithms for inference from a given knowledge base represented in some formalization. These two areas of study are closely related.

Logic-based formalism is perhaps the most commonly used knowledge representation form in AI systems. See Genesereth and Nilsson[18] for more discussions of logical foundations of AI. Under a logic-based knowledge representation scheme, an intelligent agent's knowledge base is a set Δ of logical sentences in the representation logic language, and the inference problem faced by the agent becomes the problem of deriving logical consequences of Δ using valid inference rules of the logic. Propositional logic and first-order logic are most frequently used in practice for knowledge representation and reasoning.

When using logic as a tool for knowledge representation, one has to first define the syntax of the language for the logic, which specifies the basic symbols, logical connectives, and rules to formulate well-formed expressions (well-formed formulas) in that logic. For example, in propositional logic, the basic symbols are propositional symbols (typically represented by uppercase letters such as P, Q, R) each representing a proposition that can be true or false. The logical connectives in propositional logic includes $\wedge$, $\vee$, $\neg$, etc. So if P, Q are propositional symbols, then $P \wedge Q$, $P \vee Q$, and $\neg P$ are all well-formed formulas in the logic.

A logic must define the semantics of the language. Intuitively, semantics defines the "meaning" of well-formed formulas. The semantics of a logic defines the truth value of each formula for each possible world. The truth value for any well-formed formula in a possible world is obtained compositionally from the truth value of the basic proposition symbols in that possible world. For example, consider a propositional logic with two proposition symbols P and Q, where P stands for the proposition "John is a professor at Harvard" and Q denotes the proposition "John lives in Boston." Here we have totally 4 possible worlds: {TT, TF, FT, FF}, where each possible world spells out the truth value assignment for P and Q, in that world. For example, the possible world "TF" tells us that P is true and Q is false in this world. Thus in this world, the well-formed formula "$P \wedge Q$" will be assigned the truth value "false (F)" because "$P \wedge Q$" is true in a possible world if and only if both "P" and "Q" are true in that world.

Once the semantics of a logic is defined, we can use logic for the purpose of reasoning, namely, we can ask the question, "can we derive conclusion ϕ given our knowledge base Δ"? This is the problem of checking whether ϕ is a logical consequence of Δ—whether ϕ is true in all possible worlds in which Δ is true. Automated reasoning is responsible for this task. Automated reasoning research aims at finding efficient, valid inference algorithms to support derivation of logical consequences. For automated reasoning in propositional logic and first order logic, AI researchers pioneered by Alan Robinson have developed the resolution inference rule[19] and many of its variants. Since the logical reasoning problem in propositional logic is essentially reduced to the satisfiability (SAT) problem, which is known to be computationally hard, many heuristic methods have been developed which aim to find efficient solvers for the SAT problem.[20–22] The development of resolution-based inference in first-order logic and the drive for a unified language for declarative knowledge representation and automated reasoning have led to the logic programming technique, hallmarked by the language PROLOG.[23] In using PROLOG, one represents knowledge by a PROLOG program, and automated reasoning is carried out by the PROLOG interpreter which essentially performs resolution.

Researchers have developed a plethora of non-monotonic logics for representing commonsense knowledge in the 1980s and 1990s. The idea is based on the observation that human commonsense knowledge is not well represented by propositional or first-order logic, so something new needs to be developed. Among the various frameworks proposed, we have Reiter's Default Logic,[24] McCarthy's circumscription,[25] etc. The studies

on non-monotonic logics and reasoning are closely related to the study of logic programming. Logics for dealing with time, events, knowledge, and belief have also been developed by AI researchers in order to more accurately model the real world. For example, situation calculus[26] and event calculus,[27] and temporal logic[28] as well as logic about actions[29] deal with time-event-action related representation issues. Various logics on knowledge and beliefs[30] handle problems of representing (and reasoning about) beliefs and knowledge.

Knowledge representation studies involve not only developing formalisms (logic, etc.) for representing the real world, but also methodologies as to how to model the real world and represent the model within the chosen formalism. Generally speaking, decision on how to represent the world would require the identification of an ontology, which specifies the concepts (categories) for modeling the world and the taxonomy (inheritance hierarchy) relating the concepts. Other semantical relationships among concepts can also be included in an ontology. For example, when building an ontology for a university, we would identify concepts such as students, professors, courses, departments, employees, staff, etc. We can also organize the people in the university into a taxonomy (a tree) T with top node labeled as "person." The two children nodes below the root would be labeled by "student" and "employee," indicating a student is a person, and an employee is a person too. We could also identify other semantical relationships among concepts in this domain: for example, the relationship "enrolled-in" can be identified between "student" and "course," indicating that students take courses. This kind of ontology specification bears close similarity with Semantic Networks,[31] a representation scheme developed in early years of AI research. Clearly, tools supporting the construction and maintenance of ontologies are highly desirable. Current research on knowledge representation appears to focus on developing formal systems and tools for representing and processing ontologies with applications in the Semantic Web.[32] This includes the studies of a unified knowledge representation framework based on XML, RDF, OWL (Web Ontology Language),[33] etc. and development of tools for extracting/editing ontologies using the unified representation, and studies of inference procedures for query-answering with such representations. Some current works investigate the problem of knowledge acquisition by agent with commonsense knowledge formalized by some logic.[34]

Machine Learning

An intelligent system must have the ability to learn new knowledge so as to adapt in an ever-changing world around it. Machine Learning[35,36] study focuses on developing computational theories and algorithms that enable computers to learn. Since the early years of AI development, many researchers have pursued the ideas of a learning machine, and the field of Machine Learning is now a quite matured subfield within AI. Machine learning is closely related to the fields of data mining[37] and pattern classification.[38]

A typical intelligent agent with learning capability could be modeled as consisting of a learning element, a knowledge base, and a performance element. The agent interacts with the outside environment by performing some tasks (by the performance element) in the environment, and getting experience through observing the environment and its feedbacks to the agent. The learning element of the agent learns useful knowledge from the experiences, such that the learned knowledge will enable the performance element to do better on the task in the future. For example, consider a computer program that learns to play the game of checkers. The performance element here is a component that plays the game by using an evaluation function f on board configuration features to choose the next move. The outside environment is another copy of the program itself, and the experience gained by the computer will be a sequence of games between the computer and its opponent, as well as the game outcomes (win, loss, or draw). The learning element of the system could be a least-mean square-based linear function learning algorithm, if we define the evaluation function f to be a linear function of the game board features. Arthur Samuel's Checkers program[8] has tested such set-ups.

Machine learning tasks can be classified as supervised, unsupervised, and reinforcement learning dependent on the kind of experience available to the learning agent. In supervised learning, the task is learning (an approximation of) a function f from a set of input–output pairs for f: $\{x_1, f(x_1), \ldots, x_m, f(x_m)\}$. Here the experience is encoded in the supervision: the function values at the points $x_1, \ldots, x_m$. In the case of learning to classify Web pages as "interesting" or "uninteresting," the function value $f(x_i)$ for a Web page x_i will be just binary: 1 or 0 (denoting whether a Web page is interesting or uninteresting). In unsupervised learning, we do not have a beneficial teacher providing the labels ($f(x_i)$) for each observed x_i, we can only identify the patterns present in the observed data $\{x_1, \ldots, x_m\}$. In some sense, unsupervised learning basically amounts to forming clusters from the data and thus identifying the inherent structures in the data. For reinforcement learning, the agent does get some feedbacks from the environment, but not in the form of direct supervision $f(x)$ for each observed instance x. Instead, the agent would perform a sequence of actions in the environment and then receive a "reinforcement" signal after performing the action sequence. For example, consider a robot exploring an open area with obstacles and trying to reach a specific goal location without bumping into the obstacles. Here we do not give specific supervisions as to what is the best move for each location—because such supervision may not be available anyway in practice. Instead, reinforcement signals could be assigned to reward or punish a sequence of actions. If the robot reached the target location through several moves without bumping into obstacles, it would get a positive reward. It would get a negative reward (punishment) when stumping into an obstacle.

Symbolic learning approaches represent the knowledge to be learned in symbolic forms such as decision trees, formulas in propositional logic, logic programs, etc., and learning often takes place in some form of symbolic manipulation/inference, loosely speaking. One popular learning algorithm is Quinlan's Decision Tree learning algorithm,[39] which constructs a decision tree from a set of training examples, in a top-down fashion. Each example is represented by a vector of attribute-value pairs, together with a class label for the example. In each step of the tree construction, the algorithm checks to see if the examples associated with the current node are of the same class. If so, the node is a leaf node, and marked by the class name. Otherwise, the algorithm chooses the "most discriminating attribute" A to subdivide the examples associated with the node into disjoint subsets, and thus growing the tree. Then the tree construction process is applied recursively until the resulting subsets are "pure," namely, consisting of examples from one class. Various works have been done on learning Boolean functions, learning decision lists, and learning logic programs.

Artificial Neural Networks (ANN)[40,41] follow a different approach to the learning task. ANN research was motivated by the desire to construct simplified mathematical, computational models that mimic the way human brain works, and hoping to achieve better performance on tasks requiring human intelligence. It is observed that human brains consist of large number of biological neurons, which are massively connected, each with relatively low switching speed in communications compared with the switching speed of electronic circuits. However, humans can perform, with amazing speed, complex cognitive tasks such as recognizing a familiar face, which is still a difficult task for computers in spite of their speed advantages. This suggests that the processing power of the human brain may come from its highly parallel mode of information processing, and the connections patterns among the neurons are crucial in making such massively parallel processing possible. The study of ANN models represents efforts in trying to simulate this model of human brains. An ANN consists of a number of simple processing units, called neurons, each capable of computing simple functions such as linear functions and threshold functions, and sigmoid functions. The neurons are interconnected with real-valued weights. Neural networks can be used to do predictions, to perform classification tasks, to approximate functions, and to find clusters in input datasets. Learning in ANN amounts to adjusting the numerical-valued weights that connect the neurons. Such learning could be supervised, unsupervised, or a hybrid of supervised and unsupervised. In the supervised learning, perhaps the most well-known learning algorithms are perceptron training algorithm for a single linear threshold unit, and the backpropagation algorithm for training multilayer feedforward networks. Neural networks have been used widely in many successful applications.

Genetic Algorithms (GA)[42] are another distinctive family of methods for learning. GA are search algorithms that patterned after natural evolution. In using GA for learning, we are interested in searching for good solutions for a problem by selecting candidate solutions and recombining parts of candidate solutions guided by the mechanics of natural selection (survival of the fittest) and natural genetics (children inherit good traits from parents, with occasional mutations). GA maintains a current population of strings, each encoding a candidate solution to the problem. A fitness function f is defined that measures the merit of a string as a solution to the problem. The objective of GA is to search for the best string which maximizes the fitness value. GA applies the genetic operators reproduction, crossover, mutation to the current population in generating the next population of candidate solutions. In the reproduction process, strings from the current population are sampled with probabilities proportional to their fitness values. Crossover operations will produce two new strings from two parent strings by exchanging segments of the parents. And finally mutations may be applied to randomly alter one bit in a string. Through evolutions of strings from one generation to the next, GA perform structured yet randomized search of the space of all possible strings, often efficiently, in looking for the optimal or near optimal solutions. Koza's genetic programming[43] further extends the idea of GA by evolving computer programs for problem solving. GA research is closely related to studies of Artificial Life and evolutionary computing.

Another type of machine learning is statistical learning,[44] utilizing probabilities and Bayesian theories for learning. In particular, graphical models aim at generating models represented as directed or undirected graphs with (conditional) probability tables attached to each node in the graph, and the entire graph captures joint distributions of a set of random variables. This includes learning Bayes Belief Networks,[45] learning (conditional) Markov Networks, etc. In recent years Probabilistic Relational Models (PRM) and related learning models have also been developed.

Natural Language and Speech Processing

An intelligent system must have the capability to communicate and interact effectively with the outside world. Effective communications include receiving information (in various forms) from the world, understanding such information, and sending out information in suitable forms understandable to the outside world. Natural language processing and speech processing address the problems involved for an intelligent computer to communicate with humans using natural (written or spoken) language such as English.

Natural language processing[46] research mainly handles the task of communicating with written natural language. The main topics studied include language understanding, language generation, and machine translation. The inputs to a natural language understanding system are written texts (articles, or paragraphs or sentences) of some language, and the desired outputs are semantical structures represented in some form, which capture the semantic meanings of the inputs. Language generation handles the opposite side of the problem: Given semantic meanings to be communicated to the outside world, a natural language generator produces correct natural language sentences (paragraphs, articles) that convey the meanings accurately. Machine translation tackles the task of automated translating texts from the source language to the target language, say, from English to French.

In speech processing,[47] the tasks are speech understanding and speech generation. Clearly, the apparatus of natural language processing techniques can be used as components of a speech processing system. For speech understanding, the main hurdle is speech recognition, which requires the capability of converting the spoken language inputs into written texts (so that natural language understanding tools can be utilized subsequently). Similarly, for speech generation, the major task is to map written texts to speech utterance. Converting continuous speech signals to written text requires multiple steps, from the initial step of signal sequence segmentation, to the step of phoneme recognition, followed by the step of mapping phonemes to texts. Signal processing techniques are needed to handle speech signal noise removal and segmentation. Neural Networks and Hidden Markov Models are commonly used techniques for speech recognition and generation.[48] Speech recognition and generation techniques[49] are widely used in day to day applications such as automated information systems in airlines, banks, etc.

Natural language processing requires several important techniques. First, syntactical analysis tools such as parsers are necessary for analyzing the syntactical structures of sentences according to the language grammar—to find the subject, predicate, and the object in a sentence. Semantical analysis tools are needed to give semantic interpretation to the sentences. Contextual information and pragmatic background knowledge are also essential for semantic disambiguation (word meaning disambiguation, reference resolution, etc.). Thus knowledge representation is also an important topic related to natural language processing.

Natural language processing is closely related to text-mining, which is an active area of study involving computer science and LIS. Text-mining aims to discover useful knowledge from large collections of textual documents, which can be seen as a generalization of natural language understanding. The studies in text-mining include text summarization, concept extraction, and ontology extraction.

Signal, Image Processing, and Robotics

The communications between an intelligent agent and the outside world can take various forms, such as visual and audio signals, in addition to utterances in natural language. Moreover an intelligent agent should be able to act in the world and thus effecting changes to the world around it as well. Signal and image processing research develops techniques that support computer perception and understanding of information in image, audio, and other sensory forms (such as radio signals, infrared, GPS signals). Robotics put together the techniques of AI study and build robots that can act intelligently, change the world, and achieve desired goals.

Although signal processing has been mostly studied by researchers in Electrical Engineering (EE), it has close connection to building fully autonomous intelligent agents. Image processing[50] and computer vision[51] are important topics in AI and EE. In image processing, the main task is image understanding, namely, to build a semantic model of a given imagery; and in computer vision, the main task is visual scene understanding, i.e., to building a model of the world (the perceived visual scene). Further extension of visual scene understanding would include understanding video streams (sequence of scenes). The "understanding" of a visual scene/image involves recognition of the objects present, the relevant photometry/geometry features of the objects, and the (spatial or other) relationships among the objects. To achieve the objectives of image/scene understanding, several stages of image processing operations are needed. Initial processing of images includes low-level operations such as smoothing to filter out noise in the image signals, edge detection, and image segmentation to decompose the image into groups with similar characteristics. These low-level processing operations are local computations and require no prior knowledge about the images for the particular application. The next stage processing involves object recognition, which requires isolating each distinctive object, determining the object's position and orientation (relative to the observer), and identifying the object shape. Objects are outlined by edges and described by a set of features, which are chosen by the designer of the image processing system. The feature could be shape-based (geometric features) or photometric features (such as textures, brightness, shading). For this processing stage, the computations are not necessarily local, features for characterizing different objects could require computation involving the pixels of the entire image. Supervised learning or pattern classification[38] methods are typically used for object recognition. The problem of object recognition from images is still highly challenging: a good object recognition system must perform well in spite of variations in the input image. The variations include changing illumination of the image, different pose and orientation of the objects, and translation, scaling of the objects. We humans are very good at recognizing, for example, familiar faces even if the faces are varied by wearing eye glasses, putting on a hat, having a different facial expression, or being illuminated differently. But such variations are still very hard to handle by computers.

Robotics[52] studies the techniques for building robots, i.e., intelligent agents with capabilities to act in the physical world. The research in Robotics concerns with both the hardware and software aspects of robots. A robot possessed a set of sensors for perceiving its surrounding environment and a set of effectors (also called actuators) for effecting actions in the environment. For example, for a mobile robot such as the planetary rovers that explore the surface of Mars, it has range sensors for measuring distance to close-by obstacles and image sensors (cameras) for getting images of surrounding environment. It also has effectors such as wheels/legs, joints for moving around. Robots can be classified into three categories: (1) Manipulators, which are robotic arms physically anchored at a fixed location, for example, garbage collection robot arms on the garbage van; (2) Mobile robots that move around using wheels, legs, etc.; and (3) Hybrid—mobile robots with manipulators. In particular, recent years have witnessed an increasing interest in building the so-called humanoid robots which, resemble humans in physical design and physical appearance.

The research problems studied in robotics call for utilization of all major AI techniques. A robot must be able to perceive its environment and represent the state of its environment in some knowledge representation form, it must be able to learn from its past experiences, it must be able to perform inference in making decisions about the correct move, and it must be able to plan and act intelligently, it must be able to handle uncertainty, and it must be able to communicate effectively to teammates and human users. Robotic perception addresses the problem of constructing internal models/representations of the environment from the sensory signals of the robot. This includes the study on localization, i.e., locating the position of specific objects, on environment mapping which allows the robot to construct a map of its environment by observation and exploration. Robotic motion research concerns with the planning and control of robot moves by the effectors. Various control architectures have been proposed in the literature.

Robotics has found wide range of successful applications in the real world. We will present some in the next entry on AI applications.

AI APPLICATIONS

AI has found many successful applications in various sectors of the real world. Here we sample some of them. The online resource from Wikipedia[53] gives more samples.

Artificial–Association

Game playing. Since early days of AI, researchers have studied the problem of computer game playing using heuristic search methods and machine learning. Arthur Samuel's Checkers playing program pioneered the studied in this aspect. Along with the advances in computing power and AI research, many successful computer game playing systems have been developed that can compete at human master levels. TD-Gammon[54] is a neural network based program that plays the game of Backgammon very well. The most well-known computer game player is perhaps the Deep Blue[9] Chess program from IBM. In a 6-game match against world Chess champion Kasparov, Deep Blue achieved 2 wins, 3 draws, and 1 loss, thus overall it has won the match. Today there are many online computer game playing programs (chess, go, checkers, backgammon, etc.) that people can play with and have fun. Almost all such game programs utilize ideas from AI in one way or the other.

Financial applications. Prominent financial firms in Wall Street have employed proprietary software systems for predicting stock-market trends and predicting stock prices for assisting mutual fund managers to boost investment returns. Although the details of such proprietary systems are held secret, it is known that at least a number of them used neural networks.

Medicine and health care. In more than 100 hospitals across the United States, nurses receive help from robotic "tugs"[55] that tow carts that deliver everything from meals to linens. Miniature robots have been used in surgery procedures for a number of diseases.[56] Data mining and machine learning techniques have been applied to find patterns of diseases and treatment effects of various medications from huge amounts of medical data. Intelligent medical imaging tools have been widely used to identify tumors/nodules from x-ray/CT-scan images for early detection and diagnosis of cancers. Moreover, computational biology combined with microarray technology in biological sciences has enabled the medical scientists to quickly identify or pin-point the genes responsible for certain diseases.[57] The construction of large online medical knowledge bases and the availability of such medical knowledge to ordinary people contribute significantly to boost preventive care in public health.

Engineering and manufacturing applications. The ideas of heuristic search and GA have been widely used in solving optimization problems commonly seen in engineering applications such as job-shop scheduling and air traffic scheduling. In manufacturing, utilization of robotic arms at assembly lines is quite common, and such application enhances the productivity tremendously.

Environment protection. Remote-sensing techniques have been widely used for gathering information about the oceans, the atmosphere, the space, and the earth. It is a difficult task to process the huge amount of environment data and find trends in environment change so as to meet the challenge of climate change and global warming. AI methods such as image processing, pattern classification, and data clustering have been applied successfully for analyzing environment data to assist scientists in environment-related research.[58]

Space science explorations. Mobil robots have been used to explore the unknown terrains on Mars. According to the Mars rover's web page,[59]

NASA's twin robot geologists, the Mars Exploration Rovers, launched toward Mars on June 10 and July 7, 2003, in search of answers about the history of water on Mars. They landed on Mars January 3 and January 24 PST, 2004 (January 4 and January 25 UTC, 2004). The Mars Exploration Rover mission is part of NASA's Mars Exploration Program, a long-term effort of robotic exploration of the red planet.

After more than four years of geological surveying the Mars Exploration Rovers robots have ceased to communicate (November 11, 2008).

Intelligent information systems. AI research has significantly impacted the studies in LISs. Ideas in AI have been widely applied in information technology to build smart information systems. On the other hand, the explosive development of the Internet and the Web has fueled AI with many interesting and challenging research problems. Along with the challenges are the great opportunities to bring AI closer to ordinary people's day-to-day lives. Nowadays we take it for granted that we can find information about anything by using online search engines such as Google, Yahoo, or Microsoft Live. Millions of consumers utilize online shopping tools to buy services and products. Digital libraries are a commonplace accessible to a much larger audience than before. What people probably did not realize is that behind all these nice and fascinating online tools and services (such as search engine, online shopping tools, etc.) there are important contributions of AI. For example, association rule mining and other AI methods are routinely used in many major online shopping Web sites so that related products can be recommended to consumers.

Intelligent information systems studies have developed a number of AI-based approaches in information extraction, indexing, and smart information retrieval. In information extraction, text-mining and natural language processing methods are developed to obtain semantical information from texts in the form of concepts and their relationships. Such information is then used for indexing the source texts to facilitate retrieval.[60] User profiles can be constructed by fuzzy clustering on user information-seeking behaviors (Web-click streams, etc.) to personalize the information service to individual users.[61] User information-seeking behavior includes not only current session Web-click streams of the user, but also previously logged Web search activities that help to model the user. Fuzzy rule, Neural Networks, and GA have been applied to adapt user queries for better retrieval performance.

There are continuous efforts in building large-scale commonsense knowledge bases and making information/knowledge in the collection accessible to ordinary people. The Wikipedia is one of such knowledge bases.[62] On the other hand, the studies in Semantic Web[32] aims at building large knowledge bases in formats such that the semantics (contents) of the information can be interpreted and processed by computers across the Web. Clearly Semantic Web would promote knowledge sharing and intelligent query-answering beyond what the current Web search engines would support. Along this line, the CYC project is another notable example.[63]

Many multimedia information retrieval systems have been constructed, resulting in various interesting applications such as music retrieval system,[64] video-clip retrieval system,[65] etc. Image retrieval for security surveillance has been in practical use for quite some time.

Machine translation. Machine translation is one special type of intelligent information system service, which supports automatic translation of texts from a source language to a target language. Today one can use machine translators at various online search engines, for example, Google. Although the performance is still not as good as that of human translators, machine translators are very useful in several ways. For one thing, human translators are highly specialized professionals and thus expensive to hire. Secondly, human translators would get tired and could not work as fast as computers. The common practice in using machine translation is to let the machines do the first (quick) cut of the translation and then let human translators polish the results produced by machines. This would greatly enhance the productivity of translation.

CONCLUSIONS

AI is an exciting research area. AI research is multidisciplinary in nature, drawing on advances in mathematics, philosophy, logic, computer science, information theory, control, cognitive science, and linguistics. The objective of AI is to understand the nature of intelligence and to build computer systems that behave intelligently. AI research covers a wide range of topics, many of which are briefly discussed in this entry. AI has found many successful applications that impact our daily life significantly. The entry samples some AI applications. AI and LISs have close connections, and cross-fertilization of research efforts between the two fields has been fruitful. Looking forward, we see great opportunities as well as challenges in realizing the dream of AI, which we embrace wholeheartedly.

ACKNOWLEDGMENT

This work was partially supported by NSF grant ITR-0326387.

REFERENCES

1. Russell, S.J. Norvig, P. *Artificial Intelligence: A Modern Approach*, 2nd Ed. Prentice Hall: Upper Saddle River, NJ, 2003.
2. Luger, G.F. *Artificial Intelligence: Structures and Strategies for Complex Problem Solving*, 6th Ed. Addison Wesley: New York, 2008.
3. Nilsson, N. *Artificial Intelligence: A New Synthesis*, Morgan Kaufmann Publishers: San Mateo, CA, 1998.
4. http://en.wikipedia.org/wiki/Artificial_intelligence.
5. Jenkinson, A.J. (translator). 2007, written at The University of Adelaide, Prior Analytics, eBooks @ Adelaide, South Australia, Australia.
6. Pascal, B. *Pensees de M. Pascal sur la Religion et sur quelques autre sujts*, Chez Guillaume Desprez: Paris, France, 1670.
7. Leibniz, G.W. *The Art of Discovery*, 1685; W 51.
8. Samuel, A. Some studies in machine learning using the game of checkers. IBM J. **1959**, (3), 210–229.
9. Hsu, F.-H. *Behind Deep Blue: Building the Computer that Defeated the World Chess Champion*, Princeton University Press: Princeton, NJ, 2002.
10. Turing, A.A. Computing machinery and intelligence. Mind. **1950**, *59*, 443–460.
11. Buchanan, B.G. A (very) brief history of artificial intelligence. AI Mag. **2005**, 53–60.
12. Newell, A.; Shaw, J.C.; Simon, H.A. Report on a general problem-solving program. Proc. Int. Conf. Inform. Process. **1959**, 256–264.
13. Pearl, J. *Heuristics: Intelligent Search Strategies for Computer Problem Solving*, Addison-Wesley: New York, 1984.
14. Hart, P.E.; Nilsson, N.J.; Raphael, B. A formal basis for the heuristic determination of minimum cost paths. IEEE Trans. Syst. Sci. Cybernet SSC4. **1968**, *2*, 100–107.
15. Dechter, R.; Judea, P. Generalized best-first search strategies and the optimality of *A**. J. ACM. **1985**, *32*(3), 505–536.
16. Sowa, J.F. *Knowledge Representation: Logical, Philosophical, and Computational Foundations*, Brooks/Cole Publishing Co.: Pacific Grove, CA, 2000.
17. *Handbook of Automated Reasoning*; Robinson, A., Voronkov, A., Eds.; Elsevier Science B. V. and MIT Press: Amsterdam, the Netherlands, 2001.
18. Genesereth, M. Nilsson, N. *Logical Foundations of Artificial Intelligence*, Morgan Kaufmann Publishers: San Mateo, CA, 1988.
19. Robinson, J.; Alan, A. Machine-oriented logic based on the resolution principle. J. ACM. **1965**, *12* (1), 23–41.
20. Marques-Silva, J.P. Sakallah, K.A. GRASP: A new search algorithm for satisfiability Proceedings of International Conference on Computer-Aided Design Santa Clara, CA, 1996; 220–227.
21. Zhang, H. SATO: An efficient propositional prover Proceedings of International Conference on Automated Deduction (CADE-97); Springer-Verlag: London, U.K.
22. Zhang, L. Solving QBF with combined conjunctive and disjunctive normal form Proceedings of Twenty-First National Conference on Artificial Intelligence (AAAI 2006) Boston, MA, 2006; July.

Artificial–Association

23. Kowalski, R. Predicate Logic as a Programming Language Proceedings IFIP Congress North Holland Publishing Co.: Stockholm, Sweden, 1974; 569–574.

24. Reiter, R. A logic for default reasoning. Artif. Intell. **1980**, *13*(1–2), 81–132.

25. McCarthy, J. Applications of circumscription to formalizing common-sense knowledge. Artif. Intell. **1986**, *28*(1), 89–116.

26. Pirri, F.; Reiter, R. Some contributions to the metatheory of the situation calculus. J. ACM. **1999**, *46*(3), 325–361.

27. Kowalski, R.; Sergot, M. A logic-based calculus of events. New Gen. Comput. **1986**, *4*, 67–95.

28. Emerson, E.A. *Temporal and Modal Logic. Handbook of Theoretical Computer Science*, MIT Press: Cambridge, MA, 1990; Chapter 16.

29. Gelfond, M.; Lifschitz, V. Representing action and change by logic programs. J. Logic Program. **1993**, *17*, 301–322.

30. Chen, J. The generalized logic of only knowing that covers the notion of epistemic specifications. J. Logic Comput. **1997**, *7*(2), 159–174.

31. Sowa, J.F. Semantic networks. In *Encyclopedia of Artificial Intelligence*; Shapiro, S.C., Ed.; Wiley: New York, 1987; revised and extended for the second edition, 1992.

32. Berners-Lee, T.; James, H.; Ora, L. The semantic web. Sci. Am. Mag. **2001**, May 17 *284*(5), 34–43.

33. http://www.w3.org/TR/owl-features/.

34. Kandefer, M. Shapiro, S.C. Knowledge acquisition by an intelligent acting agent Logical Formalizations of Commonsense Reasoning, Papers from the AAAI Spring Symposium Technical Report SS-07-05; Amir, E., Lifschitz, V., Miller, R., Eds.; AAAI Press: Menlo Park, CA, 2007; 77–82.

35. Mitchell, T. *Machine Learning*, McGraw-Hill: New York, 1997.

36. Bishop, C.M. *Pattern Recognition and Machine Learning*, Springer: New York, 2007.

37. Tan, P. Steinbach, M. Kumar, V. *Introduction to Data Mining*, Addison-Wesley: New York, 2006.

38. Duda, R.O. Hart, P.E. Stork, D.G. *Pattern classification*, 2nd Ed.; Wiley: New York, 2001.

39. Quinlan, J.R. Induction of decision trees. Mach. Learn. **1986**, *1*(1), 81–106.

40. Bishop, C.M. *Neural Networks for Pattern Recognition*, Oxford University Press: Oxford, U.K., 1995.

41. Jain, A.K.; Mao, J. Artificial neural networks: A tutorial. IEEE Comput. **1996**, March 31–44.

42. Goldberg, D.E. *Genetic algorithms for search, optimization, and machine learning*, Addison-Wesley: New York, 1989.

43. Koza, J.R. *Genetic Programming: On the Programming of Computers by Means of Natural Selection*, MIT Press: Cambridge, MA, 1994.

44. Vapnik, V. *Statistical Learning Theory*, Wiley-Interscience: New York, 1998.

45. Pearl, J. *Probabilistic Reasoning in Intelligent Systems*, Morgan Kaufmann Publishers: San Mateo, CA, 1988.

46. Allen, J.F. *Natural Language Understanding*, 2nd Ed. Benjamin Cummings: Melano Park, CA, 1987, 1994.

47. Deng, L. O'Shaughnessy, D. *Speech Processing: A Dynamic and Optimization-Oriented Approach*, Marcel Dekker Inc.: New York, 2003.

48. Rabiner, L.R. A tutorial on hidden Markov models and selected applications in speech recognition. Proc. IEEE. **1989**, *77*(2), 257–286.

49. http://www.nuance.com/naturallyspeaking/legal/.

50. Chan, T.F. Shen, J. (Jianhong). Image Processing and Analysis - Variational, PDE, Wavelet, and Stochastic Methods, SIAM Publisher: Philadelphia, PA, 2005.

51. Davies, E.R. *Machine Vision: Theory, Algorithms, Practicalities*, Morgan Kaufmann Publishers: San Mateo, CA, 2004.

52. Bekey, G. *Autonomous Robots*, Massachusetts Institute of Technology Press: Cambridge, MA, 2005.

53. http://en.wikipedia.org/wiki/Category:Artificial_intelligence_applications.

54. Tesauro, G.; Sejnowski, T. J. A parallel network that learns to play backgammon. Artif. Intell. **1989**, *39*(3), 357–390.

55. Peter, T.A. Robot set to overhaul service industry. The Christian Science Monitor. 2008, February 28, http://www.csmonitor.com/2008/0225/p01s01-usgn.html.

56. http://en.wikipedia.org/wiki/Robotic_surgery.

57. Schena, M.; Shalon, D.; Davis, R.W.; Brown, P.O. Quantitative monitoring of gene expression patterns with a complementary DNA microarray. Science **1995**, *270*, 467–470.

58. Erotoz, L. Steinbach, M. Kumar, V. Finding clusters of different sizes, shapes, and densities in noisy, high dimensional data Proceedings of SIAM Int. Conf. on Data Mining San Francisco, CA, 2003; May.

59. http://marsrovers.nasa.gov/overview/.

60. Punuru, J. Chen, J. Extraction of non-hierarchical relations from domain texts Proceedings of IEEE International Symposium on Computational Intelligence and Data Mining. Honolulu, HI, 2007; April.

61. Martin-Bautista, M.J.; Kraft, D.H.; Vila, M.A.; Chen, J.; Cruz, J. User profiles and fuzzy logic for web retrieval issues. J. Soft Comput. **2002**, *6*, 365–372.

62. http://www.wikipedia.org.

63. Matuszek, C. Witbrock, M. Kahlert, R. Cabral, J. Schneider, D. Shah, P. Lenat, D. Searching for common sense: Populating Cyc from the web Proceedings of the Twentieth National Conference on Artificial Intelligence Pittsburgh, PA, 2005; July.

64. Zhang, X.; Ras, Z.W. Sound isolation by harmonic peak partition for music instrument recognition. Fund. Inform. **2007**, *78*(4), 613–628.

65. Proceedings of the 6th ACM International Conference on Image and Video Retrieval, CIVR 2007; Sebe, N., Worring, M. Amsterdam, the Netherlands, 2007; July 9–11.

Artificial Neural Networks and Natural Language Processing

Hermann Moisl
Center for Research in Linguistics, University of Newcastle upon Tyne, Newcastle upon Tyne, U.K.

Abstract

This entry gives an overview of work to date on natural language processing (NLP) using artificial neural networks (ANN). It is in three main parts: the first gives a brief introduction to ANNs, the second outlines some of the main issues in ANN-based NLP, and the third surveys specific application areas. Each part cites a representative selection of research literature that itself contains pointers to further reading.

INTRODUCTION

Two preliminary notes:

- NLP is here regarded as language engineering—i.e., the design and implementation of physical devices that process human linguistic input in some application and not as cognitive modeling.
- The discussion is concerned solely with text. This is not meant to imply that speech processing is any sense less important in NLP than text processing—far from it. It is, rather, simply a consequence of space limits on this entry, which preclude consideration of what has become a research discipline in its own right.

ARTIFICIAL NEURAL NETWORKS

An ANN is an artificial device that emulates the physical structure and dynamics of biological brains. There are two broad approaches to such emulation. One approach regards ANNs as biological models that capture aspects of brain structure and dynamics as accurately as possible. The other sees ANNs as computational systems, which, although neurally inspired, can be developed in line with required computational properties without regard to biology. Historically, these two approaches have proceeded more or less independently, but recently, with rapid advances in brain-imaging techniques and consequent improved understanding of the brain, there are clear signs of convergence. This discussion focuses on ANNs as computational systems.

ANNs were first proposed by McCulloch and Pitts[1] in 1943 and were developed throughout the 1950s and 1960s.[2,3] But, in 1969, Minsky and Papert[4] showed that there were computable functions that the ANN architectures of the time could not compute,[5] as a consequence of which ANN-based research activity diminished significantly. Some researchers persevered, however, and by the early 1980s interest in them had begun to revive.[6] In 1986, Rumelhart and McClelland published their now-classic *Parallel Distributed Processing*[7] volumes. Among other things, these proposed the backpropagation learning algorithm, which made it possible to train multi-layer nets and thereby to overcome the computational limitations that Minsky and Papert had demonstrated. The effect was immediate. An explosion of interest in ANNs ensued, which continues today. This has generated a considerable variety of network architectures; details are available from numerous textbooks.[8–10]

ISSUES IN ANN-BASED NLP

Motivation

In general, a new technology is adopted by a research community when it offers substantial advantages over what is currently available. NLP has from the outset been dominated by a technology based on explicit design of algorithms for computing functions of interest and implementation of those algorithms using serial computers, an approach widely known as "symbolic NLP." It is only since the early 1980s that alternative technologies have been extensively used, among them ANNs. In what follows, we look briefly at the main advantages and disadvantages of ANNs relative to symbolic NLP.

Advantages

Of the various advantages referred to in the literature, the following two are arguably the most important:

- Function Approximation
  It has been shown that certain ANN architectures, including widely used ones, can approximate any computable function arbitrarily closely.[11–14] The function *f* which a given ANN approximates is determined by

Encyclopedia of Library and Information Sciences, Fourth Edition DOI: 10.1081/E-ELIS4-120008648

Artificial–Association

its parameter values, or weights, and these parameter values are learned from a data set D⊂f. In principle, therefore, NLP functions can be approximated from data using ANNs, thereby bypassing the explicit design of algorithms for computing those functions.

Why should one want to dispense with explicit design? Looking back on several decades' work on symbolic NLP, and artificial intelligence (AI) more generally, some researchers have come to believe that a variety of problems are too difficult to be solved by algorithm design given the current state of software technology[15] and have instead turned to function approximation techniques like ANNs. A, and probably the, chief advantage of ANN technology for NLP, therefore, is that it offers an alternative way of implementing NLP functions which symbolic NLP has thus far found it difficult to implement.

- Noise Tolerance
  Practical NLP systems must operate in real-world environments characterized by noise, which can for present purposes be taken as the presence of probabilistic errors in the data (e.g., spelling or syntax errors). A frequent criticism of symbolic AI/NLP systems is that they are brittle in the sense that noisy input, which the designer has not taken into account, can cause a degree of malfunction out of all proportion to the severity of the input corruption. The standard claim is that ANNs are far less brittle, so that the performance of an ANN-based system will "degrade gracefully" in some reasonable proportion to the degree of corruption.

Noise tolerance is a by-product of ANN function approximation. ANNs approximate a function from data by fitting a regression curve to data points, and the best approximation—the one which generalizes best—is not the curve that passes through the data points, but the one that best captures the general shape of the data distribution.[16,17] Most of the data points in a noisy environment will be at some distance from the regression curve; if the input corruption is not too severe, the regression model will place the corresponding output in or near the training data distribution.

Disadvantages

Scaling. Given some ANN architecture, the aim of learning is to find a set of connection strengths that will allow the net to compute a function of interest. Empirical results have repeatedly shown that the time required to train an ANN increases rapidly with the number of network connections, or network complexity, and fairly quickly becomes impractically long. The question of how ANN learning scales with network complexity is therefore crucial to the development of large nets for real-world

applications. The answer is that learning is an NP complete or intractable problem—"We cannot hope to build connectionist networks that will reliably learn simple supervised learning tasks."[18]

This result appears to bode ill for prospects of scaling ANNs to large, real-world applications. It is all very well to know that, in theory, ANNs with suitable architectures and sufficient complexity can implement any computable function, but this is of little help if it takes impractically long to train them. The situation is not nearly as bad as it seems, however. The intractability result is maximally general in that it holds for all data sets and all ANN architectures. Although theoretically interesting, such generality is unnecessary in practice[19] and a variety of measures exist, which constrain the learning problem such that intractability is either delayed or circumvented. These include restricting the range of ANN architectures used,[19] developing mechanisms for determining optimal network complexity for any given learning problem such as network growing and pruning algorithms,[18] biasing the net toward data to be learned,[20,21] explicit compilation of knowledge into network weights,[22] preprocessing of inputs by feature extraction, where the features extracted reflect the designer's knowledge of their importance relative to the problem,[16] incremental training,[20] and transfer of weights from a net that has successfully learned a problem similar to the one of interest.[23,24]

Inscrutability. A network that has learned its training data and generalizes well realizes a desired behavior, but it is not immediately obvious how it does so; the set of connection strengths determine the behavior, but direct examination of them does not tell one very much. Because of this, ANNs acquired a reputation as "black box" solutions soon after their resurgence in the early to mid-1980s and have consequently been viewed with some suspicion, particularly in critical application areas like medical diagnosis expert systems where unpredictable behavior, or even the possibility of unpredictable behavior, is unacceptable. ANNs are, however, no longer the black boxes they used to be. Mathematical tools for understanding them have been developed,[13,14] and inscrutability has become less of a disadvantage in the application of ANNs.

Discussion

It needs to be stressed that the intention here is not to argue that ANN-based NLP is necessarily "better" than its symbolic counterpart in a partisan sense. They are alternative technologies, each with its strengths and weaknesses, and, in an NLP context, can be used pragmatically in line with one's aims. The current position with respect to NLP is that ANN-based systems, "while becoming ever more powerful and sophisticated, have not yet been able to provide equivalent (let alone alternative superior) capabilities to those exhibited by symbolic systems."[25]

ANN-Based NLP

NLP is historically intertwined with several other language-oriented disciplines, including cognitive science, AI, generative linguistics, and computational linguistics. In general, the interaction of these disciplines has been and continues to be important. It remains, however, that each discipline has its own research agenda and methodology, and it is possible to waste time engaging with issues that are simply irrelevant to NLP. Now, it happens that ANN-based research into natural language has been strongly cognitive science oriented. For present purposes, it is important to be clear about the significance of this for NLP.

The formalisms invented in the 1930s and 1940s to define computable functions were soon applied to modeling of aspects of human intelligence including, of course, language. There have been two main strands of development. One is based on automata and formal language theory and has come to be known as the "symbolic" paradigm. The other is based on ANNs and is known as the "connectionist" or "subsymbolic" paradigm. Until fairly recently, the symbolic paradigm dominated thinking about natural language in linguistics, cognitive science, and AI. It reached its apotheosis in the late 1970s, when Newell and Simon[26] proposed the Physical Symbol System Hypothesis (PSSH), in which "physical symbol system" is understood as a physical implementation of a mathematically stated effective procedure, the prime example of which is the Turing Machine. The PSSH was widely accepted as the agenda for future work in linguistics, cognitive science, and AI: in essence, the first two proposed cognitive virtual architectures, and the third implemented them. At this time, however, interest in ANNs was being revived by cognitive scientists who saw them as an alternative to the dominant symbolic paradigm in cognitive modeling, and a debate soon arose on the relative merits of the PSSH- and ANN-based approaches to cognitive modeling. The PSSH case was first put in 1988 by Fodor and Pylyshyn,[27] who labeled what we are here calling the symbolic as the "classical" position, and Smolensky[28] argued the ANN-based case;[29–32] between them they set the parameters for subsequent discussion.[33–43] The essentials of the debate are as follows.

The symbolist position (all quotations are from Fodor and Pylyshyn)[27] descends directly from the PSSH and thus proposes cognitive architectures, which compute by algorithmic manipulation of symbol structures; the mind is taken to be a symbol manipulation machine. Specifically, there are representational primitives—symbols. Being representational, symbols have semantic content (i.e., each symbol denotes some aspect of the world). A representational state consists of one or more symbols, each with an associated semantics,

> in which 1) there is a distinction between structurally atomic and structurally molecular representations, 2)

structurally molecular representations have syntactic constituents that are themselves either structurally molecular or structurally atomic, and 3) the semantic content of a representation is a function of the semantic contents of its syntactic parts, together with the syntactic structure.

Transformations of mental states "are defined over the structural properties of mental representations. Because these have combinatorial structure, mental processes apply to them by virtue of their form." Therefore,

> if in principle syntactic relations can be made to parallel semantic relations, and if in principle you have a mechanism whose operation on expressions are sensitive to syntax, then it is in principle possible to construct a syntactically driven machine whose state transitions satisfy semantic criteria of coherence. The idea that the brain is such a machine is the foundational hypothesis of classical cognitive science.

The subsymbolic position (all quotations are from Fodor and McLaughlin)[28] defines cognitive models that are "massively parallel computational systems that are a kind of dynamical system." The primitives are subsymbols, which are like classical symbols in being representational, but unlike them in being finer-grained: they "correspond to constituents of the symbols used in the symbolic paradigm...Entities that are typically represented in the symbolic paradigm as symbols are typically represented in the subsymbolic paradigm as a large number of subsymbols." The difference between symbolic and subsymbolic models lies in the nature of the semantic content. The subsymbolic position distinguishes two semantic levels, the conceptual and the subconceptual: "The conceptual level is populated by consciously accessible concepts, whereas the subconceptual one is comprised of finer-grained entities beneath the level of conscious concepts." In classical models, symbols typically have conceptual semantics (i.e., semantics, which correspond directly to the concepts that the modeler uses to analyze the task domain, whereas subsymbols in subsymbolic models have subconceptual semantics; the semantic content of a subsymbol in a subsymbolic ANN model corresponds directly to the activity level of a single processing unit in an ANN. Subsymbolic representations are, moreover, not operated on by processes, which manipulate symbol structures in a way that is sensitive to their combinatorial form because subsymbolic representations do not have combinatorial form. Instead, they are operated on by numeric computation. Specifically, a subsymbolic ANN model is a dynamical system whose state is a numerical vector of the activation values of the units comprising the net at any instant t. The evolution of the state vector is determined by the interaction of the current input, the current state of the system at t, and a set of numerical parameters corresponding to the relative strengths of the connections among units.

Artificial–Association

How do the symbolic and subsymbolic paradigms relate to one another as cognitive models? In the symbolist view,

> classical and connectionist [= subsymbolic] theories disagree about the nature of mental representations. For the former but not the latter, representations characteristically exhibit combinatorial constituent structure and combinatorial semantics. Classical and connectionist theories also disagree about the nature of mental processes: for the former but not the latter, mental processes are characteristically sensitive to the combinatorial structure of the representations on which they operate. These two issues define the dispute about the nature of cognitive architecture.

Now, any adequate cognitive model must explain the productivity and systematicity[44–46] of cognitive capacities. Symbolic models appeal to the combinatorial structure of mental representations to do this, but subsymbolists cannot: "Because it acknowledges neither syntactic nor semantic structure in mental representations, it treats cognitive states not as a generated set but as a list," and among other things lists lack explanatory utility. Because they cannot explain cognitive productivity and systematicity, subsymbolic models are inadequate as cognitive models; they may be useful as implementations of symbolically defined cognitive architectures, but this has no implications for cognitive science.

In the subsymbolist view, on the other hand, subsymbolic models refine symbolic ones rather than, as the symbolists maintain, implementing them. In a now-famous analogy, the relationship between symbolic and subsymbolic paradigms is likened to that which is obtained between the macrophysics of Newtonian mechanics and the microphysics of quantum theory. Newtonian mechanics is not literally instantiated in the world according to the microtheory, because fundamental elements in the ontology of the macrotheory, such as rigid bodies, cannot literally exist according to the microtheory. In short, "in a strictly literal sense, if the microtheory is right, then the macrotheory is wrong." This does not, however, mean that Newtonian mechanics has to be eliminated, for it has an explanatory capacity, which is crucial in a range of sciences and branches of engineering and which a (strictly correct) quantum mechanical account lacks; such explanatory capacity is crucial in the subsymbolic view. Thus, cognitive systems "are explained in the symbolic paradigm as approximate higher-level regularities that emerge from quantitative laws operating on a more fundamental level"—the subconceptual—"with different semantics." Or, put another way, symbolic models are competence models that idealize aspects of physical system behavior, whereas subsymbolic models are performance models that attempt to describe physical systems as accurately as possible.

What are the implications of this debate for ANN-based NLP? First, the debate has forced a reexamination

of fundamental ideas in cognitive science and AI, and, because much of the ANN-based research on NL is done within a cognitive science framework, ANN-based NLP cannot afford to ignore developments in the corresponding cognitive science work. This is not bland ecumenism, but a simple fact of life.

Second, notwithstanding what has just been said, it remains that the cognitive science focus of the debate can easily mislead the NLP researcher who is considering ANNs as a possible technology and wants to assess their suitability. The debate centers on the nature of cognitive theories and on the appropriateness of symbolist and subsymbolist paradigms for articulation of such theories. These issues, although intrinsically interesting, are orthogonal to the concerns of NLP as understood here. Cognitive science is concerned with scientific explanation of human cognition, including the language faculty, whereas NLP construed as language engineering has no commitment to explanation of any aspect of human cognition, and NLP systems have no necessary interpretation as cognitive models. The symbolist argument that the ANN paradigm is inadequate in principle for framing cognitive theories is, therefore, irrelevant to NLP, as are criticisms of particular ANN language-processing architectures in the literature on the grounds that they are "cognitively implausible," or fail to "capture generalizations," or do not accord with psycholinguistic data.

Third, once the need for cognitive explanation is factored out, the debate reduces to a comparison of standard automata theory and ANNs as computational technologies.[38] So construed, the relationship is straightforward. We have taken the aim of NLP to be design and construction of physical devices that have specific behaviors in response to text input. For design purposes, the stimulus-response behavior of any required device can be described as a mathematical function (i.e., as a mapping from an input to an output set). Moreover, because the stimulus-response behavior of any physical device is necessarily finite, the corresponding input-output sets can also be finite, with the consequence that every NLP function is computable; in fact, the sizes of the I/O sets are specifiable by the designer and can, therefore, be defined in such a way as to make the function not only finite and therefore theoretically computable but also computationally tractable. As such, for any NLP mapping, there will be a Turing Machine—a PSS—which computes it. But we have seen that certain classes of ANN are Turing equivalent, so there is no theoretical computability basis for a choice between the two technologies. The choice hinges, rather, on practical considerations such as ease of applicability to the problem in hand, processing efficiency, noise and damage tolerance, and so on.

And, finally, the debate has set the agenda for ANN-based language-oriented research in two major respects: the paradigm within the research is conducted, and the ability of ANNs to represent compositional structure. These are discussed in the following sections.

Research paradigms

The symbolist/subsymbolist debate has resulted in a trifurcation of ANN-based natural language oriented research, based on the perceived relationship between PSSH- and ANN-based cognitive science and AI:

- The symbolic paradigm[47,48] accepts the symbolist view of the position of ANNs relative to cognitive science. It regards ANNs as an implementation technology for explicitly specified PSS virtual machines and studies ways in which such implementation can be accomplished.
- The subsymbolic paradigm[49–51] subdivides into what is sometimes called "radical connectionism," which assumes no prior PSSH analysis of the problem domain but relies on inference of the appropriate processing dynamics from data, and a position which in essence regards prior PSSH analysis of the problem domain as a guide to system design and/or as an approximate or competence description of the behavior of the implemented system. *Neural Networks and a New Artificial Intelligence*[42] exemplifies the radical position, and Smolensky, in a manner of speaking the father of subsymbolism, has in various of his writings[28,52–54] taken the second.
- The hybrid paradigm,[38,55–60] as its name indicates, is a combination of the symbolic and the subsymbolic. It uses symbolic and subsymbolic modules as components in systems opportunistically, according to what works best for any given purpose. A subsymbolic module might, for example, be used as a preprocessor able to respond resiliently to noisy input, whereas the data structures and control processes are conventional PSS designs.

Interest in the hybrid paradigm has grown rapidly in recent years, and, to judge by relative volumes of research literature, it is now the most often used of the above three alternatives in engineering-oriented applications like NLP. It is not hard to see why this should be so. The hybrid paradigm makes full use of theoretical results and practical techniques developed over several decades of PSSH-based AI and NLP work and supplements it with the function approximation and noise resistance advantages of ANNs where appropriate. By contrast, the symbolic and subsymbolic paradigms are in competition with established PSSH-based theory and/or methodology. On the one hand, the symbolic paradigm has yet to demonstrate that it will ever be superior to conventional computer technology as an implementation medium for PSS virtual machines. On the other, the subsymbolic paradigm essentially disregards existing PSSH-based NLP theory and practice and starts afresh. Of the three paradigms, therefore, it is the least likely to generate commercially exploitable systems in the near future, although it is the most intriguing in pure research terms.

Representation

The most fundamental requirement of any NLP system is that it be able to represent the ontology of the problem domain.[61–63] One might, for example, want to map words to meanings or strings to structural descriptions: words, meanings, strings, and structures have to be represented in such a way that the system can operate on them to implement the required mapping. Most ANN-based NL work has been directly or indirectly concerned with this issue, and this section deals with it in outline.

There are two fundamentally different approaches to ANN-based representation:

- Local representation: Given some set E of objects to be represented and a set N of network units available for representational use, a local representation scheme (or "local scheme" for short) allocates a unique unit or group of units $\in$ N for each object $e \in$ E.
- Distributed representation: Given the same sets E and N, a distributed representational scheme uses all the units $n \in$ N to represent each $e \in$ E.

The difference is exemplified in the pair of representational schemes for the integers 0–7 shown in Fig. 1. In the local scheme, each bit represents a different integer, whereas in the distributed one all the bits are used to represent each integer, with a different pattern for each. Because, in the local scheme, each bit stands for one and only one integer, it can be appropriately labeled, but in the distributed scheme, no bit stands for anything on its own; no bit can be individually labeled, and each is interpretable only on relation to all the others.

Local and distributed schemes both have advantages.[41,61,64,65] Much of the earlier work used localist representation, and although the balance has now shifted to the distributed approach, significant localist activity remains.[47,66–69] In what follows, local and distributed approaches to representation of primitive objects and of compositional structure in ANNs are discussed separately.

	Local	**Distrib**
1	0000001	0000001
2	0000010	0000010
3	0000100	0000011
4	0001000	0000100
5	0010000	0000101
6	0100000	0000110
7	1000000	0000111

Fig. 1 Local and distributed representational schemes.

Local representation. *Representation of primitives.* Local representation of primitive objects is identical to that in the PSSH approach: in a PSS, each object to be represented is assigned a symbol, and in local ANN representation, each object is assigned a unit in the net.

Representation of structure. Local representation of primitive objects is straightforward, but representation of compositional structure is not. The difficulty emerges from the following example.[70] Assume a standard AI blocks world consisting of red and blue triangles and squares. A localist ANN has to represent the possible combinations. One unit in the net is allocated to represent "red," another for "blue," another for "triangle," and yet another for "square." If one wants to represent "red triangle and blue square," the obvious solution is to activate all four units. Now represent "blue triangle and red square" using the same procedure. How can the two representations be distinguished? The answer is that they cannot, because there is no way to represent the different color-to-shape bindings in the two cases. In other words, there is no way to represent constituency in a net like this. Localists have developed a variety of binding mechanisms to overcome this problem.[25,64,70,71]

Distributed representation. *Representation of primitives.* In distributed ANNs, each primitive object is represented as a pattern of activity over some fixed-size group g of n units, or, abstractly, as a vector v of length n in which the value in any vector element v_i represents the activation of unit g_i, for $1 \leq i \leq n$. Such representation has significant advantages over the localist approach.[17]

Representation of structure. The symbolist/subsymbolist debate made representation of compositional structure a major issue in ANN-based cognitive science on account of its explanatory capacity. The symbolists claimed that ANNs were incapable of representing structure, and for that reason dismissed them as inadequate for cognitive modeling. In response, adherents of ANN-based cognitive science have developed a variety of structure-representing mechanisms. But, as noted, NLP as construed here is not primarily interested in explanation, but in implementing mappings of interest, and many ANN architectures can implement any computable mapping without—as the symbolists observe—any recourse to compositional structure. Whatever its importance for cognitive science, therefore, is representation of compositional structure an issue for ANN-based NLP? The answer is that, in principle, compositional structure is not necessary for NLP, although it may be useful in practice. A discrete-time, continuous space dynamic system, such as a two-layer feedforward ANN with sigmoid activation function, may theoretically be capable of implementing any computable function, but, for some particular function, is finding the required weight parameters computationally tractable, and will network

complexity have reasonable space requirements? It may well turn out that compositional structure makes implementation of certain NLP functions easier or indeed tractable; the need for compositional structure in ANN-based NLP is an empirical matter, and for that reason researchers need to be aware of the structuring mechanisms developed by ANN-based cognitive science.

Using a distributed ANN to represent compositional structure is difficult because arbitrarily complex structures have to be represented with a fixed-size resource (i.e., over some specific group of units). To see this, assume that the primitive objects in a given domain are represented as feature vectors. An ANN that uses distributed representations by definition uses all the available units for each vector. There would be no difficulty about individually representing "man," for example, or "horse." But how would the net represent "man" and "horse" at the same time? Even more difficult is representation of relations, such as "man on horse." The problem, therefore, has been to find ways of overcoming this difficulty.

The crucial insight came from van Gelder in 1990.[72] He argued that distributed representations can be compositional but not necessarily in the sense intended within the PSSH paradigm. Van Gelder's article has become very influential in ANN-based cognitive science generally; its importance for present purposes lies in the distributed compositional representation, which it proposes and which underlies several distributed ANN-based representational mechanisms. Space limitations preclude a detailed account of it here. Suffice it to say that, as an alternative to the concatenative compositional representation used by symbolic cognitive science, where the symbol tokens of an expression's constituents are spatially present in the compositional expression, van Gelder proposes nonconcatenative representation, where this is not the case. The importance of such nonconcatenative representation is that it breaks the link between abstract complexity and spatial representation size: because it does not require constituent tokens to be physically present in an expression token, it becomes possible in principle to represent abstract constituency relations over a fixed-size resource. What is needed in ANN terms are "general, effective, and reliable procedures"[80] to compose constituent tokens into and to decompose them from expression tokens represented over the representational units of the net. Several such nonconcatenative mechanisms have been proposed, chief among them tensor products,[73,74] recursive auto associative memories (RAAM),[75–78] and holographic reduced descriptions.[79–81]

Sequential processing

Text processing is inherently sequential in the sense that word tokens arrive at the processor over time. ANN-based work on NL has addressed this sequentiality[82,83] by using three main types of network architecture:

- Multilayer perceptrons

  MLPs are a feedforward architecture in which time is not a parameter: they map inputs to outputs instantaneously. MLPs consequently appear to be inappropriate for sequential processing. Nevertheless, the earlier ANN-based NL work used them for this purpose by spatializing time. Given a set of symbol strings to be processed, the MLP is given an input layer large enough to accommodate the longest string in the set, as in Fig. 2. This was unwieldy both because, depending on the input encoding scheme, it could result in large nets that take a long time to train and because of the inherent variability in the length of NL strings. It is now rarely used.

- Time-delay neural networks

  A time-delay neural network (TDNN) is an MLP whose input layer is a buffer k elements wide, as shown in Fig. 3. It processes input dynamically over time $t_0, t_1 \ldots t_n$ by updating the input buffer at each t_i and propagating the current input values through the net to generate an output. The problem here is buffer size. For example, any dependencies in a string whose lexical distance is greater than the buffer size will be lost. In the limiting case, a buffer size equal to the input string length reduces to an MLP. TDNNs have been successfully used for finite state machine induction and in NLP applications.[84–86]

- Recurrent networks

  Recurrent networks (RANN) use a fixed-size input layer to process strings dynamically. RANNs used for NL work are discrete-time, and input successive symbols in a string at time steps $t_0, t_1 \ldots t_n$, as in Fig. 4. The net's memory of sequential symbol ordering at any point in the input string is maintained in the current state, which is fed back at each time step.

  RANNs are dynamic systems that can be understood in terms of standard automata theory. A RANN that is driven through a trajectory in continuous-state space by a sequence of input signals is interpretable as an automaton driven through a trajectory in discrete space by an input symbol string, and the response of that RANN to a set of signal sequences as an automaton that defines a string set. Moreover, if the dynamics of the RANN are learned from input-output data rather than explicitly compiled into the net, then the RANN can be taken to have inferred an automaton which defines some language L or, equivalently, to have inferred the corresponding grammar.

There has been a good deal of work on the use of RANNs for grammatical inference, both for formal and for natural languages, and using a variety of RANN architectures.[82,87–100] The training of a Simple Recurrent Network (SRN) as a finite state acceptor is paradigmatic:

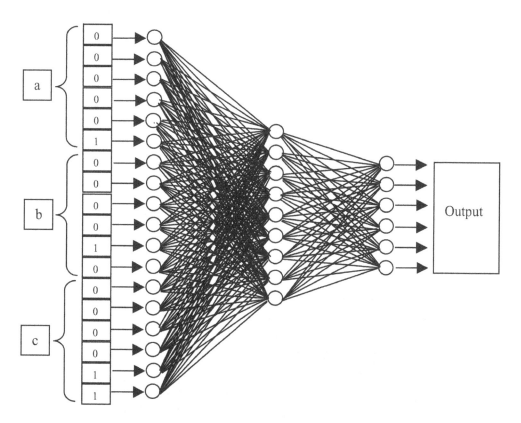

Fig. 2   A feedforward ANN for string processing.

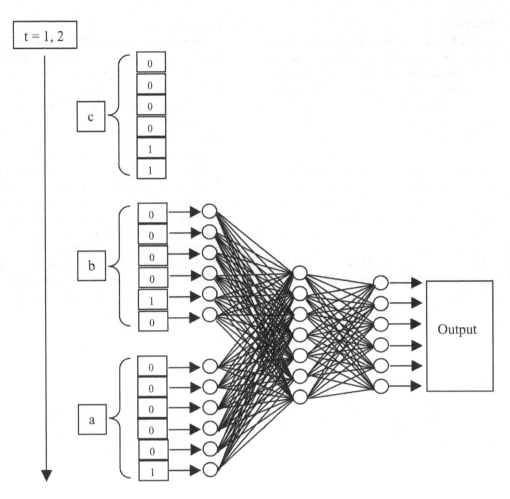

Fig. 3 A TDNN for string processing.

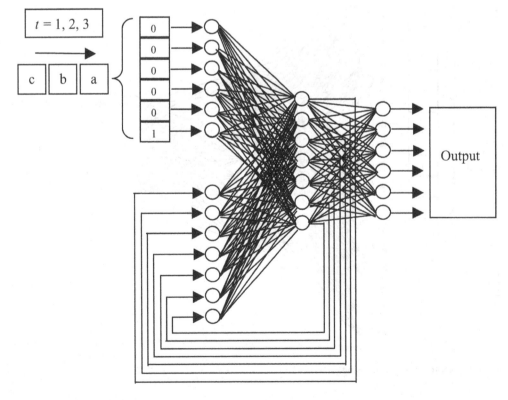

Fig. 4 A RANN for string processing.

given a language L and a finite set T of pairs (a, b), where a is a symbol string and b is a boolean which is true if $a \in L$ and false otherwise, train an SRN to approximate a finite state acceptor for L from a proper subset T′ of T. The SRN, like various other RANN architectures used for grammatical inference, is a discrete-time, continuous-space dynamic system. To extract discrete computational states, the continuous ANN state space is partitioned into equivalence classes by using, for example, statistical clustering algorithms based on vector distance, and each cluster is interpreted as a single computational state. Any finite state machine extracted in this way is a possible computational interpretation of the RANN, but it is not unique, because the number of states extracted depends on the granularity of the continuous-space partitioning and on the partitioning algorithm used.

Meaning representation

There are some NLP applications, like document search, where the meaning of the text being processed is not an issue. In others, semantic interpretation of text is necessary but reasonably straightforward; an example would be an NL command interpreter for a database front end, where both the syntax of input strings and the semantic interpretations to which they are mapped are both well defined and severely restricted relative to normal linguistic usage. However, when one moves to AI-oriented applications, such as (more or less) unrestricted NL understanding systems, semantic interpretation becomes a difficult and still largely unresolved problem. ANNs do not provide an easy solution, but they do offer a promising alternative to existing PSSH-based approaches.

Meaning is variously understood by different disciplines and by researchers within them. It does, however, seem uncontroversial to say that meaning of NL linguistic expressions has to do with denotation of states of the world and that semantic interpretation is a mapping from strings to denotations. That, in any case, is what is assumed here. PSSH-based AI and NLP systems have implemented the mapping by constructing system-internal representations of some aspect of the world—the "domain of discourse"—and then relating input strings to the representation.[101] How best to represent the world has become a research discipline in its own right—knowledge representation—and numerous formalisms exist. Some ANN-based work on semantic interpretation continues in the PSSH tradition in the sense that they use explicitly designed domain representations. Other work takes a radically different approach, however: input strings are mapped not to explicitly designed representations of the world, which, inevitably, reflect a designer's analysis of what is significant in the task domain, but to representations that are learned from the world via transducers without designer intervention. At its most ambitious task, this line of research aims to embed NLP systems in robotic agents that not only receive inputs from an environment via, say, visual, acoustic, and tactile transducers, but also interact with and change the environment by means of effectors. The aim is for such agents to develop internal world representations by integrating inputs and internal states via self-organization based on adaptive interaction with the environment: "Concepts are thus the 'system's own,' and their meaning is no longer parasitic on the concepts of others (the system designer)."[102] In particular, agents would learn to represent the meanings of words and expressions from their use in specific environment-interactive situations. Work on this is proceeding,[42,43,102–112] although it must be said that, to keep experimental simulations tractable, the goal of real-world interaction is often reduced to explicitly designed microworlds reminiscent of ones like the famous SHRDLU in the PSSH tradition.

ANN-BASED NLP: AN OVERVIEW

ANN-based NL research[17,25,113–115] began, fairly slowly, in the early 1980s with articles on implementing semantic networks in ANNs,[116] visual word recognition,[117–119] word sense disambiguation,[120] anaphora resolution,[121] and syntactic parsing.[122–124] In 1986, Lehnert published an article[125] on the implications of ANN technology for NLP, an indication that this early work had by then attracted the attention of mainstream work in the field. Also in 1986 the *Parallel Distributed Processing* volumes[7] appeared, and these contained several chapters on language: McClelland and Kawamoto on case role assignment, McClelland on word recognition, and Rumelhart and McClelland on English past tense acquisition. All of these were to be influential, but the latter had an effect out of all proportion to the intrinsic important of the linguistic issue it dealt with. Rumelhart and McClelland presented an ANN that learned English past-tense morphology from a training set of (past tense/present tense) verb form pairs, including both regular ("-ed") and irregular formations. They regarded their net as a cognitive model of past tense morphology acquisition on the grounds that its learning dynamics were in close agreement with psycholinguistic data on past tense acquisition in children, and, because it was able to generalize the regular tense formation to previously unseen present tense forms after training, that it had learned was an aspect of English morphology. Crucially, though, the net did this without reference to any explicit or implicit PSS architecture. This was quickly perceived as a challenge by symbolist cognitive scientists and became a test case in the symbolist versus subsymbolist debate outlined above. Pinker and Prince[126] made a long and detailed critique of the Rumelhart and McClelland model, in response to which the model was refined by a succession of researchers.[51,127–132]

From an NLP point of view, the chief importance of Rumelhart and McClelland's work and its successive refinements is not in its validity as a cognitive model but in the impetus that it gave to ANN-based NL research. It made 1986 a watershed year in the sense that the number of language-oriented articles has increased dramatically since then: disregarding speech and phonology on account of this article's focus on text processing, there has been further work on a wide variety of topics, a representative selection of which follows:

General language acquisition: see also the above discussion of sequential processing.[41,89,90,133–135]

Morphology: see also the above discussion of past tense formation.[23,136,137]

Lexical category learning[89,90]

Lexical semantics[49,64,68,102–111,138–143]

Syntax and parsing[41,49,89,90,115,144–154]

Sentence semantics[49,155–159]

Metaphor interpretation[160,161]

Text understanding[49,162–164]

Language production[49,50,140,165–168]

Dialog systems[49,50]

Character recognition[169–172]

Text classification[173–176]

Data mining and text summarization[177–181]

Text compression[182]

REFERENCES

1. McCulloch, W.; Pitts, W. A logical calculus of the ideas immanent in nervous activity. Bull. Math. Biophys. **1943**, *5*, 115–133.

2. Medler, D. A brief history of connectionism. Neural Comput. Surv. **1998**, *1*, 61–101.

3. Taylor, J. The historical background. In *Handbook of Neural Computation*; Fiesler, E., Beale, R., Eds.; Institute of Physics Press, 1997.

4. Minsky, M.; Papert, S. *Perceptrons*, MIT Press, 1969.

5. Pollack, J. No harm intended: a review of the *perceptrons* expanded edition. J. Math. Psychol. **1989**, *33*, 358–365.

6. Feldman, J.; Ballard, D. Connectionist models and their properties. Cogn. Sci. **1982**, *6*, 205–254.

7. Rumelhart, D.; McClelland, J., Eds. *Parallel Distributed Processing*; MIT Press, 1986.

8. Ellacott, S.; Bose, D. *Neural Networks*, International Thomson Publishing, 1996.

9. Rojas, R. *Neural Networks: A Systematic Introduction*, Springer, 1996.

10. Haykin, S. *Neural Networks: A Comprehensive Foundation*, 2nd Ed. Prentice Hall, 1998.

11. White, H.; Hornik, K.; Stinchcombe, M. Multilayer feedforward networks are universal approximators. Neural Netw. **1989**, *2*, 359–366.

12. White, H. *Artificial Neural Networks: Approximation and Learning Theory*, Blackwell, 1992.

13. Smolensky, P.; Mozer, M. Rumelhart, D. *Mathematical Perspectives on Neural Networks*, Lawrence Erlbaum, 1996.

14. Ellacott, S.; Bose, D. *Neural Networks: Deterministic Methods of Analysis*, International Thomson Computer Press, 1996.

15. Arbib, M.; Ed. *The Handbook of Brain Theory and Neural Networks;* MIT Press, 1995; 20.

16. Bishop, C. *Neural Networks for Pattern Recognition*, Clarendon Press, 1995.

17. Moisl, M. NLP based on artificial neural networks. In *Handbook of Natural Language Processing*; Dale, R., Moisl, H., Somers, R., Eds.; Marcel Dekker, 2000.

18. Judd, J. Complexity of learning. In *Mathematical Perspectives on Neural Networks*; Smolensky, P., Mozer, M., Rumelhart, D., Eds.; Lawrence Erlbaum, 1996.

19. Baum, E. When are *k*-nearest neighbor and back-propagation accurate for feasible-sized sets of examples? In *Computational Learning Theory and Natural Learning Systems*; Hanson, S., Drastal, G., Rivest, R., Eds.; MIT Press, 1994.

20. Elman, J. Learning and development in neural networks: the importance of starting small. Cognition **1993**, *48*, 71–99.

21. Frasconi, P.; Gori, M.; Soda, G. Recurrent neural networks and prior knowledge for sequence processing: a constrained nondeterministic approach. Knowl.-Based Syst. **1995**, *8*, 313–332.

22. Omlin, C. Stable encoding of large finite state automata in recurrent neural networks with sigmoid discriminants. Neural Comput. **1996**, *8*, 675–696.

23. Gasser, M. Transfer in a connectionist model of the acquisition of morphology. In *Yearbook of Morphology 1996*; Baayen, H., Schroeder, R., Eds.; Foris, 1997.

24. Thrun, S.; Pratt, L. *Learning to Learn*, Kluwer, 1997.

25. Dyer, M. Connectionist natural language processing: a status report. In *Computational Architectures Integrating Neural and Symbolic Processes*; Sun, R., Bookman, L., Eds.; Kluwer, 1995; 391.

26. Newell, A. Physical symbol systems. Cogn. Sci. **1980**, *4*, 135–183.

27. Fodor, J.; Pylyshyn, Z. Connectionism and cognitive architecture: a critical analysis. Cognition **1988**, *28*, 3–71.

28. Smolensky, P. On the proper treatment of connectionism. Behav. Brain Sci. **1988**, *11*, 1–74.

29. Fodor, J.; McLaughlin, B. Connectionism and the problem of systematicity: why Smolensky's solution doesn't work. In *Connectionism and the Philosophy of Mind*; Horgan, T., Tienson, J., Eds.; Kluwer, 1991.

30. Fodor, J. Connectionism and the problem of systematicity (continued): why Smolensky's solution still doesn't work. Cognition **1997**, *62*, 109–119.

31. Smolensky, P. The constituent structure of connectionist mental states: a reply to Fodor and Pylyshyn. In *Connectionism and the Philosophy of Mind*; Horgan, T., Tienson, J., Eds.; Kluwer, 1991.

32. Smolensky, P. Connectionism, constituency, and the language of thought. In *Meaning and Mind. Fodor and His Critics*; Loewer, B., Rey, G., Eds.; Blackwell, 1991.

33. Clark, A. *Microcognition: Philosophy, Cognitive Science, and Parallel Distributed Processing*, MIT Press, 1989.

34. Horgan, T.; Tienson, J., Eds. *Connectionism and the Philosophy of Mind*; Kluwer, 1991.

35. Churchland, P.; Sejnowski, T. *The Computational Brain*, MIT Press, 1992.

36. Dinsmore, J.; Ed. *The Symbolic and Connectionist Paradigms: Closing the Gap;* Lawrence Erlbaum, 1992.

37. Clark, A. *Associative Engines. Connectionism, Concepts, and Representational Change*, MIT Press, 1993.

38. Honavar, V.; Uhr, L. *Artificial Intelligence and Neural Networks: Steps Toward Principled Integration*, Academic Press, 1994.

39. Port, R.; van Gelder, T., Eds.; *Mind as Motion. Explorations in the Dynamics of Cognition;* MIT Press, 1995.

40. Macdonald, C.; Macdonald, G., Eds.; *Connectionism: Debates on Psychological Explanation*; Basil Blackwell, 1995.

41. Elman, J.; Bates, E.; Johnson, M.; Karmiloff-Smith, A.; Parisi, D.; Plunkett, K. *Rethinking Innateness. A Connectionist Perspective on Development*, MIT Press, 1996.

42. Dorffner, G.; Ed.; *Neural Networks and a New Artificial Intelligence*; Thomson Computer Press, 1997.

43. Clark, A. *Being There. Putting Brain, Body, and World Together Again*, MIT Press, 1997.

44. Hadley, R. Systematicity in connectionist language learning. Mind Lang. **1994**, *9*, 247–273.

45. Niklasson, L.; van Gelder, T. On being systematically connectionist. Mind Lang. **1994**, *9*, 288–302.

46. Niklasson, L.; Sharkey, N. Systematicity and generalization in compositional connectionist representations. In *Neural Networks and a New Artificial Intelligence*; Dorffner, G., Ed.; Thomson Computer Press, 1997.

47. Jagota, A.; Plate, T.; Shastri, L.; Sun, R., Eds. *Connectionist Symbol Processing: Dead or Alive?*; Neural Computing Surveys, 1999; Vol. 2, 1–40.

48. Witbrock, M. The symbolic approach to ANN-based natural language processing. In *Handbook of Natural Language Processing*; Dale, R., Moisl, H., Somers, R., Eds.; Marcel Dekker, 2000.

49. Miikkulainen, R. *Subsymbolic Natural Language Processing: An Integrated Model of Scripts, Lexicon, and Memory*, MIT Press, 1993.

50. Miikkulainen, R. Text and discourse understanding: the DISCERN system. In *Handbook of Natural Language Processing*; Dale, R., Moisl, H., Somers, R., Eds.; Marcel Dekker, 2000.

51. Dorffner, G. The subsymbolic approach to ANN-based natural language processing. In *Handbook of Natural Language Processing*; Dale, R., Moisl, H., Somers, R., Eds.; Marcel Dekker, 2000.

52. Smolensky, P.; Legendre, G.; Miyata, Y. Integrating connectionist and symbolic computation for the theory of Language. In *Artificial Intelligence and Neural Networks: Steps Toward Principled Integration*; Honavar, V., Uhr, L., Eds.; Academic Press, 1994.

53. Prince, A.; Smolensky, P. Optimality: from neural networks to universal grammar. Science **1997**, *275*, 1604–1610.

54. Smolensky, P.; Legendre, G. *Toward a Calculus of the Mind/Brain: Neural Network Theory, Optimality, and Universal Grammar*; MIT Press.

55. Wermter, S. *Hybrid Connectionist Natural Language Processing*, Chapman & Hall, 1995.

56. Sun, R.; Bookman, L. *Computational Architectures Integrating Neural and Symbolic Processes*, Kluwer, 1995.

57. Sun, R.; Alexandre, F. *Connectionist-Symbolic Integration. From Unified to Hybrid Approaches*, Lawrence Erlbaum, 1997.

58. McGarry, K.; Wermter, S.; MacIntyre, J. Hybrid neural systems: from simple coupling to fully integrated neural networks. Neural Comput. Surv. **1999**, *2*, 62–93.

59. Wermter, S.; Sun, R., Eds. *Hybrid Neural Systems*; Springer, 2000.

60. Wermter, S. The hybrid approach to artificial neural network-based language processing. In *Handbook of Natural Language Processing*; Dale, R., Moisl, H., Somers, R., Eds.; Marcel Dekker, 2000.

61. Sharkey, N. Connectionist representation techniques. Artif. Intell. Rev. **1991**, *5*, 143–167.

62. Sharkey, N.; Jackson, S. Three horns of the representational trilemma. In *Artificial Intelligence and Neural Networks: Steps Toward Principled Integration*; Honavar, V., Uhr, L., Eds.; Academic Press, 1994.

63. Sharkey, N.; Sharkey, A. Separating learning and representation. In *Connectionist, Statistical, and Symbolic Approaches to Learning for Natural Language Processing*; Wermter, S., Riloff, E., Scheler, G., Eds.; Springer, 1996.

64. Shastri, L. Structured connectionist models. In *The Handbook of Brain Theory and Neural Networks*; Arbib, M., Ed.; MIT Press, 1995.

65. Thorpe, S. Localized versus distributed representations. In *The Handbook of Brain Theory and Neural Networks*; Arbib, M., Ed.; MIT Press, 1995.

66. Feldman, J. Structured connectionist models and language learning. Artif. Intell. Rev. **1993**, *7*, 301–312.

67. Bookman, L. *Trajectories through Knowledge Space. A Dynamic Framework for Machine Comprehension*, Kluwer, 1994.

68. Regier, T. *The Human Semantic Potential*, MIT Press, 1996.

69. Feldman, J.; Lakoff, G.; Bailey, D.; Narayanan, S.; Regier, T.; Stolcke, A. L0: the first five years. Artif. Intell. Rev. **1996**, *10*, 103–129.

70. Bienenstock, D. Geman, S. Compositionality in neural systems. In *The Handbook of Brain Theory and Neural Networks*; Arbib, M., Ed.; MIT Press, 1995.

71. Sun, R. Logics and variables in connectionist models: a brief overview. In *Artificial Intelligence and Neural Networks: Steps Toward Principled Integration*; Honavar, V., Uhr, L., Eds.; Academic Press, 1994.

72. van Gelder, T. Compositionality: a connectionist variation on a classical theme. Cogn. Sci. **1990**, *14*, 355–384.

73. Smolensky, P. Tensor product variable binding and the representation of symbolic structures in connectionist systems. Artif. Intell. **1990**, *46*, 159–216.

74. Legendre, G.; Miyata, Y. Smolensky, P. Distributed recursive structure processing. In *Advances in Neural Information Processing Systems 3*; Touretzky, D., Lippman, R., Eds.; Morgan Kaufmann, 1991.

75. Pollack, J. Recursive distributed representations. Artif. Intell. **1990**, *46*, 77–105.

76. Chalmers, D. Syntactic transformations on distributed representations. Connect. Sci. **1990**, *2*, 53–62.

77. Callan, R.; Palmer-Brown, D. (S)RAAM: an analytical technique for fast and reliable derivation of connectionist symbol structure representations. Connect. Sci. **1997**, *9*, 139–160.

78. Adamson, M.; Damper, R. B_RAAM: a connectionist model which develops holistic internal representations of symbolic structures. Connect. Sci. **1999**, *11*, 41–71.

79. Plate, T. Holographic recurrent networks. In *Advances in Neural Information Processing Systems 5*; Giles, C., Hanson, S., Cowan, J., Eds.; Morgan Kaufmann, 1993.

80. Plate, T. Holographic reduced representations. IEEE Trans. Neural Netw. **1995**, *6*, 623–641.

81. Plate, T. A common framework for distributed representation schemes for compositional structure. In *Connectionist Systems for Knowledge Representation and Deduction*; Maire, F., Hayward, R., Diederich, J., Eds.; Queensland University of Technology, 1997.

82. Bengio, Y. *Neural Networks for Speech and Sequence Recognition*, International Thomson Computer Press, 1996.

83. Sun, R.; Giles, C. L., Eds. *Sequence Learning: Paradigms, Algorithms, and Applications*; Springer, 2001.

84. Clouse, D.; Giles, C.; Horne, B.; Cottrell, G. Time-delay neural networks: representation and induction of finite-state machines. IEEE Trans. Neural Netw. **1997**, *8*, 1065–1070.

85. Bodenhausen, U. Geutner, P. Waibel, A. Flexibility through incremental learning: neural networks for text categorization. *Proceedings of the World Congress on Neural Networks (WCNN)*; Lawrence Erlbaum Associates, 1993.

86. Bodenhausen, U.; Manke, S. A connectionist recognizer for on-line cursive handwriting recognition. *Proceedings of the IEEE, International Conference on Acoustics, Speech, and Signal Processing*, IEEE, 1994.

87. Jordan, M. Attractor dynamics and parallelism in a connectionist sequential machine. *Proceedings of the Eighth Conference of the Cognitive Science Society*; Lawrence Erlbaum Associates, 1986.

88. Servan-Schreiber, D. Cleeremans, A. McClelland, J. Learning sequential structure in simple recurrent networks. In *Advances in Neural Information Processing Systems 1*; Touretzky, D., Ed.; Morgan Kaufmann, 1989.

89. Elman, J. Finding structure in time. Cogn. Sci. **1990**, *14*, 179–211.

90. Elman, J. Distributed representation, simple recurrent networks, and grammatical structure. Mach. Learn. **1991**, *7*, 195–225.

91. Servan-Schreiber, D.; Cleeremans, A.; McClelland, J. Graded state machines: the representation of temporal contingencies in simple recurrent networks. Mach. Learn. **1991**, *7*, 161–193.

92. Pollack, J. The induction of dynamical recognizers. Mach. Learn. **1991**, *7*, 227–252.

93. Watrous, R.; Kuhn, G. Induction of finite-state languages using second-order recurrent networks. Neural Comput. **1992**, *4*, 406–414.

94. Zeng, Z.; Goodman, R.; Smyth, P. Discrete recurrent neural networks for grammatical inference. IEEE Trans. Neural Netw. **1994**, *5*, 320–330.

95. Horne, B.; Giles, C.; Lin, T. Learning a class of large finite state machines with a recurrent neural network. Neural Netw. **1995**, *8*, 1359–1365.

96. Frasconi, P.; Gori, M.; Maggini, M.; Soda, G. Representation of finite state automata in recurrent radial basis function networks. Mach. Learn. **1996**, *23*, 5–32.

97. Lawrence, S.; Fong, S.; Giles, C. Natural language grammatical inference: a comparison of recurrent neural networks and machine learning methods. In *Connectionist, Statistical, and Symbolic Approaches to Learning for Natural Language Processing*; Wermter, S., Riloff, E., Scheler, G., Eds.; Springer, 1996.

98. Tino, P.; Horne, B.; Giles, C.; Collingwood, P. Finite state machines and recurrent neural networks—automata and dynamical systems approaches. In *Neural Networks and Pattern Recognition*; Dayhoff, J., Omidvar, O., Eds.; Academic Press, 1998.

99. Parekh, R.; Honavar, V. Grammar inference, automata induction, and language acquisition. In *Handbook of Natural Language Processing*; Dale, R., Moisl, H., Somers, R., Eds.; Marcel Dekker, 2000.

100. Omlin, C.; Giles, C. *Symbolic Knowledge Representation and Acquisition in Recurrent Neural Networks: Foundations, Algorithms, and Applications*; World Scientific Publishing.

101. Poesio, M. Semantic analysis. In *Handbook of Natural Language Processing*; Dale, R., Moisl, H., Somers, R., Eds.; Marcel Dekker, 2000.

102. Dorffner, G.; Prem, E. Connectionism, symbol grounding, and autonomous agents. *Proceedings of the 15th Annual Conference of the Cognitive Science Society*; Lawrence Erlbaum Associates, 1993.

103. Nenov, V.; Dyer, M. Perceptually grounded language learning: part 1—A neural network architecture for robust sequence association. Connect. Sci. **1993**, *5*, 115–138.

104. Nenov, V.; Dyer, M. Perceptually grounded language learning: part 2—DETE: a neural/procedural model. Connect. Sci. **1994**, *6*, 3–41.

105. Harnad, S. Grounding symbols in the analog world with neural nets. Think **1993**, *2*, 12–78.

106. Dyer, M. Grounding language in perception. In *Artificial Intelligence and Neural Networks: Steps Toward Principled Integration*; Honavar, V., Uhr, L., Eds.; Academic Press, 1994.

107. Srihari, R. Computational models for integrating linguistic and visual information: a survey. Artif. Intell. Rev. **1994–1995**, *8*, 349–369.

108. Harnad, S. Grounding symbolic capacity in robotic capacity. In *The Artificial Life Route to Artificial Intelligence: Building Embodied, Situated Agents*; Steels, L., Brooks, R., Eds.; Lawrence Erlbaum, 1995.

109. Jackson, S.; Sharkey, N. Grounding computational engines. Artif. Intell. Rev. **1996**, *10*, 65–82.

110. Sales, N.; Evans, R.; Aleksander, I. Successful naive representation grounding. Artif. Intell. Rev. **1996**, *10*, 83–102.

111. Bailey, D.; Feldman, J.; Narayanan, S.; Lakoff, G. Embodied lexical development. *Proceedings of the 19th Annual Meeting of the Cognitive Science Society*; Stanford University Press, 1997.

112. Pfeifer, R.; Verschure, P. Complete autonomous systems: a research strategy for cognitive science. In *Neural Networks and a New Artificial Intelligence*; Dorffner, G., Ed.; Thomson International Computer Press, 1997.

113. Selman, B. Connectionist systems for natural language understanding. Artif. Intell. Rev. **1989**, *3*, 23–31.

114. Sharkey, N.; Reilly, R. Connectionist natural language processing. In *Connectionist Approaches to Natural Language Processing*; Reilly, R., Sharkey, N., Eds.; Lawrence Erlbaum, 1992.

115. Christiansen, M.; Chater, N. Connectionist natural language processing: the state of the art. Cogn. Sci. **1999**, *23*, 417–437.

116. Hinton, G. Implementing semantic networks in parallel hardware. In *Parallel Models of Associative Memory*; Hinton, G., Anderson, J., Eds.; Lawrence Erlbaum, 1981.

117. McClelland, J.; Rumelhart, D. An interactive activation model of context effects in letter perception: Part 1. An account of basic findings. Psychol. Rev. **1981**, *88*, 375–407.

118. Rumelhart, D.; McClelland, J. An interactive activation model of context effects in letter perception: part 2. The contextual enhancement effects and some tests and enhancements of the model. Psychol. Rev. **1982**, *89*, 60–94.

119. Golden, R. A developmental neural model of visual word perception. Cogn. Sci. **1986**, *10*, 241–276.

120. Cottrell, G.; Small, S. A connectionist scheme for modeling word sense disambiguation. Cogn. Brain Theory **1983**, *6*.

121. Reilly, R. A connectionist model of some aspects of anaphor resolution. *Proceedings of the Tenth Annual Conference on Computational Linguistics*; ACL, 1984.

122. Small, S.; Cottrell, G.; Shastri, L. Towards connectionist parsing. *Proceedings of the National Conference on Artificial Intelligence*; AAAI, 1982.

123. Fanty, M. *Context-Free Parsing in Connectionist Networks*, TR 174, Department of Computer Science, University of Rochester, 1985.

124. Selman, B.; Hirst, G. A rule-based connectionist parsing system. *Proceedings of the Seventh Annual Meeting of the Cognitive Science Society*; Lawrence Erlbaum, 1985.

125. Lehnert, W. Possible implications of connectionism. *Theoretical Issues in Natural Language Processing*; University of Mexico, 1986.

126. Pinker, S.; Prince, A. On language and connectionism: analysis of a parallel distributed processing model of language acquisition. Cognition **1988**, *28*, 73–193.

127. MacWhinney, B.; Leinbach, J. Implementations are not conceptualizations: revising the verb learning model. Cognition **1991**, *40*, 121–157.

128. Plunkett, K.; Marchman, V. U-shaped learning and frequency effects in a multi-layered perceptron: implications for child language acquisition. Cognition **1991**, *38*, 43–102.

129. Daugherty, K.; Seidenberg, M. Rules or connections? The past tense revisited. *Proceedings of the Fourteenth Annual Meeting of the Cognitive Science Society*; Lawrence Erlbaum, 1992.

130. Daugherty, K.; Hare, M. What's in a rule? The past tense by some other name might be called a connectionist net. *Proceedings of the 1993 Connectionist Models Summer School*; Lawrence Erlbaum, 1993.

131. Plunkett, K.; Marchman, V. From rote learning to system building: acquiring verb morphology in children and connectionist nets. Cognition **1993**, *48*, 21–69.

132. Marcus, G. The acquisition of the English past tense in children and multilayered connectionist networks. Cognition **1997**, *56*, 271–279.

133. Plunkett, K. Language acquisition. In *The Handbook of Brain Theory and Neural Networks*; Arbib, M., Ed.; MIT Press, 1995.

134. Seidenberg, M.; Allen, J.; Christiansen, M. Language acquisition: learning and applying probabilistic constraints. In *Proceedings of GALA 1997: Language Acquisition: Knowledge Representation and Processing*; Sorace, A., Heycock, C., Shillcock, R., Eds.; University of Edinburgh, 1997.

135. Bates, E.; Elman, J.; Johnson, M.; Karmiloff-Smith, A.; Parisi, D.; Plunkett, K. Innateness and emergentism. In *A Companion to Cognitive Science*; Bechtel, W., Graham, G., Eds.; Basil Blackwell, 1998.

136. Hare, M.; Elman, J. Learning and morphological change. Cognition **1995**, *56*, 61–98.

137. Plunkett, K.; Juola, P. A connectionist model of English past tense and plural morphology. Cogn. Sci. **1999**, *23*(4).

138. Dorffner, G.; Taxonomies and part–whole hierarchies in the acquisition of word meaning—a connectionist model. *Proceedings of the 14th Annual Conference of the Cognitive Science Society*; Lawrence Erlbaum, 1992.

139. Dorffner, G.; Hentze, M.; Thurner, G. A Connectionist model of categorization and grounded word learning. In *Proceedings of the Groningen Assembly on Language Acquisition (GALA'95)*; Koster, C., Wijnen, F., Eds.; Center for the Study of Cognition, 1996.

140. Scheler, G. Generating English plural determiners from semantic representations: a neural network learning approach. In *Connectionist, Statistical, and Symbolic Approaches to Learning for Natural Language Processing*; Wermter, S., Riloff, E., Scheler, G., Eds.; Springer, 1996.

141. Scheler, G. Learning the semantics of aspect. In *New Methods in Language Processing*; Somers, H., Jones, D., Eds.; University College London Press, 1996.

142. Narayanan, S. Talking the talk *is* like walking the walk: a computational model of verbal aspect. *Proceedings of the Annual Conference of the Cognitive Science Society*; Stanford, 1997.

143. Clouse, D.; Cottrell, G. Regularities in a random mapping from orthography to semantics. *Proceedings of the Twentieth Annual Cognitive Science Conference*; Lawrence Erlbaum, 1998.

144. Waltz, D.; Pollack, J. Massively parallel parsing: a strongly interactive model of natural language interpretation. Cogn. Sci. **1985**, *95*, 1–74.

145. Hanson, S.; Kegl, J. PARSNIP: a connectionist network that learns natural language grammar from exposure to natural language sentences. *Proceedings of the Ninth Annual Conference of the Cognitive Science Society*; Lawrence Erlbaum, 1987.

146. Howells, T. VITAL—a connectionist parser. *Proceedings of the Tenth Annual Conference of the Cognitive Science Society*; Lawrence Erlbaum Associates, 1988.

147. Jain, A.; Waibel, A. Parsing with connectionist networks. In *Current Issues in Parsing Technology*; Tomita, M., Ed.; Kluwer, 1991.

148. Wermter, S.; Lehnert, W. Noun phrase analysis with connectionist networks. In *Connectionist Approaches to*

Natural Language Processing; Reilly, R., Sharkey, N., Eds.; Lawrence Erlbaum, 1992.

149. Kwasny, S.; Johnson, S.; Kalman, B. Recurrent natural language parsing. *Proceedings of the Sixteenth Annual Meeting of the Cognitive Science Society*; Lawrence Erlbaum, 1994.

150. Miikkulainen, R. Subsymbolic parsing of embedded structures. In *Computational Architectures Integrating Neural and Symbolic Processes*; Sun, R., Bookman, L., Eds.; Kluwer, 1995.

151. Tabor, W.; Juliano, C.; Tanenhaus, M. Parsing in a dynamical system: an attractor-based account of the interaction of lexical and structural constraints in sentence processing. Lang. Cogn. Processes **1997**, *12*, 211–271.

152. Henderson, J.; Lane, P. A connectionist architecture for learning to parse. *Proceedings of 17th International Conference on Computational Linguistics and the 36th Annual Meeting of the Association for Computational Linguistics (COLING-ACL'98)*; University of Montreal: Canada, 1998.

153. Tabor, W.; Tanenhaus, M. Dynamical models of sentence processing. Cogn. Sci. **1999**, *23*(4).

154. Henderson, J.; Constituency, context, and connectionism in syntactic parsing. In *Architectures and Mechanisms for Language Processing*; Crocker, M., Pickering, M., Clifton, C., Eds.; Cambridge University Press.

155. McClelland, J.; Kawamoto, A. Mechanisms of sentence processing: assigning roles to constituents. In *Parallel Distributed Processing*; Rumelhart, D., McClelland, J., Eds.; MIT Press, 1986.

156. St. John, M.; McClelland, J. Learning and applying contextual constraints in sentence comprehension. Artif. Intell. **1990**, *46*, 217–257.

157. St. John, M.; McClelland, J. Parallel constraint satisfaction as a comprehension mechanism. In *Connectionist Approaches to Natural Language Processing*; Reilly, R., Sharkey, N., Eds.; Lawrence Erlbaum, 1992.

158. Wermter, S. A hybrid symbolic/connectionist model for noun phrase understanding. In *Connectionist Natural Language Processing*; Sharkey, N., Ed.; Kluwer, 1992.

159. Miikkulainen, R. Subsymbolic case–role analysis of sentences with embedded clauses. Cogn. Sci. **1996**, *20*, 47–73.

160. Wermter, S.; Hannuschka, R. A connectionist model for the interpretation of metaphors. In *Neural Networks and a New Artificial Intelligence*; Dorffner, G., Ed.; Thomson International Computer Press, 1997.

161. Narayanan, S. Moving right along: a computational model of metaphoric reasoning about events. *Proceedings of the National Conference on Artificial Intelligence AAAI-99*, AAAI, 1999, Orlando, FL.

162. Miikkulainen, R. Integrated connectionist models: building AI systems on subsymbolic foundations. In *Artificial Intelligence and Neural Networks: Steps Toward Principled Integration*; Honavar, V., Uhr, L., Eds.; Academic Press, 1994.

163. Wermter, S.; Weber, V. SCREEN: Learning a flat syntactic and semantic spoken language analysis using artificial neural networks. J. Artif. Intell. Res. **1997**, *6*, 35–85.

164. Narayanan, S. Reasoning about actions in narrative understanding. *Proceedings of the International Joint Conference on Artificial Intelligence IJCAI-99*, AAAI, 1999 Stockholm.

165. Ward, N. *A Connectionist Language Generator*, Ablex, 1994.

166. Cottrell, G.; Plunkett, K. Acquiring the mapping from meaning to sounds. Connect. Sci. **1995**, *6*, 379–412.

167. Dell, G.; Burger, L.; Svec, W. Language production and serial order: a functional analysis and a model. Psychol. Rev. **1997**, *104*, 123–147.

168. Aretoulaki, M. Towards a hybrid abstract generation system. In *New Methods in Language Processing*; Somers, H., Jones, D., Eds.; UCL Press, 1997.

169. Basak, J.; Nikhil, R.; Pal, S. A connectionist system for learning and recognition of structures: application to handwritten characters. Neural Netw. **1995**, *8*, 643–657.

170. Shustorovich, A.; Thrasher, C. Neural network positioning and classification of handwritten characters. Neural Netw. **1996**, *9*, 685–693.

171. Chiang, J.-H. A hybrid neural network model in handwritten word recognition. Neural Netw. **1998**, *11*, 337–346.

172. Lucas, S. Character recognition with syntactic neural networks. In *Handbook of Natural Language Processing*; Dale, R., Moisl, H., Somers, R., Eds.; Marcel Dekker, 2000.

173. Merkl, D. Text classification with self-organizing maps: some lessons learned. Neurocomputing **1998**, *21*, 61–77.

174. Kaski, S.; Honkela, T.; Lagus, K.; Kohonen, T. WEBSOM—self-organizing maps of document collections. Neurocomputing **1998**, *21*, 101–117.

175. Wermter, S. Neural network agents for learning semantic text classification. Inf. Retriev. **2000**, *3*, 87–103.

176. Kohonen, T.; Kaski, S.; Lagus, K.; Salojarvi, J.; Paatero, V.; Saarela, A. Organization of a massive document collection. IEEE Trans. Neural Netw. **2000**, *11*, 574–585.

177. Merkl, D. Text data mining. In *Handbook of Natural Language Processing*; Dale, R., Moisl, H., Somers, R., Eds.; Marcel Dekker, 2000.

178. Winiwarter, W.; Schweighofer, E.; Merkl, D. Knowledge acquisition in concept and document spaces by using self-organizing neural networks. In *Connectionist, Statistical, and Symbolic Approaches to Learning for Natural Language Processing*; Wermter, S., Riloff, E., Scheler, G., Eds.; Springer, 1996.

179. Aretoulaki, M.; Scheler, G.; Brauer, W. Connectionist modeling of human event memorization with application to automatic text summarization. *AAAI Spring Symposium on Intelligent Text Summarization*; Stanford, 1998.

180. König, A. Interactive visualization and analysis of hierarchical neural projections for data mining. IEEE Trans. Neural Netw. **2000**, *11*, 615–624.

181. Chen, C.; Honavar, V. Neural architectures for information retrieval and database query. In *Handbook of Natural Language Processing*; Dale, R., Moisl, H., Somers, R., Eds.; Marcel Dekker, 2000.

182. Schmidhuber, J.; Heil, S. Compressing texts with neural nets. In *Handbook of Natural Language Processing*; Dale, R., Moisl, H., Somers, R., Eds.; Marcel Dekker, 2000.

Arts Literatures and Their Users

Lisl Zach
iSchool, Drexel University, Philadelphia, Pennsylvania, U.S.A.

Abstract
The various arts disciplines such as dance, painting, and sculpture are often considered to belong to the larger category of the humanities. User studies in the humanities have focused primarily on the needs of scholars and academic researchers. User studies in the arts have also investigated the information needs of practicing artists and others involved in arts-related activities. While these studies are often exploratory in nature, they do point out that the needs of practicing artists go well beyond the information services traditionally provided by academic libraries. With the advent of the World Wide Web, artists and other users of information about the arts have access to some of the greatest libraries and museum collections in the world without ever leaving their homes. However, retrieval of textual, image, and sound data still poses significant challenges for the user.

INTRODUCTION

Arts literatures, or works about the arts, are as broad a field of materials as the various topics they cover. Until recently, information about the arts would have been found predominantly by seeking out primary materials or by consulting books or journal articles. However, due in large measure to the advent of the World Wide Web, the wealth of digital information sources has increased dramatically. These sources have changed the ways in which people look for and interact with information about the arts.

While users of information about the arts range from children looking for material to include in school assignments to retirees looking for information about museums they plan to visit, the main focus of this entry will be on the users most closely involved with the arts—scholars working in the various disciplines of the arts and the creative artists themselves. These user groups have specialized information needs and have been the subject of several previous research studies. In addition to examining what we know about art scholars' and artists' use of information, this entry will also address some of the current issues in retrieving information in the arts.

DEFINITION OF THE ARTS

Aristotle was the first to make the distinction between the arts and the sciences—a distinction that has been preserved in the academic traditions of today. For Aristotle, the sciences were concerned with the natural world while the arts focused on the "artificial." During the Middle Ages, the term "liberal arts" came to be applied to the academic fields of grammar, rhetoric, logic, arithmetic, geometry, music, and astronomy, while many of the disciplines that we now think of as comprising the arts (such as dance, painting, and sculpture) were considered to have little or no intellectual value. In modern times, the arts are often regarded as a subcategory of the humanities—those disciplines that are concerned with the study of human culture compared to the more empirical disciplines of the natural and social sciences. Blazek and Aversa, for example, include the visual and performing arts in the list of fields covered by their definition of the humanities.[1]

One definition of the arts is offered by the National Foundation on the Arts and the Humanities Act, which established the National Endowment for the Arts and the National Endowment for the Humanities in 1965. In it, "the arts" were decoupled from the humanities and defined to include, but not be limited to "music (instrumental and vocal), dance, drama, folk art, creative writing, architecture and allied fields, painting, sculpture, photography, graphic and craft arts, industrial design, costume and fashion design, motion pictures, television, radio, film, video, tape and sound recording, the arts related to the presentation, performance, execution, and exhibition of such major art forms, all those traditional arts practiced by the diverse peoples of this country. [sic] and the study and application of the arts to the human environment." National Foundation on the Arts and the Humanities Act (20 U.S.C. 952(c)). See http://www.arts.gov/about/Legislation/Legislation.pdf.

ARTS OCCUPATIONS

According to data released by the National Endowment for the Arts' Office of Research and Analysis, more than 2.1 million U.S. workers have artist occupations as their primary jobs. This number represents more than 7% of the total workers in all professional occupations. In addition, more than 300,000 workers hold secondary jobs as artists.[2]

Encyclopedia of Library and Information Sciences, Fourth Edition DOI: 10.1081/E-ELIS4-120043466

Artificial–Association

Table 1 Artist occupations.

	Primary jobs	Secondary jobs	Self-employed (%)	Postsecondary education or training
Designers	803,000	55,000	27	Bachelor's or associate's degree
Architects	239,000	6,000	20	Bachelor's degree
Fine artists and animators	245,000	32,000	62	long-term on-the-job experience
Musicians and singers	223,000	96,000	41	long-term on-the-job experience
Writers and authors	185,000	33,000	68	Bachelor's degree
Photographers	158,000	37,000	69	long-term on-the-job experience
Producers and directors	129,000	13,000	30	degree plus work experience
Announcers	63,000	21,000	25	long-term on-the-job experience
Actors	55,000	5,000	17	long-term on-the-job experience
Dancers and choreographers	35,000	3,000	19	long-term on-the-job experience
Other entertainers and performers	29,000	3,000	n/a	not available

As shown in Table 1, almost half of the total number of primary jobs belong to designers (a term that covers commercial and industrial designers as well as window dressers) and architects; the remaining categories include fine artists (i.e., painters, sculptors, and illustrators) and animators, musicians and singers, writers and authors, photographers, producers and directors, announcers, actors, dancers and choreographers, and other entertainers and performers.

Approximately one-third of artists are self-employed. This percentage is highest among photographers, writers and authors, and fine artists and animators. With the exception of designers and architects, most artists rely on on-the-job experience rather than formal academic education or training. These characteristics support a popular image of artists as solitary, eccentric, and preoccupied with their own creative processes that originated in the sixteenth century and has persisted ever since.[3] However, most artists today "live and work in the same mundane world of dollars and deadlines as their professional colleagues" (p. 344)[4] and so have many information needs not associated with their creative processes.

The numbers shown in Table 1 do not include academics and scholars who teach or do research in these areas unless they are also practicing artists. According to the Bureau of Labor Statistics, there are 72,100 postsecondary teachers of art, drama, and music. Data extracted from the Bureau of Labor Statistics Occupational Statistics homepage: http://www.bls.gov/oes/home.htm for Standard Occupational Classification code 251121 on March 30, 2008. In addition, the arts world includes many professionals who are neither practicing artists nor academic scholars but who do have specific job-related information needs.[5] The diversity of user groups in the arts and the variety of their information needs make a simple definition difficult, but the following review provides some insights into the characteristics of these users.

USER STUDIES IN THE ARTS

Users in the humanities have been studied systematically by researchers for over 30 years. Early projects to identify information needs in the humanities were carried out by the Centre for Research on User Studies at the University of Sheffield;[6,7] further descriptions of the information needs of humanities scholars appeared in the early 1980s with the publication of articles by Weintraub[8] and Stone.[9] While the Weintraub article focused primarily on the implications for library services, the Stone study identified certain characteristics of humanities scholars that have been confirmed repeatedly by subsequent research. These include the tendency to work alone, to browse extensively, and to rely heavily on primary materials in addition to books and journals. Since the publication of the Stone study, researchers have discussed and expanded on the characteristics of humanities scholars[10–14] and tracked changes in their information behavior in response to technology.[15–18] A more thorough consideration of the humanities literatures and their users may be found elsewhere in this work, and it is not the intention of this entry to do more than indicate the context in which research on user groups in the arts has been conducted.

The overlap among the subject areas considered to be part of the humanities in general and those that are more strictly defined as belonging to the arts specifically is reflected in the studies of users in these areas. The distinction between the two areas is further complicated by the fact that often users of information about the arts are not themselves artists, but rather scholars such as art or music historians. These arts scholars may share more characteristics in common with other humanities scholars than with the visual or performing artists whose works they study. Furthermore, user studies point out that "scholars in the humanities do not have a homogeneous information-seeking behavior or homogeneous information needs" (p. 195).[11] That this is also true of users in the arts becomes apparent when looking at studies of users in specific arts disciplines.

A further characteristic of user studies in the arts is that they, like the studies of users in the humanities, often take a library-centric perspective. That is, the focus of many studies has been on users' interactions with libraries and how library services might be better designed to serve these user groups' information needs.[19–22] This perspective may not fully reflect either the information needs or the information-seeking behavior of users in the arts, especially practicing artists. Users in the arts, whether scholars or practicing artists, often look beyond traditional academic library resources when searching for information or inspiration.

The concept of serendipity, and the role that it plays in the creative process, is addressed in several studies related to users in the arts.[4,23] This concept is also well documented in studies of humanities and interdisciplinary scholars.[24,25]

Art Historians

Among the earliest user studies specifically addressing a particular group in the arts was Stam's 1984 dissertation summarized in the article "How Art Historians Look for Information."[26] Drawing on Allen's technology transfer model,[27] Stam "attempted to establish some normative data about these scholars as a first step" towards providing better library and information services for these users.[27] Stam's study found that art historians, like other humanities scholars, relied heavily on institutional and personal libraries. Art historians who were also college faculty augmented their personal libraries with a significant number of slides. Stam further found that art historians visited museums an average of eight times per year. This behavior was consistent with Stone's earlier finding on the range of materials needed by humanities scholars including "primary materials such as … works of art …" (p. 296).[9] Stam's study also found that art historians working in museums used informal channels of communication (telephone calls, social/professional events, and correspondence) more frequently than did their counterparts working in academic environments. Both groups, however, interacted more with institutional colleagues than they did with librarians. Stam's conclusion from her data was that "information seeking among art historians seems to be a fairly private activity…. The process … appears to be a contemplative undertaking involving objects of art, reproductions of those objects and related objects, and written descriptions and observations about works of art" (pp. 29–30).[27] Even those arts scholars working in interdisciplinary areas, where collaboration is often the norm, are "definitely loners" (p. 358).[28]

The art historians' need for visual information, whether from primary or secondary sources, is a defining characteristic of this particular user group. The art object, or at least a visual representation of it, is essential to the art historians' research activities.[29,30] This characteristic is shared by art students and artists in both the fine and the applied arts.[20,23,31] A study of visual arts students confirmed that extensive browsing is another characteristic that this user group shares with more experienced art historians. However, as might be expected, student browsing was more general in nature than that of arts historians. Students in the study "claimed they looked for images that broadened their knowledge about art or a specific area of art, that addressed specific creative problems and inspired or offered relief from creative block" (p. 448).[20] Students were also interested in survey material on the history of Western art and in material that described specific works of art in the artist's own words. However, when asked whether they sought help in finding appropriate library resources to meet their information needs, only a few students in the study admitted that they had asked a librarian for assistance. Most students worked on their own or relied on other students for help and advice on where to find information. Further studies have confirmed that while library resources appear to be important in the development of student artists, finding ways to support their information seeking remains a challenge for academic librarians.[19,32]

The question of how art historians have responded to the advent of online information and digital images has been addressed by several researchers. In 1995, results of a 2 year project to study the use of online databases at the Getty Center for the History of Art showed that "scholars' response to online searching was positive, if fraught with some anxiety" (p. 36).[15] However, scholars were not comfortable with the new technology and did not use it as a routine part of their information-seeking process. In 1999, a further Getty report on the information needs and seeking of scholars and artists in relation to multimedia materials[33] predicted that the speed of adoption of new technology by users of the Getty Institute would be slower than that shown by social scientists and natural scientists. This apparently limited use of technology by art historians is particularly surprising given the dependence on visual images as a source of information mentioned in many of the user studies cited above. Results of a survey conducted in 1999/2000 of 121 art historians based in British institutions of higher education indicated that less than half of the respondents (47.1%) felt that access to digital images had affected their work methods, while only 24.8% felt that digital images had affected their actual research interests.[34] Further studies have continued to confirm this impression of art historians[35,36] and conclude that "many have a rather limited awareness of electronic resources and haven't fully developed the skills to utilize them to their fullest potential" (p. 6).[37]

Music Scholars

In contrast to art historians, scholars working in the area of music have been the subject of only a few formal studies. Most frequently they are mentioned, if at all, in the context of other scholars working in the humanities[11,14,15,38] or as a member of the fine arts faculty.[22] Those instrumental instructors who are members of the fine arts faculties may also have secondary jobs as practicing musicians. Other areas of music scholarship include historical musicology, the study of music theory and composition, music education, ethnomusicology, and electronic music.[39] Like other humanities scholars, those working in the area of music are frequently interdisciplinary. The information resources they need include, but are not limited to "recordings and music scores including study scores, performance scores, and the scholarly collected editions" (p. 229).[22] Especially for music scholars who are not instrumental instructors, the information resources will also include material from other disciplines, both within and outside of the humanities.

A useful model of the research process used by music scholars has been developed by Brown.[40] Her model, based on interviews with 30 full-time noninstrumental music faculty was further tested through a survey of 175 music scholars in the United States and Canada. The model defines six stages in the research process, from idea generation to dissemination, and identifies information resources associated with each stage. The model, which built on research into the scholarly process of literary critics,[41] shares many characteristics in common with those describing the work of art historians and confirms several of the major findings already identified in the literature such as the predominately solitary nature of the scholarly process, the reliance on personal as well as institutional collections for information, and the importance of browsing in the information-seeking process.[26] Like Stam's art historians, these music scholars also used informal channels of communication to keep up with current trends in their fields. Although these scholars did use e-mail and discussion groups to stay in touch with colleagues, they preferred face-to-face communication and personal interactions at conferences.[42] Use of technology in general seems to have followed the distribution curve identified by Bates,[43] although this in itself is surprising given that the study was conducted 5 years after the report of the Getty End-User Online Searching Project was released. However, no more recent data are available on the use of technology by scholars in the arts.

Other Arts Scholars

Research on other academic user groups in the arts (dance, film, theatre, etc.) is difficult to find. Faculty from dance and theatre were included in the study by Reed and Tanner focused on improving library services for fine arts faculty.[22] Other insights into the information-seeking process of these scholars may be gleaned from the types of materials that are made available to them in specialized library collections.[44] Librarians in art, music, and other specialized collections have long been intimately involved in the scholarly activities of their users and strive hard to understand their needs.[45,46] However, an absence of user studies in this area has left a gap in our knowledge of how these users' information needs and information-seeking behavior is changing as a result of the proliferation of online information resources.

Fine Artists

Very few studies have been done on practicing artists as a user group,[47,48] and what exists can only be considered suggestive because of the small numbers that have been studied. A review of what literature there is about fine artists, including an annotated bibliography, may be found in the online version of Cowan's 2004 article.[49] As with the studies of scholars in the arts, what has been written about practicing artists has looked primarily at artists' use of libraries, both actual and potential, and has not focused on their broader information needs. Studies of this type include Cobbledick's interviews with four artists—a sculptor, a painter, a fiber artist, and a metalsmith—and Powell's interviews with nine studio artists.[4,50] Cowan's own study reports the results of a series of interviews with a single artist. While none of these studies produced generalizable results, they do point to some common characteristics among artists. Like other humanities user groups, fine artists tend to work alone and to browse extensively in search of inspiration or useful information. However, Cobbledick points out that, "the extent and importance of browsing can be overstated," since the artists she interviewed looked for information with a specific need in mind and browsed only within a limited range of material (p. 362).[4]

Musicians

User studies of practicing musicians are at least equally scarce and, like other studies of users in the arts, have focused primarily on the information needs that can be addressed by library services.[51,52] A 2006 study of composers of electroacoustic music notes that, not only are there no user studies of this "idiosyncratic group," but also "the larger community of composers of music" has not been addressed (p. 1 and 3).[53] Similar to the studies of fine artists, the findings about electroacoustic composers are based on a very small sample—five composers who were graduate students in American universities. However, an important characteristic of these users may have relevance for other groups of artists as well. While their creative output situates these composers in an arts discipline, their information needs include resources from computer science, engineering, and mathematics. Even though these types of interdisciplinary needs were not identified specifically in the studies of fine artists, the increasing use of computer technology in all forms of art makes it likely that other types of practicing artists would have similar information needs. In addition, artists of the generation just now entering into its most productive years are individuals who grew up with technology. The tools for their creative processes will most likely include a higher number of interdisciplinary resources than ever before.

Another significant characteristic of this user group is its dependence on Internet-facilitated information sources such as e-mail, Listservs, and Web sites. The composers in the study reported using Listservs to find information on technical questions, while Web sites of professional organizations provided information on conferences and festivals. Personal Web sites were used by the composers to find information on other composers and to communicate information about themselves. While communication with colleagues was seen as an important source of information for both scholars and artists,[4,12,42] the artists interviewed by Cobbledick in 1995 did not embrace the use of the available technology to support their information-seeking or communication activities. Whether this is a characteristic of fine artists compared to musicians or whether this is a generational effect requires further research to determine. An interesting insight into the creative process of musicians can be found in the discussion by Eaglestone, Ford, Brown, and Moore.[54]

Other Artists

Reports about the information needs of practitioners in other arts disciplines are largely anecdotal. National service organizations such as the American Composers Forum, American Music Center, Chorus America, Dance/ U.S.A., League of American Orchestras, Meet the Composer, National Performance Network, OPERA America, and Theatre Communications Group provide information services to both organizational and individual members as well as hosting conferences and meetings at which information is exchanged. Much of the information provided by these organizations is practical in nature, addressing the mundane needs of the users, such as employment opportunities, rather than supporting their creative activities. However, these organizations also provide fora for discussions of artistic issues, information about trends in the disciplines, and opportunities for networking with colleagues. Since these organizations serve a national membership, much of the information they provide is disseminated electronically through e-mail, Listservs, discussion boards, and organizational Web sites. Other organizations such as the Performing Arts Data Service (PADS) in the United Kingdom maintain Web sites that offer a wide variety of digital resources to researchers and other users in the areas of music, film, broadcast arts, theatre, and dance.[55] Most of these organizations keep statistics on user requests and make some effort to identify user needs. However, systematic studies are yet to be done in these areas.

IDENTIFYING AND RETRIEVING ARTS INFORMATION

Although simple text retrieval has long been the norm for traditional library literature, information retrieval in the arts is no longer primarily text-based. Other sources of information, especially digital images and sound, play important roles in satisfying the information needs of scholars and practicing artists. Each type of information— textual, graphic, and sound—presents different challenges to users looking for arts information.

Subject Retrieval

Subject retrieval (e.g., by a particular artist, work, or theme) has traditionally been one of the common ways in which users access material about the arts. Results from the Getty End-User Online Searching Project show that 49% of the scholars in the study used individual names as subjects in their searches and 57% used some other common term.[43] However, a persistent obstacle to building large-scale subject retrieval systems in the humanities has been the "diversity and complexity of textual, graphic, aural and artifactual sources used by scholars in their research" (p. 608).[56] According to Tibbo, "the interdisciplinary and unique research interests of humanistic scholars make convenient subject coverage with appropriate index terms challenging" (p. 616).[56] With the increase in full-text databases, especially of secondary material, there "seems to be an assumption that full-text searching is the solution to all information retrieval problems" (p. 228).[57] Unfortunately, full-text searching is not a panacea, and users often become discouraged or frustrated by the number and range of the items that are retrieved. This leads East to predict that, even though vast amounts of secondary material is now being made available digitally, "it seems all too likely that they [scholars] will use these digital collections to look for documents that they have identified form other sources rather than performing subject searches to discover further resources" (p. 239).[57] This is consistent with earlier findings that humanities scholars do not make regular use of abstracting and indexing tools except when "venturing into unfamiliar areas," but prefer to follow bibliographic references from sources already know to them (p. 204).[58]

Image Retrieval

The advent of searchable databases of digital images has dramatically changed the way users look for graphic materials. In 2000, all of the major information sources for the visual arts listed in Blazek and Aversa's guide were print materials.[1] Today, few students or scholars would rely on print material but would look instead for digital images.[21] There are now numerous large image collections available to researchers all over the world. For example, in 2001, The Andrew W. Mellon Foundation launched ARTstor, a collection of digital images available on the Web for use by scholars and educators.[59] The collection includes approximately 700,000 images in the areas of art, architecture, the humanities, and social sciences. ARTstor: Images for Education and Scholarship, see: http://www. artstor.org/index.shtml. Other major image databases exist, providing access to some of the most extensive library and museum collections in the world.[60] One result of the shift to digital images from print material is that "more and more users perform searches on their own," rather than relying on the help of human intermediaries such as art librarians (p. 696).[61] This trend has made research into user issues such as relevance criteria essential. How users look for images and evaluate what they find continues to be an area for fruitful investigation. Two studies funded by the British Institute for Image Data Research (IIDR) and Resource: The Council for Museum, Archives and Libraries have provided a foundation for further research.[62,63] Additional issues arise in the area of multimedia retrieval, which involves the retrieval of two or more media, i.e., text, image, or, sound.[62,64] In multimedia retrieval, as in image retrieval, the success of the process depends on the appropriate representation of the item content and the indexing tools available to describe it.

Melodic Retrieval

In the past, users looking for a particular piece of music needed to know the title, and preferably the composer, of the work they wanted. The difficulty of such a task may be imagined from the fact that the Music Division of the Library of Congress has several million pieces of published sheet music, most of which is not fully cataloged. The Library of Congress, Performing Arts Reading Room, see: http://www.loc.gov/rr/perform/askalib/aboutsheetmusic.html. Although music scholars and practicing musicians may have the expert knowledge to find the music sources they need, the general public often asks a librarian to identify a piece of music based on a few hummed bars. Since the 1990s, with the increase of digital music collections, information retrieval researchers have been studying the possibility of electronically searching musical databases by means of direct acoustical input.[65] An important characteristic of the "query-by-humming" approach is that the user of the system does not need to have expert knowledge of music notation or theory.[66] In its simplest form, query-by-humming allows the user to "sing a few bars and have all melodies containing that sequence of notes retrieved and displayed" (p. 11).[65] Approaches to melodic retrieval have been investigated by a number of researchers. Results of several studies may be found at the University of Illinois' music-ire Web site http://www.music-ir.org/. These results show that mismatches between the query and the desired retrieval target are still frequent.[67] Difficulties occur both because of the "limitations of amateurs in the vocal production of melody," (p. 688),[66] and because of the technical aspects of the retrieval process. However, work in this area is sure to become more sophisticated due to the increasing popularity of searching for music to download from the Web.

CONCLUSION

In the conclusion to her study on fine artists, Cobbledick wrote:

> Artists need to have access to the universe of knowledge, not merely to some of its parts, and libraries that would meet their information needs must become access points to that universe. The development of computer networking, full-text databases and electronically stored images can help make this access a possibility, but it remains to be seen how the artist community as a whole will respond to these new sources of information once they become more readily available (p. 365).[4]

It is unlikely that in 1996, Cobbledick could have foreseen how soon this prophesy would come true. Over the past 15 years, the impact of the World Wide Web as a growing source of information has been felt in the arts as in every other aspect of life. With an ease that was unthinkable when the early studies of users in the humanities were conducted, and was still unfamiliar when the 1995 report from the Getty Online Searching Project was released, scholars and practicing artists can now search for information from the greatest libraries and museum collections without ever leaving their homes. However, retrieval of textual, image, and sound data still poses significant challenges for the user. Research into various types of retrieval systems as well as the broad range of users' information needs is still necessary. Traditional user studies in the arts have focused primarily on scholarly research, with some exploratory studies into the information-seeking behavior of practicing artists. While these have provided insights, research is needed to investigate more fully the contexts in which users search for information as well as the barriers that exist to meeting their information needs.

REFERENCES

1. Blazek, R. Aversa, E. *The Humanities: A Selective Guide to Information Resources*, 5th Ed. Libraries Unlimited: Littleton, CO, 2000.
2. National Endowment for the Arts, O. o. R. &. A, National Endowment for the Arts: Washington, DC, 2006; Report no. Note #90.
3. Vasari, G. *The Lives of the Most Excellent Painters, Sculptors, and Architects*, Modern Library: New York, 2005; Notes: Original work published 1550.
4. Cobbledick, S. The information-seeking behavior of artists: Exploratory interviews. Libr. Quart. **1996**, *66*(4), 343–372.
5. Zach, L. When is 'enough' enough? Modeling the information-seeking and stopping behavior of senior arts administrators. J. Am. Soc. Inform. Sci. Technol **2005**, *56*(1), 23–35.
6. Corkill, C. Mann, M. *Information Needs in the Humanities: Two Postal Surveys*, Centre for Research on User Studies: Sheffield, U.K., 1978; BLR&DD Report no. 5455.

7. Corkill, C. Mann, M. Stone, S. *Doctoral Students in the Humanities: A Small-Scale Panel Study of Information Needs and Uses 1976-79*, Centre for Research on User Studies: Sheffield, U.K., 1981; BLR&DD Report no. 5637.

8. Weintraub, K.J. The humanistic scholar and the library. Libr. Quart. **1980**, *50*(1), 22–39.

9. Stone, S. Humanities scholars: Information needs and uses. J. Doc. **1982**, *38*(4), 292–313.

10. Fulton, C. Humanists as information users: A review of the literature. Aust. Acad. Res. Libr. **1991**, *22*(3), 188–197.

11. Lonnqvist, H. Scholars seek information: Information-seeking behavior and information needs of humanities scholars. Int. J. Inform. Libr. Res. **1990**, *2*(3), 195–203.

12. Palmer, C.L.; Neuman, L.J. The information work of inter-disciplinary humanities scholars. Libr. Quart. **2002**, *72*(1), 85–117.

13. Watson-Boone, R. The information need and habits of humanities scholars. Res. Quart. **1994**, *34*(2), 203–216.

14. Wiberley, S.E.; Jones, W.G. Patterns of information seeking in the humanities. Coll. Res. Libr. **1989**, *50* (November), 638–345.

15. Bates, M.J.; Wilde, J.N.; Siegfried, S. Research practices of humanities scholars in an online environment: The Getty Online Searching Project Report no. 3. Libr. Inform. Sci. Res. **1995**, *17*, 5–40.

16. Ellis, D.; Oldman, H. The English literature researcher in the age of the internet. J. Inform. Sci. **2005**, *31*, 29–36.

17. Reynolds, J. A brave new world: User studies in the humanities enter the electronic age. Ref. Libr. **1995**, *49–50*, 61–68.

18. Wiberley, S.E.; Jones, W.G. Time and technology: A decade-long look at humanists' use of electronic technology. Coll. Res. Libr. **2000**, *61*(5), 421–431.

19. Bennett, H. Bringing the studio into the library: Addressing the research needs of studio art and architecture students. Art Doc. **2006**, *25*(1), 38–42.

20. Frank, P. Student artists in the library: An investigation of how they use general academic libraries for their creative needs. J. Acad. Libr. **1999**, *25*(6), 445–55.

21. Lorenzen, E.A. Selecting and acquiring art materials in the academic library: Meeting the needs of the studio artist. In *Selecting Materials for Library Collections*; Fenner, A., Ed.; Haworth Information Press: Binghamton, NY, 2004; 27–39.

22. Reed, B.; Tanner, D.R. Information needs and library services for fine arts faculty. J. Acad. Libr. **2001**, *27*(3), 229–233.

23. Layne, S.S. Artists, art historians, and visual art information. Ref. Libr. **1994**, *47*, 23–36.

24. Duff, W.M.; Johnson, C.A. Accidentally found on purpose: Information seeking behavior of historians in archives. Libr. Quart. **2002**, *72*(4), 472–496.

25. Foster, A.; Ford, N. Serendipity and information seeking: An empirical study. J. Doc. **2003**, *59*(3), 321–340.

26. Stam, D.C. How art historians look for information. Art Doc. **1997**, *16*(2), 27–30.

27. Allen, T.J. *Managing the Flow of Technology: Technology Transfer and the Dissemination of Technological Information with the R&D Organization*, The MIT Press: Cambridge, MA, 1977.

28. Spanner, D. Border crossings: Understanding the cultural and information dilemmas of interdisciplinary scholars. J. Acad. Libr. **2001**, *27*(5), 352–360.

29. Bakewell, E. Beeman, W.O. Reese, C.M. *Object, Image, Inquiry: The Art Historian at Work*, The Getty Art History Program: Santa Monica, CA, 1988; Notes: Report on a collaborative study by the Getty Art History Information Program (AHIP) and the Institute for Research in Information and Scholarship (IRIS) Brown University.

30. Brilliant, R. How an art historian connects art objects and information. Libr. Trends **1988**, *37*(3), 120–9.

31. Van Zijl, C.; Gericke, E.M. Information-seeking patterns of artists and art scholars at the Vaal Triangle Technikon. S. Afr. J. Libr. Inform. Sci. **1998**, *66*(1), 23–33.

32. Latimer, K. SOS (self-help or spoonfeeding): teaching students the art of retrieving architectural information. Art Libr. J. **2002**, *27*(1), 5–8.

33. Bates, M.J. Information needs and seeking of scholars and artists in relation to multimedia materials. University of California, 2001.

34. Bailey, C. Graham, M.E. *The corpus and the art historian*, 2000; In paper presented at the Thirtieth International Congress of the History of Art. Art History for the Millennium: Time. Section 23 Digital art history time.

35. Beaudoin, J. Image and text: A review of the literature concerning the information needs and research behaviors of art historians. Art Doc. **2005**, *22*(2), 34–37.

36. Rose, T. Technology's impact on the information-seeking behavior of art historians. Art Doc. **2002**, *21*(2), 35–42.

37. Elam, B. Readiness or avoidance: E-resources and the art historian. Collect. Build. **2007**, *26*(1), 4–6.

38. Case, D.O. Collection and organization of written information by social scientists and humanities: A review and exploratory study. J. Inform. Sci. **1986**, *12*(3), 97–104.

39. Rebman, E. Music. *The Humanities and the Library*, 2nd Ed.; Couch, N., Allen, N., Eds.; American Library Association: Chicago, IL, 1993; 132–172.

40. Brown, C.D. Straddling the humanities and social sciences: The research process of music scholars. Libr. Inform. Sci. Res. **2002**, *24*, 73–94.

41. Chu, C.M. *The Scholarly Process and the Nature of the Information Needs of the Literary Critic: A Descriptive Model*, University of Western Ontario: Canada, 1992; Unpublished doctoral dissertation.

42. Brown, C.D. The role of computer-mediated communication in the research process of music scholars: An exploratory investigation. Inform. Res. **2001**, *6*(2).

43. Bates, M.J. The Getty End-User Online Searching Project in the humanities: Report no6: Overview and conclusions. Coll. Res. Libr. **1996**, *57*(6), 514–523.

44. Palazzola, B. American modern dance: Primary sources in book form at the University of Michigan Music Library. Collect. Build. **1991**, *11*(3), 20–24.

45. Clegg, S.M. User surveys and statistics: The opportunities for music libraries. Fontes Artis Musicae **1985**, *32*, 69–75.

46. Stam, D.C. Tracking art historians: On information needs and information-seeking behavior. Art Libr. J. **1989**, *14*(3), 13–16.

47. Bates, M.J. Learning about the information seeking of interdisciplinary scholars and students. Libr. Trends **1996**, *44*(2), 155–164.

48. Case, D.O. *Looking for Information: A Survey of Research on Information Seeking, Needs, and Behavior*, Academic Press: Amsterdam/New York, 2002.

49. Cowan, S. Informing visual poetry: Information needs and sources of artists. Art Doc. **2004**, *23*(2), 14–20 Notes: The full version of the paper, with a transcript of the artist interview and an annotated bibliography, may be found at http://www.geocities.com/xelazulu/.

50. Powell, E.F. *Information Seeking Behaviors of Studio Artists*, University of North Carolina at Chapel Hill: Chapel Hill, NC, 1995; Unpublished doctoral dissertation.

51. Gottlieb, J. Reference service for performing musicians: Understanding and meeting their needs. Ref. Libr. **1994**, *47*, 47–59.

52. Narveson, L. *The Information Needs and Seeking Behaviors of Amateur Musicians: A Qualitative Study*, University of North Carolina at Chapel Hill: Chapel Hill, NC, 1999; Unpublished doctoral dissertation.

53. Hunter, B. A new breed of musicians: The information-seeking needs and behaviors of composers of electroacoustic music. Music Ref. Serv. Quart. **2006**, *10*(1), 1–15.

54. Eaglestone, B.; Ford, N.; Brown, G.; Moore, A. Information systems and creativity: An empirical study. J. Doc. **2007**, *63*(4), 443–464.

55. Duffy, C.; Owen, C. The view from the performing arts: Workshop on scholarly use of digital resources in the performing arts. New Rev. Acad. Libr. **1998**, *4*, 182–184.

56. Tibbo, H.R. Indexing for the humanities. J. Am. Soc. Inform. Sci. **1994**, *45*(8), 607–619.

57. East, J.W. Subject retrieval from full-text databases in the humanities. Libr. Acad. **2007**, *7*(2), 227–241.

58. Green, R. Locating sources in humanities scholarship: The efficacy of following bibliographic references. Libr. Quart. **2000**, April *70*, 201–29.

59. Marmor, M. ArtSTOR: A digital library for the history of art. J. Libr. Admin. **2002**, *27*(3), 26–29.

60. Chen, H.L. A socio-technical perspective of museum practitioners' image-using behaviors. Emerald Libr. **2007**, *25*(1), 18–35.

61. Choi, Y.; Rasmussen, E.M. Users' relevance criteria in image retrieval in American history. Inform. Process. Manage. **2002**, *38*, 695–726.

62. Cooniss, L.R. Ashford, A.J. Graham, M.E. *Information Seeking Behaviour in Image Retrieval*, British Library: London, 2000.

63. Cooniss, L.R. Davis, J.E. Graham, M.E. *A User-Oriented Evaluation Framework for the Development of Electronic Image Retrieval Systems in the Workplace*, British Library: London, 2003.

64. Auffret, G.a.P.Y. Managing full-indexed audiovisual documents: A new perspective for the humanities. Comput. Hum. **1999**, *33*, 319–344.

65. McNab, R.J. Smith, L.A. Witten, J.H. Henderson, C.L. Cunningham, S.J. Towards the digital music library: Tune retrieval from acoustic input Proceedings of the first ACM International Conference on Digital Libraries; Fox, E., Marchionini, G., Eds.; ACM: New York, 1996; 11–18.

66. Dannenberg, R.B.; Birmingham, W.P.; Pardo, B.; Hu, N.; Meek, C.; Tzanetakis, G. A comparative evaluation of search techniques for query-by-humming using the MUSART testbed. J. Am. Soc. Inform. Sci. Technol. **2007**, *58*(5), 687–701.

67. Pardo, B.; Shifrin, J.; Birmingham, W. Name that tune: A pilot study in finding a melody from a sung query. J. Am. Soc. Inform. Sci. Technol. **2004**, *55*(4), 283–300.

ASLIB

David Bawden
Lyn Robinson
City, University of London, London, U.K.

Abstract

ASLIB was, from 1924 to 2010, an independent membership organization for special librarianship, technical and commercial information work, and latterly for information management. It was highly influential in the development of documentation and information science, in the United Kingdom and worldwide. Its activities included research and consultancy, training, professional development, publishing, and technology development. ASLIB was for many years the *de facto* UK center for information research, especially information and library management, information organization, and computer applications. It has had several names, being at times the Association of Special Libraries and Information Bureaux, ASLIB, the Association for Information Management, and ASLIB. In 2010, ASLIB became a part of Emerald Group Publishing, and activities ceased in 2016.

INTRODUCTION

ASLIB was, from 1924 to 2010, an independent membership organization for special librarianship, technical information work, and information management. It was highly influential in the development of documentation and information science. Since 2010, it has been a part of Emerald Group Publishing. It has had several names, and name variants, over the years, and for convenience it is referred to as ASLIB throughout this entry. This entry is structured into five chronological sections, reflecting major changes in ASLIB's status.

ORIGINS

ASLIB was founded as a direct result of a conference held at Hoddesdon, Hertfordshire, England, in September 1924, at which delegates from British special libraries and information bureaux met to discuss the desirability of an association to represent their interests. For a full account of its origins, see Muddiman[1,2] and Hutton,[3] and for personal recollections, see Pearce.[4]

Its origins, like much of the information environment emerging at that time, can be traced to the "second industrial revolution," beginning about 1870, in which new industries developed, based on the rapid developments in chemistry, materials science, metallurgy, electricity, and precision engineering. These industries were "information intensive," needing access to knowledge from scientific research, to data of all kinds, and to commercial and technical intelligence. Experience in the 1914–1918 war had shown the necessity for unimpeded access to such information, coordinated at national level.[4,5]

This led to the need for individuals and departments who could provide such information, actively and in detail, in a way different from traditional libraries. Inspiration for such work was provided by the documentation movement, pioneered in Continental Europe by Paul Otlet and Henri Lafontaine.[6,7] This used novel technical methods, particularly index cards, and intellectual tools, especially classification, to provide access to information at much finer levels of granularity then heretofore. (It may be noted that information technology and information organization have remained major themes throughout ASLIB's activities.) Companies, government-supported research associations for specific areas of science-based industry, universities, and larger public libraries throughout Britain began to set up special libraries, intelligence bureaux, and technical information sections. These efforts were largely uncoordinated and without any cooperation. Some initiatives had been undertaken, including a Conjoint Board of Scientific Societies founded by the Royal Society, which had advised on library/information matters between 1916 and 1923, and conferences organized by the government Department of Scientific and Industrial Research (DSIR) on these themes, but there was little to show for these efforts.

It was to attempt to improve this situation that the Hoddesdon conference was convened from September 5 to 8, 1924, at the initiative of staff from metallurgy research associations.

The conference was stated to be "open to all men and women who need to utilize information systematically, or who are interested in the conduct of information bureaux, intelligence services and special libraries." It is interesting to note that from this earliest stage ASLIB was open to information users, and those with any sort of interest in information matters; it was never envisaged as an organization solely for what would come to be termed "information professionals." Eighty-four delegates attended and resolved to appoint a standing committee of 16 members who had the task of creating an organization to promote cooperation and mutual assistance among information departments, to come into being within two or three years. Discussions with the (UK) Library Association about possible affiliation came to nothing; a story that was to be repeated 70 years later. A second conference in 1925 with over 200 delegates, addressed by Paul Otlet, showed continuing enthusiasm. At a meeting in London on March 29, 1926, the Association of Special Libraries and Information Bureaux, with the acronym ASLIB, was created, as an independent voluntary association. The association was formally incorporated on November 30, 1927. Its main objective was stated to be "To facilitate the co-ordination and systematic use of sources of knowledge and information in all public affairs and in industry and commerce and in all the arts and sciences . . . ."

1924–1948

This period in ASLIB's history is examined in detail by Muddiman[1,2]; see also Ditmas.[8]

The new association established itself in the Bloomsbury area of London, initially in Bloomsbury Square, and from 1928 in larger premises in Bedford Square, with a small office staff. In 1931, with the economy faltering and income dropping, there was a move to smaller offices in Russell Square. Its secretary at this point was Edith Ditmas, a formidable woman to whom Muddiman[1] attributes a major role in ASLIB's survival in these difficult times. (Muddiman notes that the *Daily Mirror* newspaper dubbed her the "Woman Oracle of Russell Square".) Ditmas[8] (p. 269) contrasted ASLIB's position at "the centre of a spider's web of contacts, each with a vast range of sources of information" with its situation as "an organisation whose only visible assets were one small room on the top floor of a tall Bloomsbury house, some ill-assorted office equipment which so filled the room that the wall-cupboards could only be opened when some of the furniture was moved to the landing, two typewriters and a typist." The association lived in these straitened circumstances throughout the 1930s, moving to slightly larger premises in Museum Street in 1936, while managing to develop and extend its activities. Further discussions were held with the Library Association between 1928 and 1930 as to whether some form of amalgamation might be possible but in the end failed, due to partly ASLIB's insistence that its interests were in the organization, management, and use of information, rather than in libraries *per se*. The nature of the ASLIB membership, institutional rather individual, and catering for users of information as much as for those directly involved in its management, also made integration problematic.

Its funding came largely from membership subscriptions, with limited support from grants from the Carnegie United Kingdom Trust. Membership, almost entirely institutional rather than individual, was 220 in 1927: the largest sections of membership were private sector companies (53) and scientific or technical associations (50) but also included were research associations (18), higher education institutions (18), and public libraries (17). Membership rose to over 400 by 1932, dropped to 280 in 1934 under the influence of economic depression, and had risen to 1043 by 1949, but the general balance of the membership categories remained largely constant throughout. Although ASLIB's office remained London-based, regional branches were formed in the 1930s, two for the north of England, in Lancashire and Cheshire and in Yorkshire, and one for London and the Home Counties in southeast England. Internationally, ASLIB became closely tied into the pan-European efforts to develop the documentation movement into what became in 1937 the Federation International de Documentation (FID). From its earliest days, it had close links with the British Society for International Bibliography, an affiliate member of Otlet's International Institute of Bibliography. The two organizations worked closely together on issues of classification and indexing, and on promoting documentation principles. There were also links with the United States: Hutton[3] notes that a representative from the U.S. National Research Council was present at the Hoddesdon conference, and thereafter there were few ASLIB meetings without an American representative or contribution.

ASLIB's main objectives throughout this period were to foster cooperation within the special library sector, to enable interaction and exchange of experience between those working in the area, to act as a clearinghouse for specialized information, and to promote, and lobby for, the wider dissemination and better use of published information. While there was never any intention of ASLIB's creating its own central collection of information, its first chairman, J.G. Pearce of the British Cast-Iron Research Association, suggested that it should promote a "free trade in non-confidential information" and thereby "unlock the intellectual capital of Britain."[4]

The ASLIB annual conferences, initially attracting 200 delegates from the United Kingdom and overseas, rapidly became a fixture, known for their successful mix of the theoretical/technical and the practical/vocational. Their topics covered developments in bibliography, documentation and classification, technical advances such as microforms and reprographic techniques, and planning and management processes. Most notable was the conference of 1938, held jointly with the FID with large numbers of international delegates, meeting under the shadow of the Munich crisis, and which may be seen as foreshadowing the end of the internationalist documentation movement.[9]

An enquiry service was launched at an early stage, offering advice by telephone on the best sources for specialized information. A register of specialist translators was established in 1931, an indicator of an area in which ASLIB was to be heavily involved for several decades.

Publications was also an early area of activity: the first *ASLIB Directory: a guide to sources of specialised information in Great Britain* appeared in 1928; the *Aslib Information* newsletter in 1929, initially quarterly, and later monthly; and in 1935, the quarterly *ASLIB Book List*, an alerting service of significant new scientific and technical books. From the mid-1930s, publication of reference works began, typically annual listings of technical books, directories, and yearbooks.

Education and training for special libraries was also a concern for ASLIB, and attempts were made from 1929 to persuade educators to give this greater attention. As a result of ASLIB's lobbying, the library school at University College London introduced an elective special libraries course that ran intermittently throughout the 1930s, while the Library Association included the topic in the syllabus for its Fellowship examinations, though not for the more widely followed Associate examinations. This was not regarded as satisfactory, and education and training for the special library and technical information sector was to become a major theme for ASLIB in the following decades.

The 1939–1945 conflict initially caused predicable problems for ASLIB, as its regular activities and membership decreased due to wartime conditions. However, in the longer term, its contributions to the war effort greatly increased its profile, and established it on its postwar path. There was a need to keep up, and increase, the flow of information for industrial and military purposes, in face of the disruption to the normal publication and communication channels, due to shortages of materials and personnel and the cessation of communication with continental Europe, as well as potential and actual destruction of collections through bombing. In 1941, ASLIB was funded by the Royal Society and the Rockefeller Foundation to research the problems and propose solutions. This led to ASLIB designing and managing a service to obtain difficult-to-obtain materials of all kinds, generally those originating in hostile or occupied countries, obtained through neutral nations, reproduce them on microfilm, and distribute them to Allied scientific, industrial, and military recipients worldwide. Funded by the UK and U.S. governments, and based at the Science Museum in London, the service, kept secret throughout the period of its operation, made a considerable contribution to the war effort and was maintained after the war's end, management being passed on to the Royal Society of Medicine.[10,11] Other contributions included a series of *War Time Guides to Specialized Sources of Information* in important technical and industrial areas (essentially sections of the *ASLIB Directory*, revised and updated to reflect the wartime situation), and an expanded enquiry service. A series of short training courses in special librarianship were also run from 1943 onward and proved very popular. ASLIB's first textbook, a *Manual of Special Library Technique*, was published to supplement these courses. The value to the nation of these activities was recognized by the British government, which, in 1944, made a substantial grant to ASLIB over a five-year period, via the DSIR.

In the immediate postwar period, and with its newly found success, ASLIB reflected on its future course.[1] Should it become an individual member professional association, or focus on a role as a national documentation center? If the latter, should it remain independent, or seek to become a part of a monolithic state-run information institute? Eventually, it settled upon development as a partly state-supported independent entity, focusing on a continuation of its role as a coordinator and facilitator of the national system of special libraries and technical information centers. Its high status as a result of its wartime contributions saw ASLIB participating in numerous committees and official bodies relating to information and documentation, and it took a significant role in the highly influential Scientific Information Conference organized by the Royal Society in 1948.

With this aim made possible by an increased membership, and a government grant linked to membership income, ASLIB expanded its activities considerably in the late 1940s, with its membership rising to over 1000 and its staff to 16. Its main areas of focus were enquiries (with over 3,000 telephone enquiries and over 30,000 written enquiries *per annum* by the end of the decade); conferences and education; expansion of international activities; promotion of, and lobbying for, the importance of special libraries and technical information; and publications.

In respect of publications, in 1947, ASLIB launched an academic and theoretical journal, *Journal of Documentation*, to complement its more ephemeral and practical serial publications. The following year, *Aslib Proceedings* was launched, for the publication of papers given at conferences of ASLIB and related organizations.

In the education area, the 1948 ASLIB annual general meeting passed a resolution agreeing to develop a postgraduate qualification in information work. This ambitious proposal was never followed up, and instead ASLIB embarked on creating an extensive program of short training courses.

Muddiman[1] sums up the situation at the end of this phase of ASLIB's life by agreeing with its claim that it was generally recognized as the main British clearinghouse for specialized information and had scope for expanding this role in several ways in an increasingly information-conscious environment. It was at the center of the emerging field of documentation, which would develop into information science.

Artificial–Association

1949–1980

Issues in the period have been reviewed in part by Muddiman,[2] and by Wilson,[12] who draws attention to the way in which ASLIB reached out to those beyond the library world with an interest in information and focused on information management as an integral part of management as a whole.

As a response to its new situation, ASLIB transformed itself into a new organization with a new corporate structure, and a new name. Proposals for a "new ASLIB" had been announced in 1948, and on May 24, 1949, the association formally combined with the British Society for International Bibliography, with which it had long had close links. The new association was simply titled "ASLIB." The joining of the two organizations, one with a largely institutional membership and one with exclusively individual members, gave a unique flavor to the organization. So too did the mix of funding: partly direct government support, partly membership subscriptions, and partly fees for the increasingly wide range of services provided. The 1977 annual report comments that, because of this mixed membership, people had different ideas about what ASLIB was, decades after its restructuring.

Edith Ditmas retired as director at this point, to devote herself to study and research, and also to do editorial work for ASLIB. She was succeeded as Director by Lesley Wilson who saw himself as a modernizing manager, and who steered ASLIB through the growth of the next three decades.

With the security of a further five-year grant-in-aid from the DSIR, matching membership subscriptions, ASLIB moved to larger premises in Palace Gate in west London, to cater for the increased staff numbers: 16, serving a membership of just over 1000. The new building gave the director an individual office for the first time, together with a glass-enclosed veranda used for meetings. The membership at that time comprised 32% research associations and learned societies, 20% industrial and commercial, 20% overseas members (uncategorized), 9% UK government departments, and 19% individuals, and it grew steadily year on year, reaching a highpoint of nearly 3000 by 1966. Staff numbers increased to match, reaching 29 by 1955, 51 by 1964, and attaining a maximum of 69 in 1971, reflecting the much wider range of activities being undertaken by then. The director always had a major influence on the organization's direction, advised by a council largely comprising senior figures from the information world, from the scientific establishment, and from industry. The first chair of the council in the new ASLIB structure was Harry Hyams of Shell Petroleum; three decades later, his nephew, Montagu Hyams of Derwent Publications, was to be appointed vice president of the ASLIB Council. In a manner typical of a British organization of the time, its council always included a proportion of Honorary Members, representing the upper strata of society.

In 1958, it could count an earl, a baroness, and four knights among its members; 10 years later, it had a viscount, two lords, three knights, and a vice admiral.

Although ASLIB remained headquartered in London throughout the period, regional branches for Northern England and for Scotland were established in the new structure, joined by a branch serving the English Midlands in 1951. These regional branches remained unchanged throughout the life of ASLIB as an independent organization. An ASLIB office opened in Birmingham, in the Midlands region, in 1968, sharing facilities with other organization and staffed by one liaison officer. Overseas membership remained significant, at least 20% of the total, throughout the period, and there were proposals for an overseas branch structure. Nothing came of this, perhaps due to the wide geographical spread of membership, with over 70 countries generally represented.

A new initiative was the setting up of special interest groups among the membership: in 1951, groups were set up for aerospace, textiles, economics, and food and agriculture, and in the following two years these were joined by engineering, fuel and power, and chemicals. This set of groups was fluid through this period, as groups formed, merged, renamed themselves, and disbanded. Most common were groups, liked those named above, devoted to information in a particular industry or sector; they were later joined by groups for the furniture industry, for film libraries, for biosciences, for social sciences, for transport, and for electronics. Other groups focused on processes and techniques, such as the technical translation, mechanization (later computer applications), and coordinate indexing (later informatics) groups. Only one, the One-Man Band group, brought together those working in similar environments (a note in the February 1990 issue of *Aslib Information* pointed out that although women were in a majority of the membership and the committee, the group liked the imagery of the male name, rather than the gender-neutral One-Person Library). Some of the larger groups, with memberships of several hundred, had activities and reach equivalent to that of professional bodies in their own right, and some undertook original research; see, for example, the Informatics group's study of the indexing process.[13] The branches and groups contributed greatly to ASLIB's activity, with extensive programs of events and publications complementing the central provision.[14]

Conferences and meetings were a major preoccupation. The ASLIB annual conference became one of the major events of the year for the information professions, and a great variety of both regular and *ad hoc* conferences and meetings were organized, both centrally and by groups and branches, and covering the whole gamut of topics relevant to the wide interests of the ASLIB membership. A few themes are constant throughout the period: management processes and costs; demonstrating the value of information; planning and control of operations; document handling and reproduction; subject-specific resources;

classification and indexing; dealing with publishers and other suppliers; new technologies and research results; translations and the language barrier; and reports of international developments. A particularly notable venture was the organization, on behalf of FID, of the influential Dorking conference on classification in information retrieval.

Training programs were developed further, the main offerings being courses for newcomers to special library and information work at either junior or senior levels, soon joined by a course of intermediate-level staff.

In all, there were typically at least 10 ASLIB events, including training courses, per month throughout this period, rising at times to 20.

The publications program continued, based around the four periodicals—*Aslib Information*, *Aslib Proceedings*, *Journal of Documentation*, and *Aslib Book List*—regularly updated directories, and bibliographies and guides to information sources in specific subjects. An *Index to Theses* from British universities commenced annual publication in 1953. The register of specialist translators, which in 1950 included 126 names, was complemented by an *Index to Scientific and Technical Translations*, later renamed the *Commonwealth Translations Index*. A similar register of specialist indexers was established later but was never so successful as that for translators; this was attributed to the competing services of the Society of Indexers, whereas ASLIB was alone in providing access to translators.

A series of ASLIB manuals, providing detailed and practical advice on special library topics, was established. In 1956, the first issue of the highly regarded and influential *Aslib Handbook of Special Librarianship and Information Work* was published. In subsequent years, the publication program expanded to encompass textbooks and monographs on a variety of information-related topics.

The enquiry service, renamed later as the information service, continued its work of providing answers and referral to technical queries. It was complemented by a new in-house ASLIB library, with a collection of journals, books and pamphlets, and facilities for document reproduction. This dealt with queries on library/information issues. To give an idea of the early scale of operations, in 1950, 100 periodicals were taken in the library and 271 books and 228 pamphlets added to the stock. There were 138 personal enquiries, 868 by post, 1310 by telephone, 718 loans of material, and 94 requests for translations. As an extension of this kind of work, a consultancy service was established in 1952, to deal with enquiries needing considerable work to solve. The use made of the service grew dramatically, so that in 1956 more than 30,000 requests were made, attributable in part to a large number of new members from small or one-person information departments.

Lobbying and promotion activities, as throughout all of ASLIB's existence, involved a continual process of direct contacts with government ministries and agencies, and with international organizations, on many information-related topics. ASLIB were also represented on numerous national and international committees and working groups of many organizations: the British National Committee on Documentation, British Standards Institute, International Standards Organization, National Book League, British National Bibliography, UNESCO, the British National Committee of UNISIST, and IFLA, to name but a few. ASLIB also served as the British representative to FID and in a joint panel with the Confederation of British Industry considering industrial information issues.

In 1956, a "staff employment register" was established by the membership department. This later developed into a full recruitment agency activity.

The last of ASLIB's major activities was initiated in 1957, when a research committee was formed, and produced an outline for a research program into aspects of library and information work; this covered a wide area, including library procedures, information retrieval, costs and effectiveness of services, indexing, bibliographic control, abstracting services, and access to non-English material. From the start, it had an emphasis on rigorous research with clear practical application that became the hallmark of ASLIB's research in later years. The research department was set up in the following year, its first remit being the oversight of the Cranfield retrieval experiments on behalf of the funders, the U.S. National Science Foundation. Projects over the next few years included studies of the language barrier, on interlibrary loan in special libraries, on abstracting journals and on gray literature, and a major study of information users in science and technology.

As early as 1955, with 29 staff and 1187 members, there had been complaints that the ASLIB headquarters was cramped, so that, for example, members could not read in the library. Two years later, with 32 staff and a membership doubled to 2254, the situation became impossible, and in 1958 ASLIB moved to its own buildings in the imposing surroundings of Belgrave Square, in London's West End. The move was supported by a government grant and donations by members to a building fund. The effort was worth it, as the following year's annual report concluded that the building provided, for the first time in ASLIB's history, fully satisfactory conditions for staff, committee meetings, courses (which could now be held in-house), the library, and other activities.

The new building, together with a further DSIR five-year grant and a steadily increasing membership, led to a general expansion of activities. To the existing training courses, which provided a general introduction for new special library staff, were added specialized courses on a range of topics including patents information, classification, Eastern European material, and document production. The courses typically lasted between three and five days, indicative of a less time-pressured age. The space in the new building allowed courses to be complemented by demonstrations and exhibitions of equipment and procedures, such as photocopying and journal binding. Training was now seen as so significant that in 1960 the post of education officer was established to plan new courses. ASLIB also joined in discussions with the UK Institute of Information Scientists on formal training for those engaged in technical information work and endorsed the syllabus of the two-year evening course at Northampton College of Advanced Technology (later City University London), which later became the MSc Information Science.[15] In 1963, ASLIB produced a leaflet on "Information work as a career," aimed at school leavers and college students, which was effective for many years in bringing more publicity to what had been a little known profession.

By the time ASLIB's 40th anniversary was celebrated in 1964, it was approaching its peak in term of size, with a full staff complement of 70 serving a membership of nearly 3000. It was a major force on the national and international information scene, in particular in respect of its conferences, publications, and research, as well as its lobbying and influencing.

Peak member numbers were reached in 1966. The annual report covering that year recorded 2887 members: 1014 industrial and commercial members, 205 from UK government departments and agencies, 272 public and national libraries, 459 universities and colleges, 403 other nonprofit organizations, and 534 individuals. Thereafter, a gradual decline in membership set in; by the end of this period, in 1980, the total membership was 2082.

The nature of government support for ASLIB changed around this time. A new government department, the Office for Scientific and Technical Information (OSTI), was formed in 1965, partly a result of ASLIB's representations for the need for long-term government policies toward information services. One of OSTI's first initiatives was a joint investigation with ASLIB on ways of strengthening ASLIB's research, consultancy, and training activities. One result was a decision to focus on computer applications. Another was a strong steer toward applied rather than basic research, and to make a closer link between research and consultancy. The report recognized ASLIB as a prime mover in information research.

Thereafter, government support for ASLIB came from OSTI, and later from the British Library's Research and Development Department, and was focused on supporting specific activities and programs, rather than the previous open grant-in-aid arrangements. As a result, the research department expanded considerably, and produced a remarkably wide body of results,[16] becoming the *de facto* national center for research in information science and documentation. In 1966, it had a staff of 15, divided into teams focusing on mechanization, operations, and surveys. It also offered consultancy and advice services, which brought in a significant income, and contributed

books and reports to ASLIB's publishing program. The following year it moved into its own premises in Hobart Place, close to the main ASLIB building, with new projects on cost-effectiveness, on bibliographic records, on computer typesetting, on clerical processes in libraries, and on the use of metals information. In 1970, it changed its name to the Research, Development, and Consultancy Division (in 1979, this would become simply Research and Consultancy), reflecting a renewed focus on helping members to solve their information problems, with a particular focus on cost-effectiveness and on new technology. In 1975, the department, now with 17 research staff and 5 full-time consultants, moved to still larger premises in Bedford Row, Bloomsbury.

The training program also expanded along the lines recommended in the OSTI report, launching courses in mechanization, microforms, dissemination of information, systems design, information retrieval, and thesaurus construction, as well as information for subject-specific areas such as business, finance, food and agriculture, pharmacy, and medicine. The training program took on, for the first time, an international dimension: a course was held in Nairobi for participants from several East African countries, and other courses were organized in London for visiting groups from overseas.

The events program diversified further. The annual conference was held outside the United Kingdom for the first time, in the Netherlands in 1966 and in Germany in 1971. An annual ASLIB lecture series was started.

The publications program also diversified, with a number of new monographs on topics such as mechanization, classification, and thesaurus construction, a new *ASLIB Reader* series, bringing together significant writings of topics of central importance, and a range of directories and bibliographies. A third, and much expanded, edition of the *ASLIB Handbook* was produced. The contents list for this handbook give a good indication of what ASLIB considered to be core elements of its area of operation at this stage:[17] the special library and information service, administration, selection and acquisition, classification, information retrieval, filing and storing, technical report literature, library planning, service routine, subject enquiries and literature searching, abstracting, publications of the information department, mechanical aids in library work, and organizations in the special library field.

A new *Aslib Occasional Publications* series was started to disseminate reports from the research department: early examples dealt with the changing nature of the scientific journal, the use made of technical libraries, and the library/information literature. A new journal, *Program*, dealing with library automation and formally published by the School of Library Studies at the Queen's University Belfast, was incorporated into ASLIB's periodicals portfolio, as was the *Technical Translators Bulletin*, formerly published by the special interest group, and *Audiovisual Librarian*, previously published by the Audiovisual group.

Publication of *Forthcoming International Scientific and Technical Conferences* was taken over from the UK government Department of Education and Science. In 1969, ASLIB partnered with the Library Association in producing the new *Library and Information Science Abstracts* secondary publication, this arrangement continuing until 1980.[18]

The employment register, now established as a licensed recruitment agency, was renamed the Professional Appointments Register and became one of the few agencies specializing in library and information jobs.

The library took on the role of systematically collecting thesauri, subject heading lists, classification schemes, and bibliographies, and regularly listing them.[19,20] In 1979, this was formalized by ASLIB being nominated by the British Standards Institute as the official clearinghouse for British thesauri, including those planned or in preparation.

By the mid-1970s ASLIB was still at the peak of its activities and influence, a small decline in membership having been offset by income from other sources, particularly consultancy to large national and international bodies. However, changes in the economic, social, and technical environment began to cause concern for the future. One immediate response was for ASLIB to rationalize its idiosyncratic membership structure. Individual members were offered a reduced fee as affiliates, without the voting rights of full, institutional, members, and the anomaly by which any UK government department could have free membership by virtue of the grants supplied by the government was removed. Student membership was introduced to attract the new generation.

At an ASLIB conference on "The information worker: identity, image and potential" held in London in November 1976, Dennis Lewis of ICI Plastics Division gave a talk entitled "There won't be an information profession in 2000 AD"; Lewis's "Doomsday Scenario" came to have a major influence on the future of ASLIB. In the following year, ASLIB suffered its first major cutback, when rising costs forced the reduction of the research department from 17 to 11. In 1980, it was further reduced by three, and its separate premises were given up.

Other trends at the time were positive, with ASLIB becoming a European center for the promotion of the new technology of online searching, which led to a focus on this topic in the library, which hosted an online information center, research projects, training courses, and publications, including monographs, particularly J.L. Hall's influential *Online retrieval sourcebook*, a "European user" series, how-to guides, and a new monthly newsletter, *Online Notes*. In 1978, ASLIB's long-standing interests in technical translation and in applications of technology came together in the organization of the first of the *Translating and the computer* conferences, which were to continue for over three decades.

In an environment starting to give concern to all information-related professional organizations, there was a renewed interest in cooperation. A Joint Consultative Committee, involving ASLIB, the Library Association, the Institute of Information Scientists, the Society of Archivists, the British Library, and the SCONUL academic library group, was revised after some years of dormancy, and a tripartite conference between the first three organizations was held in Sheffield in September 1980.

1980–2010

This next stage of ASLIB's story is one of an attempt to adapt to a changing environment, which saw some innovation and success, but overall a decline and retrenchment, which, at the end of period, saw ASLIB lose its independent status.

The reasons for this somewhat depressing state of affairs are multiple and affected many other information-related organizations in similar ways. Most immediately, the UK government ceased to offer direct financial support, in the way that it had for nearly four decades. This not only affected ASLIB directly but also the research associations and similar bodies that made up a significant part of the membership. The decline of British science-based industry over the period had a direct effect on ASLIB's membership, as did changes in the information environment that led to the merging, downsizing, or closure of many special libraries (appropriate enough, one the 1980 conferences addressed the topic of "cuts and the special library"). Organizations of diverse type, including consultancies, publishers, and universities, were increasingly involved in the activities that previously had been left to organizations like ASLIB. Finally, social changes posed changes for very many membership-based professional bodies, and ASLIB, despite its unique position and generally loyal membership base, was far from immune to this.

Dennis Lewis, who had predicted the demise of the information professional, took over in 1981 as Director from Basil Saunders, who had served for a short period following the retirement of Leslie Wilson. He produced a development plan for ASLIB, which, in view of the deteriorating financial situation, was put into action in 1985, earlier than anticipated, moving toward a "slimmer and trimmer" organization, run on fully commercial lines.[21] ASLIB left the fading splendor of its Belgrave Square location for premises in the Holborn area of central London. The research and consultancy division was closed, the research staff continuing to work on British Library projects, and the consultants forming an independent "Information Partnership." ASLIB still offered a limited consultancy, associated with the information service and the training program, as it does to the present day. The cessation of ASLIB research caused dismay to many in the information world and is still seen by many as marking the end of ASLIB's unique position in that world. Small survey research was still carried out at ASLIB; see, for example, a survey of IT use by the information service.[22]

The new vision was for a focus on membership services, publications, and professional development, with ASLIB undertaking a positive leadership role in the development of information management in the United Kingdom. At the time, the byline "The Association for Information Management" was added to the ASLIB name.

The branches and groups continued active programs, with some new groupings added, most notably the Information Resources Management Network.[23] The Informatics group was involved in organizing a series of Information Retrieval Package fairs. The nature of the conference program changed, with the long-standing annual conference supplanted by less regular multipartite events with other information organizations. The staff register was spun off as "ASLIB Professional Recruitment," an independent wholly owned subsidiary of ASLIB.

Publications expanded, with an ambitious program of new books, and new journals in local area networking and records management. More emphasis was placed on shorter publications, particularly addressing topical issues: these included a series of monthly newsletters on topics such as information technology, library automation, and business information, and a series of short "Know How Guides" on a variety of information technology and information management topics.

The training program, now offering almost exclusively one-day courses, was further expanded, particularly in areas of technology and online searching. More than 50 courses were run in 1985, more than half dealing with some aspect of online searching. Ten years later, the ASLIB training program was similarly responsive to the needs of moment in focusing on Internet-related courses.

In 1989, Dennis Lewis was succeeded as Director by Roger Bowes, a former newspaper executive, who continued Lewis's policy of downsizing to profitable areas, and on corporate information management as ASLIB's focus. The organization moved premises again, to Old Street in the eastern part of Central London. Advantage was taken of the layout of the new building to create a "One-Stop Information Shop," with services including quick reference, desk research, library, consultancy, and recruitment visible and accessible from the street. The Society of Archivists took up space in the building, and it was hoped to make it a focus for other information organizations. However, this did not happen, and ASLIB moved again to occupy two floors of an office building in the City of London.

The Saunders report of 1989 recommended the merger of the independent UK information-related associations, to form a single body to represent the information professions. ASLIB took an active part in the discussions, particularly with the Library Association and the Institute of Information Scientists, but concluded that the association was too different in nature and purpose for a merger to be feasible. ASLIB therefore stood apart from the merger of the other two organizations that formed CILIP; many considered this a missed opportunity to carry ASLIB's values and perspectives further.

In 2002, a further major downsizing took place, as ASLIB abandoned their publication activity, selling their books business to Taylor & Francis, and their journals to Emerald Group Publishing, with ASLIB members given favorable terms as purchasers and subscribers. Only a monthly *Managing Information* newsletter, the successor to *Aslib Information*, remained. Thereafter, ASLIB and Emerald worked increasingly closely together in providing services to the membership.

Shortly thereafter, the ASLIB library and information service closed, and its stock and the organization moved to smaller premises near Victoria Embankment. By 2005, membership had halved in three years, to less than 800, and in the same period staff numbers had fallen from 40 to 10. In December 2005, ASLIB went into voluntary liquidation.

It reemerged in early 2006 as a private company, having been bought by the former director, Roger Bowes. Operating from premises in Shoreditch, East London, it continued to offer professional development and recruitment services. The training program remained active, offering about 50 one-day courses, one of a wide variety of topics in information management, library/information skills, information organization, management, information governance, information technology, and subject-specific information. Onsite courses, on the customer's premises, were offered, and the first ASLIB distance-learning course, in thesaurus construction, was launched. There was still activity in some of the groups and branches, though most were becoming inactive, but the only regular conference was the long-standing *Translating and the Computer* meeting; ASLIB otherwise supported the meetings and conferences of other groups.

At the end of 2008, ASLIB moved to its final London location, an office suite in Goswell Road, Finsbury. With no in-house training facilities, courses were run, as in ASLIB's early days, in hired venues. The recruitment agency ceased operating during the course of the year, and the training courses, *Managing Information* magazine, and the *Translating and the Computer* conference were ASLIB's only remaining activities. The only active group was the combined engineering and technology special interest group; all the other groups and branches were dormant. Clearly, the organization was no longer viable as an independent entity.

2010–DATE

In April 2010, ASLIB was acquired by MCB Group, the holding company for Emerald Group Publishing Ltd. The

April issue of *Managing Information* noted that "Aslib will operate as a sister organization to Emerald, drawing on Emerald staff and resources, but still retaining its independence." The London offices were closed, and ASLIB activity centered on Emerald's headquarters in Bingley, Yorkshire. In the next issue, it was noted that "Our original ethos, which still applies today, was 'to serve those engaged in the collection, treatment and dissemination of information in many departments of human activity'."

Membership arrangements continued largely unchanged, though with an increased emphasis on Emerald's general management information benefits. Training and group activity was initially suspended, though the annual *Translating and the Computer* conference was held, and the ASLIB yearbook and membership directory published online, and the *Managing Information* newsletter continued to appear, with a new emphasis on research reports drawn from Emerald publications. A new design and branding saw the uppercase "ASLIB" restored as the organization name, still with the "Association for Information Management" byline.

ASLIB was relaunched with a member meeting at the London Online Information Meeting in December 2010. Five Emerald staff were initially named as assigned to the association's activities. An advisory council was formed, though sadly without any peers, knights, or admirals. A survey of the membership showed that training was the most highly regarded service, and the training program resumed in 2011, with public courses in London, onsite courses, and an increased distance-learning provision. The courses were categorized as business and official information sources, cataloging and classification, copyright and intellectual property, general management and communication skills, library and information management skills, and web and Internet. Most were based on courses offered formerly, but there were new offerings in social media and in web profiles and privacy. The engineering and technology group restarted activity, and a business information community of practice was formed.

Over the next three years, the program was slowly expanded. New training courses were introduced, in areas familiar from old ASLIB days, including information governance, information management, general management, and intellectual property. Some were in a half-day format, some ran outside London, and two-day "masterclasses" ran in Malaysia and Hong Kong. The 70th anniversary of ASLIB training activities was celebrated with a feeling that this element of the old association was still vibrant.

A program of member evening meetings and joint meetings with other bodies was started, and a two-day ASLIB Knowledge and Information Strategy Seminar at the British Library was held in December 2013, with the aim, again familiar from former days, of bringing together researchers and practitioners. Publishing activities under the ASLIB name restarted, with a series of monographs published by Emerald "in association with Aslib," and a new quarterly newsletter, *Privacy and Data Protection*, was launched.

However, Emerald proved no more able to sustain a membership-based organization than had the old ASLIB. It was announced at the end of 2014 that ASLIB would focus its efforts on expanding its professional development program. ASLIB membership and associated benefits, including group membership, ended, and the two newsletters ceased publication.

As from January 1, 2015, ASLIB was integrated with the wider Emerald organization, providing, in effect, a library/information professional development arm. No longer an association, it is known simply as "ASLIB." The name also lives on in Emerald's *Aslib Journal of Information Management* (formerly *Aslib Proceedings*). At that time, 34 courses were offered, as public courses in London and Manchester and as onsite training, with some via e-learning, in eight main areas: business and official information sources, information organization, copyright and intellectual property, customer services skills, knowledge management, general management, marketing and communication, and writing and editing.

In July 2015, Emerald announced that, in view of changing professional development requirements, ASLIB training courses would cease. All activities with the ASLIB name, other than publication of *Aslib Journal of Information Management* ceased in May 2016.

CONCLUSIONS

Richards wrote that the history of documentation, or information science, in Britain is largely synonymous with that of ASLIB, its members, and their activities.[11] Muddiman[1] suggested that ASLIB could fairly be regarded as an innovative organization, which actively shaped the development of the "informational turn" in society, and, in its earlier days, gave impetus to the involvement of the sciences of information and documentation in an increasingly information-intensive world. Both are correct, and we might also reflect on the number of the discipline's thought leaders who have been involved with ASLIB: Brian Vickery, Blaise Cronin, Steve Robertson, and Alan Gilchrist to name but four. So multifaceted have its activities been that it is difficult to single out the most significant achievements, but they must certainly include the promotion of the idea of rigorous, applied research, support for the effective introduction of several generations of information technology into library/information settings, and a continual insistence on the importance of the information disciplines and professions.

The influence of ASLIB has been very great throughout its history, as it has reinvented itself through changing times, and we should remember that ASLIB's story may not yet be over.

310

ASLIB

Artificial–Association

REFERENCES

1. Muddiman, D. A new history of ASLIB. J. Document. **2005**, *61* (3), 402–428.

2. Muddiman, D. Black, A.; Muddiman, D.; Plant, H. The history and development of ASLIB, 1924–1960. In *The Early Information Society: Information Management in Britain before the Computer*. Ashgate: Aldershot, U.K., 2007; 79–102.

3. Hutton, R.S. The origins and history of ASLIB. J. Document. **1945**, *1* (1), 6–20.

4. Pearce, J.G. The origin of ASLIB. Aslib Proc. **1967**, *19* (2), 64–65.

5. Black, A. "Arsenals of scientific and technical information": Public technical libraries in Britain during and immediately after World War 1. Libr. Trends **2007**, *55* (3), 474–489.

6. Rayward, W.B. In *European Modernism and the Information Society: Informing the Present, Understanding the Past*; Ashgate: Aldershot, U.K., 2008.

7. Wright, A. In *Cataloging the World: Paul Otlet and the Birth of the Information Age*; Oxford University Press: New York, 2014.

8. Ditmas, E.M.R. Looking back on ASLIB. Libr. Rev. **1961**, *18* (4), 268–274.

9. Muddiman, D. Documentation under duress: The joint conference of the International Federation for Documentation (FID) and the Association of Special Libraries and Information Bureaux (ASLIB), Oxford-London, 1938. Libr. Trends **2013**, *62* (2), 378–401.

10. Moholy, L. The ASLIB microfilm service: The story of its wartime activities. J. Document. **1946**, *2* (3), 147–173.

11. Richards, P.S. Aslib at war: The brief but intrepid career of a library organization as a hub of allied scientific intelligence 1942–1945. J. Educ. Libr. Inform. Sci. **1989**, *29* (4), 279–296.

12. Wilson, L. Saunders, W.L. Aslib and the development of information management. In *British Librarianship Today*; Library Association: London, U.K., 1976; 15–29.

13. Jones, K.P. How do we index? A report of some ASLIB informatics group activity. J. Document. **1983**, *39* (1), 1–23.

14. Wilson, L. Group activity in Aslib. Aslib Proc. **1960**, *12* (7), 264–268.

15. Robinson, L.; Bawden, D. Information (and library) science at City University London: Fifty years of educational development. J. Inform. Sci. **2010**, *36* (5), 618–630.

16. Taylor, P.J. Research at ASLIB 1959–1977. Aslib Proc. **1978**, *30* (3), 104–114.

17. Ashworth, W. *Handbook of Special Librarianship and Information Work*;; Aslib: London, U.K., 1967.

18. Gilchrist, A.; Presanis, A. Library and information science abstracts: The first two years. Aslib Proc. **1971**, *23* (5), 251–256.

19. Walkley, J.; Hay, B. Annotated list of thesauri held in the Aslib library. Aslib Proc. **1971**, *23* (6), 292–300.

20. Gilbert, V. A list of thesauri and subject headings held in the Aslib library. Aslib Proc. **1979**, *31* (6), 264–274.

21. Lewis, D.A. Director's report on Aslib. Aslib Proc. **1986**, *38* (8), 239–247.

22. Sippings, G. The use of information technology by information services: The Aslib information technology survey, 1987. Electron. Libr. **1987**, *5* (6), 354–357.

23. Montgomery, S.; Robertson, G. Aslib information resources management network. Aslib Inform. **1992**, *20* (9), 324–326.

Association for Information Science and Technology

Diane H. Sonnenwald
University College Dublin, Dublin, Ireland

Charles H. Davis
Debora Shaw
Indiana University, Bloomington, IN, U.S.A.

Abstract

The Association for Information Science and Technology (ASIS&T) was founded as the American Documentation Institute in 1937. In tandem with the evolution of the field, the association has continued to grow and change its structure and focus. This article describes the Association's history, purpose, structure (regional chapters plus topical special interest groups [SIGs]), governance, meetings, and publications.

INTRODUCTION

ASIS&T is an association of researchers and information professionals leading the search for new and better theories, techniques, and technologies to improve access to information. Among its membership are some 2000 individuals and organizations in over 50 countries. They have expertise in a variety of fields, such as information science, computer science, linguistics, management, psychology, library science, engineering, law, medicine, chemistry, physics, and education. They share a common interest in improving how individuals, organizations, and society store, retrieve, analyze, manage, archive, share, and disseminate information. Members come together for mutual benefit.

ASIS&T provides online and face-to-face educational and conference programs, a highly regarded journal and other publications, and local and international networking opportunities. These activities bridge the gap between research and practice and between theory and products. The Association's flagship publication is the highly regarded *JASIST*, the *Journal of the Association for Information Science and Technology* (formerly the *Journal of the American Society for Information Science and Technology*). ASIS&T is a nonprofit 501(c)3 professional association organized for scientific, literary, and educational purposes.

HISTORY

1937— Beginnings in Documentation

ASIS&T was founded on March 13, 1937, as the American Documentation Institute (ADI), a service organization made up of individuals nominated by and representing affiliated scientific and professional societies, foundations, and government agencies. Its initial interest was in the development of microfilm as an aid to learning. Watson Davis, the Institute's first president (1937–1943), was the editor of *Science News Letter* and director of Science Service, a nonprofit organization founded to promote public appreciation of science. ADI compiled an impressive record of achievement in its early years: the development of microfilm readers, cameras, and services; fostering negotiations and research that resulted in the so-called gentleman's agreement covering the photo duplication of copyrighted materials; the establishment of programs for the storage and reproduction of auxiliary publications in support of journal editors; the operation of an Asian scientific literature service during World War II; support of Interlingua, an early rival of the artificial language Esperanto, to foster international science communications; and cosponsorship of the 1958 International Conference on Scientific Information.[1–4]

1950s— Transition to Modern Information Science

As the number of people engaged in developing new principles and techniques in the many areas of documentation and information services increased, the Bylaws were amended in 1952 to admit individual as well as institutional members. Thus, ADI became the national professional society for those concerned with all elements and problems of information science. With the 1950s came increasing awareness of the potential of automatic devices for literature searching and information storage and retrieval. As these concepts grew in magnitude and potential, so did the variety of professional interests.[1,4,5]

Encyclopedia of Library and Information Sciences, Fourth Edition DOI: 10.1081/E-ELIS4-120051442

1960s—The Information Explosion

During the 1960s, the problems created by the "information explosion" gained national attention and ADI membership increased sevenfold. Reflecting this change in its total range of activities, as well as the emergence of information science as an identifiable configuration of disciplines, the membership voted to change the name of the ADI to the American Society for Information Science (ASIS). The name change took effect January 1, 1968 and emphasized the fact that the membership of ASIS was uniquely concerned with all aspects of the information transfer process. The new Society was a national professional organization with regional chapters for those concerned with designing, managing, and using information systems and technology. Harold Borko's frequently cited article, *What Is Information Science?*, helped to define the field: "Information science is that discipline that investigates the properties and behavior of information, the forces governing the flow of information, and the means of processing information for optimum accessibility and usability. It is concerned with the body of knowledge relating to the origination, collection, organization, storage, retrieval, interpretation, transmission, and utilization of information."[6] One of the Society's flagship publications, the *Annual Review of Information Science and Technology (ARIST)*, was launched in 1966, under the editorship of Carlos A. Cuadra and with support from the National Science Foundation. Contributors to *ARIST* provided overviews and syntheses of research on contemporary topics.

1970s—The Move to Online Information

The transitions from batch processing to interactive computing and from mainframe to mini- and microcomputers accelerated in the 1970s. Traditional boundaries among disciplines began to fade; many library schools and some departments of computer science added "information" to their titles. In 1974, the Society's *Newsletter* became the *Bulletin of the American Society for Information Science*, a mainstay membership publication of the Society. During this decade, ASIS made the midyear meeting an annual event that focused on a single topic of current interest. In 1976, the Society sponsored a bicentennial conference on the role of information in the development of the United States. From 1970 to 1974, ASIS administered the ERIC Clearinghouse on Library and Information Sciences; the Society was also an active participant in the planning and implementation of the 1979 White House Conference on Library and Information Services.

1980s—Personal Computers Change the Market

By the 1980s large databases, such as MEDLINE at the National Library of Medicine, and user-oriented services, such as Dialog, SDC, and CompuServe, were for the first time accessible by individuals from their personal computers. ASIS added to and revised its SIGs to respond to the changes, establishing groups on office information systems and personal computers as well as international information issues and rural information services. By the end of the decade, SIGs had been formed and reformed to represent various interests, including nonprint media, social sciences, energy and the environment, and community information systems, and ASIS had its first regional chapters outside of North America.

1990s and 2000s—The Human/Social Perspective

The social implications of information and communication technologies became increasingly apparent in the 1990s and ASIS responded. Ann Prentice's "environmental scan" identified key topics that were expected to reshape work and the workforce, including public access to and ownership of information, internationalization, and increasing expectations of information technology. Themes that appeared regularly in the Society's *Bulletin* and at national meetings included information management, globalization, intellectual property, computer networks (National Research and Education Network, the Internet), the digitization of text and visual information, bibliometrics, information retrieval, and human–information interaction. Interest in the representation of information found new applications in, for example, markup languages, taxonomy, metadata, and information architecture. The understanding of information retrieval was broadened and deepened by such developments as hypertext, the World Wide Web, search engines and browsers, and the addition of new domains to the Text REtrieval Conferences.

ASIS experienced changes in its own use of information technology. Merri Beth Lavagnino introduced the ASIS-L LISTSERV in 1992. The Society created its first website in 1995;[7] it was chosen as a Yahoo! "Pick of the Day" in 1996. After considerable deliberation about making ASIS's publications available electronically, a digital library was unveiled in 2006. In 2007, the *Bulletin of the American Society for Information Science and Technology* became an electronic-only publication.

The Society's work reflected the increasing interest in social and historical aspects of information science in other ways as well. In 1991, it renamed and reoriented one of the SIGs to focus on the History and Foundations of Information Science, and in 1992, ASIS adopted its first professional guidelines. Conferences cosponsored with the Chemical Heritage Foundation on the History and Heritage of Scientific and Technical Information Systems were held in 1998 and 2002.

Reflecting both internal and external influences, the members changed the name in 2000, becoming the American Society for Information Science and

Technology (ASIS&T). President Eugene Garfield championed the change, calling attention to the importance of practical, information technology–focused programs. ASIS&T continued its annual meetings but discontinued the midyear meeting in favor of "summits" featuring topics such as information architecture, knowledge management, and digital archives for engineering and scientific research.

2010s—Connecting across Boundaries in the Digital World

New possibilities emerged with digital technologies, along with new socioeconomic challenges, and presentations at the Annual Meeting and papers published in *JASIST* and the *Bulletin* reflected this. Topics such as data analytics, research data archiving and sharing, information visualization, health informatics, crisis informatics, mobile information seeking, and collaborative information behavior emerged as new themes at the Annual Meeting and in *JASIST* and the *Bulletin*.

ASIS&T-sponsored webinars (digital audio and visual seminars) were introduced in 2010. Free to all members, the webinars provide synchronous and asynchronous learning opportunities. Typically 60 minutes in length, they address current topics and trends in information science such as metadata, search engine optimization, digital preservation, health informatics, research data archiving, web analytics, information architecture, and information visualization. Other webinars provide career coaching, such as advice on publishing in international journals, conducting blended learning courses, and mentoring. The webinars are conducted by individual members, affiliated groups, and SIGs.

In 2012, the membership voted overwhelmingly to change the association's name from the *American Society for Information Science and Technology* to the *Association for Information Science and Technology*, keeping the acronym ASIS&T. The name change reflected the increasing global challenges that information science addresses and the increasing benefits from addressing these challenges collaboratively across sociopolitical and geographic boundaries. ASIS&T also increased its international activities, providing reduced membership fees based on gross national income per capita, establishing an Asia-Pacific Chapter, and holding its first annual meeting outside North America in 2016.

PURPOSE

The mission of the ASIS&T is to advance the information sciences and related applications of information technology by providing focus, opportunity, and support to information professionals and organizations.

The Association seeks to stimulate participation and interaction among its members by affording them an environment for substantive professional exchange. It encourages and supports personal and professional growth through opportunities for members to extend their knowledge and skills, develop and use professional networks, pursue career development goals, and assume leadership roles in the Association and in the broader information community. ASIS&T increases the influence of information professionals among decision makers by focusing attention on the importance of information as a vital resource in a high-technology age; in this way the organization seeks to promote informed policy on national and international information issues. It supports the advancement of the state-of-the-art and practice by taking a leadership position in the advocacy of research and development in basic and applied information science.

To accomplish these goals, ASIS&T edits, publishes, and disseminates publications concerning research and development; convenes annual meetings providing a forum for papers, discussions, and major policy statements; holds smaller local and special interest meetings, as well as special summits; and acts as a sounding board for the promotion of research and development and for the education of information professionals.

CHAPTERS AND SIGS

To give scope and focus to the diverse interests of its members, SIGs were established in 1966. These groups are, in effect, small professional organizations within ASIS&T, which encourage the discussion and development of both theoretical and practical issues. As of 2017, the Association's 16 SIGs were Arts and Humanities; Classification Research; Digital Libraries; Education for Information Science; Health Informatics; History and Foundations of Information Science; Information and Learning Services, Information Needs, Seeking and Use; Information Ethics and Policy; International Information Issues; Knowledge Management; Management; Metrics; Scientific and Technical Information; Social Informatics; and Visualization, Images and Sound.

All ASIS&T members are automatically enrolled as members of the humorous and ironic SIG/CON, which seeks to "explore the fundamental notions of information science, and expose them for what they are."[8] The official SIG/CON website describes the initial meeting: "Those who stumbled into the first SIG/CON session found it to be one of the most informative and intellectually challenging sessions of the conference, a tradition which continues to this day."[8] Dr. Llewellyn C. Puppybreath, III, founder and perpetual chair of the group, has yet to appear at any of its meetings. The breadth of topics covered is evident in this list of titles from presentations: "Baloonean Logic" by Ev Brenner, "Conical Classifications" by C. David Batty, "Titular Colonicity" by Candy Schwartz, and "Early Binary Separation: Implications for Information Retrieval (relevance judgments by identical twins separated at birth)" by Joseph Janes.

At a local and regional level, ASIS&T had, in 2017, 13 chapters in major cities and regions in the United States and around the world: Arizona, the Asia-Pacific Region, the Carolinas, Central Ohio, Europe, Indiana, Los Angeles, New England, New Jersey, Northern Ohio, the Pacific Northwest, Potomac Valley (Washington D.C. area), and Taipei. These chapters provide local forums for discussion and information exchange for members in a given locality.

In addition, 39 student chapters channel student interest and engagement in information science: the Catholic University of America, Drexel University, Europe, Florida State University, Indiana University, Long Island University, Louisiana State University, McGill University, North Carolina Central University, Pratt Institute, Rutgers University, San Jose State University, Simmons College, St. John's University, State University of New York (Albany), State University of New York (Buffalo), State University of New York (Oswego), Taipei University, University of Alabama, University of British Columbia, University of California at Los Angeles, University of Denver, University of Hawaii, University of Illinois at Urbana-Champaign, University of Iowa, University of Maryland, University of Michigan, University of Missouri, University of North Carolina at Chapel Hill, University of North Texas, University of Pittsburgh, University of South Carolina, University of South Florida, University of Tennessee (Knoxville), University of Texas (Austin), University of Toronto, University of Washington, University of Wisconsin-Milwaukee, and Wayne State University. Additional regional and student chapters in South America and Africa are being established.

GOVERNANCE

The ASIS&T Board of Directors consists of the president, president-elect, immediate past president, treasurer, a parliamentarian, directors chosen by the SIGs and chapters, and six directors-at-large. ASIS&T maintains its headquarters under an executive director in the Silver Spring area in Maryland. The Association's web address is www. asist.org.

The Executive and Budget and Finance committees are standing committees of the Board of Directors; the standing committees of the Society are Awards and Honors, Constitution and Bylaws, Information Science Education, International Relations, Leadership, Membership, Nominations, Publications and Scholarly

Communications, and Standards. An annual business meeting of the membership is held during the ASIS&T annual meeting.

The presidents of the Society have been the following:

American Documentation Institute

Watson Davis 1937–1943
Keyes D. Metcalf 1944
Waldo G. Leland 1945
Watson Davis 1946
Waldo G. Leland 1947
Vernon D. Tate 1948–1949
Luther H. Evans 1950–1952
E. Eugene Miller 1953
Milton O. Lee 1954
Scott Adams 1955
Joseph Hilsenrath 1956
James W. Perry 1957
Herman H. Henkle 1958
Karl F. Heumann 1959
Cloyd Dake Gull 1960
Gerald J. Sophar 1961
Claire K. Schultz 1962
Robert M. Hayes 1963
Laurence B. Heilprin 1964–1965
Harold Borko 1966
Bernard M. Fry 1967

American Society for Information Science

Robert S. Taylor 1968
Joseph Becker 1969
Charles P. Bourne 1970
Pauline A. Atherton 1971
Robert J. Kyle 1972
John Sherrod 1973
Herbert S. White 1974
Dale B. Baker 1975
Melvin S. Day 1976
Margaret T. Fischer 1977
Audrey N. Grosch 1978
James M. Cretsos 1979
Herbert B. Landau 1980
Mary C. Berger 1981
Ruth L. Tighe 1982
Charles H. Davis 1982–1983
Donald W. King 1984
Bonnie C. Carroll 1985
Julie A. C. Virgo 1986
Thomas H. Hogan 1987
Martha E. Williams 1988
W. David Penniman 1989
Toni Carbo 1990
Tefko Saracevic 1991
Ann E. Prentice 1992
José-Marie Griffiths 1993
Marjorie M.K. Hlava 1994

James E. Rush 1995
Clifford Lynch 1996
Debora Shaw 1997
Michael Buckland 1998
Candy Schwartz 1999
Eugene Garfield 2000

American Society for Information Science and Technology

Joseph A. Busch 2001
Donald H. Kraft 2002
Trudi Bellardo Hahn 2003
Samantha Hastings 2004
Nicholas Belkin 2005
Michael Leach 2006
Edie Rasmussen 2007
Nancy Roderer 2008
Donald O. Case 2009
Gary Marchionini 2010
Linda C. Smith 2011
Diane H. Sonnenwald 2012

Association for Information Science and Technology

Andrew Dillon 2013
Harry Bruce 2014
Sandra Hirsch 2015

PUBLICATIONS

Publication affords a major channel to communicate current significant research reports, development work, and professional activities, both to the membership and to the interested public. It also serves as a historical record of achievement. In addition to a substantial monograph series, ASIS&T produces three continuing publications.

The *Journal of the Association for Information Science and Technology* (*JASIST*, formerly the *Journal of the American Society for Information Science and Technology*) is a highly ranked and frequently cited source for reports of research and development in information science. The *Journal of Documentary Reproduction* (edited by Vernon D. Tate and published by the American Library Association, 1938–1942) is considered the predecessor of the Association's journal. In 1950, it was reborn as *American Documentation*. With the Association's changes of name, it became the *Journal of the American Society for Information Science* (in 1970), the *Journal of the American Society for Information Science and Technology* (in 2001), and the *Journal of the Association for Information Science and Technology* (in 2014). The journal, also known as *JASIST*, serves as a forum for new research in information transfer and communication processes in general, and in the context of recorded knowledge in particular. Concerns include the generation, recording, distribution, storage, representation, retrieval, and dissemination of information, as well as its social impact and management of information agencies. *JASIST* consistently ranks at or near the top in assessments of influence in the field. It gradually increased from four issues per year (1950–1969) to its current monthly publication. The journal has been edited by Vernon D. Tate (1950–1951), Mortimer Taube (1952), Jesse H. Shera (1953–1960), Luther H. Evans (1961), James D. Mack (1962–1963), Arthur W. Elias (1964–1976), Charles T. Meadow (1977–1984), Donald H. Kraft (1985–2008), Blaise Cronin (2009 to date), Blaise Cronin (2009–2015), and Javed Mostafa (2016 to date).

The proceedings of the annual meetings collect significant contributed, peer-reviewed papers. Since 2002, these proceedings are available to meeting registrants and digitally through the ASIS&T Digital Library and the Wiley Online Library. All *JASIST* issues are also available in the ASIS&T Digital Library.

The *Bulletin of the American Society for Information Science* (formerly the Society's *Newsletter*) began publication in 1974. It is a bimonthly news magazine focused on developments and issues affecting the field, pragmatic management reports, opinion, and news of people and events in the information science community. Irene Travis was named editor in 1997; in 2007, the *Bulletin* became a digital-only publication. It is currently available free of charge via the ASIS&T website.

A fourth publication, the highly regarded *ARIST* was published from 1966 to 2011. *ARIST* surveyed the landscape of information science and technology, providing the reader with an analytical, authoritative, and accessible overview of recent trends and significant developments. The initial volumes of this series were supported by a grant from the National Science Foundation. *ARIST* was edited by Carlos A. Cuadra (1966–1975), Martha E. Williams (1976–2001), and Blaise Cronin (2002–2011). Many *ARIST* volumes are available in the ASIS&T Digital Library. The *Advances in Information Science* series published in *JASIST* continue the *ARIST* tradition of in-depth reviews of the field.

MEETINGS

Both the annual meeting held in the fall of the year and the special summits provide important forums for communication among people with diverse interests. The annual meetings bring together researchers and professionals from virtually every field: attorneys, bankers, electronic and print publishers, government officials, businesspersons, librarians, educators, engineers, social scientists, information managers, computer scientists, and many others. Activities specially held for students include a student design competition, student paper awards and presentations, a doctoral colloquium, and a newcomer's

Table 1 Recipients of the Award of Merit and Watson Davis Award

Year	Award of Merit Recipients	Watson Davis Award Recipients
1964	Hans Peter Luhn	
1965	Charles P. Bourne	
1966	Mortimer Taube	
1967	Robert Fairthorne	
1968	Carlos A. Cuadra	
1969	[not awarded]	
1970	Cyril W. Cleverdon	
1971	Jerrold Orne	
1972	Phyllis Richmond	
1973	Jesse H. Shera	
1974	Manfred Kochen	
1975	Eugene Garfield	
1976	Laurence Heilprin	James M. Cretsos
		Laurence B. Heilprin
		Lois F. Lunin
		Gerald J. Sophar
		Herbert S. White
1977	Allen Kent	Arthur W. Elias
		Irene Farkas-Conn
		Simon M. Newman
		Roy D. Tally
1978	Calvin N. Mooers	Mary C. Berger
		Joe Ann Clifton
		Charles H. Davis
1979	Frederick Kilgour	Frank Slater
1980	Claire K. Schultz	Gerard O. Platau
1981	Herbert S. White	Jan Krcmar
1982	Andrew A. Aines	Edmond Sawyer
1983	Dale B. Baker	Toni Carbo
		Margaret T. Fischer
1984	Joseph Becker	Robert Tannehill, Jr.
	Martha Williams	
1985	Robert L. Chartrand	Lawrence Woods
		Michel Menou
1986	Bernard M. Fry	N. Bernard Basch
1987	Donald W. King	George Abbott
		Barbara Flood
1988	F. Wilfrid Lancaster	Bonnie C. Carroll
1989	Gerard Salton	Madeline Henderson
		G. Daniel Robbins
1990	Pauline Atherton Cochrane	Julie Virgo
1991	Roger K. Summit	Mickie Voges-Piatt
1992	Robert S. Taylor	Marianne Cooper
1993	Robert M. Hayes	Debora Shaw
1994	Harold Borko	Audrey Grosch
1995	Tefko Saracevic	Martha Williams
		Donald Kraft
1996	Jean Tague-Sutcliffe	Marjorie M.K. Hlava
1997	Dagobert Soergel	Karla Petersen
1998	Henry Small	Judy Watson
1999	José Marie Griffiths	Jessica L. Milstead
2000	Donald R. Swanson	Candy Schwartz
2001	Patrick G. Wilson	Julie Hurd
2002	Karen Spärck Jones	Thomas Hogan
2003	Nicholas J. Belkin	Nancy Roderer
2004	Howard D. White	Joseph Busch

(Continued)

Table 1 Recipients of the Award of Merit and Watson Davis Award *(Continued)*

Year	Award of Merit Recipients	Watson Davis Award Recipients
2005	Marcia Bates	Michael Buckland
2006	Blaise Cronin	Trudi Bellardo Hahn
		Steve Hardin
2007	Donald H. Kraft	Paula Galbraith
2008	Clifford Lynch	Samantha Hastings
2009	Carol Tenopir	Edie Rasmussen
2010	Linda C. Smith	Barbara Wildemuth
2011	Gary Marchionini	Robert Williams
2012	Michael Buckland	K.T. Vaughan
2013	Carol C. Kuhlthau	Beata Panagopoulos
2014	Marjorie M.K. Hlava	Vicki Gregory
2015	Michael E.D. Koenig	Michael Leach
2016	Peter Ingwersen	Donald O. Case
		Diane H. Sonnenwald

brunch. The topical summits feature presentations from research and industry leaders, opportunities for extensive personal interaction, and awards recognizing excellence in all aspects of the field.

SIGs also organize and present specialized meetings. The SIG on Classification Research has conducted the highly regarded Workshop of Classification Research each year since 1990. The SIG on Information Needs, Seeking and Use holds a research symposium in conjunction with ASIS&T annual meetings. Other SIGs that periodically hold seminars and workshops include the SIG on Metrics, the SIG on Social Informatics, and the SIG on Knowledge Management. In 2012, a one-day seminar on international perspectives on the history of information science and technology was held to help celebrate ASIS&T's 75th anniversary. (Seminar co-chairs were Toni Carbo and Robert Williams.)

AWARDS

ASIS&T's Award of Merit is the Association's highest honor. It is bestowed on individuals who have made noteworthy contributions to the field of information science, including the expression of new ideas, the creation of new devices, the development of better techniques, and outstanding service to the profession of information science. The Watson Davis Award commemorates the Society's founder. It is given to members who have shown outstanding continuous contributions and dedicated service to the ASIS&T. The recipients of these awards are listed in Table 1.

Twelve additional awards are given annually to recognize students, academics, and practitioners for their contributions to information science education, research, and the Association. There are six chapters and three SIG awards presented annually as well.

CONCLUSION

ASIS&T has responded to more than seven decades of change in our understanding and use of information technologies. It has evolved from an organization of institutions to an individual membership society. As ASIS&T advances into the twenty-first century, its meetings reach academic, industry, and professional audiences, and its long-standing publications retain their high impact as they expand to digital formats.

The text describing the history of ASIS&T through the 1980s can be found on the ASIS&T website and is used with permission from ASIS&T. An earlier version of this article appeared in Davis, C.H.; Shaw, D. *American Society for Information Science and Technology (ASIS&T). Encyclopedia of Library and Information Sciences*; Taylor & Francis, 2009, DOI: 10.1081/E-ELIS3-120044410.

REFERENCES

1. Farkas-Conn, I.S. *From Documentation to Information Science: The Beginnings and Early Development of the American Documentation Institute-American Society for Information Science*; Greenwood Press: New York, 1990. [downloaded on 14 October 2013 from http://www.asis.org/Farkas-Conn-FDTIS.html].

2. Richards, P.S. *Scientific Information in Wartime: The Allied-German Rivalry 1939–1945*; Greenwood Press: Westport, CT, 1994.

3. Schultz, C.K. ASIS: Notes on its founding and development. Bull. Am. Soc. Inf. Sci. **1976**, *2* (8), 49–51.

4. Schultz, C.K.; Garwig, P.L. History of the American Documentation Institute: A sketch. Am. Doc. **1969**, *20* (2), 152–160.

5. Hahn, T.B.; Buckland, M.K., Eds. *Historical Studies in Information Science*; Information Today: Medford, NJ, 1998.

6. Borko, H. What is information science? Am. Doc. **1968**, *19* (1), 3.

7. Association for Information Science and Technology [Homepage of the Association for Information Science and Technology], ASIS&T Association for Information Science and Technology, http://www.asist.org/ (accessed May 2017).

8. Schwartz, C. SIG/CON [what? how? why?], 2001. http://web.simmons.edu/~schwartz/con/sig-con.html (accessed May 2017).

Association for Information Systems (AIS)

William R. King
University of Pittsburgh, Pittsburgh, Pennsylvania, U.S.A.

Dennis Galletta
Katz Graduate School of Business, University of Pittsburgh, Pittsburgh, Pennsylvania, U.S.A.

Abstract

The Association for Information Systems (AIS) is a professional organization whose membership is made up primarily of business school academics who specialize in information systems(IS) development, implementation, and evaluation. The current membership is 3965, which is over 50% saturation of the overall population of an estimated 7000 IS academics worldwide.

The Association for Information Systems (AIS) is a professional organization whose membership is made up primarily of academics who specialize in information systems (IS) development, implementation, and evaluation. As the majority of early academics in the field were housed in business schools, much of our research still has a focus on the "bottom line"—the business consequences of IS choices. This special perspective is reflected in the mission statement of AIS:

> To advance knowledge in the use of information technology to improve organizational performance and individual quality of work life.

Worldwide, most recently, there are also many members housed in computer science and information science programs as well, but the organizational perspective remains.

OBJECTIVES OF AIS

The purpose of AIS is to serve as the premier global organization for specialists in IS in order to:

- Create and maintain a professional identity for IS educators, researchers and professionals, researchers, and educators;
- Promote communications and interaction among members;
- Provide a focal point for contact and relations with bodies in government, the private sector, and in education that influence and/or control the nature of IS;
- Improve curricula, pedagogy, and other aspects of IS education;
- Create a vision for the future of the IS field and profession;
- Create and implement a modern, technologically sophisticated professional society;
- Establish standards of practice, ethics, and education where appropriate; and
- Include professionals worldwide.

To these ends, AIS conducts conferences and meetings, publishes books, journals, and other materials; cooperates with other organizations interested in the advancement and practice of IS, stimulates research; promotes high professional standards and promotes the growth of IS and the profession's quality throughout the world.

BACKGROUND ON IS IN BUSINESS SCHOOLS

The IS area developed in business schools in the late 1960s with an initial focus on programming business applications of computers, often through the use of COBOL, a business-oriented computer programming language. This addition to business curricula was controversial with some faculty arguing that business students would never become computer programmers and others arguing that managers needed to be informed "end users," to understand the nature of the tasks that analysts and programmers perform, and to develop a realistic awareness of the capabilities and limitations of computer "solutions."

This programming emphasis rapidly shifted to one that emphasized the "fit" between computer systems and the business enterprise as illustrated by the development of planning processes for deciding how computer systems could best be used to aid in the achievement of business goals.

HISTORY OF AIS

In the early days of IS in business schools, most academics in the area came from other disciplines such as economics,

Encyclopedia of Library and Information Sciences, Fourth Edition DOI: 10.1081/E-ELIS4-120044815

accounting, organizational behavior, operations research, and management science. Because of this, most IS academics had professional affiliations in other underlying disciplines. Thus, some did not see IS as a distinct professional field even though they were teaching and researching IS topics.

Nonetheless, as the field grew explosively in the 1970s and 1980s with ever-greater demands for new IS programs and class offerings, the notion of IS as a professional field of study and practice grew. Many IS academics saw the need for an organization that could represent the professional values and aspirations of IS business-school academics. Although the focus differs somewhat from region to region, the predominant approach was, and still is, to recognize the different needs of IS academics from those of faculty in computer and information science.

Interestingly, the first major effort in this direction came in 1980 with the creation of a major international research conference—the annual International Conference on IS (ICIS)—a nonprofit organization with a governing executive committee that was responsible for site selection and choosing the conference chair and other key positions for upcoming conferences. This conference was created through a grass-roots effort by senior IS academics, primarily from North America. It rapidly became a major focal point for the research interests of academics across the world.

As ICIS grew and prospered, various informal groups met there to discuss the need for a professional organization to more broadly represent the interests of IS academics. Several studies and surveys were conducted with mixed results and little action.

Finally, a study was planned by a group of senior people who met informally and commissioned Dr. William R. King, University Professor at the University of Pittsburgh to organize a task force to comprehensively study the issue of creating a professional organization, and to assess the level of support for the idea. King contacted numerous senior people to get their ideas; he found that they almost unanimously favored the creation of such an organization. So, rather than leading a study, King formed an organizing committee of about 40 senior academics from around the globe. While this group was creating the general design for a new organization, King attended academic conferences held by related professional organizations and regional IS conferences that had recently been initiated. At those conferences, he held information sessions to discuss the objectives of the proposed organization and to solicit ideas.

The organizing committee conducted an electronic constitutional convention to agree on a constitution for the new organization and appointed King to be its first Executive Director. Operating out of his university office with the help of his assistant and a doctoral student, he proceeded to solicit members and had a membership roster of 1800 charter members within 6 months. The charter members elected King as the first President of AIS in an election that also filled various officer and council slots.

GOVERNANCE OF AIS

The Association for IS is led by a president who is annually elected from one of three world regions—the Americas, Europe and Africa, and Asia-Pacific—on a rotating basis. The governing Council is made up of elected functional vice-presidents and other officers and council members who are elected in the three world regions. There are also some appointed positions, such as secretary and treasurer. Thus, AIS is truly a global organization that can reflect the diverse interests and needs of business-school academics all over the world.

EVOLUTION OF AIS

The Association for IS was initially operated out of a single office at the University of Pittsburgh with only one part-time paid employee, King's administrative assistant, and many activities conducted by member-volunteers. In 1997, it became apparent that more professional management was necessary, and the Council circulated a request for proposals for a permanent site and staff. Georgia State University submitted the winning proposal and the organization operated from there, with Dr. Ephraim R. McLean of Georgia State as Executive Director for 10 years. Currently, AIS has a paid Executive Director and a small number of staff members and contractors to handle membership, conferences, technology, and finances. Some degree of outsourcing allows AIS to reduce its dependence on volunteers.

In spite of professionalizing many operations, AIS still involves a great deal of volunteer work. There are several elected positions, including that of the President (and Past-President and President-Elect), Regional Representatives, and VPs. The VPs cover technology, communications, chapters, meetings and conferences, Special Interest Groups (SIGs), member services, and accreditation. Holding one of these positions carries with it an expectation of traveling to one or two meetings per year, responding to communications throughout the office-holder's tenure, and, most importantly, being innovative, and responsive to emerging issues and challenges.

Special Interest Groups began to be created in 2001. In November 2000, VP of Member Services Dennis Galletta conducted a survey of 11 other professional associations, and AIS fortunately compared favorably to those associations when considering dues amount and services to members. One of the most striking services in many of those associations was in the area of SIGs, not yet offered by AIS at that time. After evaluating 12 proposals, a first slate of six were created. As of early 2009, there are 33 AIS

SIGs. Many SIGs serve as valuable resources to members, as many of them provide valuable research and teaching resources, as well as newsletters, pre-conference activities, and conference tracks. With funding (actual and in-kind) from AIS, SIGHCI (Human–Computer Interaction), Syracuse University, and the University of Pittsburgh, Council has just created the first *Transactions* journals: *AIS Transactions of HCI* (*TOHCI*), edited by Dennis Galletta and Ping Zhang, and *AIS Transactions on Enterprise Systems* (*TES*), edited by Norbert Gronau. If these journals are a success, Council will entertain proposals from other SIGs for further *Transactions*.

Given the costs inherent in AIS initiatives and operations, it is important for AIS to generate revenue to cover a wide range of activities. Fortuitously, one of the most important AIS activities is that of conferences, which both provide benefits to members and generate funds for the Association. Currently, 70% of the $1.6 million AIS budget is funded by conferences. Based on strategic planning exercises initiated by President Michael Myers in 2006, Council has begun to explore ways in which the organization can use its intellectual capital to fund more of the costs of AIS. The recent move to expand memberships to practitioners will also broaden our base of revenue.

Milestones in the Evolution of AIS

The details of the evolution of AIS, as well as its truly global nature, may best be characterized by describing the foci of each president's term. As well, such a description may provide a useful case-study of the creation and development of a successful academic professional organization. The accomplishments are not necessarily reflective of a particular president but of the timing of the milestone. In many cases, the initiatives were accomplished over several presidential terms. Also, the affiliations of the President at that time are provided; in some cases they have either moved to other universities or retired.

1995 (Bill King, University of Pittsburgh, United States of America)

Bill King's administration represented the culmination of a history of imagining and planning. As described earlier, years of discussion about creating the Association were put into place at that time. Initial meetings were held to determine what the name of the organization would be (which was surprisingly difficult because of the diverse language structures and cultures), what services it would provide, what its structure would be, and to develop a strategic plan for its future development.

King arranged for AIS to support existing nascent conferences in Europe and Asia (ECIS—European Conference on Information Systems; PACIS—Pacific Asia Conference on Information Systems) and asked Dennis Galletta to organize the Inaugural Americas Conference

on IS (AMCIS) in Pittsburgh in August of 1995. The strategy for the conference was for it be complementary to ICIS in being inclusive, with a higher acceptance rate than that of ICIS (which is normally significantly lower than 20%), and with more of a focus on teaching. The first AMCIS generated 239 paper submissions and 149 were accepted by the program committee (chaired by Hugh Watson of the University of Georgia), representing a 62% acceptance rate. The 587 attendees in that first Americas regional conference enjoyed keynote talks by Tom Davenport, Herbert Simon, and James Wetherbe.

Attendees enjoyed a dinner/dance riverboat ride on the three rivers of Pittsburgh to kick off the Inaugural Conference. Such amenities were common at ICIS and AMCIS carried on this tradition. The conference was innovative in several ways: it featured electronic submissions and reviewing, electronic registration via Web site, teaching workshops, technology briefings, tutorials, and software demonstrations. Since 1995, the conference has been held each year in August, and attendance has risen to around 1000 attendees as membership in AIS has risen to nearly 4000.

Before the Inaugural AMCIS conference was held, there was some controversy as some Council members feared that regional conferences would erode the popularity of ICIS, which was generally perceived to be of such high quality that it constituted a "crown jewel" of the field. The fears proved to be unfounded as ICIS continued to attract 1000–1500 participants, and as all regional conferences have prospered.

The Association for IS began by making arrangements with several existing journals to provide a discounted subscription to a journal of choice to each member. Currently, members deal directly with any of 32 journals and provide their membership number to obtain the discount.

1996 (Niels Bjorn-Andersen, Copenhagen Business School, Denmark)

President Bjorn-Andersen worked to make sure that AIS was established as a true global organization. When he took office, his main efforts were aimed at stimulating support in the various regions through traveling to the major regional conferences. His message was that the field needed a global organization, and that AIS was not exclusively run by Americans. He garnered more members from the non-American regions and worked to make sure that officers represented all regions.

Council established membership for researchers from developing countries at a very affordable level, and Bjorn-Andersen initiated a series of negotiations on enlisting already existing local IS faculty organizations as chapters.

He stimulated improvements in the faculty directory service and integrated a European directory of faculty into the global directory done so well by Dave Naumann and his colleagues at the University of Minnesota.

1997 (Ron Weber, University of Queensland, Australia)

Several key milestones were reached in 1997. The seeds for a merger of AIS, ICIS, and ISWorld were planted. Discussions were held about making the *MIS Quarterly* available to all AIS members electronically. Also, two electronic AIS journals were begun: *Journal of the AIS* and *Communications of the AIS*. The first editor of the *Journal of the AIS* was Philip Ein-Dor, and the first editor of the *Communications of the AIS* was Paul Gray, both of whom developed the journals into the highly-ranked publications that they are today. Key decisions that year include the formation of an audit review committee, keeping the new journals in an electronic-only format, the development of strategic planning for AIS, and the genesis of the first AIS chapter (Southeastern United States of America).

1998 (Gordon Davis, University of Minnesota, United States of America)

Much of the complexity involved in building a global association was exhibited in this time period, as many initiatives took several years to come to fruition. For example, merger discussions between AIS and ICIS reached a crescendo during 1998. Also, trying to normalize many conference procedures as AMCIS began to mature, and ICIS continued in its 19th year. Significant attention was paid to conference and AIS finances, as there were some difficulties with an outside contractor who handled funds and conference implementation. During this period, relationships were cultivated with organizations worldwide for potential alliances.

1999 (Bob Galliers, University of Warwick, United Kingdom)

During 1999, plans for new Lyons Electronic Office (LEO) and Fellow awards were drafted to honor contributors to the field. Named after the world's first commercial application of computing (The LEO), the purpose of the LEO Award is to recognize truly outstanding individuals in the IS community, both academics and practitioners, who have made exceptional lifetime contributions to research in and/or the practice of IS. As of 2009, 21 LEO awards have been conferred.

The AIS Fellow award recognizes individuals who have made outstanding contributions to the IS discipline in terms of research, teaching, and service. A Fellow need not have excelled in all three categories. Nonetheless, she or he is expected to have made exceptional contributions in at least one of these categories and to have made significant contributions in the other two categories. A Fellow is also expected to have made significant global contributions to the IS discipline as well as outstanding local contributions in the context of their country and region. As of 2009, 48 Fellow awards have been conferred.

2000 (Mike Vitale, University of New South Wales, Australia)

During 2000, final consolidation of AIS and ICIS was planned. Attorneys drew up the necessary paperwork, and the AIS constitution and other legal documents were thoroughly reviewed and updated to reflect the new nonprofit corporate structure. Also, functions and procedures were refined for the office in Atlanta. Also that year, it was decided that a periodic member survey would be conducted and that AIS would become involved in curriculum and accreditation efforts.

2001 (Blake Ives, Louisiana State University, United States of America)

The consolidation of AIS and ICIS was finally completed and celebrated in New Orleans, Louisiana, in December 2001, complete with a wedding cake inscribed with "AIS + ICIS." Also that year, AIS became involved in *MIS Quarterly's* Policy Committee by having its VP of Publications as a member of the Committee, by appointing a member, and by choosing the Committee's chair every other term. Also, a contract was signed with *MIS Quarterly* and *MISQ Executive* to offer them both free to AIS members electronically. The Minnesota IS Faculty directory was integrated into the "ISWorld" online portal. That year, AIS responded formally to Association to Advance Collegiate Schools of Business, the premiere accreditation agency for academic business programs, which had omitted MIS from its business program accreditation guidelines; AACSB subsequently "fixed" its guidelines and added a requirement for business programs to include MIS. This one initiative has led several schools to create a required MIS course, or to make sure that the required course was retained despite any curriculum innovations. Heavy membership discounts were offered to IS academics in "non-rich" countries. Finally, AIS began studying accreditation as a possible avenue for program development.

2002 (Phillip Ein-Dor, Tel Aviv University, Israel)

Budgetary integration of AIS and ICIS was accomplished for the first time in 2002, which was largely accomplished by AIS Treasurer Richard Elnicki. Also, global geographical integration was encouraged. New affiliations were struck with German and French IS Societies. An agreement was reached with *MISQ Executive* to provide the journal electronically to all members. The electronic version of the *MIS Quarterly* was built. Previous ICIS and AMCIS proceedings were also provided online.

2003 (K.K. Wei, National University of Singapore)

In 2003, memberships from Region 3 achieved a greater than twofold increase, thus making AIS into a truly global community. A large number of chapters were created in 2003, including new chapters representing Chinese-speaking individuals and also those spanning different continents including Europe (Ireland, Italy, Slovenia, and United Kingdom), Middle East (Israel and Morocco), and Asia (Pakistan). Wei initiated a new strategic planning function for AIS. A new Code of Research Conduct was adopted by the AIS Council. A member survey was undertaken, which indicated a high degree of member satisfaction. Particular areas were noted for future decision making.

2004 (Richard Watson, University of Georgia, United States of America)

In 2004, two successive losses exceeding $200,000 underscored the importance of cost cutting. Establishing the financial viability of AIS monopolized action during this term. The budget was scrutinized, and the cycle of losses was turned into more than a $100,000 surplus. A number of actions were taken to establish the long-term financial health of AIS, including an analysis of the operations of AIS by the Finance Committee, chaired by Joe Valacich, and a review of the financial contributions of AMCIS and ICIS by a committee chaired by Malcolm Munro, VP of Meetings and Conferences.

Also, AIS replaced its in-house Web site system with an open-source content management system to reduce costs and improve the currency of the site.

2005 (Claudia Loebbecke, University of Cologne, Germany)

The e-library was enhanced in 2005 by allowing cross-searching of all AIS conferences. Also, new conferences were added to the e-library (those from conferences in Australia/Asia (ACIS), Bled, and Europe (ECIS). Internationalization efforts continued, and hundreds of new members from China marked the success of that initiative. Also, membership fees in developing countries were lowered again to broaden the membership of AIS. The open source Web site was implemented. Model curricula were approved in conjunction with IEEE and ACM. The Distinguished Member award was initiated for honoring deceased members. As of 2008, there are three such members (Claudio Ciborra, Gerardine DeSanctis, and Heinz Klein). Finally, a new conference management firm was appointed to help with conference negotiation, planning, and execution.

2006 (Michael Myers, University of Auckland, New Zealand)

In 2006, a key event was a strategic planning meeting for AIS Council to provide a clear vision and direction for the future growth and development of AIS. This meeting marked the inauguration of a firm strategic planning function within AIS. Strategies were developed by AIS Council for internationalization, communications, journals, chapters and SIGs, conferences, accreditation, corporate sponsorship, and membership growth, amongst other things. Council developed a vision to transform AIS into a more professional association. An orientation program was established, whereby all new Council members are introduced to the principles of association leadership, the state of the association, the strategic plan, and the opportunities and challenges of their particular role whether it be vice-president or region representative. The AIS Technology Committee met and had its own strategic planning session. Recommendations that were adopted by AIS Council were the following: 1) Scholar One was adopted as the electronic review system for all AIS journals and conferences; 2) A revamped AIS Web site was launched, including regularly updated content on AIS SIGs and chapters; and 3) AIS Council agreed to fund an AIS association management system, an AIS e-library system and to bring together into one unified portal the AIS Web site and AISWorld. An affiliation of *Information Systems Journal* with AIS was approved. Two new journals were launched: an AIS Journal in Spanish and the Pacific Asia Journal of the AIS. Finally, a VP of Technology was introduced and the position filled by presidential appointment.

2007 (Dennis Galletta, University of Pittsburgh, United States of America)

During Galletta's term, a task force explored how to increase membership value and another worked on issues of member diversity. In a bold move to increase membership, Council decided to include practitioners as an important additional membership category. This move enables student chapters to provide a "membership path" for future members, and for graduates to serve as contacts for us and our current students. We also adopted a German language journal into the e-library, and Council insisted that full German text be included rather than only incorporate English abstracts. To address the enrollment downturns that had been experienced in the early twenty-first century, a committee maintained an AIS wiki for storing materials and strategies for student recruitment. A brochure "Picture yourself studying IS" was created for co-branding with interested schools. Further, a student video competition was held, and a Second Life half-island was purchased. To increase value to members, an AIS discussion forum was created entitled "What in the World can AIS do for you?" AIS adopted the Senior Scholars' "Basket" of six top IS journals and featured it prominently on the AIS Web site. Including six journals in the basket was seen as a solution to enhance the promotability of academics in a relatively new field in which the quality assessments of journals was previously highly variable across universities. The basket includes our own *Journal of AIS*, providing welcome recognition of the journal's quality. Finally, the AISWorld and AIS Web sites were integrated based on a design contest that was won by Dave Haseman and Craig Claybaugh of University of Wisconsin at Milwaukee.

THE FUTURE OF AIS

The next two Presidents as of 2009 are David Avison (2008) of ESSEC Business School in Paris, and Bernard Tan (2009) of National University of Singapore. The member initiative, run by David Avison during 2007, and continued by Bernard Tan after Avison took over as President in July 2008, is a key component in determining the future of AIS. He has found that there is no shortage of ideas for extending the impact of AIS and broadening its reach. As of early 2009, the current membership is 3965, which is estimated to be over 50% saturation of the overall population of an estimated 7000 IS academics worldwide. Tables 1–3 provide details of the current AIS membership.

Conferences, journals, and associations are contacting AIS on an accelerating basis to explore partnering or merger. The international organization is taking on more regular roles in providing AIS plenary presidential addresses to all three regional conferences (AMCIS, ECIS and PACIS) in addition to ICIS. The Association for IS'

Table 1   Membership by category.

Academic	2609
Student	1273
Professional	59
Retired Academic	24
Total	3965

Table 2   Membership by region.

The Americas	2038
Europe and Africa	1289
Asia Pacific	638
Total	3965

Table 3   Top ten membership levels by country.

United States	1768
Canada	170
Germany	165
United Kingdom	161
Australia	156
France	132
China	98
Taiwan	90
Sweden	79
Italy	74
Total top ten	2893

intellectual capital is recognized to be one key to improving the ability of AIS to become more self-sustaining and to reduce fund-raising pressure on conferences.

The Association for IS has come a long way from having nearly all operations completed by volunteers to having a professional staff and bank of consultants. The "professionalization" of the office has been gradual but dramatic. Memberships have grown over the years on a healthy basis. The decision to include practitioner members and student chapters should take AIS to a new level.

Artificial–Association

Association for Library Collections and Technical Services

Charles Wilt
Association for Library Collections and Technical Services (ALCTS), Chicago, Illinois, U.S.A.

Abstract

The Association for Library Collections and Technical Services (ALCTS) is one of the 11 divisions of the American Library Association (ALA). In 1957, this division was established from the merger of several ALA units with common interests. It was called the Resources and Technical Services Division until 1989 when the present name was approved by the membership. ALCTS focuses on six areas: standards, best practices, publications, continuing education, professional development, and information exchange.

INTRODUCTION

The Association for Library Collections and Technical Services (ALCTS) is responsible for the following activities: acquisition, identification, cataloging, classification, and preservation of library materials, the development and coordination of the country's library resources, and those areas of selection and evaluation involved in the acquisition of library materials and pertinent to the development of library resources. ALCTS has specific responsibility for

1. Continuous study and review of the activities assigned to the division
2. Conduct of activities and projects within its area of responsibility
3. Synthesis of activities of all units within the American Library Association (ALA) that have a bearing on the type of activity represented
4. Representation and interpretation of its type of activity in contacts outside the profession
5. Stimulation of the development of librarians engaged in its type of activity, and stimulation of participation by members in appropriate type-of-library divisions
6. Planning and development of programs of study and research for the type of activity for the total profession

DESCRIPTION

Vision

ALCTS is the recognized dynamic leader and authority for principles, standards, best practices, continuing education, and new developments in the selection, management, and preservation of all information resources.

Mission

To shape and respond nimbly to all matters related to the selection, identification, acquisition, organization, management, retrieval, and preservation of recorded knowledge through education, publication, and collaboration.

ALCTS is one of 11 divisions of ALA. With about 3500 personal, organizational, and international members, ALCTS represents a wide range of the library community, public, academic, and special. Members of ALCTS have the opportunity of joining one of five sections: Acquisition (AS), Cataloging and Metadata Management (CaMMS), Collection Management (CMS), Preservation and Reformatting (PARS), and Continuing Resources (CRS). Each of these sections represents specific areas of interest within ALCTS and offers members an opportunity to meet and discuss issues, trends, and developments with others with similar interests.

A staff of three, who specialize in specific areas, manages ALCTS, which supports the ongoing work of the association. The Executive Director is responsible for the overall management and fiscal health of the association. Other staff oversee the website, continuing education, programming, meeting management, publications, and membership.

Products, Services, and Member Benefits

The association focuses on six areas: standards, best practices, publications, continuing education, professional development, and information exchange. Within these six areas, ALCTS offers products and services for librarians throughout their careers.

Standards

The purpose of ALCTS is to develop, evaluate, revise, and promote standards. Standards are vital to the organization

Encyclopedia of Library and Information Sciences, Fourth Edition DOI: 10.1081/E-ELIS4-120053686

of information, such as metadata standards, cataloging revisions, resource description and access (RDA), and preservation. Through its own committee structure and liaison work with outside organizations, ALCTS members make a significant impact. ALCTS appoints the ALA representative to the National Information Standards Organization (NISO), who informs the association of critical changes in U.S. and international standards.

Best Practices

The association is the authority on many vital issues confronting libraries today. Its members provide input to national bodies such as ALA on practices and programs relating to technical services.

Publications

The association publishes outright or has significant input into many of the crucial publications in technical services including RDA and the Library of Congress subject headings. The ALCTS publications cover the range of our interests from e-publications to monographs to directories to guides. *Library Resources & Technical Services* (*LRTS*), ALCTS quarterly research journal, consistently ranks as one of the premier research journals in library technical services. The *ALCTS News*, a timely news source, keeps the membership informed of the activities of the association.

Continuing Education

Substantial and high-quality conference programming including outstanding preconferences is offered by ALCTS. The association offers nearly 12–15 programs on a variety of topics at each ALA Annual Conference. ALCTS offers web-based continuing education on a broad array of topics. The ALCTS e-Forum brings a monthly discussion of current topics to over 4000 subscribers.

Professional Development

ALCTS offers great opportunities for members to increase their knowledge and further their professional careers. Besides serving on committees, there are opportunities to work with outside groups and organizations in developing important policies for the library community. Members represent the library community on the RDA Steering Committee, the Program for Cooperative Cataloging, and the NISO.

Information Exchange

There are forums for the exchange of information among members and nonmembers. Twice a year, at the Midwinter Meeting and the Annual Conference, over 40 topical interest groups meet to address some important issues facing the library community. There are numerous electronic discussion lists available for subscription.

HISTORY

ALCTS has been a division of ALA since 1957, when the Resources and Technical Services Division (RTSD) was formed from the merger of several ALA units with common interests. In 1989, the membership of RTSD voted to change the name of the division to ALCTS. It was during this same period that ALCTS began a cooperative shared staffing arrangement with the Library Administration and Management Association. This shared staffing arrangement lasted over a decade. In January 2001, the boards of each division approved the dissolution of the arrangement effective September 1, 2001. The first bylaws for RTSD provided for four sections: Acquisitions, Cataloging and Classification, Copying Methods, and Serials. In 2007, ALCTS celebrated its 50th anniversary as a division of ALA. In 2014, ALCTS celebrated the 25th anniversary of the name change from RTSD to ALCTS.

The Acquisitions Section was formed from the Board on Acquisition of Library Materials, created in 1951. In 1973, the Resources Section was formed by the merger of the RTSD Resources Committee and the AS. In 1991, the Resources Section changed its name to Acquisition of Library Materials Section, and some of its committees were assigned to a newly formed Collection Management and Development Section (CMDS). In 1992, the section again adopted the name Acquisitions Section.

CaMMS dates from 1900 when the Cataloging Section of ALA was founded. In 1940, the section became the Division of Cataloging and Classification (DCC). The DCC included a Council of Regional Groups, formed by Margaret Mann in 1923.

The Copying Methods Section became the Reproduction of Library Materials Section in 1967. This section dates to 1936, when the Committee on Photographic Reproduction of Library Materials was created. In 1948, the committee changed its name to Committee on Photo-Duplication and Multiple Copying Methods. This committee was replaced by the Copying Methods Committee, which joined RTSD as the Reproduction of Library Materials Section in 1957. In 1994, the section merged with the Preservation of Library Materials Section (PLMS) to form the PARS.

The CRS has its origins in the Serials Round Table formed in 1929.

In 1980, the original four sections were joined by the PLMS, in recognition of the growing awareness and knowledge of preservation issues. In 1994, PLMS merged with the Reproduction of Library Materials Section to form the PARS.

GOVERNANCE

Bylaws

The ALCTS bylaws govern and guide the association. In its 17 articles and numerous subsections, the bylaws lay out the essence of ALCTS.

Board of Directors

A 19-member Board of Directors governs ALCTS. The board has authority over the affairs of ALCTS. It sets the policies and programs of the association in relationship to the goals and objectives identified by the membership and by ALA. The Board of Directors is composed of voting and nonvoting officers and members. These officers and members have particular duties as set forth in the ALCTS bylaws. The voting members of the board include the president, president-elect, past-president, ALCTS councilor to the ALA Council, three directors-at-large, the chairs of the sections of ALCTS, and the chairs of the Budget and Finance Committee, the Organization and Bylaws Committee, the Planning Committee, and the Affiliate Relations Committee. The ALCTS Executive Director, the editor of the ALCTS News, and the board Intern are ex officio nonvoting members.

The Executive Committee of the board meets to plan or carry out actions resulting from previous board action, to review financial matters of the association, and to plan future board action. The Executive Committee is composed of the president, president-elect, past-president, councilor, and the Executive Director.

Each of the five ALCTS sections is governed by its own executive committee, which is elected by its membership. The chair of the section serves on the ALCTS Board of Directors.

STRATEGIC PLAN

In 2015, the ALCTS Board reviewed and approved a revised strategic plan. This plan guides the association in the development and implementation of a wide range of activities, policies, and goals.

ALCTS Strategic Plan 2015

Adopted June 26, 2015

Preamble

This 3 yr strategic plan was developed by the ALCTS Planning Committee and approved by the Board of Directors at the 2015 ALA Annual Conference. The plan, similar in design to its 2011 predecessor, is not meant to encompass the whole of ALCTS activities, but rather those areas where increased focus is needed. The plan is fluid, that is, achieved objectives may roll off the plan prior to its expiration, while emerging areas of strategic importance may be added as warranted. The Planning Committee will make such recommendations to the Board of Directors annually, and the plan will be updated accordingly. The strategic plan in its entirety will be reviewed and updated during the 2017–2018 term with the expectation that a new plan will be approved at the 2018 ALA Annual Conference.

1. Increase awareness of the ALCTS mission and activities to outside groups.
   a. Raise awareness of our mission and our contributions to the profession.
   b. Strengthen advocacy for user access to resources and information.
2. Increase participation in ALCTS activities.
   a. Encourage a culture of year-round participation in meetings in all formats.
   b. Expand opportunities for virtual members to participate in the association's activities.
   c. Reach out to underrepresented and underinvolved groups (support staff and students) to encourage their participation in webinars and online meetings.
3. Develop ALCTS as a vibrant, relevant organization.
   a. Give special attention to member retention.
   b. Develop a culture of continuous review and examination of programs and services to maintain relevance.
   c. Recruit new members, particularly students and faculty in iSchools and library programs, and public and special librarians.
   d. Identify and address requirements for the financial sustainability of ALCTS, particularly fund-raising.

Financial Plan

A sound and responsible financial base that is derived from a broad range of revenue streams supports ALCTS' vision and mission. Planning and management of ALCTS' financial resources ensures the association's fiscal health.

The environment in which ALCTS operates is distinguished by a rapidly changing technology; a level-declining and changing membership base; increased opportunities in nontraditional library fields; changing member expectations and types of involvement; increased competition; increasing the number of initiatives requiring sophisticated technology; changing attitudes of leaders and potential members about how the association operates; evolving staffing patterns, levels, and responsibilities within acquisitions, cataloging, serials, preservation, and collection development areas; an increased need for practical staff training; and an increased international influence.

In response, ALCTS recognizes the need to ensure efficient use of association resources through a continued sound and broad-based financial infrastructure to support its broad-based activities and initiatives. Providing the necessary revenue and staff to support important standards-driven programs and community-based programs is crucial in supporting ALCTS interests in relevant standards-based activities in the United States, internally within ALA and externally in cooperation with other library organizations and internationally working with the International Federation of Library Associations and Institutions (IFLA). Membership recruitment and retention is the foundation upon which any association is based. ALCTS strives to increase its membership revenue and membership base by providing diverse opportunities, creating services that appeal to staff who have paraprofessional library positions, exploring ways to attract nonlibrary information management professionals, and focusing membership campaigns on ALCTS new and existing programmatic areas. Continuing education and professional development programs offer reliable and consistent revenue, in addition to supplementing the career advancement of library professionals and paraprofessionals. A key to any financial plan is a well-rounded publishing endeavor. This informs members and nonmembers of current trends and issues in librarianship and enables ALCTS to maintain a high visibility.

Revenue growth and fiscal responsibility ensure that ALCTS has the resources to continue to develop programs and services for its members in the long term.

THE DIVISION

Committees

Committees of ALCTS support the operation of the association, produce revenue through publications and continuing education, support the advocacy of the association in areas of interest, and provide a forum for members and nonmembers on a wide variety of topics. Association-wide committees give operational support to ALCTS, produce the products and services ALCTS is known for, provide a forum for the discussion of trends and issues important to the library community in ALCTS areas of interest, and help to promote and develop ALCTS members and leaders. These committees fall into four broad areas: operational, revenue, advisory, and topical.

Operational Committees

The operational committees of ALCTS advise, recommend, implement, and guide the association. The Budget and Finance Committee, which is primarily responsible for the fiscal health of the association, reviews, approves, and recommends acceptance of the ALCTS annual budget. This committee performs budget analyses and fiscal planning based on the ALCTS financial plan. It advises all ALCTS groups on fiscal matters, including fiscal implications of all division publications and programs. The Organization and Bylaws Committee advises the Board of Directors and through it the association on the establishment, functions, and discontinuance of sections, committees, and other groups as the needs of the association may require. It reviews documents and practices and advises the officers of the association and its sections on the bylaws, policies, and procedures of the association. It has authority on questions of the bylaws and any amendments needed to that document. With the adoption of the ALCTS Strategic Plan and the continuing planning efforts, the Planning Committee is a key contributor to the operations of ALCTS. It oversees this planning effort by continually updating the strategic issues, by coordinating the planning initiatives of the association and its sections, and by informing the ALCTS leadership through workshops and announcements on planning in general.

Revenue Committees

The ALCTS Revenue Committees are Continuing Education, Fund-raising, Membership, Program, and Publications. Each has a specific charge and area of responsibility. The Continuing Education Committee oversees the association's programs outside the conference setting. It develops criteria for continuing education, reviews continuing education plans, recommends approval of continuing educational offerings, and researches the educational needs of library staff at all levels in the areas supported by ALCTS. The Fund-raising Committee establishes fund-raising goals and priorities that support the association's strategic and financial plans. It also secures financial and in-kind support for ALCTS activities and further develops relationships with the commercial and private sectors. The Membership Committee is charged as the primary resource to the development and implementation of a continuous campaign to recruit and retain members for ALCTS (see Membership section). The Program Committee oversees the conference and preconference program planning, recommending a slate of programs to the board, and provides advice to those planning programs and preconferences. It is the work of the Publications Committee that produces the range of monographic and serials publications. This committee coordinates the association's publication program and provides a forum for discussion of the editorial policy for ALCTS' serial publications.

Advisory Committees

There are a number of committees whose purpose is to monitor trends and issues outside ALCTS, provide for the professional development of ALCTS members, and offer a forum to discuss research: Advocacy and Policy, Standards, Affiliate Relations, International Relations, Leadership Development, and *LRTS* Editorial Board. International Relations recommends members to serve on international committees, particularly within IFLA, and to monitor international developments. Leadership Development is responsible for the professional development of ALCTS member leadership through orientation and leadership training. The *LRTS* Editorial Board advises the editor on matters relating to editorial policies and journal content and assists in the selection of contributors and the evaluation of manuscripts. The Standards coordinates ALCTS input, development, and information on the wide range of standard-producing organizations. Affiliate Relations is the primary contact group with state and regional technical services organizations. Advocacy and Policy purpose to monitor state, national, and international advocacy and public policy developments and issues.

INTEREST GROUPS

The informal discussion of developments, trends, and issues is supported by ALCTS through its interest groups that are managed by the Interest Group Coordinator. Any group of ten or more members of ALCTS, for divisional groups, or section, for those interested in discussing common problems, which fall within the interests of ALCTS or the section, may form an interest group upon written petition, describing the purpose of the group, and upon approval of the Section Executive Committee or ALCTS Board. The petition shall include the requirements for membership. The ALCTS interest groups offer the library community a forum to explore a variety of topics and that addresses the needs of technical services professionals at different levels and in different types of libraries. The ALCTS interest groups include

• Creative Ideas in Technical Services
• Electronic Resources
• Electronic Resources Management
• Functional Requirements for Bibliographic Records
• Linked Library Data
• MARC Formats Transition
• Metadata
• New Members
• Newspaper
• Public Libraries Technical Services
• Publisher–Vendor–Library Relations
• Role of the Professional Librarian in Technical Services
• Scholarly Communication

• Technical Services Managers in Academic Libraries
• Technical Services Directors of Large Research Libraries
• Technical Services Workflow Efficiency

For information on section level discussion groups, visit each section's website listed under "Sections."

SECTIONS

The five ALCTS sections, AS, CaMMS, CMS, CRS, and PARS, are important contributors to the association. Through the sections, members of ALCTS have an opportunity to interact with other members who share common interests and concerns. The sections consider issues, trends, and policies related to their specific area of expertise. The sections produce programs and publications and in many cases serve as the voice of ALCTS on a variety of matters that go beyond ALCTS. Any group of 50 or more members of ALCTS or of ALA, whose interest falls within ALCTS but is distinct from that of any existing section, may be established as a section upon written petition, and upon approval by the association. The name of the new section must clearly indicate its field of activity. Each section defines its own functions and manages its own affairs, provided, however, that no section shall adopt bylaws or other rules for the transaction of its business that are inconsistent with those of ALCTS or engage in any activity in conflict with the activities of ALCTS. Each section is governed by an executive committee, which is elected yearly by its members. The chair of the section serves on the ALCTS Board of Directors.

Acquisitions Section

This section is dedicated to acquisition and associated bibliographic control for all formats of information resources through purchase, lease, and other access methods and in all types of libraries.

The AS does this through programs and publications specifically targeted to library staff whose primary responsibility is acquisitions and related areas. Members of AS can contribute through a number of committees and discussion groups within the section. Each year, the AS presents the HARRASSOWITZ Leadership in Acquisitions Award to a person who has shown outstanding leadership in the field of acquisitions. The committees of AS are Acquisitions Organization and Management, Education, Leadership in Library Acquisitions Award, Policy and Planning, Publications, Research and Statistics, and Technology.

Cataloging and Metadata Management Section

CaMMS' primary focus is on providing leadership to the library community on cataloging, metadata and related discovery, and access issues through encouragement and promotion of activities relating to all formats and to all types of institutions. CaMMS does this not only through programs and publications but also as a significant contributor to the revisions of cataloging rules and metadata standard policies nationally and internationally. The Committee on Cataloging: Description and Access and the Subject Analysis Committee, representing a broad range of ALCTS members, ALA units, and national and international organizations, advise regularly on the structure and function of the foundation tools the library community utilizes to make information accessible to its users. The prestigious Margaret Mann Citation, given yearly to the person who has made superlative contributions to cataloging and classification, is presented by CaMMS. Its members have the opportunity of serving on these other CaMMS committees, Cataloging: Asian and African Materials; Cataloging of Children's Materials; Continuing Education; Research and Publications; Recruitment and Mentoring; and Policy and Planning. In addition to the committees, CaMMS supports numerous task forces, subcommittees, and six discussion groups.

Collection Management Section

The CMS contributes to the library community in all matters relating to collection management and development, selection, and evaluation of library materials in all types of institutions. It does this through interest groups, which address administration, electronic resources, practical issues, and evaluation and assessment for collection management. Other interest groups aim at collection management professionals at varying levels of management in three distinct library organizations: Chief Collection Development Officers of Large Research Libraries, Collection Development Librarians of Academic Libraries, and Collection Management in Public Libraries. The CMS gives one award: the ProQuest Coutts Award for Innovation. Members can gain valuable experience by serving on one of CMS' committees: Continuing Education, Planning, and Publications.

Continuing Resources Section

The CRS has its role in fostering the importance of serials through the distribution of information concerning serials, through open discussion of the trends and issues of serials, and through publications. The CRS seeks to encourage specialized training for librarians and other library staff in the field of continuing resources and supports the need for a serials curriculum. The CRS gives two awards: the Ulrich's Serials Librarianship Award for superior serials librarianship and the First Step Award, a Wiley Professional Development Grant, to new librarians committed to serials librarianship. Members of the CRS have many opportunities to participate in the section, including

committees and discussion groups. The CRS committees are Acquisitions, Education, Research and Publications, Continuing Resources Cataloging, Policy and Planning, Standards, and Holdings Information. The two interest groups are College and Research Libraries and Access to Continuing Resources.

Preservation and Reformatting Section

The PARS provides leadership in the areas of librarianship relating to the preservation and reformatting of library materials in all types of institutions and in the application of new technologies to assure continued access to library collections. It is most noted for the Preservation Administrators Interest Group (PAIG), which serves as one of the most prominent communication vehicles for preservation staff in the country. The PARS has three awards: Paul Banks and Carolyn Harris Preservation Award, the George Cunha and Susan Swartzburg Award, and the Jan Merrill-Oldham Professional Development Grant. PARS members are responsible for revising the ALA Preservation Policy, one of ALA's major policy statements for the library community, adding digital preservation to the policy in 2008. The PARS provides members with many opportunities to serve and influence the preservation community. Its interest groups include Books and Paper, Intellectual Access, Digital Conversion, Digital Preservation, and Promoting Preservation. Its committees include Program, Planning, and Publications, and Preservation Standards and Practices. It has 1000 members.

PUBLISHING

ALCTS publishing is a diverse effort from serials to monographs, from web publications to "best practices." Publications cover all the areas of ALCTS interest: acquisition, cataloging and classification, collection management and development, preservation and reformatting, and serials.

Library Resources & Technical Services

Established in 1957, *LRTS* is the official electronic quarterly journal of ALCTS supporting the theoretical, intellectual, practical, and scholarly aspects of collection management, acquisitions, cataloging and metadata, preservation and reformatting, and continuing resources, by publishing articles (subject to double-blind peer review), book reviews, editorials, and correspondence. Each issue includes reports of current, ongoing research on technical services, and related issues. The articles serve as the record of scholarly communication in the field. *LRTS* is one of the most frequently cited journals in the field. *LRTS* includes shorter notes reporting unique and evolving technical processes and research methods and substantive book reviews on new publications. *LRTS* is read by librarians in all types of libraries worldwide.

ALCTS News

The *ALCTS News* is the primary public communications voice of the association. Published as a continually updated news source, *ALCTS News* features the association's news, committee and discussion group reports, reports of programs and preconferences, election results, informative editorials, and related news. *ALCTS News* is free to anyone interested in ALCTS activities, products, services, and news.

Guides Series

The Guides series, acquisitions and collection management, offers "best practices" and "hands-on" information on a variety of topics across ALCTS.

Sudden Selector's Guides

This series helps library workers get familiar with new subject areas by covering the tools, resources, people, and organizations they will need to keep collections relevant.

Monographs Collection

The Monographs Collection publishes relevant monographs on highly relevant topics of interest to the library community.

z687

*z*687 is an online collection of white papers and think pieces by library technical services professionals for their peers. This publication series has been discontinued.

MEMBERSHIP

ALCTS has 3500 members representing all types of libraries and library staff. Although personal members account for most of the membership, ALCTS also has 420 organizational and corporate members. Many members belong to at least one section. Section membership is free and members can join as many sections as they wish.

CONTINUING EDUCATION AND PROGRAMMING

In this era of rapid change, the need for many quality venues for Continuing Education and Programming (CE) is critical. Targeted, affordable, and accessible CE is one of the most requested benefits of members. This requires CE offerings to be both practical and theoretical,

providing basic information on a variety of topics for the beginner, but also providing advanced information for the seasoned librarian.

ALCTS offers CE opportunities to its members and the library community in general in a variety of formats: Annual Conference preconferences, Annual Conference programs, web-based courses, webinars, Midwinter symposium, virtual symposia and preconferences, and e-Forums.

Programs offered at the ALA Annual Conference bring to one place topics of particular importance in the areas in which ALCTS has an interest. These programs represent topics that are of immediate concern; offer updates to new developments, issues, and trends; and in general educate the library community as a whole.

Preconferences and symposia bring ALCTS expertise to a setting at the ALA Annual Conference and Midwinter Meeting in which the entire library community can benefit from timely topics and knowledgeable speakers. These preconferences and symposia address a specific topic over one or two days at both the practical and theoretical level.

The development of web-based CE continues as a high priority for ALCTS. Through web-based learning, library staff from across the country and internationally can participate in a meaningful educational experience. ALCTS web-based courses are intended to address the continuing educational need of those who cannot attend workshops in person. These courses address the need for basic, practical information.

The Continuing Education Committee is charged with providing leadership and coordinating the overall development of the continuing education initiatives. The Program Committee is charged with providing leadership and coordinating the overall development of programs and preconferences presented at the ALA Annual Conference and Midwinter Meeting.

AWARDS

The ALCTS Awards Program recognizes significant contributions by those working in libraries in ALCTS areas of interest. ALCTS presents 13 awards for lifetime achievement, excellence in publishing, excellence in acquisitions, cataloging, preservation, serials, innovation, collaboration, and leadership, for librarians new to the profession who show particular promise, and for support staff travel to the ALA Annual Conference. ALCTS also awards the Presidential Citation to members who have contributed significantly in recent years to the association but who do not qualify for any of the other ALCTS awards.

Ross Atkinson Lifetime Achievement Award

This award was established in 2007 to honor the legacy of Ross Atkinson, distinguished library leader, author, and scholar, whose extraordinary service to ALCTS and the library community at large serves as a model for those who follow.

The award, sponsored by EBSCO, is given to recognize the contribution of a library leader through demonstrated exceptional service to ALCTS and its areas of interest (acquisitions, cataloging and classification, collection management and development, preservation and reformatting, and serials). The award consists of a citation and $3000.

Edward Swanson Memorial Best of *LRTS* Award

The Best of *LRTS* Award is given to the author(s) of the best paper published each year in *LRTS*. Each of the papers published in the volume year is eligible for consideration with the exception of official reports and documents, obituaries, letters to the editor, and biographies of award winners. The award is given to the paper whose content is a significant contribution about one or more issues addressed by ALCTS and its sections. In 2011, it was named in memory of Edward Swanson in recognition of his numerous contributions to making *LRTS* an outstanding professional journal.

ALCTS Outstanding Publications Award

This award honors the author or authors of the year's outstanding monograph, article, or original paper in the field of technical services, including acquisitions, cataloging, collection management, preservation, continuing resources, and related areas in the library field. Papers published in *Library Resources & Technical Services*, the ALCTS journal, are ineligible.

Ulrich's Serials Librarianship Award

The Ulrich's Serials Librarianship Award is presented for distinguished contributions to serials librarianship demonstrated by such activities as leadership in serials-related activities through participation in professional associations and/or library education programs, contributions to the body of serials literature, conduct of research in the area of serials, development of tools or methods to enhance access to or management of serials, and other advances leading to a better understanding of the field of serials. ProQuest sponsors the Ulrich's Award.

First Step Award, a Wiley Professional Development Grant

The First Step Award, offered by John Wiley & Sons, provides a librarian new to the serials field with the opportunity to broaden perspectives and supports professional development by funding travel to the ALA Conference and by participating in Serials Section activities. All ALA members with five or fewer years of professional experience in the serials field, who have not previously attended an ALA Annual Conference, are eligible. This award is presented to the librarian who shows a commitment to professional development in the serials field as evidenced by participation in continuing education activities, workshops, previous participation in professional activities, and a commitment to or interest in serials-related work.

HARRASSOWITZ Leadership in Library Acquisitions Award

The Leadership in Library Acquisitions Award, sponsored by HARRASSOWITZ in honor of Dr. Knut Dorn, is given to recognize the contributions by and outstanding leadership of an individual to the field of acquisitions librarianship. This recognition is made for individual achievement of the highest order.

Margaret Mann Citation

The Margaret Mann Citation is awarded by CaMMS for outstanding professional achievement in cataloging or classification through either publication of significant professional literature, participation in professional cataloging associations, or valuable contributions to practice in individual libraries. A $2000 scholarship is donated by OCLC to the U.S. or Canadian library school of the winner's choice. The Mann Citation represents the highest recognition of a professional in the cataloging or classification area.

Esther J. Piercy Award

The Esther J. Piercy Award was established by the RTSD of ALA in 1968 in memory of Esther J. Piercy, editor of the *Journal of Cataloging and Classification* from 1950 to 1956 and of *LRTS* from 1957 to 1967. This award, sponsored by YBP, Inc., recognizes the contribution of a librarian with not more than 10 yr of professional experience who has shown outstanding promise for continuing contribution and leadership to those areas of librarianship included in library collections and technical services.

Paul Banks and Carolyn Harris Preservation Award

The Paul Banks and Carolyn Harris Preservation Award, sponsored by Preservation Technologies, L.P., recognizes the contribution of a professional preservation specialist who has been active in the field of preservation and/or conservation for library and/or archival materials and who has shown superb leadership in professional associations at local, state, regional, or national level; has made significant contributions to the development, application or utilization of new or improved methods, techniques, and routines and to the professional literature; and has been a leader in training and mentoring in the field of preservation.

Artificial-Association

George Cunha and Susan Swartzburg Award

Established in 2007 by the PARS, the award honors the memory of George Cunha and Susan Swartzburg, early leaders in cooperative preservation programming and strong advocates for collaboration in the field of preservation.

The award, sponsored by the Library Binding Institute ($1250), acknowledges and supports cooperative preservation projects and/or rewards individuals or groups that foster collaboration for preservation goals. Recipients of the award demonstrate vision, endorse cooperation, and advocate for the preservation of published and primary source resources that capture the richness of our cultural patrimony. The award recognizes the leadership and initiative required to build collaborative networks designed to achieve specific preservation goals. Since collaboration, cooperation, advocacy, and outreach are key strategies that epitomize preservation, the award promotes cooperative efforts and supports equitable preservation among all libraries, archives, and historical institutions.

ProQuest Coutts Award for Innovation

The ProQuest Coutts Award for Innovation is given to recognize the contribution of an individual who has demonstrated innovation and excellence in the practice of electronic collection management and development. This new award has been established by the CMDS to recognize significant and innovative contributions to electronic collections management and development practice. The award consists of a citation, a listing on the ALCTS Awards website, and $2000.

Outstanding Collaboration Citation

This Outstanding Collaboration Citation recognizes and encourages collaborative problem-solving efforts in the areas of acquisition, access, management, preservation, or archiving of library materials. It recognizes a demonstrated benefit from actions, services, or products that improve and benefit in providing and managing library collections.

Jan Merrill-Oldham Professional Development Grant

The Jan Merrill-Oldham Professional Development Grant is awarded by the ALCTS PARS to provide librarians and paraprofessionals new to the preservation field with the opportunity to attend a professional conference and encourages professional development through active participation at the national level. The grant is to be used for airfare, lodging, and registration fees to attend the ALA Annual Conference.

Presidential Citations

These very special awards honor ALCTS members who make significant contributions to the association and to the profession but whose accomplishments do not fall within the criteria for ALCTS' other awards. They are awarded by the current ALCTS President. The Presidential Citation is intended to recognize distinguished achievement by a member or members.

ALCTS Honors

ALCTS Honors is intended to recognize ALCTS members who have dedicated themselves to ALCTS through their service. That service can be holding an office, being a committee member, a committee chair, task force member, or other service as deemed applicable. It is the totality of service that this award seeks to acknowledge.

CONCLUSION

ALCTS, as one of the "functional" divisions of ALA, represents the voice of the library community in matters that reflect its interests, issues, and concerns: acquisitions, cataloging and metadata, collection management, preservation and reformatting, and continuing resources. In this role, ALCTS is seen as the "expert" by a vast array of constituent organizations and individuals, nationally and internationally. ALCTS advocates for creating, collecting, organizing, delivering, and preserving information resources in all forms to the library and information communities through the development and promulgation of principles, standards, and best practices. ALCTS leads through its members by fostering educational, research, and professional service opportunities. ALCTS is committed to quality information, universal access, collaboration, and lifelong learning.

Association for Library Service to Children (ALSC)

Virginia A. Walter
Department of Information Studies, University of California, Los Angeles, Los Angeles, California, U.S.A.

Abstract

The Association for Library Service to Children (ALSC) is one of 11 divisions of the American Library Association (ALA). Its members are drawn from the ranks of children's and school librarians, library educators, commercial vendors, and representatives of the children's book publishing community. Starting as the Club of Children's Librarians in 1900, it now boasts more than 4000 members and a broad-based program of activities focusing on advocacy, education, and partnerships. While best known outside of the library community for its implementation of the Newbery and Caldecott awards for distinguished contributions to American children's literature, the Association today engages in many activities and initiatives aimed at improving library services for children through the identification and promotion of best practices, the professional development of its members, and a strong program of advocacy.

INTRODUCTION

The Association for Library Service to Children (ALSC), one of 11 divisions of the American Library Association (ALA), is the primary American professional association for those who are concerned with the advancement of library service to children. Its members include children's librarians, school librarians, library educators, commercial vendors, and members of the children's book publishing community.

This entry looks briefly at the historical development of ALSC and discusses its current mission and activities. Among its more significant contributions to library service to children are the implementation of prestigious book awards, initiatives aimed at improving practice, continuing education and professional development, and advocacy.

HISTORY

Several of the pioneers of library service to children saw the need for both networking among their ranks and for acceptance within the larger library community. To this end, Caroline Hewins, Alice Jordan, and Anne Carroll Moore brought together a group of like-minded colleagues at the ALA Conference in Montreal in 1900; they called themselves the Club of Children's Librarians.[1] A year later, in 1901, the Section for Library Work for Children was recognized by the parent organization, the ALA. Jacalyn Eddy notes that this was an important event in the recognition of library work for children as a bona fide specialization within the library profession.[2] That section evolved into the American Association of School Librarians in 1915 and the Children's Library Association and the Young People's Reading Roundtable in 1930. In 1941, those two associations and the roundtable merged to become the Division for Children and Young People. When ALA went through a restructuring in the mid-1950s, that division split into the Children's Library Association and the Association of Young People's Librarians, the latter being concerned with the relatively new specialization of teen services. In 1958, the Children's Library Association was renamed the Children's Services Division of the ALA and announced that it had "responsibility to speak for the ALA on those matters which concern children's books and other library materials and their use in libraries in any type which serve children."[3] In 1976, the association changed its name to the Association for Library Service to Children, and it has retained this name since then.

ALSC TODAY

There are currently more than 4000 members of ALSC, making it the largest organization in the world dedicated to library services for children. Its 2006–2011 Strategic Plan claims as its core purpose: Creating a better future for children through libraries. It aims to accomplish that purpose through three primary goals involving advocacy, education, and collaboration.[4]

Governance

ALSC accomplishes its goals and objectives through a complex governance structure. An elected Board of Directors provides leadership to the Association as a whole. Members participate directly in the work of the Association through committees. Membership on most committees

Encyclopedia of Library and Information Sciences, Fourth Edition DOI: 10.1081/E-ELIS4-120043639

is by appointment of the vice president or president; but a few, specifically the more prestigious awards committees, are comprised of a mixture of elected and appointed representatives. Traditionally, ALSC committee members were required to attend both the midwinter and annual conferences. However, the development of new forms of electronic communication has made it possible recently for some committees to accomplish their work virtually without actually being present at the conferences.

ALSC committees are divided into eight priority groups: Child Advocacy, Evaluation of Media, Professional Awards and Scholarships, Organizational Support, Projects and Research, Awards, Partnerships, and Professional Development. A Priority Group Consultant, ordinarily an ALSC veteran with much experience serving on committees, is assigned to each of these committee clusters and is available to troubleshoot, answer questions, and provide leadership and guidance where needed.[5]

ALSC maintains collegial relationships with the other two ALA divisions that promote library services to young people: The Young Adult Services Association (YALSA) and the Association of School Librarians (AASL). Their Boards meet together regularly to discuss matters of mutual concern, and their representatives on the ALA Council, along with other youth services librarians serving on that body, meet informally as a Youth Caucus.

The work of the association is funded through membership dues, grants, merchandise sales, and endowments. An Executive Director and office staff located at the ALA Headquarters in Chicago support the volunteer efforts of the membership.

Activities

Awards

Probably the activity for which ALSC is best known to the general public is its implementation of the Newbery and Caldecott Medals, the oldest and most prestigious awards given to children's literature in the United States. The idea for the Newbery Medal, an award given to the most distinguished contribution to American children's literature in a given year, came not from a librarian, but from the secretary of the American Bookseller's Association, Fredric G. Melcher. In 1919, he had launched Children's Book Week, a national initiative designed to promote the output of the emerging children's book publishing industry. Invited to speak about Children's Book Week at the 1921 ALA conference, he saw the possibilities within the library community for wider organized support for children's books. Writing in 1965 as he looked back on the evolution of the award, he noted many benefits that had occurred. Children's librarians had become more interested in the new books being published each year. Their status as a profession had increased. Creative people were more likely to write for children because award-

winning books had a longer life. He also paid tribute to the children's librarians in ALA who had cultivated the strict standards that gave the medal its luster.[6]

Seventeen years later, children's book publishing had proliferated sufficiently to warrant a separate award for distinguished illustration. Again, Melcher was responsible for proposing the Caldecott Medal, which has been awarded every year since 1938. Originally, the Caldecott Medal book and runners-up were chosen by the same committee that named the Newbery Medal. In 1980, a separate Caldecott Committee was established.[7] The prestige of these two awards has grown immensely. They are now announced, along with other ALA awards given to significant contribution to children's literature and film, at a press conference at the annual Midwinter Conference in the largest auditorium at the conference site. The award-winning author and illustrator and the ALA President are traditionally invited to appear the next morning on the Today show, and the awards are announced in the national news media. The gold seals on the books have become symbols of quality in literature for children.

Over the years, the ALSC leadership, with ALA's approval, has added many additional awards to highlight and promote contributions to different aspects of children's literature and related media. The Laura Ingalls Wilder Award was first given to its namesake in 1957; it honors an author or illustrator whose books published in the United States have made a lasting contribution to literature for children. Between 1960 and 1980, the Wilder Award was given every 5 years. From 1980 to 2001, it was awarded every 3 years. Beginning in 2001, it has been awarded every 2 years.[8] The Mildred L. Batchelder Award honors a former executive director of the Association for Library Service to Children. Established in 1966, it is a citation awarded to an American publisher for a children's book for the most outstanding book originally published in a foreign language in a foreign country and subsequently translated into English and published in the United States.[9] The Andrew Carnegie Medal for Excellence in Children's Video was awarded for the first time in 1991.[10] In 1996, ALSC and Reforma, the National Association to Promote Library and Information Services to Latinos and the Spaniish-Speaking, established the Pura Belpre Award, presented to a Latino/Latina writer and illustrator whose work best portrays, affirms, and celebrates the Latino cultural experience in an outstanding work of literature for children and youth. Originally given every other year, it becomes an annual award in 2009.[11] Nonfiction, while eligible for both the Newbery and Caldecott awards, is singled out for special attention with the Robert F. Sibert Informational Book Medal, established in 2001.[12] Awards that honor contributions by African-American authors and illustrators are not given by ALSC; they are an activity of the Coretta Scott King Task Force of the Ethnic and Multicultural Materials Information Exchange Roundtable (EMIERT) of ALA.[13] More recently, the American Indian

Library Association, an ALA affiliate association, has launched its own awards honoring American Indian authors and illustrators of books for children.[14] The Theodor Seuss Geisel Award, established by ALSC in 2004 and first presented in 2006, is given to the author and illustrator of the most distinguished American book for beginning readers.[15] The newest award is the Odyssey Award for Excellence in Audiobook Production. It is jointly administered by ALSC and the Young Adult Library Services Division.[16]

The awards administered by ALSC have done a great deal to publicize quality children's literature and have helped to legitimize the role of children's librarians in promoting the best in children's books. However, ALSC has also worked to improve the practice of children's librarianship in areas other than book evaluation and the related area of collection development.

Improving Practice

This entry will focus on just three activities undertaken by ALSC that have as their objectives the improvement of library service to children. The first of these is the development of a set of core competencies for children's librarians. Established in 1999, the competencies outline a broad range of skills, knowledge, and aptitudes required for those professionals who practice the specialization of children's librarianship. The competencies are divided into seven broad categories: knowledge of client group; administrative and management skills; materials and collection development; programming skills; advocacy; public relations, and networking skills; and professionalism and professional development. As the authors of a book that provides resources and examples of best practices related to the competencies point out, children's librarians are "the original multitaskers of the library world."[17] The ALSC competencies are intended to guide children's librarians as they develop their professional skills and also to inform public library administrators about what they can and should expect from these specialists.

Every Child Ready to Read (ECRR) is a collaborative initiative sponsored by ALSC and the Public Library Association (PLA). ECRR began with a partnership between PLA and the National Institute of Child Health and Human Development (NICHD). In 2000, NICHD had released the National Reading Panel's Report outlining research-based findings on reading development in America's children. PLA partnered with NICHD to disseminate the report through public libraries. The next step in the collaboration was to develop model public library programs that incorporated the research. PLA contracted with two emergent literacy researchers, Dr. Grover C. Whitehurst and Dr. Christopher Lonigan, to develop a model program for parents and caregivers that would give them the skills for preparing young children to read.

ALSC joined PLA as a partner in this endeavor. They selected 20 demonstration sites representing a broad range of library size and population demographics. The sites tested and evaluated the materials. In 2002, a second round of evaluation testing was conducted at 14 sites. Finally, the materials that were developed by Whitehurst and Lonigan were packaged as scripted workshops and made available for purchase or through free downloads on the ALA Web site.[18] In addition to the research-based curriculum that is aimed at encouraging early literacy skills in infants, toddlers, and preschoolers, there is an emphasis on modeling techniques such as dialogic reading during regular library storytimes.[19] Anecdotal evidence suggests that the ECRR approach to library-sponsored parent and caregiver education has been widely adopted, but an evaluation study is currently being conducted to see just how broadly the materials are being used.

Finally, El dia de los ninos/El dia de los libros (Children's Day/Book Day) is an activity that calls attention to the growing Latino and Spanish-speaking population of the United States. Dia, as the initiative is known, is a celebration of children, families, and reading that emphasizes the importance of literacy for children of all linguistic and cultural backgrounds. Children's book author Pat Mora proposed linking the already existing observation of Children's Day with literacy. ALSC has joined with REFORMA to sponsor the annual event, now celebrated on April 30. Funding from the W.K. Kellogg Foundation and Target has helped to generate promotional materials and national interest. Public libraries have observed Dia in a number of ways, from community festivals to author visits and booklists celebrating Latino culture and heritage.[20]

Continuing Education and Professional Development

Programs such as ECRR and Dia are initiatives designed to generate best practices in particular areas of library service to children. Through a wide program of continuing education, ALSC aims to provide multiple forums for children's librarians to develop professionally and presumably to then provide increasingly effective service to children.

There are three primary vehicles for providing continuing education opportunities to the ALSC membership. These are conference programs, regional institutes, and publications. The ALA holds two conferences each year. The Midwinter Conference is intended as a working meeting for the leadership of the Association and its divisions. Boards, divisions, roundtables, task forces, and committees meet, but association policy dictates that there be no programs for the general membership.

The Summer Conference, on the other hand, is a three-ring circus with program offerings from morning to night from Saturday through Monday. Many ALSC committees organize programs designed to educate and update members

Artificial–Association

on emerging trends and current thinking about best practices. The preliminary program for the 2008 summer conference in Anaheim indicates a wide array of continuing education offerings for ALSC members on topics ranging from readers' theater to programming for non-English-speaking preschoolers and their families to Library 2.0 for children's librarians and gaming for elementary school children. The summer conference is also a venue for showcasing authors, illustrators, and distinguished speakers from related fields. Distinguished pediatrician, Dr. Barry Brazelton, was the keynote speaker at the ALSC President's Program in 2008.

In addition to the programming offered through the annual summer conferences, ALSC has made a practice of holding a 2-day national institute every 2 years. These institutes, held in different parts of the country, have tended to focus on two or three current issues. In September, 2008, for example, the Institute features three tracks: technology and children's services, programming in the new millennium, and reading promotion with an emphasis on "tweens."[21]

Advocacy

ALSC organizes its working committees into priority groups, and Priority Group 1 is concerned with child advocacy. The function statement for this cluster of activities reads: "To identify, evaluate and make recommendations on issues, legislation and services concerning children on the local, state and national levels."[22] The Committees charged with this general function include Intellectual Freedom, International Relations, Legislation, Library Service to Special Population Children and their Caregivers, Early Childhood Services and Programs, Public Awareness, and School-Age Programs and Services. The International Relations committee has been active recently, calling attention to books that foster global perspectives and international understanding. Both the Intellectual Freedom and Legislation committees also meet with their counterparts from other ALA divisions in larger assemblies of interest. Together, these committees help to ensure that ALSC meets its responsibilities to advocate for the well-being of children in the larger society as well as within the library profession.

CONCLUSION

As ALSC embarks on its second century of service, it faces new challenges and opportunities. Its existing leadership is drawn from the ranks of librarians who are facing retirement; in fact, both its 2008–2009 president and vice president/president-elect have already retired. There is some danger that the Association, with its many years of tradition, may be perceived as being a little stodgy. It must find a way to attract more young librarians to join, become active, and be judged ready to assume responsible positions in the very near future. These younger professionals promise to bring fresh perspectives that may revitalize current activities of the association and suggest important new directions. One of the strategies being used to draw in younger professionals is an increased use of digital media to communicate with members—the ALSC wiki and blog as well as its discussion list—and the opportunity to participate in some committees virtually without the need to be physically present at conferences.

While the hundred-plus years of history may sometimes feel like an albatross around the neck of those members who would like the Association to be more innovative, this sense of tradition and stability also attracts many potential partners to work with its staff and leaders. Head Start, NASA, and producers of public television programs such as *Between the Lions* and *Dora the Explorer* have all initiated partnerships with ALSC.

ALSC has made a permanent imprint on children's book publishing and children's reading through the prestigious awards it bestows on works of distinction. It has yet to make the same impact on digital media for children, in spite of its efforts to promote "great Web sites for kids." Perhaps this will be a contribution of the second 100 years.

REFERENCES

1. *A Brief History of the Association for Library Service to Children*, ALSC Wiki.
2. Eddy, J. *Bookwomen: Creating an Empire in Children's Book Publishing, 1919–1939*, University of Wisconsin Press: Madison, WI, 2006.
3. CLA Becomes CSD. *Top of the News*, March 12, 1958.
4. ALSC Strategic Plan. Available at http://www.ala.org/ala/alsc/boaradcomm/alscstratplan/alscstrategic.htm.
5. *ALSC Handbook of Organization*, ALSC: Chicago, IL, 2008; 51–58.
6. Melcher, F.G. The origin of the Newbery and Caldecott Medals. In *Newbery and Caldecott Medal Books: 1956–1965*; Lee, Kingman, Ed.; The Horn Book, Inc.: Boston, MA, 1965; 1–2.
7. Randolph Caldecott Medal. Available at http://www.ala.org/ala/alsc/awardsscholarships/literaryawds/caldecottmedal/aboutcaldecott/aboutcaldecott.htm.
8. Laura Ingalls Wilder Award. Available at http://www.ala.org/ala/alsc/awardsscholarships/literaryawds/wildermedal/wildermedal.htm.
9. Mildred, L. Batchelder Award. Available at http://www.ala.org/ala/alsc/awardsscholarships/literaryawds/batchelderaward/batchelderaward.htm.
10. Andrew Carnegie Medal for Excellence in Children's Video. Available at http://www.ala.org/ala/alsc/awardsscholarships/literaryawds/carnegiemedal/carnegiemedal.htm.
11. Pura Belpre Award. Available at http://www.ala.org/ala/alsc/awardsscholarships/literaryawds/belpremedal/belpremedal.htm.

12. Robert, F. Sibert Informational Book Medal. Available at http://www.ala.org/ala/alsc/awardsscholarships/literaryawds/sibertmedal/Sibert_medal.htm.

13. Coretta Scott King Award. Available at http://www.ala.org/ala/emiert/corettascottkingbookaward/corettascott.cfm.

14. American Indian Library Association. Available at http://aila.library.sd.gov/.

15. Theodor Seuss Geisel Award. Available at http://www.ala.org/ala/alsc/awardsscholarships/literaryawds/geiselaward/GeiselAward.htm.

16. ALSC/Booklist/YALSA Odyssey Award for Excellence in Audiobook Production. Available at http://www.ala.org/ala/alsc/awardsscholarships/literaryawds/odysseyaward/Odysseyaward.htm.

17. Cerny, R. Markey, P. Williams, A. *Outstanding Library Service to Children: Putting the Core Competencies to Work*, American Library Association: Chicago, IL, 2006; v.

18. Every Child Ready to Read. Available at http://www.ala.org/everychild.

19. Ghoting, S.R. Martin-Diaz, P. *Early Literacy Storytime @ Your Library: Partnering With Caregivers for Success*, American Library Association: Chicago, IL, 2006.

20. http://www.ala.org/dia About Dia.

21. ALSC National Institute. Available at http://www.ala.org/ala/alsc/alscevents/nationalinstitute2008.cfm 2008.

22. *ALSC Handbook of Organization*; ALSC: Chicago, IL, 2008; 53.

Association of College and Research Libraries (ACRL)

Mary Ellen Davis
Mary Jane Petrowski
American Library Association, Chicago, Illinois, U.S.A.

Abstract

The Association of College and Research Libraries (ACRL) is the higher education association for librarians. Representing more than 10,500 academic and research librarians and interested individuals, ACRL (the largest division of the American Library Association) is the only individual membership organization in North America that develops programs, products, and services to help academic and research librarians learn, innovate, and lead within the academic community. Founded in 1940, ACRL is committed to advancing learning and transforming scholarship. ACRL represents librarians working with all types of academic libraries—community and junior college, college, and university—as well as comprehensive and specialized research libraries and their professional staffs. ACRL activities are guided by the core values, vision, and goals in ACRL's strategic plan, the Plan for Excellence. The core purpose of ACRL is to lead academic and research librarians and libraries in advancing learning and transforming scholarship. ACRL advances its work by serving as a channel of communication among academic librarians, faculty, students, administrators, other information professionals, higher education organizations, federal, state, and local governments, and the larger society. It is the leading professional organization of choice for promoting, supporting, and advancing the values of academic libraries to the higher education community. ACRL and, indeed, the American Library Association itself, were founded to establish regular channels for communication among librarians. Today ACRL is a dynamic, inclusive organization that has grown into a large association encompassing all types of positions in all types of academic and research libraries.

INTRODUCTION

The Association of College and Research Libraries (ACRL) is the higher education association for librarians. Representing more than 10,500 academic and research librarians and interested individuals, ACRL (the largest division of the American Library Association) is the only individual membership organization in North America that develops programs, products, and services to help academic and research librarians learn, innovate, and lead within the academic community. Founded in 1940, ACRL is committed to advancing learning and transforming scholarship. ACRL represents librarians working with all types of academic libraries—community and junior college, college, and university—as well as comprehensive and specialized research libraries and their professional staffs. In July 2015, ACRL had a total of 10,598 members (accounting for 18.9% of ALA's membership): 9,941 personal members, 639 organizational members, and 18 corporate members. Approximately 43% of the personal members work in research/doctoral-granting institutions, 24% in comprehensive institutions, 15% in four-year colleges, 12% in two-year/technical institutions, 1% in independent research libraries, and 1% in information-related organizations. ACRL activities are guided by the core values, vision, and goals in ACRL's strategic plan, the Plan for Excellence. The core purpose of ACRL is to lead academic and research librarians and libraries in advancing learning and transforming scholarship.

ACRL advances its work by serving as a channel of communication among academic librarians, faculty, students, administrators, other information professionals, higher education organizations, federal, state, and local governments, and the larger society. It is the leading professional organization of choice for promoting, supporting, and advancing the values of academic libraries to the higher education community. ACRL and, indeed, the American Library Association itself, were founded to establish regular channels for communication among librarians. Today ACRL is a dynamic, inclusive organization that has grown from its early origins of college and reference librarians to a large association encompassing all types of positions in all types of academic and research libraries. ACRL members hold a variety of positions and responsibilities in the areas of management, public and information services, technical services, online services, assessment, information literacy, data curation and management, collection development, rare books and special collections, nonprint media, and distributed education.

ORIGINS OF ACRL

Since the late nineteenth century, conferences and meetings of professional groups have been an American institution. They reflect our penchant for association and our passion for professional self-improvement. In 1853 American librarians held their first convention in New York City.

About one-fifth of the 81 librarians who attended the meeting were college librarians.[1] Not until a generation had passed, however, and the crisis surrounding the Civil War was over, did American librarians hold a second national meeting. In the spring of 1876, Melvil Dewey and Frederick Leypoldt sent out their famous call for a conference of librarians to promote "efficiency and economy in library work."[2] Of the 103 librarians present when the conference convened in Philadelphia on September, 10 were college librarians.[3] The focal point of the 1876 meeting was the reading of papers on practical library subjects such as cooperative cataloging, indexing, and public relations. The response to the program was apparently positive because the conference participants voted on the final day of the meeting to establish the American Library Association and to hold Annual Conferences.[4]

From the beginning, the American Library Association was a predominantly public library organization. But, the areas of common interest between public and academic libraries are extensive, and for the first dozen years of the association's existence the college librarians attending ALA conferences did not hold separate meetings. Finally, in 1889, a group of 13 college librarians caucused at the Annual Conference in St. Louis and recommended that a college library section be formed. The following year at the 1890 Annual Conference in the White Mountains of New Hampshire, 15 librarians representing most of the major colleges of the Eastern seaboard, including Harvard, Yale, Columbia, the Massachusetts Institute of Technology, and Brown, held the first meeting of the College Library Section.[5] The new section was a small, relatively informal discussion group attended for the most part by administrators who could afford long distance travel. The annual meetings of the section provided a forum for the presentation and discussion of papers on such topics as reference work, cataloging, departmental collections, union lists, and the like.[6]

In 1897, the section acquired a new name, the College and Reference Library Section (to recognize the participation of reference librarians) and, after the turn of the century, began to select officers to plan annual meetings. Not until 1923, however, did the section adopt its own bylaws and thereby cross the line that separates a discussion group from a section within ACRL today. The 1923 bylaws regularized the existence of the section by establishing a Board of Management with three officers to conduct the business of the section between conferences and provided for the levying of annual membership dues of 50 cents.[7] During the course of the 1920s, attendance at section meetings grew from 90 in 1923 to 240 in 1926 and peaked at 800 in 1928 before dropping off to 600 in 1929. The meeting program of the section during the twenties and thirties included general sessions for the whole section as well as separate roundtables for college and reference librarians. The topics discussed at the early section meetings are issues that still confront academic librarians today: faculty status and personnel classification, teaching students, interlibrary loan, library standards, etc.[8]

From 1890 to 1938, the College and Reference Library Section served primarily as a forum for discussion. But, beginning in the 1920s, pressure began to build in the academic library profession for the creation of a stronger professional organization capable of undertaking a broad range of activities, programs, research, and publications. The occasion for a radical restructuring of the section came in the mid-1930s when ALA roundtables representing teachers, college librarians, and junior college librarians expressed the desire to affiliate with the College and Reference Library Section. In 1936, the chair of the section appointed a Committee on Reorganization to develop plans for restructuring the section. The final report of the committee in 1938 recommended the adoption of new bylaws that would transform the section into an Association of College and Reference Libraries with full autonomy over its own affairs. The new bylaws provided for the creation of subsections within the association for college libraries, junior college libraries, teachers college libraries, university libraries, and other groups that might wish to affiliate.

ACRL BECOMES A DIVISION

The section approved the proposed bylaws in June 1938 and officially became the Association of College and Reference Libraries (ACRL) by the end of the year. The ALA Council responded by ratifying a new ALA constitution that made provision for the creation of self-governing divisions within ALA, entitled to receive a share of ALA dues.

ACRL swiftly prepared a new constitution to meet the conditions for division status, and the ALA Council recognized ACRL as ALA's first division on May 31, 1940.[9] The Association of College and Reference Libraries started its new life with six nearly formed subsections of its own: Agricultural Libraries Section, College Libraries Section, Junior College Libraries Section, Librarians of Teacher Training Institutions Section, Reference Libraries Section, and University Libraries Section. When the Reference Libraries Section departed to join the newly formed Library Reference Services Division in 1956, ACRL substituted "Research" for "Reference" in its name and became the Association of College and Research Libraries.[10] With its sections, chapters, and discussion groups, ACRL grew rapidly after its beginnings in 1938: membership jumped from 737 in 1939 to 2,215 in 1941, rose to 4,623 in 1950, increased to 9,324 in 1975, reached 11,524 in 2000, and dropped back to 10,598 in July 2017.[11]

ACRL Office

First executive secretary

ACRL and its network of sections and committees grew so rapidly after 1938 that by the end of World War II the association could no longer, as A. H. Kuhlman put it, "be expected to run of its own accord."[12] The elected leaders of ACRL were convinced that it was now essential to have a professional executive secretary, working under the direction of the president and Board of Directors, to integrate the activities and services of the association. As early as 1931, the ALA Council, recognizing that the interests of academic libraries had not always received adequate attention at ALA headquarters, authorized the appointment of a College Library Advisory Board (CLAB) to advise the ALA Board of Directors on academic library questions. One of the first recommendations of CLAB was that a full-time academic library specialist be employed at ALA headquarters to provide information and advisory services for college librarians. The ALA Council approved this recommendation for a college library specialist in principle, but throughout the rest of the 1930s and the war period, ALA never found the money to fill the position.[13] The issue came to a head in 1946 when ACRL, with its growing membership and pressing need for professional staff, made clear that it would seriously consider withdrawal from ALA if the question of funds for a paid executive was not resolved satisfactorily. ALA responded within the year by appropriating funds to finance an ACRL headquarters staff.[14]

Orwin Rush, the librarian of Clark University, came to ALA headquarters in the spring of 1947 as ACRL's first executive secretary. After launching the new ACRL office and clearing the way for its future, Rush departed for the University of Wyoming in 1949. In his place came "young Arthur Hamlin, fresh from the University of Pennsylvania." Hamlin described the ACRL office in the early fifties this way: "Physically, the ACRL headquarters office is a second floor front room, complete with fireplace, in the large, old-fashioned, reconverted mansion which is ALA headquarters at 50 East Huron Street in Chicago. Here an active staff of four, the executive secretary, the publications officer, a secretary and a clerk-typist, with their typewriters, telephones, file cabinets, and visitors hold forth. Like many a library staff area, ACRL headquarters is a noisy, crowded, active place."[15] In 1961, a modern headquarters building replaced the old mansion.

Communities of Practice

One of ACRL's primary strengths is the effectiveness of its communities of practice, including committees, discussion groups, editorial boards, interest groups, sections, and chapter affiliates in meeting the interests of ACRL's diverse membership. Membership in ACRL provides opportunities to become involved with communities of practice that focus on specializations within the profession. The 44 chapter affiliates provide members with networking opportunities at the local level throughout the North America.

Chapters

In 1952, ACRL took the first step toward encouraging participation at the local level by recognizing its first local chapter—the Philadelphia Area Chapter. ACRL currently has 44 chapters, two of which include Canadian provinces. In 2017, two new chapters formed: a second Michigan chapter affiliated with the Michigan Academic Library Association, and the ACRL Idaho chapter. The purpose of the chapters is to bring the national organization closer to individual members and to provide programs beneficial to members at the local level.

Discussion Groups

In the 1970s, ACRL added a new community of practice to its national organization—the discussion group. As of September 2017, ACRL had 22 discussion groups. In a sense, the discussion groups are a reincarnation of the original College Library Section. They provide a relatively informal framework for librarians with similar interests to gather to exchange ideas and information.

Interest Groups

In 2008, ACRL members approved a bylaws change allowing for the creation of Interest Groups. As of September 2014, ACRL members can affiliate with an unlimited number of sections and interest groups with no additional charge. By October 2009, seven new interest groups were established including Academic Library Services to International Students, Health Sciences, Image Resources, Numeric and Geospatial Data Services in Academic Libraries, Residency Programs, Universal Accessibility, and Virtual Worlds. In May 2011, the Digital Curation Interest Group was approved by the Board. In June 2012, the Librarianship in For-Profit Educational Institutions Interest Group and the Library and Information Science (LIS) Education Interest Group were approved by the Board. In June 2013, the Technical Services Interest Group was approved by the Board. In 2017 the Board approved the 5 new interest groups including the Academic Library Services for Graduate Students Interest Group, the Asian, African, and Middle Eastern Studies Interest Group, History Librarians Interest Group, Institutional Research Interest Group, and Library Services for Graduate Students Interest Group. As of 2017, ACRL had 21 interest groups.

Sections

By 1979, the association had 13 sections: the three "types-of-libraries" sections (College, Community College, and University) plus the Arts Section, Asian, African, and Middle Eastern Section (which became an interest group in 2017), Anthropology Section, Instruction Section (name changed from the Bibliographic Instruction Section in 1995), Education and Behavioral Sciences Section (into which the old Teachers Training Section was incorporated), Public Policy and International Relations Section (name changed from Law and Political Science Section in 2016), Rare Books and Manuscripts Section, Science and Technology Section (with which the Agricultural Section was merged), European Studies Section (created by a merger of the Slavic and East European Section and the Western European Studies Section in 2017). Between 1987 and 1990, three more sections were formed: Women's Studies Section in 1987 (name changed to Women and Gender Studies in 2011); African-American Studies Librarians Section in 1989 (which became an interest group in 2016); and the Distance Learning Section in 1990 (name changed from Extended Campus Libraries Services Section in 1998). In 1994, the Literatures in English Section (name changed from English and American Literature Section in 2000) was formed. By 1997, ACRL had 17 sections, and this number remained static until 2017 when the number dropped to 15 sections following the merger of SEES and WESS as well as the transition of AAMES and AFAS to interest groups. The Digital Scholarship Section was created in 2017.

Coordination and Oversight

It is the responsibility of the ACRL executive director and staff to coordinate the work of ACRL's 55 committees, 16 sections, 22 division-level discussion groups, 9 editorial boards, 21 interest groups, 22 discussion groups, and 44 chapters. To ensure the smooth operation of this complex structure, the headquarters staff monitors the many procedural details associated with appointments, archiving, awards, budgets, elections, meetings, programs, reports, and so on. ACRL currently has 15.75 FTE approved positions for its Chicago office (housed in the ALA headquarters) and 22.4 positions for its *CHOICE* office in Middletown, Connecticut. The ACRL office works closely with committees and sections to plan stimulating meetings at ALA conferences, and also manages the arrangements for ACRL preconferences and conferences. Planning for these conferences begins years in advance as detailed arrangements are worked out for hotel space, meeting times, exhibits, programs, publicity, and finances.

The ACRL office supports ACRL's publication program by providing assistance to the editors of *C&RL*, *RBM*, and *Publications in Librarianship*, by working closely with the editor of *CHOICE* who reports to the executive director, by publishing and distributing the many publications of ACRL committees and sections, and by writing, editing, and publishing *C&RL News*, the association's monthly news publication. With the exception of *CHOICE*, the ACRL staff based in Chicago manages the production of all ACRL books and journals.

Clearinghouse

Together with the ALA Headquarters Information Center, the ACRL office serves as a clearinghouse for information on academic library concerns and issues. The office handles inquiries regarding policies and practice. It also offers information about ALA activities and services and is in daily contact with the staff of other ALA divisions and offices, including the Washington Office.

Ambassador

ACRL serves as the ambassador for academic libraries and librarians at ALA headquarters. The ACRL Executive Director plays a key role in representing the association to other library and information associations as well as to higher education and government communities. In this role, the executive director attends meetings and gives presentations in many parts of the country each year. In doing so, the director strives to maintain and establish lines of communication between the academic library profession and other communities. In 1984, a new standing committee, the Professional Liaison Committee, was established to further cooperative efforts and to put stronger emphasis on ACRL's liaison efforts with other associations. To build upon this work, in 1995, the ACRL Board developed the Professional Liaison Committee as the Council of Liaisons, identifying an initial nine important higher education associations to which it will send a liaison. The Council and the Board annually review the list of liaison organizations. In 2009, the Board expanded its liaison work and reconstituted the Council of Liaisons as the Liaison Coordinating Committee.

Strategic Planning

Guiding all association activity is the strategic planning process adopted by ACRL. This process relies on member input to articulate the direction of the professional organization, and to identify areas of highest priority for association activity. Since 1981, ACRL has updated its strategic plan, mission, and vision on a regular basis. Each year the ACRL Board of Directors sets the priorities and performance indicators for the association. At the 2003 Midwinter Meeting, the Board authorized contracting with Tecker Consultants to lead a strategic planning process. Extensive data gathering took place including telephone interviews, focus groups, leadership sessions, and an all-member web-based survey. The Board reviewed the data, drafted a plan, tested its thinking with the members, and made revisions. The new strategic plan, "Charting Our Future: ACRL Strategic Plan 2020," was approved by the Board at the 2004 Annual Conference and is reviewed annually with minor changes being made each fall at the Fall Executive Committee Meeting. The Board is now working to encourage ACRL's units to align their work with the strategic plan. The Board began a new strategic planning cycle at the 2010 Midwinter Meeting which culminated in a new "Plan for Excellence" which was approved in June 2011. The Board reviews progress on the plan at its annual Strategic Planning Session and makes adjustments as needed. In September 2015, the Board began discussion on the revision of the Plan for Excellence and in March 2016 approved updates to the goals and objectives.

IMPACT ON HIGHER EDUCATION, SCHOLARLY COMMUNICATION, AND CIVIC DEVELOPMENT

ACRL is ALA's key link to the higher education community, and one of ACRL's strategic directions is to ensure that the contributions of academic and research libraries and librarians to higher education, scholarly communication, and civic development are recognized by society. To this end, ACRL has undertaken several initiatives.

Working with Higher Education Associations

In 1995, ACRL identified a number of higher education organizations with which to share ideas and implement programs in areas of mutual interest. ACRL assigned member liaisons to these organizations and these individuals comprised the ACRL Council of Liaisons. The ACRL Board expanded the work of its liaison program in 2010 under the umbrella of a Liaison Coordinating Committee that encouraged ACRL units to explore liaisons and work with other associations. These organizations currently include: the Academy of Criminal Justice Sciences (ACJS), American Anthropological Association (AAA), American Association for the Advancement of Science (AAAS), American Association of Community Colleges (AACC), American Chemical Society (ACS), American Institute of Biological Sciences (AIBS), American Physical Society (APS), American Political Science Association (APSA), American Society for Engineering Education (ASEE), American Sociological Association (ASA), Association for Information Science and Technology (AIST), Association for Library and Information Science Education (ALISE), Association for the Study of African American Life and History (ASAALF), Council of Botanical and Horticultural Libraries, Council of Independent Colleges (CIC), EDUCAUSE Learning Initiative (ELI), International Association of Aquatic and Marine Science Libraries and Information Centers, Medical Library Association (MLA), Modern Language Association (MLA), National Resource Center for the First-Year Experience, National Women's Studies Association (NWSA), Society for College and University Planning (SCUP), Society for Information Technology and Teacher Education (SITE), Special Libraries Association (SLA), and the United States Agricultural Information Network. One of ACRL's early successful liaisons was with the American Association for Higher Education (AAHE), which ceased operations in mid-2005. Sample collaborative activities included joint sponsorship of a provosts' luncheon at the AAHE annual conference and AAHE participation in developing the Information Literacy

Competency Standards for Higher Education. ACRL has also developed subject-specific information literacy standards in collaboration with subject discipline organizations. ACRL also belongs to the Council for Higher Education Management Association (CHEMA), and the Coalition for Networked Information (CNI), which has led to increased collaboration. In 2011, ACRL partnered with the Association for Institutional Research (AIR), the Association of Public and Land-grant Universities (APLU), and the Council of Independent Colleges to convene two national summits as part of the IMLS-funded project "Building Capacity for Demonstrating the Value of Academic Libraries." In September 2012, ACRL continued its partnership with AIR and APLU on the three-year "Assessment in Action: Academic Libraries and Student Success" program, made possible by IMLS. ACRL has also collaborated with CIC through participating and supporting CIC's Information Literacy and Fluency workshops for campus teams of librarians, subject-discipline faculty, and administrators. Beginning in 2010, ACRL increased its presence with the American Council of Learned Societies, and the executive director has been invited to make several presentations on scholarly communication and open access.

Standards and Guidelines

Developing standards is an area where ACRL committees have made some of their most important contributions to academic librarianship. In 1957, the ACRL Committee on Standards, after two year of work, produced the "first real set of 'Standards for College Libraries' to enjoy the consensual support of the profession."[16] Since then, ACRL committees have developed standards for university libraries and two-year learning resources programs. In 2004, the ACRL Board approved the outcomes-based "Standards for Libraries in Higher Education," inclusive of all academic libraries, and this standard was revised in 2011. ACRL recruited and hired a team of trainers and have a licensed one-day workshop, "Planning, Assessing, and Communicating Library Impact: Putting the Standards for Libraries in Higher Education into Action," to a variety of campuses and chapters. Guidelines have also been developed in many specific areas including cultural competencies, personnel, instruction, branch libraries, library services for distance education, rare books and special collections, and undergraduate libraries. "Standards for Libraries in Higher Education" (2011) took the 2004 outcomes model a step further and are designed to guide academic libraries in advancing and sustaining their role as partners in educating students, achieving their institutions' missions, and positioning libraries as leaders in assessment and continuous improvement on their campuses.

In January 2000, the ACRL Board approved the "Information Literacy Competency Standards for Higher Education." The American Association of Higher Education and the Council of Independent Colleges have endorsed these standards and they are widely used on campuses across the country. There is global interest in the information literacy competency standards and they have been translated into 8 languages. Discipline-based information literacy guidelines have been developed for anthropology and sociology, journalism, literatures in English, political science, psychology, science and technology, as well as teacher education.

In June 2012, the Board approved recommendations of a review task force that ACRL's "Information Literacy Competency Standards for Higher Education" be extensively revised. In spring 2013, a second task force began the revision process, leading to the development of a draft "Framework for Information Literacy for Higher Education." Following a process that included several drafts and public comment periods, the Board took the official action of approving the Framework at the 2017 ALA Annual Conference in Orlando.

Among its guidelines in the personnel arena, the ACRL Committee on Academic Status in 1971 drew up "Standards for Faculty Status for College and University Libraries." The ACRL Board approved the Standards in June 1971, and as a corollary, ACRL drafted a "Joint Statement on Faculty Status of College and University Librarians" with the American Association of Colleges (AAC) and the American Association of University Professors (AAUP). ACRL, AAC, AAUP, and a host of other associations endorsed the statement, which laid down a clear definition of the obligations and benefits of academic status. This Joint Statement was reaffirmed by the ACRL Board of Directors at the 2007 Annual Conference. In 1992, "Standards for Faculty Status for College and University Librarians" was revised—"the first revision of this seminal document in the 21 years that had elapsed since its approval by a voice vote of the membership at Dallas in 1971."[17] At the 1993 Midwinter Meeting, the ALA Council, by consent, approved the incorporation of the revised Standards into the ALA Handbook of Organization. Council's exceptional action reaffirms "faculty status as the desired and appropriate condition of academic librarians nationally and lends the document the support of the prestigious parent body."[18] A new revision of the Standards was approved by the ACRL Board of Directors at the 2011 Annual Conference.

In 1997, the Board approved changing the name of the Academic Status Committee to the Academic Librarians Status Committee, and by 2013, the Board decided it was best to take up the revision of these guidelines with task forces as needed and dissolved the standing committee. At the 2005 Annual Conference, the ACRL Board of Directors approved an updated version of the "Guidelines for the Appointment, Promotion, and Tenure of Academic Librarians." At the 2006 Midwinter Meeting, the ACRL Board of Directors approved the "Guidelines for Media Resources in Academic Libraries" and a revision of the "Guidelines for Academic Status for College and University Librarians." The 2007 Annual Conference saw the ACRL Board of Directors reaffirm the 1989 "Statement on the Certification & Licensing of Academic Librarians"

and the "Statement on the Terminal Professional Degree for Academic Librarians," and approve the "Standards for Proficiencies for Instruction Librarians and Coordinators." 2009 through mid-2011 saw a number of standards revised and developed, including "ALA-SAA Joint Statement on Access to Research Materials in Archives and Special Collections Libraries," "Guidelines for Curriculum Materials Centers," "A Guideline for Appointment, Promotion and Tenure of Academic Librarians," "A Guideline for the Screening and Appointment of Academic Librarians," "Information Literacy Standards for Teacher Education," "Psychology Information Literacy Standards," "Standards for Proficiencies for Instruction Librarians and Coordinators: A Practical Guide," "Statement on the Certification & Licensing of Academic Librarians," and "Statement on the Terminal Professional Degree for Academic Librarians."

New and revised standards approved from 2012 through 2017 include "ACRL Statement on Academic Freedom," "ACRL/RBMS Guidelines For Interlibrary And Exhibition Loan of Special Collections Materials," "Characteristics of Programs of Information Literacy that Illustrate Best Practice: a Guideline," "Competencies for Special Collections Professionals," "Guidelines for Recruiting Academic Librarians," "Diversity Standards: Cultural Competency for Academic Libraries," "Framework for Information Literacy for Higher Education," "Guidelines for University Library Services to Undergraduate Students," "Joint Statement on Faculty Status of College and University Librarians," "Proficiencies for Assessment Librarians and Coordinators," "Roles and Strengths of Teaching Librarians," and "Standards for Distance Learning Library Services."

Awards

The ACRL awards program honors the best and brightest stars of academic librarianship. Sixteen awards recognize and honor the professional contributions and achievements of ACRL members. This special recognition by ACRL enhances the sense of personal growth and accomplishment of its members, provides its membership with role models, and strengthens the image of its membership in the eyes of employers, leadership, and the academic community as a whole. Among its most prestigious achievement awards are the Academic/Research Librarian of the Year Award and the Excellence in Academic Libraries Awards, both sponsored by GOBI Library Solutions from EBSCO.

There are three basic types of ACRL awards: achievement and distinguished service, research/travel grants, and publications. Achievement and distinguished service awards are intended to honor academic and research librarians for significant past achievements, such as publications, program development, or general leadership in the profession. Such awards include a plaque and may also involve a cash award. Research and travel awards, normally in the form of grants, can also recognize past achievements, but their main purpose is to assist academic and research librarians in completing a research project, usually relating to some aspect of academic or research librarianship. Publication awards are given for outstanding articles, bibliographies, catalogs, etc. ACRL awards are given to either individuals or groups. Depending upon the terms of the award, recipients do not necessarily need to be members of ACRL.

Value of Academic Libraries

As librarians are increasingly called upon to demonstrate the value of academic libraries and their contribution to institutional goals, ACRL is responding with research and resources to support the profession in meeting this challenge. ACRL's Value of Academic Libraries Initiative, part of the 2011 Plan for Excellence, is a multiyear project designed to provide academic librarians with competencies and methods for demonstrating library value relative to the mission and goals of postsecondary institutions.

ACRL has long been concerned with accountability, assessment, and student learning. In the early 1980s, ACRL was on the cutting edge of these issues with a publication on assessment to "stimulate librarians' interest in performance measures and to provide practical assistance so that librarians could conduct meaningful measurements of effectiveness with minimum expense and difficulty." The association is a national authority that the higher education community looks to for standards and guidelines to enhance library effectiveness.

Building on this work and given the emphasis on assessment issues, ACRL leadership and staff held an invitational meeting in 2009 with members of the library research community to discuss actions related to assessing and documenting library value. ACRL then commissioned a report on existing research and literature on the topic: The Value of Academic Libraries: A Comprehensive Research Review and Report. This report recommends that ACRL create a professional development program to build the profession's capacity to document, demonstrate, and communicate library value in alignment with institutional goals.

In response, ACRL, together with three partners – the Association for Institutional Research, Association of Public and Land-grant Universities, and the Council of Independent Colleges – was awarded an IMLS grant in 2011 from the Federal Institute of Museum and Library Services for a National Leadership Collaborative Planning Grant Level II. The grant proposal, "Building Capacity for Demonstrating the Value of Academic Libraries," submitted in February 2011, covered costs to convene two national summits in late fall 2011 to determine the professional competencies that librarians need. College and university chief academic officers, senior institutional researchers, representatives from accreditation commissions and higher education organizations, and academic librarians attended the summits to share their best thinking. A white paper, "Connect, Collaborate, and Communicate: A Report from the Value of Academic Libraries Summits" summarizing the findings and setting a framework for future action was published in June 2012.

As a direct result of the IMLS collaborative planning grant, ACRL submitted a follow-up proposal in early 2012and was awarded a National Leadership Demonstration Grant of $249,330 by the IMLS for the program "Assessment in Action: Academic Libraries and Student Success" (AiA) in September 2012. A professional development program to strengthen the competencies of librarians in campus leadership and data-informed advocacy has been designed, implemented, and evaluated. The project goals and anticipated outcomes consider the experiences of students in academic libraries and the impact of the library on student learning and success. Over 200 institutions have participated in the three-year project. Each participating institution identified a team, consisting of a librarian and at least two additional team members as determined by the campus (e.g., faculty member, student affairs representative, institutional researchers, academic administrator). The librarians participated as cohorts in a 14-month long professional development program that included team-based activities carried out on their campuses. Supported by a blended learning environment and a peer-to-peer network, the librarians led their campus team in the development and implementation of an action-learning project, examining the impact of the library on student success, and contributing to assessment activities on their campus. In January 2015, ACRL released a report synthesizing the project reports of the first 75 teams to participate in the AiA program. A report synthesizing the second-year projects was issued in fall 2015, "Academic Library Contributions to Student Success: Documented Practices from the Field." Additional reports were published in 2016 ("Documented Library Contributions to Student Learning and Success: Building Evidence with Team-Based Assessment in Action Campus Projects") and 2017 ("Academic Library Impact on Student Learning and Success: Findings from Assessment in Action Team Projects").

In 2011, the ACRL Plan for Excellence identified "Value of Academic Libraries" as one of three areas for strategic focus for 2011–2016. The goal is to have academic libraries demonstrate alignment with and impact on institutional outcomes.

Additionally, in 2011, ACRL established a standing committee on the Value of Academic Libraries to continue the association's commitment to and focus on this work. The Committee's first years have produced an active blog and a value bibliography. In 2015, ACRL began planning a research project in which a small number of questions would be replicated using the same research approach at a variety of academic institutions.

Research and Scholarly Environment

In 2002, ACRL embarked on a three-year scholarly communications initiative as one of its highest strategic priorities; this initiative continues today through the work of ACRL's Research and Scholarly Environment Committee (which had been named the Scholarly Communications Committee until July 2012), established as a standing division level-committee. The committee coordinates the association's scholarly communications activities and hosts a discussion group for further exploration of these issues beyond the initial three-year launch of the program.

Addressing issues critical to the future of all academic libraries, the association works to reshape the current system of scholarly communications, focusing on education, advocacy, coalition building, and research. Broad goals of the initiative include creating increased access to scholarly information; fostering cost-effective alternative means of publishing, especially those that take advantage of electronic information technologies; and encouraging scholars to assert greater control over scholarly communications.

ACRL has partnered with the Association of Research Libraries (ARL) and the Scholarly Publishing and Academic Resources Coalition (SPARC) signing on to letters in support of a number of calls for broader access to scholarly works, including several regarding the National Institute of Health's (NIH) policies on open access for federally funded research. ACRL and SPARC partnered to offer an invitational only webcast to prepare committed grassroots advocates to take action in 2007. ACRL has participated in the Information Access Alliance (IAA) and the Open Access Working Group (OAWG), cosponsoring a symposium on anti-trust issues.

ACRL partnered with ARL to create the ARL/ACRL Institute on Scholarly Communication, which offered its signature event in July 2006, December 2006, and July 2007. This immersive learning experience prepares participants as local experts within their libraries and equips them with tools for developing campus outreach strategies. ACRL and ARL formalized their agreement and offered the first licensed regional institute for December 2007 and a second in December 2008. A regional event allows the ACRL to support institutions that would not, for whatever reason, attend a national event. This move to a regional event as a natural evolution in the life of the institute will enable the ACRL to refocus its national efforts to meet the changing needs of the library community. Under the ARL/ACRL Institute on Scholarly Communication name, ARL and ACRL also jointly offered an in-person workshop in conjunction with the ACRL 2009 Conference, an 8-part webinar series in 2010, and in-person workshops in conjunction with ACRL 2013, ACRL 2015 and the Library Assessment Conference.

In May 2009, the Board identified six as strategic priorities for 2009–2013, one of which was to, "enhance ACRL members' understanding of how scholars work and the systems, tools, and technology to support the evolving work of the creation, personal organization, aggregation, discovery, preservation, access and exchange of information in all formats." Given this focus and the perceived need to help members of the profession develop a basic awareness and understanding, ACRL invested in this workshop to extend the reach from a one-off workshop to a subsidized travelling program, now called "the road show." In its seventh year, and with the 2014 workshops completed, the road show will have visited 26 different states, the District of Columbia, 1 U.S. territory, and 1 Canadian province. The 36 workshops offered over the program's six year will have reached over 2200 participants from over 600 colleges and universities. Additionally, in summer 2014, the presenters began offering webinars through ACRL eLearning that allow for deeper exploration of specific topics.

ACRL and ARL partnered with SPARC to offer 4 webcasts on author rights during Fall 2006–Spring 2007. ACRL again partnered with SPARC to offer a full-day institute in conjunction with ALA MW 2015 on Tackling Textbook Costs Through Open Educational Resources: A Primer. ACRL convened an invitational meeting in July 2007 to collectively brainstorm the evidence needed to inform strategic planning for scholarly communication programs. In November 2007, ACRL issued a resulting white paper, "Establishing a Research Agenda for Scholarly Communication: A Call for Community Engagement."

In 2009, ACRL joined the Library Copyright Alliance (LCA) and submitted briefs and comments for the courts and the Department of Justice regarding the proposed Google Book Settlement. LCA has submitted other amicus briefs of the court, as appropriate, to support the doctrine of fair use. Other action through the LCA include communicating concerns to the U.S. Trade Representative on the Anti-Counterfeiting Trade Agreement, releasing issue briefs on international copyright and the legal status of streaming films, supporting the U.S. statement on copyright exceptions for the blind and visually impaired before the World Intellectual Property Organization. In November 2012, LCA convened an invitational meeting in Washington, D.C., with approximately 20 library leaders in order to explore, articulate, and prioritize anticipated copyright-related issues facing the library community. The purpose of the meeting was to develop an appropriate strategy with articulated priorities to help inform the library associations' work.

In 2011, ACRL appointed a visiting program officer to develop a robust and sustainable model for the Scholarly Communications 101 road show workshop, support the work of the Scholarly Communications Committee, and develop other initiatives to advance ACRL's work in this area.

In 2011, the ACRL Plan for Excellence identified "Research and Scholarly Environment" as one of three areas for strategic focus for 2011–2016. The goal is to have academic librarians accelerate the transition to a more open system of scholarship. The objectives were revised in 2017 as follows:

1. Increase the ways ACRL is an advocate and model for open dissemination and evaluation practices.
2. Enhance members' capacity to address issues related to scholarly communication, including but not limited to data management, library publishing, open access, and digital scholarship.
3. Increase ACRL's efforts to influence scholarly publishing policies and practices toward a more open system.

In March 2013, ACRL published the white paper, "Intersections of Scholarly Communication and Information Literacy: Creating Strategic Collaborations for a Changing Academic Environment," written by a working group of leaders from many areas of the association. A formal task force was charged in October to promote its use. Also in 2013, members of the Research and Scholarly Environment Committee (formerly Scholarly Communications Committee) began to update content in the popular scholarly communication toolkit and migrated from Drupal to WordPress. It is routinely promoted in September in advance of October's Open Access Week.

In June 2015, the ACRL Board approved funding for curriculum design of two new ACRL licensed workshops to follow the popular "road show" model and ACRL now offers workshops on data management and the intersections of scholarly communication and information literacy.

Student Learning

ACRL has undertaken a number of initiatives related to information literacy. For those working in information literacy, ACRL supports programs in the areas of professional development, assessment, and instructional development. Spearheading many of these programs was the Institute for Information Literacy (IIL). IIL was charged with preparing librarians to become effective teachers in information literacy programs; supporting librarians, other educators, and administrators in taking leadership roles in the development of information literacy programs; and forging new partnerships within the educational community to work toward information literacy curriculum development. Conceptualized by Cerise Oberman, dean of libraries at SUNY Plattsburgh, the Institute for Information Literacy (IIL), was established by the ACRL Board at the 1997 ALA Annual Conference under the name National Information Literacy Institute. IIL is now known as the Student Learning and Information Literacy Committee (SLILC) and the Immersion program is now coordinated by the Immersion Program Committee.

ACRL approved the "Information Literacy Competency Standards for Higher Education" at Midwinter 2000. In response to inquiries from members about their use on campuses, a training program was developed and was made available to members during Annual Conferences and Midwinter Meetings. In 2016, the Board approved the "Framework for Information Literacy for Higher Education" and rescinded the "Information Literacy Competency Standards for Higher Education."

A visiting program officer for information literacy was hired to develop educational content to help the profession use the Information Literacy Framework and those outside the profession to understand its application to the curriculum.

Many ACRL sections have developed discipline-specific standards for information literacy including "Information Literacy Standards for Science and Technology," (approved in 2006), "Research Competency Guidelines for Literatures in English" (approved in 2007), "Information Literacy Competency Standards for Anthropology and Sociology," "Political Science Research Competency Guidelines" (approved in 2008), "Psychology Information Literacy Standards" (approved 2010), "Information Literacy Standards for Teacher Education" (approved in 2011), "Visual Literacy Competency Standards for Higher Education" (approved in 2011), "Information Literacy Competency Standards for Journalism Students and Professionals" (approved in 2011), "Information Literacy Standards for Teacher Education" (approved 2011), "Information Literacy Standards for Nursing" (approved 2013). ACRL has also reviewed information literacy standards for music developed by the Music Library Association.

ACRL has also developed a number of standards to assist practitioners in developing and maintaining strong educational programs including: "Guidelines for Instruction Programs in Academic Libraries" (approved 2011), and "Characteristics of Programs of Information Literacy that Illustrate Best Practices: A Guideline" (first approved in June 2003 and revised in 2012).

Grant-funded projects

In 2000, ACRL received a $150,000 National Leadership grant from the Federal Institute of Museum and Library Services to develop tools and training to help librarians better assess student learning outcomes in information literacy courses. The work of 30 librarians and their campus teams in implementing and assessing information literacy courses was widely disseminated through presentations and publications.

In 2006, ACRL's Rare Book and Manuscript Section (RBMS) received funding from IMLS to offer scholarships to newer and aspiring professionals in these fields to its annual conference, which in 2006, has a special focus on issues of mutual interest to special collections librarians, archivists, and museum professionals. A total of 33 attendance scholarships were offered with preference given to applicants from professionally underrepresented backgrounds. The purpose was to increase the conference participants' knowledge of allied fields and increase their interest in communicating and collaborating with allied professionals. In addition, the aim was to stimulate scholarship recipients to show greater interest in pursuing or continuing a career as special collections librarians, archivists, or museum professionals.

In 2011, ACRL's RBMS members were also instrumental in getting a small ($3000) grant from the Gladys Krieble Delmas Foundation to support the mounting of back files of *Rare Books and Manuscripts Librarianship* on the *RBM*/HighWire website. The Gladys Krieble Delmas Foundation generously awarded ACRL/RBMS $5000 in 2013 and $8000 in 2017 to support scholarships to attend the annual RBMS conference.

In 2011, ACRL, together with three partners—the Association for Institutional Research, Association of Public and Land-grant Universities, and the Council of Independent Colleges—was awarded an IMLS grant from the Federal Institute of Museum and Library Services for a National Leadership Collaborative Planning Grant Level II. The project entitled "Building Capacity for Demonstrating the Value of Academic Libraries" had a budget total (direct and indirect) of: $162,076, with the Grant subtotal: $99,985 and Cost Share subtotal: $62,091. The grant covered costs to convene two national summits in late fall 2011, to determine the professional competencies that librarians need. College and university chief academic officers, senior institutional researchers, representatives from accreditation commissions and higher education organizations, and academic librarians attended the summits to share their best thinking.

In 2012, ACRL was awarded a follow-up National Leadership Demonstration Grant of $249,330 by the IMLS for the three-year program "Assessment in Action: Academic Libraries and Student Success" (AiA). The grant funded a professional development program to strengthen the competencies of librarians in campus leadership and data-informed advocacy.

Public Policy Advocacy

In September of 1997, the Association of College and Research Libraries (ACRL) in concert with the ALA Washington Office embarked upon its public policy initiative. The goal was to educate academic librarians about legislative/public policy issues pertinent to academic libraries and higher education. ACRL staff, the ACRL Government Relations Committee, the ACRL Research and Scholarly Environment Committee, the *ACRL Legislative Network*, the ACRL Board, the ALA, Washington Office, and other appropriate ACRL entities carry out the advocacy work. These groups work together to develop a legislative agenda that identifies ACRL policy priorities.

ACRL communicates information on its policy priorities via many means. A Legislative Network consisting of a representative from each of the 42 ACRL chapters and an electronic distribution list (LEGNET) was established

to share the legislative agenda with other ACRL members, their institution's administration and their congressional representative. Information on issues is also disseminated using the ACRLeads electronic distribution list, *C&RL News*, flyers, letters, and the ACRL Legislative web site (http://www.ala.org/acrl/issues/washingtonwatch). When the ALA conference takes place in Washington, D.C., ACRL has offered advocacy preconferences designed to acquaint academic librarians with federal legislative issues and to equip them with the skills needed to deliver effective messages to congressional representatives. ACRL also encourages participation in ALA Legislative Day, and has, in the past, provided travel grants for first time attendees and hosted a special luncheon to highlight issues of importance to academic librarians.

The Board approved the recommendations of ACRL's Task Force on National Advocacy at the 2005 Midwinter Meeting. To supplement the existing ACRL Legislative Network, ACRL created a new position of Legislative Advocate. Legislative Advocates work as much as possible with other library legislative efforts in the state or region. 45 Legislative Advocates served in the inaugural group, appointed in Spring 2007. ACRL hired a short-term Visiting Program Officer Michael McLane in late 2007 to expand the program, organize training opportunities, and undertake assessment. This program has since been phased out as sophisticated web-enabled "action alerts" have made it easier to communicate directly with librarians in specific congressional districts.

In 2009, ACRL joined the Library Copyright Alliance and has been active in submitting briefs and comments to courts and agencies as appropriate. ACRL is also active in the Open Access Working Group, a coalition to promote legislative and policy that would expand access to results of publicly funded research. ACRL regularly releases a legislative agenda in conjunction with National Library Legislative Day. The 2017 agenda includes federal funding for libraries, network neutrality, privacy and government surveillance.

@Your Library Campaign and Grassroots Advocacy

Although generally viewed positively, libraries are often taken for granted. Recognizing this challenge, in 2001, the American Library Association launched The Campaign for America's Libraries, a five-year commitment, to speak loudly and clearly about the value of libraries and librarians to our communities, schools, academic institutions, and businesses, as well as to our society, democracy, and the new digital age. Based on research and crafted to target key audiences, The Campaign worked to raise public understanding that libraries are dynamic, modern community centers for learning, information, and entertainment. The campaign was designed to heighten awareness regarding the vibrancy, vitality, and real value of today's libraries, to galvanize public support, and influence public policy. Working under the umbrella of the American Library Association @Your library campaign, ACRL led the effort to develop a public relations campaign for academic libraries. Consumer research was undertaken in order to develop promotional materials. A toolkit was developed, mailed to all ACRL members, distributed at the 2003 ACRL Conference.

ACRL also launched a multiyear effort to emphasize the importance of academic libraries and librarians to the higher education community. A series of ads, focusing on the exciting things happening @Your library, was placed in *The Chronicle of Higher Education* beginning in 2001. Testimonials from faculty, students, and administrators were an important component of this campaign. ACRL's Excellence in Academic Libraries award winners have also been recognized in the ads. In 2003, the Board established the Marketing Academic and Research Libraries (MARL) Committee to continue developing @Your library campaign tools for academic and research libraries and to eventually update the toolkit. In 2005 and 2007, MARL presented a Best Practices in Marketing @Your library Award. In early 2010, MARL launched the ACRL Marketing Minute, a biweekly series of quick tips and insights on marketing research, trends, and data delivered through Facebook and Twitter.

ACRL President Camila Alire continued efforts to emphasize the value of academic and research libraries and librarians during her presidential year in 2006. A grassroots advocacy toolkit focusing on "The Power of Personal Persuasion" was developed and mailed to every ACRL member. ACRL President Pamela Snelson continued the advocacy role by commissioning research on what senior academic administrators expect from librarians, and subsequent ACRL Presidents have also focused on advocacy.

ALA ended the Campaign for America's Libraries in 2015, and ACRL continues its advocacy for academic libraries through the ALA Libraries Transform campaign.

Recruitment to the Profession

Professional associations such as the American Library Association, the Association of College & Research Libraries, the Association of Research Libraries (ARL), and state-based associations are key stakeholders in recruitment and retention efforts. The work of the profession is changing and association can help in recruiting talented individuals and in providing training to those already in the profession. Over the years, ACRL has undertaken a number of initiatives. The Personnel Administrators & Staff Development Officers Discussion Group of the Association of College & Research Libraries (ACRL) established the Ad Hoc Task Force on Recruitment & Retention Issues in early 2001 to examine how academic libraries can successfully recruit

and retain professionals in an increasingly competitive environment. In 2002, ACRL and ARL formed a joint task force to work on recruitment issues. A short video, "Faces of a Profession @Your Library" highlighting the benefits of careers in academic and research libraries, was developed under the Task Force's leadership. This video is freely available for downloading from the web.

In 2007, ACRL released "Achieving Racial and Ethnic Diversity among Academic and Research Librarians: The Recruitment, Retention, and Advancement of Librarians of Color," a white paper commissioned by the ACRL Board of Directors working group on diversity. Building on the 2002 ACRL white paper, "Recruitment, Retention & Restructuring: Human Resource in Academic Libraries," the authors discuss efforts to promote, develop, and foster workplaces that are representative of a diverse population, along with addressing the development of a workplace climate that supports and encourages the advancement of librarians from underrepresented groups.

In 2003, ACRL founded the ACRL Spectrum Mentor Program and committee to encourage ALA Spectrum LIS students to pursue academic librarianship. To date, ACRL has contributed $102,500 to the ALA Spectrum program, sponsoring 12 Spectrum Scholars, and has also matched more than 120 Spectrum Scholars with an ACRL member mentor. More than 20 Spectrum Travel Grants consisting of registration and a travel stipend have also been awarded to help scholars attend every ACRL Conference since 2005, with a total value of more than $20,000.

Summits

- *Summit on Technology and Change in Academic Libraries*—In November 2006, ACRL convened a roundtable of librarians, higher education administrators, publishing, and information industry leaders to address how technologies, on the one hand, and the changing climate for teaching, learning, and scholarship, on the other, will likely recast the roles, responsibilities, and resources of academic libraries over the next decade. One of the outcomes of the roundtable was a roadmap for ACRL to help its members deal with the ongoing changes in the profession and the academy. A white paper entitled "Changing Roles of Academic and Research Libraries" was posted to the ACRLog. Then Vice-President Julie Todaro provided a detailed response to the essay and invited others to comment.
- *Stepping Through the Open Door: A Forum on New Modes of Information Delivery in Higher Education*—In March 2007, ACRL joined EDUCAUSE and the National Association of College Stores to jointly sponsor an invitation-only forum focused on changing roles within higher education. An article from a participant appeared in *C&RL News* and the report was issued by the conveners in September 2007.

- *Building Capacity for Demonstrating the Value of Academic Libraries*—The summits convened in 2011 as part of ACRL's Value of Academic Libraries Initiative, are described in the "Value of Academic Libraries" section of this chapter.
- *Forum on Data Management*—In January 2014, ACRL hosted a forum on data management with Drexel University Libraries. The goal of this forum was to have a dialogue between disciplinary faculty and academic librarians on the management, curation, and sharing of research data. The faculty agreed that they would want to share most of their data, although they allowed for the possibility that some data might need to remain private for a period of time, depending on the nature of the data and whether or not the scholar was still working with it. They saw strong roles for librarians including providing consultation on data management plans, curation, and long-term preservation of data. ACRL offered two workshops on data management in 2015 and now offers a licensed workshop.

COMMITMENT TO PROFESSIONAL DEVELOPMENT AND GROWTH

ACRL supports and enhances the professional development and growth of academic and research librarians through its numerous professional development activities.

ACRL Conferences

On its 40th birthday in 1978, ACRL took a giant step forward by convening its first National Conference, distinct from ALA, in Boston. The conference featured a 3-day program of major addresses and research papers that attracted 2625 participants. Participants praised the conference for focusing on academic librarianship, for stimulating research on the issues facing academic librarianship, and for bringing together librarians with a common professional interest in academic libraries.

The conference, now held every odd year and begun primarily as an outlet for presenting formal research papers, has responded to member interests and changes in the profession by diversifying its content over the years and attracting a more global audience. ACRL dropped the "National" from the Conference name in 2011, as the conference now welcomes attendees from over 26 countries. New programming has been developed by adding panel sessions, by inviting noted leaders to write papers on current topics, by adding more opportunities for networking and informal dialogue (e.g., roundtable discussions, dinner with colleagues), and by engaging attendees more actively in the learning process with a wider range of session formats. The 1999 National Conference offered first ever "conference-within-a-conference" on the topic of student learning

and the first live Web broadcast. The 2005 conference saw the first full-fledged Virtual Conference offered in conjunction with a National Conference. The 2007 conference held the first Cyber Zed Shed, now known as TechConnect Presentations, which demonstrate technology-related innovations. IdeaPower and THATCamp formats were added to the 2011 and 2013 conferences, respectively, to allow for more late-breaking and spontaneous generation of content.

Interest in presenting at the conference has steadily increased since 1978. Typically the conference will feature about 84 contributed papers, 60 panel sessions, 6–7 preconferences, 16 workshops, 30 TechConnect presentations, 200 poster sessions, and 100 roundtable discussions. Acceptance rate, on average, has been around 20–30%, depending on the proposal type. Attendance at the conference has climbed over the past few years with about 3000 attendees, 1000 exhibitors, and 500 virtual participants for each biennial conference. Slidecast recordings have been available on demand to broaden the reach to those unable to attend or those who want to review F2F content again after the conference. ACRL also launched its first Virtual Conference online community in 2005, and has continued the practice since to allow attendees the opportunity connect and access the conference content for up to one year after the event.

Since its first Conference in 1978, ACRL has gone on to hold successful conferences in Minneapolis (1981, 1881 participants), Seattle (1984, 1754 participants), Baltimore (1986, 2309 participants), Cincinnati (1989, 2735 participants), Salt Lake City (1992, 2241 participants), Pittsburgh (1995, 2721 participants), Nashville (1997, 2973 participants), Detroit (1999, 3080 participants), Denver (2001, 3388 participants), Charlotte (2003, 3427 participants), Minneapolis (2005, 3946 participants), Baltimore (2007, 4784 participants), Seattle (2009, 4321participants), Philadelphia (2011, 5312 participants), Indianapolis (2013, 4824 participants), Portland (2015, 5072 participants) and Baltimore (2017, 5200 participants).

The ACRL 2019 Conference, "Recasting the Narrative," will be held in Cleveland, Ohio, April 10–13, 2019.

E-Learning

ACRL offers online courses that provide low-cost continuing education opportunities for librarians in a format that allows them to work around their busy schedules. Courses are typically 3 weeks long and include real-time and asynchronous activities. Current offerings include, "Designing Curriculum and Developing Educators for the Information Literacy Courses of Tomorrow."

In 2004–2005, ACRL also began offering live webcasts, providing participants with interactive learning experiences on topics such as information literacy, standards, scholarly communication, and distance learning. Webcasts typically last 60–90 min and feature a live audio presentation accompanied by PowerPoint slides, web sites, and other resources. Participants can interact with instructors through text chat, asking questions using a microphone, and responding to polls. ACRL established a group registration rate for webcasts in April 2006. The group rate allows an institution to register multiple people and project the session to participants in the same location. ACRL introduced e-learning scholarships in 2010 and were supported again in 2011 and 2012. In FY12, five scholarships that covered the cost of a webcast registration fee ($50 each) and five scholarships that covered the cost of a multi-week course ($135) were awarded. In partnership with TLT Group in the 2000s, ACRL offered a three-part online information literacy seminar series, featuring webcasts on assessment, collaboration, best practices, and information literacy resources.

For the fourth time, ACRL offered a multi-day online-only professional development institute in the spring of a nonnational conference year. The 2012 Spring Virtual Institute, "Extending Reach, Proving Value: Collaborations Strengthen Communities," featured feature a keynote presentation, concurrent live webcasts, and asynchronous lightning talks. The SVI program explored how libraries of all types are capitalizing on community collaborations. As the webcasts increased, interest in a multiday, themed virtual conference decreased, so additional virtual institutes have not been planned.

In spring 2008, ACRL offered its first Springboard Event. The 90-min Springboard webcast, now an annual program, is available for free to the full ACRL membership, with the archived content available afterwards to the world via the web. Approximately 230 ACRL members participated in ACRL's 3rd annual Springboard webcast featuring John Palfrey, Henry N. Ess III Professor of Law, Vice Dean, Library and Information Resources, Harvard Law School, and Faculty Co-Director, Berkman Center for Internet & Society, and co-author of "Born Digital: Understanding the First Generation of Digital Natives." The topic of privacy was chosen in conjunction with ALA's Office of Intellectual Freedom's inaugural "Choose Privacy Week."

ACRL and Choice have teamed up to offer sponsored webinars which have proven to be an attractive vehicle for communication between librarians and vendors. The ACRL/CHOICE webinar series have proven popular with librarians, too, with hundreds of librarians registering for each event.

Leadership

With the quantity of information growing exponentially and the expansion of technology into all aspects of the educational process, higher education is looking for those who can lead the way in utilizing these new tools wisely and navigate the numerous challenges facing us all. It is more important than ever for academic librarians to step up and guide administrators, faculty, and students through the minefields of this new information environment.

ACRL takes its responsibilities for developing leaders seriously. Since 1999, it has partnered with the Harvard University Graduate School of Education to offer a 5-day institute designed to increase the ability of library directors to lead and manage. The Institute helps participants to assess their own leadership capabilities and to analyze how well their own organizations are positioned to meet current and future challenges. Sessions focus on such topics as managing change, human resources, applying technology, team building, and staff motivation. Learning is through the case study method. ACRL held reunions for the alumni of this program at the Midwinter Meeting during the early/mid-2000s and reoffered a reunion at the 2017 Midwinter Meeting. In addition, ACRL partnered with Harvard to offer the Advanced Leadership Institute for Senior Academic Librarians. This new institute was designed exclusively for senior library leaders and alumni of the Leadership Institute for Academic Librarians and addressed critical leadership issues including collaboration and alliances, influence and leadership, managing expectations of presidents and provosts, and the future role of the academic library. The program was offered in March 2008. While programmatically successful, the recessionary economy dampened registration for the March 2010 session and it was cancelled, but was held in March 2012.

In addition, ACRL collaborated with six other higher education associations to offer the Women's Leadership Institute in December 2008. This unique program brought together mid-level administrators from across campus functions to share experiences, develop a better understanding of the campus as a workplace and culture, and create new networks and networking skills. The success of this endeavor led the associations to offer it the program again and ACRL has participated annually since the initial offering. The next Women's Leadership Institute will be offered on December 3–6, 2017.

Mentoring and Training Programs

Recognizing the importance of training and mentoring, ACRL developed the Academic Library Internship for Administrators of Black College Libraries, an internship program for librarians of predominantly black institutions. In December 1973, the Andrew W. Mellon Foundation agreed to underwrite the program with grants totaling $350,000. During the four-year period of the program (1974–1978), 25 librarians from predominantly black institutions of higher education served as management interns for periods of 3–9 months at nationally known academic libraries. The evaluation conducted at the end of the program suggests that the interns carried back to their home institutions a broad understanding of the management techniques and styles employed in large academic libraries. In 1987, ACRL received another grant from the Mellon Foundation to conduct a planning project to assist staff in libraries of historically black colleges and universities.

Under the leadership of Larry Hardesty, the College Libraries Section developed a mentoring program for new college library directors, which is now in its twentieth year. In 2012, Irene Herold succeeded Hardesty as the President of the Board and a new Board was created, which consisted of seven former participants in the program. The faculty for the seminar is now led by Susan Barnes Whyte and Mellissa Jadlos. The program had received funding from the Council on Library Resources, is now primarily self-supporting with participants bearing the costs, and is now a separate 501(c)3 program and is not an official ACRL project. Since its inception, over 333 first-year college library directors and over 152 experienced library directors have participated in the program.

The ACRL Board established the (ACRL Dr. E.J.) Josey Spectrum Scholar Mentor Committee in 2003, to provide conference programs on mentoring, recruit and maintain a pool of academic and research librarians to serve as mentors to Spectrum Scholars. In 2011, the E.J. Josey Spectrum Scholar Mentor Committee revised its mentor recruitment, requirements, and training to more effectively manage the mentor/Scholar matching process. The Committee has forged a strong relationship with the ALA Spectrum Scholar Program staff at ALA.

"Your Research Coach" was established in 2004–2005 by the CLS Research for College Librarianship Committee to help academic librarians with research and scholarly projects. The program provides mentoring to librarians whose institutions cannot offer such support, and who need to publish or present to attain tenure. As of 2015, there are 13 coaches, with 12 assigned, taking part in the program. The committee will open up the program to other sections, starting with the Instruction (IS) and Distance Learning (DLS) sections, once they are able to find more coaches.

Information Literacy Immersion Programs

The ACRL Information Literacy Immersion Program provides information literacy training and education for librarians in the areas of pedagogy and leadership. The popular 4.5 day "Classic" Immersion program provides two tracks for intensive training and education of librarians: 1) the teacher track, focusing on individual development for those who are interested in enhancing, refreshing, or extending their individual instruction skills; and 2) the program track, focusing on developing, integrating, and managing institutional and programmatic information literacy programs. A national faculty of 15 outstanding instruction librarians has been assembled to design, write, and teach the immersion program. The first immersion program was held in July 1999, at Plattsburgh State University of New York and was geared toward academic librarians. Since then, the program has been offered annually with a total of 26 immersion programs training more than 1900 people have been held or licensed in California, Canada, Colorado, Florida, Illinois, Iowa, Massachusetts, Ohio, Rhode Island, Texas, Vermont, Washington, and Wisconsin. While most immersion programs have been held for librarians only, some have been developed to include both librarians and teaching faculty. In addition, ACRL licensed its first institutional-based Immersion program to Cornell University in FY12. The model was very successful and ACRL licensed a similar program to a Hong Kong academic library consortium in FY13, Yale University in FY14, and the Singapore academic library community in FY15.

In fall 2006, the Immersion faculty offered an addition to the Immersion program: "Intentional Teaching: Teaching: Reflective Practice to Improve Student Learning." The Intentional Teaching program provides 3.5 days of learning and reflection for academic librarians and offers a mixture of structured and co-constructed learning segments such as peer discussions, individual reading, and reflection times, and participant-led communities of practice.

The ACRL Institute for Information Literacy announced another new addition to the Immersion program in 2008. The *Assessment: Demonstrating the Educational Value of the Academic Library* is intended for librarians active in teaching and learning and those with leadership roles for information literacy program development who want to improve their knowledge and practice of both classroom and program assessment. The inaugural Assessment Immersion Track was offered in December 2008.

A new structure to the Immersion Program was introduced in 2009. Intentional Teacher and Assessment tracks have been offered simultaneously since fall 2009, while Teacher Track and Program tracks continue to be offered together in summer. This new schedule blends immersion programs with similar formats and lengths to offer a more cohesive immersion experience.

Two new Immersion Programs were offered in FY13, "Practical Management for the Instruction Coordinator" and "Teaching with Technology & Instructional Design Program." The Practical Management was reoffered in FY15.

Workshops and Preconferences

ACRL sponsors workshops, seminars, and preconferences at ALA conferences. Most notable in this area are the conferences developed by the ACRL Rare Books & Manuscripts (RBMS). For 55 year, RBMS has provided 3 days of programming for rare books, special collection, archives, and manuscripts librarians at its annual conference.

ACRL also offers a variety of programs through its extensive chapter network. Local and regional chapters typically offer annual conference programming. To support these efforts, the ACRL Board of Directors has allocated funding for the ACRL president, vice-president/president-elect, and the executive director to visit ACRL chapters. Currently, funds to support 10 visits per year are budgeted. The ACRL Chapters Speakers Bureau fosters closer relations between the Association and its members by creating opportunities for leaders to share perspectives and concerns at the regional and national level.

Annual Conference Programs

As a means of addressing issues of concern and to increase the knowledge of academic librarians, ACRL units and members develop programs to present at the ALA Annual Conferences. ACRL Sections, interest groups, committees, and personal members submit program proposals to the ACRL office 9 months prior to the Annual Conference at which the program is to be presented. This includes programs that are not requesting funding, as well as those that are asking for funds. ACRL encourages its units to cosponsor programs with other ACRL or ALA units and outside organizations. The ACRL Professional Development Committee (PDC) reviews and selects programs for presentation.

The ACRL Board of Directors provides $27,500 from ACRL's budget to support Annual Conference programs (including the ACRL President's Program and excluding cost of audiovisual equipment, which is largely paid for by ALA). The PDC determines how these funds are allocated among the program proposals. How well proposals meet the program criteria is one of the determining factors in whether it is funded or not.

Career and Job Services

ACRL offers three ways for academic librarians to find out about career opportunities and for employers to build a pool of highly qualified individuals from which to recruit for vacant positions: 1) as the only magazine targeted specifically to academic/research librarians, *College & Research Libraries News* (*C&RL News*) is a print vehicle to advertise in print academic library job listings; 2) the ALA JobLIST online career center (joblist.ala.org), launched in 2006, is a joint project of *American Libraries* (*AL*) magazine, *C&RL News*, and ALA's Office for Human Resource Development and Recruitment (HRDR), which incorporates the formerly separate *AL* and ACRL job sites and many services of HRDR, including placement services at the ALA Midwinter Meeting and Annual Conference; and 3) at each ACRL Conference, the association works with HRDR to offer a job placement service. Since 2009, JobLIST has also maintained an active presence on social media sites, including Facebook, Twitter, and LinkedIn. Content shared on these sites includes notifications of new job listings, news items specifically about

the job market for librarianship, and general job searching tips and strategies. In 2011, a free biweekly e-newsletter, *ALA JobLIST Direct*, was also launched to reach yet another target audience with a selection of job search and career advice content, as well as promoting the online service.

ACRL Consulting Services

ACRL has offered library consulting services to member and non-member libraries since 2007. ACRL consultants have worked to provide services including external reviews, staff retreats, and strategic planning facilitation for Appalachian State University, Arkansas State University, Association of Academic Health Sciences Libraries (AAHSL), Buffalo State University, California State University – San Marcos, Center for Research Libraries, Central Michigan University, Chadron State College, Community College of Philadelphia, Dominican University, Duquesne University, Fairleigh-Dickinson University, Georgia College, Gonzaga University, Hampshire College, Kean University, Lamar University, Lewis University, Louisiana Tech, Loyola University, MIT, Mount Holyoke, New Mexico Institute of Mining and Technology, New Mexico State University, North Park University, Northeastern Illinois University, Our Lady of the Lake University, Purdue University, St. Xavier University, SUNY Council of Library Directors, Texas A&M, Triangle Research Libraries Network (TRLN), Tufts University, University of California – Merced, University of California – Riverside, University of California – San Diego, University of Illinois at Chicago, University of Kansas, University of Maryland Baltimore County, University of North Carolina - Charlotte, University of Puerto Rico, University of Saint Francis, University of Wisconsin – Milwaukee, University of Wisconsin – Stout, Western Michigan University, and Wheaton College. ACRL staff consultants, Kathryn Deiss and Kara Malenfant, authored a chapter, "Successful external reviews: Process and practicalities," for *Reviewing the Academic Library: A Guide to Self-Study and External Review*, which guides library leaders and external reviewers in planning and conducting an external review and effectively using the results for future planning.

Building the Knowledge Base

ACRL strives to be a national and international interactive leader in creating, expanding, and transferring the body of knowledge of academic librarianship. One of the principle motives for creating a separate unit for academic librarians in 1938 was to stimulate research and publication in academic librarianship. The ALA First Activities Committee, a body appointed in the 1920s to review the activities and structures of ALA, reported in 1928 that the ALA publishing program had neglected scholarly and bibliographic publication, the areas of greatest interest to academic librarians. This neglect, said the committee's report, had been so extensive "as to threaten at times actual withdrawal of the College and Reference Section from A.L.A."[19]

C&RL and C&RL News

A year after its creation in 1938, ACRL established an official journal called *College & Research Libraries* (*C&RL*). The first issue of the new quarterly publication appeared in December 1939. It was at one and the same time a professional journal, an official organ of ACRL, and a vehicle for the exchange of news about libraries and librarians. A. F. Kuhlman, the first editor of *College & Research Libraries*, believed that "the absence of a professional journal devoted specifically to the interests of college, university, and reference libraries . . . no doubt accounts to a large extent for the lack of a definitive literature dealing with these institutions."[20] Under a series of able editors, from Kuhlman in the 1940s to Wendi Arant Kaspar who assumed the editorship in 2016, *C&RL* established itself as a premier scholarly journal for the publication of empirical research in academic librarianship and helped to build a body of knowledge and intellectual technique for the academic library profession. In 1950, Arthur Hamlin, then ACRL's executive secretary, called *C&RL* "the principal jewel in the Association crown."[21]

The ACRL Board of Directors decided in 1951 to make *College & Research Libraries* a membership benefit so that all members would receive the journal without charge. This far-reaching decision made it possible for *C&RL* to play a key role in unifying the association and the profession. In light of the growing quantity and quality of research about academic librarianship, the Association decided in 1956 to publish *C&RL* on a bimonthly rather than a quarterly basis. The Board approved making *C&RL* an open access publication starting in April 2011.

At the 2012 ALA Annual Conference, the Board approved transitioning *C&RL* to an online-only publication beginning with the January 2014 issue.

In November 2011, the University of Illinois at Urbana-Champaign began a volunteer project to digitize the remaining backfiles of *C&RL*. The back content was added to the main *C&RL* website during the 2012–2013 fiscal year. All current content and complete back issues, beginning in 1939, are freely available to the public on publication website at http://crl.acrl.org. The ACRL Board of Directors expressed their thanks to Illinois for this generous contribution through a formal resolution.

In 1967, the people and news portions of *C&RL* were separately published, allowing the journal to focus on its role as a scholarly journal. Since 1967, *College & Research Libraries News* has served as the official magazine of record of the association and as a clearinghouse for news about academic libraries, librarians, and higher education. A history of the first 30 year of *C&RL News* appeared in the September 1996 issue as part of an anniversary celebration. In 1993, *C&RL News* became the first ALA print publication available through the Internet. In 2002, *C&RL News* began offering an electronic contents service and, also in 2002, *ACRL Update*, an electronic biweekly news publication, was launched to provide more current information and news. RSS feeds of *C&RL News* contents were launched in 2008. The full text of all *C&RL News* articles from 2004 to the present is freely available to all on the web at http://crln.acrl.org.

In January 2010, the online versions of ACRL's three serial publications (*C&RL*, *C&RL News*, and *RBM*) moved to a new home through a partnership with HighWire Press. The move of ACRL's publications from the association website to the HighWire platform provided a number of benefits, including improved search capabilities both within and across publications, increased Web 2.0 functionality, and online access for individual and institutional non-member subscribers.

ACRL Publications in Librarianship. In 1952, ACRL began the ACRL Monographs series. By 2014, the series had grown to 66 titles and is now called ACRL Publications in Librarianship. The first volume in the series was Joe W. Kraus's *William Beer and the New Orleans Public Libraries, 1891–1927.* PIL #58, *Centers for Learning* (Elmborg & Hook) won the 2007 Ilene E. Rockman Publication of the Year Award. Recent titles include *Interdisciplinarity and Academic Libraries* (PIL #66, Mack & Gibson 2012), and *The Changing Academic Library,* 2nd Edition (PIL #65, Budd 2012: *Not Just Where to Click: Teaching Students How to Think About* Information (Swanson and Jagman, Editors), *Assessing Liaison Librarians: Documenting Impact for Positive Change* (Mack and White, Editors), and in Fall 2015, ACRL issued what could be the final PIL volume:, PIL #69, *Leadership Development Programs for Academic Libraries* (I.M.H. Herold, Editor).

Choice

Founded in 1963 with a grant from the Council on Library Resources, Choice is the Middletown, Connecticut–based publishing unit of ACRL with an annual operating budget of more than three million dollars. In 2008 the ACRL and ALA Boards approved the purchase of office space for Choice in the form of a condominium located on Main Street in Liberty Square. In 2009 Choice moved into its top floor space. The design is LEED-certified by the U.S. Green Building Council. Choice is the publisher of Choice magazine, the premier U.S. academic review journal; *Choice Reviews*, a continuously updated database of our reviews; and other publications, described below.

Choice360

The Choice website, www.choice360.org organizes our content and provides librarians with an easy way to view our webinars, download bibliographic essays, listen to our podcasts, get free trials for our subscription products, or contact our editorial staff. It also features the "The Open Stacks" blog, which publishes a wide variety of Choice content.

Choice Reviews. *Choice's* concise, critical, and authoritative reviews have been called "the best short critical evaluations of new titles available anywhere." By a wide margin, subscribers rate *Choice* #1 among the review sources they use to select materials for academic libraries. *Choice Reviews* (www.choicereviews.org) contains every review published by the Choice staff since September 1988, with some 6000 records added each year. Updated daily, the database functions both as a collection-development resource for academic libraries and as a discovery tool for bibliographic research by students and researchers.

Choice. *Choice* magazine has been in continuous publication since 1964. In addition to its reviews, *Choice* is home to "Ask an Archivist," a series of interviews with curators of special collections; guest editorials; lists of forthcoming titles in various disciplines; an annual guide to academic book pricing; the "University Press Forum"; and the "Digital Resources Buying Guide," published each August. Choice's highly esteemed Outstanding Academic Titles list, representing the "best of the best" new titles reviewed during the preceding year, is published each year in January and is widely regarded as comprising the essential titles for purchase by undergraduate libraries. *Choice* reviews are also available under license in other widely used products and services.

ccAdvisor. A collaboration between Choice and The Charleston Company, *ccAdvisor* (www.ccadvisor.org) is a new review service dedicated to the identification and assessment of academic databases, web sites, and tools. Each review features a scoring matrix that rates resources by the quality of their content, user experience, price, and purchase and contract options.

Resources for College Libraries. Released in 2006, *Resources for College Libraries* (www.rclinfo.net) is the digital successor to *Books for College Libraries*, 3rd edition. The product of a collaborative effort between Choice and ProQuest, RCL is published under the direction of more than seventy subject-area experts, who update the work on an annual basis, and lists almost 90,000 works across 61 curricular areas. With the publication of its companion product, *Career Resources*, in 2009, RCL now includes over 4500 core works for vocational-technical subjects as well.

Choice Bibliographic Essays

Written by subject area experts, these comprehensive essays contain discussion and annotation on hundreds of titles. Essays from recent years are available on the LibGuides platform (www.ala-choice.libguides.com).

The Authority File

In 2017 Choice launched, "The Authority File: A Choice podcast," which features weekly episodes of author/editor conversations about new books and products, thought-leadership interviews and discussions, and technology and product case studies.

ACRL-Choice Mobile App

The ACRL-Choice app allows users to interact with ACRL content in a new and more convenient way, with articles, blog posts, webinars, and other content conveniently organized around membership and community, professional development, and content development.

RBM: A Journal of Rare Books, Manuscripts, and Cultural Heritage

RBM, a semiannual publication, began in the spring of 1986 as *Rare Books and Manuscripts Librarianship* (*RBML*) on a trial basis under the leadership of the Rare Books and Manuscripts Section. The journal was incorporated into the ACRL publishing program in 1988. In 2000, the journal underwent a major revision, including a new name, a new graphic treatment, and a new editorial focus. The editorial focus was broadened to include all types of special collections in a variety of media in order to address the broad range of issues and concerns of professionals who work with such collections. Full contents of the past 12 months of *RBM* are available online to subscribers at http://rbm.acrl.org. Contents older than the past 12 months are freely available to all. Thanks to a generous grant from the Gladys Krieble Delmas Foundation, the complete backfiles of *RBML* were digitized and added to the publication website in spring 2011.

ACRL Nonserial Publications

Monographic works pertinent to academic and research librarianship are published by ACRL. ACRL publishes books on a broad spectrum of topics. ACRL has worked to challenge standard publishing models and has experimented with open access and digital publications as well as traditional publications. Typically, ACRL publishes 8–12 titles annually, each with an intended audience of 500–1500. The publications program is intended to be self-supporting.

Titles published in 2016-2017 include: *2016 Academic Library Trends and Statistics, 2015 Academic Library Trends and Statistics, Bridging Worlds: Emerging Models and Practices of U.S. Academic Libraries Around the Globe* (Pun, Collard, and Parrott, Editors), *Choosing to Lead: The Motivational Factors of Underrepresented Librarians in Higher Education* (Olivas, Editor), *Critical Library Pedagogy Handbooks—Vol. 1: Essays and Workbook Activities and Vol. 2: Lesson Plans* (Pagowsky and McElroy, Editors), *Collaborating for Impact: Special Collections and Liaison Librarian Partnerships* (Totleben and Birrell, Editors), *Creative Instructional Design: Practical Applications for Librarians* (West, Hoffman, and Costello, Editors), *Curating*

Research Data—Vol. 1: Practical Strategies for Your Digital Repository and Vol. 2: A The First-Year Experience Cookbook (Pun and Houlihan, Editors), *Handbook of Current Practice* (Johnston, Editor), *Librarians: Exploring Selves, Cultures, and Autoethnography* (Deitering, Stoddart, and Schroeder, Editors), *Mobile Technology and Academic Libraries: Innovative Services for Research and Learning* (Canuel and Crichton, Editors), *Reading, Research, and Writing: Teaching Information Literacy with Process-Based Research Assignments* (Broussard), *Rewired: Research-Writing Partnerships within the Frameworks* (McClure, Editor), *The Small and Rural Academic Library: Leveraging Resources and Overcoming Limitations* (Kendrick and Tritt, Editors), *Students Lead the Library: The Importance of Student Contributions to the Academic Library* (Arnold-Garza and Tomlinson, Editors).

CLIPP (Formerly ClipNotes)

The popular ClipNotes series, developed by the ClipNotes Committee of the College Library Section (CLS), has undergone a significant change: it will henceforth be known as simply CLIPP (College Library Information on Policies and Procedures). The change is not merely cosmetic; there will be a substantive change to the way content is treated. Essays in CLIPP volumes will be more analytical of best practices while retaining the survey method used in the ClipNotes series. The CLS CLIPP Committee is currently seeking proposals for this new series.

Library Statistics

ACRL's involvement in library statistics goes back to 1906 when James T. Gerould read a paper to the College and Reference Library Section on comparative statistics. Gerould himself started an annual compilation of Statistics for Academic Libraries. Known in the 1920s as "Princeton Statistics," the compilation later became *ARL Statistics*.[22] In 1941, ACRL began to collect statistics for college and university libraries and continued to do so until the late 1950s when the service was discontinued in order to avoid duplicating the efforts of the National Center for Educational Statistics. In 1979, however, the ACRL University Library Section, citing the need for up-to-date comparative library statistics in a usable format, proposed that ACRL collect comparative statistics for the university libraries not covered by *ARL Statistics*. This led to the publication of *ACRL University Library Statistics 1978–1979* in 1980. Additional statistical studies of university libraries were published in 1983 and 1985. In 1984 and 1985, a special Task Force on Library Statistics worked to define the statistical needs of academic libraries. Its work served

as the basis for the new standing committee on Academic Library Statistics (now known as the ACRL Academic Library Trends and Statistics Editorial Board). This committee recommended expanding the survey universe to include the "ACRL 100 libraries" and also revised the survey form to match that used by the federal government. These survey results were published in 1987. In 1989, the survey returned to the non-ARL university libraries. Also in 1989, ACRL issued a compilation of the data from 1979 to 1989 in machine-readable form. Since 1989, the ARL-like survey was administered and published covering the years 1990–1991, 1992–1993, 1994–1995, and 1996–1997.

In 1998, ACRL published the final edition of *University Library Statistics*, covering 1996–1997, and initiated a new statistics project, *Academic Library Trends and Statistics*. This annual comprehensive data gathering effort includes libraries at all institutions of higher learning in the United States and Canada. Until 2015 the ARL survey instrument was used with permission. A core set of data, intended for comparative analysis over time, consists of four major categories: Collections, Expenditures, Personnel and Public Services, and Networked Resources and Services. Additional questions are used to gather data on a variety of topics of interest to the profession and to identify trends and other changes that are having an impact on library operations. Results, arranged by Carnegie classifications, are published, as well as made available on an annual subscription basis through ACRL Metrics.

In June 2006, based on information provided by the University of Illinois Graduate School of Library and Information Science Library Research Center (LRC) and committee discussion, the ACRL concluded that some libraries are burdened by a subset of the ARL survey questions and the effort necessary to gather data in support of these questions, have resulted in diminishing response rates. Beginning with the 2006 survey, ACRL produced two shortened versions of the existing ARL survey and tailored it to better meet the unique needs of member libraries. The following versions of the existing ARL instrument were offered to ACRL member libraries: ARL survey (existing instrument); four-year college survey (shortened version of existing ARL instrument); community college survey (shortened version of existing ARL instrument). Four-year colleges and community colleges had the option of submitting the complete ARL survey (including supplementary e-metrics questions). Beginning in 2007, 4-yr colleges, community colleges, and non-ARL doctoral libraries had the option of completing the new ARL supplementary statistics in addition to the annual survey. In 2010, ACRL partnered with Counting Opinions to develop a longitudinal database, providing access to annual trends and statistics data collected since 1999. ACRL Metrics was launched in June 2010, and provides access to ACRL data from 1999–2014 as well as the NCES Academic Library Survey data from 2000–2012. In 2015, ACRL began using the current IPEDS Academic Libraries (AL) component augmented with key questions from the 2012 NCES Academic Library Survey instrument.

Section Newsletters

Most of the 16 ACRL sections publish a semi-annual newsletter. These newsletters provide information about the section's activities. A few sections, such as the Western European Studies Specialists and the Slavic and East European Section (SEES) produce in-depth newsletters. The SEES newsletter, published since 1985, averages 75–80 pages and serves as the official record of the section, reporting on section activities and on relevant activities in the field of Slavic and East European librarianship. Minutes from SEES mid-winter and annual committee meetings are included, along with the minutes of AAASS (American Association for the Advancement of Slavic Studies) Bibliography and Documentation, a reports section, information on new grants and significant acquisitions, new professional appointments, and a bibliography of recent publications. Profiles of special library collections and papers from the annual SEES program are frequently included. The newsletter is distributed internationally and serves as an archival record of Slavic and East European librarianship in North America. In FY2010, sections transitioned from print to electronic only publishing.

ACRL Website

The web provides a powerful opportunity for ACRL units to share information with their membership and with the academic community at large. The URL of the ACRL webpage is http://www.acrl.org. The site maintains comprehensive information about ACRL and its programs, including mission and goals, strategic plan, advocacy, information literacy, conferences, institutes, preconferences, publications, standards, membership and links to sections and chapters. ALA launched a new information architecture in September 2008. After conducting user studies, ACRL redesigned its website in 2009, using the ALA design as a guide. ACRL Sections and committees are encouraged to mount their pages on the ALA server, but are also free to choose non-ALA servers. In 2011, ALA began using Drupal for its content management system and ACRL's web pages were converted to the new systems. An alternative to ALA Connect where members can work, share documents, vote, and have threaded discussions will be implemented in the fall of 2017. The Web Management Group is also working on Responsive Redesign of the current website and all associated microsites.

ACRL Insider

In January 2008, ACRL launched a new weblog titled ACRL Insider. Primarily written by staff and ACRL officers, ACRL Insider provides updates on ACRL activities, services, and programs, including publications, events, conferences, and eLearning opportunities. The blog also keeps readers up-to-date on ACRL operations, including updates from members of the Board of Directors. Through ACRL Insider, ACRL fosters openness and transparency by providing an outlet for connection between members and the Board and staff. ACRL Insider complements ACRLog to provide a big picture view of the association. A weekly Member of the Week post spotlights the diversity and accomplishments of the ACRL membership, and is one of the blog's most popular features. ACRL Insider is available online at www.acrl.ala.org/acrlinsider/.

ACRLog

To provide a forum for major issues, the ACRLog, ACRL's issues blog, was launched in October 2005. On an average day, the ACRLog is visited 4000–4500 times and more than 4600 individual posts are read. ACRLog is the work of a team of regular contributors in addition to guest posts featuring a variety of voices, including first-year librarians. ACRLog is available at acrlog.org/.

Keeping Up With...

In April 2013, ACRL launched Keeping Up With..., an online current awareness publication featuring concise briefs on trends in academic librarianship and higher education. Each edition focuses on a single issue including an introduction to the topic and summaries of key points, including implications for academic libraries. Keeping Up With... has featured information on digital humanities, flipped classrooms, altmetrics, MOOCs, critical librarianship, competency-based education, net neutrality, and more on the ACRL website at http://www.ala.org/acrl/publications/keeping_up_with.

ACRL Podcasts

ACRL Podcasts have provided fresh dimensions on the issues and events in academic librarianship since January 2007. Podcasts include discussions of ACRL publications, interviews with *C&RL News* article authors, recordings of ACRL Candidate Forums, interviews with ACRL

Conference invited speakers, a series of podcasts on the Value of Academic Libraries initiative, and an interview with filmmaker John Waters at the ACRL Conference in Baltimore. ACRL Podcasts are available on the ACRL Insider blog and through iTunes. In addition, several ACRL groups, most notably the Residency Interest Group, produce podcasts on topics specific to their constituencies.

ACRL Social Media

Over the past several years, the association has been on the cutting edge of new technologies to communicate and connect with members. As of December 2015, ACRL's Facebook page (http://www.facebook.com/ala.acrl) boasts 6,429 fans, while 13,692 interested parties receive daily updates on association activities on Twitter (http://www.twitter.com/ala_acrl), and 385 follow the association on Pinterest. These social media tools allow ACRL to create community with members in new and exciting ways.

CONCLUSION

Academic libraries are moving into a century of change that calls for strengthening our collaborations and community relationships if we are to succeed. Through its publications, professional development programs, public policy advocacy, and work with higher education associations, ACRL will continue to enhance the effectiveness of academic and research librarians to advance learning, teaching, and research in higher education. Scholarly communication, student learning, and articulating the value of academic libraries will be of particular concern to the profession and the association in coming years. ACRL initiatives in these areas will help academic librarians learn from one another, become more effective in their work, advance the quality of academic library service, and to promote a better understanding of the role of libraries in academic and research institutions.

ACRL Presidents (beginning with 1938)

1938–1939 Frank K. Walter	1947–1948 William H. Carlson
1939–1940 Phineas L. Windsor	1948–1949 Benjamin E. Powell
1940–1941 Robert B. Downs	1949–1950 Wyllis E. Wright
1941–1942 Donald Coney	1950–1951 Charles M. Adams
1942–1943 Mabel L. Conat	1951–1952 Ralph E. Ellsworth
1943–1944 Charles B. Shaw	1952–1953 Robert W. Severance
1944–1945 Winifred Ver Nooy	1953–1954 Harriet D. MacPherson
1945–1946 Blanche Prichard McCrum	1954–1955 Guy R. Lyle
1946–1947 Errett Weir McDiarmid	1955–1956 Robert Vosper

(Continued)

ACRL Presidents (beginning with 1938) (Continued)

1956–1957 Robert W. Orr	1987–1988 Joanne R. Euster
1957–1958 Eileen Thornton	1988–1989 Joseph A. Boissé
1958–1959 Lewis C. Branscomb	1989–1990 William A. Moffett
1959–1960 Wyman W. Parker	1990–1991 Barbara J. Ford
1960–1961 Edmon Low	1991–1992 Anne K. Beaubien
1961–1962 Ralph E. Ellsworth	1992–1993 Jacquelyn McCoy
1962–1963 Katherine M. Stokes	1993–1994 Thomas Kirk
1963–1964 Neal R. Harlow	1994–1995 Susan K. Martin
1964–1965 Archie L. McNeal	1995–1996 Patricia Senn Breivik
1965–1966 Helen Margaret Brown	1996–1997 William Miller
1966–1967 Ralph E. McCoy	1997–1998 W. Lee Hisle
1967–1968 James Humphry III	1998–1999 Maureen Sullivan
1968–1969 David Kaser	1999–2000 Larry Hardesty
1969–1970 Philip J. McNiff	2000–2001 Lizabeth (Betsy) Wilson
1970–1971 Anne C. Edmonds	2001–2002 Mary Reichel
1971–1972 Joseph H. Reason	2002–2003 Helen H. Spalding
1972–1973 Russell Shank	2003–2004 Tyrone H. Cannon
1973–1974 Norman E. Tanis	2004–2005 Frances J. Maloy
1974–1975 H. William Axford	2005–2006 Camila Alire
1975–1976 Louise Giles	2006–2007 Pamela Snelson
1976–1977 Connie R. Dunlap	2007–2008 Julie B. Todaro
1977–1978 Eldred R. Smith	2008–2009 Erika C. Linke
1978–1979 Evan I. Farber	2009–2010 Lori A. Gotsch
1979–1980 Le Moyne W. Anderson	2010–2011 Lisa Janicke Hinchliffe
1980–1981 Millicent D. Abell	2011–2012 Joyce L. Ogburn
1981–1982 David C. Weber	2012–2013 Steven J. Bell
1982–1983 Carla J. Stoffle	2013–2014 Trevor A. Dawes
1983–1984 Joyce Ball	2014–2015 Karen A. Williams
1984–1985 Sharon J. Rogers	2015–2016 Ann Campion Riley
1985–1986 Sharon Anne Hogan	2016–2017 Irene M.H. Herold
1986–1987 Hannelore B. Rader	2017–2018 Cheryl A. Middleton

CHOICE Editor & Publishers

1963–1966 Richard Gardner (founding editor)	1982–1984 Rebecca Dixon
	1984–1995 Patricia Sabosik
1966–1972 Peter Doiron	1995–2013 Irving Rockwood
1972–1978 Richard Gardner	2013–present Mark Cummings
1978–1979 Louis Sasso	
1979–1982 Jay Poole	

Publications in Librarianship Editors

1952–1953 Lawrence S. Thompson	1977–1982 Joe W. Kraus
	1982–1988 Arthur P. Young
1953–1956 David K. Maxfield	
1956–1960 Rolland E. Stevens	1988–1993 Jonathan A. Lindsey
1960–1966 William V. Jackson	1993–1998 Stephen E. Wiberley Jr.
1966–1970 David W. Heron	1998–2003 John M. Budd
1970–1972 Edward G. Holley	2003–2008 Charles Schwartz
1972–1977 Kenneth G. Peterson	2009–2014 Craig Gibson

(Continued)

Artificial–Association

ACRL Presidents (beginning with 1938) *(Continued)*

C&RL Editors	1984–1990 Charles Martell
1939–1941 A. F. Kuhlman	1990–1996 Gloriana St. Clair
1941–1948 Carl M. White	1996–2002 Donald E. Riggs
1948–1962 Maurice F. Tauber	2002–2008 William Gray Potter
1962–1963 Richard Harwell	2008–2012 Joseph Branin
1963–1969 David Kaser	2013–2016 Scott Walter
1969–1974 Richard M. Dougherty	2016–2019 Wendi Arant Kaspar
1974–1980 Richard D. Johnson	
1980–1984 C. James Schmidt	

C&RL News Editors	1980–1990 George M. Eberhart (staff)
1967–1979 member editors include:	1991–2001 Mary Ellen K. Davis (staff)
David Kaser, David Doerrer, Michael Herbison, Alan Dyson, Susana Hinojosa, Mary Frances Collins, Anne Dowling, John V. Crowley	2001–2002 Maureen Gleason, acting editor (staff)
1979 Jeffrey T. Schwedes (First staff editor)	2002–2007 Stephanie Orphan (staff)
	2007–present David Free (staff)

RBM (formerly RBML) Editors	2003–2008 Richard Clement
1986–1989 Ann S. Gwyn	2008–2013 Beth Whittaker
1989–1993 Alice D. Schreyer	2014–2017 Jennifer Karr Sheehan
1993–1999 Sidney E. Berger	
1999–2003 Lisa Browar/Marvin Taylor	

ACRL Executive Directors	1968–1972 J. Donald Thomas
1947–1949 N. Orwin Rush	1972–1977 Beverly P. Lynch
1949–1956 Arthur T. Hamlin	1977–1984 Julie Carroll Virgo
1956–1957 Vacant	1984–1990 JoAn S. Segal
1957–1961 Richard Harwell	1990 Cathleen Bourdon (Acting)
1961–1962 Mark M. Gormley	1990–2001 Althea Jenkins
1962–1963 Joseph H. Reason	2001–present Mary Ellen K. Davis
1962–1968 George M. Bailey	

ACKNOWLEDGMENTS

The authors wish to acknowledge former ACRL staff whose history of ACRL (Chapter 15 in the *ACRL Guide to Policies and Procedures*) provides the basis for this entry.

Name of the Association

"The ACRL Planning Committee discussed changing the name of the division from the Association of College and Research Libraries to the Association of College and Research Librarians. The ACRL Planning Committee did not recommend such a change. Since this issue comes up from time to time the Planning Committee set down some of the reasons for the recommendation to provide a record for the future.

1. The term 'libraries' is used generically to include librarians and the mission and goals of librarianship. ACRL supports the needs of libraries and their goals through supporting legislation and promoting access to information as well as a wide variety of other ways. These activities relate to libraries more directly than to librarians.
2. The term 'libraries' rather than 'librarians' includes users and their needs.
3. The time involved in the bureaucracy of changing the name would take away from the business of the association. That time expenditure is not justified since the name change would not affect the direction of the association.
4. There are costs involved in changing the name on printed materials such as stationary and brochures."

Source: ACRL Board, January 1986.

REFERENCES

1. Hale, C.E. *The Origin and Development of the Association of College and Research Libraries, 1889–1960.* Xerox University Microfilms: Ann Arbor, MI, 1976; 24.
2. Thomison, D. *A History of the American Library Association, 1876–1972*; American Library Association: Chicago, IL, 1978; 6.
3. Hale, 25.
4. Thomison, 8, 9.
5. Hale, 33–36.
6. Ibid., 36–37, 49, 52, 69.
7. Ibid., 40–42, 46–48, 66–68.
8. Ibid., 75–76, 82.
9. Ibid., 106–107, 109–112, 119, 121–124, 136–138; Association of college and reference libraries: report of the committee on reorganization. ALA Bull. **1938**, *32*, 810–15; Reorganization of the college and reference section. ALA Bull. **1937**, *31*, 591, 593–598.
10. Hale, 190, 198–199, 235.
11. Ibid., 83, 179.
12. Kuhlman A. F. Can the association of college and reference libraries achieve professional status? College Res. Libr. **1946**, *7*, 151.
13. Hale, 156–161.
14. Thomison, 168–169.
15. ACRL Organizational Manual. Association of College and Reference Libraries: Chicago, IL, 1956; 10–11.
16. Kaser, D. A century of american librarianship as reflected in its literature. College Res. Libr. **1976**, *37*, 116.

17. Kroll, S. *Academic Status: Statements and Resources*, 2nd Ed.; Association of College and Research Libraries: Chicago, IL, 1994, iii.
18. Ibid.
19. Thomison, 116.
20. Kuhlman, A.F. Introducing "college and research libraries" College Res. Libr. **1939**, *1*, 8.
21. Hamlin A.T. Annual report of the ACRL executive secretary, 1949–1950. College Res. Libr. **1950**, *11*, 272.
22. Kroll, S. ed., *Academic Status: Statements and Resources*, 2nd Ed.; Association of College and Research Libraries: Chicago, IL, 1994, 54, 55.

Artificial–Association

Association of Library Trustees, Advocates, Friends and Foundations (ALTAFF)

Sally Gardner Reed
Association of Library Trustees, Advocates, Friends and Foundations (ALTAFF), Philadelphia, Pennsylvania, U.S.A.

Abstract

The Association of Library Trustees and Advocates (ALTA) was a division of the American Library Association (ALA) and the only division developed for lay supporters and citizens who govern libraries. Founded as the Trustees Section in 1890, this division was one of the oldest units within the ALA. The mission of ALTA was to provide programs and publications designed to help America's library trustees increase their effectiveness both in terms of library governance and, more recently, in terms of library advocacy. On February 1, 2009 Friends of Libraries U.S.A. (FOLUSA), an independent NGO founded to support library friends groups, merged with the ALTA. This merger resulted in an expanded division of ALA that is now called the Association of Library Trustees, Advocates, Friends and Foundations (ALTAFF).

HISTORY

Founded in 1890 as the Trustees Section, this unit was authorized by the Council as the Trustees Division in 1940. In 1951, it merged with the Library Extension Division and the Division of Public Libraries, becoming part of the Public Libraries Division. In 1961, it again became an independent division, the American Library Trustee Association. In 1999, the Association of Library Trustees and Advocates (ALTA) undertook a name change from the American Library Trustee Association to the Association for Library Trustees and Advocates, retaining the acronym ALTA, in order that the American Library Association (ALA) members, other divisions and units of ALA, the profession at large, and the general public will understand the important role of advocacy for libraries that ALTA plays. The ALTA is the only division of the ALA dedicated to providing resources, programs, publications, and services to America's public library trustees and advocates.

MISSION AND RESPONSIBILITIES AND MEMBERSHIP

The mission of ALTA is to support the development of effective library service for all people in all types of communities and in all types of libraries. In the discharge of this mission ALTA assists the efforts of all those who govern libraries, promote libraries, and advocate for libraries.

ALTA has the specific responsibility for:

- Education through a continuing and comprehensive program for Library Trustees to enable them to discharge their responsibilities in a manner best fitted to benefit the public and the libraries they represent.
- Providing a means for Trustees to have access to information and ideas that will prove useful to them in the governance of their libraries.
- Promoting strong state and regional Trustee organizations.
- Providing to all who value libraries the materials and support they need to be effective advocates for their libraries on the local, state, and national levels.

Membership

With a membership of over 5,000 ALTA represents library trustees and advocates throughout the United States and Canada. Members receive a quarterly newsletter, discounts on publications, reduced conference registration fees, opportunities to network with trustees from across the country, and a subscription to *American Libraries*, the magazine of ALA.

GOVERNANCE AND STRUCTURE

ALTA is governed by a Board of Directors comprising the President, the Vice-President, the Second Vice-President Elect, the immediate Past President, and committee cluster chairs. The work of the organization is done through its committee structure which is, in turn, organized under clusters.

The committee cluster structure and charges are as follows.

Encyclopedia of Library and Information Sciences, Fourth Edition DOI: 10.1081/E-ELIS4-120044817

Advocacy Committees Cluster

- *Advocacy Committee*—To identify and encourage library advocates; provide information to support library advocacy; to identify, explore, and publicize current and emerging advocacy issues; to provide ongoing education about advocacy issues; and to create strategic alliances with other organizations promoting library advocacy.
- *Awards Committee*—To concentrate on the administration, promotion, and development of awards; to evaluate nominations and select worthy recipients for citations; to evaluate each award every three years; to serve as hosts during the Annual Conference to award recipients who are first-time attendees and/or who are not actively involved in ALTA activities; to approve and recommend to the ALTA Board of Directors all new awards.
- *Intellectual Freedom Committee*—To foster the principles of intellectual freedom among public library trustees; to advise the ALTA Board on matters relating to intellectual freedom; and to serve as liaison to the ALA Intellectual Freedom Round Table, and the ALA Intellectual Freedom Committee.
- *Legislative Committee*—To monitor national legislation for its impact on public library service; to encourage trustee participation in ALA and state legislative activities; to represent the interests of the ALTA division with the ALA Legislation Committee and with other appropriated groups having similar goals.

Business Committees Cluster

- *Development*—To encourage and develop relationships between ALTA and the private sector for the support of ALTA activities and services; to develop a recognition program for those vendors or others who contribute to the ALTA development program; and to develop procedures for working with the ALTA office for coordination of these activities.
- *Leadership Development*—To identify and develop ALTA member leadership; to prepare and present an annual orientation and leadership program for all ALTA volunteers that will provide an essential understanding of the division and its policies and procedures.
- *Membership Committee*—To develop and pursue an aggressive and continuous campaign to recruit and retain members for ALTA from existing and potential member groups such as library trustees, librarians, and friends of libraries.
- *Nominations Committee*—To prepare an ALTA slate of candidate for inclusion in the ALA Election Ballot.

Education Committees Cluster

- *Annual Conference Programs*—To coordinate conference programs sponsored by ALTA at Annual Conferences; to ensure broad coverage; diversity, and limited duplication of programs; to receive conference program requests and review them for adherence to established procedures and budgets. To encourage the presentation of successful programs; to submit program requests along with the committee's recommendations to the ALTA Board for its action.
- *Regional Programming*—To develop or identify programs or speakers who can present programs around the country on behalf of ALTA; to identify successful local and state program agendas that can be shared with the state representatives and replicated in other areas of the country.
- *President's Events*—To plan the ALTA President's events including the President's Reception at the Midwinter Meeting and the President's Program at the Annual Conference.
- *PLA National Conference*—To work with Public Library Association (PLA)'s national conference committee to develop and identify programs of interest to trustees and advocates for the PLA National Conference.

Publications Committee Cluster

- *Newsletter*—To solicit or write articles for the *VOICE* in conjunction with ALTA and ALA staff.
- *Publication Development*—To initiate and coordinate the publication of books and pamphlets pertinent to the role of public library trustee service; evaluate the need for revision of published materials; to aid in originating new publication, and in defining the purpose, audience, and scope of each publication. To identify existing publications throughout the country that might be brought to ALTA for national exposure and sale.
- *Web Site Development*—To advise office staff on the development of the ALTA Web site and to represent the interests of ALTA to the ALA Web site Advisory Committee; to contribute to the development and management of Web 2.0 technologies including online communities, blogs, wikis, photo-sharing sites, etc.

THE MERGER OF ALTA AND FOLUSA

In 2008, the ALTA Board in conjunction with the Board of Friends of Libraries U.S.A. (FOLUSA) established a joint task force to determine whether and how these two organizations could combine to create a new, larger division for trustees, advocates, friends, and foundation members. FOLUSA was founded in 1989 as an independent NGO whose mission was to support library friends groups and encourage library advocacy.

The ALTA/FOLUSA joint task force developed a new set of bylaws that would effectively combine these organizations into a new, expanded division of ALA. The ALTA

Board unanimously approved sending the new bylaws to ALTA members for their approval in September 2008.

On February 1, 2009 FOLUSA officially merged with ALTA to become the Association of Library Trustees, Advocates, Friends and Foundations (ALTAFF). According to its bylaws, "The purpose of ALTAFF is to support the development of effective library service for all people in all types of communities and in all types of libraries." The new division will have a minimum of four sections: 1) Trustee Section; 2) Friends Section; 3) Foundation Section; and 4) Corporate Friends Section.

The bylaws also define the role of the new division: http://www.ala.org/ala/mgrps/divs/alta/Final%20ALT AFF%20Bylaws%20.pdf.

Article II. Purposes and Responsibilities.

Section 3. ALTAFF has the specific responsibility for:

1. Educating through a continuing and comprehensive program for Library Trustees to enable them to discharge their responsibilities in a manner best fitted to benefit the public and the libraries they represent.

2. Encouraging and assisting the formation of and the development of Friends of Library groups and Library Foundations.

3. Providing a means for Trustees to have access to information and ideas that will prove useful to them in the governance of their libraries.

4. Providing Friends of Library groups and Library Foundations access to information and ideas that will prove useful to them in fundraising, library promotion, and the operation of their organizations.

5. Promoting strong state and regional Trustee and Friends of Library organizations.

6. Providing to all who value libraries the materials and support they need to be effective advocates for their libraries on the local, state, and national levels.

7. Making the public aware of the existence of formalized citizen groups such as Trustees, Friends of Library groups, and Library Foundations, and the services they perform to encourage and develop expanded citizen participation in the support of libraries across the country.

Association of Research Libraries (ARL)

Lee Anne George
Julia Blixrud
Association of Research Libraries, Washington, District of Columbia, U.S.A.

Abstract

The Association of Research Libraries (ARL) is a not-for-profit membership organization comprising over 120 libraries of North American research institutions. ARL influences the changing environment of scholarly communication and the public policies that affect research libraries and the diverse communities they serve. ARL pursues this mission by advancing the goals of its member research libraries, providing leadership in public and information policy to the scholarly and higher education communities, fostering the exchange of ideas and expertise, and shaping a future environment that leverages its interests with those of allied organizations. This account of association priorities and activities updates the entries in the first and second editions of the encyclopedia and focuses on the years 2001–2007.

FOUNDING AND FOUNDATION

For seventy-five years, ARL has addressed issues of concern to the library, research, higher education, and scholarly communities. The Association was established at a meeting in Chicago on December 29, 1932, by the directors of several major university and research libraries who recognized the need for coordinated action and desired a forum to address common problems. Forty-two libraries adopted a constitution that stated, "the object shall be, by cooperative effort, to develop and increase the resources and usefulness of the research collections in American libraries." On December 5, 1961, the Association was incorporated under the laws of the District of Columbia and certified that "the particular business and objects of the society shall be: Exclusively for literary, educational and scientific purposes by strengthening research libraries." A grant from the National Science Foundation in 1962 enabled the Association to establish a full-time secretariat in Washington, D.C., with a paid executive director and staff.

In 1987, a series of member discussions led to the construction of a vision statement that portrayed the future aspirations for the Association. Following the statement's review by the membership and adoption by the Board of Directors, ARL undertook a planning process to develop the Association's strategy for the 1990s. These efforts culminated in a new mission statement, a values statement, revised programmatic objectives, and a set of financial principles to guide the ARL leadership. In 1988, a new Executive Director was hired and charged with building the Association's capacity to implement the strategic plan.

ARL's mission and objectives were reviewed and updated in 1994. The ARL Board annually adopted priorities to guide the ARL program for the current year and developed a statement of priorities to guide the Association programs for the next 3–5 years. In 2001, the membership undertook a review of core ARL programs. In February 2004, the ARL Board recognized it was time for a comprehensive membership-wide review and assessment of the ARL agenda. This plan was developed by the Strategic Planning Task Force, based on member input and with guidance from the ARL Board.

Guiding Principles

The following principles guided the Task Force in its work. (The "we" in these statements refers to the Association.)
Distinctive mission

- We complement and build on the strengths of other organizations.
- We rethink historic assumptions.
- Our policy positions guide our strategies.

Community

- We are a member-driven organization.
- We are accountable to our members.
- We provide opportunity for full engagement by all member representatives.
- We respect the diversity of our membership.

Intellectual freedom and scholarly communication

- We promote and advocate barrier-free access to research and educational information resources.

Collaboration

- We build relationships with other higher education societies and associations.
- We work closely with other library-related associations, councils, federations, etc.

Diversity

- We encourage and support our members as they strive to reflect society's diversity in their staffing, collections, leadership, and programs.
- We strive to employ a diverse staff.

Operational effectiveness

- We are focused on the needs of our member libraries.
- We allocate our resources wisely and practice sound fiscal management.
- We promote continuing staff development and growth.

The resulting strategic plan identified key priorities for ARL for 2005–2009, areas where the members agreed ARL should play a leadership role at this point in time. The three strategic directions identified in the plan relate to scholarly communication; information and public policy; and research, teaching, and learning. http://www.arl.org/arl/governance/stratplan.shtml.

Summary of Strategic Directions (2005–2009)

Strategic Direction I

ARL will be a leader in the development of effective, extensible, sustainable, and economically viable models of scholarly communication that provide barrier-free access to quality information in support of teaching, learning, research, and service to the community.

Strategic Direction II

ARL will influence information and other public policies, both nationally and internationally, that govern the way information is managed and made available.

Strategic Direction III

ARL will promote and facilitate new and expanding roles for ARL libraries to engage in the transformations affecting research and undergraduate and graduate education.

SCHOLARLY COMMUNICATION

ARL members have had a long-standing interest in issues of scholarly communication, initially focusing on the increasing costs of journal subscriptions. In 1989, the membership voted to establish a formal office to address their concerns. The Office of Scientific and Academic Publishing (OSAP) was created in 1990 to understand and influence the forces affecting the production, dissemination, and use of scientific and technical information. The Office's agenda expanded to address all forms of scholarly information and in 1995 the name was changed to the Office of Scholarly Communication (OSC) to reflect the broadened scope. The OSC promotes Strategic Direction I by working to create new models for scholarly exchange that build on the widespread adoption of digital technologies and networking for research, teaching, and learning; improve the traditional systems of scholarly exchange; and increase the purchasing power of libraries and the terms and conditions under which content is made available. http://www.arl.org/sc/index.shtml.

New Publishing Models

ARL has been a leader in advocating the development of innovative systems that offer barrier-free access to research and educational resources. As libraries, research institutions, scholarly societies, commercial publishers, and others experiment with a variety of models to provide digital, online, unfettered access to scholarly information, a number of business models have emerged utilizing different approaches to handling publication costs, managing collections, and providing user access.

In early 2002, an ad hoc task force met to review ARL's strategy for managing intellectual property in the best interests of the academic community and the public. The task force recommended that ARL promote "open access to quality information in support of learning and scholarship." Open access, in this context, refers to works created with no expectation of financial remuneration and available at no cost to the reader on the public Internet for purposes of education and research. The task force developed a 5-year action agenda to promote open access. Activities were identified in seven major areas: education, advocacy, legal, legislative, new funding models, global alliances, and research. The task force also identified essential partners to engage in these efforts, including scholars and scientists, the higher education and library associations, university counsels, scholarly societies, and numerous others.

In February 2002, ARL signed on to the Budapest Open Access Initiative (BOAI), a movement to accelerate progress in the international effort to make research articles in all academic fields available on the public Internet at no cost to the user. Hundreds of individuals and organizations around the world, including scientists and researchers, universities, laboratories, libraries and library organizations, foundations, journals, publishers, and learned societies, have signed the initiative.

ARL published *Open Access Bibliography: Liberating Scholarly Literature with E-Prints and Open Access Journals* in early 2005. This compilation presents over 1300 selected English-language books, conference papers (including some digital video presentations), debates, editorials, e-prints, journal and magazine articles, news articles, technical reports, and other printed and electronic sources that are useful in understanding the open access movement. It also includes a concise overview of key concepts that are central to the movement. http://www.escholarlypub.com/oab/oab.htm.

Since 2002, when DSpace and other institutional repository software began to be available, an increasing number of research libraries and their parent institutions have established institutional repositories to collect and provide access to diverse locally produced digital materials. A 2006 survey of ARL member libraries found that they are rapidly implementing this approach to asset management. Their repositories house a wide range of content, including theses and dissertations, preprints, postprints, and many other formats. http://www.arl.org/bm~doc/spec292web.pdf.

In 2006, in concert with other members of the Open Access Working Group, a coalition committed to collective advocacy of open access to research, ARL offered a statement of support in response to the Research Councils U.K. policy that mandated open access in digital repositories for funded works. The statement notes,

> We believe that open access research dissemination is an indispensable part of the overall remedy to the serious problems now facing the system of scholarly communication. Moreover, open access is a necessary ingredient in any plan to fully realize the social benefits of scientific advances. While these advantages are important no matter the source of the funding, it is particularly critical when the research is publicly funded and the resulting output is a public good.

http://www.arl.org/sparc/advocacy/oawg.html.

A growing number of research libraries have entered into agreements that carry the collections they have built and nurtured beyond their institutions and into the world in digital format. In response to concerns that these agreements, particularly for mass digitization projects, provide responsible management of these collections, ARL engaged a consultant to provide guidance when negotiating such agreements. The consultant identified seven core library interests in evaluating digitization partnerships. He advised that "strategic, community-wide, and societal interests...must be an explicit part of the negotiation" and that, as stewards of their collections, libraries should consider carefully issues of exclusivity, uses of the digital files, respect for the public domain and copyright, preservation, use of standards, the quid pro quo, and transparency. Developed at the request of ARL's Scholarly Communication Steering Committee, a checklist suggested some of the questions negotiators should ask themselves as they define their objectives in entering an agreement. http://www.arl.org/resources/pubs/br/br250.shtml.

Scholarly Publishing Market

OSC collects data that provides libraries with information on changes and trends in the market for scholarly publications. Given the significant and increasing investment research libraries are making in electronic journals, ARL undertook a series of surveys of its members' electronic journal subscriptions to better understand the issues libraries are facing in ensuring that electronic resources can be used effectively on campus.

ARL developed the first survey in 2002 to gather information about members' subscriptions, expenditures, and licensing terms for a set of 16 commercial and not-for-profit publishers to determine how many titles libraries were subscribing to from each publisher, what they were spending, and under what pricing model. Another survey sent out in 2003 gathered information on additional issues in licensing, particularly how libraries were thinking about their upcoming negotiations for many of their electronic journal packages. A summary of the results is available at http://www.arl.org/resources/pubs/br/asit.shtml.

In 2005, ARL surveyed member libraries on their experiences with large bundled collections of journals to build understanding of the market practices for these bundles. The survey responses provided a broad picture of journal bundling practices and journal collecting as members provided insights into their acquisition rates, satisfaction with pricing terms, the frequency of nondisclosure agreements, contract lengths, cancellation terms, protection of bundled titles in cancellation projects, and other topics.

Information Access Alliance

OSC tracks mergers and acquisitions in the scholarly publishing arena and endeavors to raise awareness of library concerns about the increased consolidation of the publishing industry with antitrust authorities. The Information Access Alliance (IAA), formed by ARL and six other library organizations, advocates a revised analysis of publisher mergers. The IAA urged U.S. Justice Department scrutiny of two mergers of large journal publishers, the Candover and Cinvens acquisition of Kluwer and Springer and the John Wiley & Sons acquisition of Blackwell Publishing.

In 2005, the IAA and the American Antitrust Institute hosted an invitational symposium on "Antitrust Issues in Scholarly and Legal Publishing." The meeting presented perspectives from the library community, economists, and antitrust experts in the legal community to an audience of federal and state regulators, economics and antitrust scholars, and librarians. Participants explored issues surrounding consolidation in the scholarly and legal publishing industry and related issues arising from the development of bundling as a pricing strategy.

ARL offered comments on the "Study on the Economic and Technical Evolution of the Scientific Publication Markets in Europe" from the European Commission (the executive branch of the European Union). The study provides a detailed analysis of the journal publishing market and makes a number of policy recommendations encouraging broader support for public access and open access approaches, regulation of journal bundling, and more vigorous antitrust review of mergers. ARL commented on antitrust and journal bundling issues through the IAA. http://www.informationaccess.org/.

Institute on Scholarly Communication

Strategic Direction I also focuses on "the development of library professionals who have the expertise and knowledge to contribute to enhanced and transformed systems of scholarly communication." To this end, ARL and the Association of College & Research Libraries (ACRL) jointly sponsor the Institute on Scholarly Communication. The institute provides an immersive learning experience to prepare participants as local experts within their libraries and equip them with tools for developing campus outreach strategies. With the number of participants growing, the institute is developing additional resources, such as webcasts and survey instruments, to support library outreach efforts. http://www.arl.org/sc/institute/index.shtml.

SPARC

OSC was instrumental in establishing the Scholarly Publishing and Academic Resources Coalition (SPARC), an international alliance of academic and research libraries working to correct imbalances in the scholarly publishing system, in 1998. SPARC has become a catalyst for change. Its pragmatic focus is to stimulate the emergence of new scholarly communication models that expand the dissemination of scholarly research and reduce financial pressures on libraries.

SPARC's role in stimulating change focuses on educating stakeholders, including authors, publishers, and libraries, about the problems facing scholarly communication and the opportunities for change; advocating policy changes that advance the potential of technology to advance scholarly communication and that explicitly recognize that dissemination is an essential, inseparable component of the research process; and incubating real-world demonstrations of business and publishing models that advance changes benefiting scholarship and academe. SPARC has advanced this agenda by:

- Stimulating the development of increased publishing capacity in the not-for-profit sector and encouraging new players to enter the market;
- Providing help and guidance to scientists and librarians interested in creating change;
- Creating an environment in which editors and editorial board members claim more prominent roles in the business aspects of their journals;
- Demonstrating that new journals can successfully compete for authors and quickly establish quality;

- Effectively driving down the cost of journals; and
- Carrying the methods and message of change to international stakeholders.

In 2007, membership in SPARC numbered nearly 800 institutions in North America, Europe, Japan, China, and Australia. SPARC worked with the Ligue des Bibliothèques Européennes de Recherche (LIBER) and other European organizations to establish SPARC Europe in 2001. SPARC also is affiliated with major library organizations in Australia, Canada, Denmark, New Zealand, the United Kingdom and Ireland, and North America. http://www.arl.org/sparc/.

Create Change

Originally designed in 2000, the Create Change Web site has been a significant resource for librarians and faculty members seeking to improve the scholarly communication system. In 2006, ARL and SPARC partnered to update the site to reflect new developments. The redesigned site highlights the perspective of faculty members to provide both an understanding of key issues in scholarly communication and ideas for taking action to promote change. One of the most popular features of the site, the "Cases in Point" section, highlights interviews of faculty members actively working with new scholarly communication models. http://www.createchange.org/.

PRESERVATION

The nature of library collections is changing and with change come new challenges for preservation. Paper-based books and manuscripts have been the mainstay of scholarly communications and library collections for hundreds of years. But in less than two decades digital information has become a mainstay of research in all disciplines. Web documents, moving images, sound recordings, and data sets are growing more important and mainstream for scholarship. The continued work of developing preservation strategies in research libraries requires a new level of intensity to succeed in an information landscape that is more complex and less stable than ever.

Recognizing research librarians' fundamental role and responsibility for preservation, the ARL Committee on the Preservation of Research Library Materials held retreats in 2000 and 2001 to refocus its agenda and develop a new action plan. A statement developed by the committee and approved by the board in 2002 reaffirmed the commitment of ARL members to preserving collections as basic to an understanding of intellectual and cultural heritage through an active stewardship that enables current and future consultation and use of library resources. The statement also acknowledged the difficulties faced by libraries in trying to effectively balance preservation needs and available resources. http://www. arl.org/preserv/presresources/responsibility_preservation. shtml.

Approximately 150 people attended the 2002 conference "Redefining Preservation, Shaping New Solutions, Forging New Partnerships" that was cosponsored by ARL and the University of Michigan Libraries. Fifteen high-priority actions were identified as needing national attention. Representatives of ARL, the Library of Congress, the Council on Library and Information Resources (CLIR), and the National Endowment for the Humanities subsequently met to review the recommendations and suggest individual and collective strategies for action. The preservation of audio–visual materials was identified as a key area needing attention. As a first step, the Preservation and Conservation Studies program of the School of Information at the University of Texas at Austin, the Library of Congress Preservation Directorate, the National Recording Preservation Board, and ARL cosponsored "Sound Savings: Preserving Audio Collections" in 2003. The program covered topics ranging from assessing the preservation needs of audio collections to creating, preserving, and making publicly available digitally reformatted audio recordings. Conference attendees articulated seven areas for future action to move the field effectively forward. http://www.arl.org/preserv/sound_savings_proceedings/.

In 2004, the ARL Board endorsed digitization as an acceptable preservation reformatting option and released the paper "Recognizing Digitization as a Preservation Reformatting Method" as a first step in building community support and facilitating the development of policies, standards, guidelines, and best practices. CLIR, the Coalition for Networked Information (CNI), OCLC, and RLG joined ARL in supporting digitization as a viable preservation reformatting strategy. http://www.arl.org/bm~doc/ digi_preserv.pdf.

In 2005, ARL endorsed the statement "Urgent Action Needed to Preserve Scholarly Electronic Journals." It reflected ARL's recognition that it was a crucial time for the library community to act in support of initiatives that will ensure enduring access to scholarly e-journals. The statement arose out of a meeting of library leaders hosted by The Andrew W. Mellon Foundation and articulated four actions needed to support the development of qualified preservation archives for scholarly e-journals. http:// www.arl.org/bm~doc/ejournalpreservation_final.pdf.

During a two-day invitational workshop hosted by the University of North Carolina at Chapel Hill in 2006, 30 leaders in preservation and research libraries discussed the future of preservation programs and activities within research libraries as well as at ARL. Participants included program leaders from organizations active in the preservation realm, including CLIR, CNI, the Library of Congress, The Andrew W. Mellon Foundation, the Preservation and Reformatting Section (PARS) of ALA/ALCTS and RLG-OCLC. Discussion focused on which organizations could

take various roles and what the responsibilities of research libraries are in a new era of preservation needs. Informed by the conversations and ideas generated, the Task Force on the Future of Preservation in ARL Libraries developed recommendations for an action agenda relating to ARL's three strategic program areas. A summary of the recommendations is available at http://www.arl.org/bm~doc/ arlbr251preserv.pdf.

PUBLIC POLICIES AFFECTING RESEARCH LIBRARIES

It was no accident that the ARL offices were established in Washington, D.C. Research libraries are part of a larger community of higher education and scholarly societies that tended, in the 1960s, to congregate offices in Washington to influence federal policy. Since one of the key roles member libraries look to ARL to perform is to represent their interests before Congress and other federal agencies, the Association has a history of engagement in federal legislation affecting information policies and appropriations. Initially, it was exclusively the role of the executive director to monitor developments and look for occasions when the interests of research libraries should be articulated before the Federal Government. In some years, trying to influence legislation was an all-consuming activity for the Association's executive director and elected officers. By the early 1980s, ARL members concluded that the range of federal relations issues was expanding enough to warrant a dedicated capacity to engage these developments. In 1984, ARL added a federal relations program officer to bring focus to the increasing array of issues. In 1996, the program officer registered as a lobbyist.

A primary goal of Strategic Direction II is to influence legislative action that is favorable to the research library and higher education communities. To achieve this goal, the public policies program helps ARL members keep abreast of the legislative landscape, as well as rapidly changing issues, players, regulations, and community priorities. Program staff track the activities of state and federal legislatures as well as regulatory and government agencies in North America and abroad. Staff analyze, respond to, and seek to influence public initiatives on information, intellectual property, and telecommunications policies. In addition, the program promotes funding for numerous agencies and national institutions and advances ARL members' interests on these issues. The program works with a variety of agencies and offices on public policy issues including the National Science Foundation, United States Geological Survey, Government Printing Office, Office of Science and Technology Policy, Institute of Museum and Library Services, and the Office of Management and Budget, among others. http://www. arl.org/pp/.

Copyright and Intellectual Property Policies

Copyright and related intellectual property laws have important and substantial effects on the nature and extent of information services libraries provide to their users. As a key focus for Congress, the courts, and state legislatures became updating copyright and intellectual property laws to meet the challenges of the networked environment, the ARL Board identified intellectual property and copyright as a defining set of issues for the future of scholarly communications. The Digital Millennium Copyright Act, the Sonny Bono Copyright Term Extension Act, peer-to-peer file sharing and digital rights management, and legislation to create additional protections for databases have dominated the public policies agenda. Other areas of concern include orphan works—those works whose owners are difficult or even impossible to locate; Internet neutrality—the concept of keeping the Internet open to all lawful content, information, applications, and equipment; and fair use legislation. http://www.arl.org/pp/ppcopyright/.

The public policies program participates in a number of collaborative efforts with a diverse constituency of library, education, legal, scholarly, consumer, and public interest associations; hardware and software manufacturers; and telecommunications providers to advance its agenda and raise library and scholarly community awareness of issues associated with copyright and intellectual property management. Through these partnerships the program has represented ARL interests in a number of *amici curiae* briefs that were filed in copyright and intellectual property court cases. http://www.arl.org/pp/ppcopyright/copyresources/intlcourt.shtml.

Changes to copyright laws extend around the globe and the program has a growing emphasis on international copyright treaties. To help address national and international copyright issues, ARL and four other library associations created the Library Copyright Alliance (formerly named the Shared Legal Capability). The purpose of the alliance is to work toward a unified voice and common strategy for the library community in responding to and developing proposals to amend national and international copyright law and policy for the digital environment. Its mission is to foster global access and fair use of information for creativity, research, and education. Intellectual property laws are currently undergoing major changes in response to the growth in the use of digital formats for works. The alliance is principally concerned that these changes do not harm, but rather enhance, the ability of libraries and information professionals to serve the needs of people to access, use, and preserve digital information. http://www.librarycopyrightalliance.org/.

To assist member libraries with their copyright education activities, ARL engaged a Visiting Scholar for Campus Copyright and Intellectual Property projects. Working closely with the Public Policies Steering Committee and key ARL and SPARC staff, the visiting scholar leads the planning and development of a multiphase ARL Copyright Education Initiative to offer information, resources, and tools that are reflective of library principles and goals and are specifically targeted to major campus constituent groups. The Know Your Copy Rights® Web site, which provides resources for librarians working on positive copyright educational programs for academic users of copyrighted materials, was launched in 2007. The site offers a range of tools to help librarians view copyright education from the perspectives of key academic stakeholders. The new resources will help librarians develop messages that are targeted to different campus groups such as faculty, students, legal counsel, academic leadership, and library staff. http://www.knowyourcopyrights.org/.

Public Access Policies

The U.S. government funds research with the expectation that new ideas and discoveries from the research will propel science, stimulate the economy, and improve the lives and welfare of Americans. ARL has promoted legislation and selected agency initiatives that would make federally funded research publicly available under certain circumstances. Enhancing access to federally funded research is a priority for the library community as such initiatives improve access by the public, provide for effective archiving strategies for these resources, and ensure accountability of the federal investment. http://www.arl.org/pp/access/index.shtml.

ARL has been a strong supporter of the National Institutes of Health (NIH) Public Access Policy. The NIH policy requested that, beginning May 2, 2005, all NIH-funded investigators submit to PubMed Central an electronic version of their final manuscripts upon acceptance for publication in peer-reviewed journals. http://www.arl.org/pp/access/accessfunded/nihaccess.shtml.

The Federal Depository Library Program is an important program that enables public access to federal government information. ARL and others in the library community are engaged in reexamining the role of the Federal Depository Library Program in the networked environment. http://www.arl.org/pp/access/fdlp/index.shtml.

Privacy, Security, and Civil Liberties

One of the key concerns of the library community has been legislation intended to promote homeland security but that may restrict access to information and have a negative impact on civil liberties. There is a need to balance the interests of public access to information, privacy, and security concerns. The USA Patriot Act and related antiterrorism measures broadened the surveillance capabilities of law enforcement and contained new provisions governing criminal and foreign intelligence investigations. ARL, with others in the library and higher education communities, worked extensively with House and Senate staff and met with representatives of the FBI, law enforcement, and the Office of Management and Budget to discuss these measures. The program continues to monitor new legislation in this arena.

Cyberinfrastructure

An NSF-funded ARL workshop in 2006 examined the role of research and academic libraries in the stewardship of scientific and engineering digital data. Participants explored issues concerning the need for the new partnerships and collaborations among domain scientists, librarians, and data scientists to better manage digital data collections, necessary infrastructure development to support digital data, and the need for sustainable economic models to support long-term stewardship of scientific and engineering digital data for the nation's cyberinfrastructure. The workshop report reflects the recognition that digital data stewardship is fundamental to the future of scientific and engineering research and the education enterprise and hence to innovation and competitiveness. http://www.arl.org/bm~doc/digdatarpt.pdf.

Federal Funding

ARL, with others in the library and education communities, supports the annual appropriations of a number of federal and congressional agencies, national libraries, and agency programs and initiatives. These include the Institute of Museum and Library Services, the Library of Congress, the National Agricultural Library, the National Archives and Records Administration, the National Endowment for the Humanities, the National Science Foundation, and the U.S. Government Printing Office.

RESEARCH, TEACHING, AND LEARNING

The transformation of research libraries mirrors to a large degree the ongoing evolution of research institutions, especially the changes underway in the very processes of research, teaching, and learning. The 2005–2009 strategic planning process identified the need for Strategic Direction III, which focuses on new and expanding roles for ARL libraries to engage in the transformations affecting research and undergraduate and graduate education.

In 2005, the webcast "Teaching, Learning & Research: Libraries and Their Role in the Academic Institution" convened library staff to discuss the role libraries play in academia and highlighted the way one library is moving to more fully engage with faculty and students in this process.

In 2006, the Research Teaching and Learning (RTL) Steering Committee focused on establishing subgroups to pursue ARL's newest strategic direction. The RTL Steering Committee established a Task Force on Library

Roles in Enhanced Environments for Teaching and Learning to advise on a strategic agenda to advance research library roles in teaching and learning. The task force recommended undertaking a broad environmental scan of member activities in support of teaching and learning; securing professional assistance in defining a public relations campaign that addresses the roles research libraries play in the teaching and learning enterprise; strengthening partnerships that leverage common interests, particularly with the EDUCAUSE Learning Initiative and CNI; enabling professional development opportunities for library staff; and identifying best practices for library facilities. The RTL Steering Committee and the Scholarly Communication Steering Committee together established a Joint Task Force on Library Support for E-Science to recommend and initiate strategies to address emerging issues in the role of research libraries in e-science. Other subgroups within Strategic Direction III address special collections and diversity. In 2007, a Program Director for Research, Teaching, and Learning was hired to support these activities.

DIVERSITY INITIATIVES

The need for support in the recruitment and training of librarians, especially minority librarians, was expressed as early as the 1960s and 1970s in funding provisions of the Higher Education Act Title II-B. For many years, ARL worked with members to address their growing concerns about recruiting and retaining a diverse workforce in research libraries. Two grants from the H. W. Wilson Foundation in 1990 and 1991 enabled ARL to establish the project "Meeting the Challenges of a Culturally Diverse Workforce" and hire a part-time Diversity Consultant. Demand for seminars, resource materials, and consulting services on diversity topics continued to grow. By 1993, the ARL membership recognized the need for a full-time program to address minority recruitment and retention. A grant from the Gladys Krieble Delmas Foundation in 1994 assured a stable beginning for the program.

ARL's Diversity Initiatives encapsulate a suite of efforts implemented across the strategic directions that aid the Association in defining and addressing diversity issues in ARL libraries. This program seeks to encourage exploration of the rich gifts and talents that diverse individuals bring to the library. ARL staff work closely with a broad range of libraries, graduate library education programs, and other library associations to promote awareness of career opportunities in research libraries and support the academic success of students from groups currently underrepresented in the profession.

ARL launched the Leadership and Career Development Program (LCDP) in 1997. The 18-month program prepares mid-career librarians from underrepresented racial and ethnic groups to take on increasingly demanding leadership roles in ARL libraries and addresses the needs of research libraries to develop a more diverse professional workforce that can contribute to library success in serving increasingly diverse scholarly and learning communities. Over the course of four LCDP offerings, 80 librarians completed the program and a large percentage of them have either been promoted within their libraries or have taken new positions with significantly expanded responsibility.

In 2005, ARL conducted an evaluation of the program's effectiveness. The feedback confirmed the continued need for this program and recommended that its instructional design be tied directly to ARL's strategic directions. The goal of the redesigned LCDP is to provide meaningful exposure to and experience with the strategic issues that are shaping the future of research libraries and to prepare professionals of color for increasingly demanding leadership roles in ARL libraries.

The Initiative to Recruit a Diverse Workforce began in 2000 with support from ARL member libraries. The Institute of Museum and Library Services (IMLS) provided additional support in 2003. The program offers a stipend of up to $10,000 over two years, a mentoring relationship with an experienced librarian, and a leadership training curriculum to MLIS students from underrepresented groups who are interested in careers in research libraries. This multi-year initiative reflects the commitment of ARL members to create a diverse academic and research library community that will better meet the new challenges of global competition and changing demographics. In 2006, the program was awarded a three-year grant through the IMLS Laura Bush 21st Century Librarian Program. These additional funds allow ARL to address the growing need for specialized librarians to help users who create and need access to digital resources and scientific data. The primary goal of the project is to educate, develop, and hire new librarians from underrepresented racial and ethnic groups, especially those with a background in applied and natural sciences and information technology. Forty-five graduate students will receive stipends, mentoring, and leadership development experiences to launch their careers in an ARL library.

The ARL Academy: Careers in Academic and Research Libraries was designed to recruit, educate, and promote visibility within the profession of MLIS students who bring previously gained educational and professional experiences to academic and research libraries. The Academy was a partnership between ARL and three library and information science schools—Catholic University of America, Simmons College, and the University of North Carolina at Chapel Hill. The program was generously supported through an IMLS Recruiting and Education Librarians for the 21st Century grant.

Each fall, from 2004 through 2006, five students from each of the three schools were selected as ARL Academy Fellows. Fellows brought Ph.D.'s or specialized educational accomplishments to the program. As part of the program, Fellows were immersed through their classes in the core philosophical and theoretical context necessary to successfully contribute within libraries at professional levels. Simultaneously, they gained mentored work experience in an ARL library to significantly increase their exposure, competence, and marketability upon graduation.

LEADERSHIP DEVELOPMENT

For over 30 years, the Office of Leadership and Management Services designed and facilitated effective and well-attended library staff development programs and offered services that helped research libraries serve their clientele through the training and strategic deployment of talented individuals. During the 2004 strategic planning process neither professional development and training nor organizational development emerged as top priority issues for ARL members.

In 2005, the membership expressed a range of views about ARL's ongoing involvement in leadership development, especially about the appropriate level of investment of dues to address issues in this arena. The Board established a Task Force on Leadership Development to provide advice on ARL's future approach to addressing executive leadership development issues on behalf of its member libraries. The task force was asked specifically to identify and clarify needs for executive leadership development and succession planning in research libraries and to develop recommendations for strategies to be adopted by the Association for addressing those needs. The responses to a survey conducted by the task force indicated a convergence of views on the need for leadership development resources, especially for executive leadership roles, yet a range of views on how ARL should contribute to meeting those needs. The task force recommended that ARL make a targeted investment to address a set of leadership development needs in the evolving research library environment.

ARL and five member libraries piloted the ARL Research Library Leadership Fellows Program in 2005. This new executive leadership program identifies the unique demands facing directors of large research libraries and prepares participants to develop the skills and professional networks to move into those positions. The pilot program was sponsored by the University of California at Los Angeles; Columbia University; University of Illinois at Urbana-Champaign; University of Texas at Austin; and University of Washington. The second offering began in 2007 and was sponsored by the University of California, Berkeley and the California Digital Library; Harvard University; University of Minnesota; North Carolina State University; Pennsylvania State University; and the University of Toronto. The program offers an opportunity for development of future senior level leaders in large research libraries. It exposes and engages library staff who have the desire and potential for leadership at ARL libraries to themes and institutions that will enhance their preparedness.

COLLECTIONS AND ACCESS

Global Resources

The importance of foreign materials to research libraries was a concern to the Association almost from its beginning. The Farmington Plan, proposed in 1942 by a committee of the Librarian's Council of the Library of Congress, was sponsored by ARL in 1944 and began operation in 1948. The plan was a voluntary agreement under which some 60 libraries attempted to bring to the United States at least one copy of each new foreign monograph of research value. In 1968, with a grant from the Ford Foundation, ARL established its Center for Chinese Research Materials to help bring rare and scattered Chinese materials to libraries at a reasonable cost. The center became a separate organization in 1986.

The ARL Foreign Acquisitions Program, begun in 1991 with support from The Andrew W. Mellon Foundation, assessed the state of global resources in research libraries in North America. The project found a pattern of retrenchment across most collecting areas and an aggregate reduction on the number of unique titles acquired from overseas. A variety of strategies to monitor and respond to this situation were recommended. In 1995, the ARL Board approved the strategic plan for the AAU/ARL Global Resources Program (GRP) as part of the Association's collections activities. The program began in early 1997 with funding from The Andrew W. Mellon Foundation. Originally intended to be a three-year grant, the funding enabled over five years of activity focused on improving access to international research materials through cooperative structures and the use of new technologies, and on generating increased communication with the scholarly community regarding future information needs. In addition, the GRP funding served as seed money for the regional projects, two of which received significant additional funding from the U.S. Department of Education's Title VI Program for Technological Cooperation and Innovation for Foreign Information Access. The regional projects sponsored by the GRP addressed seven countries or world areas: Africa, Eastern Europe, Germany, Japan, Latin America, South Asia, and Southeast Asia. Each developed differently, based on the needs of scholars who use materials from the area and on the perceptions of area specialist librarians of the most pressing challenges for information access.

In 2003, the GRP Advisory Committee, comprised of leaders from both libraries and the academy, reaffirmed the program's importance as a sustainable, broad-gauged

vehicle through which AAU and ARL could combine cooperation with technology in order to expand access to international information. The report, "Scholars, Libraries, and the AAU/ARL Global Resources Program," framed a discussion of several different organizational scenarios and the associated financial options. As a result, the GRP was reframed as the Global Resources Network (GRN).

On January 1, 2006, the Center for Research Libraries (CRL) assumed leadership, governance, coordination, and services of the GRN. During the transition year, the existing GRN Advisory Committee continued to provide intellectual leadership and direction to the network, ensure a smooth transition, and provide ARL oversight while ARL members contributed to GRN funding. CRL now provides administrative, technical, legal, financial management, and communications support for the regular activities of the GRN and its related projects.

Special Collections

For research libraries, special collections are a point of considerable pride. Indeed, these collections are what distinguish and differentiate research libraries. ARL conducted surveys of special collections in member libraries in 1979 and 1998. The goals of the 1998 survey were to equip members to protect and promote special collections as an essential element of research libraries; to articulate the role of special collections within the library program; and to visibly integrate special collections with the goals of the library and the university. The survey results found that special collections constitute a vast and varied resource that is growing not only in size but in scope. http://www.arl.org/bm~doc/spec_colls_in_arl.pdf.

In 2001, a symposium was held at Brown University to explore the prospects and promise of special collections in the expanding electronic environment. "Building on Strength: Developing an ARL Agenda for Special Collections" brought together ARL directors, heads of special collections, invited guests, and speakers to articulate a long-term programmatic agenda for special collections in research libraries. http://www.arl.org/rtl/speccoll/spcollres/

An ARL Special Collections Task Force was formed in late 2001 and charged to engage and advance the agenda that emerged from the symposium. This group brought together ARL directors and special collections librarians, including representatives of the ALA Rare Books and Manuscripts Section (RBMS) and the Society of American Archivists (SAA). They developed a report and an action plan that addresses key points in the symposium agenda, including: enhancing access to collections and backlogs; coordinating planning for collecting nineteenth and twentieth century materials and those in new formats; defining core competencies among special collections librarians and creating training opportunities; and incorporating special collections topics into the agenda of ARL standing committees. http://www.arl.org/rtl/speccoll/spcolltf/.

Over 190 participants from libraries, archives, and funding agencies attended the "Exposing Hidden Collections" conference in 2003 to explore the challenges of providing access to uncataloged and unprocessed archival, manuscript, and rare book materials. One of the major recommendations from the conference was identification and promotion of a shared commitment to certain themes and subjects to encourage cooperative action among libraries and archives to process this material. A survey assessed the interest of libraries and archives in cooperative projects on the themes and also asked for brief descriptions of the top three hidden collections a library or archive was most eager to process, regardless of theme, format or vintage. The task force recommended the development of a position statement to encourage libraries and archives to expose hidden collections though some form of expedited access; a recommendation for a technical strategy for an inventory of unprocessed collections that includes establishing cooperative ties to the Program for Cooperative Cataloging and with the group revising the rare book cataloging standard; and an ACRL/RBMS pre-conference on using collection level records to deal with backlogs of unprocessed special collections. The white paper and a summary of the conference discussions are available at http://www.arl.org/rtl/speccoll/hidden/.

Throughout the work of the Special Collections Task Force, the urgent need to develop the next generation of special collections librarians and administrators has been a recurring theme. A small working meeting was held in 2003 that brought together library directors, special collections librarians and archivists, and library and information science educators to discuss possible responses to this critical situation. The white paper "Education and Training for Careers in Special Collections" found a pressing need for recruitment, education, and training of special collections professionals at all career levels. It calls for the articulation of professional competencies needed for positions in special collections, the development of a shared culture among librarians and archivists, and encouragement of the provision of opportunities for development and professional growth for mid-career practitioners. http://www.arl.org/rtl/speccoll/spcolled/.

A statement of principles, "Research Libraries and the Commitment to Special Collections," contained the key message that

> Special Collections represent not only the heart of an ARL library's mission, but one of the critical identifiers of a research library.... The development, preservation, support, stewardship, and dissemination of major special collections is both a characteristic of the true research library, and an obligation assumed by all members of the Association of Research Libraries.

The statement also articulates the kind of actions that member libraries of ARL should take to support special collections, including providing reliable funding for the support, staffing, and preservation of special collections; building special collections in keeping with institutional collection development policies, existing strengths, and regional or national commitments; and entering a new collection area only if there is a firm commitment to develop the collection and make it accessible to users. http://www.arl.org/rtl/speccoll/speccollprinciples.shtml.

In 2006, the RTL Steering Committee reviewed the final report of the Special Collections Task Force and identified priority elements for ARL's future agenda on special collections as encouraging concerted action and coordinated planning for collecting and exposing nineteenth and twentieth century materials in all formats (rare books, archives and manuscripts, audio, and video); and identifying criteria and strategies for collecting digital and other new media material that currently lack a recognized and responsible structure for stewardship. The steering committee established a new Special Collections Working Group in 2007 to pursue this new agenda. http://www.arl.org/rtl/speccoll/spcolltf/status0706.shtml.

Resource Sharing

For many years the access services program undertook activities to support resource sharing among research libraries in the electronic environment and to improve access to research information resources while minimizing costs for libraries. This program worked to strengthen interlibrary loan and document delivery performance, interoperability among library systems, cooperative cataloging programs, and policies that increased user access to information both onsite and remotely.

A centerpiece of the program was the North American Interlibrary Loan and Document Delivery (NAILDD) Project, initiated by ARL in 1993 to facilitate the development of standards, software, and system design capabilities to improve interlibrary loan and document delivery (ILL/DD) services for users, and to make them more cost effective for research libraries. The NAILDD Project involved the collaboration of over 40 key ILL/DD vendors and system suppliers.

The first Directors Forum on Managing ILL/DD Operations was held in 1995 and marked the beginning of concerted efforts to understand and improve interlibrary loan and document delivery services. The Interlibrary Loan/Document Delivery Performance Measures Study was a two-year effort to measure the performance of ILL departments in 119 North American research and college libraries. The study, funded by The Andrew W. Mellon Foundation, examined four performance measures: direct cost, fill rate, turnaround time, and user satisfaction. This study highlighted the characteristics of high-performing borrowing and lending operations in research libraries.

Techniques to implement these "best practices" were the basis for the "From Data to Action" workshops. Over 400 librarians and representatives from the commercial community attended the twelve workshops offered between October 1998 and March 2001.

The Assessing ILL/DD Services Study in 2002 and 2003 was ARL's third effort in a decade to measure the performance of interlibrary loan operations in North American libraries. The study tracked the performance of mediated and user-initiated ILL/DD operations in 72 research, academic, and special libraries including unit cost, fill rate, and turnaround time for mediated borrowing and lending services. These same measures were also taken for seven user-initiated services. The final report of the study, "Assessing ILL/DD Services: New Cost-Effective Alternatives," confirmed that, in general, user-initiated services have lower unit costs, higher fill rates, and faster turnaround times than mediated ILL/DD services. The report also identified high-performing borrowing and lending operations and laid out strategies for libraries seeking to improve local services.

STATISTICS AND MEASUREMENT

The ARL Statistics and Measurement Program focuses on describing and measuring the performance of research libraries and their contributions to research, scholarship, and community service. ARL serves a leadership role in the development, testing, and application of academic library performance measures, statistics, and management tools. Grounded in the tradition of the North American research library environment, the program provides analysis and reports of quantitative and qualitative indicators of library collections, personnel, and services by using a variety of evidence gathering mechanisms, and tools. http://www.arl.org/stats/.

ARL Statistics

ARL Statistics is a series of annual publications that describe the collections, expenditures, staffing, and service activities for the member libraries of the Association of Research Libraries. The annual statistics series also includes the *ARL Academic Health Sciences Library Statistics*, the *ARL Academic Law Library Statistics*, and the *ARL Preservation Statistics*.

The *ARL Annual Salary Survey* is a compilation of data covering over 12,000 professional positions in ARL libraries. Tables display average, median, and beginning salaries; salaries by position and experience, sex, and race/ethnic background; and salaries in different geographic regions and sizes of libraries. Additional tables cover law, medical, Canadian, and nonuniversity research libraries. http://www.arl.org/stats/annualsurveys/.

Performance Measures

New measures that address issues of service quality, electronic resource usage and value, and outcomes assessment are also being developed. StatsQUAL®, is a gateway to library assessment tools that describe the role, character, and impact of physical and digital libraries. Through StatsQUAL®, libraries gain access to a number of resources that are used to assess library's effectiveness and contributions to teaching, learning, and research. StatsQUAL®, presents these tools in a single powerful interactive framework that integrates and enhances data mining and presentation both within and across institutions. StatsQUAL®, includes instruments and data such as LibQUAL+®, DigiQUAL®, and MINES for Libraries®, as well as a growing dataset of survey results.

LibQUAL+® is a rigorously tested Web-based survey that libraries use to solicit, track, understand, and act upon users' opinions of service quality. Results have been used to develop a better understanding of perceptions of library service quality, interpret user feedback systematically over time, and identify best practices across institutions. http://www.libqual.org/.

The DigiQUAL® project is modifying and repurposing the existing LibQUAL+® protocol to assess the services provided by digital libraries. DigiQUAL® has identified 180+ items around twelve themes related to digital library service quality. The first phase of DigiQUAL® involved testing subsets of these elements with five pilot sites.

MINES for Libraries®™ is an online transaction-based survey that collects data on the purpose of use of electronic resources and the demographics of users. As libraries implement access to electronic resources through portals, collaborations, and consortium arrangements, the Measuring the Impact of Networked Electronic Services (MINES) protocol offers a convenient way to collect information from users in an environment where they no longer need to physically enter the library in order to access resources.

To address the issues of learning outcomes, ARL supported Project SAILS (Standardized Assessment of Information Literacy Skills). This 2003–2006 project was funded through an IMLS grant to Kent State University to develop a Web-based standardized test that allowed cohorts of students to be evaluated against the ACRL Information Literacy Competency Standards for Higher Education. Kent State made the project operational at the end of 2006. http://www.projectsails.org.

ARL is helping libraries develop effective, sustainable, and practical assessment activities that demonstrate the libraries' contributions to teaching, learning, and research. "Effective, Sustainable, and Practical Library Assessment," grew out of a two-year project called "Making Library Assessment Work," which involved 25 libraries during 2005–2006. The service involves a site visit to each participating library, a report to each library with recommendations on practical and sustainable assessment, and follow-up assistance in implementing the recommendations. It is now open to both ARL member and nonmember libraries. http://www.arl.org/stats/initiatives/espassessment/.

TECHNOLOGY

As a strategic response to the realization that telecommunications networks would play a major role in the reform and enrichment of teaching, learning, and education in the twenty-first century, ARL, CAUSE, and EDUCOM (now EDUCAUSE) formed the Coalition for Networked Information (CNI) in 1990. CNI is an organization dedicated to advancing the transformative promise of networked information technology for the advancement of scholarly communication and the enrichment of intellectual productivity. In establishing CNI, the sponsor organizations recognized the need to broaden the community's thinking beyond issues of network connectivity and bandwidth to encompass networked information content and applications. Reaping the benefits of the Internet for scholarship, research, and education demands new partnerships, new institutional roles, and new technologies and infrastructure. The Coalition seeks to further these collaborations, to explore new roles, and to catalyze the development and deployment of the necessary technology base. http://www.cni.org (See the entry on "Coalition for Networked Information (CNI)," p. 1080.)

ARL's Web site was established in 1994 and by the end of 2006 had grown to be extremely large and complex, consisting of over 50,000 pages. In January 2007, ARL launched a newly designed Web site. The updated, streamlined design was developed to improve navigation within the site. The new site focuses on current activities of the Association and many legacy files were not migrated. These were archived and are made available upon request.

MEMBERSHIP

Membership in the Association of Research Libraries is necessarily limited to research institutions sharing common values, goals, interests, and needs. The members of ARL are research libraries distinguished by the breadth and quality of their collections and services. Each member also makes distinctive contributions to the aggregation of research resources and services in North America. Membership is based on the research nature of the library and the parent institution's aspirations and achievements as a research institution. Membership is by invitation upon the recommendation of the Board of Directors and approval of the membership. Once achieved, membership is presumed

to be continuing. The criteria for ARL membership derive from efforts to define a universe of similar institutions that share a commitment to providing the materials and services needed for serious study and research. The principles and procedures of membership are explained at http://www.arl.org/arl/membership/qualprin.shtml. A list of member institutions is at http://www.arl.org/arl/membership/members.shtml.

GOVERNANCE

ARL's Board of Directors is composed of 12 member library representatives who are elected by the membership to serve 3-year terms. ARL committees, task forces, and working groups are also composed of member representatives who wish to work on specific issues important to the Association membership. There are steering committees for each of the three strategic directions: Scholarly Communication, Public Policies Affecting Research Libraries, and Research, Teaching, and Learning. The chairs of these committees serve as nonvoting ex officio members of the ARL Board. Other standing committees include the Membership Committee, Nominating Committee, and Statistics and Assessment Committee. A number of Advisory Committees, Working Groups, and Task Forces work on specific projects and programs of the Association. They are ongoing or temporary as appropriate. http://www.arl.org/arl/governance/cmte.shtml.

There are two membership meetings each year to transact business, provide a forum for discussion of emerging issues, and build the Association's agenda. Since the 1970s, member representatives and invited guests have gathered at an October meeting in Washington, D.C., and a May meeting hosted by and near a member library. During the fall meeting, in addition to committee meetings and group discussion sessions, the membership approves the dues for the coming year and elects new Board members. The spring meeting agenda is built around a specific topic of concern to research libraries. Minutes, and later Proceedings, of meetings 1–133 are available in print. Proceedings since meeting 124 are available at http://new.arl.org/resources/pubs/mmproceedings/.

BIBLIOGRAPHY

1. An engaging overview of the Association's first 60 years, originally a speech by David H. Stam, can be found online at http://www.arl.org/bm~doc/pluscachange.pdf. A more detailed history is Frank M. McGowan's doctoral dissertation, *The Association of Research Libraries 1932–1962* (University of Pittsburgh, 1972).

Association of Specialized and Cooperative Library Agencies (ASCLA)

Barbara A. Macikas
American Library Association, Chicago, Illinois, U.S.A.

Abstract

The Association of Specialized and Cooperative Library Agencies (ASCLA) is a division of the American Library Association (ALA). ASCLA is the professional home for state library agency personnel, librarians serving special populations, multitype librarians, and consultants and other independent librarians. The ASCLA members were strong advocates of the Library Services and Technology Act (LSTA) and will be active in urging reauthorization of the act. The ASCLA also sponsors the Americans with Disabilities Act (ADA) Assembly, an ALA-wide group designed to foster coordination of efforts to meet the challenges and opportunities presented by the ADA legislation.

INTRODUCTION

The Association of Specialized and Cooperative Library Agencies (ASCLA), a division of the American Library Association (ALA), enhances the effectiveness of library service by providing networking, enrichment and educational opportunities for its diverse members, who represent state library agencies, libraries serving special populations, multitype library organizations and independent librarians. It is a community that focuses on evolving issues that cut across library boundaries. Emphasizing the future, ASCLA supports individuals in their personal development and career growth.

DEFINITIONS

State library agencies are those organizations created or authorized by the state government to promote library services in the state through the organization and coordination of a variety of library services.

Specialized library agencies are those organizations that provide materials and services to meet the information needs of persons whose access to library services and materials is limited because of confinement, sensory, mental, physical, health, or behavioral conditions.

Multitype library cooperatives are combinations, mergers, or contractual associations of two or more types of libraries (academic, public, special, or school) crossing jurisdictional, institutional, or political boundaries, working together to achieve maximum effective use of funds to provide library and informational services to all persons above and beyond those that can be provided through one institution. Such cooperative agencies may be designed to serve a community, a metropolitan area, a region within a region, or may serve a statewide or multistate area.

Independent librarians provide services outside of the traditional library setting.

MEMBERSHIP

The ASCLA has the distinction of being the ALA's smallest division yet it has the most diverse membership. With just over 1000 members, ASCLA is the professional home for state library agency personnel, librarians serving special populations, multitype librarians, and consultants and other independent librarians. These four groups joined together over the years when various other ALA divisions and roundtables merged. For example, in 1977, two ALA divisions, the Association of State Library Agencies and the Health and Rehabilitative Library Services Division merged and took on the name ASCLA. The independent librarians joined ASCLA in 1998 when their ALA roundtable, Independent Librarians' Exchange Round Table voted to merge with the division.

ORGANIZATION

The ASCLA is organized into four membership sections: Independent Librarians' Exchange Section, Interlibrary Cooperation and Networking Section (ICAN), Libraries Serving Special Populations Section (LSSPS), and State Library Agency Section (SLAS). Each section offers members committee and discussion group opportunities. The ICAN offers discussion groups on interlibrary cooperation, the virtual library, and network management. The LSSPS offers the following forums: Library Service to People with Visual or Physical Disabilities, Library Service to the Deaf, Library Service to the Impaired Elderly, and Library Service to Prisoners. The SLAS offers

Encyclopedia of Library and Information Sciences, Fourth Edition DOI: 10.1081/E-ELIS4-120044401

discussion groups for Library Services and Technology Act (LSTA) coordinators, consultants for services to children and young adults, consultants to institutional libraries, and state library consultants.

The ASCLA also sponsors the Americans with Disabilities Act (ADA) Assembly, an ALA-wide group designed to foster coordination of efforts to meet the challenges and opportunities presented by the ADA legislation.

PUBLICATIONS

The ASCLA publishes *Interface*, a quarterly newsletter containing news of division activities. The association also publishes monographs; recent titles include *Guidelines for Library Services for People with Mental Illnesses; Revised Standards and Guidelines of Service for the Library of Congress Network of Libraries for the Blind and Physically Handicapped; The Functions and Roles of State Library Agencies* and *Library Networks in the New Millennium: Top Ten Trends*.

AWARDS AND SCHOLARSHIPS

The ASCLA sponsors the following awards and scholarships:

- The ASCLA Exceptional Service Award: A citation presented to recognize exceptional service to patients; to the homebound; to medical, nursing, and other professional staff in hospitals; and to inmates.
- The ASCLA Leadership and Professional Achievement Award: A citation presented to recognize leadership and achievement to one or more ASCLA members in the following areas of activity: consulting, multitype library cooperation, networking, statewide service and programs and state library development.
- The Cathleen Bourdon Service Award: A citation presented to recognize an ASCLA personal member for outstanding service and leadership to the division.
- The ASCLA/National Organization on Disability Award: A $1000 award and certificate for a library organization that has provided outstanding services for people with disabilities.
- The Francis Joseph Campbell Award: A citation and medal presented to a person who has made an outstanding contribution to the advancement of library service for the blind and physically handicapped.
- The Century Scholarship: Up to $2500 annually to be used for accommodations or services not provided by law or the university to a student or students with a disability pursuing a degree in library and information science at an ALA-accredited library school. The award supports the ALA mission of improving service at the local level through the development of a representative workforce that reflects the communities served by all libraries in the new millennium.

CONTINUING EDUCATION OPPORTUNITIES

During the ALA Annual Conference, the ASCLA offers programs on topics such as planning and implementation of open source software for cooperatives and consortia, universal accessibility of physical and virtual collections and of the physical plant, outreach and leadership. The association also presents full-day workshops before the ALA Annual Conference on topics such as accessibility, strategic planning, and sustainability. Finally, the ASCLA has developed an online continuing education course on the topic of selecting Spanish-language materials for adults, an introductory course to help public library staff learn the basics of developing a Spanish-language collection.

STANDARDS

The ASCLA sets standards on recommended performance for the profession. The association has adopted the following standards:

- The Functions and Roles of State Library Agencies, 2000.
- Guidelines for Library and Information Services for the American Deaf Community, 1995.
- Library Standards for Adult Correctional Institutions, 1992.
- Library Standards for Juvenile Correctional Institutions, 1999.
- Library Services for People with Mental Retardation, 1999.
- Revised Standards of Service for the Library of Congress Network of Libraries for the Blind and Physically Handicapped, 1995.

PROBLEMS AND ISSUES

One of ASCLA's challenges is to increase membership rates. Although there are advantages to being a small organization with a diverse membership, it is critical that ASCLA recruit more members. Building the membership will help provide a better financial footing for the organization. Another challenge is to find effective and economical ways of providing continuing education to members, particularly those members who are unable to attend ALA conferences. ASCLA has begun a program on online continuing education and has plans to increase the number of topics in this area.

Artificial–Association

CONCLUSION

The ASCLA members were strong advocates of the LSTA and will be active in urging reauthorization of the act. The ASCLA members closely worked with the ALA Washington Office on the development of a website database that provides stories of the impact LSTA funds have had on local communities and ASCLA's Legislative Committee works to strengthen ALA's advocacy efforts in this arena.

Members of the ADA Assembly will continue to assist libraries in making their services and collections assessable to people with disabilities through such policy statements and development of new guidelines for libraries.

In 2007, the ASCLA ICAN members and other librarians from networks, cooperatives and consortia, participated in the development of the first-ever database of information about how networks and cooperatives operate and the many ways in which these collaborative organizations help to advance learning communities. The ASCLA will encourage participants from the base study to regularly update directory information and the data derived from the surveys to keep the data as up-to-date as possible. The final report database is housed on the ASCLA web. Data was collected from over 200 participating library networks, cooperatives and consortia. The data was collected through a project administered by the ALA Office for Research and Statistics (ORS) and funded in part by a grant from the Institute of Museum and Library Services.

BIBLIOGRAPHY

1. More information on the Association of Specialized and Cooperative Library Agencies, including full-text of policies and links to disability resources is available on the ASCLA website, http://ascla.ala.org.

Australia: Libraries, Archives, and Museums

Alex Byrne
University of Technology, Sydney—Sydney, New South Wales, Australia

Abstract

Over the two centuries since European settlement, Australia has developed fine archives, libraries and museums which are guided by strong curatorial professions. Globally linked and technologically proficient, their standards conform to international models of good practice and sometimes lead the way. The particular challenges posed by distance within a very large country and distance from major centers of curatorial practice and the opportunities offered by an advanced economy have enabled Australia's museologists, archivists and librarians to engage with distinctive issues including those relating to the development of a new multicultural national identity which celebrates both the ancient Indigenous knowledge systems and those brought by visitors and settlers since 1788.

INTRODUCTION

Occupied by humans for some 60,000 years, Australia has long traditions of preserving and transmitting knowledge both orally and through iconographic art as well as the development of technology. However, it was with the arrival of British colonizers from 1788 that European modes of keeping records, gathering collections and publishing observations and opinions as well as the habits of reading and book-based study came to the southern continent. The colonizers brought with them the emerging concepts of modern libraries, museums and archives that were being developed in northern Europe and North America and which were significantly shaped by the Enlightenment. And the explorers and colonizers displayed a need both to share their new found knowledge about the vast land which they encountered and to stay in touch with developments at "home" on the other side of the world.[1]

As settlement proceeded through the nineteenth century, the establishment of museums and libraries marked the growing maturity of the developing colonial cities of Sydney, Hobart, Melbourne, Brisbane, Adelaide and Perth and some significant ports and regional centers. The creation of those institutions confirmed to the European sensibilities of both settlers and visitors that the new cities were truly becoming civilized and that the frontier of settlement was receding from them. In postgold rush "Marvelous Melbourne," for example, the establishment and construction of the imposing Museum and Library of Victoria on one of the grand avenues, Swanston Street, signaled that the city's aspiration to be considered one of the great cities of the British Empire.

After protracted negotiations, the colonies agreed to federate to create a new nation, the Commonwealth of Australia, on January 1, 1901. The words of the first prime minister of Australia, Sir Edmund Barton, "For the first time in history, we have a nation for a continent and a continent for a nation,"[2] expressed the greatest challenge to all institutions in Australia, the "tyranny of distance."[3] While the nation spreads across an area (7,682,300 km² or 2,966,153 mi²) which is almost the same as the total size of the contiguous United States of America, much of the country is arid and the northern part is tropical and mostly extremely remote. The population (now approximately 21 million) was, and continues to be, concentrated in cities and major towns along the coastline, especially in the southeastern quarter. With the exception of the national capital which is located 147 km (92 mi) inland in the purpose built city of Canberra, both governments and major civic and private institutions have been located in the coastal capitals. With vast distances and thin populations it has been difficult to provide services across the nation to smaller cities, towns and remote communities (Fig. 1).

The other key aspect of distance, the distance from the primary centers of professional validation in Europe and North America, constrained Australia's libraries, archives and museums in the early years but has been a lesser concern since the spread of modern transport and telecommunications technologies which place Australians less than 24 hr flight or a few mouse clicks from those centers. Although Australia lacks any borders with adjoining nations, its peoples have long had links to other countries. They extended from the trade and cultural links that the Yolgnu of northeastern Arnhem Land enjoyed for centuries with the Macassan trepang (*Holothuria* sp. also known as *bêche de mer* or sea cucumber) fishermen from their north to today's complex of familial, cultural, trade and professional linkages. That diverse globalism was vastly enriched through the large-scale mass migration programs initiated after World War II. A continuing process, it has brought people from some 200 nations to Australia, changing it and its institutions from the

Encyclopedia of Library and Information Sciences, Fourth Edition DOI: 10.1081/E-ELIS3-120043843

Fig. 1 Map of Australia from *CIA World Factbook*.

formerly predominant orientation towards British institutions and models to a confident transnational outlook.

One of the first nations to introduce free, secular and compulsory elementary education with the pioneering Education Acts in Victoria and the other colonies from 1872, Australia has long enjoyed high standards of education and high levels of literacy. A culture of reading was well established even in nineteenth century rural life and Australia continues to rank highly in international comparisons of literacy and numeracy. Its researchers produce some 2% of the recognized international scholarly literature, a proportion much higher than its population would indicate. Science was inspired by the unique flora and fauna (first noted by early visiting explorers and collectors, notably Sir Joseph Banks on James Cook's *Endeavour* in 1770), the need to develop agriculture appropriately for the various climatic regions and soils, and the identification of ore bodies. Research and technological development has tended to follow those priorities with marked international standing in biomedicine, agriculture and the extractive industries.

Those strengths echo the orientation of Australia's economy to global trade essentially in agricultural, mining and energy commodities and, more recently, financial, professional and educational services. A domestic manufacturing sector has been reshaped by the progressive removal of trade barriers since the 1980s so that most manufactured goods are imported and a fairly small number of companies produce for export. Publishing and other creative industries in Australia face a considerable challenge from imported media and consequently focus on

generating Australian cultural products with low levels of governmental support.

As in the United States, the Australian states have considerable autonomy within the federation so the early differences among them and their institutions have persisted to a significant extent. The national institutions have lacked the authority to direct state and local institutions and their status as the leading national institutions has in some cases been contested by longer established state institutions. Only the National Library of Australia has had legislated authority to provide leadership within its sector but even it has had to attempt to lead by persuasion rather than direction and the legitimacy of that role has sometimes been questioned.

However, the combination of relatively small professional communities with quite strong professional associations has led to good communication and sharing across the memory professions. This has resulted in consistent standards of practice which have been shaped by well-regarded professional education programs and which rate highly in international comparisons. English speaking, welcoming to expatriates from other countries and willing to travel, Australian professionals have been well informed about developments in other countries, especially the British Isles and North America and have been known as innovators and early adopters of innovations. An increasing number have practiced in other countries or have provided leadership in their professions internationally.

LIBRARY AND INFORMATION PROFESSIONS, SYSTEMS AND SERVICES

Origins and Legislation

Following examples imported from Britain, the earliest libraries in Australia were subscription libraries and mechanics' institutes.[4] The Australian Subscription Library and Reading Room, established in Sydney in 1826, replaced a cooperative lending system created by some of the town's leading citizens who had made a shared catalog of their private libraries in 1821. By fits and starts, this led to the opening of a free public reference library in 1869 and the establishment of a lending branch in 1877 and a country-circulating department in 1883. The lending branch was subsequently transferred to the corporation of the city of Sydney in 1909 and added a children's section in 1915.

Melbourne's experience was happier. A mechanics' institute had been initiated in 1840 but Judge Redmond Barry's establishment of a free circulating library in 1842 was the genesis of the first free public library in Australia. Its construction and the purchase of books were approved by the parliament of the colony, newly separated from New South Wales and rich from gold discoveries, with

Barry as chairman of the board of trustees. The other colonies (which became states following their federation for forming the Australian nation in 1901) followed similar patterns during the late nineteenth century (Fig. 2).

However, the development of libraries in Australia was somewhat slow and patchy until the middle of the twentieth century. The landmark investigation by Ralph Munn and Ernest Pitt which was supported by the Carnegie Corporation of New York and whose report, known eponymously as the Munn-Pitt Report,[5] was published in 1935 demonstrated that much had to be done to provide adequate library services across the continent. Implementation of its recommendations was delayed by World War II but the report laid the foundations for the well-regarded libraries of modern Australia.

In common with other institutions in Australia, libraries open to the public have been largely initiated in response to public pressure but then funded and administered by government. Public library provision is a responsibility of local authorities in urban municipalities and rural shires sometimes through collaborative regional services. State libraries are established by state legislation, such as the Library Board of Western Australia Act 1951. Together with the state libraries and the University of Sydney, the National Library benefits from legal deposit legislation applying to printed publications, although its scope has yet to be extended to digital publications. Other legislation of particular relevance to libraries includes the Copyright Act 1968, Classification (Publications, Films and Computer Games) Act 1995 and complementary state laws, Disability Discrimination Act 1992, national and state privacy legislation, and laws relating to online services. Australia lacks a national charter of rights but freedom of expression and freedom of access to information have been substantially protected under the common law system.

National Library and Information Services

In 1901, the first federal parliament recognized the need to establish a collection and services to support its work. The Commonwealth Parliamentary Library was transferred to the new capital, Canberra, with the parliament in 1926. Earlier, the growing collection had been significantly augmented when the Australiana collection of E.A. Petherick was acquired in 1909, laying the foundation for the Commonwealth National Library. That function was eventually separated from the Parliamentary Library by the National Library Act 1960 which gave it several responsibilities: the development of a national collection which would be comprehensive in relation to Australia and the Australian people, provision of bibliographical services, and cooperation in library matters including "the advancement of library science."

The National Library has vigorously pursued the creation of the national collection which now extends to digital

Fig. 2 The statue of the founder of the State Library of Victoria, Sir Redmond Barry, surveys the passing throng before its imposing building on Swanston Street, Melbourne (Photographer Alex Byrne).

resources as well as the traditional printed, pictorial, cartographic and audiovisual resources. In spite of some competition from state heritage collections, especially the rich resources held by the Mitchell Library at the State Library of New South Wales, its collections have achieved the status which was required by the Act. Establishment of the Australian Bibliographic Network in 1981 as a vehicle for collaborative development of bibliographic services demonstrated the foresight and innovation of Australian librarians. Now known as Libraries Australia, the combination of the Australian National Bibliographic Database and various shared services have positioned Australian libraries well to take advantage of new technologies including the emerging Web 2.0 capabilities. It is an excellent example of collaboration which has involved libraries of all types and returned benefits to all and to their clients. In a related area, the National Library coordinates the national cooperative interlibrary loan service, so vital for a very large nation.

In other aspects, the National Library's attempts to show leadership have been more contested. The Australian Libraries Summit convened in October 1988 under the leadership of Warren Horton, then Director-General of the National Library, brought together leaders of the profession from all sectors but few of its recommendations had far reaching results. More recently, his successor as Director-General, Jan Fullerton, has encouraged less publicized but more strategically focused collaboration as a member of National and State Libraries Australasia (NSLA which includes the National Library of New Zealand/Aotearoa). Among the innovations which that collaboration has delivered is PictureAustralia that has exposed the invaluable image collections of NSLA members and other participants, resources which were previously very difficult to access. It has been followed by other collaborative services such as MusicAustralia and PeopleAustralia. Partnership also underpins the operation of the PANDORA Web Archive which was initiated in 1996 and aims to preserve the Australian Web domain. The collaborative traveling exhibition, *National Treasures from Australia's Great Libraries*,[6] which followed the 2002 blockbuster *Treasures from the World's Great Libraries*,[7] demonstrated the value of library heritage collections to a broad public as it was shown across Australia to enthusiastic crowds of visitors.

Various attempts have been made to develop national library and information plans but have been frustrated by the constitutional division of power and authority in the nation's federal structure, a factor which has similarly bedeviled many attempts at national planning in other spheres. For example, following the Scientific and Technological Information Services Enquiry Committee's report,[8] the National Library attempted to establish a national science and technology library and information service in cooperation with the national scientific research organization, the Commonwealth Scientific and Industrial

Research Organisation (CSIRO). This failed to gain adequate financial support and the National Library eventually largely vacated the field, leaving the delivery of scientific and technological information to CSIRO, university and special libraries. However, the National Library maintained an important focus on collecting international publications in the social sciences and humanities, thereby fulfilling a role to inform Australians about the world.

State and Public Libraries

The state and territory library systems operate as the peak libraries for the public within their jurisdictions. They provide major reference libraries, maintain documentary heritage collections and provide support and coordination to public libraries. Some have additional responsibilities for the provision of services to government ministries or to schools. Because they were the first substantial libraries within their jurisdictions and are centrally located within the capital cities, the state libraries fulfill major roles as reference and research libraries and have substantial collections, especially on their home state or territory. Preeminent among those heritage collections is the Mitchell Library within the State Library of New South Wales. Founded by the gift of David Scott Mitchell which was announced in 1898, the collection is unsurpassed in the strength of its collections of Australiana and particularly in items from or relating to the early settlement of Australia.[9] His collection joins with the State Reference Library, the Dixson Collection and many other specialist collections to form a quasinational library operating at the state level to provide a rich resource for consultation by the general public as well as researchers. It is not a lending library but coordinates and supports the state's public library network under a regime which requires adherence to minimum standards in return for financial grants—a model broadly followed by all states and territories.

Although early library services in Australia were provided by private subscription libraries, mechanics' institutes and schools of arts, few private lending libraries survive. They were supplanted by the developing public library network. A notable exception is the Athenaeum Library in Melbourne which continues to offer a fine collection, comfortable reading room and a lively program of speakers.

Funding for each state's public library network is provided by state and local governments. Federal Government interest in public libraries was evident briefly when the Whitlam Labor Government commissioned an enquiry into public libraries under the chairmanship of Allan Horton, the University Librarian at the University of New South Wales and a forceful advocate for libraries in Australia.[10] Its recommendations, unfortunately, were sidelined due to a change in government and economic difficulties in the late 1970s. Nevertheless, services have continued to be expanded so that in Western Australia, for example, there are now 239 registered public libraries throughout a very large state.

The public libraries are lending libraries with relatively small reference collections and specialized services to assist such segments of their populations as the house-bound, young children and business. Many operate mobile libraries and services to institutions such as nursing homes for the elderly. All provide Internet access which is usually filtered due to federal government law to protect children from exposure to undesirable materials. The libraries are sometimes operated through regional consortia spanning a number of local government areas particularly outside the major cities where they wrestle with the challenges of providing services to thinly dispersed populations. Other services of benefit to the public include those provided in hospitals and prisons and in remote mining communities (Fig. 3).

The particular challenges faced by Indigenous communities in remote areas have let to the development of the Indigenous Knowledge and Library Centre models, which have been pursued by the Northern Territory Library Service and the State Library of Queensland. Recognized by receiving the Access to Learning Award of the Bill and Melinda Gates Foundation in 2007, the Northern Territory approach seeks to offer Indigenous peoples a means of recording and holding traditional knowledge in culturally appropriate ways as well as access to the mainstream information required to support education and community activity.

School and College Libraries

There are some 20,000 school libraries in Australia at both primary and secondary levels and located in the state and Roman Catholic diocesan school systems as well as in independent schools. Generally poorly housed and equipped until the late 1960s, except in the richer independent schools, they developed quickly after the Federal Government responded in 1968 to a library association campaign by providing earmarked funding to support their establishment.[4] Staffing, however, remains variable with the best staffed by teacher librarians (who hold combined qualifications as teachers and librarians) but others staffed only by library technicians or supervised part time by a teacher.

Combined school community models provide a means of offering satisfactory library services in many areas outside major cities and towns, especially in South Australia and the Northern Territory. Although they can be challenging to manage because of the different needs of the student and general populations, the models have been successful and have been emulated in other countries.

Fig. 3 The striking Brisbane City Library defines a new public space across the river from the State Library of Queensland (Photographer Alex Byrne).

Australia's well-developed system of technical and further education is supported by good library services in the state operated technical colleges but little library support is evident from the private providers of technical education. The same must be said of private tertiary colleges, including those which offer courses to international students particularly in such fields as business, information technology, design and English language.

Academic and Research Libraries

Established in 1859, the University of Sydney was Australia's first university. It was soon followed by the University of Melbourne and at intervals by universities in each of the other colonial (state following federation) capitals. Originally modeled on U.K. universities, Australian universities tended to import academic staff and did not offer doctoral degrees until 1948. They were all established and to a considerable degree funded by government, benefiting only little from philanthropy.

After World War II, the drive to industrialize the economy and the commencement of very large infrastructure projects—notably the Snowy Mountains Scheme—created a demand for technologists. With the support of the federal Menzies Coalition Government, universities began to expand and the Australian National University (ANU) was established. It was followed by others until, by 1988, there were twenty-one universities in Australia as well as a network of state established colleges of advanced education and institutes of technology both of which were accredited to offer degrees but had extremely limited capacity to undertake research. In 1989, the federal Labor Government initiated a process by which the colleges and institutes merged with existing universities or with each other to form new universities. It also began to cap budgetary allocations to higher education by reinstating tuition fees with a system which permitted deferred payment via taxation surcharges. Extension of this policy by the succeeding Federal Coalition Government continued to reduce the proportion of government funding to higher education with the share at some universities declining from over 90% to around 30% by 2007. All now supplement their revenues by attracting large numbers of international students, enrolling almost 80,000 from more than 12 nations in 2002. However, the university system remained essentially government sponsored and regulated with 37 public but only three private universities.

At first, university libraries developed slowly. By the time of the Munn-Pitt report, their collections remained inferior to the state libraries and were almost all poorly housed and staffed. But, like their universities, they developed quickly during the decades following World War II. Quick to adopt new technologies, they benefited from close professional contact with colleagues in the United Kingdom and the United States and, increasingly, other nations. To support the expanding range of degree programs and increasing emphasis on research, collections ballooned during the 1960s and 1970s and continued to grow steadily in spite of the difficulties caused by the increasing cost of serials and unfavorable exchange rates. Perhaps for that reason, Australian university libraries rapidly introduced policies to prefer digital journals. The Council of Australian University Librarians (CAUL), established as a committee of the universities' chief librarians who began meeting from 1927, became a much more businesslike vehicle for consortial purchasing as well as other collaborative activities (Fig. 4).[11]

As the new millennium was reached, Australian university libraries were very well regarded internationally. They were considered to be leaders in a number of areas including their long-standing leadership in supporting distance education, the rapid adoption and application of information technologies, and the development of the focus on information literacy which has accompanied the emergence of the information society and the rapid spread of the Internet. Many university librarians assisted colleagues in universities elsewhere, particularly Southeast Asia and China. Among many acknowledged leaders were: Harrison Bryan who repositioned the libraries of the Universities of Queensland and Sydney as well as the National Library of Australia; John Shipp who drove the development of CAUL and promoted international relations as well as serving as University Librarian at the Universities of Wollongong and Sydney; Meg Cameron who promoted high standards of service to distance education students during her long service as University Librarian at Deakin University; and Arthur Ellis who promoted the development of the Australian Bibliographic Network during his service at the National Library of Australia and as University Librarian at the University of Western Australia.

Special Libraries

As the Australian economy matured, government departments, hospitals and business and industrial enterprises saw the need to establish libraries and information services to support their activities. Fine special libraries developed in companies such as the mining and steel giant BHP, publishing companies such as Fairfax and legal firms. Major acute hospitals were expected to have adequate library services which were often linked to university library services in support of medical and, later, nursing education. Government departments in particular frequently gave their libraries a broad ambit to collect widely. Improvements in the professional education of librarians created new expectations of professional service to their organizations and collaborative resource-sharing networks were established.

Extensive agricultural extension services were an early response to the need for development across a vast but thinly populated landmass. They led to the establishment

Artificial–Back

Fig. 4 The bronze-clad Fisher Library constructed in the 1960s contrasts with original neogothic sandstone buildings of the University of Sydney (Photographer Alex Byrne).

of agricultural colleges and many government research organizations that were established to investigate and provide advice on the development of the agriculture, mining and manufacturing sectors and, more recently, environmental protection. They included such bodies as the Sugar Research Institute, state departments of primary industries, the Australian Institute of Marine Science and the Great Barrier Reef Marine Park Authority and almost all developed good library services. Foremost among them was the national research and development body, the CSIRO, which provided in effect the national science and technology library and information services. Among the leaders of those services was Des Tellis who pioneered the Australian Earth Sciences Information System (AESIS).

However, changing priorities during the 1990s led to the closure or curtailment of many of these library and information services. Major research collections were transferred to university or other libraries or dispersed when tighter fiscal policies demanded narrower focus. The libraries and information services that survived demonstrated a capacity to visibly add value to their enterprises. Legal and medical services continue to be strong but many services in business and government have been eliminated or severely reduced.

Librarianship as a Discipline and Profession

Preeminent among Australian librarians and library educators must be John Metcalfe, long-term Principal Librarian of the Public (now State) Library of New South Wales, University Librarian at the University of New

South Wales and founder of what became the School of Information, Library and Archive Studies, University of New South Wales.[4] His achievements as a library director, educator and innovator brought him international recognition.

Metcalfe was also instrumental in the establishment of the Australian Institute of Librarians in 1937, one of the immediate consequences of the Munn-Pitt report. It would become the Library Association of Australia (LAA) and subsequently the Australian Library and Information Association (ALIA).[12] It replaced the LAA which had been established with representatives from all the Australian and New Zealand colonies in 1896 but had become defunct in 1928. Nevertheless, the creation of that first "LAA." within two decades of the start of the first professional library associations in the United Kingdom and the United States demonstrated the commitment and ambition of Australian librarians to advance their field.[13] Other specialist and regional associations have been established, including notably the Australian School Library Association, but ALIA remains the national professional organization which seeks to influence governments while fostering professional discourse and setting standards for professional education.

Professional education for librarians was introduced into universities and colleges of advanced education only from the late 1960s. Before that time it had been offered through a system of mentoring with admission to the profession regulated through examinations conducted by the professional association, the LAA. In university libraries, assistant librarians were expected to have a degree and to then sit for the registration examinations while public librarians often took only the registration examinations. Many studied by correspondence with the assistance of colleagues in the workplace, some attended the three library schools which operated in at the state libraries in Sydney and Melbourne and the National Library in Canberra.

One of the major achievements of the national association lay in the transfer of professional education into universities and colleges because that step enhanced professional recognition, generated a stream of well-qualified librarians, provided opportunities for further professional development and expanded research and publishing in the field. A dual system of qualifications replaced the registration process: three- or four-year-long undergraduate degrees and generally one year long postgraduate diplomas. The latter have largely been replaced by master's degrees in recent years. Fine library and information science schools were established in many universities and were led by distinguished library educators such as Jean Whyte, John Balnaves and Margaret Trask. However, most of these schools found themselves in considerable difficulty and many closed in the face of changing university priorities from the late 1990s. Their reconstruction so that they may contribute to the renewal of an ageing professional workforce is now a major issue for the association and its members.

The profession in Australia has also benefited enormously from the association's championing of the education and recognition of library technicians, a paraprofessional element of the library workforce. Trained in technical colleges, library technicians have proven to be an innovation which has significantly enhanced the operation of libraries and information services.

Although the transfer of professional education into universities and colleges was a key stage in professionalization, the battle for professional recognition was not entirely won since salaries for librarians lagged behind those in other fields with comparable qualifications. The principle of wage parity was recognized when librarians at the State Library of New South Wales successfully argued their case although the resultant increase in salaries did not automatically flow onto those employed in other libraries.[14]

International Influences

Professional leadership extended to many aspects of practice and participation in a wide range of international initiatives as well as those at a national level. They have included the technological innovations pioneered by such organizations as OCLC in the United States and the Joint Information Systems Committee in the United Kingdom. Frequent travelers, Australian librarians have participated in many international organizations and often visit overseas colleagues. Reciprocal visits to Australia are encouraged

and Australia has hosted all of the major international library congresses including those of the International Federation of Library Associations and Institutions, the International Association of Music Libraries and the International Association of Technological University Libraries. Australian librarians have served as presidents and senior officials in many such bodies as well as contributing to the work of specialist committees and projects.

Perhaps one of the most significant Australian contributions will prove to lie in the attempts of the profession to engage with the library, information and knowledge issues relating to Indigenous peoples. Work which started in Australia in the early 1990s[15] has been both locally and internationally influential.[16]

ARCHIVES AND ARCHIVAL SCIENCE

In the military and naval traditions to which they were accustomed, the authorities of the first settlements in New South Wales and Van Diemen's Land (now Tasmania) kept records of the progress of the new colony, including births and deaths, marriages, law cases, stores, arrival of ships and land grants. Because they were penal colonies, many of the early records concern the assignment of convicts, keeping of order and punishments. Similar patterns were followed as the other Australian colonies were established including the convict free settlements in South Australia and Western Australia. The records were transmitted to London so the principal early Australian records were located in repositories in the United Kingdom. They had to be copied in the twentieth century through the Australian Joint Copying Project, a major project of repatriating records which began in 1945.[17] Nevertheless, the practice of record keeping was established early and maintained continuously with records offices established as each of the Australian colonies achieved responsible self-government.

National and State Archives

As for libraries, official archives in Australia are split between levels of government. The National Archives of Australia manages records at the federal level.[18] However, the formalization of that responsibility came fairly late in Australia and its achievement was protracted. The first national archivist, Ian Maclean, was appointed in 1944 and served to 1968. During his term, the Director of Archival Management at the U.S. National Archives in Washington, T.R. Schellenberg, visited to propose in 1954 the establishment of national archives. Two decades later in 1974, W. Kaye Lamb, former Dominion Archivist of Canada, reaffirmed the need to take responsibility for government archives. His report was followed by the 1975 appointment of Robert Neale as the first Director-General of the Australian Archives. He served until the Archives Act 1983 came into force, finally providing legislative assurance that Australia's national archives would be protected. Thanks to the commitment of his successors, including the long-serving George Nichols, the National Archives of Australia became a recognized national cultural institution and provided leadership especially in electronic recordkeeping.

Despite its long delayed establishment, the National Archives of Australia now enjoys major recognition for its high standards and has been able to guide Australian Government agencies to adopt good record keeping practice and to select and preserve the national archival collection. Under the Archives Act records in that collection that are over 30 years old are made publicly available, offering valuable insights to historians and the general public.

State records offices perform similar functions at the state and territory level. In South Australia, for instance, the State Records Act 1997, Freedom of Information Act 1991, and Information Privacy Principles provide the legal and regulatory framework for government recordkeeping including both state and local government records.[19] The Director of State Records develops standards and issues guidelines relating to the management of official records, including their destruction with the approval of the State Records Council. Although provisions are similar, the ambit of legislation varies among the jurisdictions. In some states, local government authorities maintain their own archives.

The nexus between record keeping and transparency in government is well established in Australia although many, especially media organizations, object to the charges levied and restrictions imposed on access under freedom of information legislation.[20] In most jurisdictions, ministers and sometimes public servants have wide powers to apply "conclusive certificates" which prevent the release of documents sought by investigators. In the absence of a constitutional right to information, Australians can only appeal to courts to order disclosure on the basis of public interest.

One of the most important achievements of the national and state archives in Australia resulted from the tragic events uncovered by the national inquiry into the forced removal of Indigenous children from their parents under eugenics policies.[21] Included among its recommendations was the need for those who had been removed to access files relating to them and their families so that they could recover their history and reconnect with their families. Indigenous staff members have been employed by the archives to assist enquirers in obtaining access to sensitive and sometimes restricted records. The Indigenous access teams liaise with Indigenous support groups such as Link-up to ensure that counseling and support will be available to those who undergo such emotional experiences.[22]

Similar, although usually much less traumatic, needs to discover "roots" drive much of the growing interest in genealogical inquiry. Government archives in Australia are vital to such inquiry because they hold official migration and shipping records as well as records relating to births, marriages and deaths. Of particular interest are the records relating to transportation of convicts to some of the Australian colonies by British authorities. Once a source of considerable shame, today the identification of a convict ancestor can be welcomed because it identifies the family as early, if involuntary, settlers. Many of the records have been copied to microform or directories and sometimes digitized so that they may be consulted in state libraries and some larger public libraries as well as the libraries of genealogical associations.

Nongovernmental Archives

Universities, colleges and other public and private organizations including businesses, universities and associations generally maintain their own archives although those established by legislation, including universities and public corporations, are usually subject to the state records requirements of their jurisdiction which may include transfer of important records to the state authority. The Noel Butlin Archives Centre is a nationally significant collection of primary source material relating to business and labor which was established in 1953 at the ANU in Canberra. Holding the records of Australian companies, trade unions, employer and professional associations and industry bodies from the 1820s to the present, the Centre is named after Noel Butlin, the first Professor of Economic History at the ANU who began the collection.[23]

Other significant archives include the Bob Hawke and John Curtin Prime Ministerial Libraries (modeled on the presidential libraries of the United States), the Reserve Bank of Australia Archives, the National Library of Australia manuscript collection and the very important records of Australians at war in the Australian War Memorial. Of particular interest is the Archives of the Benedictine Community of New Norcia which holds records of the missionary work of the monks at New Norcia near Perth and in the north of Western Australia.[24] The material—in Spanish, Catalan, French, English, Italian and Latin—includes official correspondence with Rome, records of the abbey and town, and lexicographical work in the various Aboriginal languages of the districts in which the monks worked.

Another important initiative, especially because of Australia's strong record in science and technology, is the Australian Science and Technology Heritage Centre.[25] It was established by the University of Melbourne in 1999 to continue the academic, research and heritage activities of the Australian Science Archives Project, started in 1985, to foster the preservation and promotion of the heritage of Australian science, technology and medicine. Its central Internet accessible resources are Bright Sparcs, which incorporates biographical and bibliographical details about

Australian scientists, and the Virtual Library for the History of Science, Technology and Medicine.

Among the technologies adopted early in Australia were photography, film, sound recording and radio. The National Film and Sound Archive (NFSA) was established in 1984 following a disagreement on the management of audiovisual resources within the National Library. Although the National Library continues to hold valuable audiovisual resources, responsibility for preserving those elements of the national documentary heritage rests with the NFSA which has been an independent statutory authority since July 1, 2008 and works in collaboration with producers such as Film Australia. *australianscreen* delivers online clips from Australian feature films, documentaries, TV programs, shorts, home movies, newsreels, advertisements, and sponsored films produced over the last 100 years.[26]

Archival Science as a Discipline and Profession

The Australian Society of Archivists is the national professional association for archivists[27] and the Records Management Association of Australia is the peak organization for records managers.[28] Both provide opportunities for professional interaction through conferences and publications and guide standards and professional education. Several Australian universities offer courses in archives and records management usually via one year full-time graduate diploma. A master's degree normally requires an additional year of full-time study or original research. Interest in doctoral study is increasing.

At the institutional level, the Council of Australasian Archives and Records Authorities comprises the heads of the government archives authorities of Australia, New Zealand and each of the Australian states and territories.[29] A key undertaking is its Australasian Digital Recordkeeping Initiative launched in 2004 to promote a single Australasian approach to digital public record-keeping across all jurisdictions and to pool resources and expertise to find better ways to ensure that digital records are preserved and made accessible for the future.[30] It was sparked by a paper by the influential head of the Department of Prime Minister and Cabinet, Dr. Peter Shergold, "Digital amnesia: the danger in forgetting the future."[31]

The Register of Australian Archives and Manuscripts is a guide to collections of personal papers and nongovernmental organizational records held by Australian libraries and archives.[32] It succeeds the *Guide to Collections of Manuscripts Relating to Australia* which was published by the National Library of Australia from 1965 to 1995. It includes more than 37,000 records of collections of manuscript and archival material held in Australia, whether of Australian origin or not; photocopies or microfilms of Australian manuscript or archival material, including nongovernment overseas material relating directly to Australia; personal records held in government

archives; and single unpublished primary sources such as a letter, deed or diary. But it does not include government records, unpublished secondary sources, or material that is entirely in pictorial, video, cartographic, musical or sound formats.

MUSEUMS AND MUSEOLOGY

To an even greater extent than libraries, a central role for museums in Australia has been to present the cultures and heritage of the world to the citizens of a far-off land. The major galleries attempted to build representative, "study" collections of artists and art movements recognized to be significant in the metropolitan centers Europe. Many of the museums similarly presented world cultures and manufactures but many also strove from early times to build collections which would represent and enable study of the new land. Collection of biological and ethnographic objects began before colonization as navigators and naturalists took curiosities back to their metropolitan centers. However, Aboriginal peoples had already established keeping places in which important and sacred objects, such as the *tjurungas* of Central Australia were preserved.

For a nation established through colonization, the frontier of cultural contact is in many ways commemorated through material culture as Philip Jones has shown in his landmark study, *Ochre and Rust.*[1] The adoption and adaptation of European materials and tools by Aboriginal peoples and sometimes the reciprocal use of Aboriginal technologies demonstrated the moveable and contested nature of the frontier. But it also highlights the changing perspectives of museology as artifacts, which were once considered to exemplify contamination of traditional Indigenous technology, are now seen to be evocative of cultures in collision. The embracing of Australia's Indigenous heritage and the contemporary vitality of Indigenous art and culture has been a major theme in the development of Australian museums and galleries.

National and State Museums and Galleries

The state art gallery of Victoria, entitled the National Gallery of Victoria (NGV) was established in 1861 and continues under that name, operating two major galleries, the NGV International which presents art from Europe, North America and Asia and the Ian Potter Centre: NGV Australia which shows Australian art, including many works by Indigenous artists. Benefiting enormously from the generous Felton Bequest in 1904, the NGV was enabled to construct an outstanding collection of Western art which fulfilled the mission of presenting world culture to the residents of the antipodes.[33] The Art Gallery of New South Wales, in contrast, had to wait until the last decades of the twentieth century to develop its collections to a noteworthy standard, following the appointment of

Edmund Capon as Director in 1978.[34] As an expert in Asian art, he has developed a major collection of Asian art as well as building the Indigenous and modern art collections. His flair for publicity has generated record attendances especially for major exhibitions including the populist annual Archibald Prize and visiting "blockbuster" exhibitions for which visitors queue for long periods. The Queensland Art Gallery's has gained reputation and built its collections similarly by exploiting its flagship international art event, the "Asia–Pacific Triennial of Contemporary Art," a series of exhibitions uniquely focusing on the contemporary art of Asia and the Pacific, including Australia.[35]

Although the original city plan for Canberra included two galleries, typically for most of the museums and galleries at the truly national, federal level, the National Gallery of Australia, was only established in 1973, long after its state rivals.[36] With federal government funding it built a significant study collection of international works, including controversially Jason Pollock's *Blue Poles*, and a major collection of Australian Indigenous art.[37] It developed a fully professional curatorial capacity which enabled it to solicit major exhibitions from overseas galleries with the risk underwritten through a federal government program now known as Art Indemnity Australia. Now complemented by other programs backed by state governments and extended to state and regional galleries and museums, this is a very necessary provision if precious and sometimes fragile works are to be entrusted to the rigors of touring to and around a far-off continent.

In 1998, the National Gallery was complemented by the National Portrait Gallery, also in Canberra. It aims "to increase the understanding of the Australian people – their identity, history, creativity and culture – through portraiture."[38] Together with the major pictorial collections held by the National Library of Australia, a significant collection at the ANU, the works of war artists at the Australian War Memorial, a number of fine smaller private and public galleries and many practicing artists, these major galleries have made Australia's "bush capital" a city of art which is promoted as destination for tourists with cultural interests. In doing that, Canberra has finally overcome the initial advantage of the state capitals although they—especially Melbourne, Sydney and Brisbane—continue to be major centers for art exhibition (Fig. 5).

Similar tardiness at the national level was evident in the creation of museums with the National Museum of Australia opening in Canberra only in 2001 although its enabling Act had been passed in 1980, it had been building collections for many decades and it had occupied temporary exhibition spaces for some years.[39] The Australian National Maritime Museum, located in Sydney and opened in 1991, celebrates Australia's maritime history and includes one of the world's largest collections of floating vessels which can be visited.[40]

Fig. 5 The innovative architecture and enticing public spaces at the Ian Potter Centre: NGV Australia at Federation Square, Melbourne (Photographer Alex Byrne).

The Australian War Memorial is an exception to the belated creation of museums and galleries at the national, federal level.[41] Conceived in 1916 by the writer of the official history of Australia's participation in World War I, C.E.W. Bean, it was shaped by its first Director, Lieutenant Colonel J.L. Treloar, who occupied the position from 1920 to 1952. A wealth of resources for the researcher and lay person—records of military engagements, collections of war relics and depictions of battle—began to be collected in 1917. A temporary exhibition opened in Melbourne in 1922, moved to Sydney in 1925 and closed in 1935 with the permanent Memorial opening in Canberra in 1941 during World War II. The almost unique combination of memorial, museum, gallery and documentation centre commemorates Australia's long history of participation in international conflicts and peacekeeping operations and provides a major research facility.

The major state museums have tended to have a natural history and ethnographic focus. The oldest, the Australian Museum in Sydney, is an important research institution with specializations in taxonomy, anthropology and mineralogy. It was founded in 1827 when Earl Bathurst, Secretary of State for the Colonies, wrote on 30 March, to Lieutenant-General Darling, Governor of New South Wales, offering up to £200 per annum for "... the Government to afford its aid towards the formation of a Public Museum at New South Wales where it is stated that many rare and curious specimens of Natural History are to be

procured ...."[42] It opened to the public in 1857 in its current location on College Street, Sydney and continues to attract strong attendances to visit both its continuing displays and special exhibitions including some imported "blockbusters" (Fig. 6).

Similar models were followed in other colonies and subsequently states and territories, some combining museum and art gallery, including Tasmania and the Northern Territory. Perhaps reflecting Australian pragmatism and the pressing need to understand the nature of the settlers' new country, curatorial traditions developed more quickly in regard to scientific priorities than art. In keeping with the belief that Australia's Indigenous peoples would disappear, ethnographic interest emphasized traditional cultures and technologies rather than those which were developing across the frontier of contact. The South Australian Museum in Adelaide has an especially rich collection of Aboriginal artifacts which have been represented in the outstanding Australian Aboriginal Cultures Gallery.[43] Curated in consultation with traditional owners and with over 3000 items on display in an interactive setting, the Gallery exhibits the largest collection of Aboriginal artifacts and archival material in the world. It endeavors to tell the story of one of the world's oldest and most continuous cultures in a context which recognizes contemporary Indigenous life as well as traditional cultures and technologies. Although it retains the commitment to scientific rigor and public education, today's

Fig. 6 The floating collection of the Australian National Maritime Museum in Darling Harbour, Sydney (Photographer Alex Byrne).

South Australian Museum is a far cry from the original, somewhat dilettante, conception of a museum to satisfy intellectual pursuits such as literature, arts, history and natural science that had been proposed with the foundation of the South Australian Literary Association in London on August 29, 1834.

Other Museums, Galleries and Curated Environments

Museums and galleries proliferated during the nineteenth century as the Australian colonies developed and the immigrant population swelled. Especially after the discovery of gold in Victoria and later other areas, towns and cities grew and soon wanted a museum, gallery or combination. Many survive today and have been joined by new institutions established through local effort or specialist interest, often with government support in the form of establishment grants and, more rarely, continuing subventions. Australia has largely lacked the tradition of munificent philanthropy found in the United States because of its later settlement, much smaller population and lesser industrialization. Galleries and museums have consequently been unable to acquire overseas works on the scale of the major collections to be found in the United States. Many have nevertheless developed strong collections which increasingly focus on Australia and its broader region, including East and Southeast Asia and the Southwest Pacific.

Among the regional museums and galleries, the Bendigo Art Gallery has built on its colonial heritage to become a highly recognized gallery featuring a strong permanent collection and a stimulating program of locally curated and touring temporary exhibitions.[44] Close by is the Bendigo Joss House which is managed by the local council in association the National Trust of Australia and continues to be used as a place of worship.[45] The only surviving building of its kind in regional Victoria, it commemorates the influx of many Chinese sojourners during the gold rushes. Other properties classified and sometimes managed by the National Trust in and around Bendigo—including the Shamrock Hotel, Post Office and Castlemaine Market—reflect the wealth of the Victorian goldfields. Together with other museums, botanic gardens, national parks and many other curated attractions, this one regional city illustrates not only the widespread interest in and support for preservation and presentation but also the depth of curatorial capacity and expertise that is evident across Australia.

Specialist museums and galleries abound including the Powerhouse Museum in Sydney which was developed from the Museum of Arts and Sciences and has rich technological collections.[46] The Museum of Sydney presents the history and archaeology of the first European settlement in Australia[47] while the Sydney Jewish Museum commemorates the Holocaust as well as displaying the culture and experiences of the Jewish migrants to Australia.[48] Others extend from toys and photography to public transport and militaria. The large museums are complemented by smaller specialist museums, a variety of art galleries, many historic houses which can be visited, others that are heritage listed for preservation, and historic sites and precincts. Although as Australia's largest city Sydney is more richly endowed than many other localities, the variety is typical of that to be found in other cities and towns.

Many have taken advantage of new technologies to engage visitors and stimulate understanding interactively including the South Australian Museum presentation of Indigenous culture and technology. The Questacon, located in Canberra, is Australia's National Science & Technology Centre which promotes interest in science, especially by children, through quirky and exciting exhibits and the use of young explainers who take touring exhibitions and outreach programs across Australia.[49]

The museums and galleries are joined by a range of other curated environments that includes zoos and botanical gardens, aquaria and national parks. The major botanical gardens trace their histories back to the colonial era when they were conceived partly as pleasure grounds and partly as scientific facilities which experimented with the acclimatization of exotic plants to the very different biogeography of Australia as well as the growing of native plants in controlled conditions and preparation of specimens for transmission to European herbaria, especially the famous Kew Gardens in London.

Zoos and aquaria also followed the European models until recent decades when they were influenced by the new open style developed at ZSL Whipsnade Zoo in the United Kingdom and elsewhere. Many are now focused on sustainable practices and the preservation of endangered indigenous and exotic species with strong scientific programs. Although most of the major facilities are to be found in capital cities, some such as the Taronga Western Plains Zoo near Dubbo, New South Wales, are located in rural areas (Fig. 7).

Scientific work and conservation is advanced through the extensive network of national parks across the vast landmass of Australia and stretching northwards to Christmas Island off Java and southwards to the Australian Antarctic territories.[50] They include national parks within metropolitan boundaries, the Sydney Harbour National Park being an outstanding example, as well as relatively small parks surrounding geographical features, such as the Naracoorte Caves Conservation Park in South Australia, large tracts of significant landforms and vegetation, including the impressive Kakadu National Park in the Northern Territory, and marine parks, notably the Great Barrier Reef Marine Park which extends more than 2300 km from the tip of Cape York down the Queensland coast. Australia was early to recognize the importance of national parks with the establishment in 1879 of the Royal

Fig. 7   Powerful rock art at Nourlangie Rock in Kakadu National Park which has been added to the World Heritage List for reasons of both natural and cultural richness (Photographer Alex Byrne).

National Park just beyond the southern suburban fringe of Sydney, the world's second oldest national park after Yellowstone in the United States. However, widespread proclamation and the scientific management of national parks only occurred a century later, in the last decades of the twentieth century. Many are vital to the preservation of endangered native species of fauna and flora as well as sensitive landforms and a growing number is managed in collaboration with the traditional Indigenous owners of the land areas.

As in other developed countries, there are very many commercial, specialist and local museums, galleries and collections across Australia. Some are well curated, others amateurish. A vibrant art sector is reflected in the commercial galleries as well as private and institutional collections. Universities have built art collections, often including an Indigenous component, and some have museums that reflect their research fields such as the Nicholson and Macleay Museums at the University of Sydney and the Ronald and Catherine Berndt Museum at the University of Western Australia.

Museology

Early collectors included visiting scientists, explorers and pastoralists who often collected for donation or sale to museums in Europe or elsewhere. Their personal collections frequently passed to the major public museums or were used to establish local museums and, in some cases university museums. Some found their way to the market for curiosities and many continue to be sold through the major auction houses.

From a museum perspective, almost certainly the most influential Australian collector was the botanist Baron Sir Ferdinand Jakob Heinrich von Mueller.[51] His story is, in many ways, emblematic of a settler society bringing diverse traditions and seeking to understand a very different continent. It explains much about the development of Australian science and museology. Fiercely patriotic to his new country, Mueller was driven by scientific curiosity to investigate the flora of the colonies of South Australia, Victoria and other regions including remote North Australia and Papua. But he was also conscious of the need to develop viable industries and consequently looked for applications of Australian flora as well as the acclimatization of exotic plants. At a time in which settlers tended to look to known European agronomy, Mueller identified uses of Australian plants for medicinal and commercial purposes and was responsible for exporting eucalyptus seeds to California, India, Algeria, Hong Kong and elsewhere as a measure to combat malaria, an initiative which has had global impact. In his later years, his preoccupations extended to the promotion of the exploration of New Guinea and Antarctica.

After emigrating from Germany to Adelaide at the end of 1847, Mueller investigated the South Australian flora and contributed papers to the Linnean Society in London, the German *Linnea* and Adelaide newspapers. Moving to Melbourne in 1852, he was appointed Victorian government botanist in 1853. His initial explorations and collection of specimens added new genera to the flora of Australia. As botanist with A. C. Gregory's North Australian Exploring Expedition in 1855–1857, he observed nearly 2000 species, including some 800 not previously identified in Australia. Appointed director of the Botanical Gardens in Melbourne, Mueller developed a herbarium and prepared his *Fragmenta Phytographiae Australiae* published in twelve parts 1858–1882. He had long hoped to write a flora of Australia and had compiled much material towards it, but with extreme reluctance was compelled to assist George Bentham in preparing *Flora Australiensis* 1863–1878. The tension between science and the creation of public pleasure grounds led to Mueller's dismissal from the directorship of the botanic garden in 1873 since "no foundations exist . . . neither are statues erected . . . works of art we can call forth at pleasure, while time lost in forming the plantations cannot be regained." Despite that major disappointment, he continued his scientific investigations as government botanist, dying in 1896 just before the Australian colonies federated.

As Mueller's career and extraordinary achievements demonstrate, Australian museology was scientifically focused and internationally linked from the outset. Although some of the motivation for establishment of the early museums derived from the desire to demonstrate that the colonies were becoming civilized, as it did for the colonial public libraries, a major impetus came from curiosity about the settlers' new land. Unlike the galleries which tended to emphasize education and the fostering of intellectual life through the presentation of civilization to the residents of distant colonies, the museums collected and exhibited the flora, fauna, geology, ethnography and developing manufactures of those colonies and surrounding regions, especially in the Southwest Pacific.

Museology is well developed as a discipline and profession in Australia with active professional associations and a range of professional educational options. Museums Australia is the national organization for the museums sector.[52] It promotes development of the museum sector, articulates ethical standards, facilitates training and raises public awareness. Many other associations represent more specific interests. At the institutional level, the major organizations are the Cultural Collections Council of Australia, Council of Australian Museum Directors and Council of Australian Art Museum Directors.

Courses range from those more focused on the technical aspects to those that emerge from scientific disciplines or ethnographic and cultural studies.[53] Australian universities which provide cultural and museum studies courses include Deakin, Macquarie, Western Sydney, Sydney, Curtin and the ANU.

Key Contemporary Issues

The international links forged in the early days of Australian curatorial practice continue. Australian museums and their staff are active participants in the major international organizations dealing with curatorial issues, especially the International Council of Museums and the International Council on Museums and Sites. They have vigorously supported UNESCO sponsored initiatives such as the World Heritage List on which are inscribed many natural areas—such as the Australian Fossil Mammal Sites at Riversleigh and Naracoorte (1994), Gondwana Rainforests (1986) and Great Barrier Reef (1981), some cultural areas—Melbourne's Royal Exhibition Building and Carlton Gardens (2004) and the Sydney Opera House (2007)—and areas meeting both criteria—such as Kakadu National Park (1981).

The handling of Indigenous material culture and art has long been a major issue for Australian museums and galleries.[54] Curators have developed considerable understanding of the proper ways to manage both the collections built up since the early days of settlement and contemporary cultural expression. Most work closely with relevant Indigenous organizations and communities to ensure effective consultation and, where possible, engagement with curation. Some have established "keeping places" to which cultural objects, particularly those of a secret or sacred nature, may be returned for safekeeping under appropriate cultural conditions while still being preserved from deterioration. Concerns about the treatment of human remains no longer surface in Australia but Australian curators continue to work with Indigenous peoples to recover those still held in overseas museums, resulting in periodic successful negotiation for return for burial with appropriate ceremony.

International recognition of contemporary Australian Indigenous art over recent decades has led to its inclusion in all major Australian galleries, many galleries in other countries and many other public and private collections in Australia and overseas. The opening of the Musée du Quai Branly in Paris in 2006 was a signal of that recognition: not only were Australian Aboriginal artworks included in the collection but Aboriginal artists contributed to the decoration of the museum itself.[55] While this has been criticized as a manifestation of exoticism, it has forcefully demonstrated that Indigenous art is no longer considered to be of ethnographic interest only but it is now seen as expression of continuing cultural traditions.

CONCLUSION

Over the century since Australia became a nation, its memory institutions and their practitioners have drawn on

Artificial–Back

the professional dispositions brought by settlers and the much older Indigenous traditional knowledge to develop distinctive patterns of practice. Globally linked and technologically proficient, their standards conform to international models of good practice and sometimes lead the way.

The particular challenges posed by distance within a very large country and distance from major centers of curatorial practice and the opportunities offered by an advanced economy have enabled Australia's museologists, archivists and librarians to gain reputations as "early adopters" and innovators. The development of the institutions and the modes of practice have been shaped by those challenges and by the competitiveness inherent in a federation. That has now largely been replaced by cooperation and a degree of convergence which is expressed through the Cultural Collections Council of Australia and in other forums.

The memory institutions have been central to the development of a new multicultural national identity which celebrates both the ancient Indigenous knowledge systems and those brought by visitors and settlers since 1788. Not uncontested, that identity is developing its own shape through reconciliation with Indigenous Australians, celebration of the diversity of modern Australia and recognition of the particular needs and challenges of the "wide brown land."

REFERENCES

1. Jones, P. *Ochre and Rust: Artefacts and Encounters on Australian Frontiers*, Wakefield Press: Kent Town, SA, Australia, 2007.

2. Ward, R.A. *A Nation for a Continent: The History of Australia, 1901–1975*, Heinemann: Melbourne, VIC, Australia, 1983.

3. Blainey, G. *The Tyranny of Distance*, Sun Books: South Melbourne, VIC, Australia, 1966.

4. Biskup, P. *Libraries in Australia*, Centre for Information Studies: Wagga Wagga, NSW, Australia, 1994.

5. Munn, R. Pitt, E.R. *Australian Libraries; a Survey of Conditions and Suggestions for Their Improvement*, ACER: Melbourne, VIC, Australia, 1935.

6. *National Treasures from Australia's Great Libraries*; National Library of Australia: Canberra, ACT, Australia, 2005.

7. *Treasures from the World's Great Libraries*; National Library of Australia: Canberra, ACT, Australia, 2001.

8. *The STISEC Report: Report to the Council of the National Library of Australia by the Scientific and Technological Information Services Enquiry Committee, May 1973*; National Library of Australia: Canberra, ACT, Australia, 1975.

9. Fletcher, B.H. *Magnificent Obsession : The Story of the Mitchell Library, Sydney*, Allen & Unwin: Crows Nest, NSW, Australia, 2007.

10. *Public Libraries in Australia: Report of the Committee of Inquiry into Public Libraries*; Government Printer: Canberra, ACT, Australia, 1976.

11. http://www.caul.edu.au.

12. http://www.alia.org.au.

13. Whyte, J.P. Jones, D.J. *Uniting a Profession: The Australian Institute of Librarians 1937–1949*, Australian Library and Information Association: Kingston, ACT, Australia, 2007.

14. Willard, P. Public sector reform in Australia and its impact on libraries. Int. Inform. Libr. Rev. **1995**, *27* (4), 359–373.

15. Byrne, A. Garwood, A. Moorcroft, H. Barnes, A. Comps. *Aboriginal and Torres Strait Islander Protocols for Libraries; Archives and Information Services*, ALIA for ATSILIRN: Canberra, ACT, Australia, 1995.

16. Nakata, M., Langton, M., Eds. *Australian Indigenous Knowledge & Libraries*; Australian Academic and Research Libraries: Canberra, ACT, Australia, 2005.

17. http://www.nla.gov.au/collect/ajcp.html.

18. http://www.naa.gov.au.

19. http://www.archives.sa.gov.au.

20. Bettington, J., Ed.; *Keeping Archives*, 3rd Ed.; Australian Society of Archivists: Dickson, ACT, Australia, 2008.

21. *Bringing Them Home: Report of the National Inquiry into the Separation of Aboriginal and Torres Strait Islander Children from Their Families*; Commonwealth of Australia: Canberra, ACT, Australia, 1997. Available at http://www.humanrights.gov.au/pdf/social_justice/bringing_them_home_report.pdf (accessed August 17, 2008).

22. Read, P. *A Rape of the Soul So Profound: The Return of the Stolen Generations*, Allen & Unwin: St Leonards, ACT, Australia, 1999.

23. http://www.archives.anu.edu.au/nbac/html/history.html.

24. Strong, R. *Report on the Research Potential of the Archives of New Norcia Abbey for the Archives; Research and Publications Committee of New Norcia Abbey, March 1995*. Available at http://www.newnorcia.wa.edu.au/archives_report.htm (accessed August 17, 2008).

25. http://www.austehc.unimelb.edu.au.

26. http://www.nfsa.gov.au.

27. http://www.archivists.org.au/about-asa.

28. http://www.rmaa.com.au/docs/about/index.cfm.

29. Council of Australasian Archives and Records Authorities. Available at http://www.caara.org.au.

30. Australasian Digital Recordkeeping Initiative. Available at http://www.adri.gov.au.

31. Shergold, P. Digital amnesia: the danger in forgetting the future. Canberra, May 26, 2004. Available at http://pandora.nla.gov.au/pan/53903/20051109-0000/www.pmc.gov.au/speeches/shergold/digital_recordkeeping_2004-05-26.html (accessed August 17, 2008).

32. http://www.nla.gov.au/raam.

33. Poynter, J. R. *Mr Felton's Bequests*, Miegunyah Press: Carlton, VIC, Australia, 2003.

34. http://www.artgallery.nsw.gov.au.

35. http://www.qag.qld.gov.au/exhibitions/archive/2006/apt5.

36. Mollison, J., Murray, L., Eds. *Australian National Gallery: An Introduction*; Australian National Gallery: Canberra, ACT, Australia, 1982.

37. Caruana, W., Ed. *Windows on the Dreaming: Aboriginal Paintings in the Australian National Gallery*; Ellsyd Press: Chippendale, NSW, Australia, 1989.

38. http://www.portrait.gov.au.
39. http://www.nma.gov.au/about_us/history.
40. http://www.anmm.gov.au.
41. http://www.awm.gov.au/aboutus/origins.asp.
42. http://www.austmus.gov.au/archives/fact01.htm.
43. http://www.samuseum.sa.gov.au/page/default.asp?site=1.
44. http://www.bendigoartgallery.com.au.
45. http://www.nattrust.com.au/trust_properties/affiliated_pro-perties/bendigo_joss_house.
46. http://www.powerhousemuseum.com.
47. http://www.hht.net.au/museums/museum_of_sydney.
48. http://www.sydneyjewishmuseum.com.au.
49. http://www.questacon.edu.au.
50. http://www.environment.gov.au/parks/index.html.
51. Mueller, Sir Ferdinand Jakob Heinrich von [Baron von Mueller] (1825–1896), in Australian Dictionary of Biography Online Edition. Available at http://www.adb.online.anu.edu.au/biogs/A050353b.htm.
52. http://www.museumsaustralia.org.au.
53. http://www.collectionsaustralia.net/sector_info_item/46.
54. Lynda, K. Gordon, P. Developing a community of practice: museums and reconciliation in Australia. In *Museums, Society, Inequality*; Sandell, R., Ed.; Routledge: London, U.K., 2002; 153–174.
55. Van den Bosch, A. Cultural memory re-presented at the Quai Branly Museum. In *Museum Marketing: Competing in the Global Marketplace*; Rentschler, R., Hede, A.-M., Eds.; Butterworth-Heinemann: New York, 2007; 3–10.

Australian Library and Information Association (ALIA)

Sue Hutley
Jane Hardy
Judy Brooker
Australian Library and Information Association, Deakin, Australian Capital Territory, Australia

Abstract

The Australian Library and Information Association (ALIA) is the professional association for the Australian library and information services sector. It seeks to empower the profession in the development, promotion and delivery of quality library and information services to the nation, through leadership, advocacy, and mutual support. The ALIA represents the interest of 6000 members, the profession and Australia's 12 million library users.

The objects of the Association are listed in its constitution. They are

- To promote the free flow of information and ideas in the interest of all Australians and a thriving culture, economy, and democracy.
- To promote and improve the services provided by all kinds of library and information agencies.
- To ensure the high standard of personnel engaged in information provision and foster their professional interests and aspirations.
- To represent the interests of members to governments, other organizations, and the community.
- To encourage people to contribute to the improvement of library and information services through support and membership of the association.

INTRODUCTION

The Australian Library and Information Association (ALIA) is the professional organization for the Australian library and information services sector. It seeks to empower the profession in the development, promotion and delivery of quality library, and information services to the nation, through leadership, advocacy, and mutual support (Fig. 1).

The Association is governed by a Constitution and is guided by a vision, mission, objects, and values. Our policy statements are developed by an elected Board of Directors and implemented by the ALIA National Office.

The Association's Web site is http://www.alia.org.au.

HISTORY OF THE ASSOCIATION

In 1937, the Australian Institute of Librarians (AIL) was formed as a result of the Munn-Pitt survey of public libraries, the work of John Metcalfe, principal librarian of the Public Library (now the State Library) of New South Wales, and the support of the Carnegie Corporation, in the United States, which was also providing financial support for the Free Library Movement in Australia. Metcalfe introduced a national system of examining and certifying librarians, under the auspices of the new organization. This met one of the principle aims of the AIL: improvement in the professional education and status of librarians.

The Institute supported at every opportunity the appointment of librarians to professional positions in libraries.

Carnegie and British Council grants helped the postwar development of the Institute. A Carnegie grant enabled the launching of the Institute's professional periodical, the *Australian Library Journal*, in 1951. The Institute expanded its professional activities to include conferences, more publications, and surveys of library needs, particularly the expansion of public library services. In 1949, the Institute changed its name to the Library Association of Australia (LAA).

From 1944 until 1980, the Association's board of examiners set the syllabus and examined candidates for the LAA's professional registration examinations. From 1961, when the LAA recognized the course of the first Australian School of Librarianship at the University of New South Wales, the Association gradually moved from providing professional education to setting and monitoring its standards. The LAA developed a system for the recognition of tertiary-level courses, which gradually superseded the Association's registration qualifications.

In the 1960s, the LAA also articulated the need for specialist training for archivists, teacher-librarians, and library technicians. Courses for archivists and teacher-librarians were established through the 1960s and 1970s in universities and colleges of advanced education, which later became universities. The Association passed responsibility for course recognition for archivists to the Australian Society of Archivists in the 1980s.

Encyclopedia of Library and Information Sciences, Fourth Edition DOI: 10.1081/E-ELIS4-120044957

Artificial-Back

Australian Library and Information Association

Fig. 1   ALIA logo.

The Association played a major part, in the 1970s, in formalizing an educational and career structure for library technicians within the library and information sector. Library technicians deal with library processes, such as acquisition, cataloging, circulation, interlibrary lending, and simple reference queries. At a senior level, they review operating procedures, evaluate equipment, assist in promotional programs, and supervise staff. The Association established a benchmark for standards that enabled all members of the sector to compare library technician qualifications nationally. The Association also created a section for the technicians and an annual award for achievement.

Features of the Association's work in the 1970s included the promotion of federal government funding for libraries and the encouragement of special interest groups within the Association, including a strong interest in the growing computerization of and networking opportunities for library and information services.

In 1989, recognizing the development of librarians as information providers and managers, the Association changed its name to the Australian Library and Information Association. After much discussion the Association created a permanent office in Canberra, the national capital of Australia. The ALIA House was opened in November 1990.

In 2000, the Association was incorporated as a public company limited by guarantee.

The Australian Library and Information Association has joined with other like-minded member organizations, such as National and State Libraries Australasia, the Australian School Library Association, Council of Australian University Librarians and Public Libraries Australia to present a united front in advocating on behalf of our profession. Developing international relations is also part of ALIA's agenda, especially in light of Brisbane being selected to host the International Federation of Library Associations and Institutions 2010 World Library and Information Congress.

The year 2007 marked ALIA's 70th anniversary since the establishment of the AIL. As a membership organization, ALIA continues to focus on high level customer service in every part of its business. With close to 6000 members, ALIA provides a wide range of services for those in the library and information services sector.

CONSTITUTION

The Australian Library and Information Association is incorporated under Australian Corporation Law and operates predominantly for the promotion, development, and attainment of its Objects. It is a not-for-profit organization. The Objects as set down in the constitution are

1. To promote the free flow of information and ideas in the interest of all Australians and a thriving culture, economy, and democracy.
2. to promote and improve the services provided by all kinds of library and information agencies.
3. to ensure the high standard of personnel engaged in information provision and foster their professional interests and aspirations.
4. to represent the interests of members to governments, other organizations and the community.
5. to encourage people to contribute to the improvement of library and information services through support and membership of the Association.

ALIA CORE VALUES

A thriving culture, economy, and democracy requires the free flow of information and ideas. Fundamental to that free flow of information and ideas are Australia's library and information services. They are a legacy to each generation, conveying the knowledge of the past and the promise of the future.

Library and information service professionals therefore commit themselves to the following core values of their profession: promotion of the free flow of information and ideas through open access to recorded knowledge, information, and creative works; connection of people to ideas; commitment to literacy, information literacy and learning; respect for the diversity and individuality of all people; preservation of the human record; excellence in professional service to our communities; and partnerships to advance these values.

GOVERNANCE AND STRUCTURE

The Association is governed by a Board of Directors. The ALIA Board of Directors is the elected policy-making body of the Association. There are eight members of the Board, seven of whom are elected: the president, the vice-president, and five elected directors.

The President, Vice-President and four Directors are elected by the membership at large; one Director is elected by institutional members. The ALIA Executive Director is the eighth (non-voting) member of the Board.

All voting directors must be personal members and represent the interests of the organization as a whole, rather than those of a particular constituency. Terms of office are for 2 years with a maximum of two consecutive terms.

Standing Committees are formed by the Board to advise it on key priority areas. Current Committees include: the Governance Standing Committee; Finance, Audit and Risk Management Standing Committee; Education and Professional Development Standing Committee; Membership and Awards Standing Committee; and the Research and Publishing Standing Committee.

There are 10 Advisory Committees whose members are endorsed by the Board. These committees provide expert advice in areas such as copyright and intellectual property, online content regulation, interlibrary lending, research, public libraries, special libraries, new generation issues, and TAFE (Technical and Further Education).

A National Advisory Congress is convened each year to provide members with the opportunity to participate in discussion on policy, planning and any topical issues relevant to the Association. An ALIA Board Director facilitates each of the regional meetings, with a national meeting convened that includes regional representatives.

Groups and committees within the Association provide a focus for communication and participation for members. There are over 50 endorsed self-nominating ALIA groups that are based on states, regions, sectors, and special interests. Groups run a wide variety of activities ranging from e-lists, networking, and social activities, through to professional development events, workshops, and symposiums.

The Association is supported by a national secretariat based in ALIA House in the nation's capital, Canberra.

ADVOCACY

The Australian Library and Information Association undertakes an active advocacy role supporting the objects and core values of the Association in a wide range of forums both nationally and internationally. The Australian Library and Information Association contributes to and participates in Commonwealth, State, and Territory government committees and inquiries, writing submissions, media releases, and lobbying. The main issues for advocacy include information access, online content regulation, national broadband strategy, copyright, privacy, and government information.

The Australian Library and Information Association supports programs encouraging wide ranging community engagement, and providing the opportunity to promote the role and value of libraries, and library and information professionals. The public campaigns include Library Lovers Day, Library and Information Week, Information Awareness Month, and National Simultaneous Storytime. The Australian Library and Information Association also organizes the annual Summer Reading Club, with libraries across the nation holding a range of events aimed at enticing the public, especially children, to make use of the services available in libraries over the summer months.

EDUCATION AND TRAINING: EDUCATION STANDARDS

The Australian Library and Information Association sets and maintains standards for entry into the library and information profession in Australia through ALIA's course recognition program. It plays a vital role in ensuring that education for the profession produces graduates who have the ability to provide excellent library and information services to benefit the nation and individual clients and who can respond to and meet the ever-changing information needs of a dynamic society.

The Australian Library and Information Association works collaboratively with educators and training providers, employers, and practitioners to promote the development and continuous improvement of courses in library and information management. Initial education should prepare library and information graduates who can achieve excellence in practice.

Library and information professionals have a responsibility to commit to professional development and career-long learning. Similarly, their employers and the ALIA have a responsibility to provide opportunities which enable library and information professionals to maintain excellent service delivery.

The Association's commitment to its members achieving their career goals is demonstrated by providing mechanisms which enable members to plan and undertake learning and other development activities; offering a mentoring program to assist members in their ongoing professional development; organizing partnerships with other professional and training organizations to make available learning activities and opportunities in library and information management and other disciplinary studies; formally recognizing members who participate in ALIA's professional development scheme; facilitating forums where knowledge can be created, shared, and disseminated to enable members to better understand the dynamic environment in which they, as library and information professionals, and their clients operate.

The ALIA Professional Development Scheme enables ALIA members to accumulate Professional Development (PD) points which provides recognition of ongoing learning through the use of the Certified Practitioner (CP) post-nominal.

MEMBERSHIP

Membership of ALIA is open to individuals and organizations. The only membership requirement is an interest in the sector.

Membership of ALIA entitles members to a wide range of benefits, free publications, online services, and discounts on conferences. Members of ALIA can belong to as many ALIA interest groups as they wish. The membership year runs from 1 July to 30 June.

Associate members hold an ALIA-recognized library and information science qualification at under-graduate or post-graduate levels. They are entitled to use the post-nominal letters AALIA while a financial member of ALIA. They are also entitled to membership of the professional development scheme. Graduates in the allied fields of records management, archives, and computing studies whose qualifications are recognized by their professional body may be admitted to associate membership. They may then use the post-nominals: AALIA(RecMan), AALIA(Arch) or AALIA(CS), respectively.

Library technician members hold an ALIA-recognized library technician qualification. A library technician member is entitled to use the post-nominal letters ALIATec while a financial member of the Association. They are also entitled to membership of the professional development scheme.

Associate fellow members have to demonstrate current professional or technical knowledge and skills. To qualify for Associate fellow, the applicant must demonstrate a minimum of 5 years associate or library technician membership of ALIA; continuity of participation in the professional development scheme and compliance for at least the previous 5 years; and current employment in the Australian library and information sector or an allied sector. Associate fellows may use the post-nominal AFALIA or AFALIATec.

Students and retirees also have membership categories.

Institutional members of ALIA have the right to vote in ALIA elections and to take advantage of the many benefits offered—both core services and "value-adding". A nominated institutional representative holds the formal right to vote. There are six grades of institutional membership based on the non-salary budget of the library and information service within the institution.

AWARDS AND SCHOLARSHIPS

The Australian Library and Information Association awards, at the national and group level, are presented each year to reward members of the library and information community and to celebrate their achievements.

Premier peer-nominated awards include: the HCL Anderson Award honoring outstanding achievement within the library and information sector by an Associate member; Redmond Barry Award honoring outstanding contribution to the library and information sector by an individual not eligible for associate membership; and the Ellinor Archer Pioneer award recognizing pioneering in new areas of library and information science by an individual or institutional member.

The ALIA Fellowships recognize the attainment of an exceptionally high standard of proficiency in library and information science and a distinguished contribution to the theory and practice of library and information science.

Research awards include: Study Grant; Dunn & Wilson Scholarship: Ray Choate Scholarship; YBP/Lindsay & Croft Research Award for Collection Services; and Twila Ann Janssen Herr Research Award for Disability Service. Excellence is recognized by the ALIA Excellence Award and the Metcalfe Award. Other awards include Aurora Scholarship, student awards, merit awards, Letters of Recognition, group-based awards, and a silver pin for recognizing volunteers who serve five terms on ALIA committees.

COPYRIGHT ADVICE SERVICE

The Australian Library and Information Association supports balanced copyright and related laws that advance the interests of society as a whole. The ALIA's Copyright and Intellectual Property Advisory Committee monitors copyright and intellectual property policy and legislative developments of interest to members and advises the Board of Directors on these matters. The Australian Library and Information Association is also a member of the Australian Libraries' Copyright Committee and the Australian Digital Alliance.

The ALIA's Copyright Service provides members with up-to-date information on copyright and how it affects the library and information sector. The service responds to members' queries about implementing their obligations under copyright and intellectual property law.

INDUSTRIAL RELATIONS ADVICE SERVICE

The Australian Library and Information Association provides a range of industrial and employment services and support including an online list of job vacancies, current salary rates, information on changes to industrial relations laws, library workforce data, rights and obligations when facing redundancy, etc.

INTERLIBRARY LOAN VOUCHER SCHEME

The Australian Library and Information Association operates an interlibrary lending voucher system, with vouchers used by libraries as payment for interlibrary loans, photocopies or microform copies made for retention by another library. The scheme was introduced to provide a simple and secure form of interlibrary currency with a minimum of record-keeping.

CONFERENCES

The Australian Library and Information Association's conferences provide professional development opportunities, promote excellence in practice, provide networking opportunities, promote research and innovative use of technologies, and build relationships with sector trade organizations.

Major conferences include the ALIA Biennial Conference, ALIA Information Online Conference, National Library and Information Technicians Conference, and the New Librarians Symposium. A number of annual symposiums, conferences, and workshops are organized by ALIA groups.

PUBLICATIONS

The Australian Library and Information Association's publishing program relates directly to ALIA's mission and values. The publishing program facilitates the flow of information and ideas as part of ALIA's engagement with members, the wider library sector, and the Australian community.

aliaNEWS is a free monthly e-newsletter available to any person interested in ALIA and the library and information sector.

The Australian Library Journal began in 1951. Published quarterly, it contains a wide coverage of Australian library issues ranging from ongoing research to day-to-day news from the workplace. It is the acknowledged flagship publication of the ALIA, and an appropriate vehicle for publishers, suppliers, and agencies wishing to reach the decision-makers. The journal is available through subscription.

Australian Academic and Research Libraries is a quarterly journal devoted to all aspects of librarianship in university and college libraries, including the TAFE sector, and in research libraries of all types. It publishes original, refereed contributions on all aspects of librarianship past and present, pure and applied bibliography, publishing, information science, and related subjects. All articles appearing in *AARL* are fully peer-refereed.

inCite is the premier news magazine for the library and information sector in Australia. Published monthly, it is distributed to all ALIA members and to subscribers around the world.

Other ALIA publications include ALIA's *Annual report*, *Australian Dictionary of Acronyms and Abbreviations*, and *Uniting a Profession*: *The Australian Institute of Librarians 1937–1949*.

CONCLUSION

The Australian Library and Information Association is the professional organization for the Australian library and information services sector. The Australian Library and Information Association celebrated its 70th anniversary in 2007. The Association is governed by a Constitution and an elected Board of Directors. The National Office is based in the nation's capital, Canberra.

The Australian Library and Information Association focuses on providing high quality services for its members and for the library and information profession in Australia, including national conferences, publications, copyright and industrial relations advice, a professional development scheme, an awards program, e-lists to facilitate communication, and support for over 50 self-nominating groups across the country. Through its course recognition process, ALIA ensures high educational standards for graduates of ALIA-recognized courses.

The Australian Library and Information Association's advocacy focuses on promoting the value of libraries and information services in ensuring the free flow of information and ideas in the interest of a thriving democracy, culture, and economy.

Authentication and Authorization

David Millman
Academic Information Systems, Columbia University, New York, U.S.A.

Abstract
Information systems security, access management, and privacy are normally understood through the processes of authentication and authorization. This entry provides a brief historical context and then describes several methods of authentication, such as passwords, digital signatures, network topology, smart cards, biometrics, and public key cryptography. It also discusses the most popular methods of authorization, including mandatory, discretionary, and role-based access control.

AUTHENTICATION

Authentication is the process of validating an assertion of identity. It normally involves the assertion of identity by an entity to a computer system and the procedures followed by that system to assess the truth of the assertion (i.e., whether the entity is who it claims to be). An entity may be an individual person, or it may be an organization, an account, a document, a device, or another computer system. Typically, an individual supplies an identity code and a password, as confirmation of the identity, to a computer system to gain access to services associated with that system. The establishment of identity does not necessarily result in access to services. Decisions concerning what level of access is permitted, and to which services, is made by an authorization process, described below, which may be entirely independent from authentication.

Authentication mechanisms must validate identity, protect against disclosure of secure information used to establish identity, and provide nonrepudiation—that is, preventing retraction or denial of an identity assertion. The level of confidence in an identity depends on the type of authentication and on the procedures used to implement it. These vary widely. Higher levels of trust usually require additional administrative and technical complexity. Ideally, authentication methods and implementation procedures are chosen to guarantee a level of confidence appropriate to the situation. For example, impersonation of identity may pose a higher risk in military or medical systems than it would in other cases.[1]

Well into the 1980s, most computer systems served a small community exclusively. Establishment of individual identity was often a direct personal transaction with a system administrator. The principle methods for accessing a computer system were either through a device directly connected to the computer, such as a punch card reader, or over a communications network in which each device had a unique wiring path to the computer (i.e., wires were not shared). Compromising someone's identity by tapping the correct set of wires was relatively inconvenient. Subsequent developments in communications technology and systems architecture have had an important impact on the requirements and characteristics of authentication infrastructures.

The introduction of shared networks and distributed system architectures, beginning in the 1970s, encouraged the decentralization of services and administration. Authentication became a distinct network service, an intermediary between a community of individuals and an assortment of other services on the network.[2] Services such as e-mail, printing, and file sharing could be obtained independently from numerous distinct devices attached to the network. Individuals' workstations were attached to the network in the same way. Authentication transactions occurred frequently among an individual's workstation, the authentication service, and the destination services. The shared network architecture also made it possible for any device attached to the network to examine all traffic that passed across the network, making spying much easier. Today's Internet uses a similar distributed architecture.

An authentication service is an arrangement between a community of individuals and an administrative organization in a position to verify their identities. The scope and characteristics of the community served by an authentication authority therefore becomes an important consideration in distributed architectures generally, and in the Internet in particular. Individuals may perform a wide variety of activities on the Internet and may be considered members of many separate communities of use, such as those who are employed by a particular company or those who purchase products from a particular retailer. It is not currently practical or desirable for a single authentication service to manage the identities of all communities and as independent services proliferate it will become increasingly cumbersome for an individual to authenticate his or her identity separately with each distinct service to which access is desired.

Encyclopedia of Library and Information Sciences, Fourth Edition DOI: 10.1081/E-ELIS4-120008659

Methods of Authentication

Authentication methods for individuals fall into three broad classes: those based on what one knows (e.g., a password); those based on what one possesses (e.g., a "smart card"); and those based on what one is (e.g., a fingerprint).

Shared secrets

Authentication that depends on the use of shared secrets is by far the most widespread. In this type of authentication, the individual and the system responsible for validating the individual's identity share a common piece of information, the secret, often a string of characters commonly called a password. In the simplest systems, after someone transmits an identity code and password, the system checks an internal table and, if a match is found, the system authenticates the person for whatever access is deemed appropriate by the authorization process.

Such password-based schemes have several difficulties. Passwords may be stolen by observing the traffic on the communications lines or by simply looking over someone's shoulder. Passwords may be shared explicitly. Passwords may be discovered by, for example, programmatically trying every word in a dictionary. Choosing passwords well (e.g., selecting nondictionary words) and changing them often improves security somewhat.

The Kerberos authentication service was an important improvement in password-based shared secret methods.[3] Kerberos was developed at the Massachusetts Institute of Technology and first used there in 1986. It takes advantage of the ability of an individual's workstation to perform cryptographic processing. When using Kerberos, a password is transmitted across the network encrypted, only once, to retrieve what is referred to as a ticket. The ticket is itself largely encrypted and can be used as proof of authentication for a fixed interval—perhaps for several hours—to obtain access to other services on the network.

Kerberos uses a central administrative authority service to perform the initial authentication and also to establish secondary tickets for each network service the individual uses. The set of services and individuals listed in the central administrative database are called a Kerberos realm. Kerberos first made common the use of a single authority over a set of highly distributed entities and remains an important contribution.

One-time passwords

The security of passwords is substantially improved when a password is changed after each time it is used. In one such scheme, the individual and the authenticating system arrange for a quantity of perhaps 1000 passwords in advance.[4] Each time a password is used it is discarded, the next password on the list is used for the next authentication, and so on.

In a variant method, the correct password changes as a function of the time, changing each minute, for example.[5] The individual possesses a small device, the size of a credit card, that continually calculates and displays the current correct password. The authenticating system validates identity through software that performs the same calculations.

Another method is based on the individual and the authenticating system having knowledge of numerous shared secrets, which are ideally chosen so that only the individual is likely to know them all, such as a favorite color, childhood address, friend's birthday, etc.[6] For each authentication, the system challenges the individual to confirm a random subset of the secrets, asked in a random order. This method is interesting because it combines properties of both shared secrets and one-time passwords. Each authentication is somewhat unpredictable, and the individual does not need to possess a list or a device to answer the questions.

Public key encryption

Public key refers to an encryption technology based on the mathematical properties of two carefully chosen numbers. The numbers are chosen so that they are inverses of each other: information encrypted with one number may be decrypted only with the other, and vice versa. The numbers are also chosen so that knowing one of them gives no clue to what the other is. One number is designated the private (or secret) key, the other the public key. The private key is never divulged; the public key is distributed widely.

An individual may encrypt information with their private key. Anyone who possesses the corresponding public key may decrypt the encrypted information and may further infer that the information must have originated with the individual. Similarly, information encrypted by anyone using the individual's public key can only be decrypted by the individual, using the private key. This elegant mechanism reduces the task of authentication to simply decrypting an arbitrary encrypted transmission with the individual's public key, because only that individual could have created it.[7]

Authentication systems built on public key methods are largely concerned with the administrative procedures needed to ensure that the system possesses the correct public key in the first place, and that the private key has not been divulged. Some of these procedures are discussed below.

Smart cards

A smart card, or smart token, is a small computer with low power requirements, making it appropriate for uses that require portability.[8,9] The computer is usually embedded in a plastic card the size of a credit card, but objects such as rings and buttons are also used. They are designed to protect the information stored within them both by performing only specific, well-defined communications and by employing safeguards against physical tampering.

For authentication, smart cards are especially useful for storing a private key and for performing the computations needed to encrypt or decrypt information using that key. This enables public/private key authentication to be performed with the private key never leaving the smart card.

Often, smart cards have security features that enable it to verify the identity of the possessor of the card. The card itself thus engages in an authentication process with the possessor—most often using a simple password or "PIN"—before performing further operations.

Smart cards may communicate through physical contact with another device, such as a conventional computer, a telephone, or a satellite receiver, or they may communicate by radio contact within a short distance (a few inches) of a device, as do the proximity cards used to gain access to a building or a mass transit system. It is interesting to note that possession of a smart card does not imply possession of the information stored within it. In most existing implementations, the issuer of the card controls this information.

Biometric

A biometric method of authentication identifies an individual through personal physical characteristics. Biometric systems exist for analysis of fingerprints, handprints, voices, faces, and patterns in the iris and retinal blood vessels of the eye. Details of these methods vary widely, with a corresponding variance in accuracy and efficiency. There are few standards in place.[10]

These systems compare the characteristics presented by an individual with a database of those previously stored. In the context of authentication, an individual asserts a claim of identity and uses the biometric as supporting evidence. Biometrics are also used to discover an unknown identity by searching through the database of characteristics. The latter method is common in, for example, law enforcement.

Related to biometric methods are those that measure the characteristics of an individual's performance of an action, for example, systems that measure the pressure, direction, and rate of a stylus when a person signs their name; such systems are considered to be relatively reliable. In another method, identification can be determined by the latency pattern of keystrokes, the speed and timing with which the keys are struck, when an individual types a particular phrase on a keyboard.[11] Robust biometric systems can prove the most reliable, nonforgeable means of authentication, assuming the individual has the body part being measured, but they have come under criticism for their invasive nature and for privacy concerns.

Other kinds of authentication

The most effective forms of authentication probably use a combination of several of the methods described above. For example, using a biometric means to authenticate an individual's identity to a smart card, and then using the card to perform further authentication for some fixed period of time is a relatively secure and convenient scheme;[12] smart cards with fingerprint sensors are currently available. Choosing the appropriate combination of authentication methods requires a careful analysis of the requirements and risks of the environment.

Authentication increasingly happens between entities in which neither party is an individual person. All the methods above, except for biometrics, may be used for an organization's accounts, and for hardware and software systems.

Between computer systems on the Internet, a popular form of mutual authentication relies on the correspondence of identification with network addresses (Internet Protocol or IP addresses). In other words, one system authenticates to another only by virtue of its topological location in the network. Reliance on particular network addresses is increasingly unreliable as networks are growing and changing quickly and as services are becoming increasingly distributed.[13]

Digital Signatures

Authentication as discussed so far has been in the context of a transaction between two entities. Another important form of authentication is with respect to digital documents or other digital objects. The goals are the same—to ensure identity, integrity, and nonrepudiation—but in this case it applies to a set of fixed data. Authentication of this type is commonly called a digital signature.

Public key methods provide a natural foundation for digital signatures. Data encrypted with someone's private key are in some sense "signed" by that person. Although anyone may decrypt these data using the corresponding public key, only the holder of the private key can have created it.

More typically, however, it is desirable to separate the data from the signature so that decryption is only required when signature verification is required. In this case, a *characteristic function* (also called a *message digest function*) is used. A characteristic function of a particular block of data has two properties: it is significantly smaller than the original data, and no other data will generate the same characteristic function. Thus, it can be considered a unique "fingerprint" of the original data. Encrypting this fingerprint with the private key creates a signature that may be attached to the original data. The data may be verified by recomputing its characteristic function, decrypting the signature with the public key to obtain the original characteristic function and comparing them to see if they match.

Process and Authority

In many of the methods discussed above, the individual or entity presents some form of authenticating information to a system that performs the authentication by comparing the presented information with information stored internally, perhaps after some computation. The internally stored information must be established by an initialization process.

Initialization processes vary widely, becoming more complex depending on the degree of trust required in the identity thus established and on the risk involved in impersonation. If the risk is considered relatively low, as it is with many free services on the Internet, initial identity may be established simply by requesting it in the course of first using the service. If the risk is considered to be greater, independent confirmation of identity, as by transmitting a copy of a government-issued identity document or by appearing in person with such documents before an official of the authenticating organization, may be required.

These kinds of procedures parallel conventional methods for confirming identity. Once identity is established, the authentication methods described above are used for ongoing access. When identity is compromised, for example, by a stolen password or private key, all or part of the initialization process normally needs to be repeated.

Public key systems present an additional challenge to the initialization process. Public keys may be distributed widely, to any entity to which one wants to send encrypted information, but it is impractical to engage in an identity initialization transaction with every other entity who may want to decrypt or authenticate using the public key. There are two popular methods employed to solve the problem of distributing public keys: ITU-T international standard X.509 certificates[14] and PGP (Pretty Good Privacy).[15–17]

X.509 relies on formalized Certificate Authorities (CA), who are responsible for asserting that a particular public key corresponds to a particular entity. In this scheme, an entity engages in an identity initialization process with a CA, resulting in a digital document containing the name of the entity and its public key; this document is digitally signed, as above, by the CA. Such a document is known as a digital certificate. Subsequently, the entity uses this certificate, rather than the public key alone, for authentication transactions because the certificate carries the assertion and trust-risk value of the CA. In other words, the CA's digital signature represents a measure of due diligence in confirming the correspondence between the individual's identity and the public key. Additional information may also be carried in the certificate, usually relating to the nature of the original identity initialization,

the purposes for which identity has been assured, and the level of trust in the identity.

The CAs themselves possess certificates, which may be signed by a higher order CA. The higher order CA assures the correspondence of the lower order CA and its public key, as above, and also asserts that the lower order CA has appropriate policies and procedures to issue trusted certificates. The higher order CA may also have a certificate signed by an even higher CA; this process can continue through many levels of authority until reaching an ultimate or "root" authority, who signs its own certificate. In current practice, there exist many root authorities, and the number of levels between a root and an end entity (e.g., an individual) is usually small (less than five).

In addition to policy relationships that may be established by hierarchies of certificate authority, there may also exist "bridge" CAs, who specify formal agreements and mappings among the policies of two or more CAs that are not part of the same hierarchy.[18–20]

Although digital certificates are in increasing use today, there is less experience with the potentially complex interactions of the policies of the supporting CAs. An alternative to using a certificate with an intricate set of policies behind it is to use a number of different certificates, each with a relatively simple policy, for different purposes.

In the PGP architecture, certificate authorities are not used to invest trust in the correspondence between an identity and a public key. Instead, PGP certificates contain any number of digital signatures, with each signature investing a degree of trust in the identity-key correspondence asserted by the certificate. The validity of a PGP certificate is determined by the sum of trust in the signers. The validity of each signer is determined by the same process, establishing a network or "web of trust" among certificates.

Public key systems must also have special procedures when someone's private key becomes compromised or is revoked for cause (e.g., termination of employment). Current designs recommend that the authenticating system consult a list of revoked certificates, which is normally maintained by the CA. In PGP, individual signers may revoke their own signature, and maintaining a centralized list of revoked certificates is similarly suggested.

It is also recommended that these systems should establish procedures for recovering a private key if it is lost. Traditional backup methods may not provide adequate security protection. In cases where a key is associated with a role, such as Chief Financial Officer, rather than an individual, a lost private key may have far-reaching impact. In business, conventional escrow procedures may suffice. More generally, schemes have been developed to divide a key into segments, or "shares," that are distributed to trusted parties so that more than one party must cooperate to reconstruct the key.

Processes to support longer-term digital signatures are not well developed. After a private key is revoked or expires, documents signed with that key should remain valid if they were signed while the key was still valid; this suggests an archive of private keys. The policies and procedures for creating, maintaining, and accessing such an archive have yet to be adequately specified or implemented.

Traditional signatures are used, for example, to assert authorship of a document, agreement with a document, and witnessing of an event, but there is no current method to express these semantics with digital signatures.[21] Traditional methods for verifying the authenticity of a document, such as corroboration by an expert, forensic analysis, consistency of dating, provenance, and common sense, are not addressed by the purely computational nature of digital signatures. Additional research is necessary to better understand these issues and to suggest procedures and infrastructures to support them. None of the authentication methods discussed, and the processes to support them, adequately address historical expectations of privacy.

One method for improving privacy protection is to introduce an intermediary authenticating service, which performs authentication with an individual and then acts as a proxy for further transactions on behalf of that individual.[22] There is little consensus on the behavior, responsibilities, policies, and trust relationships of such intermediaries, and the intermediary itself may willingly or, if compromised, unwillingly disclose private information, tieing individuals to their proxied activities.

Privacy concerns are an active area of social and legal discourse as well as of technical research. Traditional libraries have historically acted as intermediaries for access to information, and have carefully implemented policies and procedures to guarantee individual privacy. The library community has taken an active role in translating those methods to digital information.[23,24]

AUTHORIZATION

Authorization, sometimes called Access Control or Access Management, is the process by which an individual or other entity is permitted to perform some operation on a computer system. Authorization may apply to an organization as well as to individuals, or to automated processes acting on their behalf. In the technical literature, the individual, organization, or process is referred to as the "subject." Authorization happens in the context of (i.e., subsequent to) authentication. The authorization decision is, in other words, given someone's identity, what may they do? What information may they see; what may they create or destroy; what may they change?

The earliest authorization process was implicit: if one had physical access to the computer, one could do anything with it. (To some extent, this holds true today for personal computers.) When computing and information resources, or "objects," as they are called, are shared among many subjects, more elaborate methods are necessary to ensure that appropriate access is granted to each.

Current authorization systems divide access rules into three broad categories: whether a subject may retrieve an object ("read"); whether a subject may create, change, or destroy an object ("write"); and the extent to which a subject may alter the authorization rules themselves. These categories are called "operations." Thus, authorization is the process of determining which operations are permitted between a given subject and object.

Operations are performed by computer software applications. In early systems, the authorization decisions were made directly in the application, and the behavior of the program logic depended explicitly on the subject's authenticated identity. This method has been superseded by independent authorization systems and infrastructures that interact with applications in a range of ways. For example, in some cases, an application explicitly consults the authorization service to determine if an operation is permitted. Or the authorization service may be part of the underlying operating system, automatically enforcing permissions on operations requested by applications. Analysis of application and operating system logic and their interactions can offer a relative measure of security or trust in the authorization service.[25]

Authorization may be considered an implementation of policy,[26] and an important goal of authorization systems is flexibility, accuracy, and efficiency in implementing policies. Actual policies can be highly complex, involving classifications, hierarchies, interrelationships, exceptions, degrees, and emergencies. They may be motivated by legal, business, or political needs. They may change frequently. Not surprisingly, the development of effective authorization systems is an active area of research. Three principle system models are currently most common: mandatory, discretionary, and role based.

Mandatory Access Control

In mandatory access control (MAC) or "lattice-based" access control systems, a security administrator assigns classification labels to all subjects and all objects. These labels are ordered to reflect the levels of security required, for example, "unclassified," "confidential," "secret," and "top secret." Objects may be further categorized within a security level. Subjects are assigned labels corresponding to their maximum security clearance, and to the categories within that clearance that they may access.

A principal concern of MAC systems is the control of the information flow between levels. A typical implementation, for example, may permit subjects to read objects only at their own or lower levels and to write objects only at a higher level. The strict controls on information classification and flow, coupled with the central authority of the security administrator, make MAC systems attractive in military applications. The policy known in the financial community as the "Chinese Wall,"[27] which is designed to prevent financial analysts from reading objects that could create a conflict of interest with respect to other objects they have already seen, may be enforced with a variant of a traditional MAC implementation.

Discretionary Access Control

Discretionary access control (DAC) relies less on a central security administrator, putting more of the burden on an object's owner. The owner is usually the subject that created the object. In DAC, the owner may change the access permissions of an object, granting, for example, read or write access to other subjects. In some implementations, the owner may delegate this granting ability itself to other subjects or may transfer ownership of an object to another subject.

Although DAC is much more flexible than MAC, the security is weaker. For example, in DAC, it is possible to make a copy of an object that has fewer restrictions than the original. Because security policy is not governed by a single authority but is distributed across all owners, it is difficult to assess the overall security of a DAC implementation. Nonetheless, DAC is widely popular in many operating systems and databases.

Role-Based Access Control

Role-based access control (RBAC) has gained significant attention for its power and flexibility.[28] Within RBAC, subjects are assigned one or more roles, with each role embodying one or more permissions. Permissions may be read, write, and the ability to adjust the authorization system itself.

Roles typically are identified with a business or organizational capacity, such as "Chief Financial Officer" or "enrolled student," and thus simplify making changes in permissions when a subject's role changes (because you just change the person's role instead of revamping all the permissions that person has). Roles may be organized in hierarchies, so that certain roles imply others through inheritance. Constraints may be placed on the system to implement desired policies (e.g., multiple signature requirements, performance of tasks in a particular order, enforcement of the separation of responsibilities).

RBAC systems are also considered flexible enough to implement both MAC and DAC policies.[29]

There is much ongoing research on RBAC devoted to such goals as improving its performance in highly dynamic settings[30] and improving its ability to express complex policies.

Other Kinds of Access Control

The methods above play a predominant role in existing implementations and in research. But the goal of enforcing policy is broad and may be approached in many ways. For example, the "trust management" approach combines the trust relationships of digital signature authentication with the expressions of policy.[31] Generally, the digital rights management community has been highly active in developing mechanisms to express trusted policy assertions on the Internet.[32–35]

REFERENCES

1. U.S. Department of the Treasury. Use of electronic authentication techniques for federal payment, collection, and collateral transactions. Fed. Regist. **2001**, *66*(2), 394–397.
2. Israel, J.; Linden, T. Authentication in office system internetworks. ACM Trans. Off. Inf. Sys. **1983**, *1*(3), 193–210.
3. Steiner, J.G.; Neuman, B.C.; Schiller, J.I. Kerberos: an authentication service for open network systems. Proceedings of the Winter 1998 Usenix Conference, February 1998.
4. Lamport, L. Password authentication with insecure communication. Commun. ACM **1981**, *24*(11), 770–774.
5. RSA Security, Inc. *Strong Enterprise User Authentification: RSA ACE/SERVER Solution White Paper*. Available at http://www.rsasecurity.com/products/securid (accessed January 2001).
6. Ellison, C.; Hall, C.; Milbert, R.; Schneier, B. *Protecting Secret Keys with Personal Entropy*, Future Generation Computer Systems; Elsevier, 2000; Vol. 16, 311–318.
7. Needham, R.; Schroeder, M. Using encryption for authentication in large networks of computers. Commun. ACM **1978**, *21*(12), 993–999.
8. Husemann, D. The smart card: Don't leave home without it. IEEE Concurr. (April–June) **1999**, *2*, 24–27.
9. Chadwick, D. Smart cards aren't always the smart choice. Computer **1999**, *32*(12), 142–143.
10. Phillips, P.; Martin, A.; Wilson, C.; Przbocki, M. An introduction to evaluating biometric systems. Computer **2000**, *33*(2), 56–63.
11. Joyce, R.; Gupta, G. Identity authentication based on keystroke latencies. Commun. ACM **1990**, *33*(2), 168–176.
12. Sanchez-Reillo, R. Securing, information and operations in a smart card through biometrics, Proceedings of IEEE 34th Annual 2000 International Carnahan Conference on Security Technology; IEEE, 2000; 52–55.
13. Lynch, C., Ed.; *A White Paper on Authentication and Access Management Issues in Cross-Organizational Use of Networked Information Resources; Revised Discussion Draft Of April 14, 1998*; Coalition for Networked Information: Washington, DC., 1998. http://www.cni.org/projects/authentication/authentication-wp.html (accessed November 2000).
14. International Telecommunication Union. *ITU-T Recommendation X.509: Information Technology—Open Systems Interconnections—the Directory, Authentication Framework*; International Telecommunication Union: Geneva, 1996.

15. Zimmermann, P. *The Official PGP User's Guide*; MIT Press: Cambridge, MA, 1995.

16. Garfinkel, S. *PGP: Pretty Good Privacy*; O'Reilly & Associates, 1994.

17. PGPi Project. Available at http://www.pgpi.org (accessed January 2001).

18. Guida, R. *Applying for and Interoperating with the Federal Bridge Certification Authority*; U.S. Department of the Treasury, 2000. Available at http://csrc.nist.gov/pki/fbca/fbcaguide_20001207.pdf (accessed February 2001).

19. *Federal Bridge Certification Authority*. National Institute of Standards and Technology, Computer Security Resource Center, Eds.; Available at http://csrc.nist.gov/pki/fbca/ (accessed February 2001).

20. *NIST PKI Program*; National Institute of Standards and Technology, Computer Security Resource Center, Eds.; Available at http://csrc.nist.gov/pki (accessed January 2001).

21. Lynch, C. Authenticity and integrity in the digital environment: an Exploratory analysis of the central role of trust. In *Authenticity in a Digital Environment*, Council on Library and Information Resources: Washington, D.C., 2000.

22. Givens, B. Infomediaries and Negotiated Privacy: Resources. Proceedings of the Tenth Conference on Computers, Freedom and Privacy: Challenging the Assumptions: Toronto, ON, Canada, 2000. 165–166.

23. Arms, C.; Klavans, J.; Waters, D. *Enabling Access in Digital Libraries: A Report on a Workshop on Access Management, Washington, D.C., April 6, 1998*; Arms, C., Ed.; Council on Library and Information Resources: Washington, DC, 1999.

24. Millman, D. Cross-organizational access management: A digital library authentication and authorization architecture. D-Lib Mag. **1999**, *5*(11), Available at http://www.dlib.org/dlib/november99/millman/11millman.html (accessed December 2000).

25. Joshi, J.; Aref, W.; Ghafoor, A.; Spafford, E. Security models for web-based applications. Commun. ACM **2001**, *44*(2), 38–44.

26. Arms, W. Implementing policies for access management. D-Lib Mag. **1998**. Available at http://www.dlib.org/dlib/february98/arms/02arms.html (accessed November 2000).

27. Brewer, D.; Nash, M. The Chinese Wall Security Policy. Proceedings of IEEE Symposium on Security and Privacy, IEEE, 1989; 206–214.

28. Sandhu, R.; Coyne, E.; Feinstein, H.; Youman, C. Role-based access control models. Computer **1996**, *29*(2), 38–47.

29. Osborn, S.; Sandhu, R.; Munawer, Q. Configuring role-Based access control to enforce mandatory and discretionary access control policies. ACM Trans. Inf. Syst. Secur. **2000**, *3*(2), 85–106.

30. Barkley, J.; Beznosov, K.; Uppal, J. Supporting relationships in access control using role based access control. Proceedings of the Fourth ACM Workshop on Role-Based Access Control, ACM, 1999; 55–65.

31. Blaze, M.; Feigenbaum, J.; Ioannidis, J.; Keromytis, A. *The Keynote Trust Management System Version 2*; IETF Network Working Group RFC 2704, September 1999. Available at http://www.ietf.org/rfc/rfc2704.txt?number=2704 (accessed January 2001).

32. Dalal, M. Tractable deduction in knowledge representation systems .Proceedings of the Third International Conference on Principles of Knowledge and Reasoning KR92, Cambridge, MA, 1992. 393–402.

33. Erickson, J.; Williamson, M.; Reynolds, D.; Vora, P.; Rodgers, P. Principles for standardization and interoperability in web-based digital rights management.In *W3C Workshop on Digital Rights Management*; 2001. Available at http://www.w3.org/2000/12/drm-ws/pp/hp-erickson.html (accessed March 2001).

34. Roscheisen, M.; Winograd, T. The Stanford FIRM framework for interoperable rights management. In *Forum on Technology-Based Intellectual Property Management*; Interactive Media Association, White House Economic Council and White House Office of Science and Technology: Washington, D.C., 1997. http://mjosa.stanford.edu/~roscheis/IMA (accessed January 2001).

35. Stefik, M.; Silverman, A. *The Bit and the Pendulum*; Xerox Palo Alto Research Center, 1997. Available at http://www.XrML.org/PDFs/Pendulum97Jul29.pdf (accessed January 2001).

Automated Acquisitions

Patricia A. Smith
Allison V. Level
Colorado State University, Fort Collins, Colorado, U.S.A.

Abstract

"Automated Acquisitions" describes the automation of acquisitions work, which began with the development of simple ordering and receiving systems. Over the years, these systems have expanded to automate acquisitions work generally and to integrate acquisitions with other major library functions such as cataloging, circulation, and collection development. Automation has also promoted electronic interfaces with monograph and serial vendors and with institutional accounting systems. Increasing growth of electronic collections requires additional system development to accommodate the various components of this complex workflow. Workflows in acquisitions are enhanced by online cataloging services, such as OCLC's Promptcat or vendor-provided MARC records, which allow libraries to receive books cataloged and shelf-ready; and electronic books have led to many interesting new models of acquisitions, including demand-driven acquisitions. To promote operability among various systems, standards adopted by publishers, vendors, and libraries are increasingly important and a focus of development in the field.

INTRODUCTION

Libraries have automated acquisitions for a variety of reasons, including lowering unit costs, improving access and service to users, speeding the ordering–cataloging process, accessing ordering, and in-process information, collecting and organizing acquisition data, gathering data for assessing collections, and linking with other systems both inside and outside the library to assure efficient workflows. The complexity of electronic resource management (ERM) has added new challenges for workflows, data collection, and system linking.

When selecting a system, librarians should consider 1) the capacity of the system, that is, whether it will handle the required number of orders, funds, and the like; 2) the cost of installation and maintenance; 3) compatibility with other systems; and 4) functionality, that is, whether it offers the functions required and whether staff can easily perform them. If a system appears to meet these criteria, then the library should verify performance by requesting written documentation from the vendor and securing references from institutions using the system. Most libraries also prepare a formal request for proposal (RFP) from several vendors, a proposal that lists all the library's requirements.

As a minimum, automated acquisition systems should perform some or all of the same activities required in a manual acquisitions operation but with the added requirement that the system perform them more efficiently and effectively. Automated systems can achieve these ends by making data transfer more accurate, by reducing costs through economy of scale, and by collecting new data or manipulating it in new ways.

THE EVOLUTION OF SYSTEMS

Automation of acquisitions functions began in the 1930s and 1940s, before the computer age, with the use of punch-card systems for in-process control and ordering. Because of the complexity of acquisitions processes and the difficulty of integrating them with local accounting policies and procedures, these early efforts relied on single-function systems.

Great impetus to automated acquisitions came in the 1960s when research libraries and many medium-sized libraries, especially in the United States, were growing rapidly. Federal grants, the PL480 program, and the expansion of higher education all contributed to a phenomenal increase in acquisition budgets for both academic and public libraries, but with no or limited increase in staffing. Coinciding with the dawning of the computer age, the pressure generated by these large budgets and significant gifts led individual libraries in the 1960s and 1970s to rely increasingly on computerized single-function or multiple single-function systems. Developed in-house, these simple, typically machine-dependent systems operated in a batch mode; that is, the data were collected and processed periodically, and order forms, reports, or other printed products or electronic files were the result. Generally, libraries automated only parts of the acquisitions process, usually order/receiving, in-process files, or accounting functions. These systems typically operated on a general purpose mainframe computer housed and managed outside the library.

Other libraries using the same hardware sometimes borrowed or purchased these systems but doing so typically required significant local programming to run

Encyclopedia of Library and Information Sciences, Fourth Edition DOI: 10.1081/E-ELIS4-120053312

effectively. Because these programs were machine dependent, library literature from this period contains close descriptions of the computer hardware required. Few libraries, of course, had the resources to develop such systems, nor were early commercial vendors able to create successful generic programs for acquisitions because practices from library to library vary so much more than, for example, those in circulation.

In 1969, Baker and Taylor released BATAB, the first jobber-supplied, commercial acquisitions system. BATAB was a batch software package that ran on a local mainframe and featured a range of acquisitions functions, including selection lists and purchase orders; claims and cancellation lists; detailed fund accounting; and historical, open-order, and in-process reports. BATAB focused on relations between libraries and book jobbers, Baker and Taylor in particular, where orders could be placed by output via magnetic tapes. Other book vendors followed suit, and in less than a decade, Brodart offered the first jobber-supplied online order service, Instant Response Order System (IROS). Released in 1978, IROS gave customers online files and online ordering. Marketed primarily to public libraries, IROS did not offer fiscal or processing management, but online access to jobbers' publication databases, nonetheless, remains today a constant of book and serial vendors.[1]

The advent of online capability in the later 1970s meant that in 1982 Washington Library Network (WLN), a bibliographic utility, brought out an acquisitions system as part of its networked services to libraries. Other major utilities—Online Computer Library Center (OCLC), Research Libraries Information Network (RLIN), and University of Toronto Library Automation Systems (UTLAS)—soon followed.[2] These networked systems brought an important advance over then-current systems for acquisitions by offering integration with other library processes including circulation and cataloging and provided full-function acquisitions services. As a result, the bibliographic utilities gave birth to the first integrated library systems (ILSs).

ILSs proliferated in the 1980s as commercial designers joined the marketplace to create multifunction, multiuser systems combining cataloging, circulation, serials, and acquisitions modules. Acquisitions data could now be shared among all modules and displayed for users in the online catalog. These commercial systems continued to evolve throughout the 1990s although fewer than half of the academic libraries implementing new systems opted to include acquisitions or serials components during this period.[3] A decade later, however, most academic and larger public libraries had fully functioning ILSs and could choose among several types: 1) turnkey systems where the library is responsible for little more than turning the key to the system, that is, the commercial vendor installs the software and hardware for the library, connects the library to the network, and provides most technical support; 2) standalone systems where the library essentially stands alone, that is, the library purchases or is free to purchase the hardware and software separately, installs both, and independently administers the system except for such limited technical support as hardware and software commercial vendors normally provide; 3) hosted systems where the library buys the commercial vendor's software but the vendor hosts the software on its server, which library staff access through the Internet; and 4) service-as-a-subscription (SaaS) systems where, rather than buying the software, the library pays a subscription for online access to its own version of the commercial vendor's ILS software.[4]

ILSs provided by commercial vendors were typically proprietary systems that guaranteed performance but restricted access to the source code, thus preventing customization for local needs. Responding to rising library dissatisfaction with this lack of flexibility, ILS commercial vendors have made their systems more "open," especially by offering application programming interfaces (APIs). APIs allow libraries to do such local programming as establishing interfaces with a parent institution's financial system.[5] Since vendors charge for additional features, costs have risen as well, leading to more library frustration with traditional ILS commercial vendors.

As an alternative to proprietary systems, libraries continue to explore open source software in hopes of finding more affordable and customizable systems. Open source systems strive for the same functionality as commercial ILSs, but members of an open source community share in the development of the source code, which is free for anyone to download and update.[6] The success of these efforts varies, and users have found important functionality lacking in some.[7] Two systems working primarily with public libraries and consortia are Koha (www.koha.org), which first appeared in 1999, and Evergreen (http://evergreen-ils.org). Open Library Environment (OLE) by the Kuali Foundation (www.kuali.org/ole) is an important initiative by academic libraries.

Into this mix of systems has come yet another potentially revolutionary force. As electronic collections began to appear in the early 2000s, librarians realized that the automated acquisitions and serials modules of the traditional integrated library system were not capable of tracking all the necessary information or of accommodating workflows that ERM requires.[8] Many libraries developed home-grown databases, spreadsheets, or standalone systems to fill the gaps. Recognizing the disorganization of this approach, the Digital Library Foundation (DLF) and the National Information Standards Organization (NISO) established a steering group to guide the development of standards, specific data elements, and functionality for ERM to foster interoperability (ERMI) among various applications.[9]

Encouraged by library interest and guided by this ground-breaking DLF ERMI report, automation vendors began to produce standalone ERM systems or to incorporate ERM as modules of their ILS. Open source ERM systems have also appeared. CUFTS, an extension of a suite of open source discovery tools developed at Simon Fraser University,[12] and the University of Notre Dame Libraries' Centralized Online Resources Acquisitions and Licensing (CORAL) are standalone open source ERM systems. Librarians have also tried to address the obvious need for ERM and ILS interoperability, especially with the acquisitions module, and in 2008 another DLF subcommittee studied this issue and identified key acquisitions elements requiring interoperability. This report also saw the need for an identifier unique to each resource for accurate interfacing.[10]

Ongoing, critical analysis by librarians of changing workflows will continue to be an indispensable part of achieving an effective system for ERM.[11] Essential elements that have been identified include tracking acquisitions and budget information, license terms and conditions, and administrative and contact information for individual electronic resources. Giving users ready access to electronic materials is also important. Providing full texts, for example, requires such tools as OPEN URL link resolvers and knowledge bases. Reports and data analysis going beyond current ILS capabilities, too, must be developed for collection assessment and selection. Per-use cost for individual electronic journals, particularly those in serial publisher's "big deal," for example, is another primary need. Although standards such as Counting Online Usage of Networked Electronic Resources (COUNTER) and products like the Standardized Usage Statistics Harvesting Initiative (SUSHI) are available to facilitate compiling and reviewing statistics on the use of a library's electronic resources, such tools are still incomplete in their coverage and application and vendor data are inconsistent.[12] Additional statistics and data often have to be gathered from individual vendor webpages and tracked and manipulated in local spreadsheets. In short, tools and methods for assessment and evaluation of electronic usage still require research and development.[13]

Responding to current needs, major automation developers introduced new systems in 2012 variously referred to as library management systems, web-scale management solutions, or library service platforms (LSPs). These new systems seek to consolidate in a single interface the essential functions of a traditional ILS for print materials with the multiple tools needed for ERM. These platforms or services combine acquisitions, ERM, serials, reporting capability, link resolvers, discovery programs, and other relevant library software under a comprehensive knowledge base or information center.[14] The Global Online Knowledge bases (GOKb) project being funded by a Mellon Foundation grant and led by the University of North Carolina Libraries (http://gokb.org/) is a current effort to provide a comprehensive knowledge base and will be adopted for Kuali/OLE's open source library platform— freely available to anyone whether a librarian, a commercial system vendor, or a publisher. Since major institutions in Europe and the United States are contributors, GOKb promises to be a significant source of information and metadata needed to manage electronic resources.[15] Unlike the traditional ILS that comes with a prescribed user interface, these new systems permit libraries to implement the discovery tools they choose, even a product from another commercial vendor. These systems also seek to improve ERM by simplifying other processes such as input on trial databases, adding holdings for titles, and coping with licenses. To address the need for improved data to aid decision making regarding electronic resources, LSPs also propose to support more sophisticated reporting, using analytics and data mining across various components of collections.[16]

Another potential advantage of these new systems is that, based on cloud-computing with its reduction in both upfront costs for equipment and for local maintenance staff, libraries may be able to achieve important savings. Depending on the annual subscription fees, these services may also be options for small libraries that have little or no automation.[17] Major commercial vendors offering LSP services include Alma by Ex Libris, Sierra by Innovative Interfaces, Intota by Serials Solutions, and WorldShare by OCLC, and OLE by Kuali Open Library Environment.[18] In 2014, the University of Chicago launched OLE; a number of other universities are scheduled to become early adopters of OLE in 2015.

As libraries consider the emerging options for improving or replacing their current systems, much—as always in automated library acquisitions—remains to be learned about impacts on costs, staff, and workflow.

BASIC COMPONENTS OF AN AUTOMATED ACQUISITIONS SYSTEM

A 1999 model RFP in *Library Technology Reports*[19] and the ongoing experience of acquisitions librarians make it possible to list the basic features desirable in an automated system. The elements in this RFP are standard features in the acquisitions modules that commercial vendors provide in an ILS and are primarily based on the acquisitions of print materials. Acquiring electronic publications also involves similar ordering information and budget control, but additional tools will be required to manage other aspects of electronic collections.

> *Verification of bibliographic data for new orders*: An automated system for acquisitions should include a link to a source of bibliographic data in MARC format that can be downloaded into the new order file. The source of these data can be a bibliographic utility such as OCLC, Library of Congress MARC

records, or a database of MARC or MARC-like bibliographic records from a materials vendor. Vendor records usually have the added advantage of including price and invoice data that can be loaded into the system, thus reducing posting time by staff.

Centralization and integration of on-order and in-process information: Even the most basic systems maintain an automated file of titles on order or in process. When acquisition is part of an integrated library system or has links to other systems in the library, this file, at the discretion of the librarian, can be accessed through the online catalog. In this case, the file should be interactive so that changes made in the acquisitions record are also reflected in other library files. Patrons and selectors should be able to access the data in standard ways: author, title, series, ISBN, LCCN, keyword. Acquisitions staff and searchers should be able to access information with appropriate local keys, such as purchase order number, and be able to find titles cataloged or on order so that duplicate orders can be rejected and relationships to other titles in the collection can be identified. If the librarian is ordering an added copy or added volume, the system should permit borrowing of a bibliographic record from the online catalog or from an acquisitions file.

Generation of printed and/or electronic purchase orders and quotations requests: The system should handle different formats—print, nonprint, and hybrid monographs and serials that have associated electronic components. It should accommodate a full range of order types, including serials, standing orders, rush orders, prepaid orders, deposit account titles, and gifts. Once a purchase order has been entered, the system must not only store all bibliographic and purchasing information but also index and link critical elements so that the record can be retrieved by all appropriate search keys. The system should produce orders in print or send them electronically to vendors that have such capability.

Online vendor files: Each vendor record should store and index a variety of information about itself, including ordering and pay-to addresses, claiming intervals, names of contacts and phone numbers, and notes such as discounts. If the order system links vendors to an institutional accounting system, the record should contain a field for vendor numbers matching the number in the institution's file. Vendor records should also provide the link needed to generate an electronic order to that vendor.

Automated review of order records for claims or cancellations: The system should offer file management options such as alerts for claiming or canceling, but ideally the library should have the option of setting the system to claim or cancel automatically if the vendor permits automatic cancellations. The system should produce electronic and printed claim/cancellation requests or offer a list for librarians to review and track the date and type of requests generated.

Recording of receipt information: The system should be able to handle titles that are received in full and titles where only part of the order is received. When partial receipts are recorded, the system should disencumber only the cost of the part or the volume received and keep remaining parts of the order open.

Payment of invoices: When invoices are received, staff should be able to enter invoice data automatically into each individual purchase-order record associated with that invoice. If the system does fund accounting, the system should create an electronic invoice record that can be approved online. Approval of the invoice should produce, electronically when possible, a voucher or a check.

Fund accounting: Basic fund accounting should include a fund record file with the capability of handling all the funds needed by the library. The system should record encumbrances, disencumbrances, and expenditures against each fund as transactions occur. Users should be able to view online the fund allocation, the amount encumbered, the amount expended, and the uncommitted and cash balances. The system should likewise validate fund codes entered during ordering and disallow incorrect codes. Each fiscal year, the system must provide the end-of-fiscal-year processing needed to move to a new fiscal year. Since requirements vary by library, the system should be flexible enough to establish new fund records for the new fiscal year or to leave a purchase order linked to a previous fiscal year so that it is paid with funds from the appropriate fiscal year.

Management reports: Automated systems should include a report generator adaptable to local needs in the type and timing of reports and offer options for online or printed output. The data should be compatible with standard database or spreadsheet software. Basic reports include fund reports, vendor-performance statistics, and staff performance statistics. Other reports should be available for collection development, for example, a list of titles acquired by fund or selector linked with data in other library modules, such as circulation or cataloging.

Security: Automated systems should provide for a variety of password levels, with two being an absolute minimum. Many functions such as fund control require secondary password levels so that unauthorized staff cannot create or change records. Ideally, passwords should be controlled at the local level by authorized staff and be constructed so that one log-on unique to an individual staff member serves to authorize all the functions that the person is permitted to perform. Histories of transactions with associated passwords should also be maintained by the system so that supervisors can review actions to see who has performed them. The system should, finally, perform validity checks for funds and other elements deemed critical by local auditors.

Archiving: Since most libraries periodically archive completed acquisitions and fund data, the system should provide common archiving options: tape, microform, or disk. Another desirable option for easy retrievability would be to have the archive reside on a separate file as part of the integrated library system. Should a library migrate to another system, the system should be able to provide a copy of the library's acquisition files in an electronic format.

Other features that augment this list should include OPAC searching options, links to other databases or modules in an ILS, Z39.50 clients and servers, and web-based clients. Librarians now expect systems to include ERM components, which allow them to track individual holdings, licenses, and usage statistics associated with electronic products. The acquisitions system should have the qualities of good library systems, such as system integrity, file output for data sharing or migration, logical screen and function layouts, and user support.

Finally, fully automated systems should support interoperability with other local and external systems related to acquisitions of library collections and adhere to industry standards that enable effective exchange of data.

INTERFACES WITH EXTERNAL ACCOUNTING SYSTEMS

An important interface that many libraries require as part of its process for paying vendors is a linkage with the parent institution's automated financial system where the check is generated. Automating this accounting interface between the two accounting systems can be beneficial to both the library and the larger institution.[20] Such an interface should permit vendor and voucher information to be sent to the parent system and must provide for consistency and security of data between the two systems. Since financial systems, security requirements, and audit controls vary widely from institution to institution, most links to parent accounting systems have been locally developed in cooperation with an institution's auditors. The emergence of APIs as an ILS function has assisted these local efforts to customize financial interfaces.

Before implementing an automated payment system, librarians must acquaint themselves both with the institution's audit requirements and with the automated system's internal controls. Traditional auditors still require some parts of the old paper trail for invoices, but use of electronic file transfer protocol (FTP) and reliance on electronic passwords for authorizing transactions will eventually make most paper records obsolete. Generally, too, auditors require that the functions of ordering, receiving, and payment be separated and that records of who performed these transactions be maintained for verification with lists of authorized signatures. As automation advances, even signatures will be replaced with authorized lists of electronic passwords. Electronic fund transfer (EFT) from the account payable unit of an institution to the bank of the vendor is becoming increasingly common.

INTERFACES WITH THE BOOK AND SERIAL INDUSTRY

By the 1990s, book and serial vendors had automated the majority of their internal operations and could offer libraries a broad range of automated services, including everything from order processing, claiming, and invoicing to approval-plan profiling and management.[21] The sophistication of these processes will vary among vendors depending on the library market they serve. Smaller vendors may not offer the range of services a larger academic vendor can; therefore, libraries need to explore with their vendors what they can provide. In many cases, vendors are willing to work with a customer to develop a service.

Various online ordering models are available. Serial vendors, for example, offer password access to their databases to allow customers to verify, order, and claim serials within the serial vendor's database. Book vendors have also automated their publication databases and offer online access to libraries. The result is that before placing a firm order, libraries can now call up specific titles to see if the title will come as part of the library's approval plan or as a standing order and, in some cases, check the jobber's inventory to see if a copy of the book is in stock. While online with the jobber, library staff can also place orders or request that a title be sent as part of the approval plan. The jobber's system then creates a file that acquisitions staff can use to download electronically into the library's acquisitions system, creating an order record in the process. This download requires specially programmed loaders to map and link the data from the book jobber's system to the formats required by the library's acquisitions system. After library staff members determine that the title is not a duplicate and that funding is available, they can confirm the order with the vendor by returning the order electronically to the book jobber. MARC cataloging records can also be provided by the jobber in this model or the vendor can send the title to OCLC's Promptcat service for cataloging before it is sent to the library. The Promptcat service has been merged with OCLC's WorldCat Cataloging Partners program, which also includes the former Cataloging Partners program. OCLC provides a list of the print and e-book vendors that participate. As another alternative for expediting firm orders, staff can set up in their local acquisitions module an address for participating vendors and send electronic orders directly from the ILS without going online to the vendor. Other services include electronic invoicing via FTP.

Approval plan vendors, such as Coutts Information Service or YBP Library Services, routinely supply books shelf-ready. For each shipment, they will provide the library an electronic file of bibliographic records with attached invoices that can be loaded into the library's acquisitions system to create an order record facilitating payment. Working with the WorldCat Cataloging Partners model, libraries can use these records to also serve as the basis for the library's cataloging record. The vendor sends a manifest of the libraries' approval titles to OCLC's cataloging service. While the books are being shipped to the library, OCLC uses automated search algorithms to check OCLC's bibliographic files for a matching cataloging record and puts that record in a file for the library to load into its local online system. An advantage of using OCLC's service for cataloging records for those libraries that maintain their holdings symbol in OCLC is that the WorldCat Cataloging Partners' program automatically sets the library's holdings symbol in OCLC according to a time frame predetermined by the participating library, eliminating the need for staff to upload records to OCLC once the book is available for check out.

The advent of electronic books has brought new interfaces incorporating vendor-provided discovery records which may not meet the standards of a full MARC record. As a result, patrons can find available titles in the library's online catalog and request a short-term loan (STL) at some fraction of the list price. After a specified number of such requests, the library purchases the title. When the library purchases these titles, however, the discovery records can be upgraded to a full catalog record by the vendor or the library or sent to OCLC's cataloging services, using the capabilities of automated systems at the e-book vendor. If the item comes via an approval plan, the record exchange involves yet one more system in this complex array of interfaces.

STANDARDS

Standards are the key factor for successful interfaces between vendors and library acquisitions/ERM systems. This exchange of electronic data in a standardized form from one computer to another is called electronic data interchange (EDI). For EDI to work, two standards are necessary, one for the contents of the message and a second for the packaging of the transmitted information.

The creation of standards has fallen to voluntary professional and industry associations. One of these, the Accredited Standards Committee (ASC) (http://web.ansi. org/), part of the American National Standard Institute

(ANSI), has taken responsibility for establishing standardized formats for ordering, shipping, receiving, payment, and claims. The specific standard for electronic data interchange is X12, and the ANSI X12 formats were created for transmitting and receiving information. Another group, the (NISO, http://www.niso.org/), which is accredited by ANSI, has taken responsibility for standards relating to metadata and retrieving, storing, repurposing, and archiving information.

The Book Industry Standards and Communications Committee (BISAC, http://www.bisg.org/index.html) has established fixed formats for transmitting orders and claims for monographs. A parallel organization, the Serial Industry Systems Advisory Committee (SISAC), has promoted voluntary standards for electronic transmission of serials information. In 1998, these committees merged into a group named Book and Serial Industry Communications (BASIC, http://www.bisg.org/basic.html) to continue to foster standards for EDI formats for books and serials based on the international EDI standards. This group, which is officially a working group of the Book Industry Study Group (BISG), coordinates its efforts in creating standards with EDItEUR, a nonprofit agency with members from various countries around the world, including the United States (http://www.editeur.org). EDItEUR supports the EDIFACT guidelines for transactions used in the book trade and also manages important identifiers such as the international ISBN and ISSN as well as maintaining the ONLINE INFORMATION EXCHANGE (ONIX) standard for publishers to use for data fields (e.g., ISBN, title, author, reviews, etc.) when sending information describing books to libraries or book stores. ONIX is also extending to journal subscriptions and licenses.

In 2007, the 10-digit ISBN was replaced by the 13-digit ISBN and required everyone in the book supply chain from publishers, jobbers, and libraries to make certain their systems would support the full 13-digit ISBN. As a result, many systems have dropped BISG/BISAC and require the EDIFACT formats for transmitting orders electronically.

The transition to electronic collections has led to major efforts to develop standards or guidelines for exchanging data for usage statistics, license information, and other facets of electronic publications. NISO is sponsoring an initiative called Standardized Usage Statistics Harvesting Initiative (SUSHI, http://www.niso.org/committees/ SUSHI/) using a standard data container for transferring COUNTER statistics into a digital repository. Integrated library systems are now employing SUSHI to gather usage statistics for a library. Project COUNTER is an effort to produce guidelines for counting electronic use statistics and is widely supported by publishers and libraries. The Digital Library Federation (DLF, http://www.diglib.org/) supported the Electronic Resource Management Initiative (ERMI), which has developed guidelines for defining elements of electronic resources and their associated licenses.

An extension of this effort is the joint NISO/EDItEUR ONIX-PL Working Group developing a standard for exchanging license information between publishers and libraries. The Knowledge Base and Recommended Practice (KBART) project (www.niso.org/workrooms/kbart) is an ambitious effort to provide guidance to content providers and knowledge base creators for metadata formatting so that knowledge bases and link resolvers get users to the desired content. Many additional standards exist and many more will continue to emerge.

THE INTERNET AND E-COMMERCE

Acquisitions librarians have found a variety of uses for the Internet that add convenience to daily routines and increase access to resources. The listserv and social media have become popular forums for discussion of acquisitions and serials issues and an excellent way to stay abreast of major trends. ACQNET-L and SERIALIST-L are two of the common forums for monograph and serial acquisitions. LIBLICENSE-L is important for obtaining information regarding the licensing of digital publications and other legal issues. ERIL-L promotes discussion of issues related to managing electronic resources. Techniques in Electronic Resource Management (TERMS) is an ongoing project to develop best practices for ERM and uses Facebook, Twitter, and a blog (http://www/6terms.tumblr.com) to share ideas.

Websites offer a wealth of sources for collection development and acquisitions. Borrowing ideas or procedures from the web pages of other acquisitions departments has become common practice. Staff can also access other library catalogs to find citations to aid in bibliographic verification. Many libraries also order titles from Amazon.com or Barnes & Noble's online store as a quick source for rush items.

And nowhere has using the Internet been more gratifying to librarians than in helping them find out-of-print materials. The formerly idiosyncratic and disorganized world of used and antiquarian book dealers has seen new markets develop as buyers and hundreds of dealers embrace the Internet as a distribution center.

The use of credit cards for e-commerce is now common. Among the advantages are ease in dealing with e-commerce dealers and reduction of check-writing for petty sums. On the down side, credit card transactions require extra steps in tracking and do not interface directly with library systems.

IMPACTS ON WORKFLOW AND ORGANIZATION

Automation of acquisitions data and integration of that data with other library systems has led to major reexamination of workflows and organization within technical services. Traditionally, acquisitions departments did preorder searching and ordered materials, while cataloging departments performed the bibliographic and authority searching necessary for cataloging a book. When the online bibliographic utilities such as WLN, RLIN, and OCLC became available, acquisitions and cataloging librarians realized that the order verification information could also be used by the catalogers if staff agreed on policy and procedures. The recognition that the knowledge, skill, and tools employed by acquisitions and cataloging staff are similar led many technical services librarians to consolidate preorder and precataloging searching into a single unit that might report to either acquisitions or cataloging departments. As the integration between acquisitions and cataloging modules on automated systems evolved, acquisitions and cataloging librarians gave greater attention to generating accurate bibliographic records at the earliest possible stage of processing. In particular, staff could eliminate several searching and keying steps, thereby reducing the costs of searching bibliographic databases and shortening the time between receipt and cataloging for a book. As a result, some libraries adopted cataloging at the point of receipt for selected materials so that the staff member who searched and found acceptable bibliographic copy for a book when it was received could also catalog that book. This combination of tasks has led to the creation of fast cataloging units in many libraries. Increasingly, acquisitions department staff members now perform cataloging using Library of Congress MARC records retrieved from bibliographic utilities or from jobber-supplied MARC records received with the book. Discovery records provided by e-book vendors to help libraries and patrons find titles for short-term loans in demand-driven acquisitions (DDA) products can also be upgraded to full cataloging records by acquisitions staff.

Automated acquisitions has also brought important benefits for collection development librarians, and the traditional lines separating acquisitions from selecting have begun to show signs of blurring as well. Budget reports now can be drawn with greater frequency and timeliness, while the increased ability of automated systems to correlate acquisitions data with circulation and cataloging data has improved the management reports that can be compiled for collection analysis. Selectors, moreover, can now access the centralized file of orders to review the status of their orders without relying on the acquisitions department.

Because selection has also become more automated, many libraries have moved away from collection bibliographers toward subject/reference librarians who use the time gained for instruction and reference as well as selection of print and electronic materials. Certainly, the automation of a large portion of the publishing universe allows the possibility of more sophisticated purchase plans replacing the costly title-by-title examination selectors once practiced with traditional approval plans. The possibility of receiving approval books shelf-ready and the increased use of OCLC's WorldCat Cataloging Partners service for cataloging has led to further implementation of plans where the books are not returned to vendors. As these purchase plans replace approval plans, collection developers and acquisitions librarians will increasingly focus on becoming expert designers and managers of profiles and systems.

Another area of staffing change has come with the creation of ERM librarian positions within acquisitions or serials units. The more traditional serials librarian roles have also changed to include greater focus on electronic acquisitions and require a greater need for expertise with licensing, authenticated access/proxy servers, open uniform resource locator (URL) linking, and creation of A to Z database and journals lists. ERM librarians often help gather and organize usage statistics and provide licensing and cost data to collection development librarians who then incorporate the metrics into the selection and evaluation process for the library.[22]

The ease of sharing data from automated acquisition systems means also that collection developers operating within consortia can now remove significant obstacles to resource sharing and cooperative acquisitions. Sharing acquisition data allows members to tag titles automatically for cancellation or purchase and so to expedite cooperative ventures.

ELECTRONIC PUBLICATIONS AND PATRON-DRIVEN ACQUISITIONS

Electronic book products have raised new questions of distribution, pricing, ownership, and contracting for the acquisitions librarian. E-books are promoted to academic libraries in several ways including 1) publishers offering direct marketing of content to the library, 2) specialist aggregators packaging and presenting books in particular subject areas, 3) large aggregators offering e-books in many disciplines from multiple publishers, and 4) library vendors offering e-books and print books from both the aggregators and from individual publishers.[23]

1. Direct purchases of individual e-books from publishers usually allow libraries more permissive digital rights management (DRM) and give patrons greater freedom to download or share. Some publishers, however, opt to sell only an entire "package" of titles and do not make individual titles available for purchase. Other publishers offer an evidence-based model (EBM) allowing access to a specific list of titles for a set period of time. In this model, a library agrees to pay a specified dollar amount at the beginning of the time period, and then at the end of the period, the library selects the titles to purchase, up to the defined dollar amount. The pool of titles available to the patrons during the open period is quite extensive and represents a dollar amount several times greater than the amount paid upfront by the library. At the end of the EBM specified period, the library reviews the "evidence" provided by usage statistics to guide final title selections.

2. Specialist aggregators often bundle titles around related subject areas or types of books such as reference materials. Bundling can be a benefit for users since the e-books are collected together by subject or type, and the specialist aggregators often have a search feature that allows searching across all content in the specialist package. But like database/journal aggregators, titles can change and move into and out of the specialist packages.

3. Large aggregators offering e-books from multiple publishers permit librarians the significant advantage of automating and streamlining the process for discovery records, book purchase, and use of e-books by users. The advantage for users is that they see their books through the same platform.

4. Library vendors selling e-books and print books from aggregators and individual publishers offer a variety of different purchasing models. The standard model permits libraries to purchase books title by title and own the book outright. Another model is common for approval plan vendors. Here libraries add electronic books via their approval plan profiles, which then link the approval plan with e-book aggregators or e-book publishers. Finally, some vendors offer patron-driven acquisitions (PDA) or DDA by which libraries acquire books only as patrons ask for them, often with an option for an STL first, before repeated uses trigger purchase of a book. In 2014, NISO announced a recommended practice on DDA that covers these various options and will serve to guide libraries, publishers, aggregators, and vendors on management of DDA services.[24]

The PDA or DDA model is also moving into use by library consortia. The Orbis Cascade Alliance and the Colorado Alliance of Research Libraries are two groups operating a DDA selection model cooperatively among several libraries. Formerly, the choices for "ownership" were limited to purchase or lease; now there are options for the number of simultaneous users of a particular title, limits on the status of users, and sometimes frequent changes in what access a particular publisher will allow. As Rick Anderson describes, "There are many PDA models currently under development, and we can expect that they will proliferate and be refined over time. While some libraries will (and should) still be building monumental collections in 2021, I am confident that the standard approach for most research libraries will be patron-driven."[25] Whether to acquire e-books as collections or singly or as subscriptions or with perpetual access are just some of the issues acquisitions librarians now face.

For electronic journal articles, acquisitions librarians have yet another patron-driven model now poised for further development. In the past, requests for articles not available from an academic library's current subscriptions went to an interlibrary loan (ILL) department and could be sent electronically by direct document delivery to patrons. A vendor- or publisher-supplied model for journal articles has emerged by which some academic libraries now fill requests on an article-by-article or pay-by-the-drink basis. A library, working through a publisher or vendor, provides money on deposit so that affiliated researchers can request the needed article directly from the publisher or vendor. Expanded use of patron-driven direct document delivery from vendors and publishers permits delivery with little or no staff intervention and may eventually offer an alternative to conventional e-journal subscriptions.

The e-book and direct article delivery models are in flux as publishers and vendors are still testing the market with a variety of ways to offer e-books and journal articles. Spending on electronic materials will continue to outpace spending on print.[26] Libraries are still pondering effective models for acquisitions and making changes to workflows and budgets accordingly.

THE FUTURE

As library technology and electronic resources become more pervasive and more complex, demands for automated acquisitions systems meeting a variety of needs will continue. System vendors and libraries will continue to improve functions that assist in comprehensively managing electronic resources. The concept of a publisher will be pushed to new limits by electronic information. Exploration of patron-driven acquisitions will continue. Article databases and document delivery will supplement and may replace current subscriptions models. Consortial purchasing will grow and open new opportunities for development in both library and vendor systems. Librarians, publishers, and vendors will develop new standards to speed sharing and accuracy. Yet unknown forms of online resources and services will be required.

As electronic resources replace print materials, the need for traditional automated acquisitions may diminish as acquisitions becomes nothing more—nor less—than providing users fast and convenient ways of accessing the world's vast body of information in the cloud.

ACKNOWLEDGMENTS

The authors are indebted to Margo Sasse, formerly of Colorado State University Libraries, for her original contributions to the first edition of this work.

REFERENCES

1. Grosch, A.N. *Library Information Technology and Networks*; Marcel Dekker, Inc.: New York, 1995; 1–55.
2. Boss, R. *Automating Library Acquisitions: Issues and Outlook*; Knowledge Industry Publications, Inc.: White Plains, NY, 1982, 5–8, 79–98.
3. Boss, R. Options for acquisitions and serials control automation in libraries. Libr. Technol. Rep. **1997**, *33* (4), 407–409.
4. Peters, A. *Integrated Library Systems: Planning, Selecting and Implementing*; Libraries Unlimited: Santa Barbara, CA, 2010; 5–8.
5. Breeding, M. Opening up library systems through web services and SOA, hype or reality? Libr. Technol. Rep. **2009**, *45* (8), 4–42.
6. Molyneux, R.E. An open source ILS glossary: version 2. Publ. Libr. Q. **2011**, *30*, 65–165.
7. Rapp, D. Open source reality check. Libr. J. **2011**, *136* (13), 14–36.
8. Anderson, E. Electronic resource management systems: a workflow approach. Libr. Technol. Rep. **2014**, *5* (3), 1–47.
9. Jewell, T.D.; Anderson, I.; Chandler, A. *Electronic Resource Management Report of the DLF ERM Initiative*; Digital Library Federation: Washington, DC, 2004. http://old.diglib.org/pubs/dlf102/ERMfinal.pdf (accessed October 2014).
10. Medeiros, N.; Chandler, A.; Miller, L. *White Paper on Interoperability between Acquisitions Modules of Integrated Library Systems and Electronic Resource Management Systems*; Digital Library Federation: Washington, DC, 2008; 1–28. http://old.diglib.org/standards/ERMI_Interop_Report_20080108.pdf (accessed October 2014).
11. Emery, J.; Stone, G. Techniques for electronic resource management. Libr. Technol. Rep. **2013**, *49* (2), 1–29.
12. Welker, J. Counting on counter. Comput. Libr. **2012**, *32* (9), 6–11.
13. Grogg, J.; Fleming-May, R. The concept of electronic resource usage and libraries. Libr. Technol. Rep. **2010**, *46* (6), 5–16.
14. Yang, S. From integrated library systems to library management services: time for change? Libr. High Tech News **2013**, *30* (2), 1–8.
15. Hill, K.; Collins, M. Building a better knowledgebase: a community perspective. Ser. Libr. **2014**, *66* (1–4), 106–114.
16. Breeding, M. Mining data for library decision support. Comput. Libr. **2013**, *33* (5), 23–25.
17. Breeding, M. Lowering the threshold for automation in small libraries. Comput. Libr. **2012**, *32* (3), 23–26.
18. Breeding, M. Library systems report 2014: competition and strategic cooperation. Am. Libr. **2014**, *45* (5), 21–33.
19. Boss, R.W. A model RFP for an automated library system. Libr. Technol. Rep. **1999**, *35* (6), 717–820.
20. Lamborn, J.L.; Smith, P.A. Institutional ties: developing an interface between a library acquisitions system and a parent institution accounting system. Libr. Collect. Acquis. Tech. Serv. **2001**, *25* (3), 247–261.
21. Boss, R.W.; McQueen, J. The uses of automation and related technologies by domestic book and serials jobbers. Libr. Technol. Rep. **1989**, *25* (2), 125–251.

22. Schmidt, K.; Korytnyk Dulaneym, C. From print to online: revamping technical services with centralized and distributed workflow models. Ser. Libr. **2014**, *66* (1–4), 65–75.

23. Walters, W.H. E-books in academic libraries: challenges for acquisition and collection management. Portal: Libr. Acad. **2013**, *13* (2), 187–211.

24. North American Standards Organization. *Demand Driven Acquisition of Monographs: A Recommended Practice of the National Information Standards Organization*; North American Standards Organization: Baltimore, MD, 2014, 1–37. http://www.niso.org/apps/group-public/download.php/13373/rp-20-2014_DDA.pdf (accessed October 2014).

25. Anderson, R. Collections 2021: the future of the library collection is not a collection. Serials **2011**, *24* (3), 211–215.

26. Breeding, M. Balancing the management of electronic and print resources. Inf. Today **2014**, *34* (5). http://www.infotoday.com/cilmag/jun14/Breeding–Balancing-the-Management-of-Electronic-and-Print-Resources.shtml (accessed September 2015).

Automatic Abstracting and Summarization

Daniel Marcu
Information Sciences Institute, University of Southern California, Marina del Rey, California, U.S.A.

Abstract

After lying dormant for a few decades, the field of automated text summarization has experienced a tremendous resurgence of interest. Many new algorithms and techniques have been proposed for identifying important information in single documents and document collections, and for mapping this information into grammatical, cohesive, and coherent abstracts. Since 1997, annual workshops, conferences, and large-scale comparative evaluations have provided a rich environment for exchanging ideas between researchers in Asia, Europe, and North America. This entry reviews the main developments in the field and provides a guiding map to those interested in understanding the strengths and weaknesses of an increasingly ubiquitous technology.

GENRES AND TYPES OF SUMMARIES

It is almost impossible to conceive of a world without summaries. Daily, we skim over headlines to decide what news to read. We go over scientific abstracts to decide what papers to study. We read reviews to decide what books to buy. And we browse television guides to decide what movies to watch.

Most of the summaries that we use today are produced manually by trained abstractors. However, an increasingly large number of researchers is now tackling the problem of automatic summarization. This turns out to be quite difficult because even defining the scope of the problem is not an easy feat. Consider, for example, the text fragment in Fig. 1a, which is taken from *Time* (October 2, 1997).

What is a good summary of the text in Fig. 1a? A *headline* such as that in Fig. 1b may convince a layperson who is hit annually by a flu to read the whole article. However, a molecular biologist interested in the function of various enzymes may be more interested in an abstract such as that shown in Fig. 1c.

As the examples in Fig. 1b and 1c illustrate, the communicative goal with which a summary is written and the background knowledge of the intended audience clearly influence the information that one may consider to be important.

The intended communicative goal and the background of the audience are only two of the facets that are useful when characterizing summaries. Other facets that have received significant attention are listed below [see Hovy,[1] Mani,[2] and Jones[3] for a list of dimensions along which summaries can be characterized].

- *Usage*. When they can be used only for quick categorization, summaries are called *indicative*. When they can be used as substitutes of the original documents to access significant content, summaries are called *informative*.
- *Relation to source*. When summaries contain sentences/clauses that are "lifted" verbatim from the input source, they are called *extracts*. When the important information in a text is paraphrased to ensure cohesion, coherence, and a higher degree of compression, summaries are called *abstracts*. A particular type of abstract, which is extremely short and uses a peculiar syntax is the *headline*. For example, the texts in Fig. 1d and Fig. 1e are extractive summaries of the text in Fig. 1a, the text in Fig. 1b is a headline, and the text in Fig. 1c is a short abstract.
- *Purpose*. When summaries provide the author's view, they are called *generic*. When they reflect the user's interest, often expressed in the form of a query, they are called *query oriented*.
- *Source type*. When they are based on only one text, they are called *single-document summaries*. When they fuse information provided in multiple texts, they are called multidocument or multidoc summaries.

The dimensions above are not always clearly cut. A summary can be at the same time indicative and informative. Or it can be indicative for one audience and informative for another. A summary that is obtained by selecting important sentences and then simplifying them by deleting unimportant words and phrases is not really an extract because it contains clauses/sentences that are not part of the input. But it is not an abstract either because paraphrasing is not employed extensively. By the same token, one may argue that it is not appropriate to call "abstract" and extract in which only the pronominal dangling references have been replaced by the corresponding noun phrases.

Encyclopedia of Library and Information Sciences, Fourth Edition DOI: 10.1081/E-ELIS4-120008882

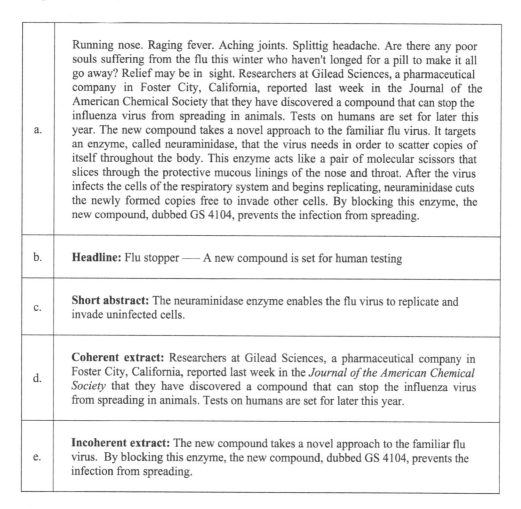

a.	Running nose. Raging fever. Aching joints. Splittig headache. Are there any poor souls suffering from the flu this winter who haven't longed for a pill to make it all go away? Relief may be in sight. Researchers at Gilead Sciences, a pharmaceutical company in Foster City, California, reported last week in the Journal of the American Chemical Society that they have discovered a compound that can stop the influenza virus from spreading in animals. Tests on humans are set for later this year. The new compound takes a novel approach to the familiar flu virus. It targets an enzyme, called neuraminidase, that the virus needs in order to scatter copies of itself throughout the body. This enzyme acts like a pair of molecular scissors that slices through the protective mucous linings of the nose and throat. After the virus infects the cells of the respiratory system and begins replicating, neuraminidase cuts the newly formed copies free to invade other cells. By blocking this enzyme, the new compound, dubbed GS 4104, prevents the infection from spreading.
b.	**Headline:** Flu stopper —— A new compound is set for human testing
c.	**Short abstract:** The neuraminidase enzyme enables the flu virus to replicate and invade uninfected cells.
d.	**Coherent extract:** Researchers at Gilead Sciences, a pharmaceutical company in Foster City, California, reported last week in the *Journal of the American Chemical Society* that they have discovered a compound that can stop the influenza virus from spreading in animals. Tests on humans are set for later this year.
e.	**Incoherent extract:** The new compound takes a novel approach to the familiar flu virus. By blocking this enzyme, the new compound, dubbed GS 4104, prevents the infection from spreading.

Fig. 1 Sample text and various types of summaries.

SUMMARY EVALUATION

The discussion in the previous section is much more important than first meets the eye. The facets and summary types enumerated above do not merely provide a common vocabulary to natural language researchers. Rather, they set the foundation for addressing a serious concern: If there are so many facets and factors that influence the way summaries are constructed and presented, how can one measure the quality of a summary? What makes a summary good and another one bad? Unless these questions can be answered adequately, it is impossible to assess progress in the field of automatic abstracting.

The first metrics for measuring the quality of a summary have been proposed in the context of producing generic extracts. Given a document, one or more human judges are asked to mark the sentences that are important. The sentences on which a majority of the judges agree to be important are taken to be the "gold standard." The quality of automatic summarizers is measured in terms of recall—the number of important sentences correctly identified by the summarizer divided by the total number of sentences in the gold standard—and precision—the number of important sentences correctly identified by the summarizer divided by the total number of sentences selected by the summarizer. Note that both recall and precision have to be taken into consideration as one can easily rig one of the metrics. By marking as important all sentences in a text, one can easily obtain a 100% recall. Clearly though, the precision is going to be much smaller. A metric that summarizes the recall and precision metric is the F-value metric, which is a harmonic average of precision and recall ($F = 2 R P/(R+P)$). The F-value is a number between recall and precision that it is higher when recall and precision are closer.

The F-value cannot be rigged the same way recall and precision can when considered in isolation. Unfortunately, the use of the F-value metric can also pose problems. Let's say that five human judges mark as important sentences in a document that has 20 sentences. Let's say that each judge marks 10 sentences as important but a majority of the judges agree on only two sentences. Is this two-sentence gold summary of any good? If five judges randomly choose sentences from a document and mark them as important, just by the laws of odds it is likely that two sentences will be marked as important by a majority of them. This is clearly a problem as it corresponds to a case where there is no agreement between human judges with respect to what is important in a text (see Carletta,[4] Krippendorff,[5] and Siegel[6] for methods of computing agreement between multiple judges). If humans cannot agree on what is important, it means that a summarizer that chooses sentences randomly is as good as a human.

Some of the summarization studies that are referred to in this entry were carried out before statistical methods for measuring interannotator agreement and adequate experimental design techniques became part of the tools used by natural language researchers. For example, a classic summarization paper published by Edmundson[7] measured the performance of various summarization methods in terms of recall with respect to a gold standard built by only one human. Forty years after the experiment of Edmundson was carried out, it is impossible to tell how difficult the texts given as input were to summarize—no human agreement figures are reported in the paper. Still, as the reader can see in this entry, we still take many such results as indicative of the performance of various methods. Understanding how to measure the quality of a summary is by itself a research area[8] that evolves concurrently with research aimed at developing increasingly sophisticated summarization techniques.

The recall, precision, and F-value metrics represent only a small fraction of the metrics proposed to measure the quality of a summary. Alternative metrics have been proposed to address their limitations. For example, Radev et al.[9] proposed an evaluation metric in which sentences are not judged on a boolean scale as important or nonimportant, but are given a utility score. A summarizer that selects sentences of high utility is given a higher reward than a summarizer that selects sentences of medium or low utility.

To enable the evaluation and encourage the development of systems that produce not extracts but abstracts, the Document Understanding Conferences[10,11] use human judges to measure the degree of overlap between a model summary and abstracts/extracts produced by various systems. In addition to the informational content, systems participating in DUC are also evaluated in terms of output grammaticality, cohesion, and coherence.

The metrics discussed so far pertain to the intrinsic quality of a summary. However, in many scenarios, summaries are also evaluated extrinsically, with respect to their utility for solving specific tasks. For example, systems that participated in the SUMMAC evaluation[12] were evaluated by measuring how often their outputs could be used by humans to determine whether the documents for which the summaries were generated were relevant to a given topic. In another task, summarization systems were judged by measuring how often their outputs could be used to successfully carry out a document categorization task.

Reviewing all research in summary evaluation is beyond the scope of this entry.[1,2,8,10–12] In the rest of this entry, whenever the "performance" of a summarization system is referred to without using any additional qualifiers, the F-value of that system, as measured against a gold standard, is meant. When the performance of a system was measured using a different metric, this is made explicit. Because the science of evaluating summaries is not stable and because a generally accepted evaluation standard is yet to emerge, comparing the performance of various summarization algorithms and techniques is challenging.

SUMMARIZATION-SPECIFIC COMPONENTS

Overview

In dealing with the facets discussed previously, researchers in summarization developed numerous components/techniques that are now part of many summarization systems. Importance identifiers are used to determine the most important sentences and clauses in a document or collection of documents. Sentence compression algorithms are used to compress extracts further, by deleting nonimportant phrases and clauses in long sentences. Paraphrasing components are used to increase the compression and readability of the output. Information extraction and natural language generation systems are used to generate abstracts via semantic-based representations. Headline generators use specialized techniques to determine the most representative concepts in a text and to render them using a language that is often different from that used in typical English.

Fig. 2 shows the components that have received significant attention (the components are represented as ellipses) and the summary types these components are meant to produce (the summary outputs are represented in bold fonts surrounded by boxes). Each summarization pipeline in Fig. 2 takes as input a document and some optional parameters, such as the desired compression, a user query, etc. Additional components that were developed to generate multidocument summaries are discussed later in this entry.

Importance Identifiers

Arguably, most of the effort in the summarization research field has gone into the development of algorithms and methods for identifying important sentences/clauses in single document texts.

Position-based method

Many of the summarization systems assume that sentences that occur at the beginning of documents are more likely

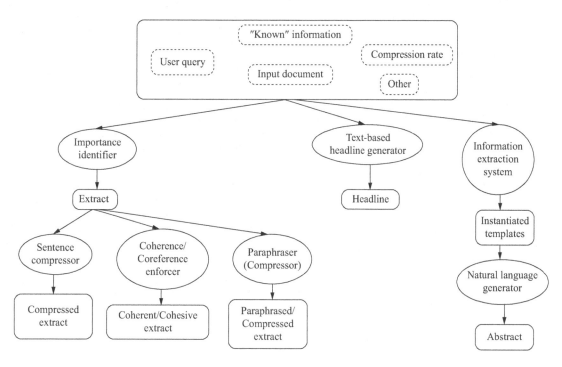

Fig. 2 Components (ellipses) and sought outputs (bold rectangles) of single document summarization systems.

to be important than sentences that occur at the end. The simplest way to operationalize this assumption is to build a summarizer that always selects the first sentence in a text; or the first k sentences in a text, when a smaller degree of compression is required.

Empirical evaluations of algorithms that implement this assumption have revealed that the position-based method is the best[7,13] or second best[14] single method for identifying important sentences in text. Although the performance of this method varies significantly with text genre and compression rate, this method is usually capable of identifying around 33% of the important sentences in a text.[7,13,14]

An extensive study aimed at deriving a genre-dependent optimum position policy have been carried out by Hovy and Lin[15] who showed that in technical articles that announce computer-related products, the sentence that is most likely to be important is the first sentence of the second paragraph, then the first sentence of the third paragraph, then the second sentence of the second paragraph, and so on; in news articles, the sentence that is most likely to be important is the first sentence of the first paragraph, then the first sentence of the second paragraph, and so on.

Title-based method

Edmundson[7] was the first to show that the words in titles and headings are more likely to be used in important sentences in a text than in nonimportant sentences.

Systems that employ this heuristic in combination with others tend to yield summaries of higher quality. For example, the title-based method increased the performance of Edmundson's position-based summarizer by 8% and the performance of Teufel and Moens's[14] cue-phrase-based summarizer by 3%.

Cue-phrase method

Cue-phrase-based systems capitalize on the observation that important sentences contain "bonus phrases," such as *significantly, in this paper we show*, and *in conclusion*, whereas nonimportant sentences contain "stigma phrases," such as *hardly* and *impossible*. The cue-phrase method yielded the best results when used to identify important sentences in scientific articles[14]—it identified 55% of the important sentences in a text—and was shown to increase the performance of summarization systems by 7% to 9% on other genres, when applied in conjunction with other methods.[7,13]

Word-frequency method

To my knowledge, the claim of Luhn[16,17] that important sentences in a text are those that contain words that occur "somewhat" frequently has not been validated empirically in any summarization system. In Edmundson's experiments,[7] this method decreased the performance by 7% when combined with other methods. In Kupiec et al.'s experiments,[13] this method decreased performance by 2%.

The method appeared to mildly help only in Teufel and Moens's system,[14] where it increased performance by 0.2% when combined with other methods. In spite of not being properly validated, the word-frequency method continues to be used in many implemented systems.

Cohesion-based methods

The working hypothesis that constitutes the foundation of all cohesion-based methods is that important sentences/paragraphs are the highest connected entities in more or less elaborate semantic structures. There are several approaches to characterizing the level of connectedness. They are based on word co-occurrences, local salience and grammatical relations, coreference, lexical similarity, and combinations of the above.

Word-based. The most straightforward approaches apply Information Retrieval techniques to compute the similarity between the paragraphs in a text.[18–20] The paragraphs that have the highest collective similarity to the other paragraphs are assumed to be central/important to the document in which they belong. When evaluated against a collection of encyclopedia articles, this method did not improve significantly over a position baseline.[20] The assumption that important sentences are those that contain important concepts have been also explored in a different representation space, using methods specific to latent semantic analysis.[21]

Lexical chains-based. Lexical chains are successions of semantically related words that create a context and contribute to the continuity of meaning.[22] For example, a lexical chain defined over the text in Fig. 1a, may contain the semantically related words (nose, fever, headache, flu, pill). Lexical chains are automatically built[23–25] using large lexical databases, such as thesauri and the Wordnet.[26] Lexical-chain-based approaches to text summarization[23,25] assume that important sentences are those that are "traversed" by "strong" chains, where the strength of a chain is defined by its length and number of distinct words in it. Empirical evaluations have shown promising results but have not provided yet uncontroversial evidence that this summarization method yields higher performance than others.

Connectedness-based. The connectedness-based method[27] fuses many of the elements specific to cohesion-based methods. Systems that use this method first map the words in a text into nodes in a graph and then create arcs between the nodes whenever an adjacency, grammatical, coreference, or lexical-similarity-based relation holds between the words corresponding to the nodes. The sentences that contain the most connected words are considered important. The method has been shown to produce good results when evaluated extrinsically, on a document categorization

task.[12] The author is not aware of intrinsic evaluations of this approach.

Discourse-based

The working hypothesis of the discourse-based method is that the hierarchical discourse structure of texts[28] can be used to determine the important sentences/clauses in a text.[29–32] Intuitively, if a paragraph p in a text elaborates on a sentence i in the same text, it is reasonable to assume that a summary should contain only the sentence i as the information in paragraph p is subsidiary to that in i.

Marcu[30] presented various algorithms for automatically deriving the hierarchical discourse structure of texts, and Marcu[30] and Carlson et al.[33] discussed methods for exploiting the discourse structure to determine the most important sentences/clauses in a text. For short texts, the performance of discourse-based summarization systems approaches that of humans.

Integration of the methods

None of the methods above has proved to provide consistent, high-performance results across all text genres, document lengths, and compression rates. In fact, experiments carried out by Jing et al.[34] suggested that many of these methods are quite unstable. Systems that perform well in a given summarization setting may perform poorly in a different one. To increase robustness, many summarization systems do not rely on only one of the above methods. Rather, they use supervised machine learning techniques, such as naïve Bayes,[13,14] decision trees,[35,36] inductive learning,[35] and rhetorical parsing tuning,[30] to adjust the contribution of each summarization method in determining the most important sentences in a text. In stable experimental conditions, summarization systems that use machine learning to integrate various methods/techniques always outperform systems that use only one summarization method.

Sentence Simplifiers/Compressors

Professional abstractors do not create abstracts by simply copying the most important sentences in a text. Often, they delete nonimportant segments in the important sentences to yield summaries of higher compression.[37–39] For example, a sentence such as that in Fig. 3a, can be simplified as shown in Fig. 3b.

To simplify sentences, some researchers[40–45] manually developed collections of rules that posit, for example, that during compression all adjectives and relative clauses in a sentence should be dropped. As one may expect, this approach can often yield ungrammatical outputs. To address this problem, other researchers[39,46] tackled the sentence simplification problem in a probabilistic framework, using a noisy-channel approach similar to that

a.	Although the modules themselves may be physically and/or electrically incompatible, the *cable-specific jacks* on them *provide industry-standard connections.*
b.	*Cable-specific jacks provide industry-standard connections.*

Fig. 3 Sentence compression example.

employed in speech recognition,[47] machine translation,[48] part-of-speech tagging,[49] and information retrieval.[50]

Using an automatically collected corpus of sentences and their human-generated compressions, Knight and Marcu[39,46] developed generative probabilistic models that explain how short sentences can be rewritten as long sentences through a sequence of syntax-based stochastic operations. Mathematically, any long sentence L can be generated from a short sentence S with a probability $P(L|S)$. A language model component assigns a probability $P(S)$ to any possible sentence: grammatical sentences have high probability, whereas nongrammatical sentences have low probability. Once the probability distributions $P(L|S)$ and $P(S)$ are estimated, given a sentence l, sentence compressions of arbitrary length can be generated by looking for sentences s that maximize the product $P(l|s) P(s)$ [i.e., sentences that are both good compressions ($P(l|s)$ is high) and grammatical ($P(s)$ is high)]. Empirical evaluations of Knight and Marcu's sentence compression algorithms showed that the noisy-channel approach outperforms two baseline algorithms that compress sentences by dropping words randomly or by choosing subsequences of maximal n-gram probability. However, Knight and Marcu's compression system still does not come close to human performance levels.

Sentence Paraphrasers

The simple deletion of syntactic constituents in long sentences is not the only way professionals create abstracts. Often, information from multiple sentences is fused together or, for coherence, cohesive, and stylistic effects, information in one or multiple sentences is paraphrased. Numerous researchers studied in detail the types of paraphrase operations employed during summarization[38,51,52] and applied them in implemented systems.[38,52] However, systems capable of generating multiple paraphrases of arbitrary sentences are yet to be developed.

For example, it is not clear at all what kind of computational processes one should use to produce the short abstract in Fig. 1c, which paraphrases some of the information in the text in Fig. 1a, while making explicit information that is not rendered in the original text. Or consider the examples of compression/paraphrasing in Figs. 4 and 5, in which two sentences were compressed/paraphrased into one (the text in italics has been preserved or modified slightly during compression).

Both examples of compression/paraphrasing in Figs. 4 and 5 were produced by professional abstractors. Mimicking this process automatically is not within the capability of many summarization systems. Fortunately, the recognition and generation of paraphrases is currently on the radar of many research communities. Researchers in natural language generation and summarization,[53] data mining and question answering,[54] and information extraction[55] methods for automatically learning paraphrase-specific information by exploiting bilingual, parallel texts,[54] or large corpora of monolingual text.[54,55] It is likely that progress in this area will have a significant impact on the next generation of automatic abstraction systems.

a.	He *holds a bachelor's degree* in *chemistry.*
b.	"Maintaining an organization like ISFUG is like building a castle in the sand; it just requires constant work to keep it trim," said *Berkman, an economist with the Commerce department's* Bureau of Economic Analysis.
	Compression with paraphrasing: B*erkman, who holds a bachelor's degree in chemistry, is an economist with the* U.S. *Department of commerce.*

Fig. 4 Compression/paraphasing example 1.

a.	*Nonetheless, policies on its use vary.*
b.	*Agencies* such as the Internal Revenue Service and the Farm Credit Administration promote *shareware use, but* one NASA center shuns it.
	Compression with paraphrasing: Federal agencies generally use shareware, but policies on its use vary.

Fig. 5   Compression/paraphrasing example 2.

Coherence/Coreference Enforcers

Although extraction-based summaries contain, by construction, only well-formed grammatical sentences, they often do not sound right. For example, the summary extract in Fig. 1e of the text in Fig. 1a sounds odd because it is not clear to what entities "the new compound," "this enzyme," "the new compound," and "the infection" refer. Also, even if we could felicitously solve these references, the two sentences still do not seem to go well together because the discourse relation between the two is not clear.

Numerous techniques have been proposed to rewrite extracts so as to yield more readable summaries. Many of the systems tested as part of the large-scale summary evaluations carried out in conjunction with the DUC[10,11] use anaphora resolution techniques to replace with name entities some of the pronouns in the extracts they produce. More sophisticated summary revision operations, such as aggregation of constituents of two sentences on the basis of referential identity and reduction of coordinated constituents were proposed by Mani et al.[56] Aggregation operations specific to events have been proposed by Maybury.[57] Sentence-reordering algorithms have been proposed by Barzilay et al.[58]

In an attempt to deal with sentence and document compression in a uniform framework, Daumé and Marcu[59,60] showed how the noisy-channel approach to sentence compression of Knight and Marcu[39] can be extended to the text level. Their empirical evaluations show that a document compression system that integrates the deletion of syntactic constituents with the deletion of clauses and sentences outperforms a compression system that sequentially simplifies the sentences in a text.

HEADLINE GENERATORS

The vast majority of the work on headline generation has been carried out in a statistical-based noisy-channel framework.[61–64] Given large collections of «text, headline» tuples, which can be easily collected from the Web,

one can estimate probability distributions, $P(w_d|w_h)$, that reflect the likelihood of a word w_d occurring in a document when another word w_h occurs in a headline. By treating documents and headlines as bags of words, one can easily estimate the probability, $P(D|H)$, of a document given a headline. A classic trigram language model, P(H), can be trained to differentiate between well- and ill-formed headlines. Once the parameters of these models are estimated, one can construct document headlines by searching for sequences of words h that maximize the product, $P(d|h) P(h)$. The headlines generated in this manner tend to contain the most representative words in a document; however, the n-gram language models that were employed so far in headline generators seem to be too weak to enforce that the generated headlines are grammatical.

A different approach to headline generation is taken by Daumé et al.,[65] who first identified the input document type. This type can fall into one of these four classes: single event, multiple event, biography, and natural disaster. In addition to the type, Daumé et al. also automatically determined the most salient entities and relations in the input. Depending on the input type, they use a set of predefined templates to produce grammatical headlines. For example, when a multievent template of the form "MainEvents in Location1, Location2, and Location3" is instantiated using entities specific to a collection of documents about eclipses, it yields the headline "Eclipses in Hawaii, Mexico, and Bay Area." The method produces impressive results when the document type and salient entity identifiers work correctly, but produces odd results when these components fail.

SUMMARIZATION VIA SEMANTIC REPRESENTATIONS

Most of the light semantic-based summarization systems developed to date implement the pipeline architecture in Fig. 2. An information extraction system specialized in dealing with certain events, such as "terrorism" or "natural disasters," first extracts from the document(s) to be

SOURCE:	Reuters
SOURCE: DATE	Febr 27, 1993
INCIDENT: DATE	Febr 26, 1993
LOCATION:	World Trade Center
TYPE:	Bombing
NUMBER OF VICTIMS:	at least 5

Fig. 6 Example of template from the "terrorism" domain.

summarized text fragments that fit a predefined template (see Fig. 6 for an example). Once the template has been filled, a natural language generation component is used to map the information in the template into well-formed natural language sentences.[66,67]

Because the types of entries associated with information extraction templates are limited, before generation, one can apply additional operations to modify, delete, and aggregate information extracted from text.[67] This is particularly useful in the context of multidocument summarization, where the information extracted from multiple articles may be inconsistent.

This approach produces good summaries in the limited domains for which the information extraction and natural language generation systems have been tuned. Unfortunately, the approach does not generalize easily to arbitrary domains. In general, the role of semantics in summarization is still to be determined. What constitutes an appropriate, computable representation for the semantics of texts that can be exploited in the context of summarization applications remains a controversial issue. Some researchers[68] map texts into description logics and perform condensation operations on formal representations. Other researchers[65,66,69] tend to use intermediary representations that are closer to the surface and/or syntactic forms. No researcher in the field claims that high-performance summarization systems can be built without a deeper understanding of the semantics of texts than current technology supports. But where the middle ground between formal elegance and coverage is remains an open question.

MULTIDOC SUMMARIZATION

A multidocument summarization system takes as input a collection of documents and produces a summary of the entire collection. In some instances, producing generic multidocument summaries is easier than producing generic single document summaries. If a collection of documents is about the same event, let's say an earthquake, it is likely that many of the input documents contain sentences describing the number of victims, location, time, and strength of the earthquake. Clustering techniques that measure the word overlap between sentences are usually sufficient for choosing from the set of all sentences in a collection one that is similar to many others. If a sentence and its variants were produced by many

authors, it is likely to be important. In such instances, important information can be, hence, detected using simple word overlap. In other instances, producing multidocument summaries is more difficult than producing single document summaries. When the collection given as input is heterogeneous and describes multiple events of the same kind, for example, earthquakes that happened in various locations over a time interval, it is no longer easy to determine what information is most important.

Besides the challenges specific to single document summarization, multidocument summarizers have to address a whole range of additional problems. When one summarizes document collections, it often happens that the information given in them is inconsistent. One document may claim that a car crash has produced six victims, whereas another document may claim seven. Which document should one believe? How can one determine that the two documents talk about the same car crash to begin with?

Determining the absolute time of the events reported in the input documents becomes an important factor in deciding what should go in a summary. If a collection reports the saga of a military conflict, what should go in a short summary? The cause of the conflict? The major steps in the conflict? Or only the outcome?

Equally important is the explicit mentioning of the dates/times at which the events occurred.[70] If the dates/times associated with the reported events are not explicitly mentioned, there is a serious danger for human readers to make incorrect inferences. The simple juxtaposition of some sentences may mislead readers to believe that, for example, an event in one sentence caused the event in the subsequent sentence, if the reader cannot infer from the summary that the events took place 10 years apart.

Collections about similar events tend to contain significant amounts of redundant information. But recognizing where redundancies occur is by no means a trivial problem because the relation between sentences in arbitrary documents is difficult to characterize:[71] sentences can be semantically equivalent, they can subsume each other, or the information they convey can partially overlap. Recognizing the exact relation that holds between two arbitrary sentences is still an open problem.

The field of multidocument summarization has created the context and need for research in various areas. It led to the development of methods that are well suited for generating nonredundant summaries;[72,73] methods that can be used to detect and track events over time while producing "evolving," temporal summaries;[74] methods for fusing sentence fragments[66,69] to produce grammatical sentences; and in-depth studies of specific summary types (biographical).[2]

Since 2001, the DUC[10,11] in North America and the NTCIR Workshop on Evaluation of Chinese & Japanese Text Retrieval and Text Summarization[75] in Asia provided an excellent forum for evaluating and comparing various multidocument summarization algorithms. Participants in large-scale evaluations worry not only about building useful systems, but also about figuring out how to measure their performance.[76,77] It appears that a significant gap exists between the summarization systems that participate in DUC and NTCIR evaluations, which are capable of handling unrestricted document collections using simple surface-based techniques, and the systems developed to advance the state of the art in information fusion,[66,69] paraphrasing,[53] or text compression.[59] It would be interesting to see how and when the latter work gets integrated into robust summarization systems capable of handling arbitrary text types.

OTHER KINDS OF SUMMARIES

Throughout this entry, the focus has primarily been on techniques and algorithms developed for producing single document and multidocument generic summaries. By no means does this cover all the work in text summarization. For example, Teufel and Moens[78] went beyond generic summarization to devise algorithms for classifying sentences in scientific articles according to their rhetorical goal (Background, Topic/Aboutness, Related Work, Purpose/Problem, Solution/Method, Result, Conclusion/Claim). A system capable of recognizing these roles can be used to generate summaries tailored to specific communicative goals.

An increasingly significant body of work has been focusing on summarizing dialogues and meeting transcripts,[79–81] video[82,83] diagrams,[84] soccer games,[85] and web pages for personal digital assistant access.[86] As we make progress in developing new methods for representing and accessing nontextual media types, it is likely that we will develop additional techniques for summarizing nontextual information that exploits peculiarities specific to various media types. For example, Amitay and Paris[87] proposed a method for summarizing Web pages that makes use of the text in the links that point to a page.

CONCLUSION

The field of natural language summarization has come a long way since the initial experiments of Luhn.[16] Although we still struggle to define summaries and attempt to evaluate them, there is much reason for optimism. Summarization systems are no longer the appanage of a small research community. Summarization technology is used daily to summarize news (http://www.cnn.com), to consolidate and summarize news published by multiple sources (http://www.cs.columbia.edu/nlp/newsblaster/), to provide handheld device access to information (http://www-diglib.stanford.edu/~testbed/doc2/

PowerBrowsing/index.html), to enable voice-based navigation of the Web (http://www.voxera.com), and to provide feedback to students on their written essays (http://www.etstechnologies.com). Given the increasing success of the technology, there is a certain danger to believe that summarization is a solved problem. It is certainly not. The success of the technology is not determined so much by the quality of the output as by the willingness of the users to tolerate anything that enables them to deal easier with the deluge of information to which they are subjected. As the expectations of the users increase, the need for developing new techniques for automatic abstracting will also increase. Building a summarization system that is better than one that selects the first n sentences in a news article continues to be a significant intellectual challenge.

REFERENCES

1. Hovy, E.H. Automated text summarization. In *The Oxford University Handbook of Computational Linguistics*; Mitkov, R., Ed.; Oxford University Press: Oxford, 2002.
2. Mani, I. *Automatic Summarization*, Natural Language Processing Series John Benjamins Publishers Co.: Amsterdam, 2001; Vol. 3.
3. Jones, K.S. Introduction to text summarization. In *Advances in Automatic Text Summarization*; Mani, I., Maybury, M., Eds.; The MIT Press: Cambridge, MA, 1999; 1–12.
4. Carletta, J. Assessing agreement on classification tasks: The kappa statistic. Comput. Linguist. **1996**, *22*(2), 249–254.
5. Krippendorff, K. *Content Analysis: An Introduction to Its Methodology*, Sage Publications: Beverly Hills, CA, 1980.
6. Siegel, S. Castellan, N.J. *Nonparametric Statistics for the Behavioral Sciences*, 2nd Ed. McGraw-Hill, 1988.
7. Edmundson, H.P. New methods in automatic extracting. J. Assoc. Comput. Mach. **1968**, *16*(2), 264–285.
8. Baldwin, B.; Donaway, R.; Hovy, E.; Liddy, E.; Mani, I.; Marcu, D.; McKeown, K.; Mittal, V.; Moens, M.; Radev, D.; Jones, K.S.; Sundheim, B.; Teufel, S.; Weischedel, R.; White, M. *An Evaluation Roadmap for Summarization Research*. Available at http://www-nlpir.nist.gov/projects/duc/roadmapping.html.
9. Radev, D.; Jing, H.; Budzikowska, M. Centroid-Based Summarization of Multiple Documents: Sentence Extraction, Utility-Based Evaluation, and User Studies Proceedings of the ANLP/NAACL-2000 Workshop on Automatic Summarization Seattle, WA May 4, 2000; 21–30.
10. Proceedings of the 1st Document Understanding Conference (DUC-2001) New Orleans, LA September 2001.
11. Proceedings of the 2nd Document Understanding Conference (DUC-2002) Philadelphia, PA July 2002.
12. Firmin, H.T.; Sundeim, B. TIPSTER-SUMMAC Summarization Evaluation Proceedings of the TIPSTER Text Phase III Workshop Washington, DC, 1998.
13. Kupiec, J.; Pedersen, J.; Chen, F. A Trainable Document Summarizer Proceedings of the 18th ACM/SIGIR Annual

Conference on Research and Development in Information Retrieval Seattle, WA 1995; 68–73.

14. Teufel, S.; Moens, M. Sentence Extraction as a Classification Task Proceedings of the ACL'97/EACL'97 Workshop on Intelligent Scalable Text Summarization Madrid, July 11, 1997; 58–65.

15. Hovy, E.H.; Lin, C.-Y. Automated text summarization in summarist. In *Advances in Automatic Text Summarization*; Mani, I., Maybury, M., Eds.; The MIT Press: Cambridge, MA, 2000; 81–94.

16. Luhn, H.P. A statistical approach to mechanized encoding and searching of literary information. IBM J. Res. Develop. **1957**, *1*(4), 309–317.

17. Luhn, H.P. The automatic creation of literature abstracts. IBM J. Res. Develop. **1958**, *2*(2), 159–165.

18. Salton, G.; Buckley, C.; Singhal, A. Automatic analysis: Theme generation and summarization of machine-readable texts. Science **1994**, *264*, 1421–1426.

19. Salton, G.; Allan, J. Selective text utilization and text traversal. Int. J. Human-Comput. Stud. **1995**, *43*, 483–497.

20. Mitra, M.; Singhal, A.; Buckley, C. Automatic Text Summarization by Paragraph Extraction Proceedings of the ACL'97/EACL'97 Workshop on Intelligent Scalable Text Summarization Madrid, July 11, 1997; 39–46.

21. Gong, Y.; Liu, X. Generic Text Summarization Using Relevance Measure and Latent Semantic Analysis Proceedings of the 24th Annual Conference on Research and Development in Information Retrieval (SIGIR-01) New Orleans, LA September 2001; 19–25.

22. Morris, J.; Hirst, G. Lexical cohesion computed by thesaural relations as an indicator of the structure of text. Comput. Linguist. **1991**, *17*(1), 21–48.

23. Barzilay, R.; Elhadad, M. Using Lexical Chains for Text Summarization Proceedings of the ACL'97/EACL'97 Workshop on Intelligent Scalable Text Summarization Madrid, July 11, 1997; 10–17.

24. Hirst, G.; St-Onge, D. Lexical chains as representations of context for the detection and correction of malapropisms. In *Wordnet: An Electronic Lexical Database and Some of Its Applications*; Fellbaum, C., Ed.; The MIT Press: Cambridge, MA, 1997.

25. Silber, G.H.; McCoy, K.F. Efficient Text Summarization Using Lexical Chains Proceedings of the ACM Conference on Intelligent User Interfaces New Orleans, LA January 2000.

26. Fellbaum, C., Ed. *Wordnet: An Electronic Lexical Database;* The MIT Press: Cambridge, MA, 1998.

27. Mani, I.; Bloedorn, E. Multi-document Summarization by Graph Search and Matching Proceedings of the 14th National Conference on Artificial Intelligence (AAAI-97) Providence, RI July 27–31, 1997; 622–628.

28. Mann, W.C.; Thompson, S.A. Rhetorical structure theory: Toward a functional theory of text organization. Text **1988**, *8*(3), 243–281.

29. Hobbs, J.R. Summaries from Structure Working Notes of the Dagstuhl Seminar on Summarizing Text for Intelligent Communication Dagstuhl, December 13–17, 1993.

30. Marcu, D. *The Theory and Practice of Discourse Parsing and Summarization*, The MIT Press: Cambridge, MA, 2000.

31. Ono, K.; Sumita, K.; Miike, S. Abstract Generation Based on Rhetorical Structure Extraction Proceedings of the International Conference on Computational Linguistics (COLING-94) Tokyo, 1994; 334–348.

32. Jones, K.S. What might be in a summary?. *Information Retrieval 93: Von der Modellierung zur Anwendung*; Universitatsverlag Konstanz: Konstanz, 1993; 9–26.

33. Carlson, L.; Conroy, J.M.; Marcu, D.; O'Leary, D.P.; Okurowski, M.E.; Taylor, A.; Wong, W. An Empirical Study of the Relation Between Abstracts, Extracts, and the Discourse Structure of Texts Proceedings of the Document Understanding Conference New Orleans, LA September, 13–14, 2001; 11–18.

34. Jing, H.; Barzilay, R.; McKeown, K.R.; Elhadad, M. Summarization Evaluation Methods: Experiments and Analysis Proceedings of the AAAI-98 Spring Symposium on Intelligent Text Summarization Stanford, CA March, 23–25, 1998; 60–68.

35. Mani, I.; Bloedorn, E. Machine Learning of Generic and User-Focused Summarization Proceedings of 15th National Conference on Artificial Intelligence (AAAI-98) Madison, WI July 26–30, 1998.

36. Lin, C.-Y. Training a Selection Function for Extraction Proceedings of the 8th International Conference on Information and Knowledge Management (CIKM'99) Kansas City, MO November 2–6, 1998; 55–62.

37. Endres-Niggemeyer, B. *Summarizing Information*, Springer-Verlag: Berlin, 1998.

38. Jing, H. Sentence Reduction for Automatic Text Summarization Proceedings of the 1st Annual Meeting of the North American Chapter of the Association for Computational Linguistics NAACL-2000 Seattle, WA 2000; 310–315.

39. Knight, K.; Marcu, D. Summarization beyond sentence extraction: A probabilistic approach to sentence compression. Artif. Intell. **2002**.

40. Jing, H.; McKeown, K.R. The Decomposition of Human-Written Summary Sentences Proceedings of the 22nd Conference on Research and Development in Information Retrieval (SIGIR-99) Berkeley, CA August, 15–19, 1999.

41. Mahesh, K. Hypertext Summary Extraction for Fast Document Browsing Working Notes of the AAAI-97 Spring Symposium on Natural Language Processing Tools for the World-Wide-Web Stanford, CA March 1997; 95–103.

42. Carroll, J.; Minnen, G.; Canning, Y.; Devlin, S.; Tait, J. Practical Simplification of English Newspaper Text to Assist Aphasic Readers Proceedings of the AAAI-98 Workshop on Integrating Artificial Intelligence and Assistive Technology Madison, WI July 26–30, 1998.

43. Canning, Y.; Tait, J.; Archibald, J.; Crawle, R. Cohesive Page Regeneration of Syntactically Simplified Newspaper Text Workshop on Robust Methods in Analysis of Natural Language Data (ROMAND 2000) Laussane, October 19–20, 2000.

44. Chandrasekar, R.; Doran, C.; Bangalore, S.B. Motivations and Methods for Text Simplification Proceedings of the 16th International Conference on Computational Linguistics (COLING '96) Copenhagen, August 5–9, 1996.

45. Grefenstette, G. Producing Intelligent Telegraphic Text Reduction to Provide an Audio Scanning Service for the Blind Working Notes of the AAAI Spring Symposium on

Artificial–Back

Intelligent Text Summarization Stanford, CA March 23–25, 1998; 111–118.

46. Knight, K.; Marcu, D. Statistics-Based Summarization—Step One: Sentence Compression Proceedings of the 17th National Conference on Artificial Intelligence (AAAI-2000) Austin, TX July 30–August 3, 2000; 703–710.

47. Jelinek, F. *Statistical Methods for Speech Recognition*, The MIT Press: Cambridge, MA, 1997.

48. Brown, P.F.; Della Pietra, S.A.; Della Pietra, V.J.; Mercer, R.L. The mathematics of statistical machine translation: Parameter estimation. Comput. Linguist. **1993**, *19*(2), 263–311.

49. Church, K.W. A Stochastic Parts Program and Noun Phrase Parser for Unrestricted Text Proceedings of the 2nd Conference on Applied Natural Language Processing Austin, TX 1988; 136–143.

50. Berger, A.; Lafferty, J. Information Retrieval as Statistical Translation Proceedings of the 22nd Conference on Research and Development in Information Retrieval (SIGIR-99) Berkeley, CA August 15–19, 1999; 222–229.

51. Chuah, C.-K. Linguistic Processes for Content Condensation in Abstracting Scientific Texts, Ph.D. thesis; Université de Montréal: Montréal, 2001.

52. Jing, H.; McKeown, K.R. Cut and Paste Based Text Summarization Proceedings of the 1st Annual Meeting of the North American Chapter of the Association for Computational Linguistics NAACL-2000 Seattle, WA April 29–May 3, 2000; 178–185.

53. Barzilay, R.; McKeown, K. Extracting Paraphrases from a Parallel Corpus Proceedings of the 39th Annual Meeting of the Association for Computational Linguistics Toulouse, July 9–11, 2001; 50–57.

54. Lin, D.; Pantel, P. Discovery of inference rules for question answering. J. Nat. Lang. Eng. **2001**, *Fall-Winter*.

55. Shinyama, Y.; Sekine, S.; Sudo, K.; Grishman, R. Automatic Paraphrase Acquisition from News Articles Proceedings of the Human Language Technology Conference (HLT-02), Poster Presentation San Diego, CA March 24–27, 2002.

56. Mani, I.; Gates, B.; Bloedorn, E. Improving Summaries by Revising Them Proceedings of the 37th Annual Meeting of the Association for Computational Linguistics Baltimore, MD 1999; 558–565.

57. Maybury, M. Generating summaries from event data. Inf. Process. Manag. **1995**, *31*(5), 733–751.

58. Barzilay, R.; Elhadad, N.; McKeown, K. Sentence Ordering in Multidocument Summarization Proceedings of the First International Conference on Human Language Technology Research (HLT'01) San Diego, CA March 18–21, 2001; 149–156.

59. Daumé, H.; III Marcu, D. A Noisy-Channel Model for Document Compression Proceedings of the 40th Annual Meeting of the Association for Computational Linguistics (ACL-02) Philadelphia, PA July 7–12, 2002.

60. Daumé, H.; III Knight, K.; Langkilde-Geary, I.; Marcu, D.; Yamada, K. The Importance of Lexicalized Syntax Models for Natural Language Generation Tasks Proceedings of the 2nd International Conference on Natural Language Generation Harriman, NY July 1–3, 2002.

61. Banko, M.; Mittal, V.; Witbrock, M. Headline Generation Based on Statistical Translation Proceedings of the 38th Annual Meeting of the Association for Computational

Linguistics (ACL-2000) Hong Kong, October 1–8, 2000; 318–325.

62. Berger, A.; Mittal, V. Query-Relevant Summarization Using FAQs Proceedings of the 38th Annual Meeting of the Association for Computational Linguistics (ACL-2000) Hong Kong, October 1–8, 2000; 294–301.

63. Jing, R.; Hauptmann, A. Title Generation for Machine-Translated Documents Proceedings of the 17th International Conference on Artificial Intelligence (IJCAI'01) Seattle, WA August 4–10, 2001.

64. Witbrock, M.J.; Mittal, V.O. Ultra-summarization: A Statistical Approach to Generating Highly Condensed Non-extractive Summaries Proceedings of the 22nd International Conference on Research and Development in Information Retrieval (SIGIR'99), Poster Session Berkeley, CA August 1999; 315–316.

65. Daumé, H.; III Echihabi, A.; Marcu, D.; Munteanu, D.S.; Soricut, R. GLEANS: A Generator of Logical Extracts and Abstracts for Nice Summaries Proceedings of the 2nd Document Understanding Conference (DUC-02) Philadelphia, PA July 11–13, 2002.

66. McKeown, K.R.; Klavans, J.L.; Hatzivassiloglou, V.; Barzilay, R.; Eskin, E. Towards Multidocument Summarization by Reformulation: Progress and Prospects Proceedings of the 16th National Conference on Artificial Intelligence (AAAI-99) Orlando, FL July 18–22, 1999.

67. Radev, D.; McKeown, K.R. Generating natural language summaries from multiple on-line sources. Comput. Linguist. **1998**, *24*(3), 469–500.

68. Reimer, U.; Hahn, U. A Formal Model of Text Summarization Based on Condensation Operators of a Terminological Logic Proceedings of the ACL'97/EACL'97 Workshop on Intelligent Scalable Text Summarization Madrid, July 11, 1997; 97–104.

69. Barzilay, R.; McKeown, K.R.; Elhadad, M. Information Fusion in the Context of Multi-document Summarization Proceedings of the 37th Annual Meeting of the Association for Computational Linguistics (ACL-99), University of Maryland June 20–26, 1999; 550–557.

70. Lin, C.-Y.; Hovy, E. From Single to Multi-document Summarization: A Prototype System and Its Evaluation Proceedings of the 40th Anniversary Meeting of the Association for Computational Linguistics (ACL'02) Philadelphia, PA July 7–12, 2002.

71. Radev, D. A Common Theory of Information Fusion from Multiple Sources, Step One: Cross-Document Structure Proceedings of the 1st ACL SIGDIAL Workshop on Discourse and Dialogue Hong Kong, August 2000; 74–83.

72. Carbonell, J.G.; Goldstein, J. The Use of MMR, Diversity-Based Reranking for Reordering Documents and Producing Summaries Proceedings of the 21st Annual International ACM SIGIR Conference on Research and Development in Information Retrieval Melbourne, August 24–28, 1998; 335–336.

73. Goldstein, J.; Mittal, V.; Carbonell, J.; Kantrowitz, M. Multi-document Summarization by Sentence Extraction Proceedings of the ANLP/NAACL-2000 Workshop on Automatic Summarization Seattle, WA May 4, 2000; 40–48.

74. Allan, J.; Gupta, R.; Khandelwal, V. Temporal Summaries of News Topics Proceedings of the 24th Annual Conference on Research and Development in Information

Retrieval (SIGIR-01) New Orleans, LA September 2001 10–18.

75. Proceedings of the Workshop on Evaluation of Chinese & Japanese Text Retrieval and Text Summarization Tokyo, March 2001.

76. McKeown, K.; Barzilay, R.; Evans, D.; Hatzivassiloglou, V.; Kan, M.Y.; Shiffman, B.; Teufel, S. Columbia Multidocument Summarization: Approach and Evaluation Document Understanding Conference (DUC-2001) New Orleans, LA 2001; 43–64.

77. Marcu, D. Discourse-Based Summarization Proceedings of the Document Understanding Conference (DUC-2001) New Orleans, LA September, 13–14, 2001; 109–116.

78. Teufel, S. Moens, M. Argumentative classification of extracted sentences as a first step towards flexible abstracting. In *Advances in Automatic Text Summarization*; Mani, I., Maybury, M., Eds.; The MIT Press: Cambridge, MA, 1999; 155–175.

79. Reithinger, N.; Kipp, M.; Engel, R.; Alexandersson, J. Summarizing Multilingual Spoken Negotiation Dialogs Proceedings of the 38th Annual Meeting of the Association for Computational Linguistics (ACL'00) Hong Kong, October 2000; 310–317.

80. Zechner, K.; Waibel, A. DiaSumm: Flexible Summarization of Spontaneous Dialogues in Unrestricted Domains Proceedings of the 18th International Conference on Computational Linguistics (COLING'00) Saarbruecken, July 31–August 4, 2000; 968–974.

81. Zechner, K. Automatic Summarization of Spoken Dialogues in Unrestricted Domains. *Ph.D. thesis;* Carnegie Mellon University: Pittsburgh, PA, 2001.

82. Merlino, A.; Maybury, M. An empirical study of the optimal presentation of multimedia summaries of broadcast news. In *Advances in Automatic Text Summarization*; Mani, I., Maybury, M., Eds.; The MIT Press: Cambridge, MA, 1999; 391–401.

83. Takeshita, A.; Inoue, T.; Tanaka, K. Topic-based multimedia structuring. In *Intelligent Multimedia Information*; Maybury, M., Ed.; The MIT Press: Cambridge, MA, 1997; 259–277.

84. Futrelle, R.P. Summarization of diagrams in documents. In *Advances in Automatic Text Summarization*; Mani, I., Maybury, M., Eds.; The MIT Press, 1999; 403–421.

85. Raines, T.; Tambe, M.; Marsella, S. Automated Assistants to Aid Humans in Understanding Team Behaviors Proceedings of the 4th International Conference on Autonomous Agents Barcelona, June 3–7, 2000; 419–426.

86. Buyukkokten, O.; Garcia-Molina, H.; Paepcke, A. Seeing the Whole in Parts: Text Summarization for Web Browsing on Handheld Devices Proceedings of the 10th International WWW Conference Hong Kong, May 1–5, 2001.

87. Amitay, E.; Paris, C. Automatically Summarising Web Sites—Is There a Way Around It? Proceedings of the 9th International Conference on Information and Knowledge Management (CIKM'00) Washington, DC 2000; 173–179.

Automatic Discourse Generation

John A. Bateman
University of Bremen, Bremen, Germany

Abstract

Automated discourse generation, also known as natural language generation (NLG), is a subfield of computational linguistics focusing on theories and methods for creating natural "texts" of all kinds automatically by machine. This entry provides an overview of the main application areas of natural language generation, motivations for adopting natural language technology, the theoretical problems addressed, and the technical solutions adopted.

INTRODUCTION

Automated discourse generation (also commonly known as "text generation" or "natural language generation": hereafter: NLG) is a branch of computational linguistics, or natural language processing, concerned with the question of how "texts" in natural human languages can be created automatically by machine. Very many NLG systems have now been created, ranging from fielded application systems to research prototypes aimed at investigating particular aspects of the language generation problem. In general, an NLG system has the task of converting representations of information maintained in some computationally accessible form (e.g., in databases, numerical sources, knowledge representations, or ontologies) into appropriate textual renditions of that information expressed in natural human languages. NLG as a field therefore concerns itself with the construction of such systems, with the identification of appropriate system architectures, their component modules, interconnections between modules, suitable processing schemes, reliable algorithms, and evaluation methods.

The range of texts addressed in the field of NLG is very broad and continues to expand. Generated texts may be written (still the majority) or (increasingly) spoken; they may be extended in length, be single paragraphs (the majority), or consist instead of single phrases or sentences—as is often the case in database responses, diagram caption generation, machine translation, question-answering, etc. The texts produced may consist of simple character strings or invoke more sophisticated text formatting, punctuation, or page layout; they may exhibit a linear structure (the majority), be part of a dialogue or interactive setting (increasingly), or be (partially) organized as hypertext. The texts may also be explicitly tailored for audiences differing in expertise, knowledge, interest, or cognitive load; they may be produced in a variety of natural languages; they increasingly combine natural language (sentences, paragraphs) with nonlinguistic material such as graphs, pictures, or diagrams; and they may even target nonverbal languages such as sign language. In short, wherever we find natural human language being employed, we are dealing with a potential field of application for NLG. Established NLG applications include the generation of weather reports from meteorological data in multiple languages, customer response letters, health information leaflets and letters, environmental reports, software documentation, dialogic human–computer interfaces, and many more.

NLG also provides a significant set of methods for linguistic research, particularly overlapping with text linguistics, discourse analysis, pragmatics, and semantics. Moreover, most theories of syntactic description developed within linguistics have also been explored within NLG systems, providing additional valuable input concerning the formal and computational properties of those theories. This applies also to functional theories of linguistics, an area underrepresented in computational linguistics otherwise. This is due to the basic orientation within NLG to uncovering the language *choices* that are available when constructing texts and the *motivations* for those choices; this makes clear connections with the general goals of functional linguistics. In all these areas, a computational account of the human linguistic system that is sufficiently well specified to allow the automatic generation of natural texts needs to combine a breadth and detail of linguistic description unusual in noncomputational text linguistics, semantics, and syntactic theory and thus serves to push the articulation of linguistic theory further along a broad front of issues.

A variety of techniques are employed when constructing NLG systems, ranging as we shall see below from simple "fill-in-the-gaps" to deep intention-based reasoning. However, whereas systems employing simpler techniques are increasingly finding general use, as in dynamically generated Web pages or mail-merge, more sophisticated NLG systems are still rarely deployed. This is due partly to the inherent complexity of the task of language production and

Encyclopedia of Library and Information Sciences, Fourth Edition DOI: 10.1081/E-ELIS4-120043875

the overheads that constructing usable systems necessarily bring on the one hand, and partly to a lack of awareness of NLG technology and approaches on the other. As NLG becomes more well known and the supporting technology more streamlined, we can expect increasingly rapid transfer of research results into challenging applications. Research on applied NLG is therefore a particularly dynamic and active area.

The kinds of applications where it is worthwhile investigating the use of NLG can be characterized as follows. First, when there is a high volume of constantly changing but highly restricted text, as in weather reports or other numerically based summaries, then an NLG system specifically tailored to the demands of the required text type can offer significant benefits. This is further multiplied if, on the basis of the same underlying data, versions of the text are required in multiple languages (as in the early weather report systems employed in Canada for parallel English/French versions). Second, when some combination of variables means that very many possible texts might be required (as in "comparison" texts that describe on demand differences between two objects, products, situations, etc.), although each single text is relatively simple, automatically generating the required text can be a better option. Third, when the particular text required depends on some user-specific variable, such as level of expertise concerning a subject (as in medical reports from a single medical record for patients and doctors), or on the time available to the user or the cognitive load that the user is under (as in car navigation systems where the driver might at some point need to deal with a difficult traffic situation or be traveling at speed), or on particular difficulties (such as hearing, restricted vision, etc.), then the flexibility inherent in full automatic NLG might be the only option for producing appropriate natural language output. This is particularly the case when the effects of any of these variables needs to be *combined*.

In general, the more flexible, context-dependent, and user-adaptive a piece of language needs to be, then the more appropriate sophisticated NLG techniques become. In dialogue situations, e.g., a system needs to respond quickly and effectively to user input, selecting appropriate levels of complexity, detail, and styles for its answers. But simpler techniques can only reproduce more straightforward interactions. We are already seeing a marked increase both in the requirements being placed on such systems and in the expectations of their users and so more sophisticated techniques will certainly be necessary.

NATURAL LANGUAGE GENERATION: BASIC COMPONENTS

We can characterize NLG in two ways: first, there are the kinds of problems that need to be solved when producing natural language and, second, there are the kinds of

computational techniques and components that are being developed to provide those solutions. There is a close relationship between these views. The more flexible and situation-dependent the language to be produced is, the more complex the computational techniques necessary become. Simple, inflexible language does not require advanced techniques. For example, the simplest method of generation appears in every automated bank teller machine and is responsible for printing out Welcome. Please insert your card (and all the other messages). Obviously, the text can be made arbitrarily sophisticated, with as many sentences as required. However, the system has no way of exploiting commonalities across sentences, and hence no way of recombining parts of existing sentences to produce new ones—not even to change Good morning! into Good afternoon!. "Canned text" systems of this kind illustrate how simplicity comes at the expense of power.

The obvious generalization is to create templates instead of canned sentences or paragraphs. Every form letter that contains spaces into which your name and address and the date are automatically entered is an example of such a text template. Though enabling some saving of space, template systems by and large suffer from the same lack of combinatory power that hampers canned text systems. As a consequence, template systems are made progressively more complex by including various conditions on which particular template is to be used. For example, changing the grammatical subject of a sentence to be filled by a plural noun phrase rather than a singular noun phrase will also require a change to the verb, otherwise a nongrammatical sentence results. For languages with more morphology than English (i.e., the majority of languages), the number of templates multiples explosively. Nevertheless some quite successful generation systems have been built on sophisticated notions of templates that avoid the pitfalls of simple template combination.[1]

The kinds and degrees of flexibility required in any particular application or research project in NLG vary widely, however, and so there is an accompanying diversity of opinion concerning the components and architectures best adopted for building NLG systems. Attempts to provide overall systematicity or standardization in the field are still of arguable generality.[2] Nevertheless, one commonly adopted structure is Reiter's so-called *pipeline* architecture,[3] which sees generation as a process whereby data is progressively transformed, structured, and augmented in order to produce the final text. This is shown graphically in Fig. 1.

Historically, development has moved from the bottom of this architecture to the top. The first NLG systems were very much concerned with organizing grammatical sentences so that their production could be controlled. Here various views of grammar and syntax from linguistics were employed, often augmented with conditions so that particular structures could be selected according to the precise textual information to be expressed. NLG was

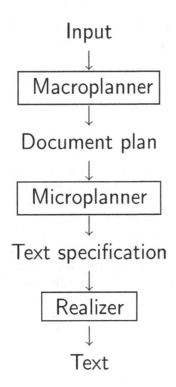

Fig. 1 Reiter's pipeline architecture for NLG.

therefore seen from very early on as a problem of choice under a variety of communicative constraints. Deciding on appropriate grammatical structures and lexical material is called *tactical generation* and answers the basic question of *how* information is to be presented. The corresponding computational components are responsible for *realization* and are therefore called *realizers*. Such realizers can either be targeted toward specific, narrowly defined areas of grammatical competence for particular kinds of texts, or can aim for general grammatical coverage. In the latter case, the grammatical resources drawn upon can become extremely extensive, rivalling standard reference grammars.[4,5] Some realizers aim for ease of use rather than broad coverage.[6] The input to realizers also varies: some require substantial grammatical information, thereby pushing grammatical decisions "upstream" in the entire generation process,[6] others abstract away from grammatical information so that input can already be semantic.[4] In the latter case, the realizer takes over responsibility for the grammatical decisions to be made.

In both cases, there are a range of phenomena to be covered to do with the fine-detailed organization of the sentences to be produced: Are simple sentences to be grouped together into more complex sentences? Are pronouns to be used to refer to repeated entities? Which words are to be selected when there are several similar in meaning available? All these issues are referred to as *microplanning*. The output of a microplanning component needs to be a specification suitable for passing to some realiser. Within this specification, decisions are recorded concerning which option is taken from a range of options

that are all grammatical but which differ in their textual import. They cannot therefore be made by a grammar alone but need to be responsive to the overall communicative plans guiding the generation process.

Microplanning then contrasts with *macroplanning* or, *text planning*, which is responsible for deciding on the overall organization of a generated text. It is the text planner that decides on *what* is to be expressed and how that content is to be organized—argumentatively, narratively, rhetorically, etc.—in order to make its point. This area of the generation task is termed *strategic generation*. Commonly, a text or document plan is created either by kinds of templates for entire text structures or by planning methods developed from Artificial Intelligence. In the latter case, the plans rely on the specification and satisfaction of communicative goals which, as a by-product, select particular "chunks" of content to be expressed. A completed text plan then includes structured chunks of content plus specified relations holding between those chunks. These relations eventually drive the selection of linking phrases and conjunctions, such as "therefore," "because," etc. by the realizer. Text plans can also be used to select suitable formatting options, such as bulleted lists for sequences of similar information, etc.

NATURAL LANGUAGE GENERATION: AN EXAMPLE

Consider the following short biography text:

> Anni Albers is American, and she is a textile designer, a draughtsman and a printmaker. She was born in Berlin on June 12, 1899. She studied art in 1916–1919 with Brandenburg. Also, she studied art at the Kunstgewerbeschule in Hamburg in 1919–1920 and the Bauhaus at Weimar and Dessau in 1922–1925 and 1925–1929. In 1933 she settled in the USA. In 1933–1949 she taught at Black Mountain College in North Carolina.

This is not a sophisticated text; indeed, it could be improved in many ways. Its relevance for us here is that it was generated fully automatically by an NLG system as one of several hundred similar texts on the basis of information acquired by a prototype information system in the area of art history.[7] The biographical texts used here reflect the information that was held at some particular time within the experimental information system: there is no claim of historical or factual accuracy.

In order to produce the text, the generation process ran through the components of the NLG pipeline shown above. A macroplanning step produced a textual organization appropriate for a biography anchored to particular chunks of information from the system's knowledge base. This was then passed on to a general-purpose realizer containing a large-scale grammar of English. Clearly, if this text were the only possible result from the system when considering the selected artist, then there is no particular win in employing NLG technology rather than some simple template methods, or even a human writer. Consider, however, the following text:

> Anni Albers. American textile designer, draughtsman and printmaker. Born in Berlin June 12, 1899. 1916–1919 studied art with Brandenburg. 1919–1920 studied art at the Kunstgewerbeschule in Hamburg. 1922–1925 and 1925–1929 studied at the Bauhaus at Weimar and Dessau. Albers settled in the USA in 1933. 1933–1949 taught at Black Mountain College in North Carolina.

This text was also produced automatically by the same system, and what is more, it was produced from *the same set of data* using the same lexicon and grammar. It is different to the first text for the simple reason that its intended purpose is slightly different. In the latter text we are producing a biography sketch in note form; it might be more appropriate laid out as a table, or as additional information to a diagram or "info-graphic." However the text is presented, its different function is appropriately accompanied by linguistic variation. Compared to the first text we see elements in different places, different connections between sentences, and so on.

Whereas the production of an alternative text in response to differing communicative goals and user requirements would represent a major overhead for a non-NLG-based text production system, this fine-grained matching of text and function is precisely what an NLG system undertakes to provide. The variations shown in this second text could be produced by some transformation of the first version, but it should be clear that making such a transformation work reliably for a wide range of artists with differing amounts and kinds of information held about them would be an endeavor requiring very careful attention.

When we change the intended function slightly again and go on to consider a further variant, we see another different set of "transformations" again:

> Anni Albers, the American textile designer, draughtsman and printmaker who was born in Berlin on the 12 June 1899, taught at Black Mountain College in North Carolina from 1933 until 1949 after settling in the USA in 1933. Previously, she had studied art with Brandenburg in 1916–1919, at the Kunstgewerbeschule in Hamburg in 1919–1920 and at the Bauhaus at Weimar and Dessau in 1922–1925 and 1925–1929.

Any *one* of the three example texts could probably be constructed (although not particularly easily) with carefully handcrafted transformations of some "underlying" information, but this approach does not naturally lend itself to the creation of the alternatives. When variable text production is going to be an aim for some information

presentation system, for whatever reason, then it becomes appropriate to consider whether more sophisticated NLG techniques could be usefully employed. This is particularly relevant for current moves being investigated in the pursuit of message/text "personalization"; the requirement that text be produced flexibly is therefore by no means nowadays an unlikely situation.

The range of variation in these three examples illustrates a very small subset of the linguistic variations actually possible. This is a significant NLG issue in its own right because all such variations in linguistic phenomena need to be *orchestrated* so as to indicate the coherence and textual unity of the resulting product. In much NLG work, therefore, the point of interest is not what kind of variations are *possible*—large-scale generation systems will in any case be able to produce the grammatical variation exhibited here—but rather the question *when* and *why* any one variant is to be selected rather than another. Generation is fundamentally a matter of *choice* and of uncovering the reasons why one choice may be better than another in any particular context of use: this is the essential task of strategic generation and text planning.

NATURAL LANGUAGE GENERATION AND LINGUISTIC VARIATION

Above we described a generic architecture for an NLG system. We can also characterize NLG systems and their architectures from the perspective of *the range of language phenomena* that they handle. To do this requires more linguistic sophistication and so is not generally favoured within the NLG community itself. There characterizations in terms of the "standard" modules that we have seen are more common: tactical generation, microplanning, text planning, and so on. But considering the linguistic phenomena at issue provides an equally, if not more, insightful characterization because the phenomena are not shaped or restricted by architectural decisions within any particular NLG system.

To gain a theoretical hold of the linguistic phenomena and the alternatives that are relevant for building NLG systems, it is useful to employ the linguistic notion of *stratification*. Stratification refers to the differing kinds of information, at differing levels of abstraction, that contribute to the organization of any text. Thus, we can talk of lower levels of linguistic abstraction, such as phonology, morphology, syntax, or lexicogrammar (typically placed within the realiser component), and semantics and the higher levels of abstraction, such as those of text structure and style (the responsibility of macroplanning). Texts might then vary in their use of different text structures, of different syntactic structures, of different phonologies (when spoken), and so on. Regardless of an NLG component's actual inputs and its context of application, we can identify generation tasks in relation to the level of linguistic abstraction, or stratum, at which variation is being offered or controlled and consider appropriate computational mechanisms for their support.

A minimally necessary (but not sufficient) set of organizations capable of explaining how a text works—i.e., capable of explaining its "textuality"—consists of the two least abstract linguistic strata:

- Lexicogrammar: certain sequences of elements can be recognized as belonging to English (or some selected language) while others are not. The possible grammatical structures used in a language and the associated words that may (or may not) occur in them define a very constraining organization that must be adhered to in any text. Sentences are not arbitrary combinations of words.
- Semantics: in addition to knowing which structures and words are possible, there is also a clear organization that constrains the selection of knowledge in general to the particular kinds of knowledge necessary for linguistic expression. This varies across languages—for example, some languages require the textual status of definiteness to be expressed in the grammar (English) whereas some do not (Chinese, Russian).

The former area has been widely addressed within linguistics and can draw on a host of work on syntactic theory; the most commonly employed approaches for NLG include Combinatory Categorial Grammar (CCG), Head-Driven Phrase Structure Grammar (HPSG), Systemic-Functional Grammar (SFG), and Tree-Adjoining Grammar (TAG). For the latter area, a particularly useful orientation is to consider the consideration of the *semantic control* that a detailed lexicogrammar requires in order to make its selections. Early representations of such control information annotated structural rules or lexical units with "semantic" or "pragmatic" information as needed for particular effects.[8] Now there are several more sophisticated accounts available that offer a broader range of controlling/constraining semantic information for motivating lexicogrammatical generation. Even though particular NLG systems will differ in the precise input form they adopt, *linguistic semantic* representations provide a good basis for considering alternatives independent of particular implementations.

The most obvious semantic area for controlling lexicogrammatical decisions—particularly from the perspectives of logic, natural language understanding, or applications of NLG—is the "propositional content" of individual sentences: the traditional "who did what to whom (when and how)." This can be used to guide the lexicogrammatical selection of basic units (such as, often, clause *vs.* noun phrase *vs.* adverbial, etc.) and the basic grammatical structures used within these categories (e.g., intransitive, transitive, or ditransitive clause structures). For example, a "logical" form "*studyʹ (Albersʹ, artʹ, tᵢ),*"

that is intended to represent a predicate of studying that relates Anni Albers to the discipline art at some time t_i, might drive the lexicogrammar to select a clause with a transitive structure such as *Albers studies art*. A full representation of this aspect of a clause's meaning would then require at least additional information concerning temporal relations (particularly with respect to the time of speaking in order to control grammatical tense decisions), logical scopes of various quantifiers, etc. Various representations are used for this kind of information, termed *ideational*, but nowadays these show considerable commonalities. In particular, most adopt a so-called Neo-Davidsonian semantics involving two essential features: first, "events" are considered proper objects in their own right and, second, the relations between participants and events are characterised by a small set of named relations.

An example is shown in Fig. 2, couched in the widely used "Sentence Plan Language" originally developed for representing inputs to the realizer of the Penman text generation system. This semantics again represents a logical form for a sentence involving Albers studying art. Semantic-based inputs allow a system to abstract away from particular sentences as they might need to appear in the concrete texts generated and so provide more support for variation. They even allow in many cases a generalization across different languages, so that parallel versions of the same text might be produced by employing different realizers or different grammars.[9] Similar kinds of information are now provided in many modern semantics formalisms and can be expressed in description logics, minimal recursion semantics,[10] hybrid dependency logic,[11] and others. Within NLG, it is description logic (including its most recent manifestation in the Web Ontology Language, OWL) that is now the most common.

Much information necessary for controlling a realizer is, however, left unspecified in this representation. Missing information includes particular choices of words, grammatical selections such as active or passive, word orders, pronominalizations, tense, focusing constructions, etc. The constraints for selecting these are generally drawn from two further sources of control, termed *textual* and *interpersonal*, respectively.

Textual meaning is involved because NLG is concerned with producing coherent texts and not just isolated sentences. Almost all examples of texts actually generated

(e1 / study
   : actor (x1 / female :name Anni-Albers)
   : actee (x2 / art)
   : temporal-locating (t1 / time :year 1916))

Fig. 2 Example skeleton semantics for: *"Anni Albers studied art in 1916."* e1, x1 and x2 are logical variables; the terms following forward slashes are semantic types; terms introduced by colons are roles or properties.

to date (including our example texts above) have problems related to their expression of textual meaning—i.e., there are places where a human text producer would have made selections resulting in a more "fluent," or natural, text. A central research problem in NLG is then to isolate and describe the sources of constraint for lexicogrammatical decisions that influence textuality: i.e., again, how to control a lexicogrammar. The richer the lexicogrammatical resources that are represented, the more complex the problem becomes. If a lexicogrammar offers few possibilities, then issues of control are reduced; but possibilities for fluency are then also reduced. Many current approaches to NLG therefore concentrate on finding appropriate control constraints for general resources rather than on relying on artificial restrictions on the resources themselves—either by extending the semantics to include suitable statements or by weighting the selections made in the grammar probabilistically according to the desired language style.[12]

Purely probabilistic methods are problematic, however, in that textual organization can be employed strategically to improve text flow and is not simply a matter of style. Any reasonable lexicogrammatical resource of English will offer, e.g., in addition to the sentences that appeared in the example texts, sentences such as:

e. With Brandenburg, she studied art in 1916–1919.
f. In 1916–1919, Anni Albers studied art with Brandenburg.
g. She studied art with Brandenburg in 1916–1919.
h. The third artist shown above studied art with Brandenburg in 1916–1919.
i. Moreover, she studied art in 1916–1919 with Brandenburg.
j. It was in 1916–1919 that she studied art with Brandenburg.

Sentences (e)–(g) might be argued not to differ in propositional content, although they have different preferred contexts of use in texts; sentence (h) introduces semantic content to refer deictically to the artist in question, thereby bringing in issues of *referring expression* generation; sentence (i) makes its relation to the preceding text more explicit and so can also be taken as introducing more information; and sentence (j) has a similar function, but uses a particular grammatical structuring to realize that function.

Moreover, different linguistic units can include more or less of the content to be expressed. Some consequences of this for sentences appropriate to our Bauhaus texts are illustrated in the following examples:

She studied art with Brandenburg in 1916—1919, and at the Kunstgewerbeschule in Hamburg in 1919—1920, the Bauhaus in Weimar in 1922—1925 and the Bauhaus in Dessau in 1925—1929.

She studied art with Brandenburg. That was in 1916—1919.
Albers studied with Brandenburg in 1916—1919. She studied at the Kunstgewerbeschule in Hamburg in 1919—1920. She studied at the Bauhaus in 1922—1925. . . .
Anni Albers was a text designer, draftsman, and printmaker, who was born in Berlin on 12 June 1899.

Here, sentence (k) includes both the original sentence and further content closely related to that of the original, while sentence (l) divides the original content over two sentences and uses "discourse deixis" to bind the sentences together into a text fragment. Sentence (m) then goes to the extreme and separates out the events grouped together in (k), each studying period is realized in a separate sentence. Sentence (n) shows a similar decision, where the first two sentences of the example text have been combined into one.

The phenomenon at issue in all these sentences is termed *aggregation* in NLG; detailed rules and heuristics for controlling this grouping have been suggested as an important component of microplanning. A similar phenomenon applies at all linguistic levels: sentences (k)–(m) could also be expressed as (o), while the first sentence of the first example text above would read better as (p) and the similar variations found in the latter two example texts.

Anni Albers was an art student from 1916 to 1929.
Anni Albers is an American textile designer, draftsman and printmaker.

In (o) many events have been combined and covered in a single property attribution: *being an art student*; similarly in (p) the nationality and professions have been combined in a single complex property. The decisions for these presentational forms depend again on the particular granularity goals of the text being produced, which in turn depends on the situation and hearers/readers involved.

The second area of control, the *interpersonal* meaning, is important because any natural language utterance obviously does more than simply express an intended propositional content. Clauses also express the attitude of the speaker to the content being expressed. This is minimally necessary in order to motivate control of variation such as the following based on the sentence "She studied art in 1916–1919 with Brandenburg" from the example texts:

Did she study art with Brandenburg in 1916—1919?
She studied art with Brandenburg in 1916–1919, didn't she?
Actually, she studied art with Brandenburg in 1916–1919.

Here, sentence (q) changes the speech function of the sentence to a query; sentence (r) changes the speech function to request support from the hearer; and sentence (s) introduces more content, but here that information reflects the speaker's speech act force. In systems for spoken generation, there is also the further variation:

She studied art with Brandenburg in 1916–1919?

In (t), the sentence structure is as for an assertion, but the intonation selected indicates a question. There are several accounts of intonation used in generation and sources of control for lexicogrammars should then include provision for such variation. Interpersonal meanings can also have a decisive influence on other aspects of a text. For example, expressions of various speaker attitudes will partially determine the most appropriate words selected ("artist" vs. "charlatan"; "artwork" vs. "junk"). When generation systems have access to words varying in this way, they need to make sure that they do not select inappropriately. Interpersonal force is also important for the correct realization of requests, orders, statements, etc. and so can have substantial consequences for grammatical selections as well as word selections.

The appropriate control of a lexicogrammar therefore requires many additional constraints over and above the propositional content to be expressed. Providing such information enables broad coverage lexicogrammars to produce textually appropriate lexicogrammatical structures. This concerns the "how" to express what is to be expressed issue of NLG: i.e., tactical generation. We can say linguistically that any text reveals many "predispositions" of the speaker that pervasively influence both the information selected for expression and its form of expression. This consistency in selected semantic options is then a common issue across all three of the areas of meaning described here. Ideational, interpersonal, and textual selections all exhibit a unity that is characteristic of textuality. One abstract mechanism proposed to describe this is the systemic-functional notion of *register*. The theory of register states that language varies systematically according to its situation of use: consistency in situation therefore calls for consistency in the takeup of semantic options.[13] Several systems now attempt to provide stylistic control, often employing probabilistic language models to select a particular slant from a broad range of "overgenerated" possibilities.[12]

TEXT PLANNING

The next—and many researchers in NLG would consider the main—task in NLG is to guarantee that *sequences* of such constraint specifications are produced in such a way that they are mutually consistent and together combine to create recognizable *text*. Thus, as always in NLG, it is not sufficient that a full NLG system offers the means for controlling lexicogrammars to produce textually varied constructions, such a system must also be able to select

from the theoretically available possibilities those that are most appropriate for the instantial text being produced. Texts are not arbitrary combinations of sentences.

Early attempts to organize texts relied on the intrinsic organization of the knowledge being expressed; i.e., the organization of the knowledge determined the structure of the texts produced. This cannot be guaranteed to give an understandable or suitable text for any particular communicative purpose and so is inappropriately rigid. To achieve more flexibility, NLG now commonly plans out text structure independently of the content to be expressed, appealing to notions of "discourse structure." Currently the most commonly used approach here is *Rhetorical Structure Theory* (RST).[14] RST provides a general description of the relations holding among segments of a text, whether or not they are grammatically or lexically signalled.

Descriptions of texts, or texts generated using RST, are decomposed hierarchically into a nested set of related text "spans." RST defines approximately 25 relations which may hold between these spans, motivated originally on the basis of detailed descriptive linguistic analyses of some 400 texts of varying content and genres. An RST analysis of the Bauhaus biography text illustrating some of the relations defined is shown in Fig. 3. This analysis claims that the main point of the text (obtained by following the straight vertical lines downward from the top) is the fragment stating that Albers taught at Black Mountain College, thereby spreading the Bauhaus movement; other information is related to this as indicated by the identified rhetorical relations—i.e., background, circumstance, elaboration, and sequence. This particular rhetorical analysis brings out the fact that the text is intended to stand as an illustration of the spread of the Bauhaus movement to the United States and not as a neutral biography, which would typically consist of a simple sequence of events arranged chronologically.

RST definitions bring constraints to bear on the kinds of meanings that the related text spans must carry, and on the communicative effect achieved by the combined set of text spans. Constructing discourse structure in terms of RST relations has proved itself to be useful for supporting selections of linking forms and textual connectives, such as the "moreover" of sentence (d), or the deliberate *non*-selection of a conjunction: it is only because the discourse is being developed in a particular way that a particular form is appropriate. Once an NLG system has an RST-style text plan available, it has much of the additional information discussed above as necessary for motivating nonpropositional semantic choices. RST has also been used to constrain the recency relationship for anaphors, choices of theme, intonational phrasing, hyperlink generation, summarization options, and selections of focus. Several detailed overviews of the development of rhetorical relation-based NLG methods have been given,[15] as well as introductory examples showing text planning using

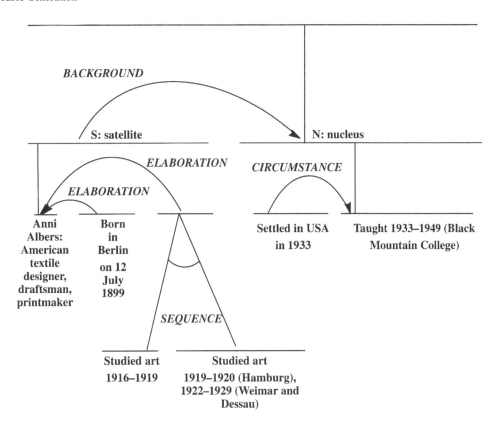

Fig. 3 Example RST analysis of a Bauhaus biography text.

RST in action.[16,17] Other commonly used approaches to text structuring are *text schemas* introduced to NLG by McKeown[8] and, more recently, discourse grammars based on tree-adjoining grammars.[18] All take on the task of selecting and structuring content to be expressed in a text flexibly responsive to communicative goals.

CONCLUSIONS: CURRENT DEVELOPMENTS AND OUTLOOK

The demand for natural language texts providing information of every conceivable kind is rocketing. Thus it is certain that NLG will form a key information technology in the future—there is scarcely an area of information presentation where beneficial applications of automatic text production cannot be imagined. Although it may still be some time before NLG techniques begin making a substantial impact on systems in everyday use, increased awareness of the potential will certainly be the most crucial aspect in hastening their adoption.

Several extensions to the field that have already occured will contribute to this. Early on NLG moved away from considering text production in a single language to parallel generation in several languages.[9] More recently, there has also been a move away from generating only texts to consider the automatic production of combinations of text, graphics, diagrams, and pictures.[19,20] Moreover,

generation techniques are now increasingly being employed in the originally rather separate field of dialogue systems. As the dialogues expected of question–answer systems, information systems, robotic systems, avatar generation (involving speech, facial expression, and gesture), etc. become more sophisticated, it is natural to turn more to the tools of NLG. There is clearly very much more to learn here concerning how these modalities can be combined appropriately and a regular conference has now been started to present research in this area.[21]

NLG systems are now beginning to appear on the World Wide Web, the pervasiveness of this new media giving fresh impetus to previously more academic explorations of automatic hypertext generation. Many aspects of NLG—including text planning, multimodal information presentation, and tactical (surface) generation are now combining in the construction of (possibly partially) synthetic hyperdocuments. This area is certain to be highly significant in increasing the acceptance and use of NLG technology, as well as in driving future NLG research. Until recently, the most significant technical bottleneck preventing the wider exploitation of NLG technology was the availability of sufficiently well-structured knowledge sources; clearly the closer we come to the Semantic Web, the easier it will be to employ NLG technology productively.

Partly as a move toward supporting applicability, the topics covered in international conferences relevant to

NLG over the past 5 years now include substantially more work on *evaluation methods* than was previously the case and there is ongoing discussion of how this can be furthered. Several "shared tasks" are being defined so that alternative approaches and systems can be compared. The most well accepted of these tasks so far is that of generating referring expressions, since this can be seen as a relatively modular component of the NLG task overall. The other topics now being addressed most include dialogue, stylistic control, and text planning in general.

FURTHER INFORMATION

Further information about natural language generation, ongoing work, and upcoming conferences, can be found on the Web site of the Association for Computational Linguistics (ACL) Special Interest Group on NLG (SIGGEN): http://www.siggen.org. There are two biannual conferences dedicated to NLG: the International Natural Language Generation Conference (running in even years) and the European Natural Language Generation Workshop (running in odd years). The proceedings of these conferences can be found in the ACL electronic paper archive.

Slightly more technical introductions to NLG can be found in several encyclopedias or handbooks,[16,17,22,23] while Reiter and Dale[3] have produced a textbook level introduction to the field. Finally, an almost complete list of natural language generation systems, with domains of application, references and theoretical foundations, is maintained in Bateman and Zock's "B-to-Z of Natural Language Generation Systems" at http://www.purl.org/net/nlg-list.

REFERENCES

1. Busemann, S. Ten years after: An update on TG/2 (and friends). In *Proceedings of the 10th European Workshop on Natural Language Generation*, Aberdeen, Scotland, 2005.
2. Cahill, L.; Carroll, J.; Evans, R.; Paiva, D.; Power, R.; Scott, D.; van Deemter, K. From RAGS to RICHES: Exploiting the potential of a flexible generation architecture. In *Proceedings of the Conference of the Association for Computational Linguistics (ACL'01)*, Toulouse, France, 2001.
3. Reiter, E.; Dale, R. *Building Natural Language Generation Systems*; Cambridge University Press: Cambridge, U.K., 2000.
4. Bateman, J.A. Enabling technology for multilingual natural language generation: The KPML development environment. J. Nat. Lang. Eng. **1997**, *3* (1), 15–55.
5. Elhadad, M.; Robin, J. A reusable comprehensive syntactic realization component. In *Demonstrations and Posters of the 1996 International Workshop on Natural Language Generation (INLG '96)*, Herstmonceux, U.K., 1996; 1–4.
6. Lavoie, B.; Rambow, O. A fast and portable realizer for text generation systems. In *Proceedings of the 5th Conference on Applied Natural Language Processing; Association for Computational Linguistics*, Washington, DC, 1997; 265–268.
7. Bateman, J.A.; Teich, E. Selective information presentation in an integrated publication system: An application of genre-driven text generation. Inform. Process. Manag. Intl. J. **1995**, *31* (5), 753–767.
8. McKeown, K.R. *Text Generation: Using Discourse Strategies and Focus Constraints to Generate Natural Language Text*; Cambridge University Press: Cambridge, U.K., 1985.
9. Bateman, J.A.; Kruijff-Korbayová, I.; Kruijff, G.-J. Multilingual resource sharing across both related and unrelated languages: An implemented, open-source framework for practical natural language generation. Res. Lang. Comput. **2005**, *3* (2), 191–219.
10. Copestake, A.; Flickinger, D.; Pollard, C.; Sag, I. Minimal recursion semantics: An introduction. Res. Lang. Comput. **2005**, *3*, 281–332.
11. Baldridge, J.; Kruijff, G.-J. Coupling CCG and hybrid logic dependency semantics. In *Proceedings of 40th Annual Meeting of the Association for Computational Linguistics*, Philadelphia, PA, 2002; 319–326.
12. Paiva, D.S.; Evans, R. A framework for stylistically controlled generation. In *Natural Language Generation: Third international Conference (INLG 2004)*, Belz, A., Evans, R., Piwek, P., Eds.; Springer: Berlin, New York, 2004; 120–129.
13. Bateman, J.A.; Paris, C.L. Phrasing a text in terms the user can understand. In *Proceedings of the Eleventh International Joint Conference on Artificial Intelligence, IJCAI'89*, Detroit, MI, 1989. 1511–1517.
14. Mann, W.C.; Thompson, S.A. Rhetorical structure theory: Toward a functional theory of text organization. Text **1988**, *8* (3), 243–281.
15. Hovy, E.H. Automated discourse generation using discourse relations. Artif. Intell. **1993**, *63* (1–2), 341–385.
16. vander Linden, K. Natural language generation. In *Speech and Language Processing: An Introduction to Speech Recognition, Computational Linguistics and Natural Language Processing*; Jurafsky, D., Martin, J., Eds.; Prentice-Hall: NJ, 2000; Chapter 20, 763–798.
17. Bateman, J.; Zock, M. Natural language generation. In *Oxford Handbook of Computational Linguistics*; Mitkov, R., Ed.; Oxford University Press: Oxford, U.K., 2003; Chapter 15, 284–304.
18. Gardent, C.; Webber, B. Describing discourse semantics. In *Proceedings of the 4th TAG+ Workshop*, University of Pennsylvania: Philadelphia, PA, 1998.
19. Geurts, J.; van Ossenbruggen, J.; Hardman, L. Application-specific constraints for multimedia presentation generation. In *Proceedings of the International Conference on Multimedia Modeling 2001 (MMM01)*, CWI: Amsterdam, the Netherlands, 2001; 247–266.
20. Bateman, J.A.; Kamps, T.; Kleinz, J.; Reichenberger, K. Constructive text, diagram and layout generation for information presentation: The DArt$_{bio}$ system. Comput. Linguist. **2001**, *27* (3), 409–449.
21. van der Sluis, I.; Theune, M.; Reiter, E.; Krahmer, E., Eds. *Proceedings of the Workshop on Multimodal Output Generation (MOG 2007)*; Centre for Telematics and

Information Technology (CTIT), University of Twente: Enschede, the Netherlands, 2007.

22. McDonald, D.D. Natural language generation. In *A Handbook of Natural Language Processing: Techniques and Applications for the Processing of Language as Text*; Dale, R., Moisl, H., Somers, H., Eds.; Marcel Dekker: New York, 2000; 147–179.

23. Horacek, H. Textgenerierung. In *Computerlinguistik und Sprachtechnologie—Eine Einführung*; Carstensen, K.-U., Ebert, C., Endriss, C., Jekat, S., Klabunde, R., Langer, H., Eds.; Spektrum Akademischer Verlag: Heidelberg, Germany, 2001; 331–360.

Back-of-the-Book Indexing

Nancy C. Mulvany
Bayside Indexing Service, Fort Collins, Colorado, U.S.A.

Abstract
The book index occupies a special niche in the information retrieval world. Each index is a unique, authored work. Each book is a closed system. The text presented in a book does not change; the material is stable and fixed. Book indexers provide readers with a nonlinear way to access information in a text. Even though closed-system indexing predates the development of the printing press, a book index can be thought of as hypertext.

INTRODUCTION

We have all used book indexes. Generally they appear in nonfiction books in the back of the book. Most people do not realize that book indexes are written by people. They are not generated by computers. A table of contents presents some of the topics discussed in the book in a linear fashion, from the book's beginning to the end. The back-of-the-book index is an information map that systematically brings together access points to every important topic discussed in a book, including concepts and nuances developed by the book author.

Although the word index is used in many diverse contexts today—index of economic indicators, consumer price index, database indexing, and search engine indexing—a book index has a particular meaning and application. Susan Klement provides a useful framework for discussing indexing in "Open-system versus closed-system indexing: A vital distinction"[1] when she writes:

> Closed-system indexing assists people finding a unit or units of relevant information *within a document*, while open-system indexing is designed to facilitate the retrieval of one or more *documents* that contain relevant information.

Internet search engines are examples of open-system indexing. An index to a periodical such as *Forbes* is also an open-system index. The Internet grows every day; it also changes every day as Web pages are edited or removed. *Forbes* will continue to add new issues as long as the magazine is published. Open systems are in a state of flux.

A book is a closed system. There is a beginning, a middle, and an end. The book and its audience is a self-contained universe. Unlike an open-system index that casts a wide, but shallow net, a book index deeply presents a systematic guide to the information contained in a text in a manner that enables readers to quickly find specific topics or concepts.

It is the insular nature of a book index that sets it apart from many other information retrieval devices or finding aids. While there may be 20 books about Web page design, each will be unique. The books will likely cover similar topics, but the presentation and writing styles of the authors will be different. The indexes for these books will also be unique. The indexes will reflect not only specific topics covered in the text, but also conceptual relationships that are presented in unique ways by each author.

Book indexing has a long history that will be reviewed briefly. It is important to recognize that a book index is not simply an alphabetic list of words that appear in the text. There are several critical elements that appear in book indexes that we do not find in computer-generated word lists. These elements and presentation formats for indexes will be discussed along with publishers' style guides and international standards. Although it is difficult to generalize about the way indexes are prepared, the basic mechanics of indexing will be presented. Lastly, indexing specialization, training, and professional associations are discussed.

A BRIEF HISTORY OF BOOK INDEXING

Closed-system indexing has a long history. It is natural to assume that the book index did not appear until books were printed. However, if we think of this particular information retrieval device as a closed-system index it is apparent that the need for an index of this type emerged before bound, printed books were produced. Hans Wellisch[2] writes:

> Indexing of books did not begin, as is commonly thought, after the invention of printing. It started with the rise of universities in the 13th century. Although no two manuscripts of the same work were exactly alike and folio or page numbers were seldom used, indexes to theological treatises, lives of the saints, medical and legal compendia

Encyclopedia of Library and Information Sciences, Fourth Edition DOI: 10.1081/E-ELIS4-120043677

Artificial–Back

and, most of all, to collections of sermons were compiled, using chapter and entry numbers instead of pagination.

Bella Hass Weinberg[3] has dated a Hebrew manuscript citation index back to the twelfth century. Hazel K. Bell in *From Flock Beds to Professionalism: A History of Indexer-Makers*[4] suggests that "perhaps the first known indexer is Ptolemy, who produced an index to his first world atlas *c*. A. D. 150."

When documents that could not be easily browsed were produced, there was a need for a finding aid. The development of the printing press made possible the mass production of bound books. Pagination of the books made possible the compilation of book indexes for each title. These indexes are referred to as "back-of-the-book indexes" because they usually appeared in the back of books. Today we are witnessing the transition of books from printed pages to electronic media. Although the media for book presentation may be changing, it is still useful to remember that need for a book index has not changed. Closed-system indexing exists independent of presentation media.

WHAT IS AN INDEX?

In the United States, the National Information Standards Organization (NISO)[5] defines an index as "a systematic guide designed to indicate topics or features of documents in order to facilitate retrieval of documents or parts of documents" (p. 39).

The International Organization of Standardization's (ISO) *ISO 999*[6] defines an index as an

> alphabetically or otherwise ordered arrangement of entries, different from the order of the document or collection indexed, designed to enable users to locate information in a document or specific documents in a collection.
>
> (Entry 3.5)

Many find these definitions too broad and imprecise. More thorough and lengthy descriptions of the purpose of an index can be found in the *British Standard's* "Function of an Index"[7] and the American Society for Indexing's (ASI) "Criteria for the H.W. Wilson Award."[8]

For general purposes, I find this definition useful: "An index is a structured sequence—resulting from a thorough and complete analysis of text—of synthesized access points to all the information contained in the text."[9]

A computer-generated list of words in the text, even arranged alphabetically, is not an index. For example, a concordance does not require analysis and synthesis of a text and its meaning. The concordance can only list words that appear in the text; it cannot include concepts or indicate relationships between topics. An alphabetical list of words does not truly qualify as the structured sequence that we associate with a proper book index.

ELEMENTS OF AN INDEX

A book index includes several major components: headings, locators, cross-references, and an arrangement scheme. Headings are composed of terms that readers are likely to look up. Locators indicate where in the text the heading can be found. Often locators are page numbers. Cross-references are linking devices that provide navigation guidance for synonyms or other related terms within the index itself. The arrangement of entries in an index is usually alphabetical. Here is an example:

nutrition, 34–40
    adults, 125–130
    children, 67–73
    infants, 53–59
    See also supplements

Headings

In the example above "nutrition" is the main heading and "adults, children, infants" are subheadings. Often headings are terms that appear verbatim in the text. For example, "Art Trading Fund, 22," will be found on page 22. Some types of entries are inverted in indexes. A common example of an inverted heading is a personal name: "Lakoff, George, 12." The nutrition example above is a conceptual heading. The word "nutrition" may or may not appear on the pages referenced. However, the indexer has determined that nutrition is discussed on these pages and has provided access to the discussion.

Locators

Each heading is followed by a locator. In the example above the locators are page numbers in a book. Locators must be precise and accurate. In some books, such as dense legal texts, a page number lacks precision. Instead, the locator may take the form of chapter number and paragraph number, such as "reserve against returns, 12.34(a)." Additionally, locators must be helpful. A main heading followed by a string of 30 separate locators is not helpful. Instead, subheadings should be provided that help guide the reader to specific information about the main heading.

Cross-References

The last line in the nutrition example is a *See also* cross-reference that directs readers to another closely related heading, in this case the "supplements" heading. Another type of cross-reference, the *See* cross-reference, performs a vocabulary control function. For example, "food. *See* nutrition," tells readers that information about food will be found at the "nutrition" heading.

The main heading and the entire block of information that follows it is referred to as an entry or entry array. The "nutrition entry" is composed of the main heading, subheadings, locators, and the cross-reference.

Arrangement of Entries

Lastly, a major element of any index is the systematic arrangement of entries. Book indexes are usually arranged in alphabetic order. There are two basic forms of alphabetic orders, word-by-word and letter-by-letter. To further complicate matters, many index entries include elements that are not letters of the alphabet:

> Lakoff, George
> San Francisco
> !ls command
> War of 1812

All four terms contain spaces. The first and third terms include punctuation marks and the last term contains a number. Any arrangement scheme must take all elements of an index entry into consideration. *NISO TR03: Guidelines for Alphabetical Arrangement of Letters and Sorting of Numerals and Other Symbols*[10] provides the most cogent discussion of alphabetical arrangement. Briefly, *NISO TR03* presents the following rules for the order of characters:

1. Space that precedes any other character (hyphen, dash, and slash are treated as a space).
2. Punctuation marks are ignored.
3. Symbols other than numerals, letters, and punctuation marks precede numerals and letters.
4. Numerals precede letters.
5. Letters follow numerals.

Word-by-word alphabetizing

If the rules above are followed, the resulting arrangement order is called word-by-word.

> type font
> type foundry
> Type/Specs Inc.
> Type4You Inc.
> typeface
> typeset

Letter-by-letter alphabetizing

The letter-by-letter form of alphabetizing eliminates *NISO TR03's* first rule, the space character (including hyphens, dashes, and slashes) is ignored.

> Type4You Inc.
> typeface
> type font
> type foundry
> typeset
> Type/Spec Inc.

These two styles of alphabetizing yield very different results, particularly in long indexes. Most users of book indexes do not think about various ways to alphabetize entries. They expect an entry to be in a particular place. If they do not find the entry, they may assume that the topic is not discussed. Choosing an arrangement order that works for users is critical. Unfortunately, as James Anderson[5] points out, there is a lack of agreement regarding arrangement:

> Two de facto standards widely used in libraries and databases in the United States are the American Library Association (ALA) and the Library of Congress (LC) filing rules. The guidelines for alphanumeric arrangement in *The Chicago Manual of Style* are used as a de facto standard by many publishers. These three codes of alphanumeric arrangement are, however, incompatible with each other (Entry 9.1).

Many book publishers follow their own alphabetizing styles. North American publishers often follow the guidelines in *The Chicago Manual of Style*, published by the University of Chicago Press, which calls for letter-by-letter alphabetization. It is important to note that Chicago's form of letter-by-letter alphabetization does not strictly follow the recommendations of *NISO TR03* discussed earlier. Style guides will be discussed in more detail later.

PRESENTATION FORMATS FOR INDEXES

There are two basic presentation formats of indexes: indented style and run-in style. Before discussing these formats it is important to keep in mind that the layout of an index is directly related to its usability.

The main goal of an index is to provide quick and easy access to information contained in the text. Unlike other types of text, indexes are not read in a linear fashion from beginning to end. Instead, readers jump into the index at various points. They scan for particular entries. The design of an index should enhance scanning by readers. They should be able to quickly determine at any place where they are in the index. Indexes that are dense and cluttered will frustrate users. The ideal index is almost transparent. Readers turn to the index, quickly find the desired entry, and locate the information in the text of the book. They are in and out of the index very quickly.

The indented and run-in formats for indexes are distinguished by the way subheadings are formatted. The indented style begins each subheading on a separate line, indented under the main heading. The run-in style separates subheadings with a semicolon and runs them all together as one block of text indented under the main heading. Here are some examples:

Indented style:
    nutrition, 34–40
        adults, 125–130
        Asian perspective on, 200–202
        children, 67–73
        European perspective on, 210–212
        infants, 53–59
        See also supplements
Run-in style:
    nutrition, 34–40
        adults, 125–130; Asian perspective on, 200–202; children, 67–73; European perspective on, 210–212; infants, 53–59. *See also* supplements

Above, the indented format uses seven lines, whereas the run-in style uses only four lines. Saving space is the main reason some publishers prefer run-in format. However, the indented format is easier to scan and use. In a complex index that includes sub-subheadings a run-in format cannot easily accommodate the extra level of detail. There are times when a hybrid style can be used with good effect. The indexes in the 15th edition of *The Chicago Manual of Style* and the 2nd edition of *Indexing Books* use a combination of the two styles. Subheadings are presented in indented format while sub-subheadings are run-in. There are many more issues related to the presentation format of indexes. For a more detailed discussion see "Format and Layout of the Index" (pp. 191–217) in *Indexing Books*.[11]

STYLE GUIDES

As mentioned earlier, most North American publishers follow the indexing style requirements of the University of Chicago Press, in particular Chapter 18 "Indexes" in *The Chicago Manual of Style*.[12] The chapter about indexes is published as a separate booklet that is often given to authors preparing their own indexes.

Some publishers use a modified form of "Chicago style" or devise their own indexing style guide that meets their unique needs. At minimum an indexing style guide will include rules for five elements:

1. Alphabetizing of main headings.
2. Arrangement of subheadings.
3. Format of entries.
4. Format of reference locators.
5. Format and placement of cross-references.

Of course, there are many other elements included in publishers' style guides. For example, a scholarly publisher will provide rules for the indexing of endnotes or footnotes. A technical publisher will have guidelines for the arrangement of numbers and symbols in index entries.

In the United Kingdom, *The Oxford Guide to Style*[13] includes a detailed discussion of the index style requirements for Oxford University Press.

STANDARDS

Currently the United States does not have a standard for the presentation of indexes. The NISO makes available a technical report, *Guidelines for Indexes and Related Information Retrieval Devices*[5] that did not garner enough votes to become a standard.

The ISO released *ISO 999: Information and Documentation; Guidelines for the Content, Organization, and Presentation of Indexes* in 1996.[6]

MECHANICS OF INDEXING

Generally book indexing does not begin until the pagination for a book is complete. Most book publishers send the indexer a set of printed, paginated pages. The pages are not bound and there is printing only on one side of the paper. Some publishers send indexers electronic copies of the book, usually as a PDF file.

One question indexers are often asked is "Do you have to read the entire book?" Yes, the indexer reads the entire book! Some indexers scan through all the book pages and mark possible index entries on each page. Others skim the table of contents and then start indexing.

Professional indexers use specialized indexing software. The programs can be set up to produce an index that conforms to the publisher's style. Index entries are typed into the program along with their locators. The software handles the alphabetizing along with many other tasks. For example, professional indexing software will perform cross-reference checking: if a cross-reference is created for an index entry that does not exist, the indexer will be warned. The three major indexing programs are Macrex (http://www.macrex.com/), SKY Index (http://www.sky-software.com/), and CINDEX (http://www.indexres.com/).

In the United States most publishers' book contracts stipulate that the book author must provide the index. Since professional software is costly, currently around $500.00, many book authors prepare indexes using their word processors. Some authors still use index cards to compile the entries. A more detailed discussion of manual methods of indexing can be found in *Indexing Books*.[14]

Regardless of the tools used to compile an index, going through the text and creating index entries is only the first step. The index must be edited and checked for accuracy. Some indexers set aside 25% of their time for editing tasks. Of course, the actual proportion of indexing time devoted to editing will vary from book to book. More time may be needed for a lengthy (500+ pages) book than for a shorter one.

Often book authors are shocked to discover that their editor expects the index to be delivered in a short span of time. Often 2 weeks is the time budgeted for indexing a 300-page book. The indexing process is demanding. Even professional indexers find it difficult to index for much more than 6 hours a day. Book authors, many without any experience or training in indexing, are often overwhelmed.

The most difficult aspect of indexing is term selection. Indexers must decide not only which terms go in the index, but also the form of the term. Some terms are obvious. For example, a book about California geography will surely include index entries for various mountain ranges such as the Sierra Nevada, Siskiyou, and Santa Lucia mountains. Unfortunately not all term selection is so straightforward. Often authors use synonymous words to describe the same topic. The indexer must make vocabulary control decisions and choose one term for the index entry and provide a *See* cross-reference from the term that is not used. If the author uses both "autos" and "cars," the indexer may decide to use "cars" as the index entry and add a cross-reference (autos. *See* cars).

As the indexer works through the text an intricate mapping structure begins to emerge. Many indexers find themselves returning to index entries written earlier and tweaking them so that they fit better into the overall index. The ability to keep the entire index structure in one's mind while working is critical. An index must always treat similar concepts and terms in a consistent manner. Users of indexes depend on this consistency to locate desired information. Using the autos/cars example, it is a disservice to readers to scatter information between two entries, as in:

autos, 23, 44–45
cars, 23, 56–58, 60

Readers looking at only one of the entries above will not get all the reference locators. They should not need to look in two places in the index. The more consistency the indexer can add to the index structure while writing the index, the less editing will be required after the initial term selection is complete.

The preferred vocabulary for an index is always the author's vocabulary. The indexer is not free to impose an external vocabulary on an index. But while indexers must remain true to the author's language, they must also anticipate the vocabulary needs of the readers. This often requires reconciling the readers' external vocabulary with that of the book through the use of cross-references. Take the example of a book about British dog shows that will be distributed in the United States. An American reader might not know that the "herding group" is called the "pastoral group" in the United Kingdom. The astute indexer will provide a cross-reference: "herding group. *See* pastoral group."

As the indexer works, the index itself is massaged into a form that will provide quick and complete access to the information contained in the book. If the indexer does a good job, the index structure will be transparent to users. Readers will not think about index structure. Rather, they will quickly locate their desired information and return to the text. Such a transparent and functional information access structure does not emerge on its own. It is the result of deliberative work.

SPECIALTIES

Many indexers are generalists. They can comfortably index on a wide variety of topics. Most nonfiction books that are mass distributed to the general public can be indexed by any competent indexer. However, there are many books that require special skills due to either the genre of book or the subject of the book.

Genre Specialties

Cookbooks and biographies are books that have unique indexing requirements. Cookbook indexes, when done well, are extremely detailed. While not every ingredient in a recipe is indexed, the main ingredients are often indexed along with the recipe name and type. In a barbeque cookbook, "Jimmy's Down Home Sauce," would likely be indexed by its name. Since it appeared in the "pork" entry of the book, it would also be indexed under "pork sauces." Likely, there is also entry under "sauces," and the type of sauce that it is, "St. Louis style," "North Carolina style," or "Memphis style" to name a few. It may also appear under "tomato-based sauces." Jimmy's secret ingredient may be bourbon, so it could also appear under "bourbon."

"Jimmy's Down Home Sauce" is just one recipe. Many cookbooks have more than one recipe per page, so the density of indexing per page can be quite high. Some cookbook indexers specialize within the genre. Some indexers may feel very comfortable with dessert and bread baking books, but they may not feel confident about indexing an Indian cookbook. More information about cookbook indexing can be found at the Culinary Indexing Special Interest Group of the American Society for Indexing (http://www.culinaryindexing.org/).

Biographies, autobiographies, and letters pose their own set of indexing issues. How does one handle the indexing of the main character? How does one arrange the subheadings: alphabetic or chronological order? What if the subject led a colorful life? How does one index scandals? Hazel K. Bell has written extensively about this type of indexing. The 3rd edition of Bell's *Indexing Biographies and Other Stories of Human Lives* is an excellent reference.[15] In this book she also briefly discusses the indexing of fiction, yet another genre specialty. Bell writes:[15]

> Although by no means standard practice, the indexing of fiction has been undertaken from time to time. Serious novels may be indexed in the same way as biographies or histories, as narratives concerning groups of people and the events in their lives.

It should be noted that Bell has written many fiction indexes, including indexes for five novels of A.S. Byatt.

Subject Specialties

Some books, because of the subject matter, require that the index be prepared by a specialist. Some indexers specialize in medical or legal indexing. Even within these two groups there are subspecialties. One medical indexer may easily work with a 600-page textbook on renal disease, while not feeling comfortable with a radiology textbook. Indexers who specialize in a particular subject area often have advanced degrees in the subject and keep in touch with the current terminology and advances in the field. Such intimate subject knowledge is needed when writing complex indexes for a technical book. The indexer must be fluent with the terminology so that appropriate cross-references can be provided and scattering information among synonymous terms can be avoided.

TRAINING

In the past many book indexers were self-taught. Formal coursework in book indexing is rarely offered in curricula of graduate schools of library and information science. Courses in "indexing and abstracting" are offered by some schools. However, it is often the case that the "indexing" part of the course does not cover book indexing in depth.

With only a few notable exceptions, most training courses are unaccredited. Some are offered through professional associations, while others are offered through continuing/adult education programs of colleges and universities. The ASI maintains a list of courses and workshops at http://www.asindexing.org/site/courses.shtml.

It is natural to wonder if training in indexing is necessary. But given the number of poorly written book indexes it is obvious that people are not born with an innate ability to index! There are two main benefits of book indexing coursework: students find out if they like the nature of indexing work and if they are able to do it well. Liking the work and doing it well are prerequisites for a successful career in book indexing. Since most book indexers work in a freelance capacity, it is also necessary that prospective indexers enjoy running their own small business. Some courses lightly cover the business aspects of book indexing.

Authors who must index their own books are often referred by their publishers to *The Chicago Manual of Style*. Unfortunately many writers find the discussion of indexing in *The Chicago Manual of Style* difficult. Another option is to contact the ASI and find a professional indexer.

PROFESSIONAL ASSOCIATIONS

Founded in 1957, the Society of Indexers (SI) in England is the original professional association for indexers. Over the years other indexing societies have affiliated with SI. Since 1958, SI has published *The Indexer: The International Journal of Indexing* (http://www.theindexer.org/). Below, the affiliated and nonaffiliated societies are listed in alphabetical order along with the current URL for their Web sites. Many of the Web sites provide a great amount of supplemental information about indexing.

ASI: http://www.asindexing.org/
Association of Southern African Indexers and Bibliographers (ASAIB): http://www.asaib.org.za/
Australian and New Zealand Society of Indexers (ANZSI): http://www.aussi.org/
China Society of Indexers (CSI): http://www.chindex.fudan.edu.cn/
Deutsches Netzwerk der Indexer (DNI): http://www.d-indexer.org/
Indexing Society of Canada/Société canadienne d'indexation (ISC/SCI): http://www.indexers.ca/
Nederlands Indexers Netwerk (NIN): http://www.indexers.nl
SI: http://www.indexers.org.uk/

SUMMARY

A back-of-the-book index is a human-authored text that provides a structured sequence of access points to the information contained in the book. A book, whether it is printed on paper and bound, or presented as pixels on a screen, is a closed system. The closed-system index is not constrained by information that may or may not be added in the future. Instead, the closed-system index focuses on a contained universe of information that is static. The very nature of the book, a human-authored text, demands a hand-crafted, human-authored index that is unique to each book.

REFERENCES

1. Klement, S. Open-system versus closed-system indexing: a vital distinction. The Indexer **2002**, *23*(1), 23–24.

Artificial–Back

2. Wellisch, H. The art of indexing and some fallacies of its automation. LOGOS **1992**, *3*(2), 70.
3. Weinberg, B.H. The earliest Hebrew citation indexes. J. Am. Soc. Inform. Sci. **1997**, *48*(4), 318–330.
4. Bell, H.B. *From Flock Beds to Professionalism: A History of Indexer-makers*; Oak Knoll Press: New Castle, DE, 2008.
5. Anderson, J.D. National Information Standards Organization. *Guidelines for Indexes and Related Information Retrieval Devices*; NISO Press: Bethesda, MD, 1997; NISO Technical Report 2.
6. International Organization for Standardization, *ISO 999: Information and Documentation; guidelines for the Content, Organization, and Presentation of Indexes*, International Standards Organization: Geneva, Switzerland, 1996.
7. *British Standard Recommendations for Preparing Indexes to Books, Periodicals, and Other Documents, BS 3700: 1988*, British Standards Institute: London, 1988.
8. http://www.asindexing.org/site/WilsonAward. shtml#awcrit (accessed July 2007).
9. Mulvany, N.C. *Indexing Books*; University of Chicago Press: Chicago, IL, 2005; 8.
10. Wellisch, H. *NISO TR03: Guidelines for Alphabetical Arrangement of Letters and Sorting of Numerals and Other Symbols*; NISO Press: Bethesda, MD, 1999; NISO Technical Report 3.
11. Mulvany, N.C. *Indexing Books*; University of Chicago Press: Chicago, IL, 2005.
12. *The Chicago Manual of Style*, University of Chicago Press: Chicago, IL, 2003.
13. Ritter, R.M.; Hart, H.; Oxford University Press. In *The Oxford Guide to Style*; Oxford University Press: Oxford and New York, 2002.
14. Mulvany, N.C. *Indexing Books*; University of Chicago Press: Chicago, IL, 2005; 245–249.
15. Bell, H.K. *Indexing Biographies and Other Stories of Human Lives*, 3rd Ed.; Society of Indexers: Sheffield, U.K., 2004; Vol. 1, 12 (Occasional Papers on Indexing).

BIBLIOGRAPHY

1. Booth, P.F. *Indexing: The Manual of Good Practice*; K.G. Saur: Munich, Germany, 2001.
2. Browne, G.; Jermey, J. *The Indexing Companion*; Cambridge University Press: Cambridge, U.K., 2007.
3. Fetters, L.K. *Handbook of Indexing Techniques: A Guide for Beginning Indexers*, 3rd Ed.; Fetters Infomanagement Co.: Corpus Christi, TX, 2001.
4. Mulvany, N.C. The human-written index: why it will survive. LOGOS. **2004**, *15*(2), 76–81.
5. Stauber, D.M. *Facing the Text: Content and Structure in Book Indexing*; Cedar Row Press: Eugene, OR, 2004.
6. Wellish, H.H. *Indexing from A to Z*, 2nd Ed.; H.W. Wilson: New York, 1995.

Artificial–Back

Bibliographic Control *[ELIS Classic]*

Robert L. Maxwell
Special Collections and Metadata Catalog Department, Brigham Young University, Provo, Utah, U.S.A.

Abstract

Bibliographic control is the process of creation, exchange, preservation, and use of data about information resources. Formal bibliographic control has been practiced for millennia, but modern techniques began to be developed and implemented in the nineteenth and twentieth centuries. A series of cataloging codes characterized this period. These codes governed the creation of library catalogs, first in book form, then on cards, and finally in electronic formats, including MAchine-Readable Cataloging (MARC). The period was also characterized by the rise of shared cataloging programs, allowing the development of resource-saving copy cataloging procedures. Such programs were assisted by the development of cataloging networks such as OCLC and RLG. The twentieth century saw progress in the theory of bibliographic control, including the 1961 Paris Principles, culminating with the early twenty-first century Statement of International Cataloguing Principles and IFLA's *Functional Requirements for Bibliographic Records* (FRBR). Toward the end of the period bibliographic control began to be applied to newly invented electronic media, as "metadata." Trends point toward continued development of collaborative and international approaches to bibliographic control.

Bibliographic control is the process of creation, exchange, preservation, and use of data about information resources. These resources may be of any type: ink on paper (e.g., books, scores, maps); recordings (e.g., music, spoken word); visual media such as photographs, films, paintings; three-dimensional objects such as sculptures or coins; electronic media, including digital reproductions of any of the previously mentioned media and original digital productions; and any other medium in existence or not yet invented, covering the whole array of intellectual creation. Such resources appear in various publication patterns—monograph, serial, or hybrids (currently known as integrating resources), and may be published or unpublished. Techniques of bibliographic control may also be extended to resources that are not the product of intellectual creation such as inanimate objects (rock specimens, chemicals), living creatures (plants and animals), or created objects that are not traditionally thought of as being subject to bibliographic control, such as clothing, buildings, or food products, indeed anything that an individual or institution might want to collect or keep track of.

Individuals and institutions wish to control these resources for a number of reasons. Collectors may want to ensure that they do not collect more than one copy of a given resource, or they may wish to learn of resources within the scope of their collection that they do not yet have. Institutions such as libraries have an interest in helping their users find materials already within their collection or that they may have access to. Publishers need to know who their authors are and at what stage their projects are, and when resources are completed they need to keep

track of their availability for sale. Retailers need to know whether a resource is available in their own inventory and at what price, or whether they can order a copy for a customer. Libraries and collectors may need to keep track of treatment needed periodically in their collections (such as conservation of deteriorating books, or "refreshing" digital collections as new platforms and systems arise). Tools created during the process of bibliographic control are necessary to fulfill all of these purposes.

There are many ways to control such resources. Many individuals practice a form of bibliographic control in their home library by a simple shelf arrangement method, perhaps grouping books by genre or author, or arranging their CD collection by composer. This allows them to easily find a wanted resource, or to verify that they already own (or do not own) a copy. This type of bibliographic control requires no record-keeping at all, except in the sense that the collection itself is the record. However, as their collection grows, individuals may begin keeping inventory lists, especially if the collection is too large to scan visually or if they intend to dispose of the collection in some way and want to record what they had.

The earliest libraries and archives appear to have used shelf arrangement systems to control their holdings. Scrolls or tablets may have been arranged on shelves in a fixed order logical to the creator of the archives, with the objects positioned so that a label or text was visible. A description of early Greek (Minoan-Mycenaean) filing systems is found in Staikos.[1] Once collections got beyond room size, however, shelving arrangement was not enough, so lists and catalogs were created. For a useful

Encyclopedia of Library and Information Sciences, Fourth Edition DOI: 10.1081/E-ELIS4-120043092

Bibliographic–Bibliothèque

summary of early catalogs, see Carpenter.[2] The most famous of these is *The Pinakes* of Callimachus, one of the keepers of the Library of Alexandria, a collection that numbered in the hundreds of thousands. Konrad Vössing estimates that the library contained 500,000 scrolls during the third century B.C.[3] Ancient estimates went as high as 700,000. Callimachus's 120-volume work was a bibliography of all of Greek literature, based on the holdings of the Library, of which it was in effect a catalog. It was arranged by broad subject, and identified individual resources in the library by citing the first line of each work and the number of its lines. The combination of first line of a scroll with its number of lines was an important identification device in early manuscripts, since they often lacked titles.[4] It is unknown whether *The Pinakes*, which have not survived, had directional aids such as location listings, but the work presumably had some use as a finding aid and possibly for inventory control.

Inventory lists such as that of Callimachus seem to have been the rule through the Middle Ages. These lists often followed the pattern of *Pinakes* in dividing collections into broad categories, frequently subdivided by size (folio, quarto, octavo). As these lists grew in size they took the form of books. Among the earliest of these book catalogs were the seventeenth century catalogs of the Bodleian Library, some of which were arranged alphabetically.[5]

The idea that the construction of tools for bibliographic control should follow explicitly formulated rules did not really take hold until the mid-nineteenth century. Antonio Panizzi's famous (and at the time controversial) 91 rules for the compilation of the British Museum's catalog was one of the first such attempts. These rules were printed before the text of the catalog itself, so that users of the catalog could understand its arrangement and manner of construction.[6] Other single-author codes followed, notably those of Jewett[7] and Cutter.[8]

These and other codes were followed (or not) based on their merits, but before the end of the nineteenth century no particular need was felt for agreement between libraries on cataloging standards. As long as a library's catalog was consistent within itself, for purposes of bibliographic control it mattered little if its compilers followed different rules from those of other libraries. However, a new development caused a hasty move toward agreement on standards: the invention of the card catalog and the availability of catalog cards for sale.

Until the mid-nineteenth century catalogs were created in book form. This caused obvious problems when libraries expanded: in a growing library, the day after the book catalog came off the press it was already out of date. Blank sheets could be interleaved within a book catalog for manuscript additions, but eventually these filled up and a new edition had to be published at great expense. Indeed, the impetus for Panizzi's formulation of his 91 rules was the need to publish a new edition of the British Museum's book catalog. An ingenious solution to this problem was the invention of the card catalog. New cards, each containing bibliographic data for a single resource, could be interfiled with the old, allowing the catalog to expand indefinitely without the need for new editions or republication.

The publication of book catalogs did not stop after the adoption of card catalogs, because they had one great advantage over card catalogs. In most cases the card catalog existed in a single instance only, usually a prominent feature of the library's reference room. Book catalogs were published in multiple copies, in some cases until the end of the twentieth century, and those of prominent libraries such as the Library of Congress, the British Museum, the Bibliothèque Nationale, or the New York Public Library were purchased by institutions around the world. Knowledge of the holdings of libraries other than one's own is clearly important to bibliographic control, allowing collectors to learn of resources outside their own institutions and allowing institutions to borrow from one another.

Such needs prompted the publication of printed union catalogs, the most prominent of which was the National Union Catalog (NUC), a compilation that attempted to include a reproduction of a catalog card for every book held in the major libraries of the United States and Canada, together with a list of all libraries that held each book. The main set, which included printed works published before 1956, ran to 754 folio-sized volumes (published 1968–1981), and was followed by several supplements. This enormous work is still not entirely superseded even in an age of massive bibliographic databases. A recent study discovered that more than 25% of the works represented in the NUC had not yet been recorded in OCLC.[9] The figure has undoubtedly shrunk somewhat since the publication of Beall and Kafadar[9] because of the merger of the RLIN database with OCLC.

A logical extension of the invention of the card catalog was the concept of sharing cataloging. If new cards created by a library's catalogers could easily be added to its catalog, so too could they be copied and added to another library's catalog. In 1898 the Library of Congress began to print its catalog cards, and began offering them for sale to other libraries in 1901. Purchasing ready-made catalog cards rather than creating them from scratch for some of the books in a library clearly made economic sense, but in order to be usable these cards had to be compatible with cards already in the library's catalog. This was particularly important in the matter of headings—resources in alphabetic (or "dictionary") card catalogs needed to group together under authors, titles, and subjects, but if different standards were followed to work out these headings, chaos could result. To a lesser extent standards of description needed to be compatible between cards as well, if for no other reason than to avoid confusing catalog users. So in order to benefit from the economic advantages of shared cataloging, shared standards needed to be agreed on.

So it is not entirely coincidental that the first of these shared codes appeared within a decade of the start of LC's card program, in 1908. As a beginning of standardization within the English-speaking world, the American Library Association and the (British) Library Association attempted to work out a set of mutually agreed-upon rules. This was unsuccessful, and two editions of the code were published.[10] Nevertheless, agreement between libraries within a single nation was a significant step forward, and the two codes were similar. It is probably also significant that these codes dealt primarily with the formation of entries. As already noted, agreement on the form of headings is probably more important in a shared environment than agreement on descriptive conventions. Indeed, it was not until the publication of the first edition of the *Anglo-American Cataloging Rules* (AACR) that rules for entry and description appeared together for the English-speaking library community within a single code.[11] As with the 1908 rules, AACR1 unfortunately was published in two editions.

AACR followed an important international agreement concluded in 1961 by the International Federation of Library Associations and Institutions (IFLA). Known as the Paris Principles, the agreement has become the foundation of subsequent standards-making work in the cataloging world. The definitive text of the Paris Principles is International Conference on Cataloguing Principles, Paris, 1961, *Statement of Principles*, annotated edition with commentary and examples by Eva Verona (London: IFLA Committee on Cataloguing, 1971). The principles began with a statement of the function of the catalog, which is also a good summary of some of the principal purposes of bibliographic control:

2. Functions of the Catalog
    The catalog should be an efficient instrument for ascertaining
2.1 whether the library contains a particular book specified by
    a. its author or title, *or*
    b. if the author is not named in the book, its title alone, *or*
    c. if the author and title are inappropriate or insufficient for identification, a suitable substitute for the title; *and*
2.2 a. which works by a particular author *and*
    b. which editions of a particular work are in the library.

The Principles established the importance of such basic concepts as uniform headings, personal and corporate authorship, and the function of main and added entries.

At the turn of the millennium the international cataloging community has created a new document intended as an updated version of the Paris Principles. The Statement of International Cataloguing Principles was developed by the IFLA Meeting of Experts on an International Cataloguing

Code during five annual meetings held in Frankfurt, Germany (2003), Buenos Aires, Argentina (2004), Cairo, Egypt (2005), Seoul, Korea (2006), and Pretoria, South Africa (2007). The final text is available in several languages at http://www.ifla.org/en/publications/statement-of-international-cataloguing-principles. It is expected that this document will become the basis for future cataloging codes throughout the world.

The Statement recites a number of objectives for the catalog, basic principles of bibliographic control that have become current in the years since the Paris Principles were promulgated. These are

*   *Convenience of the user* of the catalog: cataloging decisions should be made with the user in mind.
*   *Common usage*: standardized vocabularies should reflect the vocabulary the majority of catalog users would expect.
*   *Representation*: descriptions and controlled forms of names should be based on the way entities describe themselves.
*   *Accuracy*: the description should be faithful to the entity described.
*   *Sufficiency and necessity*: only those elements that are essential to uniquely identify an entity should be included.
*   *Significance*: elements should be bibliographically significant.
*   *Economy*: bibliographic control should be achieved by the simplest or least costly method.
*   *Consistency and standardization*: descriptions and access points should be standardized as far as possible in the interest of consistency and sharing of records.
*   *Integration*: description of and access to all types of materials should be based on a common set of rules as much as possible.
*   The rules in a cataloging code should be *defensible* and *not arbitrary*.

The formulation of the Paris Principles and the first edition of AACR in the 1960s coincided almost exactly with another significant development for bibliographic control, the use of computers as an aid to cataloging. The MAchine-Readable Cataloging (MARC) standards were first developed at the Library of Congress in the mid-1960s under the leadership of Henriette Avram. A readable introduction to the MARC format in the United States is *Understanding MARC Bibliographic: Machine-Readable Cataloging*, 7th ed.; Library of Congress: Washington, DC, 2003, also available at http://www.loc.gov/marc/umb/. There remain numerous national MARC standards but the English-speaking world generally conforms to a standard currently called MARC21. First intended as an automated method for printing catalog cards, MARC's potential for exchange, retrieval, and display of records soon became evident, and MARC became the foundation of electronic catalog databases. Just as significantly for bibliographic control, the MARC format allowed the creation of electronic union catalog databases (such as Research Libraries Information Network [RLIN] and OCLC, discussed below), which combined the advantages of card catalogs (easy extensibility and revisability) and printed book catalogs (access beyond the premises of the library) and eventually led to the near demise of both.

Another important bibliographic control standard, the International Standard Book Number (ISBN), was initiated in 1968 in Great Britain as the "Standard Book Number," a nine-digit number assigned to individual editions of a publisher's output. Within the same year the concept was extended worldwide to become the ISBN, the number expanding to 10 digits. In 2007 the number expanded again, to 13 digits. Since 1970 the ISBN system has been overseen by the International ISBN Agency in Berlin. Blocks of numbers are assigned to publishers by national agencies within the 166 member countries. The website of the International ISBN Agency is http://www.isbn-international.org/.

ISBN as an identifier of a discrete edition of a book is of obvious significance to bibliographic control, and the concept quickly expanded to other formats. There are now standard numbers for serials (ISSN) (administered by the ISSN International Centre, http://www.issn.org/), music (ISMN) (administered by the International ISMN Agency, http://www.ismn-international.org/), audiovisual materials (ISAN) (administered by the ISAN International Agency, http://www.isan.org/), recordings (ISRC, International Standard Recording Code) (administered by the International Federation of the Phonographic Industry (IFPI), http://www.ifpi.org/content/section_resources/isrc.html), and others. Any of these can be recorded within the MARC record for the appropriate resource as an identifier, a link, or finding device.

The ability to create electronic catalog records using the MARC format was as significant a leap forward for sharing cataloging as the Library of Congress's sale of its printed cards, and it allowed a much wider community of cataloging participants. Although LC was not the only library producing such cards, an institution had to be relatively large and respected in order to afford to publish, and to expect a market for, its catalog cards. On the other hand, the only thing any library needed under the MARC regime to produce catalog records of potential worth to other libraries was a computer system capable of deploying MARC and a way to distribute the records. Further, before the advent of MARC it was cumbersome, if not impossible, to revise cataloging a library might receive on cards, e.g., to note local characteristics (such as inscriptions, unique bindings, or defects in the library's copy). In an electronic environment it was not a difficult matter to enhance "copy" (the electronic catalog record a library might receive from another cataloging agency) not only for local characteristics, but also to enhance and correct the records for other

reasons. In a shared cataloging environment this benefited all who came after (and who, themselves, could continue to enhance a given record).

At about the same time as these developments the Library of Congress initiated a new program to enhance the ability of libraries in the United States and elsewhere to perform bibliographic control, the Cataloging in Publication (CIP) program. LC had conducted an experiment called "cataloging in source" in 1958 and 1959 under which its staff cataloged approximately 1000 books prior to their publication from page proofs so that cataloging information could be printed on the verso of title pages. Although the experiment showed that this could be done successfully, the Library ultimately dropped the project, feeling that the disadvantages outweighed the benefits, both to itself and to the publishers. However, the concept was revived in 1971 with the establishment of the CIP program. A good account of the beginnings of the program is found in Clapp.[12] The program's current home page is http://cip.loc.gov/. The purpose of the program was to "provide the recipient with standardized and professionally prepared cataloguing information that he cannot generate himself, but not to supply data which are readily available to him in the book which he holds."[13] Consequently, LC's CIP would not include subtitles, imprint information, or physical description. However, CIP data would include main entry in its authorized form, a brief identifying title, series statements, notes about the resource, LC subject headings, added entries in their authorized forms, LC call number, a Dewey class number, the LC card number, and any ISBNs associated with the resource. These cataloging data would be supplied based on galley proofs, and a CIP record would be returned to the publisher to reproduce on the verso of the title page of the final publication.

CIP instantly became a very popular service and is now included in the publications of most major American publishers. It was also an excellent match for the newly developed MARC record exchange format: a record might potentially become available to libraries before the resource was even published, and was easily revised when the resource became available in its final form. CIP programs now exist around the world, involving many national libraries including the British Library, Library and Archives Canada, the National Library of Australia, and the Deutsche Nationalbibliothek. LC's CIP program has recently migrated to an "eCIP" program ("electronic CIP") under which rather than paper galleys publishers send the Library electronic versions of the manuscript. This has not only resulted in faster and more efficient processing, but it has allowed the Library to consider expanding its partners. It has recently invited major university libraries to participate by producing eCIP for the output of their university presses and has also expanded the program to allow libraries with subject or other specialties to produce eCIP for specialist publishers.

The advent of electronic cataloging did not spell the immediate end of card catalogs for bibliographic control: some libraries continued to use their card catalogs well into the twenty-first century. However, most libraries in the Western world had by then retrospectively converted their cards to electronic format so that the bulk of their inventory was controlled by their Online Public Access Catalog (OPAC). But the idea of sharing electronic cataloging did give rise to the idea of Universal Bibliographic Control (UBC).

UBC was an ideal under which all the world's information resources, whether published or unpublished, would be placed under bibliographic control, i.e., cataloged.[14] Two approaches were proposed: the creation of a single worldwide bibliography, or alternately (and more realistically), the creation of national bibliographies. The promotion of UBC occupied much of the attention of the cataloging world during the second half of the twentieth century. IFLA spearheaded the movement with its IFLA Universal Bibliographic Control and International MARC Programme (later "Core Activity") (UBCIM). For background on the now-defunct IFLA UBCIM, see documentation at the IFLA website, http://archive.ifla.org/VI/3/ubcim.htm and http://archive.ifla.org/VI/3/ubcim-archive.htm. During the last 30 years of the century, UBCIM promoted not only the idea of UBC, but was also responsible for the publication of the ISBDs and UNIMARC, an IFLA-sponsored version of the MARC standard developed by the Library of Congress.

The International Standard Bibliographic Descriptions (ISBDs) were first envisaged in the late 1960s to allow a standardized form of content and exchange of bibliographic data. The value of and need for such documentation in the context of UBC is clear and the ISBDs have enjoyed a remarkably broad worldwide acceptance. The first ISBD (for monographs) was published in 1971, and other format-specific ISBDs followed in the ensuing decades. A consolidated version was published in 2007, harmonizing the format-specific ISBDs. A digital version of the consolidated edition of ISBD (Munich: Saur, 2007) is available at http://www.ifla.org/files/cataloguing/isbd/isbd-cons_2007-en.pdf. Nearly all cataloging codes developed since the publication of the ISBDs, including the second edition of AACR, have been based on them.

With the advent of electronic and digital formats and especially the rise of the Internet and the World Wide Web, a serious increase in the amount of information resources occurred, and as of the early twenty-first century, thanks to this rapid increase of these resources, many of the goals of UBC appeared unattainable. The UBCIM Core Activity was disbanded by IFLA in 2003, replaced by the IFLA-CDNL Alliance for Bibliographic Standards (ICABS), a federation of a number of national libraries charged with continuing to support goals of bibliographic control, but with less emphasis on the ideal of UBC. For information about ICABS, see http://archive.ifla.org/VI/7/icabs.htm.

Given the demise of an explicit goal of UBC on the international level, it is somewhat ironic that the notion of a single worldwide bibliography may actually be coming to fruition through the rise of OCLC to near-monopoly status as the database containing the world's cataloging data. As mentioned above, one of the obvious benefits of the invention of MARC was the ability to share and exchange cataloging electronically. However, in order to do this efficiently some exchange mechanism had to be set up. During the late 1960s and early 1970s numerous library consortia organized themselves explicitly for the purpose of creating electronic union catalog databases. This would benefit libraries beyond allowing the sharing of records and thus lowering the cost of cataloging. Such union catalogs were also efficient mechanisms for purposes such as interlibrary loan, allowing the discovery of the location of wanted items. In addition they allowed libraries to see each other in ways they had not been able to before, particularly for comparison of their collections.

Three of the principal consortia developed in the United States during this period were The Ohio College Library Center (OCLC), The Research Libraries Group (RLG), and The Washington Library Network (WLN).

The Washington Library Network was founded in 1976 as a regional union database for libraries in the state of Washington. In view of its expansion to areas of the North American west coast and northwestern Rocky Mountains it changed its name to Western Library Network in 1985, and then to WLN in 1990. Although the focus of WLN was the northwestern United States, it did support libraries from outside the region in Australia, New Zealand, and South Africa. In 1999 WLN merged with OCLC.

RLG was founded in 1974 by Columbia, Harvard, and Yale Universities and the New York Public Library. Its membership expanded across the United States to many other university and research libraries, and its union catalog became known as RLIN. Throughout its history RLG retained its character as a consortium of research libraries, and so its catalog reflected the character and needs of its members, including strong holdings in special and rare materials, and the ability to catalog using vernacular scripts such as Chinese, Japanese, Korean, and Arabic.

RLIN's structure was based on a "clustering" concept. Every member library input its own records for its own holdings (usually "deriving" a new record from some other library's record for a given resource), but records for a single edition of a resource clustered together through a set of complex algorithms. The clustered structure of the database allowed libraries to see the records of any other RLG member, which might differ from one another.

This was a particularly suitable arrangement for research libraries, which often had significant special collections holdings. Patrons of these collections were often interested in researching unique aspects of the collections, such as books that once belonged to a prominent historical figure, information that could be found on the records of individual libraries.

Aside from RLIN, RLG also developed a number of tools for archival bibliographic control, allowing better access to the holdings of the research libraries' manuscript collections, as well as, later on, tools for union access to members' new digital collections. In 2006 however RLG, like WLN, merged with OCLC. It retains a presence within OCLC as RLG Programs.

Although not the only bibliographic database on the stage today, with 100,000,000 bibliographic records (and counting) OCLC is clearly now the principal source of the world's bibliographic data. OCLC's home page is http://www.oclc.org/. A free version of the bibliographic database is available at http://worldcat.org/. OCLC was founded in 1967 as the Ohio College Library Center. Like WLN, it was originally the database of a regional library consortium. As it grew it changed its name to Online Computer Library Center, and is now officially known simply as OCLC. OCLC had a different character from RLG, attracting more public and school libraries, although a number of university and research libraries also joined (some were members of both OCLC and RLG). The structure of its database was different as well. Rather than the clustering model followed in RLIN, OCLC had a "master record" model wherein each edition of any resource would have one and only one record. Libraries added a holdings record to the master record to notify database users that they owned an item, and a select group of libraries was allowed to revise master records.

OCLC's merger with RLG precipitated a change in OCLC's database structure. RLG members were accustomed to using the clustered organization of the database and many wanted to continue to receive the benefits of being able to see each others' records after they moved to the OCLC environment. In order to accommodate this, OCLC changed its database structure to allow for "institution records." Under this structure the master record remains the principal record for a given edition of a resource, but a library may, at its discretion, link its own record (the "institution record") to the master record. When the RLIN and OCLC databases merged in 2007 a number of RLG member libraries had their RLIN records added to the OCLC database as institution records, and the number of such records has continued to grow.

In addition to maintaining its mammoth bibliographic database, OCLC is also a significant player in research into various aspects of bibliographic control through its Programs and Research division, supporting undertakings such as analysis of and data mining within the OCLC database, research into automatic classification, projects developing open and persistent URLs, and the development of a "virtual international authority file," a project that is exploring the feasibility of combining the name authority files of the Library of Congress, the Deutsche Nationalbibliothek, and the Bibliothèque Nationale de France.

There have been a number of recent developments in bibliographic control of potentially great importance to the future. The Statement of International Cataloguing Principles has been noted above. This statement was based in part on two new documents, Functional Requirements for Bibliographic Records (FRBR) and its companion document Functional Requirements for Authority Data (FRAD). FRBR is available at http://archive.ifla.org/VII/s13/frbr/frbr current toc.htm (HTML version) or http://www.ifla.org/files/cataloguing/frbr/frbr_2008.pdf (PDF version). The final version of FRAD, when published, will be available through the IFLA Cataloguing Section's home page http://www.ifla.org/en/cataloguing.

FRBR is not a cataloging code, but instead a conceptual model of the bibliographic universe. Based on the entity-relationship database model, FRBR works out how bibliographic entities interact with each other. Entity-relationship is a model introduced in the 1970s in which a given data universe is divided into specific entities linked by specific relationships, each of which have specific attributes. In FRBR and FRAD 11 entities are defined: work, expression, manifestation, item, person, corporate body, family, concept, object, event, and place.

The first four of these are called the primary entities because they have to do with resources familiar to bibliographic control, the objects of intellectual creation. "Work" is defined as "a distinct intellectual or artistic creation." When a work is set down in some form it becomes an "expression": "the intellectual or artistic realization of a work in the form of alphanumeric, musical, or choreographic notation, sound, image, object, movement, etc. or any combination of such forms." If the content or form of an expression changes (as when a writer revises a novel, or when the novel is translated into another language) this usually results in a new expression of the same work, or if the change is extensive enough, it can result in a new work. "Manifestation" is "the physical embodiment of an expression of a work." This usually means a publishing event, such as the 1936 publication of Margaret Mitchell's *Gone with the Wind*. If the same content is published again, for example, the 2006 publication of *Gone with the Wind*, this becomes a second manifestation of the same expression. "Item" is a single instance of a manifestation, i.e., a copy.

All FRBR entities have explicit relationships with each other. For example, the work entity might have a "created by" relationship with a person entity (the author). The manifestation entity might have a "produced by" relationship with a corporate body entity (the publisher). A person entity might have a "member of" relationship with a family entity.

FRBR also defines what it calls the "user tasks," i.e., the reasons users come to bibliographic databases.

According to FRBR these are four: 1) users need to *find* materials relevant to their needs; 2) once they have found a group of potential records or resources they need to *identify* the resource retrieved, i.e., confirm that the resource is what they were looking for; 3) they then need to *select* from the resulting group a resource appropriate to their needs; and 4) they need to *obtain* the selected resource.

The focus of FRBR on these user tasks is based on the principle of user convenience, one of the core principles underlying the Statement of International Cataloguing Principles discussed above.

FRBR works best within the entity-relationship database model. In such a database every instance of an entity would have one and only one record, and all entity records would be linked to other entity records by relationship links. For example, the work record for *Gone with the Wind* would be linked to the person record for Margaret Mitchell; the expression records for all expressions of the work (text, spoken word, etc.); concept entity records for the topics of the work (e.g., Scarlett O'Hara, the U.S. Civil War) and forms (e.g., Historical fiction, War stories); and so on. In turn, the person record for Margaret Mitchell would be linked to any other work records for works she created (e. g., *Lost Layson*); the concept entity records would be linked to any other related works (e.g., the concept record for Scarlett O'Hara would be linked to the work record for the film version of *Gone with the Wind* and other works in which she figures prominently such as *My Beloved Tara* by Jocelyn Mims). The expression records would be linked to their appropriate manifestation records, which would be linked in individual libraries to item records.

This exponentially expanding web of linked entity records is very different from the flat-file MARC universe, in which all details pertaining to a single publishing event (the work, the expression, the manifestation, the author, the concepts, etc.) are contained within in a single record. At first glance the complexity of entity-relationship caused many to doubt the utility of such a massive change, but there are overriding advantages both to the creator of the database (e.g., the cataloger) and to the users of the collection.

In the current MARC environment, every time a new record is created for one of the hundreds of publications of *Gone with the Wind*, including translations, revised editions, etc., everything to do with the work has to be repeated in every record. This includes information about the author (including the required standardized form of her name), subject headings, summaries, notes about awards, and anything else pertaining to the work. Ideally all this information should be identical in each MARC record, and any changes to one record (e.g., the addition of a new subject heading or a change in form of the author's name) should be made to all. This is a lot of work. In contrast, in an entity-relationship bibliographic database all this information would either be contained within the work record as attributes (e.g., a plot summary, notes about awards) or

linked to the work record by relationship links (e.g., the author's person entity record, the concept entity records). Any changes would be made only once, at the work level.

From the point of view of the user, too, there are important advantages to such a database structure. In the most common type of interfacing with a MARC database currently used, keyword searching, the user might receive hundreds of hits organized randomly when doing a search on *Gone with the Wind*. In a well-designed entity-relationship database the user might instead initially be presented only with work records, in this case possibly only two, one for the novel and the other for the film. From there the user could choose an expression (e.g., a recording, a French translation, or the original English-language expression). The user would then only see the few manifestation records associated with the chosen expression.

The first cataloging code to be explicitly based on FRBR and FRAD is *RDA: Resource Description and Access*, currently under development (see http://www. rdaonline.org/). Designed by the Joint Steering Committee for Development of RDA, it is intended as a replacement for AACR2, and is expected to be released in early 2010. As of this writing (2009) the structure of the code points toward its use in an entity-relationship database, containing instructions for recording attributes of each FRBR/FRAD entity as separate records, and instructions for linking these entities with relationship links. Although it is realized that RDA will first be implemented within a MARC environment, it is clear that its authors intend it to be used eventually within a FRBR-based entity-relationship environment. Although the database structure for this environment has not been designed yet, models within the larger database world exist, which should help development.

One of the driving forces behind the replacement of AACR2 with RDA is the changed environment in which libraries and other similar organizations find themselves at the beginning of the twenty-first century. One of the principal changes to this environment is the proliferation and wide availability of digital resources, both at individual institutions such as libraries and within the larger environment, including the World Wide Web. These digital resources are potentially subject to bibliographic control for the same reasons as physical resources are. And their proliferation has the potential to overwhelm traditional methods of bibliographic control. RDA is intended to address some of these problems, and to attract other communities, including the metadata communities, to use a common standard.

Metadata, a term coined about 1970, means data or information about other data. As such, all the techniques of bibliographic control discussed above produce metadata, since catalog records of all kinds contain data describing other data, i.e., works as embodied in a physical object such as a book. However, the common meaning of the term is data or information about digital objects.

Unlike traditional cataloging, which until recently exclusively described physical objects, and described them in a place physically separate from the objects themselves (the catalog), metadata is often attached directly to the digital object being described. This is possible because electronic metadata, itself digital, resides in the same place (e.g., the Internet, or a database) as the digital objects it describes.

This also makes possible the creation of metadata not only by professional catalogers (as was previous practice), but also by the creators of the objects themselves or their users. Indeed, the huge quantities of digital information resources demand at least some metadata creation by the creators and users of the resources. The nature of digital resources also opens the possibility of automatic harvesting of portions of the metadata from the object itself, with minimal human intervention. Because of all this a need was perceived for a metadata standard that was simpler than traditional cataloging standards. A workshop was hosted by OCLC in 1995 that gave rise to the "Dublin Core" (OCLC is located in Dublin, Ohio), a basic set of 15 elements that could be used for metadata creation. These included elements such as title, creator, subject, language, and date, which were easily understandable by anyone creating a digital object.

The Dublin Core is maintained by the Dublin Core Metadata Initiative (DCMI), which continues to refine the model and develop database architecture and modeling. DCMI's home page is http://dublincore.org/. One of its activities is a joint initiative with the creators of RDA. The DCMI/RDA Task Group, formed in April 2007, is charged with defining an RDA element vocabulary in order to make RDA's underlying bibliographic elements reusable in the "Semantic Web" (defined as a common framework that allows data to be shared and reused across applications), and citable with Uniform Resource Identifiers (URIs). Information about the Semantic Web is available at http://www.w3.org/2001/sw/. One of the hoped-for results of the DCMI/RDA Task Group is greater cooperation between the metadata and cataloging communities and broader use of RDA beyond traditional cataloging, including by the metadata communities. See information about the Task Group at http://dublincore.org/dcmirdataskgroup/.

All of these standards have been developed to facilitate the sharing of bibliographic data, so that data creators such as catalogers do not have to redo work that has already been done by others. Catalogers have long tried to organize themselves so that they could share cataloging, beginning with LC's catalog card sales and progressing to the use of each other's electronic copy through databases such as OCLC. However, the supply of reliable copy has not always been good, and a number of initiatives were attempted during the twentieth century to form alliances among cataloging agencies that followed a common standard and whose records could be used by others with little or no modification.

The most successful of these efforts was the Program for Cooperative Cataloging (PCC). Founded in 1995, the PCC built on successful existing programs for the creation of serial and name authority records. At the present time the PCC has four components: a program for the creation of name authority records called NACO; a program for the creation of subject authority records called SACO; a program for the creation of monographic bibliographic records called BIBCO; and a program for the creation of serial bibliographic records called CONSER. For information about the PCC, see http://www.loc.gov/catdir/pcc/. Participants in any of the PCC programs are given extensive training and their work may be subject to quality control by other PCC participants, ensuring a supply of high-quality cataloging data that others can use with confidence. Although much of the PCC's activity is hosted and supported by the Library of Congress, the Program is governed by its member institutions. In 2007 the 624 PCC member institutions contributed nearly 200,000 name and series authority records, about 3,000 subject authority records, over 22,000 serial records, and about 66,000 bibliographic records to the shared databases.

The beginning of the twenty-first century saw a great deal of thoughtful planning for the future of bibliographic control. In 2000 the Library of Congress sponsored the Bicentennial Conference on Bibliographic Control for the New Millennium, which produced an action plan for the Library. (The Papers are published in *Proceedings of the Bicentennial Conference on Bibliographic Control for the New Millennium* [Washington, DC Library of Congress Cataloging Distribution Service, 2001], also available at http://www.loc.gov/catdir/bibcontrol/conference.html. The action plan is available at http://www.loc.gov/catdir/bibcontrol/actionplan.html.) This was followed by three important and controversial white papers, one on the future of bibliographic services in the University of California system, another on the future of cataloging at Indiana University, and the third a paper on the future of bibliographic control in general, commissioned by the Library of Congress and written by Karen Calhoun. The publication details of these white papers were the following: University of California Libraries, Bibliographic Services Task Force, *Final Report, December 2005: Rethinking How We Provide Bibliographic Services for the University*, http://libraries.universityofcalifornia.edu/sopag/BSTF/Final.pdf; Jackie Byrd et al. *A White Paper on the Future of Cataloging at Indiana University*; January 15, 2006, http://www.iub.edu/~libtserv/pub/Future_of_Cataloging_White_Paper.pdf; and *The Changing Nature of the Catalog and its Integration with Other Discovery Tools* final report, March 17, 2006, http://www.loc.gov/catdir/calhoun-report-final.pdf., respectively. Fast on the heels of these publications, in November 2006, the Library of Congress convened the Working Group on the Future of Bibliographic Control. After broad, nationwide consultation, the Working Group produced its final report in January 2008 (*On the Record : Report of The Library of*

Congress Working Group on the Future of Bibliographic Control, January 9, 2008, http://www.loc.gov/bibliographic-future/news/lcwg-ontherecord-jan08-final.pdf). The Working Group, addressing the bibliographic control community in general and not just the Library of Congress, recommended that

1. The community increase the efficiency of its production through increased cooperation, including use of data from nontraditional sources (including publishers, foreign libraries, etc.).
2. Increase efforts to expose rare and unique materials currently "hidden" because of lack of adequate bibliographic control.
3. Position for the future by recognizing that the World Wide Web is both a technology platform and a platform for the delivery of standards and bibliographic data, and by recognizing that not only people, but also machines will interact with bibliographic data.
4. Incorporate user-supplied information such as social tagging into descriptions and work to realize the potential of the FRBR framework.
5. Strengthen the profession through education and measurements that will inform decision-making.

The Working Group concluded that

> The future of bibliographic control will be collaborative, decentralized, international in scope, and Web-based. Its realization will occur in cooperation with the private sector, and with the active collaboration of library users. Data will be gathered from multiple sources; change will happen quickly; and bibliographic control will be dynamic, not static. (*On the Record*, p. 4.)

At a time of great change in the world of bibliographic control this seems as good a statement as any of what the future might bring.

REFERENCES

1. Staikos, K.Sp. *The History of the Library in Western Civilization: From Minos to Cleopatra*; Oak Knoll Press: New Castle, DE, 2004; 42–44.
2. Carpenter, M. Catalogs and cataloging. In *Encyclopedia of Library History*; Wiegand, W.W., Davis, D.G., Eds.; Garland: New York, 1994; 107–108.
3. *Brill's New Pauly: Encyclopaedia of the Ancient World*, Brill: Leiden, the Netherlands, 2005; Vol. 7, 505.
4. Hornblower, S.; Spawforth, A., Eds. *The Oxford Classical Dictionary*, 3rd Ed.; Oxford University Press: Oxford, U.K., 1996; 277.
5. Carpenter, M. Catalogs and cataloging. In *Encyclopedia of Library History*; Wiegand, W.W., Davis, D.G., Eds.; Garland: New York, 1994; 108–109.
6. Panizzi, A. Rules for the compilation of the catalogue. In *Catalogue of Printed Books in the British Museum*; British Museum, Department of Printed Books: London, U.K., (Printed by Order of the Trustees), 1841; Vol. 1, v–ix; reprinted in *Foundations of Cataloging: A Sourcebook*; Carpenter, M.; Svenonius, E.; Eds., Libraries Unlimited: Littleton, CO, 1985; 3–14.
7. Jewett, C.C. *On the Construction of Catalogues of Libraries*; Smithsonian Institution: Washington, DC, 1852.
8. Cutter, C.A. Rules for a printed dictionary catalogue. In *Public Libraries in the United States of America Their History, Condition, and Management*; Government Printing Office: Washington, DC, 1876. pt. 2.
9. Beall, J.; Kafadar, K. The proportion of NUC Pre-56 titles represented in OCLC WorldCat. Coll. Res. Libr. September **2005**, *66* (5), 431–35.
10. *Catalog Rules: Author and Title Entries*; American Library Association: Chicago, IL, 1908 and *Cataloguing Rules: Author and Title Entries*; Library Association: London, IL, 1908.
11. *Anglo-American Cataloging Rules. North American Text*; American Library Association: Chicago, IL, 1967; *Anglo-American Cataloguing Rules. British Text*; Library Association: London, U.K., 1967.
12. Clapp, V.W. Cataloging in publication. UNESCO Bull. Libr. January–February **1973**, *27* (1), 2–11.
13. Clapp, V.W. Cataloging in Publication. UNESCO Bull. Libr. January–February **1973**, *27* (1), 5.
14. The fundamental work on UBC through the mid-1980s is Dorothy Anderson. Anderson, D. Universal bibliographic control. In *Encyclopedia of Library and Information Sciences*; Dekker: New York, 1984; Vol. 37, Suppl. 2, 366–401.

Bibliographical Society (London)

Julian Roberts
Wolfson College, University of Oxford, Oxford, U.K.

Abstract

This entry sets out the principal aims of the Society, as recorded on its Web site, and traces how these have been pursued and achieved since its foundation in 1892, through its meetings, publications, through its Transactions (*The Library*), and since 1992, the awarding of financial help for bibliographical research. Bibliography has evolved, notably under French and German influence, into a world-wide discipline of the History of the Book, and the Society's leading members have played a major part in such ventures as the as yet uncompleted *Cambridge History of the Book in Britain*.

On its Web site, the Bibliographical Society sets out its objectives:

To promote and encourage study and research in the fields of:

> Historical, analytical, descriptive, and textual bibliography, the history of printing, publishing, bookselling, bookbinding, and collecting; to hold meetings at which papers are read and discussed; to print and publish a journal and books concerned with bibliography; to maintain a bibliographical library; from time to time to award a medal for services to bibliography.

To support bibliographical research by awarding grants and bursaries.

With the exception of the last, none of these objectives would have seemed alien to most members of the Society from its foundation in 1892 to the present time. Funds for bursaries and awards were raised in celebration of the Society's centenary in 1992. The earliest members were also a little unsure of the meaning of the word "bibliography," and there are in early issues of the *Transactions* papers which discussed what would now be considered "subject" bibliography. Two commemorative volumes have been published, on the jubilee and the centenary, 1942 and 1992. Notwithstanding its early concentration on foreign books, mainly of the incunable period and collaboration with the German-based Gesamtkatalog der Wiegendrucke, the Society soon focused on the definitive recording of British books, and upon the textual problems of the Elizabethan and Jacobean drama. The issue of the *Short-Title Catalogue (STC)* in 1926, was a milestone of the former, and W.W. Greg's *Bibliography of the English printed drama to the Restoration*, of the latter. *STC* appeared in a revised edition between 1976 and 1991. The Society also played a major part in the design and completion of the *Eighteenth Century Short-Title Catalogue*. Both the earlier projects involved research in, and publication of, several books of the records of the Stationers' Company of London. The Society's concentration on pre-1640 English books was rudely interrupted by the exposure of Thomas J. Wise (a former President), but the exposure was the work of two future Presidents, John Carter and Graham Pollard, both—particularly Pollard— influential on the future direction of the Society. Ecclesiastical libraries have always been important sources of early books, and the *Cathedral Libraries Catalogue* begun in 1944, and long in gestation, was completed between 1984 and 1998. The editing and publication of such substantial works proved costly for a members' society, and the British Library offered welcome support for this and later works of scholarship, particularly those requiring extensive illustration. Keith Maslen's edition of the Bowyer ledgers was published in collaboration with the Bibliographical Society of America.

The Society's claim, also on the Web site, that it is the "senior learned society dealing with the study of the book and its history" might indeed be contested by the Edinburgh Bibliographical Society which was established in 1890.

The Society's objectives and achievements have been set out in volumes published on the occasions of its jubilee (1942) and its centenary (1992). *The Bibliographical Society, 1892–1942: Studies in Retrospect*, appeared in 1945, and was reprinted in 1949. The three-year delay was, understandably, caused by the conditions of war which had curtailed the Society's activities in many other ways. The book was edited by F.C. (later Sir Frank) Francis (President 1964), who from the relative security of the British Museum, had shouldered many of the Society's burdens—which included the Honorary Secretaryship and Editorship of *The Library*. *The Book Encompassed: Studies in Twentieth-Century Bibliography* was edited by Peter Davison (Honorary Editor of *The Library*, President 1992, and Gold Medallist 2003), and appeared more punctually. Both volumes, which differ widely in their appearance, are yet a continuing witness to the skills of Cambridge

Encyclopedia of Library and Information Sciences, Fourth Edition DOI: 10.1081/E-ELIS3-120044814

Bibliographic–Bibliothèque

University Press. Francis was uniquely qualified to write the historical introduction to the earlier volume. He detected a surprising uncertainty in the early objectives, which was perhaps mirrored in its first officials, and this was only resolved with the election in 1893 of A.W. Pollard of the British Museum as the Honorary Secretary of the Society, an office which he held solely until 1912, and jointly with R.B. McKerrow until 1934. The names of Pollard and McKerrow (both Gold Medallists in 1929) will recur in the ensuing entry. The strong connection with the British Museum was crucial to the development of the Society. The Department of Printed Books, under the direction of Panizzi, had become the greatest all-round historical collection, not only of British, but also of foreign books, and in particular of incunabula. And there was another, subtler, influence at work, that of a great museum with incomparable collections of artefacts. Thus it was natural that books should be studied, not only as texts, but as physical objects. The Society's earliest publications related to these foreign books. Despite the premature death in 1903 of Robert Proctor, a member since 1894, who had given the world the concept of "Proctor order" in arranging incunabula, the earliest British connections with the German "*Gesamtkatalog der Wiegendrucke*" were through the Museum and the Society; the Director of the Wiegendruckkommission, Dr. Konrad Haebler had been an Honorary Member since 1905. The Society has since then published both books and articles in *The Library* on fifteenth-century books. It is appropriate that the most recent (June 2008) issue of *The Library* should carry authoritative reviews of the 11th and final volume (on English incunabula) of the British Library's *Catalogue of Incunabula* (*BMC*), and of the Bodleian Library's new *Catalogue of Books Printed in the Fifteenth Century*. The former review is by Bettina Wagner of the Bayerische Staatsbibliothek in Munich, who formerly worked on the Bodleian incunable catalog, and the latter by Richard Sharpe, Professor of Diplomatic at Oxford. He draws attention to the Bodleian catalog's concentration on the texts printed; a conscious move, very proper to a university library, away from that on printing houses, which was a legacy from Proctor. Two of the Society's Gold medallists, Konrad Haebler (1929), and Victor Scholderer (1951, and President 1946) were perhaps best known as incunabulists.

It was in a paper by A.W. Pollard, delivered in 1913, that the author described the work of the Society on English printers as "meagre and scrappy." This is a slightly dismissive judgement of what had so far appeared; Gordon Duff's *A Century of the English Book Trade* in 1905 (reprinted in 1948); the *Hand-Lists of Books printed by London Printers, 1501–1556*, of 1913 (which as Pollard writes in his Preface "were begun at my suggestion as long ago as 1894"); and the series of four Printers' Dictionaries which began in 1907, and were followed by others taking the coverage up to 1775, finishing in 1932. Their abiding worth was recognized by their being reprinted by the Society in a single volume in 1977. They owed much to the archival investigations of H.R. Plomer (1856–1928), which I can personally testify, are still of value in this and related fields.

Pollard's sights were, however, by 1918, set upon a project which had at least been aired in the Society's earliest years, a "Short-Title Catalogue of English Books, 1501–1640." This owed a lot to the British Museum's own catalog of 1884. But by 1920, no bibliographer could be unaware, not only of collections of pre-1640 books in other British libraries, but of the increasing numbers moving westwards across the Atlantic. Nevertheless, the objective of Pollard and of his collaborator Gilbert R. Redgrave, was realized in the astonishingly brief period of eight years, and the "short-title catalog (STC)" of 1926 became the ancestor of a series taking the record up to 1800, and in which the Society has been, in various degrees, involved. It could do little more than bless the continuation 1641–1700 by Donald Wing of Yale University. Wing paid STC the compliment of imitating its style of entry closely, though making the obvious improvement of replacing the numeric location code with a more memorable alphabetic code, such as "LLP" instead of "L2," to signify Lambeth Palace Library. Nor did he allow himself to be deterred, as perhaps his predecessors had been, by the dramatic rise in printing output occasioned by the Civil War. Yet the revision of the 1475–1640 STC, undertaken in 1948 by F.S. Ferguson and W.A. Jackson, was always a Society project. The history of this project, which outlived both editors, and was brought to a triumphant conclusion in three volumes, 1976–1991, by Katharine F. Pantzer, originally Jackson's assistant at Harvard, has been told in the prefaces to the three volumes, of which the second (I–Z) was published first in 1976 (with an anonymous Preface by the present writer), volume 1 (A–H) in 1986, and volume 3 in 1991. The last contains several indexes, of which the first, of Printers and Publishers, encapsulates the unique knowledge of this part of the book trade which Dr. Pantzer had gained in the course of the revision, and largely, though not entirely, obviates the need for a revised Printers' Dictionary of the period. She was awarded the Society's Gold Medal in 1988. The Society remained the publisher, though the financial requirements of editing and publication proved to be far beyond the resources of a relatively small members' society, and occasioned several efforts of fundraising in Britain and the United States.

There was a determination, nevertheless, in the Society's ranks that something more than the benevolent neutrality (and war-induced impotence) that had been shown toward Donald Wing's enterprise, would have to be displayed, if a continuation to Wing's STC were called for. Preoccupation with the "New bibliography," and in particular with the textual problems of the Elizabethan and Jacobean drama (a subject which calls for extended discussion, below), had perhaps distracted attention from the bibliography of the eighteenth and later centuries. However, in 1962, consideration was given to the possibility of an eighteenth-century short-title catalog. The Sub-Committee of the Society's Council appointed to consider it was chaired by the bibliographically omniscient President, Graham Pollard, and included Sir Frank Francis, and two of the Society's most knowledgeable *dix-huitiemistes*, L.W. Hanson and D.F. Foxon. The report they produced remained on the table until 1976, when a conference was called by the newly established British Library and the American Society for Eighteenth-Century Studies. Contributions to the conference by members of the Society, including the present writer, then Honorary Secretary, were prominent. In the end the *English Short-Title Catalogue* (ESTC), as it soon became known, came into being under the editorship of Robin Alston (President 1988 and Gold Medallist 1997), not in hard copy, but in electronic form, and to it have been added the contents of the Society's 1475–1640 volume and the 1641–1700 volume (under the auspices of the Modern Language Association). The ESTC is now run, and updated, by the British Library and ESTC (North America) at Riverside, California

The enumerative bibliography of early books was seen by W.W.Greg (President 1930, Gold Medallist 1935) as "one of the first tasks of bibliography" (*Studies in Retrospect*, 1945). He began this task very early in his membership of the Society, which he joined in 1898, with *A List of English Plays written before 1643 printed before 1700* (1900) and *A List of Masques, Pageants etc. supplementary to A List* (1902). Greg's single-minded devotion to the study of English drama had begun with his friendship at Cambridge with R.B. McKerrow, and came to a culmination only in the year of his death (1959) with the final volume of his *Bibliography of the English Printed Drama to the Restoration*. The first two books were the Society's earliest contribution to English literary studies, and Greg followed them with editions of *Henslowe's Diary* and *The Henslowe Papers* (1904–1908, not published by the Society). Greg and McKerrow were, in their scholarly examination of Elizabethan and Jacobean drama, joined by A.W. Pollard, whose interest had been aroused in the course of his museum duties, by a bound copy of the "Pavier Quartos" of "1619," brought in for examination, and now in the Folger Library. Greg, in particular, challenged the reigning orthodoxy of Sir Sidney Lee on the subject of the unreliability of the Shakespeare quartos and First Folio. This formed the subject of Pollard's own *Shakespeare Folios and Quartos* of 1909, and his Sandars Lectures published in *The Library* from 1916, and his *Shakespeare's Fight with the Pirates* (1920).

F.P. Wilson's *Shakespeare and the New Bibliography* is by far the longest contribution to *Studies in Retrospect*, and solidly established the Society's claim to have "made a difference" to literary scholarship. Wilson saw the Society as "leader and inspirer of these studies," "due in the main to three of the outstanding members of the Society," Pollard, Greg, and McKerrow.

> The presence of these men in the British Museum from necessity or choice or both made that library and the neighboring restaurants—especially during the summer migration from America—the best centre for Elizabethan studies in the world.

The subheadings in Wilson's article are suggestive, as they underline the necessity of documentary evidence in addition to the study of the book as a material object: 1) copyright and the relations between author, publisher, and bookseller; 2) the interpretation of imprints; 3) entry or non-entry in the Stationers' Register; 4) licensing; and 5) piracy. These are topics which pre-date the Bibliographical Society, and are still under discussion. Robin Myers in her paper "Stationers' Company Bibliographers: the first 150 years: Ames to Arber" in *Pioneers in Bibliography* (1988) demonstrates that bookmen had for a long time been making use of the Company's records. The foundation stone for the investigation of the collateral, archival evidence called for by F.P. Wilson, was laid by Edward Arber (1836–1912). He was never, apparently, a member of the Society, but his *Transcription of the Registers of the Company of Stationers of London 1554–1640*, (1875–1895), in Robin Myers' words "heralded the 'new bibliography' of Pollard, Greg, Plomer, and others by giving scholars access to a large enough amount of the early records to enable them to work at a distance from Stationers' Hall." Where Arber had led into early records, the Society followed, with *Records of the Court of the Stationers' Company, 1576–1602*, edited by Greg and E. Boswell (1930); *Records of the Court 1602–1640*, edited by William A. Jackson (1957); *The Loan Book of the Stationers' Company...1592–1692*, edited by W. Craig Ferguson (1989); and, most recently, *Index to the Court Books...1679–1717*, edited by Alison Shell and Alison Emblow (2007). Graham Pollard proposed an edition of "Stationers' Liber A," but neither he nor D. F. McKenzie who took up the task were able to complete it. A photographic facsimile, originally prepared for Pollard, provided me with the names of "Brothers" of the Company—aliens who were engaged in the importation of books—for work on the "Latin Trade." "Liber A" remains in prospect for publication.

Greg's magnificent four-volume *Bibliography of the English Printed Drama to the Restoration* (1939–1959) was completed in the year of his death, and reprinted by the Society in 1962. A summation of Greg's unparalleled scholarship over a period of 60 years, it is also a sustained example of quasi-facsimile transcription at its most elaborate. Ending his contribution to *Studies in Retrospect*, Greg presciently named two American scholars, Fredson Bowers and Charlton Hinman as providers of valuable bibliographical analysis. Of these, Hinman had already given in *The Library* of 1940 a foretaste of his monumental *The Printing and Proof-reading of the First Folio of Shakespeare* (1963); Bowers (Gold Medallist 1969) was a frequent contributor to *The Library* (and to his own *Studies in Bibliography)* and speaker before the Society. Despite his differences with Greg over bibliographical description, he cherished a hope of continuing Greg's *Bibliography* up to 1700.

Hinman's great work was able to distinguish several different compositors at work in Jaggard's shop on the First Folio. It is thus the only begetter of a school of compositor analysis, directed largely upon the drama of the period. Bibliographical research can establish the nature and reliability of the copy underlying a printed text, but it is less good at emending a corrupt text. Or it can point out that differing texts may be equally valid, representing different stages of an author's intentions. Thus the recent Oxford *Complete Works* of Shakespeare (1986), edited by Stanley Wells and Gary Taylor, prints two versions of *King Lear*; prints the Folio version of *Hamlet*, but has an addendum of two and a half pages from the Quarto version (which appeared in the poet's lifetime); publishes the Folio *Othello*, which is 160 lines longer than the Quarto of 1622; and several other plays have briefer additions at the end, which may be perfectly authentic Shakespeare—who was not around to decide which version should go down to posterity. As for the actors and producers, they will do what they have been doing ever since at least 1600—act and produce whichever version they prefer, or have time for.

There is an ironic footnote to F.P. Wilson's words (above) about the society of friends who met at the British Museum from necessity or choice. For it was there that Pollard and McKerrow met and discussed the impact of *An Enquiry into the Nature of Certain Nineteenth Century Pamphlets* (1934) by John Carter and Graham Pollard. The book implied that Thomas James Wise had forged and sold some of the most sought-after literary pamphlets. Wise (President 1922) was seen as the Society's principal expert in the nineteenth century, a period which the Society had perhaps neglected. It was felt that the novelty and standard of bibliographical scholarship merited Gold Medals. Perhaps Wise did too; they felt an affection for him. Both Pollard and Carter became Presidents (1960 and 1968) and Gold Medallists (1969 and 1975). When Wise died in 1937, his Ashley Library was sold to the British Museum for £66,000. That A.W. Pollard's affection for him would have survived the discovery that it contained 89 of the 206 leaves stolen from the British Museum, mostly from the Garrick collection of plays, is unlikely. The discovery of the thefts was made in the course of cataloging the Ashley Library by David Foxon (President 1980, Gold Medallist 1984), and was published by the Society in 1959 as *Thomas J. Wise and the Pre-Restoration Drama*. The subject of a Presidential Address by Simon Nowell-Smith (President 1962), himself a bibliographer and collector in Wise's "period," was "T.J. Wise as Bibliographer." It seems that Wise himself did not give a Presidential Address. In 1983, Nicolas Barker (President 1983, Gold Medallist 1999) published, with John Collins, *A Sequel to An Enquiry*, which showed that Wise, had an accomplice in the creation and marketing of the forgeries, Harry Buxton Forman.

In 1943, the Society and the Pilgrim Trust were asked to support a union catalog of pre-1700 books in the Anglican cathedrals of England and Wales, and in 1944 the catalog's creator, Miss Margaret Hands, began work at Worcester. The optimism so often found at the beginning of such schemes, fuelled (if that is the right word) by a certain naivety about what work in cathedral libraries is like, led to a far greater time-span of work than had originally been envisaged; so that with the exhaustion of the Pilgrim Trust's original grant, and the retirement of Miss Hands upon marriage, led to the lapsing of the catalog. Nevertheless, the handwritten slips remained accessible in the North Library of the British Museum. Six libraries remained uncataloged, among them three of the largest. Mrs. McLeod (as Miss Hands had become) returned to catalog Carlisle. Several other factors combined to its revival and ultimate publication. One was the existence, (in the case of Durham), or foundation of universities in or near, cathedral cities. Another was a provision in the British Library Act for assistance to be given to cataloging projects outside the British Library. The *Cathedral Libraries Catalogue* was published by the Society and the British Library in three volumes from 1984 to 1998. The Editor-in-Chief was (appropriately) Dr. David Shaw of the University of Kent at Canterbury (President 2002). It was also fitting that a French specialist should edit a catalog, the bulk of which contained books printed on the Continent of Europe before 1701.

The Society displayed a related concern for lesser, parochial, libraries, in its publication in 2004 of Michael Perkin's *Directory of the Parochial Libraries of the Church of England and the Church in Wales*, itself deriving from *The Parochial Libraries of the Church of England*, edited in 1959 by N.R. Ker (Gold Medallist 1975).

The centenary in 1992 coincided with the achievement of the Society's longest and most arduous tasks, STC. It also saw the maturing, if not the final completion of the *Cathedral Libraries Catalogue*. Another publication, *The Bowyer Ledgers*, edited by Keith Maslen and John Lancaster, published in 1991 in collaboration with the Bibliographical Society of America, had its origins among the desiderata of Graham Pollard's presidency, while *John Dee's Library Catalogue* (1990), edited by Julian Roberts (President 1986) and Andrew G. Watson (Gold Medallist 2001) had a gestation period of a mere 15 or 16 years.

The increasing financial strain of producing and distributing bibliographical works of quality, sometimes requiring illustration, has led the Society into collaboration (as in the case of *The Bowyer Ledgers*, above). Collaboration has already been noted with the British Library over *The Cathedral Libraries Catalogue*, and the British Library also appears in the imprint of two recent publications which are generously illustrated, *Eloquent Witnesses: Bookbindings and their History*, (2004) edited by Mirjam M. Foot (President 2000), and *The Earliest Books of Canterbury Cathedral: Manuscripts and Fragments to c. 1200* (2008) by Richard Gameson. The Society has just announced its support for another collaborative venture in "Watermarks in English Fifteenth-Century Books," a project with strong international links. The Society also supported the conference "Collecting Revolution: the Historical Importance of the Thomason Tracts" (2008) in which several members read papers.

There now exist a great many outlets, both national and regional, for bibliographical research beyond the Society. Some of these were noted in *The Book Encompassed*; others are the Book Trade History Group, of which the late Peter Isaac (President 1994) was a founder, and the annual conferences organized by Robin Myers (President 1996) and others, whose proceedings have appeared since 1981 as the Publishing Pathways series. Particularly significant are two series, the *Cambridge History of Libraries in Britain and Ireland* (three volumes, 2006) and the ongoing *Cambridge History of the Book in Britain* (1999–, of which seven volumes are promised). If a personal example is permissible here, my Presidential Addresses of 1987–1988 appeared, not over the Society's imprint, but as a chapter "The Latin Trade" in volume 4 (2002) of the *Cambridge History of the Book in Britain*.

As the very earliest publications suggest, there has always been a powerful international concern. In 2000 the Society published *A Dictionary of Members of the Dublin Book Trade 1550–1800* by Mary Pollard. An even wider international concern dominates the reviews, in the "Recent Periodicals and Books" and "Books Received" sections of *The Library*. The most recent Gold Medallist, Dennis Rhodes (2007), is best known for his work on Italian books. Other recent honorands have been Bernhard Fabian (1999, also the Honorary Secretary for Germany), and Henri-Jean Martin (1994). The influence of the latter on British bibliography has been profound. As D.F. McKenzie somewhat mischievously noted in his contribution to *The Book Encompassed*, "'The Book' has still a ring of an imported phrase"—as in "L'Apparition du livre" or, more widely, in "L'Histoire du livre." Imported abstraction or not, it is now canonized in the title of the Cambridge series, or as one of the Society's objectives in, for example, the Society's "Publications Policy" in *The Library* for March 2008.

I now turn to the other objectives of the Society, as set out at the beginning of this entry.

Membership. After the uncertainty of the earliest years, the membership grew steadily, until the Council decided that the roll of members should be limited, as far as Great Britain was concerned, to 300. The number of American members was 62 by 1894, and an American Secretary, E. D. North, was appointed. The limit upon numbers was dropped by 1919, and none has since been imposed. The Annual Report for 2006–2007 gave a total of 885 members. "Academics" form a far higher proportion of the membership than they did a century ago. The somewhat elastic requirement that a new member should be proposed by an existing member has now been dropped in favor of a simple statement of bibliographical interests. The original subscription of "one guinea" (£1.05) has risen (2008) to £33 for U.K. members and $65 for Americans. If the impression was given early of a "gentlemen's club" this was soon dispelled by the election of women members, as was fitting and, indeed, inevitable for a society so closely linked to the profession of librarianship. Nevertheless, the Society had to wait until 1996 for its first woman President, Robin Myers. Since then, there have been two women Presidents, Mirjam Foot in 2000 and Elisabeth Leedham-Green in 2006. Katharine F. Pantzer, principal editor of the revised STC, was awarded a Gold Medal in 1988.

The Society's Officers are, a President, elected for a term of two years, Vice-Presidents, an Honorary Secretary, an Honorary Treasurer, and an Honorary Editor of *The Library*. The governing body is an elected Council. Something of the spread of the Society's influence may be gathered from the existence, in addition to an Honorary Secretary and Treasurer for America, of Secretaries for Japan, Germany, and Australia.

Meetings. There are normally six meetings during the "winter" season, which have for some time been held at University College, London, though the custom of holding the Annual Meeting at various other historic libraries in London has grown up. A summer meeting is held at such a library outside London. Two of the sessional meetings are "named" meetings; the Graham Pollard Memorial Lecture, devoted, if possible to one of Pollard's interests, and the endowed Homee and Phiroze Randeria Lecture. The latter, devoted to bookbinding, is a recent and welcome development for a Society which has been presided over by such eminent scholars of bookbinding as H.M. Nixon (1972), A.R.A. Hobson (1977), and Mirjam M. Foot (2000).

"The Library." *The Library* was begun by J.Y.W. (later Sir John) MacAlister in 1889. He was then Honorary Secretary of the Library Association, and the first three series were published independently of the Society, though members, particularly A.W. Pollard who helped edit it, contributed to it. The "official organs" of the Society were then the *Transactions* and the *News Sheet*. The two journals were amalgamated in 1920, so that the second series of the *Transactions* was also the fourth series of *The Library*. The latter has so flourished that it is now (from 2000) in its seventh series. It is now appears quarterly, and is published for the Society by the Oxford

University Press. As it is an "Oxford journal," the full text from its foundation is available online. The details of this, and of all other publications, appear regularly in *The Library*.

The Society's library. This is perhaps the least perfectly realized of the Society's objectives. It has had a number of locations of limited accessibility; the former Rooms of the British Academy in Burlington Gardens, University College, London, Stationers' Hall, and most recently in the Senate House Library of London University. Borrowing from here is not possible, though the contents are accessible online, through the University of London Research Library Services catalog.

The Gold Medal. An endowment for a Gold Medal for services to bibliography was offered by Eustace Bosanquet in 1929. The amount was then sufficient for the award of five medals, which were, as they still are, awarded on an international basis. A post-war increase in the price of gold, and a decrease in the value of the fund, made it necessary for the medal to be struck in silver-gilt. The award is made on the recommendation of a special subcommittee of Council. The recipients since 1929 are listed on the Society's Web site.

Grants and bursaries. The impetus to raise an endowment for such awards was a feature of the Centenary celebrations of 1992, and is principally due to Barry Bloomfield (President 1990). After his death in 2002, his widow made a gift from his estate to fund an award for study in his own fields of interest, connected with the India Office Library, and modern literary bibliography. Major awards of up to £2000 are made annually on the recommendation of a subcommittee of the Council, and are advertised in the national press. In addition to other "named" awards, the Society, benefiting (in common with the Bibliographical Society of America) from the will of Katharine F. Pantzer Jr., who died in 2005, can award a fellowship of up to £4000 and a scholarship of up to £1500, named after her. Details of the awards made appear in *The Library*.

BIBLIOGRAPHY

General Works

1. Bibliographical Society. Website available at http://www. bibsoc.org.uk.
2. Davison, P., Ed. *The Book Encompassed: Studies in Twentieth-Century Bibliography*; Cambridge University Press: Cambridge, U.K., 1992.
3. Duff, E.G. *Fifteenth Century English Books: A Bibliography of Books and Documents Printed in England and of Books for the English Market Printed Abroad*, Bibliographical Society: London, U.K., 1917; [Reprinted 1964.].
4. Francis, F.C. *The Bibliographical Society, 1892–1942: Studies in Retrospect*, Bibliographical Society: London, U.K., 1945; [Reprinted 1949.].
5. Pollard, A.W. Our twentieth-first birthday. Trans. Bibliogr. Soc. **1916**, *13*, 9–27.
6. Roberts, J. The Bibliographical Society as a band of pioneers. In *Pioneers in Bibliography*; Myers, R., Harris, M., Eds.; St. Paul's Bibliographies: Winchester, U.K., 1988; 86–100.
7. Sharpe, R. The present and future of incunable cataloguing, II. Library. **2008**, *9* (Series 7), 210–224.
8. Wagner, B. The present and future of incunable cataloguing, I. Library. **2008**, *9* (Series 7), 197–209.

Printers' Dictionaries

1. Duff, E.G. *A Century of the English Book Trade…From the Issue of the First Dated Book in 1457 to the Incorporation of the Company of Stationers in 1557*, Bibliographical Society: London, U.K., 1905; [Reprinted 1948.].
2. Duff, E.G.; Greg, W.W.; McKerrow, R.B.; Plomer, H.R.; Pollard, A.W.; Proctor, R. *Hand-Lists of Books Printed by London Printers, 1501–1556*, Bibliographical Society: London, U.K., 1913.
3. McKerrow, R.B., Ed. *A Dictionary of Printers and Booksellers, in England, Scotland and Ireland, and of Foreign Printers of English Books, 1557–1640*; Bibliographical Society: London, U.K., 1910, General Ed.
4. Plomer, H.R. *A Dictionary of the Booksellers and Printers who were at Work in England, Scotland and Ireland from 1641 to 1667*, Bibliographical Society: London, U.K., 1907.
5. Plomer, H.R. *A Dictionary of the Printers and Booksellers who were at Work in England, Scotland and Ireland from 1668 to 1725*, Bibliographical Society: London, U.K., 1922.
6. Plomer, H.R.; Bushnell, G.R.; Dix, E.R.M. *A Dictionary of the Printers and Booksellers who were at Work in England, Scotland and Ireland from 1726 to 1775*, Bibliographical Society: London, U.K., 1932.

The four last dictionaries were reprinted in compact form by the Society in 1977. For the purposes of the dictionaries, the compilers deemed "England" to include Wales.

7. Greg, W.W. *A Bibliography of the English Printed Drama to the Restoration*, Bibliographical Society: London, U.K., 1939–1959, 4 [Reprinted by the Society, 1962.].
8. Pollard, A.W.; Redgrave, G.R. *A Short-Title Catalogue of Books Printed in England, Scotland, and Ireland and of English Books Printed Abroad 1475–1640;* 2nd Edition.; *First Compiled by Pollard, A.W.; Redgrave, G.R.*, Ed.; Bibliographical Society: London, U.K., 1986; [first published in 1926, and later reprinted], Revised and Enlarged, begun by Jackson W.A.; Ferguson, F.S., completed by Pantzer K.F.; 1976, 1991; Vol. 3.

Stationers' Company Records, etc, mostly Published by the Society

1. Arber, E., Ed.; *A Transcript of the Registers of the Company of Stationers of London, 1554–1640*; Stationers' Company: London, U.K., 1875–1894; Vol. 5.
2. Barker, N.; Collins, J. *A Sequel to an Enquiry into the Nature of Certain Nineteenth Century Pamphlets*, Scolar: London, U.K., 1983.

3. Carter, J.; Pollard, G. *An Enquiry into the Nature of Certain Nineteenth Century Pamphlets*, Constable: London, U.K., 1934.

4. Ferguson, W.C. *The Loan Book of the Stationers' Company, with a List of Transactions 1592–1692*, Bibliographical Society: London, U.K., 1989.

5. Foot, M.M. *Eloquent Witnesses: Bookbindings and Their History: A Volume of Essays*, Bibliographical Society, British Library: London, U.K., 2004, Oak Knoll Press: Newcastle, DE.

6. Foxon, D. *Thomas J. Wise and the Pre-Restoration Drama*, Bibliographical Society: London, U.K., 1959.

7. Gameson, R. *The Earliest Books of Canterbury Cathedral: Manuscripts and Fragments to c.1200*, Bibliographical Society; British Library: London, U.K., Dean and Chapter of Canterbury, U.K., 2008.

8. Greg, W.W.; Boswell, E., Eds. *Records of the Court of the Stationers' Company: 1576 to 1602, from Register B*; Bibliographical Society: London, U.K., 1930.

9. Hinman, C. *The Printing and Proof-Reading of the First Folio of Shakespeare*, Clarendon Press: Oxford, U.K., 1963.

10. Jackson, W.A., Ed. *Records of the Court of the Stationers' Company, 1602–1640*; Bibliographical Society: London, U.K., 1957.

11. Maslen, K.; Lancaster, J. *The Bowyer Ledgers: The Printing Accounts of William Bowyer, Father and Son*, Bibliographical Society: London, U.K., 1991, Bibliographical Society of America: New York.

12. McLeod, M.S.G. *The Cathedral Libraries Catalogue: Books Printed before 1701 in the Libraries of the Anglican Cathedral Libraries of England and Wales*, Edited and Completed by James, K.I; Shaw, D.J; British Library, Bibliographical Society: London, U.K., 1984–1998; [The second volume, in two parts, bears the names of Lawrence. Le R. Dethan and others as additional editors.].

13. Myers, R. Stationers' Company bibliographers: The first 150 years: Ames to Arber. In *Pioneers in Bibliography*; Myers, R.; Harris, M., Eds.; St. Paul's Bibliographies: Winchester, U.K., 1988; 40–57.

14. Nowell-Smith, S. T.J. Wise as bibliographer. Library. **1969**, *24*(Series 5), 129–141.

15. Perkin, M. *A Directory of the Parochial Libraries of the Church of England and the Church in Wales*, Bibliographical Society: London, U.K., 2004.

16. Pollard, A.W. *Shakespeare Folios and Quartos: A Study in the Bibliography of Shakespeare's Plays, 1594–1685*, Methuen: London, U.K., 1909.

17. Pollard, A.W. *Shakespeare's Fight with the Pirates, and the Problems of the Transmission of his Text*, 2nd Ed.; University Press: Cambridge, U.K., 1920; revised.

18. Roberts, J.; Watson, A.G. *John Dee's Library Catalogue*, Bibliographical Society: London, U.K., 1990.

19. Shell, A.; Emblow, A. *Index to the Court Books of the Stationers' Company, 1679–1717*, Bibliographical Society: London, U.K., 2007.

Bibliographical Society of America (BSA)

Hope Mayo
Houghton Library, Harvard University, Cambridge, Massachusetts, U.S.A.

Abstract

The Bibliographical Society of America (BSA) is the oldest scholarly society in North America dedicated to the study of books and manuscripts as physical objects. It was organized in 1904 and incorporated in 1927 with the principal objectives of promoting bibliographical research and issuing bibliographical publications. These objectives have been and continue to be accomplished through a broad array of activities, including meetings, lectures, and fellowship programs, as well as the publishing of books and the *Papers of the Bibliographical Society of America* [*PBSA*], North America's leading bibliographical journal. The society is open to all those interested in bibliographical problems and projects, and its membership includes bibliographers, librarians, professors, students, and collectors worldwide. Libraries are welcome as institutional members.

The Bibliographical Society of America (BSA) is the oldest scholarly society in North America dedicated to the study of books and manuscripts as physical objects. It was organized in 1904 and incorporated in 1927 with the principal objectives of promoting bibliographical research and issuing bibliographical publications. These objectives have been and continue to be accomplished through a broad array of activities, including meetings, lectures, and fellowship programs, as well as the publishing of books and the *Papers of the Bibliographical Society of America* [*PBSA*], North America's leading bibliographical journal. The society is open to all those interested in bibliographical problems and projects, and its membership includes bibliographers, librarians, professors, students and collectors worldwide. Libraries are welcome as institutional members.[1]

The body of this entry has been adapted from the presidential address delivered at the 100th annual meeting of the Society, 23 January 2004, and subsequently published as: Mayo, H. The Bibliographical Society of America at 100: Past and Future. PBSA **2004**, *98* (4), 425–448. Used by permission of the Bibliographical Society of America.

FOUNDING AND OBJECTIVES

The BSA was founded on October 18, 1904. It was one of a number of scholarly and professional societies organized in the Anglo-American world during the late nineteenth and early twentieth centuries, including the American Library Association (1876), the Library Association, Great Britain (1877), the Edinburgh Bibliographical Society (1890), and the Bibliographical Society, London (1892). Organizations such as these served to create and reinforce professional identity; through meetings, publications, and correspondence they promoted scholarly communication; and they sought to organize knowledge in their fields, as well as, in most cases, codifying standards of scholarly investigation and professional conduct. In the United States, the late nineteenth and early twentieth centuries saw a kind of knowledge explosion, with the founding of research universities, such as Johns Hopkins (1876) and University of Chicago (1890), as well as the transformation of older colleges, such as Harvard and Yale, into similar institutions. These developments were accompanied by the formation of research libraries and by increased publication in all disciplines. In an age when publication meant print, and there were no union catalogs, making this abundance of information accessible to those who needed it was a priority for librarians and for learned societies. At a time when the American Historical Association and similar organizations in other disciplines formed committees on bibliography, and the American Library Association discussed creating a Bibliography Section, it is not surprising that there should have been a move to form a Bibliographical Society, reflecting a view that bibliography was a discipline with its own techniques, standards, and contributions to be made.

It is impossible to discuss the founding and the history of the BSA without giving some attention to its predecessor, the Bibliographical Society of Chicago. The individuals who formed the Chicago society were for the most part the same ones who were responsible for organizing the BSA 5 years later. These men and women, representatives of institutions such as the John Crerar Library, the Newberry Library, the Chicago Public Library, and the University of Chicago Library, were conscious that they were forming an organization that would differ deliberately from library clubs and the American Library Association on the one hand, and from book collectors' clubs, such as the Grolier Club and the Caxton Club, on the other. At the organizational meeting of the Chicago

Encyclopedia of Library and Information Sciences, Fourth Edition DOI: 10.1081/E-ELIS4-120044819

society, Aksel Josephson of the John Crerar Library, convenor and chair of the organizing committee, remarked:

> The object—the sole object—of the society should be bibliographical study and research. Our work should cover the whole range of the history of the book—printing, illustration and binding, publishing and bookselling, history and management of libraries, and last, but not least, bibliography proper, the registry of printed books and the recording of their contents. We wish to enlist the interest and cooperation of the bibliographer and the librarian, of the student of literature and literary history, of the private collector, if he is willing to make his library available to students, and of the printer and bookseller for whom his trade is something more than a means of making money.[2]

These objectives are remarkably similar to those that still inform the work and character of the BSA. From the beginning, the founders intended that the creation of the Bibliographical Society of Chicago should lead to the formation of an American Bibliographical Society, a topic proposed at their first meeting and often discussed thereafter. To that end, they held frequent meetings, both in Chicago and in conjunction with conventions of the American Library Association. They began a journal—four volumes of the *Year-book of the Bibliographical Society of Chicago* appeared between 1900 and 1903—and they published two small monographs, all in an effort to establish a record of professional activity and national visibility. With the formation of the BSA, the Chicago society transformed itself into a chapter of the BSA and continued to hold local meetings until in 1912 it voted to dissolve itself and transfer its assets to the national society.

ORGANIZATION

At its organizational meeting, the BSA adopted a constitution which stated that "The name of this society shall be the Bibliographical Society of America" and that "The object of the society shall be to promote bibliographical research and to issue bibliographical publications."[3] Officers were elected, including: William Coolidge Lane, Harvard University Library, President; Herbert Putnam, Librarian of Congress, First Vice President; Reuben Gold Thwaites, Wisconsin State Historical Society, Second Vice President; Charles Alexander Nelson, Columbia University Library, Secretary; and Carl B. Roden, Chicago Public Library, Treasurer. Wilberforce Eames was named librarian, and a council of five members was chosen. In 1938, the office of Librarian was discontinued, and the position of Permanent Secretary, subsequently called Executive Secretary, was created to handle the ongoing administration of the Society's business. At present, the BSA is governed by an elected council consisting of the officers (President, Vice President, Secretary, and Treasurer) and 12 councillors.

MEETINGS

In its early years, the BSA held general meetings twice a year, often in conjunction with conferences of the American Library Association, the American Historical Association, or other similar organizations, e.g., the American Association for the Advancement of Science, the American Historical Association, or the American Association of Law Libraries. One result was that meetings took place in various parts of the country or occasionally in Canada. Between 1904 and the end of 1941, the Society met in more than 35 different American cities in 22 states, from coast to coast, and in Ottawa, Toronto, and Montreal, Canada. The practice of meeting exclusively in the Northeast originated during World War II, and since 1960 all the official, annual meetings of the Society have been held in New York City.

Joint meetings with other scholarly and professional societies and meetings in various places produced programs reflecting bibliographical topics of local interest or the interests of the partner organization. For example, the 1908 conference in Richmond, with the American Historical Association, offered several papers on the bibliography of publications from or about the Confederate States of America; the 1911 meeting in Pasadena, with the American Library Association, heard a presentation on the lost history of Father Kino; the 1922 meeting in Ann Arbor focused on the materials for American history in libraries of the Great Lakes region; and the New Haven meeting of 1931 dealt with topics related to Connecticut printing history and to Yale University. On other occasions, the Society identified themes reflecting its own interests and organized panels to discuss them: for example, "Bibliography Today: Where Do We Stand"? (1959), "Bibliography and the Collecting of Historical Materials" (1964), "Bibliographical Techniques in the Ancient World" (1965), "Bibliographical Needs and Problems of Collectors, Librarians, and Booksellers" (1966), "Prospects for an Eighteenth-Century STC" (1977), and "Techniques of Shared Cataloging" (1981). Since the 1970s, the program at the annual meeting has usually consisted of a single address by an invited speaker. Throughout the history of the Society most of the papers presented at its meetings have subsequently been published in the *Papers of the Bibliographical Society of America*.

During its early decades the BSA relied on frequent meetings, meetings in various parts of the country, and joint meetings with kindred learned and professional societies to provide interest, variety, and broad programming appeal. After the practice of meeting annually in New York City had become well established, the Society instituted a regional or joint meetings policy, intended to encourage other organizations and institutions to host bibliographical gatherings in other parts of the country. A number of such meetings were held in the 1908s. More recent examples include "The History of Libraries in the

United States" (Philadelphia, April 2002), and, celebrating the Society's centennial, "Roughing It: Printing and the Press in the West" (St. Louis, October 2004). The Bibliographical Society of America also underwrote invitational conferences on the *Bibliography of American Literature* (*BAL*) (1992), on "Book Catalogues" (1995), and on "Marks in Book" (1997), the proceedings of which were published as special issues of *PBSA*. Since 2000, the Society has sponsored lectures or paper sessions at conferences of the American Printing History Association (APHA) and the Society for the History of Authorship, Reading, and Publishing (SHARP), at preconferences of the Rare Books and Manuscripts Section (RBMS), and at the St. Louis Conference on Manuscript Studies.

FELLOWSHIPS AND PRIZES

The Bibliographical Society of America funds several programs intended to promote bibliographical scholarship. A number of short-term fellowships of 1 or 2 months are awarded annually to support investigations of bibliographical topics, as well as research into the history of the book trades and publishing history. Eligible topics are defined as those concentrating on books and documents in any field, but focusing on the book or manuscript (the physical object) as historical evidence. The New Scholars Program funds early-career scholars who have been invited to deliver papers on bibliographical subjects at a forum immediately preceding the BSA annual meeting. The Society also awards three prizes: The William L. Mitchell Prize for Bibliography or Documentary Work on Early British Periodicals or Newspapers, the Justin G. Schiller Prize for Bibliographical Work on Pre-Twentieth-Century Children's Books, and the St. Louis Mercantile Library Prize in American Bibliography.

PUBLICATIONS

The publications sponsored by the Society include a periodical, monographs, and electronic publications. Funding for this activity central to the Society's mission has come primarily from endowed funds derived from bequests from George Watson Cole and Lathrop C. Harper, supplemented by grant funding in the case of some larger projects.

Periodical Publications

The publication of a bibliographical periodical was discussed at the founding meeting of the Society and recommended at the very first meeting of the Council. The journal itself came into being in 1906. The first number, printed in April 1906, was denominated *Bibliographical Society of America Proceedings and Papers Volume One Part One 1904–1905*. The second number, *Volume One, Part Two* was printed in 1907 and covered the years 1906–1907. Also in 1907, the Society began publishing *The Bulletin of the Bibliographical Society of America*, which included proceedings of the Society's meetings, news, and notes, and listings of American bibliographical publications. After 1912, the *Bulletin* was discontinued and its contents were merged into the journal, which, starting with volume 4 (1909), was entitled *Papers of the Bibliographical Society of America*. Volume 100 of the *Papers* appeared in 2006. The objectives of the Society's periodical publications were well expressed by William Coolidge Lane, its first president, in the first issue of the *Bulletin*:

> With this first number of a Bulletin, the BSA begins in a humble way the issue of a quarterly record of American bibliography. Whether it will prove useful, whether it is to succeed in bringing together the material it seeks, what it may eventually become, are all questions for the future. . . . Without constant help from a wide circle of correspondents, the Editors cannot make this bulletin what they wish it to be—a record of current bibliographical work in America, such a record as will be of value to all book lovers, book collectors, book workers, and book purveyors (librarians). . . . Any serious bibliographical work undertaken in America, as well as any such work relating to America undertaken abroad, is included in its scope. . . . The greater part of the titles recorded in this Bulletin will be of special bibliographies of various kinds, but works on the science and art of bibliography, works on manuscripts and public archives, accounts of library collections, both public and private, and the more important library catalogs, and contributions to the history of printing and bookselling will all find a place here.[4]

This breadth of scope and purpose still characterize *PBSA*, which has published articles relating to all branches of bibliography and what is now called book history. In the words of the current guidelines for contributors,

> Contributions to the *Papers* may deal with books and manuscripts in any field, but should involve consideration of the book or manuscript (the physical object) as historical evidence, whether for establishing a text or illuminating the history of book production, publication, distribution, or collection, or for other purposes. Studies of the printing, publishing, and allied trades are also welcome.[5]

Between 1926 and 1951 the Society also published, at irregular intervals, 71 numbers of a *News Sheet* which printed announcements of and reports of meetings, lists of new members, notes and queries, and notices of bibliographical works published or in preparation. Lists of members were published at irregular intervals from 1905 to 1979, and yearly since 1982. From 1981 through 2006, the business documents of the Society, that is, the minutes of annual meetings and the annual auditor's reports, were published in *PBSA*. Since 2007, these records have been printed in the annual list of members. Other business and administrative records exist in the Society's archives, formerly kept at the American Antiquarian Society and now hosted by the Grolier Club of New York City.

Monograph Publications

A complete list of the thirty-seven monographs published to date by the BSA may be found on the Society's Web site (http://www.bibsocamer.org). The Society's monograph publishing program was inaugurated in 1919 with the publication of the first *Census of Fifteenth-Century Books Owned in America*, edited by George Parker Winship. The second *Census of Incunabula in American Libraries* appeared in 1940, edited by Margaret B. Stillwell; the third *Census*, edited by Frederick R. Goff, in 1964; and a supplement to the third census in 1972. In 1980, the BSA granted permission to the British Library to keyboard the contents of Goff as the basis on which that library would develop its Incunabula Short-Title Catalogue, an electronic catalog that is now the most comprehensive listing of incunable editions and locations of copies. Other significant contributions to incunable studies have appeared as articles in *PBSA*.

Second in the series of the Society's monograph publications was the continuation in 1927, and the eventual completion in 1936 of Joseph Sabin's *Bibliotheca Americana: A Dictionary of Books Relating to America*. Work on this monumental compilation was first interrupted in 1881, with the death of Sabin himself. It was then taken up by Wilberforce Eames, who between 1884 and 1892 brought the alphabet from "Pennsylvania" to "Smith." After that, the project lapsed again until the mid-1920s, when the BSA, with funding from the Carnegie Corp., took on responsibility for the work. Between 1927 and 1929 parts 117–121 were completed under the supervision of Eames himself, and the remaining parts, through 172 and the end of the alphabet, were edited by R. W. G. Vail. The completion of Sabin's work was followed by the publication of a series of bibliographies of early American printing organized by state: Albert H. Allen, ed., *Arkansas Imprints, 1821–1876* (1947); Albert H. Allen, ed., *Dakota Imprints, 1858–1889* (1947); John Eliot Alden, ed., *Rhode Island Imprints, 1727–1800* (1949); and Lester Hargrett, *Oklahoma Imprints, 1835–1890* (1951). These grew out of the WPA-funded American Imprints Project of the Depression era, and were taken over by the Society after the death in 1944 of Douglas McMurtrie, who had been the director of that project. Other monographs in the area of American bibliography have included *A Bibliography of Oliver Wendell Holmes*, by Thomas F. Currier and Eleanor M. Tilton (1953); Roger P. Bristol's *Supplement to Charles Evans' American Bibliography* (1970); and Hazel A. Johnston's *Checklist of New London, Connecticut, Imprints, 1709–1800* (1978).

The Bibliographical Society of America's most significant contribution in the area of Americana has been the *Bibliography of American Literature [BAL]* published in nine volumes between 1955 and 1991. Begun by Jacob Blanck, who edited volumes 1–6, and completed after by Blanck's death by Michael Winship, the project was supported by funding from J. K. Lilly and the National Endowment for the Humanities. A comprehensive bibliography of the first editions of selected American literary authors, *BAL* pioneered in applying modern bibliographical principles and techniques to the study of nineteenth-century books and showed how analysis of their physical characteristics could help elucidate nineteenth-century printing and publishing history.[6] In 1997, a digitized version of BAL was made available on CD-ROM or by subscription on the Web.

Some other areas represented in monographs published by the BSA are dictionaries of the book trades: C. Paul Christianson, *A Directory of London Stationers and Book Artisans 1300–1500* (1990); Sidney F. Huttner and Elizabeth S. Huttner, *A Register of Artists, Engravers, Booksellers, Bookbinders, Printers and Publishers in New York City, 1821–42* (1993); and editions of sources for printing and publishing history: Warren S. Tryon and William Charvat, eds., *The Cost Books of Ticknor and Fields* (1949*)*, and K. I. D. Maslen and John Lancaster, eds., *The Bowyer Ledgers: The Printing Accounts of William Bowyer Father and Son Reproduced on Microfiche, with a Checklist of Bowyer Printing 1699–1977* (published jointly with the Bibliographical Society, London, 1991). Recent contributions have included titles as varied as: Kenneth E. Carpenter, *The Dissemination of "The Wealth of Nations" in French and in France 1776–1843* (2002); Brian Alderson and Felix de Marez Oyens, *Be Merry and Wise: Origins of Children's Book Publishing in England 1650–1850* (copublished with the Pierpont Morgan Library and the British Library, 2006); Milton McC. Gatch, *The Library of Leander van Ess and the Earliest American Collections of Reformation Pamphlets* (2007); and Andrea Krupp, *Bookcloth in England and America, 1823–50* (2008).

Electronic Publications

Since 2003, the Society has maintained Bibsite, a portion of its Web site that "functions as a service to bibliographers by offering a means for scholars to provide public access to accumulated bibliographical research material that may be useful to other researchers."[7] Materials posted include supplements to published bibliographies, as well as tables of statistics, databases of images, and bibliographical listings deemed more suitable for online than for print publication. In other areas of electronic publication, the BSA has made the full text of the *Papers* available through ProQuest and is exploring the possibility of reissuing its back list of monographs as e-books.

From the date of its founding, the BSA's reason for existence has been to promote bibliographical research and to issue bibliographical publications. The Society has never promulgated a definition of bibliography, nor attempted to choose among the several applications of that term. Rather, the topics of research encouraged by the BSA and represented in its publications have included examples of every kind of bibliographical activity, as well as studies that build on bibliographical evidence to explore the context, whether literary or historical, in which books were produced and used. It expects to continue to promote, in whatever medium, work that will demonstrate the value of material evidence for the study of books and book history.

REFERENCES

1. BSA Website. Available at http://www.bibsocamer.org/ (accessed August 2008).
2. Josephson, A.G.S. Introductory Remarks at the Organization Meeting, October 23, 1899. Year-book of the Bibliographical Society of Chicago, 1899–1900; 7.
3. Bibliographical Society of America. Constitution. PBSA **1906**, *1*(1), 7 and in all succeeding versions of the constitution or bylaws.
4. Lane, W.C. Bulletin of the Bibliographical Society of America. **1907**, *1*(1), 1–2 [Untitled headnote].
5. PBSA Web site. Available at http://www.bibsocamer.org/ Papers/default.htm (accessed August 2008).
6. Stoddard, R.E. An issue devoted to the "Bibliography of American Literature". PBSA **1992**, *86*(2), 127–210.
7. Bibsite. Available at http://www.bibsocamer.org/BibSite/ bibsite.htm (accessed August 2008).

BIBLIOGRAPHY

1. *The Bibliographical Society of America 1904–79: A Retrospective Collection*, University Press of Virginia for the Bibliographical Society of America: Charlottesville, VA, 1980.
2. Edelstein, J.M. The Bibliographical Society of America, 1904–1979. PBSA **1979**, *73*(4), 389–422.
3. Mayo, H. *The Bibliographical Society of America: A Centennial History*, Bibliographical Society of America: New York, forthcoming.
4. Van Hoesen, H.B. The Bibliographical Society of America–Its leaders and activities, 1904–1939. PBSA **1941**, *35*(3), 177–202 and one folding table.

Bibliography

D. W. Krummel

Emeritus, Graduate School of Library and Information Science, University of Illinois at Urbana-Champaign, Champaign, Illinois, U.S.A.

Abstract

The basic families of bibliography are distinguished, followed by an overview of the history of bibliography. Major works in the world of bibliography are cited in their contexts. The practices of assembling and searching bibliographical lists are summarized, and the last section, which focuses on physical bibliography, surveys the study of books as artifactual evidence.

BIBLIOGRAPHY IN GENERAL

Bibliography cites and studies books. The citing and compiling of citations into lists (i.e., bibliographies) are *reference* (or *enumerative*, or *systematic*) bibliography, the studies of printed artifacts are *physical* bibliography. The two activities go together.

Books, narrowly defined, are monographs printed on paper. There are millions of them, their numbers are growing, and they are much used. Books may also be defined broadly, to include other forms of written records: pamphlets, journals and their contents, maps, music, manuscripts, pictures, recordings, movies, texts of all kinds, tangible objects, online, or conceptual. Books are physical objects that exist because of their ideas. They are content as well as form, messages as well as media. The content can significantly determine the makeup and appearance of the object that readers are aware of. In turn, the object affects the way readers view the content. Bibliography deals with the content by citing and studying the physical sources. By specifying the physical evidence it locates the content; by suggesting the context of the evidence it reflects on the authority of the content.

Bibliography addresses the needs of readers by describing and prescribing. (Patrick Wilson's *Two Kinds of Power* explores the distinction.) Readers, however, still enjoy the freedom to use the citations as they please, selecting what they want, ignoring what they think they do not need, and reading into their selections whatever they wish. Bibliographical work is still validated by responsible citation and study of the written record, as informed by respect for readers and their intellectual freedom. The written record is implicitly historical: it deals with writings that now exist; it is also political in that it is designed for future use by readers.

Bibliographical citations are also transmitted informally and implicitly. They are passed on orally: indeed, personal communication is probably the most powerful but also the least controllable and measurable medium of bibliography. The bedrock, however, is the citation, on paper or online, ideally formulated to be concise but sufficient, and based in physical evidence.

THE HISTORY OF BIBLIOGRAPHY

The Origins of Bibliography

The story of bibliography begins not in the West with Gutenberg's invention of printing around 1450. From the ancient Near East, three millennia earlier, archival inventories survive on cuneiform tablets. Early Egyptian lists are known about as well. The first famous bibliography is the *Pinakes* ("tablets"), conceived in the fourth century B.C.E. at the great library at Alexandria, which may have included the library of Aristotle, the mentor of Alexander the Great. Its contents were thought of less as an archive than as literature, and thus they were organized by the names of authors. The tablets disappeared when the library was destroyed, but later sources testify to their use for over a millennium as references to the known and lost writings of classical civilization.

Other bibliographies served other purposes. The Greek physician Galen (A.D. 129–199) prepared a list of his authentic works to help distinguish them from other works falsely attributed to him. In his *Retractiones* (A.D. 427), an aging St. Augustine cited his earlier writings in order to revisit them. The Venerable Bede concludes his ecclesiastical history (ca. 734) with a list of several dozen brief titles of his sources, beginning with books of the Bible, and ending with his other writings. In the book world around 985, probably in Baghdad, Muhammad ibn Abi Ya'qub Ishaq al-Nadim prepared a vast index, the *Fihrist al-'Ulum*, and cites writings that can often be traced back to the *Pinakes*.

One landmark above all was to affect all later bibliographical practice: St. Jerome's bio-bibliography (ca. 392) of writings by 135 authors, mostly Church Fathers, now

Encyclopedia of Library and Information Sciences, Fourth Edition DOI: 10.1081/E-ELIS4-120044335

Bibliographic–Bibliothèque

known as *De viris illustribus*. Jerome's goal was to justify his choice of sources so as to argue for the authority of his "Vulgate" Bible. His list, and its two notable supplements, by Gennadius of Marseilles, ca. 490, with 91 added names, and Isidore of Seville, ca. 620, with 33 more, were to inform Biblical exegesis over the following centuries. Several earlier bibliographical sources, like the *Pinakes*, were also author lists, but with Jerome the names of the authors invited readers to honor their names by critically evaluating their work.

Newly founded libraries often listed their holdings, which grew from a dozen or so volumes in the early Christian era to hundreds in the late middle ages. Citation practices were casual, and bibliography, inventory, and catalog were the same. In fact, the holdings were usually better recorded in the memories of readers. A famous catalog-poem of Alcuin (ca. 735–804), for instance, recites the holdings of his library at York in a mnemonic sequence that identifies the relative location of volumes on the shelf. From the fifteenth century also come several union lists of the manuscripts in Dutch and English libraries. The latter are valuable evidence of early holdings that predate the confiscations of Henry VIII.

The Rise of Modern Bibliography

The spread of printing after 1450 introduces a modern era in bibliography. If books were to circulate and flourish commercially, more was needed than word-of-mouth promotion. As the output of the press grew, the bibliographical record expanded, both in numbers and in kinds, for use by readers and purchasers, authors and printers, merchants and rulers. The lists generally fall into four bibliographical genres: commercial booktrade lists (booktrade bibliographies, publishers' and booksellers' catalogs), topical lists (subject bibliographies), advisory lists (bibliographies of prohibited or recommended writings, whether of political, religious, or social grounds), and catalogs of collections. The four categories overlap, perhaps not always in the minds of their original compilers or in the special interests of their sponsors, but as they were used in their day and over the course of history.

Commercial Lists: The very first printed list (ca.1469) was a broadside issued by Peter Schoeffer, Gutenberg's successor, who sensed the value of listing the titles he had printed. As the book trade emerged, titles were promoted through catalogs. Beginning in 1564, citations of the titles from many publishers, exhibited at the semiannual Easter and Michelmas book fairs in Leipzig and Frankfurt, were collected and issued in lists known as *Messkataloge*. In time, these trade lists were to expand to become current national bibliographies of German books. The *Messkataloge* were succeeded by several extensive periodical lists: Wilhelm Heinsius's *Allgemeines Bücher-Lexikon* (1812–1894), Christian Gottlob Kayser's *Vollständiges Bücher-Lexikon* (1834–1912), *Hinrichs' Katalog* (1857–1913), and others leading up to today's *Deutsche Nationalbibliographie* (begun 1921).

Trade lists were needed because the German book trade, like the country itself, was decentralized: publishers needed to share their markets. In other countries, where the trade was centered in one city (like Paris or London), current lists emerged later, but still on the initiative of the booktrade. Most lists were not so much national as cultural resources: in a time when Latin was giving way to local vernaculars, the language of potential purchasers was primary. The Netherlands book trade lists (Johannes van Abkoude's *Naamregister van de bekendste en meest ungebruik zynde Nederduitsche boeken*, begun in 1788, later replaced *Brinkman's Cumulatieve catalogus van boeken*) have enjoyed an especially long life. In Great Britain, no continuing lists date from the eighteenth century. Only in the mid-nineteenth were two competing firms to emerge to share the market: Samson & Low (*Publisher's Circular and Publisher and Bookseller*, begun in 1837) and Whitaker (*The Bookseller* and *Current Literature*, begun in 1858). The Norwegian *Norsk bogfortegnelse* dates from 1848, the Danish *Dansk bogfortegnelse* from 1851, the Swedish *Svensk bokhandels-tidning* from 1863. For the Russian booktrade, the St. Petersburg firm of A. F. Bazunov prepared a *Sistematicheskii katalog* beginning in 1869. A *Bibliographie de Belgique* was begun in 1875, with extensive overlap with the Brinkman lists from the Netherlands, since Flemish publishers in both countries shared their market. In the United States, the major early trade lists were the work of Orville A. Roorbach (*Bibliotheca americana*, 1852–1861) and James Kelly (*American Catalogue of Books*, 1866–1871). A continuing succession of American trade lists began with Frederick Leypoldt (later R.R. Bowker; *American Catalogue of Books*, begun 1876) and H.W. Wilson (*United States Catalog; with Cumulative Book Index*, begun 1898).

Over time, governments replaced the booktrade in sponsoring the national bibliography. The readership was now a civil polity rather than the booktrade's customers. The most famous example was the *Bibliographie de la France* (begun in 1811), long encumbered by the arbitrary decisions of functioning bureaucracies. Booksellers and readers thus often preferred booktrade lists like Otto Lorenz's *Catalogue général de la librairie française* (1867–1945), or later lists known as *Biblio* (now subsumed in a separate section of the *Bibliographie nationale française*). Other governments assumed responsibility for listings their country's publishing output, if often selectively. The most spectacular saga of a national bibliography is that of the Russian *Knizhnaia Letopis'*, begun in 1907 and issued continuously with few delays through a century of revolutions, invasions, and other tortuous events.

Topical Lists: Other early printed subject bibliographies cover medicine (Symphorien Champier's *Index*, 1506), law (Giovanni Nevizzano's *Inventarium*, 1522), botany (Otto Brunfels' *Herbarum*, 1530), and many other subjects. These are landmarks in the history of the disciplines, but mostly for the intellectually curious. They deal with concepts and theories; the skills were to be learned not by reading books but through instruction and apprenticeship under master practitioners. The age of modern subject bibliography began with eighteenth century academic scholarship.

The most common subject lists were the early bibliographical *Acta*, lists in the spirit of St. Jerome that record the good deeds, not of the apostles of the church but of the apostles of learning, but still for critical study and veneration. The order was by authors' names, listed alphabetically or occasionally chronologically, rather than by subject subdivisions. Notable early lists are devoted to religious orders, beginning with Johannes Trithemius's *Carmelitiana bibliotheca* (1493). Later major works include Lucas Wadding's Franciscan *Scriptores ordinis minorvm* (1650) and Jacques Quétif's Dominican *Scriptores ordinis praedicatorum recensiti* (1719–1721), each of them later revised. The greatest of the genre is Augustin and Alois de Backer's vast *Bibliothèque des écrivains de la Compagnie de Jésus* (1835–1861, later revised by Carlos Sommervogel), still used to locate the early writings on the expansion of European settlement around the world from the vantage of the Jesuit missionaries who worked to convert the indigenous populations to Christianity.

Several early *Acta* celebrate the emerging national literatures. The compilers defined the criteria for inclusion; here again, the dividing lines were linguistic more often than political, since many national boundaries were not yet fixed. John Bale's *Illustrium maioris Brittanniae scriptorum* (1548), for instance, follows a chronology that begins with the Biblical Seth (!) and ends up with Bale himself. Italian literature is covered in *La libraria* (1550) and *La seconda libreria* (1551), by Anton Francesco Doni, with the brief entries and quirky coverage that one would expect of this famously erratic poet. French bibliography begins with the *Premier volume de la Bibliothèque* of François Grudé de la Croix du Maine (1584) and its competitor, *La Bibliothèque* of Antoine du Verdier (1585). Each has titles the other missed. Other early *Acta* celebrate the hero-authors of regions and cities, among them Michele Poccianti's *Catalogvs scriptorvm florentinorvm* (1589) for Florence, Leone Allacci's *Apes Vrbanae* for Rome (1633), Louis Jacob de St. Charles' *Bibliographia Parisina* (1643, complemented by a *Bibligraphia gallica*, 1651, for the rest of France), and for Oxford, Anthony à Wood's *Athenae Oxoniensis* (1691-92ff.).

Bibliographies before the nineteenth century were rarely devoted to individual authors, the famous exception being Erasmus's autobibliography at the end of his 1523 *Lucubrationum*, a forerunner of the lists that academics today prepare in applying for grants and promotion.

Lists of Recommended and Forbidden Books: The flood of printing in general, but Reformation and later the Counter-Reformation religious tracts in particular, called for bibliographies of opposite kinds. Books recommended for the faithful have their model in the *Bibliotheca selecta* (1574) of the learned Jesuit Antonio Possevino (1534–1611). Faithful readers are not always easy to separate from eager of busy readers, whose needs were served by a vast array of recommended books, advisory and critical reading lists, and bibliographical "epitomes." The genre merges with prose works like Helen Haines *Living with Books* (1935 and 1950), long a mainstay of library school classes in book selection. All such works beg the criteria for inclusion and exclusion and ultimately the cause of intellectual freedom: readers are grateful for guidance, but they need to ask where it is coming from. The classic framing of the problem is in Lester Asheim's two famous essays, "Not Censorship but Selection," *Wilson Library Bulletin*, 28 (1953), 63–67, and "Selection and Censorship: A Reappraisal," *Wilson Library Bulletin*, 58 (1983), 180–184.

More famous are the bibliographies of the opposite kind: ones that ban books. The famous example is the *Index librorum prohibitorum* of the Roman Catholic Church, begun in 1564, as fearful in its consequences as it was absurd in its execution. The set went through many editions before it was finally abolished in 1966. Listing the forbidden fruit in bibliographical citations was not a wise decision: forbidden fruit is always tastier. Political censors, in contrast, rarely publicized the apostates, preferring instead to destroy the copies and often their authors. This has left modern bibliographers with the hunt for Marprelate tracts and Mazarinades, pamphlets of the English Civil War and the American and French Revolution, booklets, diatribes, and broadsides, all of it crucial historical evidence for the study of the intellectual roots of political, social, and cultural change. The bibliography of pornography is still different. Sexually explicit material had been collected avidly, less by institutions than by a small and secretive world of private collectors, in which written citations are often privileged; and tastes in erotica have often been subjectively redefined. The genres of literature suppressed on sexual, social, political, and religious grounds is surveyed in the *Banned Books* series (1998; new eds., 2006).

Library Catalogs: The genre begins with the *Catalogus Graecorum librorum* (1575), covering 126 Greek manuscripts in the city library of Augsburg. The manuscripts had been presented to the city by the patrician Johann Jakob Fugger, and Hieronymus Wolf, his librarian, had prepared this catalog. A difference between public and private libraries then was irrelevant. Other notable early library catalogs include the *Ecloga Oxonio-Cantabrigiensis* (1600), a union list of the manuscripts in Oxford and Cambridge libraries compiled by Thomas James, who also prepared the first catalog of the Bodleian library (1605). In the third Bodleian catalog (1674), the librarian, Thomas Hyde, first suggested the need for consistency, anticipating the need for cataloging codes.

Over the nineteenth and twentieth centuries, many libraries published catalogs of their holdings. Among the great early American catalogs were those of the Astor Library in New York (1886–1888), and Charles Ammi Cutter's for the Boston Athenaeum (1874–1882). Also much respected are the catalogs of the two greatest libraries of Europe, the Bibliothèque Nationale in Paris (*Catalogue général*, 1900–1981), and the British Museum in London (1900–1905 and earlier, 1931–1954, and 1968–1979, superseded by the *British Library General Catalogue of Printed Books to 1975*, 1979–1987). The Library of Congress complemented them with its *Catalog of Books Represented by. . . Printed Cards* (1942–1946).

Universal Bibliography: The dream of one single bibliography, vast, comprehensive, and sufficient, is timeless. The first printed bibliography, Johann Trithemius's *Liber de scriptoribus ecclesiasticis* (1494), actually covers more than the writings of ecclesiastics: it nobly assumes nothing less than that all writings are the work of God, a vision of the wondrous totality of the intellectual universe. In truth, Trithemius's work is based mostly on the books in his monastery library at Sponheim. His work is still modest alongside Conrad Gessner's *Bibliotheca universalis* (1545–1555). Trithemius (1452–1516) was a learned scholar who looked inward (his best known other work is in cryptography), Gessner (1516–1565) was a brilliant polymath who looked outward to new fields involving the world at large, from natural history and geology to linguistics (in all of which fields he was historically important). Gessner's curiosity led him to end his work with a fourth volume, the famous "pandects," which provided subject access to the earlier volumes. Inevitably, his successors were lesser minds, who created selective abridgements of his work with shortened citations. Among Gessner's ambitious but sad successors is Francesco Marucelli (1625–1703), who copied out 111 volumes of citations to create his *Mare magnum*, a great bibliographical ocean still preserved but largely inaccessible in his Bibliotheca Marucelliana in Florence. Several other works have a national spin in their titles, although their purpose was not to record the nation's publications but to cite writings that were part of the national heritage, published anywhere: Christophorus Hendreich's *Pandectae Brandenburgicae* (1699; only A-B were completed) and Robert Watt's *Bibliotheca Britannica* (1824) come to mind. In the twentieth century, the world visions have been revived and begun to be implemented in the name of Universal Bibliographical Control.

Three Later Developments

Over the course of the nineteenth century, three innovations were to alter the world of bibliography profoundly: cataloging codes for libraries, serial lists of current writings, and the formal study of books as physical objects. All had precedents in earlier periods. All were justified by emerging readerships in the new democratic societies; by a burgeoning book production made possible because of the readers as well as the Industrial Revolution; and by a proliferation of subject specialties that resulted from and in turn fostered to the new agenda of academic scholarship.

Library Cataloging Codes: (*See Cataloging*.) As libraries were established—in particular the publicly financed ones beginning in the nineteenth century, where costs needed to be measured—rules and uniform citation practices began to be developed. The goal was to provide for sharing bibliographical citation practices, and in time citations themselves, based on practices of compiling lists. Compilers of bibliographies were in turn influenced by cataloging codes, whose citations were easy to adapt and whose arbitrary practices saved the compilers many decisions.

The difference between cataloging and bibliography has been often proposed and continuously redefined over the years, and a firm relationship probably should not be fixed. The two obviously have much in common, not only in the material they describe but often also in the ways they describe it. They are still different in their objectives. Cataloging describes copies for all readers, bibliographies cover delimited literatures, and serve the needs of identified audiences of readers. Many writings have long fallen outside the scope of cataloging practices—periodical articles and many other analytics among them—and often particular communities of readers make best use of citations formulated with the familiar terms they use and in the structures they know. Ideally, cataloging data are based on autopsy of copies, but the content is frequently verified in bibliographies. Bibliographers too are expected to have examined the copies they cite, but today they are greatly helped by the way catalogs provide not only locations but also citations to adapt.

Serial Bibliographies of New Writings: Scholarly journals arose with the learned societies of the late seventeenth century, along with a few short-lived bibliographies. The *Acta* were still sufficient, especially with supplements. One of the most famous of them, Christian Gottlieb Jöcher's *Allgemeines Gelehrten-Lexikon* (1750–1751), for instance, had a supplement as late as 1897. A changing academia led to lists like Jeremias Reuss's *Repertorium commentationum a societatibus litterariis editarum* (1801–1821). Periodicals of all kinds called for indexing, but the indexes were particularly important to the proliferating fields of scholarship. Slowly the biographical focus gives way to subject classification, in the interests of providing a current awareness service to the growing scientific communities. Detailed indexes were needed in fields that were evolving, and occasional annotation of the entries in time led to today's abstracting services.

Periodical indexes for general libraries began with William F. Poole's *Index to Periodical Literature* (1853, new ed., 1891 with suppls. to 1907). In 1905, H.W. Wilson began his indexes, beginning with the *Readers' Guide to Periodical Literature* (1905). The largest foreign counterpart is the *Bibliographie der deutschen Zeitschriften-literatur* (1896–1964) and counterpart *Bibliographie der fremdsprachigen Zeitschriften-literatur* (1911–1994),

Physical Bibliography: The study of books as physical objects has its roots in the work of the early printers and book craftsmen, who asked how they could make their books more attractive and easier to produce. Discerning collectors, antiquarian dealers, librarians, and inquisitive readers in time came to ask the same questions, particularly over the course of the nineteenth century. A crucial figure was the Cambridge University librarian Henry Bradshaw (1831–1886), who proposed a "natural history method" for studying the changing output of the early printers and rationalized the collational formulas that are the basis of descriptive bibliography. The founding of bibliographical societies around the turn of the century served to establish the new field, thanks in part to their journals and other publications. [See Bibliographical Society (London); Bibliographical Society of America. Their scholarship is discussed in the Section "Physical Bibliography".]

THE WORLD OF BIBLIOGRAPHIES

Today's wealth of enumerative bibliographies ranges separate monographs in book form, often in multivolume sets, to pamphlets, articles, and handouts. Many lists are also now being posted online, where they can be easily updated, but also silently reconceived, renamed, or even deleted.

National Imprint Bibliographies: Retrospective

Most retrospective lists are censuses of known copies, but a few cite unrecorded the titles of works now lost. The bibliographies fit along an axis of place (whether defined by geographical borders, by language, or by the effective market for the book trade) and time (from the earliest imprints to the present). The genres and their contexts are detailed in Friedrich Domay's *Bibliographie der nationalen Bibliographien* (1987).

United States: Annals for the Colonial and early Federal years are recorded in Charles Evans's *American Bibliography* (1903–1959) and extended by Ralph R. Shaw and Richard H. Shoemaker in *American Bibliography* (1958–1963) and the *Checklist of American Imprints*, so far through 1846 (1964). The online *North American Imprints Project* will subsume the extant entries, minus Evans' chronicle of the press. The *Bibliography of American Imprints to 1901* (1993) is particularly strong for the early years, thanks to recent cataloging at the American Antiquarian Society.

Great Britain: The landmarks are the *Short-Title Catalogue..., 1475–1640* (1926; 2nd Ed., 1976–1991, commonly called either "STC" or "Pollard and Redgrave," for the guiding spirits behind the 1926 Ed.), and Donald M. Wing's *Short-Title Catalogue of Books..., 1641–1700* (1945–1951; 2nd Ed., 1972–1998). These are now incorporated into the online *English Short-Title Catalogue*, which also subsumes the *Eighteenth Century Short Title Catalogue*, an ambitious online descriptive bibliography project. *The Nineteenth Century Short-Title Catalogue* (1984), in contrast, covers a considerably larger literature mostly by merging the card catalogs of several major British scholarly libraries. For the general literature of the later years and up to the highly selective *British National Bibliography* in 1950, booktrade lists are needed, along with the British Museum and other library catalogs. For regional writings, Harry G. Aldis's *List of Books printed in Scotland before 1700* (1916 and 1970) has brief citations, while Eiluned Rees's *Librae Walliae* (1987) cites Welsh books, 1546–1820, in fine detail.

Romance-Language Nations: No single French list covers the years before the *Bibliographie de la France* (1811) and its booktrade counterparts. Probably the most useful is still Alexandre Cioranescu's *Bibliographie de la litté-rature française du 16. [17 and 18.] siècle* (1959–1969). Italy similarly has no single sources, although the sixteenth century is being covered by the online "EDIT16" (*Censimento nazionale delle edizioni italiane del XVI secolo*), while the nineteenth is the basis for *CLIO: Catalogo dei libri italiani dell Ottocento* (1991). For these and other areas, the British Museum—long seen as having the world's second-best collection of everything—prepared short-title cataloges for its holdings. That for France, 1470–1600, was the work of Henry Thomas (1924 and 1986), that for Italy the work of A.F. Johnson and others (1958), extended in Dennis E. Rhodes's *Short-Title Catalogue of Seventeenth-Century Italian Books* (1986).

Sketchy book-trade bibliographies cover Spain and Portugal, often their New World colonies. The basic historical work is Antonio Palau y Dulcet's vast and conspicuously stenographic *Manual del librero hispano-americano (*1948–1977). The heroic bibliographer of Latin America is José Toribio Medina. His *Biblioteca hispano-americana* (1898–1907) is useful, but the most impressive of his many works is the eight volume *La Imprenta in Mexico, 1539–1821* (1908–1912; the first volume, based on work by Joaquín Garcia Icazbalceta, revised 1954, 1981). Medina also compiled lists for many other countries. His work is complemented by the work of other historical bibliographers, notably covering Cuba (Carlos Manuel Trelles y Govín, 1902–1926) and Haiti (Max Bissainthe, 1951 and 1973). There are also British Museum short-title catalogs by Henry Thomas, of Spanish (1921), and Portuguese books (1926) with a 2nd Ed., by Dennis E. Rhodes (1989, some with revised editions).

Germany and Adjacent Areas: The Leipzig and Frankfurt booktrade lists extended to cover much of the output of the German-speaking world, i.e., Austria, parts of Switzerland and of what now are many other central European nations. The two enormous "GV" sets (*Gesamtverzeichnis der deutschsprachigen Schrifttums, 1700–1910* and *1911–1965*) interfile the entries from the earlier lists of Heinsius, Kayser, Hinrichs, their successors, and supplements. The earlier periods are covered in the "VD" series: the *Verzeichnis der im deutschen Sprachbereich erschienenen Drucke des XVI. Jahrhunderts* (1983) and the online *Verzeichnis... des XVII. Jahrhunderts*. A.F. Johnson listed the British Museum's German books, 1455–1600 (1962); David A. Paisey prepared a supplement (1990), and a list for the years 1601–1700 (1994).

For the Low Countries, Wouter Nijhoff & M.E. Kronenberg's *Nederlandsche bibliographie* (1923–1961) covers the years 1500–1540. Later holdings in Belgian libraries are listed in Elly Cockx-Indestège and Geneviève Glorieux, *Belgica typographica, 1541–1600* (1968). A.F. Johnson and Victor Scholderer prepared a British Museum *Short-Title Catalogue of Books Printed in the Netherlands and of Dutch and Flemish Books Printed in Other Countries from 1470 to 1600* (1965), which was extended to 1620 by Anna E. C. Simoni (1990).

For Swedish imprints, Isak Collijn's *Sveriges bibliografi* covers the years to 1600 (1927–1938) and later (1942–1946), which Rolf du Rietz is extending in *Swedish Imprints, 1731–1833* (1977). For Denmark, Christian Walther Bruun's *Bibliotheca danica, 1482 til 1830* (1877–1931), was extended to cover the years 1831–1840 by H. Ehrencron-Müller (1943–1948). For Norway, Hjalmar Pettersen's *Bibliotheca norvegica* (1899–1924) covers the years 1643–1813.

Other Nations: The most heroic saga in all of bibliography is probably that of several generations of the family of Estreicher, librarians at the Jagiellonska Library in Krakow, who since 1872 have been compiling the *Bibliografia polska*, covering imprints up to 1900. Other notable lists from Eastern Europe include the lovely but uncompleted Czech *Knihopis ceskych a slovenskych tisku* (1925–1948); Ioan Bianu's *Bibliografia romaneasca veche 1508–1830* (1903–1936); and the online Hungarian *Magyar Nemzeti Bibliográfia* data bases, covering early imprints from what is now Hungary as well as early editions in the Hungarian language printed elsewhere.

Other notable retrospective national lists include John A. Ferguson's *Bibliography of Australia, 1784–1900* (1941–1969), Austin Bagnall's *New Zealand National Bibliography to the Year 1960* (1970), Sidney Mendelssohn's older *South African Bibliography* (1910), and Marie Tremaine's *Bibliography of Canadian Imprints, 1751–1800* (1952, with a supplement by Sandra Alston and Patricia Lockhard Fleming, 1999), as well as Fleming's superb *Upper Canada Imprints, 1801–1820* (1988) and *Atlantic Canadian Imprints, 1801–1840* (both 1991).

National Imprint Bibliographies: Current

Ideally these should extend the retrospective lists. Usually, however, there are chronological gaps and different criteria for inclusion, the exception being the *Bibliographie Nationale Français*, a direct successor to the *Bibliographie de la France*. The *British National Bibliography* (begun 1950), one of the most successful recent national bibliographies, was a totally new project. It was the model for *Canadiana* (also begun 1950), the *Australian National Bibliography* (begun 1961), and the *New Zealand National Bibliography* (begun 1967)—all government supported. In Germany, the *Deutsche Nationalbibliographie* was established in Leipzig in 1921 as successor to national and trade lists that date back to the *Messkataloge*, but with many changes. During the Iron Curtain era, a *Deutsche Bibliographie* was issued out of Frankfurt in the West. National reunification led to a *Deutsche Nationalbibliographie* based on cooperative efforts.

The mid-twentieth century also saw official national bibliographies set up in Eastern Europe, serving the cause of ideological control along with the promotion of books and reading. The quality varied from the barely competent to the superb. It is not clear how much these lists are being weakened by a less rigid enforcement of copyright deposit laws now that publishing is no longer managed by the state. In most Third World countries, current national bibliographies have been meager, fostered, and sustained mostly by a weak local book trade working with a national library and with encouragement from UNESCO and friends abroad. For several decades beginning in the 1960s, the American Libraries Book Procurement Center at the Library of Congress managed a program for acquisitions of books from countries with counterpart currencies (PL-480), and cataloging them in an *Accessions List* series. Several countries with well developed booktrades have strong national bibliographies that use their national scripts, the Japanese *Nihon Zenkoku shoshi* being a good example. Recent titles are cited in Barbara L. Bell's *Annotated Guide to Current National Bibliographies* (1998). For the latest developments, the invaluable sources now are the *Reference Resources Europe* works maintained by the Italian publisher Casalini, annually in print (RREA, 1995) and online (RREO).

The United States has no national bibliography, aside from the Bowker and Wilson trade lists. It is generally assumed that there is very little of importance that is not in today's online union catalogues. New registrations at the Copyright Office in the Library of Congress have been listed in print since 1891. The data are often useful in bibliographical work, but the *Catalog of Copyright Entries* is often not well organized for use and hard to call a national bibliography. In 1948, the Bowker Co. began issuing *Books in Print* (a French counterpart, *Les Livres disponibles*, began in 1977, the German *Verzeichnis lieferbarer Bücher* in 1978), now accessible online.

Current Online Subject Bibliographies

Vast in their size and timely in their service to research, most current subject bibliographies are now online. (See the entry on "Online Database and Information Retrieval Services Industry," p. 3963.) Constructed not as a sequence of entries (alphabetic, chronological, or classified) with indexes, the online sources are accessible through search terms, such as can be regularly updated to reflect new concepts and relationships. Some lists give entries only, others include abstracts. Many of the most respected of them grew out of sources that began in print, among them the MEDLINE/PubMed database (successor to the *Index medicus*, begun in 1879), WilsonWeb Library (incorporating many of the H.W. Wilson Co. library lists, including the *Readers Guide to Periodical Literature* (begun in 1905), *Chemical Abstracts* (begun in 1907), the Modern Language Association's *Annual Bibliography* (growing out of the *American Bibliography*, begun in 1921), *Biological Abstracts* (begun in 1926), and the *International Bibliography of the Social Sciences* (begun in 1951).

The proliferating world of online bibliography is expanding in many imaginative directions. Services like *The Web of Science* (begun in 1961 as the *Science Citation Index*, later expanded to cover writings in the social sciences, arts, and humanities) document the spread of learning in footnotes and references. Vendors often produce several services, some of which they have taken over from earlier printed sources, others of which they have originated. *America: History and Life* has *Historical Abstracts* as a counterpart; in Lexis/Nexis, the former focuses on legal research, the latter on the news media. Some provide full texts, which still require bibliographical entries for access, others are built around search-term access in ways that imply that bibliographical authority is irrelevant.

Printed bibliographies have been either subsidized or justified by likely purchasers. Most current online subject bibliographies are rather different: as commercial businesses, they need to follow different economic rules as they negotiate operational decisions that will reconcile the demands of investors with the needs of their identified readers.

Other Bibliographies

Countless other lists cover narrowly defined subject areas, whose readers come to know their strengths and weaknesses. The bibliographies describe the literatures of subject areas large and small, of academic disciplines, and of topics within and between the disciplines and subject areas, as qualified by historical periods, geographical areas, and perspectives. They serve readers at all levels, from children and amateurs to specialized scholars, and they include citations in many languages. Often their entries are classified in categories that reflect the structures of their readers' fields of interest, annotations address the specific agenda of the discipline. Lists often include periodical articles and sections of larger works, chapters of books, essays in anthologies, poems and songs in collections, or individual maps in atlases—what bibliographers call "analytics" and which are outside the purview of library cataloging. Their scope is limited to the written record as it exists when the list is being compiled: future titles need to go into a supplement or a later edition. Many lists are now published online, although the linear layout is still usually that of the printed page. Among the printed and online genres are the following:

Bibliographical Guides: Bibliographies of bibliographies—lists that cite the lists that cite the sources—are of limited but occasional usefulness. The most extensive, now four decades old, is Theodore Besterman's *World Bibliography of Bibliographies* (4th Ed., 1965–1966). It cites over 100,000 titles, out of a likely several million today. It is limited to lists published separately (in other words, bibliographies at the end of or as parts of other works are excluded). It can be very useful, particularly for its juxtapositions of lists that cover the same general topic but also for its totals of the number of entries cited in each list; but there is general consensus that bibliographies of bibliography are rarely a first place to look. Besterman himself devised his subject headings. They usually work, occasionally they are idiosyncratic, even counterintuitive. Alice E. Toomey's supplement, *A World Bibliography of Bibliographies, 1964–1974: A List of Works Represented by Library of Congress Printed Catalog Cards* (1977), is arranged under Library of Congress subject headings, which are often no less counterintuitive. Besterman's precursors are discussed in Archer Taylor's *A History of Bibliographies of Bibliography* (1955). One step beyond such lists is Aksel G.S. Josephson's *Bibliographies of Bibliographies Chronologically Arranged* (1901, with supplements to 1913). No successor to any of these works has ever appeared. Considering the magnitude of the task—the complexity of the genres, the sheer number of titles—the prospects of successors seem slim.

Bibliographical guidance is often easier to dig out of selective guides to reference sources. Most notably of these has been in the United States the *Guide to Reference Books* (11 eds., 1902–1996), long the work of a notable lineage of American reference librarians: Alice Bertha Kroeger, Isadore Gilbert Mudge, Constance Winchell, Eugene Sheehy, and Robert Balay. Their successor, the *Guide to Reference*, was recently published by the ALA in a Web-based version. Foreign counterparts include, for Great Britain, A.J. Walford's *Guide to Reference Material* (1959; new eds, under different titles); for Germany, either Georg Schneider's *Handbuch der Bibliographie* (1923; latest ed., 2005) or Wilhelm Totok and Rolf Weitzel, *Handbuch der bibliographischen Nachschlagewerke* (1954 and later eds.); and for France, Louise-Noëlle

Malclès's ancient but venerated *Les sources du travail bibliographique* (1950–1958). A selection of the most important new reference sources, many bibliographies among them, appears annually in *College and Research Libraries*, as well as in reviewing media like the *American Reference Books Annual* (1970, and as *ARBA online*); also in the H.W. Wilson *Bibliographical Index* (1937, now online at *Bibliographical Index Plus*).

Subject Lists, Monographic, and Analytic: These range from brief check-lists, often of ephemeral interest—their very existence is often best known by informal communication among specialists—over to monuments of scholarship like the multivolume Clarendon Press *Bibliography of British History* (1928–1970), the *New Cambridge Bibliography of English Literature* (1969–1977), along with T.H. Howard-Hill's *British Literary Bibliography* series (1966)—to single out a few major works. Several smaller lists serve also as basic guides to specialist reference librarians and as textbooks in historical methods courses for graduate students. Notable examples include the *Harvard Guide to American History* (Oscar Handlin, later Frank Freidel; 1954 and later eds.), and Vincent Duckles's *Music Reference and Research Materials* (1964, based on early syllabi, 1948–1957; 5th Ed., 1997). Monroe Nathan Work's *Bibliography of the Negro in America and Africa* (1928), maps out a literature that supports a richly expanding field of scholarship.

Among scholarly descriptive bibliographies, two landmarks are devoted to literary genres: W.W. Greg's *Bibliography of the English Printed Drama to the Restoration* (1939–1959) and David F. Foxon's *English Verse, 1901–1750* (1975). The output of historically significant presses is cited and discussed by Horace Hart (Oxford University Press, 1900, extended by Harry Carter, 1975), Philip Gaskell (John Baskerville, 1959; the Foulis Press, 1964), D. F. McKenzie (the early Cambridge Press, 1966), C. William Miller (Benjamin Franklin, 1974), William S. Peterson (Kelmscott Press, 1984), and Marcella Genz (Eragny Press, 2004).

Bio-Bibliography: Collected bio-bibliographies no longer predominate as in the days of the *Acta*, although several are still today among the mainstays of the bibliographical reference shelf. Many of the most respected of them cover national literatures. On thinks of Karl Goedeke's *Grundrisz zur Geschichte der deutschen Dichtung aus den Quellen* for Germany (1859, reissued and extended; new Ed. in progress); the Maurist *Histoire littéraire de la France* (1733*ff*.), first of many French guides; Thomas Erslew's *Almindeligt forfatter-lexicon* (1843–1853) for Denmark, Jens Halvorsen's *Norsk Forfatter-Lexikon, 1814–1880* (1885–1908), succeeded by Holger Ehrencron-Müller's *Forfatterlexikon omfattende Danmark, Norge og Island indtil 1814* (1924–1935) for Norway, Bengt Ahlén's *Svenskt författarlexikon* (1942) for Sweden, and Innocencio Francisco da Silva's *Diccionario bibliographico portuguez*

(1858–1923)—the latter notable for its filing of entries by the author's *first* name (those who work extensively with Portuguese surnames will know why). Two other sets of *Acta* are mainstays of the reference shelf in fields where creativity is basic: Robert Eitner's *Quellenlexikon* (1900–1904), with citations of manuscripts printed editions of composers, and Ulrich Thieme and Felix Becker, *Allgemeine Lexikon der bildenden Künstler* (1900–1907; new Ed., 1983), for artists and their work. Appropriate to their celebratory function, most bio-bibliographies are printed; online lists have been slow to emerge.

Bibliographies devoted to one person emerged only over the last century, and are essentially of two kinds. Those devoted to writings *about* the person tend to have short entries, and are essentially of biographical interest. Writings *by* the person range from brief checklists, to superbly detailed descriptive bibliographies that are essential to textual scholarship. Among the many notable examples of the latter are those on Cotton Mather and his family (Thomas J. Holmes, 1931–1940), on Samuel Johnson (J.D. Fleeman and James McLaverty, 2000), on several eighteenth century authors by Allen Hazen, nineteenth century American authors by Joel Myerson, on twentieth century authors by Donald Gallup and Matthew J. Bruccoli, and others by William B. Todd.

Commercial Lists: Trade lists have largely been subsumed in national bibliographies, although publishers still issue lists to promote their titles. Many antiquarian booksellers now cite their offerings on the Internet, with citations are often models of descriptive bibliography and annotations (i.e., "blurbs") of impressive historical scholarship. The laborious efforts are justified in the interests of finding customers, although scholars in general often learn to benefit from the record of the bibliographical facts and the lore behind the materials.

Library Publications: Library catalogs often function as bibliographies, even if their main goal is to organize and find items in the collection. Most library catalogs are now in union catalogs and merged on the Internet in the interest of document delivery. Most libraries also prepare finding aids ("pathfinders") for work in the collections, selective bibliographies of a sort organized with a view to proposing search strategies for their readers. Exhibition catalogs, as they call attention to valuable or important works in the collections, often function as bibliographies. They work like dealers' catalogs in that the historical details in their annotations serve not to entice purchasers so much as to delight visitors.

Clearly, the most important of all bibliographies today are the library union catalogs. The great *National Union Catalog, pre-1956 Imprints* (1968–1980, known either as "NUC" or often for the name of its publisher, Mansell), in 685 volumes with a 69 volume supplement, grew out of the Library of Congress *Catalog of Printed Books* (1942–1946), in 167 volumes, with later supplements. It is updated but so far not entirely duplicated in the WorldCat bibliographic database, maintained by OCLC (the Online Computer Library Center), to which over 60,000 member libraries contribute. (See the entry on "OCLC: A World-wide Library Cooperative," p. 3924.) Its over 88 million records cover books, serials, sound recordings, musical scores, maps, visual materials, mixed materials, and computer files. Recently, OCLC subsumed Research Library Information Network (the RLIN, maintained by the Research Library Group; not all RLIN bibliographical records are in WorldCat as of this writing, however). Online Computer Library Center also maintains close contact with the union catalogs in other countries and often records their holdings.

Incunabula: Books from Gutenberg's day to 1501 (the "cradle" of printing) make up a much smaller but still very special bibliographical world of their own. The books are scarce, most of them large and beautiful. They testify to a world much different from ours today, so as to call for expertise of three kinds. The content is expressed either in a Latin that is usually corrupt or in vernaculars that were just beginning to be formed. The physical books call for a deep knowledge of analytical bibliography, particularly of type forms, since most printers needed to provide their own materials and adapt their skills. Finally, modern bibliographers need to know the bibliographical history, since older sources are still cited and very useful. The first major figure in this history is Georg Panzer, whose *Annales typographici* (1793–1803) are still useful for being arranged by imprint date. Ludwig Hain's *Repertorium bibliographicum* (1826–1838) and its successors may not be held in high respect today, but scholars still consult them. More venerated is Robert Proctor, whose concept of "Proctor order" (titles are gathered first under country, then under city, then under date) is seen in the great catalog of the British Museum (begun in 1908 and finally completed in 2007). The next event was the *Gesamtkatalog der Wiegendrücke*, a unified guide begun in 1925 but still incomplete since the 1940s. Frederick Goff's one volume *Incunabula in American Libraries* (1964) is still an indispensable starting point, although the "ISTC" project (now the *Illustrated Incunabula Short Title Catalogue*, and online now as "IISTC2") promises to be the essential bibliographical record for incunabulists.

BIBLIOGRAPHICAL PRACTICE

Compiling Bibliographies

Enumerative bibliography involves *compilers* preparing *lists*, which contain *citations* (consisting of *entries* and, often, *annotations*) of written material, for the use of *readers*. Enumerative bibliography succeeds when the lists work the other way around: readers search the lists and find the citations. Readers are the unpredictable factor: they often find what they need in unexpected places and they use it in ways that can rarely be anticipated.

A bibliography, as W. W. Greg proposed and Theodore Besterman famously recalled, must be based on "some guiding principle." A transcript of citations becomes a bibliography when the principle becomes legitimized through use, although compilers commonly explain their decisions in their introduction. Admittedly, readers rarely read introductions: they impatiently turn instead to the citations. The introductions thus serve mostly to fix the list in the context of the literature. Rules for compilers follow conventions that are rarely specified but are widely understood and justified through use. The basic decisions involve:

Scope: Compilers make their topics manageable by defining their scope. All bibliographies include or exclude titles on grounds determined by the physical object or the intellectual content of the material being described. The obvious physical distinction separates books from other forms of recorded knowledge. Some forms are closely akin to books—pamphlets, government documents, and ephemera come to mind—and in lists usually fit comfortably alongside books. Continuations—serials, periodicals, journals, newspapers—usually work like books and rarely pose problems, although their component units ("analytics") can be overwhelming in their profusion. Other forms—maps, printed music, and pictures—often benefit from separate listings, as do audiovisual materials—sound recordings, films, and videos—as well as electronic forms, so as to lead to hybrids like discography, filmography, videography, and Webliography. The less the documents look and work like printed books, generally the harder they are to fit into a bibliography, although readers are often best served by including them.

Citation Style: Bibliographical citation practices are spelled out in style manuals, which arose to provide consistent practices for scholarly journals. The many style manuals differ in details and have evolved over the years, but the most important today is arguably the *Chicago Manual of Style* (latest Ed., 2003), now compatible with the dissertation style long associated with Kate L. Turabian. The Modern Language Association prefers slightly different practices in its *Style Manual* (latest Ed., 2008) and *Handbook* (new Ed. expected, 2009), as do the British university presses at Oxford Horace Hart (latest Ed., 2005) and Cambridge (Judith Butcher, latest Ed., 2006). Scientific usage, most often prescribing authors' given names as initials and the publication date before the title (once called "Harvard style"), is spelled out in manuals from the American Chemical Society (latest Ed., 2006), the American Psychological Association (latest supplement on electronic references, 2007), and the Council of Science Editors (latest Ed., 2006; formerly the Council of Biology Editors), among others. The International Organization for Standardization and the American National Standards Institute have also addressed bibliographical style. Many publishers further specify their preferences in "house style" sheets. Cataloging codes can often inform bibliographical practice, although compilers commonly adapt the rules for the materials being cited, and with a view to the convenience and background of their selected readers.

Annotation: Compilers often find it important to define the aim, scope, or thesis of the writing being described; to identify the sources and context; to single out significant features; or to praise or incriminate the work (the former best done in understatement, the latter often by quoting the text). The goals of describing and evaluating are not as different as one might think. The writing style may be either conventional prose (to catch the reader's attention) or "telegraphic" (busy readers, making quick decisions on whether or not to look at the writing, can usually do without the subject and often the predicate). Readers often benefit from comparing and contrasting entries, often done by annotating several entries together.

Organization: Most entry sequences follow one of three orders: alphabetical, chronological, or classified. Alphabetic lists are usually by main entry, most often the name of the author. The alphabet is not intrinsically intelligent but it is easy to search. Chronological lists suggest the growth of the literature, and reflect the ideal of *historia litteraria*. Classified lists display the literature under categories. Compilers are often stimulated by obvious juxtapositions but have second thoughts about titles that belong in more than one or no category. The idea of systematic bibliography is something of a counterpart to the systematic card catalogues in many libraries. Linear sequences usually ask for indexes, in printed lists as either internal cross-references or separate indexes, in computer lists through entry access.

Well designed bibliographies need to be clear, precise, and efficient. In the course of helping readers identify, tabulate, and explore, they are most effective when they lead readers think about the substance rather than the methodology, and to compare entries by moving from broad overview to fine structure.

Bibliographical Searching

Searching for specific citations is unlike subject searching, much like hunting is unlike farming. Readers still need to know the sources they work with, their scope, organization, and citation practices. But they also need to decide whether citations, once located, are actually what they are looking for: will any edition, or version, suffice; how specific must the citation be? Often the exact needs are determined only as the search progresses. Generally, searches lead to one of four outcomes: (1) success: the right citation has been located; (2) positive failure: the citation has been found but it is not what is needed; (3) negative success: the citation has not been

found, but it may be there; and (4) failure: the item has not been found, and it is certain that it is not there. Distinguishing between (3) and (4) is helpful in sizing up the bibliographical infrastructure of the literature. In general, the smaller and more transparent the list being searched, the more easily it can be determined that the result is failure rather than negative success. In many vast lists, it is simply not possible to guess where an item has been logically but ingeniously concealed. For this reason, bibliographical searching benefits from smaller lists and a knowledge of how they work.

The more important the search, the more time should be allotted to it. But many important ones turn out to be very brief: it is often the intellectual curiosity of the searcher that makes them time-consuming. Online searching is almost always faster than work with paper sources, if also less cardiovascular. But printed sources, laid out visually and deployed physically, ask to be searched for contexts. Locating entries is different from understanding them, so searching can be boring and routine or highly stimulating.

PHYSICAL BIBLIOGRAPHY

As nineteenth-century scholarship flourished and became increasingly methodical, the study of the artifactual evidence of bibliography became self-consciously analytical. Antiquarian booksellers and collectors had long known that some books were more desirable than others as physical objects, philologists and literary scholars had understood the importance of authenticity of literary writings, historians had long celebrated the bibliographical lore behind famous events, while librarians and private collectors had long taken special pride in their celebrated rarities. These groups, as they coalesced into bibliographical societies, stimulated each other in their dialogue and wrote up their findings in journals that aspired to the scholarship of academic disciplines. The goal was announced, boldly and provocatively, by W.W. Greg (in the Bibliographical Society's 1945 *Studies in Retrospect*, p. 25), thus: "bibliography is the study of books as material objects, irrespective of their contents." This study has come to be seen as several different activities.

Analytical Bibliography

The study of physical books entails an analysis of typography, paper, ink, "printing house practices" (i.e., layout and design, typesetting, imposition, and presswork), as well as illustration and binding. Philip Gaskell's *New Introduction to Bibliography* (1972) is the standard overview, but much work has been done since his day, as synthesized in many writings and notably discussed in G. Thomas Tanselle's annual essays in *Studies in Bibliography*. The graphic arts become important in evaluating the working aesthetics of particular printers, traditions, and periods, while technology is now essential to the study of paper (e.g., the "Leningrad method," also digitized watermark archives), ink ("PIXE," i.e., particle-induced X-ray emission), layout ("fingerprinting") and other evidence. Probably most important of all have been the digital images that make it possible to examine copies in distant locations side-by-side.

Evidence of tampering emerges out of a close study of the physical objects, making analytical bibliography essential to establishing their authenticity. The famous example is the legendary exposure of the forgeries of the highly respected bibliographer Thomas J. Wise, uncovered and written up by John Carter and Graham Pollard in their *Enquiry into the Nature of Certain Nineteenth Century Pamphlets* (1934).

Bibliographers recall Falconer Madan's phrase, "the duplicity of duplicates," and instinctively look for differences between copies that appear to be identical, and then seek to explain how and why the differences came about. In work with printed books of the hand-press era, they separate the differences and have conventional names for them. *Editions* are the basic units of production, that is, all the copies run off from the same printing surfaces. *Impressions* distinguish all the copies of a single press run. *Issues* involve changes overtly announced, for instance on title pages, and deal often with conditions of sale. *States* involve covert changes, made within a press run and not meant to be noticed, least of all mentioned on the title page. Often the levels of differences are hard to explain, or ambiguous, in which case the term *variant* is a useful recourse. (The conceptions are Anglo-American: terms used in continental Europe, like *Ausgabe, Auflage*, and *triage*, fit awkwardly into their hierarchy.)

Descriptive and Textual Bibliography

Descriptive bibliography is enumerative bibliography informed by analytical bibliography. The goal is to specify the exact particulars of the physical objects for scholarly use. Its rhetoric is ruled by conventions that provide for details that identify the important physical characteristics. Analytical bibliographers need to know the history of the materials and practices of printing and publication. Descriptive bibliographers will also know the grammar for formulating citations based on models that date from Henry Bradshaw in the nineteenth century, as later refined by W. W. Greg and codified in Fredson Bowers' *Principles of Bibliographical Description* (1949). Enumerative citations are assumed to apply to all copies of a particular title; any differences are irrelevant for the intended readers.

Descriptive bibliographers may identify either a specific copy, with details on all of its relevant idiosyncrasies, or they may conceive of ideal copy (which, in Bowers'

conception, is "a book which is complete in all its leaves as it ultimately left the printer's shop in perfect condition and in the complete state that he considered to represent the final and most perfect state of the book"). Citations in enumerative bibliographies are generally shorter, and their conciseness is assumed to be sufficient. Descriptive bibliographies explore physical details in order to allow readers reasons either to infer that the description is sufficient, or to record the insufficiencies.

Citations in descriptive bibliographies usually transcribe the title page and collate the gatherings. The ideal title-page transcription covers all its printed statements, often with type styles in "quasi-facsimile" presentation. The collation uses a grammar that records the gatherings by their signatures and the number of leaves in each gathering. A knowledge of early printing practices is essential, beginning with the basic practices of imposing the type within the form of the press, for purposes of accommodating sheets that were later folded once, twice, or three times (so as to produce folio, quarto, or octavo format). The directions of the paper's chain and wire lines and the placement of the watermark are most often the crucial evidence. Changes during a press run were common. Leaves were added, removed, or cancelled, unsigned gatherings were added at the beginning, or interpolated. The early printing shop was constrained in its conditions— labor was usually cheap and materials were expensive— and the physical book needs to be explained in terms of how it was assembled.

The precise details of producing a physical book often explain its content. Textual bibliography is a search for authenticity of literary works based on a probing of the printed evidence. Its roots are in classical philology, Biblical studies, and modern editorial practice. Its calls on the practices of analytical bibliography to authors whose writings exist in variant printed editions, often in the absence of manuscript sources. Locating the crucial evidence requires a thorough familiarity with the text, a close reading of many copies in search of variants, and a knowledge of printing house practices, as well as an understanding of the authors and their working relationships with their printers, editors, and publishers. Contrary to what one might suspect, there are important differences in the texts not only of earlier authors of the "hand press era" (before about 1830) but also of many twentieth century authors as well.

Historical Bibliography

The written history of the artifacts that define our civilization begins as "book appreciation," popular accounts of books and printing that recall the classic lore, of Laurens Janszoon Coster in Holland as a possible precursor of Gutenberg, of the Columbus letter and role of printing in the rapid spread of the news across Europe, of the *Depositio cornuti typographici* and Wayzgoose revelry among printing apprentices and journeyman, of the dispersal of the 1640 *Bay Psalm Book* and the forging of the 1639 "Oath of the Freeman," of Walt Whitman's *Leaves of Grass* as a reminder of his work as a printer. The early craftsmen were heroes, their editions were works of art.

Bibliographers delight in these stories, and also in the vast profusion of other names, titles, and events they work with. They were rarely collected into historical narratives, however; before the twentieth century the historiography of bibliography was rather meager. In works like Karl Schottenloher's *Das alte Buch* (1919, later eds. as *Bücher bewegten die Welt*), the account begins to be framed as cultural history. The *annales* tradition of French academic scholarship, rich in charts, statistics, and maps, characterizes *L'Apparition du livre* (1958), begun by Lucien Febvre and completed by Henri-Jean Martin, as well as the later *Histoire de l'édition française* (1981–1985), to date the most lavish of several recent national histories of the book. Two complementary exhibitions in London in 1963 further expanded the agenda of historical bibliography. At Earl's Court, a display on the history of the printing crafts and technology reflected a flourishing scholarship in the processes of book production, obvious tied to analytical bibliography. Two major periodicals emerged—the *Journal of the Printing Historical Society* (1965) and *Printing History* (1979)—along with syntheses based on the acts of production, such as Michael Giesecke's *Der Buchdruck in der frühen Neuzeit* (1991) and Adrian Johns's *The Nature of the Book* (1998). The second part of the 1963 exhibition, at the British Museum, was devoted to the first publications of the historic landmarks of Western thought, and led to the celebrated *Printing and the Mind of Man* catalogue (1967), along the way implying the crucial role of the antiquarian booktrade in preserving our cultural heritage. Two other works stand out in the rise of historical bibliography: Elizabeth Eisenstein's *The Printing Press as an Agent of Change* (1979), which surveys the views of earlier writers who evaluated the impact of Gutenberg's invention, and Robert Darnton's study of the archival evidence of the publishers of the great eighteenth century *Encyclopédie* in *The Business of Enlightenment* (1979).

Are there significant differences between historical bibliography and the new fields of study called Print Culture and Book History? (See the entry, "History of the Book," p. 2142.) If there are, they are subtle and often irrelevant. All of the families of bibliography benefit from being conceived with the needs of both readers and materials in mind. Alfred W. Pollard, one of the heroes from the last century, famously reflected these sentiments when he described bibliography as a "big umbrella."

BIBLIOGRAPHY

1. ALA, Reference and Adult Services Division. Guidelines for the Preparation of a Bibliography, 1992; revised 2001, in RUSA, Winter 2001, also online at http://www.ala.org/ala/rusa/protools/referenceguide/guidelinespreparation.cfm.
2. Balsamo, L. *Bibliography: Story of a Tradition*, Rosenthal: Berkeley, CA, 1990.
3. Barnard, C.H. *Bibliographical Citation*, 2nd Ed. Clarke: London, U.K., 1960.
4. Bates, M.J. Rigorous systematic bibliography. RQ. **1976**, *16*, 7–26.
5. Belanger, T. Descriptive bibliography. In *Book Collecting: A Modern Guide*; Jean, P., Ed.; Bowker: New York, 1977; 97–101. Available at http://www.bibsocamer.org/.
6. Berger, S. *The Design of Bibliographies*, Mansell: London, U.K., 1991.
7. Blum, R. *Bibliographia: An Inquiry into Its Definitions and Designation*, ALA: Chicago, IL, 1980.
8. Breslauer, B.H.; Folter, R. *Bibliography: Its History and Development*, Grolier Club: New York, 1984.
9. Davison, P. *The World Encompassed*, University Press: Cambridge, U.K., 1992.
10. Esdaile, A. *Student's Manual of Bibliography*, Allen and Unwin: London, U.K., 1931, later eds. by Roy Stokes.
11. Hackman, M.L. *The Practical Bibliographer*, Prentice-Hall: New York, 1970.
12. Harmon, R.B. *Elements of Bibliography: A Simplified Approach*, 3rd Ed. Scarecrow: Lanham, MD, 1981.
13. Harner, J. *On Compiling an Annotated Bibliography*, Modern Language Association: New York, 1985, 1991.
14. Harris, N. *Analytical Bibliography: An Alternative Prospectus*, L'Institut d'Histoire du Livre: Lyons, France 2004. Online at http://ihl.enssib.fr/siteihl.php.
15. Kieft, R.H. The return of the "Guide to Reference" (Books). Refer. User Serv. Q. **2008**, *48*(4), 8–10 6.
16. Krummel, D.W. *Bibliographies: Their Aims and Methods*, Mansell: London, U.K., 1984.
17. McKenzie, D.F. *Bibliography and the Sociology of Texts*, British Library: London, U.K., 1986. The Panizzi Lectures, 1985.
18. Pollard, A.W. The arrangement of bibliographies. Library, **1909**, 168–87 n.s., 10.
19. Robinson, A.M.L. *Systematic Bibliography*, 4th Ed. Bingley: London, U.K., 1979.
20. Stokes, R. *The Function of Bibliography*, André Deutsch: London, U.K., 1969.
21. *Studies in Retrospect*, The Bibliographical Society: London, U.K., 1945.
22. Tanselle, G.T. Bibliographical history as a field of study. Stud. Bibliogr. **1988**, *41*, 33–63.
23. Tanselle, G.T. *Introduction to Bibliography*, Book Arts Press: Charlottesville, VA, 2002, 19th revision.
24. Wilson, P. *Two Kinds of Power: An Essay in Bibliographical Control*, University of California Press: Berkeley, CA, 1988.

Bibliometric Overview of Information Science

Howard D. White
College of Information Science and Technology, Drexel University, Philadelphia, Pennsylvania, U.S.A.

Abstract

This entry presents an account of the core concerns of information science through such means as definitional sketches, identification of themes, historical notes, and bibliometric evidence, including a citation-based map of 121 prominent information scientists of the twentieth century. The attempt throughout is to give concrete and pithy descriptions, to provide numerous specific examples, and to take a critical view of certain received language and ideas in library and information science.

PROBLEMS OF TERMINOLOGY

"Information" and "science"—two formidable words joined. The combination has not gained wide understanding. Few persons can quickly associate it with a subject matter as they can geology with rocks or astronomy with stars. That may be because even expert definitions of information science (IS) resemble attempts to describe an elephant by describing the tent under which it is performing: some given in Zins,[1] for example, could be parodied as "Information science is the science of data, information, knowledge, and messages." Abstractions like this also cover, e.g., artificial intelligence, cognitive psychology, cognitive science, computer science, cybernetics, database management, informatics, and systems theory, as attested by *The Study of Information*,[2] an interdisciplinary volume from a generation ago. One section of that large book resembles the present encyclopedia in being devoted to "*library* and information science"—L&IS—but this qualification, useful to insiders, is not readily intelligible to ordinary readers. In the popular view, librarians check out books and put them back in order on the shelves. So where is the science? It must involve something like memorizing the Dewey decimal system. Isn't that what librarians mean by "information"? And hasn't Google ended all that?

As it happens, even L&IS experts who would laugh at such notions have not clearly established their field for nonexperts. This entry attempts to do that through such means as definitional sketches, identification of themes, historical notes, and bibliometric evidence, including a citation-based map of information scientists. As an overview, it also points to entries in this encyclopedia that develop the histories of certain subfields more elaborately and systematically. It was written wholly independently of the ELIS entry on *Information science*, p. xx by Tefko Saracevic, but turns out to complement his quite well, arriving at similar conclusions through different means (See *Information Science*, p. 2570).

First, a quick commentary on librarianship. Traditional librarianship is not a science, nor is it usually based on science. It is a range of services that many intelligent persons can carry out intuitively. This is not to deny that librarians need coursework. It is merely to say that if "science" is taken to mean novel claims rigorously tested, then "library science," which dates from the nineteenth century, is a misnomer; it really just means something like "library administration." In Pierce Butler's 1933 interpretation,[3] it connotes a vague ethical stance, as when he recommends that librarians study "function" not "process." Butler's book, once considered classic, has been deservedly debunked by contemporary writers, not least his idea of science.[4,5]

There are, of course, empirical studies of library operations that could be called "library science," were the term fresh. These studies do reveal new, nonobvious facts about the world, and some of them have been made by librarians.[6,7] But the activities most people associate with libraries are not defined by them, and libraries would not be markedly different if they had never been conducted. The bent of traditional librarianship is normative rather than positive. That is to say, librarians are not mainly concerned with *what is*, in the scientist's sense of generalizable truths uncovered through measurement. They are mainly concerned with *what ought to be* in the politics of physical and intellectual access to texts. They are most at home with qualitative standards, such as codes, and ethical pronouncements, such as being for information literacy and against censorship. Library science in this sense is a policy science, not something more rigorous. Students entering library schools have probably gathered as much. They know they already possess the mindset required for professional life, and all they really need for marketability are socialization into the field and some technical skills.

What is gained, then, by tacking "library" onto "information science" to form L&IS? Principally, differentiation from the other information sciences as described by

Encyclopedia of Library and Information Sciences, Fourth Edition DOI: 10.1081/E-ELIS4-120044527

Bibliographic–Bibliotheque

Machlup and Mansfield[2] or T.D. Wilson.[8] Does L&IS belong in this company, or is "L&IS" just another overblown title for what librarians do? The answer is, yes, it belongs, and, no, it isn't another name for librarianship. Although its various projects are far from being tightly cumulative, decades of evidence from what P. Wilson[9] calls "bibliographical R&D" compel the conclusion that L&IS is real science and technology. Parts of L&IS are as mathematical as physics. Thousands of studies involve theorizing, measurement, the development and testing of hypotheses, experimentation and other respectable forms of data-gathering, the gradual improvement of explanatory models, and the engineering of real-world systems.

Among the information sciences, moreover, the domain of L&IS is unique. The discipline's principal object of study is the age-old interface at which people communicate with bodies of related writings—broadly speaking, with literatures. At this interface, people seek to match their interests and questions with responses that would otherwise be hidden in unreadably large masses of text. At the same time, they want responses that meet the standards of their dialogues with other persons—standards of relevance, truth, novelty, timeliness and appropriate length. However, the literature side of the interface is mindless, and whatever responsiveness it has is built into it artificially through systems. Even when these systems include human intermediaries (such as reference librarians), they often function very imperfectly. It therefore makes sense to speak of the literature interface as a barrier that parts of L&IS try to reduce or remove.

L&IS DEFINED

In brief, L&IS consists of *research on literature-based answering*. Central to it are the properties of human questioners, the properties of various mediating systems (since literatures can't talk), the properties of literatures themselves, and the properties of answers. These interrelated concerns define subfields of L&IS as a scientific and technical enterprise. Taking each in turn:

1. Questioners' properties, both cognitive and sociocultural, imply what in the past were called user or usage studies and are now known as studies of information behavior. The latter title embeds literature-based answering in a broader context of behaviors, such as reliance on people rather than writings for answers. "User" connotes reliance on formal systems, and part of studying people's relationship with writings has always been exploring reasons for their nonuse. However, millions of people still seek answers from writings, and how they express their needs through questions put to literature-based systems—for example, in Internet searches—remains a perennial topic of interest. So do the uses they make of information once found. Many studies have been conducted of information seeking and use in particular groups, such as children, high school students, medical doctors, scientists, and humanists (See the entries on *Information behavior*, p. 2381; *Information behavior models*, p. 2392; *Information needs*, p. 2452; *Sense-making*, p. 4696; *Information searching and search models*, p. 2592; and numerous articles on the information behavior of various groups; *User-centered revolution, 1970-1995 [ELIS classic]*, p. 5461; and *User-centered revolution, 1995–2008*, p. 5496; among others.)

2. Properties of mediating systems—the systems that bring people and writings together—imply research on both verbal instruments and computer algorithms that serve to help people find what they want. The enterprise has been known for more than 50 years as information retrieval (IR), but it has far earlier roots. The verbal instruments have always included bibliographic metadata, such as descriptive and subject cataloging, classification and indexing schemes, and abstracts. The preestablished lists of authorized terms used in subject cataloging, classification, and indexing are called controlled vocabularies. Below these content-laden instruments are the content-neutral computer algorithms. Nowadays on the Web, the natural-language "keywords" by which people put their questions can be algorithmically matched with natural-language "keywords" in the full texts of documents, and the older mediating systems, such as controlled-vocabulary indexing or classification schemes, have been called into question, although they still have their staunch defenders (See numerous entries on information organization and description, such as *Classification theory*, p. 1045; *Descriptive cataloging principles*, p. 1481; *Metadata and digital information*, p. 3610; specific systems such as *Library of congress classification (LCC)*, p. 3383; and *Anglo-American cataloging rules (AACR)*, p. 99; and standards and protocols, such as *Encoded archival description (EAD)*, p. 1699; and *Resource description framework (RDF)*, p. 4539).

3. The properties of literatures imply the field of bibliometrics, which devises measures of the growth, connectivity, impact, scattering, and obsolescence of publications in various subject areas (See *Bibliometric research: history [ELIS classic]*, p. 546; *Citation analysis*, p. 1012; *Information scattering*, p. 2564; and *Webometrics*, p. 5634; among others). Contemporary extensions, glossed in Table 1, are called scientometrics, informetrics, and, most recently, webometrics. Research projects in bibliometrics and information retrieval have often been pursued without much contact, but one major exception involves the citation indexes published by Thomson Reuters. They are studied as retrieval tools in IR and are used to model literatures and investigate citers' behavior in bibliometrics.

Table 1 Glosses on some overlapping subfields in IS.

Bibliometrics: The quantitative study of literatures. Measures countable properties across multiple related writings and looks for empirical regularities in structure. Can be considered a variety of cliometrics in that it typically deals with historical data that can change over time. May include data on usage of publications (e.g., in library collections). Units of analysis on which counts are made include documents (e.g., articles and books in subject areas), authors' *oeuvres*, and journals; also the publications of organizations or countries, or in particular languages. Can be either descriptive or evaluative in intent. Has its own theory and methods, but is related to such fields as content analysis, text mining, corpus linguistics, and stylometrics.

Scientometrics: Includes bibliometrics, especially as applied to scientific and technical literatures and to patents, but extends quantitative study to units of analysis other than publications, such as scientists and research organizations. May be a quantitative form of the sociology of science.

Webometrics: The quantitative study of Web pages and Web sites in such aspects as their growth and connectivity. Synonymous with cybermetrics, although the latter may be construed as dealing with the entire Internet rather than just the World Wide Web.

Informetrics: As currently defined, covers much of the quantitative side of IS, but focuses more on publications and information systems than on the psychology and sociology of information seekers. Oriented toward statistics and mathematical modeling.

4. The properties of answers, such as their relevance, accuracy, completeness, and concision, imply evaluative research on the output of information retrieval systems. Both before and after the advent of the computer, IR systems can be construed as hypotheses put forward by information scientists as to what will facilitate literature-based answering. A good part of the science of the field consists in testing how well these systems work—assessing their strengths and weaknesses and proposing improvements in the quest for artificially induced relevance (See *Information retrieval experimentation [ELIS classic]*, p. 2526; *SMART system: 1961–1976 [ELIS classic]*, p. 4770; *User-oriented and cognitive models of information retrieval*, p. 5521; *XML information retrieval*, p. 5700; *Web retrieval and mining*, p. 5615; and *Text REtrieval Conference (TREC)*, p. 5182; among others.) Some researchers are now devising systems that try to answer unforeseen questions directly rather than providing documents on a topic. Others are experimenting with algorithmic summarization of documents.

This four-part division has warrant in existing organizations, all with periodic conferences and journals. Repeating the order above:

1. The American Society for Information Science and Technology (ASIST) enrolls members who represent all four parts, but ASIST is now recognizably the home of the so-called cognitive wing of IR, who specialize in information behavior and advocate user-centered design. Among its publications are the *Journal* of the Society (JASIST) and the *Annual Review of Information Science and Technology*. An independent biennial conference called Information Seeking in Context is addressed to behavioral interests.

2. Persons concerned with linguistic aspects of indexing, classification, and metadata are members of ISKO, the International Society for Knowledge Organization. They publish a journal, *Knowledge Organization*.

3. The algorithmists and evaluators of experimental computerized document retrieval systems belong to the Special Interest Group on Information Retrieval (SIGIR) within the Association of Computing Machinery, which publishes annual proceedings of the SIGIR conferences. Some of these researchers compete in the Text Retrieval Conference (TREC), an important annual contest staged by the (U.S.) National Institute of Standards for comparative evaluation of retrieval algorithms. Also for technical types, ACM and the Institute of Electrical and Electronics Engineers sponsor a Joint Conference on Digital Libraries, and IEEE by itself sponsors the Conference on Information Visualization.

4. Bibliometricians and kindred spirits have created ISSI, the International Society for Scientometrics and Informetrics. Their specialized journals are *Scientometrics* and the *Journal of Informetrics*.

There are also, of course, generalists whose interests cut across all these organizations and may extend to membership in the American Library Association or other librarians' groups. In Britain such interests meet in the Chartered Institute for Library and Information Professionals, which resulted from a merger of the [British] Library Association and the Institute for Information Scientists in 2002. International conferences intended to have cross-disciplinary appeal include Conceptions of Library and Information Science.

As noted, the phrase "L&IS" is not likely to evoke these definitions and details for most hearers or readers. It is true that libraries hold subsets of literatures and that librarians try to make them responsive to human inquiries. To certain initiates, therefore, libraries are the paradigm of the human–literature interface.[10] But in practice this interface is not limited to libraries, nor is its scientific study limited to that context. Can the matter be cleared up by substitution of a more suitable word or phrase than L&IS? Not really; if the language provided a better name for the field, it would have been adopted by now.[11] English in this case is lean in possibilities.

Leaving "library" implicit, the field becomes just "information science" or IS.[12] This name connotes "information" in at least two senses—as a synonym for textual content and as the presumed cognitive outcome of reading the texts. "Documentation," its forerunner, connoted rather less. "Literature science"—meaning in White and McCain[13] something like "literature response theory"—might suggest major parts of IS better than "library science," but the other senses of "literature" bar its acceptance; relatively few people think first of subject literatures or linked Web pages as opposed to belles lettres when they encounter the word, and some humanists might deplore the conjunction. The phrase "and technology" is sometimes added to IS to evoke the field's long-standing stake in engineered systems. All the other information sciences have distinctive names of their own (e.g., artificial intelligence), leaving IS to be principally understood as it is in names like the American Society for Information Science and Technology or the former Institute for Information Scientists in Britain.

MEDIATORS AND AUTOMATION

The founder of the latter organization, Jason Farradane, is also apparently the person who named IS as a field in 1953. He did so by articulating the concept of an "information scientist."[14] At least initially, however, he did not use the term in today's sense of someone who does research on literature-based information systems and their users. He used it in the context of British laboratory work to name a scientist, such as a bench chemist, who is adept in searching the literature for materials pertinent to research projects—so adept that this becomes a full-time occupation. (Thus, one might also see terms like "information chemist" or "literature chemist"; the older "documentalist" had not really caught on.) Dyson and Farradane[15] call this person "an expert *user* of all sources of information."

The idea of persons who mediate between inquirers and literatures is certainly not original with Farradane;[16] one obvious earlier instance of it is the reference librarian.[17] Farradane's information scientists, however, were supposed to have more substantive and technical expertise than the typical reference librarian. The same perceived shortcoming has beset librarianship for more than a century. For example, the Special Libraries Association broke away from the American Library Association in 1909 to signal its members' more particular subject knowledge and more proactive services.[18] For similar reasons (as well as a greater interest in microfilm technology), members of the American Documentation Institute left mainstream librarianship in 1937.[19] Or, from the realm of theory, take Havelock et al.,[20] a massive treatise on the role of persons who link others with potentially useful writings. In seeking to create a corps of such persons who could bring published educational research to school teachers and administrators for practical application, Havelock modeled his linkers on agricultural extension agents. Librarians of any kind are not even mentioned.

As distinct from a librarian, the expert user of literatures on behalf of others gradually became an "information officer," "information specialist," or "information professional," and Farradane himself evolved into an information scientist as the term is currently understood. At the heart of that understanding lies the idea of *automation* (See *Library automation: history*, p. 3326.) Even in the 1930s and 1940s, what is now called IS had begun its long-term project of automating the human–literature interface.[21] One well-known godfather of this project is Vannevar Bush. Describing the design of the Memex, his imaginary retrieval device from the 1930s, Bush[22] summarily dismissed the indexing schemes of librarians and documentalists, whom he also never consulted;[23] instead, he foresaw that users would thematize their own collections of microfilmed documents with personal "associative trails."[21] The implicit tendency of IS, in other words, is to eliminate human intermediaries—the indexers, classifiers, catalogers, abstractors, librarians, and professional literature searchers who stand between textual resources and their non-specialized end users.[24] Thus, information scientists in Farradane's original sense have been increasingly marginalized by a line of researchers whose inclination is to substitute algorithmic for costlier manual processes wherever possible. (The process is sometimes explicitly called "disintermediation.") This has led to the current situation in which the dominant symbol of the human–literature interface is not the traditional library—a room or building staffed by people who provide service with classified, cataloged, and indexed books and serials—but Google, which directly matches the natural language of users with the natural language of documents in cyberspace (For pre-Google antiquity, see *Internet search tools: history to 2000*, p. 2996).

While traditional libraries still have a place in the new order of things, there is much evidence that a profound change has indeed occurred.[25] Many have seen it coming. For instance, describing "The New Generation" in his book *Digital Libraries*, the computer scientist William Arms writes (ch. 14):[26]

> In 1997, a Cornell student who was asked to find information in the library reportedly said, "Please can I use the Web? I don't do libraries." More recently, a faculty member from the University of California at Berkeley mused that the term "digital library" is becoming a tautology. For the students that she sees, the Internet is the library. In the future, will they think of Berkeley's fine conventional libraries as physical substitutes for the real thing?

THE GOOGLE BOOM

Speaking of Google, it is interesting that its creators, Sergey Brin and Lawrence Page, have never been members of the American Society for Information Science and Technology and are not usually described as information scientists in the "library" sense. (They were computer science students at Stanford University.) Yet their publications draw on ideas from IS. For example, the idea of algorithmically ranking documents by their suitability as answers (as Google does Web pages) was established in IS decades ago, notably in the work of Gerard Salton. The Saltonian tradition now lives in the proceedings of the Text Retrieval Conferences, one of which Brin and Page[27] cite when they claim that TREC techniques do not scale up to handling the billions of pages on the Web. The key Google ranking algorithm PageRank, based on counts of links to particular Web pages, was influenced by the idea of citation analysis from bibliometrics, attributed in Page et al.[28] to Eugene Garfield, the founder of citation databases in science and scholarship.

But more important than Brin and Page's intellectual ancestry is their practical success in establishing a universal interface between inquirers and literatures in the widest sense. Their search engine has arguably had the greatest impact on libraries of any innovation in the past century. The old and the new are presently joined in such modules as Google Book Search (for retrieving passages from book texts in full), Google Scholar (for retrieving articles and other documents based on bibliographic citations), and OCLC's Open WorldCat (for retrieving books held by local libraries in response to queries in Google and other search engines). Most notable is the use of Google's main module—by worldwide nonspecialized publics—for answering reference questions and doing topical literature searches. These developments seem consistent with traditional library values even if they are not particularly auspicious for librarians.

During what might be called the Google boom, work recognizable as IS has become quite fashionable in academe, especially in computer science departments. Moreover, at this writing about 15 of America's best library schools have morphed into "schools of information," also known as "iSchools," with computer science professors well represented on their faculties. In the past, with notable exceptions such as Donald Kraft (Louisiana State University) and Bruce Croft (University of Massachusetts at Amherst), computer scientists have been known to find the research problems of IS unappealingly messy, involving as they do both people and natural language as subject matter. But today the de facto technical vanguard of IS includes computer scientists at many elite institutions (for example, Ricardo Baeza-Yates at Yahoo! Research, Susan Dumais at Microsoft Research, Marti Hearst at UC Berkeley, Jon Kleinberg at Cornell, Christopher Manning at Stanford, and Simone Teufel at Cambridge). Most such researchers resemble Brin and Page in not being members of ASIST. Dillon[29] similarly observes that "the term 'information science' has been taken on by philosophers and mathematicians who see themselves very much

distinct from the LIS domain..." The field is also now highly international, with numerous researchers from countries outside the traditional homes of IS in North America, Western Europe, and Russia.

One cause of the upturn in IS-related computer science was the digital libraries initiative of the 1990s. Funded in the United States and Britain by agencies that for years had largely ignored problems of the human–literature interface, this initiative made information scientists of persons who might enter a traditional library only to avoid the rain. The larger attraction was the transformative effect of the Internet and the Web on the world economy, which made new R&D on information systems not only intellectually stimulating but also potentially very lucrative. (Brin and Page are multibillionaires.)

Perhaps the foremost development in current IS is the explosion of interest in graph-theoretical work on dynamic networks—that is, the modeling of groups of entities as countable nodes and links whose counts can change over time.[30–32] Quantitative network theory is abstract enough to be applicable to many different disciplines, but in IS it has been used to model—and visualize in color—such things as the papers connected by citation in the literatures of science and technology or bundles of hyperlinked pages across the entire Web. These are bibliometric (or webometric) applications. But network theory also turns up in IR—for instance, in the PageRank algorithm and the HITS algorithm,[33] each employed in Web retrievals. The same methods and formulas can also effectively model interrelated persons, such as "invisible colleges" of communicating scientists or networks of coauthors. Interestingly, the PageRank measure, used to rank pages in Google, resembles an earlier measure, Bonacich centrality, used by sociologists to rank the centrality of people in social networks.[34] Ranks based on network ties lead to automated recommendations of documents or people as possible sources of information. Google and the online bookseller Amazon.com have shown the relevance of such recommendations to librarianship.

There has long been a mathematical tradition in IS. In bibliometrics, for example, it began with S. C. Bradford and A. J. Lotka and continued with William Goffman and Manfred Kochen. The data typically involve power-law and lognormal distributions. In exploring and possibly exploiting such data today, algorithmically minded computer scientists are joined by physicists (e.g., Albert-László Barabási, Guido Caldarelli) and applied mathematicians (e.g., Mark E. J. Newman). In the past, quantifiers like these have occasionally taken positions in schools of L&IS, but the tendency now is for such persons to "do IS" on a variety of fronts outside the discipline.

It is necessary to mention these new information scientists because existing studies of IS seldom capture them. Existing studies[35–37] characterize the discipline in terms of contributions to a limited range of home journals. However, the computer scientists and other researchers now contributing to IS (using that term expansively) publish in conference proceedings, monographs, and journals of their own. No scholar as yet has assembled and analyzed their scattered works in the bibliographic databases ordinarily used to analyze IS as a field, such as Social Scisearch, Scopus, and Google Scholar. As a result, IS is currently defined in terms of past generations of canonical authors (mostly American or British), as opposed to the younger talents who are taking the field in new directions. This is not to say that the canonical figures no longer represent major subfields of IS or that continuity is lacking between old subfields and new—Zhao and Strotmann[38,39] have made a start toward capturing the latter—merely that, in present research, brain and computer power is being upgraded with unforeseeable but probably dramatic results. Striking advances have occurred, for example, in the visualization of bibliometric data[40,41] and in text mining,[42] with implications for tomorrow's libraries. We are a long way from Pierce Butler.

A MAP OF IS

Several information scientists have visualized IS through author maps. White and McCain,[36] for example, defined the field by mapping the 120 authors with the highest citation counts in 12 IS serials during 1972–1995. These computer-produced maps group authors algorithmically on the basis of cocitation. That is, the authors frequently cited together are placed relatively close together in "intellectual space." Groups of proximate authors can usually be identified as subdisciplines or research specialties. The broadest groupings in such maps, corresponding to subdisciplines, are, first, authors mainly concerned with the design and evaluation of systems through which people seek answers from subject literatures, and, second, authors mainly concerned with the literatures that may provide those answers and whose quantitative properties are worth investigating in their own right. The first subdiscipline could be sweepingly characterized as "IR," comprising authors who are, for example, algorithmic retrievalists, computer interface designers, specialists on indexing vocabularies, conductors of user studies, and systems evaluators. The second subdiscipline could be characterized as "Domain Analysis," made up of authors who are, for example, bibliometricians, citationists, visualizers of scholarly and scientific fields, and researchers on disciplinary communication patterns.

An analysis of one such disciplinary map follows. The White and McCain[36] data, with one author added, were remapped as a Pathfinder Network (PFNET) in White.[43] A new configuration of that PFNET appears as Fig. 1. The two articles just cited give methodological and interpretive details, along with the mapped authors' full identities. (A fair number of these authors are referenced and indexed in this encyclopedia.) Here, it is enough to say

that Fig. 1 coherently organizes some major twentieth-century contributors to IS and suggests their degrees of influence. Quite a few of them, one must note, are now dead. Of those living, almost all have attained senior positions in their institutions, and many are retired or near retirement.

The map derives from a matrix in which raw cocitation counts for every pair of the 121 authors have been obtained from Social Scisearch. Most of the author pairs were cocited at least once during 1972–1995, but the PFNET algorithm links each author in the matrix only to that other author with whom he or she has the *highest* (or tied highest) cocitation count. For example, at lower left, Terry Winograd is linked only to Herbert Simon. This means that, of all Winograd's cocitation counts with the other 120 authors, his is highest with Simon. Nearby, William Paisley is linked to both Thomas J. Allen and Diana Crane. This means that Paisley has a tied highest count with both Allen and Crane. All other links are shorn away.

From the PFNET's radical simplifications, strong patterns emerge. To reinforce them, the size of each author's node has been made proportional to the number of other authors directly linked to it. The four authors dominating the linkages—stars in anyone's history of IS—are Gerard Salton, F. W. Lancaster, Eugene Garfield, and Derek de Solla Price. Salton has the most authors linked to him, followed by Garfield. Lancaster, however, is the most central author in the map. Cutting the link between him and Garfield would divide the authors into two subdisciplines of roughly equal size, the top one representing varieties of research in information retrieval and the bottom one representing varieties of domain-analytic research in scientometrics and the sociology of science. Derek Price, a polymath with doctorates in both physics and the history of science, is a key link in the latter.

Authors restricted to one position by PFNETs may in fact have published works in more than one specialty and even in both subdisciplines. An account of the multiple contributions of authors emerges from the factor analysis in White and McCain.[36] That analysis nevertheless shows these authors' *main* specialties, as perceived by citers, in their highest factor loadings. Authors who load most highly on the same factors tend to be those explicitly linked in Fig. 1.[43]

Salton and the authors around him are experimental retrievalists. Among them are Cyril Cleverdon and Michael Keen, leaders of the paradigm-setting projects on IR at the Cranfield Aeronautical College in England during the 1950s and 1960s. Researchers in the Cranfield tradition experiment with strategies of indexing and searching to improve scores of effectiveness in document retrieval. They design systems, run trials, and use human relevance judgments to assess the results. They also devise and justify measures by which system outputs can be evaluated (e.g., scores in precision and recall). Most have

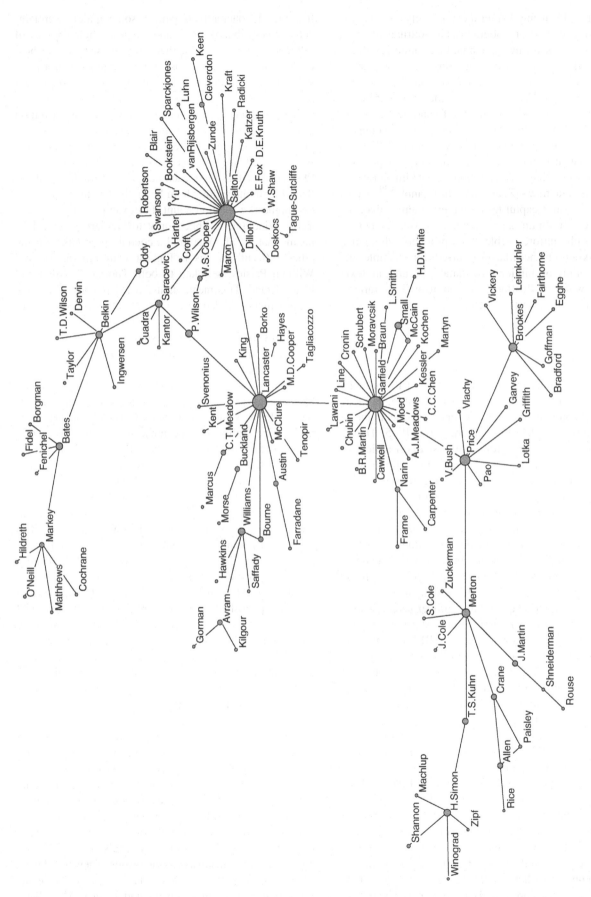

Fig. 1   A map of 121 authors in IS, 1972–1995.

mathematical backgrounds, and virtually all are oriented toward statistics and algorithms. For many years their work was identified with "test-bed" collections—relatively small sets of topically constrained documents whose properties they knew well and on which their experiments were conducted. They did not usually evaluate retrievals from large commercial bibliographic databases such as those vended by Dialog or Thomson Reuters (See Salton's own entry: *SMART system, 1961–1976*, p. 4770). In this tradition, moreover, the judges who assessed retrieved documents for relevance to queries were usually surrogates for end users, and there was little or no work on "real-world" users of systems. Largely pre-Web themselves, the Saltonians' influence lives on in the Text Retrieval Conferences. TREC members base their research on much larger collections of documents than their forerunners but still resemble them in mindset.

Another group of retrievalists branches off to the left of the Salton group. In the so-called cognitive revolution in IR, which dates from the late 1970s, they increasingly reacted against the experimental retrievalists' emphasis on systems in favor of a more user-oriented view. (Some of the authors shown around Salton—e.g., Don Swanson and Steven Harter—also did this.) The cognitive group might jointly be seen as "practical retrievalists." Key transitional figures, maintaining a foot in both camps, are Tefko Saracevic, Nicholas Belkin, and Robert Oddy. Saracevic is linked to the philosopher Patrick Wilson and the psychologist Carlos Cuadra because all are noted for their work on relevance, often said to be the central notion in IS. The names around Belkin are identified with psychological aspects of user theory; around Marcia Bates, with users as online searchers; around Karen Markey, with users of online public access catalogs (OPACs) in libraries. The practical retrievalists are relatively nonquantitative in their approach, and much of their work involves users outside science and technology, such as public library patrons, students, and humanities scholars. Their work also casts doubts on the paradigm of IR that was initiated by the Cranfield team and furthered by experimentalists such as Salton, Keith van Rijsbergen, and Karen Sparck Jones.

The simplifying assumptions of early IR were that indexers describe documents with a fixed vocabulary, that users employ the same vocabulary in searching, and that the system brings both contentedly together when query terms and document terms match in topic. A major achievement of the cognitive group has been to establish that lay users seldom know or can tell the system exactly what they want, and learned users, though more sophisticated, do not robotically interact with a system or judge documents solely by topical match. Neither kind of user is much given to looking up indexing terms in thesauri or classification schedules to assist their searches. Both kinds flounder around exploratively, follow clues to diverse sources, and change their minds. Their search strategies and relevance judgments are a creative process that goes well beyond the simplified input–output retrieval cycle of the textbooks.

The man who probably did most to popularize the simplified cycle was F. W. Lancaster, the quintessential information scientist between the Cranfield project (in which he participated) and the coming of the Internet. His book *Information Retrieval Systems*[44] is the *locus classicus* of the paradigmatic diagram in which users' "request profiles" and indexers' "document profiles," both based on controlled vocabulary, meet and sometimes match in a single system. Lancaster's main factor loading is as an experimental retrievalist, but he was always more oriented toward everyday library practice and operational bibliographic systems than most of the Saltonians. For one thing, he led a large-scale evaluation of real-world document retrieval in the National Library of Medicine's MEDLINE database; for another, he wrote textbooks on the complexities of actual indexing languages and actual Boolean searching in online systems such as Dialog. He taught a generation of librarians how to take a systems approach to evaluation[45] and wrote voluminously on many other IS subjects.

It therefore seems fitting that the authors around Lancaster in Fig. 1 are themselves more diverse in their specialties than the authors around either Salton or Garfield. Most have their main factor loadings in general library systems (e.g., Allen Kent, Donald King, Robert Hayes, and Fred Kilgour); others have them in online retrieval (e.g., Martha Williams), OPACs (e.g., Michael Gorman), indexing theory (e.g., Derek Austen), user theory (e.g., Patrick Wilson—a bridging figure with the cognitive retrievalists), and even bibliometrics (e.g., Philip Morse), an area in which Lancaster also published. Many of the "Lancastrian" authors resemble him in being generalists who exhibit a wide range of interests across their publications (e.g., Michael Buckland).

In contrast to this mixed lot, the authors around Eugene Garfield all load most highly on citation analysis, the largest specialty in bibliometrics. Garfield founded the Institute for Scientific Information (ISI)—now Thomson Reuters—in Philadelphia in 1960. At ISI he and his staff developed citation indexes commercially as a new tool for retrieving documents, complementing subject indexes. (Lawyers and judges had long had Shepard's citation indexes for discovering chains of precedent in legal proceedings, and one of Garfield's motives for his indexes was to give scientists a way of tracking the chains of corrections, addenda, and retractions that affect the claims made in scientific papers; see p. 162).[46] From the beginning, however, it was evident that ISI's large-scale citation databases could be used for more than document retrieval. They could be used to operationalize the notion of scientific specialties as corpora of citation-linked writings—corpora whose composition changed over time, yielding mappable data for intellectual histories.[41] The databases could also be used to operationalize the notion of scientific impact or reputation, because they showed the varying citedness of individual papers, authors, journals, organizations, and countries.[47]

Led by the prolific Garfield and the comparably influential Price, the citation analysts found many new applications of ISI data for descriptive or evaluative purposes. One of Garfield's contributions was the well-known journal impact factor, still widely used for assessing the prestige of journals, although various replacements have been proposed.[48] Price virtually created the field of scientometrics by using citation data to study, e.g., author productivity (acknowledging Lotka as a forerunner), the growth of literatures, networks of scientific papers, invisible colleges, differentiation of the hard and soft sciences from the humanities, and cumulative advantage as a model of differences in scientific reputations. Among other innovators, Francis Narin, Mark Carpenter, and J. D. Frame did pioneering work on evaluative bibliometrics, notably on networks of journals that cite each other. Henk Moed and B. R. Martin are known for their uses of citation analysis in evaluating research teams at institutes and universities. Maurice Line gathered citation data to study the ageing (i.e., decline in use) of literatures; so did Belver Griffith. M. M. Kessler came up with the static associative measure called bibliographic coupling—the number of references documents share in their endnotes. Henry Small (Thomson Reuters's director of research) topped this with a dynamic associative measure, cocitation—the number of later papers that cite any particular pair of earlier ones. Howard White and Katherine McCain generalized Small's cocitation measure to authors' *oeuvres* and journals, respectively. Besides having uses in retrieval, bibliographic coupling and cocitation measures underlie the maps of science (such as Fig. 1) that have been appearing since the 1970s.

Linked to Price at right are B. C. Brookes's group of theoreticians who contributed significantly to the mathematical side of bibliometrics. Their work is concerned with fundamental processes of literature dynamics rather than with citation analyses of particular fields. Famous early students of these processes, all with distributions of bibliometric data named after them, are S. C. Bradford (connected to Brookes), A. J. Lotka (connected to Price), and George Zipf (connected to Herbert Simon at far left). Brookes himself, Ferdinand Leimkuhler, Robert Fairthorne, and Brian Vickery are all known for their work on "Bradford's law" (which pertains to regularities in the distribution of articles on a subject over journals). Leo Egghe, who was Brookes's doctoral student, is currently a leader in mathematical IS and the founding editor of the *Journal of Informetrics*.

From Price leftward, most of the authors represent disciplines other than IS. However, they share his interest in communication patterns in science and technology (which include citation networks and invisible colleges). Robert

Merton's group are largely identified with one of his specialties, the history and sociology of science. Merton, Harriet Zuckerman, and Thomas Kuhn (of paradigm fame) helped to legitimize citation analysis theoretically. Stephen and Jonathan Cole, Diana Crane, William Paisley, and Ronald Rice have all published citation analyses themselves, although their research topics are considerably broader. Herbert Simon (the Nobel laureate) and his group are sources of ideas imported into IS. Quantitative work by Simon and Zipf, for example, has been used in bibliometrics, and Claude Shannon's way of measuring "information" has had a diffuse influence.

TOO NARROW A VIEW?

As stated at the outset, many experts define IS at a high level of generality. This entry attempts to define IS more concretely by describing its center. It attempts to state the field's main scientific and technological concerns—literature-based answering, the human–literature interface—with many specific examples. When that is done, readers can assess not only what has been included but what has been excluded, and the exclusions can be judged. On grounds of clarity, that seems preferable to formulations of IS that imply anything and everything, from cloth books for preschoolers to the genetic code. (The same vague imperialism softens the field's name to "information studies.") Two earlier pieces by White and McCain[36,49] took a similar line, and both have been criticized as leading to unduly narrow conceptions of IS. A bit more exposition thus seems called for, one that also extends earlier definitional themes.

Fig. 1 is a core-and-periphery model of IS; it depicts specialties as central authors surrounded by less central ones. It also depicts Lancaster's group as being the most central. Through Lancaster, the retrievalists above and scientometricians below are joined to a diverse group of generalists with applied interests in real-world libraries and information systems. However, if the center of the field is conceived more widely, the entire map of 121 authors can be taken as modeling the core of IS. On analogy with a target, it would be the bull's-eye in a set of concentric circles representing adjoining specialties and disciplines. (Something like an outer circle of authors now radiates left from Price.)

Does this mean, however, that just these 121 authors and no others fully characterize IS? Certainly not. First, they represent the period 1972–1995, and, as stated earlier, a whole new generation of information scientists has risen since then. (Mapping approximately the same intellectual space, Zhao and Strotmann[38,39] show both new names and new specialties joining the older ones.) Second, while the 121 are an elite of sorts—they are the top citees in a set of 12 serials oriented toward IS and library automation—the cutoff was determined mainly by computational and presentational limits. The elimination of lower-ranked contributors to IS or librarianship (or both) was not meant to deny them a place in the field or to impugn what they write about. It simply means that, across 12 representative serials, the work of certain authors has been the most prized by their citing colleagues. Revealing the nature of this work is an evidence-based way of bringing a field into focus.

Not seeing their own interests represented in Fig. 1, Danton[50] and Webber[51] questioned the set of serials from which the top authors were chosen. White and McCain[36] acknowledged that the names they mapped were far from exhausting the field. A larger sample from the same or different serials would indeed add both authors and specialties. Areas such as library management and library education, for example, could be induced in Fig. 1 by expanding the pool of citees to include authors like Richard de Gennaro and Herbert S. White, both highly cited during 1972–1995 in library-oriented serials not among the original 12.

The subtext here, however, is that what appears on a citation map is never wholly value-free. It has a political dimension in which invidious comparisons are made. People who do not find themselves or their specialties on a map may feel slighted. Or they may fear that the maps will be used prejudicially, to rule out creativity in new directions.

In the latter vein, P. Wilson[52] argued that IS is both messier and far richer than White and McCain's[49] "restrictive" characterization of it as IR and bibliometrics, which is essentially repeated in Fig. 1. Cronin[53] argued that the "exclusionary" maps in White and McCain[36] may have revealed something of classical IS, but miss the many other contributory literatures being explored in the *Annual Review of Information Science and Technology* (ARIST), whose editor he had just become.

Obviously the White and McCain papers do not stifle creativity, or there would be no new literatures for ARIST to review. But in one sense Wilson and Cronin are correct: cocitation analysis drastically simplifies. The maps are high-altitude views of the consensus hidden in the bibliographic record. This consensus has been evolving over years or decades, and so the maps inherently look toward the past rather than the future. The best that can be said for them is that they provide a check on the overinclusive definitions of IS that many experts seem to favor. The *Online Dictionary for Library and Information Science*, for example, defines IS thus: "The systematic study and analysis of the sources, development, collection, organization, dissemination, evaluation, use, and management of information in all its forms, including the channels (formal and informal) and technology used in its communication." This phrasing covers all of the more than 500 sprawling, interdisciplinary ARIST chapters ever written, but also all of interpersonal communication, all of mass communication, all of education, and all of remote sensing (to go no further)—enormous areas of study that already have plentiful researchers of their own. At the same time, it misses what is both central and unique to IS, namely the human–literature interface.

Another check on the nature of IS appears in Table 2, which shows the 20 ARIST chapters most highly cited in the Web of Science (WoS) and Google Scholar (GS) as of early 2008. Most of these chapter titles concisely imply what is now the human–computer–literature interface as an object of scientific study. The top 20 are taken from the entire run of ARIST, 1966–2007, and are a merger of the top 15 titles in each database. In each set of 15, 10 titles overlapped and 5 were unique. The unique chapters appear in Table 2 with abbreviations of their sources; the overlapping chapters are unmarked. The WoS counts reflect citations in the journal literatures covered by Thomson Reuters. The GS counts reflect many of the same citations but draw on a wider range of sources (e.g., conference proceedings, books) and so are usually higher. The WoS counts are higher for relatively old chapters.

The topics that come to the fore accord quite well with the divisions of IS given here. Information behavior studies are well represented. So is IR, in both its algorithmic and cognitive wings; so is the bibliometric study of literatures, including the recent emphasis on visualization. The one area not immediately apparent is the linguistic side of bibliography, as seen, for example, in discussions of subject indexing, classification, search strategies, metadata. That area is indeed present at the level of chapter titles in ARIST; it is just further down in the rankings.

Several titles in Table 2 imply that IS includes studies of communication between people. According to Webber,[51] this is an area that White and McCain[36] explicitly leave to other fields. More accurately, the latter paper claims that the *main* dialogue studied in IS is that between people and literature-based information systems. But interpersonal dialogues oriented toward that center have always been and will remain important topics in IS. For example, no one would rule out research on how reference librarians interact with their customers, or on how colleagues consult as a means of avoiding literature searches, or even on how people inform themselves when, for whatever reason, they have little or no access to publications of any kind. There are, however, countless other studies of interpersonal communication that address matters information scientists never will. Does this mean that all such studies are irrelevant to IS? Not if the right context is set up; IS borrows from other fields all the time. Yet it also seems likely that most would enter IS in a very limited context. They would be cited once or twice, rather than repeatedly.

The lesson to take away is that virtually all definitions of IS need elaboration with examples. The present account has advocated a way of getting examples that is distinctive to IS—by citation counts, using the ranking principle that is fundamental to both IR and bibliometrics. The ranked citation frequency distributions underlying both Fig. 1 and

Table 2   Major topics in IS as shown by the most-cited ARIST chapters, 1966–2007.

Year	Titles	Authors	WoS	GS
1986	Information needs and uses	B. Dervin, M. Nilan	317	383
1994	Relevance and information behavior	L. Schamber	144	212
1989	Bibliometrics	H. D. White, K. W. McCain	132	215
1968	Information needs and uses	W. J. Paisley	101	76
1987	Retrieval techniques	N. J. Belkin, W. B. Croft	92	189
2002	Scholarly communication and bibliometrics	C. L. Borgman, J. Furner	74	194
1996	Query expansion	E. N. Efthimiadis	63	161
1997	Evaluation of information retrieval systems	S. P. Harter, C. A. Hert	62	82
1985	Probability and fuzzy set applications to information retrieval (WoS)	A. Bookstein	59	59
1996	Social informatics of digital library use and infrastructure (WoS)	A. P. Bishop, S. L. Star	56	55
1980	The impacts of computer-mediated organization and interpersonnel communication (WoS)	R. E. Rice	54	20
1977	Bibliometrics (WoS)	F. Narin, J. Moll	53	40
1997	Visualization of literatures	H. D White, K. W. McCain	52	63
1993	Browsing—A multidimensional framework	S. J. Chang, R. E. Rice	48	80
1991	Cognitive research in information science (WoS)	B. L. Allen	44	60
2003	Visualizing knowledge domains (GS)	K. Börner, C. Chen, K. W. Boyack	42	153
2001	Conceptual frameworks in information behavior (GS)	K. E. Pettigrew, R. Fidel, H. Bruce	40	64
1990	Information need and use studies (GS)	E. T. Hewins	38	70
1998	Cross-language information retrieval (GS)	D. W. Oard, A. Diekema	32	92
2002	Collaboratories (GS)	T. A. Finholt	23	112

Table 2 are highly typical of bibliometric data. They consist of a relatively small core of highly cited items, a larger number of decreasingly cited items, and a still longer tail of items with little or no citation. Ranking by citedness raises some items above a certain threshold, as if by balloting in an election. Those, in the judgment of all citing authors, are the items most relevant to the field. So how to prioritize one's examples for greatest relevance? Look to the cores.

REFERENCES

1. Zins, C. Conceptions of information science. J. Am. Soc. Inform. Sci. Technol. **2007**, *58*(3), 335–350.
2. Machlup, F., Mansfield, U., Eds. *The Study of Information: Interdisciplinary Messages;* Wiley: New York, 1983.
3. Butler, P. *An Introduction to Library Science*, University of Chicago Press: Chicago, IL, 1933.
4. Buckland, M.K. Documentation, information science, and library science in the U.S.A. Inform. Process. Manage. **1996**, *32*(1), 63–76.
5. Cronin, B. Pierce Butler's an introduction to library science: a tract for our times?. J. Libr. Inform. Sci. **2004**, *36*, 183–188.
6. Bensman, S.J. Bradford's law and fuzzy sets: statistical implications for library analysis. Int. Fed. Lib. Assoc. J. **2001**, *27*(4), 238–246.
7. Bensman, S.J. Urquhart and probability: the transition from librarianship to library and information science. J. Am. Soc. Inform. Sci. Technol. **2005**, *56*(2), 189–214.
8. Wilson, T.D. Information behavior: an interdisciplinary perspective. Inform. Process. Manage. **1997**, *33*(4), 551–572.
9. Wilson, P. Bibliographical R&D. In *The Study of Information: Interdisciplinary Messages*; Machlup, F., Mansfield, U., Eds.; Wiley: New York, 1983; 389–397.
10. Miksa, F. Library and information science: two paradigms. In *Conceptions of Library and Information Science: Historical, Empirical and Theoretical Perspectives*; Vakkari, P. Cronin, B., Eds.; Taylor Graham: London, U.K., 1992; 229–252.
11. Schrader, A.M. In search of a name: information science and its conceptual antecedents. Libr. Inform. Sci. Res. **1984**, *6*(3), 227–271.
12. Vakkari, P. Library and information science: content and scope. In *Information Science: From the Development of the Discipline to Social Interaction*; Olaisen, J. Munch-Petersen, E. Wilson, P., Eds.; Scandinavian University Press: Oslo, Norway, 1996; 169–231.
13. White, H.D.; McCain, K.W. Visualization of literatures. Annu. Rev. Inform. Sci. Technol. **1997**, *32*, 99–168.
14. Shapiro, F.R. Coinage of the term information science. J. Am. Soc. Inform. Sci. **1995**, *46*(5), 384–385.
15. Dyson, G.M.; Farradane, J.E.L. Education in information work: the syllabus and present curriculum of the Institute for Information Scientists Ltd. J. Chem. Doc. **1962**, *2*, 74–76 [Reprinted in J. Inform. Sci. **2002**, 28, 79–81].
16. Learned, W.S. *The American Public Library and the Diffusion of Knowledge*, Harcourt, Brace: New York, 1924.
17. Wyer, J.I. *Reference Work: A Textbook for Students of Library Work and Librarians*, American Library Association: Chicago, IL, 1930.
18. Ball, R. Future trends in special library services. Int. J. Spec. Libr. **2000**, *34*(3/4), 133–140.

19. Williams, R.V. The documentation and special libraries movements in the United States, 1910–1960. J. Am. Soc. Inform. Sci. **1997**, *48*(9), 775–781.

20. Havelock, R.G. et al. *Planning for Innovation through Dissemination and Utilization of Knowledge*, Center for Research on the Utilization of Scientific Knowledge: Ann Arbor, MI, 1970.

21. Buckland, M.K. Emanuel Goldberg, electronic document retrieval, and Vannevar Bush's Memex. J. Am. Soc. Inform. Sci. **1992**, *43*(4), 284–294.

22. Bush, V. As we may think. Atl. Mon. **1945**, *176*, 101–108.

23. Burke, C. The other Memex: the tangled career of Vannevar Bush's information machine, the Rapid Selector. J. Am. Soc. Inform. Sci. **1992**, *43*(10), 648–657.

24. Arms, W. How effectively can computers be used for the skilled tasks of professional librarianship?. D-Lib Mag. **2000**, *6*(7/8). Available at http://www.dlib.org/dlib/july00/arms/07arms.html (accessed November 2008).

25. Hemminger, B.H.; Lu, D.; Vaughan, K.T.L.; Adams, S.J. Information seeking behavior of academic scientists. J. Am. Soc. Inform. Sci. Technol. **2007**, *58*(14), 2205–2225.

26. Arms, W. *Digital Libraries*, MIT Press: Cambridge, MA, 2000. Available at http://www.cs.cornell.edu/wya/DigLib/ (accessed November 2008).

27. Brin, S.; Page, L. The anatomy of a large-scale hypertext Web search engine. Stanford InfoLab Publication Server, **1998**. Available at http://ilpubs.stanford.edu:8090/361/ (accessed November 2008).

28. Page, L.; Brin, S.; Motwani, R.; Winograd, T. The PageRank citation ranking: bringing order to the Web. Stanford InfoLab Publication Server, **1999**. Available at http://ilpubs.stanford.edu:8090/422 (accessed November 2008).

29. Dillon, A. Library and information science as a research domain: problems and prospects. Inform. Res. **2007**, *12*(4), paper colis03. Available at http://InformationR.net/ir/12–4/colis/colis03.html (accessed November 2008).

30. Newman, M.E.J. The structure and function of complex networks. SIAM Rev. **2003**, *45*(2), 167–256.

31. Chakrabarti, D.; Faloutsos, C. Graph mining: laws, generators, and algorithms. ACM Comput. Surv. **2006**, *38*(1). Available at http://www.cs.cmu.edu/~deepay/ (accessed November 2008).

32. Börner, K.; Sanyal, S.; Vespignani, A. Network science. Annu. Rev. Inform. Sci. Technol. **2007**, *41*, 537–607.

33. Kleinberg, J. Authoritative sources in a hyperlinked environment. J. ACM **1999**, *46*(5), 604–632.

34. Bollen, J.; Luce, R.; Vemulapalli, S.S.; Xu, W. Usage analysis for the identification of research trends in digital libraries. D-Lib. Mag. **2003**, *9*(5). Available at http://www.dlib.org/dlib/may03/bollen/05bollen.html (accessed November 2008).

35. Persson, O. The intellectual base and research fronts of JASIS 1986–1990. J. Am. Soc. Inform. Sci. **1994**, *45*(1), 31–38.

36. White, H.D.; McCain, K.W. Visualizing a discipline: an author co-citation analysis of information science, 1972–1995. J. Am. Soc. Inform. Sci. **1998**, *49*(4), 327–355.

37. Bates, M.J. A tour of information science through the pages of JASIS. J. Am. Soc. Inform. Sci. **1999**, *50*(11), 975–993.

38. Zhao, D.; Strotmann, A. All-author vs. first-author co-citation of the information science field using Scopus. Proceedings of the 71st Annual Meeting of American Society for Information Science and Technology, **2007**, *Vol. 44*, unpaginated CD-ROM.

39. Zhao, D.; Strotmann, A. Information science during the first decade of the Web: an enriched author cocitation analysis. J. Am. Soc. Inform. Sci. Technol. **2008**, *59*(6), 916–937.

40. Börner, K.; Chen, C.M.; Boyack, K.W. Visualizing knowledge domains. Annu. Rev. Inform. Sci. Technol. **2003**, *37*, 179–255.

41. Morris, S.A.; Martens, B.V. Mapping research specialties. Annu. Rev. Inform. Sci. Technol. **2008**, *42*, 213–295.

42. Hearst, M. Untangling text data mining Proceedings of the 37th Annual Meeting of the Association for Computational Linguistics, 1999. Available at http://people.ischool.berkeley.edu/~hearst/papers/acl99/acl99-tdm.html (accessed November 2008).

43. White, H.D. Pathfinder networks and author cocitation analysis: a remapping of paradigmatic information scientists. J. Am. Soc. Inform. Sci. Technol. **2003**, *54*(5), 423–434.

44. Lancaster, F.W. *Information Retrieval Systems: Characteristics, Testing, and Evaluation*, Wiley: New York, 1968.

45. Lancaster, F.W. *The Measurement and Evaluation of Library Services*, 1st Ed. Information Resources Press: Washington, DC, 1977.

46. Hertzel, D.H. History and development of ideas in bibliometrics. Encycl Lib. Inf. Sci. **1987**, *42*(suppl. 7), 144–218.

47. Moed, H.F. *Citation Analysis in Research Evaluation*, Springer: Dordrecht, the Netherlands, 2005.

48. Bensman, S.J. Garfield and the impact factor. Annu. Rev. Inform. Sci. Technol. **2007**, *41*, 93–155.

49. White, H.D.; McCain, K.W. Bibliometrics. Annu. Rev. Inform. Sci. Technol. **1989**, *24*, 119–186.

50. Danton, J.P. Authors of information science [letter]. J. Am. Soc. Inform. Sci. **2000**, *51*(9), 882.

51. Webber, S. Information science in 2003: a critique. J. Inform. Sci. **2003**, *29*(4), 311–330.

52. Wilson, P. The future of research in our field. In *Information Science: From the Development of the Discipline to Social Interaction*; Olaisen, J., Munch-Petersen, E., Wilson, P., Eds.; Scandinavian University Press: Oslo, Norway, 1996.

53. Cronin, B. Preface and introduction. Annu. Rev. Inform. Sci. Technol. **2002**, *36*, vii–xiii.

Bibliometric Research: History [ELIS Classic]

Dorothy H. Hertzel
Case Western Reserve University, Cleveland, Ohio, U.S.A.

Abstract

Hertzel marshals a vast amount of information on the origins and development of one of the core areas of information science research—bibliometrics, or, as it is also known, informetrics. The study of the statistical properties of the domain of recorded information is a large field with an extensive body of research results.

—*ELIS Classic, from 1987*

INTRODUCTION

The word "bibliometrics" first appeared in print in 1969 in Alan Pritchard's article "Statistical Bibliography or Bibliometrics?" in the December issue of the *Journal of Documentation*.[1] Pritchard's article was the result of his judgment that the expression "statistical bibliography" should be replaced with a better term. He used "statistical bibliography" in his unpublished "Computers, Statistical Bibliography and Abstracting Services"[2] and again in his *Statistical Bibliography: An Interim Bibliography*, published in May 1969.[3] In December 1969, in "Statistical Bibliography or Bibliometrics?" he stated, "The term [statistical bibliography] is clumsy, not very descriptive, and can be confused with statistics itself or bibliographies on statistics."[1] As a result of the prompting of his friend, M.G. Kendall, Pritchard suggested that the word "BIBLIOMETRICS, i.e., the application of mathematics and statistical methods to books and other media of communication" be substituted for "statistical bibliography."[1]

In the same issue of *Journal of Documentation* appeared Robert A. Fairthorne's classic article "Empirical Hyperbolic Distributions (Bradford–Zipf–Mandelbrot) for Bibliometric Description and Prediction,"[4] in which the author used the word "bibliometric" and acknowledged Alan Pritchard as the donor of the term.[4] Fairthorne, a close friend of Pritchard, admitted that a phrase in the article—"This term [bibliometrics] resuscitated by Alan Pritchard"[4]—incorrectly suggested previous use, but in personal correspondence, he definitely verified that "Alan Pritchard *did* coin the word 'Bibliometrics'."[5]

Because the relation between the terms "bibliometrics" and "statistical bibliography" has now been established, it seems imperative that a history of bibliometrics should begin with its predecessor, statistical bibliography and its components: statistics and bibliography.

BIBLIOGRAPHY

"Bibliography" was derived from two roots: biblion, book; and graphos from graphein, to write.[6] Webster defined it as a "history of books, an account of manuscripts...and information illustrating the history of literature," as well as "a list of an author's writings; or the literature dealing with a certain subject or author."[6]

"In post-classical Greek times, when the word originated, it meant the writing or copying, i.e., the production of books; as late as the eighteenth century it was understood as the study of these ancient manuscript books."[7] Through the centuries, the meaning of "bibliography" has undergone a series of changes, some of which are considered here.

Monks, in copying manuscripts, also made lists of the books being copied; these lists, catalogs, or inventories are considered early bibliographies. At first, there were only limited numbers of lists of books or catalogs. Georg Schneider suggested this was because of lack of titles, limited need, and a limited area. He stated that, in the narrow sense, bibliography is a "study of lists of books" and further suggested that "true [i.e., modern] bibliography" began in 1564 when Georg Willer of Augsburg published his catalog of books, a listing for sale by him at the Frankfurt fair.[8]

From individual listings, bibliography progressed to the attempt to produce a "world bibliography." Konrad Gesner (1545) made the "first attempt to collect in one list all the scholarly publications of the world in bibliographic form,"[8] resulting in his having been acclaimed the "Father of Bibliography,"[8,9] but here, there is a difference of opinion. Archer Taylor claimed, "we may conveniently date bibliography from the activities of Johannes Tritheim, a practical and theoretical bibliographer of the late fifteenth century," but he agreed that "another man who was to display an even greater ability as a bibliographer" was Conrad Gesner (Murray acclaims "The

Encyclopedia of Library and Information Sciences, Fourth Edition DOI: 10.1081/E-ELIS4-120009034

Bibliographic–Bibliotheque

Bibliotheca of Conrad Gesner, a monument of human industry and one of the triumphs of bibliography."[9]).[10]

Between 1600 and 1700, bibliographies on particular subjects were published; these "are descriptive accounts of the literature of the subjects they deal with rather than catalogues of books."[9]

Taylor's investigations led him to the conclusion that "By 1700 bibliography was a highly developed art. And it was more than that: It was a deliberate and conscious maintainer of an intellectual and learned tradition descending from the beginning of time...."[10] This tradition can be traced to the beginning of the sixth century with the founding by Cassiodorus of an Italian monastery, one of the rules of which was the service of God "by diligent study and carefully copying of texts,"[11] and in the establishment in 529 of the Benedictine Order with its "very great emphasis on reading."[11]

The transition of meaning of bibliography was probably due to numerous reasons. These include the greater availability of books; rise of a middle class; development of "public" libraries; more people becoming literate; the passing of the Renaissance and advent of the Reformation; continued interest in humanism; the time of Galileo, Descartes, Kepler, and Newton; and the beginning of scientific progress.

By the eighteenth century in France, bibliography was "writing about books;"[8] later, it was called "the science that deals with literary production"[8] and finally, "the science of books."[8]

According to Thomas Hartwell Horne,

GENERAL BIBLIOGRAPHY...is, in strict language, a science; which consists in the knowledge of books, of their different editions and degree of rarity and curiosity, their real and reputed value, and the rank which they ought respectively to hold in a system of classification. General bibliography comprises works or catalogues, whose design is to give us a knowledge of every kind of book whatsoever....[12]

Bibliography is one of the oldest and yet one of the most modern of the sciences. It is old, because at all times scholars had lists of works of authors, catalogues of libraries, and other similar aids; it is modern, because it is only within a comparatively recent period that bibliography has been developed systematically.[9]

In 1935, Oxford University Press published *The Beginnings of Systematic Bibliography* by Theodore Besterman. Besterman divided bibliography into two parts, "the enumeration and classification of books, and the comparative and historical study of their make-up."[13] His research indicated that "bibliographies started as biographies in which the writings of the subjects were referred to as facts in their lives."[13] Eventually this slant changed, with the emphasis on the writings becoming the more important.

Current bibliography may be said to be the result of demands of research, development of numerous journals, and the formation of many societies, literary and scientific.[14] It can be concluded that the most important change in the meaning of bibliography was caused by change in its purpose, which at first was to preserve or record past items (lists or inventories), whereas its primary purpose now is to aid in dispensing knowledge (guides).[14]

This supports E. Wyndham Hulme's *Statistical Bibliography*, in which he defined bibliography as "the science of the organization of recorded knowledge."[15]

STATISTICS

"Statistics" was derived from the German "Statistik," which came from the medieval Latin "statisticus," which in turn was derived from the Latin "status" meaning state, position, standing (Latin for "to stand").[16]

Statistics

as a plural noun means collections or sets of facts that are related...In practice, it is customary to restrict its meaning to facts that are numerical or can in some way be related to numbers. As a singular, collective noun, statistics means the science of collecting or selecting statistical facts, sorting and classifying them, and drawing from them whatever conclusions may lie buried among them.[17]

In other words, "statistics" means facts or data of a numerical kind, assembled, classified, and tabulated to present information about a subject; originally, it referred to the physical, social, economic, intellectual, political, and industrial state of a country or its people as indicated in tables or numerical statements.

Yule and Kendall, in their *An Introduction to the Theory of Statistics*, first published in 1911, wrote that the now obsolete term "statist," meaning a statesman or politician, was used as early as 1602 in Hamlet,[18] and that the

earliest occurrence of the word "statistics" yet noted is in *The Elements of Universal Erudition* by Baron J.F. von Bielfeld (1770). One of its chapters is entitled "Statistics" and contains a definition of the subject as "The science that teaches us what is the political arrangement of all the modern states of the known world."[18]

Studies in the History of Statistical Method, by Helen Walker, asserted "The term "Statistik" occurs for what is probably the first time in the writings of Achenwall who is usually hailed as the father of statistical science."[19] His first statistical work appeared in 1748 and occurred in 1749 as the introduction to another of his works. The term "Statistik" was used in the preface.[19] Helen Walker's research led her to state that Achenwall and others "did not devote themselves so much to enumeration and computation as to verbal descriptions of the political situation and of all facts of interest in their countries."[19]

Although the word did not appear until the eighteenth century, the rudiments of the subject, statistics, were found in very early ancient times and at first were only approximate counts. As early as 2000 B.C., there was a population census in Judea;[20] in ancient Rome and Greece, the census was taken for real estate taxes; and in the Middle Ages, there were taxes, tithes, and land registers of various countries.[20] As early as "762 Charlemagne ordered the detailed descriptions of church lands"[21] in France, and by the first half of the ninth century, "the serfs and peasants being attached to the land were included in the land census."[21] Eventually, it became necessary for those in power, for their own protection, to know their military and financial standings and that of other powers. Thus,

> In the beginning, statistics considered the most urgent need of the period, namely the knowledge of the state, and as its method comparison, without, however, possessing an exact sense of proportion nor a distinct consciousness of the necessity of measurement.[20]

The gradual change in the meaning of statistics was a result of the era and the conditions of the time, starting with a listing of land ownership and progressing to a description of states, but at first, statistics was not an exact enumeration. Further development was due not only to the change from feudal or medieval system to the "modern state" but also to the need to know of the comparison status for an evaluation of power which necessitated more accurate counting.

Gradually, statistical facts became "public" knowledge, for as early as 1536–1544 the *Cosmographia* of Sebastian Muenster described the world, giving maps of known countries, history, organization, laws, customs "and in detail the chief cities with their wealth and trade."[20] This may be called "descriptive statistics." Various other works followed; these included public laws, estimates of taxes, military strength, and commerce. In 1626, "the publication of the so-called *Respublicae Elzeviranae*," which gave descriptions of single states written by distinguished statesmen of the period,[20] was begun. In 1660, statistical studies were introduced into the curriculum of the University of Helmstedt, and from there, the giving of similar lectures spread to other areas.[20]

It was during the "17th century [that] the games of hazard grew into prominence"—lotteries, probabilities in games of chance, and annuity life insurance,[20] and this is thought to be the true beginning of inductive statistics, explained as follows by Walpole: "Any treatment of data leading to predictions or inferences concerning a large group of data is known as *inductive statistics*."[22] Based on the theory of probability, which had its beginning in

gambling or games of chance, inductive statistics projects or determines a possible or probable outcome.

History showed that Galileo was consulted for solutions to some problems on gambling, solved the problems, and wrote a paper concerning games of chance. Not long after in France, Blaise Pascal was approached by his friend Chavalier de Mere with similar questions.[17] "Chevalier de Mere proposed to B. Pascal the fundamental "Problem of Points" to determine the probability which each player has, at any given stage of the game, of winning the game."[23] (See V. Sanford, who claimed this problem was in evidence as early as 1494.)[24] Pascal shared the problem with his friend, Fermat. Both solved the problems, but by different methods. Pascal, as a result of his work, developed the theory of probability.[17] In 1657, Huygens developed a treatise on probability which was later included in Jakob Bernoulli's *Ars conjectandi*.[23] Bernoulli, elaborating on the subject, made the innovative suggestion that the "ideas [of probability] might be applied to civil, moral, and economic affairs."[24]

Shortly after, Abraham de Moivre published *The Doctrine of Chances*, the outcome of which study led to the development of the normal curve of distribution.[24] This same bell-shaped distribution is "often referred to as the *Gaussian distribution*, in honor of Gauss (1777–1855) who also derived its equation from a study of errors."[22] The ideas of probability also led to the further development of insurance which was based on "mortality." During the seventeenth century, some investigators had used vital statistics to make predictions and draw some very interesting conclusions.[20] The "first attempt to interpret mass biological phenomena and social behavior from numerical data—in this case, fairly crude figures of births and deaths in London from 1604 to 1661"—was in *Natural and Political Observations Made upon the Bills of Mortality*, by John Graunt of London in 1662.[25]

John Graunt (1620–1674) was born in London and was a shopkeeper by trade.

> *The Bills of Mortality* which attracted Graunt's attention were issued weekly by the company of parish clerks and listed the number of deaths in each parish, the causes, and also an "Account of all the Burials and Christnings, hapning that Week."[25]

Impressed with Graunt's work, Charles II fostered his membership in the Royal Society of London. The influence of Graunt's work spread and soon affected "the gathering and study of vital statistics on the Continent."[25]

Thirty years later, Edmund Halley, of comet fame, at the request of the Royal Society, wrote the mortality rates report "An Estimate of the Degrees of the Mortality of Mankind, drawn from curious Tables of the Births and Funerals at the City of Breslaw; with an Attempt to ascertain the Prices of Annuities upon Lives."[20]

This was followed by "Some further Considerations on the Breslaw Bills of Mortality." Together, the papers are the foundation for all later work on life expectancy, indispensable, of course, to the solvency of life-insurance companies.[25]

"During the eighteenth century it [subject of probability] developed rapidly, and in the nineteenth it found application in the mathematical study of statistics."[24]

On the other hand, the approach of Gottfried Achenwall, the father of statistical science, was descriptive. In 1746 he began his statistical lectures and soon had some of his related writings published; in the introduction of one of these appeared for the first time the word "statistik."

> Up to this time the word had been merely suggested. . .by statista, one versed in the knowledge of the state, or by the adjective use, rationes statisticae, by Oldenburg, 1668, bibliotheca statistica, by Thurmann in 1701, and the collegium statisticum of Schmeitzel. Achenwall derives the word from the Italian, ragione di stato, practical politics, and statista, statesman.[20]

He alleged the "final object [of statistik] is to gain political wisdom by means of knowledge of the various states" both internal and external.[20]

In 1749, Achenwall's work which later became *Reiche and Völker* was published.[20] This covered "Spain, Portugal, France, Great Britain, the Netherlands, Russia, Denmark, and Sweden," giving a careful picture of each, of land and peoples, in seven distinct groups of questions:

1. The literature and sources of information.
2. The state, its territory and the changes of the same.
3. The land, its climate, rivers, topography, divisions, abundance or scarcity of products.
4. The inhabitants, numbers and character.
5. The rights of the rulers, the estates, the nobility and the classes of the inhabitants.
6. The constitution of the court and the government, laws, and administration of churches, schools, and justice, industry, home and foreign commerce, currency, finances, debt, army and navy.
7. The interests of national life and politics, as well as the outlook for the future.[20]

"The work found such general recognition that it was translated into all languages," with the result that Achenwall's concept of statistics in addition to the word was widely distributed and accepted.[20]

Running parallel with Achenwall's idea was the need for lists of "official statistics"—tables of vital statistics, census, and demographics. By 1782, the scope of the tables had been expanded considerably, and "the statistical details [were] collected in general outline tables."[20] The information was not only descriptive, it was exact enumeration.

The origin of the publication of details of official statistics has been attributed to Anton Friedrich Büsching.[20] He is also credited with the founding of *Das Magazin für Historiographie und Geographie*, claimed to be the first statistical periodical because of its statistics from many German and other countries.[20] Büsching's importance resulted from his interest in details and not in generalities as in Achenwall's case, and this interest in details "led to a careful examination as to the completeness and correctness of the data, and promoted, by the scrutiny of their origin and arrangement, the progress of critical methods."[20]

Throughout the 1800s, numerous countries developed commissions, departments, or bureaus of statistics with publications of various reports and annuals. Undoubtedly, the developing spirit of nationalism, free trade movement, and world exhibitions were contributing factors to the ever-rising interest in statistics.[23]

Adolph Quetelet, a Belgian statistician and sometimes referred to as the founder of modern statistics, "developed the conception of statistics as a general method of research applicable to any science of observation,"[19] and his theory widened the application of statistics when he applied statistical methods to education and sociology. Quetelet, responsible for the First International Statistical Congress which was held in Brussels in 1853, was also "one of the first statisticians to demonstrate that statistical techniques derived in one area of research are also applicable in most other areas."[22]

By 1872, M. Haushofer by stating "statistics is a method and a science. . .designates statistics as an essentially methodical science,"[20] and this was supported by August Meitzen, a German professor, who in 1886 evaluated statistics as "a method of scientific investigation by means of enumeration of objects and the numerical comparison of the results of such enumeration."[20] His concept of statistics was that it is a "science of method" based on counting plus a comparison of the results.

This information leads to the conclusion that statistics may be divided into two areas, descriptive and inductive, or inferential. Descriptive statistics is really a description or compilation of data in a form that is clear and usable, whereas in inductive statistics, data pertaining to a sample of a population are used to arrive at a probable conclusion or prediction concerning the whole. Blalock called this "generalizing on the basis of limited information."[26] The conclusions of inductive statistics, based on incomplete data, being uncertain or probable, depend on the theory of probability,[22] and though the original meaning of statistics dealt with counting, the modern statistical method "deals more with the design, execution, data analysis, and interpretation of studies."[27]

> Originating as the mathematical tool of the gambler, the science of probability has become fundamental in the

knowledge of the physicist, the biologist, the technologist, the industrialist, the businessman, and the philosopher.[17]

It seems logical that the science of probability and statistics would also have been used by the librarian or the bibliographer and eventually applied to the field of bibliography. So, statistics, using the theory of probability (counting, analyzing, and interpreting) combined with bibliography (knowledge dispensers) became statistical bibliography, as Pritchard defined it.

STATISTICAL BIBLIOGRAPHY

In the papers using the expression "statistical bibliography," there are only a few explanations of what is meant. Researchers claim E. Wyndham Hulme (1922) was the first to use the expression,[15] and according to Alan Pritchard, from then until 1969, only two other authors, Charles F. Gosnell and L. Miles Raisig, used the phrase. Antecedent to Hulme was Cole and Eales's use of "statistical analysis" in "The History of Comparative Anatomy, Part I (1917)."[28] After Hulme, "statistical bibliography" was used by Gosnell in 1943 in his dissertation, "The Rate of Obsolescence in College Library Book Collections, as Determined by an Analysis of Three Select Lists of Books for College Libraries"[29] and in his article of 1944, "Obsolescence of Books in College Libraries;"[30] and by Raisig in 1962 in his "Statistical Bibliography in the Health Sciences."[31]

Hulme believed "statistical treatment must show the existence of phases of activity and retardation in each science and might even indicate approximately the period when its ultimate boundaries would be reached."[15] "Bibliographical statistics, employed with the requisite qualifications, are without question able to reveal the shape and period of such movements."[15]

Cole and Eales, calling their work a statistical analysis of the literature, graphically represented the "activities of comparative anatomists for a period of time." As they stated,

> In other words it seemed possible to reduce to geometrical form the activities of the corporate body of anatomical research, and the relative importance from time to time of each country and division of the subject.[28]

Here we have tabulations, charted and analyzed as to the important periods in the development of anatomy study or progress.

Gosnell's doctoral thesis, "The Rate of Obsolescence," purports "to discover lines of trend or curves of distribution by means of which this rate of obsolescence may be expressed in mathematical form."[29] "The degree of such obsolescence may be discovered by statistical analysis."[29] His Chapter II, Statistical Bibliography, was divided into several parts, the first of which was a definition.

> In bibliography, which is the science of history and description of books major emphasis has always been qualitative...Very little attention has been given to the significance per se of the number or bulk of publications in a group or series. *Statistical bibliography* implies emphasis or the quantitative aspect rather than the qualitative.[29]

Using "three select lists of books for college libraries, the Shaw *List*, issued in 1931, the Mohrhardt *List*, 1937, and the Shaw *Supplement*, 1940,"[29] Gosnell proceeded to make an analysis, developing a curve of obsolescence after first preparing tables of the distribution of imprint or publications dates. He also made the observation that "The curve of obsolescence in a sense represents the reverse of compound interest."[29]

In "Obsolescence of Books," an article based on his thesis, Gosnell used two assumptions; the first suggests that principles of choice may be discovered by the documentary analysis of the list.[30] (Here Gosnell is qualifying his use of the specific three lists.) "The second [under the section heading, Statistical Bibliography] is that mere masses of books (or titles) may be analyzed for certain characteristics without reference to their individual titles."[30] Here using "collections of books as populations," he suggested that certain characteristics or trends be noted to make a reliable prediction of obsolescence.[30]

Gosnell's idea that "Books represent one of the higher forms of culture and the rate at which they are discarded and replaced may give some suggestion as to the rate of evolution of the general culture of which they form a part" seems to be a variation of the Hulme theme.[30]

Raisig, claiming there was a "potential utility of statistical bibliography as a method of analyzing information needs,"[31] stated that his article was "a review of investigative methods and results in the health sciences."[31] He gave the following as definition of "statistical bibliography:" "the assembling and interpretation of statistics relating to books and periodicals; it may be used in a variety of situations for an almost unlimited number of measurements."[31] He continued,

> bibliographical statistics have been collected... for these [three] main purposes: to demonstrate historical movements, to determine the national or universal research use of books and journals, and to ascertain in many local situations the general use of books and journals.[31]

"The citation analysis of the research usage of periodicals has yielded the greatest number of published results and the greatest variety of interpretations."[31] Raisig's article leaned to a discussion of citation literature with particular critical emphasis on the Gross and Gross analysis (see the next section, Seminal Bibliometric Papers).

Glenn Wittig, in his extensive research, found several more users of "statistical bibliography:"[32] H.H. Henkle in 1938 in the article "Periodical Literature of Biochemistry;"[33] Herman H. Fussler in 1948 in his dissertation, "Characteristics of the Research Literature used by Chemists and Physicists in the Soviet Union,"[34] which was also published in *Library Quarterly* in 1949;[35] and Dale Barker in his dissertation of 1966, "Characteristics of the Scientific Literature cited by Chemists of the Soviet Union."[36]

Henkle's paper used the expression "statistical bibliography" to report "the results of a study in which the methods of statistical bibliography were applied to an evaluation of the periodical literature of biochemistry."[33] Although he did not "define" statistical bibliography, he did state, "the bibliographies [he used] were analyzed statistically and a list of journals arranged in the order of the frequency with which they were cited."[33] He believed his research would help in the "problem of selecting periodicals for the library."[33] Henkle listed many papers of biomedical interest which have been patterned after the Gross and Gross method. His interest is a research study of citations in the biochemistry field and he made comparisons with the results of other scientists, concluding "one of the important facts shown by these bibliographical studies is the extent to which the journal literature demonstrates the interrelationships of the sciences."[33]

Fussler's research, published as two papers in the *Library Quarterly*, examined the characteristics of the literature of chemistry and physics. Although Fussler wrote, "This study is a form of statistical bibliography,"[35] he did not define statistical bibliography. However, he did make this explanatory remark:

A number of analyses of research literature have been made by listing the references contained in one or a group of important journals in the field to be surveyed. Various considerations suggested that the adoption of this technique or some modification of it would yield the most useful sample of the general research literature for chemistry and physics.[35]

Dale Lockard Barker's doctoral thesis, "Characteristics of the Scientific Literature," was quite an ambitious endeavor. The objectives of his paper were:

to identify and describe in terms of selected characteristics, the scientific literature used by Soviet chemists and to determine in part the nature of that use. Specifically, the study sought a) to establish the outstanding properties of Soviet chemical literature; b) to provide greater insight into the characteristics of Western chemical literature as reflected in Soviet use; c) to extend our knowledge of some basic properties of world scientific literature without respect to its national origins; and, d) to determine patterns of literature used by Soviet chemists.[36]

Although Barker did not define statistical bibliography, he described what it included. In the section headed Statistical Bibliography, the author pointed out that:

the knowledge of scientific literature afforded us by descriptive and historical studies has been augmented by a considerable body of statistical research. This has been directed toward various phases of scientific communication, but has sought principally to describe the characteristics of the literature available to scientists, to identify the literature required by them, or to determine how they use the literature.[36]

"Statistical bibliography" was used by Pritchard when he wrote "Computers, Statistical Bibliography and Abstracting Services"[2] and again in his *Statistical Bibliography: An Interim Bibliography*, which:

contains a 700 item bibliography of literature on statistical bibliography, arranged in author order. The areas covered are: citation studies, abstracts, journals studies, direct literature studies with additional relevant material on user surveys, the history and sociology of science and citation structures. The period covered is from 1881–1969.[3]

Alan Pritchard's definition in the introduction to *Statistical Bibliography* is:

The statistical analysis of the means of communication in order to illuminate the processes of communication, the factors which influence them and the inter-relationships between the history and sociology of a science and the literature of the science.[3]

In December 1969, Pritchard's "Statistical Bibliography or Bibliometrics," where he cited his preference of terms, appeared in print.

About a year later, in September 1970, *Research in Librarianship* published Pritchard's document, "Computers, Bibliometrics and Abstracting Services;"[37] Pritchard said that it and his unpublished paper, "Computers, Statistical Bibliography and Abstracting Services," "were very similar and for practical purposes you can take it that they were the same."[38]

"Bibliometrics" as a term was now accepted in print (Table 1).

BIBLIOMETRICS

Alan Pritchard, who first used the word "bibliometrics," described it as the "application of mathematics and statistical methods to books and other media of communication."[1] This was paraphrased by Robert A. Fairthorne as "quantitative treatment of the properties of recorded discourse and behaviour appertaining to it."[4] In a later

Table 1 Chronological list—statistical bibliography to bibliometrics.

No.	Year	Author and title	Publication	Cites these works on	Is cited in these works[a]	Expression and location
1.	1917	Cole, F. J., and Eales, N. B. "The History of Comparative Anatomy Part 1. A statistical analysis of the literature"	*Science Progress*, Vol. 11, April 1917, pp. 578–596	—	2, 8, 9, 11, 12	Statistical analysis subtitle, p. 578
2.	1922	Hulme, E. Wyndham *Statistical Bibliography in Relation to the Growth of Modern Civilization*	Lectures, May 1922, Book: Butler and Tanner Grafton, London, 1923	1	7, 8, 9, 11, 12	Bibliographical statistics, p. 9 Statistical bibliography, preface, title, pp. 5, 21
3	1938	Henkle, H. H. "The Periodical Literature of Biochemistry"	*Bulletin of the Medical Library Assoc.* Vol. 27, 1938, pp. 139–147	—	7, 8, 9, 11	Statistical bibliography, pp. 139, 140 Statistical analysis p. 139
4.	1943	Gosnell, Chas. F. "The Rate of Obsolescence in College Library Book Collections as Determined by an Analysis of Three Select Lists of Books for College Libraries"	Dissertation: New York University; 1943	1, 2	11, 12	Statistical bibliography, p. 16 Statistical method, p. 158 Statistical analysis, p. 158
5.	1944	Gosnell, Chas. F. "Obsolescence of Books in College Libraries"	*College and Research Lib.* Vol. 5; March 1944; pp. 115–125	—	9, 11, 12	Statistical bibliography, pp. 115, 116
6.	1948	Fussler, Herman H. "Characteristics of the Research Literature Used by Chemists and Physicists in the United States"	Dissertation: University of Chicago, 1948; p. 115	—	11	Statistical bibliography, pp. 13, 240
7.	1949	Fussler, Herman H. "Characteristics of the Research Literature Used by Chemists and Physicists in the United States"	*Library Quarterly*, Vol. 19, 1949; pp. 19–35	2, 3	9, 11	Statistical analysis, p. 21 Statistical bibliography, p. 27
8.	1962	Raisig, L. Miles "Statistical Bibliography in the Health Sciences"	*Bulletin of the Medical Library Assoc.*, Vol. 50, July 1962; pp. 450–461	1, 2, 3	9, 11, 12, 14	Bibliographical statistics, p. 450 Statistical analysis, p. 450 Statistical bibliography, pp. 450, 451
9.	1966	Barker, Dale Lockard "Characteristics of the Scientific Literature Cited by Chemists of the Soviet Union"	Dissertation: University of Illinois, 1966, p. 297	1, 2, 3, 5, 7, 8	11	Statistical bibliography, p. 10, 17 Statistical analysis, p. 208 Bibliographic statistics, p. 99
10.	1968	Pritchard, Alan "Computers, Statistical Bibliography and Abstracting Services"	Unpublished: 1968	—	12	Statistical bibliography, title
11.	1969	Pritchard, Alan *Statistical Bibliography: An Interim Bibliography*	North-Western Polytechnic School of Librarianship, May 1969; 69 pages	1, 2, 3, 4, 5, 6, 7, 8, 9	12, 14	Statistical bibliography, title, abstract, introduction
12.	1969	Pritchard, Alan "Statistical Bibliography or Bibliometrics?"	*Journal of Documentation*, Vol. 25, December 1969; pp. 348–349	1, 2, 4, 5, 8, 10, 11		Statistical bibliography, p. 348 Bibliometrics, title, p. 349
13.	1969	Fairthorne, Robert A. "Empirical Hyperbolic Distributions for Bibliometric Description and Prediction"	*Journal of Documentation*, Vol. 25, December 1969; pp. 319–343			Bibliometrics, title, pp. 319, 325, 341
14.	1970	Pritchard, Alan "Computers, Bibliometrics and Abstracting Services"	*Research in Librarianship*, September 1970; pp. 94–99	8, 11		Bibliometrics, title, pp. 94, 95, 97

[a]Numbers refer to number listing of articles in this table.

article, "Bibliometrics and Information Transfer," Pritchard explained bibliometrics as the "'metrology' of the information transfer process and its purpose is analysis and control of the process."[39] He based his interpretation upon the fact that measurement is "the common theme through definitions and purposes of bibliometrics" and "the things that we are measuring when we carry out a bibliometric study are the process variables in the information transfer process."[39]

The *British Standard Glossary of Documentation of Terms* explained bibliometrics as the study of the use of documents and patterns of publication in which mathematical and statistical methods have been applied,[40] which is basically the same as Pritchard's original definition.

William Gray Potter, editor of the issue of *Library Trends* devoted to bibliometrics, followed suit with "Bibliometrics is, simply put, the study and measurement of the publication patterns of all forms of written communication and their authors."[41] In the same issue, Alvin M. Schrader said it even more simply, "Bibliometrics, [is] the scientific study of recorded discourse."[42]

Bibliometrics, called a quantitative science, is divided into two areas, descriptive and evaluative.

> In one of these classes is included the study of the number of publications in a given field, or productivity of literature in the field for the purpose of comparing the amount of research in different countries, the amount produced during different periods, or the amount produced in different subdivisions of the field. This kind of study is made by a count of the papers, books and other writings in the field, or often by a count of those writings which have been abstracted in a specialized abstracting journal. The other...includes the study of the literature used by research workers in a given field. Such a study is often made by counting the references cited by a large number of research workers in their papers.[43]

The two areas may also be divided as follows:

1. Productivity Count (descriptive)
   a. Geographic (Countries)
   b. Time periods (Eras)
   c. Disciplines (Subjects)
2. Literature Usage Count (evaluative)
   a. Reference
   b. Citation[43]

Nicholas and Ritchie divided the two groups as "those describing the characteristics or features of a literature [descriptive studies] and those examining the relationships formed between components of a literature [behavioral studies]."[44] Although all the descriptive studies are not evaluations, all the evaluative analyses are first descriptive with the evaluative taking the data one step further, providing "data on the condition or character of the literature as a whole."[44]

Bibliometrics, the "science of recorded discourse"—which uses specific methodologies, mathematical and scientific, in its research—is a controlled study of communication. It is the body of a literature, a bibliography quantitatively or numerically or statistically analyzed—a statistical bibliography; a bibliography in which measurements are used to document and explain the regularity of communication phenomena.

Seminal Bibliometric Papers

For the development of a list of seminal bibliometric papers, two criteria were used: 1) did the author indicate the publication an original or; 2) did peer consensus indicate the paper "a first?"

Daniel O'Connor and Henry Voos, of the Graduate School of Library and Information Studies of Rutgers University, in the *Library Trends* 1981 Summer Issue, "Bibliometrics," claimed "The scope of bibliometrics includes studying the relationship within a literature (e.g., citation studies) or describing a literature"[45] focusing "on consistent patterns involving authors, monographs, journals, or subject/language."[45]

The paper by Cole and Eales was probably the first to give a description of a literature, using publication counts and utilizing graphic illustrations to them by year and country. "The History of Comparative Anatomy—A Statistical Analysis of the Literature"[28] (1917), by F.J. Cole and Nellie B. Eales, has the qualifications that Hulme set forth for a statistical bibliography. It is considered a classic, being "one of the first papers based on significant statistical data."[46] Both Cole and Eales were scientists; F.J. Cole was a professor of zoology, and Nellie Eales was a museum curator.[28]

Cole first attempted to "apply graphic methods to an historical study of anatomical museums" but "the number of such museums...was too small to admit of satisfactory treatment by statistical methods."[28] As a result, "a similar attempt on the literature of comparative anatomy" was made: "it seemed possible to reduce to geometrical form the activities of the corporate body of anatomical research, and the relative importance from time to time of each country and division of the subject."[28]

The report recorded publications (6,436) dealing with animal anatomy for the period 1543 to 1860,[28] in chronological charts; the fluctuations shown were explained by miscellaneous influences. The authors admitted that no evaluation of papers was made before the counting; thus, a short paper counted the same as a lengthy one.[28]

The Cole and Eales objectives were:

> 1) to represent by a curve the activities of comparative anatomists as a whole from the sixteenth century to 1860; 2) to detach from this general scheme and plot separately the performances of each European country; 3) to determine in a similar way which groups of animals and what

aspects of the subject engaged the attention of workers from time to time; 4) to trace the influence of contemporary events, public bodies and individuals on the history of anatomical thought.[28]

A particularly interesting statement from the Cole and Eales publication is included here because of the noting of the occurrence by other scientists:

> The steep decline after 1835 can only be explained as an admirable example of that rhythm which underlies all the activities of the living world. This primitive abhorrence of the fixed level, which finds its expression in advance or retreat, but never in stability, is just as characteristic of the work of a community as of the internal economy of the individual.[28]

It would be very interesting to see an update of this paper from 1850 to check on the possible continuance of the Cole and Eales findings.

The second paper of importance is that of E. Wyndham Hulme;[15] it too is an historical publication count and is considered "one of the first analytical accounts of the growth of the literature."[44]

Hulme used 13 annual issues of *The International Catalogue of Scientific Literature*, from 1901 to 1913, counting author entries for various subjects and graphing the information. Because of declines and advances evidenced in the chart he did a further analysis resulting in the tabulation of the "Number of Journals Indexed arranged by Countries"[15] and listing of countries in order of total productivity of the number of journals indexed.[15]

Wyndham Hulme wrote "Bibliography is the science which collects, preserves, describes and classifies."[15] He was an advocate of books being universally classified, then put in chronological order in each class, the result of which, he believed, "presents for each period a bibliographical counterpart of the corresponding growth of the activities of the human mind."[15] In other words, Hulme believed that the chronological order of classified books would show the growth or development of a subject science, or important movements in history.

The purpose of his paper was to "ascertain and illustrate by bibliographical data, various stages in the development of the mechanics of civilization."[15] "I deal with it [civilization] as an organic growth so far as this growth can be correlated with the recorded intellectual activities of the several periods."[15] He was not as concerned with numbers as with specialization, for he was certain a sign of subject division was a sign of growth. Hulme wished to emphasize that "phases of activity and retardation in each science" could be shown by statistical treatment and that the time a science would reach its boundaries might also be indicated.[15] He gave three requirements for such statistics (which he believed were practically nonexistent):

1) the statistics must be international in scope and sufficiently extended for the purpose in view; 2) original work must be distinguished from educational literature; 3) the statistician must possess a competent knowledge of his subject matter.[15]

Hulme also stated "increased activity in the library output of a science can invariably be associated with pre-existent causes,"[15] concluding that declines and peaks are influenced by population change, political, and economic movements.[15]

Cole and Eales said something of the same when they decided that an increase of anatomical publications from 1650 to 1700 and a decline from 1700 to 1750 were due to the number of anatomists being born between 1600 and 1650 followed by a birth decline, to the founding of important scientific societies from about 1652 to 1666, and to the beginning of periodical publications.[28]

One item of note is the suggestion that a sharp decline after 1835 may have been due to "the rise of histology and embryology subsequent to the enunciation of the Cell Theory in 1838–9."[28] This sounds like Hulme's idea mentioned earlier, that specialization shows growth.

Hulsey Cason and Marcella Lubotsky published a study, "The Influence and Dependence of Psychological Journals on Each Other," in 1936.[47] It was cited by Francis Narin in his study *Evaluative Bibliometrics*[48] for its "Journal to Journal Cross Citation Influence for Journals and Fields."[48]

The purpose of the paper was "to secure a quantitative measure of the extent to which each psychological field influences and is influenced by each of the other psychological fields,"[47] with the study being limited to journals published in the English language in 1933.[47] The method used was a listing of 61 tentative journals in a survey letter sent to 54 psychologists; they received replies from 45, of which 39 were usable. Ultimately, only 28 journals were used; others were eliminated for various reasons. The specifically different references in the bibliographies, references, and footnotes of each article in each periodical were used.[47]

> For the 1933 part of each journal, we counted the number of references 1) to the same journal; 2) to each of the other journals listed in Table 1 [list of 28 journals]; and 3) to all other references including books, monographs, and journals not listed in Table 1.[47]

Narin believed Cason and Lubotsky's:

> real advance lies in the idea of constructing a cross-citing network. They summarized their data by constructing a 28 × 28 element table of the percent of references from each of the most significant journals to the others, with a summarized count of references to other publications. This seems to be the first time that the idea of a cross-citing network appears in the literature.[48]

The Cason and Lubotsky article should be mentioned even though the suggestions presented in it did not flourish.

In 1926 the *Journal of the Washington Academy of Sciences* printed the paper "Statistics—The Frequency Distribution of Scientific Productivity."[49] It was especially important because it developed a productivity formula which became known as Lotka's Law.

Alfred J. Lotka, a statistician with the Metropolitan Life Insurance Company, thinking it "would be of interest to determine, if possible, the part which men of different calibre contribute to the progress of science,"[49] developed a listing of A and B names (6891) from the *Chemical Abstracts Index* of 1907–1916 and the corresponding number of papers each produced. The same procedure, but using complete coverage, was applied to Auerbach's *Geschichtstafeln der Physik* through the year 1900.[49] From these data, Lotka formulated a general equation: "the relation. . .between the frequency y of persons making x contributions is $x^n y = $ const."[49] He then found the value of the constant when $n = 2$; this ultimately became known as the inverse square law of scientific productivity, explained as follows:

> In the cases examined it is found that the number of persons making 2 contributions is about one-fourth of those making one; the number making 3 contributions is about one-ninth, etc.; the number making n contributions is about $\frac{1}{n^2}$ of those making one; and the proportion, of all contributors, that make a single contribution, is about 60 per cent.[49]

The results were publicized in June of 1926 and have had quite an impact on a number of investigations by stimulating further analysis, controversy and research. (See further discussion of Lotka's Law in the next section.)

> The major step of analyzing citations, as opposed to publications, was taken by Gross and Gross in 1927, when they discussed the purchase of journals for a chemical library in terms of a tabulation of citations from the *Journal of the American Chemical Society*.[46]

Paul L.K. Gross and E.M. Gross, a husband and wife team, wrote a paper tackling the problem of colleges presenting "broad cultural education" without neglecting preparation for the increasing competitive demands of graduate schools. The Grosses were concerned about the preparation of graduates from small colleges who wanted to enter graduate competition and believed that some periodicals were more conducive to success than others in preparing students for advanced work.[50] On the basis of the idea that lists or files of scientific periodicals were of utmost importance in solving such a problem, Gross and Gross "decided to tabulate the references in a single volume of the *Journal of the American Chemical Society*"

chosen "as the most representative of American chemistry."[50] The result was 3633 references for 247 different periodicals or journals.[50] The tabulations were made for 5-year periods from 1871 to 1925 as found in the 1926 volume, and trends were traced from the tabulations. From these trends, need for specific periodicals was decided, which included not only specific subject journals but also some from related fields.[50] "This is the first recorded study based on counting and analyzing citations, i.e., citation analysis"[51] and has been a model for many investigations.

In late 1927, P.L.K. Gross wrote "Fundamental Science and War,"[52] in which he claimed, "the process of laying this foundation [science progresses only after the foundation stones of pure science have been firmly laid] consists in searching out, correlating and classifying knowledge."[52] Gross again used citations to support his conclusions, this time as to the effect of war on the science of chemistry. In Gross's words:

> Consideration of the method of investigation [citation counts] here employed will show that we are concerned not merely with the quantity of work published during this period (1912–1923), but that in reality we are concerned only with the good work, the work which has survived and which has proved of value to the investigators who followed. The method, therefore, has a distinct advantage over any method which counts pages or number of papers published in various journals for its basis of comparison.[52]

In other words, papers which are continually cited are the most valuable for researchers, according to Gross.

Samuel Clement Bradford, another pioneer of bibliometrics, should be considered for his special article "Sources of Information on Specific Subjects,"[53] which is the first paper published on observations on scattering. This important paper contains Bradford's law, which many scientists have used in research. Upon investigating "300 abstracting and indexing journals,"[53] Bradford found that of 750,000 articles, "only 250,000 different articles are dealt with and 500,000 are missed."[53] This disturbed him greatly, true librarian that he was, and he believed repetition would be eliminated if there were a standard classification adopted internationally to bring references to the same subject together. This would be a cooperative endeavor.

Based on data collected by E. Lancaster Jones of the Science Library from the bibliographies of *Applied Geophysics* (1928–1931) and *Lubrication* (1931–1932 plus),[53] a hypothesis was established: "to a considerable extent, the references are scattered throughout all periodicals with a frequency approximately related inversely to the scope."[53] Using this observation, Bradford proceeded to analyze his data to develop a means by which librarians could select the most usable periodicals. He found when the journals were divided into zones, each containing the same number of articles as a nucleus, the relation of the number of journals was 1: *n*: n^2.[53] In his words:

> the law of distribution of papers on a given subject in scientific periodicals may thus be stated: if scientific journals are arranged in order of decreasing productivity of articles on a given subject, they may be divided into a nucleus of periodicals more particularly devoted to the subject and several groups of zones containing the same number of articles as the nucleus, when the numbers of periodicals in the nucleus and succeeding zones will be as 1: *n*: n^2.[53]

The next scientist to incite thought and controversy in the area of "bibliometrics" is George Kingsley Zipf. Zipf's law is considered a bibliometric law and has been used in many bibliometric papers.

Zipf's book *Psycho-Biology of Language*,[54] published in 1935, has in its Preface the kernel of what later was called Zipf's law.

> If the number of different words occurring once in a given sample is taken as *x*, the number of different words occurring twice, three times, four times, *n* times, in the same sample, is respectively $\frac{1}{2^2}$, $\frac{1}{3^2}$, $\frac{1}{4^2}$, $\ldots \frac{1}{n^2}$ of *x*, up to, though not including, the few most frequently used words; that is, we find an unmistakable progression according to the inverse square, valid for well over 95% of all the different words used in the sample.[54]

From this beginning, Zipf developed a formula, $ab^2 = k$, where *a* is the number of words occurring *b* times. He also concluded the "$ab^2 = k$ relationship is valid only for the less frequently occurring words which, however, represent the greater part of the vocabulary in use, though not always a great majority of occurrences,"[54] but "it is in the frequency distribution of the rarer words that we find the closest approximation to the $ab^2 = k$ relationship."[54]

It is important to note here that Zipf, in his book *Human Behavior and the Principle of Least Effort*,[55] cited J.B. Estoup for word usage frequency by stating,

> The first person (to my knowledge) to note the hyperbolic nature of the frequency of word usage was the French stenographer J.B. Estoup who made statistical studies of French, cf. his *Gammes Stenographiques*, Paris 4 ed., 1916 (have not seen his earlier editions).[55]

Fussler's paper, as noted earlier, is another classic in the history of bibliometrics[35] and is composed of two parts: the first discusses "the importance of the literature of various subject fields to research in chemistry and physics;"[35] and the second, "the temporal span of the literature, the principal forms of the literature, the national origins of the literature used in the United States, and some attention is devoted to the more important serial titles."[35] Because of various problems facing research

libraries, Fussler believed it important to do a study to find the "more fundamental characteristics of the literature used by research personnel of the United States in two related subject fields—chemistry and physics."[35]

He planned a modification of the technique of reference listing. Using the *Physical Review* and the *Journal of the American Chemical Society* as key journals of the individual disciplines, for the years 1899, 1919, and 1939, frequency lists of samples of cited titles were made, and lists of source journals were developed.[35] The idea of using "key" journals to determine a frequency citation list which in turn gave a group of source journals was new,[35] although Fussler admitted his general procedure, that of listing and analyzing references, was not new.[35] Fussler's study differed mainly in the method of selecting the source journals.[48]

His investigations led him to conclude:

> the distribution of literature, as revealed in the study, indicates that provision of a substantial proportion of the references in a research field is possible through a small number of journals if they are properly selected but that the provision of the entire necessary research literature for a field will require a very large number of titles, many of which will be needed elsewhere in a university or in a large research organization.[35]

Because of the interrelationship of various disciplines, there is a "network of joint use of research materials"[35] not necessarily based on the classification scheme.

One item of special note is Fussler's comment about literature characteristics wherein he suggested that those of a particular sector of a discipline may differ noticeably from those of a whole field.[35]

The first presentation of "impact factor" was in 1955 in "Citation Indexes for Science" by Eugene Garfield.[56] Garfield was aroused by the quote of P. Thomasson and J. C. Stanley:

> The uncritical citation of disputed data by a writer, whether it be deliberate or not, is a serious matter. Of course, knowingly propagandizing unsubstantiated claims is particularly abhorrent, but just as many naive students may be swayed by unfounded assertions presented by a writer who is unaware of the criticisms. Buried in scholarly journals, critical notes are increasingly likely to be overlooked with the passage of time, while the studies to which they pertain, having been reported more widely, are apt to be rediscovered.[57]

Hoping to "eliminate the uncritical citation of fraudulent, incomplete, or obsolete data by making it possible for the conscientious scholar to be aware of criticisms of earlier papers," Garfield developed a plan for "a citation index that offers a new approach to subject control of the literature of science...best described as an association-of-ideas index."[56] He concluded that a basic difficulty was to build a subject index which included all possible synonyms, and that "bibliographic tools" were needed to help bridge the subject gap between source and receiver, authors and information-seeking scientists.[56]

Garfield invented "a citation code for science."[56] This code was made up of two parts, "a serial number...to identify each periodical" and a "serial number, assigned to each article in a particular publication."[56] The citation index would have these characteristics:

> A complete alphabetic listing of all periodicals covered, in addition to the code number for each periodical...would list in straight numerical order the code numbers for all the articles covered...code numbers representing articles that had *referred* to the article in question would be listed under each code number.[56]

"The system would provide a complete listing, for the publications covered, of all the original articles that had referred to the article in question," being especially "useful in historical research, when one is trying to evaluate the significance of a particular work and its impact on the literature and thinking of the period."[56] The influence of an article, Garfield named "its 'impact factor'."[56]

Garfield's "Citation Analysis as a Tool in Journal Evaluation," subtitled, "journals can be ranked by frequency and impact of citations for science policy studies" appeared in 1972.[58] Narin claimed it "is a milestone in the field"[48] for this reason. "There are more data covering more citations to more journals in that one paper than there had been in all the scientific literature up to that time."[48] Problems, of course, exist: abbreviated titles standing for more than one journal title, changes in format of journals such as mergers or splits, and changes in journal titles for special issues.[56] But Garfield claimed:

> It is apparent...that a good multidisciplinary journal collection need contain no more than a few hundred titles. That is not to say that larger collections cannot be justified, but it does say something indisputable, in terms of cost and benefit, about how large a journal collection need be (or how small it can be) if it is to provide effective coverage of the literature most used by research scientists.[56]

Using much data from his *Science Citation Index*, Garfield made numerous observations: "majority of all references cite relatively few journals,"[56] "the average paper is cited only 1.7 times a year;"[56] "predominance of a small group of journals in the citation network has been confirmed,"[56] "half of all articles published cite at least one of the 25 most frequently cited journals at least once,"[56] "predominance of cores of journals is ubiquitous,"[56] and "most frequently cited journals...are (usually) most productive in terms of articles published."[56] All of this led him to state, "I can with confidence

generalize Bradford's bibliographic law concerning the concentration and dispersion of the literature of individual disciplines and specialties."[56]

What about journals which rank low on a "times-cited" list? According to Garfield, although "citation frequency reflects a journal's value and the use made of it,"[56] low-ranking journals may also be used "for some purpose other than the communication of original research findings."[56]

In this article, Garfield also explained how his relative impact factor was determined:

> by dividing the number of times a journal has been cited by the number of articles it has published during some specific period of time. The journal impact factor will thus reflect an average citation rate per published article.[56]

In 1961, M.M. Kessler did a report on "bibliographic coupling" for MIT. This report, revised in 1962, was later published in *IEEE Transactions on Information Theory* of January 1963.[59] The paper, "An Experimental Study of Bibliographic Coupling Between Technical Papers," was described by Kessler as "one of several pilot studies designed to test an hypothesis."[59] Having been suggested three years earlier, the hypothesis was basically "that a number of scientific papers bear a meaningful relation to each other (they are coupled) when they have one or more references in common."[59]

In August 1962, Kessler sent his article "Bibliographic Coupling between Scientific Papers"[60] to the publishers of *American Documentation*; it was printed in January 1965. In 1974, Bella Hass Weinberg reviewed the subject and referred to this paper by Kessler as "the classic paper in bibliographic coupling."[61]

Kessler stated that bibliographic coupling is "a new method for grouping technical and scientific papers,"[59] with the essential facts of the method being:

a. A single item of reference used by two papers is called one unit of coupling between them.
b. A number of papers constitute a related group, G_A, if each member of the group has at least one coupling unit to a given test paper P_0.
c. The coupling strength between P_0 and any member of G_A is measured by the number of coupling units (n) between them.[62]

He also explained that:

> This method was called "Bibliographic Coupling" because it originated in the hypothesis that the bibliography of technical papers is one way by which the author can indicate the intellectual environment within which he operates, and if two papers show similar bibliographies, there is an implied relation between them.[62]

Based on the unit of coupling (a single item of reference used by two papers), two criteria of coupling, *A* and

B, are defined. Criterion A includes b) and c) in the foregoing list. An example of G_A would be:

$$A - B \overset{P_0}{\diagup} C$$

where P_0 is the test paper, and A, B, and C are members of G_A.

Criterion B contains related group G_B if each member is connected with every other member by at least one coupling unit. An example of G_B would be:

$$B \overset{A}{\diagup} C$$

where A, B, and C are members of G_B.[60]

Although, at the time of his report, Kessler was further investigating his method, he listed five bibliographic coupling properties:

1. It is independent of words and language. All the processing is done in terms of numbers. We thus avoid all the difficulties of language, syntax and word habits.
2. No expert reading or judgment is required. Indeed, the text need not be available.
3. The group of papers associated with a given test paper extends into the past as well as the future.
4. The method does not produce a static classification or permanent index number for a given paper. The grouping will undergo changes that reflect the current usages and interests of the scientific community.
5. The member papers of a G_A may be considered as the "logical references" of P_0. A population of papers, P_m may thus be produced in such a way that for each paper in P_m we substitute the members of its G_A for the actual references. (That is, replace the actual references of the paper by its logical references.) This population of papers may then be further processed according to Criterion A. The significance of this logic is being investigated.[60]

To secure data, Kessler had a computer program written, processing 137,000 references obtained from 8,521 articles in 36 volumes of the *Physical Review*. Because of the overabundance of material, it was decided to use Volume 97 only, creating 265 G_As.[60] The variance of the number of articles in each G_A was 0 to 27, with the variance of coupling units from 1 to 7.[60] Bibliographic coupling, at the present time, is recognized as a technique "effective in identifying related papers,"[48] but more analysis is needed to interpret the relationship.

Although the "epidemic theory" itself is not bibliometric, it can be in its application. Robert Kenneth Dikeman used the epidemic theory model in his

dissertation, "On the Relationship between the Epidemic Theory and the Bradford Law of Dispersion,"[63] J.J. Hubert included it in his "Bibliometric Models for Journal Productivity,"[64] and Dennis B. Worthen made a comparison of it in his "The Epidemic Process and the Contagion Model."[65]

The original report, "A Generalization of the Theory of Epidemics: An Application to the Transmission of Ideas," by William Goffman and Vaun A. Newill,[66] is included in this group of original idea papers because it is antecedent to a large portion of the bibliometric research of Goffman and his colleagues[67–70] and can be considered fundamental.

Publicized ideas have been the forerunner of many scientific developments, and Goffman and Newill's innovative suggestion of the model of the theory of epidemics is really basic to the study of idea development. It is not a technique such as bibliometric coupling and cocitation which are used in the analysis of the structure of literature. The epidemic theory is a technique explaining the transmission of ideas by means of the literature. In other words, an "infectious disease epidemic" is compared to an "intellectual epidemic," where the germ or virus is analogous to an idea, the "case of disease" analogous to the "author of paper" or the "paper containing useful ideas," the susceptible person to the "reader of paper" or "all papers containing potentially useful ideas," "death or immunity" to "death or loss of interest," or "deletion or loss."[66]

The purpose of this "epidemic theory report" was "to help to describe the publication activity within a given discipline"[66] to determine the necessity for an information retrieval system.

As explained by the authors, an epidemic process needs: "1) A specified population; [and] 2) An exposure to infectious material,"[66] with the population divided into three categories at a given time:

(a) *Infectives*: Those members of the population who are host to the infectious material.
(b) *Susceptibles*: Those members of the population who can become infectives given effective contact with infectious material.
(c) *Removals*: Those members of the population who have been removed from circulation for one of a variety of reasons.[66]

A possible example would be: If a third paper C cites papers A and B (which are then bibliographically coupled), it can be said that C was susceptible to the infectious material (idea) from infectives A and B, and because A and B are coupled, A could be considered a catalyst of B, and B the same of A. Example:

This analogy can also be applied to clusters of papers: an original paper is the infective, and papers that cite the original are susceptibles. For example:

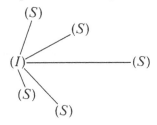

Cluster of papers

One last note: If "in general, the epidemic process can be characterized as one of transition from one state…to another,[66] a scientific front and an epidemic are closely related, for a scientific front is also an indication of a change."[66][65]

The next article in the chronological arrangement of papers is that of Alan Pritchard who advocates the use of the new word "bibliometrics" in place of "statistical bibliography," already mentioned in the first section.

Another innovative paper, this one on analysis of citations, was presented by Henry Small. Small proposed a new method of analyzing citations in 1973 in the first of a group of articles bearing on the subject cocitation. In this article, "Co-Citation in the Scientific Literature: A New Measure of the Relationship between Two Documents,"[71] Small explained his subject as follows:

> Bibliographic citations in scientific papers have been used by a variety of researchers to establish relationships among documents. Both direct citation—the citing of an earlier document by a new document—and bibliographic coupling—the sharing of one or more references by two documents—have received considerable attention. A related measure, which has been overlooked in earlier discussions, is co-citation. Unlike bibliographic coupling which links source documents, co-citation links cited documents and is, therefore, analogous to a measure of descriptor or word association. The purpose of this paper is to define this new kind of "coupling" and to distinguish it from bibliographic coupling, using an actual example from the literature of physics. Co-citation patterns are found to differ significantly from bibliographic coupling patterns, but to agree generally with patterns of direct citation.[71]

Because "Co-citation is the frequency with which two items of earlier literature are cited together by the later literature,"[71] the number of times that papers are cited together by a new or later document gives the strength of the cocitation. For comparison, bibliographic coupling, depending "on references contained in the coupled documents" is a set or permanent relation, whereas in the case of cocitation, "patterns change as the interests and intellectual patterns of the field change."[71]

Assuming that the main ideas and the experiments supporting them are in the frequently cited papers, then cocitation "can be used to map out in great detail the relationships between these key ideas."[71] Cocitation can also be used to depict a literature specialty core or cluster.[71]

Small concluded:

> The pattern of linkages among key papers established a structure or map for the specialty which may then be observed to change through time. Through the study of these changing structures, co-citation provides a tool for monitoring the development of scientific fields, and for assessing the degree of interrelationship among specialties.[71]

The preceeding information led to the listings found in Table 2. Historically, the Hulme paper was first in the presentation and application of the term "statistical bibliography." Hulme cited the Cole and Eales publication which was the first to use a publication count by country. Gross and Gross, first to consider the structure of literature, used a citation count technique to make an evaluation. Fussler was the first to develop a different method of selecting source journals for data. Garfield was the first to use an "impact factor" in citation analysis. Kessler cited his paper as the introduction of bibliographic coupling. Goffman and Newill were the first to apply the theory of epidemics to the field of literature. Small introduced the technique of cocitation. Pritchard coined the word "bibliometrics." The identification of the first papers of Bradford and Lotka to present their laws was easy; the identification of the first paper to present Zipf's law required more searching; the law was found in the preface of one of his books.

Since the 1973 paper of Small, there does not seem to have been any basically new idea presented; the many publications since then are variations, applications and/or extensions of the original hypotheses, laws, or techniques.

Empirical Laws of Bibliometrics

> One of the principal aims of Science is to trace, amidst the tangled complex of the external world, the operation of what are called "laws"—to interpret a multiplicity of natural phenomena in terms of a few fundamental principles.[18]

> Natural laws describe patterns which are regular and recurring. The scientific point of a law is twofold. First, a concrete statement of a law may give us the ability to better predict events or to shape our reactions to them. Second, a physical law may help in the development of theories which explains why a particular pattern occurs.[72]

Table 2 Chronological list of seminal papers.

Year	Author(s)	Title	Historical significance
1917	Cole, F.J., and Nellie B. Eales	"History of Comparative Anatomy"	First publication count, by country, 1843–1860
1923	Hulme, E. Wyndham	*Statistical Bibliography in Relation to the Growth of Modern Civilization*	Term "statistical bibliography" first used
1926	Lotka, Alfred J.	"Frequency Distribution of Scientific Productivity"	Inverse square law, relation of authors to papers
1927	Gross, Paul L.K., and E.M. Gross	"College Libraries and Chemical Education"	Citation count technique
1934	Bradford, Samuel C.	"Sources of Information on Specific Subjects"	Scattering observations published
1935	Zipf, George K.	*Psycho-Biology of Language*	Word rank-frequency distribution
1949	Fussler, Herman H.	"Characteristics of the Research Literature used by Chemists and Physicists in the U.S."	First to use journal references to develop a core of journals in a field
1955	Garfield, Eugene	"Citation Indexes for Science"	Term "impact factor" first used
1963	Kessler, M.M.	"Bibliographic Coupling between Scientific Papers"	Introduced bibliographic coupling
1964	Goffman, William, and Vaun Newill	"Generalization of Epidemic Theory: An Application to the Transmission of Ideas"	Epidemic theory applied to literature growth
1969	Pritchard, Alan	"Statistical Bibliography or Bibliometrics?"	Word "bibliometrics" introduced
1973	Small, Henry	"Co-Citation in the Scientific Literature: A Measure of the Relationship between Two Documents"	Introduced cocitation

This was emphasized in the work of Pranas Zunde and John Gehl,[73] where they wrote that it is important not only to describe the phenomena of a science but to try to establish tenets to explain and predict the phenomena.[73]

Abraham Bookstein has set specifications for laws.

> When a new law is proposed, it may be reasonable to demand of it, on intuitive grounds, that it remain true under a variety of circumstances differing from those in which it was discovered. Should it satisfy these demands, we may increase the validity of the law; if it does not, and yet we still wish to maintain the validity of the law, we ought to feel an obligation to explain the discrepancy.[74]

Three regularities occur in bibliometrics to which have been given the name "law": Lotka's Law of Scientific Productivity (authors publishing in a certain discipline), Bradford's Law of Scattering (distribution of publications), and Zipf's Law of Word Occurrence (ranking of word frequency).[75] Although claims have been made that the three laws are basically the same, one of their differences lies in the type of data. Lotka's law dealt with authors publishing and the number of papers published; Bradford observed the scattering of articles on specific subjects in various journals; Zipf counted frequencies of words. "Each of these distributions was empirically derived," and they "are similar to each other as special cases of a hyperbolic distribution."[76] Although many and various proofs have been offered by important scientists supporting the theory of relatedness and similarity of these "laws,"[4,77] the reader is advised to keep John Hubert's comment in mind: "literature" contains many models, and some are erroneously referred to as "laws" as if they predicted occurrences without error."[78]

In this section, each law is analyzed from the viewpoint of its literature, and some of the literature of the three empirical laws explaining each one's importance is noted.

LOTKA'S LAW OF SCIENTIFIC PRODUCTIVITY

Alfred J. Lotka was a mathematician, supervisor of mathematical research in the Statistical Bureau of the Metropolitan Life Insurance Company from 1924 to 1933.[79] It was during this time, 1926, that his definitive work, later called Lotka's law was produced.[49] His investigation was a productivity analysis (described in the preceding section). Counting names and the number of publications listed for each, the coverage was for only A and B names in *Chemical Abstracts* for 1907 to 1916 and for Auerbach's *Geschichtstafeln der Physik* from its beginning through 1900.[49] The data were tabulated and plotted, from which Lotka developed a "general formula for the relation...between the frequency y of persons making x contributions" as "$x^n y = $ const."[49]

Finding the value of the constant when $n = 2$, he observed that:

> the number of persons making 2 contributions is about one-fourth of those making one; the number making 3 contributions is about one-ninth, etc.; the number making n contributions is about $\frac{1}{n^2}$ of those making one, and the proportion, of all contributors, that make a single contribution, is about 60 per cent.[49]

Notice that Lotka's observation deals with the least number of productions.

Since the publication of Lotka's original article in 1926, much research has been done on author productivity in various subject fields. The publications arising from this research have come to be associated with Lotka's work and are often cited as proving or supporting his findings. However, a review of this literature reveals that Lotka's article was not cited until 1941, that his distribution was not termed "Lotka's law" until 1949, and that no attempts were made to test the applicability of Lotka's law to other disciplines until 1973.[80]

In that year Murphy did his study of "Lotka's Law in the Humanities."[81]

Larry J. Murphy chose for his work the first decade of the publication (on history of technology) *Technology and Culture*[81] and found "the actual number of single authors" to be 130.[81] He rounded off his numbers and plotted the results, leading him to conclude, without any statistical test for goodness of fit, that "The tabulation essentially verifies that Lotka's Law applies to this field within the humanities."[81] There was an error in Murphy's graph which was noted and corrected in a letter to the editor.[82]

In 1977 John J. Hubert, in a letter to the editor of the *Journal of the American Society for Information Science*, evaluated Murphy's work with the χ^2 goodness of fit test.[83] His conclusion was, "By using an exact interpretation of the law and by applying a valid statistical test, it can be shown that this data does not follow Lotka's law."[83] Russell C. Coile also did an evaluation of the Murphy results,[84] and he too found that Lotka's law did not apply to the particular sample of authors.[84]

Henry Voos, in 1974, measured the productivity of authors in the field of information science and compared his results with Lotka's observation.[85] Using all the articles indexed in *Information Science Abstracts* from 1966 to 1970 for his data, he made a rank-frequency listing of all the authors and their articles by year.[85] Voos "found that the relationship for information science seems to be $1/n^{3.5}$ instead of Lotka's $1/n^2$,"[85] and instead of Lotka's 60% producing one paper, "in information science the figure tends to be 88% writing only one paper."[85] The results of using the χ^2 test show "one can be more than 95% certain that the actual population fits the theoretical distribution."[85]

Coile wrote a criticism of the Voos work, making suggestions and corrections.[86] Voos responded with a revised chart, a promise for further research, and an apology.[85]

Next to apply Lotka's law to various subjects was Alan Edward Schorr.[87] He published his first application study in the fall of 1974 in conjunction with library science and used for data *Library Quarterly* and *College and Research Libraries* from 1963 to 1972.[87] His analysis indicated that "It is clear that Lotka's Law does not apply to the field of library science, where four-fifths of all papers represent the only contribution of an individual."[87] Schorr suggested:

for library science literature, scholarly production would follow an inverse quadruple power law $\left(\frac{1}{n^4}\right)$ whereby for each 100 contributors of single articles, about 6 will contribute two articles, about 1 will contribute three articles, and almost no writers would provide four articles or more.[87]

A scathing letter by Dean Tudor, chairman of the Library Arts Department of Ryerson Polytechnical Institute regarding Schorr's "Lotka's Law and Library Science" was soon received by the editor of *RQ*.[88] Tudor called it "frivolous bagatelle occupying two pages of space that could be better put elsewhere,"[88] but he did pose some interesting questions, including "is Lotka's Law still valid almost fifty years later?"[88] Alan Schorr retaliated, but with a very carefully worded and explanatory letter."[89]

In the spring of 1975, Schorr's "Lotka's Law and Map Librarianship,"[90] based on his own map librarianship bibliography,[91] was published. "Because it appears that the results do not support the inverse square law, a chi-square test was applied,"[90] and conclusions were drawn that "Lotka's Law holds true for the field of map librarianship."[90]

This time Russell Coile reacted to Schorr's work, writing "this conclusion is not true when the correct data is tested with an appropriate goodness-of-fit test"[84] and suggested, after his own calculations were made, that "it would appear that Lotka's law does not apply to this field of map librarianship."[84]

Schorr did one more test with Lotka's law, producing the paper "Lotka's law and the History of Legal Medicine,"[92] the data being secured from *International Bibiliography of the History of Legal Medicine*.[93] Schorr found:

Authors with multiple contributions fall below the numbers expected by Lotka's Law (e.g., authors with two publications account for 18.1% of single author entries instead of 25%; authors with five publications comprise 1.8% instead of 4%)...The results do not support the inverse square law and thus Lotka's Law does not hold true for the history of legal medicine.[92]

"To test for validity a chi-square test was applied... Lotka's Law is not applicable to this discipline."[92]

T. Radhakrishnan and R. Kernizan reported their "experiment to verify the satisfaction of Lotka's law, with the papers published in the area of computer science,"[94] in the article, "Lotka's Law and Computer Science Literature."

In a convenient machine-readable form we had with us the bibliographic details of the papers published during the years 1968 to 1972 in *Communications of the Association for Computing Machinery* (CACM) and in the *Journal of the ACM* (JACM).[94]

It was decided to use this information in an attempt to verify Lotka's law. In their first experiment, the assumption was made that "an author publishes exclusively through one scientific journal," but it was found that this assumption was not valid. "From the chi-squared values it is clear that Lotka's original suggestion of x/n^2 does not apply very well. However, the prediction by x/n^3 is quite close."[94] "In the second experiment we considered all the papers published by the authors irrespective of the journal."[94] A random selection of authors was made using the cumulative author of *Computer and Control Abstracts* and was repeated with JACM. The deviation from Lotka's Law was high.[94]

In the judgment of the authors:

> 1) Lotka's law, in its generalized form, seems applicable when we consider the publications of authors in one periodical; and 2) when we consider all the publications of the authors in various journals, the observed values deviate considerably from the predictions of the law. This may be due to the differences between pure and applied science.[94]

They suggested further testing the law using "the service tapes such as Engineering Index, INSPEC, etc."[94]

Two more papers testing the law in 1981 are here mentioned. K. Subramanyan did a study with library literature,[95] and Valerie L. Richardson[96] collected data from the "catalogue at State College of Victoria at Frankston."[96] Subramanyan found that the pattern did not conform to Lotka's results and suggested more work be done.[95]

Richardson stated that from her investigation, "It can be seen that the Frankston data do fit Lotka's law in its general form, but since the slope of the Frankston line is greater than 2, the data do not correspond to the inverse square law."[96]

The compilation "Frequency Distributions of Scientific Performance: A Bibliography of Lotka's Law and Related Phenomena," by Jan Vlachy,[97] although not a substantive work, is included here because of its excellence. This bibliography of 437 papers[97] is international for it lists articles in English, from England, Canada, and the United States; and in Czech, Russian, Danish, French, and German. Lotka, Bradford, and Zipf are included as well as many other areas of bibliometrics. (An interesting note: Here was found the second reference to the Estoup book.)[97]

Attention should be drawn to several items of importance: In Lotka's investigation, "Joint contributions have in all cases been credited to the senior author only,"[49] whereas some of the studies counted each contributing author separately. This could make a difference in the resultant statistics. The number of names used may have influenced the outcome; Lotka use 6819 and 1325;[49] other studies, many less; Murphy, for instance, used 130.

The time period is another variable that may have affected outcomes. Lotka's *Auerbach* data covered a complete history—and the data fit; his *Chemical Abstracts* statistics covered a 10-year period—and the data did not fit.[49]

Miranda Lee Pao summed up the prevalent problems neatly:

> Several studies have assumed the inverse square relation as the basis for testing. Others derived the value of the constant c from the percentage of single paper contributors. None of these assumptions can be traced back to Lotka.[98]

Her deduction is well taken,

> Therefore, a uniform method should be agreed upon by those attempting such a test. Comparison and generalization on author productivity may be possible only if compatible data are available and results are significant.[98]

An added note about Lotka. At Johns Hopkins University at the invitation of the Department of Biometry and Vital Statistics, Alfred J. Lotka finished his book *Elements of Physical Biology*, started in 1902 and published in 1925.[99] In the book, Lotka discussed the work of C.J. Willis, who developed a biological theory of age and area,[100] and the extract following seems to be especially interesting.

> Turning from the theoretical and debatable portion of Dr. Willis' contribution, to the factual material presented by him, we are confronted by a number of very remarkable relations...[The] ordinates represent the number of genera in the several natural families of plants and animals indicated, and the corresponding abscissae represent the number of species in each genus plotted. So...of the family Compositae 1143 genera were noted as follows:
>
> 446 genera of 1 species
> 140 genera of 2 species
> 97 genera of 3 species
> 43 genera of 4 species
> 55 genera of 5 species
> etc.

> ...the monotypic genera, with one species each, are always the most numerous, commonly forming about one third of the whole group; the ditypics, with two species each, are next in frequency, genera with higher numbers of species becoming successively fewer...Perhaps more striking still is the relation established in figures 64, 65, and 66. [These figures are graphs wherein relations between number and size are plotted logarithmically.] The quantities plotted here are of the same character as in figure 63, [these are hyperbolic curves] but they are plotted on a double logarithmic scale, with the remarkable result that the graphs obtained are in close approximation straight lines. Thus if x denotes the number of species in a

genus, and y denotes the number of genera comprising x species, we have $\log x + a \log y - b = 0$ or

$$xy^a = \text{const.}$$

that is to say, the variables x and y are connected by a hyperbolic relation. It should be noted that this relation covers a wide variety of cases, including both plants and animals.[99]

This seems very familiar when we realize Lotka's "general formula for the relation...between the frequency y of persons making x contributions is $x^n y = \text{const}$,"[49] and when his data are arranged as the Willis genera and species information:[49]

Frequency of persons (Genera)	Contributions (Species)
3991	1
1059	2
493	3
287	4
184	5

Is it possible that Lotka's idea for his "productivity law" was suggested by Willis's investigations?

BRADFORD'S LAW OF SCATTERING

Part I

That Samuel Clement Bradford was a very determined and dedicated person is very evident in his writing. His thinking was undoubtedly influenced by Paul Otlet and LaFontaine, who organized the First International Conference on Bibliography held in Brussels in 1895.[101] The theme of the conference was the need for international cooperation to develop a universal index which would recognize the requirement of a standard subject classification to be backed by a central universal library. This idea was staunchly supported by Bradford, as shown in the following.

In 1927 "Bibliography by Cooperation" appeared in *Library Association Record* (LAR). Here Bradford indicated his concern for the accelerating accumulation of "useful scientific and technical literature"[102] and reiterated the need for "bibliographical work...[to] be brought together"[102] by the use of one system of classification "as an index to all the papers that relate to a particular subject of study, no matter when they were written."[102] Because "Every bureau indexes or abstracts papers that are done by other bureaux, and only a portion of the literature is covered,"[102] he believed a universal classification system would eliminate repetition or classifying the same articles, use items not cataloged, and thereby save time, information, and money, and give better service.

He included in his paper a list of science subjects and the number of bibliographic references for each[102] to alert readers as to what the Science Library, of which he was deputy keeper, was doing to develop into an "information service covering the whole field of Science and Technology."[102] At that time, Bradford claimed the number of references assembled, covering many aspects of science and technology, as 1,212,700.[102]

This article was followed by Bradford's "The Necessity for the Standardisation of Bibliographical Methods"[103] in which Bradford explained the Science Library's method of classification and again stressed the need for cooperation. By the time of the introduction of the paper, a form of interlibrary loan already had been practiced for 2 years by the Science Library, with requests having been received from a number of other countries.[103]

In 1934, Bradford's classic article[53] containing his law was printed. Again he wrote about the number of articles abstracted and indexed by a number of abstracting and indexing journals showing his continued concern about wasted money and skills brought about by duplication of effort. He stated that "although the 300 abstracting and indexing journals notice 750,000 articles each year...only 250,000 different articles are dealt with and 500,000 are missed."[53] His concern was shared by Ernest Lancaster-Jones, assistant keeper in the Science Museum. Consequently, "A statistical analysis was made of the references in two quarterly bibliographies compiled in the Science Library, i.e., the *Current Bibliography of Applied Geophysics* and the *Quarterly Bibliography of Lubrication*"[104] by Lancaster-Jones, a trained mathematician who also "had received recognition as an expert in applied geophysics."[105] Bradford, using the data, evolved his "law."[105]

The method Bradford used was this: For each of the two subjects, geophysics and lubrication, a table was made, listing "number of journals producing a corresponding given number of references"[53] and each accumulation. (It seems natural that such a list of subjects and the number of references for each, plus a very limited budget and a sincere desire to give exceptionally good library service would cause an active and inquisitive mind, such as Bradford's, to wonder about the actual number of journals needed for coverage.)

One of Bradford's hypotheses was that

references are scattered throughout all periodicals with a frequency approximately related inversely to the scope. On this hypothesis, the aggregate of periodicals can be divided into classes according to relevance of scope to the subject concerned, but the more remote classes will, in the aggregate, produce as many references as the more related classes.[53]

Observations of the tables evinced three "rough" zones or groupings which Bradford graded as "1) Those producing more than 4 references a year; 2) Those producing more than 1 and not more than 4 a year; 3) Those producing 1 or less a year."[53] Bradford found

> The groups thus produce about the same proportion of references in each case, and the number of constituents increases from group to group, by a multiplier which, though by no means constant, approximates fairly closely to the number 5, especially for the two larger groups.[53]

From his data, Bradford constructed two graphs, plotting the logarithms of cumulated number of journals in relation to the cumulated number of references for each, geophysics and lubrication. He noted that "the later portion of each curve is remarkably close to a straight line," and observed that

> the aggregate of references in a given subject, apart from those produced by the first group of large producers, is proportional to the logarithm of number of sources concerned, when these are arranged in order of productivity.[53]

With this observation in mind, Bradford constructed a second graph or diagram. This diagram he used to develop an algebraic relation, but only for the "straight" part of the curve noted originally.[53] From this, he deduced his "law":

> Therefore, the law of distribution of papers on a given subject in scientific periodicals may thus be stated: if scientific journals are arranged in order of decreasing productivity of articles on a given subject they may be divided into a nucleus of periodicals more particularly devoted to the subject and several groups or zones containing the same number of articles as the nucleus, when the numbers of periodicals in the nucleus and succeeding zones will be as 1: n: n^2....[53]

This was the first statement of what was later to be called the verbal part of Bradford's Law. Usually referred to as Bradford's Law of Distribution of Scattering, "sometimes this regularity is also called the law of dispersal of publications."[106]

Bradford came to the conclusion that

> A standard classification must be adopted, so that references to the same subject would be brought together by the classification, irrespective of source or abstracting bureau, when, without increase of labour, a complete index to scientific literature would be achieved.[53]

In a paper on abstracting and indexing periodicals, presented in 1937 at the ASLIB Proceedings 14th Conference, Bradford told that the details of the analysis "to determine the extent to which scientific papers are scattered in periodicals devoted to other subjects,"[104] published in *Engineering* in 1934, were reprinted "in

Publication No. 1 of the British Society for International Bibliography, 1934,"[104] and added, "Only the results need be quoted here."[104]

> The law of scattering may thus be stated. If periodicals containing articles on the given subject are arranged in decreasing order of the number of such articles they contain and divided into a nucleus of journals more specifically devoted to the subject and zones of periodicals containing the same number of special articles as the nucleus, the numbers of journals in the nucleus and succeeding zones are as 1:N:N^2...where N is about 5 or 6.[104]

In his first article, Bradford used the value of 5.

In the discussion which followed his presentation, Bradford admitted:

> The whole object of the elaborate statistical investigations reported in this paper is to prove beyond any question that quantities of important scientific papers are neither abstracted nor indexed. This is one of the main reasons why so much time and expense is being wasted in duplicating previous work. The other main reason is the adoption of archaic alphabetical methods of indexing, which hide the information and prevent it being found. The adoption of standard methods would go far to remove both these defects.[104]

The International Federation for Documentation in 1946 published Bradford's "Complete Documentation in Science and Technology,"[107] which Bradford presented at the 16th Conference in Paris.[107] Again, Bradford presented the statistics he used in "Sources of Information on Specific Subjects." He repeated his description of the "law of distribution" as given in his previous papers and again supported the use of the Universal Decimal Classification. A synopsis of this paper was also printed in *Nature* of January 1947.[108]

Two more of his publications are merely mentioned here: "Complete Documentation,"[109] presented at the 1946 Royal Society Empire Scientific Conference when Bradford was president of the British Society for International Bibliography;[109] and his book *Documentation*,[110] first published in 1948. In both cases, Bradford used the same statistics he had used in his paper which first presented his law. The statistics were primarily used to support his ideas about cooperative, universal classification; the law was a natural outcome of Bradford's desire to give added proof and credence to his unswerving conviction.

Part II

Bradford's law has been the main topic of many articles in literature. A general sampling has been used here.

The discussions of Bradford's law take several directions: analysis of the law itself, attempts to refine it, comparison with other laws, and applications.

The first notable paper on Bradford's Law of Scattering was that of Brian C. Vickery.[111] When results of an analysis of some 1600 periodical references, borrowed from various libraries by the Butterwick Research Laboratories Library, were compared with Bradford's work, an inconsistency was found.[111] Vickery discovered that the law as Bradford stated it was not in total agreement with his algebraic expression.[111] After methodically examining Bradford's algebraic analysis, Vickery proceeded with his own analysis, arriving at this conclusion.

We can...regard the theoretical distribution of papers on a given subject in scientific periodicals, as derived by Bradford, as fully corroborated by the distributions observed in the sample investigations. The rectilinear relation...incorrectly assumed by Bradford to be identical with his theoretically derived relation, fits only the upper portion of the observed curve. The theoretical relation itself, however, enables us to predict the whole curve.[111]

The next paper to be noted is that of M.G. Kendall.[112] Kendall's report is an analysis of 370 journals of 1763 references in operational research. For comparison, "1465 references to statistical methodology (covering the period 1925–39)"[112] were used. Kendall suggested:

There is an obvious resemblance between this type of distribution and that of income. If we equate journals to persons and number of references to size of income we have the characteristic pattern in which a lot of people (journals) have a low income (few references) proceeding regularly to a few millionaires (journals with a large number of references).[112]

Kendall supported the idea that the Bradford type of distribution is really a Zipf distribution,[112] and he also provided "a more refined statistical explanation of the straight line observed by Bradford."[113] (See page 35 of the Kendall article[112] for Kendall's explanation.)

P.F. Cole in 1962 took "A New Look at Reference Scattering" whereby he plotted the cumulative fraction of references against the logarithm of the cumulative fraction of titles and the resultant slope he named "reference-scattering coefficient."[113]

Cole gave three possible applications of the coefficient:

a. For identifying those groups of references which, though of different sizes, are scattered to the same extent. This facilitates the establishment of further relationships between numbers of journal titles.
b. As a criterion for detecting abnormalities in the results of literature usage surveys.
c. As a concise means of expressing quantitatively the distribution of references among journal titles. The

extensive reporting in detail which is such a characteristic feature of literature usage surveys would then become entirely unnecessary..., which is probably the most important of the three, is self-explanatory.[113]

Cole came to the conclusion that the reference-scattering coefficient "may be characteristic of the subject field"[113] and found "A comparison of reference-scattering coefficients derived from data obtained in three published petroleum literature-usage surveys shows that the reference-count method gives untrustworthy results."[113]

Ferdinand Leimkuhler did a further analysis of the Bradford law and concluded that

Bradford's "law of scattering" is the inverse function for the Bradford distribution, i.e. while the latter predicts the number of references for a given portion of the journals, the former speaks of the number of journals required to obtain a given portion of the references.[114]

In a later paper (1980), Leimkuhler developed what he said was:

An exact, discrete formulation of Bradford's law describing the distribution of articles in journals (is derived) by showing that Bradford's law is a special case of the Zipf–Mandelbrot "rank frequency" law. A relatively simple method is presented for fitting the model to empirical data and estimating the number of journals and articles in a subject collection.[115]

In 1981, Brookes did a criticism of Leimkuhler's exact formulation, where he wrote:

An elementary theorem of Shannon information theory shows that his [Leimkuhler] new function is applied to only 2.3% of the statistical information inherent in the bibliography he chooses and that Bradford's original simple formulation not only fits this segment but also the whole bibliography more closely than the new formulation. As every loss of statistical information can be measured, it can be shown that sophisticated mathematical techniques cannot compensate for the information they squander.[116]

So I question Leimkuhler's present claim. More seriously, however, I question his methodology.[116]

Brookes concluded:

The results of applying Bradford's law in its original form, i.e. applied to the ranked journals with the ranks marked on a logarithmic scale with the cumulative sums of papers as the ordinates, shows a better fit over any range than Leimkuhler's new function applied to the journals only.[116]

Of course, Leimkuhler wrote an answer, making necessary corrections and adding, "I share Brookes' concern for empirical validity, but this concern should lead to more, not less, theoretical analysis."[117]

In 1969 Brookes presented a paper, "Bradford's Law and the Bibliography of Science," in which he stated, "In the management of special libraries the bibliograph offers a check on the completeness of any allegedly complete bibliography—a check that has hitherto been missing."[118] He made this claim on the basis of the application of Bradford's law in which he found, "All such bibliographs of periodical literature so far examined have either corroborated Bradford's law or else the discrepancies have been plausibly accounted for."[118]

S. Naranan, in his "Bradford's Law of Bibliography of Science: An Interpretation,"[119] called attention to the Brookes paper: "An interesting discussion of the law and its application in library systems is given by Brookes."[119] For the conclusion of his own paper, he wrote:

In summary, it is shown here that the essential features of Bradford's law of bibliography of scientific literature can be explained in terms of an underlying power law distribution of the number of articles in scientific journals. It is suggested that the law emerges as a natural consequence of exponential growth of scientific literature and journals at comparable rates. Such a model predicts a strong correlation between the age of a journal and the number of articles it carries. The proposed mechanism is likely to find wider application in many other fields of science.[119]

Brookes responded with an interesting letter. He showed that Naranan's analysis, while not valid for Bradford, "(with suitable verbal amendments) provides a plausible model of Lotka's law."[120]

This inverse square law of scientific authorship has hitherto been regarded as an inexplicable and useless statistical oddity. Naranan's model of it is therefore welcome. And, together with other measures of scientific productivity, Lotka's law has been applied by Dobrov and Korennoi in determining the optimum size of research institute in the USSR.[120]

William Goffman and Thomas Morris applied Bradford's Law to library acquisitions.[121] The authors stated,

Bradford's law should apply to the use of periodicals in a library as well as to the dispersion of articles among journals. These are both acquisitions processes; namely, processes of obtaining relevant items by means of selection.[121]

Goffman and Morris summarized their findings in the following paragraph:

It has been shown that the distributions of both circulating periodicals and their users in a medical library seem to

obey Bradford's law. Hence, a smallest core of journals which must belong to the library's collection can be defined. This core should consist of the minimal nucleus of periodicals circulating in the library plus the minimal nuclei of journals devoted to the subjects of most interest to the library's nucleus of users. As the budget allows, successive zones of periodicals corresponding to circulation and user interest can be added. As a result, the library collection can be maintained in an orderly and viable state, thus providing its patrons with the most potentially usable materials for the funds at its disposal.[121]

There was also a note of correspondence regarding the Goffman – Morris suggested "use of Bradford's law of dispersion as an aid to selection decision."[122] A. Fasler of the Department of Research of Beckett and Colman of Norwich wrote, "this appears to be a most promising method."[122] He added a warning, however, "Before it is possible to discontinue a journal subscription, it is necessary to make sure that such action will not cause great inconvenience."[122]

In 1972 Elizabeth Wilkinson wrote "The Ambiguity of Bradford's Law."[123] As Wilkinson pointed out, it is not necessary to completely understand an empirical law before using it.[123] "Despite the limitations of our understanding of such laws, then, Bradford promises useful application in the design of more rational and economic information systems."[123] She listed a number of scientists who have written papers about the Bradford law: "Vickery, Barrett, Leimkuhler, Brookes, Fairthorne, Goffman and Warren, and Naranan;"[123] and added this observation, "Yet the most remarkable feature that emerges from a study of these papers is that no two of these contributors interpret the law in mathematically identical terms."[123] She also remarked on the ambiguity of the law itself, for "Bradford formulated his law in two ways: firstly by means of a graph and secondly in words. But these two formulations can be shown not to be mathematically equivalent."[123]

Vickery has pointed out the disparity between these two formulations. He showed that if n_m journals contribute a cummulative m papers, where n_m is larger than the nucleus, the verbal formulation is equivalent to the expression

$$n_m : n_{2m}n_m; n_{3m}n_{2m} : \ldots :: 1 : a_m : a_m^2 : \ldots$$

while the graphical formulation is equivalent to the expression[123]

$$n_m : n_{2m} : n_{3m} : \ldots :: 1 : b_m : b_m^2 : \ldots$$

Wilkinson continued:

Leimkuhler has expressed the same relation in terms of the proportion of total productivity (papers)…contained in the fraction. . .of journals. . .Brookes, on the other hand, has developed the graphical formulation of the law.[123]

In summary then, the distribution of papers on a particular topic among contributing journals is expressed by

$R(n) = j \log (n/t + 1)$ for the verbal formulation, and
$R(n) = k \log n/s$ for the graphical formulation.[123]

the constants of these equations are not equal; i.e. $j \neq k$ and $t \neq s$. The verbal and graphical formulations are not, therefore, mathematically equivalent, nor do they converge for large n.[123]

$R(n)$ is the cumulative total of papers published by the first n journals;
n is the rank of a set of journals contributing papers on a particular topic constant;
j, t, k, s are constants.[123]

Commenting on "The Ambiguity of Bradford's Law" in Letters to the Editor, David A. Smith stated that it "showed clearly the empirical nature of Bradford's Law and the uncertainty about its derivation."[124]

In 1973, *Library Trends* published an issue, "Analyses of Bibliographies," to which B.C. Brookes contributed the lengthy but excellent "Numerical Methods of Bibliographic Analysis."[125] He listed five main objectives of numerical analyses,

based on the belief that quantification is a necessary component of the design of economic information systems and that measurement of the key processes of an information system is a necessary component of management control.[125]

These are:

1. The design of more economic information systems and networks;
2. The improvement of the efficiencies of information-handling processes;
3. The identification and measurement of deficiencies in present bibliographical services;
4. The prediction of publishing trends; and
5. The discovery and elucidation of empirical laws which could form the basis for developing a theory of information science.[125]

Most of the faults about the application of the technique [Bradford's Law to provide reliable estimates] that have been reported arise from lack of appreciation of the fact that the initiating data must be bibliographically well-defined in subject relevance and in period of publication and also be complete as far as they need to go. And the reliability of the estimates is naturally critically dependent on the precision of the initiating data.[125]

Other areas in which Brookes believed the Bradford law to be evident and usable were: in relation to "items borrowed from a library;" "contributors to the discussions [at a conference], ranked according to the frequency of their questions or contributions;" "index terms assigned to documents; publication of books by publishers."[125] Brookes also included a section on obsolescence[125] which subject is mentioned in the next section.

In 1977, Wilson O. Aiyepeku of Nigeria, in an extensive "analysis of the periodical literature in geography for the whole world and for the USA, British Isles, France, and Germany,"[126] found evidence "that data conforming with the verbal formulation of Bradford's distribution theory may not necessarily conform with its graphical representation and *vice versa*,"[126] which is an agreement with other investigations mentioned earlier.

From here on, most of the papers regarding Bradford were surveys of literature pertaining to the law, refinements of the law, collection of data, and application of the Bradford distribution or reiteration of some previously noted facts.

I.M. Sengupta's study fostered

an extension of Bradford's law of scattering: that during phases of rapid and vigorous growth of knowledge in a scientific discipline, articles of interest to that discipline appear in increasing numbers in periodicals distant from that field.[127]

Pope claimed "The area where the Bradford distribution has the greatest potential is in collections development, both in individual libraries and in library systems."[128] A.B. Worthen's findings supported the statement that "Bradford's law of distribution can also be applied to the monograph-publisher relationship for a subject literature."[129]

In 1977, Brookes contributed his "Theory of the Bradford Law"[130] to the Bradford commemorative issue of the *Journal of Documentation*. In this superb paper, Brookes did a complete reevaluation of the Bradford Law and concluded, "Bradford was therefore a pioneer in social mathematics."[130] Within the paper Brookes stated:

I describe how Bradford's law can be regarded as a particular example of an empirical law of social behaviour which pervades all social activities. Bradford, without knowing what he was doing in this respect, may have initiated a new branch of statistics—the statistics of individuality—which may also lead to an extension of fundamental mathematical ideas which would make mathematics more useful in the social sciences.

However, this paper will focus on the task of explaining Bradford's law in this wider context and on its applications to documentation.[130]

The analysis of the Bradford law has hitherto been applied to theoretical models which are too static, too deterministic and too physical. All Bradford data are derived by observing the *activities* of a set of sources over some appropriate period and by noting these activities, as measured in terms of the numbers of items each source accounts for in that time.

Thus, the Bradford law is concerned with:

a. A finite set of active sources (*an ensemble*) whose activities are made manifest by the generation or consumption of a specified type of item.
b. Observation of those activities over a specific *sampling* period.
c. Items of some *homogeneous* kind which are discrete and countable.

One of the points that has so far escaped analysis is the fact that the statistical distributions of such an ensemble of activity must depend on relationships between the number of active sources, the range and intensity of their activities, and the period of observation which provides the sample data. All Bradford distributions are *samples* of some on-going activity but, all too often, the sample data have been regarded as constituting a total population.[130]

ZIPF'S LAW OF WORD OCCURRENCE

The third of the triumverate of bibliometric laws was that of George Kingsley Zipf. While studying linguistics at the University of Berlin,[54] Zipf had an idea that "speech as a natural phenomena" is really "a series of communicative gestures" and after extensive research found that "the length of a word, far from being a random matter, is closely related to the frequency of its usage—the greater the frequency, the shorter the word."[54] He also discovered that

the distribution of words in English approximates with remarkable precision a harmonic series. . .an unmistakable progression according to the inverse square, valid for well over 95% of all the different words used in the sample.[54]

Zipf wrote several books "on the theory and application of his principle of relative frequency in the structure and development of language."[131] In his first thesis, *Relative Frequency: A Determinant of Phonetic Change*, published in the Harvard Studies in Classical Philology in 1929,[132] Zipf wrote:

Observing the speech of many hundreds of millions of people, we have demonstrated, in part actually, in part by induction, that the conspicuousness or intensity of any element of language is inversely proportitnate to its frequency. Using *X* for frequency, and *Y* for conspicuousness (rank) we can express our thesis thus:

$$Y = \frac{n}{X} \qquad \text{or} \qquad XY = n$$

where *n* is some constant, the actual size or value of which need not be our immediate concern now.[132]

This was undoubtedly the foundation on which his other works were based.

His next study, *The Psycho-Biology of Language*, he called "An Introduction to Dynamic Philology," whose goal was to put language study on a par with exact sciences, by use of "statistical principles."[54]

The thesis, very briefly stated, is that the key to the explanation of all synchronic and diachronic language-phenomena has been found in a statistically established tendency to maintain equilibrium between size and frequency.[131]

The "primary aim is the observation, measurement, and, as far as it is possible, the formulation into tentative laws of the underlying forces which impel and direct linguistic expression."[54]

Zipf recognized that there had been accurate investigative studies of language for about 100 years but "nothing has ever been found in the nature of speech in any of its manifestations which is not completely comprised in the statement that speech is but a form of human behavior."[54]

His method was "to analyze samples of the stream of speech of many languages into their component parts, [and]. . .study the frequency distributions of these parts."[54] By using data from three different languages (German, Chinese, and Latin) arranged in tables:

it is clear: 1) that the magnitude of words tends, on the whole, to stand in an inverse (not necessarily proportionate) relationship to the number of occurrences; and 2) that the number of different words (i.e. variety) seems to be ever larger as the frequency of occurrences becomes ever smaller.[54]

Yet the significant feature in the diminution of variety which attends upon an increase in frequency of usage is the orderliness with which the one decreases as the other increases.[54]

Zipf plotted his observations on double logarithmic paper and after drawing a line "approximately through the center of the points" developed:

the formula $ab^2 = k$ in which *a* represents the number of words of a given occurrence and *b* the number of occurrences. That is, the product of the number of words of a given occurrence, when multiplied by the square of their occurrences remains constant, for the great majority of the different words of the vocabulary in use, though not for those of highest frequency.[54]

Of particularly important note was, "One has the feeling that the exponent of *b* may well differ with differences in the size of the bulk examined,"[54] for "It is conceivable that an exponent of *b* which is larger than the square may represent the frequency distribution in bulks that may be viewed as incomplete discussions of a topic."[54]

About 14 years after Zipf's *The Psycho-Biology of Language* (1935) was printed, his next book, *Human Behavior and the Principle of Least Effort* (1949) appeared.[55] The purpose of this book, which was an introduction to human ecology, "is to establish the Principle of Least Effort as the primary principle that governs our entire individual and collective behavior of all sorts."[55] Zipf stated that, "An investigator who undertakes to propound any such primary scientific principal of human behavior must discharge three major obligations," that is, have a large verifiable number of observations, be consistent, and have an understandable presentation.[55] As to the observations:

> we may claim in all modesty to have increased the number of our observations to such a point that they may be viewed as empiric natural laws...[for] by means of the accepted methods of the exact sciences, we have established an orderliness, or natural law, that governs human behavior.[55]

What is the "Principle of Least Effort?" "In simplest terms, the Principle of Least Effort means, for example, that a person in solving his immediate problems will view these against the background of his probable future problems, as estimated by himself."[55] In other words, a person will tend "to minimize the probable average rate of his work-expenditure (over time), meaning use the least amount of effort."[55] "The connection between this principle and the rank-size law is by no means clear, and Zipf's theoretical arguments now have at most only historical interest."[133]

In writing the chapter "On the Economy of Words," Zipf applied his principle to speech development. Using Miles L. Hanley's *Index of Words for James Joyce's Ulysses*, he found in the rank frequency word distribution "a clearcut correlation" between *r* (rank) and *f* (frequency) of the words "in the sense that they approximate the simple equation of an equilateral hyperbola: $rxf = C$."[55] This, he believed, gave "clear evidence of the existence of a vocabulary balance."[55]

More analysis showed

> *N* number of different words of the same *f*-integral frequency of occurrence (under the conditions of the equation, $rxf = C$) will be inversely proportionate to the square of their frequency (approximately)—or, stated somewhat more precisely in equation form, that $N(f^2 - 1/4) = C$.[55]

This was sometimes considered Zipf's second law and has been called his "weak" law.

Zipf, applying his "Principle of Least Effort" to many different areas of language and to other disciplines, cited many noted scientists, including Lotka, Davis, Willis, and Pareto, all of whom worked with distributions, but in various areas.[55] Zipf's very wordy, philosophical, and detailed explanations are beyond the scope of this paper.

One last observation by Zipf is noteworthy:

> The greater the prestige of a person, the *ever greater* will be his power of attraction both for students and for grants of research money for the employment of technicians and for the purchase of expensive apparatus, with the result that his probable opportunities for making and reporting new "important" and "interesting" observations will tend to increase exponentially (i.e., "nothing succeeds like success").[55]

Many scientists have analyzed, refined, and evaluated Zipf's endeavors, but Ronald E. Wyllys, who had made a special study of Zipf's Law, called it "One of the most puzzling phenomena in bibliometrics"[134] and noted:

> that Zipf's law only approximates the relationship between rank *r* and frequency *f* for any actual corpus. Zipf's work shows that the approximation is much better for the middle ranks that for the very lowest and the very highest ranks, and his work with samples of various sizes suggests that the corpus should consist of at least 5000 words in order for the product *rf* to be reasonably constant, even in the middle ranks.[134]

In Volume 2 of the *World of Mathematics*, the author called Zipf's *Human Behavior and the Principle of Least Effort* a rather remarkable book; however, "that the law of least effort should be manifested everywhere in human conduct, as Zipf holds—indeed, "in all living process"— is difficult to believe."[25] Although Zipf offered much empirical data,

> It does not show, as he claims, the existence of "natural social laws," but, at best, only certain regularities. Brute empiricism is not yet science. Unless observed regularities can be brought into logical relation with other regularities previously observed, we remain at the level of description rather than explanation; and without explanation, we cannot attach much predictive weight to description.[25]

Anatole Rapaport, in writing about the "Principle of Least Effort," stated "His [Zipf's] arguments are vague appeals to the recognition of the Principle in a great variety of situations simply on the basis of its plausiblity. And even these appeals are often stretched far beyond ordinary credibility."[135] However, other serious researchers, Simon and Mandelbrot in particular, endeavoring to put Zipf's investigations on a more scientific basis, have also done some very serious evaluating of his work.[135]

Benoit Mandelbrot has published several studies of generalizations of Zipf's law [one of which is in French], dealing both with the question of whether the slope is − 1 and with the deeper problem of explaining why the *rf* products should be relatively constant.[134]

The purpose of Herbert A. Simon's paper, "On a Class of Skew Distribution Functions," published in 1955 was:

to analyze a class of distribution functions that appears in a wide range of empirical data—particularly data describing sociological, biological, and economic phenomena. Its appearance is so frequent, and the phenomena in which it appears so diverse, that one is led to the conjecture that if these phenomena have any property in common it can only be a similarity in the structure of the underlying probability mechanisms. The empirical distributions to which we shall refer specifically are: 1) distributions of words in prose samples by their frequency of occurrence; 2) distributions of scientists by number of papers published; 3) distributions by cities by population; 4) distributions of income by size; and 5) distributions of biological genera by number of species.[77]

This paper included mathematical and technical explanations of a stochastic model primarily attributed to D.G. Champernowne[77] and "the observed fit of the Yule distribution to a number of different sets of empirical data"[77] including word frequency (Zipf). Simon's "Some Further Notes on a Class of Skew Distribution Functions," which is primarily in answer to a note written by Mandelbrot, followed in 1960.[136] In Simon's words, "This note takes issue with a criticism by Dr. B. Mandelbrot of a certain stochastic model to explain word-frequency data."[136]

Bruce M. Hill, a professor of statistics, was responsible for a group of papers regarding Zipf's law. Hill's first paper was "Zipf's Law and Prior Distributions for the Composition of a Population" (1970) in which he cited Willis for first noting the law and discussed the fact that the "theoretical models yielding such a law have been proposed by Yule and Simon."[137] "Rank-Frequency Form of Zipf's Law" (December 1974), published next, "presents a theoretical derivation of the rank-frequency form of Zipf's law based on a Bose-Einstein form of the classical occupancy model."[138] "Of course, other models have been formulated, and other approaches taken, in regard to justifying the Zipf (or Pareto) law, most notable those of Mandelbrot and Simon."[138] Hill's next paper was "Stronger Forms of Zipf's Law" (March 1975), which is really an extension of his 1970 paper.[139]

In September 1975, Sichel wrote "On a Distribution Law for Word Frequencies," for as he stated,

In the past, several attempts were made to represent word frequency counts by statistical distribution laws. Of the models suggested, none was singularly successful when applied to a variety of data over the entire length of the observed word distribution.[140]

Therefore, Sichel developed a new model to which "twenty observed distributions quoted in the literature were fitted and the results look most encouraging."[140]

Also in 1975, R.E. Wyllys presented a paper at the ASIS 38th Annual Meeting in which he included an excellent summation.

Inclined toward mysticism, Zipf not only leaped to the conclusion that the "true" slope of rank-frequency curves was − 1, but also claimed that this regular slope resulted from some fundamental force of nature. In the broad sense, this claim had to be correct; but Zipf variously described the force as that of the struggle between the "life tendency" and the "death tendency" or the "Force of Diversification" and the "Force of Unification," and finally as the " Principle of Least Effort," *for none of which did he furnish an operable definition*. However, in work summarized. . .Zipf did show that an astonishingly wide range of phenomena. . .exhibited distributional behavior that could be approximated by his "law."[141]

Wyllys claimed his report and

study established that significant, detectable variations in rank-frequency curve slopes do exist, that they appear to have the potential for subject-field discrimination, and that they may perhaps also be usable for the identification of subfields of especially rapid development.[141]

BRADFORD-ZIPF PHENOMENA

A number of researchers have attempted to explain the relation between the Bradford and the Zipf laws. One of the first papers to do so was "The Bibliography of Operational Research" by M.G. Kendall.[112] Kendall suggested looking at Bradford's data through the eyes of a statistician and, contrary to Bradford, cumulate data from the bottom of the listing (see the sample chart given in Table 3). From this, he developed a relation which he reasoned indicated Bradford's Law would be true.[112] According to his analysis, this would make the Bradford and Zipf laws almost equivalent.[112] (See Kendall's paper for mathematical explanation.) Kendall called the results of Yz (see Table 3) a constant.

In 1967, Ferdinand Leimkuhler wrote his paper "The Bradford Distribution," in which he stated,

Bradford's "Law of Scattering" is the inverse function for the Bradford distribution, i.e., while the latter predicts the number of references for a given portion of the journals, the former speaks of the number of journals required to obtain a given portion of the references.[114]

Leimkuhler also commented that "Kendall has suggested an alternative explanation of the linear approximation formula, in which he considers the number of titles with a

Table 3 Using Bradford's data from lubrication to illustrate the comparison of Bradford's and Kendall's methods.

Number of journals	Number of references		
Bradford			
X	Y	ΣX	ΣXY
1	22	1	22
1	18	2	40
1	15	3	55
2	13	5	81
2	10	7	101
1	9	8	110
3	8	11	134
3	7	14	155
1	6	15	161
7	5	22	196
2	4	24	204
13	3	37	243
25	2	62	293
102	1	164	395
Kendall			
X	Y	ΣX from bottom up	YZ
102	1	164	164
25	2	62	124
13	3	37	111
2	4	24	96
7	5	22	110
1	6	15	90
3	7	14	98
3	8	11	88
1	9	8	72
2	10	7	70
2	13	5	65
1	15	3	45
1	18	2	36
1	22	1	22

reference count greater than some given value"[114] (column Z in Table 3). He showed his compliance with Kendall's mathematical manipulations by agreeing that "Thus Bradford's law and Zipf's law are essentially just two different ways of looking at the same thing."[114]

Brookes, in one of the first of a group of papers, had this to say about the Leimkuhler paper:

Leimkuler's analysis therefore provided documentalists for the first time with a general formula which expresses the distribution of papers relevant to a given topic over the range of productive journals. Up to the present, information systems have tended to evolve, documentalists using their experience and making *ad hoc* adjustments to satisfy the demands for services made on the system. But the discovery and formulation of statistical laws of wide generality would enable documentalists to *design* information services...

Unfortunately, though Leimkuhler's formulation can be used theoretically without difficulty, it has some disadvantages for the practical documentalists.[142]

An outstanding disadvantage was the need for tedious statistical computation;[142] "the exasperation evoked by an attempted practical application of Leimkuhler's formulae [that] led the author of this paper to seek a simpler formulation of the Bradford distribution."[142]

However, the near identity of Zipf's and Bradford's laws is not immediately obvious because, in practice, the most marked deviations of empirical data from the mathematical expectations of the Bradford law are likely to occur among the most productive journals of the nucleus... The Bradford law is not reliable in predicting the productivity of individual journals: it is a statistical law which relates only to large collections of journals or to major subsets of such collections.[142]

Brookes then proceeded to develop the relationship "*R* (*n*) = *k* log *n*,"[142] which he believed was "an exact formulation of the distribution which satisfies the Bradford law."[142]

The Groos Droop was also mentioned in this Brookes paper. "When empirical Bradford data are plotted in the expectation of yielding a straight line corresponding to *R* (*n*) = *k* log *n*, the ideal straight line is attained only in part."[142] Groos, after whom the droop is named, noticed when he plotted the Keenan–Atherton[143] data according to Bradford that "the plotted points eventually form a curve which droops below the linear predictions of the Bradford law."[142]

The upward curving bottom of the curve (of the bibliograph) represents the small nuclear zone of the most relevant journals. The upper end of the curve, usually termed the Groos Droop, represents the peripheral zone where relevant references are widely scattered among a great number of journals.[144]

It has been suggested but not proven that the droop indicates that the data are incomplete.[142] E.T. O'Neill, on the other hand, claimed "the pronounced droops have occurred most frequently in large and therefore presumably relatively complete sets of data."[145]

In March 1969, "The Complete Bradford-Zipf 'Bibliograph'" by Brookes was published.[146] This paper "summarizes the outcome of recent analyses of empirical data which have enabled the general form of the Bradford-Zipf distribution to be elucidated."[146] Brookes explained "Bibliograph" as follows:

In general, the complete "bibliograph" has two components...an initial rising curve...running smoothly into...a straight line...Since a source, to be recognized and counted as a source, must contribute at least one item, it is possible to estimate the end-point...of the line. It has been shown that an estimate of the total number of *contributing* sources is given by the slope of the linearity.[146]

The initial curvature indicates a situation in which restraints are imposed on the productivity of the contributing sources; the linearity indicates a true Zipf situation in which there are no restraints on productivity.[146]

The complete "Bibliograph" was soon followed by two other papers by Brookes: "Bradford's Law and the Bibliography of Science" and "Theory of the Bradford Law." In "Bibliography of Science," Brookes specified "the requirements for conformity with the law (Bradford-Zipf) in more general terms. . .:"

The Bradford-Zipf distribution can be expected to arise when selection is made of items, characterized by some common element, which are all equally open to selection for an equal period and subject to the "success-breeds-success" mechanism, but when selection of a most popular group is also, but to a weaker extent, subject to restriction. It is thus a general law of concentration over an unrestricted range of items on which is superimposed a weaker law of dispersion over a restricted range of the most frequently selected items.

The Bradford law remains empirical until it is better understood. But if it can be demonstrated to be widely applicable, reliable and useful in practical operations, there is no need to wait until the underlying theory is completely established.[118]

Bradford's law can be used to help in the designing of information systems, "in rationalizing library services, and in making more economic and fruitful use of the 29,000 scientific periodicals estimated to be the current production."[118]

Brookes gave as applications of Bradford's law:

1. Use in computerized bibliographic search systems (Medlars).[118]
2. "Management of special libraries."[118]
3. Discarding of "aged" periodicals (obsolescence).[118]
4. "Planning of special library systems."[118]
5. "Subject bibliographies can be resolved or composed without difficulty."[118]

"Theory of the Bradford Law" was printed in the *Journal of Documentation* in 1977 and included in *Key Papers in Information Science*, published in 1980. In Brooke's words, "In this paper I describe how Bradford's law can be regarded as a particular example of an empirical law of social behavior which pervades all social activities."[147]

"The analysis of the Bradford law has hitherto been applied to theoretical models which are too static, too deterministic and too physical."[147] "All Bradford distributions are *samples* of some on-going activity but, all too often, the sample data have been regarded as constituting a total population."[147] "The law as Bradford formulated it is a hybrid."[147]

Brookes then concluded:

So as Zipf abandoned the use of ranks in favour of frequencies, whereas Bradford's law can be expressed simply only in terms of ranks, I see no further reason to continue to use Zipf's name descriptively in the context of Bradford's law. When as many ambiguities surrounded the Zipf law as once surrounded the Bradford law, the use of Zipf's name in the present context serves no useful purpose.[147]

In 1969, M.K. Buckland and A. Hindle wrote a survey article, "Library Zipf" for the *Journal of Documentation*,[148] in which they mention a publication by Kozachkov and Khursin who "propose a basic model, called the 'hyperbolic ladder,' and relate it, in particular, to work in linguistics by Zipf, in documentation by Bradford, and in the science of science by Lotka."[148]

According to Buckland and Hindle,

The earliest recognition that Bradford's law of scattering was a form of Zipf's law seems not to have come until a paper by Kendall in 1960—a conclusion since endorsed by Leimkuhler, Kozachkov and Khursin, and Brookes.[148]

A number of letters resulted from Buckland and Hindle's survey. Fairthorne claimed that the relation of the Bradford and Zipf Laws was suggested as early as 1953 and may have been referred to as early 1948 in a discussion of a paper by Vickery.[149]

D.A. Smith also wrote a letter, not so much critical of the Buckland-Hindle article as critical of the theory discussed. He noted "Other authors have followed this well-worn path of references cited by Buckland and Hindle."[150]

The search for unifying principles is doubtless a worthwhile aim but will not justify the use of analytical techniques in libraries. Only attention to practical applications will do this and a study of the methods used in business management may well be of more value than the uncritical acceptance of ideas put forward by those who have a particular concern with theories of society.[150]

Back came a reply from the authors of the original article. They apologized to Fairthorne for their oversight, but to Mr. Smith they addressed this,

We chose to stress Zipf mainly because, as we wrote, 'In the field of libraries, where apparent manifestations of Zipf's laws are evident, there is independent evidence that convenience is a dominant factor in determining library use. . . .'[151]

In our own experience of attempting to apply the mathematical methods used in business management to practical problems in libraries, we have found that librarians recognize well enough that mathematical predictions can only be estimates based on assumptions and require some check on their accuracy.[151]

Brookes also wrote concerning Smith's letter:

Pareto's law [Smith seemed to be pushing for Pareto] does arise in documentation—in the data relating to the productivity of the periodicals in Bradford "nuclear" zones, where I interpret its occurrence as evidence of some control of publication... Incidentally, Pareto has no monopoly of the distribution Mr. Smith claims for him and for economics. Biologists too have found applications of the same distribution in statistical taxonomy though they chose to refer to the same (Pareto) empirical distribution as the Willis distribution. And I am at present engaged in applying the Bradford-Zipf distribution to a problem in paleontology....[52]

My own view of empirical laws in documentation is that they provide us with prospects of making considerable economies in the running of special libraries and information services.[152]

The use of the Bradford-Zipf relation was suggested by Stephen Bulick in the "interaction between book users and books available for use in a library."[153] The author found:

the verbal formulation consistently produces a lower error percentage than the graphical and is, in fact, a good overall predictor. Book use, then, is a Bradford-Zipf phenomenon...

This offers support for the strategy of shortening loan periods for frequently circulated items. Other potential uses for a function that describes the distribution of book use over a collection might include core collection determination or the derivation of a marginal utility function describing the effect on total use of adding to a collection.[153]

In October 1976, Abraham Bookstein's "The Bibliometric Distributions" was published.[154] The premise of this paper was:

One of the more surprising findings in the information sciences is the recurrence of a small number of frequency distributions. In this paper, these distributions are described, and a point of view is adopted that allows us to understand them as being different versions of a single distribution. The empirical distributions are shown to be special cases of a single theoretic distribution.[154]

All of these distributions are almost equivalent...In each case we have a set of entities (for example, chemist, words) producing events (publications, occurrences) over some dimension of extension (time, length of text); and in each case the distribution describes the number of occurrences of events over a fixed interval of that dimension. Under these conditions it is possible to describe the same distribution in at least four distinct ways: these modes of description are represented above by the distributions of Lotka, Zipf, Bradford, and Leimkuhler.[154]

The four distinct ways are noted in Bookstein:

The four ways are: 1) One may rank the entities according to the values of the attribute, and express the value of the attribute as a function of the rank. This is the form the original expression of Zipf's law takes; 2) Conversely, we can define a unit of the attribute, and indicate how much the rank must increase if we are to gain an additional unit quantity of that attribute. Bradford chose to define his law in these terms; 3) The third form is to cumulate the attribute and indicate how many entities (e.g. what rank of journal) is needed to produce, collectively, a given amount of the attribute (e.g., a given number of articles). This is a form of Bradford's law as defined by Leimkuhler; 4) Finally, one can express, for any possible value of the attribute, how many entities take that value. This, of course, is the form taken by Lotka's law.[155]

Other writers...have shown that Lotka's law and Bradford's law are different approximate descriptions of the same basic distribution, and that, at least for larger values of r, both agree with Zipf... The distributions are approximately the same; it is only the entities and events that differ.[154]

"Both Zipf's and Bradford's laws emphasize the higher-ranking entities. Lotka, on the other hand, begins with the low end."[154]

An analysis then led Bookstein to conclude, "Thus the Bradford, Leimkuhler, and Lotka laws, which are virtually equivalent to each other, and also the Zipf law, are also special cases of the forms derived by Mandelbrot and Bookstein on theoretical grounds."[154]

Citation Analysis

Since the inception of bibliometrics, citations have played a very important role in bibliometric research. At first, there were citation counts which were used to indicate importance of journals; these counts being statistical did not consider the quality of a work. Ultimately, the citations themselves were closely examined as indicated in many research papers. Citation analysis "is a method based on the principle that articles citing the same references also have much of their content in common," stated M. Osinga in "Some Fundamental Aspects of Information Science."[156] Elliot Noma followed the same thinking, shown by the comment "The similarity between an article and the articles on its reference list is one of the cornerstones of citation analysis."[157] The supposition that articles and their references are related has led to many studies, including citation counts (Gross and Gross), impact factor (Garfield), bibliographic coupling (Kessler), cocitation (Small), and citation indexes.

In discussing citation study in his thesis, Dale Lockhard Barker noted there were "two principal types of citation studies... These are 1) studies bearing on productivity in all or part of science literature and; 2) those reflecting the use of all or part of the literature."[36] Studies based on entry counts in source materials are productivity studies.[36] The source materials may be "major abstracting or indexing services, review journals, or comprehensive monographic bibliographies."[36] "The citations...*are not to literature used but to literature published*."[36] An example of such a study would be the work of Samuel Clement Bradford.

"The second type of citation study" was based on the literature used by an author.[36] An example would be the citation count presented in the Gross and Gross paper.

Linda C. Smith, in her overview of citation analysis for the Bibliometrics Issue of *Library Trends*,[158] explained the two kinds of citation: "A reference is the acknowledgment that one document gives to another; a citation is the acknowledgment that one document receives from another."[158] A relationship is implied between the cited document and the citing document, all or part. "Citation analysis is that area of bibliometrics which deals with the study of these relationships."[158]

Studies have found that reasons for citations are numerous and greatly varied. Ina Spiegel-Rosing discovered 13 different categories for use of references in *Science Studies*;[159] Charles Oppenheim and Susan Renn list seven reasons for citations to some highly cited old papers:

A. Historical background
B. Description of other relevant work
C. Supplying information or data, other than for comparison
D. Supplying information or data for comparison
E. Use of theoretical equation
F. Use of methodology
G. Theory or method not applicable or not the best one.[160]

The amount of literature about citation analysis is truly extensive, if not actually overwhelming. It is evident that whenever a subject "grows" (research fronts), new views or facts are presented and more studies then are accomplished validating or questioning those views or facts. When the views are analyzed from different aspects, articles are written and printed, and following a logical sequence, usually a bibliography and/or index of the subject is then compiled, such as Pritchard's *Statistical Bibliography: An Interim Report*, Pritchard and Wittig's "Bibliometrics: A Bibliography and Index," and Vlachy's "Bibliography of Lotka's Law and Related Phenomena."

In the case of citation analysis, Garfield established and organized his extensive *Science Citation Index*; Renata Tagliacozzo wrote her "Citations and Citation Indexes: A Review;"[161] and Hjerppe developed his report, "An Outline of Bibliometrics and Citation Analysis,"[162] followed by "A Bibliography of Bibliometrics and Citation Indexing and Analysis."[163]

Some techniques using citation analysis are obsolescence, clustering, and citation indexes; information about these is given in the following subsections.

HALF-LIFE AND OBSOLESCENCE

When Gross and Gross determined what materials were vital for the maintenance of a dynamic college and library, they were projecting needs. The flip side of this positive thinking is what is no longer needed or of use. This can be interpreted as "obsolescene." Obsolescence is the process whereby materials become no longer useful or reliable.[29] Gosnell, whose dissertation was on obsolescence, stated "the causes of book mortality or obsolescence are many, varying from pure fad through extension of scientific knowledge, technological changes, to fundamental changes in our civilization."[30] The rate of obsolescence varies with the discipline. Books according to Gosnell "represent one of the higher forms of culture and the rate at which they are discarded and replaced may give some suggestion as to the rate of evolution of the general culture of which they form a part."[29]

In 1960 Robert E. Burton and R.W. Kebler wrote "The 'Half-Life' of some Scientific and Technical Literatures,"[164] in which they compared the rate of obsolescence of scientific literature with that of radioactive substances.[164] The authors noted upon their research "A short half-life, which is equivalent to rapid obsolescence, is the result of rapidly changing techniques or interest within a subject field."[164]

Half-life as explained by Maurice B. Line was "half the active life,"[165] and this was usually interpreted "as meaning the time during which one-half of the currently active literature was published."[165]

The results of many of the half-life studies varied so greatly that they did not project a possible general application. There also were discrepancies as to meanings. This was Maurice Line's explanation of the process. If the average number of articles in each journal remained constant, and the number of journals doubled, "the probability of citation is twice as great."[165] "The so-called 'half-life' of a literature is (therefore) compounded of its obsolescence rate and its growth rate."[165] Line then proceeded to develop mathematical expressions for determining a half-life corrected because of the two rates."[165]

In 1970, Brookes analyzed Line's technique and concluded that it was "both questionable and impractical."[166] He stressed that the "rate of obsolescence is a function of both the subject literature and of the local usage of that literature."[166] A much more accurate measure of obsolescence would be in "sampling the actual usage of the literature in the local library context"[166] rather than using Line's correction method. In other words, librarians interested in the obsolescence of holdings should make their own direct measurements. Brookes believed there had been sufficient theorizing; the time had come for proper testing of the basic assumptions which he was in the process of doing, which work, "relying mainly on citation analysis, is tedious but straightforward."[166]

The next paper to deal with obsolescence was that of Sandison, who stated "some aspects of the theoretical approaches and terminology need clarification."[167] The crux of the papers on this controversial subject was in two frequently observed facts,

> first that, in any collection of citations and of items used in a library, the numbers decrease with the time since publication; and secondly that the numbers of items published and available for citation have increased year by year.[167]

Age does not completely determine obsolescence; "it is possible for older literature to be in greater demand than newer."[167]

Michael K. Buckland—in posing an interesting question, "Are obsolescence and scattering related?"—explained obsolescence as the "relative decrease in use of material as it ages" and scattering as "extent to which the use of material tends to be concentrated in a few titles."[168] Results of research "were indicative but not conclusive."[168] Line, in Letters to the Editor, commented on Buckland's hypothesis and claimed, "The reasons why scatter and obsolescence occur are numerous and complex, and vary from discipline to discipline."[169]

Sandison also wrote to the editor claiming Buckland fell into the same trap to which most citation analysis work has succumbed by "assuming that every citation can be treated as of equal weight as a parameter of a literature."[170] "But before further progress is possible in the interpretation of citation analysis, reliable data are required for the growth of the literature studied so that citation available per item can be calculated."[170]

Then, in 1973, J. Michael Brittain and Maurice B. Line coauthored "Sources of Citations and References for Analysis Purposes,"[171] categorizing "the uses to which analysis of bibliographical references and citations can be put."[171] The authors identified five sources of references and citations and enumerated advantages and disadvantages of each. Included as one of the listed items was whether obsolescence patterns could be traced.

The "measurement of obsolescence" and "obsolescence, scatter, and growth" were sections in Brookes's next publication.[125] Brookes was truly interested in libraries, as was indicated in his observations of use of a library and in his remark "It is, however, simpler to

recognized that any literature ages at a uniform rate, but that some libraries, especially new ones, can hope to attract usage at a rate which exceeds the rate of obsolescence."[125]

Although there was some evidence of a positive correlation of scatter and obsolescence, it was not wholly convincing. "It may be so, although at the present time there is no general agreement on how scatter should be defined or measured."[125] Both scatter and obsolescence seem to be determined by the rate of growth; "the faster the rate of growth, the less is the scatter and the more rapid the obsolescence."[125]

Obsolescence has been the concern of librarians for some time because of overabundance of materials and lack of housing. However, interest in obsolescence was not imperative during the period of money availability which fostered more building and greater collections. Present lack of funds has made it necessary for librarians to reevaluate expenditures and has resulted in a recurrence of interest in purchasing and weeding practices. No doubt, this has aroused the attention of researchers again, but Maurice Line has emphasized, "no measure of journal use other than one derived from a local-use study is of any significant practical value to librarians."[172]

Kaye Gapen and Sigrid P. Milner accomplished an extensive study of obsolescence research in 1981[173] and observed several interesting and important facts; there are many needs: much more investigation of the obsolescence hypothesis; agreement on definitions; proof of the "obsolescence concept" validity; and justification for costs acquired in making use studies. There are also problems: unfamiliarity of librarians with mathematical manipulations; controversial results of past research; and inherent local library "use peculiarities."[173]

CITATION INDEXES

Samuel Clement Bradford would be elated if he were to see Eugen Garfield's work. Bradford, as stated earlier, advocated cooperation in indexing science articles for a universal index, and Garfield seems to have found a solution. In his "Citation Indexes in Sociological and Historical Research" of 1963,[174] he reminded us that in his original paper of 1955, "Citation Indexes for Science,"[56] he "proposed the compilation of comprehensive citation indexes primarily as an effective means of disseminating and/or retrieving scientific literature."[174] ("Citation Indexes for Science" is discussed in the section headed Seminal Bibliometric Papers.) His second paper was a review of his efforts since the first article appeared and was an emphasis on the fact that "Impact is not the same as importance or significance. There is no specific correlation between the number of papers published by an individual and the quality or importance of his work,"[174] although there have been indications that high publication rates usually go with high quality work.[174]

Citation indexes have been called facilitators by Eugene Garfield; they facilitate personnel and fellowship evaluation, historical research, and computer programs.[174] But what is a citation index?

> A citation index is an ordered list of cited articles each of which is accompanied by a list of citing articles. The citing article is identified by a source citation, the cited article by a reference citation. The index is arranged by reference citations. Any source citation may subsequently become a reference citation.[175]

"Arrangement by author is favored in the citation index and the source index because the research scientist usually approaches the literature first by author."[175] "A complete source index containing full source-article titles and certain additional data...is similar to an upgraded conventional author index covering all disciplines."[175] It was assumed that the source paper bibliographies were true indications of previous information; otherwise, the citation index would not be valid.[175]

In his "Citation Indexing: A Natural Science Literature Retrieval System for the Social Sciences,"[176] Garfield differentiated between the dissemination and retrieval of information. Information which was current was disseminated but information which had been stored was retrieved.[176] The problems associated with both dissemination and retrieval, he believed, "are largely overcome by citation indexing."[176]

Since citation indexes deal with citations, it is important to explain the different kinds.

> In an explicit citation, the source or citing document will identify the cited works by use of formal reference citations, which enable the reader to locate the document in question. In an implicit citation, one recognizes that some other work has been drawn upon or alluded to, but the citing author does not consider it important enough for a formal citation.[176]

In other words, one is definite; the other is implied.

In the case of conventional indexes, it is necessary to use very highly trained indexers; in citation indexes, the author of a paper provides the indexing by providing the citations.[176]

In another article, Garfield remarked that clearly visible linkages are "ordinarily provided by authors in the forms of explicit citations. Less clearly seen are implicit references."[177] However, in traditional indexing systems, treating each document as an independent unit "results in the loss of important informational links;"[177] thus, by showing the linkages, citation indexes are more efficient.

In 1965, J.E. Terry, in a critical review of *Science Citation Index: An International Interdisciplinary Index to the Literature of Science*, reminded the reader that

"The whole concept of citation indexing in science should be regarded as still at an experimental stage,"[178] a fact that Garfield supported by continually stating that improvements are constantly being made.

In his "Citation Indexing for Studying Science" of 1970,[179] Garfield explained what the *Science Citation Index* does.

> Basically...the SCI does two things. First, it tells what has been published. Each annual cumulation cites between 25 and 50 per cent of the 5–10 million papers and books estimated to have been published during the entire history of science. Second, because a citation indicates a relationship between a part or the whole of a cited paper and a part or the whole of the citing paper, the SCI tells how each brick in the edifice of science is linked to all the others.[179]

"Important applications for the SCI have been found in three major areas: library and information science, history of science, and the sociology of science."[179] The first purpose of the SCI was as a retrieval tool which would be used in library and information science.[179] In historical research, citation indexing aided in identifying "key events, their chronology, their interrelationships, and their relative importance."[179] In sociological processes, by means of citation networks, the papers that have had the most impact (most cited by others) can be recognized (The reader's attention is called to item (15) of Conclusions, p. 33. "It is believed that citation analysis has been demonstrated to be a valid and valuable means of creating accurate historical descriptions of scientific fields, especially beyond the first quarter of the twentieth century when bibliographic citation had become well established as part of scientific publication."[180]). There are also different kinds of citation indexes: pertinent to one field or to one journal, or interdisciplinary.[181]

There have been difficulties with citation indexes—costs, various spellings of authors' names, authors with the same name, incorrect citing information, and other human errors—but Garfield believed the advantages far outweigh the disadvantages.[179]

An excellent summary on citation indexes can be found in *On Documentation of Scientific Literature*:[182]

A citation index consists of two different parts:

1. An index of all articles published in a selected group of periodicals in a given year (so-called sources index).
2. An index, arranged by author, of all articles cited in the articles of group (1) (so-called citation index).

The use of a citation index is: locating a known author in (2) and searching the sources, quoting his articles, in (1). If no author is known, a subject index is available leading to author names.[182]

The same writer also listed three principal uses of citation indexes:

(a) Searching for the history of an idea (patent).
(b) Searching for the use and expansion of a certain method after its first publication.
(c) Searching in multidisciplinary fields.[182]

CLUSTERING

The topic "clustering" as part of citation analysis should not be overlooked. Mark P. Carpenter and Francis Narin reported their result of a cluster analysis in "Clustering of Scientific Journals"[46] based on the *Journal Citation Index* (JCI) citations from 1,821 different journals.[46]

> The process [clustering process] used to divide sets of journals into subject areas has two underlying asumptions: first, that journals which deal with the same subject area will have similar journal referencing patterns; and second, that journals which deal with the same subject area will refer to each other.[183]

The "first experiment using a new computer-based technique to identify clusters of highly interactive documents in science"[183] was reported in Henry Small and Belver C. Griffith's study "The Structure of Scientific Literatures I: Identifying and Graphing Specialties." The paper was based on the idea that "scientific specialties" exhibiting "high levels of activity" will be represented by clusters.[183] Clusters of cited documents are groups made evident by cocitation and cocitation strengths. Cocitation is the citing of two documents together creating a link, the strength of which is indicated by the number of times the pair is cited together.

Another important study undertaken by Small was based on the hypothesis that "highly cited and co-cited papers in a cluster is a concrete representation of the cognitive structure…that the authors of the highly cited papers constitute the elite, or 'leading' scientists of the specialty."[184] Small made three suggestions regarding clusters: 1) "…these reflect the social and cognitive structures of research specialties;"[184] 2) "The authors of citing papers then comprise a subgroup of the current practitioners in the specialty;"[184] and 3) "By using the same thresholds each year the growth of the specialty can be gauged in terms of the growth of the cluster."[184]

Sullivan, White, and Barboni did an evaluation study of citation analyses and agreed that citation analysis seemed to them to be a useful technique,[185] but of Small's three claims, the second and third were not supported by their data.[185]

Trudi Bellardo also did an evaluation of the use of cocitations and made this summation:>

> The primary utility of co-citation analysis is as a research tool for studying the sociology and history of science and scientists. More specifically, co-citation analysis is a tool for understanding the specialty structure of science. The clusters of works which are connected by being co-cited are indirect indicators of the birth, growth, and death of scientific specialities and their social and cognitive structures.[186]

Since

> the observed relationships are in substance those which have been established by the collective efforts and perceptions of the community of publishing scientists…our task is to depict these relationships in ways that shed light on the structure of science,[183]

wrote Small and Griffith. This, they believed, was accomplished by the clustering method.

CONCLUSION

The idea of chronological arrangement of periodicals to show trends is a very good theory, but because of the increasing number of bibliometric papers, any research concerning the theory should use technology in the gathering of data. Alan Pritchard listed 700 items in his interim bibliography of 1969,[3] Roland Hjerppe cited 2032 items for bibliometrics and citation analysis as of 1979,[163] and Pritchard stated in the introduction of his *Bibliometrics: A Bibliography and Index* that the literature at that writing (1981?) contained 5000–6000 items.[187] This increase in the number of publications may have been due to the use of bibliometric techniques in various disciplines, to the many grants under which a number of studies have been accomplished, and to the greater interest in the theory of literature structure.

Studies have indicated that "bibliometric measures have been applied to evaluation of scientists, academic departments, and scientific publications."[48] Alan Pritchard and Glenn Wittig advised "some of the uses to which bibliometrics may be put are:"

1. The use as a visible sign of an underlying problem of social structure relating to individuals, e.g. sex differences, promotion policies, creativity.
2. The use for the evaluation of organisations—research sponsorship, government policies, standing in the academic community etc.
3. The use for the evaluation of countries, i.e. science policy studies, either looking at the situation within a country or comparing countries.
4. The use for the examination of the general growth and development of the social structure within a subject or discipline.
5. The use for evaluating individual journals or groups of journals, especially for acquisition decisions.
6. The use as raw data for operations research and other mathematical models.

7. The study of bibliometric distributions in their own right or as members of larger families which are of interest to social science generally.[187]

Citation counts have been used to indicate linkages of papers as suggested by Henry Small,[71] of individuals and their special fields mentioned by Broadus,[188] and of institutions, programs, fields, and subfields as suggested by Narin.[48] Citations are a valid measure of quality stated Lawani in his discussion regarding Nobel prize winners;[189] "Derek de Solla Price of Yale [also argued] that citations are an accurate measure of individual quality."[190] Others believed that citations were not a true "measure of eminence."[188]

Several instances of valid practical applications of citation analysis have been cited by Wade, including a court case contesting the promotion of two men over a woman who had been denied tenure, and the National Science Foundation's use of citation analysis to help determine the distribution of grants to chemistry departments.[190]

Garfield used his *Citation Index* to point out key journals in disciplines,[58] but Carlos Cuadra stressed that "no amount of quantitative or mechanical manipulation of texts, bibliographies, or other tools will enable one to identify the key contributions to information science."[191]

One of the most interesting studies was that in textiles, "The Examination of Research Trends by Analysis of Publication Numbers,"[192] and although no clear-cut trend in total textile research could be concluded, basic research was decreasing; applied research (i.e., production and properties of yarns and fabrics) was increasing.[192] It was hinted that such information could help channel careers.[192] This examination was done for total research effort rather than interest in research significance, although it was recognized that counting publications did not indicate the importance of the research.

Wade also presented a possible use as volunteered by Morton V. Malin, "to identify the gaps in a country's published research."[190]

Lawani had such strong feelings about scientists and their productivity being evaluated through citation analysis more and more, that he admonished scientists to "become familiar with the method of citation analysis" and the uses to which it is put "whether or not they consider the basis for such applications well-founded."[189]

These are isolated instances and do not have the practicability one would desire. As Nicholas Wade so aptly stated, "the impact of citation analysis on the scientific community cannot yet be assessed because all that has really been demonstrated so far is promise, not practicability."[190]

As applied to libraries, bibliometric analyses have been instrumental for book and periodical acquisitions, library use analysis, and weeding of obsolete materials. Carol Tenopir noted Bradford's law had been shown "to apply to the items borrowed from a library, the users of a library,

contributors to a discussion at a conference, the index terms assigned to documents, and the publication of books by publishers."[193] Marcia Sprules informed us of an attempt of an academic library "to evaluate its periodical holdings in all disciplines" to identify candidates for cancellation.[194]

Maurice Line wrote that "most studies of journal citations and library use are of little if any practical use to librarians...because of inadequate data collection and analysis."[195] On the other hand, Narin, in the foreword and summary of McAllister's report, claimed "Citation analysis is steadily growing as a technique for analyzing scientific productivity."[196] Librarians, who were the first to use bibliometric procedures, were not primarily mathematicians or statisticians, nor were they bibliometricians.

Although Schmidmaier observed that "Bibliometrics has its...place in library science, bibliography and informatics," he believed it needed promoting and that this should and "must begin with the popularization of the contents and teaching of basic knowledge,"[197] which is a logical assumption.

Bibliometrics is the analysis of the structure of literature using various tools, counting, rank-frequency distributions, and citation analysis; and although the structure of literature is basic to all disciplines, it is particularly important in the area of information retrieval.

Hulme's original idea was, in reality, a theory to show a possible trend or development in science by means of literature, but it was not a structure study. Gross and Gross, making a supposition about the structure of literature, developed a rank-frequency distribution from citation counts to indicate relative quality of journals; but this was a very primitive study of literature structure. Fussler's study was an extension of the Gross and Gross idea. Kessler and Small developed excellent bibliometric tools for analyzing literature structure, and based on the same supposition as the Gross idea, these tools for studying relationships have far-reaching implications. No distinctly different technique has been produced since cocitation by Small in 1973, clustering and graphing being procedures using bibliographic coupling and cocitation. The many other articles since 1973 seem to be evaluations, applications, or analyses of investigative processes in attempts to describe more fully or support or duplicate previous research.

It seems evident that bibliometrics, which was first used in attempts to evaluate journals for collection development in libraries, has been recognized and expanded to the study of the structure of literature in the larger encompassing field of information science. At the present, it is composed of methods and techniques without a coordinating theory; but it seems possible that as more theoretical information is converted to fact, and behavioral patterns are established for the assessment and evaluation of the structural components of literature (i.e., authors, publications, words, and laws) and their relationships, the more probable causal explanations will be evident and bibliometrics will be closer to being recognized universally as a science.

REFERENCES

1. Pritchard, A. Statistical bibliography or bibliometrics?. J. Doc. **1969**, *25*(4), 348–349.
2. Pritchard, A. *Computers, Statistical Bibliography and Abstracting Services*, 1968. unpublished.
3. Pritchard, A. *Statistical Bibliography: An Interim Bibliography*, North-Western Polytechnic School of Librarianship: London, May 1969; Abstract, introduction.
4. Fairthorne, R. A. Empirical hyperbolic distributions (Bradford–Zipf–Mandelbrot) for bibliometric description and prediction. J. Doc. **1969**, *25*(4), 319–343.
5. Fairthorne, R. A. Personal correspondence.
6. Webster, N. *Webster's New Twentieth Century Dictionary of the English Language*, Standard Reference Works Publishing Company: New York, 1956; 172.
7. Van Hoesen, H. B. Walter, F. K. *Bibliography: Practical, Enumerative, Historical, An Introductory Manual*, Chas. Scribner's Sons: New York, 1928; 3.
8. Schneider, G. *Theory and History of Bibliography*, Columbia University Press: New York, 1934. Translated by Ralph Robert Shaw, theoretical-historical portion of Handbuch der Bibliographie, 3rd ed., 1926; 13, 41, 272, 273.
9. Murray, D. *Bibliography: Its Scope and Method*, James Maclehose and Sons: Glasgow, 1917; 2, 7, 8, 13. Samuel Peterson, a bookseller and auctioneer, is the "Father of English Bibliography".
10. Taylor, A. *Renaissance Guides to Books: An Inventory and Some Conclusions*, University of California Press: Berkeley, CA, 1945; 1, 54.
11. Dahl, S. *History of the Book*, Scarecrow Press: New York, 1958; 40–41.
12. Horne, T. H. An introduction to the study of bibliography. *A Memoir on the Public Libraries of the Ancients* 1814; Vol. 1, 364. Printed by G. Woodfall for T. Cadell and W. Davies, Strand: London.
13. Besterman, T. *The Beginnings of Systematic Bibliography*, Oxford University Press: Humphrey Milford, London, 1935; 1, 3.
14. Malcles, L. N. *Bibliography*, Scarecrow Press: New York, 1961. Translated by Theodore Christian Hines, first published in Paris, 1956; 93, 84.
15. Hulme, E. W. *Statistical Bibliography in Relation to the Growth of Modern Civilization Lectures, May 1922*, Butler and Tanner Grafton: London, 1923; 5, 7, 9, 30, 31, 33, 39, 43, preface.
16. *Webster's New World Dictionary of the American Language*, College Ed. World Publishing Company: Cleveland, OH, 1960; 1425.
17. Levinson, H. C. *The Science of Chance from Probability to Statistics*, Rinehart and Company: New York, 1950; 5, 11, v, foreword.
18. Yule, G. U. Kendall, M. G. *An Introduction to the Theory of Statistics*, 13th Ed. Rev. Charles Griffin and Company: London, 1949; 4.
19. Walker, H. M. *Studies in the History of Statistical Method*, Williams & Wilkins: Baltimore, MD, 1929; 32, 41.
20. Meitzen, A. *History, Theory, and Technique of Statistics*, American Academy of Political and Social Science: Philadelphia, PA, 1891; Translated by Roland P. Falkner, Part First: History of Statistics; 4, 15, 17, 19, 20–22, 24, 28, 29, 30, 31, 96.
21. Koren, J. *The History of Statistics: Their Development and Progress in Many Countries*, Macmillan, for the American Statistical Association: New York, 1918; 221–222.
22. Walpole, R. E. *Introduction to Statistics*, Macmillan: New York, 1968; 1–3.
23. Cajori, F. *A History of Mathematics*, 2nd Ed. Macmillan: New York, 1919; 170, 171, 380.
24. Sanford, V. *A Short History of Mathematics*, Houghton Mifflin: Cambridge, MA, 1930; 200–202 Pacioli (1949) was among the first to introduce the "Problem of the Points" into a work on mathematics.
25. Newman, J. R. *The World of Mathematics*, Simon and Schuster: New York, 1956; Vols. 2, 3 1303, 1416, 1417.
26. Blalock, H. M., Jr. *Social Statistics*, Rev. 2nd Ed. McGraw-Hill: New York, 1979; 6.
27. Mosteller, F. Evaluation: requirements for scientific proof. In *Coping with the Biomedical Literature*; Warren, K. S., Ed.; Praeger: New York, 1981; 104.
28. Cole, F. J.; Eales, N. B. The history of comparative anatomy, Part I, a statistical analysis of the literature. Sci. Progr. **1917**, *11*, 528–596.
29. Gosnell, C. F. The Rate of Obsolescence in College Library Book Collections as Determined by an Analysis of Three Select Lists of Books for College Libraries, Ph.D. Dissertation; New York University School of Education, 1943; 1, 2, 16, 125, 159, 162.
30. Gosnell, C. F. Obsolescence of books in college libraries. Coll. Res. Libr. **1944**, *5*(2), 115–125.
31. Raisig, L. M. Statistical bibliography in the health sciences. Bull. Med. Libr. Assoc. **1962**, *50*, 151, 450–461.
32. Wittig, G. Statistical bibliography—A historical footnote. J. Doc. **1978**, *34*, 240–241.
33. Henkle, H. H. The periodical literature of biochemistry. Bull. Med. Libr. Assoc. **1938**, *27*, 139–147.
34. Fussler, H. H. *Characteristics of the Research Literature Used by Chemists and Physicists in the United States*. Ph. D. Dissertation; University of Chicago, 1948.
35. Fussler, H. H. Characteristics of the research literature used by chemists and physicists in the United States. Libr. Q. *19*, Part I, January 1949, Part II, April 2, 1949; 19–35, 119–143.
36. Barker, D. L. *Characteristics of the Scientific Literature Cited by Chemists of the Soviet Union*. Ph.D. Dissertation; University of Illinois, Chicago, 1966; 5, 10–12, microfilm.
37. Pritchard, A. Computers, bibliometrics and abstracting services. Res. Librariansh. **1970**, *15*, 94–99.
38. Pritchard, A. Personal correspondence.
39. Pritchard, A. Bibliometrics and information transfer. Res. Librariansh. **1972**, *4*, 37–46.
40. British Standards Institution. *British Standard Glossary of Documentation Terms* 1976; 30 November 7 Published under the authority of the Executive Board, Prepared under the direction of the Documentation Standards Committee.
41. Potter, W. G. Introduction. Libr. Trends **1981**, *30*(1), 5.

42. Schrader, A. M. Teaching bibliometrics. Libr. Trends **1981**, *30*(6), 151.

43. Stevens, R. E. *Characteristics of Subject Literatures*, American College and Research Library Monography Series 7, January 1953; 10–16 Chicago.

44. Nicholas, D. Ritchie, M. *Literature and Bibliometrics*, Clive Bingley: London, 1978; 9–11.

45. O'Connor, D.; Voos, H. Empirical laws, theory construction and bibliometrics. Libr. Trends **1981**, *30*, 10.

46. Carpenter, M. P.; Narin, F. Clustering of scientific journals. J. Am. Soc. Inf. Sci. **1973**, *24*(6), 425–436.

47. Cason, H.; Lubotsky, M. The influence and dependence of psychological journals on each other. Psychol. Bull. **1936**, *33*, 95–103.

48. Narin, F. *Evaluative Bibliometrics: The Use of Publication and Citation Analysis in the Evaluation of Scientific Activity*, Computer Horizons: Cherry Hill, NJ, 1976; 41, 44, 48, 50, 82, 129, 130PB 252 339.

49. Lotka, A. J. Statistics—The frequency distribution of scientific productivity. J. Wash. Acad. Sci. **1926**, *16*, 317–325.

50. Gross, P. L.K.; Gross, E. M. College libraries and chemical education. Science. **1927**, 27 *66*, 385–389.

51. Lawani, S. M. Bibliometrics: Its theoretical foundations, methods and applications. Libri. **1981**, Vols. 31, 32, 295.

52. Gross, P. L.K. Fundamental science and war. Science December 20, **1927**, *66*, 640–645.

53. Bradford, S. C. Sources of information on specific subjects. Engineering **1934**, 26, *137*, 85–86.

54. Zipf, G. K. *The Psycho-Biology of Language: An Introduction to Dynamic Philology*, MIT Press: Cambridge, MA, 1935; 3, 7, 18, 25, 41–44, 47, v, vi, preface, reprint ed., 1965.

55. Zipf, G. K. *Human Behavior and the Principle of Least Effort: An Introduction to Human Ecology*, Addison-Wesley: Reading, MA, 1949.

56. Garfield, E. Citation indexes for science: A new dimension in documentation through association of ideas. Science **1955**, *122*, 108–111, 473–476.

57. Thomasson, P.; Stanley, J. C. Science **1955**, *122*, 610.

58. Garfield, E. Citation analysis as a tool in journal evaluation. Science **1972**, *178*, 471–478.

59. Kessler, M. M. An experimental study of bibliographic coupling between technical papers. IEEE Trans. Inf. Theory **1963**, 10, 49–51, PTG I9-9.

60. Kessler, M. M. Bibliographic coupling between scientific papers. Am. Doc. **1963**, *14*, 10–25.

61. Weinberg, B. H. Bibliographic coupling: A review. Inf. Storage Retr. **1974**, *10*(5,6), 189–196.

62. Kessler, M. M. Comparison of the results of bibliographic coupling and analytic subject indexing. Am. Doc. **1965**, *16*, 223.

63. Dikeman, R. K. *On the Relationship between the Epidemic Theory and the Bradford Law of Dispersion*Ph.D. Dissertation; Case Western Reserve University, 1974.

64. Hubert, J. J. Bibliometric models for journal productivity. Soc. Indic. Res. **1977**, *4*, 441–473.

65. Worthen, D. B. The epidemic process and the contagion model. J. Am. Soc. Inf. Sci. **1973**, *24*(5), 343–346.

66. Goffman, W. Newill, V. A. *A Generalization of the Theory of Epidemics: An Application to the Transmission of Ideas*, An original report. Also published in Nature, 1964, *204*, 225–228 with slight changes. 1, 2, 13, 14, abstract.

67. Goffman, W. Stability of epidemic processes. Nature **1966**, *210*, 786–787.

68. Goffman, W. A mathematical method for analyzing the growth of a scientific discipline. J. Assoc. Comput. Mach. **1971**, *18*, 173–185.

69. Goffman, W.; Harmon, G. Mathematical approach to the prediction of scientific discovery. Nature **1971**, *229*, 103–104.

70. Goffman, W. Mathematical approach to the spread of scientific ideas—The history of mast cell research. Nature **1966**, *212*, 449–452.

71. Small, H. Co-citation in the scientific literature: A new measure of the relationship between two documents. J. Am. Soc. Inf. Sci. **1973**, *24*(4), 265–269.

72. Drott, M. C. Bradford's law: Theory, empiricism and the gaps between. Lib. Trends **1981**, *30*(1), 41.

73. Zunde, P. Gehl, J. Empirical foundations of information science. *Annual Review of Information Science and Technology*, Knowledge Industry Publications, for the American Society for Information Science, 1979; Vol. 14, 67–92, Chap. 3.

74. Bookstein, A. Patterns of scientific productivity and social change: A discussion of Lotka's law and bibliometric symmetry. J. Am. Soc. Inf. Sci. **1977**, *28*, 209.

75. Bookstein, A. Explanations of the bibliometric distributions. Collect. Manage. **1979**, *3*(2–3), 151–162.

76. O'Conner, D.; Voos, H. Laws, theory and bibliometrics. Libr. Trends **1981**, *30*(1), 10.

77. Simon, H. A. On a class of skew distribution functions. Biometrika **1955**, *42*, 425–440.

78. Hubert, J. General bibliometric models. Libr. Trends **1981**, *30*(1), 67.

79. 79.Debus, A. G., Ed. *World Who's Who in Science*; A Biographical Dictionary of Notable Scientists from Antiquity to the Present, Marquis Who's Who Western Publishing Company: Hannibal, MO, 1968; 1069.

80. 80.Potter, W. G., Ed. *Lotka's Law Revisited*; Libr. Trends, 1981, *30*(1), *21* "*Zipf is the first to call the inverse square rule 'Lotka's Law' and discusses it as an approximation, not a rigid distribution,*" 26. *Zipf's comment found in his "Principle of Least Effort,*" 514.

81. Murphy, L. J. Lotka's law in the humanities. J. Am. Soc. Inf. Sci. **1973**, *24*(6), 461–462.

82. Murphy, L. J. Letters to the editor. J. Am. Soc. Inf. Sci. **1974**, *25*, 134.

83. Hubert, J. J. Letters to the editor. J. Am. Soc. Inf. Sci. **1977**, *28*, 66.

84. Coile, R. C. Lotka's frequency distribution of scientific productivity. J. Am. Soc. Inf. Sci. **1977**, *28*(6), 360, 366, 370.

85. Voos, H. Lotka and information science. J. Am. Soc. Inf. Sci. **1974**, *25*(4), 134, 270, 271.

86. Coile, R. C. Letters to the editor. J. Am. Soc. Inf. Sci. **1975**, *26*(2), 133.

87. Schorr, A. E. Lotka's law and library science. RQ **1974**, *14*(1), 32–33.

88. Tudor, D. Letters. RQ **1974**, *14*(2), 29, 30, 187.

89. Schorr, A. E. Letters. RQ **1975**, *15*(1), 90.

90. Schorr, A. E. Lotka's law and map librarianship. J. Am. Soc. Inf. Sci. **1975**, *26*, 189–190.

Bibliographic-Bibliotheque

91. Schorr, A. E. Map librarianship, map libraries and maps: A bibliography, 1921–1973. Bulletin **1974**, *95*, 2–35. *Special Libraries Association (Geography and Map Division)*.

92. Schoor, Lotka's law and the history of legal medicine. Res. Librariansh. **1975**, *30*, 205–209.

93. Nemec, J. *International Bibliography of the History of Legal Medicine*, National Library of Medicine: Bethesda, MD, 1974.

94. Radhakrishnan, T.; Kernizan, R. Lotka's law and computer science literature. J. Am. Soc. Inf. Sci. **1979**, *30*(1), 51–54.

95. Subramanyam, K. Lotka's law and library literature. Libr. Res. **1981**, *3*, 167, 170.

96. Richardson, V. L. Lotka's law and the catalogue?. AARL **1981**, *12*, 185, 186, 188.

97. Vlachy, J. Frequency distributions of scientific performance: A bibliography of Lotka's law and related phenomena. Scientometrics **1978**, *1*, 109–130.

98. Pao, M. L. Lotka's test. Collect. Manage. **1982**, *4*, 111–112.

99. Lotka, A. J. *Elements of Physical Biology*, Williams and Wilkins: Baltimore, MD, 1925, 313, 314, vii.

100. Willis, C. J. *Age and Area: A Study in Geographical Distribution and Origin of Species*, University Press: Cambridge, MA, 1922.

101. W.C.B.S., Obituary. Libr. World **1948**, *5*, 107.

102. Bradford, S. C. Bibliography by cooperation. Libr. Assoc. Rec. **1927**, *5*, 253, 254, 258.

103. Bradford, S. C. The necessity for the standardisation of bibliographical methods. ASLIB Proc. **1928**, *5*, 104–113.

104. Bradford, S. C. The Extent to Which Scientific and Technical Literature Is Covered by Present Abstracting and Indexing Periodicals ASLIB Proceedings, 14th Conference 1937 64, 65, 70.

105. Ditmas, E. M.R. A chapter closes: Bradford, Pollard and Lancaster-Jones. Coll. Res. Libr. **1949**, 334.

106. Yablonsky, A. I. On fundamental regularities of the distribution of scientific productivity. Scientometrics **1980**, *2*, 3–34.

107. Bradford, S. C. Complete documentation in science and technology. F.I.D. Commun. **1946**, *13*(2), C1–C5.

108. Bradford, S. C. Complete documentation. Nature **1947**, *159*(4029), 105–106.

109. Bradford, S. C. Complete documentation. Report of the Royal Society Empire Scientific Conference, June–July 1946; Vol. 1, 729–749.

110. Bradford, S. C. *Documentation*, Crosby Lockwood and Son: London, 1948.

111. Vickery, B. C. Bradford's law of scattering. J. Doc. **1948**, *4*, 198–203.

112. Kendall, M. G. The bibliography of operational research. Oper. Res. Q. **1960**, *11*(1,2), 31–36.

113. Cole, P. F. A new look at reference scattering. J. Doc. **1962**, *18*(2), 58–64.

114. Leimkuhler, F. F. The Bradford distribution. J. Doc. **1967**, *23*(3), 197–207.

115. Leimkuhler, F. F. An exact formulation of Bradford's law. J. Doc. **1980**, *36*(4), 285–292.

116. Brookes, B. C. A critical commentary on Leimkuhler's 'exact' formulation of the Bradford law. J. Doc. **1981**, *37*(2), 77–88.

117. Leimkuhler, F. F. Bradford's law. J. Doc. **1982**, *38*, 126. Erratum.

118. Brookes, B. C. Bradford's law and the bibliography of science. Nature **1969**, *224*, 953–956.

119. Naranan, S. Bradford's law of bibliography of science: An interpretation. Nature **1970**, *227*, 631–632.

120. Brookes, B. C. Scientific bibliography. Nature **1970**, *227*, 1377, Correspondence.

121. Goffman, W.; Morris, T. G. Bradford's law and library acquisitions. Nature **1970**, *226*, 922–923.

122. Fasler, A. Exceptions to Bradford's law. Nature 4 **1970**, *227*, 101, Correspondence.

123. Wilkinson, E. A. The ambiguity of Bradford's law. J. Doc. **1972**, *28*(2), 122–130.

124. Smith, D. A. The ambiguity of Bradford's law. J. Doc. **1972**, *28*, 262 Letters to the Editor.

125. Brookes, B. C. Numerical methods of bibliographic analysis. Libr. Trends **1973**, *22*(1), 18–43.

126. Aiyepeku, W. O. The Bradford distribution theory: The compounding of Bradford periodical literatures in geography. J. Doc. **1977**, *33*(3), 218.

127. Sengupta, I. N. Recent growth of the literature of biochemistry and changes in ranking of periodicals. J. Doc. **1973**, *29*(2), 210–211.

128. Pope, A. Bradford's law and the periodical literature of information science. J. Am. Soc., Inf. Sci. **1975**, *26*(4), 212.

129. Worthern, D. B. The application of Bradford's law to monographs. J. Doc. **1975**, *31*(1), 19–25.

130. Brookes, B. C. Theory of the Bradford law. J. Doc. **1977**, *33*(3), 173–250.

131. Joos, M. Book reviews. *George K. Zipf's The Psycho-Biology of Language (Boston: Houghton Mifflin Company, 1935)*, Language. 1936; Vol. 12, 196–197.

132. Zipf, G. K. *Relative Frequency: A Determinant of Phonetic Change*, Harvard Studies in Classical Philology Harvard University Press: Cambridge, MA, 1929; Vol. 40, 1–95.

133. Kruskal, W. H., Tanur, J. M., Eds.; *International Encyclopedia of Statistics*; The Free Press: New York, 1978; Vol. 2. Anatol Rapoport, "Rank–Size Relations Zipf's Law," 848.

134. Wyllys, R. E. Empirical and theoretical bases of Zipf's law. Libr. Trends **1981**, *30*(1), 53, 55, 56, 58.

135. Rapaport, A. The stochastic and the 'teleological' rationales of certain distributions and the so-called principle of least effort. Behav. Sci. **1957**, *2*, 150–151.

136. Simon, H. A. Some further notes on a class of skew distribution functions. Inf. Control **1960**, *3*(1), 80–88.

137. Hill, B. M. Zipf's law and prior distributions for the composition of a population. J. Am. Stat. Assoc. **1970**, *65*(331), 1220.

138. Hill, B. M. The rank-frequency form of Zipf's law. J. Am. Stat. Assoc. **1974**, *69*(348), 1017, 1025.

139. Hill, B. M.; Woodroofe, M. Stronger forms of Zipf's law. J. Am. Stat. Assoc. **1975**, *70*(349), 212–219.

140. Sichel, H. S. On a distribution law for word frequencies. J. Am. Stat. Assoc. **1975**, *70*(351), 542–547.

141. Wyllys, R. E. Measuring Scientific Prose with Rank-Frequency ("Zipf") Curves: A New Use for an Old

Phenomenon Proceedings ASIS 38th Annual Meeting, 1975; Vol. 1230, 30. *Inf. Revolution.*

142. Brookes, B. C. The derivation and application of the Bradford–Zipf distribution. J. Doc. **1968**, *24*(4), 247–265.

143. Groos, O. V. Bradford's law and the Keenan–Atherton data. Am. Doc. **1967**, *18*, 46.

144. Narin, F.; Moll, J. K. Bibliometrics. Ann. Rev. Inf. Sci. Technol. **1977**, *12*, 35–57.

145. O'Neill, E. T. Limitations of the Bradford Distributions Proceedings of the American Society for Information Science Greenwood: Westport, CT, 1973; Vol. 10, 177–178.

146. Brookes, B. C. The complete Bradford–Zipf 'bibliograph'. J. Doc. **1969**, *25*(1), 58–60.

147. Brookes, B. C. Theory of the Bradford law. In *Key Papers in Information Science*; Griffith, B. C., Ed.; Knowledge Industry Publication: White Plains, NY, 1980.

148. Buckland, M. K.; Hindle, A. Documentation notes, 'library Zipf'. J. Doc. **1969**, *25*(1), 52.

149. Fairthorne, R. Library Zipf. J. Doc. *25*, 152. Letters to the Editor.

150. Smith, D. A. Library Zipf. J. Doc. **1986**, *25*, 153–154 Letters to the editor.

151. Buckland; Hindle Messrs. Buckland and Hindle reply. J. Doc. **1969**, *25*, 154. Letters to the Editor.

152. Brookes, B. C. Mr. Brookes replies. J. Doc. **1969**, *25*, 155. Letters to the Editor.

153. Bulick, S. Book use as a Bradford–Zipf phenomenon. Coll. Res. Libr. **1978**, *39*(3), 215, 218.

154. Bookstein, A. The bibliometric distributions. Libr. Q. **1976**, *46*(4), 416–423.

155. Bookstein, A. *Bibliometric Symmetry and the Bradford–Zipf Laws, 6–7, in preparation.*

156. Osinga, M. Some fundamental aspects of information science. Int. Forum Inf. Doc. **1979**, *4*(3), 31.

157. Noma, E. Untangling citation networks. Inf. Process. Manag. **1982**, *18*(2), 43–53.

158. Smith, L. C. Citation analysis. Libr. Trends. **1981**, *30*(1), 83–106.

159. Spiegel-Rosing, I. Science studies: Bibliometric and content analysis. Soc. Stud. Sci. **1977**, *7*, 97–113.

160. Oppenheim, C.; Renn, S. P. Highly cited old papers and the reasons why they continue to be cited. J. Am. Soc. Inf. Sci. **1978**, *29*, 225–231.

161. Tagliacozzo, R. Citations and citation indexes: A review. Meth. Inf. Med. **1967**, *6*(3), 136–142.

162. Hjerppe, R. *An Outline of Bibliometrics and Citation Analysis*, Royal Institute of Technology Library: Sweden, October, 1978; Report TRITA-LIB-6014.

163. Hjerppe, R. *A Bibliography of Bibliometrics and Citation Indexing and Analysis*, Royal Institute of Technology Library: Stockholm, Sweden, December 1980. Report TRITA-LIB-2013.

164. Burton, R. E.; Kebler, R. W. The 'half-life' of some scientific and technical literature. Am. Doc. **1960**, *11*, 18–22.

165. Line, M. B. The 'half-life' of periodical literature apparent and real obsolescence. J. Doc. **1970**, *26*(1), 46–47.

166. Brookes, B. C. The growth, utility, and obsolescence of scientific periodical literature. J. Doc. **1970**, *26*(4), 283, 286, 291, 294.

167. Sandison, A. Use of older literature and its obsolescence. J. Doc. **1971**, *27*(3), 184, 199.

168. Buckland, M. K. Are obsolescence and scattering related?. J. Doc. **1972**, *28*(3), 242.

169. Line, M. B. Letters to the editor. J. Doc. **1973**, *29*(1), 107.

170. Sandison, A. Letters to the editor. J. Doc. **1973**, *29*(1), 107–108.

171. Brittain, J. M.; Line, M. B. Sources of citations and references for analysis purposes: A comparative assessment. J. Doc. **1973**, *29*(1), 72–83.

172. Line, M. B. Rank lists based on citations and library uses as indicators of journal usage in individual libraries. Collect. Manage. **1978**, *2*(4), 313, 315.

173. Gapen, D. K.; Milner, S. P. Obsolescence. Libr. Trends **1981**, *30*(1), 107–124.

174. Garfield, E. Citation indexes in sociological and historical research. Am. Doc. **1963**, *14*(4), 289–291.

175. Garfield, E. Science citation index—A new dimension in indexing. Science May 8, **1964**, *144*, 650–652.

176. Garfield, E. Citation indexing: A natural science literature retrieval system for the social sciences. Am. Behav. Scientist **1964**, *7*(10), 58–61.

177. Garfield, E. Primordial concepts, citation indexing, and historio-bibliography. J. Libr. Hist. **1967**, *2*, 238–239.

178. Terry, J. E. Science citation index: An international, interdisciplinary index to the literature of science. J. Doc. **1965**, *21*(2), 139–141 Review.

179. Garfield, E. Citation indexing for studying science. Nature August15, **1970**, *227*, 669–671.

180. Garfield, E. Sher, I. H. Torpie, R. J. *The Use of Citation Data in Writing the History of Science*, Institute for Scientific Information, Inc.: Philadelphia, PA, December 31,1964.

181. Martyn, J. An examination of citation indexes. Aslib Proc. **1965**, *17*(6), 186–196.

182. Loosjes, P. *On Documentation of Scientific Literature*, Butterworths: London, 1973; 84–85.

183. Small, H.; Griffith, B. C. The structure of scientific literatures I: Identifying and graphing specialties. Sci. Stud. **1974**, *4*, 17–40.

184. Small, H. G. A co-citation model of a scientific specialty: A longitudinal study of collagen research. Soc. Stud. Sci. **1977**, *7*, 139–166.

185. Sullivan, D.; White, D. H.; Barboni, E. J. Co-citation analyses of science: An evaluation. Soc. Stud. Sci. **1977**, *7*, 223–240, 324.

186. Bellardo, T. The use of co-citations to study science. Libr. Res. **1980–1981**, *2*(3), 231–237.

187. Pritchard, A. Wittig, G. R. *Bibliometrics: A Bibliography and Index*, ALLM Books: Watford, 1981; Vol. 1, 3–5.

188. Broadus, R. N. The applications of citation analyses to library collection building. Adv. Librariansh. **1977**, *7*, 310, 313.

189. Lawani, S. W. Citation analysis and the quality of scientific productivity. Bioscience **1977**, *27*(1), 26–31.

190. Wade, N. Citation analysis: A new tool. Science **1975**, *188*(4187), 429–432.

191. Cuadra, C. A. Identifying key contributions to information science. Am. Doc. **1964**, *15*, 289–295.

192. David, H. G.; Piip, L.; Haly, A. R. The examination of research trends by analysis of publication numbers. J. Inf. Sci. **1981**, *3*, 283–288.

193. Tenopir, C. Distribution of citations in databases in a multidisciplinary field. Online Rev. **1982**, *6*(5), 402.

194. Sprules, M. L. Online bibliometrics in an academic library. Online **1983**, *7*(1), 26.
195. Line, M. B.; Sandison, A. Practical interpretation of citation and library use studies. Coll. Res. Libr. **1975**, *36*, 393.
196. McAllister, P. R. *Review and Analysis of Importance and Utilization Measures Contained in 'Evaluative Bibliometrics'*, 1978; *January 19. iii Report prepared for National Science Foundation by Computer Horizons, Inc., Cherry Hill, NJ.*
197. Schmidmaier, D. *Application of Bibliometrics in Technical University Libraries,* Proceedings of the 7th Meeting of IATUL, Leuven, May, 16–21, 1977 129–135.

SELECTED BIBLIOGRAPHY

Bibliography

1. Ferguson, J. *Some Aspects of Bibliography*, George P. Johnston: Edinburgh, 1900.
2. Taylor, A. *A History of Bibliographies of Bibliographies*, Scarecrow Press: New Brunswick, NJ, 1955.
3. Van Hoesen, H. B. Walter, F. K. *Bibliography: Practical, Enumerative, Historical*, Charles Scribner's Sons: New York, 1928. 2nd printing, 1937.

Bibliometrics/Statistical Bibliography

1. Aiyepeku, W. O. Towards a methodology of bibliometrics. Niger. Libr. **1974**, *10*(2–3), 85–90.
2. Potter, W. G., Ed. *Bibliometrics*. Libr. Trends, 1981; Vol. 30.
3. Donohue, J. C. *Understanding Scientific Literatures: A Bibliometric Approach*, MIT Press: Cambridge, MA, 1973.
4. Ferrante, B. Bibliometrics: Access in the library literature. Collect. Manage. **1978**, *2*, 199.
5. Meadows, A. J.; O'Conner, J. G. Bibliographic statistics as a guide to growth points in science. Sci. Stud. **1971**, *1*(1), 95–99.
6. Pritchard, A. Statistical bibliography of bibliometrics. J. Doc. **1969**, *25*(4), 348–349.

Citation Analysis

1. Bensman, S. J. Bibliometric laws and library usage as social phenomena. Libr. Res. **1982**, *4*(3), 279–312.
2. Bonzi, S. Characteristics of a literature as predictors of relatedness between cited and citing works. J. Am. Soc. Inf. Sci. **1982**, *33*, 208–216.
3. Brookes, B. C. Numerical methods of analysis. Libr. Trends. **1973**, *22*, 18–43.
4. Cole, J. R. Ortega hypothesis. Science. October 27, **1972**, *178*, 368–375.
5. Force, R. W. A bibliometric analysis of literature of environmental education. J. Environ. Educ. **1978**, *9*(3), 29–34.
6. Garfield, E. *Citation Indexing — Its Theory and Application in Science, Technology, and Humanities*, Wiley: New York, 1979.
7. Garfield, E. Is citation analysis a legitimate evaluation tool?. Scientometrics. **1979**, *1*, 359–375.
8. Marshakova, I. Citation networks in information science. Scientometrics. **1981**, *3*, 13–25.
9. Martyn, J. Bibliographic coupling. J. Doc. **1964**, *20*, 236.
10. McAllister, P. *Review and Analysis of Importance and Utilization Measure Contained in Evaluative Bibliometrics*, Computer Horizons: Cherry Hill, NJ, 1978. Contract No. NSF PRM-7682854, PB-278 744/8 SL.
11. Price, D. de S. General theory of bibliometric and other cumulative advantage processes. J. Am. Soc. Inf. Sci. **1976**, *27*, 292–306.
12. Price, D. de S. Networks of scientific papers. Science 30, **1965**, *149*, 510–515.
13. Pritchard, A. Citation analysis versus use data. J. Doc. **1980**, *36*, 268–269.
14. Small, H. G. Cited documents as concept symbols. Soc. Stud. Sci. **1978**, *8*, 327–340.
15. Small, H. G. Co-citation context analysis: The relationship between bibliometric structure and knowledge. Proc. Am. Soc. Inf. Sci. **1979**, *16*, 270–275.
16. Wiberley, S. E., Jr. Journal rankings from citation studies: A comparison of national and local data from social work. Libr. Q. **1982**, *52*, 348.

Laws

Lotka

1. Allison, P. et al. Lotka's law: a problem in its interpretation and application. Soc. Stud. Sci. **1976**, *6*, 269–276.
2. Bookstein, A. Patterns of scientific productivity. J. Am. Soc. Inf. Sci. **1977**, *28*, 206–210.
3. Coile, R. Bibliometric studies of scientific productivity. Proc. ASIS 39 Annu. Meet. **1976**, *13*, 90.
4. Lotka, A. J. Frequency distribution of scientific productivity. J. Wash. Acad. Sci. **1926**, *16*(12), 317–325.
5. Meadows, A. J. *Communication in Science*, Butterworths: London, 1974.
6. Murphy, L. J. Erratum. J. Am. Soc. Inf. Sci. **1974**, *25*, 134.
7. Schorr, A. E. Lotka's law. RQ 1975, *15*, 90.
8. Schorr, A. E. Lotka's law and the history of legal medicine. Res. Librariansh. **1975**, *30*, 205–209.

Bradford

1. Avramescu, A. Theoretical foundation of B's law. Int. Forum Inf. Doc. **1980**, *5*, 15–22.
2. Bookstein, A. *Symmetry Properties and Discoverability of Bibliometric Laws,* Presented at Operations Research of America, the Institute of Management Sciences Joint Society National Meeting Miami, FL November 3–5, 1976.
3. Bradford, S. C. *Documentation*, 2nd Ed. Crosby Lockwood: London, 1953.
4. Bradford, S. C. Bradford issue. J. Doc. **1977**, *33*, 173–250.
5. Braga, G. M. Some aspects of the Bradford distribution. Pro. ASIS Annu. Meet. **1978**, *15*, 51–54.

6. Goffman, W. Dispersion of papers among journals based on mathematical analysis of two diverse medical literatures. Nature **1969**, *221*, 1205–1207.

7. Naranan, S. Power law relations. J. Doc. **1971**, *27*, 83–97.

8. Praunlich, P.; Kroll, M. Bradford's distribution—New formulation. J. Am. Soc. Inf. Sci. **1978**, *29*(2), 51–55.

9. Wilkinson, E. A. Erratum. J. Doc. **1972**, *28*(3), 232.

Zipf

1. Booth, A. A law of occurences for words of low frequency. Inf. Control **1967**, *10*, 386–393.

2. Good, I. J. Distribution of word frequencies. Nature 16, **1957**, *179*(4559), 595–596.

3. Hill, B. M. Zipf's law and prior distributions for the composition of a population. J. Am. Soc. Inf. Sci. **1970**, *65* (331), 1220–1232.

4. Hill, B. M. Rank frequency form of Zipf's law. J. Am. Soc. Inf. Sci. **1974**, *69*, 1017–1026.

5. Hill, B. M.; Woodroofe, M. Stronger forms of Zipf's law. J. Am. Soc. Inf. Sci. **1975**, *70*, 212–219.

6. Mayes, P. B. Use of the B-Z distribution to estimate efficiency values for a journal circulation system. J. Doc. **1975**, *31*, 287–289.

7. Parker Rhodes, A. F.; Joyce, T. A theory word frequency distribution. Nature **1956**, *178*, 1308.

8. Rapoport, A. Rank–size relations. *International Encyclopedia of Statistics*, 1978; Vol. 2, 847–854 New York.

9. Scarrott, G. Will Zipf join Gauss?. New Sci. **1974**, 402–404.

10. Zipf, G. K. *Human Behavior and the Principle of Least Effort*, Hafner Publishing: New York, 1965.

Statistics

1. Levy, H. Roth, L. *Elements of Probability*, Clarendon Press: Oxford, 1936; 1–11 Chap. I, Historical Introduction.

2. Parl, B. *Basic Statistics*, Doubleday: New York, 1967.

3. Reichmann, W. J. *Use and Abuse of Statistics*, Oxford University Press: New York, 1961.

4. Runyon, R. P. Haber, A. *Fundamentals of Behavioral Statistics*, 4th Ed. Addison-Wesley: Reading, MA, 1980.

5. Shirey, D. L. Statistical Methods and Analysis. *Encyclopedia of Library and Information Science*; Dekker: New York, 1980; Vol. 29, 78.

6. Warren, K. S. *Coping with the Biomedical Literature: A Primer for the Scientist and the Clinician*, Praeger Special Studies/Praeger: New York, 1981.

7. Westergaard, H. *Contributions to the History of Statistics*, 1st Ed. P. S. King and Son: London, 1932.

8. Yule, G. U. The introduction of the words "statistics, statistical" into the English language. J. R. Stat. Soc. (Lond.) **1905**, *68*, 391–396.

Bibliothèque Nationale de France

Noémie Lesquins
Scientific Mission (DSR), National Library of France, Paris, France

Abstract

With more than 10 million volumes and an annual increase of about 50,000, the Bibliothèque Nationale de France is one of the biggest libraries in the world. It is also one of the oldest and since the sixteenth century, it has been entrusted with the mission of collecting, cataloging, preserving, and providing access to the French print heritage. Although the library's history consists of several turning points, the last decade of the twentieth century has brought an unprecedented change in the life of the institution: new information technologies, new buildings, new collection management politics, and new services. More than ever, today the library is part of a national and international network of libraries and other cultural institutions whose goal is both to share the wealth of their resources and assert their identities.

A SHORT HISTORY OF THE PAST: FROM THE BIBLIOTHÈQUE ROYALE TO THE BIBLIOTHÈQUE NATIONALE

The Royal Library

The origin of what is now known as the Bibliothèque Nationale de France is the French kings' own private libraries. The most famous one is King Charles V (1364–1380) own collection, composed of 917 manuscripts and cataloged in 1343. In 1368, a specific space in the Louvre was built for this collection. Yet, according to the tradition, it was dispersed in 1380 after its owner's death. The first step to the creation of a royal library was made by Louis XI (1461–1483) who donated his private library to his heir. The new principle was respected by his followers. As the consequence of the advent of printing, the library expanded in the fifteenth century, and a 1622 inventory reports a volume of 4712 manuscripts and prints.

It is under François the First's reign (1515–1547) that a major decision was made to order any printers and booksellers to deposit copies of all books printed or sold in the kingdom. Although this law was not fully enforced in the beginning, the 1537 "Ordonnance de Montpellier" marks the birth of the legal deposit (dépôt légal) that remains one of the major missions of the Bibliothèque nationale de France to this day.

From the fifteenth to the seventeenth century, the library was housed in various royal palaces. Transferred from Paris to Amboise, from Amboise to Blois, from Blois to Fontainebleau, it was finally moved back to Paris at the end of the sixteenth century. In 1666, Colbert, Louis XIV Prime Minister, decided to move it into two of his own private residences located on rue Vivienne. In the following decades, that location was extended to the entire block on the rue Richelieu, rue des Petits-Champs and rue Colbert, where parts of the collection are still stored.

The constant expansion of the library in the seventeenth and eighteenth century was mainly due to a determined acquisition and gift policy. Nicolas Clément, librarian of the king in 1670, invented a classification based on 23 letters corresponding to subjects, from religion to literature, including history, arts, sciences, law, and foreign languages. It was used until 1996. The collections were enriched not only with books and manuscripts but also with medals and engravings, hence the creation by the Abbé Bignon, librarian of Louis XV, of the first departments in 1720: Manuscripts, Printed Books, Titles and Genealogy, Engraved Plates and Prints, and Medals and Stone Engravings. Parts of the collections were already open to scholars, but in 1720, the Abbé Bignon extended the access to the general public one day a week, from 11 A.M. to 1 P.M.

The French Revolution changed the contents and structure of the Library tremendously: around 250,000 books, 14,000 manuscripts and 80,000 engravings were confiscated from religious congregations and aristocratic families to be added to the collections, and a 1794 decree changed the Bibliothèque du Roi into the Bibliothèque Nationale.

The Bibliothèque Nationale Throughout the Nineteenth and Twentieth Century

Although its name was changed to Bibliothèque Impériale under Napoleon the First (1804–1815) and Napoleon the Third (1852–1870) and to Bibliothèque Royale under the Restoration (1814–1830) and the Monarchy of July (1830–1848), the library structure and mission kept developing in the same direction whatever the political system was.

The nineteenth century was for the Bibliothèque Nationale, as well as for many other great libraries worldwide, an important time of development in building and

Bibliographic–Bibliothèque

cataloging. With the constant increase of the collections, the lack of space became one of the major issues in the middle of the nineteenth century. In 1862, Napoleon III commissioned the French architect Henry Labrouste (1801–1875) to renovate some of the original buildings, the Hôtel Tubeuf and the Galerie Mazarine, and to build new premises on the rue de Richelieu and rue des Petits-Champs. His most famous work for the library is the Printed Books Reading Room, inaugurated in 1868 and well known as an avant-garde masterpiece of cast-iron architecture.

At the same time, making inventories of the collections became a priority. Head librarians Joseph Naudet (1840–1852), Jules Taschereau (1852–1874), and Léopold Delisle (1874–1905) devoted most of their time and energy to cataloging. Catalogs of medical sciences and of French, British, Spanish and Portuguese, Asian, American, and Oceanic history were first compiled, followed in 1897 by Léopold Delisle's General Catalog of Printed Books (Catalogue Général des Livres Imprimés) that was continued throughout the twentieth century until the computerization of the catalog in the 1980s.

Expansions continued throughout the twentieth century, especially under the governance of Julien Cain from 1930 to 1964, to house current holdings as well as rare and precious pieces and to provide access to the public. A new catalog and reference room was built in 1935–1937, and a periodicals room in 1936; special departments (Engravings and Maps and Plans) were moved within the Richelieu block to free up space for the Printed Books Department; new departments were created to preserve and collect newly formed collections, such as oriental manuscripts and music and sound archives; decentralized annexes were created in Versailles in 1934, 1954, and 1971, in Sablé (200 km west of Paris) in 1980, and in Provins (east of Paris) in 1981.

Several laws and decrees managed to improve the legal deposit process and extended it to new materials. Since 1943, printers have the obligation to deposit in regional libraries copies of any material (books, periodicals, flyers, postcards, posters, etc.) printed in the country. The scope of the obligation was extended to imported books; photographic, cinematographic, and sound materials in 1925 and 1943; audiovisual items in 1975; and computer resources in 1992. In 1880, 12,414 prints were deposited; in 1993, the number reached 45,000 volumes.

With the computerization, the increasing readers demand, the growth of published material, and the pressure of new conservation requirements, among which is the urgent treatment of brittle and acid paper and the conservation of new media, the Bibliothèque Nationale had to face the issues of equipment, staff, and space at the dawn of the twenty-first century. A great debate in the 1980s finally ended with the decision by President François Mitterrand (1981–1995) to build a new library.

THE BIBLIOTHÈQUE NATIONALE DE FRANCE

History of the Project

The official launch of a new national library project dates back to the traditional Bastille Day presidential speech made by François Mitterrand in 1988, in which he announced "the construction and development of one of the, or maybe the largest and most modern library in the world." The ambitious project was defined as such: "the role of this great library will be to cover every field of knowledge, to be accessible to all, to make use of the very latest data transmission technologies, to provide for remote document consultation, and to forge links with other European libraries." A planning committee, the Établissement Public de la Bibliothèque de France, was commissioned for the planning for the new institution, including the international architectural contest, the construction schedule, and the collection and staff management. Although it was first planned to transfer into the new building only the collections printed after 1945, the final decision was made to transfer all prints into the new building, when the special departments would remain in the old site.

In April 1989, the location for the new building was chosen by the river Seine in a southeast area of Paris called Tolbiac. The international architectural contest was chaired by the American architect I. M. Pei. Four projects were selected by the jury and presented to the French President who finally chose the French architect Dominique Perrault on August 21, 1989. Dominique Perrault's project consists of four glass towers symbolizing four open books built on a wooden esplanade and surrounding a deep patio with a garden. On the east and west sides of the esplanade, two escalators provide access to entrance halls from which readers can reach either the reference library reading rooms on the upper level or the research library on the lower level. The project and its final result triggered and is still the topic of a great polemic, especially on the storage of books in glass towers.

On January 3, 1994, a decree merged the Bibliothèque Nationale and the Établissement Public de la Bibliothèque de France into a single institution, the Bibliothèque Nationale de France. The keys to the building were officially given by the architect on March 23, 1995, and the reference library was opened on December 20, 1996. After 2 years, during which most of the collections held on the Richelieu site were transferred to the new Tolbiac site, the research library was finally opened on October 8, 1998. A new site was also built by Dominique Perrault in Marne-la-Vallée (20 km east of Paris) for conservation, preservation, and restoration.

This great project presented an opportunity to modernize a century-old library that could not afford to miss the technological turn that revolutionized the information world in the last decades of the twentieth century. It helped the library to improve its services to the public and to the national and international library community.

Mission and Organization

The Bibliothèque Nationale de France is in charge of five major national missions: acquiring and cataloging collections (through legal deposit, acquisitions, exchanges, gifts, and donations), preserving the collections (through preventive conservation, restoration, duplication), providing the public with access to the collections (inside the library within its 24 reading rooms as well as outside through the Digital Library and remote duplicated document delivery), highlighting the collections (through education and exhibitions programs, publishing, and other cultural activities), and last but not least, cooperating with national and international libraries and institutions (through conservation sharing, acquisition sharing, information and expertise sharing, and the creating of union catalogs).

As most cultural institutions in France, the Bibliothèque Nationale de France is a public institution. Both the President and General Director are named by the government, and most important decisions are made under the control both of the Ministry of Finances and of the Ministry of Cultural Affairs (Ministère de la Culture et de la Communication), more precisely of the Book and Library administration (Direction du Livre et de la Lecture). Funding is also mainly public: in 2001, the French state provided the library with a 105.8 million Euro budget, to which 23.7 million Euro from grants, donations, and internal resources were added. Yet, private funding is an essential and a major resource for the acquisition of special collections such as the manuscripts of Chateaubriant's *Mémoire d'Outre-tombe* bought in 2000 or of *Voyage au bout de la nuit* by Louis-Ferdinand Céline bought in 2001 for 11 million francs (1.67 million Euro).

Today, the Bibliothèque Nationale de France has more than 2800 employees. Three main directors rule the Library: the Staff Director, the Collection Director, and the Services and Networks Director. Furthermore, four delegates to the General Director are, respectively, in charge of a Development Office, a Bureau for International Affairs a Bureau for Communication, and a Bureau for Cultural Activities. Two committees assist the President of the library with all administrative and scientific decisions. A financial service linked to the Ministry of Finance centralizes, distributes, and controls the resources and expenses of the establishment.

Although the collection was formerly divided according to the material type (prints, periodicals, audiovisual, photographs, etc.), it is now divided into four print collection departments based on disciplines (Philosophy, History, and Human Sciences; Law, Economics, and Political Sciences; Science and Technology; Art and Literature), one special department for Rare Books (Réserve

des livres rares), seven special departments based on material criteria and partly inherited from older departments (Manuscripts; Prints and Photographs; Maps and Plans; Coins and Medals; Music and Opera; Performing Arts; Audiovisual Collections) and two departments that are special libraries onto themselves with their own site (Bibliothèque-Musée de l'Opéra and Bibliothèque de l'Arsenal). A specific Department for Bibliographic Research is also under the supervision of the Collection Director.

Print and audiovisual collections were transferred to the Tolbiac site; the rest was left at their original sites.

Public and Collections

Tolbiac site

One of the key features of the new library site, also known as site François-Mitterrand, is opening its doors to the general public, when the Bibliothèque Nationale was traditionally for researchers only. Because it is not possible to provide the large number of people with access to the whole national print heritage, among which are rare, sometimes unique, and fragile pieces, the Tolbiac site is composed of two reading rooms levels: the upper level with a reference library open to any person over 16 and the lower level with a research library open to only authorized readers. Among the average 3000–3500 readers who come to the library everyday, about two thirds, mostly students, use the reference library and more than one third use the research library.

The reference library is composed of 10 thematic reading rooms providing 1700 seats. The 280,000 volumes of books and periodicals with an encyclopedic coverage of a first academic degree level are stored on open stacks and classified with the Dewey system as in many public libraries. Most of them are in French, but a wide range of foreign languages, such as English, German, Spanish, Italian, Portuguese, and also oriental languages is well represented. In every reading room, computers provide access to the library catalog (up to 8 million records), about 250 CD-ROMs and databases (bibliographic indexes, directories, dissertation abstracts, union catalogs, dictionaries, encyclopedias, etc.), a growing number of electronic periodicals, and the Internet. Admission to the research library is given only after an interview with a librarian who evaluates the needs of the user. Dealing with both of its missions of preserving and providing access with the national printed heritage, the Bibliothèque Nationale de France is considered a library of "last resort." The collection of print materials is composed of 11 million volumes of books, among which 200,000 are kept in the Rare Books Department, and of about 350,000 periodicals titles, among which 32,000 are current ones. They have mostly entered the library through legal deposit and exchanges and are kept in closed stacks, either on the top

11 floors of the towers or on the basement level near the reading rooms. Every reading room of the research library also has open stacks holding about 330,000 volumes bought by librarians to compose the lower level open access reference library.

The library information system oversees the catalog and the documents and reading seats reservation system that is linked to an automated document delivery system made of 8 km of tracks and 150 delivery stations. Documents can be sent from any storage to any of the 14 reading rooms of the lower level in about 45 minutes.

Digitization has become an opportunity for libraries to duplicate and protect the original collection. Besides the 100,000 printed works and 300,000 fixed images that have been digitized, the library owns about 76,000 microforms and 950,000 microfiches that are substitutes for the documents.

The Department of Audiovisual Collections is the only special department on Tolbiac site. It covers all disciplines on audio, visual, and electronic material deposited by publishers. This collection of 50,000 multimedia documents, 900,000 sound recordings, and 90,000 video recordings is complementary to the Institut National de l'Audiovisuel (National Institute for Audiovisual) holdings that are also housed on the Tolbiac site and mainly consist of TV and radio legal deposit.

Richelieu site and the other libraries

Five special departments are still housed in the old Bibliothèque Nationale site, rue de Richelieu: Manuscripts (with a collection of 225,000 Western and Oriental volumes of manuscripts and xylographs), Engravings (with a collection of 12 million engravings, photographs, posters, and postcards), Maps and Plans (with a collection of 890,000 maps, atlases, reference documents, and globes), Coins and Medals (with a collection of 530,000 coins, medals, and antiques), Music (with a collection of 2 million music documents and recordings).

These departments are associated with other libraries that joined the national library. The Music Department is in charge not only of the collection housed in Richelieu but also of the Bibliothèque-Musée de l'Opéra housed in the Opéra Garnier building and composed of 130,000 prints and 240,000 special materials. The Bibliothèque de l'Arsenal (1 million prints and 120,000 special documents), which opened in 1797 and was originally composed of two precious encyclopedic collections owned by two members of the royal family, was annexed to the national library in 1934 and established as a department in 1977. It is well known among bibliophiles for its rare and precious editions and bindings. The Arsenal site also houses the Performing Arts Department (2.4 million objects and documents), which was created in 1976 after collector Auguste Rondel's donation and opened an annex in the Maison Jean Vilar in Avignon in 1979. A project for

the National Institute for the History of Art (Institut national d'histoire de l'art) composed of a reference and research library and of an academic research section has been in development since the late 1990s on the Richelieu site. It consists of the merging of three major art libraries (Bibliothèque d'Art et d'Archéologie, created after Jacques Doucet's private collection; Bibliothèque Centrale des Musées Nationaux, and Bibliothèque de l'École Nationale Supérieure des Beaux-Arts), providing free access to 550,000 volumes in the famous salle Labrouste. The École nationale des Chartes and its library, mainly specialized in medieval history, is also associated with the project. This institute is meant to be complementary with the Bibliothèque Nationale de France precious collection, which is being reorganized while the buildings and preservation conditions are being modernized. The former periodicals reading room of the Richelieu site, also called salle ovale, is meant to become in the coming years another reference library providing the large public with access to all the library digitized and microfilmed collections.

A LIBRARY FOR THE TWENTY-FIRST CENTURY

The Library on the Web

In the 1980s, the development of new information technologies enhanced a major evolution in the library world. Remote services are part of all contemporary libraries' missions and duties, especially those libraries with a national and international role. The new project for the French national library was an opportunity to define new missions, complementary to the traditional ones and based on technological progress. Twenty years later, the Bibliothèque Nationale de France has developed two major information tools that will continue to develop and grow: the Web site and the Digital Library.

The Bibliothèque Nationale de France Web site, http://www.bnf.fr, was created in 1996. It contains about 100,000 pages and provides information, services, and bibliographic products for the Internet public, the actual library's users, and the national and international community of librarians and publishers.

Information to the general audience mainly takes the form of a repertory of Internet resources called "Les Signets de la Bibliothèque Nationale de France" (the Bibliothèque Nationale de France Internet bookmarks) that are classified, enriched, and updated by specialized librarians. Service to the readers and researchers is based on the integrated information system that gives them the possibility to search in catalogs (BN-OPALE PLUS for printed material and BN-OPALINE for special collections), reserve a seat on Tolbiac site, order documents, or ask for duplication on various material (color or black and white photocopies, microfiches, microfilms, facsimiles, slides, photographs, digitized images on CD-ROM or on disk, etc.). Information and services to the library and publisher community cover various activities: bibliographic standards (ISBD and ISSN recommendations, MARC formats); legal texts (legal deposit laws and decrees); bibliographic products (subject headings lists, catalogs, French National Bibliography online with the possibility to download records); and general information about the library (annual reports, legal deposit activity, and national and international cooperation and partnership).

In 2001, the Bibliothèque Nationale de France Web site was visited around 14,300 times every day by 11,700 different users (41% are located in France, 17% in Canada and the United States, 3.5% in Italy, and 38.5% in the rest of the world). An average number of 90,700 pages were looked at every day, mostly the catalogs and the Digital Library.

The Bibliothèque Nationale de France Digital Library, Gallica, was launched in 1997. It was reorganized in 2000 and includes in 2002 a free of right collection of 80,000 fixed images, several dozens hours of sound recordings, and more than 60,000 printed volumes. Most prints are digitized in image format, but all tables of contents can be accessed in text format as well as about 1250 documents, such as reference books, classical literature, and rare editions. In 2001, about 4500 Internet users entered Gallica every day, and 10 million pages were downloaded every month.

Gallica complements the virtual exhibitions and the Internet education programs also available on the Library's Web site. The Digital Library growth is based on 3-year digitization programs of 20,000 prints and 80,000 fixed images selected according to two major directions: completeness of the cultural heritage and encyclopedic library (new authors, dictionaries, and periodicals) and development of thematic projects such as Traveling (in France, in Africa, in Italy, etc.), Memories of the Bibliothèque Nationale de France (a selection of material of an outstanding research and art value), and special collections (illuminated manuscripts, incunabula, Philidor collection in partnership with the Library of Versailles, Gaignières collection in partnership with the Bodleian Library, Dunhuang collection in partnership with several foreign institutions coordinated by the Mellon Foundation).

National and International Cooperation and Partnership

Institutional and human networks based on expertise, financial resources, and documentation sharing are key priorities of the Bibliothèque Nationale de France.

Since 1994 a partnership policy based on long- or short-term programs has been created under the name of "réseau des pôles associés" (partner institutions network).

Every partner institution, mostly libraries or groups of libraries, signs an agreement with the national library displaying the cooperation patterns. In 2002, there are 64 «pôles associés», representing 71 institutions. Among them, 20 are city libraries, 18 are academic libraries, and 33 are research institutions, museums, or record offices.

Collecting and preserving the national heritage is one of the partnership's purposes: in 1996, agreements with 25 regional libraries and record offices were drawn up to reinforce the effectiveness of printer legal deposit and to enhance links between national and local copyright systems. Cross-control helps all partners to improve their collection depth.

Developing specialized collections of a high research level is a second purpose: seeking to ensure a comprehensive national coverage of the world's knowledge, the Bibliothèque Nationale de France shares its collection development policy with 39 libraries (23 located in Paris and its vicinity, 16 are spread out all over the country), mostly academic and research libraries that specialize in various disciplines or documents, from medieval art to biomedical engineering, from comics to posters. The library partly helps its partners financially.

Improving access to resources is the other action led by the Bibliothèque Nationale de France in cooperation with other libraries for the benefit of scholarship and research: The French Union Catalogue (Catalogue Collectif de France) was inaugurated on the Web in May 2001 and is composed of a National directory of libraries and document centers (Répertoire National des Bibliothèques et Centres de Documentation, RNBCD) providing detailed practical and technical information on more than 3500 French institutions and of 13 million bibliographic records from three major databases: the Bibliothèque Nationale de France's print collection catalog (BN-OPALE PLUS), the academic libraries union catalog (SUDOC), and a reservoir of bibliographic records resulting from a retrospective conversion program for handpress books held by city libraries. This recon program was also partly funded by the Bibliothèque Nationale de France.

Other services, such as shared digitization programs or remote document supply, are also developed.

Dealing on a large scale with most of librarianship issues, the Bibliothèque Nationale de France also shares resources and expertise with other similar institutions within international organizations. It is a member of various organizations promoting French language around the world and helping French-speaking countries to do so. Through a continuing professional education program, it provides French-speaking librarians from developing countries with training in particular domains of librarianship and in new fields of technology. An active cooperation network was also built in the past decades with French-speaking Canadian librarians, mostly from the Bibliothèque Nationale du Québec (BNQ) and the University of Laval to share information, documentation, internships, and bibliographic records. Translating English-speaking library reference tools such as subject headings, ISBD, Dewey classification, and editing a French-speaking national libraries directory (Le Répertoire des Bibliothèques Nationales de la Francophonie) providing information on national libraries or equivalent institutions in not less than 43 countries, are some of the main actions led in that field of cooperation.

Before it was even built, the Bibliothèque Nationale de France had a mission to "forge links with other European libraries." Many collaborative projects are initiated by the Conference of European National Librarians (CENL) and funded by the European Commission. Among their programs, the Bibliothèque Nationale de France plays an active part in Biblink (Linking Publishers and National Bibliographic Services), Bibliotheca Universalis (a G-7 global information society pilot project), CD-BIB (National Libraries Project on CD-ROM), CoBRA (Computerized Bibliographic Record Actions), EROMM (European Register of Microform Masters), NEDLIB (Networked European Deposit Library), and RENARDUS (a project that aims at improving access to existing Internet-accessible services across Europe) and participates with various other organizations (The Ligue des Bibliothèques Européennes de Recherches, LIBER; the Consortium of European Research Libraries, CERL; the European Bureau of Library, Information, and Documentation Associations, EBLIDA) in actions promoting the interests of research libraries and academic users and mainly consisting in the making of shared catalogs and bibliographies.

Last, but not least, collaborating with libraries worldwide is another role of a national library. Within various specialized organizations, the Bibliothèque Nationale de France makes its contribution to international research on specific librarianship issues: computer files ISBD, permanent UniMARC, preservation and conservation within the IFLA, Russian and Mediterranean libraries projects in the UNESCO, bibliographic standards within the International Standardization Organization (ISO), and the International Standard Serial Number (ISSN), and on more specific collection issues within (e.g., the International Federation of Film Archives (FIAF)), the International Association for musical libraries (AIBM/IAML), or the International Council of Museums (ICOM).

The last decade of the twentieth century has been an important turn in the life of the French national library. The Tolbiac project played a pivotal role in the institution's growth and helped it to make up for lost time. The impact of new technologies in knowledge and collection management mainly consisted of the improvement of services, both inside the library and outside its physical frontiers. As any other national library, the Bibliothèque Nationale de France has now both dimensions of a keeper of the past and an initiator of innovating projects.

BIBLIOGRAPHY

From The Bibliothèque Royale to the Bibliothèque Nationale

1. Balayé, S. *La Bibliothèque Nationale des Origines à 1800*; Droz: Genève, Switzerland, 1988.
2. Balayé, S.; Foucaud, J.-F. La Bibliothèque Nationale (1800–1914). In *Histoire des Bibliothèques Françaises*; Editions du Cercle de la Librairie, Promodis: Paris, France, 1991; Vol. 3, 296–355.
3. Blasselle, B. *La Bibliothèque Nationale*; Presses Universitaires de France: Paris, 1993. Collection "Que sais-je?" 2496.
4. Blasselle, B.; Melet-Sanson, J. *La Bibliothèque Nationale de France, Mémoire de l'Avenir*, Updated Ed.; Gallimard: Paris, France, 1996. Collection «Découvertes Gallimard».
5. Cain, J. *Les Transformations de la Bibliothèque Nationale et le dépôt annexe de Versailles*; Editions des Bibliothèques Nationales: Paris, France, 1936.
6. Cain, J. *Les Transformations de la Bibliothèque Nationale de 1936 à 1959*; Editions de la Déesse: Paris, France, 1960.
7. Dennery, E. Bibliothèque Nationale de France. In *Encyclopedia of Library and Information Science*; Marcel Dekker: New York, 1969; Vol. 2, 435–448.
8. Duchemin, P.-Y. La Bibliothèque Nationale (1945–1975). In *Histoire des Bibliothèques Françaises*; Editions du Cercle de la Librairie, Promodis: Paris, France, 1992; Vol. 4, 366–379.
9. Duchemin, P.-Y. La Bibliothèque Nationale (1975–1990). In *Histoire des Bibliothèques Françaises*; Editions du Cercle de la Librairie, Promodis: Paris, France, 1992; Vol. 4, 682–697.
10. Dupuigrenet-Desroussilles, F. *Trésors de la Bibliothèque Nationale*; Nathan: Paris, 1986.
11. *Etude sur la Bibliothèque Nationale et Témoignages Réunis. Hommage à Thérèse Kleindienst*; Bibliothèque Nationale: Paris, France, 1985.
12. Foucaud, J.-F. *La Bibliothèque Royale sous la Monarchie de Juillet*; Bibliothèque Nationale: Paris, France, 1978.
13. Kleindienst, T. *La Bibliothèque Nationale. Histoire, organisation, fonctions. Cours destiné aux Élèves de l'École Nationale Supérieure des Bibliothèques*; Paris, France, 1970.
14. Kleindienst, T. Les Transformations de la Bibliothèque Nationale (1914–1945). In *Histoire des Bibliothèques Françaises*; Editions du Cercle de la Librairie, Promodis: Paris, France, 1992; Vol. 4, 84–113.
15. La Bibliothèque du Roi, Première Bibliothèque du Monde (1664–1789). In *Histoire des Bibliothèques Françaises*; Editions du Cercle de la Librairie, Promodis: Paris, France, 1988; Vol. 2, 209–233.
16. La Bibliothèque Nationale. Beaux-Arts Mag. **1993**. special issue.
17. Mortreuil, T. *La Bibliothèque Nationale, son Origine et ses Accroissements jusqu'à nos jours*; Champion: Paris, France, 1878.
18. Pastoureau, M. *Bibliothèque Nationale, Paris*; Albin Michel, Bibliothèque Nationale: Paris, France, 1992. Collection «Musées et monuments de France».

19. *Revue de la Bibliothèque Nationale*; Bibliothèque Nationale: Paris, France, 1981–1994.
20. Vallée, L. *La Bibliothèque Nationale. Choix de Documents pour Servir à l'Histoire de l'Établissement et de ses Collections*; E. Terquem: Paris, France, 1894.

The Bibliothèque Nationale de France

1. Bélaval, P. *Rapport du groupe de travail sur l'avenir du site Richelieu-Vivienne, remis le 30 Juin 1993 à Monsieur Jacques Toubon, ministre de la Culture et de la Francophonie*; Paris, France, 1993; 23–25.
2. Bélaval, P. *Rapport du groupe de travail sur la mise en service de la Bibliothèque de France à Tolbiac, remis à Monsieur Jacques Toubon, ministre de la Culture et de la Francophonie, le 30 juin 1993*; Paris, France, 1993; 24-3–24.20.
3. Bibliothèque Nationale–Bibliothèque de France: Où en sont les grands chantiers? Bull. Bibl. Fr. **1993**, *38*, 3.
4. Bibliothèque Nationale de France. In *Encyclopedia of Library and Information Science*; Marcel Dekker: New York, 1969; Vol. 2, 435–447.
5. Jamet, D., Ed. *Bibliothèque Nationale de France, Premiers Volumes*; Institut français d'architecture: Paris, France; Carte Segrete: Rome, Italy, 1989.
6. *Bibliothèque Nationale de France 1989–1995: Dominique Perrault*; Arc en Rêve Centre d'architecture: Bordeaux, 1995.
7. Bibliothèque Nationale de France. In *Au Seuil du Vingt-et-unième Siècle*; Bibliothèque Nationale de France: Paris, France, 1998; 69.
8. Bibliothèque Nationale de France. In *Into the Twenty-First Century*; Bibliothèque Nationale de France: Paris, France, 1998.
9. Bibliothèque Nationale de France. In *Rapport d'activité 2000*; Bibliothèque Nationale de France: Paris, France, 2000.
10. Bibliothèque Nationale de France. In *Projet d'établissement 2001/2003*; Bibliothèque Nationale de France: Paris, France, 2001.
11. Cahart, P.; Melot, M. *Proposition pour une grande bibliothèque, rapport du Premier ministre*; La Documentation française: Paris, France, 1989.
12. Conseil Supérieur des Bibliothèques. In *Rapport à Monsieur le Président de la République sur la Bibliothèque de France*; Conseil supérieur des bibliothèques: Paris, France, 1992.
13. Établissement Public de la Bibliothèque de France. In *Le projet*; Établissement Public de la Bibliothèque de France: Paris, France, June 1990.
14. Établissement Public de la Bibliothèque de France. In *Rapports des Groupes de Travail 1991*; Établissement Public de la Bibliothèque de France: Paris, France, November 1991.
15. Gattégno, J. *La Bibliothèque de France à mi-parcours: de la TGB à la BN bis?* Editions Cercle de la Librairie: Paris, France, 1992.
16. Jackson, W.V. The French National Library, a special report. World Libr. **1999**, *9* (1), 3–30.
17. Kessler, J. The Bibliothèque François Mitterand of the Bibliothèque Nationale de France: Books, Information and Monuments. In *Building Libraries for the 21st Century, the Shape of Information*; T. Webb, Ed.; Mc Farland and Company: London, U.K., 2000; 197–230.
18. La Bibliothèque Nationale de France Vis à Vis **1995**. special issue 2.
19. La Bibliothèque Nationale de France Connaiss. Arts **1996**, special issue 99.
20. La Bibliothèque Nationale de France, première étape. Bull. Bibl. Fr. **1997**, *42*, 6.
21. *L'Avenir des Grandes Bibliothèques, Colloque International Organisé par la Bibliothèque Nationale, Paris, January 30–February 2, 1990*; Bibliothèque Nationale: Paris, France, 1991.
22. *Les Grandes Bibliothèques de l'Avenir, Actes du Colloque International des Vaux de Cernay, June 25–26, 1991*; La Documentation française: Paris, France, 1992.
23. Mandosio, J.-M. *L'Effondrement de la très Grande Bibliothèque Nationale de France: ses Causes, ses Conséquences*; Edition de l'Encyclopédie des nuisances: Paris, France, 1999.
24. Renoult, D. The Bibliothèque Nationale de France: A National Library for the 21st Century. In *Building Libraries for the 21st Century, the Shape of Information*; T. Webb, Ed.; Mc Farland and Company: London, U.K., 2000; 231–264.

Binding [ELIS Classic]

Lawrence S. Thompson
University of Kentucky, Lexington, Kentucky, U.S.A.

Abstract

Lawrence Thompson has written extensively about bibliography and the history of books and printing. He contributed many articles to the first edition of *ELIS*, including this one on Binding. The binding of books is one of several book arts, as Thompson's discussion reveals.

—*ELIS Classic*, *from 1969*

INTRODUCTION

The binding of books is both a craft and an art. The forwarding of books (all the processes of binding before lettering and decoration) is a craft requiring a high degree of technical skill and a considerable fund of information about the materials and processes used in book production during all periods of history. Finishing (decoration of the cover in various styles) is an art in which some gifted binders of the past have attained substantial heights of achievement, comparable to artistic creativity in other media.

OVERVIEW

A binder's equipment and tools must be kept in an orderly manner, for a job can easily be ruined in a poorly organized shop. Minimum equipment consists of the bone folding stick (in constant use for many purposes), a pair of 6-inch scissors, a paste pot (earthenware is customary), a glue pot, which will hold about a pint, a glue brush (round, about 1 inch), a flat paint brush for the paste pot (about 1 1/2 inch), a cobbler's knife, a package of bookbinder's needles, a wooden sewing frame, a flat piece of marble or stone about 10 × 15 inches for doing any work which requires a firm surface (old toilet partitions are ideal), a lying press made of beechwood blocks and moved with wooden screws, a plough for cutting edges in the lying press, a backing hammer, a cheap board cutter, and, if possible, a standing press (in which to press books for drying, thus freeing the lying press for routine work in progress). A power cutter, two or three "finishing" presses, a tooling stove (for the finisher), an electrical paring machine, and a power-driven device for sharpening knives and shears are examples of other equipment that the more experienced hand binder may need.

When a binder receives a volume, his first task is collation, and if he deals with printed books from the fifteenth through the eighteenth centuries, he is likely to need to know something about descriptive bibliography. Further, the folding of printed sheets is a process that requires meticulous care and competence in collation. If a book is to be rebound, it must be "pulled" (i.e., separate cover from text and remove old cords and tapes and the old glue, often a tedious process). If folds of old books are damaged, a "guard" (a narrow strip of paper used to repair a damaged fold, or to attach a plate or a single leaf) may be necessary. Special techniques are involved in guarding and mounting maps and folded sheets. A similar problem may arise in the case of pages of new books that are incorrectly printed. These pages must be cut out from the entire sheet, or "cancelled," and the corrected page supplied by the printer must be tipped in.

Before sewing, the pages of a book must be pressed firmly, both to expel air and to crease the folds of the sheets. If there is to be no edge-decoration or "rough-gilding," books may be trimmed before sewing. However, a book must be "cut in boards" if there is to be a smooth gilt edge, a marbled edge, gauffering, or a fore-edge painting (marbling is largely passé, gauffering and fore-edge painting are now essentially *tours de force*). Gilding smooth edges (see below) is a highly specialized technique, and application of gold leaf is a tedious and delicate operation.

There are various styles of sewing. Most primitive of these is stabbing, and the earliest codices were held together in this manner. In Western Europe and America small items were sold without covers and stab-stitched through the eighteenth century; and some modern American textbooks with wide inner margins are still sewn in this manner. With the rise of parchment as a writing material in the early Middle Ages, a much tougher material, which could be folded, was available. Threads could be put through holes in the folds. This process of sewing on raised thongs or cords goes back to the tenth century in northern Europe. Today large books (e.g., pulpit bibles) are generally the only works sewn in this manner, although imitation raised bands are commonly used to lend the illusion of luxury. Recessed thongs or cords were

Encyclopedia of Library and Information Sciences, Fourth Edition DOI: 10.1081/E-ELIS4-120008968

first used in the sixteenth century as a step toward automation, since it is quicker and cheaper than raised-cord sewing (not for the aesthetic qualities of a smooth spine). "Two-sheets-on" sewing, a device in which each length of thread holds two sections instead of one, was used from the beginning of the seventeenth century until the latter part of the nineteenth, when sewing machines began to be used widely in commercial binderies. It is to be contrasted with the method of passing the thread continuously through all sections, from beginning to end, without a break. Oversewing (also called "overcasting" or "whipstitching") was developed in the eighteenth century to hold heavy plate books together firmly. It is a method for making a series of sections, then sewing all along them. Oversewing machines are the mainstays of modern library binders. They have produced many an expendable book that has been circulated far more times than a hand-sewn book would have been used; but it has also resulted in the virtual destruction of many fine books sent out for rebinding by injudicious library binding assistants.

Regardless of the sewing method used, the sewn book must be "knocked-up" to be square at head and spine, and the expansion resulting from the sewing can be reduced by hammering along the back edges. The spine is glued with a thin hot glue (still preferable to the yet untested plastic product) diluted with some insecticide such as DDT, the book again "knocked-up" to round the spine, and the spine lined. Preparation of cover boards for the case requires exact measurements, lining-up (to prevent the pigmentation of the boards from showing through the end papers), marking the boards for sewing, and marking the boards for tapes.

End papers (sheets before and after the text of a book) may or may not be decorated. The classic type of decorated end papers is marbled, presumably originating in Japan in the twelfth century A.D., and reaching Europe via Persia and Turkey. By the end of the sixteenth century marbled paper was used in the Netherlands, and its use spread eastward and across the Channel in the next century. The oriental *serefsan* (gold-flecked) papers, used for end papers as well as texts, is a variation of marbling that has been given little attention, although *serefsan* has been used imaginatively by many oriental calligraphers and binders. "Dutch gilt" (or "Dutch flowered") papers were produced in Germany and Italy in the eighteenth century, imported into the Netherlands, then exported to France and England. They were not marbled, rather printed by engraved rolls, wood-blocks, or incised metal. In the twentieth century, Douglas Cockerell[1] and a few Parisian and Scandinavian binders (especially Ingeborg Börjesson) have produced marbled papers of distinction.

Commercial binders often use maps, engravings, or pertinent photographs (generally reproduced by offset) for end-paper illustration. A classic variation of the end-paper is the doublure or lining of silk, leather, or other material in lieu of end sheets. The Islamic doublure of the fifteenth century had a strong influence in Europe and was well known on the continent beginning in the sixteenth century. Leather doublures were common in England and Western Europe from the latter part of the eighteenth century on (see Boyet, below). Most doublures are plain, although fairly elaborate borders are common. Since the first and last sections of a book are the weakest, and since elaborate doublures need added protection, leather hinges, either sewn in with the end papers or pasted down on the end papers, are usual in bindings with doublures. Linen hinges are often used for more common bindings.

Trimmed edges may be handled in different ways. In the middle ages each leaf was trimmed separately. In the sixteenth century the proliferation of books through printing required smooth trimming of edges. At this time we have the beginning of edge-painting (e.g., the Pilone Library; not fore-edge painting as yet, simply edge-painting), and gauffering (decoration of the edge with a tool on the gilded edge). In the middle of the seventeenth century the fore-edge painting (an image under a smooth gilt edge, visible only when fanned out) came into vogue. Floral, heraldic, and landscape themes were popular with fore-edge painters of the eighteenth and early nineteenth centuries. The art has become less popular in the twentieth century, but it is still practiced with vigor and imagination by such artists as the elder Joseph Ruzicka and Kenneth Hobson. Edge marbling was common in the eighteenth century, by the same method as the one used for marbling flat paper, by transfer via solution in muriatic acid, or by rollers. Sprinkling, most frequently with Venetian red, has been used in Europe since the sixteenth century and is still a characteristic of law books. "Rice marbling" is a similar process.

Traces of headbands may be found as early as the seventh or eighth century in the well-preserved Stonyhurst Gospel. The headband as we know it in modern books is in colored silk or thread. Beading shows at the bottom, and turned-in leather of the spine forms a hood. Sixteenth-century binders used this style, and it is still in vogue. In the nineteenth century the more sought-after hand binders used fine silks and elaborate patterns, and some quite handsome work was produced. Stuck-on headbands have been used since the sixteenth century, but the best hand binders of all periods have sewn them on to the spine.

Up to the nineteenth century most books were bound with "tight backs" (covering material attached firmly to the spine, or sandwiched between cover and spine when there are additional layers of material). "Hollow backs" permit leaves to throw up in the middle, and the outer spine is not folded or broken. Hollow backs began to be used widely in Europe at the end of the eighteenth century and became the order of the day after about 1820. "Spring backs," "capable of retaining a firm situation" (patent granted to John and Joseph Williams, ca. 1799), are used in ledger bindings.

Leather was the primary covering material for books until the third decade of the nineteenth century, when commercial publishers began to use cloth on a production-line basis. Goatskin, sheepskin, and calfskin have been the usual materials, although other sources of leather are known, ranging from weasel skin for a set of Lenin's collected works, to human skin (e.g., from Confederate soldiers, used by a Unionist Army surgeon to bind his texts and manuals, now in the College of Physicians and Surgeons in Philadelphia). Calf was popular in the sixteenth century, sheep in the seventeenth and eighteenth. Careless tanning methods have resulted in the virtually complete deterioration of hundreds of thousands of bindings. Straight-grained morocco was widely used in the eighteenth and early nineteenth centuries and is still in use today, since it does not show scratches and other damage as plainly as other leathers. Russian leather, originally prepared in and imported from Russia, is made from calf or cowhide; it was widely used in England, North America, and Western Europe since about 1730.

Embroidered covers were used for fine bindings in the Tudor and Stuart periods in England, possibly even as early as the late fourteenth century (Anne de Felbrigge's putative binding of her manuscript Psalter). Jeweled bindings, based on metal plates covered with enamel or with plaques in ivory or other metals, and in-laid with jewels, were the deluxe book covers in France and Germany from the sixth to the fourteenth centuries. In the twentieth century Sangorski and Sutcliffe and a few other hand binders have produced jeweled bindings, but tax laws and customs "finkishness" have discouraged hand binders from working in this area.

Clasps are a practical device for binders who use vellum, a material likely to curl in almost any climate. However, all books and leathers may curl, and clasps are practical for all types of bindings. Far Eastern and Islamic binders have used an ivory pin on a strong ribbon, to be inserted into a loop on the front cover. Some surviving Coptic bindings have a simple strap that can be fitted to the fore-edge of the front cover. Holes and pins were used in Europe until the fourteenth century, when metal clasps and catches began to be used. A significant key to bibliographic description is the position of clasps and catches: in the Low Countries and the Germanies the clasps were fitted to the bottom cover, the catch on the front cover. In France and England these positions were reversed.

Bosses were put on covers in the Middle Ages to save the decorations from wear and tear resulting from pushing over tables or shelves, or possibly from an overflow of sacramental wine (German student *Kommersbücher* of the nineteenth century were provided with bosses to elevate the covers a millimeter or so above the beer spilled on the *Stammtisch*). A related practical device was the chain, and *catenati* are still to be seen at Wimborne and Hereford. The length of chains ranged from three to five feet. Chains served less to prevent theft than to avoid abuse of books.

The earliest technique of finishing is blind tooling, impression of tools on leather (generally wet), without use of gold or color. Coptic bindings of the seventh century show evidence of widened blind tooling. Blind tooling was always in use; and all surviving romanesque bindings of the twelfth and thirteenth centuries are in blind, with the exception of a French binding of the early thirteenth century on which there are traces of gold. Gothic bindings of the fifteenth and sixteenth centuries reveal much greater variety of stamps and rolls. A peculiar laborsaving technique developed at this time was the large panel stamp, a device that often provided a convenient medium for imaginative artistic expression. The panel stamp originated in the Netherlands and was used there throughout the fourteenth century, and during the fifteenth and first half of the sixteenth centuries it was used widely by binders all over Northern Europe. Impressions were almost invariably in blind. A related device was developed in commercial binderies of the late nineteenth century, the "butter stamp," which could be applied to the whole spine of a book and decorate it with one impression.

Related to the panel stamps as a decorative device is the supralibros, or superexlibris, stamped in the center of the front cover to indicate ownership. It was used in France from the sixteenth century on, and it spread from that country to the rest of Europe. Generally the supralibros is the armorial device of the owner, but it may also be a monogram or some other device indicating ownership. Unfortunately, some armorial bindings are forgeries, since a binder might retain the stamp and use it on books not belonging to the rightful owner of the device. When provenance lends value to a book, binders with a low level of scrupulousness have been tempted to make improper use of a supralibros stamp. The supralibros is generally tooled in gold leaf.

Blind tooling has had spurts of popularity with certain binders or schools (e.g., the "ecclesiastical" or "divinity" style in Victorian England), but after the middle of the sixteenth century it was largely restricted to cheap leather bindings on trade books. Another finishing technique, practiced mainly in Germany during the Gothic period, but still used occasionally, is *cuir ciselé* (also called *cuirbouilli;* German, *Lederschnitt*; incised leather). Under this method no elaborate dies were needed, for the leather was simply cut and then worked so that it stood out in relief.

Gold tooling *à petits fers* seems to have been a Moorish invention developed in Córdoba and was probably imported into Christendom via Naples in the late fifteenth century. Applying gold leaf with a hot tool was not a Venetian art, for binders in Venice simply followed the usual oriental manner of painting the impressions made by blind tooling. In any event, the art of gold tooling was used by Italian binders in the early sixteenth century and quickly spread to France. In both countries, bindings produced in this period show a high level of artistic achievement. Gold tooling did not spread to the Germanies and other parts of Europe until somewhat later. Much early gold tooling was rather crude, but techniques improved gradually. Today virtually every design is drawn or tooled on thin paper, which is then positioned on the cover as a guide for the heated tools. It is said that the great English binder Roger Payne (1738–1797) first used paper patterns, and the accuracy of his tooling would tend to confirm this attribution. However, they did not come into common use until a generation after Payne's death, and we cannot safely give him certain credit for this invention.

Gold leaf is applied after the leather has been neatly smeared with glaire (or glair, white of egg). The dry leather is greased and covered with gold leaf, and the pressure of the hot tool causes the gold to adhere to the leather by the coagulation of the glaire. Pure egg white is generally altered with various other liquids, ranging from spirits of wine to vinegar, ammonia, salt, and camphor to avoid putridity. Today dried albumen is most generally used, since it works as well as fresh eggs and is far more convenient. Among oils used for binding, hog lard, palm oil, and Vaseline have been used. (But neither Vaseline nor any other mineral oil should ever be used for rubbing bindings.) Some soft unabrasive material such as flannel lightly sprinkled with olive oil should be used to clean off the leather after tooling.

"Smooth gilding" (in contrast to "rough gilding," see above) is an operation much like gold tooling and appeared on the European scene about the same time or a little later. It requires consummate patience and exactness of technique, and in large shops there are gilders who do nothing else. An essential step in edge-gilding is careful burnishing with an agate burnisher once the gilding is thoroughly dry. The purpose of gilding is protection of a book from the penetration of dust, and it is essential for the protection of fore-edge paintings.

There are devices other than tooling in gold or blind for decorating book covers, and we even have such bizarre examples of decoration as the Argentine *nonato* binding (skin from the foetus ["not born"] of an Argentine pampas calf, with hair left on). There are highly sophisticated uses of intarsia (inlay), a technique used as early as the 1560s by a French binder for achieving polychromistic effects with various colors of leather. In the twentieth century inlay has been used with grace and imagination by a number of master binders. Much "inlay" is actually "onlay," since the base leather is not cut, but rather impressed with the design and the impressions filled with cutout pieces of the desired color.

Certain other devices for decorating leather should be noted (these are mostly of British origin). "Tree calf" is a marbled calf on which the pattern resembles the trunk and branches of a tree, and the earliest example comes from England around 1775. Tree calf has not been used extensively in the twentieth century, perhaps, in part, because the marbling process is a cause of deterioration of leather. "Etruscan calf" received its name from the resemblance to the "black figure" pottery of the Etruscans with contrasting shades or colors of the decoration. Most of the Etruscan bindings have a classical urn in the center surrounded by Greek palmated leaves and outer borders reminiscent of Hellenic entablature. They were popular in the quarter of a century on either side of 1800. "Landscape bindings," used in the same period, show landscapes on the covers, generally freehand drawing with India ink or acid, or possibly stenciled or otherwise printed, also come from this period.

"Illuminated bindings" originated in France and were popular in the mid-nineteenth century. They show designs in both gold and colors. Another process for showing designs in color is the "Sutherland tooling process," patented in 1896 by G.T. Bagguley of Newcastle-under-Tyne, and while it achieves rather striking effects, it is too delicate to be handled and is generally restricted to doublures. It was used originally on books bound for the Duchess of Sutherland. Richard Cosway (ca. 1747–1821), the noted British miniaturist, lent his name to "Cosway bindings," leather bindings with miniature paintings inserted in their covers. Actually Cosway bindings were introduced only in the beginning of the present century by Rivière and Son. A talented Miss Currie copied the miniatures, but the bindings were designed by Mr. Stonehouse.

Binding materials other than leather offer special problems to the finisher. Vellum cannot be tooled in blind or in gold with much success. Silk, actually too fragile for a practical binding, may be tooled, but care must be taken not to stain the material with glaire or to scorch it with a hot tool. Velvet can be decorated only with large tools, since it has such a deep pile. When unusual or exotic materials are used (e.g., tapa cloth for a James Michener book, snake skin for a Raymond Ditmars book), decoration should be restricted to pasted-on spine labels, since the natural design lends the desired effect.

Lettering is a late development, for in the Middle Ages and even into the sixteenth century, most books were shelved on their backs. The fore-edge may have been lettered, or a piece of vellum inscribed with the author and title was glued on the back (the upper side when the book was shelved with the fore-edge to the wall). Late in the sixteenth century the increased size of libraries required that books be shelved upright to economize on space, and the more elaborate decoration of spines suggested that the books could also be decorative household furnishing. In the first half of the sixteenth century gold-tooled spine lettering came from Italy and soon thereafter from Paris, then spread to England and the Continent. Tooled lettering did not become common until the latter part of the seventeenth century. Lettering was done with large individual tools (sometimes resulting in curious abbreviations), but in the late eighteenth century the typeholder and lettering pallet began to be used. About the same time printed paper labels were used widely, but commercial binders abandoned them when gold blocking on cloth came into general use in the 1830s.

Binding–British

Restoration and preservation of bindings is an entire discipline in itself. Such official or quasi-official agencies as the Institute for Restoration and Preservation of the Akademiia Nauk in Leningrad or the Istituto de Patologia del Libro in Rome, private investigators such as W.J. Barrow of Richmond, Virginia., libraries such as the New York Public Library (John Archer and H.M. Lydenberg), organizations such as the Library Binding Institute, and many individual binders have conducted productive studies in this area. The definitive manual on preservation and restoration is still lacking. At this point it is sufficient to issue a few caveats: Never attempt homemade restoration with miscellaneous adhesives and adhesive tapes. Never use any binder or restorer who is not recommended by one of the major rare book libraries. In general, do not rebind or attempt restoration unless the book is subject to further deterioration in its present condition. If a noble book suggests the need of a noble binding, consider a chemise inserted into a tastefully decorated slip case, leaving the book in its original state.

At this point a word about slip cases (sometimes called "thumb cases"), Solander cases, and chemises is in order. These devices are far and away the best means of preservation, for they do no injury to the book as it has been handed down through the years, and they also protect it from further damage and offer an opportunity to the binder to pay homage to a fine volume. Meticulous care must be exercised in measurement for slip cases, and it is essential that they be constructed to permit easy removal of fragile books. They may be in full leather, half leather, paper (plain or decorated), cloth, or most any other material. Leather spines—but not others—may have imitation raised bands. Leather and half-leather slip cases may be tooled in the same way as a leather binding. The Solander case, named for its inventor, Dr. Daniel Charles Solander, Anglo-Swedish naturalist on the staff of the British Museum in the 1760s and 1770s, differs from the slip case in that books are inserted from one end, not from the foreedge. It may be covered in the same manner as a slip case and decorated if covered in a suitable material. A chemise, or a protective cover, lined with muslin or some other soft material, should be fitted over any book worthy of a slip case or Solander case before the case is designed. Slip cases or Solander cases are useful for vellum bindings, susceptible to curling even under the most careful temperature and humidity control.

HISTORY OF HAND BINDING

The history of bookbinding as we know it begins with the Roman diptych, two facing receptacles for wax tablets, or *pugillares* connected at the back with rings. Diptychs were made of ebony, boxwood, ivory, or some other hard material, and the outside covers were sometimes elaborately carved. The text proper was incised with a stylus on the wax surface. There were also triptychs, including three *pugillares*, and some containing even more.

The codex form of the book is associated with the rise of Christianity. More specifically, the Coptic bindings from early Christian communities in Egypt reveal most of the basic characteristics of the binding of codices, and possible future discoveries may tell us much more than we know at present about the genesis of binding practices. Only fragments of the earliest bindings used by the Coptic Church in Egypt survive, but they are sufficient to tell us a great deal about the origins of the binding of the codex. Surviving examples of Coptic binding, mainly from the ninth, tenth, and eleventh centuries and quite fragmentary, show the full heritage of Hellenistic ornament in blind tooling, *cuir ciselé*, and even rudimentary *pointillé*—the basic styles of the previous 1000 years, and the basic techniques of the next millenium.

Greek art as well as Greek science owes much to the Christian Syrians and Egyptians, who were tutors of the Moslem invaders of the seventeenth century. The sophisticated Coptic leathercraft was picked up at once by the Mohammedans and taken to Sicily and Spain along with other skills inherited from the orient (e.g., papermaking, marbling, gold tooling). The full story of Europe's artistic and intellectual legacies from Islam is still imperfectly known; but when it is told, the provenance of the technology and the artistic traditions of Western European binding will be much better understood.

Western Christendom had a tradition that went back to late Roman antiquity. In Europe as well as Africa the codex book was a necessary development, but techniques and decorative traditions were different. There were leather bindings on boards in the early Middle Ages, but we have few examples before the eleventh century. The Stonyhurst Gospel of the seventh or eight century, with rudimentary blind tooling on leather or birch boards, and the Victor Codex at Fulda are well-known examples of medieval binding in the conventional manner in Europe. A very small group of decorated Carolingian leather bindings has survived.

As early as the seventh century there is an example of a jeweled binding, the Gospel for the Empress Theodelinda in Monza. The jeweled binding, with gold filigree, pearls, carved ivory, and enamel was frequently produced from the ninth through the twelfth centuries for emperors and secular and ecclesiastical dignitaries, but more as an act of piety than as a display of artistic skill. Noteworthy examples are the Codex Aureus (presented to St. Emmeran by Emperor Arnulf in the ninth century; now in the Bayerische Staatsbibliothek) and the lectionary from the Bamberg Cathedral (a gift of Henry II, also in the Bayerische Staatsbibliothek).

More important are the romanesque bindings of the twelfth and thirteenth centuries, of which some 200 survive. The stamps represented biblical characters such as David, Samson, and the Virgin, as well as themes from classical mythology, such as centaurs and mermaids. A characteristic of the romanesque bindings is successive rectangular borders around a central square, rectangle, or circle. Romanesque bindings were produced in Germany, England, and France. There were two especially important shops in Paris in the neighborhood of the Sorbonne between 1135 and 1146.

Blind tooling, not only with individual dies but also with rolls and panel stamps (see above) reached a high level of artistic achievement in the fifteenth and sixteenth centuries when Gothic styles prevailed. Several unusual features of bindings during this period should be noted. Signed bindings [e.g., by the noted binderpriest, Johannes Rychenbach (Richenbach), chaplain of Geislingen] began to appear. Lettered inscriptions, such as a *Laus Deo* or an *Ave Maria*, on bindings suggest the rapidly developing concept of typographic printing. During the Gothic period we have quite elaborate inscriptions, giving even the full name and status of the binder and the year and the date when the work was completed.

The single stamps used in the Gothic period have been analyzed in meticulous detail by Ernst Kyriss,[2] and thanks to his labors, styles and trends of this period are better known than any other before the Renaissance. Two major types of arrangement of the stamps can be identified. One is characterized by compartmentalization of the surface of the binding by means of lines, with the small stamps arranged within the compartments. The other type showed a large central panel surrounded by smaller stamps arranged in a geometric design. The octavo book, which became quite common in the early period of typographic printing (in contrast to the quarto and folio volumes of the manuscript period) were especially well adapted to the use of panel stamps (of necessity, relatively small). A large proportion of Gothic bindings are of monastic or of some other type of clerical provenance.

The late period of the Italian Renaissance was the source of the seminal traditions of modern hand binding. The Venetians had intimate contact with the Levant, before and after the fall of Constantinople in 1453; and the Neapolitans had equal intimacy with the Moors of Spain and North Africa. Both traditions were related, and undoubtedly both had their effects in Italy, with interlacing, reticulation, narrow fillets, cable patterns, and tiny roundels, which characterized Italian bindings of the late fifteenth century. The introduction of gold tooling to Italy (see above) was a major step forward. The vigor and imagination of Aldus Manutius as a manufacturer of books, regardless of how much or how little he personally innovated, cannot be overemphasized. He and his successors used every idea that drifted into Italy with full effect. The relationship of typographic design and the finishing of bindings in the Aldine tradition is especially obvious in the Aristotle of 1495 and the famous *Hypnerotomachia* of 1499, in which knobs and borders in the bindings have a striking similarity to the decorations on initial letters in the text.

The styles, which we know from Venice, are inseparable from the bindings commissioned or collected by Jean Grolier de Servin (1479–1565), French diplomat and bursar in Italy and friend of Aldus Manutius. Most of his some 3000 books, many bound in an apogee of the early Venetian style, with exquisite gold-tooling, went to his son-in-law, Méry de Vic, Seigneur d'Erménouville, but they were later dispersed, and those authentic items with the inscription, *Io. Grolierii et amicorum*, are special treasures of any collector or library lucky enough to acquire them. Grolier was addressed by Geoffroy Tory as "amateur de bonnes lettres, & de tous personnages sauans," and some of his *amici* are listed in a famous quatrain:

Vatibus, historicis, addit calcaria; cunctos
    Scriptores fovet, his fertque patrocinium.
Musurus, Stephanus Niger, atque Thylesius, Aldus
    Lascaris, Arpinus, sunt mihi firma fides.

A successor of Jean Grolier in Italy was Thomas Mahieu (Maiolus), for whom bindings similar to Grolier's were executed. Some 90 are known and identified. The styles are quite similar to those of Grolier's bindings, with the characteristic interlaced strapwork, with impeccable gold tooling. In the same stylistic tradition are the so-called "Canevari bindings," once ascribed to the papal physician, Demetrio Canevari. Actually these bindings were most probably ordered for books presented by Cardinal Alexander Farnese (son of Pier Luigi Farnese, duke of Parma and secretary to Pope Paul III) for his nephew, Alexander Farnese, about 1545–1547 as a reference collection. Regardless of genealogical and political problems, the "Maioli" and "Canevari" bindings are superb examples of finishing techniques of the sixteenth century.

The heritage of Marc Lauwrin, or Laurin, Seigneur de Watervliet, cannot be overlooked. Classicist, antiquarian, and patrician of Bruges, he has left two or three examples of work in the Grolier–Mahieu–Canevari style. In the Louvre there is a binding inscribed "Io. Grolierus M. Laurino D.D."[3] "M. Lavrini et amicorum" appears on his few surviving bindings.

A method of decoration used in Italy in the sixteenth century was a sort of cameo binding. Relief impressions, like ancient cameos, were produced by stamping leather with dies cut in intaglio. A noble example may be found in the British Museum's Grenville Library: Celsus, *De Medicina* (Venice, 1497), inscribed by Grolier: "Est mei Io. Grolier Lugd. & amicorum." The workmanship on the cameos is almost flawless. One of the Canevari (*vice*, Farnese) bindings has a superb cameo of Apollo driving a chariot hitched to two horses toward a crag on which Pegasus is perched, the whole oval cameo being encircled by the inscription:

$$\text{OP}\theta\Omega\Sigma \quad \text{KAI} \quad \text{MH} \quad \Lambda \text{O}\Xi\Omega\Sigma$$

It is likely that some of the important surviving Italian bindings of the sixteenth century were produced in Lyon as well as in Venice and possibly Rome, Milan, and Naples. Toward the end of that century Italian binding deteriorated, along with other aspects of Italian art.

In France the Guild of St. Jean Latran, under the protection of the University and settled in its precincts, had a strong influence on the production and sale of books, not only in the Quartier de l'Université, but also throughout Paris and the rest of France. The guild was founded with a membership "en grant nombreur, riches et oppulenz" but it conservatively followed traditional styles, the Gothic in the fifteenth and early sixteenth centuries, the Italianate in the sixteenth. It was dissolved along with other Parisian guilds on March 17, 1791, although its influence had been almost nonexistent for over a century.

The early gold-tooled bindings of France date from the second decade of the sixteenth century [e.g., Thomas Linacre's copy of his translation into Latin of Galen's *Methodus Medendi* (Paris, 1519) presented to Henry VIII].—By the time of Francis I the influence of Italian binding traditions becomes obvious in France. A singularly fine example of this tradition is the binding of the *Vulgata* (Paris, 1538–1540; plate I in Marius-Michel,[4] *La reliure française*), of which the main centerpiece is Italianate, yet the main design is Gallic. A binder for Francis I was Pierre Roffet, le Faucheur, "relieur du Roy." Perhaps the most famous binder, illustrator, and printer of this period was Geoffroy Tory (ca. 1480–1533), educated in Italy and strongly influenced by Italian traditions. Tory was probably not a finisher, but he undoubtedly drew designs both for his illustrations and his bindings. It is doubtful that any other stationer of Western Europe in the sixteenth and seventeenth centuries conceived and produced work of comparable quality.

The bindings executed for Henri II and Diane de Poitiers bring into sharp focus the peculiar characteristics of French bookbinding (and French art in general) of the mid-sixteenth century. The crowned 'H' and the cipher of two 'D''s interlaced with the 'H,' surrounded by carefully arranged interlaced borders, is a high point of Renaissance art. The subtle contrasts of curved and angular forms, the intricate designs of the borders with interlaced fillets, and the suggestive devices and ciphers (H + D) of the king and his noble mistress are precursors of the baroque. There is a sharp contrast with the relative simplicity of the Grolier bindings. The collection of Henri II is largely preserved in the Bibliothèque Nationale; but that of Diane (Duchess of Valentinois after Henry succeeded his father in 1547) was dispersed, and only some 30–odd have been identified (Bauchart).

At the turn of the century a remarkable but short-lived dynasty of binders began with Nicolas Éve. He was binder in ordinary to Henri III and used the "semi" (strewn style, e.g., his binding of the 1581 Horace, an armorial panel surrounded by a field of fleurs de lis). Clovis Éve, possibly the nephew of Nicolas, worked for later noble patrons, above all Marguerite de Valois, Queen of Henri IV, in essentially the same style. His typical design consisted of a field broken by a series of ovals, with flowers and sprigs of oak or pomegranate in the middle (the so-called style "a la fanfare"). Generally there would be borders of palm branches. The front cover usually shows a shield with three fleurs de lis. The legend, *EXPLICATA NON ELUDET*, is on the lower cover [e.g., the *Caesar* (Paris, 1564) in the British Museum, c. 19. a. 15].

In the early seventeenth century a still unidentified compatriot of Cyrano, "Le Gascon," produced some exceptionally fine work, tooled with singular care, in geometrical designs. His "pointillé" style (tool forms in dotted outline) was characteristic of his work. Some have identified Le Gascon with Florimond Badier, but the evidence is not convincing. Badier's name appears first in 1630, and he was accepted as a master in the Guild of St. Jean Latran in 1645. Many of Badier's bindings survive in the Bibliothèque Nationale. Another seventeenth-century French binder whose name should be noted is Macé Ruette, to whom the invention of marbled paper is often incorrectly attributed. Some of the great collectors of the day for whom French craftsmen worked are Henri IV, Jacques-Auguste de Thou, Nicolas Claude Fabri de Peiresc, and Sir Kenelm Digby.

The great French tradition of binding established in the Renaissance has endured to the present time, and even today Paris is still the world's most important center of fine binding. Only a few of the more important figures may be mentioned. Luc-Antoine Boyet (or Boyer), active from about 1680 until his death in 1733, used handsome dentelle borders (small contiguous stamps at the edge of the book cover, forming a lacy pattern) and may have introduced the decorated leather doublure (see below). His special patrons were the Marquise de Chamillart and Count Hoym. His son, Ètienne, was binder and librarian of Prince Eugene in Vienna and also worked for Baron von Hohendorff. Probably a pupil of the Boyets was Augustin Duseuil, also a devotee of dentelle and doublure, but he was especially noted for his corner ornaments. Jean Charles Henri Lemonnier (d. 1782), binder for the Duke of Orléans, created some finely executed inlays in the style of *chinoiseries*. Antoine Michel Padeloup ("le jeune," 1685–1758) bound for Louis XV and the kings of Portugal. He used dentelle borders and inlay but also elaborate tooling such as feeding birds, not always in the best of taste, but technically well-nigh perfect. Mme. Pompadour was one of his best customers. Nicolas-Denis Derome ("le jeune," 1731–1788) was perhaps the greatest master of the *relieure à dentelle*. His bindings are often identified by a small bird, but this method should be used with greatest care, since Derome le jeune and Dubisson also used a bird tool. Both Padeloup and Derome were members of families whose other members won distinction as finishers.

During the Romantic period Joseph Thouvenin (1790–1834), a student of the brothers Bozérian (binders for Napoleon, among others), was the best-known artist of the period. His technique was almost flawless, his ideas about decoration sensitive and pertinent to content. He was the master of the "cathedral" style, which used Gothic architectural motifs for cover decoration. G. Trautz, a German, married the daughter of the binder Antoine Bauzonnet ("master of the fillet" or "the Boyet of the nineteenth century"). He continued the historical trends of the Thouvenin school, and he produced singularly handsome bindings (often hard to open) for such noted collectors of the period as Baron Rothschild. They are signed Trautz–Bauzonnet. Toward the latter part of the century and into our own time we have such personalities as Chambolle, Duru, Cuzin, Gruel, Lortic, and Mercier, but they exhibited greater skill as forwarders than as finishers. There are signs of a renaissance of decorated bindings in France, and another Le Gascon, Padeloup, or Derome may well be serving apprenticeship today in some Parisian atelier.

After the great age of the Grolier and Mahieu binders, Italian binding suffered an eclipse from which it has never fully emerged. Italian binding of the seventeenth and later centuries has been strongly influenced by the French and, in general, has shown little originality. However, there are some young binders in Rome and Florence who are showing much promise.

Spain has always been a deviant country as a result of the Moorish domination, and her binding traditions reflect this situation. Arabic influence is revealed in the predominance of interlacing strapwork, rings, and dots, with a few supplementing stamps (some Islamic, some Christian in theme). The "mudéjar" style of medieval Spain has many imaginative features that represent a significant period of art history. Just as in Italy, Spanish binding styles from the *siglo de oro* on have been strongly influenced by the French. In Ibero-America there have been some outstanding shops in the twentieth century in São Paulo, Buenos Aires, Santiago de Chile, and Mexico City, mainly staffed by binders trained in Germany and France.

After the Gothic period at least one truly distinguished binder emerged in the Germanies, Jakob Krause (1531–1586). For the last twenty years of his life he was court binder in Dresden and created handsomely designed bindings in pigskin with decorated rolls and panels impressed in blind, a few gilded vellum bindings (a difficult art in every respect), and monumental decorated bindings on calf, the latter influenced by French and Venetian styles. His stamps (many cut in Augsburg, where he worked for a lustrum before settling in Dresden) were designed with fantasy and imagination, cut with flawless technology. Among his apprentices were Daniel Wachsler, Gregor Schenk the Younger, Urban Köblitz, Hans Hermann, and, perhaps the most competent, certainly the most famous, Caspar Meuser (d. 1593), successor of Krause as Saxon court binder. The latter's son, Moritz, used tools in the family stock for bindings for the Saxon archives. In April 1945 an air raid by the U.S. Air Force on Dresden destroyed a large proportion of the Krause bindings in the Japanisches Palais (home of the Königliche Sächsische Landesbibliothek).

It would be an error of omission not to mention Peter Flötner (ca. 1485–1546) of Ansbach and Nuremburg, whose remarkable wood-engravings are known from many sixteenth-century sources (e.g., *Spruchgedichte* of Hans Sachs). In addition to his contributions as an illustrator, he made plaquettes for German cameo bindings, not of the quality of those of the Italian school, but of considerable merit from the standpoint of imaginative design. In addition to Augsburg, Munich and Vienna were also major centers of the binding craft, for the kings and emperors were zealous to assemble libraries bound in the best contemporary styles.

In the seventeenth century there were remarkable naturalistic flower stamps (Praque, Gotha, Augsburg) and dentelle borders (Heidelberg). Far too little is known of eighteenth-century German bindings, and we need more detailed investigations of bindings for Frederick the Great (Kraft and Roch) and Goethe (Lehmann in Berlin). In the nineteenth century some of the ablest German binders emigrated, e.g., Trautz and Bradel to Paris, Zahn to the United States, Zaehnsdorf to London.

In the twentieth century German hand binding has attained new distinction. Best known abroad, perhaps, is Ignatz Wiemeler, whose stern insistence on the essentials of individual craftsmanship and basic elements of design is reflected in the work of numerous students. Otto Dorfner, theorist, teacher, and creative binder, has had a similar influence; and Paul Adam[5] and Joseph Hoffmann belong in the same category as teachers and original artists. Fritz Helmuth Ehmcke (1978–1965), formerly professor in the Hochschule für bildende Künste in Munich, has had a strong influence on all aspects of German book arts. With F.W. Kleukens and G. Belwe he founded the Steglitzer Werkstatt. He has designed bindings, decorated papers and many typefaces in wide use. Franz Weisse (1878–1952), whose most productive years were spent as professor in the Kunstgewerkeschule in Elberfeld, was dedicated to the ideals of absolute simplicity in design. His relatively few bindings are among the best done in twentieth century Germany, but his influence in continental Europe is abiding. His home and his shop and most of his best work were destroyed one night in 1943 by an unprogrammed aerial raid. Despite the temporary setback of the period 1932–1945, hand binding and other arts, which advanced so rapidly in the Germanies from about 1890, on are now on the rebound.

English binding reached no special distinction before the eighteenth century, although there are high points. The Gothic binding traditions of the Continent crossed the channel, and many engraved fillets and stamps comparable or similar to those in use in the Germanies and the Low

Countries reappeared in England. Few pictorial panels were used, but heraldic devices appear frequently.

In the latter part of the sixteenth century, England had such distinguished collectors as Lord Lumley, the Earl of Arundel, the Earl of Leicester, and Archbishop Cranmer, and the tradition has been continued to the present day. Many employed the best domestic and continental binders, and some of their collectanea have been preserved in the British Museum and a few other great English libraries. Such collections in the British Museum as the Harleian or that of the Rev. Mr. C.M. Crachecode, both intimately associated with the binding traditions of their period, explain in part the wealth of the museum in historical hand bindings.

Gold tooling is believed to have come to England in the work of Thomas Berthelet, or Bartlet, about 1540, a full half-century after gold tooling was used on the Continent. Berthelet showed some imagination, but the Italian influence is obvious in all of his work. Berthelet was also a printer, succeeding Richard Pynson in 1530 as royal printer. In Europe, England, and the Americas there was an intimate connection between the printing and the binding trades, and both were practiced in many of the major printing houses. The book business was monolithic in the sixteenth and seventeenth centuries: all processes, from type founding through retailing of the final product, were often done by one firm.

An Elizabethan collector often compared with Jean Grolier is Thomas Wotton, whose bindings are inscribed, "Thomae Wottoni et amicorum." His books were bound either in France, or by French craftsmen imported to England. Some of the work executed for Wotton is decorated only by a central armorial design, but others have considerably more elaborate devices, often decorated with strapwork resembling the designs of Grolier's binders. Through the Civil War and the Commonwealth, English binding, like English art (save literature) in general, was undistinguished. Samuel Mearne (d. 1683), binder for Charles II, and his son, Charles, produced rich and elaborately ornamented bindings, largely based on French models, but marred by heavy hands of relatively unskilled finishers. The Mearnes themselves were probably administrators of a book production firm, not binders.

In the eighteenth century English hand binding came into its own. There was James Edwards of Halifax, who invented transparent vellum bindings; John Whitaker, master of the Etruscan binding; and, above all, that eccentric genius, Roger Payne. Payne's remarkable tooling with delicate small dies, his almost unbelievable precision, and his exact gilding may have been partially inspired by indulgence in the Bacchic tradition. He inscribed his handsome binding for Barry's *Wines of the Ancients* with an amusing bit of doggerel:

Falernian gave Horace, Vergil fire,
And barley-wine my British muse inspire.

Payne designed and cut his own tools and did most of his own work, assisted by his brother, Thomas, and later by a partner named Weir. He bound for a few distinguished collectors, notably Rev. Cracherode, Lord Spencer, and Hon. Thomas Grenville (Lord Spencer's collections being largely preserved in the John Rylands Library in Manchester, the Cracherode and Grenville in the British Museum). Payne's use of straight-grain morocco (perhaps the first example among English binders and his "diced" russia ["diced" by ruling the leather in diagonal lines]) are characteristic features of his work. His meticulous tooling, much in pointillé, has never been excelled. He used heraldic devices as center panels on order from customers.

English and North American binding of the mid-nineteenth century and later is artistically unimportant, but much solid work was done by Francis Bedford (1799–1883) and his associates, Charles Lewis and John Clarke. They produced flawless forwarding, but their finishing was a weak imitation of Renaissance styles. Robert Rivière (1802–1882), scion of a family of French artists, taught himself binding and settled in London in 1840, where he bound for the royal family, the Duke of Devonshire, and other collectors, and restored the Domesday Book. His nephew, P. Calkin (Rivière and Son since 1880), carried on the business and did much solid but highly conservative work for British collectors. Joseph Zaehnsdorf (1816–1886) learned binding in Stuttgart and Vienna and settled in London in 1842. He attracted a substantial following among English collectors, and the reputation was maintained by his son, Joseph William (1853–1930), who continued a tradition of dignified but stolid finishing. He was court binder for Edward VII and George V. Francis Sangorski (1875–1912) and George Sutcliffe (1878–1943) formed a firm in 1901 in a Bloomsbury attic and have produced some usual inlaid and jeweled bindings (see above) along with much basic routine work with simple decorations. The Sangorski and Sutcliffe bindings of the *Rubáiyát* of Omar Khayyám with the peacock design are particularly well known.

Thomas James Cobden-Sanderson (1840–1922), unhappy at the bar, took seriously the suggestion of Mrs. William Morris that he go into bookbinding, and he learned the rudiments of the craft in the shop of Roger de Coverly. Cobden-Sanderson shared the belief of William Morris that his age was one of pointless imitation, however skillful, of older styles (e.g., by Bedford, Rivière, and Zaehnsdorf). Imaginative and inventive, he did much to establish the concept of the innate quality of individual craftsmanship in contrast with mass production. While Cobden-Sanderson was certainly mistaken in many of his ideas about modern industrial tyranny, he did contribute a great deal to the craft by insisting on sound workmanship, the use of good materials, and originality in design. An apprentice at Cobden-Sanderson's Doves Bindery, Douglas Cockerell (1870–1945), put many of his master's ideals into practice, but Cockerell showed a vast fund of common sense which was almost a closed book to Cobden-Sanderson.

In the United States, English traditions of hand binding have always been the major influence. When the Club Bindery (see Figs. 1, 2, and 3) was flourishing in New York around the turn of the century, both English and French craftsmen were on the staff. Some of the best binding in America has been done at the great printing houses which maintain extra-binding departments (e.g., The Cuneo Press with George Baer, or R.R. Donnelley and Sons with Harold Tribolet) and at certain libraries which maintain hand binders [e.g., Margaret Duprez Lahey at the Pierpont Morgan Library, Ferdinand Zach at Catholic University (see Fig. 4)]. German emigré binders, who have come to both hemispheres of the Americas since the 1930s, have also made a substantial contribution to the craft in the New World.

The medieval binding styles in the Low Countries were close to those of the Germanies. In the seventeenth century one Magnus of Amsterdam was an exceptionally successful imitator of Le Gascon. John B. Smits of Haarlem and S.H. de Roos of Amsterdam executed some unusual designs around the turn of the century, and there are a number of promising young binders in Holland today. Among modern Belgian binders, René Laurent deserves special mention for the originality of his designs.

In Scandinavia styles were largely imitative up to the end of the nineteenthcentury, although there were occasional examples of superior workmanship and design, especially in Copenhagen. Hans Christian Lerche (1807–1876) was the founder of a Copenhagen hand-binding establishment, which his sons sold as late as 1921. D.L. Clément (d. 1877) catered to commercial publishers by using English cloth bindings, and when Immanuel Petersen (1836–1903), an accomplished gilder, took over his shop, both hand bindings of considerable originality in finishing and machine bindings were produced. Anker Kyster (1864–1939) started in Petersen's shop as a young gilder and developed rapidly in technical and professional skills. For most of the first part of the present century Kyster was the main personality in Danish bookbinding, skilled in forwarding technique and finishing, learned in the history of binding. His younger collaborators were Joakim Skovgaard, Hans Tegner, Johan Rohde, and, above all, Thorvald Bindesbøll. Around 1950 Henrik Park, a skilled and talented designer, became the heir to the Kyster establishment. Danish bookbinding in the twentieth century would have had a totally different tradition if August Sandgren (1893–1934) had not had a little shop in which he devoted all his energies to the ideal of individual craftsmanship, much in the tradition of Cobden-Sanderson and Marius Michel. His work is basically simple in

Fig. 1 From the Club Bindery.

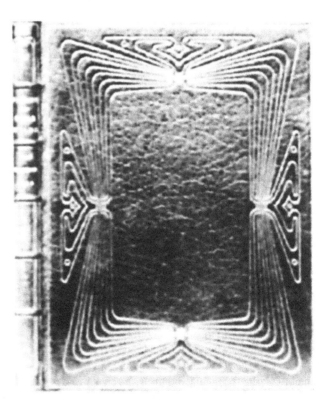

Fig. 2 "Salome," by Oscar Wilde, bound for Willis Vickery at Rowfant Bindery. Designed by Henri Hardy, and finished by Gaston Pilon. Cf. Progressive Printer, 19, 357 (1909).

Fig. 3 Anacreontis Teii Odae (Paris, Henricus Stephanus, 1554) (Club Bindery, 1898).

Fig. 4 Binding of an edition of the Council of Trent, edited by Dr. S. Kuttner of Catholic University in 1942. The original manuscript is in the Pierpont Morgan Library Binding by Dr. Ferdinand Zach and presented by Catholic University to the Pope.

construction and design, but it represents the highest standards of personal achievement. His marbled papers and lithographic papers (with Axel Salgo) are exemplary.

In Sweden most bindings before 1850 were imitations of French, and, to some extent, of English and German styles. Johan Frederick Stangenberg, who flourished between 1821 and 1841, produced simple tooled bindings on long-grain leathers. After about 1850 there were a good many solid orthodox bindings by Gustaf Hedberg (also a historian of binding). During the 1920s and 1930s Akke Kumlien, master of most of the book arts, designed some publishers' bindings for P.A. Norstedt and söner, which have had an enduring influence in the Scandinavian bookmaking tradition. Two lady binders, Countess Eva Sparre and Greta Morssing, have produced work that is a major contribution to Swedish binding tradition.

The leading Norwegian bookbinder of modern times is M.M. Refsum, who established his firm in 1887. He worked closely with many outstanding Norwegian artists, notably Gerhard Munthe and his own son, Tor Refsum.

During the 1950s significant traveling exhibits of English, West German, Swedish, and Danish hand bookbindings were sponsored by the University of Kentucky Libraries. The viability of artistic inspiration in all four

shows attracted considerable attention, and there is every reason to believe that some of the best book design and bookbinding will come from countries north of the Scheldt in the second half of the twentieth century.

CONCLUSION

Serious scholarship in the field of the history of bookbinding is largely a labor of love, and foundation support is rarely forthcoming. Nevertheless, scholars such as the late G.D. Hobson,[6] Prosper Verheyden, Luc Indestege, Ernst Kyriss, and Ilse Schunke[7] have made substantial contributions to the history of the book and to Western European cultural history with their investigations. The history of hand binding offers many opportunities for the bibliographer that have not been fully appreciated.

GLOSSARY

Antique binding. See Divinity binding.
Armenian bole. See Bole.

Basil. Sheepskin, generally poorly tanned, used for inexpensive bindings.

Beating stone. A stone set in a pan filled with sand, on which books are beaten to make the paper more compact.

Blind tooling. Impression of tools on leather by hand, to form an engraved design, not to be painted or covered with gold leaf.

Bole. A reddish mineral (friable clay) similar to chalk used to prepare edges for gilding; also known as Armenian bole.

Butter stamp. Stamp applied to the whole spine of a book, decorating it with a single impression; generally cut for collected works of single authors and used widely by nineteenth-century edition binders.

Cameo binding. A binding decorated with stamps cut intaglio (similar to a seal) and impressed on the center of the book as a medallion; background is generally painted in gold or colors; see also Plaquette binding.

Cancel. A corrected leaf, tipped in on the stub left from an incorrectly printed leaf which has been removed.

Canevari binding. Binding allegedly in the library of Demetrio Canevari (1539–1625), an Italian physician and collector of the sixteenth and early seventeenth century, whose books were probably those commissioned for binding by Cardinal Alexander Farnese for his nephew, Alexander Farnese, Duke of Parma, between 1545 and 1546; see also Farnese bindings.

Case. Simple cover of book, not yet covered with cloth or leather.

Catch stitch. A stitch which fastens one section of a book to the other, also called chain stitch or kettle stitch.

Catenati. Chained books.

Cathedral style. The early nineteenth-century style which employed Gothic architectural motifs for decoration of covers.

Chain stitch. See Catch stitch.

Champlevé. Embedding enamel in canals in metal plates used for book covers.

Chemise. Cover or jacket used to protect a binding, usually inserted in a slip case or a Solander case.

Cheveril. See Chevrotain.

Chevrotain. Leather made from the skin of small guinea deer; also called cheveril.

Cloisonné. Enamel intarsia between strips of metal placed on a porcelain or metal base.

Cosway bindings. Leather bindings with miniature paintings inserted in their covers; named for Richard Cosway (1742–1821), the English miniaturist, but who had nothing to do with the development of this style (an early twentieth-century fad).

Cottage style. Decoration associated with the Mearnes, characterized by the outline of a cottage gable on the cover.

Cropped. A bound book, of which the edges have been trimmed radically, sometimes even cutting into the text, is said to be cropped.

Cuirbouilli. A design in relief on a leather cover, formed by modeling and hammering.

Cuir ciselé. Decoration by incising the design into the leather; also known as *Lederschnitt* or incised leather.

Cumdach. A box for a book in the shape of a casket, used in Ireland up into the eleventh century.

Cusped-edged stamp. A stamp used in the south Germanies to provide the effect of a leaf design.

Cut in boards. A book whose edges are cut after the boards are laced on is said to be cut in boards.

Deckle edge. Rough edge left on a sheet of paper in the manufacturing process.

Dentelle borders. Small contiguous forms at the edge of a book cover (or doublure), forming a lacy pattern with ends pointing toward the center of the cover.

Diaper design. Repetition of lozenge forms at formal intervals.

Diced. Ruled leather, usually in crisscrossing diagonal lines to form diamonds.

Diptych. A Roman device for letter writing, consisting of two facing wooden tablets (or tablets of metal or ivory), of which the inner surfaces are covered with wax to be incised by a stylus.

Divinity binding. Books covered in thick bevelled boards with Oxford corners, often with a heavily rounded spine and with Dutch marbled end papers; popular in the period from 1840 through 1870 in England; also known as ecclesiastical or monastic bindings.

Doublure. Lining of leather, silk, or other material instead of customary end sheets; leather doublures are often decorated.

Ducali bindings. Bindings for collections of official documents of the Venetian doges, often showing both occidental and oriental techniques in a single pattern.

Dutch flowered. See Dutch gilt.

Dutch gilt. Papers decorated with wooden or metal blocks, or produced by engraved rolls; first produced in Germany and Italy, then imported into the Low Countries and exported from there to France and England (hence "Dutch").

Ecclesiastical bindings. See Divinity bindings.

End papers. Sheets not forming part of the text pasted down inside of front and back covers; often decorated, sometimes illustrated (e.g., with maps or thematic illustrations in commercial books).

Etruscan bindings. Decorated light brown or terra cotta covers with designs similar to the "black figure" pottery of the Etruscans.

Extra binder. A hand binder who uses the best manual techniques and the highest quality of material to satisfy orders for special work.

Fanfare. A cover design broken into a series of geometrically arranged ovals, bounded by fillets, and elaborately decorated with flowers, sprigs of oak, or pomegranates.

Farnese bindings. Bindings with polychrome oval plaquettes, of which some 105 survive, originally the property of the papal physician Demetrio Canevari, or of Pier Luigi Canevari; see also Canevari bindings.

Fillet. A round finishing tool which makes an impression as it is rolled on a cover to form a running line or a series of designs on a book cover; also called a roll or a roulette.

Finishing. Decoration of a book after the boards have been covered.

Fly-leaves. All the leaves of an end-paper section.

Fore-edge painting. A painting on the fore-edge of a book visible only when the edge of the book is fanned out; generally covered by gold leaf when the book is solid (or not fanned out).

Forwarding. Sewing and covering a book.

French chalk. A soft chalk used to remove grease spots.

French shell. Marbled papers with a conchological pattern popular in late eighteenth-century France.

Full-gilt. An adjective applied to books of which all the edges are gilded.

Gauffering. Decorating a book edge with a tool over the gilt edge, often with elaborate designs.

Girdle book. A book whose binding had an extra protective cover, elongated so that one end could be attached to a belt or a girdle; common in Western Europe and Ethiopia; frequently used by wandering missionaries.

Glaire. White of egg, plus vinegar, water, and occasionally other substances, used to size tooled impressions on a leather cover (or, sometimes, cover of other material) or a smooth edge of a book, before gilding.

Gothic binding. A style of decoration widely used in Western Europe in the fifteenth and six-teenth centuries, characterized by numerous stamps and rolls in blind.

Grooved boards. Boards cut out to receive slips of a book, or cover boards with grooved edges (characteristic of Greek bindings).

Guard. A narrow strip of paper, used to repair a damaged fold, or to attach a plate or a single leaf.

Harleian style. An English style of the eighteenth century, with a center of small dies in a lozenge form and a border decorated with a roulette; characteristic of the bindings of the Lords Harley, now mainly in the British Museum.

Headband. A band across the top edges of the book block, usually of silk, linen, or cotton (often in motley colors), and fitted to the contour of the back of the spine; see Tailband.

Hollow backs. Deceptively hollow construction affixed to the back of a book block before it is covered.

Imprint. Place of publication, publisher or printer, and date; on last leaf of early printed books, in sixteenth century and thereafter generally on title-page.

Incised leather. See Cuir ciselé.

Inlay. Insertion of pieces of colored leather or other materials into incised patterns on a leather cover; often called intarsia; sometimes actually onlay, since pieces of colored leather are applied to blind-stamped patterns.

Intarsia. See Inlay.

Jansenist bindings. French bindings of the second half of the seventeenth century, named for the highly orthodox Jansenists, and characterized by decorations only with blind-tooled lines.

Kettle-stitche. See Catch stitch.

Knocking-up. The process of squaring and making compact the back of the sections of a book.

Kutch. A container for a heavy piece of gold while it is being formed or beaten into thin pieces.

Laid paper. Paper manufactured in a mold in which heavy lines of wire are attached to the bottom, usually to form a pattern.

Landscape bindings. Covers decorated with a watercolor of a landscape, frequent in England in the 1820s.

Lederschnitt. See Cuir ciselé.

Lozenge. A tooled form in the shape of a diamond, or a square figure placed from one of its points; often gilded.

Lyonnaise bindings. Bindings of the sixteenth century with interlaced strapwork, often painted and with large cover ornaments, with a background filled in with dots; other French and Italian bindings from different localities show the same characteristic decoration.

Marbling. A process by which oil colors are floated on size to form a design, then transferred to leather, paper, other covering material, or book-edges.

Mosaic. Inlaid designs on book covers; also called Intarsia or Inlay.

Monastic bindings. See Divinity bindings; also applied to any binding executed in a monastery.

Morocco. Goatskins originally imported from Morocco, now mainly "Cape goat" (from South Africa), Niger (see *below*), and Turkish goatskin.

Mudéjar style. Strapwork design on a cover by medieval Spanish Moslem craftsmen who worked in Christian jurisdictions.

Niger. A "morocco" leather from the skin of small goats found in Nigeria and the Mediterranean coast of Africa.

Nonato bindings. Covers in the skin of a foetus of a stillborn Argentine pampas calf, with the hair left on (*nonato* = not born).

Onlay. Pieces of colored leather or other material affixed over an outlined tool form, border, or panel; see also Inlay.

Overcasting. A sewing process in which all the leaves of a section are bound together with thread.

Oversewing. See Overcasting.

Panel stamp. A large stamp frequently pictorial, stamped in the middle of a cover of a book; common decoration of Gothic bindings.

Plaquette binding. Binding with medallions in the center of the cover as the main decoration; see also Cameo binding.

Plough. A wooden device into which a cutting edge is fitted to cut edges of a book in a lying press.

Pointillé. Decoration of covers by tool forms in dotted outline (*fers pointillés*); used in seventeenth-century France by Le Gascon and his contemporaries.

Polaire. Book satchel used in medieval Ireland.

Pounce. Adhesive used under gold or colors to make them stick firmly to the surface which they decorate.

Polyptych. Three or more wooden tablets used for letter writing, with inner surfaces covered with wax to be incised by stylus; see Diptych.

Provenance. Record of previous ownership.

Pugillares. Tablets whitened to receive ink, or covered with wax to be engraved with a *stilus;* also called *tabellae;* Ovid, Am. i. 12, I.

Pull. To separate quires of a book from old cords and tapes and to remove the old glue.

Raised bands. Cords on the spine of a book over which the threads are passed.

Roan. Sheepskin tanned in sumac.

Roll. See fillet.

Romanesque bindings. Bindings of the twelfth and thirteenth centuries richly decorated with single stamps, produced in England, Germany, and France.

Rough gilding. Gilding of edges before a book is sewn.

Roulette. See Fillet.

Roundel. A stamp forming a double ring, usually with a dot in the center.

Russia. Calf or cowhide originally imported from Russia, tanned with willow bark, dyed with sandalwood, and soaked in birch oil.

Saddle.	Part of book-sewing machine on which quires are placed so that folds may be directly below sewing needles and loopers.
Sawn-in.	The spine of a book is sawn in to provide a canal into which cords may be placed.
Semis (semée).	A cover design in which small stamps are located at regular intervals to form a design; originally a heraldic term.
Serefsan.	Gold-flecked paper from Persia or Turkey.
Slip case.	Book container opening at the front edge.
Solander case.	Book container opening at the top, so-called after Daniel Solander, a librarian in the British Museum, 1765–1782.
Spring backs.	Backs capable of retaining a firm position.
Sprinkled calf.	Speckled surfaces on calfskin produced by sprinkling of acid.
Sprinkled edge.	An edge sprinkled with gold, red ink, or other colors.
Stabbing.	Method of sewing books by which the inner edge of the entire book-block is perforated.
Stamp.	A tool-engraved intaglio to be impressed on a leather binding; impressed unheated in medieval and early modern times; today, impressed heated with an arming or blocking press.
Strapwork.	Interlaced double lines forming a geometrical pattern; a special characteristic of Grolier bindings.
Super-exlibris.	Panel stamp in center of front cover to indicate ownership.
Supralibros.	See Super-exlibris.
Tabellae.	See *Pugillares*.
Sutherland tooling process.	A method of tooling in color on different materials used around 1900; name is from books bound for the Duchess of Sutherland by G.T. Bagguley of Newcastle-under-Tyne.
Tailband.	A band on the tail edge of a book (opposite the head) usually of silk, linen, or cotton, and fitted to the contour of the book; see Headband.
Tree-calf.	Marbled calfskin with the pattern of the trunk and branches of a tree.
Triptych.	Hinged metal tables like the diptych, but with three tables instead of two.
Two-sheets-on sewing.	A method of sewing in which each length of thread holds two sections instead of one.
Vellum.	Thin scraped calfskin prepared with lime, not tanned.
Wire marks.	Marks left in a piece of paper by the wires in the bottom of the mold.

REFERENCES

1. Cockerell, D. *Bookbinding and the Care of Books*, 5th Ed. Pitman: London, U.K.,1953.
2. Kyriss, E. *Verzierte gotische Einbände im alten deutschen Sprachgebiet*, M. Hettler: Stuttgart, Germany, 1951; 4 volumes.
3. Le Roux de Lincy, A.J.V. *Researches Concerning Jean Grolier*, The Grolier Club: New York, 1907; 100.
4. Michel, M. *La reliure française. Depuis l'invention de l'imprimerie jusqu'à la fin du XVIII siècle*, 1880; Paris.
5. Adam, P. *Der Bucheinband, seine Technik und seine Geschichte*, Seemann: Leipzig, Germany, 1890.
6. Hobson, G. D. *Maioli, Canevari and Others*, Little, Brown: Boston, MA, 1926.
7. Schunke, I. *Leben und Werk Jakob Krauses*, Insel-Verlag: Leipzig, Germany, 1943.

SELECTED BIBLIOGRAPHY

1. Arnold, T. W.; Grohmann, A. *The Islamic Book*, Pegasus Press: Leipzig, Germany, 1929.
2. Brassington, W. S. *A History of the Art of Bookbinding*, Elliott Stock: London, U.K., 1894.
3. Brunet, G. *La Reliure Ancienne et Moderne*, Paul Daffis: Paris, France, 1878.
4. In *Bookbinding in America*; Lehmann-Haupt, H., Ed.; The Southworth-Anthoensen Press: Portland, ME, 1941.
5. de Crauzat, E. *La Reliure Française de 1900 à 1925*, Kieffer: Paris, France, 1932; 2 volumes.
6. Cundall, J. *On Bookbindings, Ancient and Modern*, George Bell: London, U.K., 1882.
7. Davenport, C. *The Book, Its History and Development*, Van Nostrand: Princeton, NJ, 1908.
8. Davenport, C. *Roger Payne, English Bookbinder of the Eighteenth Century*, Printer for the Caxton Club: Chicago, IL, 1992.
9. Davenport, C. *Samuel Mearne, Binder to King Charles II*, The Caxton Club: Chicago, IL, 1906.
10. Davenport, C. *Thomas Berthelet, Royal Printer and Bookbinder to Henry VIII, King of England*, The Caxton Club: Chicago, IL, 1901.
11. Diehl, E. *Bookbinding, its Background and Technique*, Holt Rinehart and Winston: New York, 1946; 2 volumes.
12. Goldschmidt, E. P. *Gothic and Renaissance Bookbindings*, Houghton Mifflin: New York, 1928.
13. Gruel, L. *Manuel historique et bibliographique de l'amateur de reliure*, Gruel & Engelmann: Paris, France, 1887.
14. Haebler, K. *Rollen-und Plattenstempel des XVI. Jahrhunderts*, O. Harrassowitz: Leipzig, Germany, 1928–1929; 2 volumes.
15. Helwig, H. *Handbuch der Einbandkunde*, Maximilian-Gesellschaft: Hamburg, Germany, 1953–1955; 3 volumes.

16. Horne, H. P. *The Binding of Books; an Essay in the History of Gold-tooled Bindings*, Kegan Paul, Trench, Trübner and Company: London, U.K., 1894.

17. Loubier, H. *Der Bucheinband in alter und neuer Zeit*, Herman Seemann: Berlin, Germany, 1926.

18. Matthews, B. *Bookbindings Old and New*, Macmillan: New York, 1895.

19. Mejer, W. *Bibliographie der Buchbinderei-Literatur*, Hiersemann: Leipzig, Germany, 1925.

20. Mejer, W. Ergänzungsband, 1924–1932. *Bibliographie der Buchbinderei-Literatur*; Hiersemann: Leipzig, Germany, 1925.

21. Middleton, B. C. *A History of English Craft Bookbinding Technique*, Hafner Publishing Company: New York, 1963.

22. Prideaux, S. T. *An Historical Sketch of Bookbinding*, Archibald Constable: London, U.K., 1906; (with a chapter on early stamped bindings by E. Gordon Duff).

23. Prideaux, S. T. *Modern Bookbindings, Their Design and Decoration*, E.P. Dutton: New York, 1906.

24. Sarre, F. *Islamic Bookbindings*, Kegan Paul, Trench, Trübner and Company: London, U.K., 1923.

25. Thoinan, E. *Les relieurs français (1500–1800)*, Paul, Huard, Guillemin: Paris, France, 1893.

26. Thompson, E. A.; Thompson, L. S. *Fine Binding in America*, Beta Phi Mu: Urbana, IL, 1956.

27. Thompson, L. S. *Kurze Geschichte der Handbuchbinderei in den Vereinigten Staaten von Amerika*, Max Hettler Verlag: Stuttgart, Germany, 1955.

28. Uzanne, O. *The French Bookbinders of the Eighteenth Century*, The Caxton Club: Chicago, IL, 1904.

29. Weale, W.H.J. *Bookbindings and Rubbings of Bindings*, Eyre and Spottiswoode: London, U.K., 1898.

30. Weber, C. J. *Fore-Edge Painting; an Historical Survey of a Curious Art in Book Decoration*, Harvey House: Irvington-on-Hudson, NY, 1966.

Biological Information and Its Users

Kalpana Shankar
School of Informatics, Indiana University, Bloomington, Indiana, U.S.A.

Abstract
The purpose of this entry is to introduce the information science reader to the wide range of data and other resources that constitute "biological information." Attention is paid to both paper and digital sources and the use of digital libraries and cyberinfrastructure for the creation, use, and reuse of information. The entry discusses various user communities and their needs, including scientists, educators and students, policy makers, and other secondary users. The entry concludes with challenges for data sharing, preservation, and access.

INTRODUCTION

The term "biological information" covers an enormous range of content, format, and uses. The creators and users of it are just as heterogeneous. The range of disciplines encompassed within the term "biology" is itself broad, including descriptive fields that rely on fieldwork and observation to those that are exclusively computer-based, or computationally intensive. Not surprisingly, the data generated from the results of scientific activity within these fields (which is what most biologists would call "biological information") generates a variety of information formats and storage, from paper-based laboratory notebooks and anatomical drawings to petabytes of numeric data generated from high throughput screening, sensors, genomics, and proteomics. There is also a great deal of information generated by what might be termed the context of scientific practice. Information that can provide insight into the processes that generated scientific data may include grant reports, patent applications, correspondence, personnel records, and other texts and data that are not necessarily direct products of research practices. This does not even include medical information, which comes with a different and attendant set of users, privacy and other concerns, and legal ramifications. This encyclopedia piece, however, will focus on the realm of data products from the scientific research process and on contextualizing information from the conduct and organization of science within institutions.

For the information scientist, biological information presents fascinating challenges for understanding and facilitating information policy, organization, sharing, and access. This entry will first discuss some data-intensive current practices in biological research. Next, it will describe some of the variety of information that encompasses "biological information," focusing on both the data products of scientific activity (this includes results of experiments and analysis) and the documents that are created in the context of scientific research (correspondence, grant applications, and similar materials). Following that, the categories of users—scientists, educators, policy makers, and activists, will be discussed. Lastly, the entry will outline some challenges that exist for understanding biological information and its uses—highly collaborative environments, data-sharing and intellectual property regimes, proprietary formats. The conclusion will discuss potential challenges for information studies research and practice.

INFORMATION PRACTICES ACROSS THE BIOLOGICAL SCIENCES

The pedagogy of research in the biological sciences has been compared to an artisanal model.[1] In universities around the world, junior students of the biological sciences trained with more senior scientists, then set up laboratories in other academic, independent, and government scientific institutions, often introduced through networks to their future employers. While scientists used various instruments, the primary mechanism for recording information was the laboratory notebook which contained drawings, handwritten notes, and served as a daily aide-memoire for the scientist. To some extent, this model of research training and scientific data gathering persists and continues to be taught in the science classroom, although many digital tools exist for capturing and storing data.

However, the biological sciences underwent radical changes in the latter half of the twentieth century and continue to change. The widespread use of information and communication technologies (ICTs) has made the biological sciences (indeed, all sciences) globalized and highly data intensive in nature. As a result, one emergent trend that has influenced the nature of biological

Encyclopedia of Library and Information Sciences, Fourth Edition DOI: 10.1081/E-ELIS4-120043747

information is the advent of research teams can span countries and continents as expensive equipment is shared remotely and data sets analyzed in parallel. There have been efforts in numerous countries to develop information infrastructures that centralize data sharing and collection, standardize descriptive data (metadata), and make the use of remote tools more readily available. The emergent sciences of genomics (the study of genes), proteomics (the study of proteins), and chemical informatics (the study of molecular and submolecular entities in living systems) belong to this data-intensive category of science. Sensor-based biological arrays, from the particulate level to terrestrial grids, are producing terabytes of data on a daily basis. Examples include microorganism analysis,[2] climate change, and land use.[3] Systems such as the proposed Global Earth Observation System of Systems (GEOSS) integrates other information infrastructures to provide multidisciplinary perspectives on biological problems that are planetary in scale and scope of data.[4]

A second important trend in the biological sciences that has been responsible for the generation of most of the data (and related and derivative information products) is the increase in the quantity and size of collaboration, often in response to emerging problems or even crisis situations (such as the SARS epidemic). Many researchers have noted that the need to gain access to instruments, unique data, and funding for increasingly complex projects requires the assemblage of individuals and teams with diverse skills.[5] As many scientists become more and more highly specialized, these teams are necessary to solve complex problems. Teams can also minimize impact on politically, environmentally, and socially sensitive regions and areas. Governmental funding agencies such as the United States National Science Foundation (NSF) and funding programs in the European Union often require that collaborations be formed in order to use natural and data resources in efficient ways. Collaborations also cut across geographical and institutional boundaries. Nonprofits, academic, government laboratories, and industries can work together to mobilize around problems and issues of mutual interest, hopefully with benefit to all concerned parties.

A third trend that has been instrumental in generating the explosion of biological data is the increasingly data-intensive nature of the biological sciences, aided by the introduction of information technologies with the capability to generate, manipulate, and store vast quantities of data at a high rate. Benoit[6] notes that the collaboration of computer scientists and molecular biologists has generated an impressive collection of tools to manage data, conduct analyses, and present results that would not be possible without the integration of these disciplines. While the term bioinformatics has been reserved for the application of information technology, data mining, visualization, and other computational tools to questions in molecular biology, the other biological sciences (i.e., those whose research purviews reside above the level of molecule, from the cell to whole systems) have also been aided by "informatics" and in turn have generated vast quantities of data. The resulting data (often the results of extremely intensive string manipulations[6]) is stored in databases that may be local, publicly available, or commercial/proprietary.

The contribution of high performance computing to the scientific enterprise has been subsumed under the term "cyberinfrastructure" coined by the NSF to coordinate and further its sponsored research efforts in data acquisition, storage, management, and other Internet-based services around scientific information.[7] In the United Kingdom and the rest of Europe, similar efforts are usually called e-Science.[4] The term is meant to include the human resources needed to make the technical infrastructure function, but the discourse has tended to focus on a technologically deterministic vision for Web services in science.[7,8] Cyberinfrastructure grants have proven to be technological drivers in furthering the computationally intensive sciences. Challenges include making data formats mutually intelligible, the creation and implementation of standards, and the creation of tools that make data resources potentially useful to multiple communities of experts and nonexperts. An important commonality across these cyberinfrastructure projects is the need for efficient storage and access to information.[9] Many in the biological communities have relied on cyberinfrastructure projects to minimize bottlenecks to data sharing, since data derived in NSF-funded projects are generally considered open access and nonproprietary. Nevertheless, because cyberinfrastructure is built upon the current Internet, it has inherited all of the technical, legal, cultural, and policy problems of the Internet, including security, language barriers, intellectual property regimes, and others. However, cyberinfrastructure is not the only mechanism by which the biological sciences have been transformed. Traditional biological fields, such as anatomy, animal biology, and taxonomy, have also undergone dramatic changes, aided by the use of various kinds of technologies. Mapping technologies such as Geographical Information Systems (GIS), satellite imagery, and earth-based sensor networks allow scientists to use data that has changed the nature of these "traditional" biological disciplines.[10,11] The ability to analyze legacy and archival data sets with new tools can yield insights into new problems. For example, analysis of tissue samples and doctors' notes from the 1918 flu epidemic resulted in findings that an opportunistic bacterial infection may have been more responsible for the number of deaths, rather than the original flu virus itself,[12] an important consideration for future pandemic planning.

New ICTs, increase in the size of collaborations, and data-intensive practices: these three trends, when taken together, probably account for the "information explosion" in the biological sciences more than any other factors. Collaboration across disciplines and informatics together have made the biological sciences extremely data intensive. The former has enabled scientists to use equipment and resources in distant locations and answer increasingly complex problems with the help of colleagues with very different disciplinary interests. Informatics is entwined with this collaboration, since it implies the integration of biological sciences with computational power. Of course, policy initiatives and financial resources to back them can spur such efforts to greater success.

VARIETIES AND FORMATS OF INFORMATION

An early model by which biological information is organized is the UNISIST model, developed by the United Nations Educational, Scientific, and Cultural Organization (UNESCO) and the International Council of Scientific Unions (ICSU), and more recently updated by Fjordback Søndergaard et al.[13] In this model, three channels of information: tabular (or data), formal, and informal channels are defined, with three levels of information (primary, secondary, such as abstracting and indexing services, and tertiary, such as encyclopedias). This makes a good departure point, but is hardly complete.

Much biological data is still recorded on paper. Many researchers in the biological sciences still use bound laboratory notebooks in the conduct of their daily work. Paper notebooks may seem antiquated because they are not easy to search in or integrate data, but they are still prevalent and mainstream in science classrooms and many academic laboratories, which provide the fundamental education for scientists.[14] Paper notebooks are a form of data capture that are easy to understand and use, transcend computer technology that breaks or becomes outdated, are fairly simple to maintain for short- and long-term use, and do not require any proprietary systems. They are also effective mechanisms for integrating printouts, drawings, spreadsheets, and other date formats.

Digital data, not surprisingly, is far more diverse in scope. Specialized instruments usually generate data files in proprietary formats; often, these data are further manipulated using computer-based analytical tools or integrated with other data formats. Common data formats, such as those from Microsoft and Adobe products are prevalent, but others are more specialized from manufacturers of scientific equipment. Other data, although technical, are in formats that have become standard in scientific fields: GIS, for example, generate data in a range of formats that are universally accepted and used by scientists. Still other data are often generated within and managed by "collaboratories,"[15] closed groupware or collaborative systems that are built for or around a particular project or group that is heavily dependent on digital technologies.

These cover the general scope of primary resources in the biological sciences, but the data is analyzed and synthesized into various formats for dissemination outside of the immediate laboratory or research group. The range of publication formats is also numerous, and the reliance on different kinds of materials varies greatly by discipline and by individual preference. For example, to circumvent publishers' embargos on current research and the high costs of publications, many scientific communities created their own systems for circulating preprints and obtaining comments. While these have been extremely successful in the physical sciences, there has been significantly less use of preprint or e-print servers in the biological sciences.[16] Conference papers, conference posters, monographs, reference manuals, and publications in peer-reviewed journals are all important sources of biological information. Conferences and workshops will often maintain Web sites where publications from previous years can be found. In addition to general journals that may cover a wide range of scientific disciplines (such as *Nature*, one of its related journals, or *Science*), many scientists use more specialized peer-reviewed journals.

Although they may not be considered biological information per se, it is important to note another major category of documents that are important to the biological sciences: administrative materials. Scientists are responsible for obtaining grants and administering them, advising students, dealing with personnel matters, serving on peer review committees for journals and granting agencies, and depending on the status of the scientist, serving on conference and institutional, national, and other committees and boards. Many scientists may serve in multiple capacities as educators, consultants, researchers, and the like. In the context of these roles, scientists generate an enormous amount of information that may not be seen as important to the generation of biological knowledge, but are of incalculable value to other stakeholders with an interest in matters in the processes by which science is conducted.[17]

INFORMATION AND DATA REPOSITORIES IN THE BIOLOGICAL SCIENCES

Data and other information in the biological sciences is, not surprisingly, stored in distinct and various locations, physical, virtual, and both. Laboratory notebooks and digital files tend to stay within the academic laboratories that created them, just as information in the commercial sector or government laboratories tend to stay within those institutions. In the latter case, digital data sets are not retained by the National Archives and Records Administration (NARA), but, in the best of circumstances, in the archives and records centers of these laboratories. More often, they are left in the custody of the scientists whose laboratories generated them.

Because their creators often consider these documents useless once they have been used for formal publications, they tend to be regarded without much interest for long-term maintenance and preservation.[17–19] There have been many institutional efforts to organize and contextualize data sets of long-term value as there has been some interest in repurposing them for other projects or having an easily accessible historical trail. Documents have been useful to historians and others with interest in scientific activity, so there is still interest in their long-term preservation. However, inactive documents and records often do not end up in archives of science for permanent retention, unless the director of the laboratory is extremely prominent. Even then, many archivists of science are reluctant to accept "raw" data, which is impossible to interpret without the aid of context in the form of the publications that resulted from them.[17,20] Nevertheless, there are many archives of science around the world, often focused around specific individuals, institutions, disciplines or subdisciplines, or topics. These archives may include the personal and scientific papers of individuals and professional/academic societies, important instruments, and key artifacts. There are far too many to enumerate, but some interesting examples include the John C. Liebeskind Collection at the University of California at Los Angeles, the only history of pain collection in the world. BrightSparcs, a resource for finding collections about Australian scientists serves as a pointer to the physical location of archival collections. Like the BrightSparcs collection, there have been many efforts to make the finding aids of these collections available online, but digitizing the materials themselves is difficult and expensive, so few archival scientific collections are available online in their entirety.

Scientific publications are a crucial resource for obtaining peer-reviewed research results. Publicly funded and proprietary databases index and if necessary create abstracts for this literature, making it accessible to individual and institutional subscribers, and in the case of publicly funded databases, the broader public. Examples include the National Library of Medicine's PubMed-Central, and proprietary databases from other commercial publishers such as the Web of Science. For reasons similar to the impetus for e-print and preprint repositories, there have been some institutional efforts to circumvent the traditional scientific publishing model, which many critics argue favors publishers at the expense of scientists and the institutions at which they work.[21,22] These critics charge publishers with creating a model in which universities and similar institutions subsidize research publishing, and then in turn pay for subscription rights to journals and secondary databases. However, given the imprimatur these systems lend to individual scientists in establishing their professional reputations, there has been little leeway and few incentives for them to break away from this system.

There have been some efforts to create open access models for more immediate and free access to scientific publications. One example is the Public Library of Science, championed and cofounded by the former director of the National Institutes of Health Harold E. Varmus in late 2003. In addition to being a repository for publications and references, the Public Library of Science also encompasses several peer-reviewed open access journals in the biological sciences on a market-driven model in which scientists pay for publication, peer review, and online archiving.

The digital libraries movement has probably been the most influential in making biological data and other information broadly accessible and usable, particularly to nonexperts.[23] The creation of digital libraries has been instrumental in allowing educators and other nonexpert communities with interest in scientific data to minimize problems associated with information overload, poor structure, and the variable quality of scientific information available on the World Wide Web.[24] Resources in digital libraries are organized and maintained, usually around a focused topic, and can encompass a variety of information types and formats that can support searching, reuse, and other interesting reuses of data and other scientific documents.[25]

USERS OF BIOLOGICAL INFORMATION

Most evidently, the primary users of biological information are biologists themselves. It is impossible to characterize the biological research community in any general way. It is international, multidisciplinary, and collaborative. Biological scientists receive academic training in academic settings, but may work in universities, the private sector, in government laboratories, for nonprofits and civil society, and even as independent scholars and researchers (although this limits the kind of work they can do). As noted earlier, scientists are trained in an apprentice-based model in which they conduct research with senior scientists, working on projects of interest to the principal investigator. These researchers can be funded through a number of mechanisms, including government and foundation grants. Researchers can be found in other kinds of institutions as well. Although there has been significantly less biological research being conducted in the private sector in recent years, researchers in pharmaceutical companies and other biotechnology firms generate data and related information and are dependent upon multiple publication streams for disseminating results and keeping up with the current research, just like their academic counterparts.

Lastly, government laboratories and independent organizations throughout the world also employ these researchers. In the United States, the federal government administrates few biological laboratories of its own; the vast majority of its resources are disseminated in the form of grants to scientists in other institutions. Nevertheless, government laboratories often conduct research in topics that may be of little financial interest to the private sector, such as therapies for rare illnesses, or clinical research.

Of course, biologists also work in non-Western countries; their information needs and challenges are different and often fraught with difficulties. While their needs for information may be as great as their Western counterparts, many characteristics of their situations make finding that information more difficult. Subscriptions to journals and secondary databases are often prohibitively expensive, even for institutional libraries and consortia of libraries. Most of these databases are no longer published in paper format, which would be the most useful for situations with uncertain power supplies and limited Internet access. Scholarly communication at conferences and workshops is of great importance to the biological community as informal networks are important to all kinds of researchers; however, the costs of travel and often the difficulties of obtaining foreign visas make these options difficult in many situations and often marginalize scientific researchers outside main research centers.[26,27] To give one concrete example, the American Association for the Advancement of Science (AAAS), a nonprofit organization in Washington, District of Columbia, has argued that U.S. travel policies have made it extremely difficult for Cuban scientists to enter the United States to conduct legitimate research or attend important conferences.[28] These scientists, much like junior researchers and those at institutions that are considered less research oriented, are generally deemed peripheral to the core research networks; this in turn, further marginalizes their work and thus they find it harder to get and share important information.

There are other scholarly communities with interests in studying science from social science, behavioral, and humanistic perspectives. These transdisciplinary areas of scholarship are often grouped under terms like "science, technology, and society," or "science studies"; these terms are used to denote research interests that generally focus on science as a practice and epistemology with methodologies drawn from the social and behavioral sciences and humanities. History of science is one obvious area in which researchers rely on documents and artifacts produced by and about scientists. Philosophers, anthropologists, and sociologists of science also use similar artifacts and documents, such as journal articles, to make and situate their claims about science itself.[29] There are also other areas of research that may require use of biological information, such as bioethics. In general, there has not been much research into the specific ways in which this community uses biological (or for that matter, any) information, so it is difficult to state categorically what kinds of information are useful to this category of researchers. However, it is likely that their practices mirror those of other interdisciplinary fields in which attendance at workshops and conferences, collaborations with biological researchers and other scholars, and wide literature searching are important resources.[30]

The introduction of digital libraries has been extremely useful for the educational community. There are a number of reasons why educators have collaborated with computer scientists, biological scientists, and information scientists to create digital libraries. Although the biological sciences have been less affected than other scientific disciplines, concerns in the United States regarding the low interest of students in pursuing advanced degrees and subsequent careers in the sciences has led to new models of engaging students at younger ages. One way in which this challenge is being approached is through the use of real data to address real scientific problems and learn the processes of science as scientists themselves conduct it, one approach to what is often called "inquiry-based learning."[31,32] Much of the impetus to move toward more inquiry-based approaches for teaching science stemmed from the 1996 release of the National Science Education Standards (NSES) for grades K-12, which articulated numerous disconnects between classroom teaching and the ability of science students to connect their learning to science in the "real world." Other studies have found that these problems persist in undergraduate science education. Furthermore, the sciences have found a decline in the number of women and underrepresented minorities in the sciences (though the biological sciences have again been less affected). Some reports have suggested that one way to engage these students might be through exposing them to real-world problems that can be addressed through scientific inquiry.[33] Biological digital libraries have been used to support these efforts, since materials can easily be added to them to contextualize scientific data sets with age and level appropriate information.

However, researchers have noted that without "pedagogical alignment," the potential of digital libraries to foster sophisticated inquiry cannot be fully realized. Instead, they serve as yet another "information silo" in which students conduct artificial projects and teachers then move on to other lessons. In short, the science curriculum and specific learning activities must align appropriately with the content and organization of the digital library itself.

More specifically for the biological sciences, research and implementation has suggested that participation in research is important for life science education,[34] but that this is not necessarily feasible for all undergraduates in life science undergraduate programs. The use of biological data in the classroom and inquiry-based approaches may provide substitutes, however. Reading primary papers, designing "real-life experiments," analyzing data, and reading primary papers have all been employed to further information competence and fluency, although this material can be enormously challenging to the undergraduate student.[35,36]

Another important category of users is organizations and individuals involved in foundations, policy making, and advocacy (these communities are not necessarily the same, although there may be some overlap). Examples include data on disease prevalence, land use planning, and environmental information, though there are many others. These users may or may not have scientific training, but may be working with others who do to get help in interpreting the literature. There is also an enormous range of "policy makers," from local officials and planners[37] to members of Congress, as well as scientists and bureaucrats within government agencies. Transgovernmental organizations are also significant users of scientific information. Relying on primary evidence is considered necessary to making science-based policy in environmental affairs, health care, new biomedical technologies, as well as many other areas of science and technology policy. These users are not likely to delve into the primary literature of biology, much less the primary data sources. Therefore, task forces and expert panels may be convened to assemble the state of the art on important topics; policy makers may also rely on information from the Congressional Research Service nongovernmental "think tanks," some of which have partisan agendas and others that do not. In the United States, the National Research Council (NRC) and Institute of Medicine (IOM) are two quasi-governmental organizations that are tasked with providing expert opinion on various issues through the form of reports to the U.S. Congress and executive branch agencies; their reports are available to the public. Policy making, however, is both a political and a scientific process, and that complicates the ways in which it is used. Some argue that scientific information with policy implications is not useful when specifically targeted to this population; instead, synthesized consensus on such topics should be targeted to those who have the power to influence policy makers.[38]

Another important group of users that rely on and even contribute to biological knowledge are interested amateurs. Some of them may be interested in data-driven approaches to social and political change. Again, these users represent a diverse set of interests, abilities, and needs. They may be geographically localized groups with an interest in environmental issues in their community,[39] activist organizations that mobilize around a particular illness for personal, policy, and social change (AIDS activists are among the most well-known group, but there are many others),[40] or global movements interested in effecting change on transnational issues. Individuals may be interested in obtaining the latest scientific information for personal knowledge. They may be involved in what is often called "citizen-science" for personal reasons; these users are often involved in research efforts in ornithology, biodiversity, and other observational fields.[41,42] Many of the problems that plague other communities outside of the sciences persist: vast quantities of information that may or may not be valid, highly technical information, and few vetted pathways through that maze.

CHALLENGES AND CONCERNS

There are numerous open challenges to understanding the nature of biological information and making it useful. One central issue for practice and research is the issue of data sharing. The National Institutes of Health, for example, argues that data sharing is necessary to develop new analytical tools, encourage supporting studies, enable interdisciplinary research, and prevent duplication of efforts. To this end, NSF and some other grantmaking agencies require a data-sharing plan in grant proposals. In spite of an ethos of information sharing, data sets in the sciences, even those with wide import and use, are often "siloed" or not preserved in useful formats. There are many reasons for this—lack of time, resources, interest, and incentives for data sharing are all cited as barriers to true information exchange.[43,44] This varies from scientific community to community, though there have been extensive calls for data sharing in the broader scientific literature. It is often argued by scientists themselves that most information sharing is done in the form of peer-reviewed workshop papers, conference presentations, journal articles, and informal networks; therefore, sharing "raw data" is not feasible, since it would require setting up a dedicated data repository, responding to many individual requests, and other mechanisms that may be time consuming.[45] Committing long-term resources to the maintenance of data only seems feasible in fields that are highly data intensive and "cutting edge."

The fields of genomics and protein chemistry have publicly available databases and extensive toolkits for data analysis (examples include GenBank and RefSeq); these efforts are usually developed and implemented by various branches of the federal government, such as the National Library of Medicine's National Center for Biotechnology Information (NCBI). Data deposited to these databases is centrally maintained and curated. Many journals in molecular biology and related fields require submission of novel genetic or protein sequences before publication so that the accession number can appear in it. Europe and Japan have similar databases (EMBL and DNA Database of Japan, respectively), but work in collaboration with NCBI. Increasingly, many data are proprietary, as they are produced in the commercial sector, or in collaboration with commercial firms. The formal results of analyses of these data sets may be made public in publications, but the raw data is not made available for reuse.

Another concern is preservation and curation, not just of data, but also of the contextual information that is often of greater use to those outside of the biological communities. These are a complex set of concerns, both social and technical. Issues of digital preservation that haunt other arenas include problems with media, differing approaches, and metadata. These are issues that most laboratories do not routinely consider. Making information and data available to the scientific community requires appropriate architectures, integration of data formats, and agreements on the funding and labor required to maintain such systems.[46]

But to make scientific information accessible to the nonscientist, even more challenges ensue. There is certainly interest in access to such data, since at least in the United States, data sets that are funded by public resources are expected to be made public in some format. Digital libraries have been one approach to mitigating the access problem, but these are designed for educational users, not activists or policy making. For the researcher who needs access to the material that is produced as a by-product of research, such as grant applications and letters, the problems may be one of too much information in too many places, all of it ephemeral, much of it not easily integrated with other data sources.[47] Much of the work that has been done on mapping formal and informal communication networks, history of science, and similar fields was only possible with the preservation of letters and documents in archives, or at least in other accessible settings and conditions. However, much of the current practice of science is done online: in e-mail, word processors, and proprietary software or groupware. The transient nature of these systems requires more intensive preservation efforts than paper-based data sets, and if insufficient efforts are made to preserve this material, institutions end up creating what might be termed a "memory crisis"[48] for contemporary science. This problem will only get worse, as the kinds of technologies that scientists use become less and less amenable to digital preservation. It is still an open question, though, as to what needs to be saved at all for future use.

CONCLUSIONS

In the research community, current focus is, not surprisingly, on the vast array of digital data being created and used, and a suite of pressing concerns with access, storage, analysis, and policy around it. But it is worth remembering that a significant amount of information is still generated on paper, in laboratories, by researchers and their students, and that this information is just as difficult to access. The ways in which these different kinds of information are generated are, not surprisingly, not always mutually exclusive. Furthermore, the kinds of information that are often of greatest interest to the nonexpert community is information that is increasingly difficult to access, or ephemeral. The users themselves are an extremely diverse group with differing needs; making biological information useful and accessible for all potential user groups is a daunting task and one that is probably impossible. Lastly, it is important to note that access to biological information is not just a set of technical problems; there are deep reasons of practice and policy that also shape storage, access, and use patterns. Most of these areas pose emerging questions of research and practice.

ACKNOWLEDGMENTS

The author wishes to thank Jacob Warner for his research assistance, Barbara Andrews and Emily Maguire for reading preliminary drafts of this work, and the anonymous reviewer for extremely helpful suggestions for revision.

REFERENCES

1. Gibbons, M.; Limoges, C.; Nowotny, H.; Schwartzman, S.; Scott, P.; Trow, M. *The New Production of Knowledge;* SAGE Publications: London and Thousand Oaks, CA, 1994.
2. Goldman, J.; Ramanathan, N.; Ambrose, R.F.; Caron, C.; Estrin, D.; Fisher, J.; Gilbert, R.; Hansen, M.; Harmon, T.C.; Jay, J.A.; Kaiser, W.J.; Sukhatme, G.; Tai, Y.-C. *Distributed Sensing Systems for Water Quality Assessment and Management*; Center for Embedded Network Sensing Papers, 2007. Available at http://www.cens.vela.edu/pub/sensingwhitepage/sensor_whitepaper_lr.pdf (accessed January 12, 2009).
3. Smith, W.; Kelly, S. Science, technical expertise and the human environment. Prog. Plann. **2003**, *60*(4), 321–394.
4. Lautenbacher, L.C. The Global Earth Observation System of Systems: science Serving Society. Space Policy **2006**, *22*(1), 8–11.
5. Hara, N.; Solomon, P.; Kim, S.L.; Sonnenwald, D.H. An emerging view of scientific collaboration: scientists' perspectives on collaboration and factors that impact collaboration. J. Am. Soc. Inform. Sci. Technol. **2003**, *54*(10), 952–965.
6. Benoit, G. Bioinformatics. In *Annual Review of Information Science and Technology*; Cronin, B., Ed.; Information Today: Medford, NJ, 2002; Vol. 39, 179–218.
7. Atkins, D.E. *Revolutionizing Science and Engineering Through Cyberinfrastructure: Report of the National Science Foundation Blue-Ribbon Advisory Panel on Cyberinfrastructure*, National Science Foundation, 2003. Available at http://www.nsf.gov/od/oci/reports/atkins.pdf (accessed September 28, 2007).
8. Gold, A. Cyberinfrastructure, data, and libraries, Part 1. D-Lib Mag. **2007**, *13*(9/10).
9. Almes, G.; Birnholtz, J.P.; Hey, T.; Cummings, J.; Foster, I.; Spencer, B. *CSCW and Cyberinfrastructure: Opportunities and Challenges*, Chicago, IL, 2004, Proceedings of the 2004 ACM Conference as Computer Supported Co-Operative Work, Chicago, IL, 2004; 270–273.
10. Heidorn, P.B.; Palmer, C.L.; Wright, D. Biological information specialists for biological informatics. J. Biomed. Discov. Collab. **2003**, *2*(1).
11. Frame, M.T. The national biological information infrastructure: building information and geospatial technologies for biological community Proceedings of the 2004 Annual National Conference on Digital Government Research Seattle, WA May, 24–26, 2004; Association for Computing Machinery: Washington, DC, 1–4.
12. Morens, D.M.; Taubenberger, J.K.; Fauci, A.S. Predominant role of bacterial pneumonia as a cause of death in

pandemic influenza: implications for pandemic influenza preparedness. J. Infect. Dis. **2008**, *198*(7), 962–970.

13. Fjordback Søndergaard, T.; Andersen, J.; Hjørland, B. Documents and the communication of scientific and scholarly information: revising and updating the UNISIST model. J. Doc. **2003**, *59*(39), 278–320.

14. Kanare, H.M. *Writing the Laboratory Notebook;* American Chemical Society Press: Washington, DC, 1998.

15. Finholt, T.A.; Olson, G.M. From laboratories to collaboratories: a new organizational form for scientific collaboration. Psychol. Sci. **1997**, *8*(1), 28–36.

16. Lawal, I. Scholarly communication: the use and non-use of e-print archives for the dissemination of scientific information [electronic version]. Issues Sci. Technol. Librarianship **2002**, Fall. Available at http://istl.org/02-fall/article3.html (accessed September 28, 2007).

17. Haas, J.K.; Samuels, H.W.; Simmons, B.T. *Appraising the Records of Modern Science and Technology: A Guide;* Massachusetts Institute of Technology: Cambridge, MA, 1985.

18. National Science Board. *Long-lived Data Collections: Enabling Research and Education in the 21st Century*, 2005. Available at http://www.nsf.gov/nsb/documents/2005/LLDDC_report.pdf (accessed September 28, 2007).

19. Palmer, C.L.; Cragin, M.H.; Heidorn, B.P. Supporting biological information work: research and education for digital resources and long-lived data Proceedings of the 6th ACM/IEEE-CS Joint Conference on Digital Libraries Chapel Hill, NC June 11–15, 2006; Association for Computing Machinery: Washington, DC, 2006; 353.

20. Elliott, C.A., Ed. *Understanding Progress as Process: Documentation of the History of Post-War Science and Technology in the United States*; Society of American Archivists: Chicago, IL, 1983.

21. Dewatripoint, M.; Ginsburgh, V.; Legros, P.; Walckiers, A.; Devroey, J.-P.; Dujardin, M.; Vandooren, F.; Dubois, P.; Foncel, J.; Ivaldi, M.; Heusse, M.-D. *Study on the Economic and Technical Evolution of the Scientific Publication Markets in Europe*, 2006. Available at http://ec.europa.eu/research/science-society/pdf/scientific-publication-study_en.pdf (accessed September 28, 2007).

22. Correia, A.M.R.; Teixeira, J.C. Reforming scholarly publishing and knowledge communication—from the advent of the scholarly journal to the challenges of open access. Online Inform. Rev. **2005**, *29*(4), 349–364.

23. Bishop, A.P.; Neumann, L.J.; Star, S.L.; Merkel, C.; Ignacio, E.; Sandusky, R.J. Digital libraries: situating use in changing information infrastructure. J. Am. Soc. Inform. Sci. **2000**, *51*(4), 394–413.

24. Recker, M. Perspectives on teachers as digital libraries users. D-Lib. Mag. **2006**, *12*(9).

25. Fraser, S.P.; Deane, E.M. Educating tomorrow's scientists: IT as a tool, not an educator. Teach. High. Educ. **1999**, *4*(1), 91–106.

26. Kavulya, J.M. Digital libraries and development in Sub-Saharan Africa. Electron. Libr. **2007**, *25*(3), 299–315.

27. Patil, D.J.; Simon, J.; Cumberledge, S. Digital science libraries: practical approaches to supporting science in developing countries. UN Chronicle **2006**, *43*(3), 60–61.

28. Munoz, E. *The Right to Travel: The Effect of Travel Restrictions on Scientific Collaboration between American and Cuban Scientists*; American Association for the Advancement of Science: Washington, DC, 1998. Availale at http://shr.aaas.org/rtt/report/report.htm (accessed September 28, 2007).

29. Frohmann, B. The role of the scientific paper in science information systems Proceedings of History and Heritage of Science Information Systems Philadelphia, PA October, 23–25, 1998; Bowden, M.E., Bellardo, T.H., Williams, R.V., Eds.; Information Today: Medford, NJ, 63–73.

30. Palmer, C.L. *Work at the Boundaries of Science: Information and the Interdisciplinary Research Process;* Kluwer Academic Publishers: Dordrecht,the Netherlands, 1996.

31. Apedoe, X.S.; Reeves, T.C. Inquiry-based learning and digital libraries in undergraduate science education. J. Sci. Educ. Technol. **2006**, *15*(5–6), 321–330.

32. Dumouchel, B.; Demaine, J. Knowledge discovery in the digital library: access tools for mining science. Inform. Services Use **2006**, *26*(1), 39–44.

33. Goldey, E. Disciplinary integration: the sciences and humanities in learning. *Invention and Impact Building Excellence in Undergraduate Science, Technology, Mathematics, and Engineering Education*; American Association for the Advancement of Science: Washington, DC. Available at http://shr.aaas.org/rtt/report/report.htm (accessed September 28, 2007).

34. Wood, W.B. Inquiry-based undergraduate teaching in the life sciences at large research universities: a perspective on the Boyer report. Cell. Biol. Educ. **2003**, *2*(2), 112–116.

35. Blatt, R.J.R. OECD workshop on genetic testing, Vienna 2000. Banking biological collections: data warehousing, data mining, and data dilemmas in genomics and global health policy. Commun. Genet. **2000**, *3*(4), 204–211.

36. Theobald, D.M.; Hobbs, N.T.; Bearly, T.; Zack, J.A.; Shenk, T.; Riebsame, W.E. Incorporating biological information in local land-use decision making: designing a system for conservation planning. Landscape Ecol. **2000**, *15*(1), 35–45.

37. Castillo, A.; García-Ruvalcaba, S.; Martinez, R.; Luis, M.M. Environmental education as facilitator of the use of ecological information: a case study in Mexico. Environ. Educ. Res. **2002**, *8*(4), 395–411.

38. DebBurman, S.K. Learning how scientists work: experiential research projects to promote cell biology learning and scientific process skills. Cell Biol. Educ. **2002**, *1*(2), 154–172.

39. Glasson, G.E.; McKenzie, W.L. Investigative learning in undergraduate freshman biology laboratories. J. College Sci. Teach. **1998**, *27*(2), 189–193.

40. Brashers, D.E.; Stephen, M.H.; Neidig, J.L.; Rintamaki, L.S. Social activism, self-advocacy, and coping with HIV illness. J. Soc. Pers. Relat. **2002**, *19*(1), 113–133.

41. Heidorn, P.B.; Mehra, B.; Lokhaiser, M.F. Complementary user-centered methodologies for information seeking and use: systems design in the Biological Information Browsing Environment (BIBE). J. Am. Soc. Inform. Sci. Technol. **2002**, *53*(14), 1251–1258.

42. Card, J.J.; Shapiro, L.; Amarillas, A.; McKean, E.; Kuhn, T. Broadening public access to data through the development of tools for data novices. Soc. Sci. Comput. Rev. **2003**, *21*(3), 352–359.

43. Dawyndt, P.; Dedeurwaerdere, T.; Swings, J. Contributions of bioinformatics and intellectual property rights in

sharing biological information. Int. Soc. Sci. J. **2006**, *58* (188), 249–258.

44. Epstein, S. *Impure Science: AIDS, Activism, and the Politics of Knowledge;* University of California Press: Berkeley, CA, 1996.

45. Fienberg, S.E. Sharing statistical data in the biomedical sciences. Ann. Rev. Publ. Health **1994**, *15*, 1–18.

46. Birnholtz, J.P.B.M. Data at work: supporting sharing in science and engineering International ACM SIGGROUP Conference on Supporting Group Work, Sanibel Island, FL Nov, 9–12, 2003; Association of Computing Machinery: Washington, DC, 2003; 339–348.

47. Shabo, A.; Rabinovici-Cohen, S.; Vortman, P. Revolutionary impact of XML on biomedical information interoperability. IBM Syst. J. **2006**, *45*(2), 12.

48. Sherratt, T.; McCarthy, G. Mapping scientific memory: understanding the role of recordkeeping in scientific practice. Archives and Manuscripts **1996**, *24*(1), 78–85.

BIBLIOGRAPHY

1. Abbas, J. Finding science resources online with the ARTEMIS digital library. Know. Quest **2003**, *31*(3), 12.

2. Allen, D.; Tanner, K. Approaches to cell biology teaching: questions about questions. Cell Biol. Educ. **2002**, *1*(1), 63–67.

3. Bishop, A.P. Document structure and digital libraries: how researchers mobilize information in journal articles. Inform. Process. Manage. **1999**, *35*(3), 255–279.

4. BrightSparcs. Available at http://www.asap.unimelb.edu.au/bsparcs/.

5. Dedeurwaerdere, T. The institutional economics of sharing biological information. Int. Soc. Sci. J. **2006**, *58*(188), 351–368.

6. Gehring, K.M.; Eastman, D.A. Information fluency for undergraduate biology majors: applications of inquiry-based learning in a developmental biology course. CBE Life Sci. Educ. **2008**, *7*(1), 54–63.

7. Heidorn, P.B. Publishing digital floras and faunas. Bull. Am. Soc. Inform. Sci. Technol. **2003**, *30*(2), 8–11.

8. Janse, G. Communication between forest scientists and forest policy-makers in Europe—a survey on both sides of the science/policy interface. Forest Pol. Econ. **2008**, *10*(3), 183–194.

9. John, C. *Liebeskind History of Pain Collection*, University of California, Los Angeles History of Pain Project. Available at http://www.library.ucla.edu/biomed/his/pain.html.

10. Lundmark, C. BEN: the biology branch of the National Science Digital Library. Bioscience **2003**, *53*(7), 631.

11. Mardis, M.A.; Payo, R.P. Making the school library sticky: digital libraries build teacher-librarians' strategic implementation content knowledge in science. Teach. Libr. **2007**, *34*(5), 8–14.

12. Mischo, W.; Norman, M.; Shelburne, W.A.; Schlembach, M. The growth of electronic journals in libraries: access and management issues and solutions. Sci. Technol. Libr. **2006**, *26*(3/4), 31.

13. Murphy, J. Information-seeking habits of environmental scientists: a study of interdisciplinary scientists at the Environmental Protection Agency in Research Triangle Park, North Carolina. Issues Sci. Technol. Librarianship **2003**, *38*. Available at http://www.istl.org/03-summer/refereed.html.

14. National Research Council. *BIO 2010, Transforming Undergraduate Education for Future Research Biologists*; National Academy of Sciences Press: Washington, DC, 2003.

15. Palmer, C.L. Information work at the boundaries of science: linking library services to research practices. Libr. Trend. **1996**, *45*(2), 27.

16. Saylor, J.M.; Minton-Morris, C. The National Science Digital Library: an update on systems, services and collection development. Sci. Technol. Libr. **2006**, *26*(3/4), 61–78.

Blind and Physically Disabled: Library Services

Jane Rosetta Virginia Caulton
Stephen Prine
Library of Congress, Washington, District of Columbia, U.S.A.

Abstract

Library service for individuals who are blind or have a physical disability is an ever-expanding phenomenon, thanks to both the innovative spirit of the users and advances in technology. This entry discusses the 1931 establishment and subsequent growth of the National Library Service for the Blind and Physically Handicapped, Library of Congress. The importance of the network of cooperating agencies at the state and local levels and of the volunteers who support the program is also discussed. The entry presents the history of special-format reading materials in the United States. Specifically covered are the emergence of electronic braille and the development of digital audiobooks, both of which are accessible by smartphone.

INTRODUCTION

The path to providing quality library service to people who cannot read regular print with corrective lenses or who cannot handle printed materials because of a physical disability has experienced many advances, twists, and setbacks. The commitment and dedication of individuals and organizations and developments in braille and recording technology have combined to significantly improve access to written materials for all. The National Library Service for the Blind and Physically Handicapped (NLS)—a program of the Library of Congress—and its network of cooperating libraries have evolved from events dating back to the 1860s and are part of this expanding landscape. Together they form a unique organization of cooperative federal, state, and local programs that serve eligible residents of the United States and its territories and U.S. citizens living abroad.

This network, one of the oldest in the country, has its origins in the Act to Provide Books for the Blind—commonly called the Pratt–Smoot Act—which was signed into law in 1931. In the cooperative arrangement, NLS serves as the administrative agency responsible for selecting print books and magazines to be added to the national collections, for contracting with various entities to produce them in braille and audio formats, and for ensuring the availability of playback equipment. Cooperating libraries handle the daily interface with patrons; ensure that the books, magazines, and playback equipment are circulated to registered readers (see Fig. 1); and maintain necessary records on readership and inventory. This relationship between NLS and its affiliates is based on an informal agreement, which has been acceptable for the establishment of distribution centers (which now include regional libraries, subregional libraries, and advisory and outreach centers) since the program's inception.

Several other elements played vital roles in the effectiveness of the program: Free Matter postal legislation,

volunteers, the goodwill of authors and publishers, and later P.L.104-197 (known as the Chafee Amendment), which granted copyright permissions at no cost. The combination of these diverse elements made it possible for the program, in 2013, to serve more than 929,939 eligible readers. In that year, NLS circulated more than 24 million copies of the 217,896 titles included in its *Union Catalog*.

HISTORY

In the nineteenth century, library service for blind people was virtually nonexistent. In 1868, the Boston Public Library acquired a collection of books in an embossed format and began serving local residents. The Chicago Public Library followed in 1894, the New York State Library in 1895, the Library of Congress in 1897, and the Free Library of Philadelphia in 1899. Other collections were established, but most served only local populations. Several different systems of embossed formats were used by various library providers. Readers who learned one system could not necessarily read others. In addition, libraries could purchase only a few titles because of the high cost of production. These issues severely limited the ability of people who were blind to enjoy the same access to information as their sighted neighbors.

In 1917, the United States adopted the Revised Braille (known then as grade 1-1/2 braille) code as the standard for embossed materials. The system was invented in France in 1821 by Louis Braille, who used a military code as a model. The braille code is based on a six-dot cell, three rows high and two dots wide. Dots in the cell are raised according to the letters or numbers the dots represent (Fig. 2). Since its introduction, the code has been revised to accommodate changes in language and usage. Grade 1-1/2 braille integrates elements of the letter-for-

Encyclopedia of Library and Information Sciences, Fourth Edition DOI: 10.1081/E-ELIS4-120053406

Binding-British

Fig. 1 A government worker teaches a blind veteran to use a digital talking-book player. Talking books enable individuals who are blind, are visually impaired, or have a physical disability that prevents handling print materials to access books and magazines. The National Library Service for the Blind and Physically Handicapped, Library of Congress, collections include bestsellers, biographies, and self-help materials.

letter transliteration of uncontracted braille and the abbreviated form of contracted braille. By 1925, the Library of Congress Reading Room for the Blind had expanded its readership from about 150 local readers to 2400 national patrons, and the collection had grown to 2400 volumes. Although improved, library service for blind people was still fairly sparse and the standard braille used today had not yet been adopted.

Recognizing the need for a centralized national library service, blind citizens and their advocates combined their efforts to acquire transcribed materials with coordinating economies of scale. Organizations serving blind individuals, including the American Foundation for the Blind in New York, the American Printing House for the Blind in Louisville, and the Braille Institute of America in Los Angeles, began to lobby Congress. In 1930, three bills addressing the issue were introduced in the House of Representatives and one in the Senate. The bills introduced by Congresswoman Ruth Pratt of New York and Senator Reed Smoot of Utah survived.

NETWORK OF COOPERATING LIBRARIES

The passage of the Act to Provide Books for the Blind (the Pratt–Smoot Act) and its signing into law on March 3, 1931, by President Herbert Hoover, created the first national free library service for people who were blind. The Act reads:

> That there is hereby authorized to be appropriated annually to the Library of Congress, in addition to appropriations otherwise made to said Library, the sum of $100,000, which sum shall be expended under the direction of the Librarian of Congress to provide books for the use of adult blind residents of the United States, including the several States, Territories, insular possessions, and the District of Columbia.

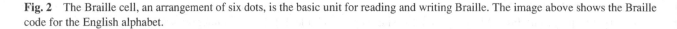

Fig. 2 The Braille cell, an arrangement of six dots, is the basic unit for reading and writing Braille. The image above shows the Braille code for the English alphabet.

The Pratt–Smoot Act also acknowledged the necessity of localized service:

> The Librarian of Congress may arrange with such libraries as he may judge appropriate to serve as local or regional centers for the circulation of such books, under such conditions and regulations as he may prescribe. In the lending of such books preference at all times shall be given to the needs of blind persons who have been honorably discharged from the United States military or naval service. (Sec. 2)

The year the law was signed, 19 libraries from around the country, including the Library of Congress, circulated books in embossed formats. Two more joined the following year. These libraries, which formed the nucleus of the modern network of cooperating libraries, were called regional distribution centers.

In 1933 the Pratt–Smoot Act was amended to include books "published either in raised characters, or on sound reproduction records, or in any other form." This change sparked an expansion of the network from 21 to 26 regional libraries by 1934. A second growth spurt began in 1950 and extended through 1976, ending with a total of 56 regional libraries.

During this latter period, the legislation received additional amendments. In 1952, the word "adult" was removed, permitting the braille and talking-book program to serve blind children; in 1962, a library of "musical scores, instructional texts, and other specialized materials" was established; and in 1966, Congress provided for the reading needs of "blind and other physically handicapped residents." All of these changes were codified in Public Law 89–522 and approved on July 30, 1966.

With the growth in responsibility, the Library of Congress expanded the Service to the Blind project to Books for the Adult Blind. The unit became the Division of Books for the Adult Blind in 1943, later the Division for the Blind (1952), then the Division for the Blind and Physically Handicapped (1966), and finally the National Library Service for the Blind and Physically Handicapped (1978).

The 57th regional library was established in Bismarck, North Dakota, in 1995, leaving only Wyoming without its own special-format library. Wyoming's patrons are served under contract by the Utah regional library. The network organized into four regions—the Midlands, North, South, and West—to facilitate interlibrary support, help librarians keep abreast of industry issues, and share information. Since 2009, two regional libraries in the Midlands have discontinued service. In each instance, however, a second regional library in the state assumed service for all eligible residents.

Subregional libraries: In the late 1960s, the concept of providing reading service through local entities emerged. Subregional libraries are extensions of the regional libraries, although they typically are funded separately and may report to a different parent organization. The NLS annual report for fiscal year 1971 shows records that the network had four subregionals in Alabama, six in Kansas, and one in Nebraska. By the mid-1980s, these had been joined by 93 others for a total of 104. Since then, however, many subregionals have fallen victim to budget cuts. In July 2014, the network had 36 subregional libraries: 6 in the Midlands, 4 in the North, 23 in the South, and 3 in the West.

Advisory and outreach centers: The economic crisis that hit America starting in 2008 resulted in the constriction of the NLS network of cooperating libraries. As funding for public libraries decreased, many subregional libraries closed. In 2011, advisory and outreach centers were introduced to the network. Georgia and Kansas replaced their subregional libraries with advisory and outreach centers, which provide reader advisors to assist patrons with service issues and conduct outreach activities. Other states have followed suit and the network now includes a total of 15 advisory and outreach centers: 12 in the Midlands, 1 in the North, and 2 in the South.

Machine-lending agencies and multistate centers: When Congress authorized the Library of Congress to issue talking-book machines (TBMs) to eligible individuals for the purpose of reading, the agency established distribution centers. The centers were important to NLS's ability to track machine placement. Today most regional libraries provide machine services to their patrons. Separate machine-lending agencies (MLAs) in Minnesota, North Dakota, and Ohio support regional libraries that do not loan playback equipment, and one in Wyoming provides playback equipment to eligible residents in the state who receive library service from Utah.

The current network is also supported by two warehouse facilities called multistate centers. The Multistate Center East (MSCE) in Cincinnati, Ohio, handles the storage and distribution needs of regional libraries east of the Mississippi River, including those in the territories of Puerto Rico, and the U.S. Virgin Islands. It manages the audio and braille foreign-language collection and, since the early 1980s, has conducted the network's Quality Assurance Program to ensure the consistency and quality of locally produced audiobooks and magazines. The second facility, Multistate Center West, in Salt Lake City, Utah, serves libraries west of the Mississippi, including those in Alaska, Hawaii, and Guam. It also is the repository for special braille collections (Fig. 3) and for NLS playback equipment.

POPULATION SERVED

The free library service administered by NLS is available to any resident of the United States who is unable to read standard print because of blindness, visual impairment, or

Fig. 3 Print-Braille books from the National Library Service for the Blind and Physically Handicapped, Library of Congress, enable blind parents to share reading with their sighted children.

a physical disability that prevents holding a book or turning pages. Service is available on a temporary or long-term basis.

The majority of program users fit into one of two broad categories. First are individuals who are born blind, visually impaired, or with a physical limitation. Once individuals from this group learn about the program and apply for service, they may use it throughout their lives. The second group—the larger of the two—are individuals who become eligible later in life because of an age-related illness or disease. Arthritis, diabetes, glaucoma, and macular degeneration are the most common causes for eligibility. A subcategory of this group includes those who lose the ability to use print because of an accidental injury.

Currently, 69% of active readers in the program are 60 yr of age or older. The onset of a condition, later in life, that makes them eligible for the program has tended to make readers who are in their 1960s or older the primary users of this program. In fact, more than 3600 readers are more than 100 yr old.

It is important to note that veterans are given primary preference by law. As stated in Section 2 of An Act to Provide Books for the Adult Blind,

In the lending of such books preference shall at all times be given to the needs of blind persons who have been

honorably discharged from the United States military or naval service.

According to the Blinded Veterans Association, 7000 aging veterans become blind each year as a result of macular degeneration, retinitis pigmentosa, glaucoma, or diabetic retinopathy. In addition, approximately 13% of the wounded soldiers returning from the Middle East have sustained visual impairment of some type. The free library service that NLS provides assures that they will be able to continue to access the written word.

THE LIBRARY COLLECTION

The braille and talking-book collection of titles was established on the ideology that eligible readers should have access to the same type of materials found in public libraries for the sighted population. The collection emphasizes recreational and informational reading material. Among the nonfiction and fiction titles, including classics and bestsellers, patrons may select from a wide range of genres. Biographies, political commentaries, religion, and science titles may be found among the nonfiction offerings, while stories about mysteries, romances, science fiction, and westerns are popular fiction selections. Age-appropriate titles in similar categories are also available for younger readers. Though most of the offerings are English, the collection includes foreign-language materials, principally Spanish. The collection also offers subscriptions to popular magazines, selected for special format according to patron interest. The *NLS Collection Building Policy* explains that the interests of "the aged, the young, professional people, and other specific groups [are to be] reflected proportionally in the collections in relation to the overall readership served."

As patron consideration is an important aspect of service, NLS is aided in its collection-building activities by a group that meets biennially. The National Collection Advisory Group includes patrons, librarians, and representatives of constituency organizations who gather information from their peers to recommend strategies for NLS to consider when making selections for the titles. Principles guiding selection are to ensure that titles are available to meet the interests of a population with broad and diverse interests, tastes, and reading levels. The *NLS Collection Building Policy* bases consideration of nonfiction titles on the following:

1. Broad trends in public interests, knowledge of developing theories and practices in the various subject areas, and availability of appropriate titles
2. Materials in areas of particular interest to readers, classic and standard materials, contemporary works, specialized works of interest to educated laymen, and works with potential informational reading reference

Annually, NLS seeks to add 3000 titles in audiobook format and about 500 titles in braille. Ratios of titles selected for audiobooks in each subject area are 60% fiction to 40% nonfiction, 70% current to 30% retrospective, and 80% adult to 20% juvenile. Braille, however, is selected based on cost: the larger and/or more complex the title, the higher the production costs. "For example: Stephen King's novel *11/22/63* contains 1576 transcribed pages in eleven volumes. NLS produced forty-three copies of 7,768 braille pages each, which cost $23,300" (*The Future of Braille*, 2014).

MUSIC LIBRARY SERVICES

A special music collection for blind persons was authorized by Congress as part of the national free library program in 1962. The collection became the basis for the music program administered through the NLS Music Section. These services, extended to individuals with disabilities in 1966, are provided directly to eligible individuals from Washington, DC. In effect, the Music Section is the national music library for people who are blind and visually impaired or have a disability.

The collection offers braille, large-print, and recorded instructional materials about music and musicians. More than 30,000 titles in audio, braille, and large-print music scores, texts, and other instructional materials (in excess of 100,000 volumes) are available. Some materials are purchased from national and international commercial sources. Other items are selected and produced in audio and braille formats by NLS with the permission of composers, authors, and publishers. Some titles are developed solely for the NLS program. Patrons may also obtain free subscriptions to several magazines produced in audio, braille, and large print. Musical recordings intended solely for listening are not included in the music collection as they are readily available commercially and from public libraries.

BRAILLE AND TALKING BOOKS

Within 2 yr of the Pratt–Smoot Act becoming law, braille had become the standard for embossed reading materials in the United States. In 1931, the Division of Books for the Adult Blind selected 157 braille titles for the collection; and starting in 1934, the national program also offered sound recordings of books on 33-1/3 rotations per minute (rpm) rigid discs (records). The addition of talking books was especially well received because many individuals who were blind had never learned to read braille.

In 1935, the Works Progress Administration was allocated funding to build and repair record players for blind adults to read Library of Congress–produced recorded books. After World War II, Congress provided an appropriation for the purchase of TBMs, which were distributed to regional libraries and MLAs for loan to eligible readers. By the mid-1960s, the Library of Congress began looking at ways to reduce the size and weight of the materials being handled by patrons, network library staffs, and the United States Postal Service (USPS). The first experiment slowed the speed of the 33-1/3 rpm record to 16-2/3 rpm, then to 8-1/3 rpm. Simultaneously, NLS began experimenting with providing books on magnetic tape. Initially, titles were produced on open-reel tape and loaned to readers who had access to open-reel tape players. While the tape technology proved effective, the open-reel format was inefficient. The experience, however, led NLS to the Phillips cassette, which was becoming the commercial recording industry standard. A C-90 Phillips cassette played for 45 min per side or 90 min overall at 1-7/8 inches per/second (ips).

NLS experimented with this format, at first reducing the play speed from 1-7/8 to 15/16 ips, which doubled the play time from 90 to 180 min. A second change was recording on four tracks instead of two, which added another 180 min. Unlike commercial cassettes that play for 1-1/2 hr, talking-book cassettes play for 6 hr. This format meant that readers needed to handle only a single cassette instead of four. The noncommercial speed prohibited talking books from being played on commercial machines, thereby ensuring copyright protection. The new format also qualified the books for Free Matter mailing.

TALKING-BOOK MACHINES

In 1947, the Library of Congress received its first appropriation for the purchase of special record players. From its introduction in 1935 through the transition to the cassette machine in 1971, the TBM was continuously modified to achieve peak performance. In the first decade, attention was given to the motor, pickup arm, and amplifier. In the 1950s, motor speeds began increasing to accommodate 33-1/3, 16-2/3 rpm, and, by 1965, 8-1/3 rpm. A lightweight model with a transistorized amplifier and plastic case was introduced in 1968. In the early 1970s, the TBM was further refined with an automatic cut-off switch, two-sided needle, and a detachable lid that held the speaker. The last model was joined by a cassette player and called the combination machine. It allowed patrons to play magazines on discs and audiobooks on cassettes, which were introduced in 1971.

Cassette machines debuted in two models: a playback-only and a record-and-play designed for students. The playback-only held a rechargeable nickel–cadmium battery and the latter used size C dry-cell batteries. Both operated at 1-7/8 and 15/16 ips. Cassette machines were produced to accommodate the four-track cassette beginning in 1974, and in 1976 an automatic shutoff, tape

Binding–British

motion sensor, and variable speed control were added. The 1980 model included a pitch restoration model that retained the pitch even at higher speeds. The Easy Machine, a cassette machine operated by two controls, was introduced to assist readers with low dexterity. The machine started and stopped with a sliding door and had a button to rewind the tape.

QUALITY IMPROVEMENTS

Recognizing that analog cassette tape would become obsolete by the early twenty-first century, NLS began exploring alternative formats for recorded books in the late 1980s. A review of the market at the time indicated that the recording and distribution of books and magazines should be moved from an analog to a digital environment and that the output of digital books required a format robust enough to stand up to the extensive handling and mailing that circulating machines and audiobooks between libraries and patrons entails.

By the mid-1990s, it was clear that digital technology would become the accepted medium. Digital recordings had emerged in several formats: digital audiotape (DAT), digital compact cassette (DCC), compact disc (CD), digital video disc (DVD), and mini discs (a smaller compact disc developed by Sony). Each of these applications utilized digital recordings, but all were susceptible to the same weaknesses as their analog counterparts. DAT and DCC relied on magnetic tape for storage, and CD and DVD encountered fragility problems not recognized at their inception. The NLS search for a durable medium led to flash memory. It was costly, but market research indicated that by 2008 the cost would drop enough to make the medium affordable.

Under the National Information Standards Organization (NISO), part of the American National Standards Institute (ANSI), an international team of experts and interested parties released the ANSI/NISO Z39.86, the first standard for digital talking books, in 2002. With the NISO standard in hand and flash memory designated as the format, NLS selected a company to design a proprietary digital TBM, a cartridge, and a cartridge container. The design phase was completed and approved in the spring of 2007, and manufacturing began in 2008. NLS launched its digital talking-book system in 2009.

The transition from analog to digital service was well received by users of the NLS program. The digital talking-book system includes a player—available in two models, a flash-memory cartridge, and the cartridge container (Fig. 4). Patrons enjoy the higher-quality sound, easy navigation, and convenience of an incremental shutoff button in the basic machine. The advanced machine offers additional features such as chapter-to-chapter navigation. Both machines weigh less than the cassette player and are easier to carry. The larger buttons accommodate individuals who have dexterity challenges.

BRAILLE AND TALKING-BOOKS SERVICE TO GO

Prior to the emergence of the digital audiobook system, NLS launched the Braille and Audio Reading Download (BARD) Internet service. BARD offers patrons the ability to have their reading material in a matter of minutes as opposed to the week or two required for regular circulation activities. Books and magazines may be downloaded from BARD and played on NLS or authorized commercial players.

Interest in developing the BARD service was spawned by the success of Web-Braille, the first NLS exploration into providing online access to its collections. Established in 1999, Web-Braille, a password-protected website, allowed patrons to access thousands of complete e-braille books and magazines in contracted (formerly known as

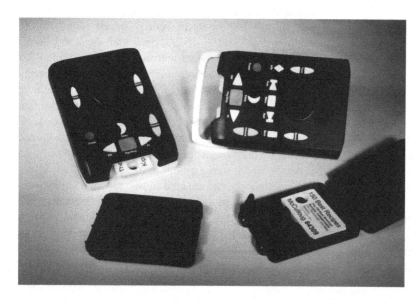

Fig. 4 The digital talking-book system issued by the National Library Service for the Blind and Physically Handicapped, Library of Congress, provides an improved reading experience. Patrons may choose a basic player (left) or an advanced player (right). A single cartridge, circulated in a lightweight plastic container, will hold most books.

grade 2) braille. Books were added to the Web-Braille site after production in press-braille (paper) distribution to braille-lending libraries. Patrons received log-on IDs through their network libraries and then accessed the site to download braille books and magazines using braille-embossing hardware.

In 2006, NLS used Web-Braille as a working model, as several elements had emerged to create an environment for establishing an audiobook download website. First, Adaptive Multi-Rate Wideband Plus, an audio compression algorithm, had been selected as the best code-decode (codec) for ensuring high-quality human-voice recording for digital production. Second, a scheme for digital rights management to ensure copyright protection had been developed; and third, a compatible digital machine had been produced. Together they presented an opportunity for NLS to make the large collection of digital audio files it had amassed in preparation for the digital talking-book system available online.

In October 2006, NLS began testing the system's usability by launching its downloadable book pilot. A year later the pilot was expanded to the present BARD system. The password-protected website provides access to thousands of recorded and braille books and magazines, with weekly additions. In 2013, Web-Braille merged with BARD, and in 2014, network libraries were invited to submit their locally produced books and magazines for inclusion on BARD. More than 60,000 patrons are registered users of the website, which has more than 50,000 book titles and 94 magazine titles. Patrons may also download instructional music materials and some foreign-language books and magazines.

BARD Mobile: Always scanning the technological environment for opportunities to improve reading accessibility for individuals who cannot see to read regular print or handle printed materials, NLS turned its attention to smartphones. In September 2013, NLS successfully launched BARD Mobile, a software application that enables registered patrons to download audio and braille books and magazines from the BARD website using an iPhone, iPad, or iPod touch. All aspects of the app were designed and developed to ensure complete accessibility and a premium patron experience, including navigability and download speed. Users must register for access through their network libraries.

Audiobook readers are able to use the device's built-in speaker, and braille readers are able to read e-braille books, magazines, and music scores using a refreshable braille display connected via Bluetooth to their iOS devices. The new app increased the portability of the free library service and patrons expressed delight that it did not require special equipment; they can use the same type of devices as their sighted peers. About 5000 patrons signed up for BARD Mobile within the first week of release. In the first year, more than 10,000 users had registered for the app and registered 17,200 devices.

EQUIPMENT REPAIR AND THE ROLE OF VOLUNTEERS

Because talking-book equipment is loaned, users had to return broken or defective equipment to their MLAs and wait to receive replacements. At first, machines that needed repair were handled by centralized repair facilities, but that often left readers waiting for weeks or months to get them back. Consequently, the Division of Books for the Adult Blind changed its policy and required MLAs to replace broken machines immediately. This requirement ensured that the reader would quickly receive a replacement player, but it also created a backlog of machines needing service. Repair at the local level was sporadic, mostly because the staff did not have the necessary technical background or training. This ongoing problem was resolved in 1960 when the program, by then the Division for the Blind, formed a partnership with a volunteer organization whose members had both technical background and the time to devote to repairing the playback equipment.

Machine-repair volunteers: The Telephone Pioneers of America, formed by long-tenured or retired telephone company employees, was a nationwide organization looking for public service projects. In 1959, the Division and the Pioneers launched a pilot machine-repair project. The success of this project proved mutually beneficial, and the Pioneers adopted the TBM-repair program as a project on June 6, 1960. This group, which in 2008 changed its name to Pioneers—A Volunteer Network, has provided decades of significant community service by repairing TBMs. In 1989, the General Electric (GE) Elfun Society, a group of retired GE employees who were also looking for a community service project, joined the NLS volunteer machine-repair program. Its repair facilities are in Fort Wayne, Indiana; Louisville, Kentucky; Lynn, Massachusetts; Schenectady, New York; Cincinnati, Ohio; Philadelphia, Pennsylvania; and Salem, Virginia. In 2006, this organization changed its name to GE Volunteers. Together Pioneers and GE Volunteers have provided millions of dollars in service to the talking-book program, network libraries, and their patrons. More than 288 repair sites still operate around the country, a tribute to the dedication of the men and women who volunteer.

Other volunteers: Volunteers have played a crucial role in ensuring that individuals who are blind or have a physically disability gained and retained access to reading materials. Even before the national program was set in place, volunteers began organizing to provide hand-copied braille materials to blinded veterans after World War I. The formation of such organizations required a method for ensuring the quality of braille transcription. The Library of Congress and the American Red Cross collaborated to produce a manual to guide the sighted transcribers and to establish a certification program. This program operated under the American Red Cross until 1943, when it was transferred to the Library of Congress.

Many braille transcriber groups operate across the United States. Some network libraries have braille-transcription programs that are supported by these volunteers. Others contribute to the network library system by donating materials or handling special assignments. Braille transcribers and proofreaders still must be certified by NLS, now handled under contract with the National Federation of the Blind in Baltimore, Maryland. Their efforts are guided by the *English Braille American Edition* and *Braille Formats: Principles of Print to Braille Transcription*, under the watchful eyes of such groups as the Braille Authority of North America (BANA) and the National Braille Association (NBA). BANA's mission statement reads in part that it "publishes rules, interprets and renders opinions pertaining to braille in all existing and future code" (http://www.brailleauthority.org), while the NBA provides professional development and training for transcribers.

Most positions in the audiobook recording industry are held by paid professionals; however, some network libraries and volunteer studios engage the help of volunteers to narrate, monitor, and review audiobooks. Books narrated by volunteers must meet the standards of narration established by NLS. In addition, volunteer-produced books nominated to the NLS *Union Catalog* must pass inspection by the NLS Quality Assurance Program at the MSCE.

Countless other volunteers help maintain library service by performing a variety of functions, from greeting visitors to inspecting and shelving returned books. These individuals contribute numerous hours to ensuring that patrons have access to reading materials. The network of cooperating libraries recognizes their work through commendations, awards, and/or special events.

COPYRIGHT

Copyright protection became an issue when talking books were introduced to the Books for the Adult Blind. At the program's inception, only embossed books, which could be used by only a small percentage of the population, were provided. Publishers did not perceive this special embossed format as a threat to their profit margins. Recorded books, however, were produced in a commercial format, so anyone who had a phonograph could use them. To address this issue, the Library of Congress set controls on recorded books. First, Books for the Adult Blind secured permission from the copyright owner to publish recorded books. Then, every recorded book contained an announcement acknowledging copyright permission and noting that the format was for blind users only.

In spite of these safeguards, publishers became more reluctant to grant permission for the special-format publications as demand for talking books grew. By 1936, the Library of Congress could only add books published before 1880, most of which were already available in braille or had been read by patrons who came to the program after losing their sight as adults. The Book Publisher's Bureau proposed that publishers grant permission for a nominal fee or at no cost to the Library of Congress, which became a clearinghouse for the special-format permissions. It tracked the number of permissions requested, the number granted, and the number of recordings made. The American Foundation for the Blind and the American Printing House for the Blind, which held government contracts to produce the books, labeled talking books "Solely for the Use of the Blind." They also ensured that this message was narrated, along with the name of the publisher.

Stakeholders felt that maintaining tight controls on TBMs was the key to copyright protection. TBMs were strategically placed in central locations because only one machine was provided for every six blind individuals in the United States. The Books for the Adult Blind program was required to track the location of the players and to issue regulations on their loan, maintenance, and repair.

This program worked for decades, although even with the controls in place it was a slow process. Then in 1996, at the request of representatives from organizations of individuals who are blind or who provided services to them, Senator John Chafee of Rhode Island introduced Legislative Branch Appropriations Bill, H.R. 3754, to amend Chapter 1, Title 7 of the U.S. Code as follows:

> It is not an infringement of copyright for an authorized entity to reproduce or to distribute copies or phonorecords of a previously published, nondramatic literary work if such copies or phonorecords are reproduced or distributed in specialized formats exclusively for use by blind or other persons with disabilities.

P.L.104-197, signed by President William Jefferson Clinton, effectively removed the need to request copyright permission for any U.S. nondramatic work intended for production in braille or recorded format specifically for readers who are blind and visually impaired or have a physical disability.

In 2013, NLS met with commercial audio publishers to discuss acquisitions of commercially recorded audio titles. Hachette Book Group of New York, New York, agreed to provide NLS with audio files at no cost. The relationship marked a major benchmark in the NLS effort to provide an ever-increasing selection of audio and braille titles for program users. Other publishers joined, and as a result, NLS increased its annual audio title offerings from 2000 to 3000. It must be noted that while the audio files have been donated, NLS must cover the costs of duplication and distribution.

FREE MATTER MAIL FOR BLIND AND HANDICAPPED INDIVIDUALS

The Free Matter for the Blind and Handicapped mailing privilege offered by the USPS predates the legislation that established NLS. Originally passed by Congress in 1904, it allowed public institutions serving blind individuals, public libraries, and blind readers to mail embossed materials free of charge. Two requirements for Free Matter mailing are noted in the *Domestic Mail Manual* Section 703.5, "Free Matter for the Blind and Other Handicapped Persons." First, the material being mailed must be in a special format intended for blind users such as braille; sight-saving type (i.e., 14-point or larger); recorded materials on cassette, records, or compact discs; or descriptive videos with audio narration. Second, the recipient of the material must be eligible to receive (and return) materials as Free Matter. The criteria specified include being an NLS patron, being blind or having a physical disability that prevents using printed materials, and being an American citizen or resident of the United States or its territories.

Residents of the United States who are eligible for braille or talking-book service are a microcosm of the general population, representing every age, gender, and socioeconomic status. The only generalization that can be made about these individuals is that their most common problem is mobility. As a consequence of their disability, they do not drive. Therefore being able to receive and return library materials through the USPS affords them a considerable degree of freedom and independence.

Free Matter postal service is vital to NLS and the network of cooperating libraries as it supports postage-free book circulation. Libraries mail books, magazines, and playback equipment to patrons free of charge. Once readers complete a book, they merely return it to the mail stream, and it is returned to the appropriate library. Without the Free Matter mailing privilege, the NLS free library service would not be nearly as successful. For the past decade, the NLS network of cooperating libraries has circulated approximately 23 million books and magazines annually, and approximately 20 million books are returned. Additionally, an untold number of playback machines, catalogs, newsletters, and order forms travel between patrons and libraries. Postal service employees who process and deliver these materials have been honored by individual network libraries for the dedication and concern NLS patrons report.

DEVELOPMENTS IN THE FIELD OF BRAILLE

The technological landscape of the twenty-first century flourished with opportunities for accessible options. Yet Braille has remained the primary literacy tool for individuals who are blind or visually impaired. The tactile format enables these people to engage in their environment by gaining the fundamental skills that make literacy possible. Amidst all of the change, however, Braille seemed to be at a standstill. In June 2013, NLS partnered with Perkins of Watertown, Massachusetts, to convene a conference exploring the status of braille, with a mind to shaping policies to expand and support this important literacy tool.

Called "The Future of Braille," the braille summit drew more than 100 librarians, educators, producers, engineers, and others to the Perkins campus to discuss issues and to formulate recommendations. Their major concerns were reducing the high cost of braille production, acquiring better technology and more training for instructors, and the need to expand knowledge about the medium. Stakeholders also heard from a representative of the U.S. Department of Education, which had just released a Dear Colleague letter advising educators to include braille in the individual education plans of qualifying students; the National Braille Press Center for Braille Innovation and the Daisy Consortium Transforming Braille Group on their efforts to produce a low-cost braille display; and the Braille Authority of North America on its adoption of the Unified English Braille code, an update of the braille code that has a set of rules applicable across the English language.

Participants recommended that NLS provide a refreshable braille display at no cost to patrons, vary the quality and/or publication medium of books depending on their use and expected shelf life, and work with publishers to acquire source texts. They also asked NLS to expand the use of tactile graphics in its books and to build support for efforts to update braille technology, specifications, and methods for selection, production (including production on demand), and distribution of braille materials. These recommendations will serve as a guide as NLS develops policy to better serve its braille readers.

FROM READER TO LEADER

On March 26, 2012, NLS marked another milestone in its development. For the first time since its establishment, the organization would be led by one of its patrons. Karen Keninger, who had lost her sight at age seven, was appointed director of the National Library Service for the Blind and Physically Handicapped. From 2008 until her appointment, she had served as the director of the Iowa Department for the Blind, where she oversaw statewide vocational rehabilitation and independent living programs and library services. Prior to that—from 2000 to 2008— she had served as director of the Iowa Library for the Blind and Physically Handicapped, where she planned and implemented a new, in-house digital recording program with a state-of-the-art recording studio.

Taking the helm of NLS, Keninger set five priorities for the program: 1) maintain the highest quality standards for all NLS products; 2) enhance the reading experience for all NLS patrons by leveraging current and future technologies to improve the reading and delivery systems; 3) expand the scope and quantity of titles available in alternative formats through the NLS system; 4) take a leading role in positioning braille as a viable, practical, and achievable literacy medium for all blind Americans; and 5) increase readership. Under her leadership, NLS has released the BARD Mobile app, engaged support from commercial publishers, held the first braille summit, and made digital audio materials produced by network libraries available on BARD.

CONCLUSION

Library service for eligible residents of the United States and its territories and for its citizens living abroad continues to expand as technology changes and new needs are identified. Within the NLS network, many libraries, both regional and subregional, offer services beyond those provided by the Library of Congress, such as large-print books and descriptive videos. Many also host community activities, such as summer reading programs for children and book discussion groups. They house recording studios that record books and magazines, normally of local significance, supplementing the NLS *Union Catalog*.

The development of library service to individuals who are blind, are visually impaired, or have a physical disability represents the combined efforts of many people, organizations, and institutions. Among these, the most important have been the readers. Their contributions have been evident from Louis Braille's use of a military code to create a tactile alphabet through contemporary demands for accessible technology, including braille displays and smartphone apps, to the rise of a patron to leadership of the national service. As noted in *That All May Read: Library Service for Blind and Physically Handicapped People*, "This history reflects the growing recognition that [people who are blind or have a physically disability] have the same interests, intellectual capacity, and ambitions as other members of society and the determination that they enjoy the same benefits."

BIBLIOGRAPHY

1. Bernstein, N.; Dixon, J. *The Future of Braille: NLS Braille Summit Presentations and Outcomes*; National Library Service for the Blind and Physically Handicapped, Library of Congress: Washington, DC, 2014. http://www.loc.gov/nls/other/futureofbraille.html.
2. *Current Strategic Business Plan for the Implementation of Digital Systems.* National Library Service for the Blind and Physically Handicapped, Library of Congress: Washington, DC, 2006.
3. Eldridge, L. R is for Reading: Library Services to Blind and Physically Handicapped Children, National Library Service for the Blind and Physically Handicapped, Library of Congress: Washington, DC, 1985.
4. Eldridge, L. Speaking Out: Personal and Professional Views on Library Service for the Blind and Physically Handicapped, National Library Service for the Blind and Physically Handicapped, Library of Congress: Washington, DC, 1982.
5. *Facts: Books for the Blind and Physically Handicapped.* National Library Service for the Blind and Physically Handicapped, Library of Congress: Washington, DC, 2007.
6. *Facts: Web-Braille.* National Library Service for the Blind and Physically Handicapped, Library of Congress: Washington, DC, 2003.
7. Majeska, M. *Talking Books: Pioneering and Beyond*; National Library Service for the Blind and Physically Handicapped, Library of Congress: Washington, DC, 1988.
8. National Library Service for the Blind and Physically Handicapped, Library of Congress; http://www.loc.gov/nls.
9. Nieland, R.A.; Thuronyi, G. *Answering the Call: Telephone Pioneer Talking-Book Repair Program 1960–1993*; National Library Service for the Blind and Physically Handicapped, Library of Congress: Washington, DC, 1994.
10. *NLS Collection Building Policy.* National Library Service for the Blind and Physically Handicapped, Library of Congress: Washington, DC, 2007.
11. *That All May Read: Library Service for Blind and Physically Handicapped People.* National Library Service for the Blind and Physically Handicapped, Library of Congress: Washington, DC, 1997.

Bliss Bibliographic Classification First Edition *[ELIS Classic]*

Jack Mills
North-Western Polytechnic, London, U.K.

Abstract
Henry Evelyn Bliss was probably the greatest American contributor to the theory of the classification of library materials. The first edition of the classification scheme he developed is discussed here by a major British thinker in this area, Jack Mills. After some years of use, the scheme was extensively revised, under the Editorship of the same Jack Mills, and the resulting second edition of the Bliss Bibliographic Classification is discussed in the entry by that name.

—ELIS Classic, from 1969

INTRODUCTION

A substantial outline of the system of library classification[1] known as the Bibliographic Classification (or Bliss Classification, or, simply, BC) was first published in 1935, and the full edition[2] appeared in 1940–1953, making it the latest of the major general classifications for libraries to appear. It was the culmination of a life-long study of the problems of library classification by its author, Henry Evelyn Bliss (1870–1955). Bliss, who was associated for almost 50 years with the College of the City of New York, in fact, applied his system in an earlier form as long ago as 1908,[3] when he reorganized the City College library. Subsequently, he undertook a very thorough investigation of the problems of organizing knowledge and the relevance to them of classification. He prepared the way for the publication of his full system using two works, both of them influential among librarians in the 1930s and 1940s, particularly in Great Britain. The first, *The Organization of Knowledge and the System of the Sciences*[4] (described by John Dewey in its Introduction as a "monumental work"), appeared in 1929, and the second, *The Organization of Knowledge in Libraries*[5] in 1933.

These two works set the scene for his classification in that they developed, after a thorough review of historical systems, both a comprehensive theoretical basis for the classification of knowledge and a set of pragmatic principles reflecting the functional requirements of documentary collections, the two combining to give the foundations of a bibliographic classification, i.e., one for documentary collections.

HISTORY

Extent of Use

Libraries began to adopt the BC as soon as the 1935 "System" appeared. By 1954, a year after the completion of the full scheme, some 50 libraries had adopted it and the first issue of the *Bliss Classification Bulletin* appeared[6] as a maintenance service for users. At first somewhat intermittent, it now appears regularly once a year; details of the service and of proposals for a new edition are given later in this entry. By 1967 more than 80 libraries had adopted the BC, and each year sees several more added to the total. These users are predominantly academic and learned libraries and government and special libraries. Most of them are in the British Commonwealth and only two are in the United States of America itself.[7] In 1967 an abridged version of the scheme[8] for use in schools was published by the (British) Schools Library Association.

Principles Underlying the BC

In the second edition of *The Organization of Knowledge in Libraries*, Bliss summarized the principles of classification for libraries under 32 headings. From these, five may be selected as of major significance in determining the characteristic features of the BC.

COLLOCATION

A library classification is a device for defining specific classes of information and for showing the relations existing between them. In this way it performs its basic function, which is to assist the retrieval of information from stores (libraries, bibliographies, indexes) by allowing a searcher first to locate a specific class in which relevant material is believed likely to be and then to make adjustments to the search (if necessary—and it usually is) by expanding or contracting the initial class according to whether too little relevant material or too much nonrelevant material is found in it. The second function is

Encyclopedia of Library and Information Sciences, Fourth Edition DOI: 10.1081/E-ELIS4-120008966

Binding–British

achieved by *collocation*, i.e., bringing together in propinquity those classes that are closely related.

Bliss attacked the alphabetical organization of information (which has been, throughout this century, the dominant form in American libraries) as serving adequately only the first function (of locating specific classes) when the needs of bibliographical searching make the second every bit as necessary. He argued that only systematic collocation could meet these needs and that this was the fundamental feature of a good classification, and it was only to be achieved by observance of certain basic principles governing the order of classes. The first of these was "consensus."

CONSENSUS

Bliss believed that knowledge should be organized consistently with the "consensus of scientific and educational opinion," since it is through the processes of science (in its broadest sense) and of education that we find that this organization is ultimately effected and its results systematized. This consensus, or agreement, he claimed, "is relatively stable and tends to become more so as theory and system become more definitely and permanently established in general and increasingly in detail."[5] The stability, Bliss argued, was a reflection of the fact that the system of knowledge was "correlative to the persistent order of nature."

At its simplest, this notion was a conscious refinement of the basic assumption made, but never stated explicitly, by other major general classifications—the Decimal Classification, the Universal Decimal Classification (UDC), the Library of Congress, Colon—all based on the accepted "disciplines" of late-nineteenth-century Western society (although in the case of the Colon scheme, the framework of main classes and divisions reflecting this is its least significant feature). But Bliss went far beyond this almost intuitive assumption in his attempt to establish an acceptable consensus, which would at the same time display fundamental relations between classes with maximum efficiency.

This consensus is, to be sure, relative (to different viewpoints) and temporary. Bliss admitted this, but claimed for it an impressive weight of historical evidence, from the Stoic "triad" (logic, physics, ethics) to the more detailed structures of the nineteenth century philosophers, and particularly of Comte, Spencer, Ostwald, and Wundt.

GRADATION

A bibliographic classification must be presented as a serial or linear order, however complex the differences used to distinguish its classes. (At least, this is true of its use for the arrangements of documents on shelves and of entries in a precoordinate index; it is not quite true of a classification used as an authority list or thesaurus for a post-coordinate index.) The two central relations underlying this serial order are subordination (of the special to the general) and coordination (of classes of the same rank, or order of division). While observing these two principles, Bliss recognized a further principle which seemed to incorporate elements of both. This was the principle of gradation in speciality:

> A very important instance is that of the natural sciences arranged in order of *speciality*, each science being in one sense individual and coordinate with its fellow sciences, yet in another sense subordinate to that on which it is mainly dependent for concepts and principles and from which it is largely derived by specialization.[4]

By this theory (advanced in various forms by a number of philosophers, particularly Comte, Spencer, and Ostwald) it is argued, for example, that physics, dealing as it does with the fundamental nature of matter and energy itself, is more "general" than chemistry, which studies organizations of matter and energy at a more specialized level. This "dependence" of chemistry on physics is seen in such critical areas as that of the theory of valency, in which the combining powers of chemical elements are explicable finally in terms of atomic structure.

It may be noted here that such a theory (which implies, for example, that all biological phenomena are ultimately explicable in terms of physics and chemistry) has obvious philosophical implications, as does the argument for consensus, which assumes that classes are correlative to a persistent order of an objectively verifiable, external nature. Bliss considered these implications at numerous points in his work, especially Chapter 10 of *The Organization of Knowledge and the System of the Sciences.*

Another interesting point is the close parallels that can be drawn between the order produced by gradation and that produced by applying the theory of integrative levels[9] in which entities are organized in a sequence of increasing complexity, from fundamental particles to atoms, to molecular aggregates, to cells, and so forth.

ADAPTABILITY

The "adaptation of logical order to practical uses and to convenience thru collocation. . ." (*A Bibliographic Classification*, Vols. I–II. p. 23) recognizes the practical implications of the "relativity of knowledge and classes" and is another major principle in the BC. It takes the form of extensive provision for alternative locations and treatments, e.g., some technologies may be collocated with others in the technology class (U) or subordinated to the science that they largely reflect (as aeronautics, say, reflects aerodynamics); or, mineralogy may go with its science crystallography, within chemistry, or under geochemistry, within geology; International law may go with international relations within political science or with law.

To implement this, extensive provision must be prepared in the notation for different decisions, and the BC provides this to a degree far excelling that of the other major schemes.

NOTATION

Bliss made valuable contributions to the theory and practice of notation, not least in his emphasis on notation being "correlative and subsidiary," i.e., that it should not determine the display of relations achieved by the order of classes. In the BC, two principles of notation are particularly prominent. First, the use of "composite" notation to provide economically for the notational specification of different classes by recurrent concepts; this is the old principle of "divide like..." used by Dewey, and, while it is not taken as far as the provision of a fully faceted notation, it does allow notational synthesis on a scale comparable with that of the UDC. Second, the notation is quite the briefest of all the major schemes. Bliss considered brevity a major element in the acceptability of a notation and sought it for the BC in a number of ways.

The Classification

The basic sequence, its structure predominantly reflecting the principle of subordination of the general to the special and of gradation in speciality, is notably coherent and mnemonically clear:

Philosophy, Logic, Mathematics (i.e., "tool" studies, and
    Methodology)
Physical sciences
    Physics
    Chemistry
    Astronomical systems and bodies
      Earth
Biological sciences
    Botany
    Zoology
      Man: Human sciences and studies
         Physical anthropology (in widest sense)
         Social anthropology (in widest sense)
            Social sciences
            Arts

To put some flesh onto this skeleton, an amplification shows some of the additional features necessitated by the practical demands of a working bibliographic classification, including notation. The latter, it should be noted, does not necessarily reflect the relations (of subordination or coordination) between the classes.

This simple linear presentation (which, however, is quite essential to a bibliographic classification) does not fully display the care with which Bliss charted the complexities of the relations involved, and the consistency with which certain principles of subordination have been applied.

The principle of gradation clearly dominates the sequence down to class H, which introduces the human sciences and studies. It is then reflected somewhat differently in the placing of generalizing theoretical studies just ahead of the concrete activities they illuminate. For example, sociology, as the fundamental, generalizing study of society and social institutions, is located ahead of the special social sciences and studies in which particular aspects of human society—its religions, political, and economic structures and its technological and imaginative artifacts—are considered. Bliss firmly rejected the bifurcation of subjects into "abstract" and "concrete" and consistently locates the "abstract" studies (the philosophy and methodology) relating to a given activity with the associated concrete ones. Similarly, Bliss secures a valuable element of predictability by following an old, pragmatic indexing principle that is rarely explicitly stated—that of forming compound subjects, in which both general and special classes are reflected, by citing the special element first, e.g., physical chemistry under chemistry, biochemistry under biology, medical psychology under psychology, educational psychology under education. Generally speaking, such compounding follows a "retroactive" principle, whereby such compounds are located by citing first the "special" activity and then going "backwards" in the sequence to the preceding and more "general" class in order to cite the "general" viewpoint. In the human studies this principle tends to be obscured by the allowances made for consensus, e.g., in allowing the various aspects of a country's history and description to be collected under that country (in classes M–O) rather than subordinating the place (the society) to the social activity (e.g., legal, political); or, in allowing the economic organization of particular industries to be subordinated to the class economics (T) rather than to place it with that industry in the neighboring useful arts class (U).

By and large, the "modulation" in the BC, whereby each major class leads onto the next in a manner clearly determined by adherence to the basic and consistent theory, as remarkably sustained when we remember the artificiality imposed on a classification by the need to linearize the multidimensional pattern of relations. Only in a few instances does the sequence raise a doubt as to why it is thus, and in these cases it is usually due to Bliss's anxiety to provide a practical order agreeable to the current "consensus." For example, education (J) is theoretically a special social science and should not precede sociology (J); Bliss argued that education, as the "training and development of mind and of mental abilities," is inseparable from psychology and therefore placed it as a bridge between education and sociology. The practical utility of this is indisputable and would seem to be a major reason for the adoption of the BC by nearly all the university Institutes of Education in the United Kingdom.

Another example of the very practical approach to collocation when the demands of theory are not overwhelming is in his interpretation of philology. Theoretically, language is the fundamental agent of social communication and by gradation should be collocated with sociology, and only its *application* to the "special" human activity of imaginative creation should be subsumed under literature. But there is a strong educational consensus manifested in the study and teaching of the two subjects to justify their collocation, and this Bliss does.

Class Q, social welfare, and so forth (previously called "applied social science"), is a composite class for which no accepted short name is available but which nevertheless collocates reasonably a number of broad social movements and problems (social pathology, women, socialism, internationalism, etc.) in which a number of special social sciences (politics, law, economics, etc.) intermingle.

One more example is the treatment of technology. By gradation, this should be regarded as a special social science (giving the "material culture" of society), and this is by and large how Bliss treats it. However, the demands of the literature obviously suggest collocation of some technologies with the science from which they spring, as in the case of electronics, nuclear engineering, much of chemical technology, and so forth. Bliss acknowledges this, and all these technologies are to be found with physics and chemistry in B and C, representing as they do what Bliss calls "the more scientific technologies." However, we have noted already that alternatives are provided for some technologies.

Provision for Alternatives

Allowing some technologies to be collocated either with a "pure" science or with the useful arts class is typical of the regard for adjustability which makes the BC by far the most flexible of the major general schemes. This regard takes two forms: first, simple "alternative location," in which a given class may be located in one of two (or more) different places but without any alteration of its *internal* arrangement; examples here are the moving of theology from Class P (religion) to AJ in order to collocate it with philosophy, or placing constitutional law in law instead of in political science (where Bliss prefers it).

The second feature of this regard for adjustability is in "alternative treatments," which are more elaborate; they involve altering the internal structure of a class by varying its "citation order," i.e., the order in which the elements of a compound class are cited (reflecting a different order of application of principles of division). A prominent example here is the rich provision under literature whereby the literature of a given language may be divided first by period and then by form, or vice-versa; or, the literary texts may be kept separate from the works *about* literature and each major group then divided in various ways, of the division into texts and literary history may be made after prior division by period and/or form. A similar rich provision of alternatives is to be found under social–political history.

Notation

1. This uses predominantly capital letters A–Z; the number 1–9 (but not zero, which would conflict with the letter O) and lower-case letters a–z are used fairly often, as shown below. The comma is used extensively as a facet indicator and the hyphen is used as a phase indicator (i.e., linking elements derived from different subject areas rather than from different facets of the same class). Although Bliss unwisely introduced some ungainly symbols (&, %, and so forth) in the final volume of the BC these were used very rarely and have since been withdrawn (Table 1).

Table 1

1–9	Anterior numeral classes (for special collections of various kinds, e.g., 6, Periodicals) Class 2 here (Bibliology and libraries) is an alternative to its preferred position in Z.
A	Philosophy and general science (including Logic, Mathematics, Metrology, Statistics)
B	Physics (including special physical technology, e.g., Radio)
C	Chemistry (including Mineralogy, Chemical technology)
D	Astronomy, Geology, Geography (General and Physical)
E	Biology (including Paleontology, Biogeography)
F	Botany (including Bacteriology)
G	Zoology (including Zoogeography and Economic Zoology)
H	Anthropology (General and Physical) (including Medicine, Hygiene, Physical training, and recreation)
I	Psychology (including Psychiatry)
J	Education
K	Social sciences Sociology, Ethnology, Anthropogeography (including Travel and description in general)
L–O	Social–political history M Europe, N America, O Australia, Asia, Africa
P	Religion, Theology, Ethics
Q	Applied social sciences and ethics; Social welfare
R	Political science
S	Law
T	Economics
U	Arts in general, Useful arts (including the less scientific technologies)
V	Aesthetic arts, Recreative arts and pastimes
W–Y	Philology: language and literature W Non-Indo–European, X Indo–European, Y English
Z	Bibliology, Bibliography, Libraries

2. Bliss sought simplicity for his notation in two main ways: first, by using for the greater part symbols whose ordinal value are very widely known (i.e., Roman letters and Arabic numerals) with a minimum use of "arbitrary" symbols; second, by maximum brevity. This latter is secured by using a wide base for the radix fraction principle of division (effectively, 35 factors (1–9, A–Z) available for dividing any and every class); by careful allocation of the notation to the classes to reflect the weight of literature to be accommodated (although his assignment of notation to the physical sciences and technologies is somewhat small by this measure); and by the frequent and deliberate abandonment of "expressiveness" (i.e., the quality of conveying hierarchical relations by length of number). An extreme example of the last item would be Bliss's use of a shorter number for a subclass than for the containing class where the weight of literature justified this (Table 2).

   A third way in which Bliss sought easily grasped notation is his likened use of literal mnemonics where these are obtainable without sacrifice of desirable order; e.g., AL, Logic; AM, Mathematics; BD, Dynamics.

3. The hospitality of the notation (its ability to accommodate subdivisions wherever called for) is achieved by the radix fraction principle (cf. the arithmetical notation of the Library of Congress), by the deliberate refusal to seek expressiveness (so that growth is not inhibited by what the class number *looks* like), and by extensive use of the principle of faceted notation (i.e., using distinctive symbols to show the conjunction of different facets, thereby allowing the independent expansion of each facet). This latter is seen in the provision for "composite specification," or "synthesis," as it is perhaps better known.

4. Composite specification[10] is achieved by a number *systematic schedules* (over 50, if the adaptations of particular schedules are counted as separate ones). Like the Auxiliary Tables of the UDC, which perform the same function, they are of two kinds—general (i.e., applicable to any class) and special (i.e., applicable only to a limited number of classes).

5. Of the former, *Schedule 1* provides for *form divisions*, using numerals, e.g., F1 dictionary of botany, F2 bibliography of botany. *Schedule 2* provides for *place divisions*, using lower-case letters, e.g., HOd hospitals in Europe, HOe hospitals in the U.K., HOed hospitals in London. It should be noted, however, that in some classes (e.g., social–political history L–O) special provision is made for Place and Schedule 2 does not apply. Also, numerals may be used as an alternative to lower-case letters.

   Schedule 3 provides for *language divisions*, using capital letters, e.g., XRY, M translations of Portuguese literature into English, XRY,P into Polish. Strictly speaking, this schedule is not general in that it is normally used only in conjunction with areas such as literature and philosophy, but theoretically it may be used to indicate the language of the document in any class. *Schedule 4* provides for *period divisions*, using capital letters, e.g., TU3A taxation in the ancient world, TUJ3N income tax in the nineteenth century. Again, in some classes (e.g., literature) special provision is made for period divisions, and these Schedule 4 divisions do not apply.

6. Of the special systematic schedules, Schedule 13 (for subdivision under any disease) is fairly typical. It includes, D diagnosis, ,e Etiology, ,n Therapeutics; these, added to the class number for a disease (e.g., HPSM epilepsy) give, say, HPSM,E etiology of epilepsy; HPSM,N therapy of epilepsy, and so on. In most cases, further detail may be obtained by drawing on the general tables, e.g., HN in therapeutics (general) and HNY is psychotherapy; the terminal "Y" may be added to the "N" from the systemic schedule to give HPSM,NY psychotherapy of epilepsy.

7. Some of the systematic schedules have extensive adaptations for particular needs, e.g., in Class J education one schedule has special adaptations for a number of different types of school. Or, in philology, schedules provide for division by linguistic factors (e.g., ,G grammar), by literary form (e.g., ,Q drama) and for arranging material under any given author; also, adaptations are made to allow fuller detail under the major literary languages.

8. If necessary, elements from several different systematic schedules can be added to a class-number. Confusion as to what schedule a given element comes from is avoided by following Bliss's advice to introduce capitals from Schedule 3 by the number "4" (from Schedule 1) and capitals from Schedule 4 by the number "3" (from Schedule 1). Schedule 1 and 2 cannot be confused since they use distinctive symbols (numbers and lower-case letters). This allows the letters from special systematic schedules to be added immediately after the comma (and sometimes even this comma may be dispensed with; e.g., MU social–political history of United Kingdom (U.K.); MUE economic history of U.K.).

As an example of what a sequence in a catalog or bibliography organized by BC looks like, using these various forms of composite specification, below is a selection

Table 2

AZ	Physical science in general
B	Physics

Table 3

UVP	Paper technology
UVP2	Bibliography
UVP3	History
UVP3K	18th century
UVP4e	United Kingdom
UVP4eK	18th century
UVP–UTT	*effect of* Printing trade
UVP,B	Research
UVP,C	Chemical/Physical studies
UVP,K	Manufacture of paper
UVP,R	Marketing
UVP,R,B	Research
UVPA	Rag papers
UVPI	Newsprint paper
UVPI,C	Chemical/Physical studies
UVPI,C,B	Research
UVPQ	Paper boards: Cardboard, etc.

of headings from Class UVP paper technology (for which a special Schedule 21 is available, as it is for all other industries in classes US–UW). This also demonstrates the *filing order* of the different symbols used in the BC notation (which is: 1–9; a–z; –; A–Z) (Table 3).

A–Z Index

Bliss provided an alphabetic relative index to the BC, printed conveniently in a separate volume and comparable in detail to the fullest relative indexes to the Dewey classification, containing about 45,000 subjects. While it is obviously the product of great care and industry, it suffers (as do the alphabetic indexes to the Library of Congress classification) from a failure to appreciate fully the dual role of the A–Z index, which is both a rapid key to the location of classes (using the *names* of subjects as entry) and as a complement to the systematic order in displaying relations between classes (i.e., indicating "distributed relatives"). Consequently, the direct alphabetic approach is constantly undermined by needless classification; for example, under the entry

Railroads, railways   TNP : UHL

there are some 80 terms listed, each with its separate number(s), e.g., Accidents UHV; Transportation TNP. Yet all but four of these are subclasses of TNP or UHL (the economic and engineering aspects of railroads, respectively) and would be located very quickly by scanning the *systematic* display of them in the schedules or classified catalog. Moreover, not all of them are repeated under their direct alphabetic form (e.g., soils), so that a searcher for other contexts of these classes (e.g., other contexts of soils) would not find them.

To put the matter in another way, Bliss was unfortunately not aware of the substantial economies and

improved efficiency of access possible here by the use of chain procedure.

Weakness of the BC

The BC was largely the work of one man. Although Bliss received assistance from a number of librarians and subject specialists, the BC was overwhelmingly the product of his own exertions (even physically, in that the schedules are reproduced by photolithography from his own typescript). It is not surprising that there are some flaws of detail in the BC; what is surprising is that there are relatively so few.

The most serious weakness is undoubtedly the same one that marks every other general system apart from Colon, and that is the failure to observe strictly the fundamental rule of classification, which is to apply one principle of division only at a time and to exhaust it before applying another. The modern name for this is, of course, facet analysis, and it is rather sad that Bliss developed his great work just a little too soon to allow it to benefit from the rigorous methods of analysis and synthesis that Ranganathan was developing in his Colon classification. In a number of classes in the BC the facet analysis is clearly imperfect, e.g., in the systematic schedule 21 (applied to paper technology above) may be found (Table 4).

Although this reflects roughly the order in which a technology might operate, it produces an unhelpful mixing of technical and economic problem. The operations facet, for example, begins at ,E with the general technical operation of engineering, changes to a sequence of economic and business operations, is dropped to introduce another facet—that of product—and then returned to at ,R marketing, and so forth. The enumeration of terms such as deterioration, repair (under ,E) means that an implicit facet (processes and operations on agents of production) is being tied to a particular class (here, the agent, plant and machinery) when in fact it might easily occur in the

Table 4

,D	Materials, Raw or Crude Acquisition….Import….Properties..; Analysis, Testing, etc.
,E	Technology (general), Engineering, and Equipment Plant….Power, Fuels; Machinery…; Deterioration, Repair
,F–,J	Business: Organization….Management….Personnel…. Safety….Insurance….Finances….
,K–,N	Production: Manufacture….Special methods….Patents….
,O–,Q	Products and By-products Properties….Testing….Substitutes….
,R–,S	Markets….Merchandising….

literature compounded with other classes (e.g., deterioration of raw materials). Other classes belonging to this facet (processes and operations on agents of reproduction) are analysis and testing; these are enumerated under ,D and ,O—but not under ,E, say, or ,Q, where they might easily occur.

Lack of thorough facet analysis not only reduces the hospitality of the system but also introduces a lack of predictability for the indexers. Where would marketing personnel go—in ,R marketing or ,G personnel? Where would safety regulations for plant go—in, plant or ,H safety, Also, because of unnecessary repetition (as testing above) the economics in schedule display implicit in composite specification are not fully utilized.

Bliss's failure to appreciate fully the need for stringent facet analysis is also apparent in his discussions of, say, subordination. He tended to see this vital principle primarily as a problem of general-before-special sequence and of collocation between major areas (e.g., the subordination of psychology to anthropology). But every bit as important is the collocation *within* classes (large or small), which is determined largely by the citation order of facets, e.g., whether, in the subject of librarianship itself, college library cataloging is subordinated to the library service (college) or the library operation (cataloging). Consistency, to give predictability in searching, in this matter is very difficult without explicit recognition of facet structure.

However, this weakness in BC is one common to all the major "enumerative" systems, and, what is more significant, is the degree to which Bliss did in fact achieve consistent and thorough analysis. In many classes the demands of composite specification led to just this, e.g., the whole of the large philology class reflects it very clearly indeed.

A weakness of another kind in the BC is the lack of detail in certain class, notably some of the physical technologies. While it is rarely less detailed than the Decimal Classification of the Library of Congress (and the large synthetic element of BC must by remembered here since it greatly enlarges its range of specification), the BC falls well short of the UDC in these areas, despite its extensive provision for synthesis. However, this shortcoming is being remedied by the current program of the BC *Bulletin*.

Present Position

The BC is maintained via the *Bliss Classification Bulletin*.[6] The first issues were edited by Bliss himself. On Bliss's death, in 1955, British librarians using his scheme cooperated to form the British Committee for the Bliss Classification, and Dr. D.J. Campbell took over the Hon. Editorship of the *Bulletin*, which the H.W. Wilson Company continued to publish and distribute free to all users of the scheme until the stocks of the present edition of the BC were exhausted. In 1967 the H.W. Wilson Company handed over the publication of the *Bulletin* to the British Committee (which reconstituted itself as the Bliss Classification Association), vesting full copyright of the classification in them.

The BC is now (1968) out of print, although available in xerographic reprint form from University Microfilm, Inc. (313 North First Street, Ann Arbor, Michigan).

Meanwhile, the current program of the annual *BC Bulletin* is to provide, together with simple amendment by addition of recently developed terms, substantial revisions of complete classes, particularly in those areas of science and technology in which the BC has so far lacked detail. In issues of the *Bulletin* detailed schedules have appeared for such classes as electronics, oceanography, automatic control and control devices, nuclear reactor engineering, sound reproduction and recording, astronautics, microchemical analysis, physics and chemistry of the atmosphere, food preservation, operative surgery, child hygiene and care, gardening and fruitgrowing, solid state physics, physical and chemical metallurgy, printing, and so forth.

CONCLUSION

To their reader, most libraries serve primarily as self-service stores of information, in which the *order* of the material is the major index to its content. The efficiency of the day-to-day retrieval is closely affected by the quality of the collocation in that order. The same holds, though to a somewhat lesser degree, with library catalogs and bibliographies. On this score, the BC is without a doubt the best general classification we have. Not merely is its basic pattern superior, but the flexibility to special needs, which is one of its main features, ensures a particularly appropriate order if a local variant is called for. Moreover, the notation which maintains its order is of exceptional brevity and is often highly mnemonic, too. As a system for organizing catalogs and bibliographies, the relative lack of detail in some classes is a drawback, but this is now being remedied.

The main handicap to the wider use of the BC has been, on the one hand, the inertia inevitably associated with existing collections, in which the cost of changing systems is considerable, and, on the other, the administrative advantages enjoyed by the Decimal Classification and the Library of Congress by virtue of their facilities for centralized classification and substantial day-to-day maintenance. These are real disadvantages the BC offers for many kinds of libraries, and it is fitting to end with a categorical assertion by a former editor of the *BC Bulletin*, Dr. D.J. Campbell: "...it is the best scheme for any but very large general libraries, and for many special libraries, while it lends itself to the modifications which every special library needs."[11]

REFERENCES

1. Bliss, H.E. *A System of Bibliographic Classification*, Wilson: New York, 1935; (2nd Ed., 1936).
2. Bliss, H.E. *A Bibliographic Classification, Extended by Auxiliary Schedules for Composite Specification and Notation*, Wilson: New York, 1940–1953; (4 vols. in 3): Vol. 1 (Classes 1–9, A–G) 1940; Vol. 2 (Classes H–K) 1947 (a 2nd edition of Vols. 1 and 2 appeared in 1952, in one volume); Vol. 3 (Classes L–Z), and Vol. 4 (General Index) 1953.
3. Bliss, H.E. A modern classification for libraries, with simple notation, mnemonics and alternatives. Libr. J. **1910**, *35*, 351–358.
4. Bliss, H.E. *The Organization of Knowledge and the System of the Sciences*, Holt: New York, 1929; 153.
5. Bliss, E. *The Organization of Knowledge in Libraries and the Subject Approach to Books*, Wilson: New York, 1933; 42–43 (2nd Ed., 1939).
6. *Bliss Classification Bulletin*, Wilson: New York, 1954–1966; Vols. 1–3, Vol. 4, No. 1, Bliss Classification Association, London (c/o Commonwealth Institute, Kensington High Street, W8), 1967–.
7. Shell, E.E. The use of Henry E. Bliss's bibliographic classification at the Southern California School of theology. Libr. Res. **1961**, *5*, 290–299.
8. Bliss, H.E. *Abridged Bliss Classification*, School Library Association: London, 1967 (150 Southampton Row, WC1).
9. Foskett, D.J. Classification and integrative levels. *Sayers Memorial Volume: Essays in Librarianship in Memory of W.C. Berwick Sayers*, Library Association: London, 1962.
10. Mills, J. Composite specification in the BC. Bliss Class. Bull. **1967**, *2*, 6–15.
11. Campbell, D.J. Bibliographic classification. *Encyclopaedia of Librarianship*, 3rd Ed.; Landau, T., Ed.; Bowes & Bowes: London, 1966.

Bliss Bibliographic Classification Second Edition

Vanda Broughton
School of Library, Archive and Information Studies, University College London, London, U.K.

Abstract

This entry looks at the origins of the *Bliss Bibliographic Classification* 2nd edition and the theory on which it is built. The reasons for the decision to revise the classification are examined, as are the influences on classification theory of the mid-twentieth century. The process of revision and construction of schedules using facet analysis is described. The use of BC2 is considered along with some development work on thesaural and digital formats.

INTRODUCTION

The Bibliographic Classification of Henry Evelyn Bliss (BC1)[1] was the last of the significant general classifications to appear, and was hailed by many as the best constructed and most scholarly of them all, "the fine flower of the enumerative period."[2] Despite that label, it included many facilities and devices for synthetic classification, what Bliss called "composite specification," or the expression of complex subject content by the building of compound classmarks. Conceptually and structurally BC1 forms a bridge between the great enumerative library classifications of the nineteenth and early twentieth century, the Dewey Decimal Classification and the Library of Congress Classification, and the scientifically structured retrieval tools of the later twentieth century, such as Colon Classification.

BC1 was adopted by many academic and special libraries, particularly in the United Kingdom and the British Commonwealth, although it was not implemented anywhere in the United States, other than Bliss's own library at the College of the City of New York, where it had originally been developed and tested. When the time came to produce a substantial revision of the classification it was BC1's British user base which took up the task, and utilized the opportunity to make the 2nd edition a fully faceted classification scheme, the only general scheme of its kind other than Ranganathan's own Colon Classification. The way in which faceted classification theory is married with the scholarly structure of Bliss's original scheme is described below under the following heads:

Background and context of the 2nd edition.
The relationship of BC2 to the original Bibliographic Classification.
Facet analysis and information retrieval.
Structural principles of BC2.
The use of BC2 today and developments for the future.

BACKGROUND AND CONTEXT OF THE SECOND EDITION

Bliss was virtually the first theoretician of bibliographic classification, writing in the period immediately before Ranganathan, and establishing very many principles of knowledge organization that are still considered significant today. These include the need to relate documentary classification to the consensus of informed opinion about the nature and structure of subject fields (scientific or educational consensus), and the way in which the whole of knowledge can be reduced to linear order through philosophic or logical method (principles of general-before-special and gradation in specialty). Features of the original classification, its main class order, the presence of systematic and special auxiliaries, the use of systematic and literal mnemonics, the facility for synthesis, the flexibility of its alternative arrangements, and its inherent hospitality to new subjects, are all highly desirable features in a classification and are still so regarded today.

Nevertheless the classification, which was a long time in the gestation period, lacked any mechanism or financial support for the revision and maintenance so clearly required, even by the time of the publication of the final volumes. The first outline of the scheme had been published in 1910 (*Library Journal* 35(8), 351–358), and an abridged version in 1935 (*A System of Bibliographic Classification*; H. W. Wilson: New York). In the years immediately following the publication of BC1, updates and amendments were published in the *Bliss Classification Bulletin*, then under the editorship of H. E. Bliss himself, but after Bliss's death in 1955 there were no obvious authorities appropriate to the intellectual work of maintenance, and the outlook for the long-term continuation of the system appeared bleak.

British librarians had long been supportive of the BC, and Bliss, in correspondence with a number of them during the period of its creation, acknowledged the part that

Encyclopedia of Library and Information Sciences, Fourth Edition DOI: 10.1081/E-ELIS4-120043531

Binding–British

they had played in the development of the scheme. BC1 was warmly received in the United Kingdom, and its British advocates did much to promote its uptake both in the United Kingdom and overseas. In 1956 a new British Committee had been elected, and in 1960 Jack Mills became Chair of that Committee and Editor of the *Bliss Classification Bulletin*. In 1967 the Bliss Classification Association (BCA) was formed, again with Jack Mills at the helm, and, as at the time there were no users of BC1 in North America, H. W. Wilson, the publishers of the original scheme, passed all rights in the classification and the *Bulletin* to the BCA, who have remained the copyright holders ever since. The BCA also accepted responsibility for the maintenance of the scheme, and under the guidance of an executive committee, many small-scale revisions and amendments were published during the 1960s, via the medium of the *Bliss Classification Bulletin*.

By the late 1960s it was clear that this piecemeal revision (mainly consisting of the addition of new concepts) was not adequate for the continuation of BC as a viable system, and that more wide-ranging and more radical revision was required. A program was drawn up for such substantial revision, under the editorship of Jack Mills, and based at the School of Librarianship, Polytechnic (later University) of North London, work beginning on the revision in 1967. The revised classification was to be published by Butterworths in Great Britain as the *Bliss Bibliographic Classification* 2nd Edition (BC2), and although the original plan was for a single three-volume work, it was subsequently decided to publish in parts, each consisting of a discrete subject area. The first revised schedules appeared in 1977 (Introduction and Systematic Auxiliaries) and were closely followed by classes for Religion, Social Welfare, Psychology, Education, and Health Sciences. Since the 1970s further volumes have been published for Mathematics, Philosophy, Sociology, Politics, Law, Economics and Business, General Science and Physics, and most recently, the Fine Arts. Social welfare and Education are now in the 3rd edition. Current work is in the areas of Chemistry, Technology, and Music.

RELATIONSHIP OF BC2 TO THE ORIGINAL CLASSIFICATION

Since BC1 was in use in a number of libraries, the extent of the revision was a significant issue. A memorandum was prepared for consideration by the members of the BCA, and published in the *Bulletin* of 1969. Two aspects of the classification were in question: the level of detail in the vocabulary, and the internal structure of classes. It was felt that the vocabulary of BC2 should be at least equivalent to that of the Dewey Decimal Classification and the Library of Congress Classification. Since some specialist libraries used BC1 there was a potential need for a vocabulary comparable to the Universal Decimal Classification,

but this was not considered an achievable objective; there was an expectation that the facilities for synthesis would go some way toward implicit provision for complex topics.

Altering the structure of the classification was a greater problem since the consequences of such changes were more serious. It was agreed that Bliss had intuitively applied much of the new classification theory, and that the scheme largely met the requirements of a modern system. The overall order was excellent, and the internal facet structure was often correct; auxiliary schedules and other devices for synthesis allowed for the expression of complex subjects and the scheme was hence hospitable to new topics. However, it was also the case that in some classes the facet structure was inconsistent and the analysis imperfect. It was decided that, despite the undoubted inconvenience to users, correcting these faults would give the classification greater long-term stability, and reduce the need for frequent revision in the future.

The overall broad structure of the classification (much admired and generally agreed to be the best of any general scheme) would remain intact; new (particularly) scientific subjects, such as space science and cybernetics, were to be added at the appropriate place in the order, and the anterior numeral classes, previously used mainly as form classes, would accommodate a discrete classification of phenomena, allowing for the easier management of cross-disciplinary and interdisciplinary works. Bliss's provision of alternative locations would be built upon and extended, to allow, for example, the location of all technologies in Class U, and a unitary class for geography in Class D. The internal structure of classes was to be made consistent with the methods of faceted classification, which would import a degree of rigor in the analysis which was in parts lacking in BC1, and would ensure uniformity in the application of citation order, filing order, and the procedure for synthesizing classmarks. This process is described in detail below.

FACET ANALYSIS AND INFORMATION RETRIEVAL

Facet analysis was conceived by S. R. Ranganathan during the 1920s and 1930s, and formally disseminated in his works of classification theory,[3] and the Colon Classification.[4] While the Colon Classification was virtually never applied in Britain, Ranganathan's ideas profoundly influenced British librarians (The library of Christ's College Cambridge is a notable exception but in 2007 it is now almost completely replaced by the Library of Congress classification.) During the 1950s and 1960s much research and development work was carried out in the area of faceted classification and information retrieval, and many significant works in the field were published. The focus of most of this activity was the Classification

Research Group (CRG), founded in 1952, to which many of the leading lights of the age belonged, and which served as a forum for the discussion of these ideas. The CRG was committed to the concept of faceted classification as a unifying theory for information retrieval and the design of information retrieval tools, and in 1955 published what may be regarded as a manifesto for facet analysis.[5]

During the 1960s work by the CRG on a new British general classification scheme funded by a grant from NATO[6] failed to come to fruition, although much of the conceptual work can be discerned in the PRECIS system of indexing developed for the British National Bibliography by the principal researcher, Derek Austin.[7] Nevertheless, the Group retained its interest in a general classification, and in 1963 held a conference to consider various aspects of this.[8] Within this context the revision project of BC2 was seen as an ideal opportunity for the incorporation of many of the theoretical principles of the facet-analytical school in a large general classification. As a consequence BC2 became identified to some extent with the new general classification scheme long envisaged by the CRG, and embodying theoretical principles and structure in a way that no universal system of classification (with the exception of Colon Classification) had ever done before. While maintaining and preserving the central features of the original classification, and the general infrastructure of that system—the main class order, broad order within classes, literal notation, and principal mnemonics—the new schedules incorporated all the aspects of modern classificatory theory—rigorous and logical analysis into facets and their constituent arrays, the imposition of a consistent and predictable citation order via the medium of an inverted schedule, a theoretically limitless capacity for synthesizing compound subjects, and the use of a retroactive notation dispensing with the need for facet indicators and greatly simplifying the process of classmark construction.

Since the beginning of the BC2 project CRG meetings have provided a major forum for the discussion of draft schedules of BC2 and debate about the implementation of facet theory in them. Individual members of the CRG have also made a significant contribution to the creation of schedules, notably Eric Coates, Douglas Foskett, and Derek Langridge.

STRUCTURAL PRINCIPLES OF BC2

Main Class Order

In most of its particulars BC2 conforms to the model of a modern general classification scheme on the faceted pattern. The main features of a bibliographic classification scheme in general and those of BC2 in particular are set out in much greater detail in the introduction to BC2,[9] but a resumé of the core principles is provided here.

It is often said of traditional library classifications, in contrast with automatically generated vocabularies, or with the recent phenomenon of social classification, that they are "top-down" classifications, since they are generated by dividing and subdividing knowledge as a whole. In all existing general schemes the primary step of division is into disciplines, or major subject fields, known as main classes.

The order of the main classes in BC2 follows that of Bliss's original scheme; this has been widely regarded as the best and most helpful main class order of all the general systems of classification, based as it is on both philosophical principles and educational consensus (that is the perception of interdisciplinary and intersubject relations from the academic viewpoint, and reflecting academic study). The order follows a logical developmental sequence often regarded as consistent with integrative level theory.[10] This was a scientific theory concerned with the basis of natural order, which was subsequently much used in the social sciences and as part of organization theory; it proposed that entities could be ordered in terms of a series of levels each of which was structurally dependent on the previous level. From a disciplinary point of view, physics, for example, should precede chemistry, since chemical substances were "dependent" on the forces and fundamental particles of physics; similarly chemistry should precede mineralogy and geology where the "simple" substances of the former are aggregated into more complex materials. Such a theory provides an excellent basis for the ordering of disciplines and produces an almost identical result to Bliss's gradation in specialty order of main classes.

In comparison, BC2 starts with mathematics and the physical sciences, progressing through the biological sciences, with increasing complexity of organisms, to mankind, as represented by anthropology, medicine, and psychology. The study of society, sociology, is collocated with various aspects of human social activity, education, religion, politics, law, commerce and technology, and the whole is rounded off with the higher intellectual activities of man, music, the arts, and literature, culminating in bibliography and library science. In addition to the natural progression of this sequence, the order gives rise to natural academic groupings of subjects, the physical, and biological, sciences, medical sciences, social sciences, commerce and industry, and the arts.

Main Class Order in BC2

A–AL	Philosophy
AM–AX	Mathematics
AY	General science
B/D	Physical sciences
B	Physics
C	Chemistry
D	Astronomy and the earth sciences

(Continued)

Main Class Order in BC2 *(Continued)*

E/G	Biological sciences
E	General and microbiology
F	Botany
G	Zoology
H/I	Human sciences
HA	Anthropology
HB/HZ	Medicine
I	Psychology and psychiatry
J	Education
K	Sociology
L/O	Historical ancillaries and history
P	Religion
Q	Social welfare
R	Politics and administration
S	Law
T	Economics, business, management
U	Technology, industry, manufacturing
W	Fine arts: art, architecture, music
X/Y	Language and literature
Z	Bibliography

While BC2 shares with the other schemes this initial division into disciplines, subsequent organization of the subject proceeds on a very different basis from the traditional serial subdivision into smaller and more specific classes on more or less pragmatic grounds.

Facet Analysis and Fundamental Categories

A fundamental feature of a faceted classification is that its classes consist of individual concepts, rather like elementary particles in Chemistry, which are combined to express the more complicated subjects of actual documents, as elementary particles are combined to make molecules. Ranganathan himself compared the faceted classification to Meccano, an engineering toy consisting of individual struts, blocks, and wheels which can be combined to create simple machines; its modern equivalent is the Danish building block toy, Lego.

The internal structure of a class is determined by the process of facet analysis, or the allocation of the individual concepts to a set of categories or facets. The conventional definition of a *facet* is the total set of concepts, within a given class, produced by the application of one broad principle of division. Therefore, if a class such as agriculture is divided by the principle of *place*, concepts (or subclasses) such as England, Montana, Bolivia, uplands, forest, coastal regions, etc., are generated. This process can also be viewed as happening in the reverse manner, in that appropriate concepts are aggregated into facets labeled with the principles of division. For that reason, a faceted classification can be regarded as a "bottom-up" classification, that is, one in which individual concepts are grouped or categorized to build the classificatory structure from its component concepts.

Ranganathan identified five *fundamental categories* into which the concepts within a class would be sorted: personality, matter, energy, space, and time. In the Colon Classification these categories often have to do double duty, so that there may be two energy facets, or two (or three) personality facets, in a given subject. Work by the CRG in the 1960s made finer distinctions between categories, so that a set of 13 were eventually agreed on, as can be seen in the work of Vickery,[11] Foskett,[12] Langridge,[13] Mills,[14] and others. They form the basis of all analysis of concepts in BC2, and are as follows:

Thing: sometimes called **entity**, this is equivalent to Ranganathan's personality category; it applies to the principal focus or object of interest of any discipline (plants in botany, substances in chemistry, nations in history).

Kind: this accommodates concepts in a genus–species relationship with **thing**, and contains general broad groupings of concepts (e.g., wind instruments, as **kind** of **thing** musical instruments) as well as specific subclasses (e.g., oboe as **kind** of **thing** wind instrument).

Part: constituents and subsystems of concepts in the **thing** category (e.g., **thing** vehicle has **parts** chassis, wheels, brakes; **thing** animal has **parts** legs, brain, skin). A separate **part** category only be considered legitimate if its members are in a part–whole relationship to all (or most) members of the **thing** category.

Material: this category is equivalent to Ranganathan's matter category, and is represented by gross materials, constituents and elements; it is more fundamental than **part** (e.g., a **thing** house, has **parts** roof, walls, windows, basement, consisting of **materials** timber, brick, tiles, glass).

Property: properties and attributes of concepts in the **thing** category.

Process: the first of two energy, or activity categories, **process** refers to inherent and self-initiating actions within **things**—actions that "happen by themselves." Examples are usually intransitive verbs (or noun equivalents), such as growth, change, disease, and flow.

Operation: these are actions which are performed on a **thing** by an external **agent**. Examples are usually transitive verbs, such as testing, building, mining, and harvesting.

Product: outcomes or results of processes within, or operations on, **things**; they usually occur as actual physical products in technologies; for example, food, drugs, or textiles in agriculture and horticulture.

By-product: another similar, self-explanatory, technological category.

Patient: the object acted on, when this is other than members of the **thing** or **entity** category, as is most usual. Instances are again largely technological; for example, in engineering, blanks (**patient**) may be stamped (**operation**) into components (**part**) for machines (**thing**).

Agent: the means by which **operations** are performed, **agents** may be either persons (or organizations of persons), or tools/equipment. Both categories of agents may occur simultaneously; for example, in medicine a surgeon (**person agent**) may excise tissue using a laser (**tool agent**).

Space: any region defined by political, administrative, physiographic, or other spatial properties (e.g., Hong Kong, Australasia, city, mountains, interior).

Time: any historical period, or concept defined by temporal properties (for example, Medieval, summer, nocturnal, permanent).

These 13 categories have been found to deal adequately with the vocabulary of a wide range of subjects, although it is usually the case that for a particular subject, only a smaller subset of categories is actually required. The complete range is probably only utilized within technological subjects, and modifications to the standard ones are found in some disciplines (notably the arts and humanities). When a particular *category* is identified within a subject, its members constitute a *facet* in that subject for example, the category of *things* or *entities* in zoology is known as the *organisms* facet, and the category of *processes* in pathology is the *disease* facet.

Organization within Facets

In the majority of subjects the number of concepts within a facet will be large, and some further means of organizing them must be found. This is done by identifying groups of concepts that share a particular defining attribute. For example, within sociology concepts in the persons facet may be divided into groups characterized by age (children, adolescents, old people), by gender (male, female), by ethnic origin (Afro-Caribbean, Caucasian, aboriginal), or indeed by a range of other properties. These groupings are called *arrays* and the defining attribute is called a *principle, or characteristic, of division*.

For example, in Class Q, Social Welfare, the facet "Persons in need" is organized into arrays by several principles of division, as shown in the (abridged) section below.

QG	Persons in need, Causes of need
	(Types of persons)
	(Defined by events, conditions)
QGN	Emergencies, disasters

QGN G	Victims, refugees
QGN J	Evacuees
QGO B	Prisoners of war
	(By social conditions)
QGP	Deprivation
	Poor
	Unemployed
	(Persons by age)
QL	Minors, children
QLV	Elderly, aged
	(Persons by disadvantage)
QMT	Physically disabled
	Blind

Order in Array

The sequence of the different arrays within a facet, and the sequence of the classes in individual arrays (order within array) are not usually based on any specific principles, although sometimes there are obvious orders to be employed (such as chronological, developmental, or spatial sequences). Ordinarily, order within arrays, and also between arrays, is not subject to any theoretical rules, but is decided pragmatically, as in this list of prisons from BC2 Class Q, which uses the degree of security as the basis of arrangement:

QQS K	Types of prisons
QQS M	Maximum security
N	Medium security
P	Minimum security, open prisons
R	With semiliberty (living in institution, working outside)
S	With restricted liberty (living at home, working at institution)

Order by degree of security in BC2 Class Q

Citation Order and the Combination of Concepts

A second fundamental characteristic of the faceted scheme, after its structural organization, is the system syntax, that is, the rules for combining basic concepts to express the content of documents, which is usually more complex. In any indexing or classification system that allows the building of classmarks or headings, the order in which concepts are combined is known as the citation order. In BC2 this is linked to the standard categories used to create the facets, and is known as standard citation order. A compound subject consisting of terms from more than one category combines them in the order of the standard categories as discussed above, that is, thing–kind–part–material–property–. . .–agent–place–time. For example:

Hospitals–emergency–fittings–procurement–UK
(thing)   (kind)   (part)   (operation)   (place)

The order of fundamental categories in the citation order is determined by Ranganathan's principle of decreasing concreteness, and also by the principle of dependency of succeeding facets. This proposes that a thing must exist in order for its kinds and parts to exist, and that an operation must have something to be performed on, and that an agent must have an action to be agent of.

In BC2 the structure does not require that all things are located in a single facet, or all operations, or all agents (a pure faceted structure). In practice the citation order can be introduced at any point. For example, in aeronautical engineering, the wing is part of the aircraft, but can itself have kinds for example, delta, and parts, for example, ailerons; likewise a process may have stages, or parts, for example, digestion has parts ingestion and peristalsis. The structure of BC2 can therefore become quite complex as this example shows:

HUH	Heart	=(Part)
HUH FP	(Treatment)	=(Part + op.)
HUH FV	Drug therapy	=(Part + op. + agent)
	(Disorders)	=(Part + process)
	(Physiology)	=(Part + process)
HUH OX	(Disorders of heart physiology)	=(Part + pro.1 + pro.2)
HUH OXJ	Arrhythmia	=(Part + pro. + pro. + kind)
HUH OXJ FV	(Drug therapy)	=(Part + pro. + pro. + kind + op. + agent)

This example also demonstrates the practice in BC2 of incorporating many compounds into the facet structure to ensure that sought terms will appear in the schedule, and hence in the index.

Filing Order and Schedule Inversion

In addition to the organization of concepts into facets and arrays, and the order of concepts within the facet, it is also necessary to provide rules for the order of facets within the class as a whole. This order is known as the filing order, since it determines the linear sequence of classes on the library shelves or in the classified file.

In BC2 the filing order begins with the most abstract facet (time) and progresses toward the most concrete facet (thing). This concept of increasing concreteness produces an order of classes which looks natural and which appears to be intuitive to users. Because this is the inverse of the citation order, such a classification schedule is known as an inverted schedule.

J	Education	
J7	History of education	(Time)
J8	Education by place	(Place)
JE	Psychology of education	(Process)
JF	Educational performance	(Process)
JG	Students, pupils	(Patients)
JH	Teachers	(Agent)
JI	Teaching methods	(Operation)
JK	Curriculum	
JL/JV	Educands and educational institutions	(Thing/kind)

Inverted schedule in BC2 Class J

The principle of inversion is not essential to a faceted classification, but it makes the filing order much more logical when compound terms are added. We can see this in a very simple classification for history consisting of two facets (place and time) where the citation order is place–time:

History
    England   (Facet 1)
    France
    Germany

    Fifteenth century   (Facet 2)
    Sixteenth century
    Seventeenth century

If some compound classes are now added, as will happen when the classification is applied to real documents, and the results filed, the following will occur:

History
  England
    Sixteenth century England
  France
    Fifteenth century France
  Germany
    Seventeenth century Germany
  Fifteenth century
  Sixteenth century
  Seventeenth century

In this arrangement the specific classes of, for example, sixteenth century England and fifteenth century France, precede the general classes fifteenth century and sixteenth century. This is counterintuitive for the majority of readers who expect general classes to come before subdivisions of those classes, and, as the complexity of subjects is increased the confusion and illogicality of the filing order would also increase.

The situation can easily be remedied by inverting the facet order, so that the first cited facet files last:

History
    Fifteenth century   (Facet 2)
    Sixteenth century
    Seventeenth century

England (Facet 1)
    Sixteenth century
France
    Fifteenth century
Germany
    Seventeenth century

It is not essential to invert the schedule, and in a digital environment where there is no linear ordering it may not offer any advantage, but where items are to be organized in a sequence it makes for a much more logical and natural arrangement. Because in BC2 the citation order is built into the schedule, this has the added advantage of making analysis and classmark building much easier for the classifier, since it is necessary only to combine the elements of a compound in the reverse order of their appearance in the schedule.

Notation

The original literal notation of BC1 has been preserved in the broad structure, since letter notations have several advantages. The notational base is large, and shorter classmarks can be achieved using letters. Because there are 26 letters, a two-character combination provides 26×26 classes ($= 676$) as opposed to 100 for numerals; at the four-character level, which is most often used for a broad arrangement of books, the difference is 456,976 to 10,000. Obviously a much greater depth of classification can be provided for within the same notational provision. Literal mnemonics, the creation of correspondences between letters in the terminology and letters in the notation, can also be used to advantage, for example, C for chemistry, CE for electrochemistry. As in BC1, the BC2 notation is nonexpressive and neither reflects the degree of hierarchy in the length of the notation, nor indicates the subordination of one class to another.

BTK	B	Bodies in gas flow
BTK	BBG	Control forces
		(parts)
BTK	D	Surfaces
	DR	Holes
		(types of bodies)
BTK	F	Streamlined bodies
	G	1-dimensional bodies
	H	2-dimensional bodies
	J	Manifolds
	L	Conduits, ducts
	N	Channels
	P	Closed conduits

In order to build compound classmarks BC1 uses the comma (to introduce auxiliary schedules peculiar to a subject), and the hyphen (to join classmarks between different classes). In BC2 the notation is used in a more sophisticated way to effect classmark synthesis without the need for such facet indicators. This device, namely retroactive notation, is, like the citation order, incorporated into the schedule, and it further reinforces the mechanical process of classmark building.

The device operates by "reserving" blocks of notation under each class for the direct addition of any preceding classes. No facet indicators or linking devices are required, since notational conflict is avoided by the appropriate assignment of notation by the schedule compiler. The indexer simply adds together the notational codes for the constituent elements of his concept analysis; since this must be, by definition, done in the reverse of schedule that is, alphabetical order, this also serves to correct any errors in citation order made by the user.

Practical Classification in BC2

Classmark synthesis is achieved in BC2 by two methods. In most cases direct retroactive number building is used, as demonstrated in example 1 below. On occasion, where there is great pressure on the notation (because of the sheer size of the vocabulary), this may be eased by the introduction of an intercalator; this is a "peg" on which to hang synthesized numbers; an example of the use of the intercalator is given at example 2.

1. Normal retroactive building of classmarks within a main class involves the direct addition of notation for a previously scheduled term, dropping the main class letter from the added term.

B	Physics
B7H	Visualization and imaging
BAF	Energy interactions and forces
BF	Waves
BFT	Scattering
BJ	Magnetism
BM	Particles

All classes preceding BM can be directly added to that class, dropping the initial letter B.

BM7 H	Imaging of particles
BMF T	Scattering of particles

2. In some places, intercalators are employed to extend the capacity of the notation. In such cases, notation is added to a specified base letter, rather than directly. For example:

BB	Mechanics
BBB T	Kinetic energy
BDC	Velocity
BDM	Circular motion

BF	Waves, wave motion
	*Add to BFA letters A/DR following B
BFA BBT	Kinetic energy of waves
BFA DC	Velocity of waves
BFA DM	Circular motion of waves

Here previous classes are added to the intercalator A, rather than directly to the class BF. Use of the intercalator is always made explicit by instructions in the schedule, and should not be used other than in compliance with these.

THE USE OF BC2 TODAY AND DEVELOPMENTS FOR THE FUTURE

Although the scheme is envisaged as a general scheme suitable for large document collections, the take-up of BC2 has been largely among special libraries. The new scheme offers a level of detail in its terminology, and a capacity for synthesis that is particularly appropriate for collections of highly specialized documents, and in some disciplines there are no comparable indexing languages, either general or special. This is particularly true for Social welfare and related subjects where adoption of BC2 has been most widespread. Of recent times it has also been applied in a number of medium sized academic libraries in the University of Cambridge.[15] It finds favor in these because of its sound theoretical principles, the visible logic and predictability of its structure, and the close match between its broad structure and the academic understanding of the constitution of subjects.

Most users of BC2 are new to the scheme, and BC1 libraries have not on the whole reclassified to the new edition, probably because the modification of the internal structure and the changes in notation would require both substantial reprocessing of stock and amendment of catalog records. The complexity of BC2 may also be seen as a disadvantage in an age when most UK librarians receive little formal education in cataloging. Over a period of time most BC1 libraries have now reclassified (or are in the process of reclassifying) to either the Library of Congress Classification or the Dewey Decimal Classification. The current emphasis on copy cataloging and the downloading of records in preference to local cataloging effects considerable savings for the majority of general libraries, making it difficult for schemes with a smaller user base to compete. Attempts in the 1970s to persuade the British National Bibliography to include BC2 classmarks were unsuccessful, and it is unlikely that this situation will change in the future.

BC2 also suffers from many of the same disadvantages as the original classification, and has been criticized on that basis. The lack of strong institutional and financial support for the scheme is a major shortcoming, and the dependence on voluntary input from BC2 enthusiasts makes consistent and systematic maintenance and updating hard to achieve. The lack of salaried staff means that schedule production is a slow process, and this has been exacerbated by the need to reengineer the software that manages the physical generation of the schedules and alphabetical index.

To date, about two-thirds of the classification is published, and the remaining classes exist in draft format. Current plans are to disseminate these drafts on the Association's Website (http://www.blissclassification.org.uk) in the hope of providing for the gaps in the available classes, and attracting input from a wider constituency. It is also intended to produce in the near future an abridged version of the scheme (based in part on the unpublished classes) which may have greater appeal, particularly for small libraries, than the full classification, and will also address the problem of the missing subjects. Although completion of the scheme may as yet be some way off, where the necessary expertise is available, some classes (J, Education and Q, Social Welfare) have already been published in a 3rd edition.

BC2 in its inception was intended, like its predecessor, primarily as a bibliographic classification, and an organizing tool for print collections. Nevertheless the physical arrangement of books (and the related ordering of items in catalogs and bibliographies), important as this is, is not the only potential application of the system.

BC2 in a Digital World

There are numerous examples of applications of classifications and other conventional indexing languages to the organization of Web-based resources, and a number of schemes are used to arrange resources on managed sites, in the same way that those same schemes are used to order books on shelves. *EU telematics for research project DESIRE* surveyed the use of a number of existing library classifications for the management of Internet resources, particularly in managed hubs and portals. The results of their investigations can be viewed at http://www.surfnet.nl/innovatie/desire1/deliver/WP3/D32–3.html. However, more recent projects have begun to look at the knowledge structures embedded in classifications as tools for handling structural and semantic relations in searching and retrieval. It would appear that the highly structured, rigorously logical and predictable BC2, with its detailed vocabulary, and sophisticated semantic and syntactic relations, should be a frontline contender in any such enterprise. A project funded by the U.K. Arts & Humanities Research Council at University College London looked at the way faceted classification data can be structured and manipulated to deliver some degree of automatic indexing for digital resources.[16] BC2 vocabularies provided the core subject terminologies, and the results were very encouraging. The regular structure of the classification lends itself readily to incorporation in a database, where many of the relationships, including the facet status of a concept, can be recorded. This can allow the mapping of a document content analysis or a search query onto the database, and the automatic generation of structured strings using controlled vocabulary and with inbuilt citation order. It did, however, demonstrate that some features of the scheme (such as the alternative locations, the flexibility of the citation order, and the lack of facet indicators) must be sacrificed in order to achieve the regularity of structure that machine manipulation demands.[17]

Thesaurus Format for BC2

As long ago as the 1980s the potential of BC2's faceted terminology as a source for a thesaurus had been realized and exploited. Jean Aitchison, known for her pioneering work on *Thesaurofacet*,[18] had written about the process of converting BC2 to a thesaurus,[19] and had developed a faceted thesaurus for international affairs from the BC2 Politics terminology.[20] Current work on the classification is investigating the feasibility of publishing future volumes in the form of a joint classification and thesaurus, rather on the model of the *Thesaurofacet*. Music is a suitably straightforward vocabulary, of relatively modest size, and has been chosen as a trial class. To date the computer programs for the classification layout and alphabetical index generation have been extended to support the thesaurus format, and work has proceeded on guidelines for the form of entry of class headings to ensure that these can be converted to thesaurus entries with the minimum of human intervention and editing.[21] Some difficulties occur as a result of the lack of vocabulary control in the classified structure, but otherwise trials of the software show that structural relationships are relatively straightforward to manage automatically, and that BC2 provides a sound basis for automatic thesaurus generation.[22]

Influence of BC2 on Other Schemes

It is also clear that the major contribution of BC2 to date lies not so much in its adoption by individual institutions, as in the influence it has had on the development of classificatory theory, and the consequent impact of this on the other major schemes. Particularly this is true of the UDC. McIlwaine, I. and Williamson, N. *A feasibility study on the restructuring of the UDC into a full faceted classification system* International Society for Knowledge Organization Conference 1994 Copenhagen, examines the use of BC2 as a source of facet analytical structure and vocabulary for incorporation into new revisions of the UDC. This theme is further developed in McIlwaine, I. UDC centenary; the present state and future prospects. Knowledge organization **1995** 22(2), 64–69. Most other systems now regard some degree of facet structure as necessary in the development of new schedules, and the advantages in

retrieval performance of highly structured and logical systems on the model of BC2 can hardly be disregarded. The gradual spread of the faceted structure throughout an increasing number of retrieval languages must be due at least in some measure to the perceived superiority of this approach, and it seems likely that the published schedules of BC2 have provided a pattern for modern retrieval language construction that is now widely regarded as the archetype of the twenty-first century scheme.

REFERENCES

1. Bliss, H.E. *A Bibliographic Classification Extended by Systematic Auxiliary Schedules for Composite Specification and Notation*, H. W. Wilson: New York, 1940/1953A1.
2. Langridge, D.W. *Approach to Classification for Students of Librarianship*, Bingley: London, U.K., 1973; Vol. 93.
3. Ranganathan, S.R. *Prolegomena to Library Classification*, 3rd Ed. Asia Publishing House: London, U.K., 1967; (Ranganathan Series in Library Science, 20).
4. Ranganathan, S.R. *Colon Classification. Basic Classification*: 6th Ed. Asia Publishing House: New York, 1960.
5. Classification Research Group. The need for a faceted classification as the basis of all information retrieval. Libr. Assoc. Rec. **1955**, *57*(7), 262–268.
6. CRG Bulletin No. 8. J. Doc. **1964**, *20*(3), 158–159.
7. Austin, D.; Butcher, P. *PRECIS: A Rotated Subject Index*, The British Library Bibliographic Services Division: London, U.K., 1969.
8. Library Association. *Some Problems of a General Classification Scheme: Report of a Conference held in London in June 1963*; Library Association: London, U.K., 1964.
9. Mills, J.; Broughton, V. *Bliss Bibliographic Classification: Introduction and Auxiliary Schedules*, 2nd ed.; Butterworths: London, U.K., 1977; 29–61.
10. Spiteri, L.F. The classification research group and the theory of integrative levels. Katharine Sharp Rev. **1995**, *1* (Summer). Available at http://wotan.liu.edu/dois/data/Articles/juljuljasy:1995:i:1:p:10952.html (accessed May 2007).
11. Vickery, B.C. *Classification and Indexing in Science*, Butterworths: London, U.K., 1958.
12. Foskett, D.J. *Classification and Indexing in the Social Sciences*, Butterworths: London, U.K., 1963.
13. Langridge, D.W. *Classification and Indexing in the Humanities*, Butterworths: London, U.K., 1976.
14. Mills, J. *A Modern Outline of Library Classification*, Chapman & Hall: London, U.K., 1960.
15. Broughton, V. Classification and subject organization and retrieval. In *British Librarianship and Information Work 1991–2000*; Bowman, J.H., Ed.; Ashgate: Aldershot, U.K., 2006; 494–516.
16. Facet Analytical Theory in Knowledge Structures. Project documentation available at http://www.ucl.ac.uk/fatks/ (accessed June 2007).
17. Broughton, V.; Lane, H. The Bliss Bibliographic Classification in action: Moving from a special to a universal faceted classification via a digital platform. In *Knowledge*

Organization and the Global Information Society; McIlwaine, I.C., Ed.; Ergon: Wurzburg, Germany, 2004; 73–78 Proceedings of the 8th International ISKO Conference, July 13–16, 2004, London, U.K.

18. Aitchison, J.; Gomershall, A.; Ireland, R. *Thesaurofacet: A Thesaurus and Faceted Classification for Engineering and Related Subjects*, English Electric Company: Whetstone, Leicester, U.K., 1969; (Compilers).

19. Aitchison, J. A classification as a source for a thesaurus: The Bibliographic Classification of H. E. Bliss as a source of thesaurus terms and structure. J. Doc. **1986**, *42*(3), 160–181.

20. Aitchison, J.; Gallimore, N.; Boyde, S. *Royal Institute of International Affairs Library Thesaurus*, RIIA: London, U.K., 1992.

21. Aitchison, J. Thesauri from BC2: Problems and possibilities revealed in an experimental thesaurus derived from the Bliss Music schedule. Bliss Classif. Bull. **2004**, *46*, 20–26.

22. Broughton, V. A faceted classification as the basis of a faceted terminology. Axiomathes. **2008**, *18*(2), 193–210. Available at http://www.springerlink.com/content/6mm3r57j1r44k5u5/.

BIBLIOGRAPHY

1. Bliss Classification Association, website Available at http://www.blissclassification.org.uk.

2. Broughton, V. Faceted classification. *Essential Classification*, Facet: London, U.K., 2004; 257–283.

3. Maltby, A. Gill, L. *The Case for Bliss: Modern Classification Practice and Principles in the Context of the Bibliographic Classification*, Saur: New York, 1979.

4. Mills, J. Attention please for BC2. Intl. Classif. **1983**, *10*, 24–28.

5. Mills, J. The Bibliographic Classification. In *Classification in the 1970s: A Discussion of Development and Prospects for the Major Schemes*; Maltby, A., Ed.; Bingley: London, U.K., 1972; 25–52.

6. Mills, J. Faceted classification and logical division in information retrieval. Lib. Trends. **2004**, *52*(3), 541–570.

7. Thomas, A.R. Bliss Bibliographic Classification 2nd edition: Principal features and applications. Catalog. Classif. Quart. **1992**, *15*(4), 3–17.

Boolean Algebras *[ELIS Classic]*

A. R. Bednarek
University of Florida, Gainesville, Florida, U.S.A.

Abstract

Boolean algebra, named after the 19th century mathematician and logician, George Boole, has contributed to many aspects of computer science and information science. In information science, Boolean logic forms the basis of most end-user search systems, from searches in online databases and catalogs, to uses of search engines in information seeking on the World Wide Web.

—**ELIS Classic**, from 1970

INTRODUCTION

In this entry attention is focused on mathematical models of proven utility in the area of information handling, namely, Boolean algebras. Following some general comments concerning mathematical models, particular examples of Boolean algebras, serving as motivation for the subsequent axiomatization, are presented. Some elementary theorems are cited, particularly the very important representation theorem that justifies, in some sense, the focusing of attention on a particular Boolean algebra, namely, the algebra of classes, and applications more directly related to the information sciences are given.

Running the risk of redundancy, attention will be called to an often-repeated observation, but one of extreme importance in applications of mathematics to physical problems. Referring to Fig. 1, it is important to realize that when one constructs a mathematical model as a representation of a physical phenomenon, one is abstracting and, as a consequence, the model formulated is doomed to imperfection. That is, one can never formally mirror the physical phenomenon, and must always be satisfied with an imperfect copy. However, following the initial commitment to a model, the logic that one appeals to dictates the resultant theorems derived within the framework of the model. Of course, the depth of the theorems realized is limited by the sophistication of the model as well as the ingenuity of those who attempt to formulate the propositions within it. After theorems are derived within the framework of the model, they are interpreted relative to the physical situation that motivated the model.

It is not necessary to go very deeply into mathematics before facing the necessity of examining, in some detail, this cycle and developing a feeling for its power as well as its limitations. By way of example, almost any student of calculus encounters, in one form or another, the following problem:

The deceleration of a ship in still water is proportional to its velocity. If the velocity is v_0 feet per second at the time the power is shut off, show that the distance S the ship travels in the next t seconds is $S = (v_0/k)[1 - e^{-kt}]$ where k is the constant of proportionality.

HISTORY

The desired equation relating the distance traveled to the time is easily arrived at by means of the calculus. However, a close look at the solution reveals a few puzzling aspects. When does the ship stop? The conclusion is that it never stops. How far does it go? The conclusion is that it goes no further than v_0/k, that is, the distance it travels is bounded. Sympathy is due the beginning student of the calculus who is puzzled by these observations, but, too often, we neglect to focus our attention on the source of the puzzlement. It really has nothing to do with the limit process that plays such an integral role in analysis, nor must we drag poor Zeno into the picture. This disturbing conclusion is not the consequence of any faulty mathematics, but is more directly related to the naïveté of the original model. If we say that the deceleration of a ship in still water (an idealization in itself) is proportional only to its velocity, then the conclusion that asserts itself is that the ship never stops but only goes a finite distance.

The usual remedy applied in such cases as the ship problem is to construct a more sophisticated model, that is, a model that takes into account more of the phenomena observed. For example, in the ship problem, the assertion that the deceleration is proportional only to the velocity might be amended to include friction in some way, resulting in an equation of greater complexity, the formulation and solution of which require a more general mathematical model. We might extend the preceding model to look like Fig. 2.

The great power of mathematics lies in its ability to reflect several different phenomena at one time, and the theorems derived within the framework of a single

Binding–British

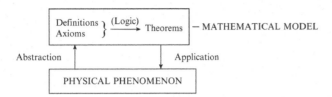

Fig. 1

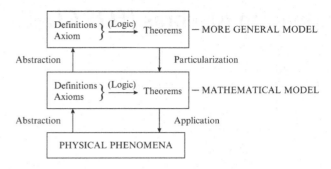

Fig. 2

axiomatization of these varied phenomena will, in turn, be applicable to each of them. However, the trade-off that exists between generalization and depth must constantly be kept in mind. That is, it should be remembered that it is difficult to prove deep theorems in very general models. But when axioms are added to the model, the phenomena which the model reflects begin to be delimited, and certainly one does not wish to undermine the real power of mathematics, that is, its ability to treat a variety of situations at the same time.

Examples

We now turn our attention to an examination of some of the particular examples of the model that is the principal concern of this entry, Boolean algebras. One must keep in mind that the common characteristics of these models are precisely those that will constitute the elements of our later axiomatization. To avoid infinite regress a certain level of sophistication on the part of the reader, if not actual mathematical experience, is assumed.

Example 1 (A finite algebra)

The system considered in this example consists of the two digits, 0 and 1, and two binary operations of multiplication, "$\cdot$," and addition, "$+$." The operations are defined by the multiplication and addition tables shown.

$\cdot$	0	1
0	0	0
1	0	1

$+$	0	1
0	0	1
1	1	1

Multiplication   Addition

If x, y, and z are any *variables* that are allowed to assume one of the two values 0 or 1, then the structure defined above has an algebra possessing (among others) the following properties:

$$x \cdot x = x \qquad x + x = x \tag{1a}$$

$$x \cdot y = y \cdot x \qquad x + y = y + x \tag{1b}$$

$$x \cdot (y \cdot z) = (x \cdot y) \cdot z \qquad x + (y + z) = (x + y) + z \tag{1c}$$

$$(x \cdot y) + x = x \qquad (x + y) \cdot x = x \tag{1d}$$

$$x \cdot (y + z) = (x \cdot y) + (x \cdot z) \qquad x + (y \cdot z) = (x + y) \cdot (y + z) \tag{1e}$$

Each of the above can be verified by a consideration of all the possible values of the variables.

If B is the collection consisting of the elements 0 and 1, and for each x in B x' is defined by $x' = 1$ if $x = 0$, and $x' = 0$ if $x = 1$, then

$$\begin{array}{llll} x \cdot x' = 0 & & \text{and} & x + x' = 1 \\ 0 \cdot x = 0 & 1 \cdot x = x & \text{and} & 0 + x = x \qquad 1 + x = x \end{array} \tag{1f}$$

We let $[B; \cdot, +, ']$ denote the system described in Example 1.

Example 2 (Algebra of propositions)

The elements in this example are *propositions*, that is, statements to which it is possible to assign one of the truth values "true" or "false." Two propositions p and q are defined to be equal if and only if they have the *same* truth value. We consider the two logical binary operations of *conjunction* and *disjunction* as well as the unary (operating on a single proposition as contrasted with a binary operation, which operates on pairs of propositions) operation of *negation*. The conjunction of the propositions p and q is denoted by pq and is the proposition corresponding to that obtained by applying the logical connective "and." The conjunction is defined to be true only if both p and q are true. Otherwise it is false. The disjunction of p and q, denoted by $p + q$, is the proposition corresponding to that obtained by applying the logical connective "or." The proposition $p + q$ is false if and only if both p and q are false. The negation of p, denoted by $\bar{p}$, is the proposition having truth values opposite those of p. It corresponds to the logical statement, "It is false that p."

All of the above can be summarized nicely by employing "truth tables" that give the truth values of compound statements, realized by applying the operations discussed to the truth values of the component propositions.

p	q	pq
T	T	T
T	F	F
F	T	F
F	F	F

p	q	$p + q$
T	T	T
T	F	T
F	T	T
F	F	F

p	$\bar{q}$
T	F
F	T

Conjunction   Disjunction   Negation

One can verify that, in view of the definitions above, if p, q, and r are any propositions, the following statements hold.

$$pp = p \qquad p + p = p \tag{2a}$$

$$pq = qp \qquad p + q = q + p \tag{2b}$$

$$p(qr) = (pq)r \qquad p + (q + r) = (p + q) + r \tag{2c}$$

$$(pq) + p = p \qquad (p + q)p = p \tag{2d}$$

$$p(q + r) = pq + pr \qquad p + (qr) = (p + q)(q + r) \tag{2e}$$

$$p\bar{p} \text{ is always false and } p + \bar{p} \text{ is always true} \tag{2f}$$

Denoting $p\bar{p}$ by 0 and $p + \bar{p}$ by 1 we have

$$0q = 0 \quad 1q = q \quad 0 + q = q \quad 1 + q = q$$

(*Note*: It is easy to see that $p\bar{p} = q\bar{q}$ and $p + \bar{p} = q + \bar{q}$ for any propositions p and q.)

We illustrate the employment of the truth table technique in the verification of part of the assertion in Eqs. 2e, namely, that $p + (qr) = (p + q)(p + r)$.

p	q	r	qr	$p + qr$	$p + q$	$p + r$	$(p + q)(p + r)$
T	T	T	T	T	T	T	T
T	T	F	F	T	T	T	T
T	F	T	F	T	T	T	T
T	F	F	F	T	T	T	T
F	T	T	T	T	T	T	T
F	T	F	F	F	T	F	F
F	F	T	F	F	F	T	F
F	F	F	F	F	F	F	F

Since the columns headed "$p + qr$" and "$(p + q)(p + r)$" have identical entries the propositions $p + qr$ and $(p + q)(p + r)$ have the same truth value and are therefore equal.

The structure described in Example 2 is denoted in the sequel by $[\wp; ,+, -]$.

Example 3 (Algebra of sets)

The term *set* is taken as undefined and used synonymously with class, aggregate, and collection. The objects that constitute a set E are called the *elements* of E. To denote the logical relation of "being an element of E" we use the notation $x \in E$. This is read: "x is an element of E." The denial of this relation is symbolized by $x \notin E$.

The notation of $E = \{xP(x)\}$ denotes the set E consisting of all x for which the proposition $P(x)$ is true. When the set under consideration is finite, it is often denoted by a simple listing of its elements; thus, $E = \{a,b\}$ is the set consisting of the elements a and b.

If $A = \{xP(x)\}$ and there are no elements which satisfy the proposition $P(x)$, A is said to be the *empty set*. The empty (or *null*) set is denoted by ϕ.

If the sets E and F have the property that every element of E is an element of F, E is called a *subset* of F; this is denoted by $E \subset F$. If the set E is a subset of F, but F is not a subset of E, then E is said to be a *proper subset* of F, or F *properly contains* E. The empty set ϕ is a subset of every set.

Two sets E and F are *equal*, written $E = F$, if and only if $E \subset F$ and $F \subset E$.

Given two sets E and F, we define the *union*, denoted by $E \cap F$, by the set equation

$$E \cup F = \{x | x \in E \text{ or } x \in F\}$$

Similarly, the *intersection* of E and F, denoted by $E \cap F$, is defined by

$$E \cap F = \{x | x \in E \text{ and } x \in F\}$$

In general, consideration centers on subsets of a fixed set often referred to as the *universal set*. In particular, if X is the universal set, we let $\wp(X)$ denote the set of all subsets of X. The set $\wp(X)$ is often called the *power set* of X. If $E \in \wp(X)$, then the *complement* of E, denoted by E', is defined as $E' = \{xx \in X \text{ and } x \notin E\}$. If E and F are elements of $\wp(X)$, that is, subsets of X, then the *difference* of the sets E and F, denoted by $E - F$, is the set defined by

$$E - F = \{x | x \in E \text{ and } x \notin F\}$$

It should be noted that $E - F = E \cap F'$.

It is often helpful to employ the schematics shown in Fig. 3 in visualizing the set-theoretic relations and operations defined above. The rectangular area represents the

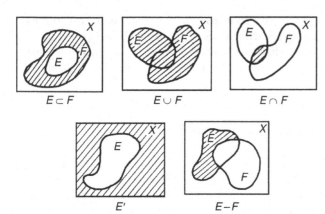

Fig. 3

universal set X; subsets of X are denoted by areas within the rectangle.

Focusing our attention on a particular universal set X and its power set $\wp(X)$, it is easy to verify the following properties (in no sense exhaustive) of the algebra of sets, where E, F, and G are arbitrary subsets of X.

$$E \cap E = E \quad E \cup E = E \tag{3a}$$

$$E \cap F = F \cap E \quad E \cup F = F \cup E \tag{3b}$$

$$E \cap (F \cap G) = (E \cap F) \cap G \quad E \cup (F \cup G) = (E \cup F) \cup G \tag{3c}$$

$$(E \cap F) \cup E = E \quad (E \cup F) \cap E = E \tag{3d}$$

$$E \cap (F \cup G) = (E \cap F) \cup (E \cap G)$$
$$E \cup (F \cap G) = (E \cup F) \cap (E \cup G) \tag{3e}$$

$$E \cap E' = \phi \qquad\qquad E \cup E' = X$$
$$\phi \cap E = \phi \quad X \cap E = E \quad \phi \cup E = E \quad X \cup E = E \tag{3f}$$

In the sequel we denote the above algebra of sets by $[\wp(X); \cap, \cup, {}']$.

Axiomatization

In the preceding section we examined three structures possessing some common properties, namely, (1i), (2i), and (3i), where $i = 1, 2, 3, 4, 5, 6$. We abstract to construct the important mathematical (see Fig. 1) model called a *Boolean algebra*, named in honor of G. Boole who first studied it in 1847.[1,2]

A *Boolean algebra* is a set B with two binary operations $\wedge$ (cap) and $\vee$ (cup) and a unary operation (complementation) satisfying the following axioms:

$$x \wedge x = x \quad \text{and} \quad x \vee x = x \tag{Ia}$$

$$x \wedge y = y \wedge x \quad \text{and} \quad x \vee y = y \vee x \tag{Ib}$$

$$x \wedge (y \wedge z) = (x \wedge y) \wedge z \text{ and } x \vee (y \vee z)$$
$$= (x \vee y) \vee z \tag{Ic}$$

$$(x \wedge y) \vee x = x \text{ and } (x \vee y) \wedge x = x \tag{Id}$$

$$x \wedge (y \vee z) = (x \wedge y) \vee (x \wedge z) \text{ and}$$
$$x \vee (y \wedge z) = (x \vee y) \wedge (x \vee z) \tag{Ie}$$

B contains *distinct* elements 0 and 1 such that

$$x \wedge x' = 0 \qquad\qquad \text{and} \quad x \vee x' = 1$$
$$0 \wedge x = 0 \quad 1 \wedge x = x \quad \text{and} \quad 0 \vee x = x \quad 1 \vee x = 1 \tag{If}$$

This is by no means the only axiomatization possible,[3] but it is probably the one most commonly used.

To emphasize the relationship between the above axiomatization and the preceding particularizations

Boolean algebra	Example 1	Example 2	Example 3
B	set $\{0,1\}$	set $\wp$ of all propositions	set $\wp(X)$ of all subsets of a fixed set X
$\wedge$	$\cdot$	conjunction	intersection
$\vee$	$+$	disjunction	union
${}'$	${}'$	negation	complementation
0	0	$p\bar{p}$	empty set ϕ
1	1	$p + \bar{p}$	universal set X

(Examples 1, 2, and 3) we identify in tabular form the corresponding structural elements.

We now prove a particular theorem to illustrate the generation of results within the framework of the model and their subsequent application.

Theorem. If $(B; \wedge, \vee, {}')$ is a Boolean algebra, then for any x and y in B we have

$$(i) \, x'' = x$$

$$(ii) \, (x \wedge y)' = x' \vee y'$$

Proof. First of all we prove that every element has only one complement. Suppose $\bar{x}$ is an element such that $x \wedge \bar{x} = 0$ and $x \vee \bar{x} = 0$. Then

$$\bar{x} = \bar{x} \wedge 1 = \bar{x} \wedge (x \vee x') = (\bar{x} \wedge x) \vee (\bar{x} \wedge x')$$

$$= 0 \vee (\bar{x} \wedge x') = \bar{x} \wedge x'$$

but

$$x' = x' \wedge 1 = x' \wedge (x \vee \bar{x}) = (x' \wedge x) \vee (x' \wedge \bar{x})$$

$$= 0 \vee (x' \wedge \bar{x}) = (x' \wedge \bar{x})$$

and since $\bar{x} \wedge x' = x' \wedge \bar{x}$, we have $\bar{x} \wedge x'$. We then apply the preceding by demonstrating that $(x \wedge y) \wedge (x' \vee y') = 0$ and $(x \wedge y) \vee (x' \vee y') = 1$, so that $(x \wedge y)' = x' \vee y'$.

$$(x \wedge y) \wedge (x' \vee y') = [(x \wedge y) \wedge x'] \vee [(x \wedge y) \wedge y']$$

$$= [(y \wedge x) \wedge x'] \vee [(x \wedge y) \wedge y']$$

$$= [y \wedge (x \wedge x')] \vee [x \wedge (y \wedge y')]$$

$$= [y \wedge 0] \vee [x \wedge 0] = 0 \vee 0 = 0$$

$$(x \wedge y) \vee (x' \vee y') = [(x \wedge y) \vee x'] \vee y'$$

$$= [(x \vee x') \wedge (y \vee x')] \vee y' = [1 \wedge (y \vee x')] \vee y'$$

$$= (y \vee x') \vee y' = (x' \vee y) \vee y'$$

$$= x' \vee (y \vee y') = x' \vee 1 = 1$$

An interpretation (application) of this theorem in Example 2 yields the fact that the negation of the conjunction of two propositions is the disjunction of the negations of each of them. For example, "it is false that x is a positive integer and x is greater than or equal to 5" is logically equivalent to the proposition "x is not a positive integer or x is less than 5."

An interpretation of the above in Example 2 yields the set-theoretic equation

$$(E \cap F)' = E' \cup F'$$

that is, the complement of the intersection of two sets is equal to the union of their complements, as shown in Fig. 4.

We describe very briefly some of the more significant results and developments in the theory of Boolean algebras. For a comprehensive treatment of the subject, see Birkhoff[3] and Halmos[4] and their bibliographies.

Every Boolean algebra can be made into a ring with identity in which every element is multiplicatively idempotent; that is, $x^2 = x$ for every x. This is accomplished by defining addition and multiplication as follows

$$x + y = (x \wedge y') \vee (x' \wedge y) \text{ and } xy = x \wedge y$$

Because rings are more familiar and more carefully studied many of the useful concepts can be translated into the context of Boolean algebras.

Conversely, if one starts with a ring with identity in which every element is idempotent (usually called a Boolean ring), defining $\wedge$ and $\vee$ by

$$x \wedge y = xy \quad x \vee y = x + y + xy$$

the Boolean ring is converted into a Boolean algebra.

Two Boolean algebras, B_1 and B_2, are said to be *isomorphic* if there exists a function $h: B_1 \rightarrow B_2$ that maps B_1 onto B_2 in such a way that distinct elements of B_1 are mapped onto distinct elements of B_2, and h preserves the operations; that is, $h(x \wedge y) = h(x) \wedge h(y)$; $h(x \vee y) = h(x) \vee h(y)$; and $h(x') = h(x)'$.

If X is a compact Hausdorff space, then the class of sets that are both open and closed forms a Boolean algebra. A topological space is totally disconnected if the only components (maximal connected sets) are points. There is a very important *representation theorem* in the theory of Boolean algebras, the Stone Representation Theorem (M.H. Stone[5]). If B is a Boolean algebra, then a compact totally disconnected Hausdorff space S exists such that B is isomorphic to the Boolean algebra of all open-closed subsets of S.

An Application

We consider here one modest application of the preceding to switching theory. Switching theory is concerned with circuits composed of elements that can assume a finite number of discrete states, most commonly two states. These circuits are modeled as described earlier, and the models are analyzed. This is an idealization; the models neglect such characteristics as stability, temperature effects, and transition times. The theory of Boolean algebras has played an important role in the analysis of these models for circuits made of binary (two-state) devices.

A *switching function* is a rule by which the output of a composite circuit can be ascertained from the states of its components. If the variables x, y, and z denote switches and each switch can assume one of the states, open or closed (0 or 1), then the function $w = x \wedge y$ describes the output of a series circuit containing the switches x and y. Similarly, $t = x \vee y$ is a function describing a parallel circuit

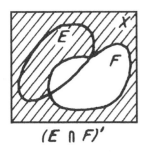
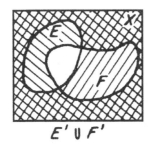

(E ∩ F)'   E' ∪ F'

Fig. 4

$x \wedge y$ $x \vee y$

Fig. 5

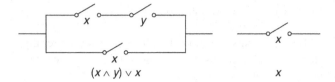

Fig. 6

containing the switches x and y. These components, along with the negation function (x' is a switch that is open whenever x is closed and closed whenever x is open), allow the construction and analysis of complex circuits. This analysis can be carried out by the use of truth tables, and the circuits can be indicated by a diagram, as shown in Fig. 5.

With this interpretation it is readily seen that the above is a Boolean algebra. For example, the verification of axiom (Id) involves the observation that the circuits in Fig. 6 are equivalent.

After observing that it is indeed a Boolean algebra, the machinery of that algebra may be used to synthesize circuits, consider questions of realizability, minimize circuitry, and so forth. We can only hint at the possible applications.[6]

REFERENCES

1. Boole, G. *The Mathematical Analysis of Logic*, Cambridge, 1847.
2. Boole, G. *An Investigation of the Laws of Thought*, London, 1854.
3. Birkhoff, G. *Lattice Theory*; Providence: RI, 1961; 155.
4. Halmos, P.R. *Lectures on Boolean Algebras*; Van Nostrand: Princeton, NJ, 1963.
5. Stone, M.H. The theory of representations for boolean algebras. Trans. Am. Math. Soc. **1936**, *40*, 37–111.
6. Flegg, H.G. *Boolean Algebra and Its Application*; Wiley: New York, 1964.

Brazil: Library Science

Mariza Russo
Faculty of Administration and Accounting Sciences (FACC), Federal University of Rio de Janeiro, Rio de Janeiro, Brazil

Abstract

This entry presents the history of Library Science in Brazil, from the creation of the first Library Science course in 1911, in the National Library, Rio de Janeiro, to the current courses offered today. The various levels of education for the professional librarian are described as well as the evolution of the profession from the early recognition of it as such through the twenty-first century including both undergraduate and graduate courses. It also focuses on the paradigmatic changes in information demands and discusses the professional job market highlighting the opportunities open to potential entrants.

INTRODUCTION

For many years in Brazil, Library Science as an avocation, had as its place in the library, which was mainly viewed from a static perspective as "a warehouse." It was more for storing a collection rather than for exploiting it, and the more extensive it was, the more valuable it was perceived to be. For the profession to grow and broaden its functions, a radical change in its outlook was required. The result was a new trend toward an interest in facilitating access to the collections—which resulted in a growth in research and, in turn, a growth in knowledge.

The practice of the profession, at this time, was to find studious individuals, lovers of books and belles lettres, or historians—almost always highly cultured individuals—who would work with two main objectives: "to guard the collection zealously" and "to preserve it for the future."

With the advent of the Industrial Revolution several events stand out that had world-wide influence for the practice of Library Science: increased production of the printed book; new scientific discoveries; the proliferation of scientific societies and the accompanying growth of scientific periodicals. These events brought about substantial changes leading to the information age and the changing role of the book keepers. Library professionals had to take on the responsibilities of acquiring, conserving, organizing, and distributing information and acquire the skills to do all of these.

DEVELOPMENT OF LIBRARY SCIENCE

There are several indispensable indicators for an area of study to be considered scientific: that it has developed theories, accompanied with practices that prove those theories; that there is a scientific infrastructure based on scientific societies, channels of communication, educational institutions and research; and most importantly, on qualified staff.

The formation of such staff represents a major component for the development of any field of study. For Library Science, it is achieved through courses at the undergraduate level in Latin American countries, and at the postgraduate level in Europe and in the United States.[1]

In Brazil, courses of Library Science are offered mainly as undergraduate courses and postgraduate specializations, resulting in students having the legal right to the title "librarian" upon graduation. Graduate courses for master's and doctorate degrees are completed in programs of various fields of study, with a major in Information Science.

The undergraduate and specialization courses result in the basic training needed to practice the profession. To become a professor or researcher, completion of a master's or doctorate program is required and sometimes both.

Undergraduate Courses

In Brazil, the evolution of the professional course of Library Science is characterized by three distinct phases.[2] The first one, from 1879 through 1929, followed the French traditions under the aegis of the National Library of Rio de Janeiro. The second, from 1929 through 1962, was based on American models, with the Mackenzie College, in São Paulo leading the way. The third, from 1962 until the present, begins with the establishment of the first official minimum curricula requirements (1962, revised in 1982) approved by the Federal Council of Education, and by the directives for curricula decreed in the Law of Directives and Educational Foundation (1996).

First phase

Library Science in Brazil, as a profession, closely follows the creation to the National Library. This library was formerly the Royal Library of Portugal, brought to Brazil by the Portuguese Crown, as it took refuge from the

Encyclopedia of Library and Information Sciences, Fourth Edition DOI: 10.1081/E-ELIS4-120045421

597

Binding–British

invading French army in its more prosperous colony in 1808. Officially established in 1810, it was opened to the public in 1814. Headed by clerics in the early days, its first secular director was appointed in 1846. Subsequently, the directors began to voice their concern for the lack of qualified staff and, with the objective to ameliorate these circumstances, the first course of Library Science was created in 1911 within the National Library. The first group of students in 1915 studied four disciplines: Bibliography (subdivided into Administration of Libraries and Cataloging); Paleography and Diplomatics; Iconography and Numismatics. This course was later revised following the model of the École Nationale de Chartes, in Paris—the first school in the world (1821) for the formation of library professionals. In this phase, the predominance of general culture reigned to the detriment of Library Science practices.

Second phase

The second phase began with the pragmatic model of education for Library Science and the Elementary Course of Library Science, created in the late 1920s at the Mackenzie College of São Paulo. This school was soon followed by another, the Course of Library Science of the Department of Culture of the City of São Paulo, in 1936 directed by the city's Municipal Librarian Rubens Borba de Moraes. Both schools adopted the American model for the courses of Library Science, favoring its technical emphasis over the study of general humanities.

These two courses brought significant changes to the library scene limiting the term "Modern Libraries," to those whose bookshelves and catalogs were open to the public with free access, in ample and comfortable installations, accommodation, with good illumination, and trained librarians to assist the readers.

In 1944, a new reform occurred in the National Library (see Fig. 1), incorporating changes in the Course of Library Science, that de-emphasized the humanist and scholarly aspects and emphasized American technical practices. This change was brought about by Brazilian librarians who had studied for specializations in the United States. This reform and the creation of the Course of the Free School of Sociological Politics of São Paulo, with funding from the Rockefeller Foundation in 1940, opened up new opportunities for the field. Professionals from other Brazilian states that participated in these courses in Rio de Janeiro and São Paulo, upon returning to their homes, reorganized old libraries, created new ones and started new Library Science courses in their respective home states. For example, the Schools of Library Science of Bahia, of Belo Horizonte and Paraná, among others were founded soon afterwards.

Third phase

The proliferation of Library Science courses in the country—each one with a different grading system—started to cause concern among library professionals. This gave rise to a national movement with the goal to standardize education to improve the performance of the average Brazilian librarian. The first "Minimum Curriculum" for the course of Library Science was established by the Federal Council of Education (CFE), by means of a Resolution of November 16, 1962. Decree no. 326/62, authored by Council member, Josue Montello, presented the following basic characteristics:

1. *Course offerings*: History of the Book and Libraries; History of Literature; History of Art; Introduction to Historical and Social Studies; Evolution of Philosophical and Scientific Thought; Organization and Administration of Libraries: Cataloging and Classification; Bibliography and Reference; Documentation; Paleography.
2. The duration of the course changed from two to three academic years.

Fig. 1   The National Library of Brazil.

This act had the immediate consequence of elevating the profession to a "profession of superior level," and was regulated by means of the Law no. 4.084/62 and Decree no. 56.725/65.

By this time there were 10 courses of Library Science in the country and by the end of 1969, eight new courses had been created. Between 1970 and 1977, 11 new additional courses were created. In the structure of their curricula, these courses followed the national model "Minimum Curriculum," which began to be criticized for being overly bureaucratic and an impediment to timely changes in the curricular structure.

In 1971, the Brazilian Association of Education for Library Science and Documentation (ABEBD) announced that it agreed that a revision of the curriculum was necessary. In 1980, the Secretariat of Higher Education of the Ministry of Education and Culture organized a task force, MEC/ABEBD. A year later, they presented a proposal to the Federal Education Council for the restructuring of the Minimum Curriculum for Library Science courses.

The second minimum curriculum for undergraduate coursework in Library Science was approved in 1982. It established a minimum course duration of 4 years and maximum of 7. The new curriculum established core courses for Communication, Society, Economic and Political Studies of Contemporary Brazil, and Cultural History; Basic courses for Logic, Portuguese Language, a Modern Foreign Language, Methods and Techniques of Research and Professional Development courses in Applied Information for Library Science, Production of Knowledge Registers, Collection Development, Bibliographical Control of Knowledge, Dissemination of the Information, and Administration of Libraries.

All previous legislation concerning Library Science education was revoked by the new decree no. 9.394, passed on December 20, 1996. The new curriculum directives allowed for more flexibility with course structures addressing new issues relevant to Brazilian society.

Today there are 43 Library Science courses in the country following these curriculum directives; 70% are established in public institutions and 30% in private institutions. Library schools are mostly located in the Southeastern region (47%).

One of the newest courses in the country was established by the Federal University of Rio de Janeiro in July of 2005. This Course of Library Science and Management of Information Units (CBG) aims to produce library managers and administrators qualified to function in the twenty-first century, where flexibility, speed, autonomy, innovation, and competitiveness are the best words. This course, whose first group started in August, 2006, includes in its curricular package, the same disciplines of Library Science and Administration in the same proportion, but goes beyond the general matrix necessary to form professionals attuned to the demands of the current marketplace.

Graduate Courses

The creation of the Brazilian Institute of Bibliography and Documentation (IBBD) in 1954, had great importance for the formation of Library Science professionals in Brazil as it offered the first courses of graduate level study in the field.

Particularly noteworthy was their first Course in Scientific Documentation (CDC) for—initially—librarians, with the goal of enabling them to work with technical and scientific literature. This course trained innumerable librarians from Brazil and elsewhere in Latin America until the end of the 1960s. In the 1970s, the IBBD became part of the new Brazilian Institute of Science and Technical Information (IBICT) which, nevertheless, maintained the IBBD's concern for access to information and interest in the formation of human resources in this area.

Strongly influenced by the international movement to create a new program, the first graduate course at the master's level, entitled the Information Science Course was established in 1970 for IBICT. Until 2000 it was affiliated with the Federal University of Rio de Janeiro (UFRJ). During 2003–2006 it was affiliated with the Federal Fluminense University (UFF). Currently, it is renewing ties with the UFRJ, as part of the same University's Center of the CBG—Center of Legal and Economic Sciences (CCJE).

According to Christovão,[3] Brazil had no qualified individuals to teach at the master's level in Information Science; consequently professors were imported from England and the United States by IBICT to teach the graduate level courses. Over time, as Brazilian students graduated and became qualified teachers and researchers, they obviated the need to import professors. Steadily they began to take over the advising of theses and dissertations and directing new lines of inquiry.

As of 2008, there are eight graduate programs of Information Sciences in Brazil and the discipline figures prominently in two others: in the Graduate Program in Communication, at the School of Communication of the University of São Paulo (ECA/USP), and in the Graduate Program in Communication and Information of the Federal University of Rio Grande do Sul (UFRGS).

THE JOB MARKET FOR LIBRARIANS

The accelerated transformations occurring world-wide have radically influenced work processes in all professions. Amongst other factors, the revolution in the world of microelectronics and telecommunications (CD-ROM, robots, Internet, email) and the shift from the industrial age to the information age has profoundly changed educational policies. This visible transformation of the labor market requires a new model for academic preparation, one that responds to the need for increasing professional competencies that can address unpredictable situations.[4]

As the profession seeks to maintain relevancy and renewal, the teaching of Library Science, structured systematically, must continually identify the characteristics and conditions of the labor market and fields of research in relation to the profession of the librarian.

Historically, the labor market for librarians was limited, as Fontoura notes, to the preservation of information in the form of clay tablets, parchment scrolls, or papyrus sheets.[5]

In Brazil, according to Stumpf,[6] the librarian became an integral part of the work force after of the creation of the first course for Library Science, in the National Library. From this point forward, the professional was absorbed by the budding labor market, and primarily by public service institutions.

In the 1960s with the recognition of the profession as being of "superior level" and the promulgation of Law 4.084/62 that regulated the profession, the market expanded with new opportunities. With this growth and expansion, however, arose many new avenues of research.

Studies on the Labor Market for Librarians

Among the many studies on the labor market for librarians expanding in the 1970s, that of Souza and Nastri[7] stands out for its view that the major problem of the librarians' labor market stems from the lack of any effective, serious national policy addressing or regulating the profession. Corroborating this position, Souza[8] highlights the varying conditions and inequalities of working environments in Brazilian libraries that librarians face. An exit strategy for this situation came with planning and coordination at the national level of professionals, interested politicians, and appointees, which enabled and legitimized the performance of library professionals and enhanced their value in the eyes of the general public.

Again, referring to Souza and Nastri,[7] the course curriculum for Library Science did not reflect a consensus about the type of professional the courses would turn out because too much emphasis was being given to technical practices. They also indicated that too many professors were out of touch and not keeping up with current research; the quality of students entering the undergraduate courses declined each year and seeming to lack enthusiasm for the course or even interest in it.

Other studies on the labor market of the librarian carried out in the 1990s emphasized the diversity of professional opportunities: documentation services, communication and information, culture and leisure, education, research, information, technology, planning, and policymaking. Mueller and Baptista[9] analyzed the performance of Library Science students in the local market of the Federal District of Brasilia before and after graduation. The study by Tarapanoff[10] sought to identify the profile of the information professional facing the challenges presented by the new technologies for interlibrary exchanges. Finally,

Maranhão[11] researched the labor market of Rio de Janeiro with the intention of revising the Library Science curriculum prevalent in the 1980s. "Change" is the ever-present keyword in these analyses coupled with a call for adaptability during times of continual evolution. In the studies, the labor market described consists of several types of institutions with different foci. At the same time, they characterize the typical librarian as a professional who invests little in continuing education and seems unprepared to face the challenges presented by the evolution of technology and increasing globalization.

Tarapanoff[10] argues that from the decade of 1990 there were many changes to the social and economic environments that impacted the strategies, structure, and the management of information centers or units. She observes that society is increasingly more dependent on information which she asserts will give rise to a broader labor market for the librarian and more recognition of the role libraries and library professionals play.

Although the dependence of society on information and of its ability to broaden the horizons of the librarian's labor market, it is evident that this will only happen if the library professional is able to adjust to the new demands. To become a modern professional it is necessary to continuously adjust to new paradigms as they relate to the services and concepts of information. Courses for professional training must consider the dynamic, ongoing changes, injecting new content and practices into its programs to address these new situations.

Segmentation of the Labor Market for Librarians

While analyzing the labor market for librarians, Valentim[12] was the first to identify three major sectors: the traditional information market; the "unoccupied" information market; and the "information-trends" market.

The first group—the traditional information market—represents the one most sought after by professionals and the one most highly recognized by society.

Within this sector, the public library is most familiar and attracts a great number of professionals. Unfortunately in Brazil, due to the dearth of school libraries, the public library tends to function more as a school library and this distorts the role its professionals perform. Consequently public library services may take a back seat to the school library services demanded by the users. Among the major public libraries in Brazil are the National Library located in Rio de Janeiro, the Municipal Library of Sao Paulo, the State Library of Rio Grande do Sul, and the State Library of Bahia.

School libraries represent another type of library in this first group. Most suffer from inadequate funding resulting in a myriad of deficiencies. On the whole, school libraries are staffed by persons with limited or no professional library training and users, not knowing the educational background of the attending person, may inappropriately disparage the image of the librarian for inadequate rendering of services. Some private schools do have adequate libraries with trained personnel such as the American Schools in Rio de Janeiro and in Sao Paulo, Colégio de São Bento in Rio, as well as the many binational English language schools located in most major cities.

University or academic libraries also employ a great number of professionals. Although chronically underfunded, they stand out for having seemingly endless users that require efficient library products and services. Nearly all public universities and many private ones have one or more libraries for their students and faculty members. Some notable libraries with online services are at the University of Campinas, the Federal University of Rio de Janeiro, University of Sao Paulo, Federal University of Minas Gerais, University of Brasilia, and the Catholic Pontifical Universities in Rio de Janeiro, Sao Paulo, and Porto Alegre.

Also in this sector, specialized libraries include the libraries of research institutions, think tanks, or corporations, both private and public. These are almost exclusively located in metropolitan areas and attract a considerable number of professionals. It is worth noting that in the case of private companies, any financial turbulence directly impacts library funding, particularly staffing patterns and collection development. This instability demonstrates the questionable importance that entrepreneurs give to their information services. On the other hand, research institutions and think tanks, besides having better qualified staff (often with doctorates) than might be found in corporations, tend to value the services offered and usually invest heavily in resources for its library. Some of the outstanding specialized libraries are the Library of the Federal Senate in Brasilia, specializing in social sciences, law, and politics; the Pan American Health Organization Specialized Center BIREME offering medical and health information online and the Instituto Brasileiro de Informação em Ciência e Tecnologia in Brasilia which offers a wide variety of information resources in their library and online.

Cultural centers, which may also be considered as modern public libraries, are slightly different in that they usually function in partnership with another institution, and are located next to a museum, theater, auditorium, or cinema and often close to one or more of such facilities offering a variety of cultural activities for leisure and entertainment. However, although they are in larger urban centers where the demand for cultural activities is greater, these libraries tend to hire few professionals.

Archives and museums located in urban areas are also part of this sector and usually include a library space in their facilities; however, they also tend to hire few librarians.

The second sector, existing information markets but occupied by non-librarians, mainly pertains to school libraries. Few school library staff are trained librarians, in spite of the constant vigilance of the Federal Council of Library Science and 15 regional councils which are authorized to levy fines on unqualified or undocumented practitioners and their employers.

Publishing houses and bookstores also hire very few Library Science professionals, although they could be very useful in this sector providing specialized expertise in cataloging and scientific editing for publishers and in bookstores, collection development, and publicity or public relations services. With the creation of the position of "literary consultant," this market sector is experiencing growth. The consultant is a type of reference librarian who provides clients with information on titles of specific interest.

Private companies, even those without a physical library, can use librarians as information purveyors or analysts, supplying relevant information to make the company more competitive in its business.

Internet service providers and Web-based purveyors represent another under-explored opportunity. They have the need to organize, process, and disseminate the information they manage—tasks that are basic to the profession. Databases are also fertile grounds that use few librarians tending to prefer systems analysts. Some of the major database content providers are SciELO, the Scientific Electronic Library Online covering a selected collection of Brazilian scientific journals; BIREME mentioned above and the IBICT online resources.

Librarians are finding that work as contractees or consultants is becoming an alternative to full-time, steady employment but this usually requires a great deal of experience, a recognized name, and the capital to invest in one's own business.

The third sector, that of, the "informational-trends" market, consists of a prospective market for librarians that demands performance based on the paradigm of access to information, when information is considered fundamental to the development of an organization.

The librarian who is prepared to offer excellent information services will have a place of prominence in this market; however, to act in this niche with quality and ability, the librarian must be sufficiently observant; act in an enterprising fashion; be flexible, dynamic, bold, a team player, pro-active and, mainly, to be future-oriented and interested in continuing education.[12]

CONCLUSION

The evolution of Library Science in Brazil now points to a paradigm shift toward emphasizing ease of access to documents and the information contained within, and the abandonment of excessive concern with the conservation of collections. With this new focus, the discipline that has benefited enormously from the technologies of information and communication now requires a professional cadre capable of addressing this new proposal: to give the right information to the right user at the right time to satisfy his/her informational needs.

It has become increasingly important that there exists a perfect harmony between the library schools and the marketplace. Schools need to provide students with the opportunity not only to learn about modern technologies related to information, but also to develop the management skills necessary to administer the full gamut of information resources, financial resources, material, and human resources. Library Science programs also need to consider the development of appropriate attitudes to manage the social responsibilities of the profession the student has chosen. In turn, the marketplace needs to monitor these abilities, ensuring that future professionals attain the levels of competency that the general public and the marketplace require.

ACKNOWLEDGEMENT

This entry was translated by Pamela Howard-Reguindin.

REFERENCES

1. Guimarães, J.A.C. Profissionais da informação: desafios e perspectivas para a sua formação. In *Profissional da informação: o espaço de trabalho*; Baptista, S.G., Mueller, S.P.M., (Org.). Thesaurus: Brasília, Brazil, 2004; 87104.

2. Castro, C.A. de. *História da Biblioteconomia brasileira: perspectiva histórica*; Thesaurus: Brasília, Brazil, 2000.

3. Christovão, H.T.A. Ciência da Informação no contexto da pós-graduação do IBICT. Ciência da Informação, Brasília, Brazil, **1995**, *24* (1), 3.

4. Bandeira, G.P.; Ohira, M.L.B. Quem é o bibliotecário no estado de Santa Catarina: mercado de trabalho. In Congresso Brasileiro de Biblioteconomia e Documentação, Porto Alegre, Brazil, 2000; 19, **Anais**... [CD-ROM].

5. Fontoura, M.T.W.T.; da, C. *Ocupação efetiva do bibliotecário e a relação desta ocupação com as atribuições formais*. 109f. Dissertação (Mestrado em Educação) Faculdade de Educação, Universidade Federal do Rio Grande do Sul, Porto Alegre, Brazil, 1980.

6. Stumpf, I.R.C. (Coord.) *Mercado de trabalho para profissionais bibliotecários na Grande Porto Alegre*; NEBI/UFRGS: Porto Alegre, Brazil, 1987.

7. Souza, M.A. de; Nastri, R.M. Análise do mercado de trabalho do bibliotecário no interior do Estado de São Paulo. Perspectivas em Ciência da Informação jul./dez. **1996**, *1* (2), 189–206.

8. Souza, F.das; de, C. Política bibliotecária no Brasil. In Congresso Brasileiro de Biblioteconomia e Documentação, 14, Recife. **Anais**... Associação Profissional de Bibliotecários de Pernambuco: Recife, Brazil, 1987; 259–276.

9. Mueller, S.P.M.; Baptista, S.G. Mercado de trabalho do bibliotecário em Brasília. Disponível em: http://dici.ibict.br/archive/00000819/01/T167.pdf (accessed September 10, 2008).

10. Tarapanoff, K. *Perfil do profissional da informação no Brasil*; IEL: Brasília, Brazil, 1997.

11. Maranhão, M.I.; de, C. *Mercado de trabalho para profissional bibliotecário no Estado do Rio de Janeiro*. Dissertação (Mestrado em Ciência da Informação) CNPq (IBICT)/UFRJ (ECO), Rio de Janeiro, Brazil, 1994.

12. Valentim, M.P. (Org.). *Profissionais da informação: formação, perfil e atuação profissional*; Polis: São Paulo, Brazil, 2000; 31–51. (Coleção Palavra-Chave, 11).

Brazil: Library Science—Distance Education

Mariza Russo
Federal University of Rio de Janeiro, Rio de Janeiro, Brazil

Abstract

This study presents an innovative approach in the field of librarianship fostered by the Federal Council of Library Science and the Open University of Brazil (UAB) that encourage the implementation of a bachelor's degree in library science via distance education (DE).

The distribution of classroom courses in the country, concentrated primarily in the big cities, is described, emphasizing the need to expand the education of librarians to meet the demands of professional positions of information, especially in smaller cities and rural areas. It relates the first steps to implementing the DE course, which is supported by the Center for Higher Distance Education in Rio de Janeiro, an institution that has extensive experience in preparing materials for this type of education. The different stages for planning the development of tasks involving the preparation of teaching materials for DE are outlined. It also points out that the Federal University of Rio de Janeiro was selected, through public bidding, to take on this challenge, which is scheduled for implementation in December 2015, when simultaneously the UAB will launch the course via the public education institutions. The innovation that this initiative brings to the country can be summed up by the possibility of expanding access to information for Brazilian citizens and the opportunity to include a large proportion of the Brazilian population in higher education.

INTRODUCTION

Since the 1970s, new technological dynamics, coupled with the focus on flexibility and on the variety of organizations, stand up to mass production unleashing intensive information technologies, which are fast becoming key competitive strategies of the twenty-first century. Such an environment is characterized by great changes, in which scientific and technological knowledge constitutes the basic inputs of the organizations where the slogans are flexibility, speed, autonomy, innovation, and competitiveness.[1]

In this context, the economy, previously based on land and capital, evolves to be based on knowledge, requiring the mastery of new knowledge bases and the ability to transform them into competitive advantage.

The society that emerges from this paradigm shift—alternatively called the Information or Knowledge Society—requires interdisciplinary collaboration and educational projects that meet the challenges of scientific and technological advances and justify their academic programs in a manner that guarantees a quality education compatible with the demands of the professional market.

Oliveira[2] states that the resources of these emerging technologies are important tools in the Knowledge Society; however, it is pointed out that the partial integration of peripheral countries to the new economy—with unequal access to information and knowledge—sets insurmountable limits to underprivileged social segments to meet the challenges of the current technoeconomic context.

Studies by the Paulo Montenegro Institute (IPM)—a Brazilian nonprofit organization that develops and executes projects in education[3]—show that even in the metropolitan regions of Brazil, where the potential access to educational resources is greater, literacy skills may still be insufficient. Data from the National Index of Functional Literacy (INAF) show that Brazil has advanced, especially in the early levels of literacy, but has failed to visibly progress in achieving full domain of literacy skills that are essential for full inclusion in a literate society. According to the IPM educators, much of this progress is due to universal access to primary and secondary schooling. They understand, however, that despite the progress, the difficulties that reach higher levels of literacy are increasingly acute for much of the population.

Other studies recorded by INAF show that the arrivals of new social strata to the higher educational stages come often accompanied by a lack of suitable conditions for these strata to reach the highest levels of literacy, thus reinforcing the need for a new quality of education. They also highlight that quality involves not only the amount of study hours or the increase of the amount of content taught, but also among other things the creation of new flexible models that allow any Brazilian to broaden their studies when desired at different stages of life.

In turn, the educator Arnaldo Niskier[4] stresses that for many years the Brazilian educational system was centered on a classic paradigm, in which the physical structure of the school was recognized as the only place from which knowledge emanated. This conventional model of

Encyclopedia of Library and Information Sciences, Fourth Edition DOI: 10.1081/E-ELIS4-120053307

Binding–British

education had a major drawback, which was translated into the homogeneous treatment of students while completely ignoring the learning pace of the individual student, resulting in losses both in the teaching and learning processes. In this learning process, there is a precondition for the existence of two parts, which should have complementary goals, that is, that one wants to learn and build new knowledge and the other wants to share and teach.[5]

However, the interaction between these parties doesn't to occur linearly; such procedures should be designed not as a mere transmission and acquisition of knowledge, but rather as a joint construction of them, in which the party in charge of teaching assumes responsibility for the guidance and knowledge construction of the party interested in learning. This process is characterized as guided, social, and communicative, resulting in a shared knowledge that includes the active participation of students in the subject matter.

It is at this conjuncture that distance education, with its concept of open education, can be considered an innovation in terms of education, given the new, added components such as flexibility of programs, agility of administrative mechanisms, and fundamentally, the emphasis on student autonomy regarding the choice of location and hour of study.

From the perspective of Formiga,[6] the modality of distance education, both as undergraduate and graduate studies, in different areas of knowledge, contributes to make available a pedagogical proposal, which aims not only to address theoretical and technical training but also to enhance the socialization of learners, which is essential to meet the needs of the current international educational scene.

Although Brazilian universities are offering undergraduate courses, specializations, and master's degrees in the classroom and by distance, thus enhancing higher education in the country, these institutions are mindful in maintaining the quality of this educational experience. It is precisely with this in mind that the Ministry of Education evaluates any course offerings. On the other hand, students envision the expansion of distance learning courses as a way to recover study time that was not completed, or an alternative to reconcile work and study simultaneously. Other points in favor of distance education are the convenience and easiness for those living far from large urban centers. Thus, the expansion of distance education occurs both by educational institutions and by students interested in this type of study.

The legislation regarding distance education in Brazil was first mentioned by Law No. 5.692, dated August 15, 1971, using correspondences, radio, and television on supplementary courses. However, the Law of Guidelines and Bases (LDB) dated December 20, 1996, allowed the higher education system to enlarge the visibility and applicability of this sort of education.

A new decree, No. 5,622, dated December 19, 2005, repealing the previous ones in some aspects, presents 37 articles with improvements to distance education.[8] The changes made by this new decree promoted greater credibility to distance education, especially the issues of compulsory attendance to classrooms, not only to perform evaluations, but also to allow participation in internships and the defense of final papers. This measure leading to the creation of centers for distance education emphasized that the results of classroom tests should prevail over other evaluation results. Another relevant issue regarding this decree was the application of the National System of Higher Education Assessment to the Higher Distance Education movement.

Besides the rapid technological developments and the new act mentioned earlier, in 2005 the creation of UAB made possible the expanding and the strengthening of distance education. UAB was established, with ties to the Coordination of Improvement of Higher Education Personnel (CAPES). The goal was to broaden entry to higher education for people from all social classes with the ultimate result being a significant, positive impact on the workforce of the country.

Using the English model of their Open University, the creation of UAB was driven by the unfavorable educational environment existing in 2004, when only 11% of young people aged 18–24 had access to higher education in the classrooms. This low percentage led the government to create policies generating growth in the number of places available in higher education and thereby promote more opportunities for education in the country.

In order to combat this scenario and to improve the Brazilian workforce, the UAB was created by the Ministry of Culture with the mission of offering higher education courses and distance programs in partnership with public universities through consortia with both municipal and state governments.

UAB's project represented a landmark for Brazilian education given its purpose of expanding access to higher education and the training of teachers for basic education. It emphasizes the articulation of public institutions of higher learning with the centers for distance education designed to support, in a decentralized manner, educational and administrative activities.

For this purpose, the UAB system was created to unite public institutions of higher education that offer on-campus courses and distance education programs. These institutions, in addition to having a well-established university campus, must have confirmed experience in research and extension required for the successful implementation of distance learning initiatives.[9]

This system was created to meet the unmet needs for higher education, especially for students who live in regions far from major urban centers. It supports the adoption and promotion of distance education through the use of ICT and various learning resources: printed material,

Binding–British

audio, video, multimedia, Internet, email, chats, forums, and videoconferences. Since 2007, this system is linked to the CAPES, Brazilian institution that plays a key role in the expansion and consolidation of higher education in the country.

Nonetheless, the usage of undergraduate distance education in Brazil could be stronger as the UAB system offers fewer than 100 bachelor degree programs, with an emphasis in management (80% of courses). Until further expansion of the program occurs, it is not possible to reach more conclusive results as to the effectiveness of the programs.[9]

In parallel to the Brazilian educational scenario, one field of study, library science, is becoming an area of training professionals to manage the vast information generated worldwide, with the mission to promote optimal recovery in benefit of the society.[10] This training has, so far, been grounded in classroom courses, heavily concentrated in the Southeast region of the country preparing professionals to work in larger cities but leaves mostly public library positions unoccupied, or occupied by laypersons, in smaller Brazilian cities.

In view of the arguments presented earlier, this study analyzes the implementation of the Bachelor in Library Science from a distance education perspective, which is an innovative alternative to meet existing demands in the professional labor market of Brazilian libraries.

Another justification for analyzing the initiative lies in the fact that the vast majority of actions undertaken so far in the implementation of distance learning in the field of librarianship cover postgraduate courses only. These situations have not produced the expected effect in the area, partly because of the small number of courses offered and, second, because almost all courses created involve costs that are beyond the ability to pay for most Brazilian professionals.

TRAINING IN THE AREA OF LIBRARIANSHIP

In Brazil, library science began when the first libraries were founded, arising from the religious orders of the Benedictines, Franciscans, and Jesuits. However, the true roots of the field of library science are attributed to the creation of the National Library (BN), which had its origins in the Royal Library brought to Brazil by the Portuguese royal court in 1808. Its official foundation occurred in 1810, and in the following year it was opened to researchers who obtained royal consent. In 1814, it was fully opened to the public.[11]

While in Rio de Janeiro the development of the field of librarianship has been linked to the trajectory of BN, in São Paulo, it was influenced by the school library of the Mackenzie College. However, in both institutions, the pressing concern at the time was the need to resolve internal matters such as unprepared personnel managing these libraries much more than training staff to work in any kind of library.[11]

The Federal Council of Library Science (CFB) considers that library science began as a field of study in Brazil in 1911, when Manuel Cicero Peregrino da Silva, director of the National Library, officiated the creation of the first librarianship course in Brazil. It was also the first such course in South America and the third in the world.[11]

Until the early 1930s, Brazilian librarianship was under the strong influence of the French humanist model of the École Nationale des Chartes, a course founded in France in 1821, which was the first school in the world created for the education of personnel to work in libraries. When creating the Elementary Course of Library Science in 1929, the Mackenzie College in São Paulo initiated the most pragmatic course using American technical standards as a basis of instruction.[12]

Based on the course taught at the Mackenzie College, in 1936 the first school of librarianship was established. It was initially part of the Sao Paulo Department of Culture and later was incorporated into the Free School of Sociology and Politics, also based in the city of São Paulo. This new school of librarianship focused heavily on the technical aspects of librarianship and strictly adhered to American principles as practiced at the course of Columbia College in the United States.

The 1940s saw the development of modern techniques of library science in Brazil and also witnessed the expansion of librarianship courses to other states. In the 1950s and 1960s, the creation of professional library associations in the country, as well as the completion of the first professional events and congresses in the field, created even more opportunities for relevant professional education and development. These events were promoted with the aim of bringing together librarians from throughout Brazil, in order to exchange experiences. They also gave rise to the creation of numerous libraries, particularly in federal agencies all the while encouraging further participation of candidates in professional courses. The various courses of librarianship in the country began to emerge in 1942 in several states, mainly linked to the federal universities, and by 1965, there were 14 official library science courses.

Since the enactment of the LDB, which repealed all previous legislation concerning undergraduate courses, professionals in the field of librarianship began discussions on the guidelines set out in the Act. The resulting document, "Curriculum Guidelines for Librarianship Courses," relaxed the curricular structure of training courses for librarians, being more directed to the needs of Brazilian society. Presently, the 43 existing courses in 2014 in the country follow these guidelines in establishing their curricula.

MODEL CLASSROOM TEACHING OF LIBRARY SCIENCE

The CFB indicates that the number of classroom-skilled library professionals working in Brazil is approximately

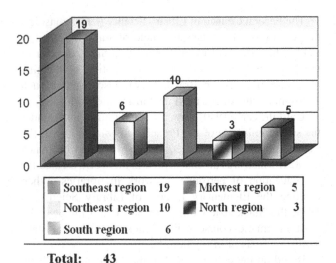

Total: 43

Fig. 1 Number of librarianship courses in Brazil.
Source: From Russo.[1]

30,000 librarians. This data may be contrasted with the vast continental dimensions of the country and its population of over 200 million inhabitants, resulting in a scenario in which information is considered a vital asset (Fig. 1).[13]

When analyzing the distribution of library science courses in Brazil, the following chart shows the highest incidence in the Southeast (44%), which is also the most economically developed region of the country.

Fig. 2 shows another aspect of the distribution of librarianship courses among Brazilian regions allowing the analysis of the relations between the number of courses and the size of Brazilian region.

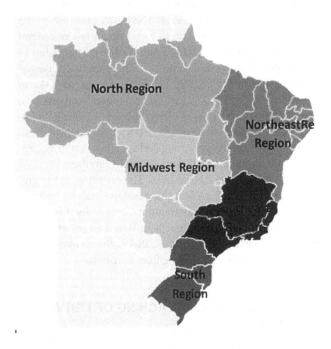

Fig. 2 Librarianship courses in Brazil regions.
Source: From Russo.[1]

In analyzing this distribution, the CFB highlights that the distribution of bachelors' training courses for librarianship is deficient in relation to the size of the country, especially with regard to the North and Midwest, which in geographical terms represent more than 50% of the national territory.[13]

The data allow us to state categorically that the distribution of trained and qualified professionals in the country reflects the need to provide alternatives for the education of librarians to meet the significant social demand for effective information services.

Thus, considering the disparities exposed with current offerings for bachelor's degree courses in library science, the alternative methodology of distance education could serve to reverse the current scenario.

Library Science and Distance Education

The involvement of librarians with distance education dates back to the late nineteenth century, when, in 1888, American librarian Melvil Dewey invited professionals from the School of Librarianship in Albany, NY, to develop correspondence courses to update librarians in small American libraries.[14]

In Brazil, distance education in library science has occurred more significantly during the last ten years, focusing on graduate-level courses. Such initiatives almost always aim at training professionals on current issues or topics in which their work activities require specific training. Many institutions are offering these courses, both in the public and private sectors, encouraging either recent graduates or more experienced librarians to embrace these opportunities for improvement or continuing education making use of one of the main facilities of distance education, which is the flexibility of time and place to study.

Libraries, in turn, are also making use of distance learning platforms to provide courses and training for users in various topics of interest such as information literacy, standards of scholarly work, access to databases, and many others, some of which already perform reference service through these virtual tools to promote real-time interaction with its users.

There are few studies analyzing the application of distance education in undergraduate courses in librarianship. Most of them present initiatives in disciplines where professors give classroom lectures complemented by distance learning sessions, which are used in virtual learning environments (AVAs). Among these environments, the most used is called Moodle, created in 2001 by Australian educator and computer scientist Martin Dougiamas, which allows open access. This software provides the instructor with tools for creating web-based courses with access control so that only enrolled students may have access to its content and facilities.

Even with the scenario described earlier, distance education in Brazil has been developing in various disciplines. Since 2006, the government has been providing greater incentives for the creation of numerous undergraduate courses in this mode by UAB.

The idea of creating an undergraduate degree in librarianship via distance education arose from the demand at the Open University in 2008. There was a great need to prepare library technical assistants to meet the needs of the classroom centers and libraries that were unable to service users taking their courses.

Motivated by this invitation and the opportunity to take higher inland education, the CFB, aware of the Brazilian library scenario and the lack of qualified personnel to develop products and services tailored to the needs of its actual and potential users, presented a counterproposal to the sponsoring agency suggesting the design of a distance course, not for the preparation of technical support personnel but instead for the training of full-fledged librarians. To support their stance, they used examples of international experiences in other countries based on current professional literature.

The acceptance of this counterproposal resulted in a partnership between the UAB and the CFB for the development of the pedagogical content of that course. This choice was considered due to the fact that the UAB system functions as the connector between higher educational institutions and state and local government in order to meet local demands for higher education. It was also reasoned at that time that the educational institutions that would be responsible for offering the courses in distance education would be exclusively public, federal, state, and municipal institutions called public institutions of higher education (IPES). This decision was based on the fact that these IPES that act with the support of classroom centers functioning as operational units with administrative and educational activities also provide the physical, technological, and educational infrastructure for the monitoring of distance courses.

Given this structure, various documents were drawn up by working groups of the two institutions involved including the pedagogical foundation to support universities that will offer training in distance education for librarianship. Using as a basis the studies of the Brazilian Association of Education in Information Science, the pedagogical design of the course was structured in eight different sections: Section 0, Basic Module; Section 1, Theoretical Foundations of Library and Information Science; Section 2, Organization and Representation of Information; Section 3, Resources and Information Services; Section 4, Policies and Management of Information Environments; Section 5, Information Technology and Communication; Section 6, Research in Library and Information Science; Section 7, Internships and Complementary Activities.

The proposal of the course curriculum covers a total of 2685 hr, distributed in eight distinct sections, lasting 4 yrs or eight semesters. In the project, it is emphasized that "[…] the details of the school year and semester modules must be in accordance with current legislation, with the school regulations of the university and the proposed project submitted and approved by the UAB" (Reference,[13] p. 22).

The counterpart of the institution that will offer the course must demonstrate the physical, material, and informational resources available as well as faculty members (teachers and tutors). For the latter, the weekly workload timetable, curricula, and other documents proving their competence to teach must also be presented.

The project also stipulates that tutors will play a fundamental role in the process and, therefore, should be described in detail by the proposing institution. They will be responsible for distance tutoring while mediating the learning process with those students geographically distant. Tutors will attend students at the classroom centers at reserved times. The tutor should therefore be well versed in the pedagogical course design; the teaching materials and the specific content under his/her responsibility, in order to answer questions, assist in individual and group activities and also participate in the evaluation process with teachers.

As for the administrative part, the IPES shall provide technical personnel to offer the support needed to complete the courses under their responsibility. This will include tasks emanated from the faculty members at the headquarters of the institution and also in the classroom centers. In these spaces, there will be technical support jobs for the provision of libraries, laboratories, maintenance of equipment, and physical facilities. This team should also operate in the academic department of the course, with the records and monitoring of enrollment and academic procedures, as well as providing academic support to tutors and teachers in classroom and distance activities.

The importance of the communication process is stressed in the project, insofar as the success of a distance course depends on the interaction between all participants in the process. In this sense, one should rely on an AVA, which should enable the following features: electronic mail, videoconferencing, discussion forum, chats, and other means of communication.

Also described in the pedagogical project is the curriculum content of the course, focusing on the sections previously mentioned, showing the options and objectives of each discipline.

After approval of this pedagogical project, CAPES launched in 2012 a Request-for-proposal (RFP) to select the Brazilian university that would manage the production of educational material to be offered by this course. Five institutions submitted proposals and the Federal University of Rio de Janeiro (UFRJ), through the Course of Library and Information Management Units (CBG), were selected to develop this innovative task.

The Notice 12/2012, which selected the UFRJ for managing the production of this course, also gave financial resources to select and train instructors for the content development of the many disciplines that make up the course curriculum. The Center for Distance Higher Education of Rio de Janeiro (CEDERJ) was chosen as UFRJ's partner institution in this project and is responsible for training the content developers, as well as for monitoring the production of the material to ensure consistency with the special characteristics of distance learning and teaching. With regard to expected results, the main points to be considered were 1) store the material developed by the project in a virtual repository, to be available to all IPES coordinators of the courses; 2) follow international standards to enable interoperability between different platforms; 3) prepare metadata that is compatible with international standards to enable the retrieval of information.

Thus, it was necessary to set up permanent teams to monitor and provide technical and operational support to the project. The following teams were created: 1) Technical Commission of Librarianship for Monitoring and Evaluation (CT) established by CAPES with the task of monitoring the management of the content development process of teaching materials, as well as other issues relating to the course offerings in library science under the UAB system. They act as an advisory committee. 2) Management Commission for the Production of Teaching Materials for the Library Science Course in Distance Education (CG) established by UFRJ, which is comprised of four members and two alternates, five of them are LS professors and one is a librarian from the CBG/UFRJ. Its function is to manage the operation of the project for production of teaching materials, monitoring the performance of all parties involved in the process. 3) Evaluation Commission of the Production of Teaching Materials for the Library Science Course in Distance Education (CA) established by UFRJ, being formed by two members nominated by the Technical Commission for Monitoring and Evaluation of Library Science, two members nominated by the Commission of Management of Production of Educational Materials of the course on librarianship, and a member of the distance education specialist, with experience in the production of educational materials for distance education. Its functions are to participate in the selection process of authors and readers, follow the process of production of teaching materials, and verify if the contents meet the guidelines of the pedagogical design of the course. This commission shall have a deliberative nature and includes CAPES participation in its decisions.

A summary of the key steps of the implementation process of the project of management of the distance education teaching materials for the course is presented here:

1. *Analysis and selection*: Selection of authors and readers and hiring technical staff; training of authors and readers; signing the Declaration of Commitment; setting deadlines for delivering the materials and Deed of Assignment of Rights of Use for the teaching materials and support within the UAB system and in projects, programs, and courses of interest to the federal government.

2. *Design and development*: Creating and augmenting the virtual repository; preparation of teaching materials by subject (author); proofreading the teaching materials by subject (reader); preparation of instructional design and graphic design; CG and CA managing the preparation of teaching materials by subject, and validation of the material by CT.

3. *Certification and printing of course materials*: Registration of the teaching material produced in the International Standard Book Number, by the National Library Foundation Registration; production of matrices/electronic files for reproduction and control of the production of the material by CG.

4. *Delivery of material*: Recording of the materials developed; submission of materials to CAPES; sending interim and final reports on the progress of the project to CAPES/UAB.

It is important to note that the steps mentioned earlier will be followed to develop each of the disciplines, while seeking to meet the goals of the teaching and learning process and the guidelines of the educational program.

Besides the mandatory printed course material, supporting materials such as videos with the presentation of the course, tutorials, and programs of specific subjects for teachers, tutors, and students will also be produced.

After the design of the project, the team at UFRJ started actions aimed at developing a methodology for its implementation. These actions describe the path from design to development of this entry.

In order to maximize efforts, formal partnerships were made between UFRJ and the following institutions:

• CEDERJ—For advice on training and instructional design of the object of this public notice process;
• Jose Bonifacio University Foundation—To support the financial management of resources allocated to the project.

For the physical execution of the project, we sought a proper installation within the university campus at UFRJ, with the ultimate decision to occupy space at the Center for Distance Education (NEAD), a newly created center at UFRJ to provide the technological infrastructure and logistics for the development of distance learning courses at the university.

The development of this course material is scheduled to be completed by the end of 2015 or early 2016, when the subject content of the first four periods will be ready. At that point, CAPES will send out another tender for the launching of the first official course in distance education for library science.

The preparation of this course content will result in a new, dynamic program of library science offered via distance education, which should meet the demands of the labor market in urban as well as rural areas. In this context, the expectations of the Brazilian government for the expansion of training opportunities at the university level should serve as a guiding point of this initiative.

CONCLUSIONS

The changes perceived in society, arising from the availability of vast amounts of information transmitted primarily by electronic networks, are omnipresent and causing a substantial impact on the socioeconomic, cultural, and educational environments of the world.

With regard to Brazil, concern about access to information and, hence, knowledge, on the part of its population also poses serious problems for the government, not only because of its territorial dimensions, but also because of persistent social inequalities observed in the regional development of the country.

The conventional educational model in effect throughout the entire country makes clear that the difficulties arising from this type of education will not be resolved in the short term, thus leaving a large part of the population on the sidelines of any formal educational activities.

In this context, the modality of distance education is considered a viable way to promote changes in an unfavorable environment leading to an education for many people who, for different reasons, have not had the chance to obtain one by traditional methods. This concept of teaching—more open, flexible, agile, dynamic, and autonomous—is being seen by the Brazilian government as a strategic tool to insert citizens into society, bringing improvements for their future and for the country in general.

In order to achieve this goal, UAB was created in 2005. Among its objectives is to offer degree courses in different fields of knowledge, expanding access to public higher education. In 2008, the discipline of library science was selected to create a new undergraduate course in this mode of teaching.

As mentioned in this study, the fact is that the choice of this area fell, first, on the issue of providing libraries to the Centers of Classroom Support (PAP), units that physically support the courses of UAB. However, negotiations with the CFB brought to light other demands of jobs for librarians, such as part of the labor market, which had professional posts in public and school libraries occupied by laypersons. Given this fact, those information units failed to provide the quality services for which librarians are trained and, thus, society lost out on that benefit.

We conclude that the provision of distance education courses in library science, with its more inclusive proposal, stands out as truly innovative, as it proposes to raise the bar for teaching in the area while increasing the possibilities for training professionals throughout the country with the goal of meeting the needs already identified.

One of the innovative aspects of this course lies in the initiative having been fulfilled by a partnership between a government agency, CAPES/UAB, and a professional union, the CFB, in order to conduct an educational course for professionals to work in the country's libraries, thereby solving the problem of a lack of qualified professionals in so many units of information such as documentation centers, special libraries, or public/academic libraries.

The increased supply of education in library science combined with opportunities for broader access to the course via distance education can lead to advantages for the candidates, such as saving time and resources, convenience of study, autonomy and flexibility in the tasks, and monitoring by tutors and more. This innovative characteristic of work and learning serves well as an enriching element of the educational process.

The incremental development of instructional materials and content produced for the course in distance education will also help classroom courses offered in the country, since this type of material has unique characteristics. They are written specifically for the student body, with the central aim of awakening their interest. The personal and conversational style facilitates active student participation in the learning process and eventual application of what is learned.

On the other hand, the production of teaching materials for courses in distance education presents major challenges to the team responsible for its preparation. The approach needs to be multidisciplinary, relying not only on experts in distance education but also on teachers responsible for the production of content, reviewers who are responsible for grammatical and editorial revisions, and experts in design and visual programming who create and develop graphic design and interactive content.

The strong presence of ICT in the teaching of library science courses via distance education, through the use of AVA, blogs, chats, wikis, and other collaborative writing tools, is considered an innovative approach in the area, a characteristic that may influence classroom courses. It is understood that ICT may serve primarily as a tool to streamline the interaction between the parties involved in the teaching process and the completion of tasks in order to build collective knowledge.

Increasing the number of library school graduates in the country, be it by classroom or distance education, serves well to promote greater visibility for the library science profession and also the body of work it encompasses.

Given these facts, the joint action by CAPES/UAB, the CFB and UFRJ will help significantly in overcoming the obstacles described. The collective construction, with the overarching purpose to provide as many Brazilians as possible with access to higher education and of ensuring a better life, will also contribute to a stronger, economically sound country.

This effort established a new landmark in the field of library science, but its success will depend on the commitment of the sponsoring institutions, UAB and CFB, and also on the IPES. Through cooperation, this groundbreaking initiative will have the chance to reshape the library science area, and enrich its educational process bringing benefits to the Brazilian society.

ACKNOWLEDGMENT

The author thanks Pamela Howard-Reguindin (Director, Library of Congress Office, Islamabad) for translating this entry.

REFERENCES

1. Russo, M. Formação em Biblioteconomia a distância: a implantação do modelo no Brasil e as perspectivas para o mercado de trabalho do bibliotecário. 244 f. Dissertation (Doctorate in Production Engineering) Universidade Federal do Rio de Janeiro: Rio de Janeiro, Brazil, 2010.
2. Oliveira, F.B. A Contribuição estratégica da Educação. Tecnologia da informação e da comunicação: desafios e propostas estratégicas para o desenvolvimento dos negócios, Pearson Prentice Hall: São Paulo, CA, 2006.
3. Instituto Paulo Montenegro, Available at: http://www.ipm.org.br/ipmb_pagina.php?mpg=4.02.01.00.00&ver=por (accessed September 13, 2014).
4. Niskier, A. *Filosofia da Educação: uma visão crítica*; Loyola: São Paulo, Brazil, 2001.
5. Mercer, N.; Estepa, F.G. A educação a distância, o conhecimento compartilhado e a criação de uma comunidade de discurso internacional. In *Educação a distância: temas para o debate de uma nova agenda educativa*; Litwin, E. (Org.), Ed.; Artmed: Porto Alegre, Brazil, 2001; 21–37.
6. Formiga, M. Educação superior, educação a distância e educação corporativa. *Desafios da educação: contribuições estratégicas para o ensino superior*; de Oliveira, F.B. (Org.), Ed.; E-Papers: Rio de Janeiro, Brazil, 2009; 53–61.
7. Niskier, A. O direito à tecnologia da esperança, 2000. Available at: http://www.cjf.gov.br/revista/numero6/artigo18.htm (accessed September 13, 2014).
8. da Gomes, C.A. C. A legislação que trata da EAD. *Educação a distância: o estado da arte*; Litto, F.M.; Formiga, M. (Org.), Eds.; Pearson Education do Brasil: São Paulo, Brazil, 2009; 21–27.
9. Sistema Universidade Aberta do Brasil. Available at: http://uab.capes.gov.br/2014 (accessed on Sept 30, 2014).
10. Shera, J.H. Sobre Biblioteconomia, Documentação e Ciência da Informação. *Ciência da Informação ou Informática?*; Gomes, H.E. (Org.), Ed.; Calunga: Rio de Janeiro, Brazil, 1980; 90–105.
11. de Castro, C.A. *História da Biblioteconomia brasileira: Perspectiva histórica*; Thesaurus: Brasília, Brazil, 2000.
12. Caldin, C.F.; Os 25 anos do ensino da Biblioteconomia na UFSC; Encontros Bibli: Revista de Biblioteconomia e Ciência da informação: Florianópolis, Brazil, 1999n. 7, abr.
13. Graduação em Biblioteconomia na modalidade a distância: projeto pedagógico. CAPES: CFB: Brasília, Brazil, 2010.
14. Sacchanand, C. Distance education in Library and Information Science in Asia and Pacific region. Proceedings of the IFLA General Conference, 64, Amsterdam, the Netherlands, 1998. Available at: http://archive.ifla.org/iv/ifla64/129-140e.htm (accessed September 9, 2014).

Brazil: Museums

Regina Dantas
Museu Nacional, HCTE, Universidade Federal do Rio de Janeiro, Brazil

Pamela Howard-Reguindin
Library of Congress Office, Nairobi, Kenya

Abstract

This entry presents a history of the National Museum as an example of the challenges to Brazilian museums facing political objectives at different times. The New Museology is examined, as a landmark for the transformation of the institutions of memory, and an example of the ever-changing activities of the National Museum (Museu Nacional) in Rio de Janeiro. The text should be considered as a contribution to museological sciences encouraging reflections between professionals of different subspecialties in various museums.

INTRODUCTION

Brazil, one of the largest economies of the world, has a land mass covering more than half of South America, and a population of more than 190 million people. Since Pedro Alvares Cabral set foot on the continent in 1500, Brazil has produced a wide range of museums and libraries to serve its diverse population. Brazilian museums have been established and supported by royalties, politicians, bureaucrats, philanthropists, banks, universities, and wealthy families and individuals. The many unique museums found all over the country range from classical art collections in Sao Paulo to snakes and butterflies in the Amazon to creative folk arts depicting the utterly weird to daily life events on the outskirts of Rio de Janeiro. Libraries too have been founded and supported by a similar cast of characters. They are also wide ranging from the massive National Library on Rio Branco Avenue to a very petite one tucked away on a winding, uphill Rio road specializing in only *literatura de cordel* or "string literature" published only in Brazil (Fig. 1).

OVERVIEW

The idea of "museum," as conceived today, was formed during the Renaissance and subsequent Age of Enlightenment, when the methods, systems, and scientific purposes characteristic of public museums were defined. From the eighteenth century, most notably with the French Revolution, when the concept of "national patrimony" became part of the ideological strategies of the modern European state, the museum would become a privileged space to protect the fragments that "materialize the collective inheritance of the nation."[1] It became a symbolic space to legitimize the State, reconciling historical continuity with the creation of new spaces of memory. Portugal was not exempt from such developments, and colonial minister Martinho de Melo e Castro was instructed by Queen Maria I to require colonial administrators such as Brazilian viceroy (from 1779 to 1790) Luis de Vasconcelos e Sousa, conde de Figueiro´, to send back to Lisbon samples of local flora and fauna—even live animals[2] and anything else representing the cultures of each colony, its mineral and agricultural potential, and the customs of its inhabitants for inclusion in Lisbon's royal museum, the Real Museu da Ajuda. The object of all this was the embellishment of the capital reflecting the scientific discoveries of the day. Within two decades, however, the capital would be moved from the old continent to the New World, together with the Museum, where it would become the Royal, and subsequently the National Museum, which is now part of the Federal University of Rio de Janeiro [MNRJ].

THE FIRST BRAZILIAN MUSEUM

According to Jose Lacerda de Araújo Feio, director of the National Museum, 1967–1971, the exact date of the inauguration of the House of Birds is unknown. It apparently opened even before the building to house it had been finished. Feio calculates that it started in 1783.[3] For Maria Margaret Lopes, professor at the Universidade Estadual de Campinas, the Museum of Natural History opened in 1784, the same year in which Viceroy D. Luis de Vasconcelos e Sousa created the Cabinet of Natural History Studies and quite probably the Museum of Natural History as well.[4] Luis de Vasconcelos e Sousa was succeeded as viceroy, 1790–1801, by Fernando José de Portugal e Castro, future third conde de Resende and second marquês de Aguiar, who was devoid of interest in the

Encyclopedia of Library and Information Sciences, Fourth Edition DOI: 10.1081/E-ELIS4-120053416

Binding–British

Fig. 1 Map of Brazil.
Source: CIA The World Factbook. https://www.cia.gov/library/publications/the-world-factbook/geos/br.html.

activities of the Museum of Natural History, and consequently, the establishment declined due to inattention and underfunding. In 1810, 20 years after its creation, the director, Francisco Xavier C. Caldeira, died and was replaced by Dr. Luiz Antonio da Costa Barradas. Sometime during 1811, the building was transformed into a workshop for diamond cutting, thus discouraging the collection of animals. On June 22, 1813, Fernando Jose de Portugal e Castro signed decree 20 of Prince Regent Dom João (1767–1826) giving an order "to extinguish the museum of this Court." All the materials were summarily packed into two large boxes comprised of more than a 1000 bird skins, many insects, and some mammals.

The boxes and the furniture were given, for 1 yr's safekeeping in a single room, to one Costa Barradas, presumably the father of the namesake Maranhão magistrate Joaquim da Costa Barradas, and the building was later razed.

In its place, a building for the Royal State Treasury was constructed, that later sheltered the National Treasury. In parallel to the decline of the Museum of Natural History, we pass now to an analysis of the situation in Europe as we identify the moment that would change the trajectory of Portugal and, consequently, of Brazil. The transference

of the Portuguese Court was a carefully planned endeavor. It was, without question, a strategic plan, conceived at the end of the eighteenth century, as an emergency solution that would save the Portuguese Crown during crisis situations. However, the transference only materialized when the threat became imminent to the integrity of the monarchy. The Prince Regent Dom João, the future King João VI, convinced that the Crown would only be guaranteed if he could preserve his possessions in the New World, whose vast natural resources were superior to those of Portugal, left Lisbon in November of 1807, with an entourage of some 15,000 courtiers and subjects.

The transference of the Lusitanian State structure to Brazil marked the end of the colonial period for Brazil.[5] The city of choice was Rio de Janeiro, which was the principal port city and since 1763 the colony's capital. Shortly after his arrival, Dom João, who intended to transform the region into the capital of the Portuguese monarchy, created a naval academy, a school of gunnery and fortification, a military archive, a gunpowder mill, the São João Theater, a government press, the Botanical Garden, the Academy of Fine Arts, the Chamber of Commerce, and the Royal Library, among others. On June 6, 1818, the Royal Museum was created,[6] by the decree of

Dom João VI implemented by Interior Minister Thomas Antonio de Villanova Portugal. As a first step, a building was acquired from João Rodrigues Pereira de Almeida, created in 1828, barão de Ubá. Father Jose da Costa Azevedo (1818–1823) was appointed director.

He had previously been responsible for the Cabinet of Minerals and Physics, in the Military Academy. The Royal Museum, known from 1822 as the National Museum, represented, as did other institutions, a "transposition of European models to the tropics, demonstrating an alignment to the analogous initiatives in Europe."[1] The first collections of the Royal Museum were indigenous artifacts and natural products of the region. Dom João VI himself offered two octagonal display cases, containing 80 models of workshops of the professions most common at the end of the eighteenth century, made at the time on the instruction of his elder brother Prince Dom Jose, 1761–1788. One example is "a vase of gilt silver, crowned with coral representing the Emperor Constantine's victory and consequent triumph of Christianity at the 312 battle of the Milvian Bridge; two keys; a carved Greek marble foot with sandal; one firearm of carved ivory from the Middle Ages and a fine collection of oil paintings"[7] (Fig. 2).

As the monarchy strengthened in Brazil, so did the museum. With João VI's return to Portugal, his elder son, Dom Pedro, became Prince Regent of Brazil and, as such, opened the Royal Museum to the public. In October of 1821, public visits to the Royal Museum began as decreed: "On Thursday of each week from ten in the morning until one in the afternoon, not being a holy day, to all persons, Foreign or National, who make worthy use of it for its knowledge and qualities."[8]

After this, the collections of the Royal Museum grew as the Brazilian kingdom became an empire with Prince Pedro now Emperor Dom Pedro I, advised by his chief minister Jose Bonifácio de Andrada e Silva to adopt a policy of encouraging visiting naturalists to donate artifacts and samples from the different regions of Brazil for the Museum.[9] Analyzing the correspondence of the Imperial and National Museum from 1822, we can see that the work of identifying and researching materials and samples sought by the Imperial Government had been added to the activities and responsibilities of the museum staff. It also becomes evident that, from the second half of the nineteenth century, increasing requests for the Museum's services points to a constant interaction between members of the Government and the administrators of the Museum. Gradually, the Imperial and National Museum became a vital resource center for the Empire (Fig. 3).

As the Imperial government explored Brazil's natural wealth, and then with the second emperor Pedro II[10] constantly encouraging scientific studies, the sciences in Brazil developed and grew throughout the nineteenth century. The Museum, with its scientific specialties—such as botany, zoology, geology, and ethnography—provided the

Fig. 2 Golden vase of silver donated by Dom João to the Royal Museum.

Fig. 3 Imperial and National Museum in Campo de Santana, Downtown, Rio de Janeiro.

foundation for great contributions to the natural sciences, heralded in the latter nineteenth century as fundamental to Brazil's progress. As the struggle for independence triumphed and internal cohesion was gradually secured, the Empire, despite its continued dependence on slave-based agriculture, joined the ranks of "civilized" countries. While in 1842, the Imperial and National Museum was recognized as Brazil's national museum, other museums were being created: the Emílio Goeldi Museum, for example, in Belém in 1861, or the Paulista Museum in São Paulo in 1890. The Goeldi Museum stands out as the largest museum of natural history in Northern Brazil. It, like the National Museum, is also dedicated to natural history. The concept of a museum for São Paulo dates back to the year of independence in 1822, but for various reasons, it was only in the Republican period that the institution was consolidated as a "place of memory,"[12] being inaugurated in 1895. Over the next 100 years, it survived many modifications in its structure and eventually became a historical and university institution connected to the University of São Paulo. In 1946, the National Museum was affiliated with the University of Brazil (currently the Federal University of Rio de Janeiro) and has adapted to the teaching, research, and extension missions of the University. In this way, it was strengthened as an institution of natural and anthropological sciences in an academic environment.

NEW MUSEOLOGY

The science of museums, for many years, was related to the analysis of practices occurring in the various institutions. During the 1970s, discussions about the concepts of museums and patrimony led to the concept of Museology and, in turn, to "a social science that studies museum objects as a knowledge source" in the 1980s.[13] In this way, the new concept of Museology adjusted to the New History[14] that had departed from the factual narrative into an alliance with narratives of the oppressed, the disenfranchised, women, Indians, and new interpretations of daily life.

Additionally, we can link studies of Social Memory[15] and its excellent argument on patrimony[16] to the exigencies of the New Museology connecting the museum with its community. Throughout the twentieth century, there was a proliferation of diverse thematic museums in different regions of Brazil—some affiliated with universities; others with schools and communities. The museums, when being used as a "laboratory" of an educational institution, provided students the opportunity to improve their knowledge of preservation and scientific analysis through the collections on display and other primary resources. For example, we highlight the current National Museum, an institution that is being used for postgraduate extension courses, research, and education in the areas of social anthropology, botany, and zoology, but facing the quandary of disconnection from education and research versus extension studies. Education and "elitist studies" done within the museum walls did not reach the community resulting in a dead end as far as outreach was concerned. At the turn of the twenty-first century, this changed with increasing activities directed toward schools and visitors to the museum's expositions. Not satisfied with short-term results, the institution's directors went outside its walls undertaking activities in the Boa Vista Park directly across the street. The anniversary of the institution's founding was chosen for the first such event, and this has continued each year ever since. Under large tents, minilaboratories, simulations, and exhibits are set up for geology, botany, and archaeology, along with presentations of vertebrates, invertebrates, or fossils: a new focus for the museum's activities and a great new interaction with the community. This gave the institution a new look changing it from a deposit for old objects into an educational laboratory presenting the results of research and education directly to the community. The Museum of Art in São Paulo (MASP) has taken outreach and continuing education seriously since the 1970s. Their educational services include children's art clubs, museum studies for students, and guided tours for school groups and adults. Through political maneuvers and policy changes, the museums of the New Museology[18] have been transformed from places that rescued popular culture only to display it for the benefit of the privileged into vibrant institutions for the education of the entire populace.

Anthropological analysis suggests that museographic representations should be viewed as fact rather than fiction; thus, objects in museums play a strategic role in this process. Objects are seen as part of distant realities in space or time, establishing a delicate continuity that connects the visible with the invisible.[19]

CONCLUSION

The first museum created in Brazil was established to keep, collect, and conduct research on the country's culture and its vast natural resources. As the years passed, the National Museum became an instrument in the Imperial regime's drive to establish, develop, and unify the nation. This drive extended into the present as many new museums were established in the late nineteenth and twentieth centuries highlighting the historical and artistic wealth of the nation such as the MASP, National Museum of Fine Arts in Rio, Historical Museum of the Republic, Museum of Modern Art, the State Pinacoteca in São Paulo, Casa do Pontal Museum, and Museum of Astronomy and Related Sciences, among many others. For the most part, the role of education in museums during the early nineteenth century, alas, did not last into the next century. Education, research, and public outreach suffered inferior roles during the twentieth century with a few notable exceptions. These factors culminated in the estrangement of the institution and a lack of interaction between these three functions. According to the precepts of the New Museology, a change was required to present the National Museum's activities to the public, in order to stimulate reflection and appreciation as a tool for education. Finally, we look forward to the day when the daily work of museums will contribute to the integration of different professions: museum specialists, curators, historians, anthropologists, sociologists, biologists, archaeologists, librarians, and archivists. Beyond preserving objects, they must all develop studies that will enhance the knowledge base, updating and disseminating information about museum holdings, thus encouraging more reflection and thought about the resources they exhibit with a multidisciplinary outlook.

REFERENCES

1. Cícero Antonio Fonseca de, A. O Colecionismo Ilustrado na Gênese dos Museus Contemporâneos. Anais do Museu Histórico Nacional **2001**, *3*, 232–254.
2. Maria Fernanda Baptista, B. A Cidade Colonial do Rio de Janeiro, entre o Mar e o Sertão. In: Ler e Escrever para Contar. Documentação, Historiografia e Formação do Historiador. Ilmar Rohloff Mattos (Org.). 1ed. Access Editora: Rio de Janeiro, Brazil, 1998, v., p. 11–32.
3. José Lacerda de Araújo, F. O Museu Nacional e o Dr. Emílio Joaquim da Silva Maia. Publicações Avulsas do Museu Nacional **1960**, *35*, 1–31.
4. Maria Margaret, L. *O Brasil descobre a Pesquisa Científica: os museus e as ciências naturais no século XIX*; Ed. HUCITEC: São Paulo, Brazil, 1997; p.27.
5. Lucia Maria Bastos Pereira das, N.; Humberto Fernandes, M. *O Império do Brasil*; Nova Fronteira: Rio de Janeiro, Brazil, 1999; 28–29.
6. BR MN. AO, pasta 1, doc.2, 6.06.1818. Decreto de criação do Museu Real—atual Museu Nacional. SEMEAR/Museu Nacional, Brazil.
7. Cícero Antonio Fonseca de, A. O Colecionismo Ilustrado na Gênese dos Museus Contemporâneos. Anais do Museu Histórico Nacional **2001**, *3*, 232–245.
8. Ladislau, N. *Investigações históricas e scientíficas sobre o Museu Imperial e Nacional do Rio de Janeiro*; Instituto Philomático: Rio de Janeiro, Brazil, 1870; 22.
9. BR MN MN.DR, CO. AO.9. Decreto de 24 de outubro de 1821 dando início as visitas públicas no Museu Real. SEMEAR/Museu Nacional, Brazil.
10. João Baptista de, L. *Fastos do Museu Nacional do Rio de Janeiro*; Imprensa Nacional: Rio de Janeiro, Brazil, 1905; 12.
11. Maria Walda de Aragão, A. Dom Pedro II e a Cultura. Arquivo Nacional: Rio de Janeiro, Brazil, 1977. (Publicações Históricas, 1a. série, 82).
12. Ilmar Rohloff de, M. *O tempo Saquarema: a formação do Estado Imperial*; ACCESS: Rio de Janeiro, Brazil, 1994.
13. Pierre, N. *Les Lieux de memoire. (trad. P. Jordan, Mary Seidman Trouille)*; University of Chicago Press: Chicago, IL, 2001.
14. Peter, Van M. O Objeto de estudo da Museologia. Tradução Débora Bolsanello e Dolores Estevam Oliveira. UNIRIO/UGF: Rio de Janeiro, Brazil, 1994.
15. Jacques, Le G. Memória. In História e Memória. Ed. da UNICAMP: São Paulo, Brazil, 2003.
16. Maurice, H. *Les cadres sociaux de la mémoire*; Felix Alcan: Paris, France, 1925.
17. José Reginaldo Santos, G. A retórica da perda: os discursos do patrimônio cultural no Brasil. Ed. UFRJ: Rio de Janeiro, Brazil, 2002.
18. José Reginaldo Santos, G. Os museus e a representação do Brasil: os museus como espaços materiais de representação social. (Museums and the representation of Brazil: museums as material spaces of social representation). Revista do Patrimônio **2005**, *31*, 254–273.
19. Krzystof, P.; Coleção.; Fernando, G. Memória-História Imprensa Nacional: Casa da Moeda: Porto, Portugal, 1984; 51–86.

British Library

Andy Stephens
OBE, Board Secretary, Head of International Engagement, The British Library, London, U.K.

Abstract

The British Library holds one of the richest and most comprehensive collections of the world's knowledge. The collection is at the heart of the unparalleled range of services that are provided to national and international users in reading rooms and exhibition galleries and through document supply. The Library is active in its use of electronic services and digitization to provide effective and ever-wider access to its holdings. One of its main tasks is to collect, preserve, and provide access to the United Kingdom's National Published Archive. Significantly, new Legal Deposit regulations, which came into force in April 2013, extended this remit into the digital domain and conferred on the British Library (and the other UK legal deposit libraries) the right to receive a copy of every UK electronic publication including blogs, e-books, and the entire UK web domain.

INTRODUCTION

The British Library (BL) is the national library of the United Kingdom and one of the world's greatest libraries. It is also most emphatically a library for the world. It has its origins in the former Library of the British Museum and it holds one of the most comprehensive collections of the world's knowledge. Its collection can be said to contain both the memory of the nation and also the DNA of civilization. The collection is at the heart of the unparalleled range of services that the BL provides to national and international users in reading rooms and exhibition galleries and through document supply and bibliographic services. The Library is active in its use of electronic services and digitization to provide effective and ever-wider access to its holdings. One of its main tasks is to collect, preserve, and provide access to the United Kingdom's National Published Archive. Significantly, new Legal Deposit regulations, which came into force on April 6, 2013, extended this remit into the digital domain and conferred on the BL (and the other UK legal deposit libraries) the right to receive a copy of every UK electronic publication including blogs, e-books, and the entire UK web domain.

HISTORY, FOUNDATION, AND FUNDING

The British Library Act 1972[1] established the BL as the "...national library for the United Kingdom... consisting of a comprehensive collection of books, manuscripts, periodicals, films and other recorded matter, whether printed or otherwise." The Act charged the British Library Board with managing the Library as "...*a national centre for reference, study and bibliographical and other information services, in relation both to scientific and technological matters and to the humanities*" and with making its services available "... in particular to institutions of education and learning, other libraries and industry."

The BL commenced operations in July 1973. While this was the first time that the United Kingdom as a whole had formally had a national library, the creation of this new body amalgamated a number of existing library institutions—each having a national remit—into a single administrative unit. Of these existing institutions, arguably the most notable was the library of the British Museum.[2,3]

The British Museum had been founded by Act of Parliament in 1753. Prompted by the decision of Sir Hans Sloane, the eminent physician and scientist, to bequeath his large collections of books, manuscripts, drawing, antiquities, and natural history specimens to the nation, the foundation collections of the British Museum Library also included the Harleian Manuscripts collected by Robert Harley, the first and Earl of Oxford and his son Edward, and the manuscripts collection of Sir Robert Bruce Cotton. The Cottonian collection included such treasures as the Lindisfarne Gospels, the Cotton Genesis, and two of the four surviving copies of Magna Carta. One phrase in the British Museum Act of 1753 has a permanent relevance: "For publick use to posterity."

Sir Anthony Panizzi is generally credited with making the British Museum Library the great treasure-house that it became in his lifetime. An Italian émigré and patriot, he held a succession of positions: first assistant librarian (1831–1837), then keeper of printed books (1837–1856), and finally principal librarian and director (1856–1866). He was knighted in 1869. He argued "I want a poor student to have the same means of indulging his learned curiosity, of following his rational pursuits, of consulting the same authorities, of fathoming the most intricate inquiry, as the richest man in the kingdom, as far as books go. . . ." It was his aim "to bring together from all quarters

Encyclopedia of Library and Information Sciences, Fourth Edition DOI: 10.1081/E-ELIS4-120053304

the useful, the elegant and the curious literature of every language; to unite with the best English Library in England, or the world, the best Russian library out of Russia... and so with every language from Italian to Icelandic, from Polish to Portuguese." Panizzi enforced the system of legal deposit, organized the production of the new catalogue, and oversaw the construction of the famous Round Reading Room of 1857 in the quadrangle of the Museum building to a design by Smirke. The current strength of the BL's collections derives in no small part from Panizzi's legacy.

The catalyst for the passage of the British Library Act had been the 1969 Report of the National Library Committee.[4] Chaired by Dr F.S. Dainton (later Lord Dainton), the Committee had been tasked with "the examination of the various National Libraries, and, in particular, to consider whether these institutions should be brought together into a unified framework." The Committee concluded that in the interest of efficiency and effectiveness the institutions should be integrated into a new authority. The other institutions that were amalgamated into the new national library in 1973/1974 were the National Reference Library of Science and Invention, the National Lending Library for Science and Technology, the National Central Library, the British National Bibliography Ltd, and the Office for Scientific and Technical Information. (Subsequently, in 1982, the India Office Library and Records also joined the BL, followed by the British Institute of Recorded Sound, later renamed the National Sound Archive, in 1983.) It was the amalgamation of these component parts that defined the compass of the new national library. As the 1971 White Paper put it, *"The objective of the British Library will be to provide the best possible central library services for the United Kingdom"*.[5]

With respect to the breadth of the BL's mission, Dame Lynne Brindley (the Library's former chief executive) has argued[6] that, as a national library, the BL is effectively "... *in a class of one, and continues to be unique, particularly in its wide scope of activities and services.*" The value that the BL brings is essentially threefold:

- It is a critical resource for UK research, it is an integral component of the national research infrastructure, and it plays a correspondingly significant role in ensuring the research excellence of the UK.
- It underpins business and enterprise through its contribution to knowledge transfer, creativity, and innovation.
- It is a world-class cultural institution with a vital role as a holder of the national memory.

Thus, the BL occupies a prominent place in the intellectual and cultural life of the nation. It is an integral component of the research infrastructure and it plays a correspondingly significant role in ensuring the research excellence of the United Kingdom and in supporting

creativity and innovation. An independent economic impact study[7] commissioned by the BL in 2003 suggested that the total value added to the UK economy by the Library each year was £363 million, or £4.40 for every £1 of public funding. An updated study undertaken in 2013 confirmed that the Library generates a net economic value of £419 million for its users and UK society as a whole. The study showed that the benefit cost ratio had increased to 4.9 from 4.4 in 2003 and, accounting for value placed on the Library internationally, the benefit cost ratio was 5.1.

The early history of the Library from its foundation, and thus from the first edition of this encyclopedia, was characterized by an ever-closer fusion of its constituent parts, developed through a number of different structures, each of which moved the Library forward.

Full integration was achieved and most symbolically demonstrated by the opening of the Library's purpose-built premises at St. Pancras in London by Her Majesty Queen Elizabeth II in June 1998 (Fig. 1). The building, designed by Sir Colin St. John Wilson, now stands as a confident symbol of the importance of *all* libraries to the nation's cultural, scientific, educational, and economic success in the twenty-first century. However, its construction was not without controversy.[8] HRH The Prince of Wales described the new BL as looking "more like the assembly hall of an academy for secret police." A campaign was mounted by the Regular Readers Group seeking to retain reader facilities in the British Museum, notably the Round Reading Room and North Library. A 1994 Parliamentary Select Committee described the building as "an edifice that resembles a Babylonian ziggurat seen through a fun-fair distorting mirror." And inquiries by the National Audit Office in both 1990 and 1996 were highly critical of the project management of the construction of the building, which was significantly delayed due to technical problems with the book shelving, with cabling damaged during installation, with shortcomings in the fire protection systems, and remedial works. The building finally opened to readers' acclaim in December 1996, and the knighthood conferred on St. John Wilson in the 1998 New Year's Honours list for services to architecture can perhaps be seen as constituting the final word on the architecture of the BL building.

At St. Pancras, the Library was able to consolidate services from 18 different buildings around London, including the move from the historic Round Reading Room at the British Museum, thus bringing humanities researchers to join their peers from the scientific disciplines in 11 new reading rooms (with 1200 reader places) covering all fields of research.

The Board has adopted a two-site estates strategy based on its headquarters building at St. Pancras and a campus site in Boston Spa, Yorkshire. In line with this strategy, construction of the BL's new state-of-the-art Centre for Conservation, to the rear of the St. Pancras building, completed in spring 2007.

Fig. 1 BL's headquarters in St. Pancras, London, is a sleek, modern complement to the Victorian Gothic railway station alongside.

Boston Spa promises to be the site of future expansion because it offers both space and relatively cheaper building costs and, at the time of writing, construction of a major new collection storage building was well under way. This low-oxygen, high-density storage unit will employ a fully automated storage and retrieval system and have a capacity for 260 km of storage (ca 7 million items). The Library's newspaper strategy, which addresses the future consolidation of the newspaper collections currently held in a separate newspaper library in North London, envisages the storage of newsprint in an additional newspaper storage facility, adopting a similar technological solution, at Boston Spa with access via digital and microfilm surrogates at St. Pancras. The Library employs broadly 2000 staff at its sites in London and Boston Spa.

The British Library Board has adopted a two-site estates strategy based on its headquarters building at St. Pancras and its campus site in Boston Spa, Yorkshire. In line with this strategy, construction of the BL's Centre for Conservation, to the rear of the St. Pancras building, completed in spring 2007. The center, which was officially opened by HRH The Princess Royal in October 2007, provides a world-leading facility for book conservation and state-of-the-art technical facilities for the preservation of the Library's recorded sound collection. A program of behind-the-scenes tours is an important adjunct to the work of the center, enabling members of the public to see: conservation techniques ranging from manuscript repair using sturgeon glue to the digital processing of archival sound recordings. The center also offers a national and international internship program, the aim of which is to give interns the opportunity to work with the Library's experienced staff to gain skills and also confidence in what they are doing.

Boston Spa promises to be the site of future expansion for the BL because it offers both space and relatively

cheaper building costs. During the last decade the Library has taken significant steps to improve the quality of its collection storage facilities for the ever-growing physical collection and the progressive concentration of the Library's collection. December 2009 saw the opening of a major new £26 million storage building for printed books at Boston Spa (and as a consequence the vacation of three leasehold outhouse storage buildings in London). This new storage facility at Boston Spa provides capacity for ca.7 million items (the equivalent of 262 linear kilometers of traditional shelving) in high density, preservation standard storage, with fully automated retrieval systems operating in a low-oxygen environment, and growth space for the next decade. This automated, high-bay, high-density solution is used frequently in warehouse management but rarely in archives and libraries, and its size and characteristics make it one of the most remarkable library buildings in the world. Collection moves of low-use stock into the new facility began in January 2009 and completed in early 2012.

At the time of writing, construction of another major new collection storage building also on the Boston Spa site—for newspapers—had just completed (Fig. 2). This new Newspaper Storage Building will hold the BL's newspaper collection in the best of archival conditions and permit the vacation of the Library's Colindale facility in North London, home of the newspaper collection since 1905. The new facility will store the equivalent of 33 linear shelf kilometers of newspaper content. Like the Additional Storage Building, the store has a fully automated retrieval system and full temperature and humidity control, and it will enable the BL's print newspaper collections to be kept in archival-standard conditions for the first time ever. Temperature in the storage void is held at a constant 14°C (+/− 1°C) and relative humidity at 52%.

Fig. 2 BL's new Newspaper Storage Building under construction at Boston Spa in 2014.

To eliminate fire risk, oxygen levels in the void are reduced to 14.9vol% by continuously introducing nitrogen. The construction of this new storage building is one element of the BL's newspaper strategy. Newspapers are an immensely rich source for research but they deteriorate quickly because of the poor quality of the paper they are printed on and it is crucial that they are properly preserved for future generations. The Library's ambition is to digitize the best of its historical collections, to open up the collections, and to make them much more widely accessible on the web.

The Library employs broadly 1600 staff at its sites in London and Boston Spa.

The Library's funding comes mainly from UK central government, through the vote of the Department for Culture, Media, and Sport. (At the time of writing the balance of the Library's incoming resources is annually in broad terms: grant-in-aid of £95 million, supplemented by the Library's trading income of £14 million, primarily from document supply services, and a further £15.6 million from donations, grants from foundations, and similar forms of support.)

STRATEGY AND USER COMMUNITIES

In September 2010, following a year of extensive research and consultation with stakeholders and experts from the fields of technology, information, publishing, libraries, the media, learning, and research, the BL published its *2020 Vision*,[9] setting out its priorities and aspirations for the next decade. The *Vision* highlighted what the Library considered was likely to be the key trends and opportunities over the next 10 years, indicating how the BL planned to take advantage of those opportunities to remain a great national library and a major hub of the global information network. And the *Vision* also provided a long-term perspective on where the information world and the Library was heading and the strong sense of direction that would be needed, particularly during the subsequent years of financial constraint.

The experts consulted by the Library provided a range of views on what the world of 2020 would look like. These predictions included the following:

- Technology will be in a constant state of beta—rapidly changing and, by 2020, resulting in a very different environment from today.
- The new generation of "digital natives" will enjoy wider access than ever to a huge range of online content in all formats; they will assume that *everything* is available on the web—an incorrect assumption, as even by 2020 a huge amount of legacy content will remain undigitized.
- The online landscape will increasingly resemble the *semantic web*—in which computers become capable of extracting, classifying, and analyzing data to create context from content.

- The business models underpinning scholarly publishing will change dramatically—more teaching, learning, and research will take place virtually and multi- and interdisciplinary research will continue to grow in importance.
- Knowledge institutions will need to reposition themselves, demonstrating the distinct value they add to the knowledge economy.

Our mission and 2020 Vision

Our mission: Advancing the world's knowledge

　Our vision: In 2020, the BL will be a leading hub in the global information network, advancing knowledge through our collections, expertise, and partnerships, for the benefit of the economy and society and the enrichment of cultural life.

　Our vision is supported by five key themes which set out the strategic priorities for the Library:

1. Guarantee access for future generations
2. Enable access for everyone who wants to do research
3. Support research communities in key areas for social and economic benefit
4. Enrich the cultural life of the nation
5. Lead and collaborate in growing the world's knowledge base

---

In February 2011, the Library published *Growing Knowledge: the British Library's Strategy 2011–2015*,[10] which set out how the Library intended incrementally, to move toward delivering its 2020 Vision over the period. The strategy recognized that all five strategic priorities were valuable in that they would enable the Library to support a wide range of users, with varied needs; however, in a period of financial constraint, efforts would need to be focused on essential activities. Therefore, in its strategy, the Library attached greatest weight to those activities relating to its statutory remit to make accessible the nation's and the world's store house of recorded knowledge, both now and in the future. Thus, its key implementation priorities for the period were to

- Achieve regulations enabling implementation of the Legal Deposit Libraries Act 2003
- Manage the ingest and storage of voluntary and legal deposit content
- Develop the Library's digital infrastructure
- Open a state-of-the-art facility for storage of physical newspapers
- Develop selective collaborative stewardship arrangements to collect and connect to content
- Digitize up to 20 million pages from the Library's newspaper collection in partnership with brightsolid

- Establish, in partnership, at least one major new large-scale digitization initiative in addition to the newspaper digitization initiative
- Obtain agreements from publishers to provide access to licensed materials off-site through a range of models

The Library's users come from diverse backgrounds, span all disciplines, and have varied needs. The Library defines its broad audience communities as follows:

- Researchers
- Business people
- The library network
- Schools and young people
- The general public

The BL is a great historic institution, with a critical mass of its own and a vibrant international brand. However, it is not an island. In caring for and managing its collection, and in providing access, it seeks to work productively with a range of partners where there is a demonstrable mutual benefit. And the absence of a unitary governance, policy, and funding structure within government for library provision nationally places a particular onus on the libraries themselves to collaborate both within and across sectors. Some notable examples of collaborative projects led by the Library follow.

The UK Research Reserve (UKRR) is a collaborative distributed national research collection managed by a partnership between the higher education sector and the BL. It allows higher education libraries to deduplicate their journal holdings of a title if two copies are held by other UKRR members, ensuring continued access to low-use journals, while allowing libraries to release space to meet the changing needs of their users. With funding from the Higher Education Funding Council for England (HEFCE), it has the target of releasing 100 km of shelf space by 2015 and to date has achieved £37 m in savings through eliminating the need for duplicated storage in the system.

Electronic Theses Online Service (EThOS) is the national aggregation service for UK PhD theses, provided by the BL. It aims to transform the visibility and availability of this previously rarely used resource: It demonstrates the quality of UK research and supports the UK Government's open-access principle (see the following text) that publications resulting from publicly funded research should be made freely available for all researchers, providing opportunities for further research. Currently, EThOS contains 350,000 records relating to theses awarded by over 120 institutions; around 120,000 of these also provide access to the full text thesis, either via download from the EThOS database or via links to the institution's own repository. Of the remaining 250,000 records dating back to at least 1800, three quarters are available to be ordered for scanning through the EThOS digitization-on-demand facility.

The British Library's Business & Intellectual Property Centre (BIPC) provides access to world-class business and intellectual property information and expertise under one roof. The Library's information resources—which include market research reports, up-to-the-minute company and financial information, and patent and trademark databases—effectively give small businesses access to the same quality and depth of resources as those of a large multinational. The information is supplemented by free or highly discounted workshops, one-to-one clinics, mentoring, and networking events all designed to inspire people and to teach them the skills they need to set up and run a business. Since it was set up in 2006, it has welcomed over 350,000 people through its doors and has helped to create over 2,700 businesses. Now, in a joint project with the UK Intellectual Property Office and six major English city public libraries, the Library is extending the successful BIPC model across the country by establishing a cobranded network with one-stop shop centers for small businesses and entrepreneurs in the city libraries of Birmingham, Leeds, Liverpool, Manchester, Newcastle, and Sheffield. Once up and running, the six centers will provide direct advice and guidance to over 25,000 people and will create over 500 businesses and 1,000 jobs each year.

COLLECTIONS AND SERVICES

The BL is unique among national libraries for the breadth and richness of its historic collections. The Library's collection is one of the largest in the world, holding over 150 million items in all known languages and formats, including books, journals, newspapers, magazines, sound and music recordings, patents, databases, maps, manuscripts, stamps, prints, drawings, and much more. Based on entries in catalogues, the Library's collection size[11] is calculated as given in the following table:

Monographs	14,768,923
Serial titles	858,414
Manuscripts	370,396
Newspapers (titles)	52,000
India Office Records	413,801
Philatelic items	8,280,939
Cartographic items	4,574,521
Music scores	1,618,761
Sound disks	1,531,844
Sound tape items	256,867
Digital audio files	74,318
Videos	39,983
Digital video files	23,638
Prints and drawings	33,210
Photographs	323,039
Patents	68,634,309
Reports in microform	10,433,593
Theses	269,442
Total	112,505,998

In the field of arts and humanities, the BL holds the world's largest collections of western and oriental manuscripts (including substantial archives) and unrivalled collections of British printed books from 1476 to the present date, of European printed books from 1455 to date, and of Asian and African printed books dating back to the origin of printing in these areas. In addition the Library holds substantial collections of sound recordings of oral history and of recorded sound relating to English language and literature. The chief historic components of these collections are the library of the British Museum, collected since its foundation in 1753. Of special importance for arts and humanities are the collections that came with the India Office Library and Records (incorporated in 1982), ranging from the foundation of the East India Company in 1600 to Indian independence and with the British Institute of Recorded Sound (incorporated in 1983). Holdings include material over 3,000 years old (Chinese oracle bones), 310,000 manuscript volumes ranging from Leonardo da Vinci's Notebook to Harold Pinter's archives, *Magna Carta*, Lindisfarne Gospels, the first edition of the Canterbury Tales, Tyndale's translation of the New Testament, the most recently published British books, and the recording of Nelson Mandela's Rivonia trial speech.

The BL houses one of the foremost collections of scientific, technical, and medical literature in the western world. Its contemporary collection of modern scientific literature includes all subject areas and disciplines and includes materials in many formats including journals, research-level monographs, conference proceedings, reports, and electronic reference materials. The Library estimates that there are approximately 3.45 million volumes of print serials and monographs in the collection.

The Library's collections in the field of social sciences reflect the full history of official publishing in the United Kingdom and its constituent parts. The Library has 12,000 volumes of House of Commons sessional papers from the nineteenth and twentieth centuries, containing more than 185,000 parliamentary papers. There are 2400 volumes of UK parliamentary debates, all available on open shelves in the Social Sciences Reading Room. Government publications from across the world include records of legislatures, censuses, and other official statistics. The Library has all print communications from the United Nations, the European Union, and the Organisation for Economic Co-operation and Development as well as other intergovernmental bodies such as the World Bank. Collecting from around the world covers economic, political, social, and cultural development and includes many rare items. The Library's collections of trade literature and market research are among the most comprehensive in the world, including over 66 million patent specifications from across the world.

The BL's acquisitions budget in the year 2013–2014 on collection items (for material of research value) was £14.6

million. Research undertaken a few years back as part of the Research Support Libraries Group (RSLG) project underlined the critical importance of the BL's reference and document supply collections in underpinning the resources of the libraries of the higher education institutions of the United Kingdom. When the budgets for acquisitions in UK institutions were ranked against those in the United States, the BL's acquisitions spend was found to be on a par with the best U.S. research collections at Harvard and Yale. However, the highest ranking UK institutions not in receipt of legal deposit (Manchester, John Rylands) were ranked at the 83rd place and, more significantly, the last 85 places out of 221 were taken up by UK universities. The RSLG report[12] observed:

"Researchers gain access to external hard copy resources in two ways: remotely via inter-library loan or document supply; and directly by visiting other institutions in person. Inter-library loan and document supply in the UK is deeply entrenched and effective..... The resources of the British Library are, we believe, one of the main reasons why UK universities are able to match, and in many cases exceed, the quality and scale of research undertaken within similar-sized universities in other parts of the world. Comparisons with North American universities, for example, show that only three UK libraries—the British Library and the university libraries at Oxford and Cambridge—acquire material on a scale comparable with the leading North American private or state universities. A large research intensive university in the UK typically acquires some 10,000 serials annually along with some 30,000 monographs—respectively about a quarter and a half of what the main Carnegie Research 1 universities in the USA, with which they would wish to be compared in the breadth and quality of their research, acquire."

Since then the information world has continued to change rapidly. Technology has improved, user expectations have increased, and publisher business models have changed. The BL's current content strategy, *From stored knowledge to smart knowledge*,[13] has responded to the unprecedented scale of these changes and has adopted a number of key principles, the most relevant of which in this context are as follows:

- The UK's publications received through Legal Deposit will underpin content development.
- We will select content in terms of its research value.
- Connecting to content will become more important.
- We will continue to invest in heritage materials.
- We will conduct a strategic review of the acquisitions budget (in the context of changing levels of provision in the UK HE and public library sectors).
- The Library will continue to make the print-to-digital transition.
- We will add value to content through curation and encourage our users to add value through community.

- The Library will develop a "without walls" approach to access.

The importance attached by the Library to heritage acquisitions is reflected by the budget it has established for purchasing such items, equivalent in broad terms to about 10% of the Library's total grant-in-aid-derived acquisitions budget. This is supplemented by a number of BL trust funds, the largest of them in terms of endowment being the Shaw Fund, although, for an institution that has been going for 250 years, the Library does not have huge endowments. Heritage assets purchased by the Library in 2013–2014 were valued in aggregate at £2.7 million. In recent years, the Library has been fortunate to acquire a number of very significant acquisitions. Notable among these have been

- The literary archives of JG Ballard, Harold Pinter, Ted Hughes, and Mervyn Peake
- The photographic collections of William Fox Talbot and Fay Godwin
- The Dering Roll, the oldest extant English coat of arms, the Macclesfield Alphabet Book, a rare medieval *pattern* book, and an illuminated medieval prayer roll that once belonged to Henry VIII
- The seventh century St. Cuthbert Gospel, the oldest intact European book

The most essential task of national libraries, and the one that more than any other makes them what they are, is that they are responsible for acquiring, preserving, and making accessible the publications—of all kinds—of the country. Therefore, it follows that legal deposit—the objective of which is to build a collection of the national literature—must be of fundamental importance. New regulations (the Legal Deposit Libraries [Non-Print Works] Regulations 2013) came into force on April 6, 2013, giving the BL and the five other UK Legal Deposit Libraries (the National Library of Scotland, the National Library of Wales, Bodleian Libraries, Cambridge University Library, and Trinity College, Dublin) the right to receive a copy of every UK electronic publication. The Regulations enable the six libraries to collect, preserve, and provide long-term access to the increasing proportion of the nation's cultural and intellectual output that appears in digital form—including blogs, e-books, and the entire UK web domain. The Legal Deposit Libraries Act 2003 had established the principle that legal deposit needed to evolve to reflect the massive shift to digital forms of publishing: the 2013 Regulations finally made digital legal deposit a reality. This had been long sought by the legal deposit libraries, concerned there was a very real danger of a digital "black hole" opening up in the nation's memory, and was a tremendously important step forward, ensuring that the fullest possible record of life in the United Kingdom in the twenty-first century is collected and preserved for future generations of researchers.

On behalf of the Legal Deposit Libraries, the British Library is archiving copies of freely accessible UK websites and web pages from the open web, using an automated crawling or harvesting process. A "snapshot" of every website within scope, currently estimated at circa 4.8 million active sites, will be archived at least once a year. Some 200–500 websites within scope will be archived on a more frequent basis—such as quarterly, monthly, weekly, or even daily—in order to ensure that rapidly changing or updated content is archived adequately. Such websites will be selected by the Legal Deposit Libraries for their importance and research value, with the crawl frequency being adapted to the circumstances and nature of the content. And in addition, the Legal Deposit Libraries will crawl other selected websites in order to develop "special collections." Perhaps four or five new collections will be developed each year for important events (which may involve crawling specific websites relatively frequently for a limited period) or important themes (which may involve crawling selected websites regularly over a longer period). By the end of March 2014, the BL had 140,000 E-journals available to access in its reading rooms and approximately 1.3 billion URLs were also available as a result of crawls of the.uk domain.

The Library's new £15 m purpose-built Centre for Conservation, funded by Government, the Library, and through fundraising, hosts state-of-the-art conservation studios, a visitor center linked to the main exhibition galleries, public demonstrations and tours, and improved conservation treatments and skills for the current cohort and new training courses for future book conservators (including a new foundation degree in book conservation with a university partner).

The Library's collection of reference materials, spanning four millennia and covering every written language, is one of the richest and the most comprehensive available to scholars. It is housed primarily in St. Pancras and serves the 11 reading rooms there (Figs. 3,4). Two other reading rooms are located in Boston Spa and in the newspaper collections building in North London. Access to the reading rooms is free—the Library welcomes everyone who wants to do research, whether for academic, personal, or commercial purposes—and, currently, Reader Passes are held by over 150,000 readers. While the reading rooms in St. Pancras are essentially interdisciplinary, they are broadly grouped into humanities on the one hand and science, technology, and business on the other. In addition, certain types of material—Manuscripts, Maps, Rare Books and Music, and Oriental and India Office Collections—each have a dedicated reading room. March 2006 saw the opening of the Library's Business & IP Centre at St. Pancras. The Business & IP Centre is designed to support small and medium-sized enterprises and entrepreneurs from that first spark of inspiration to successfully launching and developing their business.

April 2014 saw the opening of the new British Library *Newsroom*, at St. Pancras. The *Newsroom* is an integrated news and media offering—as opposed to simply a newspaper reading room—combining print and microfilm news content (transferred from Colindale) with newly created Television News Recordings from the Library's video server, with radio news, and importantly with web news collected under the Non-Print Legal Deposit Regulations introduced in 2013. The facilities of the Newsroom include

- More than 100 desks in a spacious and newly refurbished Reading Room at the heart of the BL at St. Pancras
- 40 digital microfilm viewers, offering excellent image quality and magnification, and swift, intuitive navigation—opening up microfilm to a new generation of researchers
- Microfilm of the 15 most highly used newspaper titles available instantly on open-access shelves in the Newsroom, the rest of the microfilmed archive—some 630,000 reels—available to order within 70 min
- Free access to digitized resources such as the British Newspaper Archive, which has 7.8 million scanned pages of historic newspapers, fully searchable by date, title, and keyword
- A public networking area for meeting, discussion, and collaborative research, including a large video wall displaying live news feeds and news-related Library content
- Access to the UK Web Archive, including 4.8 million archived UK domain websites, totalling more than a billion individual URLs
- The BL's extensive collection of TV and radio broadcast news—including over 40,000 programs, and growing at a rate of 60 hr every day across 22 news channels

The Library's reading room at Boston Spa was also refurbished during 2014 and now has facilities for news research broadly to match those at St. Pancras.

The Library's document supply collection is housed mainly in Boston Spa and is used in university and public libraries and top research companies both in the United Kingdom and overseas (Fig. 5). In recent years, the majority of document requests have been met by photocopies, but increasingly, researchers are being given instant access to articles and papers in electronic form as the Library develops relationships with the publishers.

In 2013–2014 the number of items supplied/consulted remotely and onsite was 10.5 million.

For the general public, the Library's building at St. Pancras provides modern facilities and galleries (Fig. 6). The centerpiece of the public space is the King's Library, housed in a magnificent glass and bronze tower (Fig. 7). The 65,000 volumes of King George III's collection were

Fig. 3   Humanities Reading Room, St. Pancras.

presented to the nation by George IV in 1823 and continue to be used by researchers every day.

Access to, and enjoyment of, the collection is also extended by a vigorous public program, both onsite and remote. At St. Pancras there is an exciting portfolio of free and charged public exhibitions, targeted at the broadest possible global audience, one that is notably diverse in culture, language, and styles of learning. The Sir John Ritblat Gallery of Treasures of the BL houses a permanent display of the jewels of the national collection. These include historic documents such as Magna Carta, icons of English literature including the Beowulf manuscript and the notebooks of the Brontës, and the sacred texts of the world's great religions. Imaginative access to a number of these, including the Lindisfarne Gospels, the Golden Haggadah, the Diamond Sutra, and Sultan Baybars' Qu'ran, is provided by Turning the Pages, a touch-screen system

developed by the Library to give visitors the opportunity to turn the pages of digital facsimiles, explore them in magnified detail, and hear or read about their production and meaning. In selecting items for exhibition, care of the collection is a paramount consideration; for example, the Library has recently introduced a rotation policy for its most iconic items, to ensure that they are rested for six months after an eighteen month period of display.

The Library plays host to two new major temporary exhibitions every year. Some recent notable examples include *Sacred: Discover what we share* (2007), which displayed some of the world's earliest and most important religious texts from the Jewish, Christian, and Muslim faiths, and *Taking Liberties: The struggle for Britain's freedoms and rights* (2008), which attracted some 100,000 visitors to see treasures including Magna Carta, the death warrant of Charles I, and the prison diary of suffragette

Fig. 4 Oriental and India Office Reading Room, St. Pancras.

Olive Wharry. Both of these exhibitions placed the visitor at the heart of current debates: in the former case exploring commonalities, crossovers, and contrasts between the three Abrahamic faiths and in the latter around rights and freedoms devolved government and free speech. Both of these exhibitions were accompanied by award-winning educational programs and enhanced by rich interactive displays. They are also accompanied by programs of public lectures, debates, readings, and performances, both in the building and on the piazza, which bring new audiences to the Library, and raise awareness of its collection.

The Folio Society Gallery in the Library's Entrance Hall plays host to four special themed exhibitions each year and attracts audiences of around 180,000 visitors per annum. Examples include the following: *Poetry in Sound* (June 2013) celebrated the centenary of the birth composer Benjamin Britten, and Beautiful Science (February 2014) explored how the visualization of scientific data is crucial

Fig. 5 The British Library site at Boston Spa, Yorkshire, is the operational center for international document supply and bibliographic services.

Binding–British

Fig. 6 The entrance hall of the Library's St. Pancras building.

Fig. 7 The tower of the King's Library, St. Pancras.

for making new discoveries and for communicating those discoveries effectively.

Other public spaces at St. Pancras include the Piazza, with sculptures by Sir Eduardo Paolozzi and Antony Gormley, and the conference center's auditorium and meeting rooms, which give the Library flexible space to mount events, conferences, and activities for a wide range of audiences.

Closely allied to the public program at St. Pancras is a program of regional outreach and cooperation with the UK public library network. Turning the Pages™ is included in the Library's complementary virtual learning environment, the Online Gallery, along with views of current and past exhibitions, and short research articles by our experts. The Library's learning program promotes the development of young people's research skills and engagement by schools and lifelong learning audiences with world culture and heritage. The Library's onsite program comprises workshops, exhibition activities, and major educational projects with schools, further education colleges, and youth and community learning groups, and public activities meet appropriate learning and access criteria.

Digitization has enabled significant numbers of collection items to be made available for a range of audiences—scholarly, school-based, commercial, or the wider public—by varying the routes into the material and the levels of interpretation provided. The Library's digitization programs, developed through a mixed approach of public and private sector funding, include major collaborations such as the International Dunhuang Project (IDP). This project is bringing together on the web a collection of thousands of Chinese manuscripts and early printed material that was dispersed a century ago from caves on the Silk Road trading route. The image bank, catalogue, and other resources promote in-depth study and establish a new paradigm for the fostering of knowledge in the digital environment by linking the collections and expertise of the BL to those of the National Library of China, the Bibliothèque nationale de France, the Institute of Oriental Studies, St. Petersburg, the National Museum, New Delhi, the Staatsbibliothek, Berlin, and other partners.

Unlike some other national libraries, the BL has not been allocated additional funding to enable it to undertake large-scale digitization. Therefore, while this has remained the case and in order to maintain the Library's core functions, the Board as a matter of policy has not allocated grant-in-aid funds toward digitization; instead, the Library has actively sought alternative forms of funding—for example, by charitable donation, sponsorship, or commercial partnership—to take this work forward on a project-by-project basis. The precise terms of funding and hence access necessarily differed between projects. Key initiatives that have significantly contributed to building critical mass include the following:

Nineteenth century books: In partnership with Microsoft, the BL commenced mass digitization of nineteenth century out-of-copyright books with an initial focus on English Literature materials. The project digitized 65,000 titles in total (25 million pages) of nineteenth century philosophy, history, poetry, and literature.

Eighteenth and nineteenth century books: In June 2011, the BL and Google announced a partnership to digitize 250,000 out-of-copyright books (40 million pages) from the Library's collections. Selected by the BL and digitized by Google, both organizations are working in partnership to deliver this content free through Google Books (http://books.google.co.uk) and the BL's website (www.bl.uk). Google are to cover all digitization costs. This project is to digitize a huge range of printed books, pamphlets, and periodicals dated 1700–1870, the period that saw the French and Industrial Revolutions, the Battle of Trafalgar and the Crimean War, the invention of rail travel and of the telegraph, the beginning of UK income tax, and the end of slavery. It will include material in a variety of major European languages and will focus on books that are not yet freely available in digital form online.

Newspapers: In 2010, the Library announced a Public–Private Partnership initiative to digitize up to 40 million pages from the BL's national newspaper collection in partnership with brightsolid online publishing, the owner of online brands including findmypast.co.uk and Friends Reunited. This 10 year agreement is set to deliver the most significant mass digitization of newspapers the United Kingdom has ever seen, making large parts of this unparalleled resource available online for the first time. There are no direct digitization costs to the BL as brightsolid is to create a commercial offering and assume the associated financial risks. The resource will be available for free to users on-site at the BL and copies of all scanned materials will be deposited with the Library to be held in the national collection in perpetuity. The Library had previously made over 4 million digitized pages of pre–twentieth century newspapers—the output of a JISC-funded digitization program—available to UK Higher and Further Education Institutions.

Sound: Also in conjunction with the JISC, the BL has digitized portions of its substantial holdings of sound material amounting to 50,000 individual recordings and 7,500 hr of sound available (music, drama and literature, oral history, wildlife, and environmental sounds).

The national library of the United Kingdom is also most emphatically a library for the world. The British Library Board's international engagement strategy places a focus on collaboration and partnership, professional leadership and engagement, and cultural diplomacy. Under the terms of the strategy, five key strands of international activity are being pursued: restoring and sustaining cultures, virtual reunification of collections, capacity building, strategic partnerships, and international professional leadership. The Library's priorities for international engagement at the current time place a focus both on India and China, and also on digital developments in its professional leadership activities. The Library has attracted £10m from the Lisbet Rausing Charitable Fund (now the Arcadia Fund) to administer an Endangered Archives Program, seeking to identify archival materials at risk in the developing world, to arrange their transfer to a safe archival haven in the country or region of origin, and to create surrogate copies for distribution by the Library. The Library continues to play its full part in the international professional bodies, including the International Federation of Library Associations (IFLA), the Conference of Directors of National Libraries (CDNL), and the Conference of European National Librarians (CENL).

Structure

Library policy and strategy is set by the British Library Board, which, under the terms of the British Library Act, comprises 13 members, under the leadership of a part-time nonexecutive chairman. In selecting members for the Board, the Act charges the Secretary of State with giving preference to "those who have knowledge and experience of library or university affairs, finance, industry, or administration". The chief executive is a member of the Board.

The Library is currently structured on the basis of four divisions: collections, operations, audiences, and finance. The directors of these divisions, together with the chief executive and the chief digital officer, form the executive leadership team of the Library, with responsibility for its day-to-day running.

Collections

The collections division is responsible for nurturing in-depth expertise and understanding of the collections and for developing policy for and managing the areas of collection development, description, and conservation. It is a major center of global excellence in collection expertise and management, uniting the Library's curatorial and conservation specialisms and also digital preservation.

Operations

The operations division is responsible for operating the services derived from the Library's collections. It is responsible for those services however they are delivered, whether in reading rooms or remotely through document supply. Divisional responsibilities also encompass IT delivery, human resources, and organizational development.

Audiences

The establishment of an audiences group reinforces both the importance of the Library's users at the heart of its strategy and also its commitment to understanding and reaching new audiences as well as serving its key constituencies. The division is responsible for the public profile of the Library, in its widest sense, and comprises the press office, public relations, exhibitions, publishing, and events, and senior staff head the marketing and communications activity to each of the Library's key audiences— researchers in higher education, business users, the library and information network, education, and the general public.

Finance

The finance division is responsible for not only the estates strategy of the Library but also the financial control of an annual spend, comprising both grant-in-aid and earned income, in the region of £120 million per annum, as well as corporate procurement, security and risk management, portfolio management, and corporate planning.

Outside the divisional structure, but within the executive leadership team, the chief digital officer is the most senior technology executive in the organization, tasked with influencing corporate strategy at the highest level to keep up the momentum and ambition of the Library's digital transformation, coordinating digital investment across the organization as a whole, setting consistent standards of architecture and metadata, and offering independent internal challenge and review.

CONCLUSION

The BL, in terms of its collections and the range of services that are derived from them, is the world's greatest research library. The Library is at one and the same time a world-class cultural institution with a vital role as holder of the national memory, a critical resource for UK research, an integral component of the national research infrastructure, and playing a correspondingly significant role in ensuring the research excellence of the United Kingdom, and it underpins business and enterprise through its contribution to knowledge transfer, creativity, and innovation. The digital information revolution has posed new challenges to the library profession and offered new opportunities. The BL is responding confidently to the new imperatives. Forty years ago the BL's founders sought to create a national library to serve the world. The Library is now achieving this in ways that its founders may not have envisaged but must certainly welcome.

FURTHER INFORMATION

The Library's website, www.bl.uk, gives details of services and collections and provides access to a wide range of digital resources.

The BL can be contacted at 96, Euston Road, London, NW1 2DB, United Kingdom.

REFERENCES

1. *The British Library Act 1972*; HMSO: London, U.K., 1972; Chapter 54.
2. Harris, P.R. *A History of the British Museum Library 1753–1973*; The British Library: London, U.K., 1998.
3. Miller, E. *That Noble Cabinet: A History of the British Museum*; Andre Deutsch: London, U.K., 1973.
4. *Report of the National Libraries Committee*, Cmnd. 4028; HMSO: London, U.K., 1969.
5. *The British Library*, Cmnd 4572; HMSO: London, U.K., 1971.
6. Brindley, L.J. The role of national libraries in the twenty-first century. Bodleian Libr. Rec. **2002**, *17* (6), 464–481.
7. Pung, C.; Clarke, A.; Patten, L. Measuring the economic impact of the British Library. New Rev. Acad. Libr. **2004**, *10* (1), 79–102.
8. Day, A. *Inside the British Library*; Library Association Publishing: London, U.K., 1998.
9. 2020 vision. British Library Board: London, U.K., 2010.
10. *Growing Knowledge: The British Library' Strategy 2011–2015*; British Library Board: London, U.K., 2010.
11. *British Library Annual Report and Accounts 2013/2014*; British Library Board: London, U.K., 2014. The Annual Reports are published on the Library's website at www.bl.uk and subsequent years' figures will be made available there.
12. Research Support Libraries Group Final Report. HEFCE: Bristol, U.K., 2003. http://www.rslg.ac.uk.
13. *From stored knowledge to smart knowledge: The British Library's content strategy 2013–2015*. British Library Board: London, U.K., 2013. http://www.bl.uk/aboutus/stratpolprog/contstrat/british_library_content_strategy_2013.pdf.

Business Informatics

Markus Helfert
School of Computing, Dublin City University, Dublin, Ireland

Abstract

The rapid changes in recent years demand constant evaluation and modification of education programs. The following contribution summarizes some aspects of current study programs in information systems (IS) and focuses on business informatics (BI). As a stream of IS, BI can be described as a method and model-centered approach focusing on business IS. The success of BI derives from the benefits that arise when business administration concepts are combined with computer science technologies and software engineering principles to form a coherent methodological approach. In addition, addressing the need for an innovative and cross-disciplinary study model to equip graduates with transformation skills we have developed a master's study program in BI. By discussing an example curriculum, this entry outlines the core elements of this program and gives direction for BI as a study domain.

INTRODUCTION

Many universities have offered various programs related to information systems (IS). However, the rapid changes in recent years demand constant evaluation and modification of education programs. Recent challenges include for instance the move toward programs, which are more applied and professionally orientated. The Bologna Declaration in Europe with its three-level study structure as well as the increasing pressure to ensure funding within most departments adds further pressure to many universities. Despite attempts being made to provide reference to curricula and guidelines, many universities and faculties struggle with the proper direction and design of the IS curricula.

Common reference curricula related to IS are, for example, the IS 2002: Curriculum Guidelines for Undergraduate Degree Programs in Information Systems; and the MSIS 2000: Model Curriculum and Guidelines for Graduate Degree Programs in Information Systems.[1,2] Although proposed for many years, several discussions and disagreement exist on the content and direction of IS curricula and IS as a discipline. Recently a joint task force of the Association for Computing Machinery and Association of Information System is aiming at revising the IS 2002 undergraduate curriculum.[3] At the same time we are experiencing decreasing enrolments in IS programs worldwide.[4]

In order to summarize aspects of current study programs in IS we focus on business informatics (BI). This entry aims to illustrate differences between BI and traditional IS study programs.[5] We structure the entry as following. First, we summarize the context of information technology (IT)- and IS-related study programs. Then we outline the main characteristics of BI. In addition, we provide a generic framework for IS study programs. Finally, we present an overview of an example study program in BI—the European M.Sc. in Business Informatics

at Dublin City University. We conclude the entry by outlining some key challenges of BI as a study domain.

CONTEXT OF IS STUDY PROGRAMS

One of the IS-related challenges at present is the increase in complexity and the growing interdependency between business aspects and IT. Organizational components like business strategy, rules, procedures, and processes, and the organization's application systems, including hardware, software, and databases, affect and influence each other. Any change in one of these components typically requires modifications in other components, whereas existing systems and structures often act as a constraint on organizations. As a consequence, the demand for graduates capable of analyzing complex information networks and project mangers managing large IT projects is expected to increase. Therefore such subjects as application integration, enterprise architecture, information management, and business process management are increasingly important.[6] In addition, claims that IT is no longer a source of strategic advantage, indicates the move from technology-orientated jobs to more business-orientated roles. Universities are expected to provide in addition to core knowledge of IS' design and implementation a broad business and real-world perspective.

Graduates should show strong analytical and critical thinking skills as well as interpersonal communication and team skills. A further necessary area of expertise for today's IT graduates is the International environment. In a decade that has seen the enlargement of the European Union and that globalized it, it is necessary for business and IT personnel to be competent to work in other countries and to work in multicultural teams. Graduates may need to understand how to manage geographically and

Encyclopedia of Library and Information Sciences, Fourth Edition DOI: 10.1081/E-ELIS4-120043667

ethnically diverse teams. Therefore, students should gain practical experience of studying and working abroad, by which they will be exposed to the business culture of another country. Spending time in another country offers many advantages: a chance to acquire new skills, to participate in multicultural teams, and to experience the benefits of cultural diversity.

BI AND IS

Traditionally universities focused on management and business studies as well as computing, software engineering, and computer science. Computing and computer science (e.g., basic informatics) addresses technical and theoretical bases of IT and software systems. Business and management provide knowledge of the principal functions of management and focus on business operations and decision making (behavioral and organizational component). The combination of both disciplines, which includes technical and social components, is generally described as IS. Terms such as management information systems (MIS), business information systems (BIS), or information systems and management (ISM) are also common.

Besides these courses, over the last decades a growing number of universities, particularly in central and northern Europe, are offering undergraduate and postgraduate degrees in *business informatics*. The term seems well established in the German-speaking countries; however is often considered controversial. Indeed, BI stems from the literal translation of the German term "Wirtschaftsinformatik," but there seems a controversy about the characteristics of BI. BI is sometimes assumed to be equal to the broad area of IS, though some indications exist that BI has a stronger emphasis on engineering principles and methods.[7] In order to provide a foundation and to frame the subject in the IS discipline, in the following we outline the characteristics of BI.

Management-orientated IS programs sometimes lack consideration of a *methodological* combination of the theoretical work of computer science with a practical orientation toward designing systems and applications. This methodological focus is the area of BI, which complements traditional areas of IS that focus on *explaining* real-world scenarios. BI aims to engage constructively to develop solutions tailored to business problems. It takes an active role in aligning business strategy, corporate goals, business processes, and IT. The core element of BI is a methodological approach to describe, explain, predict, and design information and communication systems. It involves the development of terminologies, models, and architectures that are explicit and sharable.

BI can be characterized as

- Interdisciplinary.[8]
- Focusing on business IS as socio-technical systems comprising both machines and humans.[9–12]

- Concerning the inception, development, implementation, maintenance, and utilization of business IS.[13,14]
- Describing the relationship between humans, business functions, information, and communication systems, and technology.[15]

BI can be summarized as a socio-technological and business-orientated subject with *engineering* penetration.[13] As a science discipline BI is categorized as

- Applied science that studies real-world phenomena.
- Formal science that creates and applies formal description methods and models.
- An engineering discipline that systematically designs and constructs information and communication systems.[11]

The success of the subject derives from the benefits that arise when business administration concepts are integrated with computer science technologies and software engineering principles to form a coherent methodological approach. It centers on IS architectures and business processes and provides a systematic design and construction of organizational information and communication systems.

BI Subjects

In the following we present a framework for BI and IS for which we amalgamated prominent curriculum guidelines; an undergraduate and a graduate model curriculum predominantly referred to in the Anglophone area with one frequently referred reference curriculum in the German-speaking area.

- The model curriculum and guidelines for graduate degree programs in IS (*MSIS 2000*).[1]
- The most recent version of the IS undergraduate model curriculum (*IS 2002*).[2]
- The recommendation for BI at universities (*BI recommendation*).[8]

The work of IS model curricula represent almost 30 years of experience in curriculum development. Started in the early 1970s by the Association for Computing Machinery (ACM) other organizations, including Data Processing Management Association/Association of Information Technology Professionals (DPMA/AITP), International Federation for Information Processing (IFIP), and Association for Information Systems (AIS), have aided model curriculum development. The IS 2002 model curriculum is the most recent version for an undergraduate IS curriculum, published by the ACM and AIS. IS 2002 includes detailed course descriptions and prescriptive advice on how to offer an IS undergraduate degree program. For our study we used the most recent version of the IS 2002 undergraduate curriculum, although the current version is currently reviewed by a joint ACM/AIS task force. The MSIS 2000 model curriculum was published by ACM and AIS as a guideline for master degree programs in IS. On a master's level, the curriculum is designed to accommodate students from a wide variety of backgrounds. It considers a set of interrelated building blocks including foundational skills, core subjects, integration subjects, and career tracks. Emphasizing on career development skills, the curriculum includes oral, written, and presentation skills; people and business skills; ethics and professionalism.

The model curricula are explicitly developed to include knowledge elements from three major computing disciplines: computer science, software engineering, and IS. It accumulates long experience in IS curriculum development and provides a coherent structure for a study program in IS. Thus these model curricula seem to be appropriate, even though the model curricula are primarily based on the educational system and degree structures common to the United States and Canada, with limited acceptance and use outside of this area. The two-level educational structures underlying the curricula proved to be of advantage, as many European universities are restructuring their study programs toward a two-phase curriculum with bachelor and master degrees.

The third curricula we used, the recommendation for BI, is issued by the German Society for Informatics and the Association of University Professors of Management, Germany. It is aimed at providing common directions for education in BI at universities. In contrast to the MSIS curriculum, which provides a detailed recommendation for a curriculum, the BI recommendation is intended as a guideline and is focused on key qualifications and core subjects to be taught. The BI recommendation is mainly orientated on a study program of nine semesters leading to a degree of "Diplom-Wirtschaftsinformatik" (diploma/master level in business informatics).

In order to cluster subjects and to provide a list of taught subjects, we customized the framework in an iterative process involving expert opinion from 10 academics from different countries. The structure follows the proposed curriculum building blocks in the MSIS curriculum. However, in order to accommodate particular subjects taught in some study programs, we added subject blocks of mathematics and logic, structural science, legislation, and economics, and business engineering, and included often taught business subjects, for example logistics, procurement, and supply chain management. The list of career electives and domain-specific subjects presented here illustrates just some of the possible topics. The framework is presented in Table 1.

Architectural Focus of BI

In contrast to IS, BI appears to have a stronger focus on mathematics, logic, and structural science.[7] One reason

for this could be the focus on the *systematic* construction and the application of methodological principles, which are often stated as typical for BI study programs. Indeed, mathematical principles are perceived as essential in order to systematically construct, formalize, and analyze models and architectures of IS.[16]

In this regard business informaticians are often described as IS architects (in the sense of engineers) who are actively and systematically analyzing and designing business IS. Central to this is the subject "Information System Architecture." IS Architecture describes IS through various models and refer to both a dynamic view in the form of processes and the specification of the overall structure, logical components, and the logical interrelationships of a system.[17] The conceptual description of both views builds the methodological framework for understanding the alignment of software applications and information technologies, business processes, and the corporate strategy.[18] One example of an important architectural framework for BI is ARIS—architecture of integrated information systems.[19]

STUDY PROGRAM IN BI

In keeping with the points identified above, we outline an example of a BI study program, introduced at Dublin City University—the European M.Sc. in Business Informatics. As an example, we provide an overview of the program in order to give guidelines for similar programs in BI.

The central focus of the proposed curriculum for BI is to qualify individuals to lead IS-related transformations of business. This enables them to apply technological solutions and develop IS architectures to solve business problems of organizations. With this goal in mind the curriculum focuses on an engineering perspective and the integration of cultural studies. The program is intended for students who have achieved a primary degree in computing, computer science, software engineering, or a comparable discipline. The program is designed to be completed in one calendar year of full-time study and consists of two taught semesters followed by a third term consisting of a project of practical nature.

The curriculum has an emphasis on engineering principles, and includes a module on structural science, which encompasses management science, data engineering, and data mining. It also has a strong modeling component, and includes modules on IS architecture and business process management. The integrative perspective is provided by the supply chain management module. The program also covers the more traditional IS disciplines as in the strategic management of IT module. An overview of the general program structure is provided in Table 2.

In addition to the emphasis on engineering principles and core subjects of BI, the study program also supports the building of capabilities for managing transformations

Table 1 Study framework.

Informatics and fundamentals in engineering	Business and economics	IS	Integration and enterprise engineering	Informatics in action (representative)
Information and Communication Technology (hardware, software, networks, and communication technology) Programming and algorithms, data, and object structures *Mathematics and Logic* (analysis, linear algebra, numeric, logic) *Structural Science* (decision theory and methods for strategic decision making (e.g., risk analysis), statistics and quantitative models and methods, operations research, computational modeling and simulation)	Accounting and financing Marketing, production, procurement, logistics Organization, human resources, and corporate management Legislation and economics	*Fundamentals of Information Systems* (types of IS, IS industry, IS relevant legal frameworks, management, and IS) *Principles of Business Information Systems* (principles of functional and process orientation and industry solutions) *Data Engineering* (data modeling and management, knowledge engineering, and business intelligence) *System and Software Engineering* (analysis, modeling, and design) Managing data communication and networking *Information Management* (information, knowledge, and people, project, and change management, IS/ IT policy, and strategy, ethics, and privacy)	Business Engineering and IS Architecture Integrating IS Functions, Processes and Data Integrating IS Technologies and Systems	Academia and research Academic and research libraries Biochemistry and molecular biology Consulting Consumer health information Customer relationship management Data warehousing Decision making e-Government information Electronic commerce Electronic publishing Environmental management Financing and banking Healthcare information Human factors Insurance management Knowledge management Library services Logistics Multimedia technologies Project management Techniques of IT consulting Technology management

Source: Helfert and Duncan.[7]

by aiming to expand transferable skills. In essence, transferable skills are those skills that having been learned in one context can then be applied in another. Typically these skills are based on modern teaching, learning, and assessment methods, and include

- Guided independent study and activity, with specialist input when appropriate.
- Recent or current case studies.
- Essay and report writing.

- Collaborative group work and discussions.
- Presentation of findings to the group as a whole.

In the final semester, students work on a project of practical nature. The general objective of this project is to allow students to draw on the theoretical knowledge gained over the taught element and to apply it in a practical setting in an International environment. The project of a practical nature gives students the opportunity to demonstrate their ability to analyze problems in the field of BI and draw

Table 2 Curriculum overview.

Semester 1	• Research skills/Seminar topics	• Business process management
	• IS architecture	• Strategic management of IT
	• Structural science	• Business studies
Semester 2	• Supply chain management	• Sectoral applications of IS
	• Managing and working in an intercultural environment	• Regulation in IS
		• Project management
		• Managing change
Summer	Dissertation/project of practical nature	

Source: From Helfert and Duncan.[7]

conclusions according to scientific methods within a given timeframe.

SUMMARY AND CONCLUSION

This entry presented aspects of BI and illustrated a typical curriculum. We summarized some key requirements of IS graduates. In particular, graduates need a comprehensive understanding of behavioral aspects as well as software engineering, programming, and IT. Interpersonal and communication skills as well as problem solving and critical thinking capabilities are essential for any IS and BI graduate.

The example curriculum as outlined above comprises a balanced and interdisciplinary structure, which centers on engineering principles and focuses on transformation, models, and methods. The engineering penetration throughout the program is seen as one important characteristic, which differentiates this program from management-orientated IS degrees. In contrast to business administration programs, the production of information managers, often expected in practice, is not the objective of BI. As such BI can complement the management-orientated stream of the IS discipline.

In conclusion, the BI approach appears to us not only to be innovative with regard to its interdisciplinary character, but also the engineering perspective and the integration of cultural studies and practical experience in an international setting equip graduates with essential transformation capabilities. Indeed, the focus on engineering principles in BI could play an important role in future education programs.

REFERENCES

1. Gorgone, J.; Gray, P.; Feinstein, D.L.; Kasper, G.M.; Luftman, J.; Stohr, E.A.; Valacich, J.S.; Wigand, R. Model curriculum and guidelines for graduate degree programs in information systems. Commun. Assoc. Inform. Syst. **2000**, *3*(1), 1–61.
2. Gorgone, J.; Feinstein, D.; Longenecker, H.E.; Topi, H.; Valacich, J.S.; Davis, G.B. Undergraduate information systems model curriculum update—IS 2002 Proceedings of the Eighth Americas Conference on Information Systems Dallas, TX 2002.
3. Topi, H.; Valacich, J.; Kaiser, K.; Nunamaker, J.; Sipior, J.; Vreede, G.; Wright, R. Revisiting the IS model curriculum: rethinking the approach and the process. Commun. Assoc. Inform. Syst. **2007**, *20*(11), 728–740.
4. Granger, M.; Dick, G.; Luftman, J.; Slyke, C.; Watson, R. Information systems enrollments: Can they be increased?. Commun. Assoc. Inform. Syst. **2007**, *20*(41), 649–659.
5. Helfert, M.; Duncan, H. Business informatics and information systems—Some indications of differences in study programmes. Proceedings of UKAIS Conference 2005; Newcastle: U.K., 2005.
6. Traylor, P.S. Outsourcing. CFO Mag. **2003**, *19*(15), 24–25.
7. Helfert, M.; Duncan, H. Aspects on information systems curriculum: A study program in business informatics, International Federation for Information Processing (IFIP). In *The Transfer and Diffusion of Information Technology for Organizational Resilience*; Donnellan, B., Larsen, T., Levine, L., DeGross, J., Eds.; Springer: Boston, MA, 2006; Vol. 206, 229–237.
8. Appelrath, H.J. et al. Rahmenempfehlung für die Universitätsausbildung in Wirtschaftsinformatik. Informatik Spektrum **2003**, *26*(2), 108–113.
9. Ferstl, O.K.; Sinz, E.J. *Grundlagen der Wirtschaftsinformatik*, 4th Ed.; Oldenbourg: München, Wien, Germany, 2001.
10. Retzer, S.; Fisher, J.; Lamp, J. Information systems and business informatics: An Australian German comparison Proceedings of the 14th Australasian Conference on Information Systems 2003, Delivering IT and e-Business Value in Networked Environments; Lethbridge, N., Ed.; School of Management Information Systems, Edith Cowan University: Perth, Western Australia, 2003; 1–9.
11. König, W. Mitteilungen der Wissenschaftlichen Kommission Wirtschaftsinformatik. Profil der Wirtschaftsinformatik. Wirtschaftsinformatik **1994**, *36*(1), 80–81.
12. Heinrich, L.J. *Wirtschaftsinformatik—Einführung und Grundlegung*, 2nd Ed.; Oldenbourg: München, Wien, Germany, 2001.
13. Disterer, G.; Fels, F.; Hausotter, A. *Taschenbuch der Wirtschaftsinformatik*, 2nd Ed.; Carl Hanser Verlag: München Wien, Germany, 2003.
14. Scheer, A.-W. *Wirtschaftsinformatik—Referenzmodelle für industrielle Geschäftsprozesse*, 2nd Ed.; Springer: Berlin, Germany, 1998.
15. Heinrich, L.J. *Informationsmanagement—Planung, Überwachung und Steuerung der Informationsinfrastrukutr*, 7th Ed.; Oldenbourg: München, Wien, Germany, 2002.
16. Henderson, P. Mathematical reasoning in software engineering education. Commun. ACM **2003**, *46*(9), 45–50.
17. Foegen, M.; Battenfeld, J. Die Rolle der Architektur in der Anwendungsentwicklung. Informatik Spektrum **2001**, *5* (24), 290–301.
18. Zachman, J.A. A framework for information system architecture. IBM Syst. J. **1987**, *26*(3), 276–292.
19. Scheer, A.W. ARIS. In *Handbook on Architectures of Information Systems*; Bernus, P., Mertins, K., Schmidt, G., Eds.; Springer: Berlin, Germany, 1998; pp. 541–565.

Business Information and Its Users

Eileen G. Abels
College of Information Science and Technology, Drexel University, Philadelphia, Pennsylvania, U.S.A.

Abstract

The complexity of business requires a wide array of multidisciplinary information sources to resolve business information needs. Users of business information are as varied as the discipline of business, ranging from business professionals to the lay person interested in investing to the business school student. Aspects of business touch the lives of everyone. For this reason, librarians in public, academic, and special libraries provide business information to their patrons. In order to make the most of business information, business information literacy is essential for librarians and nonbusiness professionals who seek and use business information. Business information literacy includes an understanding of business terminology as well as trends in business that have an impact on access to business information such as information overload, transparency, and the blurring of various components of business.

INTRODUCTION

Business, commercial activity involving the exchange of money for goods or services, is large and complex affecting all aspects of society. In 2006, there were approximately 26.8 million businesses in the United States. The complexity of business can be attributed to several factors: business events change at a rapid pace; business is closely tied to the economy; and business problems require multidisciplinary solutions.[1] Business consists of industries that are comprised of companies that provide products and services. Business has changed over time; we have seen a decrease in manufacturing and an increase in service industries as well as an emphasis on communication, information, and computing technologies. Technological and social changes have had an impact on business. Davis and Meyer[2] note that three factors, speed, connectivity, and intangibles, have resulted in a "new world" they refer to as "blur." In this blur, there is no longer a clear distinction between a product and a service as products now incorporate services into their offerings. The integration of new technologies will continue to blur the distinctions of once separate components of business. A trend that has had an increasingly important impact on business is the demand from the public for transparency. To encourage transparency, new laws and regulations require that more and more information be made available to everyone who interacts with business. In addition, new technologies, such as blogs, are providing more internal information to those outside of an organization.

Management is another broad area closely related to business. Managerial activities include, among other topics, the following: strategic planning, leadership, human resources, marketing, facilities management, financial management, and decision making. A variety of different types of sources provide management information, including books, videos, magazine, and journal articles.

Business information users run the widest gamut from those actively engaged in the business world to the layperson, from the entrepreneur to the student or professor in a business school. Each of these user groups has different characteristics, information needs, and information seeking behaviors. Librarians serving these various user groups in public, academic, and special libraries are called upon to provide them with business information. This entry focuses on business professionals whose various business activities relate directly to specific types of business information; business activities have influenced the forms of packaging of information to facilitate use. Business professionals include, among others, top executives, administrators, and managers. Many of the information needs and the sources used to resolve the business information needs of business professionals also apply at some level to all users of business information.

This entry will first discuss business information users and their needs. Understanding information behaviors is essential for librarians to provide appropriate library and information services. A discussion of different types of information used to resolve business information needs will be presented. A theme that will appear throughout this entry is the notion of the transparency of business activities. As will be noted, transparency is not new; however, recent events have resulted in efforts to increase transparency in business activities. Transparency has resulted in enhanced access to business information.

British–Careers

BUSINESS INFORMATION USERS AND THEIR NEEDS

The *Occupational Outlook Handbook* 2007–2008 (http://www.bls.gov/OCO/) lists a variety of managers including industrial production managers to medical and health services managers, education administrators, advertising, marketing, promotions, public relations, and sales managers. According to the *Handbook*, in 2006, administrative services managers held about 247,000 jobs; advertising, marketing, promotions, public relations, and sales managers held approximately 583,000 jobs; and financial managers held about 506,000 jobs. In addition there are related occupations that would also fall within the overall title of business professional, including, accountants and auditors, budget analysts, management analysts, and financial analysts. Every industry has its managers and for this reason, the education and training of managers varies. Business degrees at the undergraduate level or the Master of Business Administration (MBA) are common among business professionals. According to the National Center for Education Statistics (http://nces.ed.gov/programs/digest/d07/tables/dt07_290.asp), the number of degrees conferred in business in 2005–2006 was 318,042 at the bachelor level; 146,406 at the master level; and 1,711 at the doctoral level. The number of males and females were approximately equal at the bachelor's level; more males obtained the master's and doctoral degrees.

Common to business professionals is their tendency to focus on resolving the problem at hand as quickly as possible. Of major concern to the business end user is the ease of finding information, which of course means that business users will use the sources and channels most convenient to them, whether or not the sources are the most authoritative or current. However, accurate and timely information is essential for many business decisions and activities. In some cases, cost is not an issue; if critical information is needed, many business professionals will pay a premium to obtain the information in a timely fashion.

The information sources used by business professionals include informal and formal sources as well as verbal and written ones. Business professionals usually have access to "soft external information" that is available to them through personal contacts.[3] A preference for interpersonal communication among business professionals has been well documented (pp. 194–203),[4] (pp. 227–263).[5,3] Business professionals often rely on a range of people for their information: customers, consultants, colleagues, experts, and product manufacturers. It is not uncommon for business professionals to call, send e-mail, or arrange an in-person meeting when seeking information. It is not surprising then that business professionals neither use the library as a primary channel for information nor are fee-based databases among the most consulted resources.

Much information about an organization is internal to the organization and is not found through computers, in books, or in online databases. Internal information may be available through internal systems or through the knowledge of employees and others affiliated with the organization. External business information sources form only a relatively small portion of sources used by business professionals.

Businesses carry out numerous activities that require information gathering. The extent to which information seeking activities are incorporated into the workplace will depend upon the resources available. Most businesses will engage in gathering competitive intelligence to some extent. Competitive intelligence involves the systematic process of gathering, organizing, analyzing, and managing data, information, and knowledge related to the environment in which a company operates. The overall objective is to convert the data, information, and knowledge into intelligence that will help a company gain a competitive advantage.

Businesses of all sizes need to understand their customers, both active and potential. The customer base is often referred to as the market. It is through research about the market that businesses of all types strive to understand the preferences and habits of actual and potential customers as well as factors that influence buying decisions.

Other common business activities include identifying companies for mergers or acquisition, planning to offer a new product or service, or expanding to a new segment of the market. More specific business tasks related to the organization itself also require information. For example, determining pay scales, selecting a new site location, completing background checks on potential clients or suppliers, compiling lists for mailings, performing a trademark search, negotiating a contract, identifying suppliers, or filling high-level positions.

One common approach to gathering important information for decision making is environmental scanning. Environmental scanning involves the assessment of the internal strengths and weaknesses of an organization in relation to the external opportunities and threats it faces. Through a variety of sources, interpersonal, print, electronic, and broadcast, business professionals uncover information about competitors, customers, economic conditions, trends, research, products, and legal issues.

Small business owners and entrepreneurs are similar to business professionals in their preference for informal oral sources of information. Young and Welsch concluded that this was due to the preference of the small business owner for customized and personalized sources of information (pp. 42–49).[6] Franklin and Goodwin also found that small business owners relied on informal information sources that were convenient to use (p. 10).[7]

Small businesses are a key component of business in the United States. Small businesses are defined as businesses with fewer than 500 employees. The U.S. Small

Business Administration (SBA) was founded in 1953 to assist small businesses through the provision of loans and counseling. Small businesses represent over 99% of all businesses and employ 45% of the private workforce. Of the total number of businesses, over 19 million are sole proprietorships. Like business professionals in large organizations, small business owners need information about competitors. The need for competitor information by small business owners is generally associated with the initial stages of starting a business to help determine location, pricing, and products and services to offer. Once established, small business owners focus on management information including operations, finance, and marketing unless business declines or an expansion is under consideration (pp. 42–49);[6] (pp. 43–44).[8]

Product information is essential to entrepreneurs since they need to know what other similar products are on the market and what is new or different about their product. Market data are also key for the entrepreneur to determine how big the market is for the product and where a potential customer would expect to purchase the product. Legal information will also be needed to resolve questions about copyright, intellectual property, trademarks, and patents.

Various information technologies have facilitated access to business information. New forms of information technology are being used to communicate with customers and employees. In person communication and the telephone have been supplemented by e-mail, instant messaging, and Weblogs. The cell phone has become ubiquitous and in addition to its use as a phone, cell phones often serve as a telephone directory or address book, provide Internet access, and include a camera.

The amount of information available in electronic format for free on the Web has greatly increased, including some resources such as patents and financial data that were previously more difficult to obtain and expensive. Information is being formatted to be accessible from mobile communication devices. Business professionals often rely on business portals which serve as a one-stop shop for business information. To facilitate access to information, Really Simple Syndication (RSS) may be used to feed information from news Web sites, Weblogs, and podcasts to end users of information on a regular basis without requiring the information seeker to initiate the activity. New search features are making it easier to access business information. For example, dynamic query suggestion tools may provide company names and ticker symbol suggestions that eliminate the need to look up this type of information before conducting a search. Visualization tools facilitate data analysis and decision making. Managers face information overload due to the nature of their work. "By virtue of their position, managers receive more information from more sources through more channels than almost anyone else in an organization" (p. 227).[5]

Overall, the primary interest of most business information professionals in organizations of all sizes, as well as those who are self-employed, lies in resolving information needs to assist in making decisions or resolving problems. These business professionals are not concerned with the process of information seeking but rather with the results. In fact, managers do not differentiate a discrete information seeking process but rather consider information seeking to be an integrated part of the decision making process (pp. 23–35).[9] As with all information seekers, ease of access influences information seeking behaviors. Librarians working with business professionals will want to be proactive in providing information services and consider packaging information in ways that best meet the needs of business information users.

BUSINESS INFORMATION

In general, the types of information used to resolve many of the information needs described above include industry information, economic data, company information, financial data, contact information, demographic data, and legal information. Business information may be packaged in a variety of ways to meet the needs of business information users. Some business information sources have emerged from business activities themselves and are thus business specific; some sources of information are specialized or technical in nature and yet provide essential business information; some general reference tools have been adapted for use by business users; and some general information sources contain business information. Selected types of business information along with the types of sources that contain them will be described with some historical context when appropriate. Several books included in the section "For Further Information" will provide the reader with more comprehensive coverage of business information sources.

Business-Specific Sources

There is a wide range of industry information and due to the wide use of this type of information industry profiles have emerged. These industry profiles or surveys, available in print and electronic formats, provide prepackaged information about an industry including a description of the current environment; historical context of the industry; the industry structure; industry trends; key industry ratios and statistics; key players in the industry; an overview of products and services; related industry classification codes; and industry terminology.

Economic data and industry statistics are an important subset of industry information. The Economic Census dates back to the 1810 Decennial Census when questions on manufacturing were included with questions about the population. The 1905 Manufacturing Census was the first time an economic census was issued separately from the population census. In 1930, retail and wholesale trade were added. The Economic Census was initiated in 1954 and is now issued every 5 years. In 2007, the Economic Census forms were sent to more than 4 million businesses. In keeping with the notion that different groups of users have different behaviors, the Economic Census has over 600 data-gathering forms for different industries. The Economic Census is mandated by law under Title 13 of the United States Code (sections 131, 191, and 224).

The Economic Census reports data using an industry classification. Recognizing the need to develop a standard industrial classification (SIC) system for all federal statistical agencies, developmental work was begun under the auspices of the U.S. Central Statistical Board, the predecessor of the Statistical Policy Office of Office of Management and Budget (OMB). The first SIC manual was issued in 1939 by the Interdepartmental Committee on Industrial Statistics which was established by the Central Statistical Board of the United States. The classification arose from the need to collect comparable data from a variety of agencies. The SIC included all industries that made up economic activity at that time. While revised periodically over the years, it was not until 1997 that the long-standing SIC was replaced with the North American Industry Classification System (NAICS) which was developed in conjunction with Mexico and Canada to provide comparable statistical data across the three countries. The NAICS recognizes a change in the focus of the economy from manufacturing to service and among other changes, created a new Information sector that includes establishments that create and disseminate information from a broad perspective. The new Information sector brings together various industries previously included in manufacturing, utilities and transportation, or the services sectors. Publishers of newspapers, books, and periodicals, and software, broadcasting and telecommunications producers and distributors, motion picture and sound recording industries, information services, libraries and archives, and data processing services are now part of this new Information sector.

Other data gathered to foster international comparability utilize the International Standard Industrial Classification (ISIC) maintained by the United Nations Statistical Office. The first version of the ISIC was issued in 1948 with periodic revisions over the years with a fourth revision approved by the United Nations Statistical Commission in March 2006.

An industry analysis would not be complete without statistics on shipments, payroll, and consumption. In order to understand the economic situation, one must utilize employment and labor statistics as well as other economic indicators. Statistical compilations aggregate useful statistics in one resource thus facilitating access. A statistical compendium compiles statistics on a given topic or generated by a particular agency or organization. The *Statistical Abstract of the United States* has been produced since 1878 by the U.S. Census Bureau and compiles statistics

gathered by a variety of government agencies, professional associations, and other entities.

Industries consist of companies and as such company information is an essential part of business information. Because of the need for company information, prepackaged company profiles have emerged as a new document type. While the content will vary somewhat depending on the producer of the company profile, most will include contact information, key officers in the company, information about the associated industry, identification of top competitors, company affiliation information (parent and subsidiary), sales figures, number of employees, year founded, and status of the company.

Company Web sites have become a very important part of company research. In many ways, the emergence of company Web sites has changed the panorama of business research. Web sites are available for entities of all types, including companies, associations, and government agencies. Company Web sites generally provide a range of information that may include contact information, company profiles, and biographical information in addition to information about products and services. The amount and type of information provided on the company Web site varies; large public companies often provide financial information along with detailed advertising campaigns, product information in the form of videos, and opportunities for customers to communicate with staff using blogs, e-mail, and chat.

Knowledge of a company's financial state is important for a variety of information needs, including merger and acquisitions as well as investment opportunities. It is in the area of financial data that the impact of transparency can be seen the most. The Securities Act of 1933 required the provision of significant information about securities that were offered for sale to the public and prohibited misrepresentations and fraud. One year later, the Securities Exchange Act of 1934 created the Securities and Exchange Commission (SEC) with oversight of all aspects of the securities industry, including brokerage firms and stock exchanges. The SEC was authorized to require periodic reporting by publicly held companies. Among the most common financial documents are annual reports and SEC filings such as the 10-K, 10-Q, and Prospectus. These public filings contain a wealth of company information including the nature of the business, history of the company, organizational structure, company auditor, balance sheets, and income statements. A complete list of SEC filings and forms can be found on the SEC's Web site. Annual reports and other SEC filings can be obtained directly from a company or obtained for a fee either online or through a courier. The SEC's EDGAR, the Electronic Data Gathering, Analysis and Retrieval system, now provides access to real time filings for those with access to the Internet. While EDGAR does not provide access to all SEC filings, it greatly facilitates access to financial data.

British–Careers

Research departments of major investment houses have staff members who track a variety of industries and public companies within those industries. The results are published in Investment Reports that include, for example, sales and earnings forecasts, market share projections, and research and development expenditures. The primary intent of these reports is to provide information for investment decisions; however, they are useful sources of industry and company information. Some concerns have been raised about possible bias in these reports. Over the years, newspapers have reported stories about brokerage firms that have been fined for touting stocks based on the interest of the firm rather than on the accuracy of the report or for abusive trading practices.

In 2001, the Enron scandal brought to light legal and ethical issues about the conduct of businesses and cost investors billions of dollars. It is considered by many to be the most profound crisis since the Great Depression of 1929 and resulted in more transparency in business activities.[10] As a result of the Enron scandal, the Sarbanes-Oxley Act of 2002 was signed into law, introducing reforms for corporate responsibility, enhanced financial disclosures, and created the Public Company Accounting Oversight Board (PCAOB) to oversee the auditing role of accounting firms.

Financial filings are also available for nonprofits which are required to file an annual information return. The Form 990, Return of Organization Exempt from Income Tax or the Form 990-EZ (Short Form) is filed with the Internal Revenue Service. Regulations require nonprofits to make their three most recent filings available to the public. In addition, the Pension Protection Act of 2006 added a filing requirement for small nonprofits that were previously exempt from filing requirements. The purpose of the new filing requirement is to ensure transparency within the nonprofit sector.

Financial information for private companies, those companies whose securities are not sold to the public, is very difficult to obtain as there are no filing requirements. One alternative source of financial information or information related to finances for companies of all types is the company credit report. Credit reports have a long history, dating back to 1841 when the first credit service was created. Dun and Bradstreet (D&B) was established in 1841 by Lewis Tappan as the Mercantile Agency. Robert Dun assumed control over the agency in 1859 and renamed it R.G. Dun & Co. The Dun's Book was issued that same year with credit information for over 20,000 businesses. According to Hoover's company profile for Dun & Bradstreet (http://access.hoovers.com), there were over 1 million credit reports issued by 1886. R. G. Dun merged with its competitor firm John M. Bradstreet Company in 1933. The name Dun & Bradstreet was adopted in 1939. Credit reports help companies make sound credit decisions when dealing with other companies that are potential customers or suppliers.

Products and services are fundamental to businesses. Product catalogs are listings of companies' products, often with descriptions, pictures, measurements, and weights, properties, and other characteristics. Product catalogs are often available on company Web sites and are generally designed as a sales tool.

In addition to information about products and services, business professionals need information about those who purchase products or services. Off-the-shelf market research studies, prepared by market research firms, bring together market research data including sales, forecasts, market characteristics (e.g., age, income, preferences, and attitudes of consumers), new products, and competitor activities. Market research utilizes data gathered from focus groups, polls, and surveys. ACNielsen, founded in 1923 with a focus on industrial machinery, provides market research by tracking retails sales, brand loyalty, and demographics. Market research reports and data are used for general marketing activities as well as for competitive intelligence activities.

Advertising is the activity involved in bringing public attention to a product, service, or company. Advertising is available in a variety of formats: print, electronic, or broadcast (television or radio). Advertisements provide information on a company's products and services, including in many cases pricing. Advertisements also provide clues about popular products and their characteristics. For example, a review of advertisements for automobiles over time could illustrate new innovations and changing consumer tastes. Advertising may be difficult to monitor with the broad array of media through which companies advertise. There are companies, such as TNS Media Intelligence, that monitor advertising occurrences. Many company Web sites are good sources of advertising information, including current and historic advertising campaigns. In many cases, the main purpose of the company Web site is marketing so the Web site itself serves as advertising. Advertising is useful in competitive intelligence activities and is also used for price determination of new and existing products.

Specialized and Technical Nonbusiness Information

A patent is an often overlooked source of product information. A patent is a grant issued by the government giving inventors the right to exclude others from making, using, or selling the invention for 20 years. In return, the inventors fully disclose information about their invention to the public. There are three types of patents: utility patents which describe the way something works; design patents which describe the way something looks; and plant patents which are for a new variety of asexually reproduced plants. Patents are an excellent and detailed source of product information including graphics and images. They are also an important part of competitive intelligence research and general company research. Patent research is essential for research and development so that time and money are not invested in an already existing product. A trademark is a word, name, or symbol that is used to distinguish a product or service from other products and services. Trademarks apply to logos, images, and sounds.

The role of the U.S. Patent and Trademark Office (USPTO), a federal agency founded in 1802 and currently located within the U.S. Department of Commerce, is to promote the progress of science by providing investors the exclusive rights to their inventions or discoveries for a limited period of time. (Article 1, Section 8 of the United States Constitution). In addition to reviewing patent and trademark applications, the USPTO disseminates information through freely available databases on its Web site. The Patent and Trademark Depository Library Program (PTDLP) began in 1871 when a federal statute (35 USC 12) first provided for the distribution of printed patents to libraries for use by the public. Now, all of the patent and trademark information is distributed to these libraries in a variety of formats including print, microform, and DVD. New rules have been and continue to be proposed by the USPTO to improve the efficiency of the patent process; a description of these rules and their status are available on the USPTO Web site. Patents were once available in print only through the PTDLP. Then commercial database services provided access to patents for a fee. Now patents are freely available on the USPTO Web site and through other Web sites. Patent offices in several other countries also offer access to patents through databases available on their Web sites, for example, the Canadian Intellectual Property Office (http://www.cipo.ic.gc.ca/), the European Patent Office (http://www.epo.org/), and the German Patent and Trademark Office (http://www.dpma.de/).

Laws and regulations are the rules and principles governing and regulating various activities. Laws and regulations impact the way companies do business so tracking them is critical to running a business. Legislative information may be of interest to businesses that want to monitor the status of Congressional bills that affect their activities. Proposed rules are of interest to many business information users; it is during the comment period that input can be given before a rule becomes final. Individual agencies that administer the law or regulation in question may also provide access to that information. More generally, the *Congressional Record*, first published in 1873, contains legislative regulations and is the official record of the Unites States Congress. In addition, The *Federal Register*, a daily publication that was first published on March 14, 1936, contains rules, proposed rules, and notices of Federal agencies and organizations, as well as executive orders and other presidential documents. The *Federal Register* is another example of an effort to provide transparency to the public. The role of the Federal Register has grown over the years and additional information has been added for dissemination to the public. The *Code of Federal Regulations* (*CFR*) codifies the rules published in the *Federal Register*. Both the *Federal Register* and the CFR are available on the Internet for free.

Commonly Used General Reference Tools

Business professionals consult a variety of reference tools. There are encyclopedias, dictionaries, almanacs, etc. that focus specifically on business topics. Directories have emerged as a very important reference tool and business directories have evolved over the years to meet business information needs. While directories are listings of people or organizations that include contact information, the amount and type of information included in business directories varies greatly. Some directories may include only public companies or companies with earnings over a certain amount while other directories will include all types of companies, public, private, and nonprofit. Some directories provide contact information only and others will include additional company information such as financial information, detailed product information, historical information, and biographical information. Because people move from one organization to another and organizations may be acquired or go out of business, currency and accuracy is often an issue with directories. Many electronic directories offer field searching that is useful for identifying companies that meet certain characteristics.

General Information Sources

Organizations of all types provide news and press releases to announce business activities in a timely manner. Companies will provide up-to-date information about earnings; merger and acquisition activity; personnel changes; new product announcements; public offerings; and litigation. News releases from government agencies provide current information about economic indicators as well as results from economic and demographic surveys.

Newspapers are a major source of business information. Newspapers come in a variety of formats, covering different geographic areas, with broad or narrow focus. Business professionals are familiar with newspapers and look toward newspapers to answer many business questions. Daily newspapers, such as the *Washington Post*, *New York Times*, and the *Los Angeles Times* are staples for business collections. They provide news, company information, industry information, stock quotes and other financial data, biographical information, and statistics. Other newspapers with comprehensive coverage for a more targeted audience include those focused on a specific topic, such as finance, including the *Wall Street Journal*, the *Financial Times*, and the *Investor's Business Daily*. Other news publications focus on a particular industry, such as the *Cheese Market News* or *Autobeat*

Daily. In addition to national or international news coverage, regional newspapers are an often overlooked but important resource for business information. For information on private companies or people, regional newspapers can provide useful information that will not be found elsewhere.

Closely related to newspapers are business broadcasters on television or radio networks that focus on business news and business topics rather than on general news. These broadcasts are often associated with business portals that provide additional business information through blogs and videos. Business broadcasters provide one-stop shopping for business information and are therefore popular among business professionals.

Various types of magazines and journals are used by business professionals. Business magazines focus on business issues and include such titles as *Forbes*, *Fortune*, and *Business Week*. In addition to the news provided, these magazines also issue rankings such as the Forbes 500, the Fortune 500, or the Business Week's Global 1000. Trade journals are geared to a specific industry such as *Ward's AutoWorld*, *Food Processing Magazine*, or *Variety*. Popular magazines, such as *Time* and *Newsweek*, also provide useful business information as part of their more general coverage; discussions on issues and events in these popular magazines impact business. Some journals are peer-reviewed, also called refereed journals, and publish articles that have been assessed by scholars in the field. While these may be used less frequently among business professionals, they provide useful information. Probably the most popular among the academic journals is the *Harvard Business Review*. Other examples include the *American Economic Review*, a peer-reviewed journal generated by an association, and the *Monthly Labor Review*, a peer-reviewed government publication. These different types of magazines and journals are useful for different types of business information needs. Business magazines are important sources of information, providing news, company information, industry information, statistics, and biographical information. Both business and popular magazines may be read as part of environmental scanning programs to keep up on important trends and events. Business-related peer-reviewed journals are probably most useful for learning about management issues and new research findings.

Newsletters are periodic publications, usually relatively short and narrowly focused. Newsletters are typically underutilized in business research yet they may provide detailed and sophisticated coverage of industry news or subjects of interest to companies (e.g., employee benefits) or individuals (e.g., investment).

In general, monographs are not used heavily by business professionals for decision making and daily business operations. However, business books are an important part of management literature. Throughout the years, key figures have emerged in the area of management, a few of

which will be noted here. An early notable publication related to management is Frederick Winslow Taylor's work culminating in the publication of *The Principles of Scientific Management*.[11] Taylor's work focused on the scientific approach of management based on work flow and procedures. Peter Drucker was a prolific author of more than 30 books on management issues and is known as the father of modern management. Among other important contributions to management, Drucker introduced the concepts of decentralized decision making and managing for the future in his book entitled *The Concept of the Corporation* which shaped management practice. He also introduced the concept of the knowledge worker and management by objectives.[12] Michael Porter, one of the leading authorities on competitive strategy, is considered to be the father of the field of modern strategy.[13] He is often included in lists of the world's most influential thinkers on management and competitiveness.

CONCLUSION

The world of business is broad and varied as is the information available to meet the needs of business information users. Business information is needed in almost every aspect of daily life so there is a wide range of business information users. This entry has focused on business professionals as business users; understanding the information behaviors of business professionals is necessary for librarians to provide suitable business information services. Many business activities are complex, information-intensive, and time-sensitive involving a range of types of information packaged in a variety of ways. Interpersonal communication is common among business professionals who often rely on colleagues, customers, and suppliers for information. Government information, statistics, and indicators are key to business professionals in their work. Increasingly more business information is available to the public due to government demands for more transparency in business operations. New information technologies have begun to blur the distinctions between products and services and have increased information overload. Katzer and Fletcher (p. 251)[5] summarize the manager's information dilemma and challenge to librarians serving this user group:

> ... too much information, but not enough of the right information in the right format through the right channel at the right time.

REFERENCES

1. Lavin, M.R. *Business Information: How to Find It, How to Use It*; Oryx Press: Phoenix, AZ, 1992.
2. Davis, S.; Meyer, C. *BLUR: The Speed of Change in the Connected Society*; Addison-Wesley: Reading, MA, 1998.

3. Mintzberg, H. *Mintzberg on Management: Inside Our Strange World of Organizations*; Free Press: New York, 1989.

4. Auster, E.; Chun, W.C. Environmental scanning by CEOs in two Canadian industries. J. Am. Soc. Inform. Sci. **1993**, *44*(4), 194–203.

5. Katzer, J.; Fletcher, P.T. The information environment of managers. Annu. Rev. Inform. Sci. Technol. **1992**, *27*, 227–263.

6. Young, E.C.; Harold, P.W. Information source selection patterns as determined by small business problems. Am. J. Small Bus. **1983**, *7*(4), 42–49.

7. Franklin, S.G.; Goodwin, J.S. Problems of small business and sources of assistance: a survey. J. Small Bus. Manage. **1983**, *21*(2), 5–12.

8. Fann, G.L.; Larry, R.S. The use of information from and about competitors in small business management. Entrepren. Theor. Pract. **1989**, *13*(4), 35 43–44.

9. Zach, L. When is enough enough: modeling the information seeking and stopping behaviors of senior arts administrators. J. Am. Soc. Inform. Sci. Technol. **2005**, *56*(1), 23–35.

10. Tapscott, D.; Ticoll, D. *The Naked Corporation: How the Age of Transparency Will Revolutionize Business*; Free Press: New York, 2003.

11. Taylor, F.W. *The Principles of Scientific Management*; Harper and Brothers: New York, 1911.

12. Drucker, P.F. *The Concept of the Corporation*; John Day Company: New York, 1946.

13. Porter, M.E. *Competitive Strategy: Techniques for Analyzing Industries and Competitor*;, The Free Press: New York, 1998.

BIBLIOGRAPHY

1. Abels, E.G. Klein, D.P. In *Business Information: Needs and Strategies*; Bert, R.B., Ed.; Emerald Academic Press: Bingley, U.K., 2008; Vol. 31 Library and Information Science Series.

2. Esther, P. *History of the Standard Industrial Classification*, Executive Office of the President Office of Statistical Standards, U.S. Bureau of the Budget: Washington, DC. Available at http://www.census.gov/epcd/www/sichist.htm (accessed May 19, 2008).

3. Lavin, M.R. *Business Information: How to Find It, How to Use It*; Oryx Press: Phoenix, AZ, 1992.

4. Moss, R.W. *Strauss's Handbook of Business Information: A Guide for Librarians, Students and Researchers*, 2nd Ed.; Libraries Unlimited: Westport, CT, 2004.

5. Office of the Federal Register. *The Federal Register: March 14, 1936–March 14, 2006*, http://www.archives.gov/federal-register/the-federal-register/history.pdf (accessed May 19, 2008).

6. Pagell, R.A. Halperin, M. *International Business Information: How to Find It, How to Use It*, 2nd Ed. AMACON: New York, 1999.

7. U.S. Census Bureau. *Guide to the Economic Census*, available online at http://www.census.gov/econ/census02/guide/index.html (accessed May 19, 2008).

8. U.S. Census Bureau. *Statistical Abstract of the United States: 2008*, 127th Ed. Washington, DC, 2007; http://www.census.gov/statab/www/ (accessed May 19, 2008).

Business Literature: History [ELIS Classic]

Edwin T. Coman, Jr.
University of California, Riverside, California, U.S.A.

Abstract

One of the objectives of the information disciplines is to study the universe of recorded information—that is, to study the documentary products of domains of human activity—and to come to understand such bodies of literature as social and historical phenomena in and of themselves. Coman reviews the development of business from ancient times to the twentieth century, with particular attention to the history of the literatures of business and foreign trade.

—ELIS Classic, from 1970

INTRODUCTION

When the first caveman became a skilled manufacturer of arrowheads and traded them for a haunch of eohippus, the seeds of business were planted.

As the population increased, agriculture developed, civilization advanced, and villages arose that later became cities. There are no written records of very early business activities but archeologists have uncovered pottery and other artifacts that indicate trading contacts between the various settlements.

HISTORY

The Mediterranean and Asia Minor were not only the cradle of Western civilization, they were also the site at which the foundations of business were laid. The first business records were the cuneiform clay tablets found at the sites of cities in ancient Babylonia and Assyria. These date from about 2500 to 600 B.C. The tablets record contracts of sale, land titles, wills, trading ventures, lawsuits, and receipts and accounts.

The commercial activities of Babylon were codified in the *Code of Hammurabi* (ca. 2200 B.C.). This code of laws covers the relations between landlord and tenant, wages, the handling of herds and flocks, commercial contracts, wills, trading voyages, banking, and the treatment of slaves. Because it is so extensive, it supplies a large amount of information on the very active commercial life of this period.

The other seat of commercial activity was Egypt. Knowledge of manufacturing and commerce comes down to us from tomb and temple inscriptions, papyri, funerary models, and the writings of Greek authors. Most of these sources date from about 4000 B.C. to about A.D. 100. A Middle Kingdom papyrus describes the making of beer, and inscriptions on a Middle Kingdom tomb depict the making of perfume. Spinning and weaving are pictured on the walls of Twelfth-Dynasty tombs at Beni Hansan and El Bersheh.

The Greek historian Herodotus traveled extensively in Egypt during the fifth century B.C. and mentions a number of manufacturing and commercial activities in his *History*. In the first century B.C. two other Greeks—the historian Diodorus in his *Historical Library* and Strabo in his *Geography*—make similar references to the commerce and industries of Egypt.

The Greeks and their predecessors, the Mycenaeans, early turned to the sea for their livelihood. The entire Greek peninsula is intersected by bays and inlets. The mountainous character of the country made communication by land difficult and reduced the available amount of arable land. In addition, numerous islands provided harbors of refuge for small vessels and natural sites for the establishment of trading posts. Therefore, it was most natural for the Greeks to become seafarers and traders.

The Greeks were quick to locate the routes of the caravans that brought silk from China; spices from India; and gold, ivory, and slaves from Africa. Early in their history, they established trading colonies at the Dardanelles and in what are now Syria, Lebanon, and Palestine.

Clay tablets discovered near Pylos, Greece, give some indication of the trade of the Mycenaeans. Although these tablets are largely tax receipts and records of military strength, they do indicate that the Mycenaeans were doing a lively trade around the eastern Mediterranean from 1500 to 1400 B.C. This trade consisted of exchanging olives, olive oil, figs, wine, perfumes, and pottery for grain and other foodstuffs, raw materials, and a few luxury items.

When the Dorians, the progenitors of the modern Greeks, took over about 1400 B.C., much of the manufacture and trade of fine pottery lapsed because the Dorians were a vigorous but ruder people, less civilized and less interested in artistic goods.

While the Greek economy was being rebuilt, the Phoenicians began to emerge as a commercial force. They probably were the branch of the Canaanites mentioned

British-Careers

Encyclopedia of Library and Information Sciences, Fourth Edition DOI: 10.1081/E-ELIS4-120008971

repeatedly in the *Old Testament*; a seafaring people, they were centered around Tyre, Sidon, and Byblos in what is now Syria and Lebanon. Later, the Carthaginian branch extended its control to Sicily, Spain, and Cyprus—in fact, to the entire western Mediterranean. The seat of its power was Carthage, near the present city of Tunis, in Algeria. Despite various wars and invasions, the commercial and business operations of the Phoenicians flourished from about 1400 B.C. up to 146 B.C., when the Romans razed Carthage completely during the Third Punic War.

The first accounts of the activities of the Phoenicians occur in the Egyptian papyrus *Anastosi I* and the Amarana tablets from 1500 to 1400 B.C. The *Old Testament* describes the trade between Hiram I, King of Tyre, and King Solomon (1 Kings 9: 11; Ezek. 27), and Josephus mentions this same trade in his *Antiquities of the Jews* (xiv, 10, 6, and elsewhere). Herodotus makes numerous references to the seafaring and trading abilities of the Phoenicians in his *History*. Homer, in the *Iliad* (xxii, 744), speaks of the Phoenicians trading in Greek waters, and Xenophon, in his *Oeconomicus*, describes the skill of the Phoenicians as sailors and the high quality of their ships. Plato, in his *Republic* (lv, 435E), refers to the avariciousness of the Phoenicians and the hard bargains they drove.

The inhabitants of the towns on the Phoenician coast were occupied not only with trade but also with manufacturing. They had a monopoly on the production of a purple dye obtained from the gastropod *Murex*, and the manufacture of textiles and dyeing were important industries (Homer, *Iliad*, vi, 289). Metalwork in silver, gold, and electrum was also important (Homer, *Odyssey*, lv, 615 et seq.; xv, 458 et seq.). In addition, the Phoenicians were noted for their glass, and they mined iron and copper on the island of Cyprus.

During the Carthaginian period, which was the height of this civilization, the Phoenicians developed silver and copper mines in Spain and traded with the Britons for tin. The Greek writer Polemo records a special treaty regarding woven fabrics produced in Carthage that were a recognized luxury of the ancient world. In Sicily, Italy, and Greece, the Carthaginians sold black slaves, ivory, metals, precious stones, and all the products of central Africa that came to them by caravan.

The Phoenicians were the preeminent businessmen of the period from 1000 to 146 B.C. If they did not invent the alphabet and weights and measures, they improved these necessary adjuncts to business. They were the first to develop a system of accounts. Their skill and daring as seamen enabled them to be the first to venture into the Atlantic from the Mediterranean. In so doing they developed a regular trade with England and explored portions of the coast of western Africa. For a considerable length of time the Carthaginians completely controlled the trade of the western Mediterranean.

In addition to the works already cited, there are references to the business and commercial affairs of the Phoenicians and Carthaginians in the writings of Polybius, Diodorus, Livy, Appian, Justin, and Strabo.

As the Greek civilization revived from the shock of the Dorian invasion, a lively trade began in the eastern Mediterranean that competed strongly with the Phoenicians. Athens became the center of commercial activity. In addition to the earlier exports of wine, olives, olive oil, and figs, the Athenians began to export objects of art. In addition, the Laurium mines supplied them with silver as a medium of exchange. Xenophon in his *Oeconomicus* throws much light on the commerce and industry of Athens. In one passage he states that the Assembly at Athens was composed of fullers, shoemakers, blacksmiths, farmers, merchants, and shopkeepers. He also refers to the litigious tendencies of the Greeks. There is also much material on economic activities in Plato's *Republic* and in Aristotle's *Politics*, his *Constitution of Athens* and his *Nicomachean Ethics*. Aristophanes in his plays makes caustic references to the workmen and commercial customs of the Athenians. Strabo also refers to Greek business.

The conquest of Greece, first by the Macedonians and then by the Romans, shattered Greek trade and practically destroyed the industry of the country. Greece was stripped of its artistic treasures, shipping came under the control of the Romans who took over the trade from the Far East, and pirates from Crete practically destroyed the local shipping.

The Romans were interested in conquest, administration, and public works. As a matter of fact, they looked down on trade and industry. They left trade in the hands of the Greeks and other foreigners, and industry was largely performed by slave labor under the direction of foreign overseers. Cicero in his letters (*Familares; ad Atticum; ad Quintum*; and *ad Brutum*) throws some light on commercial and manufacturing activities, of which he speaks very grudgingly. Livy's *History of Rome*, Strabo's *Geography*, and Pliny's *Natural History* all contain references to business and commerce.

Agriculture did interest the Romans and was regarded as an honorable occupation. After the very early period, farming was based on the operation of large estates (latifundia system) worked by slaves under the direction of overseers. In order to keep track of this business, the Romans worked out a fairly complete accounting system. A highly romantic account of Roman agriculture is found in the *Eclogues* and *Georgics* of Vergil.

The break-up of the Roman Empire destroyed the peace and communications in the Mediterranean basin. This curtailed the large trade with the Far East. The importation of spices, perfumes, silks, ivory, precious stones, and pearls from India, Persia, and China amounted to 100 million sesterces ($4 million) annually in the time of Pliny.

As the barbarian invaders took over, commercial activities were gradually acquired by the Italians, Provençals, and Catalans. The most active in this trade were the

Venetians, the Genoese, and the Florentines. It was largely a revival of the old trade in spices and luxury goods of the Far East. The development and extent of this trade is described in detail in W. Heyd's *Histoire du Commerce du Levant au Moyen Age* (Leipzig, 1885).

There are many references to the trade of these cities in the treaties and other records in their archives. Venice and Florence became very wealthy centers of manufacturing, trade, and shipping. The Venetians profited immensely from the Crusades; they furnished supplies, funds, and transportation to the Crusaders in return for trading rights in the conquered territories as well as cash payment for their services.

As a result of these activities, the Venetians, Florentines, and Genoese accumulated large financial resources and became the bankers of Europe during the thirteenth, fourteenth, and fifteenth centuries.

The invention of printing from movable type around 1450 made it possible to make information more readily available. The first book dealing with business was that by Lucca Paccioli, *Summa de Arithmetica, Geometrica et Proportionalta* (Venice, 1494), which had a section on double-entry bookkeeping. This method of accounting was soon dubbed the Italian system and spread very rapidly. Schools were established in Venice and merchants sent their sons there to be taught this method. Soon rival schools were set up in Antwerp and Augsburg, and later in London and Paris.

In the sixteenth and seventeenth centuries, commerce and industry arose in France, England, Holland, and Germany. This was the period of the demise of the feudal system, the coming to power of the merchants, and the freeing of the artisans from the control of their feudal masters.

An early work that describes commercial developments in Antwerp and Amsterdam is Ludovico Gucciardini's *Descrezzione de Paesi Basse* (Antwerp, 1567). Sir Walter Raleigh wrote *Observations Touching on Trade and Commerce with Holland* about 1616. A work by John de Witt, which probably was the English translation of a French book by M. Delacourt, was *True Interest of Holland*, which appeared in 1667. The *Richesse de la Holland* was published anonymously in Amsterdam in 1678.

The records of treaties, suits, and business alliances of the powerful Fugger family provide insight into the operations of the Hanseatic League in the Baltic. A good résumé of this area and period can be found in Richard Ehrenberg's *Capital and Finance in the Age of the Renaissance: A Study of the Fuggers and their Connections* (New York, 1928?).

The stirrings of commercial life were very active in England at this time, and much light is thrown on commerce and industry through the various petitions, suits, and requests for charters that appear in *British State Papers—Domestic* and the records of the Privy Council, of the City of London, and of the various guilds.

Under the stimulus of developing trade, British merchants strove to improve and better control their operations. They were most interested in books that would be helpful to them. The earliest of these business books was Hugh Oldcastle's *A Profitable Treatyce: called the Instrument or Boke to learn to know the good order of kepyng of the famouse reconynge called in Latyn Dare and Habere and in Englysche, Debtor and Creditor* (London, 1543). This was followed at a considerably later date by Richard Dafforne's *The Merchants Mirror: or directions for the perfect ordering and keeping of his accounts* (London, 1635). This book went through four printings and encouraged Dafforne to publish *The Apprentice's time enterteiner accomptantly: or a methodical means to obtain the equisite art of accomptantship* (London, 1640). There is no record as to how this book was received by the apprentices but it did go through three editions.

The honors for popularity of books on accounting goes to George Fisher's *The Instructor: or young man's best companion. Containing... merchant's accompts, and a short and easy method of shop and book keeping* (Dublin, 1736). It took 31 printings to satisfy the demand.

Books on specialized accounting soon began to appear; the earliest of these was John Brown's *The Merchant's Avizo, very necessaire for their sonnes and servants when they first send them ... to Spaine and Portugal* (London, 1589). Another popular book was one by Richard Hayes entitled *The Ship and Supercargo Bookkeeper* (London, 1731). There were other books on accounting for trade with British America, India, and for ships, banks, and estates. B. S. Yaney, et al., in *Accounting in England and Scotland: 1543 to 1800* (London, 1963) give a very full history of accounting and its literature.

The late seventeenth century and all of the eighteenth century was a period of ferment. Not only were commercial rivalries bitter between England, France, Holland, and Spain, but new methods and new products were coming onto the market. This caused serious dislocation of earlier industries and was also reflected in national policies. In England, an acrimonious controversy went on concerning the extent of governmental control, whether free trade or protection was best for the country, and what was best for the national interest. This debate went on to a lesser extent in other countries, but in France it was suppressed under the mercantilistic policies of Colbert under which the state closely controlled commercial and industrial activities. This period is described in part by Germain Martin's *La Grande Industrie en France sous Règne de Louis xiv* (Paris, 1898). A very comprehensive account of the state of industry and technology appearing in the latter part of the eighteenth century was the *Encyclopédie ou Dictionnaire des Sciences, des Arts et des Métiers* by Denis Diderot (Paris, 1751–1772; 17 vols.). An earlier French publication, *Dictionnaire Universal de Commerce* by Jules Savary des Brulons (Paris, 1723–1730), was of more interest to merchants. Malachy Postlewayt published his *Universal Dictionary of Trade and Commerce* in London in 1751–1755; this was largely a translation of des Brulons's *Dictionnaire*.

There was a groping toward the development of economic theory from 1650 to 1750, especially in England. Publications, particularly tracts, appeared in increasing numbers. One of the earliest of these was *A Discourse of the Common Weal of Thys Realm of England* by Richard Hales (London, 1549). In 1581 Hales published in Cambridge an anonymous tract, *A Compendious or briefe examination of certayne ordinary complaints of diuerse of our countrymen in these our days: which although they are somewhat uniust & friuolous, yet are they all by way of dialogues thoroughly debated and discussed by W.S. Gentleman.* Hales held forth on the causes of poverty and the high cost of living, which he attributed to foreign trade and the debasement of the currency. He advocated restriction of the exportation of raw materials and of the importation of trifles (nonessential goods).

Gerald de Milynes (fl. 1568–1641), the assayer at the Royal Mint, was, as might be expected, very much concerned with monetary questions. His three books, *A Treatise on the Canker of England's Commonwealth* (London, 1601), *England's View in the Unmasking of Two Paradoxes* (London, 1603), and *Consuetedo* (London; several editions, the last in 1686), all dealt with monetary problems.

Edward Misselden (fl. 1608–1654) did battle with Hales, poking holes in his reasoning and conclusions in *Free Trade or the Means to make Trade Flourish* (London, 1621) and *Circle of Commerce* (London, 1629).

Thomas Mun (fl. 1571–1641) in his writings was an unconscious influence on Adam Smith in his works on foreign trade. These were *A Discourse on Trade from England unto East-Indies* (London, 1621) and *England's Treasure: by Forraign Trade: or the Ballance of our Forraign Trade is the Rule of our Treasure* (London, 1664), published after his death.

Nehemiah Grew (1641–1712), a British physician, pointed the way to the concept of utilitarianism and brought in ideas on consumption and distribution—in other words, he was the first to recognize the place of the consumer in economics and business. Grew's contribution was a tract, *The Meanes of a Most Ample Encrease of the Wealth and Strength of England in a Few Years Humbly represented to her Majestie in the Fifth Year of Her Reign* (London, 1707).

A highly acrimonious war was waged for a short period by two publications. The *Mercator: or Commerce Retrieved, Being Consideration on the State of British Trade* (London, 1713–1714) was one. This was a free trade sheet issued twice a week for a penny. Daniel Defoe was the editor. The opposition protectionist sheet, edited by Charles King, was *British Merchant: or Commerce Preserved; In Answer to the Mercator or Commerce Retrieved* (London, 1713–1714).

The English philosopher, Thomas Hobbes (1588–1679), held the theory that all human conduct is affected by environment, especially the drive for self-preservation. His books, *Leviathan* (London, 1651) and *De Cive* (London, 1651), influenced Mill and Bentham. The Frenchman François Quesnay (1694–1774) put forth the theory that agriculture was preeminent and land was the only source of wealth. His book, *Tableau Économique* (Versailles, 1758), epitomizes the theories of the Physiocrats. Bernard de Mandeville in his *Fable of the Bees: or Private Vices Public Benefits* (London, 1714–1729) espoused for the first time the notion that the self-interest of the individual is important to the economy.

Jeremy Bentham (1748–1832) was a leading exponent of utilitarianism, i.e., that happiness is the end of human existence. He had an influence on Mill, Adam Smith, and others. Sir James Steuart (1712–1780) was one of the first to begin to draw these strands of economic thinking together and the first to use the title "political economy" in his *An Inquiry into the Principles of Political Oeconomy: being an essay on the science of Domestic Policy in Free Nations in which are particularly Population, Trade, Industry, Money, Coin, Interest, Circulation, Banks, Exchange, Public Credit and Taxes* (London, 1767; 2 vols.). David Hume (1711–1776) added the theory of utilitarianism in international trade to the growing body of works on this subject. Mill and Ricardo were both indebted to Hume.

The time was ripe in the middle of the eighteenth century to weave all these strands of economic thought into a composite whole. Adam Smith (1723–1790) was familiar with these earlier writings and many of these men were his friends. His *An Inquiry into the Wealth of Nations and Causes of the Wealth of Nations* (London, 1776) laid the cornerstone of modern economics and became the bible of the supporters of *laissez faire*.

During this period, there were a number of articles on economics, trade, and commerce appearing in the *Philosophical Transactions of the Royal Society* (1665–). A publication of interest to businessmen was *The Annual Register* (London, 1758–), which carried much information on the political situation and the state of trade and industry in foreign countries. The Society for the Encouragement of Arts, Manufactures, and Commerce began publishing its *Transactions* in 1761. Other information of interest to businessmen was published in the *Journals of the House of Commons*, the *Journals of the House of Lords*, and the reports of committees made in response to petitions of businessmen or labor groups.

The forerunner of the British Board of Trade was formed as a committee of the Privy Council by Oliver Cromwell in 1660. This was designated as the Commission of Trade and Plantations. After being abolished in 1679, it was revived by William III in 1679 and was made a permanent body in 1786. Almost from the beginning, the Board of Trade compiled and published abstracts, memoranda, tables, and charts relating to trade and industrial conditions in the United Kingdom, its colonies, and foreign countries. It now supervises trade accounts and provides monthly and annual data on shipping and navigation; labor, cotton, and emigration statistics; and foreign and colonial customs tariffs. The Board of Trade also provides individual businessmen with accurate information, and on occasion, confidential data. The *Board of Trade Journal* (1886–) provides businessmen with current trade information.

The first businessman to write on business and economic subjects was David Ricardo (1772–1823). After making a fortune in the stock market, Ricardo retired and devoted himself to studying the economy. He studied the actual operations of business and the economy and based his observations on firsthand knowledge. He established the principle that the amount of labor expended on a product determines its worth. His major publication was *Principles of Political Economy* (London, 1817).

One of the strongest advocates of the theory of utilitarianism was John Stuart Mill (1806–1873) of the Manchester School. His views were set forth in his books, *Essays on Some Unsettled Questions of Political Economy* (London, 1843) and *Principles of Political Economy* (London, 1848; 2 vols.).

The Frenchman Jean-Baptiste Say (1767–1832) must be included here, for in his *Cours complet d'économie politique pratique* (Paris, 1828–1829), Say was the first economist to stress the importance of the entrepreneur.

The Englishman Alfred Marshall (1842–1924) was the foremost economist of his time and was largely responsible for the development of modern economics. His *Principles of Economics* (London, 1890) went through many editions, and his *Industry and Trade* (London, 1919) and *Money, Credit and Commerce* (London, 1923) are highly respected works.

As the Industrial Revolution reached full flower in the 1880s, there was a divergence between the economic theorists and the businessmen. This was partly because the businessmen were busy transforming the economy and felt that many of the older precepts were outmoded, and partly because business was developing its own literature. Businessmen were relatively unaffected by the writings of the later economists, with two notable exceptions: Karl Marx's *Das Kapital* (Hamburg, 1867) and John Maynard Keynes's *The General Theory of Employment, Interest and Money* (London, 1936). The former laid the groundwork for the formation of labor unions and emphasized the class struggle; the latter became the handbook of the inflationists.

To return for a moment to the earlier period, there are some rather interesting accounts by travelers of economic conditions and commerce in England. Among these are Daniel Defoe's *A Tour through the Whole Island of Britain, divided into circuits or Journeys* (London, 1724–1727; 3 vols.); J. Campbell's *A Political Survey of Britain,*

being a series of reflections on the Situation, Lands, Inhabitants and Commerce of this Island (London, 1774; 4 vols.). Two foreign visitors recorded their observations in books. These were Erik T. Svedenstjerne, *Resa igenom en del af England och Scotland aren 1802 och 1803* (Stockholm, 1804) and F.A. Wendeborn, *Beitrage zur Kenntniss von Grossbritannien* (Lemgo, 1780). John Smith in his *Chronicum Rusticum: or Memoirs of Wool, Woollen Manufacture and Trade* (London, 1747?) reprinted a number of rare pamphlets on this subject.

What was perhaps the first biography of a businessman appeared in 1597. This was Thomas Deloney's *The Story of John Winchcombe, commonly called Jack of Newberry* (London, 1597). This biography of a woolen manufacturer and cloth merchant ran through many printings as it was extremely popular.

Up to this point, this entry has been concerned with developments in England and on the Continent, with no references to those in the United States. During the period 1750–1870 there was little indication in the United States of the wealth of information for the businessman that was to become available during the twentieth century.

Up until the Civil War, ships' captains and newspapers were the major sources of information aside from correspondence from agents and other businessmen. When they returned from voyages to all parts of the world, the captains had firsthand information on markets and economic and political conditions. The waterfront coffeehouses of Boston, New York, Baltimore, and Charleston were much frequented by businessmen and newspapermen. News from the interior of the United States was scanty and unreliable.

During the period 1700–1860 the businessman, depending on his locality, read the following newspapers: Boston had the *Boston Gazette* (1704–1754), a new *Boston Gazette* (1755–1836), and the *Evening Transcript* (1836–1960?); in Connecticut there was the *Hartford Courant* (1764–); Pennsylvania was served by the *United States Gazette* (1789–1847), which was absorbed by the *North American* (1830–); New York City was well supplied with newspapers, which included the *New York Gazette* (1725–1840), the *New York Herald* (1835–1964), the *New York Mercury* (1752–1783), the *New York Tribune* (1841–1964), and *The New York Times* (1851–). Perhaps the first newspaper directed to the businessman was the *Journal of Commerce and Commercial Bulletin*, founded in 1827 as the *Journal of Commerce* and in 1893 consolidated with the *Commercial Bulletin* (1865–).

In 1811 an attempt was made by Hezebiah Niles to provide a national news periodical. The *Niles Weekly Register* survived until 1849. Niles, its editor, strove for accuracy and gleaned his news from the captains of ships and merchants patronizing the Baltimore Coffee-House and the Merchants' Coffee House, also in Baltimore. He also reprinted articles from other newspapers; among those from which he drew material were the

Baltimore American and the *Commercial Advertiser* of New York. Niles published the full text of major laws, regulations, and important speeches, and he interviewed travelers. While his primary aim was to provide a weekly account of news and politics, the *Register* contained much commercial information of interest to the businessman. As an example, Niles warned businessmen that upon the conclusion of the War of 1812, their prosperity would collapse in the face of renewed competition from the British.

In the 1830s, 1840s, and 1850s, the United States was embarking on the Industrial Revolution. Factories were springing up all over the country, especially in New England and the Middle Atlantic states, and railroads were reaching out to the South and Middle West and later to the Far West.

Many issues of railroad stocks and bonds were being sold, especially in the 1870s, 1880s, and 1890s. It was most difficult to obtain information on the operations and finances of these companies. John Moody in 1870 undertook to publish a compilation of data on railroads; this appeared as *Moody's Manual of Railroad Securities*. For many years this was the only volume published each year. Eventually other volumes were added. Today, *Moody's Manual on Railroads* (renamed *Moody's Manual on Transportation* to include air and steamship lines, buses, and trucking); *Moody's Manual on Industrials; Moody's Manual on Public Utilities; Moody's Manual on Banks, Insurance Companies, Real Estate and Investment Trust;* and *Moody's Manual on Governments and Municipals* are published annually. Moody was the first person to supply businessmen with accurate current information.

Credit was another troublesome factor to the businessman. This was particularly the case with wholesalers dealing with country storekeepers in the South and Middle West. In 1841, a young man in New York named Richard D. Dun undertook to supply merchants with credit information. In a large number of the smaller cities and towns, a merchant or banker whom he could trust would give him reliable information on the credit standing of individuals in their locality. R.G Dun and Company soon had competition from the Bradstreet Company, which was formed by John M. Bradstreet in Cincinnati, Ohio, in 1849. At first, both these firms supplied credit information on a fee basis for each individual inquiry. This was later supplemented by *Dun's Reference Book*, which became, on the merger of the two firms in 1933, *Dun & Bradstreet Reference Book*. This volume supplies information on the type of business and capital resources of the firm or individual and gives some indication as to how promptly bills are paid. Originally issued annually, it is now kept up-to-date with bimonthly supplements. Dun & Bradstreet will, upon request, issue individual credit reports. To provide general information on credit, business conditions, and failures, Dun started *Dun's Review* in 1893. This publication has continued as *Dun's Review and Modern Industry*.

An earlier publication that is most useful to businessmen is the *Commercial and Financial Chronicle* (New York, 1865–). This periodical provides very complete stock, bond, and commodity quotations along with articles on national finance, business conditions, and foreign trade.

The businessman has always tried to obtain up-to-date information on matters affecting his operations. This includes data on economic conditions, business trends, new improvements in methods and technology, discoveries of natural resources, and recently enacted laws and regulations. In short, statistics are highly important to his business.

The British government was quick to recognize this need of the businessman for accurate information with its Board of Trade publications, including the *Board of Trade Journal*. Its Department of Overseas Trade for many years published detailed studies of various foreign countries and colonies. These publications are supplemented by the *Royal Statistical Society Journal* (London, 1838–); the *Economic Journal* (London, 1891–) of the Royal Economic Society; and the London School of Economics and Political Science's *Economica* (London, 1921–). Although much of the material in these latter publications is concerned with methodology and theory, they do contain a number of articles of interest to the businessman.

Publications in France and Germany of a similar nature are *Journal des Économistes* (Paris, 1841–1940); *Revue Économique* (Paris, 1950–); and *Finanz-Archiv* (Stuttgart, Tubingen, 1884–1943, 1948–).

In contrast to the interest of the British government in providing information for businessmen, the U.S. government, up until 1880, provided little assistance or information to business. The first census, in 1790, was little more than an enumeration of the inhabitants. It was not until the census of 1880 that useful data on business, manufacture, finance, agriculture, and foreign commerce became available. The utility of this publication was reduced by the delay in its issuance; the last volumes did not appear until almost 10 years after the census was taken. There were earlier publications on shipping and foreign trade, but businessmen needed the bringing together of data in the decennial censuses. The situation was much improved when the Bureau of the Census began to publish *The Statistical Abstract of the United States* (1878–). This brings together a large number of statistics from governmental and other sources.

The situation was improved with the establishment of the Department of Commerce and Labor in 1903. However, the Department of Commerce did not really begin to be effective until it became a separate organization with a secretary of cabinet rank in 1913. Publications on foreign and domestic commerce were issued at this time and a modest quantity of other information was supplied to businessmen.

The Department of Commerce truly became a help to business when it was reorganized by Herbert Hoover, who was secretary of commerce from 1920 to 1927. A large number of reports, special studies, and periodicals began to be issued from that time on. Most notable of these is the *Survey of Current Business* (Washington, 1921–). This periodical with its *Weekly Supplement* supplies statistics from a wide variety of sources on every aspect of the economy, with pertinent comment.

With the establishment of the Federal Reserve System in 1915, much greater stability was given to the American banking structure. In that year the *Federal Reserve Bulletin* began to be issued and at a later date, the *Chartbook on Financial and Business Statistics* was issued on an annual basis. Another useful source for statistics is *Business Statistics*, published since the late 1940s by the U.S. Bureau of Foreign and Domestic Commerce. The Departments of Commerce and Labor and their various bureaus issue catalogs and checklists of their publications. The bulk of these are listed in the *United States Government Publications Monthly Catalog* of the U.S. Superintendent of Documents (Washington, D.C., 1895–).

Similar compilations of statistics are published in other countries, notably *Economic Trends* and the *Monthly Digest of Statistics* (Great Britain); the *Canadian Statistical Review; Economic Statistics* and the *Oriental Economist* (Japan); the *Monthly Abstract of Statistics* (India); and *Boletim Estastistico* (Rio de Janeiro, Brazil).

The advent of international organizations after World Wars I and II has resulted in the compilation of data on a worldwide basis. The League of Nations published annually the *Statistical Yearbook of the League of Nations* (Geneva, 1927–1945) and the *World Economic Survey* (Geneva, 1927–1945). These publications have been continued by the United Nations in its *Statistical Yearbook* (New York, 1948–), the *United Nations Monthly Bulletin of Statistics*, the *United Nations Economic Bulletin for Asia and the Far East*, and the *United Nations Economic Bulletin for Europe*; the two last-named periodicals appear quarterly.

Nongovernmental compilations of statistics and data on business are *Statesman's Year-book* (London, 1863–); *Whitaker's Almanac* (London, 1868–); *World Almanac and Book of Facts* (New York, 1885–); Standard and Poor's Corporation's *Trade and Securities Statistics* (New York, 1928–); and the National Industrial Conference Board's *The Economic Almanac* (New York, 1939–).

All the general periodicals directed toward the business reader, as well as trade publications, carry statistical and other data. The larger newspapers have much valuable information, especially the *London Times, The New York Times* and the *Wall Street Journal*. These three have the additional advantage of being thoroughly indexed.

The material contained in the publications listed dealing with statistics is of prime importance to businessmen and therefore has been discussed at some length.

Trade associations were formed in Great Britain early in the nineteenth century; these included associations of the cotton spinners and the woolen manufacturers. Trade associations did not appear in the United States until 1870. In 1896, the National Association of Manufacturers was established by men in manufacturing and business. The objectives of this association are 1) promotion of the industrial interests of the United States; 2) fostering of the domestic and foreign commerce of the United States; 3) the betterment of relations between employers and employees; 4) protection of the individual rights of employers and employees; 5) the dissemination of information among the public with respect to the principles of individual liberty and the ownership of property; and 6) the support of legislation to these ends.

The National Association of Manufacturers and the individual trade associations were originally concerned with protecting their interests through lobbying for legislation favorable to them, particularly protective tariffs, and defeating legislation regarded as unfavorable to business. The number of trade associations has increased enormously and there are now not only trade associations for every industry but also for subdivisions of that industry and regional associations. As an example, there are 10 associations for the leather industry, ranging from the Tanners Council of America and the National Hide Association to the Last Manufacturers Association and the National Association of Slipper and Playshoe Manufacturers. Many of these associations put out publications containing news of the trade, statistics, remarks on the general state of the economy, and references to books and articles of interest to their members. The larger associations maintain research staffs that can help members with problems. Among these are the American Iron and Steel Institute, the Copper and Brass Research Organization, the American Petroleum Institute, and the National Association of Wool Manufacturers.

The chamber of commerce has flourished in the United States as nowhere else in the world. No town is too small to have this organization beating the drums on its special economic, commercial, and cultural advantages. These developed independently until 1912 when the Chamber of Commerce of the United States of America was formed with headquarters in Washington, D.C. Its membership was made up of local chambers of commerce and some trade associations. The constitution of the Chamber of Commerce of the United States of America proclaims its purposes to be (1) to study national problems and define current questions; (2) to ascertain the views of business and to determine what business feels is the best solution of national problems; and (3) to voice and explain the views of business to government and the public. The chamber is subdivided into a number of departments that do research and publish many useful reports on special subjects and national issues. These departments are agriculture, construction and civil development, domestic distribution, finance, foreign commerce, foreign policy, insurance, manufacture, transportation and communication, economic research, education, and labor relations. The chamber publishes *Nation's Business* (Washington, D.C., 1912–).

In addition to the official publications of trade associations, there are magazines put out by commercial publishers as organs of general interest to the businessman or for specific industries. In the main, these periodicals have a broader scope and more complete coverage than do the publications of the trade associations. A very useful feature is their annual survey numbers, which sum up the state of the industry. The McGraw-Hill Book Company publishes the largest number of these trade magazines; one of general interest is *Business Week* (New York, 1929–). Other periodicals of general interest are *Nation's Business*, already mentioned, and the *U.S. News and World Report* (Washington, D.C., 1936–).

Many large companies publish magazines for their stockholders which supply information both on the operations of the company and on the industry as a whole. The larger brokerage firms put out large numbers of financial studies of firms and industries.

Schools of Business

As businesses grew in size and scope, thoughtful businessmen recognized the need for objective, scientific study of business and its operations. This thinking led to the establishment of schools of business in colleges and universities. The first of these was the Wharton School of Business at the University of Pennsylvania, which received its first students in 1881. This was followed by schools of business at the University of California and the University of Chicago in 1898. The Amos Tuck School of Administration and Finance was established at Dartmouth College in 1900, as were schools at the University of Wisconsin and New York University. The Harvard Graduate School of Business Administration came into being in 1908 and was the first to offer entirely graduate instruction. The Stanford Graduate School of Business opened its doors in 1924. Now there is scarcely a large educational institution that does not have a school of business.

This marriage between business and education has proved to be a happy one. It has provided the means to do competent and scientific research on the operations and methods of business without the pressure of day-to-day business operations and has trained a body of excellent business executives. Splendid reports, studies, and books are constantly being published by the faculties of schools of business; these are too numerous to be included. The librarians connected with the schools of business have been of immense assistance to business with their compilation and publication of bibliographies and checklists on all phases of business. A few of the excellent periodical publications of schools of business are the *Harvard Business Review* (Boston, 1922–); the *Journal of Business of the University of Chicago* (Chicago, 1928–); the *London and Cambridge Economic Services Report on Current*

Economic Conditions (London, 1923–1951); and the *Michigan Business Review* (Ann Arbor, Mich., 1949–).

Two institutions do business research independently of the academic world. These are the National Industrial Conference Board in New York and The Brookings Institution of Washington, D.C. The former is supported by firms and businessmen and supplies a large number of special reports, largely drawn from the experience of its members, to its membership. It publishes two periodicals, *The Conference Board Record* (New York, 1944–) and *The Conference Board Management Record* (New York, 1939–). The board also prepares the data for the *Economic Almanac* published by *Newsweek*. The Brookings Institution is engaged in long-range studies in depth dealing with the economy as a whole and also socioeconomic problems. The results of these studies appear at irregular intervals in the form of books. A foundation that is also on occasion concerned with research in business, with a slightly antibusiness bias, is the Twentieth Century Fund of New York.

Study of Scientific Management

After World War I, business increased greatly in scope and complexity and has continued to do so to the present day. From about 1915 on, there has been the tendency to break down business into its components, that is, finance, production, marketing, advertising and sales management, management, and personnel relations. This trend was given impetus by the studies in scientific management conducted in Europe and the United States by Oliver Sheldon (England), Henri Fayol (France), Frederick Winslow Taylor (United States), Henry Lawrence Gantt (United States), Frank B. and Lilian Gilbreth (United States), Mary Parker Follett (United States), and L. Urwick (England). This pioneering work has been continued in the United States by Wallace Clark, Harry Arthur Hopf, Ralph C. Davis, Leon P. Alford, Dexter S. Kimball, and Paul E. Holden.

Since the work of these men has been an important influence on business and industry throughout the world, it is only proper to list some of their books in this area. Frederick Winslow Taylor's books, *Shop Management* (New York, 1903) and *Principles of Scientific Management* (New York, 1911), started the trend to scientific management. Other important books in this field are Henry Lawrence Gantt's *Work, Wages and Profits* (New York, 1910) and his *Industrial Leadership* (New Haven, 1915); Harrington Emerson's *Twelve Principles of Efficiency* (New York, 1911); Frank B. Gilbreth's *Primer of Scientific Management* (New York, 1911) and his and Lilian M. Gilbreth's *Applied Motion Study* (New York, 1917). The work of these early pioneers is summarized in Frank Barkley Copley's *Frederick W. Taylor, Father of Scientific Management* (New York, 1923) and Leon P. Alford's *Henry Lawrence Gantt, Leader in Industry*

(New York, 1934). Oliver Sheldon in his *Philosophy of Management* (London, 1923) brought out the British aspects of scientific management, as did Henri Fayol in his *Industrial and General Administration* (Paris, 1920) for the French.

Later works in this field are the following: Paul E. Holden, Lounsbury Fish, and Hubert L. Smith, *Top Management, Organization and Control* (Stanford, California, 1948); A. Filippetti, *Industrial Management in Transition* (Homewood, Illinois, 1953); L. Urwick, ed., *The Golden Books of Management, An Historical Account of the Life and Work of Seventy Pioneers* (London, 1956); A. Rathe, ed., *Gantt on Management: Guidelines for Today's Executive* (New York, 1961). Other works in this important field are the following: W. B. Cornell, *Organization and Management in Industry and Business*, 4th ed. (New York, 1958); R. C. Davis, *Industrial Organization and Management* (New York, 1957); W. R. Spriegel and E. C. Davis, *Principles of Business Organization and Operation*, 3rd ed. (Englewood Cliffs, New Jersey, 1960). These books by no means exhaust the list of excellent books in this field and its subdivisions.

It was only natural that persons working in specific phases of business should join together in associations. Mechanical and civil engineers formed professional organizations at an early date. The first business professional group to organize were the accountants. The American Institute of Accountants was formed in 1887 and is now known as the American Institute of Certified Public Accountants. This American society was preceded by the Society of Accountants in Edinburgh (1854), similar societies in Glasgow (1855) and Aberdeen (1867), and the Institute of Accountants, formed in London in 1870. Scientific management was represented by the Taylor Society (1914), which in 1934 combined with the American Management Association (1923). The National Office Management Association (1919) has now changed its name to Administrative Management Society.

Other associations interested in improving business operations are the American Accounting Association (1916), the American Marketing Association (1915), and the National Association of Accountants (1919).

All these organizations publish periodicals and research collegiate schools of business, provide a most important body of business literature.

This treatment of business literature is already overlong; hence in the following list, references to the literature in specific fields of business are limited to those works that supply the greatest amount of information in the smallest compass.

Banking, Finance, and Investments

American Bankers Association, *Present Day Banking*, New York, 1954–1958; Vol. 6.

Bogen, J.I., et al. Eds. *Financial Handbook*, 3rd Ed.; New York, 1957.

Garcia, F.L.; Munn, G.G. *Encyclopedia of Banking and Finance*, 6th Ed.; New York, 1962.

Gerstenberg, C.W. *Financial Organization and Management*, 4th Ed.; Englewood Cliffs, NJ, 1959.

Jordon, D.F.; Dougall, H.E. *Investments*, 7th Ed.; Englewood Cliffs, NJ, 1960.

Prentice-Hall History of Money and Banking, Englewood Cliffs, NJ, 1962–1963; Vol. 18.

Real Estate and Insurance

Best's Insurance Reports (*Fire and Marine*) and (*Casualty and Surety*); New York, 1899–.

Dodge Statistical Service; New York, 1925?– (real estate).

Huebner, S.S.; Black, K. *Property Insurance*, 4th Ed.; New York, 1957.

Husband, W.H.; Anderson, F.R. *Real Estate*, 3rd Ed.; Homewood, IL, 1960.

The Insurance Almanac; New York, 1912–.

North, N.L.; Ring, A.A. *Real Estate Principles and Practices*, 5th Ed.; Englewood Cliffs, NJ, 1960.

Riegel, R.; Miller, J.S. *Insurance Principles and Practices*, 4th Ed.; Englewood Cliffs, NJ, 1959.

Spectator Insurance Yearbooks, Fire and Marine volume; Philadelphia, PA, 1928–.

Accounting

Barker, M., Ed., *Handbook of Modern Accounting Theory*; Englewood Cliffs, NJ, 1955.

Dickey, R.L. ed.; *Accountant's Cost Handbook*, 2nd Ed.; New York, 1960.

Lasser, J. K. *Handbook of Accounting Methods*, 2nd Ed.; New York, 1954.

Personnel and Industrial Relations

Aspley, J.C., Ed. *The Handbook of Employee Relations*; Chicago, IL, 1957.

Heyel, C., Ed. *The Foreman's Handbook*, 3rd Ed.; New York, 1955.

Yoder, D. et al., *Handbook of Personnel Management and Labor Relations*; New York, 1958.

Marketing, Sales Management, and Advertising

Aspley, J.C. Ed. *The Sales Manager's Handbook*, 8th Ed.; Chicago, IL, 1959.

Aspley, J.C. Ed. *The Sales Promotion Handbook*; Chicago, IL, 1954.

Barton, R. *Advertising Handbook*; Englewood Cliffs, NJ, 1950.

Beckman, T.N. *Marketing*, 7th Ed.; New York, 1962.

Canfield, B.R. *Sales Administration, Principles and Practices*, 4th Ed.; Englewood Cliffs, NJ, 1961.

Hepner, H.W. *Modern Advertising, Practice and Principles*, 3rd Ed.; New York, 1956.

Nystrom, P.H. Ed. *Marketing Handbook*; New York, 1948.

Philips, C.F.; Duncan, D.J.; *Marketing Principles and Methods*, 3rd Ed.; Homewood, IL, 1960.

Public Relations

Aspley, J.C.; Houten, L.L., Eds.; *The Dartnall Public Relations Handbook*, 3rd Ed.; Chicago, IL, 1961.

Cutlip, S.M.; Center, A.H.; *Effective Public Relations*, 2nd Ed.; Englewood Cliffs, NJ, 1958.

Lerley, P. Ed.; *Handbook of Public Relations*, 2nd Ed.; Englewood Cliffs, NJ, 1960.

FOREIGN TRADE

Chamber of Commerce of the United States of America. *Foreign Commerce Handbook: Basic Information and Guide to Sources*, 14th Ed.; Washington, DC, 1960.

Customs House Guide; New York (issued annually with monthly supplements).

Exporters' Encyclopedia; New York (issued annually).

This is but a sampling of the books available, not to mention the many periodicals. For more detailed information on where to locate publications, see Coman.[1]

The most pressing need of the businessman is for up-to-date information. This is most readily available in the newspapers, periodicals, and loose-leaf services. Fortunately, there are indexes that enable one to locate data in periodicals and other media. The three newspaper indexes have already been mentioned. Indexes to a broad range of publications are as follows: *Business Methods Index* (Ottawa, Canada, monthly); *Business Periodicals Index* (New York, monthly); *Applied Science and Technology Index* (New York, monthly); *Readers' Guide to Periodical Literature* (New York, monthly); *UNESCO International Bibliography of Economics* (Paris, monthly). More specialized indexes are as follows: *Index of Corporations* (Detroit, weekly); *Statistical Sources* (Detroit, monthly); *Management Index* (Ottawa, Canada, monthly); and *Accountants' Index* (New York, monthly).

The loose-leaf services that supply a stream of information daily are most popular in the United States. Commerce Clearing House, the Bureau of National Affairs, and Prentice-Hall (Englewood Cliffs, New Jersey) all have services on labor relations, taxation, and corporation law.

The tremendous amount of material being published creates a problem for the businessman as it becomes more

and more difficult for him to keep abreast of information vital to him. Various abstracting services have been devised to give the businessman a quick view of the literature in general. The most complete coverage is in *Economic Abstracts: A Semi-monthly Review of Abstracts on Economics, Finance, Trade and Industry, Management and Labour* (The Hague, 1953–). The coverage is extremely broad in that it includes abstracts of books, periodical articles, and reports of governments and international organizations published in Europe, North America, and Great Britain. The American Economic Association publishes the *Journal of Economic Abstracts* (Cambridge, Mass., 1963–), which tends to cover the more theoretical aspects of economics and business.

Specialized abstracting seems to be largely confined to personnel and industrial relations. These include *Issues & Ideas: Abstracts of the Current Literature in Management, Organization and Industrial Relations* (Chicago, 1954–); *Employment Relations Abstracts* (Detroit, 1950–); *Industrial and Labor Relations: Abstracts and Annotations of Current and Periodical Literature* (Ithaca, N.Y., 1948–1957); and *Management Abstracts* (Ann Arbor, Mich., 1955–).

One other publication, which should have been mentioned earlier, is *International Encyclopedia of the Social Sciences* (New York, 1968).

Reference has been made to the excellent bibliographies and other publications of the librarians in business libraries. This work was greatly encouraged when these librarians joined together in 1909 to form the Special Libraries Association. Various committees and chapters of this association have published many studies and very complete bibliographies on general and special subjects. The members have done much to increase the availability of business information.

Mechanization of business office operations began with the acceptance of the typewriter as a business tool in the 1890s; this was followed not too much later by the adding machine and the comptometer. Here progress rested until the advent of the computer. Now, in the larger firms, payroll and dividend checks, insurance claims, and much of the accounting for customers of banks and retail stores is done by computer. The military, the Bureau of the Census, and the Internal Revenue Service have gone heavily into the computerization of their activities. The various stock exchanges are striving to gain control of their paperwork through the use of computers.

There are still a number of problems, technical and especially economic, to be worked out before computerized retrieval of information works satisfactorily. The MEDLARS program for the retrieval of medical data and the current plans of NASA give promise of developing workable systems that will spread to other subject areas. There are two business-related computer information retrieval systems now in operation on a commercial basis. The *Standard Statistics Compustat Service*, Standard and Poor's Corporation, New York, supplies either magnetic tapes or punched cards to subscribers. Fifty-one types of basic pertinent information, updated 10 to 12 times a year, are supplied on leading industrial and utility corporations. All this material is in machine-readable language.

The *Dun and Bradstreet Sales and Marketing Identification Service* is a roster of U.S. manufacturing and commercial establishments on magnetic tape of its IBM 1401 and 1410 systems. This computerized file of 300,000 firms provides the name, address, and up to 20 "identifiers" for each company. Information includes, in addition to the name and address, the name of the chief executive, the line of business, the Standard Industrial Classification (SIC), the number of employees, information on branches, etc., and for commercial subscribers, net worth and credit rating. Data is updated quarterly and is available on computer printouts, on 3 by 5 cards, or printed on tabulating cards.

Another useful but less sophisticated service is *Thomas Register of American Manufacturers, Thomas Micro-Catalog*. The complete catalogs of all the firms listed in *Thomas Register* are available on microfiche and each subscriber is given a microfiche reader.

The literature on computer operations falls into three parts: 1) the adaptation and utilization of computers in business; 2) the impact of the computer on business and personnel; and 3) the hardware and operation of the computer.

The following books are in the first category: C.C. Barnett, et al., *The Future of Computer Utility* (New York, 1967); Alexander Blanton, et al., *Computers and Small Manufacturers* (New York, 1967); J.R. Bright, *Automation and Management* (Boston, 1958); R.G. Canning, *Electronic Data Processing for Business and Industry* (New York, 1956); and R.L. Sisson and R.G. Canning, *A Manager's Guide to Computer Processing* (New York, 1967).

Books in the second group include the following: R.A. Brady, *Organization, Automation and Society* (Berkeley, California, 1961); J. McLaughlin, *Information Technology and Survival of the Firm* (Homewood, Illinois, 1966); and L. Ricco, *The Advance against Paperwork, Computers, Systems and Personnel* (Ann Arbor, Michigan, 1967).

Works in the third category, dealing with hardware and programming, are as follows: Gille Associates, Inc., *Data Processing Handbook* (Detroit, 1960; 13 vols.); Gille Associates, Inc., *Data Processing Equipment Encyclopedia: Electro-Mechanical Systems, Punched Card, Punched Tape, Related Systems* (Detroit, 1961; 2 vols.); E.M. Grabbe, et al., eds., *Handbook of Automation, Computation and Control* (New York, 1958–1960; 3 vols.); D.B. McCracken, H. Weiss, and T.H. Lee, *Programming Business Computers* (New York, 1959); J.G. Truxal, ed., *Control Engineer's Handbook* (New York, 1958); and J.R. Ziegler, *Time-Sharing Data Processing Systems* (Englewood Cliffs, New Jersey, 1967).

Application of computers to specific segments of business operations can be found in the following books:

R.H. Brown, *Office Automation: Integrated and Electronic Data Processing* (New York, 1958, plus extra sheets to update); R.H. Brown, *Office Automation Applications* (New York, 1959; 2 vols.); E.A. Johnson, *Accounting Systems in Modern Business* (New York, 1959).

The computer industry and its application are advancing so rapidly that a bibliography is obsolete almost as soon as it is published. A good bibliography, up to its date of publication, is C.P. Bourne's *Bibliography on Mechanization of Information Retrieval* (Menlo Park, California, 1958). The best and most current bibliography is the *International Bibliography of Automatic Control* (Brussels, 1962–), which appears monthly.

Periodicals are the best source of information on this ever-changing industry. Among these are *American Documentation* (Washington, D.C., 1937–); *IBM Journal of Research and Development* (New York, 1957–); *Information Processing Journal* (Cambridge, Massachusetts, 1962–); and *Progress in Automation* (New York, 1960–).

CONCLUSION

It is difficult to compress the very large body of excellent material published on business within the compass of this entry. Many worthwhile books and other publications have been omitted, as have certain topics. Among these are directories and the large mass of material published on particular industries.

REFERENCE

1. Coman, E.T. Jr. *Sources of Business Information*, Rev. Ed.; University of California Press: Berkeley, CA, 1964.

BIBLIOGRAPHY

1. *A London Bibliography of the Social Sciences*; 1931–1932. London, U.K., (4 vols.); 1934–1960 (supplementary Vols. 5–11).
2. Grandin, A. *Bibliographie Général des Sciences Jurudiques, Politiques, Économiques et Sociales de 1800 à 1925–1926*; Paris, France, 1927–1950.
3. Johnson, H.W.; McFarland, S.W. *How to Use a Business Library with Sources of Business Information*, 2nd Ed.; Cincinnati, OH, 1957.
4. Manley, M.C. *Business Information, How to Find and Use It*; New York, 1957.
5. Mossé, R. *Bibliographie d'Économique Politique, 1945–1960; Histoire des Doctrines Statistique et Économetrie, Géographie Économique, Économie Rurale, Économie, Financière, Travail, Sociologie, Démographie*; Paris, France, 1963.
6. *Schweizerische, Bibliogrophie für Statisk und Volkswirtschaft*; Berne, Switzerland, 1937.

Canada: Libraries and Archives

Ian E. Wilson
Librarian and Archivist of Canada 2004–2009, Ottawa, Ontario, Canada

Sean F. Berrigan
Policy, Library and Archives Canada, Ottawa, Ontario, Canada

Abstract

Canada's libraries and archives have evolved from the first library in 1605 to become modern knowledge and cultural heritage institutions in the twenty-first century. Key developments in this evolution are outlined in this entry together with a profile of 13 major Canadian archives and libraries. The roles and issues of national library and archival associations and groups are examined as well as the development of professional education in Canada. It concludes with a discussion of a number of challenges and issues facing Canada's libraries and archives.

THE CANADIAN CONTEXT: AN INTRODUCTION

Canada's extensive geography, small but diverse population, federal system of government, and relatively late appreciation of the value of libraries and archives have all shaped how Canada's libraries and archives have developed. Canada is the second largest country in the world with an area of 9,984,670 km² with lengthy coastlines on the Atlantic Ocean to the east, the Arctic Ocean to the north, and the Pacific Ocean to the west. It has a population of 36.2 million largely concentrated in a ribbon close to its southern border with the United States with whom it shares close cultural and professional links. It has a G7 economy. Its diverse multicultural population now includes 4.3% aboriginal peoples and 22% who were born elsewhere with the largest number arriving from Asia and the Middle East. Its largest metropolitan area, Toronto, has more foreign-born residents than Canadian-born ones.

Canada has two official languages, English and French, but Chinese is the mother tongue of more than one million Canadians followed by Italian, German, and Polish. Algonquian languages including Cree and Ojibway together with Inuktitut are the largest language group for Canada's indigenous peoples. Canada's federal government and the Province of New Brunswick are officially bilingual (French and English); Québec is officially francophone and the remainder anglophone, but often services are available in both languages to serve substantial minorities. Canada's northernmost territory, Nunavut, created in 1999, has four official languages, Inuktitut, Inuinnaqtun, English, and French, and a population of some 29,000 occupying one-fifth the land mass of Canada. The 2011 census in fact lists over 100 maternal languages in the country. The Conference Board of Canada reported in 2013 that Canada ranked 8th out of 13 countries on the percentage of adults scoring low on adult literacy rates and that four out of 10 adults have literacy skills assessed as too low to be fully competent in most jobs in a modern economy.[1]

Canada is a federation formed in 1867 and now includes 10 provinces and three northern territories. The governments are based on British parliamentary traditions. The Government of Canada comprises the governor general as representative of HM the Queen of Canada, the Supreme Court, and Parliament with bicameral legislature: an elected House of Commons and an appointed Senate. Each of the provinces has one elected level of legislature, with a lieutenant governor representing the Crown and provincial courts. The "Constitution Act, 1982" and its predecessor, the "British North America Act 1867," divide responsibilities between the central or federal government and provincial governments. Features of the 1982 legislation include a Canadian Charter of Rights and Freedoms, provisions for the Rights of the Aboriginal Peoples of Canada, as well as a section on Equalization and Regional Disparities. The latter provisions address discrepancies among provinces with varying economies and resources. Libraries and archives are not specifically mentioned in the sets of powers ascribed to either the federal or the provincial governments. Education, however, which usually includes public libraries under its broad umbrella, is a provincial responsibility. Municipalities and school boards are created by provincial statute. All levels of government are responsible for their own administrative archives. Canada is generally acknowledged to be one of the most decentralized federations in the world.

THE FIRST NATIONS

The comments that follow reflect the imported European approach for the transmittal of authoritative knowledge across generations, emphasizing written records and print materials. There are other forms of library and archives embodied in the culture of Canada's First Nations. The

Encyclopedia of Library and Information Sciences, Fourth Edition DOI: 10.1081/E-ELIS4-120053671

observation that "When an elder dies, a whole library burns" is simply attributed as an African proverb. It describes the reality of other oral cultures as their traditional continuity in both language and memory is disrupted by modern urban culture. Oral history programs have been earnest and well intentioned over the past century, but they have largely been sporadic, reflecting short-term research interests. Much has been lost. Recently, the power of the formal oral tradition has been renewed as the Supreme Court recognized the legitimacy of aboriginal oral traditions[2] and a Truth and Reconciliation Commission has reached out to the indigenous victims of the residential school system to ensure their memories and their experience remain part of the national memory. The testimony and extensive documentation gathered by this Commission (2007–2015)[3] is now preserved and becoming accessible at the newly established National Research Centre for Truth and Reconciliation hosted at the University of Manitoba.[4]

NEW FRANCE 1605–1763

Archives and libraries have a long history in Canada beginning with the first French settlements in what is now eastern Canada. French territory grew to include the Ohio and Mississippi river valleys down through Louisiana to the Gulf of Mexico. Marc Lescarbot is credited with bringing a modest collection of books from France to what is now Annapolis Royal, Nova Scotia, then *Acadie*, in 1605. The *Collège des Jésuites* in Quebec City, established in 1635 to teach theology, classical studies, and sciences, had a library that included books from an earlier *Bibliothèque de la Mission Canadienne de Jésuites*. The extensive archives gathered by the *Séminaire de Québec*, dating from 1623 to 1800, are inscribed in UNESCO's Memory of the World Register. A law library supported the *Conseil supérieur du Québec* and was located in the Intendant's Palace in Quebec City. The Canadian archival tradition under the French regime included proposals for the appointment of a custodian of archives in 1724 and for a special archive building in 1731. When Canada became a British colony in 1763, French administrative state papers were returned to an uncertain fate in France, while the records required to support the judiciary, the church, and resident notaries were retained.

THE HUDSON'S BAY COMPANY, 1670–1870

The "Governor and Company of Adventurers of England trading into Hudson's Bay" was incorporated by Royal Charter in 1670. This granted the company a monopoly on trading for furs in all the watershed draining into the Bay. In time, as trading posts were extended inland seeking supplies of beaver pelts, the Company's lands encompassed the whole of the Northwest Territories, with settlements in the Red River Valley, and posts scattered across the prairies, up through the Peace River Country, into the mountains and down to the mouth of the Columbia River. The Royal Charter included an obligation to search for the fabled Northwest Passage but also granted the company most of the rights of the Crown, including making and enforcing laws, alienating land, and the use of armed forces. These powers continued for 200 years, until in 1870, they were surrendered back to the Crown, and the territory became part of the new Canadian Confederation.

From its earliest days, the governors of the Company, based in London, instructed its factors to maintain a careful daily record of weather, trade, exploration, and peoples. Records were sent back to London annually and were scrutinized by the governors. The resulting documents constitute an extraordinarily detailed and continuous record of the administration of a major portion of Canada over two centuries, in effect the national archives for that region. These records are essential now for the study of environmental change and have been used for legal matters involving land ownership, borders, and sovereignty as well as its unique insights into people and culture. The Hudson's Bay Company Archives were moved from London, United Kingdom, to Winnipeg in 1974 and then donated to the Government of Manitoba in 1994. The archives have received deserved international recognition being inscribed in the UNESCO Memory of the World Register.

BRITISH NORTH AMERICA 1763–1867

Governments in the various British colonies that eventually became Canada were increasingly publishing various official documents. The first press was established in Halifax, Nova Scotia, in 1752. The first issue of the *Halifax Gazette*, Canada's earliest newspaper, appeared the same year. In 1779, Governor Frederick Haldimand established a bilingual subscription library at Quebec City supported by membership and annual fees as a way of "combating the ignorance of the populations and of encouraging good relations between the former French populations and the new British one." What is now the Legislative Library of New Brunswick began when the Province of New Brunswick was created and separated from Nova Scotia in 1784. In 1787, the governor of Quebec, Lord Dorchester, appointed a commission to survey the archives of Quebec. The Legislative Council of Quebec passed an ordinance in 1790 "For the Better Preservation and Due Distribution of the Ancient French Records" to gather records concerning property and to improve access to them; Quebec Legislative Council. 30 Geo III. C8. Legislative libraries for the provinces of Upper and Lower Canada were established around 1792. Subscription libraries such as the Montreal Library were started in

1796 and the Niagara-on-the-Lake Public Library in Niagara in 1800 and at Kingston in 1804.

The development of libraries and archives in the British North American colonies prior to Confederation in 1867 was uneven, driven by different economic, administrative, social, linguistic, and educational factors. The period to the creation of Canada in 1867 witnessed waves of immigrants—from the United States and from England, Scotland, and Ireland. They brought with them increased levels of literacy and interest in books, newspapers, and other publications as well as a strong desire to better themselves in areas such as agriculture. Garrison libraries created by the large contingents of British troops stationed in Canada gave access to their collections to local residents by subscription. The libraries of legislatures in Upper Canada and Lower Canada were united in 1841 but in 1849 were destroyed when the Parliament buildings in Montreal were burned. Libraries in support of emerging medical and legal professions developed. There were also increasing levels of wealth as witnessed by the development of important private book collections. The precursors of free public libraries had their roots in a variety of community libraries. The first Mechanics' Institutes were organized in St. John's, Newfoundland, in 1827, Montreal in 1828, Toronto in 1830, and Halifax in 1831. The Atwater Library of the Montreal Mechanics' Institute is the oldest lending library in Canada. Member supported Mechanics' Institutes were located in most major population centers to provide libraries of "useful knowledge" in support of adult education including fiction, science, reference, essays, and travel accounts. Some of these received government grants in recognition of their strong support from the public. There were also school-district libraries in Nova Scotia, Ontario, and New Brunswick prior to Confederation. Association or social libraries were much in evidence. Canada's first learned society, the Literary and Historical Society of Quebec, was established in 1824 in Quebec City with a mandate to collect, preserve, and publish Canadian records. It received government support to locate and transcribe records and for publishing. Its library remains as both a research and tourist attraction.

Various governments were putting important parts of the information infrastructure in place—an "Act for the Protection of Copyright" in Lower Canada was passed in 1832 and required the deposit of a printed copy of books, charts, maps, or prints to secure copyright. In 1836, Newfoundland passed "An Act to prevent the mischief arising from the printing and publishing books, newspapers and papers of a like nature by persons unknown and to regulate the printing and publishing the same." In 1841, the United Canadas approved "An Act for the protection of Copyrights in this Province," which required the deposit of two copies of published materials with one of these copies destined for the Library of the Legislative Assembly. The year 1851 saw the first "Public Libraries and Mechanics'

Institute Act" passed by the government of the United Canadas followed shortly by the first "School Libraries' Act." In 1857, Nova Scotia appointed a records commissioner, the first provincial archivist. To support the new federal government, the Library of the Legislative Assembly, some 55,000 volumes, was moved by barge to Ottawa. By the time of Canada's Confederation in 1867, there were a number of types of libraries and archives in place in support of cultural, social, educational, and professional pursuits.

CANADA'S LIBRARIES AND ARCHIVES TO THE END OF WORLD WAR II

With Canadian Confederation in 1867, responsibility for education and, by extension, public libraries rested with the provinces. This period is one of building the library infrastructure in the provinces with very different patterns of support and development and an aspiration throughout for the establishment of a national library. At the same time, the federal government saw the need for the establishment of archival programs. The creation of Canada in 1867 with a federal government and provincial governments in place resulted in a significant scaling up of the role of the government in Canadian society. The federal Department of the Secretary of State became responsible for the keeping of "all State records and papers and papers not transferred to other Departments," and a keeper of the records was appointed. The Library of Parliament continued the functions that had earlier been put in place with the United Canadas in 1841, and legislative libraries were established in the new provinces of Ontario, Quebec, Nova Scotia, and New Brunswick. The Library of Parliament, serving the needs of federal Members of Parliament and senators, was housed in the new Parliament buildings in Ottawa in a magnificent octagonal structure designed by Thomas Fuller and Chilion Jones.

The Literary and Historical Society of Quebec had lobbied the federal government to establish a cultural archive to assist authors and others. In response, the government established an archives branch in the Department of Agriculture in 1872, and Douglas Brymner who is considered to be the first Dominion Archivist was hired. Brymner undertook to build collections and established programs to hand copy important archival materials related to the history of Canada in London and Paris. A survey and report in 1897 on the state of federal records drew attention to what the governor general described as the "most lamentable disregard for the archives of the Dominion." In response, the government consolidated responsibilities for administrative and cultural archives, and in 1904 Dr. Arthur Doughty (later Sir Arthur) was appointed Dominion Archivist and keeper of the records, a post he was to occupy for 31 years. He accelerated the program to copy Canadian records in Paris and London and had extraordinary success in acquiring the original papers of former colonial officials documenting all aspects of the Canadian experience. The first purpose-built archives building was opened in 1907, and its exhibition rooms attracted a wide range of visitors. Doughty began the national documentary art and portrait collections, systematically worked to document the war effort (1914–1918), and added maps, photographs, pamphlets, posters, portraits, documentary art, film, and even historical artifacts to create a multimedia comprehensive record of the country. Through active programs of publication, teaching, and exhibitions, the Public Archives of Canada became an active cultural presence, encouraging the universities to study and teach Canadian history. In the 1920s, faculty from universities across the country gathered in the Archives' reading room, exploring the original record often for the first time and providing leadership to the modern historical profession. Doughty remains today the only federal public servant honored by an official statue in Ottawa.[5] Several provinces followed the federal public–private model for their archives, with Ontario establishing its provincial archives in 1903 and with legislation confirming the Public Archives of Nova Scotia in 1929.

The Province of Ontario passed the "Act to Provide for the Establishment of Free Libraries" in 1882 following an investigation of Mechanics' Institutes in the province and set the pace for other jurisdictions to follow suit, which they did in British Columbia (1891 "Free Libraries Act"), Manitoba (1899 "Free Public Libraries Act"), Saskatchewan (1906), and Alberta (1907). The 1882 legislation in Ontario gave a framework for citizens to ask their municipal councils to establish and maintain public libraries. The Toronto Public Library (TPL) opened in 1884. A year later, Ontario went further with an amendment to this legislation designating Mechanics' Institutes as public libraries. By 1901, there were estimated to be over 130 public library boards in Ontario. As towns and cities across the country put their public library organizations in place, Andrew Carnegie, a Scottish-American philanthropist, began to support the construction of public library buildings. One hundred and twenty-five Canadian public libraries were constructed with Carnegie Foundation assistance, some $2.5 million, between 1901 and 1925. Archives slowly advanced in several provinces, with a new provincial archives building opened in Nova Scotia in 1931.

In collaboration with the American Library Association, the Carnegie Corporation of New York also supported a major study of the library situation in Canada, *Libraries in Canada: A Study of Library Conditions and Needs*, which was published in 1933. This Commission of Enquiry was headed by John Ridington, the librarian at the University of British Columbia (UBC), together with George Locke, the chief librarian of the TPL, and Mary Black, chief librarian of the Fort William Public Library. Their report was the result of a nationwide study, which

saw every province of the country being visited in 1930 and interviews conducted with ministers, premiers of provinces, presidents and librarians of universities, public librarians, educators, and newspaper editors—a wide net was cast. The idea for such an examination of the state of libraries was born in meetings held by Canadian librarians attending American Library Association conferences in 1925 and the conference of 1927 that was held in Toronto. The Commission concluded after some 18 months of study that "though you can find books, there is no real library service, and nearly 80% of the people of Canada have nothing that could, by any stretch of the imagination, be called library service of any kind."[6] They proposed the creation of a Dominion Library, a national library for the country, and a Dominion Librarian who would have "general charge" of the federal government's libraries whose collections should be cataloged under one common scheme. Among other duties, the Dominion Library was intended to be the recipient of the books deposited under the terms of the Copyright Act. The Commission recognized the very diverse nature of the provinces and their very different approaches to library development and support. They saw "five Canadas, each with differing backgrounds, conditions and outlooks" with only the provinces of Ontario and British Columbia recognizing "the principle that the library is an integral part of a people's welfare and educational programme." They set out some essential requirements for public library policy and legislation in the provinces and urged the formation of a national library association.[7]

While the findings of the Commission were positively received, the harsh reality of the Depression made consideration of enhanced library development very difficult. As evidence of difficult times, in 1938 an Order in Council was passed by the federal government that authorized the destruction of materials that had been received for copyright registration purposes because there was insufficient storage space. The collection had previously been offered to the Library of Parliament and the Public Archives. In 1941, the Rockefeller Foundation of New York commissioned a study of the state of Canadian libraries, which was undertaken by Charles McCombs of the New York Public Library. He, like members of the 1930 Commission of Enquiry, visited libraries and archives in all parts of the country. He made observations about the state of libraries and archives and had a number of recommendations to make for the Rockefeller Foundation to guide them in making grants to Canadian projects. These included support for the newly organized Canadian Library Council, aid for the establishment of a National Central Library, as well as support for microphotography (equipment for microfilming newspapers, manuscripts, and archival materials), and a summary guide to the manuscript collections in Canadian libraries and archives.[8] As World War II was ending, the new Canadian Library Council, with its national voice for libraries, submitted a brief to the federal Special

Committee on Reconstruction and Re-establishment that stressed the value of libraries for adult education and recommended the establishment of a Dominion Library Program "to guide, co-ordinate and encourage provincial, local and special efforts,"[9] as well as national library services, library standards, and library consultation services. The notion of national library services included the establishment of a national library.

LIBRARY AND ARCHIVAL DEVELOPMENT TO THE TWENTY-FIRST CENTURY

There are a number of milestones in the post–World War II era when Canada's archives and libraries came of age in a period of rapid economic expansion and cultural maturity. Vital leadership came with the federal government's decision to name Dr. W. Kaye Lamb as Dominion Archivist in 1948 with responsibility for preparing the way for a national library. A Canadian Bibliographic Centre was established in 1950 under his direction, which began to assume national responsibilities for the preparation of a national bibliography and for the building of a national union catalog. The Royal Commission on Arts, Letters, and Sciences reported to the federal government in 1951 on their examination of national institutions and functions, which included a general survey of the arts, letters, and sciences. This was a landmark study and led to the strengthening of a number of federal institutions including the Public Archives of Canada and the establishment of a National Library in 1953. Dr. Lamb continued as Dominion Archivist but was also named as National Librarian. Legal deposit with the new National Library also began in 1953. The Archives addressed the government's pressing need for effective records management and opened the first federal records center in 1956. In the provinces, the pace of change increased. The Saskatchewan Archives Act (1945) provided a model for archives control over the disposition of official records. Public library development was guided by the publication of the 1957 *Report on Provincial Library Service* in Ontario by Dr. William Stewart Wallace and then less than a decade later *Ontario Libraries: A Province-Wide Survey and Plan* by Francis R. St. John. "An Act Respecting Public Libraries" was passed in Quebec in 1959, as well as the establishment by statute of the Newfoundland Archives in 1959.

The expansion of universities and a growing interest in the study of Canada led to a number of influential studies at the national level. These included *The Humanities in Canada* by Watson Kirkconnell and A.S.P. Woodhouse (1947), *Resources of Canadian University Libraries for Research in the Humanities and Social Sciences* by Edwin E. Williams (1962), *Research in the Humanities and in the Social Sciences in Canada* by Bernard Ostry (1962), and F.E.L. Priestley's *The Humanities in Canada; a Report Prepared for the Humanities Research Council of Canada*

(1964). *Resources of Canadian Academic and Research Libraries* was undertaken by Robert B. Downs for the Association of Universities and Colleges of Canada in 1967, reflecting the essential role of libraries in support of the academic mission. In 1975, Professor Tom Symons' seminal *To Know Ourselves: The Report of the Commission on Canadian Studies* was published. Symons devoted a chapter to Canadian archives and the need for Canadians to be better aware of archival resources and the development of a national and regional plan to coordinate archival activity. The newly formed Social Sciences and Humanities Research Council carried forward Dr. Symons' work with the influential 1980 report of their Consultative Group on Canadian Archives. These studies reflected the professionalization of library leadership and influence in national initiatives for a society seeking to understand its own culture and experience. More recently, the journal of the Association of Canadian Archivists, Archivaria, has been recognized as one of the leading international journals in the field. The writings of Hugh Taylor, Tom Nesmith, Terry Cook and Tom Delsey have advanced theory informed by practical experience. The InterPARES Project, led by Luciana Duranti has provided a productive research center for digital recordkeeping. The three volume History of the Book in Canada will stand as a celebration of its subject and the scholarship of a generation. (Eds: P.L. Fleming, Y. Lamonde, G. Gallichan, C. Gerson, J. Michon, and F.A. Black. University of Toronto Press, 2004–2006). Through the initiative of L'association des archivistes du Quebec, the International Council of Archives and then UNESCO adopted the Universal Declaration on Archives, 2011. Two Canadians have served as President of the ICA: Jean-Pierre Wallot and Ian E. Wilson and Ingrid Parent served as President of the International Federation of Library Associations.

MAJOR LIBRARIES AND ARCHIVES

Library and Archives Canada/*Bibliothèque et Archives Canada*

Library and Archives Canada (LAC) is the country's national library and national archives. This federal knowledge institution was created by the "Library and Archives of Canada Act" in 2004, amalgamating the former National Library and National Archives. This innovative integration resulted from the close collaboration between National Librarian Roch Carrier and National Archivist Ian E. Wilson. Parliament gave the new institution an ambitious mandate, seeking to ensure that the multimedia documentary heritage of Canada, from manuscript to print to electronic, is systematically acquired, preserved, and described and is as available as possible to all citizens. It strives to be "a source of enduring knowledge accessible to all, contributing to the cultural, social, and economic advancement of Canada as a free and democratic society,"[10] in the words of the legislative preamble. The vision was the creation of an integrated knowledge institution as a strategic response to the challenges of the twenty-first century. Ian E. Wilson was named the first librarian and archivist of Canada, responsible to the Minister of Canadian Heritage.

LAC incorporates all the functions of both a national library and a national archives, plus the portrait gallery of Canada and working in the context of the networks of libraries and archives serving communities throughout the country. Through consultation and coordination, the LAC is seeking to integrate holdings, services, description, and preservation to create a new type of comprehensive knowledge-based resource, relying increasingly on electronic means to deliver services. Coordinated acquisitions build on both legal deposit of print and electronic publications; the government recordkeeping programs; donations of the papers of prime ministers, artists, writers, and business and labor leaders; occasional purchases of Canadian documentary materials; agreements with broadcast media and filmmakers; and legal authority to preserve Canadian websites. The state of the art Preservation Centre in Gatineau, Quebec, across the river from Ottawa, opened in 1997 for the National Archives, provides a shared resource for secure storage and for conservation treatment. Nitrate photographic collections have now been accommodated in a new low temperature environment in a purpose-built building opened in 2011. And the extensive military service files of the twentieth century have been moved to a newly renovated secure storage building near the Preservation Centre.

The 2004 legislation and the ambitious vision for LAC remain to be realized. Financial pressures since 2009 have resulted in the loss of experienced staff, the reduction of the public exhibition programs, the cancellation of the building for the portrait gallery, the reduction of the interlibrary loan program, and the diminution of its international leadership role, together with service reductions and severe limitations on its website. The LAC's budget was reduced by 10% between 2012 and 2015, eliminating some 215 staff positions. The Canadian Council of Archives (CCA), established in 1985, lost its funding base. The systematic digitization of significant Canadian content shifted to a consortium of research libraries through Canadiana.org. The LAC's cataloging in publication program for Canadian imprints continues as do programs for reading and literacy for youth and persons with print disabilities. Information on its current policies and activities is available on its website: http://www.collectionscanada.gc.ca.

The reductions to LAC reflected a broader government policy closing or consolidating departmental libraries. In the period 2012–2014, departmental libraries serving Fisheries and Oceans Canada, Agriculture and Agri-food Canada, Environment Canada, Canada Immigration and Citizenship, Canada Revenue Agency and Parks Canada, among others, were closed, or substantially constrained Ministers explained that holdings were offered to others, often university libraries, before disposing of materials and the remainder "recycled in a 'green way'."[11] It was expected that online services would be sufficient to replace professional library services. The full impact of these closures on research and on evidence-based governance has not yet been determined. For the first time in Canadian history, library and archives services became a partisan issue.

The changes alarmed the research community. Researchers and archivists came together in a unique Canadian Archives Summit in January 2014, with over 600 participating at the University of Toronto and across the country via live streaming. The Summit was intended as a creative participative process to engage allies and inform the two expert panels established by the Royal Society of Canada and the Council of Canadian Academies to study and make recommendations on the future of libraries and archives in Canada. Their reports were issued in November 2014 and February 2015, respectively. Both provided an eloquent and compelling defense of the continuing importance of memory institutions in a rapidly evolving content-rich, technology-enabled mobile society. Their reports stressed the changing context of information services encouraging leadership and collaboration in responding to the challenges and potential of digital technology coupled with public expectations for convenient access. Both were heavy laden with advice, with the Royal Society offering 70 specific recommendations to governments, institutions, and the professions. The Council of Canadian Academies discussed the available literature and drew on international initiatives to urge adaption to a digital environment, collaboration with others, and developing strong relationships with user communities. Both were thoughtful and strongly supportive but both were published in the final months of the then Conservative Government. Both provide useful inspiration and guidance for the future but neither organization launched an advocacy campaign and neither appears to have had a significant impact.

In both the library and archives communities, nationally and regionally, initiatives are increasingly led by consortia. Following the 2014 Canadian Archives Summit, a national working group continued the discussion, circulating a discussion paper *Canada's Archives: A vision and areas of focus for 2015–2025*, and building on that, the institutions and associations formed the Steering Committee on Canada's Archives (April 2016). The Canadian Association of Research Libraries (CARL) (www.carl-abrc.ca) has provided strong leadership in addressing the challenges of the digital era, supporting the Canadian Research Knowledge Network (www.crkn-rcdr.ca), providing university access to subscription-based scientific databases and the energetic *Canadiana.org* undertaking large-scale digitization efforts are two such partnerships. As the federal government shifted resources to advance Open Government and establishing a public–private partnership, the Open Data Exchange (2015), CARL has effectively represented the library perspective.

Space permits examination of only a representative sampling of major Canadian libraries and archives. Many of these are examples of the increasing convergence that is occurring with knowledge and documentary heritage institutions and educational programs.

Bibliothèque et Archives nationales du Québec

Under the dynamic leadership of its first president, Lise Bissonnette, the *Bibliothèque et Archives nationales du Québec* (BAnQ) has been through a major process of transformation and development over the last decade. In 2006, a new provincial legislation took effect that created the BAnQ by combining the former *Bibliothèque nationale du Québec* (BNQ) and the *Archives nationales du Québec*. This had been preceded in 2002 with legislation that combined the functions of a national library (legal deposit, national bibliography, preservation of Quebec imprint materials, and other services) with the public service functions of the municipal library in Montreal. The public face of this institution was given an enormous profile with the opening of a major new building in 2005 for public services in Montreal. The building, called *la Grande Bibliothèque*, was the result of an international architectural competition and provides an impressive new flagship for library services.

Legal deposit of materials published in Quebec started with the creation of the BNQ and the passing of regulations for legal deposit, which took effect in January 1968. The *Règlement sur le dépôt legal* required the deposit of two copies of books, pamphlets, newspapers, periodicals, artists' books, and sheet music. In 1969, the "*Bibliographie du Québec*" started as the bibliographic record of everything that was published in the province. In 1992, legal deposit was extended to prints, posters, reproductions of works of art, postal cards, sound recordings, software, and microforms.

The BAnQ is located in a number of buildings throughout the province. A preservation center was inaugurated in 1997 in Montreal. The building houses technical services staff together with a number of vaults with stringent environmental controls to support the long-term preservation of one copy of Quebec imprint materials received under the legal deposit requirements of their legislation. Archival services are headquartered in Quebec City, in a beautifully renovated heritage building adjacent to the Université Laval campus, and in Montreal in another renovated heritage building. Regional services are provided through a network of centers located in nine major population areas in the province. BAnQ offers expert and effective leadership for libraries and archives throughout Quebec and at the international level, especially in "La Francophonie."

The Provincial Archives of Quebec/Bureau des archives du Québec had been established in 1920. The BNQ had been created by the Province of Quebec in 1967. The BNQ inherited the collections and the building (*la bibliothèque Saint-Sulpice*) used to house the collections created by a religious order in 1915 and purchased by the Quebec government in 1941. The public library of the City of Montreal had been started in 1917 and included the Phileas Gagnon collection of Canadiana. The fusion of the collections and services of these diverse library and archival institutions has resulted in the creation of one organization dedicated to the acquisition, preservation and making known Quebec's published and archival heritage. For further information see the BAnQ's website at http://www.banq.qc.ca.

BAnQ and LAC have developed a close working relationship, respecting the national roles each performs. This was exemplified as they cohosted first the annual roundtable (CITRA) of the International Council on Archives in 2007 and then more than 3000 librarians attending the annual conference of the International Federation of Library Associations in 2008.

National Science Library Formerly the Canada Institute for Scientific and Technical Information (NRC-CISTI)/*Institut Canadien de l'Information Scientifique et Technique* (ICIST-CNRC)

The Canada Institute for Scientific and Technical Information (NRC-CISTI) has been recognized as Canada's National Science Library (NSL), as an integral part of the Knowledge Management Division in the National Research Council (NRC). It holds extensive print and electronic information resources in all areas of physical and life sciences, technology, engineering, and health sciences. Canada's NRC was founded in 1916 in the midst of World War I, and a librarian hired in 1917. Margaret Gill, the first chief librarian, was appointed in 1928 and through almost three decades of service developed the collections, the services, and the reputation as the basis for the NSL. The NRC Research Press started in 1929 and merged with CISTI in 1994 but in 2010 became a private company, Canadian Science Publishing.

Budget reductions and organizational changes have substantially impacted collections development and required new approaches to service delivery. While innovation and technology have always been major forces at NRC-NSL, these have been tested in the past few years. Working as a key partner with six other government's science libraries, NRC-NSL has helped establish a digital repository and coordinated acquisitions and has implemented both mobile services and a broad federated search system to facilitate public use. While headquartered in Ottawa in a purpose-built building opened in 1974, it provides services across the country linked with installations of the NRC so that in Vancouver it is part of the NRC Institute for Fuel Cell Innovation and in St. John's, Newfoundland, with the NRC Institute for Ocean Technology.

Library of Parliament/*Bibliothèque du Parlement*

The Library of Parliament traces its roots to the legislative libraries established in the 1790s for Upper and Lower Canada (now the provinces of Ontario and Quebec) and the subsequent legislative library created in 1841 for the United Province of Canada. Moves of Parliament and its Library to different Canadian cities as well as fires in 1813 (fire started in York, now Toronto, by invading American troops in the War of 1812), 1849, and 1854 meant that many materials were lost or destroyed by the time the permanent home of the Library of Parliament in Ottawa was opened in 1876. The circular building is famous for its distinctive Victorian Gothic architecture with its flying buttresses, stone exterior, and ornamental ironwork. It was designed by architects Thomas Fuller and Chilion Jones with the active involvement of the first parliamentary librarian, Alpheus Todd. Fires continued to dog the Library with the Centre Block of the Parliament Buildings burning down in 1916 (with the miraculous exception of the Library that was separated by a fire corridor) and in 1952 when thousands of books were water damaged. The magnificent Victorian structure has recently reopened after a 4-yr restoration and expansion project.

For many years until the creation of the Canadian Bibliographic Centre in 1950 and the National Library of Canada in 1953, the Library of Parliament acquired collection materials in anticipation of a national library being established. Then when the National Library was established, over 250,000 volumes and a large collection of print newspapers were transferred from the Library of Parliament. The Library of Parliament served as a deposit for materials being registered for copyright purposes in Canada. The first legislation requiring deposit for copyright registration purposes was in 1832 with the then Legislative Library in Lower Canada. Then in 1841 copyright deposit was required with the Legislative Library of the United Province of Canada and confirmed in 1868 with the new Library of Parliament of Canada. The materials required for deposit included books, maps, charts, musical compositions, photographs, prints cuts, and engravings. This provision for copyright deposit remained in place until 1921 and then returned in 1931 until 1969. From 1885 to 1956, there were two equal administrative heads of the Library, an English-speaking parliamentary librarian and a French-speaking general librarian. This arrangement ended in 1956.

In a parallel with the development of substantial collections and in the absence of a national library, the Library of Parliament was open for use by the general public until 1961 when access became limited to

parliamentarians, the governor general of Canada (Canada's head of state), members of the Privy Council, justices of the Supreme Court, and members of the Press Gallery. Since that time, the focus has been on the development of collections and specialized research and reference services in support of the House of Commons and the Senate of Canada. It has an important rare book collection including a set of John James Audubon's *Birds of America* and early Canadian, British, French, and American imprints of note. The Librarian of Parliament is an officer of Parliament. The "Parliament of Canada Act" includes provisions for the appointment of the parliamentary librarian and an associate parliamentary librarian. This legislation dating from 1988 replaces a previous standalone "Library of Parliament Act" whose roots date back to 1802.

The Library has a wide range of responsibilities. Staff from the Library support the information and research needs of individual members of the House of Commons, senators, and parliamentary committees and groups. PARLMEDIA is an important media monitoring service, among many Web services provided to parliamentarians. The Library has an extensive publishing program that is made available to the public. The former Research Branch of the Library (now the Information and Document Resource Service) was an important initiative created by parliamentary Librarian Erik J. Spicer, parliamentary librarian from 1960 to 1994. There is also a position of parliamentary budget officer within the Library. The mandate of this office is to provide objective analysis to the Senate and to the House of Commons about the state of Canada's finances and trends in the national economy, to undertake research into the nation's finances and economy, and to provide estimates of the costs of proposals contained in legislation introduced by Members of Parliament. The Parliamentary Poet Laureate with eminent poets serving for 2-yr periods is part of the Library's responsibilities. In addition, the Library is responsible for educating the general public about the institution of Parliament and for giving orientations and tours to visitors to the Parliament Buildings in Ottawa.

Archives of Ontario/*Archives Publiques de l'Ontario*

The Archives of Ontario was established in 1903 as the Bureau of Archives. It preserves extensive collections for the study of the history and people of the Province of Ontario, Canada's largest province. Holdings include official records of the Ontario Government, and original source documentation on Loyalists, African–Canadian settlement, various cultural communities, provincial politics, and business and labor organizations, as well as key sources for genealogy and records relating to indigenous peoples. It also has a large collection of vital statistics materials largely now online. There are significant collections of photographs (1,700,000), documentary art, architectural records (200,000), and cartographic records (40,000). Ontario Government records include land, court, and business registrations and the records of the Premiers of Ontario and many of their ministers. In terms of cultural archives, it holds records from some 2600 private individuals, businesses, clubs, and associations, the archives of large national businesses such as the T. Eaton department store chain and impressive architectural archives.

The Archives legislative mandate was renewed in 2006 with an emphasis on recordkeeping within the government. The Archives moved from downtown Toronto into its impressive new facility on the campus of York University in 2009. This has enabled a renewed public exhibition program and improvements to services for academic researchers and genealogists.

Bibliothèque de l'Université Laval

The University was founded by the *Séminaire de Québec* in 1852 that made its collection of some 15,000 volumes available to the new university. The two institutions continued to make their collections available together until the University moved in 1964 to its current campus at Sainte-Foy and the collections were separated. Collections exceed five million documents that include subscriptions to some 24,000 newspapers and serials of which almost 18,000 are electronic. There are 450 databases available on the Web. They have 20,000 sound recordings and 216,000 slides; 1.3 million microforms; 16,500 films, videos, and DVDs; more than 5,000 atlases; and 125,000 maps. More than half of the 2015–2016 collection budget of $12.1 million (Canadian) is dedicated to the acquisition of electronic resources. The Library has been a pioneer in the digitization of materials and making digital resources available on the Web. It has worked in collaboration with the libraries at *Université de Montréal* and the *Université du Québec à Montréal* on an important project called Érudit (www.erudit.org). The consortium's mandate is to provide a digital platform to assist in the transition and free circulation of more than 150 scholarly social science and humanities journals available and to support the world of scholarly publishing in French. They are also partners in *Nos Racines*/Our Roots, a national project to digitize local and regional history materials. This project is carried out in partnership with the University of Calgary, Dalhousie University, and the University of Toronto. They have also been an early collaborator with LAC's Theses Canada program to produce electronic theses, which then form part of the Theses Canada Portal and the Networked Digital Library of Theses and Dissertations (NDLTD). Also in partnership with LAC has been Laval's ongoing development of *Répertoire de vedettes-matière* (RVM), which provides access to more than 200,000 French subject headings and their English equivalents and is used in libraries around the world.

Bibliothèques de l'Université de Montréal

The *Direction des bibliothèques* at the *Université de Montréal* is made up of some 18 libraries and one associated library. It has collections of over four million documents and serves a number of professional faculties in areas such as library and archival science, medicine, and law, as well as humanities, social sciences, and science to students in the French language. Special collections of note include the Collection Gilles-Blain of Jean Cocteau, the Collection Louis-Melzack, and the Collection Baby, which include important early Canadian materials.

McGill University Library

McGill College was incorporated in 1821 and included a library for the Medical School. It was not until 1843 that a Faculty of Arts was in place, and in 1855 a library for the college opened with slow growth of collections primarily through donations. In the 1890s, land was donated for a library by J.H.R. Molson and funds provided for a building as well as ongoing maintenance and salaries by Peter Redpath. With a new building came centralized ordering by the library and adoption of the Cutter classification scheme under the direction of Charles H. Gould, McGill's first full-time university librarian. By 1894 there were 54,000 volumes in the main library and departmental libraries and in 1898 some 90,000 volumes. By 1898 it had collections of an almost complete set of Canadian documents dating from 1840 and publications of five provinces. An early president of the American Library Association and friend of Charles Cutter and Melvil Dewey, Charles Gould promoted the idea of the annual conference of the American Library Association being held in Montreal. This took place in 1900 and brought many Canadian and American librarians together for the first time. Gould also pioneered the publication of McGill University Papers that were used for exchange purposes by the Library and as well promoted the research carried out within the University. He is also known for the establishment of a traveling libraries project in 1901 where small collections of special materials were made available to villages, towns, and rural areas where library services were unavailable.

Today, McGill University Library is known for the quality of its rare collections including a William Blake collection, the Lawrence Lande Canadiana Collection, a collection of Napoleon-related materials, the Redpath Tracts, a collection of some 20,000 British tracts and pamphlets, and the Osler Library of the History of Medicine. It has over six million items in its collections including almost one million electronic books and 38,000 electronic journals. It operates 13 branch libraries and several specialist collections open to the public. Like other university libraries, it is fostering the growth of electronic theses, digitizing increasing amounts of its analog collections, and actively seeking ongoing financial resources for the library.

Toronto Public Library

Canada's largest public library system traces its roots to the York Mechanics' Institute founded in 1830 established "for the mutual improvement of its members in useful scientific knowledge...A library of reference and circulation will be formed." It now has some 100 branches and 11 million items in its collections. In 2015, the library recorded 18 million visits on site plus 31 million virtual visits. The significant progress of the TPL can be measured in some ways by the growth of its buildings as public spaces from a major building built in 1861 with a lecture theater, reading room, and music hall to the enormous support that Andrew Carnegie gave the TPL. In 1903, funds were received from Carnegie that paid for a new central library and three branches to be built. This was followed in later years by funds for another six branches for Toronto. In the greater Toronto area, Carnegie supported the construction of another 10 public libraries and one university library (Victoria College, University of Toronto). The Toronto Reference Library is a major landmark opened in 1977. Seventy percent of the city's population uses the library. The system now bills itself as "the world's busiest urban public library." (http://www.torontopubliclibrary.ca/media/key-facts/)

The TPL branches have research and reference libraries serving the whole city, district libraries serving large areas, and smaller neighborhood libraries. Examples of the research and reference libraries with extended hours for extensive research and reference would be the Toronto Reference Library and the North York Central Library. There are some 17 district branches in medium-sized buildings that normally have collections of more than 100,000 items and multilingual materials for the communities in the particular area. There are 78 TPL libraries designated as neighborhood branches with smaller collections. Some of the collections for which TPL is noted include the Osborne Collection of Early Children's Books; the Merril Collection of Science Fiction, Speculation and Fantasy; and the Arthur Conan Doyle Collection, all of which are supported by their own friends' organizations. The Toronto Public Library Foundation raises funds for endowments and for particular initiatives such as the Toronto Dominion (TD) Summer Reading Program and Keep Toronto Reading Program. Collections in many languages serving a very diverse population and innovation in Web services such as the Virtual Reference Library/ *Bibliothèque de Référence Virtuelle* and KidsSpace are hallmarks of this rapidly evolving institution.

The success and strong public support for the TPL is indicative of the essential role played by municipal public libraries across the country. Ontario, for example, decided in the 1980s not to build a provincial library but to link more than 1000 public library branches as a provincial library network, reinforced by support services, electronic connections, and interlibrary loan. Impressive new public libraries have been built by progressive municipalities concerned about the quality of life and access to knowledge. These cities include Vancouver, Halifax, Winnipeg, Edmonton, Calgary, and, in 2020, Ottawa.

University of Alberta Libraries

The University of Alberta celebrated its 100th anniversary in 2008. Its library is the third largest research library collection in the country. The chief librarian has the responsibility of the University of Alberta Libraries, Museums and Collections Services, as well as the University of Alberta Archives and Records Management, the Bookstore, Printing and Duplicating Services, and the University of Alberta Press. The University of Alberta Libraries include the Bruce Peel Special Collections Library, the Data Library, and a Music Library. Collections of note in the special collections library include the papers of the legendary Mountie, Sam Steele, the Alberta Folklore and Local History Collection, a Dime Novel Collection, and the Pierre Ouvrard Collection of some 300 fine bindings. There is also the Book and Record Depository (BARD). The BARD is a facility modeled on the Harvard Archives Depository, which opened in 1994 to house lesser used materials of the University and those of other institutions. With a capacity for 3.2 million books and almost 40 km of shelving, it is located in a former big-box store that was converted for collection storage purposes with constant temperature and humidity levels.

The University of Alberta Libraries has taken the lead in making regional publications available in digital form such as through the Peel's Prairie Provinces site, which makes available Bruce Peel's *Bibliography of the Prairie Provinces*, as well as the contents of many of the items from the bibliography and a wealth of other materials celebrating the history and culture of the Canadian Prairies.

University of British Columbia Library

The UBC Library has the second largest research library collection in Canada. It has more than 20 branches and divisions including the recently established UBC Okanagan Library. The main library was renovated with the addition of the Irving K. Barber Learning Center in 2009. The collections of special materials include the Wallace B. Chung and Madeleine H. Chung Collection of rare materials related to the discovery of British Columbia, the development of the Canadian Pacific Railway, and Chinese immigration to Canada. Its collections total more than 5.5 million volumes with some 260,000 e-books and more than 65,000 serial subscriptions. It has a number of important digital collections including its Bookplate

Collection, collections of photographs (Japanese–Canadian Photograph Collection, MacMillan Bloedel Photograph Collection, Peter B. Anderson Photograph collection, and the Rosetti Studios–Stanley Park Collection), and the UBC Historical Documents Collection, which includes copies of important records and publications documenting UBC's history. It has established cIRcle, UBC's information repository (www.circle.ubc.ca), which is an open-access repository for published and unpublished research, teaching, and other materials created by the UBC community.

University of Toronto Libraries

The University of Toronto Library houses the largest research library collection in Canada and third largest in North America after Harvard and Yale. The University of Toronto Library is made up of 44 libraries located on three university campuses. The Library dates from the opening of the University of King's College in the former parliament buildings in Toronto in 1841. In 1849, the university was secularized and the name changed to the University of Toronto. In 1850 the Library had approximately 4500 volumes, and as of 2015, it had holdings of 12 million books and 160,000 serials, 5.4 million microforms, and 1.9 million CD-ROMs, computer files, manuscripts, cartographic materials, graphic materials, sound recordings, film and videos, and other materials with a collection budget of some $31.8 million (Canadian). It has major partnerships with the U.S.-based Internet Archive and the Microsoft Corporation to digitize collection materials and was a founding member of the international Open Content Alliance. It has also scanned materials out of copyright from other institutions. It provides access to more than 28,000 full-text electronic journals, almost 900 indexes and abstracts, over 350,000 electronic books, and over 3000 electronic newspapers. The University of Toronto Archives and Records Management Services is a part of the Library. Its distinctive John P. Robarts Research Library for social sciences and humanities resources was originally designed by Mathers and Haldenby of Toronto and has recently been renovated to include a large Information Commons. It also houses the Thomas Fisher Rare Book Library, which includes major collections of Canadian holdings including the Northwest and the Arctic maps and atlases as well as Canadian poetry, fiction, and drama. The intention with the latter is to be comprehensive in all editions and translations as well as literary manuscripts from past and present icons of Canadian culture including Northrop Frye, Margaret Atwood, and Leonard Cohen.

Vancouver Public Library

The Vancouver Public Library (VPL) dates from 1869 with the start of the New London Mechanics Institute,

later renamed the Hastings Literary Institute. It is the third largest public library system in Canada. Over 10 million items are borrowed each year and a million reference requests answered. Its landmark central branch, Vancouver Library Square, opened in 1995 and was designed by architect Moshe Safdie. The Central Library succeeded an earlier Carnegie building opened in 1903 and a later building in 1957. The Vancouver Library Square building, which seats some 1200 patrons, includes retail space as well as a day care. The VPL has 22 branches. The VPL serves a diverse multilingual community. It maintains collections in languages other than English such as in Chinese, Farsi, German, Hindi, Italian, Japanese, Korean, Polish, Portuguese, Punjabi, Russian, Spanish, Tagalog, and Vietnamese. The use of the library as public space is important with library facilities being used for a wide variety of purposes. It has a number of innovative programs—One Book, One Vancouver, a citywide book club, is an example together with its leadership role in the AskAway program, a province-wide virtual reference service. It has a writer in residence and hosts the Vancouver's Poet Laureate.

LIBRARY AND ARCHIVAL ASSOCIATIONS AND GROUPS

Canada has a wide range of national and provincial and territorial associations and groups who support libraries, archives, librarians, library technicians, and archivists. National associations and groups include the following:

Canadian Association of Research Libraries/ *Association des Bibliothèques de Recherché du Canada*

The CARL is a national association with institutional members made up of 27 major Canadian university libraries as well as LAC, the NRC's National Science Library, and the Library of Parliament. Established in 1976, most of CARL's member universities offer doctoral programs in both the arts and the sciences. Its Mission is "to increase the capacity of individual member libraries to be effective academic partners in support of research and scholarship at the national, regional and local levels." It has a number of committees in place for copyright, effectiveness measures and statistics, government policies and legislation, and scholarly communication. It has working groups for data management, e-learning, and library education. CARL plays strong national leadership and advocacy roles in areas such as copyright, open access to digital resources, and digitizing cultural resources. *A Canadian Approach to Digital Copyright* was published in 2008 focusing on key copyright reform issues including fair dealing, damages and fair dealing, technological protection measures, and education use of the Internet. It has been instrumental in obtaining financial support from the federal government for the indirect costs of research that its members provide in support of research undertaken at their university. In the absence of a national program for library statistics, it collects important data on collections, budgets, and expenditures from its members. CARL's Education Working Group has surveyed members to develop a set of national research priorities including how to move institutional repositories to become Trusted Digital Repositories, open access and the impact of new technologies on delivering library services, and the rationalization of print collections. They have also taken the lead in the LibQUAL Canada Project to assess the quality of services provided in CARL and other academic libraries. CARL has also taken an active leadership role in advancing cooperative initiatives for the systematic digitization of unique Canadian documentary content and in providing access to digital resources internationally through the Canadian Research Knowledge Network.

Canadian Federation of Library Associations (CFLA)/*Federation Canadienne des Associations de Bibliothèques* (FCAB), Formerly the Canadian Library Association/*Association Canadienne des Bibliothèques*

The Canadian Library Association (CLA) was a national voice and advocacy organization for the Canadian library community. Founded in 1946, it was dissolved in 2016. The Association replaced the earlier Canadian Library Council, which had been established with Carnegie Corporation and Rockefeller Foundation's financial support. Its first priority was to lobby for the establishment of a national library for Canada. CLA also continued work that the Canadian Library Council had begun to microfilm significant Canadian newspapers first with funding from the Rockefeller Foundation and later from other sources such as the Canada Council. The *Canadian Index to Periodicals and Documentary Films* later the *Canadian Periodical Index* and important directories and other publications followed. Over the past decade, increasingly active provincial and other library associations and declining membership resulted in the CLA being unable to continue as a viable organization.

After months of consultation, discussions, and recommendation, a proposal was presented in the fall of 2015 to dissolve the CLA, and in May 2016 the CFLA/FCAB was formally incorporated. The new organization represents eighteen Canadian member associations and will transition a number of the CLA's activities. The CFLA/FACB brings together the provincial, territorial, and national associations across Canada and will focus on a unified and coordinated voice for the Canadian library community on both national and international issues. Its purpose is to advance library excellence, champion library values and the value of libraries and to influence national and international public policy impacting libraries and their communities. Key priorities for the new organization include continuing the building of the new federation, copyright review, and international representation.

L'Association pour l'Avancement des Sciences et des Techniques de la Documentation (ASTED)

ASTED is the national association of French language librarians and libraries. It was founded in 1973 based on the earlier *Association canadienne des bibliothécaires de la langue française* and the *Association canadienne des bibliothèques catholiques*, established in 1943. Its mandate is to promote excellence in the services and personnel of libraries and information and documentation centers as well as to promote the interests of libraries to governments and to represent the interests of North American French library science. It has an advocacy role at the provincial level primarily with the Province of Quebec and at the federal level with the Government of Canada. Federally, it speaks on issues such as copyright and the Library Book Rate. It has two periodicals, *Documentation/ et bibliothèques* issued quarterly and *Nouvelles de l'ASTED* issued five times a year. It has an important annual conference and publishes a wide range of professional literature ranging from the French editions of AACRII and Dewey to a union catalog of serials in Quebec health libraries. It also provides ongoing professional development.

Provincial and Territorial Public Library Council (PTPLC)/*Conseil Provincial et Territorial des Bibliothèques Publiques* (CPTBP)

This Council is made up of the directors of public library services in Canada's provinces and territories. Established in 1978, it was formerly known as the Provincial and Territorial Library Directors Council (PTLDC). Given the diversity of provincial and territorial legislation and programs for libraries, the agencies represented on PTPLC vary in their mandates and responsibilities. The Council exists to provide a vehicle for sharing information on policy issues and development for public library services and to act collaboratively on national issues such as literacy, services to aboriginal communities, the library book rate, persons with disabilities, and resource sharing. It facilitates cross-jurisdictional initiatives and serves as a point of contact with national library organizations and the federal government. In 2005, the PTPLC commissioned a major examination of the role of public libraries called "Public libraries in the priorities of Canada: Acting on the assets and opportunities." The PTPLC reports to a committee of provincial and territorial deputy ministers responsible for public libraries and facilitates meetings of federal, provincial, and territorial ministers responsible for public libraries.

Canadian Urban Libraries Council (CULC)/ Conseil des Bibliothèques Urbaines du Canada (CBUC)

The CULC is a national organization of Canada's largest public library systems, serving more than 40% of Canada's population. Each system serves a city of more than 100,000 people. It was formerly called the Council of Administrators of Large Urban Public Libraries (CALUPL). The organization's mission is "To work collaboratively to build vibrant urban communities by strengthening the capacity of Canada's urban libraries." Its priority issues include social inclusion of new immigrants, established multicultural groups, and youth for which it has developed an Urban Library Social Inclusion Audit for the use its members, youth and civic engagement, organizational cultural changes, and technology. It is building an Urban Library New Immigrant Programs and Services Inventory. The CULC has an advocacy role on behalf of libraries in areas such as federal copyright reforms, open access, and service standards. CULC took over the administration of the former CLA's Library Materials Shipping Tool and leads the advocacy of e-books access and pricing issues.

CULC was instrumental in the establishment of LibraryNet, a program of the federal government, to provide public Internet access with a report called "Connecting Canadians." When LibraryNet was ending in 2004, CULC responded with "Sustaining Canada's Digital Capacity: An Urban Library Strategy to Sustain Socially Inclusive ICT Networks." It collects and publishes the *Canadian Public Library Statistics* and a separate *Branch Library Activity Statistics* on an annual basis. These provide important data on expenditures, revenue sources (provincial, municipal, others), collection budgets, size of collections, number of professionals and other staff, use of electronic resources, numbers of registered users, in-person visits, number of locations, and populations served for members of CULC and other library systems that are not members. While public libraries are the responsibility of Canadian municipalities, the Council provides a forum for the discussion of the issues facing large public libraries—literacy, digital literacy, resource sharing, equity of access through public Internet facilities, social cohesion, services to aboriginal peoples, and services and meeting the information needs of an increasingly multilingual and diverse population.

Canadian Council of Archives/Conseil Canadien des Archives

The CCA was founded in 1985 to encourage and facilitate the evolution of an archival system in Canada. A national organization, it has representation from provincial and territorial archival councils, LAC, the Association of Canadian Archivists (ACA), and the *Association des archivistes du Québec*. The Council has a mandate to improve the administration and effectiveness and efficiency of the Canadian archival system and to identify national priorities. It has a number of Standing Committees and Working Groups for Standards, the Canadian Committee on Archival Description, Copyright, and Preservation. It had several funding programs supported by the federal government—including the Control of Holdings Program, the Professional Development and Training Program, the Special Projects Program, the Conservation Plan for Canadian Archival Records (CPCAR), and Preservation Management Program coupled with a training program. With support from the provincial councils, the CCA funded traveling archives advisors to encourage local institutions and their communities in preserving their archives. The CCA led the development of ArchivesCanada.ca as the national portal to access the holdings of more than 800 archives. This includes virtual exhibits, as well as digitized photographs, maps, and documents in projects supported by the former Archival Community Digitization Program. Federal funding for the CCA and the National Archival Development Program was canceled in 2012. With support from the archival community, the CCA has persevered and refocused and maintains the national database of archives holdings: http://www.archivescanada.ca.

More recently, the CCA has launched the ARCHIVESCANADA Digital Preservation Service to provide memory institutions with an affordable digital preservation service maintained in an environment to ensure sustainability, long-term accessibility, usability, and authenticity of electronic records and digital information objects.

Steering Committee on Canada's Archives (SCCA)

Following the January 2014 *Canadian Archives Summit: Towards a New Blueprint for Canada's Recorded Memory*, organizers continued to develop the ideas and deliberations raised during the Summit though the annual meetings of the CCA, the ACA, and the *Association des archivistes du Québec* (AAQ). Building on the discussions over the summer of 2014, the ACA, AAQ, CCA, and the Council of Provincial and Territorial Archivists (CPTA) and LAC came together as a working group in September 2014 to develop a renewed vision and road map to guide the Canadian Archival System. A draft Strategy entitled *Canada's Archives: A vision and areas of focus for 2015–2025* was circulated for comment in the summer of 2015 with the final version of the Strategy launched at a national event in Ottawa that November. The SCCA was established to guide the implementation of the Strategy with representatives from each of the ACA, AAQ, CCA, CPTA, and LAC.

Association of Canadian Archivists (ACA)

The Association was established in 1975 and evolved from the Archives Section of the Canadian Historical Association. It has a number of standing committees such as education, membership, ethics, and public awareness. Special interest sections provide a forum for discussion in areas such as aboriginal archives, privacy and access, government records, municipal archives, technology and archives, personal archives, religious archives, sound and moving images, and university and college archives. It seeks to provide leadership, encourage awareness of archival activities, advocate on behalf of the needs of professional archivists, and communicate to further understanding and cooperation among members of the Canadian archival system and other information and culture-based professions. In addition to organizing a major annual conference, it publishes a newsletter, the *ACA Bulletin*, as well as *Archivaria* and monographs and occasional papers. Back issues of *Archivaria*, one of the leading English language archival journals in the world, are available online.

Association des archivistes du Quebec/Association of Archivists of Quebec (AAQ)

The Association was founded in 1967 and reflects the distinctive qualities of the archival discipline, history, and profession in Quebec, Canada's francophone province. The AAQ represents more than 600 members: archivists, archival educators, and information specialists who work in both public and private sector organizations. The unique nature of the Quebec archival practice marries both European archival theory and practices with the North American methodology and profession of records management.[12] Joining management and conservation of historical archival documents with the document life cycle is reflected in both Quebec archival discipline and the pedagogy developed and taught in the curricula used in university archival education programs. The Quebec archival system is also characterized by its decentralization: archival records are managed and stored in the regions where they have been created.[13]

The AAQ publishes a biannual publication *Archives,* established in 1969, which focuses on scholarly research and reflections on the theory and practices of the archival and information management professions in Quebec.

Canadiana.org

Canadiana.org is a national not-for-profit organization that was established in 1978 with its initial funding coming from the Canada Council for the Arts. It was set up in response to concerns from librarians and scholars about the preservation of Canada's earliest printed heritage, as well as increased demands to have access to these materials for research and study purposes for Canadian Studies programs. Since its establishment, it has completed several large projects to locate and film (to preservation standards set out the *Guidelines for Preservation Microfilming in Canadian Libraries*) and produce bibliographic descriptions of these materials. This national initiative has had the effect of pulling together Canada's early published materials, some unique, some deteriorating, from scattered collections in libraries and archives in Canada and abroad and making this content available through subscribing institutions. For the past 15 yrs, Canadiana.org has focused on shifting its method of access from microform to digital. With assistance from the Andrew W. Mellon Foundation and working in partnership with the then National Library of Canada and the libraries of the University of Toronto and Université Laval, it began its Early Canadiana Online project. With the support of its members, primarily research libraries and the Department of Canadian Heritage, it has completed a major digitization project, Canada in the Making, to make pre-1900 Canadian official publications available on the Internet. It successfully completed another multiyear digitization project, Early Canadian Periodicals to 1920. In 2008 Canadiana.org merged with AlouetteCanada.ca, a national collaborative initiative on the part of Canadian research libraries and LAC. AlouetteCanada.ca, in turn, had absorbed the work of the Canadian Initiative on Digital Libraries (CIDL), a collaborative effort to coordinate digitization projects and standards in Canada. Canadiana.org will be the focus of major projects to aggregate current digitized contents and to digitize Canada's rich cultural heritage including materials from libraries, archives, and museums. With the sustained support from the CARL and a contract with the LAC, Canadiana.org is embarking on an ambitious initiative. The Héritage Project is a 10-yr commitment to digitize and make accessible online some of Canada's most popular archival collections encompassing roughly 60 million pages of primary-source documents.

DEVELOPMENT OF CANADIAN LIBRARY AND ARCHIVAL EDUCATION

Formal library and archival education takes place at a wide range of institutions across Canada. The year 2004 marked the 100th anniversary of the first formal library education program in Canada. Prior to that time, librarians and archivists learned through experience or through attendance at courses in the United States. Charles H. Gould, the then university librarian at McGill University, laid the groundwork for early librarian education in Canada. He started with an in-house training course for staff, which was also open to outside applicants. This was followed with the first McGill University Library Summer School in 1904, which was directed to people that were already in the workforce. There was a curriculum consisting of classification; cataloging; accessioning; shelflisting, charging, and ordering systems; reference work; and bibliography and included lectures, field trips, and special lectures including those delivered by Melvil Dewey and James Bain, the chief librarian of the TPL. Courses were in place from 1904 to 1911 and then in 1913 and 1914. Stimulated by the McGill experience, Ontario's Department of Education began to offer similar summer programs in 1911 in Toronto. The program in Montreal evolved to summer programs in 1927 at McGill accredited by the American Library Association then to a sessional graduate Bachelor of Library Science program at McGill University in 1930. In Toronto, a one-year academic program for university graduates leading to a Bachelor of Library Science at the Library School was jointly administered by the University of Toronto and the Ontario Department of Education and the Ontario College of Education in 1937. In 1937, the *École de bibliothéconomie* at the *Université de Montréal* was established for French language education. The University of Ottawa opened a bilingual French and English evening program for librarians in 1938 that evolved to offering a Bachelor of Library Science program in 1942.

Due to the rapid expansion of libraries in the 1960s, new graduate program library schools were created at the UBC (1961 School of Librarianship), University of Alberta (1965 School of Library Science), University of Western Ontario (1969 School of Library and Information Science), and Dalhousie University (1969 School of Library Service). The program at the University of Ottawa, *École de bibliothécaires*/Library School, closed in 1972. Carleton University began to offer a Canadian archives training course in 1959 that became a 6- to 8-week program at the Public Archives in 1969. Graduate librarian programs were accredited (and continue to be accredited) by the American Library Association. Over the last 50 years, these programs have evolved from offering Bachelor of Library Science (as a graduate degree program) to a 2-yr Master of Library Science degree program as well as doctoral programs. Changes to the programs and names of the various faculties reflect more specialization and the inclusion of other disciplines and a broader view of information management. The UBC introduced a Master of Archival Studies (MAS) program in 1981 as part of what became known as the School of Library, Archival, and Information Studies in 1984. They also introduced a First Nations concentration in 1998 and a Master of Arts in Children's Literature program in 1999. Today, the University of Toronto has a Faculty of Information with a Master of Information degree program, including eight areas of concentration together with a PhD program and a Master of Museum Studies program. Similarly, what is now the *École de bibliothéconomie et des sciences de l'information* at the *Université de Montréal* offers a Master of Information Science degree program, a doctoral program, and certificates in Archival

Studies and Digital Information Management. The program at the University of Western Ontario, now Western University, is combined with journalism as the Faculty of Information and Media Studies. The UBC was exploring a similar approach in January 2017. McGill University's program is now the School of Information Studies and offers areas of specialization in archival studies, knowledge management, and librarianship. The history department at the University of Manitoba established a master's degree program in archival studies in 1991. Finally, a new graduate program, the School of Information Studies, was launched at the University of Ottawa in 2009 as a bilingual program designed for students who want to work in the fields of archives, librarianship, and information centers.

The development of library technician programs mirrored the rapid expansion of libraries, graduate library programs, and the opening of community colleges across the country. In the late 1960s and early 1970s, library technician programs were opened at Red River Community College in 1962 in Winnipeg, Vancouver City College (now Langara College) in 1966, Southern Alberta Institute of Technology in Calgary in 1967, and Ryerson Polytechnical Institute in 1967 in Toronto. In Ontario a number of colleges of Applied Arts and Technology established programs at the same time. St. Clair College of Applied Arts and Technology and Seneca College of Applied Arts and Technology were the first in 1967. Core competencies were defined by the CLA in 1982 with their *Guidelines for the Training of Library Technicians*. There are currently programs in place in three Ontario community colleges down from a peak of eight institutions in the late 1980s, which reflect a number of closures of technician programs in the early 2000s. These colleges in Ontario were matched by a similar province-wide development of *Collèges d'enseignement général et professionnel* (CEGEPs) in the Province of Quebec intended for graduates of secondary schools. There are now some 48 CEGEPs in all regions of the province, primarily offering their courses in French with library technician programs in place in seven of these. The Nova Scotia Community College's program began in 1985. Core competencies for library technicians working in school libraries were defined in CLA's *Qualifications for Library Technicians Working in School Systems*. Library technician programs are also in place primarily to support the training of teacher–librarians at the University of Winnipeg, Lakehead University, Concordia University, Memorial University of Newfoundland's School of Continuing Education, and the University of New Brunswick's Library Assistant Program. Normally, these training programs are for 2 years.

THE CANADIAN LIBRARY AND ARCHIVAL LANDSCAPE: CHALLENGES AND ISSUES

Canada's archives and libraries share a range of issues and challenges. These include dealing with an increasingly digital environment, preservation, access, description and resource discovery, and management of government information. The continuing issues are of standards development, copyright, resource sharing, national planning, and coordination; the revitalization and replenishment of their professional ranks; convergence; working collaboratively; serving both increasingly diverse multicultural populations and indigenous peoples; and providing services for people with disabilities. Canadian libraries have some challenges and issues that archives do not share in quite the same way. These include diminishing collection budgets to address the burgeoning number of electronic resources to be acquired (while at the same time witnessing increased numbers of analog materials) and services for people with disabilities and literacy. Archives with their unique multimedia collections place high emphasis on preservation of and access to all forms of documentation. Meanwhile, they must cope with the increasingly complex legal environment for the selection and preservation of official documents, legal discovery in class action litigation together with escalating freedom of information responsibilities, and the imperatives of copyright and privacy protection.

Digital

Canada's libraries and archives have been in the forefront of adapting to an increasingly digital environment. In the 1990s, Canada's federal government established an Information Highway Advisory Council that set the stage for a range of federal initiatives including digitization programs and LibraryNet and SchoolNet programs. The National Archives launched a multiyear initiative to digitize the key records of the 660,000 Canadians who served in World War I. The National Library established a pilot project in 1993–1994 to begin the acquisition of Canadian born-digital publications. Canadian publishers, particularly publishers at the federal and provincial governments, have made a major shift to producing their publications in digital formats. Canadian libraries and archives are actively engaged in digitizing parts of their collections. Some large digitization projects such as Our Roots/*Nos Racines*, which pulls together local histories and genealogical materials from across the country, were financially supported by the Department of Canadian Heritage. This same program has also supported the transfer to digital formats of important cultural resources from the National Film Board of Canada/*Office national du film du Canada*, and Canada's public broadcaster, CBC/Radio-Canada, LAC, BAnQ, and other research libraries have shifted resources to support digitization initiatives and have been active partners internationally to harvest and preserve significant websites. Canadians have been concerned about seeking to have a critical mass of Canadian digital content in order to provide a strong Canadian voice with the Canadian experience, culture, and heritage easily available on the Internet. Mass digitization on a national scale through

Canadiana.org supported by the CARL is advancing, with minimal assistance from the federal government. Libraries, archives, museums, broadcasters, publishers, and other stakeholders came together in 2006 to develop a cohesive Canadian Digital Information Strategy, which articulated a vision:

Canada's digital information assets are created, managed and preserved to ensure that a significant Canadian digital presence and record is available to present and future generations, and that Canada's position in a global digital information economy is enhanced.[14]

> Despite many efforts, this vision for a modern knowledge society and evidence-based decision-making has not been shared by the government. Libraries and archives are advancing this agenda through reallocation of their own limited resources. In June 2016, LAC announced the launch of the National Heritage Digitization Strategy, the objective of which is to reposition Canada and provide a cohesive path toward the digitization of Canadian memory institutions' collections by encouraging the long-term viability of documentary heritage records through quality standards-based efforts and ensuring a national plan of action will be in place (www.bac-lac.gc.ca/eng/about-us/Pages/national-heritage-digitization-strategy.aspx).

Public libraries and other institutions are concerned about a growing digital divide, between those Canadians who have easy access to computers and appropriate training and computer literacy skills and those who do not. Programs such as the Bill & Melinda Gates Foundation's Library Program's gifts of computer equipment, improving Internet access, and providing computer training facilities to public libraries across Canada have made a very large contribution to addressing the digital divide. Public, government, and university libraries are in the forefront in terms of offering digital services and collections.

Preservation

In terms of preservation, the CCA released a national strategy for archives in 1994, *The Preservation Strategy for Archives in Canada*. The National Archives established an expert national task force and released an alarming report in 1995 entitled *Fading Away* to raise awareness of the need to preserve the nation's audio–visual heritage in radio, television, film, and sound. In 1993, the then National Library working in collaboration with a national committee developed a national preservation strategy for libraries, *A National Strategy for Preservation in Canadian Libraries* and *Guidelines for Preservation Microfilming in Canadian Libraries* were developed and published. The mass deacidification program of LAC has treated more than a million items in their collections. LAC also moved forward on permanent paper standards adopted by the federal government, which has largely eliminated the use of inappropriate, acidic paper by Canadian print publishers. Generally, the interest and support of preserving analog materials in paper, compact discs, tape, CD-ROMs, and other media has ebbed as digitization to ensure access has become prevalent at many custodial institutions. The LAC Preservation Centre and the new nitrate storage facility provide the best possible secure environments as do the many new library and archives buildings across the country. Several libraries have developed digital facilities meeting international standards for trusted digital repositories, but unique Canadian digital content and recordkeeping grows faster than the institutional capability. This remains a concern for institutions and the public who are coming to realize that their own digital legacy is transitory.

Access

In an environment where research is increasingly interdisciplinary and the point of access is via the Internet, researchers pay less attention to where their information needs are being met. "Access is the driver" has become a direction in many archival institutions, and while this is a natural sentiment in libraries to facilitate and support access to all collection materials, it has been less willingly embraced in many archival institutions where the training and practice has been to protect access in order to preserve materials. In Canadian libraries and other libraries, there is an increasing move to have open access for digital materials, particularly those that are publicly owned and non-exclusive licenses with private sector digitizing partners. The Simon Fraser University Library in partnership with the UBC hosts the development and provides support for the Public Knowledge Project's open-source software for scholarly publishing and communication. CARL and its members have worked with Canada's three federal research granting councils, first on endorsing open-access principles for the publication of scientific research (2010) and subsequently in encouraging the agencies to adopt an open-access policy on publications (2015). The intent is explicit: "The objective of this policy is to improve access to the results of Agency-funded research, and to increase the dissemination and exchange of research results" (http://www.science.gc.ca/eic/site/063.nsf/eng/h_F6765465.html). Similarly the three agencies adopted a *Statement of Principles on Digital Data Management*, requiring responsible stewardship with plans for metadata, preservation, and data sharing (http://www.science.gc.ca/eic/site/063.nsf/eng/h_83F7624E.html).

Archives and libraries are concerned about standards and best practices in digitization from analog and in creating born-digital materials to ensure long-term access. They are also concerned about integrity and authenticity of the originals and, given the Canadian context, how to see more materials online that speak to increasing multilingual and aboriginal audiences. To be truly useful, digital information resources have to be organized for seamless and more open access in a bilingual environment. Access for archivists also speaks to access to information and privacy concerns with the large volumes of records of federal, provincial, and municipal governments. Equitable and ubiquitous access from wherever in the country is viewed as critical to the nation's economic and cultural growth and innovation.

Description

The Canadian library community and, to some extent, the archives are implementing new standard for bibliographic description, *RDA: Resource Description and Access*, which is replacing the *Anglo-American Cataloging Rules 2nd Ed.* (AACR2). Description of published materials has been under review for some years by the Canadian Committee on Cataloguing, a national advisory committee and member of the Joint Steering Committee for Revision of AACR (JSC). The work of Dr. Tom Delsey, a Canadian, on the conceptual models Functional Requirements for Bibliographic Records (FRBR) and Functional Requirements and Numbering of Authority Records (FRNAR) prepared for the International Federation of Library Associations and Institutions (IFLA) have led the way for an overhaul of bibliographic description. He is also the editor of the new *Anglo-American Cataloguing Rules 3rd Ed.* (AACR3), which has reassessed the fundamentals of bibliographic description. The AACR3 has been developed and is being implemented jointly by LAC, the Library of Congress, the British Library, and the National Library of Australia. Archival description in Canada is carried out under the "Rules for Archival Description (RAD)" developed and updated by the Canadian Committee on Archival Description (CCAD). The RAD was revised in October 2005 and the rule revisions took effect in 2007. In 2016, CCAD convened a meeting on The Future of RAD in Ottawa and reached a consensus regarding the future of archival descriptive standard in Canada and the possible revision or replacement of RAD (http://archivescanada.ca/FutureOfRAD).

Management of Government Information

The effective management of the administrative records of all levels of government, federal, provincial, territorial, and municipal, has become a major challenge across the country. Access to information and privacy protection coupled with public expectations and political commitments to accountability and transparency has provided incentive for all levels of government to improve their recordkeeping. As in the private sector, the stringent requirements for legal discovery in class action lawsuits have lent urgency. Most jurisdictions now recognize information and records as assets, which are formally the responsibility of all managers. The federal government in particular has had a number of major inquiries and auditors' reports highlighting deficiencies in recordkeeping.

The Auditor General's Report of Fall 2014 concluded that LAC was not adequately fulfilling its responsibilities for acquiring, preserving, and providing access to government documentary heritage and had a considerable backlog of records for processing. A dedicated task force was put in place to assess and process 10 additional kilometers of archives so that they will be discoverable for users. Given the volume of government records being produced and the embrace of digital information and e-mails, government archives have been under significant pressure to put the infrastructure in place for improvements—guidelines, best practices, use of automated systems for the handling of electronic information, and more rigorous monitoring. The 2004 "Library and Archives of Canada Act" incorporated a number of new specific measures to strengthen the role of archivists in acquiring and describing Government of Canada records as well as identifying records at risk. To cope over the past two decades, the LAC pioneered risk-based functional analysis approach to the appraisal and selection of records. On a more positive note, governments have begun to make significant amounts of their published information and datasets available on the Internet to improve access by the public. The federal government is committed to the precepts of open data and in 2014 announced the formation of an Open Data Institute with the private sector.

Copyright

Canadian libraries and archives have taken a strong leadership and advocacy role as successive Canadian governments have moved to update and reform Canadian copyright legislation and to bring Canadian law into conformance with international copyright treaties. Acting collectively, library and archival groups have asserted views on the necessity of a balanced approach to copyright changes, which factors in users as a counterweight to creators. They also look to implement technology neutral approaches and technological protection measures and protection of educational use of the Internet. In 1986 Canada established the Public Lending Right Commission/ *Commission du droit de prêt public*, which compensates registered authors on an annual basis for the use of their books (print format only) held in Canadian libraries. The advent of large digitization projects has put a renewed focus on securing copyright permissions and ensuring that copyright restrictions are respected. Libraries and archives in most Canadian jurisdictions operate under copying licenses from a number of copyright collectives.

In 2004, the Supreme Court of Canada clarified and expanded the scope of a "fair dealing" defense, seeking to balance the public interest with the rights of copyright holders (2004 SCC 13: CCH Canadian Ltd vs the Law Society of Upper Canada). The *Copyright Modernization Act* of 2012 recognized education as a permitted purpose, provided it is reasonable and fair. CARL is preparing to participate in a Parliamentary review of the Act scheduled for 2017.

Resource Sharing

Equitable access by the public to collections held in libraries in Canada has been a public policy issue for many years. There is an assumption in Canada that the investment in building collections with public funds means that Canadians in principle should have access to these materials. This has manifested itself in many ways. The national union catalog maintained by LAC is one manifestation of this. With the creation of the National Library of Canada in the early 1950s, one of its first priorities was the microfilming of holdings of significant libraries across the country. Over the years, this has become a highly automated process but with the same purpose to permit library users in one library to have access via interlibrary loan to materials in other collections. Similarly, Archives Canada has brought access to the resources of participating Canadian archives and increased the ability of Canadians to know what had been collected and where it was available and possibly to seek access via interinstitutional loan. These national services and resources are bolstered by a series of agreements and provincial and regional union catalogs across the country. In 2017 LAC entered into an agreement with the Online Computer Library Center (OCLC), the massive international library cooperative, to move the Canadian National Union Catalogue to the OCLC database by the end of 2018. The Library Book Rate, a program of the Department of Canadian Heritage through Canada Post, subsidizes the transport of library materials from library to library. This is particularly important in rural and more remote parts of the country. A recent manifestation of this continued interest in resource sharing is the Resource Sharing Agreement agreed to by the Council of Prairie and Pacific University Libraries (COPPUL), the Ontario Council of University Libraries (OCUL), the *Conférence des recteurs et des principaux des universités du Québec* and the Council of Atlantic University Libraries/*Conseil des bibliothèques universitaires de l'Atlantique*. This permits students, faculty, and staff of Canadian universities standardized reciprocal interlibrary loan and document delivery privileges across Canada. Standards developments such as Z39.50 protocols and the Virtual Union Catalogue of Canada have all supported this sharing of resources.

National Planning and Coordination

Given the decentralized situation with respect to libraries and archives in Canada, it is important that there be planning and coordination mechanisms in place on a national level. The Canadian Digital Information Strategy, discussed earlier, was an important initiative of LAC in working nationally across different disciplines to achieve a common purpose. Resource sharing is another area where coordination and leadership has been critical to cut across different types of libraries (public, university, college, school, special, government) and archives (government, private, university) to ensure equitable access to information holdings for Canadians. The CCA provides a mechanism for leadership and direction for Canadian archives. The Council on Access to Information for Print-Disabled Canadians was established to give advice, identify funding requirements monitor progress and make recommendations regarding the implementation of "Fulfilling the Promise: The Report of the Task Force on Access to Information for Print-Disabled Canadians." National library associations and groups provide leadership but are more fragmented and there is no coordinated program for library development. They come together to address contentious issues but are less engaged in working collaboratively in a proactive way to support critical research, develop public policy and plans on a national scale. The lack of national library and archival statistics gathering by Canada's national statistical agency, Statistics Canada, makes coordination and information gathering and policy development much more difficult.

Human Resources

Archives and libraries in Canada share major concerns about their workforces, which will be changing dramatically over the next few years. Libraries and archives witnessed great expansion in the 1970s and grew rapidly. The demographics of Canada's workforce in archives and libraries are such that there will be many leaving their professions over the next 5–10 yrs. The Cultural Human Resources Council, a federal body, has recently completed a "Training Gaps Analysis for Librarians and Library Technicians." Competencies and training for professional and non-professional staff in libraries and archives are evolving rapidly in response to the demands of a digital environment and the needs of an increasingly diverse Canadian public. The CLA's President's Council on the 8Rs (the Rs referring to recruitment, retirement, retention, rejuvenation, repatriation, reaccreditation, remuneration, and restructuring) has collaborated with CARL and the University of Alberta to analyze library staff trends over a decade from 2004 on and has informed workforce planning (the studies are available at http://www.ls.ualberta.ca/8rs/reports.html).

Convergence

With the dramatic growth in electronic resources, there is growing recognition that Canadian researchers and users of libraries and archives are increasingly unconcerned about whether it is archival or library materials that have been used to provide information or whether it is a library or an archives that has provided the information they require. Professional education and training at graduate

programs across the country reflect an increased emphasis on offering programs where students at the same faculty can specialize in archives or library studies or more generally for information as a discipline. Similarly, trainings at postsecondary institutions offering programs for technicians are blending both library and archival training. Joint exhibits with museums and galleries have become common, evidence of the recognition that Canadian governments are giving to an integrated approach to knowledge and our collective memory.

Working Collaboratively

In the past decade, collaboration has become the rule; not the exception. In the Canadian library world there are several approaches to collaboration. Consortia are well established with research libraries having a particularly close relationship in a number of regional consortia, the COPPUL, the OCUL, *Bureau de Cooperation Interuniversitaire* (BCI) [formerly the *Conférence des recteurs et des principaux des universités du Québec* (CREPUQ)], and the Council of Atlantic University Libraries/*Conseil des bibliothèques universitaires de l'Atlantique* (CAUL/CBUA). The Partnership, which is the provincial and territorial library associations of Canada working together, offers programs and services to members of these associations. These include continuing education through the Education Institute, the Career Centre, and the electronic journal initiative, *Partnership: The Canadian Journal of Library and Information Practice and Research*. The Library Subcommittee of BCI supports cooperative collection development across the 17 French and English university libraries of Quebec, sharing cataloging resources and the development of the *Bibliothèque de recherche virtuelle québecoise* (BRVQ), the gathering of statistical data, and training and professional development. The Capital Sm@rt Library/*la BibliothèqueGéni@le de la Capitale* is a consortium of libraries, archives, and museums in the cities of Ottawa and Gatineau including the Ottawa Public Library, LAC, NRC-CISTI, and the libraries of Carleton University, the University of Ottawa, Algonquin College, and the Dominican University College as well as the Canadian Museum of History, the Canadian War Museum, and the National Gallery. All of these institutions' online catalogs have been added to the SmartLibrary Z39.50 gateway for federated searches to facilitate direct loans. Canada's archives have collaborated to establish ArchivesCanada as the search portal for access to fonds level description and increasingly to more detailed finding aids. The Toronto Public Library, the Regina Public Library, the Halifax Public Libraries, and the Vancouver Public Library have been working collaboratively on a joint project called Working Together: Library–Community Connections funded by the federal government. The project is seeking to develop better ways of serving socially excluded communities and individuals who have not traditionally used library services. In Vancouver, the focus of the project is on the poor, in Toronto on immigrants, in Regina on indigenous peoples, and in Halifax on a socially disadvantaged community. For more than a decade, the Toronto–Dominion Bank has worked with the TPL, LAC, and, now, almost 2000 public libraries to offer the successful TD Summer Reading Club to engage school children in improving reading skills each summer.

Multilingual Services

Given the diversity of Canada's population and its multilingual and multicultural character, Canada's libraries have recognized for some years the importance of building collections and providing services that better respond to information needs.

Indigenous Services and Collections

The range of services provided to Canada's indigenous peoples and the kind of collections and information they need to access are very broad. In Nunavut, the only jurisdiction in Canada with a majority of the population being of indigenous descent, a consortium of the legislative, college, public, and court libraries has worked together on a common integrated library system to provide information in a shared Web-based public access catalog with Inuit syllabics. Many of Canada's indigenous peoples live on reserves where there are a variety of services provided. School libraries play a particularly strong role. Some First Nations' territories are in the remote parts of the country, with little access to books. In 2004, Honorable James Bartleman, then lieutenant governor of Ontario, issued a call for books. The public response was overwhelming and 850,000 quality used books were delivered to remote northern communities. This initiative has continued under his successors and is now the lieutenant governor's aboriginal literacy program (see http://www.johnschofield.com/index.php/portfolio/lieutenant-governors-literacy-programs/). Given the challenges provided by a strong oral tradition and insufficient numbers of published materials for and about indigenous peoples, there has been a major attempt both to stimulate publishing and to build libraries and library infrastructure in ways most relevant to First Nations peoples. Library education has been a concern, and for many years the National Library supported a program to send indigenous students to graduate library programs. Services to indigenous youth and adults living in urban settings are also a particular challenge.

Services for People with Disabilities

Roch Carrier, National Librarian of Canada from 1999 to 2004, established the Task Force on Access for Information for Print-Disabled Canadians. Their report, "Fulfilling the Promise: the Report of the Task Force on Access to Information for Print-Disabled Canadians," made recommendations in a broad number of areas. In response to these recommendations, as well as the Federal Disability Agenda of the federal government, a number of significant achievements are being put in place including an Electronic Clearinghouse for Alternative Format Production. The pilot project for the Clearinghouse was supported by the Council on Access to Information for Print-Disabled Canadians (an advisory group to the Librarian and Archivist of Canada), the Canadian Publishers' Council, the Association of Canadian Publishers, *l'Association nationale des Éditeurs de Livres*, and leading Canadian alternative format producers with funding from Social Development Canada.

CONCLUSION

Libraries and archives in Canada have come of age in the digital era. While operating in a decentralized federation, Canada's libraries and archives and their various consortia, associations, and councils actively pursue opportunities to advance services and awareness. It is an exciting time: challenging but dynamic as technology and public expectations must be balanced with resources and capacity. New approaches thrive. The Public Service of Canada was recently encouraged to respond to this changing environment through collaboration, innovation, and courage.[15] Librarians and archivists have been doing precisely this for some time.

APPENDIX: CHRONOLOGY

1. 1635—Collège des Jésuites organized the first university library in North America.
2. 1752—First press established in Halifax, Nova Scotia.
3. 1779—First bilingual subscription library established in Quebec City.
4. 1790—Ordinance "For the Better Preservation and Due Distribution of the Ancient French Records."
5. 1832—"Act for the Protection of Copyright" required deposit of printed books, charts, maps, and prints to secure copyright.
6. 1851—First Public Libraries and Mechanics' Institute Act passed by the Government of the United Canadas.
7. 1857—Appointment of Thomas B. Akins as Nova Scotia's Commissioner of Public Records.
8. 1867—Creation of the Canadian federation with the "British North America Act."
9. 1872—Appointment of Douglas Brymner and founding of the national archives.
10. 1876—Opening of the Library of Parliament building, Ottawa.
11. 1901—Ontario Library Association organized.

12. 1912—National archives detached from the federal Department of Agriculture and recognized by statute.

13. 1920—Provincial Archives of Quebec established and "Public Libraries Act" in Ontario revised.

14. 1923—"Canadian Catalogue of Books" started.

15. 1933—Carnegie Corporation supports a national study of library conditions and needs.

16. 1941—Rockefeller Foundation examines Canadian libraries.

17. 1943—Canadian Library Council incorporated.

18. 1946—Canadian Library Association formed.

19. 1948—W. Kaye Lamb appointed Dominion Archivist and given responsibility for preparing the way for a national library.

20. 1950—Canadian Bibliographic Centre established to compile a national catalog of the holdings of Canadian libraries and to publish bibliographies of current and retrospective Canadian publications.

21. 1951—Report of the Royal commission on National Development in the Arts, Letters, and Sciences 1949–1951 has a number of recommendations with respect the archives and libraries.

22. 1952—"National Library Act" passed.

23. 1967—Opening of Public Archives National Library building in Ottawa.

24. 1967—Passage of "Bibliothèque nationale du Québec Act."

25. 1975—Publication of To Know Ourselves: the Report of the Commission on Canadian Studies.

26. 1980—"Canadian Archives"—Report to the Social Sciences and Humanities Research Council of Canada.

27. 1985—Creation of the Canadian Council of Archives.

28. 1986—Creation of Public Lending Right Commission.

29. 1987—"National Archives of Canada Act."

30. 2004—"Library and Archives of Canada Act" proclaimed merging the National Archives and National Library of Canada.

31. 2006—Establishment of the Bibliothèque et Archives nationales du Québec (BAnQ).

32. 2015—Final Report of the Truth and Reconciliation Commission of Canada: Honouring the truth, reconciling for the future.

REFERENCES

1. http://www.conferenceboard.ca/hcp/details/education/adult-literacy-rate-low-skills.aspx. Accessed Nov 3, 2014.

2. In the case: Delgamuukw v. British Columbia 1997.

3. The Final Report of the Commission http://www.trc.ca/websites/trcinstitution/index.php?p=890.

4. http://umanitoba.ca/nctr/Accessed Mar 17 2017.

5. Wilson, I.E. "A noble dream": The origins of the public archives of Canada. Archives and Libraries: Essays in Honour of W. Kaye Lamb. Archivaria **1982–1983**, *Winter*, 15.

6. Carnegie Corporation of New York *Commission of Enquiry Libraries in Canada: A Study of Library Conditions and Needs*; Ryerson Press : Toronto, Ontario, Canada, 1933; 10; Ridington, J. Chicago, IL: American Library Association.

7. Ibid. p. 137.

8. C.F. McCombsReport on Canadian Libraries New York Public Library: New York, 1941; 4Report to the Rockefeller Foundation.

9. Canadian Library Council *Canada Needs Libraries*; Canadian Library Council: Ottawa, Ontario, Canada, 1944; 4. Reprinted from Ontario Library Review, November 1944.

10. Library and Archives of Canada Act. http://laws.justice.gc.ca/en/L-7.7/index.html.

11. Canadian Association of Research Libraries. Public Affairs and Advocacy Reports, 2012 through 2016. Available at www.carl-abrc.ca accessed Mar 15, 2017. And "Time Line: The Closure of Canadian Government Libraries, Archives and Research Collections. 2014" Canadian Association of Professional Academic Librarians. www.capalibrarians.org accessed Mar 15, 2017. And "Suspicions over library consolidation" in The National Post. Jan 7, 2014.

12. Gadoury, L.; Nahuet, R.Towards an understanding of the archival discipline in Quebec in Archivaria Spring 2005 # 59 and p. 4, Couture, Carol: Taking Stock: The Evolution of Archival Science in Quebec in Archivaria Spring 2005 # 59 p. 28.

13. Ibid, p 6.

14. Canadian digital information strategy. http://www.collectionscanada.gc.ca/cdis/012033-1001-e.html.

15. 19th report to the Prime Minister on the Public Service of Canada. 2012, http://www.clerk.gc.ca/eng/feature.asp?pageId=300.

Canadian Heritage Information Network (CHIN)

Shannon Ross
Paul M. Lima
Canadian Heritage Information Network (CHIN), Gatineau, Quebec, Canada

Abstract

A national center of museum excellence, the Canadian Heritage Information Network (CHIN) enables Canada's museums to engage audiences through the use of innovative technologies.

The Canadian Heritage Information Network (CHIN) supports an active network of more than 1350 not-for-profit member heritage institutions of all sizes and disciplines. Since 1972, this special operating agency of the Department of Canadian Heritage has been connecting museums with Canadians, enabling the creation, management, presentation, and preservation of digital heritage content for current and future generations.

To achieve this, CHIN pursues three strategies: mobilize and support collaborative networks of heritage institutions and research partners; provide skills development resources for heritage professionals; and support the development and presentation of digital heritage content.

Through its programs and online services, CHIN makes it possible for heritage professionals and volunteers to:

- interact and learn with colleagues through http://www.pro.chin.gc.ca;
- explore and share collection records through a growing online national inventory titled Artefacts Canada;
- reach and captivate their audiences, including youth, through virtualmuseum.ca—Canada's virtual national museum and the Department of Canadian Heritage's most popular online destination;
- engage teachers and students through the collaborative development of interactive online spaces and accompanying digital learning resources.

OVERVIEW

With the rapid development of new technologies, the heritage community is being presented with new and important opportunities. The Internet, more than just another medium, is a new environment in which heritage institutions have an active role to play. In order to adapt to and lead by example in this new domain, they must develop the necessary skills and abilities while making themselves relevant and meaningful to the world's online audiences.

In this regard, CHIN, in close collaboration with its members, is increasing the creation of Canadian online content displaying the diversity of our heritage. As of today, more than 1350 Canadian heritage organizations are actively engaged in the network.

Through its professional Web site and related activities, CHIN helps to strengthen the knowledge and abilities of the Canadian heritage community with respect to the creation, presentation, management, and preservation of digital content. The CHIN site provides heritage professionals with easy access to extensive resources that draw from CHIN's own work and research as well as other centers of expertise around the world.

These online resources include:

- A national inventory of more than 3.5 million museum objects, including cultural and historical artifacts and natural science specimens.
- A forum in which heritage professionals and volunteers can learn, exchange knowledge and ideas, and collaborate.
- Bibliographies and data dictionaries.
- Research and reference tools.
- Publications on intellectual property, digital content, collections management, and standards.
- Courses on image digitization and automated collections management systems.
- Information on professional events and career opportunities.

In addition to its online services, CHIN works nationally and internationally, offering workshops, convening meetings on subjects of common interest to its membership, as well as sponsoring and attending conferences and events that provide opportunities for heritage professionals to increase their expertise. Underlying all of CHIN's activities is a continued, collaborative research program designed to advance our collective understanding of issues associated with new technologies and digital content, particularly in the areas of intellectual property and online audiences.

CHIN also represents the Canadian heritage community internationally in projects that facilitate access to

British-Careers

Encyclopedia of Library and Information Sciences, Fourth Edition DOI: 10.1081/E-ELIS4-120043865

heritage information, and has been recognized for its work in developing standards for museum documentation. This work encompasses a broad range of information types and the technical protocols necessary for the exchange of information in a widely distributed environment.

CHIN's internationally recognized and most commonly accessed initiative, the Virtual Museum of Canada (VMC), showcases Canadian museums and their collections to a worldwide public audience. Through online exhibits, free games, and images, the VMC integrates art, culture, and heritage from Canadian museums. Designed in part to support classroom learning, it includes interactive lesson plans and learning resources for teachers and students. At the same time, anyone can access the site to learn about and explore the treasures of Canada's museums or browse listings of museums, events, and live exhibits across Canada.

CHIN also manages the international Bibliographic Database of the Conservation Information Network (BCIN), the Web's most complete bibliographic resource for the conservation, preservation, and restoration of cultural property. While previously part of CHIN's subscription-based research and reference resources, BCIN is now offered free-of-charge.

BACKGROUND

The 1970s: Founding of the National Inventory Programme

CHIN began as the National Inventory Programme (NIP) in 1972. It was created in response to the 1970 UNESCO Convention on Means of Prohibiting and Preventing the Illicit Import, Export, and Transfer of Ownership of Cultural Property, of which Canada was an early signatory. The 1972 National Museums Policy proposed the creation of an inventory of the cultural and scientific collections held by public institutions in Canada. The mandate of NIP was to create a computerized national inventory of Canadian cultural and scientific collections, to facilitate the sharing of collections information, to conduct applied research and development on information management standards and technology, and to advise museums and the heritage community in these areas.[1] Later policies extended the mandate to include other online resources.

The establishment of the National Inventory Programme resulted in the compilation of Canadian heritage information in three National Inventories—databases of humanities and natural science collections, as well as archeological sites. At first, museums sent their paper collections catalogs to CHIN for automation, but as technology advanced the National Inventories were maintained by the participating museums through dial-up access to CHIN's mainframe computer, with support from CHIN.

The National Inventories became an important resource, used for research, comparative cataloging, and exhibition planning, but also as a tool for individual institutions to manage their collections. CHIN also provided museums with the capability to share information through electronic mail long before the Internet made this a common practice.

In support of these activities, CHIN became active in the development and implementation of documentation standards, promoting the consistent documentation of museum information at the institutional and national level, and contributing to the development of international standards initiatives (Fig. 1).

The 1980s: Expansion of Services

In 1987, CHIN added the Conservation Information Network (CIN) and a series of other reference databases to its list of resources. CHIN's resources were offered by dial-up access, free of charge to Canadian heritage institutions, and as a subscription service to others. Beginning in the late 1980s, CHIN began to work with the Canadian heritage community to assess emerging technologies for disseminating museum information. Several projects involving CD-ROM and CD-I production were completed, providing valuable experience to all participants.

The 1990s: A New Mission

In 1995, CHIN engaged in a major review of its programs, in consultation with its client community. With the improvements in desktop technology and the advent of the Internet, CHIN and Canadian museums visualized an environment where museums would create rich public information resources which would be available electronically. CHIN developed a new mission to promote the development, presentation, and preservation of Canada's digital heritage content for current and future generations of Canadians.[2]

In October 1995, all of CHIN's products and services, previously available only by dial-up access to CHIN contributors and subscribers, were launched on the Web. CHIN, working with the Canadian heritage community, also developed new collaborative products. CHIN's first virtual exhibition, "Christmas Traditions in France and Canada," released in 1995, served to bring museum content to new audiences while increasing the participants' experience in providing content for the Web.

Another new product that took advantage of the Internet to promote museums and reach new audiences was the Guide to Canadian Museums and Galleries. The Guide enabled museums to maintain their own institutional information in a searchable Internet directory—this was especially important for small museums that would not otherwise have had a presence on the Internet. CHIN also provided Internet accounts and training to museums as a

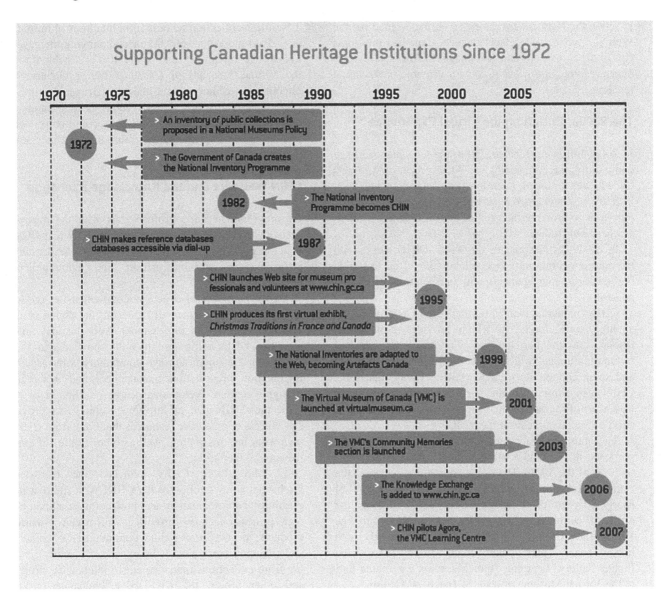

Fig. 1   A timeline of CHIN's most significant initiatives between 1972 and 2007.

means to participate in the Internet initiatives that were available, and to take advantage of the growing pool of resources for museums on the Web.

Also in 1995, as personal computers became more powerful and commercial collections management software became widely available, CHIN began assisting museums with the transition of their collections data from the CHIN mainframe to in-house collections management systems. This assistance included an assessment of a wide range of available software packages, conducted by CHIN and a team of museum professionals. The transition to in-house collections management software was accomplished by spring of 1998.

CHIN continued its participation in the development, promotion, and implementation of information standards, within the Canadian museums community and internationally. CHIN's participation in the Consortium for the Computer Interchange of Museum Information (CIMI) and the International Committee for Documentation (CIDOC) of the International Council of Museums enabled the international community to benefit from the expertise of Canadian museums while yielding important results that could be applied to the interchange and management of museum information in Canada.

As museum information became more accessible on the Internet, CHIN began producing publications to address issues of copyright and protection of museum data and images on the Web. CHIN also began to provide information to museums on image digitization and collections management through a series of publications, online resources, and workshops.

In 1996, CHIN developed the Heritage Forum, a searchable online journal for heritage professionals with content contributed online by CHIN and users alike.

In 1999, the National Inventories were redesigned for the Web and renamed Artefacts Canada. Canadian museums contribute collections data and images to Artefacts Canada over the Web, and it has become a valuable professional resource.

The 2000s: Enriching the Virtual Experience

In 2004, following feedback from the Canadian heritage community, the Archaeological Sites section of Artefacts Canada was removed. Learning With Museums, launched in 2000, was developed by CHIN with the goal of enabling museums to promote their online educational materials to teachers and students.

In 2001, CHIN launched the VMC. CHIN works with the museum community in a collaborative effort to enable the creation, management, and presentation of digital content.

CHIN maintains two Web portals: the CHIN Web site at http://www.chin.gc.ca, which provides professional resources to the heritage community and the VMC at virtualmuseum.ca, which showcases Canadian museums and their collections to a worldwide public audience. CHIN also administers the Virtual Museum of Canada Investment Programs, which support museums in the creation of digital content for the VMC.

The museum information from the Guide was re-purposed in the Find a Museum and Calendar of Events sections of the VMC. The rich resource that began as the National Inventories, which represents the results of nearly 40 years of collaboration between CHIN and the museum community, was also incorporated into the VMC—the Artefacts Canada collections records with images became part of the Virtual Museum of Canada Image Gallery. Learning With Museums was made part of the Virtual Museum of Canada Teachers' Centre.

Since the debut of the VMC, people from around the world have embraced this gateway to the exciting multimedia content created by Canadian museums. In its inaugural year, the VMC attracted almost 3 million visitors from more than 140 countries.

To coincide with the launch of the public-oriented VMC site in 2001, the focus of the CHIN Web site shifted to become exclusively a resource for heritage professionals. Artefacts Canada and the Heritage Forum remained on the CHIN site as professional resources, in addition to the publications, courses, and reference databases that CHIN offers.

PROGRAMS AND ACTIVITIES

Artefacts Canada

A growing national inventory, Artefacts Canada (also at http://www.chin.gc.ca) provides access to more than

3.5 million collection records from hundreds of museums, as well as more than 600,000 accompanying images. All images accompanying records are also accessible through the Virtual Museum of Canada's Image Gallery. The ongoing contributions of Canadian heritage institutions strengthen Artefacts Canada's value as a research and reference resource for the museum community, while increasing each participating museum's online profile through the VMC.

CHIN Web Site and the Knowledge Exchange

By providing heritage professionals with the means to share their collections and expertise, CHIN helps Canadian museums establish themselves as world leaders in the effective use of information and communication technologies.

CHIN members also access comprehensive products and services that address all aspects of the digital content process. Through its professional Web site and related activities, CHIN helps strengthen the knowledge and abilities of the Canadian heritage community with respect to the creation, presentation, management, and preservation of digital content. Welcoming nearly 4 million visits each year, the CHIN site at http://www.chin.gc.ca provides easy access to extensive resources that draw from CHIN's own work and research as well as other centres of expertise around the world.

In May 2006, CHIN launched the Knowledge Exchange, a new section of the CHIN Web site that fuels creativity, communication, and collaboration among heritage professionals by providing self-paced e-learning products, peer-to-peer online communities of practice, and engaging presentations by experts. On subjects ranging from collections management to intellectual property and other topics, the Knowledge Exchange provides museums with a wealth of knowledge on how they can better serve Canadians through emerging technologies. In 2009, the Knowledge Exchange section was expanded to include the entirety of CHIN's professional development resources. Rebranded as the Professional Exchange, the website can be freely accessed at http://www.pro.chin. gc.ca.

The VMC

For CHIN and its member institutions, the VMC is by far one of the network's proudest accomplishments. Administered by CHIN, Canada's national online museum brings a rich collection of engaging high-quality online content into Canada's homes, schools, and workplaces.

Through the VMC's digital content, heritage institutions can present their collections to a much wider audience than ever before, while showcasing their innovative use of digital technologies to the world. The Virtual Museum of Canada currently features over 500 online exhibits that bring to life Canada's treasures and stories; nearly 600,000 images of museum objects; more than 140 interactive games for children of all ages; and detailed listings for nearly 3000 Canadian heritage institutions.

In 2007, the VMC began enabling museums to further engage Canada's educators and learners in active learning through a new pilot project entitled the Agora Research Initiative. Responding to the growing use of the Internet by educators and students, the VMC is adding innovative tools and content (in the form of texts, audio and video files, and animations) that enable teachers to create their own lesson plans and projects, interact with students online, and provide interactive learning experiences.

Virtual Museum of Canada Investment Program: Virtual Exhibits Component

CHIN supports the development of multimedia online exhibits, games, and learning resources to be showcased on the VMC. Under a competitive process, heritage organizations submit proposals to the Virtual Museum of Canada Investment Program to produce bilingual virtual exhibits, interactive games, and educational tools that engage diverse audiences through quality research and appealing presentation.

Proposals are reviewed by an Editorial Board composed of representatives from the museum community, as well as other sectors such as the publishing, education, library, and new media.

Virtual Museum of Canada Investment Program: Community Memories Component

The Community Memories Program helps smaller museums to engage and involve their communities in the sharing of their local history through simple technology. The range of possible subjects for a Community Memories exhibit is endless. Examples include a town's main industry, a historical event, traditional crafts, or a way of life. These productions allow communities to rediscover the people and events that shaped their past and present, bringing otherwise dispersed records into a coherent, engaging narrative. As well, these online exhibits provide Canadians from coast to coast with insight into the values and experiences that have shaped the country's collective identity. Museums with successful proposals are supplied with user-friendly software, and museums with five or fewer full-time employees also receive a standard investment.

Research

CHIN conducts research into how museums can best approach issues such as information management, digital technologies, intellectual property, and standards. It draws on this knowledge to provide research and reference resources, as well as training opportunities for heritage professionals and volunteers, online and in-person.

CHIN is committed to staying on the leading edge of digital heritage research and technology that enhances public engagement and participation. Accordingly, it mobilizes and supports collaborative networks of heritage institutions and research partners, which leads to the publishing of original research. CHIN also works to improve heritage information management, maximize benefits for knowledge creators and distributors, and increase the ability of museums to capitalize on new business opportunities.

International Activities

CHIN works nationally and internationally, offering workshops, convening meetings on subjects of common interest to its membership, as well as sponsoring and attending conferences and events that provide opportunities for heritage professionals to increase their expertise. Underlying all of CHIN's activities is a continued, collaborative research program designed to advance our collective understanding of issues associated with new technologies and digital content, particularly in the areas of intellectual property and online audiences.

CHIN also represents the Canadian heritage community internationally in projects related to access to heritage information, and has been recognized for its work in developing standards for museum documentation. This work encompasses a broad range of information types and the technical protocols necessary for the exchange of information in a widely distributed environment.

Amongst other international initiatives, in recent years CHIN has taken a leading role in the digital Cultural Content Forum (dCCF), a partnership that brings together government agencies around the world that are dedicated to enabling heritage institutions in the fields of information and communications technologies.

MEMBERSHIP

CHIN and Canadian heritage institutions work together to strengthen our collective ability to create, present, and manage Canadian online content. This collaboration has resulted in the internationally valued Web site for heritage professionals at http://www.pro.chin.gc.ca.www.chin.gc.ca, and the highly successful VMC portal at virtualmuseum.ca.

Public, not-for-profit, Canadian museums and other heritage organizations are eligible to reap the many benefits of CHIN membership, including participation in the VMC. Membership is available at no cost to eligible institutions. Membership benefits include, but are not limited to:

- Extensive marketing support to heighten the collective profile of member institutions in schools and among the domestic and international Web surfing public (e.g., billboard campaigns, press relations, online marketing, contests, search engine optimization).

- Access to CHIN's multifaceted expertise and ongoing technical support.
- Ability to have an important voice in shaping CHIN initiatives, and to participate in the development of programs.
- Presence on the VMC.

CONCLUSION

For over 35 years, CHIN has been an international leader in the enabling of heritage institutions to engage audiences through the use of innovative technologies. CHIN has a strong and privileged relationship with hundreds of heritage institutions and organizations across Canada, and around the world; and it actively nurtures its partnerships in an effort to leverage resources and expertise for the benefit of the Canadian and international heritage community.

CHIN will continue to work closely with the heritage community in the coming years to further enhance and enrich online communication, collaboration, creativity, and access to heritage information. As emergent technologies take hold of public use and interest, CHIN will continue to assist museums and heritage institutions in finding ways to take part in these new phenomena. For instance,

the growing use of iPods and other MP3 players as educational tools enables museums to reach both wider audiences and target demographic groups. We are presently offering presentations from technical experts, as well as step-by-step instructions for transforming heritage content into effective podcasts.

CHIN is also continuing to facilitate knowledge transfer among information and heritage experts. As our members take part in new digitization projects and the development of digitization standards for museums, CHIN will continue to support the transfer of this insight to all members so that their institutions can benefit from broader knowledge.

Finally, CHIN continues to support a cultural environment where museums can create rich public heritage experiences which integrate the virtual world with real-life exhibits and experiences. The organization's longstanding mission to promote the development, presentation, and preservation of Canada's digital heritage content for current and future generations of Canadians will remain a top priority.

REFERENCES

1. Canadian Heritage Information Network, Communications Canada. *Regional Networks Policy*, CHIN: Ottawa, 1992; 3.
2. Canadian Heritage Information Network.; Network News No. 7; CHIN: Ottawa, Summer 1995.

Canadian Library Association (CLA)

Donald I. Butcher
Canadian Library Association, Ottawa, Ontario, Canada

Abstract

The Canadian Library Association/*association canadienne des bibliothèques* is a national, not-for-profit voluntary organization composed of librarians, information professionals, libraries, and information services. It is Canada's largest broad-based national organization focusing on and serving the Canadian library and information community through advocacy, continuing professional development, and the provision of direct products and services.

INTRODUCTION

The Canadian Library Association/*association canadienne des bibliothèques* (CLA) is a national, not-for-profit voluntary organization composed of librarians, information professionals, libraries, information services and suppliers to the library and information sector. It is Canada's largest broad-based national organization focusing on and serving the Canadian library and information (LIS) community. The Canadian Library Association was established in Hamilton, Ontario, Canada in 1946, and was incorporated in 1947.

Canadian Library Association's mission is to be the national voice of the LIS community and to build a stronger LIS community by providing services to its members. The Canadian Library Association provides a mix of tangible services such as continuing professional development, and intangible services such as building the Canadian LIS community by facilitating networking among members. While predominately English-language, CLA provides some services in French.

To assist in providing these services, the association has created five divisions based on type-of-library or role within the LIS community; 24 interest groups based on areas of practice; and numerous and ever-changing committees and task forces to undertake specific work. The five divisions are the Canadian Association of College and University Libraries, including the Canadian Technical and College Libraries section; the Canadian Association of Public Libraries, including the Canadian Association of Children's Librarians section; the Canadian Association of Special Libraries and Information Services, with chapters in Calgary, Edmonton, Manitoba, Ottawa, Toronto, and Atlantic Canada; the Canadian Library Trustees Association; and the Canadian Association for School Libraries.

ROLES AND ACTIVITIES

The Canadian Library Association has developed into a sectoral association, with activities combining the classic roles of professional and trade associations. The strategic plan adopted in 2004 focused on advocacy and government relations, and continuing professional development, while maintaining direct member services and building a sense of community among practitioners. A new strategic plan is to be released in 2008.

As the major national association, CLA liaises extensively with other national library associations, including the *Association pour l'avancement des sciences et des techniques de la documentation* (ASTED), the Canadian Association of Research Libraries, and the Canadian Urban Libraries Council; other associations in related heritage fields such as the Canadian Museums Association and the Canadian Council of Archives; and with other national and international associations such as the American Library Association (ALA), the United Kingdom's Chartered Institute of Library and Information Professionals (CILIP), and the International Federation of Library Associations and Institutions.

As the main national voice of the Canadian LIS community, CLA advocates for and defends the core values of librarianship such as, for example access to information or freedom of expression. The Canadian Library Association advocates on policy issues that impact library users, including federal legislation on intellectual property rights, right to privacy, and other public policy debates.

The Canadian Library Association also advocates on behalf of libraries and librarianship themselves, particularly on federal programs that directly or indirectly provide funding for library services. Examples of this are the Government of Canada's Community Access Program and Canada Post's Library Book Rate.

In its role as a provider of continuing professional development, CLA holds Canada's largest national library conference, which is moved around the country to facilitate access by those with limited professional development funding. Not wishing to duplicate the conferences of Canada's provincial and regional library associations, CLAs conference is a high-level, "pinnacle" conference targeted at the leaders and senior management in the

British-Careers

Encyclopedia of Library and Information Sciences, Fourth Edition DOI: 10.1081/E-ELIS4-120044708

Canadian Library Association (CLA)

Canadian LIS community. Also to advance the profession, CLA publishes a bimonthly magazine that provides in-depth exploration of topics in librarianship.

The Canadian Library Association has taken a leading role in fostering and disseminating research into human resources in LIS services, both by supporting a research project led by the University of Alberta's 8Rs Research Team, and by gaining funding from the Cultural Human Resources Council (a federally supported agency) for an analysis of the gaps between the sector's human resources needs and the training available. This work follows up on an earlier CHRC-funded study of human resources in the heritage community proposed by CLA, ASTED, the Canadian Museums Association, and the Canadian Council of Archives.

To help build the LIS community, CLA provides scholarships to attend library school, grants for research and education in library and information science, awards for excellence in literature for children and young adults, and awards for professional excellence and innovation in the field of library and information science.

The Canadian Library Association also provides direct services to libraries. Since 1996, CLA has administered funding under Young Canada Works in Heritage Institutions (part of the Government of Canada's Youth Employment Strategy) for summer student employment in libraries. Since 2005, CLA has hosted and administered the database of libraries eligible to access Canada Post's Library Book Rate, a set of reduced postal rates for interlibrary loan.

Along with the ALA and the CILIP, CLA copublishes the *Anglo-American Cataloguing Rules*; CLA is a member of the Committee of Principals, established in 1989 to oversee the development of the code. A new online version of the code called *RDA: Resource Description and Access* is under development for release in early 2009. Since 1993, CLA has been the exclusive Canadian distributor of all ALA Editions titles (selected titles were handled prior to that date); and CLA publishes or distributes other works.

GOVERNANCE

The Canadian Library Association is governed by a 12-person Executive Council, with six directly elected members, five division presidents sitting as voting members *ex officio*, and the Executive Director/Secretary serving as a nonvoting member *ex officio*. Each of the five divisions has its own executive boards.

EARLY HISTORY

The pre-establishment history of CLA and an account of the first 25 years could take an entry on its own. Readers

may wish to consult *The Morton Years: The Canadian Library Association, 1946–1971* by Elizabeth Hulse.[1] and several articles by Basil Stuart-Stubbs published in *Feliciter* and on the CLA Web site.[2–9] The program of activities developed in 1946 included assuming the duties of the Canadian Library Council, established in 1941. The council had an executive office in Ottawa, published the *Canadian Library Bulletin* and other publications, presented briefs to the government, began microfilming Canadian newspapers of historic importance, and coordinated a library clearinghouse.

In addition to these activities, the priorities outlined in 1946 for the CLA/*Association canadienne des bibliothèques* (the name was changed in 1969 to Canadian Library Association, but the original name was brought back in 2006) included the establishment of a national library for Canada and continuing the microfilming program.

As well, CLA established an indexing service for Canadian periodicals and documentary films (*Canadian Periodical Index*, subsequently sold), supported the publication of Canadian reference materials, developed a recommended salary scale for librarians, and created standards for libraries. A triumph of advocacy, the National Library of Canada was established by an Act of Parliament in 1953. *Feliciter*, the association's newsletter, began publication in 1956 and, following the cessation of the *Canadian Library Journal* in 1992, serves as the membership magazine. The driving force behind the successes of the first 22 years, Elizabeth Homer Morton, retired as executive secretary of CLA in 1968.

The 1970s–2000

A new constitution, adopted in 1973, provided a decisive role for representatives of the provincial library associations as full voting members of the CLA Council. Also in 1973, CLA formed five divisions to provide services on a type-of-library (academic, public, school, and special) or role (trustees) basis.

In 1989, following the CLA President's Commission on Organization (known as CLAPCO; 1987), CLA reorganized its governance. The bicameral structure (consisting of board and council) was replaced by a unicameral executive council. The position of second vice-president was eliminated, as were the provincial representatives, and the number of councillors was reduced from six to three. In 1991, the position of ASTED liaison (the chair of the CLA/ASTED Liaison Committee) was removed from the executive council and the committee disbanded.

Professional development, advocacy, library leadership, membership development, and fiscal development were the strategic priorities throughout the early 1990s. The major issues were nationwide reductions in funding to tax-supported libraries and the concomitant addition of electronic materials and revisions to the Copyright Act.

The Public Lending Right Commission, consisting of authors, librarians, and publishers, was established by the federal government in 1986 to administer a program of payments to Canadian authors for their eligible books cataloged in libraries across Canada. The Canadian Library Association is represented on the Commission.

Working with the federal government and ASTED, CLA organized the National Summit on Information Policy, in 1992, to allow Canada to maximize the benefits stemming from information resources, and, in 1995, CLA launched Library Advocacy Now! (LAN!) to train advocates and champions among the library community. In 2000, the CLA/LAN! Training Institute was established.

When Canada's Copyright Act was amended in 1988, amendments favorable to creators, specifically enabling the establishment of copyright collectives, were included. A "Phase II" set of amendments to address the concerns of users was promised immediately, but these amendments were not, in fact, introduced until 1996. Successful lobbying by the creator community resulted in reductions in the already limited library exceptions being passed in the current Copyright Act in 1997. The exceptions for libraries, archives, and museums went into effect in 1999 when the accompanying regulations were published.

The amendments to the Copyright Act passed in 1997 also established rules limiting "buying around" by libraries—importing a book from a distributor outside Canada when it is available in Canada through an exclusive distributor. The regulations governing book importation were also published in 1999.

OLAM (which originally stood for Online Account Management Service), purchased in 1987 to provide discounts to members on online database services, went into bankruptcy in 1999 and was dissolved in 2000.

Working with publishers, booksellers, and writers, CLA played a lead role in the Canadian Book Summit, held in 1999 to celebrate Canada's most successful cultural product: books.

As part of its work on copyright and in support of the continued public funding of libraries, CLA has had a watching brief on international trade initiatives: in 1998, on the Multilateral Agreement on Investment, and, since 1999, on the World Trade Organization and the World Intellectual Property Organization.

From 2000 On

The Canadian Library Association's advocacy work with the Canadian government has been impacted by government instability in the first years of the new century, with multiple federal elections and Ministerial shuffles since 2000.

In advocacy, work on "Phase III" amendments to the *Copyright Act*, which deal with copyright in a digital environment, continues. The CLA is a member of the Copyright Forum—14 national library, education, academic, museum, and archive associations representing users' concerns.

With the financial support of a number of international partners, CLA took the lead to sponsor two analyses of the impact of international trade treaties on libraries.[10,11]

Advocacy continues on both national policy issues such as intellectual property rights and right to privacy, and on funding for libraries and information services.

A joint conference with the American Library in Toronto in June 2003 was greatly impacted by incidents of Severe Acute Respiratory Syndrome (SARS) in the city. Using librarianship's core values of access to information and rational decision-making, CLA urged ALA not to relocate the conference. The American Library Association made the courageous decision to stay in Toronto, and no delegates were impacted by SARS.

The Canadian Library Association organized the second National Summit on Libraries and Literacy in 2006, to review progress since the first Summit in 1995, and to develop plans for future collaboration between libraries and adult literacy organizations.

To build the LIS community, CLA Presidents, and Executive Directors visit the universities and community colleges that teach library and information science or library technical studies, meeting the students and informing them of the role that CLA plays in the community and in their future working lives.

CURRENT TRENDS

The Canadian Library Association continues to be the national voice for the LIS community. It advances the profession through continuing professional development and the support of research; and builds community spirit through its divisions and interest groups.

The expected retirement of the "baby boom" generation of librarians within the next 10 years has put library human resources and succession planning at the top of the issues list for most libraries. The Canadian Library Association is leading efforts to publicize and operationalize the results of *The Future of Human Resources in Canadian Libraries*,[12] which provides the Canadian LIS community with the data needed to create new approaches to human resource management.

A crisis in school librarianship continues, with teacher–librarians being reassigned to classrooms and a resulting loss of service in school libraries.

Like most other North American associations, a large number of CLA members joined between the late 1960s and the mid-1970s, and many of them will be retiring in the next 5–10 years. As well, many of this "baby boomer" cohort are in senior management positions and either do not have the time or have already served in senior positions within the association. The nature of volunteering has changed, and one of CLAs biggest challenges is to adapt itself to the behaviors of potential new members so that they replace those retiring, and to encourage these new members to play an active role in CLA.

The implementation of leading-edge membership database, Web site content management software, and the collaborative tools of blogs and wikis form the operational strategy as CLA strives to serve the LIS community.

The Canadian Library Association/*Association canadienne des bibliothèques* can be reached at 328 Frank Street, Ottawa, ON K2P 0X8 Canada, or via its Web site: http://www.cla.ca/.

REFERENCES

1. Hulse, E. *The Morton Years: The Canadian Library Association 1946–1971*; Ex Libris Association: Toronto, Ontario, Canada, 1995.
2. Stuart-Stubbs, B. 1912: The ALA meets in Ottawa. Feliciter. **1997**, *43*(6), 46–51.
3. Stuart-Stubbs, B. 1912: The ALA meets in Ottawa Part 2. Feliciter. **1997**, *43*(7/8), 68–72.
4. Stuart-Stubbs, B. 1925: CLA launched . . . in Seattle? Feliciter. **1998**, *44*(5), 20–25.
5. Stuart-Stubbs, B. 1925: CLA launched . . . in Seattle? Part 2. Feliciter. **1998**, *44*(6), 26–3134.
6. Stuart-Stubbs, B. 1927: CLA born again . . . in Toronto? Feliciter. **1999**, *45*(2), 98–105122.
7. Stuart-Stubbs, B. 1927–1930: The muddle years. Feliciter. **2000**, *46*(3), 148–149.
8. Stuart-Stubbs, B. 1927–1930: The muddle years [long version]. Available at http://www.cla.ca/resources/muddle.htm (accessed March 2001).
9. Stuart-Stubbs, B. 1934–1946: The long last lap [long version] 2004. Available at http://www.cla.ca/resources/1934_1946.htm (accessed October 2007).
10. Shrybman, S. An assessment of the impact of the general agreement on trade and services on policy, programs and law concerning public sector libraries. In *Report. Prepared for the Canadian Library Association et al.* Sack Goldblatt Mitchell: Ottawa, CA, May 2001.
11. Tawfik, M.J. Is the WTO-TRIPS user-friendly?. In *Report Commissioned by the Canadian Library Association on the Impact of WTO-TRIPS on Public-Sector Libraries*, January 2005.
12. The 8Rs Research Team. *The Future of Human Resources in Canadian Libraries*; 8Rs Research Team: Edmonton, Alberta, Canada, 2005

BIBLIOGRAPHY

1. http://www.cla.ca.

Careers and Education in Archives and Records Management

Karen Anderson
Archives and Information Science, Mid Sweden University, ITM, Härnösand, Sweden

Abstract

Educators and trainers in the field of archives and records management face the challenge of providing for practitioners with widely varying educational backgrounds. At the same time, the professional environment has changed rapidly in recent years as well. Development of research programs is also crucial to the advancement of the profession and the survival of university-based professional education in the discipline. Competency standards for records management are discussed, including their appropriate role in training and education for practitioners. Examples of education and training programs from around the world are mentioned, although the coverage in this entry is not intended to be exhaustive. The role of professional associations as providers of training and in supporting and evaluating professional education programs is also addressed. The International Council on Archives Section for Archival Educators and Trainers provides a community of practice for educators and trainers. Its objectives and recent activities are described.

A wide and growing range of career choices is available to archivists and records managers. Traditional employers are national and regional archives, but there are many opportunities for specializing in aspects of the profession and the types of organizations served by the archives. Professionals with the necessary expertise required to manage digital services are in demand. Initiatives to support people of indigenous origin joining the profession and encouragement for the development of community archives are an area of growing interest.

INTRODUCTION

Education for the archives and records management profession is in tension between the need to provide an entry level qualification for new professionals and the need for generating a professional culture that is capable of finding a way forward through research while at the same time providing appropriately qualified educators for the next generation. Many countries do not have university-based programs and are reliant on short training programs for professional development, or sending prospective members of their profession overseas to gain qualifications or to further their education. At the same time, the profession must educate for a changing technological environment to ensure that sound strategies are developed for capturing, managing, and preserving born digital records. In a special issue of *Archival Science* dedicated to graduate archival education, Karsten Uhde noted: "never before in the history of archival systems has the daily work, the work material, and the whole job profile of the profession changed so quickly and radically as during the last 10–15 years."[1] Therefore, one of the current dominating interests in the profession is defining what knowledge and skills the recordkeeping professional needs, of which one manifestation is a drive to identify and document competencies for professional practice. Jeannette Bastian stated the dilemma of designing programs and developing curricula for a profession that must continue to preserve and manage the record of the past while developing strategies to capture, manage, and preserve records created in rapidly changing and evolving new technologies:

> Finding a balance between traditional archival subjects such as palaeography and new developments such as information management, insuring that students are prepared for an electronic information environment and promoting archival education as a professional discipline are all matters that deeply engage and also worry educators.[2]

Tom Nesmith also explores the issues in designing curricula for a changing archival environment,[3] while in the same issue Margaret Proctor cautions against a too-close alignment of curriculum with the shifting requirements of public policy and affirms the U.K. postgraduate programs' strategy of

> centr[ing] the educative process around the core elements of the profession—selection, preservation and provision of access; the many methods developed and developing for ensuring the effective performance of these functions; and, most importantly of all, the rationale for these functions, including the consequences for society if any aspect of them is neglected.[4]

Nevertheless, archivists and records managers must have expertise in legislation pertaining to data protection and freedom of information legislation and the ethical issues

Encyclopedia of Library and Information Sciences, Fourth Edition DOI: 10.1081/E-ELIS4-120044288

British–Careers

that arise when weighing freedom of access against protection of privacy.

While it is essential for a profession to know what it does, it must be able to validate that practice with a body of theory that is tested, confirmed, and extended through research. The tension between education for professional practice and research is an issue that preoccupies educators, most of whom constitute a small discipline interest hosted within larger university programs such as library and information science, computer science, or history. Nevertheless survival of programs within universities depends on the development of research programs that attract doctoral candidates and bring research grants to the university in order to provide the research expertise that will provide the necessary creative and investigative skills that will carry the profession forward. Furthermore, a doctoral degree is now the minimum requirement for appointment to a teaching post in professional education programs in universities. These programs must produce the next generation of professional educators.[5] This is not a new concern: Ann Pederson[6] pointed out the need to encourage archival research programs in universities in 1992, noting that among the barriers were low numbers of teaching programs and full-time professors in archival science; and Carol Couture reiterated the issue in 1996.[7] In the same issue of *Archivaria*, Barbara Craig[8] reviews Sir Hilary Jenkinson's thoughts on the need for focused archival education programs in his address at the inauguration of the first post-graduate course in England in 1947, noting that while his curriculum proposals are still relevant, education for research must be added to the archival education curriculum. In 2001, Carol Couture reported in *Archival Science*[9] on a three-year research program investigating education and research in archival science, commencing with an extensive literature review, which was followed by a questionnaire sent out to educational institutions, national archives, and researchers. Responses from more than 70 different countries were received. By 2004 Sue McKemmish and Anne Gilliland were able to say

> The past 15 years have seen unprecedented growth in the development of an archival research consciousness in the academy and in practice, as well as in scholarly awareness that the construct of *the archive*, and *recordkeeping* more generally, provides a rich locus for research and theorising. What has resulted is an unparalleled diversity of what is being studied and how.[10]

Nevertheless, Richard Cox succinctly summarizes the enormous task facing relatively small archival education programs when he says:

> it is difficult to imagine building doctoral programs, mentoring doctoral students, and guiding dissertation research if graduate educators have their time consumed

by teaching masters students, running workshops and supporting distance education.[11]

EDUCATION AND TRAINING

University programs for archives and records management are found associated with several different disciplines. Traditionally, courses have been associated with either history or library and information science programs, but it is no longer unusual to find them associated with computer science and information systems programs, reflecting the profession's need for skills and knowledge to manage and preserve records that are born digital and to provide appropriate Internet services and access for users. In the United States and Canada, although courses are taught at master level, there are only a few entire programs dedicated to archival science. Most are courses offered within library science or history programs, or taught as optional courses. University level courses devoted to records management are relatively rare in North America. Jeannette Bastian and Elizabeth Yakel[12] provide a description, based on comprehensive research, of both the availability of courses in archival science in the United States and Canada and what is being taught as "core" curriculum content for introductory archival courses. Two European institutions are dedicated entirely to the professional education of archivists for their respective national public service: The École Nationale des Chartes in Paris, France and the Archivschule in Marburg, Germany. In the United Kingdom archival education is positioned at master level, but all are full master programs dedicated to archives and/or records management, and applicants for the programs must complete a full year's appropriate work experience prior to commencement of their studies. Most European countries that have university-based archival education seat their programs at master level, but in Sweden and the Netherlands programs at both bachelor and master level are found, as is the case in China. Records management is included in the archival education programs taught in Scandinavia and the Netherlands. In Finland, the National Archives is a provider of education programs resulting in formal degrees, alongside university providers of archives and records management programs.[13]

In South America university-based archival education programs are available in Argentina, Uruguay, and in Brazil. To promote cooperation among archival educators in South America, Red Iberoamericana de Enseñanza Archivística Universitaria was formed in 2000 as a permanent Committee of the ALA (Latin American Archives Association). RIBEAU translates as "Latin American Network of Archival University Training." The purpose of RIBEAU is to promote networking and to seek common solutions to Latin American problems in the field of archival education. In Argentina, archival education is the subject of a debate between supporters of two types of degrees, on one hand, a specialized and independent course of archival education, and on the other hand, an integrated archival and library program. A third scenario considers a Professional level degree as a Librarian a pre-requisite for any archival training. In Peru the Archivo General de la Nación provides a training school that is not associated with a university and does not result in a formal degree.

In China, there is a strong tradition of archival education provided by a number of universities across the country, of which the Archives School at the Renmin University in Beijing is the largest. Programs range through vocational training courses, university-based courses at bachelor and master level, and doctoral degrees.

In Africa, university-based professional education for archivists is available in South Africa and at the Library and Archive School in Ibadan, Nigeria. An archival training program in French is also available in Dakar, Senegal. There is a strongly expressed need for assistance in developing capacity for archives and records management education in other African countries and in some countries of the Arabic-speaking world.

In Australia, university programs are also offered at both bachelor and master level, taking a recordkeeping continuum approach to archival science as described, for example, by Frank Upward and Sue McKemmish,[14] integrating what the life-cycle model separates into records management and archives. The first archival education program offered entirely by distance education without any requirement for students to attend on-campus commenced in 1994 offered by Edith Cowan University in Australia.[15,16] Caroline Williams reported on a survey of archival education programs at the International Congress on Archives in 2004,[17] at which date there were already over 30 institutions providing courses by distance education in a wide range of models, ranging from paper-based correspondence courses to full online delivery.

Competency Standards

The development of a standard set of competencies for archivists has been the subject of much interest in the profession for more than a decade, accompanied by a debate about whether or not competency models are an appropriate basis for professional education programs in higher education institutions. Marian Hoy[18] points out that there is no clear and universally accepted definition of competencies, but notes that:

> The National Training Information Service (NTIS) of the Australian Government defined competency standards as follows: "Competency Standards describe the skills and knowledge required for a person to operate effectively in the workplace. The standards have been defined by industry, are nationally recognised and form the basis of

training for that specific industry. Standards contain descriptors of outcomes to be achieved and criteria for performance."

Thus they provide benchmarks for assessment of achievement in training programs and are also useful for classifying job descriptions according to the level of performance required. Margaret Crockett[19] points out that competency standards also

> Facilitate recognition of skills across national boundaries; enable workforce mobility; improve professional knowledge and practice; help implement ARM standards; provide benchmark[s] for non-professional stakeholders.

In Australia, the *Records and Archives Competency Standard*[20] was first published in 1997, having been compiled with the active input of representatives from the Australian Society of Archivists and the Records Management Association of Australasia. Marian Hoy[21] provides a useful comparison of competency frameworks as an appendix to her article, which explores the structure, uses, and issues surrounding the Australian competency standard.

The National Training Information Service (NTIS) manages the maintenance and revision program for the Standard, which in turn forms the basis of Training Packages for records management training programs available through Innovation and Business Skills Australia[22] and provided by Technical and Further Education (TAFE) Colleges, which provide a nationally standardised system of vocational training and by other Registered Training Organizations. The NTIS maintains a system for assessing and registering trainers. Nevertheless, standards that document competencies are generally designed for and more suited as an evaluation tool for training than for education programs, which aim to provide a deeper and broader experience than the acquisition of a static set of skills.[5] Therefore the Australian Society of Archivists and the Records Management Association of Australasia developed a *Statement of Professional Knowledge* (2006)[23] as a basis for their work in accrediting and recognising university-based programs in archives and records management.

ARMA International's Education Development Group launched a project in 2005 to develop a set of competencies for records and information management professionals, which culminated in publication of the *Records and Information Management Core Competencies*[24] in 2007. The ARMA International project is a sound exemplar for approaching the development of a competency standard. It brought together an expert panel to create a first draft of the competency model, which was then validated against other models, including those from Australia and Canada. The second draft was then subjected to a survey of 2000 volunteers from the ARMA membership to validate the panel's work and contribute to the final weighting structure.[19]

In 2006, competencies for the archival profession were among the main topics for discussion at both the 7th European Conference on Archives in Warsaw and at the Society of Archivists (United Kingdom) annual conference. One outcome of the European conference was a resolution[25] to:

- "Encourage the European Branch of the ICA (EURBICA) and the Section for Professional Associations of ICA (SPA) to carry out a feasibility study for a project to develop a European competency framework for the archival profession;
- "Recognize the need to develop and sustain professional standards through certification procedures for individuals, organizations, and education programs;
- "Underline the need to advance the creation and development of virtual educational tools for promotion of online training through the use of new technologies."

The conference participants also resolved to "encourage the key educational institutions teaching archivists to:

- "Strengthen cooperation and common initiatives in order to increase the quality of archival education;
- "Improve tools for the assessment of the quality of archival education for accreditation."

The Role of Professional Associations in Education and Training

The two peak professional bodies for records managers and archivists play an important role in supporting professional education and training, as do many of their national counterparts. The International Council on Archives (ICA)[26] makes education and training one of its four strategic priorities, while ARMA International[27] is itself a provider of training which includes online courses. Provision of the ARMA certification program is contracted to the Institute of Certified Records Managers (ICRM).[28] Applicants receive the CRM designation by meeting both educational and work experience certification requirements established by the ICRM and by passing the required examinations. Two of the regional branches of ICA play an active role in the provision of training programs. EASTICA,[29] the East Asian Regional Branch, in cooperation with the Hong Kong University runs four-week intensive courses (2003, 2004, 2007, 2008) leading to the award of a Postgraduate Certificate in Archival Studies. PARBICA,[30] the Pacific Regional Branch, provides training in association with its regular conferences and has worked intensively since 1999 to encourage the development of a formal education program in the region. However, their initial plan to establish a distance education program at the University of the South Pacific stalled due to a lack of funding commitment by the university. PARBICA is now working with a wider focus, and aims to

provide a clearinghouse of information on educational opportunities for archivists in the region.

Some national professional associations also assess and accredit university-based courses within their national boundaries. The Society of Archivists (United Kingdom) does so, and provides a further program that mentors new graduates toward Registration over a minimum period of 3 years after graduation,[31] but there is no further requirement for refreshing professional knowledge.[32] The Australian Society of Archivists (ASA), Inc.[33] and the Records Management Association of Australasia[34] both have accreditation programs for university-based education programs. Graduation from a recognised program facilitates progress toward professional status in these two organizations. In the ASA, graduation from an accredited course shortens the route to professional membership, but is not an essential qualification. The Society of American Archivists (SAA) does not accredit programs, but provides a set of *Guidelines for a Graduate Program in Archival Studies* which the SAA considers "defines the minimum requirements for a graduate program in archival studies that is coherent, independent and based on core archival knowledge."[35] The Norwegian professional association Norsk Arkivråd[36] plays an active role in encouraging the development of archives and records management education and itself provides training courses and sponsors workshops for those already working in the profession. The Finnish Association of Business Archivists provides training in records management,[13] inspired by the American Certified Records Manager program sponsored by ARMA International. The German Association of Business Archivists[37] is also a provider of records management training programs.

National professional associations in general provide some form of continuing professional development programs, yet few make active participation a condition of continuing professional status, although Margaret Crockett,[32] for example, argues that this is one of the hallmarks of professionalism. The Records Management Association of Australasia requires professional members to submit an annual return as evidence of having met minimum requirements for participation in approved activities such as attendance at conferences and training courses, presentation of papers, or publications. A more rigorous approach to continuing professional development would require a system of accrediting either training providers or training courses to provide quality assurance, yet evaluation and endorsement of training programs is often seen as too difficult, or simply beyond the capacity of professional associations, although they are willing to undertake accreditation of university-based professional education programs.[5]

There are also some scholarships funded by professional associations to encourage those studying for professional qualifications. These include ARMA International's Educational Foundation,[38] the Association of Canadian Archivists Foundation,[39] the Records Management

Association of Australasia[40] and the Indigenous Special Interest Group of the Australian Society of Archivists' Loris Williams Scholarship, set up to support Aboriginal and Torres Strait Islander students in archives and records.[41] Positive action for training archivists and records managers from ethnic minorities is an important development, particularly in countries where issues surrounding the past treatment of indigenous people is in the process of being addressed, or the proportion of recent immigrants is increasing.

ICA Section for Archival Educators and Trainers

The ICA Section for Archival Educators and Trainers (ICA-SAE)[42] was formed after decisions taken at the 11th International Congress on Archives in Paris in 1988. In 1990 it succeeded the ICA Committee on Professional Training (ICA/CPT) which had been founded in 1979. It aims to provide professional support for educators and trainers worldwide and has more than 100 members spread across all continents. Its objectives are:

- To promote cooperation between archival training institutions and all those who are responsible for the education and training of archivists, and of other archives professionals.
- To assist in the development of archival research and the definition of standards, methods, and qualifications for archival education.
- To gather and disseminate information about programs of training institutions and other training activities and evaluate them.
- To cooperate with other bodies of the ICA in the preparation of training activities, training programs, and teaching for the various fields of archival work.

It has produced a range of publications and resources for teachers and trainers. Among the most notable are *What Students in Archival Science Learn. A Bibliography for Teachers*,[43] a classified bibliography of resources in a wide range of languages. Jeannette Bastian and Elizabeth Yakel[12] note that this work reports

> on the frequency with which specific readings that were used and identified which archival education programs assigned a given reading. Unfortunately, there was no analysis of these usage patterns or implied interrelationships between programs. Still, this later work represents a significant snapshot of readings assigned in different archival courses.

More recently ICA-SAE has developed and published the *Train the Trainer Resource Pack*,[44] which has been translated into Spanish, Bahasa Indonesia, Portuguese, and French, with translations into more languages proposed. The *Train the Trainer Resource Pack* is an

important means of increasing training skills among professional archivists and records managers. These training skills will enable sharing of professional knowledge and expertise more effectively in in-house training programs, courses for volunteers and in continuing professional development programs.

ICA-SAE works with educators around the globe to hold regional conferences, aiming to bring educators and trainers together to form regional networks, developing a community of practice through which knowledge is shared and capacity in education and training is increased. Since 1988, 17 international conferences have been held.

Providers of Education and Training

Many national archives provide in-house training for their own staff and those that have a legislative mandate to control and monitor recordkeeping standards in government usually also provide at least some basic training for records managers in government agencies. In some countries where there is little or no access to formal professional education, such training is crucial. However, few are able to cater for those in recordkeeping roles in nongovernment organizations. Some countries host training programs for archivists in other countries. Examples include the Les Stages Techniques Internationaux d'Archives, through which the Archives Nationales de France provides training programs for archivists in the francophone world; the Postgraduate Diploma course at the National Archives of India; the National Archives of Malaysia which provides training workshops and participates in exchange of archivists with other ASEAN countries; the Riksarkivet (National Archives) of Sweden in association with SIDA (the Swedish International Development Cooperation Agency) provides a course *Records Management in Service of Democracy* for archivists and records managers from African countries, followed up by visits by the course leaders to the participants in their own countries.[45] Russian training expertise has also historically been available to former Soviet countries. Other national archives provide active support for courses and training activities offered through ICA Regional Branches, as for example support for PARBICA programs provided by the National Archives of Australia and Archives New Zealand.

The International Records Management Trust (IRMT) was established as a nonprofit organization in 1989 "to respond to widespread and progressive deterioration in recordkeeping in the developing countries."[46] Its current goal is expressed as an aim to help "governments and organizations move to electronic environments by building trustworthy records management systems as a foundation for data integrity and management information systems."[47] One of the three major foci of the Trust is the provision of education and training programs, provided through a network of international professionals.

British–Careers

It also undertakes research on the relationship between records management and governance and provides consultancy services, all of which inform the development of its training products. The IRMT's study materials can be downloaded and used without charge.[48]

The ICA-SAE publishes a *Directory of Archival Education and Training Institutions*. The first edition was published in 1992 and the second edition in 2004.[49] The Directory is incomplete for several reasons. The first is that the information gathering used a snowball technique, commencing with members of ICA-SAE at the time and depending on them to send the survey to organizations in which those members had contacts. The second reason is that submission of entries is entirely voluntary and dependent on providers completing and returning information to the editors. Thus there are gaps in the data for the francophone world, for Africa, South America, and for the United States. However, the SAA provides an online *Directory of Archival Education*[50] which is kept up to date by education program providers. The ICA-SAE *Directory* also focused on university-based programs and only two providers of professional training courses outside formal university programs submitted an entry, yet the provision of short training courses is an important means of building capacity and improving professional skills in most countries, including those where university-based courses are available. Since the data were gathered for the last edition, new courses have begun in Japan, Switzerland, New Zealand, and South Korea, for example, and in Australia some courses have ceased and a new one has commenced. The Section therefore plans to revise the content and provide a facility for education and training providers to update their own entries online. Providers may submit a new or revised entry by contacting the ICA-SAE at http://www.ica-sae.org.

CAREER OPPORTUNITIES

Archivists and records managers have a wider range of career opportunities than ever before, and the options are growing. National, state, and regional archives remain the largest organizational employers of recordkeeping professionals in every country. National archives with a legislative mandate to play a key role in ensuring good governance and accountability have a strong role in the development of policy and standards for recordkeeping across the whole of government. They employ professionals who can research and develop jurisdiction-specific policies and standards for recordkeeping best practice, as well as developing and providing the advice, training and tools to assist compliance in government agencies' records management programs. Archival organizations now provide access to their services and collections via the World Wide Web and thus need recordkeeping professionals appropriately educated with the skills to manage the

provision of these digital services. As governments increasingly move toward providing public services via the Internet, professionals able to provide the necessary recordkeeping infrastructure that underpin e-government are in ever-increasing demand, along with expertise in data protection and privacy legislation. Large archival organizations provide opportunities for promotion through a structured career path as well as opportunities to gain wider professional expertise by moving across internal departments and services. The services of records managers are also in high demand in business and nonprofit organizations of all types. Records and information management professionals, with their systematic knowledge of organization-wide information structures, are well-qualified to take high-level positions as Chief Information Officers (CIOs). Qualified and experienced professionals who set up their own business as free-lance consultants or work for larger providers of professional consultancy services are in high demand to manage or undertake short to medium term projects that require professional expertise, or for filling temporary staffing shortages.

There is a wide range of other types of archives, as evidenced by the growing number of special interest sections of professional associations. In the International Council on Archives,[51] these include: Notarial Records, Architectural Records, Business and Labour Archives, Archivists of International Organizations, Archives of Churches and Religious Denominations, Literature and Art Archives, Municipal Archives, Sports Archives, Archives of Parliaments and Political Parties, University and Research Institution Archives as well as a Section for Archival Educators and Trainers, and a Section for Records Management and Archival Professional Associations. Many national professional associations for archivists host similar sub-groups to cater for the sectoral interests of their members. For example, the SAA hosts a similar range and also includes groups focused on particular archival functions, such as Acquisition and Appraisal, Reference, Access and Outreach, and Preservation, together with a section on Museum Archives. Museums and galleries frequently contain archives to support and extend their collections and collectively employ a considerable number of archivists, although the majority of such collections are small and often have only one archivist, or part-time volunteers. Community archives are also increasingly in evidence. The United Kingdom provides a number of examples of these, where communities and immigrant groups work to develop collections that record their memory, their experience of arrival, settlement, community development, and traditions old and new.

The Australian Society of Archivists also has special interest groups covering the usual spread of special interests, but also caters for Electronic Records, for School Archives, and for Indigenous Issues (IISIG). The IISIG has been formed for professionals with Australian and Torres Strait Island indigenous heritage and professionals who work with managing and providing access to records about indigenous people and the issues surrounding their treatment in the past. The Association of Canadian Archivists also has a Special Interest Section on Aboriginal Archives (SISAA).[52] Archivists who work with film and sound archives and in broadcast television and radio have their own professional association and the international body is the Co-ordinating Council of Audiovisual Archives Associations (CCAAA).[53] There are currently very large projects underway in a number of countries to manage the transfer of analog television and film archives to digital format in such a way as to ensure it is preserved and retrievable in the long-term.

And finally, but certainly not least, the role of educator or trainer for the recordkeeping professions is a particularly rewarding and stimulating career choice that requires sound academic achievement and well developed research, communication, and training skills.

ACKNOWLEDGMENT

The author is indebted to Dr. Helen Forde for her suggestions for improving this entry. Also, I am indebted to Professor Anna Szlejcher of the University of Cordoba for information about archival education in South America, particularly in Argentina and the development of RIBEAU.

REFERENCES

1. Uhde, K. New education in old Europe. Arch. Sci. **2006**, 6(2), 193–203.
2. Bastian, J.A. Introduction to the *Archival Science* special issue on graduate archival education. Arch. Sci. **2006**, 6(2), 131–132.
3. Nesmith, T. What is an archival education?. J. Soc. Arch. 28(1), 1–17.
4. Proctor, M. Professional education and the public policy agenda. J. Soc. Arch. **2007**, 28(1), 19–34.
5. Anderson, K. Education and training for records professionals. Rec. Manage. J. **2007**, 17(2), 94–106.
6. Pederson, A.E. Development of research programs. Archivum **1994**, 39, 312–359 *Proceedings of the 12th International Congress on Archives*, Montreal, Quebec, Canada, September 6–1, 1992.
7. Couture, C. Today's students, tomorrow's archivists: present day focus and development as determinants of archival science in the twenty-first century. Archivaria 42. http://journals.sfu.ca/archivar/index.php/archivaria/issue/view/401/showToc (accessed September 14, 2008).
8. Craig, B.L. Serving the truth: The importance of fostering archives research in education programs, including a modest proposal for partnerships with the workplace. Archivaria **1996**, 42. http://journals.sfu.ca/archivar/index.php/archivaria/issue/view/401/showToc (accessed September 14, 2008).
9. Couture, C. Education and research in archival science: general tendencies. Arch. Sci. **2001**, 1, 157–182 The
10. Gilland, A.; McKemmish, S. Building an infrastructure for archival research. Arch. Sci. **2004**, 4(3–4), 149–197.
11. Cox, R. Are there really new directions and innovations in archival education?. Arch. Sci. **2006**, 6(2), 247–261.
12. Bastian, J.A.; Yakel, E. Towards the development of an archival core curriculum: The United States and Canada. Arch. Sci. **2006**, 6(2), 133–150.
13. Lybeck, J. Archival education in Scandinavia. Arch. Sci. **2003**, 3(2), 97–116.
14. Upward, F.; McKemmish, S. Teaching recordkeeping and archiving continuum style. Arch. Sci. **2006**, 6(2), 219–230.
15. Anderson, K. Distance learning: a new approach to archival education. Arch.Manuscripts **1995**, 23(1), 48–59.
16. Anderson, K. Distance education for archival education. Janus **1998**, i(2), 37–44.
17. Williams, C. No boundaries, no limits: Distance education for archivists and records managers Paper presented at the International Congress on Archives Vienna, Austria, August, 23–29, 2004 http://www.wien2004.ica.org/imagesUpload/pres_280_WILLIAMS_C_SAE.pdf (accessed September 14, 2008).
18. Australian National Training Information Service (NTIS). Professional Development and Competency Standards: Unravelling the Contradictions and Maximising Opportunities Paper presented at the 15th International Congress on Archives Vienna, Austria, August, 23–29, 2004 definition, quoted in Hoy, M., http://www.wien2004.ica.org/imagesUpload/pres_121_HOY_SAE%2004.pdf (accessed September 14, 2008).
19. Crockett, M. What Can the Old World Learn from the New? Towards a European Competency Framework Paper presented at the 7th European Conference on Archives Warsaw May 18–20, 2006 http://archiwa.gov.pl/repository/wz/VII%20Konferencja/Papers/M_Crockett_What_can….pdf (accessed September 14, 2008).
20. National Finance Industry Training Advisory Body. *Records and Archives Competency Standards*, 2nd Ed. Commonwealth of Australia: Canberra, Australian Capital Territory, Australia, 2001.
21. Hoy, M. Record-keeping competency standards: The Australian scene. J. Soc. Archi. **2007**, 28(1), 47–65.
22. *Innovation and Business Skills Australia*. Available at http://www.ibsa.org.au/index.jsp .
23. Australian Society of Archivists, Inc. *Educational Standards*. Available at http://www.archivists.org.au/educational-standards .
24. ARMA International. Education Development Group. *Records and Information Management Core Competencies*, ARMA International, Lenexa, KS, 2007.
25. *Resolution of the 7th European Conference on Archives*. Archivists: Profession of the future in Europe. Available at http://www.archiwa.gov.pl/?CIDA=620.
26. International Council on Archives. Available at http://www.ica.org/.
27. ARMA International. Available at http://www.arma.org/.
28. Institute for Certified Records Managers. Available at http://www.icrm.org/.

29. EASTICA. Available at http://www.eastica.org/.

30. PARBICA. Available at http://www.parbica.org/.

31. Society of Archivists. Registration Scheme—FAQs. Available at http://www.archives.org.uk/careerdevelopment/registrationscheme/registrationschemefaqs.html.

32. Crockett, M. CPD and the hallmarks of professionalism: An overview of the current environment for the record-keeping profession. J. Soc. Arch. **2007**, *28*(1), 77–102.

33. Australian Society of Archivists, Inc., *Course accreditation*. Available at http://www.archivists.org.au/course-accreditation.

34. Records Management Association of Australasia. *Recognition of courses in records and information management*. Available at http://www.rmaa.com.au/docs/profdev/recognition.cfm .

35. Society of American Archivists. *Education & events*. Available at http://www.archivists.org/menu.asp?m=education.

36. Norsk, Arkivråd. Available at http://www.arkivrad.no/.

37. Vereinigung deutscher Wirtschaftsarchivare e.V. VdW (German Association of Business Archivists). Available at http://www.wirtschaftsarchive.de.

38. ARMA International Educational Foundation. Available at http://www.armaedfoundation.org/index.html.

39. The Association of Canadian Archivists Foundation. Available at http://archivists.ca/foundation/default.aspx.

40. Records Management Association of Australasia. *Peter A Smith scholarship for records management education*. Available at http://www.rmaa.com.au/docs/branches/nsw/scholarship.cfm.

41. Australian Society of Archivists, Inc. *Indigenous issues SIG*. Available at http://www.archivists.org.au/indigenous-issues-sig.

42. ICA-SAE. website. Available at http://www.ica-sae.org/.

43. Ando, M. Huiling, F. Mamczak-Gadkowska, I. Schenkolewski-Kroll, Silvia Thomassen, Theo.H.P.M. *What Students in Archival Science Learn. A Bibliography for Teachers*, 2nd Ed.; 2000; http://www.ica-sae.org/ (accessed September 8, 2008).

44. ICA-SAE. *Train the trainer resource pack*. 2007, Compiled by Crockett, M., Foster, J. 2004; Revised by Hoy, M. 2007. Available at http://www.ica-sae.org/.

45. Prager, K. Kompetens och yrkesstolthet. Tema Arkiv **2007**, *1*, 32.

46. IRMT. Available at http://www.irmt.org IRMT.

47. IRMT. About IRMT. Available at http://www.irmt.org/About_IRMT.html.

48. *The IRMT's training products are downloadable from its webpage Projects and products: Education and training materials*. Available at http://www.irmt.org/education_training_materials.html.

49. ICA-SAE. *Directory of Archival Education and Training Institutions*, 2nd Ed.; Huiling, F., Jian, W., Eds.; Compiled by http://www.ica-sae.org/ (accessed September 14, 2008).

50. Society of American Archivists. *Directory of archival education*. Available at http://www.archivists.org/prof-education/edd-index.asp.

51. International Council on Archives. *ICA's sections*. Available at http://www.ica.org/en/sections.

52. Association of Canadian Archivists. *Special interest section on aboriginal archives*. Available at http://archivists.ca/special_interest/aboriginal.aspx.

53. Co-ordinating Council of Audiovisual Archives Associations (CCAAA). Available at http://www.ccaaa.org.

Careers and Education in Information Systems

Paul Gray
Lorne Olfman
School of Information Systems and Technology, Claremont Graduate University, Claremont, California, U.S.A.

Abstract

This entry discusses the nature of information systems (IS) and the educational opportunities and careers in the information systems field within library and information science (LIS). The entry includes descriptions of the types of institutions that teach IS from a LIS point of view, standard curricula for B.A. and M.S. degrees, careers open to graduates from beginners to skilled professionals, and opportunities for continuing education. These educational possibilities are put in perspective by considering the nature of the IS industry with which professionals deal, the strategic use of IS in libraries, and the future directions of IS.

INTRODUCTION

Managing the vast stores of information and knowledge in libraries, in databases, and in knowledge repositories involves deep understanding of information systems (IS). This entry discusses the nature of IS and the educational opportunities and careers in the field.

University education in IS is centered principally in schools of business whose research objectives are to improve the application of IS in (usually profit-maximizing) organizations. Libraries are considered a specific field of application. IS studies in business schools and other departments such as computer science generally cover the fundamentals, but they do not emphasize the role of IS in library and information science (LIS). The uses of IS in LIS are taught principally in schools of information or informatics that emerged from schools of LIS. These schools usually offer full degrees or minors in IS specifically directed to libraries. Their emphasis is on the creation, processing, storing, and retrieval of information and knowledge.

A way of thinking about the use of computers in libraries is that it is similar to electronic commerce (e-commerce), the conducting of business over the Internet or over private networks. Many current library activities are forms of e-commerce. At the customer level, if you or I want a paper or a book or a question answered, we interact with the library in a way that is similar to e-commerce—ask for something and it is delivered electronically. On the supply side, libraries buy information electronically from JSTOR and other database providers such as LexisNexis, and obtain copies of books and journal articles from other libraries by e-order. E-mail is universal. All these activities depend on design and management of the IS behind them. In short, IS are at the core of how twenty-first century libraries really operate.

In keeping with the authors' backgrounds, this entry provides a view of IS in general. It covers the following topics:

- Differences among IS, computer science, computer engineering, and LIS.
- Education in IS: B.S., M.S., and Ph.D. level curricula.
- Schools of Information and Informatics.
- Training and continuing education.
- The IS industry (vendors, service providers, outsourcers, internal IS groups, and consultants).
- Specializations and job categories in IS (including total employment nationally and internationally).
- Careers in IS (in internal groups in institutions; in external vendors; working with user groups; "verticals" such as health, communications, technology vendors, education; joint projects between firms, departments; and global teams).
- Professional associations in IS.
- Strategic use of IS.
- Future of IS.

INFORMATION SYSTEMS VS. LIBRARY AND INFORMATION SCIENCE VS. COMPUTER SCIENCE VS. COMPUTER ENGINEERING

While IS, LIS, computer science, and computer engineering are all concerned with managing information and knowledge, they are studied (and consider themselves) as different disciplines. In this subsection we consider the similarities and differences among them.

Library and Information Science (LIS) is defined in Prytherch[1] as

the study and practice of professional methods in the use and exploitation of information, whether from an

Encyclopedia of Library and Information Sciences, Fourth Edition DOI: 10.1081/E-ELIS4-120043812

institutional base or not, for the benefit of users. An umbrella term, abbreviated LIS, is used to cover terms such as library science, librarianship, information science, information work etc.

In terms of IS, LIS focuses on such areas as document and records management and knowledge.

Computer Science (also called computing science) involves studying the foundations of information and computation and their application in computers. It is also concerned with computer languages and computer software.

Computer Engineering (also called electronic and computer engineering) combines computer science and electrical engineering. It focuses on software design, computer hardware design, and the integration of hardware and software.

Information Systems (IS) involves the study of the "(1) acquisition, deployment, and management of information technology resources and services ... and[2] development, operation, and evolution of infrastructure and systems for use in organizational processes."[2] Information systems differ from computer science and computer engineering in that the latter two are concerned with the creation of hardware and software rather than with their organizational implications. Information systems as a whole is a superset of IS within LIS.

EDUCATION IN IS

Table 1 summarizes the educational enterprise in IS.
    Degrees in IS are available at the

- Undergraduate
- Masters
- Ph.D.

levels. Model curricula and accreditation are available in the United States at the bachelor's and M.S. levels. Ph.D. programs are not standardized in this way. The most recent model curricula are published respectively in Gorgone, Davis et al.[2] for the undergraduate degree and in Gorgone, Gray et al.[3] for the master degree.

    In the United States, these degree programs are housed in

- Schools of Information and Schools of Informatics.
- Schools of Business.
- Schools and Departments of Computer Science.
- Schools of Liberal Arts.

Many, but not all of the Schools of Information and the Schools of Informatics are descendants of Schools of Library Science that changed their name and expanded their offerings and their faculty to include IS (see discussion below).

UNDERGRADUATE MODEL CURRICULUM

The undergraduate curriculum[2] is based on the following image of the IS professional with a bachelor's degree:

1. Has a broad real-world perspective.
2. Has strong analytical and critical thinking skills.
3. Has interpersonal communication and team skills and strong ethical principles.
4. Designs and implements information technology solutions that enhance organizational performance.

The curriculum is directed to people studying in business administration programs but is tailorable to students studying in other disciplines such as LIS. Specifically, it contains:

1. General courses in IS suitable for all students regardless of their majors or minors.
2. Specialized information technology and application design courses for both majors and minors in IS.
3. Specialized application development, deployment, and project management courses for majors in IS.

The output desired for IS majors, is shown in Fig. 1.[2] The course structure is shown in Table 2.

MSIS CURRICULUM

The M.S. model curriculum[3] is directed to create graduates who are prepared to provide leadership in the IS field. M.S. graduates should have the same skills, knowledge,

Table 1 Information systems education.

Academic level	Model curriculum	Depts/schools in universities	Delivery of offerings
Undergraduate	IS2002[2]	Schools of Business	Full time
M.S.	MSIS 2006[3]	Schools of Business, Schools of Information	Full time, part time, remote via Internet or video
Ph.D.		Schools of Business, Schools of Information	Full time, part time, remote via Internet or video
Training and continuing education		Short courses, certificates	In class, on site, and remote (e.g., via the Internet)

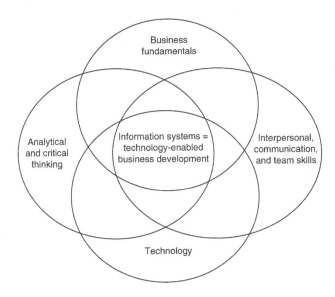

Fig. 1 High-level categorization of IS graduate exit characteristics.

and values as B.S. or B.A. graduates (Fig. 2[3]) but at a higher level.

The M.S.I.S. course set recommended for IS professionals is listed in Table 3.

To be eligible to take these courses at the M.S. level, students are expected to have completed four prerequisite courses: two in IS:

- IS fundamentals
- Programming, Data, Files, and Object Structure

and a two-course sequence in business functions and processes.

All four prerequisite courses are usually already completed by students with an undergraduate degree in IS. However, students coming from other majors, such as LIS, may be required to take some or all of these prerequisites if they did not take them as undergraduates.

The M.S.I.S. program shown in Table 3 consists of 12 courses divided into three parts, each consisting of four courses.

- IS technology.
- Managing IS.
- A coherent sequence in a specialization area.

IS technology. The IS technology courses, shown in Table 4, focus on the broad range of technology-based topics that an M.S. student needs to know. Some topics appear in several courses because they are discussed from different viewpoints. The amount of time devoted to topics varies; for example, the discussion of human–computer interaction is extensive in the analysis, modeling, and design course because of the importance of the human factor in that subject.

Managing IS. No matter what the size of the organization, successful operation requires a working, and preferably highly efficient IS. IS management turns out to be one of the more complex problems organizations face. In libraries, as in most business organizations, IS are viewed as overhead that is necessary but whose costs should be minimized. Yet minimizing cost requires a trade-off with service.

Table 5 shows the suggested content of the management courses.

Specialization track. The purpose of the specialization track is to teach students about the specific field they want to work in and to provide field experience through a practicum. IS programs, given the need for economies of scale and the available skills of their faculty, usually can offer only three to five different specialization tracks. The tracks are usually technical, organizational, human factors, or tailored to specific industries that are prevalent in an area (e.g., software in Seattle, financial institutions in New York). Typically, the specialization tracks involve selection from among elective courses. Note that specialization tracks can be lockstep or custom tailored to individual students by mixing and matching electives.

In schools of information (see next subsection) that offer both library science and IS degrees, one of the specialization tracks is based on issues paramount in LIS (e.g., courses in databases, data retrieval, information issues, knowledge management, and more specialized

Table 2 IS2002 course structure for undergraduates.

Prerequisite	IS 2002.P0 Personal Productivity with IS Technology
Information systems fundamentals	IS 2002.1 Fundamentals of Information Systems
	IS 2002.2 Electronic Business Strategy, Architecture and Design
Information systems theory and practice	IS 2002.3 Information Systems Theory and Practice
Information technology	IS 2002.4 Information Technology Hardware and Software
	IS 2002.5 Programming, Data, File and Object Structures
	IS 2002.6 Networks and Telecommunications
Information systems development	IS 2002.7 Analysis and Logical Design
	IS 2002.8 Physical Design and Implementation with DBMS
	IS 2002.9 Physical Design and Implementation in Emerging Environments
Information systems deployment and management processes	IS 2002.10 Project Management and Practice

Source: Gorgone et al.[2]

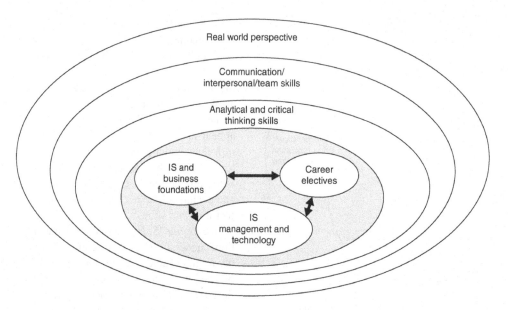

Fig. 2   M.S. level output model.

issues such as geographic IS). In M.S. programs located in schools that do not offer library science, the specialization track requires selection of appropriate electives.

The conclusion from this brief survey is that potential students and faculty with interests in both library science and IS should read school descriptions carefully if their interest spans the two disciplines.

TRAINING AND CONTINUING EDUCATION

Because of the rapid changes in information technology and software, even highly educated practitioners require updates to their skill sets. As a result, a large number of training offerings are available commercially or through universities. These offerings typically span from 3 to 5 days. Vendors offer courses at multiple locations on a weekly or monthly basis. Moreover, they produce online training (CDs, DVDs, and Internet accessible) for many software and hardware products, including the most popular ones produced by Microsoft, Oracle, and Cisco. These companies offer certifications that claim that the person who obtains them is qualified to install and maintain the

related products. Many information-related jobs call for such certifications, although the value of these certificates seems to be declining.[4] Leading-edge libraries offer training for software that supports specialist areas such as geographic IS, statistical analysis systems, databases, and bibliographic systems. Information professionals who want to manage information departments and resources often attend one of the part-time, evening M.S. programs in IS available in most urban areas. Community colleges also offer courses and associate degree programs in IS.

THE IS INDUSTRY

Information systems pervade American organizations, including libraries. To create and service these organizations, a large number of

- Equipment manufacturers
- Software firms
- Service providers and outsourcers
- Internal IS organizations
- Consultant organizations

Table 3   M.S. courses.

IS technology	Managing IS	Specialization tracks
IT infrastructure	Project and change management Strategy and policy Implication of digitization Integrated capstone	Four courses in depth in a particular subject area including, where possible, a practicum course involving experience in an organization
Analysis, modeling, Design, including Human–Computer Interaction Enterprise Systems Emerging technologies and issues		

Table 4 Contents of the technology courses.

IT infrastructure	Analysis, modeling, and design	Data and data management	Emerging technologies and issues
IT architecture	Human–computer interaction	Information content	Data mining
Enterprise Information Infrastructure	Systems development methodologies	Distribution	Sourcing
Servers and Web services	Requirements determination	Process management	Web services and business processes
Layered network architecture	Team organization and communications	Data warehousing/data marts	Technology convergence
Internet protocols	Feasibility and risk analysis	Data mining	Security
Global WAN services	Design reviews	Managing storage area networks	Organizational intelligence
Enterprise network design	System development life cycle	Large systems (ERP)	Knowledge management
Wireless technologies	Conceptual and logical data modeling	E-commerce	Mobile computing
Network management	Database implementation Data organization	Content management	Ubiquitous computing
Server architectures	Software and systems metrics		
Storage management and networks	Software package evaluation		

Source: Gorgone et al.[3]

were created over the years. The stories of vast wealth accumulated by the founders of many of these organizations are part of the folklore. In this section we describe the role of each of these types of organizations and their relation to libraries.

Equipment (hardware) manufacturers. When computers first came on the scene commercially in the 1950s and 1960s, the focus was on building computers and peripheral equipment such as displays and printers. The machines were large, horribly expensive, and designed for medium and large firms and government agencies that could afford them. Following the introduction of the IBM PC Model 51 in 1981, personal computing became generally available and is now pervasive.

The technology broadened so that it now includes telecommunications that allows computers to be networked together. Telecommunications, in turn, made it possible to connect to the Internet. Where previously connection

Table 5 Suggested content of the managing IS courses.

Project and change management	Implications of digitization	Policy and strategy	Capstone
Project life cycle	Ethics	Relationship between IS and the business	The enterprise as a system of integrated business processes
Project stakeholders	Privacy	IS and competitive position	Relationships with suppliers and customers
Project planning	Govt. regulations	Aligning IT goals and strategy	Strategic alignment of IT with the business
Software cost estimation	Globalization and sourcing	Creating IT values, vision, and mission	IT organization and governance
Work module design, assignment, version control	Intellectual property	IT strategic planning, infrastructure planning and budgeting	The value of IT
Role of repository, project library	Virtual work and telecommuting	IS implementation	Role of the CIO
Contingency planning	Implications of artificial intelligence	Inter-organizational systems	Sourcing
Reporting and controls	Security measures and planning	Outsourcing versus insourcing	Compliance
Testing and testing plans: alpha and beta	E-commerce	Globalization	Integrated enterprise architecture
Project Manager skills	Digital divide IT workforce	Risk management Virtual organization	Platform choices Impact of standards Vendor strategies

Source: Gorgone et al.[3]

to the electric power grid was required, computer equipment is moving to wireless. Miniaturization led to ever less space being required and less heat generated.

Computer prices dropped from millions of dollars in 1960 to thousands today and computer user skills are nearing universality. The result is a computer-literate population that is accustomed to the digital life. The drop in costs made it possible for libraries and information repositories to use the technology to improve service and capabilities. However, recognize that while individual personal computers are cheap (prices under $1000/unit for the hardware are common), the large number of individual units required, as well as the cost of servers, networking, and peripherals still makes the total cost of ownership considerable for an organization. The term servers, as used here, refers to dedicated computers that provide software and other services to computers on a network. In some arrangements, a server can also be used to do local computations.

Software firms. In the early years of computing, software was literally given away with the hardware, since programmers were few and programming computers was a highly complex, skilled art. By the early 1960s, firms sprang up that offered software commercially with different versions compatible with the major mainframe types. The software was high priced and was almost exclusively rented. That is, ownership did not pass to the purchaser. Rather, the software firm charged an initial price and then an annual maintenance fee, typically equal to 15% of the initial price. The result was lock-in. That is, once a particular piece of software was acquired, it became quite expensive to switch to another vendor. An annual maintenance fee was charged by the software firm to cover its cost of revising and updating the software as the hardware changed and additional features were added. For PCs, the policy is different in that its copies of software are cheap and it is possible to own the software. However, upgrades of the software must be purchased, often at or near the same price as the original. Maintenance requires separate contracts.

The range of software available expanded over the years. Typical office suites offer word processing, spreadsheets, presentation graphics, small databases, and more. Specialized software is required for large databases for data warehousing, for knowledge management and other subjects of interest to libraries.

The Internet is a separate issue. Software for access (i.e., browsers) is currently free, but hardware is required for connection. The major problems of the Internet deal with content. In addition to viruses, worms, spyware, and other "malware" directed toward ruining or interfering with content, spam is a major concern. Spam refers to unsolicited, unwanted advertising, most of it for dubious products. For libraries, be they public or in a corporation, access to pornography over the Internet needs to be avoided. Software to deal with these issues is available,

but it is not perfect. In fact, there is an arms race between providers and preventers of improper information, with one side or the other ahead at any particular time. Another content aspect is copyright protection where computer users want access to copyrighted material to which they are not entitled. Liability issues on copyright were still in contention when this entry was written in 2008.

Service providers and outsourcers. Libraries are faced with a classic "make or buy" problem. Although they can create their own computing environment, most purchase all or part of it from vendors such as catalog and database providers. Vendors include application service providers (ASPs) and outsourcers. ASPs provide computer-based services to customers over a network. A classic example is the Internet Service Provider (ISP) who offers connections to the Internet.

Outsourcers are contractors who perform computer work ranging from internal production to managing a firm's operations to providing call centers. The movement of such work is called outsourcing and offshoring. Offshoring refers to contracting beyond the boundaries of the country whereas outsourcing usually refers to moving work domestically. Domestic outsourcing, of course, simply shifts jobs from an organization to a firm that specializes in some aspect of IS but still employs domestic workers. Although newspapers and magazines are filled with stories about moving jobs from high-wage western countries to low-wage countries such as India and China, only a small portion of IS jobs have moved overseas. For example, a 2008 survey shows that approximately 1.1% of IS jobs were offshored and that the trend is toward less offshoring and more domestic outsourcing in the future.[5] Although small in percentage of jobs, offshore software outsourcing still involves large sums of money. For example, a study published in November 2007[6] estimates revenues of near $55 billion alone for India, China, Malaysia, and the Philippines.

A growing number of firms provide "Software As A Service." That is, work is sent electronically to the ASP either over the Internet or on a private network. The ASP provides the computers and software that performs contracted tasks. The advantage is that the library does not need to perform that work internally at the cost of giving up some control over the work and how it is performed.

Internal IS organizations. Where a library system (usually with multiple locations) is a government service, their information operations may either be outsourced to a centralized information group that provides services to a number of government agencies or may be served by its own IS organization. Internal organizations, like those in businesses often have the advantage that they are part of the organization and know its needs. They can be more responsive in terms of both time and resources made available. This characterization assumes that the internal organization is not dysfunctional as internal IS operations sometimes can be. With an internal IS function, the library system is locked into its organization. If the internal organization is dysfunctional, outsourcing is one way of attempting to resolve the dysfunctionality.

Consultant IS organizations. IS consultants are called in when a specific problem needs to be solved and the organization does not have the skills or the personnel needed within their internal organization. Consultants do not come cheap. They are experienced, quality people and operate in a high overhead environment since they need to cover not only their costs while employed by a client but also their costs when between engagements.

Hiring consultants is, in effect, renting knowledge. It is a speculation. It is not possible to predict whether the consultants' work will pay off. They are hired based on the reputation of their organization and people, their past success, and their potential. Knowledge rentals involve knowledge transfer from the consultant to the client. Even though consultants are temporary hires, some knowledge will stay after they leave. Some consultants offer training as part of their service. Although libraries seek their help in solving specific problems, it is wise to write the consulting contract to include knowledge transfer.

SPECIALIZATION AND JOB CATEGORIES IN IS

The United States Bureau of Labor Statistics (BLS) provides a classification and comprehensive description of information-related jobs and requirements, the number of persons (currently) employed in these jobs, and a forecast of the expected growth in the number of these jobs. The statistics presented in this section are from the 2008–2009 version of the *BLS Handbook*. Libraries typically keep only the latest annual version of the *BLS Handbook*. Thus, the data presented in Table 6 are, in some cases, obsolete in years beyond 2008. However, it is our experience that the categories and numbers change slowly. Thus, the categories and the data will remain indicative over the next several years.[7]

Categories of taxonomies almost always overlap; job taxonomies are no different. The BLS does not directly classify jobs as information-related. The BLS classifications do not map one-to-one into the educational categories discussed previously. The differences, however, are not significant. BLS lists most information-related jobs in its professional occupations. However, some are included in the Business and Financial, Administrative, and Management occupation categories. Except for Computer Programmers and Librarians (both classified as professional), job growth through 2016 is predicted to be faster or much faster than average. Table 6 lists 11 BLS information-related categories that require at least a bachelor's degree along with a job description, current and expected employment figures; and median salary.

CAREERS IN IS

While the BLS classification provides a tool for understanding the current and future employment possibilities for the main groups of information-related jobs, a more direct description of jobs available to graduates of information programs can be inferred from job advertisements. A search of jobs advertised on Monster.com[8] in eight major urban areas in the United States (within a radius of 50 mi from Los Angeles, San Francisco, Denver, Dallas, Chicago, New York, Philadelphia, and Atlanta), as of mid-July 2008 show a wide variety of job titles. Some of the titles vary by levels (e.g., Network Analyst and Network Manager), showing the natural career paths in many information-related jobs from entry level jobs (typically analysts and technicians) to senior positions (such as Manager or Chief Technology Officer). Job titles can be classified into domains (e.g., administrator, architect, analyst). Table 7 shows selected job titles by domain from the Monster.com search.

Table 8 maps the BLS occupational categories to job title domains and application areas (e.g., business systems, database, network). Salaries range from $50,000 to well over $100,000 with any specific job spanning a wide dollar amount depending on qualifications and experience. Jobs in the Information Technology field are classified into 12 different areas in Monster.com, including Computer/Network Security, Database Developer/Administrator, Enterprise Software, Information Technology/Software Development, Information Technology Project Management, Network/Server Administrator, Software/System Architect, Software/Web Developer, Systems Analysis, Telecommunications Administration/Management, and Usability. A search within these categories using a set of standard keywords for job titles provided the set of available jobs within the eight urban areas noted above. There is some overlap in available jobs using this method of search (see Table 9). While this table includes only a fraction of all jobs listed, it provides a fair distribution of jobs across different cities and job types. The highest paying jobs are more typically available in New York, Los Angeles, and San Francisco.

Within a traditional Information Systems department in a large organization, the normal hierarchy includes a senior manager (or Chief Information or Technology Officer); directors of the key IS functions, including database systems, systems development, systems maintenance, telecommunications and security, and user support. A typical career path for a specialist in database might lead from an entry level position to a database administration position, to a director of database systems, and then possibly to manager. IS managers need leadership and administration skills. They are typically experts in one of the department functions as the following discussion of career paths explains.

British–Careers

Table 6 BLS occupations in information fields.

Occupation	Job description	Employment 2006 and 2016	Salary median (2006)[a] ($)
Computer Programmers	Write, test, and maintain the detailed instructions called programs that computers follow to perform their functions. Conceive, design, and test logical structures for solving problems by computer. With the help of other computer specialists, they figure out which instructions to use to make computers do specific tasks.	435,000 declining to 417,000	65,510
Computer and Information Scientists, Research	Work as theorists, researchers, or inventors. Their jobs are distinguished by the higher level of theoretical expertise and innovation they apply to complex problems and the creation or application of new technology. The areas of computer science research range from complex theory to hardware design to programming language design. They may work on design teams with electrical engineers and other specialists.	25,000 increasing to 31,000	93,950
Computer Software Engineers, Applications	Develop, create, and modify general computer applications software or specialized utility programs. Analyze user needs and develop software solutions. Design software or customize software for client use with the aim of optimizing operational efficiency. May analyze and design databases within an application area, working individually or coordinating database development as part of a team.	507,000 increasing to 733,000	79,780
Computer Software Engineers, Systems Software	Research, design, develop, and test operating systems-level software, compilers, and network distribution software for medical, industrial, military, communications, aerospace, business, scientific, and general computing applications. Set operational specifications, formulate and analyze software requirements. Apply principles and techniques of computer science, engineering, and mathematical analysis.	350,000 increasing to 449,000	85,370
Computer Systems Analysts	Solve computer problems and use computer technology to meet the needs of an organization. They may design and develop new computer systems by choosing and configuring hardware and software. They may also devise ways to apply existing systems' resources to additional tasks. Analysts who specialize in helping an organization select the proper system software and infrastructure are often called *system architects*. Analysts who specialize in developing and fine-tuning systems are often known as *systems designers*.	504,000 increasing to 650,000	69,760
Computer Specialists	Includes a variety of jobs such as Telecommunications Specialists, who focus on the interaction between computer and communications equipment; Webmasters, who are responsible for all technical aspects of a Web site, including performance issues such as speed of access, and for approving the content of the site; and Web Designers/Developers, who are responsible for day-to-day site creation and design.	136,000 increasing to 157,000	Not available
Database Administrators	Work with database management systems software and determine ways to organize and store data. They identify user needs and set up new computer databases. In many cases, database administrators must integrate data from outdated systems into a new system. They also test and coordinate modifications to the system when needed, and troubleshoot problems when they occur. An organization's database administrator ensures the performance of the system, understands the platform on which the database runs, and adds new users to the system.	119,000 increasing to 154,000	64,670
Network Systems and Data Communications Analysts	Design, test, and evaluate systems such as local area networks (LANs), wide area networks (WANs), the Internet, intranets, and other data communications systems. Network systems and data communications analysts perform network modeling, analysis, and planning, often requiring both hardware and software solutions. Analysts also may research related products and make necessary hardware and software recommendations.	262,000 increasing to 402,000	64,600
Computer and Information Systems Managers	Includes Chief Technology Officers, who evaluate the newest and most innovative technologies and determine how these can help their organizations; MIS/IT Directors, who manage computing resources for their organizations; and Project Managers, who develop requirements, budgets, and schedules for their firms' information technology projects.	264,000 increasing to 307,000	101,580

(Continued)

Table 6 BLS occupations in information fields. *(Continued)*

Occupation	Job description	Employment 2006 and 2016	Salary median (2006)[a] ($)
Network and Computer Systems Administrators	Design, install, and support an organization's computer systems. They are responsible for LANs, WANs, network segments, and Internet and intranet systems. They work in a variety of environments, including professional offices, small businesses, government organizations, and large corporations. They maintain network hardware and software, analyze problems, and monitor networks to ensure their availability to system users. Administrators also may plan, coordinate, and implement network security measures.	309,000 increasing to 393,000	62,130
Librarians	Often called information professionals, increasingly combine traditional duties with tasks involving quickly changing technology. Help people find information and use it effectively for personal and professional purposes. Manage staff and develop and direct information programs and systems for the public and ensure that information is organized in a manner that meets users needs. Librarians with computer and IS skills can work as automated-systems librarians, planning and operating computer systems, and as information architects, designing information storage and retrieval systems and developing procedures for collecting, organizing, interpreting, and classifying information.	158,000 increasing to 164,000	49,060

[a]Note that the data are based on 2006. Add at least a 3% annual increase for each year after 2006.
Source: *BLS Handbook*.[7]

It is more likely that a career path would include lateral movement from one organization to another rather than lateral movement within organizations. For example, a database administrator may find a job at the same level in another company that may pay better or that will include use of a different database system, rather than move to a role in systems development in the same company. That is, people become specialized in their skills. Experience in particular technologies counts a great deal in developing a career path. Most job advertisements ask for specific experience. Specialization makes it difficult to begin a career in the IS field at other than the entry level. For example, a typical career can begin at a consulting firm that specializes in particular technologies. After gaining expertise in that technology, the next step might be to take a job in the organization that previously hired the individual as part of the consulting team to install its software system. Many IS jobs require certifications for using and applying specific technologies offered by companies such as Microsoft, Sun, and Cisco. Certification requires training but results in higher salaries, although the premium has been decreasing.[4]

In small organizations, however, with only a small IS staff, each staff member is expected to have the skill needed to perform multiple functions. It is possible to move laterally and vertically back and forth between small and large organizations.

PROFESSIONAL ASSOCIATIONS

Whereas the American Library Association is the principal professional association for librarians, IS people typically belong to one or both of the following two groups: The Association for Information Systems (AIS) and the Association for Computing Machinery (ACM). AIS consists almost exclusively of people who teach IS in universities and colleges whereas ACM (the much larger of the two associations) consists of both academics and practitioners with a focus on computer science rather than IS. AIS produces two electronic journals, *Communications of AIS* (*CAIS*) and the *Journal of AIS* (*JAIS*). The flagship journal of ACM is *Communications of the ACM*. ACM also publishes a large number of specialty journals that deal with specific areas of computers and computer science.

STRATEGIC USE OF IS

The concept of strategic use of IS in libraries is used in two senses:

1. The creation of specific IS that provide a way of differentiating a library system from others.
2. A multiyear plan for operations into the future.

Differentiation. The goal of a strategic information system is to differentiate the organization by improving the organization's position. A strategic information system supports and shapes the organization's strategy, often leads to innovation in the way the organization conducts its work, creates new opportunities, or develops services based on information technology.[9] In library systems, differentiation can be achieved by introducing new

Table 7 Job domains and job titles.

Job domain	Job titles
Administrator	Information Systems Administrator, Information Systems Database Administrator, Network Administrator, Senior Database Administrator, Systems Administrator
Analyst	Applications Systems Analyst, Business Analyst, Business Applications Analyst, Business Intelligence Data Analyst, Business Systems Analyst, Data Analyst, Help Desk Analyst, Information Security Analyst, Information Systems Analyst, Information Technology Analyst, Programmer Analyst, Quality Assurance Analyst, Senior Data Analyst, Senior Information Assurance Analyst, Senior Information Systems Analyst, Systems Analyst
Architect	Data Architect, Enterprise Architect, Information Architect, Information Systems Architect, Senior Information Architect, Software Architect, Web Information Architect
Auditor	Information Technology Auditor, Information Technology Senior Internal Auditor
Consultant	Business Systems Consultant, Implementation Consultant, Information Technology Consultant, Network Consultant, Senior Consultant Systems Domain, Storage Consultant
Designer	Application Designer, User Interface Designer, Web Designer
Developer	Application Developer, Database Developer, Information Technology Developer/Consultant, Senior Java Developer, Senior Software Developer, Senior Web Developer, Technical Developer, Web Developer, Web Developer/Image Librarian
Director	Director Clinical Information Systems, Director Information Systems, Director Information Technology, Global Information Technology Director, Information Technology Director
Engineer	Information Assurance System Engineer, Information Security Engineer, Information Security Network Engineer, Senior Engineer Information Systems, Senior Information Security Engineer, Senior Network Engineer, Senior Network Security Engineer, Software Engineer/Programmer, Systems Engineer, Technical Support Engineer
Librarian	Acquisitions and Collection Librarian, Corporate Librarian, Electronic Resources Librarian, Reference Librarian, Research Librarian, Technical Services Librarian
Manager	Information Security Manager, Information Services Manager, Information Systems Manager, Information Technology Manager, Manager Development, Manager Information Technology, Network Manager, Project Manager, Senior Information Security Manager, Senior Information Technology Project Manager, Senior Project Manager
Officer	Chief Technology Officer, Information Security Officer, Information Technology Disaster Recovery Officer
Professional	Information Technology Professional, System Administrator Professional
Programmer	Information Technology Programmer, Senior Application Programmer
Specialist	Application Specialist, Information Systems Specialist, Information Systems Support Specialist, Information Technology Specialist
Trainer	Business Application Trainer, Software Trainer, Web Product Support Specialist
Vice President	Chief Solutions Specialist, Chief Technology Officer, VP Business Solutions, VP Information Systems Technology
Writer	Report Writer, Senior Technical Writer, Technical Writer

Source: Monster.com[8] and *BNET Dictionary*.[9]

technologies (e.g., geographic IS, data warehouse), new services (e.g., adding an electronic data source), new analysis methods (e.g., data mining capabilities) and more.

Multiyear strategic planning. Library systems typically create their own multiyear strategic plan. A large number of such plans are found by using a search engine such as Google (e.g., Table 10 considers seven such plans). In examining these plans, the use of information technology is featured both for providing services and administrative support.

Whereas the strategic plans shown in Table 11 are concerned with the library system as a whole, IS departments also create their own strategic plan. To be effective, the department's and the system's plans should be aligned with one another. That is, they must be designed to achieve the same goals. However, in addition to supporting the system's goals, the department's work includes responsibilities that are unique to its operation, ranging from creating the needed infrastructure to

selecting new equipment and software to specifying personnel capabilities.

THE FUTURE OF INFORMATION SYSTEMS

Since the first commercial computer systems were installed in 1957, much changed in IS and even more changes can be expected. In this section, we present five of the many scenarios about the future of IS being discussed in 2008. The first four scenarios are based on a special issue on the IS Organization of the Future published in *Information Systems Management*.[10]

1. *The future organization of the IS function.* As noted above, IS organizations tend to be hierarchical, with a Chief Information Officer (CIO) at the top reporting to a senior officer of the firm and supervising departments for technology, design, internal operations,

Table 8 Occupations, application areas, and job title domains.

Occupation	Application area	Job title domain
Computer and Information Systems Managers	Director Manager	Applications Information Systems Information Technology
Computer Programmers	Internet/Web Software	Analyst Developer Programmer
Computer Security Specialists	Information Systems Information Technology Security/Data Security	Administrator Analyst Director Specialist
Computer Software Engineers, Applications	Applications Business Systems Software	Analyst Architect Consultant Developer Engineer Programmer
Computer Systems Analysts	Business Systems Information Systems Information Technology	Analyst Consultant Engineer Programmer/Analyst Specialist
Database Administrators	Database Information Systems	Administrator Analyst Manager Programmer
Network and Computer Systems Administrators	Information Systems Information Technology Network	Administrator Director Engineer Manager Specialist
Network Systems and Data Communications Analysts	Information Systems Network	Administrator Consultant Engineer Manager Programmer Specialist

Source: Monster.com.[8]

Table 10 Some library strategic systems plans.

Institution	Title
Emory University	An Overview of the Five Year Strategy for the Emory Libraries, 2007
MIT	MIT Libraries Strategic Plan 1999
New York State Library	2006–2007 Strategic Plan: Improving Library Services for All in the New Century
The Saskatchewan Multitype Library Board	Strategic Plan Update, 2003
U.S. Library of Congress	The Mission and Strategic Priorities of the Library of Congress FY 1997–2004
University of Wisconsin Libraries	Strategic Directions 2005–2007
Wisconsin Department of Public Instruction	Wisconsin Library Technology Strategic Plan, Updated 2007

Source: Google search on: strategic systems and libraries (Accessed August 2007).

computer security, and service to other departments and outsiders (such as through call centers). We use CIO to refer to the person in charge of information systems. The actual title may vary from one organization to another. With the trends to divesting many of these functions (e.g., through outsourcing and Software As A Service), the IS organization could well morph into a small group of technical and business experts managing the evolution of the organization's technology and a larger on-site group providing only maintenance-level services.[11] However, in this case, IS career paths are not likely to change in that the same roles and functions are necessary for providing IS services by either insourcing or outsourcing.

2. *The IT talent challenge.* In response to the perceived trend to offshoring and as a result of the end of the stock market bubble in technology stocks in 2001, the number of students that chose to major in IS, computer science, and related fields has declined. That decline was starting to turn around in 2008. Nonetheless, the combination of a reduced demand for IS education at the college level and an aging baby boomer IS workforce results in concern for the

Table 9 Number of available jobs by keyword in Monster.com.

	Atlanta	Chicago	Dallas	Denver	LA	NY	Phila.	SF	Total
Database	12	26	13	14	27	54	27	41	214
Librarian	4	7	5	2	8	22	6	4	58
Network	54	77	64	60	92	174	86	78	685
Programmer	35	84	56	36	122	225	105	62	725
Security	25	43	29	23	44	85	28	39	316
Software Engineer	75	72	45	54	202	157	85	346	1036
System Administrator	19	43	34	33	65	129	52	69	444
Systems Analyst	31	42	32	12	63	112	50	45	387
Total	255	394	278	234	623	958	439	684	3865

Note: LA = Los Angeles, NY = New York, Phila. = Philadelphia, SF = San Francisco.
Source: *BNET Dictionary*.[9]

British–Careers

Table 11 Effects of changing technology on IS.

From	To	From	To
Standardization	Commoditization	Data management	Knowledge management
Technological continuity	Technology integration	User friendly	Usability standards
Proliferation of hardware types	Synergy among hardware types	Access security	Protection against terrorism

recruitment and retention of IS professionals because as the demand for professionals exceeds the supply, wages escalate.[12] Therefore, IS becomes a valuable career with a long-term future.

3. *A move to Wi-Fi and telecommuting.*[13] With the proliferation of portable technology (e.g., portable PCs, cell phones) and the continuing increase in the capabilities of these devices in a world without wiring, the current computer infrastructure is replaced by a wireless infrastructure that is able to provide large masses of data, information, and knowledge. Concomitant with the technology change, the workforce also becomes mobile and telecommuting[14] is the dominant way people work. This virtual workforce is supported anytime, at anyplace, including library services. The combination of Wi-Fi and telecommuting makes it more likely that some IS professionals, particularly those involved in geographically dispersed projects do not have to move their home as their assignments change.

4. *Harnessing the Technology.*[15] Table 11, shows how changing technology leads to changes in IS.

Note that in Table 11 as well as in each of the five scenarios, the focus is much more on technology and human factors than it is on the content being managed. The strength of people trained in both library science and IS is that they view issues in terms of content as well as technology.

5. *Carr's "IT Doesn't Matter" hypothesis.* In May 2003, the *Harvard Business Review* carried an article titled "IT Doesn't Matter" by Nicholas G. Carr.[16] In this article and in a subsequent book,[17] Carr argues that IS, like railroads and electric power, is an infrastructure technology that became a commodity. Initially strategic, IS is now a commodity, no longer offering advantage. Thus, something that everyone had would no longer matter as a way of distinguishing organizations from one another. Carried to its extreme, Carr's notion implies the death of organizational computing. Developments such as outsourcing and Software As A Service are viewed as omens of Carr's vision.

Which of these five alternative scenarios will play out in the years ahead? We expect that the predominating viewpoint will be some combination of these scenarios.

The foregoing represents visions of IS in general. When we turn to library IS, the future seems bright. These systems are coming into full flower compared to their role in business in general where IS are already in an advanced

state of maturity. The general coming shortage in people skilled in IS implies that people trained in library IS will continue to be in demand and, if their interests change, they will have the skills they need to be able to transfer their talents to many other applications.

One of the key trends in libraries is convergence, streamlining, and synergy. Instead of requiring users to go through multiple steps in research, including visiting library databases and online library catalogs, libraries are investigating ways to reduce the complexity of access such as federated search engines that combine databases, the online catalog, Web sites, and Open URL link resolvers (which provide links from non-full text citation/abstract databases to full text content). Libraries are also investigating innovative ways to market services and allow library users to match their tastes and research needs with library content in the Web 2.0[18] environment. Examples include:

- Second generation catalogs with subject tag clouds and suggestions for similar reading.
- Podcasts.
- The use of social software such as blogs, Myspace, Facebook, and Second Life.

ACKNOWLEDGMENTS

We are indebted to Adam Rosencrantz of the Claremont University Center libraries for his in-depth discussions of this paper and for the example of the future of computing in libraries in the Conclusions section. We also thank the editor and an anonymous reviewer for valuable comments and suggestions.

REFERENCES

1. Prytherch, R. *Harrod's Librarians' Glossary and Reference Book: A Directory of over 9,600 Terms, Organizations, Projects, and Acronyms in the Areas of Information Management, Library Science, Publishing, and Archive Management,* Ashgate Publishing Limited: Gower, U.K., 2000; 242.

2. Gorgone, J.T.; Davis, G.B.; Valacich, J.S.; Topi, H.; Feinstein, D.L.; Longeneckr, H.E., Jr. IS 2002 model curriculum and guidelines for undergraduate degree programs in information systems. Commun. Assoc. Inform. Syst. **2002**, *11*(1), 1–53. Available at http://cais.aisnet.org/articles/11-1/default.asp?View=pdf&x=51&y=11 (accessed July 2007).

3. Gorgone, J.T.; Gray, P.; Stohr, E.A.; Valacich, J.S.; Wigand, R.T. MSIS 2006: model curriculum and guidelines for graduate degree programs in information systems. Commun. Assoc. Inform. Syst. **2006**, *17*(1), 1–63 http://cais.aisnet.org/articles/17-1/default.asp?View=pdf&x=72&y=12 (accessed July 2007).

4. Perelman, D. IT certifications declining in value. Available at http://www.eweek.com/c/a/Careers/IT-Certifications-Declining-in-Value/?kc=EWKNLCSM061008FEA.

5. Dubie, D. Outsourcing pummels offshoring in IT budget plans, Network World. Available at http://www.networkworld.com/news/2007/101507-outsourcing-tops-offshoring.html.

6. Analyst firm issues results of study on global outsourcing. Business World, November 8, **2007**, *S2*(8).

7. BLS Handbook. *Occupational Outlook Handbook (OOH);* 2008–2009 Edition, U.S. Department of Labor, Bureau of Labor Statistics: Washington, DC, 2008. Available at http://www.bls.gov/oco/ (accessed July 2008).

8. Monster.com. Available at http://www.monster.com.

9. *BNET Dictionary. Business Definition: Strategic Information Systems.* Available at http://dictionary.bnet.com/definition/strategic+information+systems.html.

10. Brown, C.V. Inform. Syst. Manag. **2007**, *24*(2), 101–102.

11. King, W.R. The IS organization of the future: impacts of global sourcing. Inform. Syst. Manag. **2007**, *24*(2), 121–128.

12. Luftman, J.; Kempaiah, R. the IS organization of the future: the IT talent challenge. Inform. Syst. Manag. **2007**, *24*(2), 129–138.

13. Scott, J.E. Mobility, business process management, software, sourcing, and maturity model trends: propositions for the is organization of the future. Inform. Syst. Manag. **2007**, *24*(2), 139–146.

14. Nilles, J. Carlson, R. Gray, P. Hanneman, G. *Substituting Communications for Transportation: Options for Tomorrow*, Wiley: New York, 1975.

15. Hoving, R. Information technology leadership challenges —past, present, and future. Inform. Syst. Manag. **2007**, *24*(2), 147–154.

16. Carr, N.G. IT doesn't matter. Harv. Bus. Rev. **2003**, *81*(5), 41–49.

17. Carr, N.G. *Does IT Matter? Information Technology and the Corrosion of Competitive Advantage*, Harvard Business School Press: Boston, MA, 2005.

18. O'Reilly, T. What is Web 2.0: Design Patterns and Business Models for the Next Generation of Software. Available at http://www.oreillynet.com/pub/a/oreilly/tim/news/2005/09/30/what-is-web-20.html.

Careers and Education in Library and Information Science

Jana Varlejs
School of Communication, Information and Library Studies, Rutgers University, New Brunswick, New Jersey, U.S.A.

Abstract

Preparation for professional careers in librarianship entails education at the master's level in many parts of the world. In North America, a degree from a program accredited by the American Library Association (ALA) is the credential preferred by most employers. While the content of library/information science education still includes traditional subjects such as reference, collection development, cataloging, and management, programs are including more courses on technology and electronic resources. Many programs are offering at least some of their courses online. Continuing learning is essential in order for librarians to adapt to rapidly changing information environments.

INTRODUCTION

Careers in librarianship may be pursued in a wide variety of settings, serve very diverse client groups, and entail quite different activities and responsibilities, ranging from storytelling to taxonomy design. There is commonality, however, which resides in the set of skills, knowledge, and attitudes that defines the profession. In much of the world, this foundation is acquired in graduate programs of library and information science. This entry deals primarily with careers and education in North America, although international developments are touched on briefly. Beginning with a general description of educational requirements, the discussion moves to the ongoing critiques of library and information science education, refers to resources on traditional and evolving career paths, and concludes with observations on trends in continuing education.

THE CONTEXT

According to the United States Bureau of Labor Statistics *Occupational Outlook Handbook*, there were about 158,000 librarians, 116,000 library assistants, and 121,000 library technicians employed in 2006. Growth in librarian positions from 2006 to 2016 is projected at 4%, but at 8% for assistants and technicians. The explanation given for the disparity is anticipated budget constraints, more work shifted to support staff, and sufficient increased enrollment in library education programs to replace retirees (http://www.bls.gov/oco/ocos068.htm#emply). Using the American Library Association's Library Fact Sheet 2, one can estimate that about 17% of librarians work in academic, 30% in public, 44% in school, and perhaps 10% in special libraries (http://www.ala.org/alalibrary/libraryfactsheet/alalibraryfactsheet2.cfm).

The *Occupational Outlook Handbook* can be recommended as a fairly reliable brief introduction to the three types of library positions and the typical educational qualifications. It should be noted, however, that the profession would question the certainty with which some pronouncements are made, particularly in regard to library education curricula. Even on the requirement of a master's degree from a program accredited by the American Library Association (ALA) as a prerequisite to a professional career in librarianship, there no longer is total agreement.

Nevertheless, this discussion proceeds on the assumption that "Education for Library and Information Science," at least in the North American context, is generally understood to refer to master's degree programs accredited by the American Library Association (ALA), of which there are currently 50 in the United States and 7 in Canada (http://www.ala.org/ala/accreditation/lisdirb/lis_dir_2007–2008_re. pdf), and that "career" implies a professional level that rests on the accredited degree, or its equivalent. Since the late 1940s, the master's degree has been accepted as the basic qualification for a professional position as a librarian.[1] "Library and Information Science" is typically shortened to LIS, and LIS education is understood to refer not only to programs that emphasize preparation for careers in library settings, but also for a broader range of information positions in industry and elsewhere. The latest *Library Journal* annual report on placements and salaries of 2006 LIS graduates states that 89.9% were placed in "some type of library agency." Employment outside of libraries, while still low, increased 37.4% over previous years.[2]

The current LIS education landscape in North America is characterized briefly below. When speaking of professional qualifications it is useful to keep a few definitions in mind. A license to practice is issued by a government body, and should not be confused with certification, which

Encyclopedia of Library and Information Sciences, Fourth Edition DOI: 10.1081/E-ELIS4-120043811

is recognition of an individual's qualifications granted by a professional organization. Accreditation, on the other hand, is given to an educational institution or program of study by an agency or professional association, based on the meeting of standards.

- A master's degree in library science (or library service or studies, or library and information science, etc.; usually designated as M.L.I.S. or M.L.S.) from a program accredited by the ALA is generally required for professional librarian positions in academic, government, and all but the smallest public libraries. According to the *Occupational Outlook Handbook*:

  In addition to an M.L.S. degree, librarians in a special library, such as a law or corporate library, usually supplement their education with knowledge of the field in which they are specializing, sometimes earning a master's, doctoral, or professional degree in the subject. Areas of specialization include medicine, law, business, engineering, and the natural and social sciences. For example, a librarian working for a law firm may hold both library science and law degrees, while medical librarians should have a strong background in the sciences. In some jobs, knowledge of a foreign language is needed (http://www.bls.gov/oco/ocos068.htm#emply).

- Qualifications for school librarianship vary by state and are different in Canada. A master's degree in librarianship from a program accredited by the ALA or a master's degree with a specialty in school library media from an educational unit accredited by the National Council for the Accreditation of Teacher Education is held to be appropriate for U.S. school librarians. Many states also require school librarians to be licensed as teachers. Regulations for specific states may generally be found on their department of education Web sites.
- Most states and provinces have licensure requirements for public and/or school librarianship, generally based on educational credentials. At the municipal or county level, some U.S. governmental units may also apply Civil Service regulations for public library positions, usually involving examinations. In the case of small public libraries where staff is not required to hold the M.L.S. degree, some form of basic training may be mandated, usually under the auspices of the state library development agency.
- For other than school and small public library positions, most employers depend on ALA accreditation of the program where the candidate obtained the master's for assurance that the person has basic qualifications. ALA accreditation is based on standards first developed in 1951 and revised in 1972, 1992, and 2008. The ALA, in its turn, receives approval from the Council for Higher Education Accreditation. Standards are designed to help programs to improve within parameters they themselves set. That is, each program declares its goals and objectives to suit its constituencies, and is judged on how well it achieves those goals and objectives. There is no single prescribed pattern that is expected or dictated by the Standards. Accreditation is granted by a committee of the ALA, based on self-studies and external review panel reports, for a period of up to 7 years. Standards are grouped under six headings: mission, goals, and objectives; curriculum; faculty; students; administration and financial support; and physical resources and facilities (http://www.ala.org/ala/educationcareers/education/accredited programs/standards/standards_2008.pdf).

- Curricula in ALA-accredited programs generally include courses on the organization of information, selection and acquisition of resources, assistance to users, application of relevant technologies, management, and what often are called "foundation" courses that deal with matters such as ethics, history, and service philosophy. Other offerings may include a range of courses on particular types of libraries, specific user groups, technical aspects of information technology, and evaluation and research methods.
- There has been a substantial shift in the delivery of M.L.I.S. programs from on-campus to the Internet. There is a consortium of schools, Web-based Information Science Education (WISE) that offer courses to each other's students on a space available basis (http://www.wiseeducation.org).
- Most programs strive to impart both theory and practical application, to introduce current issues and new technologies while continuing to cover the basics, and to allow for some level of specialization. Students are urged to get some practical experience prior to graduation, either through jobs, internships, or volunteer work.
- Most libraries prefer to hire M.L.I.S. graduates who have some experience and the specific skills required by the position to be filled. At the same time, there is an increasing interest in some institutions in hiring subject and technology experts who do not necessarily have an M.L.I.S.
- A large number of retirements were predicted some time ago, but it appears that there are not as many openings as anticipated. Some recent graduates are spending a long time searching for full time jobs, and are resorting to part time and temporary employment. This situation is especially likely to occur when the candidate limits the search to a specific location or particular preferred work environment.
- Descriptions of alternative careers (i.e., nonlibrary) for M.L.I.S. holders and advice on job searching are plentiful, both in print and online.

The above broad brush description of the current status of LIS education and employment of graduates in North

British-Careers

America masks the fact that technology for some time has been propelling change in libraries and concomitantly demanding new and different skills of librarians. It is difficult to tell how well recruitment into the profession, LIS education, job realignment, and hiring practices have responded to the changing scenario. There is evidence, however, that the profession has not been comfortable with the state of LIS education for some time.

CRITIQUES OF LIS EDUCATION FROM THE PROFESSION

There have always been complaints from librarians that educators pay too much attention to theory and not enough to practice, or conversely, that they teach technique to the neglect of theory and principles. The first formal program, launched at Columbia University by Melvil Dewey in 1886 was met with considerable suspicion, and the first graduate school to be established, by the University of Chicago in 1926, was not hailed as a necessary advancement in professional education.[3] Primarily through the forum provided by the ALA, dialogue between practitioners and educators has continued over the years. Often the accreditation process has crystallized contentious issues, although external events, such as the increased demand for librarians after World War II, the Civil Rights and Great Society movements, and—more recently—the impact of computers and the Internet, also have stimulated discussion of the adequacy of LIS programs in producing graduates prepared to meet societal needs.

The strong postwar market for librarians, which spawned a dramatic increase in the number of accredited programs, came to a halt in the 1970s. From 1978 to 1994, 17 U.S. programs were closed.[4] During that same period of retrenchment, more and more schools incorporated "information studies" or "information science" in their names and a noticeable number dropped the word library from the title. The closings, the incursion of information science (especially the hiring of faculty with doctorates in fields such as technology), the absorption of a number of independent programs into larger units in universities, and an increase in undergraduate majors in information created considerable anxiety about the direction of library education. One response was the first Congress on Professional Education (COPE), organized by ALA in 1999. Discussions revolved around the higher education context, core values, curriculum, and accreditation; a full report can be found at http://www.ifla.org/VII/s23/bulletin/vol1-2.htm#4. Two additional congresses were held, one on continuing education and the other on support staff. Some of the recommendations emanating from these meetings have been heeded, but it cannot be said that concerns have been laid to rest. Recently, an ALA presidential task force was appointed to consider the core curriculum, accreditation, and other contentious issues. A report is due in the summer of 2008.

Overlapping with the ALA congresses, a study of LIS education was carried out from 1998 to 2000, under the auspices of the Association for Library and Information Science Education (ALISE). Funded by the Kellogg Foundation, the project is known as Kellogg-ALISE Information Professions and Education Reform (KALIPER).[5] The project identified trends that confirmed critics' fears that LIS education was shifting attention and resources toward information as a major focus, but the researchers presented this finding in a positive light. At the same time, they reported that the traditional course content was still being offered, although compressed and not necessarily identified as a mandated core.

The most notable recent critique was spearheaded by Michael Gorman, 2005–2006 ALA President and Dean of Library Services at California State University-Fresno. He gave a keynote address at a conference of the European Association for Library and Information Education and Research (EUCLID) in 2003, the general thrust of which he repeated in other papers and speeches in the following years. The abstract from the EUCLID proceedings captures much of the continuing debate, and is therefore quoted at length:

> This paper advances the idea that there is a crisis in library education...Among the problems seen are that library schools have become hosts to information science and information studies faculty and curricula...peripheral to professional library work...Many of the topics regarded as central to a library education by would-be employers are no longer central to, or even required by, today's LIS curricula. Modern communications technology has led many library educators to concentrate on that technology and to dismiss anything about libraries that is not amenable to a technological solution. The gap between what is taught in many LIS schools and what is being practiced in libraries is wide and widening.[6]

Gorman goes on to raise a number of other issues, such as the impending retirement of many librarians, the uneven geographical distribution of LIS programs across the United States, the loss of a number of prestigious library schools and the resulting "diminution of the research (in quality if not quantity) into library topics." Gorman wants to see a required core curriculum taught in every school that reflects what goes on in libraries: collection development and acquisitions, cataloging, reference and library instruction, systems, management, and types of libraries. He claims not to understand why the accreditation process cannot assure this.

CRITIQUE OF LIS EDUCATION FROM WITHIN

While the KALIPER project concluded with a considerable amount of self-satisfaction, not all LIS educators are sanguine in the face of the changes of the last decades.

Expansion of course offerings with a focus on information technology and information science has made it more difficult to allow for concentration in medical, law, youth, and other specialized areas of librarianship. Courses such as the history of libraries and librarianship, reading, and outreach to underserved populations are not generally available. Florida State professor Wayne Wiegand, for example, makes a strong case for including reading studies and library-as-place in the curriculum.[7,8] LIS educators are in agreement that they have failed to recruit students who reflect the population at large, have difficulty attracting minorities to the professoriate, and are not producing enough Ph.Ds. to replace the many faculty who will soon be retiring. Some even echo a number of Gorman's criticisms about curriculum and priorities.[9,10] There is some unease about the rapidly increasing conversion of face-to-face courses to the online delivery mode. With few exceptions, master's degree programs are too short to prepare graduates who are as well qualified for today's complex library and information jobs as faculty would like. But the fear is that increasing the length of programs would cut into enrollments and therefore threaten program viability in cost-conscious higher education institutions.

IN RESPONSE TO THE CRITICS

The most serious criticism of the late 1990s concerned the perceived demise of the traditional core curriculum and the failure of the ALA's accreditation process to assure the profession that graduates of LIS master's programs are well prepared. The ALA's Office for Accreditation has acted upon the COPE I recommendation that it do a better job of preparing external review panels, by conducting training sessions in conjunction with ALISE and ALA conferences. The ALA's Committee on Accreditation (COA) has been increasingly rigorous in its reviews, and has continuously revised policies and procedures. The Office's Assistant Director, Renee McKinney, recently compared the ALA's 2005 draft of core competencies to the curricula of the LIS programs, and found that 94.6% had courses that included all of the eight that were identified in the draft (http://www.ala.org/ala/accreditation/prp/prismreports.cfm).

An earlier study by Karen Markey also identified five core courses, one each in organization, reference, foundations, and management, and one in either research or information technology. She concludes, "The curriculum remains strong in traditional coursework that seeks greater understanding of users, their information-seeking behavior, and the sources and services that libraries provide to users generally and to special populations."[11] On the other hand, a 2006 content analysis of curricula found that there was a large number of elective courses, while the number of required courses was as low as two. The four types of courses offered most frequently were organization of information, reference, introduction, and management.[12]

In response to the fusillade delivered by Michael Gorman, almost the entire Fall 2005 issue of the *Journal of Education for Library and Information Science* was devoted to articles addressing his complaints. The most pointed is the paper by Carla Stoffle and Kim Leeder.[13] Stoffle, as an academic library director with experience as a student of higher education, acting director of a library school, and member of ALA's Committee on Accreditation, takes a holistic view of the "crisis." Her main conclusion is that

> many practitioners do not understand the goals of library education, the demands under which these programs operate, or the standards to which they are held. Practitioners want to dictate a curriculum based on their interests or the hiring needs of their particular libraries, without acknowledging the tremendous range of subject matter that these schools must address in only 36 to 42 hours of coursework.[13]

OTHER PERSPECTIVES ON THE M.L.I.S.

Research libraries tend to be less concerned with the core curriculum than with qualifications that LIS graduates have obtained elsewhere during their education. Their stance is that foreign languages, area studies, and other specialized subject backgrounds are needed in research libraries, and that few M.L.I.S. holders have the necessary academic background. The Council on Library and Information Resources funds a postdoctoral fellowship at four libraries as a means of recruiting scholars to library work, bypassing the M.L.I.S. With a similar goal in mind, Yale offers a paid internship to Ph.D. students who have completed course work.

Other specialized LIS professional associations also have concerns about the products of ALA-accredited schools, and like to be involved in events such as the COPE series and to be represented on LIS schools' advisory bodies. They contribute by defining the competencies that their specialties require, providing internships, and teaching as adjuncts in LIS schools. The Special Libraries Association's (SLA) competencies document is frequently cited as a model (http://www.sla.org/content/learn/comp2003/index.cfm). It is divided into professional and personal competencies, with detailed descriptions of specific skills, and scenarios illustrating the application of professional competencies. The four major professional competencies which provide the framework for the skills lists are managing information organizations, managing information resources, managing information services, and applying information tools and technologies. While these skills are taught in LIS education programs, SLA does not have a policy that states that the M.L.I.S. is the appropriate preparation for a career in special librarianship. Instead, the last policy statement (1992) on education expresses concern that many LIS programs do not provide an adequate range of courses. The full statement is not accessible electronically, and is probably seen as no longer necessary now that the competency document exists.

The Medical Library Association (MLA), through its Academy of Health Information Professionals, provides a credentialing system for librarians who can demonstrate that they have attained specific levels of education, experience, and professional achievement. An M.L.I.S. from an ALA-accredited program is suggested as the basic academic preparation, but an equivalent graduate degree is acceptable. The American Association of Law Libraries (AALL), like the MLA, lays out a list of competencies which "may be acquired through higher education such as library and information science programs, through continuing education, and through experience" (http://www.aall.org/prodev/competencies.asp). It is fair to conclude that for many kinds of special librarianship, the M.L.I.S. from a program accredited by ALA is merely one option for meeting entry level qualifications. SLA, MLA, and AALL all place great emphasis on the importance of continuing education and professional development.

MORE ON COMPETENCIES

In the curriculum section, the 1992 accreditation standards call for taking into account competency statements promulgated by professional associations devoted to the various specializations. At the time, few such statements existed, but since then the gap has been largely closed. As previously mentioned, the ALA developed a generalized draft in 2005. The Canadian Library Association has a somewhat whimsical all-purpose list, divided into personal and professional competencies (http://www.cla.ca/infonation/skills.htm). Some of the ALA divisions devoted to types of librarianship have their own specific statements, as do a number of the nonaffiliated specialized library/information professional associations. LIS educators are expected to be familiar with those that are relevant to the courses that they design. It must be recognized, however, that the term "competency" as used in these professional pronouncements does not mean what it does when it is used by specialists who apply sophisticated measurement tools to the determination of competencies. The lists produced by LIS professionals are more like best practices descriptions, and simply represent the consensus of a group of practitioners about the ideal behavior and performance of their peers.

LIS EDUCATION IN THE INTERNATIONAL ARENA

The International Federation of Library Associations and Institutions (IFLA) is one of the best sources of information about LIS education globally. A new edition of *World Guide to Library, Archive, and Information Science Education* was scheduled for publication in late 2007. The intention is to list professional education and university level training programs worldwide. Meanwhile, there is an online directory compiled by T. D. Wilson at http://informationr.net/wl/.

IFLA approved *Guidelines for Professional Library/Information Education* in 2000 (http://www.ifla.org/VII/s23/bulletin/guidelines.htm). The introduction states that while LIS education occurs at differing levels—technical, graduate and professional, and research and doctoral—the guidelines apply mainly to the graduate and professional level. It is recommended that programs should be part of a degree-granting institution at the tertiary (university) level. Recommendations are grouped under headings similar to those used in the ALA's standards: curriculum, faculty, students, administration and financial support, and instructional resources and facilities.

In addition to IFLA publications, there are a number of English language publications that are regular sources of information about LIS education in various countries of the world, especially the journals *Education for Information, Journal of Education for Library and Information Science*, and *Libri*. In addition to ALISE and EUCLID, there are units in IFLA devoted to education and research. ALA conferences regularly attract LIS educators from abroad, and a wide array of international conferences also enable interaction among faculty from across the globe.

Perhaps the most significant development in LIS education at the international level was stimulated by the Bologna declaration in 1999, which calls for European countries to make their higher education programs more comparable and to initiate quality assurance efforts. For LIS, the likely outcome is that European preparation for library careers will look much more like that of the Anglo-American and some Asian countries. Eventually, it may be possible to tell whether librarians educated in one country have qualifications equivalent to those of another in which they want to work. Currently, it is difficult for employers in the United States, for example, to hire individuals who have received their training abroad. ALA states, however,

> The master's degree from a program accredited by the American Library Association (or from a master's level program in library and information studies accredited or recognized by the appropriate national body of another country) is the appropriate professional degree for librarians

(http://www.ala.org/ala/educationcareers/employment/foreigncredentialing/index.cfm).

CAREERS

The ALA's Office on Human Resource Development and Recruitment has general information on careers (http://www.librarycareers.org). For a more wide ranging and current snapshot of the profession, see http://www.liscareer.com, which not only collects advice on career planning and job hunting, but has a list of links to a variety of job postings. A spate of books on careers in LIS has been published, reflecting not only the concern with recruitment, but also raising awareness of alternative career paths. Since the first edition of Betty-Carol Sellen's *What Else You Can Do with a Library Degree* in 1980, there have been numerous suggestions of nontraditional jobs that benefit from the knowledge and skills acquired through LIS education. Recent examples of books with sections on alternate careers are listed at the end of this entry. The contributors to the book edited by Shontz and Murray tell how they gained the specific knowledge and skills that their jobs require, which often are not in the domain of LIS education programs. Many of them also explain how they stay up to date and continue their professional development.

The best way to learn about specific requirements for positions is to read job announcements regularly. Even if educational preparation is not mentioned explicitly, one should be able to relate the stated qualifications to the kinds of courses and training that provide the background. Learning about what employers are looking for also gives one a sense of how the field is changing, and is one way for LIS educators to see where curricula need to be adjusted. The following are excerpts from recent announcements that illustrate the range of jobs and differing requirements. Not all require an ALA-accredited M.L. I.S., and most want individuals with the kinds of personal characteristics and talents that are inherent rather than learned in a professional degree program.

Categorization Analyst

- Understanding of linguistic ambiguity and a creative approach to problem-solving.
- A background in Linguistics, Library Science, Information Sciences, or related field is preferred.
- Must have a B.A./B.S. with 1+ years of experience, or an M.L.S./M.L.I.S./M.S.
- Excellent communication and organizational skills.

We are an innovative technology company in the business of structuring and organizing massive amounts of unstructured content.

Digital Services Librarian (Assistant Professor)

- Will collaborate with library faculty and the Dean's office in developing and producing digital projects, in providing information services for students taking Web-based courses, and in the development of other Web-based applications.
- Will research, evaluate, test and recommend methods, standards, and software used in the creation of digital collections and their preservation; maintain best practice documentation for digital efforts; provide expertise to others in the creation of digital collections; and participate in grant-writing and training initiatives.
- Will collaborate with library faculty in providing reference service to distance education students and in developing tutorials and other Web-based services for off-campus and on-campus students.
- Will be expected to teach some information literacy courses and perform other duties as assigned.
- Required qualifications: Demonstrated digital project management skills; working knowledge of digital library technologies, standards, issues, and trends; ability to interact effectively with a broad range of employees within the Library, the University and in collaborative initiatives with other institutions; excellent written and oral communication skills; M.L.S. from an ALA-accredited program; a second master's degree or doctorate is required for tenure; excellent communication and interpersonal skills; demonstrated ability to work in a collegial environment.
- Desired qualifications: Demonstrated knowledge of digital conversion of materials; experience with MAchine Readable Cataloging (MARC) and Dublin Core; comprehension of relevant Internet delivery technologies for digital content; successful grant-writing activity.

Knowledge Systems Librarian/Taxonomist, Law

- Combination of knowledge and experience at the intersection of knowledge management (KM), law, and technology.... Must be able to interact effectively with attorneys and staff worldwide.
- KM Systems Librarian will maintain the quality of search results and contents and establish workflow procedures for vetting documents. Oversee workflow, reinforce and monitor data inclusion and indexing, and establish and maintain effective working relationships with all legal support departments firm wide. Develop and maintain taxonomies; both firm, office and practice group; and be able to integrate multiple language schemes. Work with Head of Technical Services to leverage Library Automation software and expand installations globally.
- Requirements: M.L.S. and/or JD (law degree) with proven expertise in building taxonomies. Must have solid IT skills including maintaining quality databases and an ability to troubleshoot difficulties with end user search and document indexing...Working knowledge of European languages would be a distinct advantage.

Staff Training and Development Coordinator

- Work collaboratively with multiple constituents including Department Heads, managers, supervisors, and library staff members to design, develop, coordinate, market, and implement training and development programs that speak to the diverse needs of the staff.
- Track staff participation in training programs...assess effectiveness.
- Lead Library Education and Training Committee; run Library's New Employee Orientation program.
- Work with University resources (i.e., University Human Resources, the Ombudsman, EEO (Equal Employment Opportunity), Employee Health and Safety, and external organizations (i.e., PALINET, the Association of Research Libraries) to achieve results.
- Maintain and update the Library HR Web site.
- Required: Bachelor's degree required. A minimum of 5 years of previous experience as a training professional in an academic library or higher education setting required. Demonstrated ability to work effectively with a variety of people at all levels in the organization. Must have superior oral and written communication skills. Previous supervisory experience required.
- Preferred: Prior experience working with diversity programs and initiatives preferred. Master's degree in Human Resource Management, Organizational Development, or Adult Education preferred.
- Desirable: Certification as a HR Professional.

Principal Librarian to Lead the Youth Services Team

- Serve an extremely active and diverse community in an urban setting. The Youth Services Department Head supervises a staff of seven, including three full-time librarians. This position will also serve as a member of the Branch Management Team; designing branch services, developing branch objectives and activities, and recommending branch procedures. The ideal candidate will demonstrate enthusiasm; excellent organization and communication skills; the ability to lead a team and work as part of a team; a collaborative and inspiring disposition; strong desire to work with and within a diverse community; and demonstrated excellence in cross cultural competencies. Evenings and weekends are required with this 35 hr per week position.
- Requirements: A master's degree in Library or Information Science in a library program from a library school accredited by the State Board of Education. Two years of professional librarian experience.

Although the positions in the above examples are from a U.S. Northeast metropolitan area, they are not especially unusual, and do reflect today's LIS environment. The Canadian Library Association' Web site has a page with hypothetical job descriptions, intended to give an idea of the range of positions and qualifications: http://www.cla.ca/infonation/jobstuff.htm.

Referring again to the excerpts from the above postings, it seems that technology expertise may not always trump management and people skills. Employers complain that LIS education gives short shrift to management, but most students resist taking courses in that area. LIS education program admissions decisions rest primarily on evidence that an individual is likely to succeed in graduate study. Admissions committees (or officers) cannot reject an otherwise well-qualified applicant merely because the applicant's motivation is confined to a lifelong love of books. The dissatisfaction that employers express with LIS curricula could at least in part be redirected to recruitment into the profession. It is well documented that most persons who decide to enter librarianship do so because they have worked in a library at some point and/or been encouraged by someone already in the field. The profession, therefore, tends toward replication rather than transformation.

CONTINUING EDUCATION

Referring again to the position announcement excerpts above, it becomes clear that very diverse skills and experience are needed across the spectrum of library/information careers, and that obtaining an M.L.I.S. may be only one step toward achieving career objectives. Other formal degree programs may actually be preferred, as in the case of the university library staff development position. All library/information personnel, however, need to participate in some form of ongoing learning, in response to continually changing technology, increasing client expectations, and pressure for accountability. Where the learning takes place can range from university classrooms, to professional conferences, to the workplace, to one's PC at home. A need may be met through offerings from a continuing education provider, workplace training, or by self-directed, informal learning.

Most LIS education programs are willing to have alumni return for additional courses as the need arises, and a few offer sixth year certificates. Some have professional development programs which range from occasional workshops, to series of webinars, to online credit-bearing courses, study tours, and other formats. Large library systems tend to have their own in-house professional development programs. There are regional consortia that play important roles in providing continuing education, and many vendors offer in-house training. The many LIS professional associations provide an increasingly diverse menu of learning opportunities, going well beyond the traditional publications and conference programs of the past. Many now regularly offer online courses, and just about all have electronic discussion lists, podcasts, newsfeeds, and blogs that encourage members to share expertise. WebJunction is a portal and clearinghouse of professional development opportunities, which makes it possible to locate a resource to meet a learning need at the moment that it is required (http://www.webjunction.org).

For some specialized areas, there are annual institutes and symposia designed to impart the latest and best practices, such as the Northeast Document Conservation Center's School for Scanning and the University of Maryland's copyright symposia. There are too many leadership development institutes to list, and several annual conferences on information literacy. Internationally, a key provider is IFLA, which holds not only an annual conference with its numerous satellite events, but also has quite a few separate programs throughout the year in different parts of the world. IFLA has a section on Continuing Professional Development and Workplace Learning (CPDWL), which has developed guidelines for quality continuing education (http://www.ifla.org/VII/s43/pub/cpdwl-qual-guide.pdf), available in a number of languages. CPDWL has held seven conferences with published proceedings and is in the process of preparing a theme issue of *Library Management* devoted to professional development.

A recent addition to the array of professional development opportunities is the Certified Library Administrator program offered by the ALA's Allied Professional Association (APA). It is currently aimed at public library directors and is an effort to make up for the lack of management education so common among LIS graduates. Nine courses are available, of which four are required: Budget and Finance, Management of Technology, Organization and Personnel Administration, and Planning and Management of Buildings. The electives are: Current Issues, Marketing, Fundraising/Grantsmanship, Politics and Networking, and Serving Diverse Populations. The APA's Web site tracks the progress of individuals seeking certification—not many at this stage, but increasing (http://www.ala-apa.org/).

CONCLUSION

Is there a crisis in LIS education? Is there going to be a shortage of library/information professionals as more librarians retire? Or, will there be fewer entry level jobs for LIS graduates as more work is passed to nondegreed staff? Will the profession succeed in diversifying its ranks to better reflect changing demographics? Will LIS educators and employers reach a better understanding of each other? Can they agree on what a core curriculum should contain? Will the accreditation process remain much as it is now, or evolve in a new direction? Will the current 31 Ph.D. programs in LIS produce a new generation of

faculty members who are well suited to teaching and conducting research in the constantly shifting higher education environment? With 18 undergraduate, as well as Ph.D. and other master's programs sharing resources with M.L.I.S. programs, many of which are increasingly using labor-intensive online delivery, will quality be maintained? Many of these questions are not new, but recur over the years. Finding answers seems to be more critical today, as LIS education programs are challenged to survive criticism from within and without.

REFERENCES

1. Holley, E.G. One hundred years of progress: the growth and development of library education. *The ALA Yearbook of Library and Information Services 1986, v. 11*, American Library Association: Chicago, IL, 1986; 23–28.
2. Maatta, S. What's an MLIS worth?. Lib. J. **2007**, *132*(17), 30–38.
3. McMullen, H. Library history: a mini-history. Am. Lib. **1986**, *17*(June), 406–408.
4. Bobinski, G. *Libraries and Librarianship: Sixty Years of Challenge and Change, 1945–2005;* Scarecrow Press: Lanham, MD, 2007.
5. Pettigrew, K.E.; Durrance, J.C. KALIPER: Introduction and overview of results. J. Educ. Lib. Inform. Sci. **2001**, *42*(3), 170–180.
6. Gorman, M. Wither library education? In *Coping with Continual Change—Change Management in SLIS*, Proceedings of the European Association for Library and Information Science Research (EUCLID) and the Association for Library and Information Science Education (ALISE) Joint Conference Potsdam, Germany, 2003; 31–August 1,2003, Ashcroft, L., Ed.; Emerald: Bradford, U.K., 2005; 1–5.
7. Wiegand, W.A. Critiquing the curriculum. Am. Lib. **2005**, *36*(January), 58–61.
8. Wiegand, W.A. MisReading LIS education. Lib. J. **1997**, *122*(June 15), 36–38.
9. Mulvaney, J.P.; O'Connor, D. The crux of our crisis. Am. Lib. **2006**, *37*(6), 38–40.
10. Cox, R.J. Why survival is not enough. Am. Lib. **2006**, *37*(6), 42–44.
11. Markey, K. Current educational trends in the information and library science curriculum. J. Educ. Lib. Inform. Sci. **2004**, *45*(4), 317–339.
12. Chu, H. Curricula of LIS programs in the USA: a content analysis School of Communication & Information, Proceedings of the Asia-Pacific Conference on Library & Information Education & Practice 2006 (A-LIEP 2006) Singapore, April 3–6,2003; Khoo, C., Singh, D., Chaudhry, A.S., Eds.; Nanyang Technological University: Singapore, 2006; 328–337. Available at http://dlist.sir.arizona.edu/1401/01/48.Heting_Chu_pp328–337_.pdf.
13. Stoffle, C.J.; Leeder, K. Practitioners and library education: a crisis of understanding. J. Educ. Lib. Inform. Sci. **2005**, *46*(4), 312–319.

BIBLIOGRAPHY

1. Dority, G.K. *Rethinking Information Work: A Career Guide for Librarians and Other Information Professionals;* Libraries Unlimited: Westport, CT, 2006.

2. In Shontz, P.K., Murray, R.A., Eds. *A Day in the Life: Career Options in Library and Information Science*; Libraries Unlimited: Westport, CT, 2007.

3. Watson-Boone, R. *A Good Match: Library Career Opportunities for Graduates of Liberal Arts Colleges;* American Library Association: Chicago, IL, 2007.

Careers and Education in Records and Information Management

Carol E. B. Choksy
School of Library and Information Science, Indiana University, Bloomington, Indiana, U.S.A.

Abstract
Careers in records management vary by organization and industry. There have been few educational-resources available other than ARMA conferences and meeting. While the career itself is interestingand challenging, the role of the records manager within organizations is often poorly understood orappreciated.

INTRODUCTION

Every type and size of organization needs records management. From the entrepreneur to the multinational, from government to nonprofit and NGOs, and from the accounting firm to the technology company, every organization needs to perform records management to find information again, to safeguard intellectual property, to comply with regulations, to be transparent, to lower costs, to serve customers better, and to respond to litigation, among other issues. The leaders of mostorganizations do not always understand the need for records management unless there are scandalslike the British intelligence report on Iraq before the war began that nearly took down Tony Blair's government, seeing paper spread far and wide in the World Trade Center collapse, or hearing about a $1.5 billion court fine against Morgan Stanley—in *Morgan Stanley v.Coleman*—for not being able to find e-mail backup tapes. Until recently, most leaders responded to these perceived risks with a hasty band-aid such as a change in procedures, a disaster recovery program, or a retention schedule. The number and severity of the scandals, from Bill Gates'e-mail in Microsoft's antitrust trial, to the deleted e-mails that brought down Arthur Andersen, from the large fines placed on broker-dealers for not managing their e-mail, to the $38 million settlement in *UBS Warburg v. Zubulake* for not handing over e-mails for which the plaintiff had clear evidence, has convinced more leaders, mostly for-profit, that they need to have a complete program. Those same leaders are now seeking records managers with education and skills torun enterprise-wide programs that are now reporting to executives in accounting, legal, and information technology. This is in sharp contrast to a decade ago when records managers reported to a director of facilities and managed boxes in off-site storage and a few records rooms.

From late 1990s to now, executives and directors managing every type and size of organization in North America have placed an increasing value on records management for finding information again for operations, disaster recovery and business continuity, and litigation response. Within those organizations generally there has also been recognition that the information technology products including, hardware, operating systems, and applications, purchased throughout the 1980s and 1990s to create, distribute, and store information actually do a poor job managing information by themselves. This increased value of records management and recognition of the current information technology situation has led to a higher value place on records managers, particularly those with the CRM (Certified Records Manager) designation. That dearth of qualified records managers has created an opportunity for professions, such as lawyers, to create retention schedules, and records management policies and procedures. This dearth has even created opportunities for library and information science to create and sell standardized taxonomies as the basis for a records retention schedule. Information technology vendors including Microsoft, Computer Associates, and EMC have created products to manage information on top of the other products these companies sell. Trade associations such as SNIA (Storage Networking Industry Association) and AIIM (Association of Imaging and Information Management, formerly the microfilm industry association) have created alliances ortried to merge with the association representing records managers, ARMA, International. These opportunistic services, products, and relationships have met with varying degrees of success depending upon the source's understanding of how complex the business objects in the records manager's portfolio are—as opposed to the cultural objects managed and studied by library science, archives, and information science. Until recently, those business objects were managed only by records managers, and were studied by the rare records manager with time for reflection.

Encyclopedia of Library and Information Sciences, Fourth Edition DOI: 10.1081/E-ELIS3-120043659

RECORDS MANAGEMENT ASSOCIATED WITH RISK

The existence of the post of records manager, a records management program, and a perceived need for a records management consultant are currently based on the risk side of records management where in policies are in adequate, procedures are lacking, information technology that did not manage records was implemented *en masse*, and the employees created messes because they thought they "know technology" or that the technology was designed to manage information. The same activities that will correct these problems can, if understood by the creator, help the organization create a more secure computing environment, manage output, and aggregate information for innovation. If the creator does not understand these problems, the organization may get a set of groups of documents and the regulations that restrain their life. The education of the creator is key to the success or failure of the organization's program. The selection of the creatoris, however, entirely in the hands of employees who do not understand records management either, hence the hiring of preformed taxonomies and attorneys to create retention schedules. Other organizations seek out persons with experience and qualifications. More and more organizations are seeking out an individual with a CRM to perform all of the tasks to create and run a records management program.

DIVERSITY OF THE ROLE

Employees who perform records management are diverse. Any person who is responsible for managing some aspect of the life cycle of documents, records, content, or information inside an organizationis performing records management. Thus, the mailroom supervisor who is capturing documents in the mailroom and the information technology (IT) director who sets the retention of e-mail in the personal mail boxes of end users to 30 days are managing records. Managing records does not mean managing only what ultimately get placed on a retention schedule, it means also ensuring that documents, content, and information that should not get placed on the retention schedule is managed (in this case, meaning formed properly and disposed of promptly).

What this means is that the titles for records managers are quite diverse. When ignorance of the existence of records management as a profession is added to that diversity, calling any organization and asking for "the records manager" can lead to some amusing circumstances. For example, in performing research for a project on organizational management and electronic records management the author made more than 150 calls to Fortune 500 and Forbes 300 companies, to identify the records manager for the study. At one Forbes 300 company the receptionist said she had worked for the company for more than 30 years and no one had ever asked that question. Only one call out of more than 150 produced the name of a person who was responsible for records management, but did not belong to ARMA.[1] (As a result of this experience, I used the ARMA member list to enroll members.) At the same time, persons who are responsible for early phases of the records life cycle, particularly those who perform information capture through scanning documents may not identify themselves as records managers because of trade advertising and analysts that cast records managers and records management in a pejorative light. This author has interviewed employees at some of the largest Fortune 500 firms that performed records management through several phases of the life cycle, including both imaging and box storage, yet those employees eschewed the title "records manager" as some how lessening what they do. Some trade association publications characterize records managers as roadblocks to successful document management implementations. In an article, Steven Good fellow, the president of Access Systems, which is a records management consultancy, writes: "When the records manager is brought into the discussion, it may be viewed as an effort to stir up more dust thus clouding the solution seen by the department and/or IT...because they can't speak in IT terms."[2] (This, of course, begs the question whether "IT" is a *linguafranca* or whether those who speak "IT" should learn to speak in the language of the employees of the organizations in which they are working.) This is a fairly typical characterization of a records manager and reflects the trade association's need to portray the solution as requiring a "tool" rather than a practice or skill set. Underlying this isthe problem that the "tools," the electronic document management systems (EDMSs) and the content management systems (CMSs), are designed to solve a tactical departmental problem where as the practice of records management is focused on information management not only throughout the lifecycle of information, but also on an enterprise-wide basis. The perspectives are thus that of 10 ft off the ground versus the 50,000-ft level. The two perspectives are not incompatible; documents, whether electronic or hard copy, must be corralled into a repository to be branded and subsequently managed. Unfortunately, the products that corral the documents manage only a tiny portion of the life cycle and often are created without the ability to "preserve" documents either in a litigation or archives sense. In addition, the policies and procedures to manage documents beyond a very narrow definition of the capture phase, such as associating a document with a retention classification are also missing.

IT AND RM

This wound of tools managing one small portion of the life cycle versus the practice of managing the entire life cycle that has existed for about 15 years is gradually healing as information technology loses its caché and as the companies that create the applications, such a Microsoft and IBM "discover" they have actually created problems that threaten the existence orat least the credibility of many organizations because of a failure to include the family planning to death certificate and document classification practices of records management. (For example, Arthur And ersen collapsed due to a lack of credibility because of the destruction and subsequent resurrection of e-mails. See Glater.[3]) This has not yet brought together all records management practitioners into one, big, happy family yet, as the article quoted above indicates. Records management is considered by most of the application companies to be an application that manages documents after the documents have been "declared a record." This relegation to less interesting phases of the life cycle has continued, so that some content management application companies have differentiated "retention management" from "records management," where in retention management is differentiating ephemeral documents—and the purging of those documents—from those documents that would need to be kept for a long period of time. Thus, software applications companies understand the practice of records management as being the management of boxes of paper documents and electronic "records," using the archivists understanding of a record as something that has passed beyond its active context to become"evidence" only.

The actual portfolio of activities performed by records managers is quite varied, so varied infact, that it is difficult to describe what all records managers do other than to say we manage the life cycle of records through the employment of policies and procedures, classification systems,risk management, legal and business requirements, and the discovery phase of litigation management. What records managers understand as the life cycle of records includes everything from the family planning phase through a death certificate. The concept of "cradle to grave" management as espoused currently by the document and content management application companies is partly what led to jury penalties as high as $38 million being applied when companies cannot pullout documents that should not have been created in the first place and that, in some instances, were deleted but resurrected because the document had not been properly destroyed. The planning phase includes creating policies, such as to whom information and the systems to create information ultimately belongs, procedures, such as ensuring that every document gets placed in an appropriate folder and does not float for eternity in the "Inbox" of an employee. The planning phase also includes a system of governance including a governance committee with enough authority to adjudicate issues, an audit program so that the organization knows whether employees are complying, a system of enforcement such as penalties for noncompliance and threats of law suit for egregious breach of policies and

procedures, and a training program that goes beyond using a software application to include why the records management program exists.

The creation phase can include everything from managing a mailroom and an imaging application there, to creating procedures to ensure the organization has draft, version, and copy or output control, to approving the purchase or intended use of a document, content, or records management application. Creation also includes limiting the content that can be placed in a document through e-mail etiquette and policies that forbid content such as pornography. No phase is hermetically sealed from other phases having created policies and procedures and introduced tools for one phase, the effects from all previous phases are felt throughout all subsequent phases. Organizations learning this currently through litigation say they are "dodging bullets" (personal conversation with general counsel of a multinational corporation) because they cannot find what they need to respond to subpoenas. Knowledge of how to write effective policies and procedures as well as how to provide training and change management are essential for implementing this phase.

RM AND LITIGATION

This leads to the question of what the records manager actually manages. There is neither "litigation phase" nor is there a limitation on what phase litigation can hit. Because of this threat, records managers now must address the management of any information captured within the organization whether that be photographs of employees dogs, internal e-mails that harass other employees, Internet caches that indicate what Web sites employees visited, in addition to those documents for which statutes, regulations, and agency bulletins as well as industry best practices and knowledge management needs. What the records manager actually manages bears little resemblance to ISO 15489 and the published definitions of archivists. Records managers are be holden to the definitions enunciated in law and the requirements of litigation—not the definitions of academic archivists. This extreme description of the "record" resembles nothing as closely as it does the current Euro-American studies of the "document" being performed by The Document Academy.[4] I have managed chunks of concrete, tools for making corporate seals, and have advised on howto track property owned by the U.S. federal government to be used in prototype engines. Unlike inventory management and supply chain management where every token of a type is managed like every other token of a type, records management requires managing processes where every token becomes its own type, can be extracted from any business process, and brought into a business process addressing litigation.

This activity requires knowledge of each business process within the organization and what purpose those business processes perform. This also includes understanding how and why employee screate and use them. Knowledge of the method of capture or creation of nearly every document involved in those business processes is required in order to ensure adequate document controls and metadata are attached to each set of documents. Control to the level of whether, for example, comments are made on documents, whether Post-It® notes, comments, or red-lines may remain attached when the document becomes idle, or whether attachments must be stripped from e-mail when filed. Another term currently used for "document control" is "document management." Document management is the phrase used to describe what activities are performed by information technology applications: EDMSs and CMS. Imaging applications do not perform document control; they create a new copy of a document in another medium. While the companies that create these products would like to claim these products do not perform records management, the activities of document control are part of the early twentieth century historical foundation for records management.

RM AND DOCUMENT OR CONTENT MANAGEMENT

Document control is the practice of ensuring the "record" is created properly using various methods of management control. It may take decades to create the "record," such as an engineering file documenting the creation of the Hoover Dam or a patient file documenting all services provided to a medical patient. The "record" may document an entire process like the creation of a single, complex document, or many different processes as with the example of the patient file. Many different skills are required to ensure appropriate controls are applied. Those skills, some of which are noted in the examples above, may require ethnographic skills to determine how the organization differentiates activities, analytical skills to determine what metadata are appropriate and what are not, managerial skills to manage change, and negotiation with upper-level executives to obtain appropriate funding for projects to implement document controls. For example, the author regularly interviews attorneys in law firms to uncover the logic of how they do their work and then negotiates with the same attorneys to help them create their own standardized list of "document types" and submatters (activities like environmental reviews for real estate deals). This requires knowledge of the technologies being used by the lawfirm, how they are implemented, and how to discuss what the attorneys do with administrators, including directors of IT, who believe the attorneys' filing activities are either chaotic or sodisparate that there can be no standardization. The records manager must also have an in-depth knowledge of security and privacy, litigation

facing the particular organization, quality controls, as well as how the organization views risk management.

Another aspect of this broader definition of "records" is the requirement of having detailed knowledge of the special content of each type of document. One university was astonished to discover the records manager knew where all the personally identifiable information was kept throughout the university and could recite the information from memory without having performed an expensive and time-consuming study. (Personal conversation with records manager of major state university.) Having worked with the documents and given extensive advice on all the university' sprocesses for so many years and with the disparate dimensions required of any records manager, this is surprising only to the nonrecords manager. Even records managers whose port folio includes only boxes or tapes in storage will have extensive knowledge of the locations of different types of documents and requirements for managing those documents appropriately throughout the end of the lifecycle.

Knowledge of the varying simultaneous dimensions required to manage documents is often confusing even to employees managing the documents, while quite natural to the records manager. The most important dimension of documents within records management is Retrievability. Documents, with the exception of the ejaculations of e-mail (which has its own set of problems), are rarely created and sent in one continuous action. Documents, even e-mails at times, are passed around, edited, approved, redrafted, abandoned, and picked up again. Being able to find the documents again iscrucial to the smooth flow of business. After documents have been completed and "filed," finding the document again can become a challenge, particularly if the person who "filed" the document is no longer with the department or even the organization. When there is a time limit on the time required to retrieve (and prepare and review) the document again, e.g., 30 days for a litigation subpoena, and there are thousands of such documents, an organization can find itself in dire straits with whom so ever needs or requires the documents. Precisely this situation held in *Morgan Stanley v. Coleman* where in Morgan Stanley kept finding more e-mail backup tapes and reporting them past the discovery phase of the trial. The judge was so exasperated with Morgan Stanley that he added a fine to the punitive damages that totaled nearly $1.5 billion. This was subsequently reversed by a higher court, but not because ofthe amount of the fine. (The jury instructions were considered in appropriately framed. Again, this had nothing to do with the total damages.)

The next dimension in importance in records management is Accessibility. This dimension has anumber of different components including speed of access, security, and privacy. Documents may be accessible, but getting hold of a specific document may take several days, many man-hours, and specialized equipment and software. Backup tapes are accessible, but depending upon the number of tapes, how the tapes were created, and what is required, finding the specific document among thousands of backups can be a struggle. In the Morgan Stanley case, documents were accessible, but could not be retrieved and restored quickly enough to respond to the subpoena. The newly revise drules of evidence do not compel the production of new types of information.[5] By using the term "electronically stored information" rather than"electronic documents," "electronic records," "electronic content," databases, etc. the federal judiciary hoped to clarify what had been interpreted poorly by responsive organizations. Records managers are now beginning to be given responsibility for part of the production of electronically stored information and are working along side general counsel and information technology to create better environments for finding information again.

RM TITLES

The actual title "records manager" has usually made reference to "records," "information," or "documents." Rather than appending "manager" to that, more and more organizations are raising there sponsibility set of these employees by making them directors and vice presidents. Sometimes the title is "assistant deputy counsel" because the person has a law degree. In multinational corporations many records managers are the global directors of records and information management. Larger organizations may have more than one records manager with duties divided according to expertise. For example, one may be an executive in charge of the entire program with responsibilities for strategy and liaising with other departments, while another with extensive records retention experience may create an organization-wide retention schedule, yet another will have had experience in imaging or microfilm and be responsible for capture and long-term storage of documents. Research indicates that records managers work closely with legal, IT, and line-of-business managers both as an advisor and as a requestor of information. Far from being locked in a basement with boxes, records managers are now part of robust organizational networks.[1]

RM EDUCATION

Until the past few years, records management was offered primarily as a community college course. Those courses were offered by practitioners with everything from a high school degree to aprofessional master's degree to a doctoral degree. Course offerings would arise with the interests of an individual practitioner in association with a college or university and often ended with that practitioner's absence. A college course or two would arise as the result of the efforts of afaculty member, but would also wane with the absence of that faculty member. There have been a few

baccalaureate degrees and minors offered, most notably at the College of Business and Public Administration at the University of North Dakota and the School of Information Science at the University of Tennessee at Knoxville. There was little effort put forth in creating an enduring program in part because the same faculty member could make additional monies through writing, consulting, and offering seminars. In other words, cooperation among faculty at different universities in creating a standardized course of study was not rewarded monetarily. Holding knowledge closely rewarded records management faculty monetarily. Faculty who did see themselves creating a course of study were few and created those courses at the ba ccalaureate level. However, as those faculty members' careers waned, so did their programs—the same as at the community college level. In addition to this, the school in which records management was taught differed greatly in focus. At the University of North Dakota, records management is taught in the context of business. At the University of Tennessee, records management is taught in the context of library and information science. An occasional dissertation would appear on records management, but no graduate positions created. When records management was taught at the graduate level, it was taught in the context of a library and information science school, usually by an archivist. Some programs have begun offering certificates in records management in MLS (Master of Library Science) courses of study, essentially a minor in records management.

On a different front, the professional association for records managers, ARMA, International, did not have staff who understood enough about university education, like the need for recognition of the profession by deans, not just individual faculty, like the need for there to be a ready market for records management graduates, like companies recruiting for records management graduates on campus and a clear research agenda for the field, to have any impact on records management education. Yet most people who count themselves as records managers, both in the public (52%) and private (64%) sectors got their education in that field from ARMA.[6]

Because no standardized course of graduate study arose, text books are all aimed at a very practical "how-to" level. Because university faculty, particularly graduate faculty, did not have practical experience and, until recently, the practitioners were primarily high school graduates, textbooks could not graduate much beyond a community college level. The content of textbooks in records management range from how to file[7] to how to manage a file room.[8–11] The very best textbook, best because of its scope, quality, and relevance to the field is the work of Robek et al.[12] (Similar but not as extensive are books by Shepherd and Yeo[13] and Penn et al.[14]) Unfortunately, this book is no longer in print. Subsequent editions attempt to deal with electronic records and fail because the scope of the editions was limited and the electronic records dealt with were primarily images on optical disk, i.e., only a fraction of the life cycle and of formats was dealt with in the text. No literature from archives was relevant to more than a tinyportion of the records life cycle, so no succor was found from that corner. Not until this author wrote *Domesticating Information: Managing Documents Inside the Organization*[15] did a book dealing with records as business objects aimed at the graduate level exist. It is my hope that this book will be replaced by a much better graduate textbook, but so far, all candidates remain written at the community college level.

A number of courses of graduate study with some records management content have arisen. University of Pittsburgh, University of Michigan, University of Texas, University of North Carolina, and University of British Columbia are the longest running. However, there are no graduate programs in records management. The programs are combined with archives education as a specialty in schools of library and information science where the student would earn an MLS degree. To restate this bluntly, there are no courses of study in records management. Whether records management is a"discipline" or not is a nineteenth century argument that we will not address here. (I believe the same could be said of library science, information science, and archives studies that borrow the methods and trappings of social science, yet ask no grand questions, like the nature of the universe or why humanity is considered a privileged subject.) The courses of study are library studies courses with electives in archives and, usually, one course in records management being taught by an archivist or a practitioner with a CRM and, occasionally, an MLS. Conceiving of adoctoral degree at the present time is difficult as there are few graduate faculty with PhDs, sufficient knowledge of records management, and a CRM, to train a student appropriately. The courses of study where records management is included, "archives and records management," have greatest emphasis on archives because the courses are taught primarily by archivists. Those courses of study emphasize a part of the life cycle most records managers never encounter: archiving for cultural purposes, i.e., handling information as a cultural object, rather than as a business object.

This difference between archives and records management is also what distinguishes records management from library and information science. Categorical statements about the subject of library science and information science declare the cultural and scientific purpose of information as *the* focus. Records management has an unembarrassed focus on business, including the business of government. Records management also focuses on the entire life cycle and the rules and governance necessary to domesticate information as thoroughly as possible. How scholars and culture will use information later is of relevance for only a tiny portion of information records managers manage.

ARMA has completed its core competencies. One of the purposes for creating the competencies is asa guide to higher education to ensure that courses on records management include the knowledge and skills required to do the job. The competencies have been embraced enthusiastically by a number of different universities that teach records management including, Wayne State University, San Jose State University, University of North Carolina, and Indiana University. Some of these programs have distance learning courses also based on the competencies. ARMA has also developed a self-assessment tool so that members of the association can identify the gaps in their knowledge and skills for further education.

UNDERSTANDING THE ROLE OF RM HISTORICALLY

The history of records management has also not been well studied. Even records management textbooks trace the history of records management to the end of World War II (WWII) when U.S. archivists were drowning in documents sent to them by the many different agencies involved in thewar. The historical record demonstrates that the basic concepts of records management—document control, life cycle management, and records retention—have been around for more than 5000 years.[16] Modern records management in the United States arose when the concepts of "scientific management" that were developed on the shop floors of manufacturing in the late nineteenth century were brought into offices in the early part of the twentieth century.[17] The term "records management" began to be used around the 1930s and was adopted by U.S. archivists immediately before and during WWII to describe what should be happening in the agencies so that the National Archives did not have to sort out the grains of wheat from the voluminous chaff. What the archivists appear to have added to the field after WWII is the term "records retention and disposition schedule" which had previously simply been called "transferring."[18]

Currently, few people with even a master's degree are teaching records management in graduate courses. Many of those classes are non degree, distance learning only, because most classrooms cannotbe filled by physically existing bodies more often than about every 5 years, except in very large markets like Houston and Los Angeles. The distance-learning methods range from the correspondence courses offered by Bill Benedon throughout the 1960s to online chat room courses like those at Wayne State University, and San Jose State University, to live video learning at Indiana University. Because of this dearth of education, the professional association for information managers, ARMA, along with its local chapters became one of the few sources of education in records management. Until a little more than a decade and a half ago, the educational sessions provided by the association and its chapters was on the latter half of the life cycle, because that was something nearly all members had in common. This meant sessions on other parts of the life cycle had to be handled by special interest groups.

The lack of graduate education also led to the development of a certification, the CRM, which requires an application, 3 years of broadly based experience in records management, the passing of five multiple-choice tests, and one essay question. This certification has been in existence for more than 35 years and has become recognized in both the public and private sectors as a measure of professional knowledge. Currently, there are more than 800 CRMs world wide, this is up from 600 about 10 years ago. This is the only certification in records management, there are many seminars with examinations that offer a certificate. Those seminars do not require experience or an application other than a fee. The CRM is currently the only way to recognize general knowledge in records management.

CONCLUSION

The diversity of the work portfolios of records managers stems from the fact that records management is needed in every organization from aeronautics through government to zoos, from the part of the life cycle that is managed and the differences in types of information managed. Not all organizations have a records manager, and most suffer from that lack when information is not created appropriately, is found by persons that should not have access to it, cannot be found again, or is not destroyed properly. While an urban legend has records managers managing boxes of paper in a warehouse, they manage evidence rooms in law firms and police stations and even DNA samples and bio-authentication data (retinal scans, fingerprints, facial scans, etc.). More than the commonly held concept of documents, records managers manage logs, caches, whole databases, database outputs, copies, drafts, versions, etc. Records managers manage information about nuclear waste, municipalities, and medical patients. The information may require machines like electron microscopes, computers, and microfilm readers to access it. The information may be in a CAD (computer-aided design) program, a laboratory testing machine's memory, or at the phone company.

The problems a records manager deals with may arise through maliciousness, laziness, or lack of direction. Identifying the source of the problem is as important as managing the process of change to reach an optimal solution. The solution nearly always requires policies and procedures. Machines do not manage information, cannot judge the level of passion a change will create, nor create a judicious statement of how information must be handled under certain circumstances.

REFERENCES

1. Choksy, C.E.B. Fortune 500, Forbes 300 and Transnational 150 Electronic Records Management Study. NHPRC Electronic Records Research Fellows 2005–2006 Final Report, 2007.
2. Goodfellow, S. What's the difference?. Is there a difference between Document and Records management?, 2007; April/May 14–17.
3. Glater, J.D. Prosecutors use notes to bolster case against Andersen. *New York Times*, June 28, 2007; May 21, 2002. Available at http://www.nytimes.com/2002/05/21/business/21AUDI.html?ex=1183003200&en=b0a1063818de5b13&ei=5070.
4. Skare, R.; Lund, N.W.; Vårheim, A. *A Document (Re)turn: Contributions from a Research Field in Transition*; Peter Lang: Frankfurt am Main, Germany, 2007.
5. U.S. *Federal Rules of Civil Procedure*, passed December 1, 2007; July 1, 2007. Available at http://www.law.cornell.edu/rules/frcp/.
6. Choksy, C.E.B. "Fortune 500" and "Government RIM Reporting". Unpublished research.
7. Stewart, J.R.; Melesco, N.M. *Professional Records and Information Management*; Glencoe McGraw-Hill: New York, 2002.
8. Read-Smith, J.; Ginn, M.L.; Kallaus, N.F. *Records Management*, 3rd Ed.; South-Western: Cincinnati, OH, 2002.
9. Saffady, W. *Records and Information Management: Fundamentals of Professional Practice*; ARMA, International: Lenexa, KS, 2004.
10. Wallace, P.E. Lee, J.A. Schubert, D.R. *Records Management: Integrated Information Systems*, 3rd Ed. Prentice Hall: Upper Saddle River, NJ, 1992.
11. Ricks, B.R.; Ann, J.; Swafford, A.J.; Gow, K.F. *Information and Image Management: A Records Systems Approach*, 3rd Ed.; South-Western Publishing Co.: Cincinnati, OH, 1992.
12. Robek, M.F.; Maedke, W.O.; Brown, G.F. *Information and Records Management*, 3rd Ed.; Glencoe: Lake Forest, IL, 1987.
13. Shepherd, E.; Yeo, G. *Managing Records: A Handbook of Principles and Practice*; Facet Publishing: London, U.K., 2003.
14. Penn, I.A.; Pennix, G.B.; Coulson, J. "Domesticating Information.". *Records Management Handbook*, 2nd Ed.; Gower: Brookfield, VT, 1994.
15. Choksy, C. E.B. Scarecrow Press: Lanham, MD, 2006.
16. Archi, A. Archival record-keeping at Ebla 2400–2350 BC. In *Ancient Archives and Archival Traditions: Concepts of Record-Keeping in the Ancient World*; Brosius, M., Ed.; Oxford University Press: Oxford, U.K., 2003; 17–36.
17. Yates, J. *Control through Communication: The Rise of System in American Management*; Johns Hopkins University Press: Baltimore, MD, 1989.
18. Choksy, C.E.B. *Domesticating Information*; Scarecrow Press: Lanham, MD, 2006; Chapter 1.

Index

A

AALL, *see* American Association of Law Libraries
AAM, *see* American Association of Museums
Aarhus Art museum, 1226
Aarhus State and University Library, 1216–1217, 1219
AASL, *see* American Association of School Librarians
AASL Hotlinks, 61
Abandoned Shipwreck Act of 1987, 1775
The Aboriginal and Torres Strait Islander Library and Information Resource Network (ATSILIRN) Protocols, 2041
Abridged WebDewey, 1259–1260
Absorption, distribution, metabolism, excretion, and toxicity (ADMET) testing, 837
Abstracts, 418–419
Academic art libraries, 251
Academic dishonesty
  definition, 3665
  faculty attitudes, 3668–3669
  individual differences, 3668–3669
  social factors, 3668
Academic e-mail messages, 2507
Academic law reviews, 2740
Academic Librarians Status Committee, 342
Academic libraries, 97, 2764, 3471–3472
  acquisitions units, organization of, 2918–2919
  administration, 9
  in Arab sector, 2548
  Armenia, 230–231
  in Australia, 384–385
  buildings, 10–11
  in China
    Peking University Library, 896
    Tsinghua University Library, 896–898
  in Croatia, 1125
  database integrators, 3472
  digital humanities (*see* Digital humanities)
  Ethiopia, 1498–1499
  external influence, 2–3
  in France, 1602–1603
  fund-raising and development
    access to donors, 2835
    annual fund, 2836–2837
    capital campaigns, 2838
    centralized *vs.* decentralized development, 2834–2835
    development activities, 2833–2834
    friends of libraries, 2836
    institutional barriers, 2834–2835
    institutionalization, 2839
    library director, role of, 2835
    literature, history of, 2832–2833
    major gifts, 2837
    planned giving, 2837–2838
    theoretical and philosophical foundations, 2839
    U.S. phenomenon, 2832
    Web communications, 2838–2839

games and gaming, 1639–1640
in Germany, 1695–1696
governance and hierarchy, 3–4
Greece, 1731–1732
history, 1–2
Hungary, 1922
in Israel, 2544–2545
Japan, 2562–2564
in Kazakhstan, 2582–2583
Kenya, 2596
Latinos, 2701–2702
library anxiety, 2785
Lithuania, 2951–2953
Mexican libraries, 3083–3086
mission, 1
in Moldova, 3125
music libraries, 3275
New Zealand libraries, 3375–3376
organizational structure, 4–5
personnel, 10
in Peru, 3608
professional associations, 3
resources and services
  expertise, 5–6
  public service, 7–8
  reference desk, 6–7
  technical services, 8–9
in Saudi Arabia, 3974
science and engineering librarians, 4009
Senegal, 4106
in Serbia, 4129–4131
Slovakia, 4177–4178
South Korea, 4310–4311
strategic planning (*see* Strategic planning, academic libraries)
Tunisia, 4628–4629
in Ukraine, 4642
in United Kingdom, 4703–4705
user privileges, 5
Venezuelan libraries, 4889–4890
Academic Library Advancement and Development Network (ALADN), 2834
Academic publications, 2826
Academic writing, 4548–4549
Academy of Beaux-Arts, 1594
Academy of Health Information Professionals (AHIP), 3035, 4356
Access control, *see* Authorization
Accessed information, 4236
Accessibility
  adaptive hardware and software, adults
    audiobooks and Playaways, 16
    audio description, 16
    Benetech, 16
    closed-circuit television, 15
    mouse challenges, 15
    outreach efforts, 16
    public meetings, signing for, 15–16
    screen reading software, 15
    talking books, 16
    TDDS and TTYS, 16

    typing and voice recognition software, 15
    virtual reference, 17
    web conferencing platform, 17
    Web sites, 16–17
  audio/recorded books, 15
  books by mail service, 14
  Braille books, 15
  building accommodations, 14
  deposit collections, 14
  homebound book delivery, 14
  large print books, 14–15
  services, 14
  symbols, 4961, 4963
Access management, *see* Authorization
Access services, 173, 373, 895, 1910, 2912, 3472, 4735
Access to Knowledge (A2K) movement, 3386
Access-to-Own model, 1211
Accountability, 2055
Accounting in England and Scotland: 1543 to 1800, 645
Accreditation
  ALA, LIS programs
    Accreditation Process, Policies, and Procedures, 18, 20
    ALISE, 19–20
    ASPA Code of Good Practice, 18–19
    BEL, 19
    COA, 18–20
    Committee on Library Training, 19
    future prospects, 20–21
    Land Grant College Act, 19
    of postsecondary education, 19
    purpose of, 18
    standards, 18, 20
  of Canadian institutions, 19, 21
  fundamental assumptions of, 18
Accreditation Board for Engineering and Technology (ABET), 1434
Accredited Standards Committee (ASC), 413
Achenwall, Gottfried, 495
ACLU v. Reno, 2396
ACM, *see* Association for Computing Machinery
Acquisitions
  in academic libraries, 2918–2919
  approval plan vendors, 2919
  automated and integrated library systems, 2919
  bibliographic networks, development of, 2921
  definition, 2918
  EDI, 2919
  in public libraries, 2919
  purchasing decisions, 2918
  shared cataloging, 2921
  in small libraries, 2919
  in special libraries, 2918
Acquisitions Institute at Timberline Lodge
  collection development, 22
  facility, 2
  history and evolution, 22–23

sections, 61
special committees, 61
Standards, 4859
standing committees, 61
task forces, 61
values, 60
vision, 59
American Association of University Professors
    (AAUP), 341
American Association of Zoological Parks and
    Aquariums (AAZPA), 5078
American Astronomical Society (AAS), 3640
American Bar Association (ABA), 2711–2712,
    2716
American Chemical Society (ACS), 3641–3642
American Civil Liberties Union (ACLU), 783,
    2402
American Committee for Devastated France
    (CARD), 1600
American Cooperative School of Tunis (ACST),
    4630
American Documentation Institute (ADI), 90,
    311, 2770
American Economic Review (AER), 3470
American Federation of Labor (AFL), 4689
American Federation of Labor-Congress of
    Industrial Organizations (AFL-CIO),
    4761
American Federation of State, County and
    Municipal Employees (AFSCME),
    4690–4691
American Federation of Teachers (AFT), 3997–
    3998, 4689
American Film Institute (AFI), 1586
American Health Information and Management
    Association (AHIMA), 1854
American Historical Association (AHA), 1788,
    3316, 4272, 4741
American Historical Review, 1791
American Indian Library Association, 334–335
American Institute for Certified Public Accoun-
    tants (AICPA), 2918–2919
American Institute for Conservation of Historic
    and Artistic Works (AIC), 1072, 3729,
    4953
American Institute of Accountants, 651
American Institute of Architects Awards,
    2845
American Institute of Certified Public Accoun-
    tants, 651
American Institute of Physics (AIP), 3639, 4763
American Journal of Mathematics, 3024, 3027
American Law Reports (ALR), 2741
American librarianship, 2890
American Library Association (ALA), 3, 60,
    229, 255, 706, 757, 1336, 1846–1847,
    2764, 2775, 2964–2965, 3728, 3775,
    3778, 4001, 4649, 4773–4774, 4777
  accreditation of LIS programs
    *Accreditation Process, Policies, and Pro-
      cedures*, 18, 20
    ALISE, 19–20
    ASPA Code of Good Practice, 18–19
    BEL, 19
    COA, 18–20, 709
    Committee on Library Training, 19
    future prospects, 20–21

  purpose of, 18
  standards, 18, 20
ACONDA/ANACONDA, 71
ACRL, 2752
affiliated organizations, 77
Allied Professional Association, 74–75, 713
ALSC, 333
awards and scholarships, 81
chapters, 77
children's literature, awards for, 852
Code of Ethics, 3917
conference
  change and controversy, 80–81
  growth and development, 79–80
  Midwinter Meeting, 80
controversy, 72–73
Council, 75
Cresap, McCormick and Paget, 70
divisions, 75–76
  self-study, 74
dues schedule transition document, 73
Executive Board, 75
GameRT, 1637
Holley Committee, OSSC, 71–72
intellectual freedom (*see* Intellectual
    freedom)
International connections, 77–78
Latinos and library services, 2699
library network, definition of, 3920
LLAMA, 2841–2845
membership and organizational change, 69
membership statistics (August 2008), 83
MIGs, 76–77
offices, 81–82
older adults, library services for, 3407
operating agreements, 73–74
organizational development
  ALA divisions and allied professional
    association, 73
  growth and democratization, 69–70
organizational self-study (1992–1995), 72
periodic scrutiny, 70
PIALA, 3548
publishing, 78–79
round tables, 76
RUSA, 3913
Science and Technology Section, 4016
standards and guidelines, 81
standing committees, 76
state library agencies, 4394
values, priorities and missions, 67–69
web-based networking, 78
American Library Association-Allied Profes-
    sional Association (ALA-APA), 3777
American Library Association conference,
    2770
*American Library Association v. U.S. Depart-
    ment of Justice*, 2396
American Library Directory, 249
American Management Association, 651
American Marketing Association, 651
American Mathematical Society, 324
American Medical Informatics Association
    (AMIA), 85–86
American Memory Project, 5027
American Museum of Natural History (AMNH),
    3252, 4944

American museums, 3234
American National Standards Institute (ANSI),
    221, 1857, 1981, 2343–2344
  conformity assessment, 88–89
  history, 87
  industry sectors and services, 87
  international standards activities, 88
  logo, 88
  NISO, 88
  process, 88
  standards panels, 89
  U.S. standardization system, 87
American National Trust for Historic Preserva-
    tion, 4169
American Psychological Association, 4246
American Radio Archives, 1565
American Records Management Association
    (ARMA), 175, 1853–1855
American Sign Language (ASL), 1184, 1187
American Society for Engineering Education,
    Engineering Libraries Division (ASEE/
    ELD), 4016
American Society for Indexing (ASI), 441
American Society for Information Science
    (ASIS), 1097, 1375
American Society for Information Science and
    Technology (ASIST), 482
  awards, 95
  chapters and SIGs, 92
  governance, 93
  history
    documentation, beginnings (1937), 90
    human/social perspective, (1990s and
      2000s), 91–92
    information explosion (1960s), 90
    modern information science transition
      (1950s), 90
    online information (1970s), 91
    personal computers (1980s), 91
  meetings, 94
  publications, 93–94
  purpose, 92
American Society for Quality National Accredi-
    tation Board (ANAB), 88
American Society of Information Science and
    Technology (ASIST), 3368
American Standard Code for Information Inter-
    change (ASCII), 5024
Americans with Disabilities Act (ADA), 10,
    377, 1530, 3575–3576, 3843
Americans with Disabilities Act Assembly, 3778
American Technology Pre-Eminence Act
    (ATPA) of 1991, 1552
American Theological Library Association
    (ATLA)
  Carver Policy Governance model, 4607
  Committee on Microphotography, 4606
  daunting problems, 4606
  Ethics index, 4607
  Executive Director, 4607–4608
  importance, 4606
  Library Development Program, 4606
  management structure, 4607
  premier professional association, 4608
  religion database, 4607
  Religion index two, 4606–4607
  Retrospective Indexing Project, 4607

Business literature
  abstracting services, 652
  accounting, 651
  banking, finance, and investments, 651
  Code of Hammurabi, 643
  on computer operations, 652–653
  cuneiform clay tablets, 643
  in Egypt, 643–644
  in England, 645–648
  foreign trade, 652
  France and Germany, publications in, 648
  in Greece, 643–644
  in late seventeenth century, 645–646
  loose-leaf services, 652
  marketing, sales management, and advertis-
    ing, 651–652
  newspaper indexes, 652
  periodicals, 653
  personnel and industrial relations, 651
  professional associations, 651
  public relations, 652
  real estate and insurance, 651
  during Romans, 644
  schools of business, 650
  scientific management studies
    in Europe and United States, 650
    list of books, 650–651
  in sixteenth and seventeenth centuries, 645
  statistics, 648–649
  trade associations, publications of, 649–650
  in United States, 647–649
  in Venice and Florence, 644–645
Business magazines, 641
Business process reengineering (BPR), 2658,
    3511
Business productivity software, 3650
Business Reference and Services Section
    (BRASS), 3909
Business Software Alliance (BSA), 3650–3651
Business Statistics, 649
Business Week, 650
Business writing, 4547–4548
Butterfly effect, 1036
Butter stamp, 540
Byzantine illumination, 1946–1947

C

Cabinets of curiosities, 1814
Cable Communications Policy Act, 1017
Cable Television Consumer Protection and
    Competition Act, 1017
Calcutta Public Library, 3326
Caldecott Committee, 334
Caldecott Medal, 334
Calendars, 133, 153, 170, 4084
California Digital Library, 1760
California Public Records Act, 1105
California's Meyers-Milias-Brown Act, 4690
California State University (CSU), 6, 9
Caliper Corporation, 1676
The Cambridge Crystallographic Data Centre,
    822
Cambridge Structural Database (CSD), 822, 834
*Cambridge Tracts in Mathematics and Mathe-
    matical Physics*, 3028
Camel library, Kenya, 2595

Cameo binding, 543
CaMMS, *see* Cataloging and Metadata Manage-
    ment Section
Campbell Collaboration, 1519
Canada
  broadcast archives, 1567
  library and archives (*see* Library and Archives
    Canada)
Canada Institute for Scientific and Technical
    Information (CISTI), 660–661
Canadiana.org, 667
Canadian Association of Research Libraries
    (CARL), 659–660, 664–665
Canadian Broadcasting Company (CBC), 1567
Canadian Committee on Archival Description
    (CCAD), 670
Canadian Council of Archives (CCA), 666
Canadian Federation of Library Associations
    (CFLA), 665
Canadian Heritage Information Network
    (CHIN)
  Artefacts Canada, 678
  BCIN, 676
  CIN, 676
  international activities, 675–676, 679
  knowledge exchange, 678
  membership, 679–680
  mission, 676–678
  NIP, 676
  online resources, 675
  research, 679
  strategies, 675
  VMC Investment Program, 676, 678–679
  Web sites, 678
Canadian Journal of Mathematics, 3027
Canadian Library Association (CLA), 19
  early history, 682
  governance, 682
  LIS community, 681
  roles and activities, 681–682
  1970s–2000, 682–683
  from 2000 On, 683
Canadian Urban Libraries Council (CULC), 666
Canberra, 186, 389, 1392–1393, 3692
Cancelbot, 785, 786
Canevari bindings, 543
Canned text systems, 431
Canstantinopolitanus, see Vienna Dioscorides
Capen, Edward, 1844
*Capital and Finance in the Age of the Renais-
    sance: A Study of the Fuggers and their
    Connections*, 645
Capitalism, 891, 1790, 1859, 1861, 4198–4199
Caracas Declaration, 4887–4890
Card sort, 4797
Careers, 706
  ALA's Office on Human Resource Develop-
    ment and Recruitment, 711
  categorization analyst, 711
  digital services librarian (assistant professor),
    711
  knowledge systems librarian/taxonomist, law,
    711
  staff training and development coordinator,
    712
  Youth Services Department Head, 712
CARIST, 4631

CARL, *see* Canadian Association of Research
    Libraries
Carlos III University, 4321
Carnegie, Andrew, 1839, 1845–1846
Carnegie Corporation, 3216–3217
Carnegie Library, 2387, 2802–2803, 3325
Carnegie United Kingdom Trust (CUKT), 1839
Carnegie–Whitney endowment, 78
Carnivore project, 2070
Carolingian illumination, 1947–1948
Carolingian leather bindings, 542
Carpenter, Nathaniel, 1685
Carthage National Museum, 4637
Carus Mathematical Monographs, 3028
CAS, *see* Chemical Abstracts Service
Cascading style sheets (CSS) specification,
    5021, 5025
Case digests, 2741–2742
Case-study model, 3814–3815
Casinos, 1123
Cason, Hulsey, 500–501
CASREACT database, 818
CAS Registry Numbers®, 824
CASS, *see* Chinese Academy of the Social
    Sciences
Cassette tapes, 3280
Catalan consortium, 2824
Catalog cards, 3450
Cataloging
  bibliographic records, 730–731
  components, 726
  forms, 724–725
  functions, 725–726
Cataloging and Metadata Management Section
    (CaMMS), 328–329
Cataloging cultural objects (CCO), 3178, 3761,
    4935, 4937–4938
  assessment of, 740–742
  elements
    authorities, 739–740
    class element, 738
    creator information, 736
    description element, 738
    location and geography, 737–738
    object naming, 736
    physical characteristics, 737
    stylistic, cultural and chronological infor-
      mation, 737
    subject element, 738
  entity relationship diagram, 735
  general guidelines
    database design, 735
    display and indexing, 735
    related works, 735
    subjective interpretation, 735
    work and image records, 735
  historical context of, 733–734
Cataloging in Publication (CIP) program, 450,
    3395
Cataloging Rules and Principles, 2921
Catalogs and cataloging
  AACR, 2920–2921
  archaeological collections, 1149
  arrangement, 744
  audiovisual archives, 1574–1575
  authority control, 2920
  BIBFRAME model, 2924

Development Officers of Research Academic
  Libraries (DORAL), 2834
Dewey decimal classification (DDC), 757, 961,
  966, 969–970, 1534–1535, 1802, 2499,
  2669–2670, 2920, 3283, 3397, 3903,
  4205, 4466, 4472, 4783, 4788
  applications and research, 1261–1263
  in China, 889
  classes, hierarchies, and relationships,
    1257–1258
  development, 1260
  general rules, 1258
  history
    development, 1258
    editions, 1259
    electronic versions, 1259–1260
  mappings, 1261
  notational synthesis, 1258
  structure and notation
    DDC Summaries, 1256–1257
    Manual and Relative Index, 1257
    schedules and tables, 1257
  translations, 1260–1261
  value of, 1256
Dewey Decimal system, 1818
Dewey for Windows, 1259
Dewey Online Catalog (DOC), 1262
DH, see Digital humanities
Dialog, 2221
Dialog generation and management system
  (DGMS), 1203
Dialogue systems, 3353
Dial-up services, 2240
Diana V. Braddom FRFDS Scholarship,
  2845
Diaries, 4531–4532
Diasporic communities, 1028
Diasporic Information Environment Model
  (DIEM), 2127
Diasporic populations
  immigrant information-seeking behavior
    research, 2125–2126
  information behavior, 2126–2127
  media studies, 2122, 2125
  research
    globalization and transnationalism, 2123
    hybridity, imagination and identity,
      2123–2124
    nation-states, 2124–2125
    transnational information networks, 2123
    voluntary economic immigrants, 2122
Diasporization, 4208
Dickinson classification system, 3283
Dictionary-based techniques, 3143
Dictionnaire Universal de Commerce, 645
Die Deutsche Nationalbibliothek, see German
  national library
Die Erde und das Leben, 1687
Differential facet, 1536
DigiQUAL®, 374
Digital Archive, 3400
Digital asset management systems (DAMS),
  4942, 4945–4946
Digital Betacam, 1568–1569
Digital books, 3472
Digital collection, 2826
Digital content, 4298

Digital content licensing
  commercial law
    choice of law and forum, 1272–1274
    contractual terms, enforceability of,
      1271–1272
    electronic self-help and denial of access,
      1275–1276
    UCC, 1270
    UCITA, 1270
    warranties, 1274–1275
    writing requirement and requirements of
      formation, 1270–1271
  copyright law, 1268
  definition, 1267
  e-rights, 1269
  moral rights, 1268–1269
  neighboring rights, 1268
Digital Cultural Content Forum (dCCF), 679
Digital divide
  Digital Inclusion Network, 1284
  educational divide, 1283–1284
  historical context, 1279–1280
  human capital development, 1284
  ICT access, 1280
  infrastructural level, programs, 1284
  international digital divide, assessments of,
    1282–1283
  social support, 1284
  in United States
    age gap, 1282
    disability, 1282
    educational attainment, 1282
    gender gap, 1281
    geography, 1282
    income, 1282
    internet users, demographics of, 1281
    NTIA reports, 1280–1281
    race/ethnicity gap, 1281–1282
Digital humanities (DH), 1295–1296
  categories, 1299
  definitions, 1286–1288, 1298–1299
  history, 1288–1290
  Index Thomisticus, 1299
  institutional location, 1290–1292
  and libraries
    collections, 1303–1304
    consultation and collaboration, 1303
    instruction, 1302–1303
    intensive commitment, 1301
    low commitment, 1301
    moderate commitment, 1301
    theory, 1299–1301
  library community, perspective of,
    1292–1293
  modularity, 1293
  pattern discovery and visualization, 1295
  representation, 1293–1295
Digital images
  create, display and share, 1307
  fundamentals
    binary image, 1309
    bit depth, 1309–1310
    bitmap image, 1308–1309
    digitization guidelines, 1309, 1311
    halftone printing, 1309
    interpolation, 1309
    lossless compression, 1310–1311

    lossy compression, 1310
    pixelation, 1309
    pixels, 1308–1310
    raster image, 1308
    resolution, 1309–1310
    vector graphics, 1308
  history of
    computing technologies, 1308
    digital fax, 1308
    image capture peripheral devices, 1308
    photographic images, 1307–1308
    visual imagery, 1307
    World Wide Web, 1308
  image file formats, 1311
  in library and archive sector
    crowdsourcing, 1313
    digitization, 1312
    technical digitization guidelines, 1312–1313
  memory institutions, surrogates for, 1307
  personal digital image collections and librar-
    ian, 1313–1314
  POD, 3733, 3735
  of provenance evidence, 3771
Digital imaging, 1414
Digital Inclusion Network, 1284
Digital library(ies), 1760, 2079, 2228–2229,
  2240, 2766, 3727
  in Denmark, 1220
  in Germany, 1694, 1698
  Greece
    Anemi, 1732–1733
    digital collections and institutional reposi-
      tory, 1733
    e-fimeris, 1733
    National Documentation Center, 1733
    Psifiothiki, 1733
  Hungary, 1923–1924
  Japan, 2566
  in Kazakhstan, 2583
  Switzerland, 4489–4490
  Venezuelan libraries, 4890
Digital Library Federation, 1312
Digital Library Initiatives, 2228–2229
Digital Library of India (DLI), 2004
Digital Library of Wielkopolska project, 2828
Digital Library System (DLS), 4311
Digitally based information art projects, 2072
Digital Media Management (DiMeMa), 3400
Digital media piracy
  economic harm, 3650
  file sharing, 3650
  legislative framework
    Copyright Act, 3656
    DMCA, 3656–3657
    e-books, 3658
    Google Book Search, 3658
    intellectual property, 3656
    libraries, 3657
    NIH Public Access Policy, 3658–3659
    Pirate Bay, 3656
    Public Law 102-561, 3656
    RIAA, 3657
    SIAA, 3656
    Software Rental Amendments Act, 3656
    Sonny Bono Copyright Term Extension Act
      (1998), 3656
    Telefonica, 3657

National Policy on Library and Information System (NAPLIS), 2001

National Program for Acquisitions and Cataloging (NPAC), 2965

National Programme for the Digitisation of Archival, Library and Museum Records, 1130

National Public Broadcasting Archives (NPBA), 1561, 1563

National-public libraries, 3321

National Register of Historic Places (NRHP), 1774

National Research and Education Network (NREN), 1551

National Research Council (NRC), 559, 660–661, 1673

National Sample Survey Organisation (NSSO), 1993

National Science Digital Library (NSDL), 2998, 3646, 4012, 4073

National Science Education Standards (NSES), 558

National Science Foundation (NSF), 555, 1029, 1392, 2217, 2517, 2772, 3646, 4263

National Science Foundation Act of 1950, 2217

National Scientific and Documentation Center (CNDST), 4106

National security, *see* Intelligence and Security Informatics

National Security Act of 1947, 2151

National Security Agency, 2151

National Security Archive (NSA), 2390

National Security Council (NSC), 2151–2152

National Security Laws, 785

National Serials Data Program (NDSP), 4139

National Society for the Study of Communication (NSSC), 2421

National Stolen Art File (NASF), 4577

National Stolen Property Act (NSPA) of 1934, 1153

National Storytelling Festival, Jonesborough, 4443

National Study of School Evaluation (NSSE), 64

National System of Public Libraries (NSPL), 4886

National System of Scientific, Technical and Organizational Information (SINTO), 3676

National Technical Information Center and Library, 1922

National Technical Information Service (NTIS), 1552, 1554

National Technical Institute for the Deaf (NTID), 1183

National Telecommunications and Information Administration (NTIA), 1279–1281, 2038

National Television and Video Preservation Foundation (NTVPF), 1579

National Terrorism Advisory Board, 4579

National Training Center and Clearinghouse (NTCC), 3035

National Training Information Service (NTIS), 687

National Trust, 4169, 4722–4723

National Union Catalog (NUC), 448, 769, 771

National War Museum of Scotland, 4719

National Zoological Park, 4194

Nation's Business, 650

Native American Graves Protection and Repatriation Act (NAGPRA) of 1990, 267, 1153, 1165–1166, 1169, 3167, 3245

Natural History, 644

Natural history museum, 1820, 3237, 4134
  Denmark, 1226
  Germany, 1704–1705
  Hungary, 1929
  India, 2021–2022
  London, 3224
  Madrid, 3152
  Tanzania, 4510
  United States, 5080

Natural language generation (NLG)
  applications, 430–431
  biography text, 432
  components, 431–432
  computational linguistics, 430
  current developments and outlook, 437–438
  dialogue situations, 431
  formal and computational properties, 430
  functional linguistics, 430
  linguistic variation
    aggregation, 435
    characterizations, 433
    discourse deixis, 435
    ideational representations, 434
    interpersonal control, 434–42
    lexicogrammar, 434–435
    linguistic abstraction, 434
    Penman text generation system, 434
    propositional content, 434
    semantic control, 434
    semantics, 434
    Sentence Plan Language, 434
    sentences, 435
    stratification, 433
    syntactic theory, 434
    textual control, 434–435
  macroplanning, 432
  message/text personalization, 433
  nonlinguistic material, 430
  non-NLG-based text production system, 433
  syntactic description, 430
  text planning, 436–437

Natural language processing (NLP), 274–275, 2201
  ANN
    advantages, 279–280
    cognitive models, 282
    connectionist/subsymbolic paradigm, 281
    disadvantages, 280
    formalisms, 281
    language-oriented disciplines, 281
    local and distributed representational schemes, 283
    meaning representation, 287
    physical symbol system, 281
    research paradigms, 283
    Rumelhart and McClelland model, 287–288
    sequential processing, 284–287
    symbolic paradigm, 281

  applications, 3353
  approaches
    connectionist approach, 3351
    hybrid approach, 3350
    similarities and differences, 3351–3353
    statistical approach, 3351
    symbolic approach, 3350–3351
  definition, 3346
  divisions, 3347
  goal, 3346–3347
  history, 3347–3348
  human-like language processing, 3346
  introspection, 3348
  levels
    discourse, 3350
    lexical, 3349
    lower *vs.* higher levels, 3350
    morphology, 3349
    phonological analysis, 3349
    pragmatic, 3350
    semantic, 3350
    syntactic, 3349
  origins, 3347
  synchronic *vs.* sequential model, 3348

Natural Resource Monitoring Partnership (NRMP), 3312

Natural SEM, *see* Search engine marketing

Nauta, Doede, 2051–2052

Navigability affordance, 1118

Nazi memorabilia, 783

NBII, *see* National Biological Information Infrastructure

Nebraska Library Association, 920

Negative feedback, 1036

Neighboring rights, 1268

NELINET, 3922, 3924

NELLI, 2825

NEMO, *see* Network of European Museum Organisations

Neoclassicism, 1742

Neo-Schumpeterian approach, 2254

NESLI2, 2825

netLibrary e-books, 1209–1211

Netscape Collabra™, 1057

Networked European Deposit Library (NEDLIB), 1333

Networked Knowledge Organization Systems/ Services (NKOS)
  aims and participants, 3366–3367
  special journal issues, 3368–3369
  workshops and special sessions, 3367–3368

"Networked Talent Model," 143–144

Network management
  activities, 3356
  ancillary support systems, 3357
  applications, 3356
  components, 3356
  dimensions, 3356
    accounting management, 3358
    configuration management, 3357–3358
    distributed computing systems, 3357
    fault management, 3357
    performance management, 3358
    security management, 3358–3359
  information and activity, 3362–3363
  LAN, 3356
  MAN, 3356

Office for Intellectual Freedom (OIF), 2387, 2391–2392
Office for Scientific and Technical Information (OSTI), 306
Office of Information and RegulatoryAffairs (OIRA), 1550–1551
Office of Management and Budget (OMB), 1550–1551, 2154
Office of Research and Development (ORD), 3562
Office of Scholarly Communication (OSC), 365
Office of Scientific and Academic Publishing (OSAP), 365
Office of Scientific and Technical Information (OSTI), 1553
Office of Scientific Research and Development (OSRD), 2811
Official Gazette, 3561–3562, 3566
The Official Museum Directory, 4379
Official statistics, 495
Off-line storage, 4942
Ohio College Library Center (OCLC), 451, 1847, 2981; *see also* Online Computer Library Center
  bibliographic network, 2981
  cataloging system, 729–730
  EMEA, 3400
  WorldShare Record Manager, 1545
OhioLink, 987
Okapi BM-25 algorithm, 2205–2206
Okapi system, 3427–3428
Older adults' information needs and behavior
  computers and Internet, 3410
  everyday life information seeking, 3408, 3411
  imperative for studying older age groups, 3407
  information literacy (Fourth Age), 3409
  information needs, 3408–3409
  information sources, 3409
  library-based research, 3407
  old, definitions of, 3406–3407
  residential communities, 3409–3410
Old Testament, 644
On2broker, 3458
OncologySTAT.com, 3472
On-demand books (ODB), 3735
One-clause-at-a time (OCAT) methodology, 1621
One Laptop Per Child (OLPC) program, 1283–1284
One-mode network, 4237
One-person librarian (OPL)
  churches and synagogues, 3416
  future, 3418–3420
  history, 3413–3414
  hospital librarians, 3415–3416
  information brokers, 3416
  law libraries, 3415
  market researchers, 3416–3417
  meaning, 3413
  nontraditional sector, 3417
  organization's goals, 3418
  prison librarians, 3415
  public librarians, 3416
  school librarians, 3416
  special libraries, 3414–3415
  tasks, categories, 3417–3418
  zoo librarians, 3416

One Thousand and One Nights, 853
One type, one printer theory, 1971
ONIX metadata upstream, 3395
Online Account Management Service, 683
Online bibliographic database, 4629
Online bibliographic retrieval, 2245
Online catalogs, 2079
Online Computer Library Center (OCLC), 2, 671, 733, 894, 984, 1390–1391, 1880, 2181, 2921, 3381, 3450–3452, 3454, 3916, 4472, 4578, 4774–4775, 4800
  advocate for libraries, 3401–3402
  Asia pacific, 3404
  Canada, 3404
  cataloging service, 3396–3397
  DDC
    Abridged WebDewey, 1259–1260
    BISAC subject headings, 1261
    Classify, 1262
    development, 1260
    Scorpion software, 1262
    Subject Headings for Children and People, Places & Things, 1261
    translations, 1260
    WebDewey, 1259
    WorldCat Collection Analysis service, 1258, 1262–1263
    XML representations, 1260
  digital collection services, 3400
  eCONTENT, 3399
  electronic books, 3399–3400
  eSerials Holdings Service, 3398
  Europe, Middle East and Africa, 3405
  finances, 3394
  Google, 3403–3404
  governance
    Board of Trustees, 3393–3394
    contribution, 3392–3393
    Global Council, 3394
    Members Council, 3392–3393
    membership, 3392
    membership participation levels, 3393
    WorldCat Principles of Cooperation, 3393
  history, 3392
  integrated library systems, 3400
  Latin American and Caribbean, 3404–3405
  outside United States, 3404
  programs and research, 3400–3401
  QuestionPoint, 3398–3399
  reference and resource sharing, 3398
  RLNs, 3921–3923
  U.S. activity, 3404
  WebJunction, 3402
  WorldCat
    bibliographic database, 3394
    collection analysis, 3398
    CONTENTdm collections, 3400
    enrichment and quality control, 3395–3396
    evolution, 3396
    FirstSearch service, 3395
    growth, 3396–3397
    and information standards, 3396
    Navigator, 3398
    Online Union Catalog, 3394
    Open WorldCat pilot program, 3402–3403
    selection, 3398
    statistics, 3395

    web services, 3403
    WorldCat Local, 3403
    WorldCat.org, 3403
Online databases, 2240–2241
Online information exchange (ONIX), 4056
Online library instruction
  assessment
    economic viability, 3444
    learner/instructor preferences, 3444
    learning outcomes, 3444
    usability, 3443
  CAI, 3444
    benefits, 3435
    computer-assisted demonstration, 3434
    independent learning tutorials, 3435
    live assistance and hand-on, computer-based learning, 3434–3435
  early days of distance education, 3434
  history
    emergence of distance education, 3432–3433
    need for teaching librarian, 3432
    online education as distance education, 3433
  instructional opportunities
    credit course, 3437–3438
    discipline-specific online library instruction, 3437
    intended audience, 3435
    librarian professional development, 3438–3440
    popular database and OPAC, 3436
    in schools of library and information studies, 3438–3439
    teaching information literacy skills, 3436–3437
    virtual tour, 3436
  internet, libraries and online learning
    assessment of, 3443–3444
    case for CAI, 3435
    instructional opportunities for online library instruction, 3435–3439
  predictors and predecessors, 3433
  technology
    chat, 3442–3443
    collaborative Web browsing, 3443
    conferencing software and courseware, 3442
    reaching online learner through electronic mail, 3440, 3442
    static Web pages, 3439
    web site interaction, 3439–3441
Online Programming for All Libraries (OPAL), 17
Online public access catalogs (OPAC), 1–2, 250, 451–452, 487, 841, 1878, 2219–2220, 2240, 2847, 2854, 2947, 3399, 3435–3436, 4159, 4467–4468, 4789, 4978–4982
  Boolean retrieval systems, 3422
  vs. card catalog, 3450
  database records, 3422
  design, improvements, 3429
    automated search heuristics, 3428
    best-match retrieval approaches, 3427–3428
    browse interface, 3426
    catalog records, enhancement of, 3427

Participatory three-dimensional modeling (P3DM), 1672–1673

Particle-induced X-ray emission (PIXE), 478

Partner institutions network, 535

Partnership libraries, 4160–4161

The Past as Prologue: The Evolution of Art Librarianship, 249

Patchwriting, 3664

Patent Abstracts of Japan (PAJ), 3569

Patent Act of 1790, 3560

Patent Act of 1793, 3561

Patent Act of 1836, 3561

Patent and Trademark Depository Libraries (PTDL), 3562

Patent and Trademark Depository Library Program (PTDLP), 640

Patent Application Information Retrieval (PAIR) system, 3566

Patent classification systems
  IPC, 3566–3567
  USPC, 3566–3567

Patent Cooperation Treaty (PCT), 815

Patent documents
  AI patents, 3565
  APC documents, 3566
  certificates of correction, 3566
  dedications, 3566
  defensive publications, 3565
  design/industrial designs, 3565
  disclaimers, 3566
  drawing, 3564
  front page, 3563–3564
  INID codes, 3564
  kind codes, 3564
  plant patents, 3565
  reexamination certificates, 3566
  reissue patents, 3565
  SIRs, 3565–3566
  specification, 3564
  TVPP publications, 3566
  utility models, 3565
  utility patents, 3564–3565

Patent information
  history
    1790–1870, 3560–3561
    1870–1970, 3561–3562
    1970–2008, 3562–3563
  monopoly right, 3560
  patentability search, 3560
  patent protection, 3560
  WIPO, 3560

Patent Lens, 3570–3571

Patent Map Guidance, 3570

Patent Office Fire of 1836, 3561

Patents, 639–640, 815, 833–834; *see also* Patent documents; Patent information

PatentScope, 3570

Patents Ordinance of 1924, 4513

PatFT patent database, 3566

Pathfinder Network (PFNET), 485, 931–932

PATRIOT Act, 2402–2403

Patron-driven acquisition (PDA), 415–416, 1209

Paul Banks and Carolyn Harris Preservation Award, 331

Paul of Aegina, 3044

Paulo Montenegro Institute (IPM), 603

Peale, Charles Willson, 1818, 4767

Pedigree chart, 1656

PeerEvaluation, 47

Peer review, version control, 4898–4899

Peer-to-peer data grids, 3000

Peer-to-peer networks, 3652

Peircean sign theory, 4098, 4101

Peking University Library (PUL), 896, 903

Penman text generation system, 434

Pennsylvania Academy of Fine Arts, 4767

Pension Protection Act of 2006, 639

People–place–information trichotomy, 1512

People with disabilities and libraries
  archives, 3575
  barriers, 3573
  cataloging and indexing, 3580–3581
  collection development, 3580
  disability
    categories of, 3574
    definition, 3573
  disability rights movement, 3574
  electronic resource accessibility
    assistive technology, 3579
    circulating equipment, 3579
    library websites, 3578–3579
    Tatomir Accessibility Checklist, 3578
    vendor database, 3578
    WCAG guidelines, 3578
  for-profit sector, 3575
  history, 3575–3576
  language, 3575
  legislation, 3575
  museums, 3575
  outreach, 3580
  physical resource accessibility
    physical books and media, 3577
    services and programming, 3577–3578
    space, 3576–3577
  print disabilities, 3574
  social model, 3573
  staff training, 3579–3580

Perceived attributes, IT adoption
  compatibility, 2296
  complexity, 2293, 2296
  observability, 2293, 2296
  relative advantage
    behavioral intention to use/actual usage, 2295
    information quality, 2293–2294
    IT quality, 2293
    service quality, 2294–2295
    user satisfaction, 2295
  triability, 2296

Performance libraries, 3276–3277

Performing arts; *see also* Visual and performing arts
  archives
    Center for Black Music Research, 4755
    Folger Shakespeare Library, 4756
    New York Public Library, 4756
  definition, 4925
  live and recorded elements, 4925
  live events, 4925

Performing Arts Data Service (PADS), 297

Perseus Project, 1290

Persian illumination (1502–1736), 1955–1956

Persistent uniform resource locator (PURL), 2155

Personal anticipated information need (PAIN) hypothesis, 2119

Personal construct theory, 2234

Personal health record (PHR), 86, 979, 3342

Personal information management, 2111
  analysis
    finding/refinding activities, 3588
    information item and form, 3586
    keeping activities map, 3588
    meta-level activities, 3588
    personal information collections, 3587
    PSI, 3587
  checkbox methodology, 3597
  convergence and integration, 3599
  e-mails, 3598–3599
  factors, 3598
  history, 3585–3586
  information fragmentation, 3585
  maintenance and organization, 3585
  observant participation, 3598
  PICs, 3584
  practical methodologies, 3597
  privacy, security and information, 3585, 3599
  research
    finding/refinding activity, 3588–3591
    GIM and PIM social fabric, 3596
    keeping activities, 3591–3593
    meta-level activity, 3593–3596
  search technology, 3599
  user-subjective approach, 3598

Personality facet
  levels of, 1536
  rounds of, 1536–1537

PERsonalized and Successive Information Search Tools (PERSIST), 1901

Personally identifiable information (PII), 1489

Personal space of information (PSI), 3587

Personnel Administration Section (PAS), 2844

Pertinence relevance judgments, 3943

Peru
  libraries
    academic libraries, 3608
    education and professional associations, 3608
    modern challenges, 3608–3609
    National Library, 3606–3608
    publications, 3607
    public libraries, 3606–3608
    school libraries, 3608
    during Spanish domination, 3606
  map of, 3606–3607

Pervasive information systems, 1488

Peschel, Oscar, 1687

Pests, 1150–1151

Peterborough library, 1839, 1843

Pew Global Attitudes project, 1061

Pew Research Center, 5028

Pew Research Institute's American Life Project, 841

PFNET algorithm, 485

Pharmacophore searching, 834–835

Phenomenography, 2754

Philadelphia Museum of Art, 1072, 3253

Philadelphia Peale Museum, 3243

Philosophical Transactions of the Royal Society, 646

Presidential libraries
  history
    Archivist report, 3715
    Claypoole, Richard, 3715
    Clinton Presidential Project, 3715
    Eisenhower Library, 3715
    George H.W. Bush Library, 3715
    Hoover, 3714
    John F. Kennedy Library, 3715
    Johnson, Lyndon B., 3715
    Lyndon Baines Johnson Foundation, 3715
    NARA, 3715
    public–private partnership, 3715
    Roosevelt, Franklin D., 3714
    Truman Grants program, 3714
  list, 3717–3718
  presidential materials
    audiovisual and photographic record, 3716
    Clinton Presidential Materials Project, 3716
    economic indicators and project, 3717
    The Foreign Gifts and Decorations Act, 3716
    library websites, 3717
    National Archives, 3716
    National Study Commission report, 3716
    Nixon Presidential Materials Staff, 3716
    personal papers and historical materials, 3716
    presidential papers, 3715
    PRMPA, 3716
    selective donation and selective destruction, 3716
  public and educational programs, 3717
Presidential Libraries Act of 1955, 3714–3715
Presidential Recordings and Materials Preservation Act (PRMPA), 3716
Presidential Records Act, 3716
Pressure ethics, 1473
Presumed credibility, 1115
Pretty Good Privacy (PGP), 404
Preventive censorship, 3676
Primary mathematics literature
  biological and behavioral sciences, 3025
  book series, 3027–3028
  journals, 3026–3027
  nonserial book, 3025
  publishers, 3026
Primary records
  card catalogs, conversions of, 3722
  definition, 3719–3721
  electronic forms, 3721
  survival and accessibility of
    artifacts, preservation of, 3728–3730
    collection-based institution, 3727
    dematerialization of information, 3726
    ownership and access, 3726
    paper facsimiles, 3727
    physical presentation of verbal texts, 3730
    primary texts, 3726
    print products, 3726
    rare books, 3726
    storage, conservation and preservation, 3727
  uses of, 3722–3725
Primos Library in Secane, Philadelphia, 1638
Principal component analysis (PCA), 3272
Principle of cumulative advantage, 4199

Print DDA programs, 1211–1212
Print disabilities, 3574
Printing
  in China, 1865
  histories, 1861
  intaglio, 1867–1868
  modern techniques, 1868
  planographic, 1868
  relief, 1865–1867
Printing Act of 1895, 2150
Printing press, 999, 3606
Print-on-demand (POD), 3736–3737, 3986, 4055–4056
  authors, opportunities for, 3736
  book publishing, impacts on, 3733–3734
    commercial and vanity publishers, 3735
    long tail, 3734
    nontraditional and traditional publishing, 3734–3735
    ODB, 3735
    suppliers, growth in, 3735
  book retailers, 3736
  digital printing
    art reproductions and artist books, 3736
    digital image, 3733, 3735
    vs. offset printing, 3733
  music publishing, 3736
Prison librarians, 3415
Pritchard, Alan, 497–499, 506
Privacy Act of 1974, 1857, 2151
Privacy Act of 1976, 1555
Privacy *vs.* information sharing, 2398–2399
Private bureaucracy, 2256
Private libraries, 748
  America, 1839
  Croatia, 1123
  Mexican libraries, 3091
  Peru, 3606
  Poland, 3674
  Saudi Arabia, 3973–3974
  Ukraine, 4642
Private press
  aristocratic plaything, 3739–3740
  author, 3740–3741
  bibliography, 3743
  clandestine, 3741
  educational press, 3739
  fine books, 3741–3742
  little, 3743
  origins, 3738
  quasi-official press, 3738–3739
  scholarly presses, 3739
Private presses, 4337
Probabilistic models, 422–423
Probabilistic Relational Models (PRM), 274
Probability, theory of, 494–496
Probate records, 1658
Problem, intervention, comparison, and outcome (PICO), 1517
Problem-solving model, 2088–2089
Procedural and descriptive markup, 4560
Procedural knowledge, 3535
Procedural markup, 3074
Proceedings of Symposia in Pure Mathematics, 3028
Proceedings of the American Mathematical Society, 3027

Proceedings of the National Academy of Sciences (PNAS), 3469
Proceedings of the Steklov Institute of Mathematics in the Academy of Sciences of the USSR, 3028
Process-based retention schedules, *see* Large aggregation retention schedules
Process knowledge, 3535
Process quality management, 1178
Producer–Archive Interface Methodology Abstract Standard (PAIMAS), 3485
Producer–Archive Interface Specification (PAIS), 3485
Product catalogs, 639
Professional associations, 303, 3377
Professional conference organizer (PCO), 2453
Professional machine bureaucracies, 3512
Professional metadata creators, 3066
Professional recognition, MLA, 3036
Professional Records and Information Management, International Association (PRISM), 1853
Professional registration, 3377
Program for Cooperative Cataloging (PCC), 454, 2871, 3395
Program for Museum Development in Africa (PMDA), *see* Centre for Heritage Development in Africa
Programme on Information and Communication Technologies (PICT), 2255
Progression of actions lifecycle model, 168
Prolegomena to Library Classification, 1534
PROLOG, 272
Promotion and Identification of Emerging Advanced Telecommunications Services (PISTA), 4322
Property rights, 2277
Prophetic Shrine Library, 3976
Propositional knowledge
  belief
    dispositional view, 2610
    state-object view, 2610–2611
  justification
    adequate indication, 2613
    contextualism, 2615–2616
    epistemic coherentism, 2614
    epistemic foundationalism, 2614–2615
    fallibilism, 2613
    inductive justification, 2613–2614
    inferential justification, 2614
    modest foundationalism, 2615
    radical foundationalism, 2615
  vs. non-propositional knowledge, 2610
  truth, 2611–2612
    coherence, 2612–2613
    correspondence, 2612
    pragmatic value, 2613
Propositions, 592–593
ProQuest Coutts Award for Innovation, 332
ProQuest Ebooks, 1211
Prospectus d'un nouveau dictionnaire de commerce, 1403
Protein Data Bank, 836
Prototypical task, 4555
Provenance, 116
Provenance information, 1364

Recurrent networks (RANN), 285–287

Recursive auto associative memories (RAAM), 284

Redarte-SP (Sao Paulo, Brazil), 256

Red de Bibliotecas Universitarias (REBIUN), 4316, 4320–4321

Red Universitaria Española de Catálogs Absys (RUECA), 4316

Redwood Library, 1841

Reference and Adult Services Division (RASD), *see* Reference and User Services Association

Reference and informational genres
  almanacs, 3898
  atlases, 3898
  bibliographies, 3898
  biological sciences, 3901
  canonical texts, 3900
  catalogs, 3898
  chronologies, 3898
  concordances, 3898
  DDC, 3903, 3905
  dictionaries, 3898
  directories, 3898–3899
  document types, 3906
  encyclopedias, 3899
  formats, 3901
  gazetteers, 3899
  handbooks and manuals, 3899
  idiosyncratic order, 3900
  "index volume" bibliographies, 3904
  LCC system, 3902, 3904–3905
  Martel's structure, 3902
  monographs, 3904
  multidisciplinary sources of information, 3901
  newsletters, 3899
  personal bibliographies, 3904
  primary literature, 3900
  publication types, 3905–3906
  secondary literature, 3900
  sourcebooks, 3899
  subject bibliographies, 3904–3905
  subject groupings, 3897
  term weighting, 3902
  tertiary literature, 3900
  union lists, 3899
  yearbooks, 3899

Reference and User Services Association (RUSA), 2699, 3913
  award, 3909–3910
  education opportunities, 3910
  guidelines, 3910
  membership, 3908
  name change, 3908
  organization, 3909
  problems and issues, 3910–3911
  publications, 3909
  reference and information professionals, 3908
  2000 ALA midwinter meeting, value statement, 3908

Reference and User Services Quarterly (*RUSQ*), 3909

Reference desk, 6–7, 3914

Reference information, 1364

Reference interview, 3912
  communication techniques, 3915–3916
  purpose of, 3915

Reference services, 1098
  bibliographic verification, 3914
  changing context of, 3913
  components of, 3912
  definition, 3912
  direct and indirect instruction, 3914
  evaluation, 3917
  history of, 3912–3913
  instruction and guidance, responsibilities for, 3914
  interlibrary loan, 3914
  readers' advisory, 3914
  ready reference, 3914
  reference collection development and maintenance, 3916–3917
  reference desk, 3914
  reference ethics, 3912, 3917
  reference interview, 3912
    communication techniques, 3915–3916
    purpose of, 3915
  reference sources, 3912, 3916
  reference transactions, definition of, 3913–3914
  reference work, definition of, 3914
  research consulting, 3914
  roving method, 3914
  virtual reference, 3915

Reference Services Section (RSS), 3909

Reference tools, 1655–1656, 1659

Reference transactions, 3913–3914

Referential markup, 3074

ReferralWeb, 3866–3867

Reflexive meta-field, 3611

REFORMA, 2699, 2701, 2703

Reformation libraries, 3954–3956

Refreezing process, 2297

Refreshing, *see* Bit preservation

Regimen Sanitatis Salernitanum, 3044

Regional Bell holding companies (RBOCs), 1018, 1020–1021

Regional Library Associations, 4328

Regional library networks (RLNs)
  Alliance of Library Service Networks, 3922
  challenges, 3924
  consortia/consortium, 3920, 3924
  definition, 3920
  educational programs, 3923
  electronic resources, 3923–3924
  federal/state agencies, 3922
  governance, 3922
  history, 3920–3921
  OCLC, 3921–3923
  services, 3923
  support/help desks, 3922
  unique projects, 3924
  web sites of, 3922

Register of Australian Archives and Manuscripts, 388

Register Plus, 3569

Registry systems, 118–119

Rehabilitation Act, 3575

Reiter's pipeline architecture, 431–432

Related term (RT), 1986

Relational databases (RDBs), 4080–4082

Relationship links, 2940

Relative index, 1257, 1262

Relative relevance (RR), 1902

Relativism, 3811–3812

RELAX NG, 1388

Relevance assessment, 1902–1903, 2172, 2174, 3708–3709, 3711, 3944, 3946, 4876, 4880

Relevance feedback, 4420

Relevance judgments, 3941–3944

Relevance measurements, 3944–3947

Relevance theory
  "berrypicking" model of literature retrieval, 3933–3934
  and citation, 3936–3937
  cognitive effects, 3934
  definition, 3926–3927
  degrees of, 3931–3933
  evidentiary relevance, 3931
  historical precedents, 3928–3929
  intermediaries and disintermediation, 3928
  intersubjective agreement, 3935
  IR evaluation tests, 3934
  literature-based systems, 3927
  objectifying subjectivity, 3935–3936
  "objective system relevance," 3933
  question relevance, 3931
  systems evaluation, 3929–3931
  topical relevance, 3931

Relief printing, 1865–1867

Religious archives
  Australia, 188–189
  China, 907
  Germany, 1701
  United Kingdom, 4737–4738

Religious publishing, 3987

Religious storytelling, 4443

Renaissance libraries
  France, 3951–3952
  Italy, 3948–3950
  Spain, 3952
  Vatican Library, 3951

Renardus Service, 1262

Renouf Press, 1724

Repertoire Bibliographique Universel (RBU), 1373

Repository-based software engineering (RBSE) spider, 2520

Representational state transfer (REST), 2187, 2347, 2895

Repressive censorship, 3676

Repterorium Bibliographicum, 1968–1969

Republic, 644

Republican Book Museum in Almaty, 2587–2588

Republican Scientific-Medical Library (RSML), 232

Republican Scientific-Technical Library (RNTB), 230, 2582

Republic of Armenia, *see* Armenia

Reputed credibility, 1115

Request-for-proposal (RFP), 607

Research Assessment Exercise (RAE), 2373, 4712–4713

Research data services (RDS), 4015

Research Libraries Group (RLG), 451–452, 2181, 3400–3401

Research Libraries Information Network (RLIN), 450, 476, 2921

Special Interest Group for Classification Research (SIG/CR), 959–960
Special Interest Group on Computer–Human Interaction (SIGCHI), 1904, 4804
Special Interest Group on Information Retrieval (SIGIR), 482, 1904, 2222
Special Interest Group on Management Information Systems (SIGMIS), 2276
Special interest groups (SIGs), 91–92, 2007, 3173–3174
Special Interest Section on Aboriginal Archives (SISAA), 691
Special interest sections (SISs), 49–50
SPECIALIST lexicon, 4673
Special librarianship
    career opportunities, 4352–4353
    characteristics, 4354–4355
    competencies, 4357–4358
    education, 4358
    employability, 4358
    end user training, 4353–4354
    evolution, 4351
    global networks, 4357
    KM, 4354
    organizations, 4355–4357
    public image, 4353
    return on investment, 4353
    technology, 4353
    titles, 4355
    Web, 4351–4352
Special libraries, 301–303, 305, 307, 1096, 2764, 3414–3415
    acquisitions units, organization of, 2918
    Armenia, 230
    in Australia, 384–385
    in China, 900–902
    in Croatia, 1125
    definition, 4361–4362
    in Denmark, 1219–1220
    digital collections, 4368
    ethics, 4365
    Ethiopia, 1498–1499
    in Germany, 1698
    globalization, 4362–4363
    Greece, 1732
    Hungary, 1923
    information technology, 4367
    in Israel, 2547
    Japan, 2565
    in Kazakhstan, 2583
    Kenya, 2596–2597
    knowledge services, 4368
    learning organization, 4367
    library associations (see Special Libraries Association)
    Lithuania, 2953
    management
        marketing, 4366
        organizations, 4365–4366
        planning and budgeting, 4366
        value evaluation, 4366–4367
    in Moldova, 3124
    New Zealand libraries, 3376
    origin, 4362
    in Peru, 3608–3609
    physical and a virtual presence, 4368
    in Saudi Arabia, 3976–3977

Senegal, 4107–4108
in Serbia, 4131
services
    acquisitions and collection development, 4363
    competitive intelligence, 4364–4365
    knowledge management, 4365
    news updating, 4364
    organization of information, 4363
    reference and research center, 4363–4364
South Korea, 4311
Tunisia, 4631
in Ukraine, 4646
in United Kingdom, 4707–4708
Venezuelan libraries, 4890
Special Libraries Association (SLA), 256, 652, 709–710, 1097, 2708, 4014, 4016, 4352, 4356–4357, 4362
    business and industry, advisory service to, 4374
    copyright legislation, 4374–4375
    core values, 4371
    corporate and technology libraries, 4374
    documentation, 4374
    Employment Committee, 4373
    Great Depression, challenges during, 4373
    information, knowledge and strategic learning, 4375–4376
    information/knowledge centers, 4370
    information professionals, 4377–4378
    knowledge services, 4372, 4376
    knowledge sharing, 4372–4373
    membership, growth of, 4374
    motto, 4373
    origin of, 4371–4372
    PAIS, creation of, 4373
    practical and utilitarian library services, 4373
    PREPS Commission, 4376
    professional knowledge workers, support to, 4370
    regional chapters, 4370
    research resources, analysis of, 4373
    research units, 4370
    responsibilities, 4374
    special libraries movement, 4374
    Vision Statement of 2004, 4370, 4377
Specialty hospital library, 1872–1873
Specialty Museums
    changing face, 4382
    examples, 4381–4382
    expertise, 4381
    number of, 4380
    organization, 4380
    types, 4379–4380
    United States, 4380
Spectrum scholarships, 81
Speculativism, 3628–3629
Speech communication, 997–998
Speech processing, 274
Speech recognition technology, 1058
SPICE (setting, perspective, intervention, comparison, and evaluation), 1517
Spiders, 2519
Spofford, Ainsworth, 2848
Spofford, Ainsworth Rand, 2881–2883
Spreading activation model, 3590
Spring backs, 539

SRW/U, 2894–2895
S-SRB, 3002
Stack management, 920–921
Standard citation order, 585
Standard Generalized Markup Language (SGML), 730–731
Standard generalized markup language (SGML), 1252, 1365, 1381, 1416, 2343, 2986, 3074–3075, 4560–4561, 5022
Standard industrial classification (SIC) system, 638, 653
Standardized Assessment of Information Literacy Skills (SAILS), 374
Standard operating procedures (SOPs), 3357, 3361–3363
Standards for Accreditation of Master's Programs in Library and Information Studies, 18, 20
Standards Institution of Israel, 2547
Standard Statistics Compustat Service, 652
Standing Committee of the National and University Libraries (SCONUL), 2545
Standing Conference of Eastern, Central and Southern African Library and Information Associations (SCECSAL), 37, 4515
Standing Conference of National and University Libraries of Eastern, Central and Southern Africa (SCANUL-ECS), 3330
Standish survey and analysis methodology, 2281
Stanford Graduate School of Business, 650
STARS, 1676
"Starvation policy," 4784
State and local government archives
    Multnomah County (Oregon) Archives, 4746
    New Orleans Notorial Archives, 4746
    San Antonio Municipal Archives Program, 4746–4747
State archives, 1222–1223
    in Dubrovnik, 1128
    establishment and administrative location, 4390–4391
    history, 4386–4389
    mission, 4384
    nature and functions, 4384–4386
    in Rijeka, 1128
    in Zadar, 1128
State Archives Bureau (SAB), 905
State Gold and Precious Metals Museum, 2587
State Historical Records Advisory Boards (SHRABs), 4743
State Historical Society of Wisconsin (SHSW), 1780–1781
State Historic Preservation Officer (SHPO), 1774
State Inspectorate of Public Libraries, 1216
State Library Agency Section (SLAS), 376
State library and state library agencies, 4392
    Arizona State Library, Archives and Public Records, 4397
    Connecticut State Library, 4397
    COSLA, 4394–4395
    definition, 4392–4393
    establishment dates, 4398–4399
    functions, 4394
    FY 2006, 4396
    GPLS, 4397
    history, 4393